The Maritime Engineering Reference Book

The Maritime Engineering Reference Book

A Guide to Ship Design, Construction and Operation

Edited by Anthony F. Molland

AMSTERDAM • BOSTON • HEIDELBERG • LONDON • OXFORD • NEW YORK
PARIS • SAN DIEGO • SAN FRANCISCO • SINGAPORE • SYDNEY • TOKYO
Butterworth-Heinemann is an imprint of Elsevier

Butterworth-Heinemann is an imprint of Elsevier
Linacre House, Jordan Hill, Oxford OX2 8DP, UK
30 Corporate Drive, Suite 400, Burlington, MA 01803, USA

Copyright © 2008 Elsevier Ltd. All rights reserved

No part of this publication may be reproduced, stored in a retrieval system or transmitted in
any form or by any means electronic, mechanical, photocopying, recording or otherwise
without the prior written permission of the publisher

Permissions may be sought directly from Elsevier's Science & Technology Rights Department
in Oxford, UK: phone (+44) (0) 1865 843830; fax (+44) (0) 1865 853333;
email: permissions@elsevier.com. Alternatively you can submit your request online by
visiting the Elsevier web site at http://elsevier.com/locate/permissions, and selecting *Obtaining
permission to use Elsevier material*

Notice
No responsibility is assumed by the publisher for any injury and/or damage to persons or
property as a matter of products liability, negligence or otherwise, or from any use or
operation of any methods, products, instructions or ideas contained in the material herein.

British Library Cataloguing in Publication Data
A catalogue record for this book is available from the British Library

Library of Congress Cataloguing in Publication Data
A catalogue record for this book is available from the Library of Congress

ISBN: 978-0-7506-8987-8

For information on all Butterworth-Heinemann publications
visit our web site at http://elsevierdirect.com

Typeset by Charon Tec Ltd., A Macmillan Company. (www.macmillansolutions.com)

Printed and bound in Hungary

08 09 10 11 12   10 9 8 7 6 5 4 3 2 1

Working together to grow
libraries in developing countries

www.elsevier.com | www.bookaid.org | www.sabre.org

ELSEVIER   BOOK AID International   Sabre Foundation

Contents

Preface — xvii

1 The marine environment — 1
1.1 The ship in the marine environment — 3
1.2 Wind — 3
1.3 Variations in level of sea surface — 4
1.4 Regular waves — 5
    1.4.1 The trochoid — 5
    1.4.2 Higher order waves. Stokes and Airy Theory — 5
1.5 The sinusoidal wave — 7
    1.5.1 Basic relationships to describe regular waves in deep water — 7
    1.5.2 Normal dispersion of a wave field — 8
    1.5.3 Orbital motion of water particles in a wave — 9
1.6 Irregular waves — 10
1.7 Spectrum formulae by Pierson/Moskowitz and Bretschneider — 12
1.8 The JONSWAP sea spectrum — 13
1.9 Maximum wave height in a stationary random sea — 13
1.10 Long-term statistics of irregular seaway — 16
1.11 Wave data from observations — 17
1.12 Wave climate — 18
1.13 Freak waves — 20
1.14 Oceanography — 21
    1.14.1 Distribution of water on earth — 22
    1.14.2 Properties of water — 22
        1.14.2.1 Chlorophyll — 22
        1.14.2.2 Circulation — 23
        1.14.2.3 Compressibility — 24
        1.14.2.4 Conductivity — 25
        1.14.2.5 Density — 25
        1.14.2.6 Depth — 27
        1.14.2.7 Dissolved gases — 28
        1.14.2.8 Fresh water — 30
        1.14.2.9 Ionic concentration — 30
        1.14.2.10 Light and other electro-magnetic transmissions through water — 30
        1.14.2.11 Pressure — 31
        1.14.2.12 Salt water and salinity — 32
        1.14.2.13 Solar radiation — 34
        1.14.2.14 Sonic velocity and sound channels — 34
        1.14.2.15 Turbidity — 35
        1.14.2.16 Viscosity — 35
        1.14.2.17 Water quality — 35
        1.14.2.18 Water temperature — 36
    1.14.3 Coastal zone classifications and bottom types — 38
1.15 Ambient air — 38
1.16 Climatic extremes — 39
1.17 Marine pollution — 40
References — 41

2 Marine vehicle types — 43
2.1 Overview — 45
2.2 Merchant ships — 45
    2.2.1 General cargo ships — 45
    2.2.2 Container ships — 45
    2.2.3 Roll-on roll-off ships (Ro-Ro ships) — 49
    2.2.4 Car carriers — 51
    2.2.5 Bulk cargo carriers — 51
        2.2.5.1 Tankers — 52
        2.2.5.2 Dry bulk carriers — 55
    2.2.6 Passenger ships — 58
    2.2.7 Tugs — 60
    2.2.8 Icebreakers and ice strengthened ships — 60
    2.2.9 Fishing vessels — 62
2.3 High speed craft — 62
    2.3.1 Monohulls — 64

		2.3.2 Surface effect ships (SESs)	64
		2.3.3 Hydrofoil craft	64
		2.3.4 Multi-hulled vessels	65
		2.3.5 Rigid inflatable boats (RIBs)	66
		2.3.6 Comparison of high speed types	66
	2.4	Yachts	66
	2.5	Warships	68
		2.5.1 Stealth	68
		2.5.2 Sensors	68
		2.5.3 Own ship weapons	68
		2.5.4 Enemy weapons	69
		2.5.5 Sustaining damage	69
		2.5.6 Vulnerability studies	69
		2.5.7 Types of warship	70
		2.5.7.1 Frigates and destroyers	70
		2.5.7.2 Mine countermeasures vessels	70
		2.5.7.3 Submarines	71
	References	73

3 Flotation and stability — 75

	3.1	Equilibrium	77
		3.1.1 Equilibrium of a body floating in still water	77
		3.1.2 Underwater volume	77
	3.2	Stability at small angles	79
		3.2.1 Concept	79
		3.2.2 Transverse metacentre	81
		3.2.3 Transverse metacentre for simple geometrical forms	82
		3.2.4 Metacentric diagrams	83
		3.2.5 Longitudinal stability	83
	3.3	Hydrostatic curves	84
		3.3.1 Surface ships	84
		3.3.2 Fully submerged bodies	84
	3.4	Problems in trim and stability	85
		3.4.1 Determination of displacement from observed draughts	85
		3.4.2 Longitudinal position of the centre of gravity	85
		3.4.3 Direct determination of displacement and position of G	86
		3.4.4 Heel due to moving weight	86
		3.4.5 Wall-sided ship	86
		3.4.6 Suspended weights	87
	3.5	Free surfaces	87
		3.5.1 Effect of liquid free surfaces	87
	3.6	The inclining experiment	88
	3.7	Stability at large angles	89
		3.7.1 Atwood's formula	89
		3.7.2 Curves of statical stability	89
		3.7.3 Metacentric height in the lolled condition	91
		3.7.4 Cross curves of stability	91
		3.7.5 Curves of statical stability from cross curves	91
		3.7.6 Features of the statical stability curve	91
	3.8	Weight movements	92
		3.8.1 Transverse movement of weight	92
	3.9	Dynamical stability	93
	3.10	Flooding and damaged stability	94
		3.10.1 Background	94
		3.10.2 Sinkage and trim when a compartment is open to the sea	94
		3.10.3 Stability in the damaged condition	96
		3.10.4 Asymmetrical flooding	96
		3.10.5 Floodable length	96
	3.11	Intact stability regulations	98
		3.11.1 Introduction	98
		3.11.2 The IMO code on intact stability	98
		3.11.2.1 Passenger and cargo ships	98
		3.11.2.2 Cargo ships carrying timber deck cargoes	101
		3.11.2.3 Fishing vessels	101
		3.11.2.4 Mobile offshore drilling units	101
		3.11.2.5 Dynamically supported craft	101
		3.11.2.6 Container ships greater than 100 m	102
		3.11.2.7 Icing	102
		3.11.2.8 Inclining and rolling tests	102
		3.11.2.9 High-speed craft	102
		3.11.3 Regulations of the US Navy	103
		3.11.4 Regulations of the UK Navy	106
		3.11.5 A criterion for sail vessels	107
		3.11.6 A Code of practice for small workboats and pilot boats	108
		3.11.7 Regulations for internal water vessels	109
		3.11.7.1 EC regulations	109
		3.11.7.2 Swiss regulations	109
		3.11.8 Summary of intact stability regulations	109
	3.12	Damage stability regulations	110
		3.12.1 SOLAS	110
		3.12.2 Probabilistic regulations	111
		3.12.3 The US Navy	112
		3.12.4 The UK Navy	113
		3.12.5 The German Navy	113
		3.12.6 A code for large commercial sailing or motor vessels	114
		3.12.7 A code for small workboats and pilot boats	114

		3.12.8	EC regulations for internal water vessels	114
References				114

4 Ship structures — 116
4.1	Main hull strength			118
	4.1.1	Introduction		118
	4.1.2	The standard calculation		119
		4.1.2.1	The wave	120
		4.1.2.2	Weight distribution	121
		4.1.2.3	Buoyancy and balance	122
		4.1.2.4	Loading, shearing force and bending moment	123
		4.1.2.5	Second moment of area	123
		4.1.2.6	Bending stresses	125
		4.1.2.7	Shear stresses	126
		4.1.2.8	Influence lines	126
		4.1.2.9	Changes to section modulus	129
		4.1.2.10	Slopes and deflections	129
		4.1.2.11	Horizontal flexure	130
		4.1.2.12	Behaviour of a hollow box girder	130
		4.1.2.13	Wave pressure correction	131
		4.1.2.14	Longitudinal strength standards by rule	132
		4.1.2.15	Full scale trials	134
		4.1.2.16	The nature of failure	134
		4.1.2.17	Realistic assessment of longitudinal strength	135
		4.1.2.18	Realistic assessment of loading longitudinally	136
		4.1.2.19	Realistic structural response	137
		4.1.2.20	Assessment of structural safety	140
		4.1.2.21	Hydroelastic analysis	142
		4.1.2.22	Slamming	142
	4.1.3	Material considerations		142
		4.1.3.1	Geometrical discontinuities	142
		4.1.3.2	Built-in stress concentrations	143
		4.1.3.3	Crack extension, brittle fracture	144
		4.1.3.4	Fatigue	146
		4.1.3.5	Discontinuities in structural design	147
		4.1.3.6	Superstructures and deckhouses	147
4.2	Structural design and analysis			149
	4.2.1	Introduction		149
		4.2.1.1	Overview	149
		4.2.1.2	Loading and failure	149
		4.2.1.3	Structural units of a ship	150
	4.2.2	Stiffened plating		151
		4.2.2.1	Simple beams	151
		4.2.2.2	Grillages	151
		4.2.2.3	Swedged plating	155
		4.2.2.4	Comprehensive treatment of stiffened plating	155
	4.2.3	Panels of plating		155
		4.2.3.1	Behaviour of panels under lateral loading	155
		4.2.3.2	Available results for flat plates under lateral pressure	156
		4.2.3.3	Buckling of panels	159
	4.2.4	Frameworks		159
		4.2.4.1	Overview	159
		4.2.4.2	Methods of analysis	161
		4.2.4.3	Elastic stability of a frame	166
		4.2.4.4	End constraint	167
	4.2.5	Finite element analysis (FEA)		168
	4.2.6	Realistic assessment of structural elements		169
	4.2.7	Composite materials		170
4.3	Ship vibration			171
	4.3.1	Overview		171
	4.3.2	Flexural vibrations		172
	4.3.3	Torsional vibrations		172
	4.3.4	Coupling		172
	4.3.5	Formulae for ship vibration		173
	4.3.6	Direct calculation of vibration		173
	4.3.7	Approximate formulae		174
	4.3.8	Amplitudes of vibration		175
	4.3.9	Checking vibration levels		175
	4.3.10	Reducing vibration		175
	4.3.11	Propeller-induced forces		175
	4.3.12	Vibration testing of equipment		178
References				178

5 Powering — 181
5.1	Resistance and propulsion			183
	5.1.1	Froude's analysis procedure		183
	5.1.2	Components of calm water resistance		184
		5.1.2.1	Wave making resistance R_W	184
		5.1.2.2	The contribution of the bulbous bow	188
		5.1.2.3	Transom immersion resistance	190
		5.1.2.4	Viscous form resistance	190
		5.1.2.5	Naked hull skin friction resistance	191

viii Contents

5.1.2.6 Appendage skin friction	192
5.1.2.7 Viscous resistance	192
5.1.3 Methods of resistance evaluation	195
5.1.3.1 Traditional and standard series analysis methods	195
5.1.3.2 Regression-based methods	197
5.1.3.3 Direct model tests	200
5.1.3.4 Computational fluid dynamics	206
5.1.4 Propulsive coefficients	210
5.1.4.1 Relative rotative efficiency	211
5.1.4.2 Thrust deduction factor	211
5.1.4.3 Hull efficiency	212
5.1.4.4 Quasi-propulsive coefficient	213
5.1.5 Influence of rough water	213
5.1.6 Restricted water effects	214
5.1.7 High-speed hull form resistance	215
5.1.7.1 Standard series data	216
5.1.7.2 Model test data	216
5.1.7.3 Summary of problems for fast and unconventional ships	217
5.1.8 Air resistance	220
5.2 Wake	220
5.2.1 General wake field characteristics	220
5.2.2 Wake field definition	222
5.2.3 The nominal wake field	223
5.2.4 Estimation of wake field parameters	225
5.2.5 Effective wake field	228
5.2.6 Wake field scaling	230
5.3 Propeller performance characteristics	233
5.3.1 General open water characteristics	233
5.3.2 Effect of cavitation on open water characteristics	239
5.3.3 Propeller scale effects	239
5.3.4 Specific propeller open water characteristics	243
5.3.4.1 Fixed pitch propellers	243
5.3.4.2 Controllable pitch propellers	243
5.3.4.3 Ducted propellers	245
5.3.4.4 High-speed propellers	246
5.3.5 Standard series data	246
5.4 Propeller theories	247
5.4.1 Early theories	247
5.4.2 Lifting surface models	248
5.4.3 Lifting–line–lifting-surface hybrid models	249
5.4.4 Vortex lattice methods	250
5.4.5 Boundary element methods	254
5.4.6 Methods for specialist propulsors	256
5.4.7 Computational fluid dynamics methods	258
5.5 Cavitation	259
5.5.1 The basic physics of cavitation	260
5.5.2 Types of cavitation experienced by propellers	265
5.5.3 Cavitation considerations in design	272
5.6 Propeller design	282
5.6.1 The design and analysis loop	282
5.6.2 Design constraints	283
5.6.3 Choice of propeller type	284
5.6.4 The propeller design basis	287
5.6.5 Use of standard series data in design	292
5.6.5.1 Determination of diameter	292
5.6.5.2 Determination of mean pitch ratio	293
5.6.5.3 Determination of open water efficiency	293
5.6.5.4 Required propeller rpm to give required P_D or P_E	293
5.6.5.5 Determination of propeller thrust at given conditions	295
5.6.5.6 Effects of cavitation	295
5.6.6 Design considerations	295
5.6.6.1 Direction of rotation	295
5.6.6.2 Blade number	297
5.6.6.3 Diameter, pitch–diameter ratio and rotational speed	298
5.6.6.4 Blade area ratio	298
5.6.6.5 Section form	299
5.6.6.6 Cavitation	299
5.6.6.7 Skew	299
5.6.6.8 Hub form	299
5.6.6.9 Shaft inclination	300
5.6.6.10 Duct form	300
5.6.6.11 The balance between propulsion efficiency and cavitation effects	300
5.6.6.12 Propeller tip considerations	301
5.6.6.13 Propellers operating in partial hull tunnels	302
5.6.6.14 Composite propeller blades	302
5.6.6.15 The propeller basic design process	303
5.6.7 The design process	303
5.7 Service performance and analysis	309
5.7.1 Effects of weather	309
5.7.2 Hull roughness and fouling	309
5.7.3 Hull drag reduction	317
5.7.4 Propeller roughness and fouling	317
5.7.5 Generalized equations for the roughness-induced power penalties in ship operation	321
5.7.6 Monitoring ship performance	324
References	334

6 Marine engines and auxiliary machinery   344

6.1 Introduction	346
6.2 Propulsion systems	346
6.2.1 Fixed pitch propellers	346

6.2.2	Ducted propellers	348
6.2.3	Podded and azimuthing propulsors	350
6.2.4	Contra-rotating propellers	351
6.2.5	Overlapping propellers	352
6.2.6	Tandem propellers	353
6.2.7	Controllable pitch propellers	353
6.2.8	Waterjet propulsion	356
6.2.9	Cycloidal propellers	357
6.2.10	Paddle wheels	357
6.2.11	Magnetohydrodynamic propulsion	359
6.2.12	Superconducting motors for marine propulsion	362

6.3 Diesel engine performance 362
  6.3.1 Rating 362
  6.3.2 Maximum rating 362
  6.3.3 Exhaust temperatures 364
  6.3.4 Derating 365
  6.3.5 Mean effective pressures 365
  6.3.6 Propeller slip 365
  6.3.7 Propeller law 366
  6.3.8 Fuel coefficient 366
  6.3.9 Admiralty coefficient 366
  6.3.10 Apparent propeller slip 367
  6.3.11 Propeller performance 367
  6.3.12 Power build-up 367
  6.3.13 Trailing and locking of propeller 368
  6.3.14 Astern running 368

6.4 Engine and plant selection 370
  6.4.1 Introduction 370
  6.4.2 Diesel–mechanical drives 373
    6.4.2.1 Overview 373
    6.4.2.2 Auxiliary power generation 373
    6.4.2.3 Geared drives 374
    6.4.2.4 Father-and-son layouts 374
  6.4.3 Diesel–electric drive 375
    6.4.3.1 Overview 375
    6.4.3.2 Flexibility of layout 377
    6.4.3.3 Load diversity 378
    6.4.3.4 Economical part load running 378
    6.4.3.5 Ease of control 379
    6.4.3.6 Low noise 379
    6.4.3.7 Environmental protection and ship safety 379
    6.4.3.8 Podded propulsors 379
    6.4.3.9 Combined systems 380

6.5 Propulsion engines 381
  6.5.1 Diesel engines 381
    6.5.1.1 Low speed engines 381
    6.5.1.2 Medium speed engines 390
    6.5.1.3 High speed engines 399
  6.5.2 Gas turbines 403
    6.5.2.1 Overview 403
    6.5.2.2 Plant configurations 404
    6.5.2.3 Cycles and efficiency 406
    6.5.2.4 Emissions 409
    6.5.2.5 Lubrication 410
    6.5.2.6 Air filtration 411
    6.5.2.7 Marine gas turbine designs 412
  6.5.3 Steam turbines 414
    6.5.3.1 Introduction 414
    6.5.3.2 Turbine types 414
    6.5.3.3 Astern arrangements 416
    6.5.3.4 Turbine construction 416

6.6 Auxiliary machinery and equipment 417
  6.6.1 Ship service systems 418
    6.6.1.1 Bilge systems 418
    6.6.1.2 Oil/water separators 420
    6.6.1.3 Ballast arrangements 425
    6.6.1.4 Domestic water systems 426
    6.6.1.5 Sewage systems 436
    6.6.1.6 Incinerators 439
  6.6.2 Shafting and propellers 439
    6.6.2.1 Overview 439
    6.6.2.2 Thrust block 440
    6.6.2.3 Shaft bearings 443
    6.6.2.4 Sterntube bearing 443
    6.6.2.5 Sterntube seals 444
    6.6.2.6 Shafting 444
    6.6.2.7 Propeller 444
    6.6.2.8 Propeller mounting 445
    6.6.2.9 Controllable-pitch propeller 445
    6.6.2.10 Cavitation 445
    6.6.2.11 Propeller maintenance 445
  6.6.3 Steering gear 446
    6.6.3.1 Overview 446
    6.6.3.2 Variable delivery pumps 448
    6.6.3.3 Telemotor control 448
    6.6.3.4 Electrical control 450
    6.6.3.5 Power units 452
    6.6.3.6 All-electric steering 458
    6.6.3.7 Twin-system steering gears 458
    6.6.3.8 Steering gear testing 461

6.7 Instrumentation and control 461
  6.7.1 Instrumentation 461
  6.7.2 Control 462
    6.7.2.1 Control theory 462
    6.7.2.2 Transmitters 463
    6.7.2.3 Controller action 465
    6.7.2.4 Controllers 467
    6.7.2.5 Correcting unit 469
    6.7.2.6 Control systems 470
    6.7.2.7 Centralized control 475
    6.7.2.8 Unattended machinery spaces 475
    6.7.2.9 Bridge control 475
    6.7.2.10 Integrated control 480

References 481

7 Seakeeping — 483

- 7.1 Seakeeping qualities — 485
  - 7.1.1 Motions — 485
  - 7.1.2 Speed and power in waves — 485
  - 7.1.3 Wetness — 485
  - 7.1.4 Slamming — 485
  - 7.1.5 Ship routing — 485
  - 7.1.6 Importance of good seakeeping — 485
- 7.2 Ship motions — 486
  - 7.2.1 Degrees of freedom — 486
  - 7.2.2 Undamped motion in still water — 486
    - 7.2.2.1 Rolling — 487
    - 7.2.2.2 Heaving — 487
  - 7.2.3 Damped motion in still water — 488
  - 7.2.4 Approximate period of roll — 488
  - 7.2.5 Motion in regular waves — 489
    - 7.2.5.1 Assumptions — 489
    - 7.2.5.2 Rolling in a beam sea — 489
    - 7.2.5.3 Pitching and heaving — 490
  - 7.2.6 Presentation of motion data — 490
  - 7.2.7 Motion in irregular seas — 492
  - 7.2.8 Motion in oblique seas — 496
  - 7.2.9 Surge, sway and yaw — 496
    - 7.2.9.1 Surge — 497
    - 7.2.9.2 Sway — 497
    - 7.2.9.3 Yaw — 497
  - 7.2.10 Large amplitude rolling — 497
  - 7.2.11 Roll excitation and influence of speed and heading — 499
    - 7.2.11.1 Motion directions of rigid body — 499
    - 7.2.11.2 Mass moment of inertia — 501
    - 7.2.11.3 Linear restoring moment — 501
    - 7.2.11.4 Natural roll period — 502
    - 7.2.11.5 Roll damping — 503
    - 7.2.11.6 GM–T_0 relationship and rolling period test — 504
    - 7.2.11.7 Modes of roll excitation in a seaway — 505
    - 7.2.11.8 Ship roll in beam seas — 506
    - 7.2.11.9 Roll in beam seas at large amplitudes — 508
    - 7.2.11.10 GZ-Variation in longitudinal waves — 509
    - 7.2.11.11 Encounter period of ship and waves — 513
    - 7.2.11.12 Encounter frequency — 515
    - 7.2.11.13 Wave group of two regular waves — 515
    - 7.2.11.14 Wave encounter of a ship in irregular seas — 521
    - 7.2.11.15 Wave energy and encounter spectra — 523
    - 7.2.11.16 Relevant frequencies of the spectrum and encounter — 523
    - 7.2.11.17 Bandwidth of the transformed sea spectrum — 526
    - 7.2.11.18 Irregular time series of wave encounter — 527
- 7.3 Limiting seakeeping criteria — 529
  - 7.3.1 Limiting critera — 529
    - 7.3.1.1 Speed and power in waves — 530
    - 7.3.1.2 Slamming — 531
    - 7.3.1.3 Wetness — 537
    - 7.3.1.4 Propeller emergence — 537
    - 7.3.1.5 Degradation of human performance — 537
- 7.4 Overall seakeeping performance — 538
- 7.5 Data for seakeeping assessments — 541
  - 7.5.1 Selection of wave data — 541
  - 7.5.2 Obtaining response amplitude operators — 543
    - 7.5.2.1 Theory — 543
    - 7.5.2.2 Model experiments — 543
    - 7.5.2.3 Ship trials — 544
- 7.6 Non-linear effects — 544
- 7.7 Numerical prediction of seakeeping — 545
  - 7.7.1 Overview of computational methods — 545
  - 7.7.2 Strip theory — 547
  - 7.7.3 Rankine singularity methods — 551
  - 7.7.4 Problems for fast and unconventional ships — 552
  - 7.7.5 Further quantities in regular waves — 554
  - 7.7.6 Ship responses in stationary seaway — 555
  - 7.7.7 Simulation methods — 556
  - 7.7.8 Long-term distributions — 557
- 7.8 Experiments and trials — 558
  - 7.8.1 Test facilities — 558
  - 7.8.2 Ship seakeeping trials — 558
  - 7.8.3 Stabilizer trials — 560
- 7.9 Improving seakeeping performance — 560
  - 7.9.1 Design and operational changes — 560
  - 7.9.2 Influence of form on seakeeping — 561
  - 7.9.3 Summary — 561
- 7.10 Ship motion control — 562
  - 7.10.1 Background — 562
  - 7.10.2 Roll stabilization — 562
    - 7.10.2.1 Stabilization systems — 562
    - 7.10.2.2 Comparison of principal systems — 564
    - 7.10.2.3 Performance of stabilizing systems — 564
    - 7.10.2.4 Fin stabilizers: Design procedure — 567

	7.10.3	Pitch damping	573	
		7.10.3.1 Pitch damping fins	573	
		7.10.3.2 Transom flaps	574	
		7.10.3.3 Interceptors	574	
References			575	

8 Manoeuvring — 578

- 8.1 General concepts — 580
- 8.2 Directional stability — 580
- 8.3 Stability and control of surface ships — 581
- 8.4 Rudder action — 583
- 8.5 Limitations of theory — 584
- 8.6 Assessment of manoeuvrability — 584
  - 8.6.1 Turning circle — 584
    - 8.6.1.1 Drift angle — 585
    - 8.6.1.2 Advance — 585
    - 8.6.1.3 Transfer — 585
    - 8.6.1.4 Tactical diameter — 585
    - 8.6.1.5 Diameter of steady turning circle — 585
    - 8.6.1.6 Pivoting point — 585
- 8.7 Loss of speed on turn — 586
- 8.8 Heel when turning — 586
- 8.9 Turning ability — 587
  - 8.9.1 Zig-zag manoeuvre — 587
  - 8.9.2 Spiral manoeuvre — 588
  - 8.9.3 Pull-out manoeuvre — 588
- 8.10 Standards for manoeuvring and directional stability — 589
- 8.11 Dynamic positioning — 590
- 8.12 Automatic control systems — 590
- 8.13 Ship interaction — 591
  - 8.13.1 Interaction — 591
  - 8.13.2 Ship to ground (squat) interaction — 592
  - 8.13.3 Ship to ship interaction — 593
  - 8.13.4 Ship to shore interaction — 598
  - 8.13.5 Summary — 598
- 8.14 Shallow water/bank effects — 598
- 8.15 Broaching — 599
- 8.16 Experimental approaches — 599
  - 8.16.1 Manoeuvring tests in sea trials — 599
  - 8.16.2 Model tests — 599
- 8.17 CFD for ship manoeuvring — 600
- 8.18 Stability and control of submarines — 603
  - 8.18.1 Control requirements and equations — 603
  - 8.18.2 Experiments and trials — 605
  - 8.18.3 Design assessment — 606
- 8.19 Rudders and control surfaces — 606
  - 8.19.1 Control surfaces and applications — 606
    - 8.19.1.1 Rudder types — 607
    - 8.19.1.2 Hydroplanes — 609
    - 8.19.1.3 Efficiency of control surfaces — 609
  - 8.19.2 Presentation of rudder data — 609
  - 8.19.3 Rudder design within the ship design process — 611
  - 8.19.4 Detailed rudder design — 612
    - 8.19.4.1 Background — 612
    - 8.19.4.2 Rudder design process — 615
  - 8.19.5 Rudder manoeuvring forces — 621
    - 8.19.5.1 Rudder forces — 621
    - 8.19.5.2 Hull upstream — 621
    - 8.19.5.3 Influence of drift angle — 621
    - 8.19.5.4 Low and zero speed and four quadrants — 622
  - 8.19.6 Numerical modelling of rudder — 627
    - 8.19.6.1 Available methods — 627
    - 8.19.6.2 Potential flow methods — 627
    - 8.19.6.3 Navier–Stokes methods — 628
    - 8.19.6.4 Rudder–propeller interaction — 629
    - 8.19.6.5 Unsteady behaviour — 630
  - 8.19.7 Guidelines for rudder design — 630
- References — 631

9 Ship design, construction and operation — 636

- 9.1 Introduction — 638
- 9.2 Ship design — 638
  - 9.2.1 Overview — 638
    - 9.2.1.1 General — 638
    - 9.2.1.2 Ship design process — 639
  - 9.2.2 Technical ship design — 639
    - 9.2.2.1 Principal requirements — 639
    - 9.2.2.2 Specification — 640
  - 9.2.3 Deadweight determined designs — 641
    - 9.2.3.1 Deadweight and dimensions — 641
    - 9.2.3.2 Cargo capacity check — 642
    - 9.2.3.3 Summary of overall model: Deadweight approach — 643
  - 9.2.4 Capacity (or space) determined designs — 643
    - 9.2.4.1 Cargo ships — 643
    - 9.2.4.2 Passenger ships — 643
    - 9.2.4.3 Container ships — 645
    - 9.2.4.4 High speed passenger/vehicle ferries — 646
  - 9.2.5 Stability check — 647
  - 9.2.6 Lightship mass estimates — 649
    - 9.2.6.1 Steel mass — 649
    - 9.2.6.2 Outfit mass — 651
    - 9.2.6.3 Machinery mass — 652
    - 9.2.6.4 Margin — 652

xii   Contents

- 9.2.6.5 Masses of fast ferries 652
- 9.2.6.6 Vertical centre of gravity (KG) 653
- 9.2.7 Design of ship lines 653
  - 9.2.7.1 Sectional area curve (SAC) – definitions: 653
  - 9.2.7.2 Modifications to sectional area curve 653
  - 9.2.7.3 Sectional area curve transformations 654
  - 9.2.7.4 Preparation of body plan 655
- 9.2.8 Statutory regulations 657
- 9.2.9 Concept design content: example 657
- 9.3 Materials 657
  - 9.3.1 Introduction 657
  - 9.3.2 Steel 659
    - 9.3.2.1 Manufacture of steel 659
    - 9.3.2.2 Heat treatment of steels 660
    - 9.3.2.3 Steel sections 660
    - 9.3.2.4 Shipbuilding steels 660
    - 9.3.2.5 High tensile steels 661
    - 9.3.2.6 Corrosion resistant steels 661
    - 9.3.2.7 Steel sandwich panels 662
    - 9.3.2.8 Steel castings 662
    - 9.3.2.9 Steel forgings 662
  - 9.3.3 Aluminium alloy 662
    - 9.3.3.1 General 662
    - 9.3.3.2 Production of aluminium 663
    - 9.3.3.3 Aluminium alloy sandwich panels 664
    - 9.3.3.4 Fire protection 665
  - 9.3.4 Composite materials 665
    - 9.3.4.1 Overview 665
    - 9.3.4.2 Introduction 665
    - 9.3.4.3 Materials selection 665
    - 9.3.4.4 Design concepts 671
    - 9.3.4.5 Design synthesis 675
    - 9.3.4.6 External issues 682
  - 9.3.5 Corrosion 683
    - 9.3.5.1 Nature and forms of corrosion 683
    - 9.3.5.2 Corrosion control 686
    - 9.3.5.3 Anti-fouling systems 688
    - 9.3.5.4 Painting ships 689
- 9.4 Ship construction 691
  - 9.4.1 Introduction 691
  - 9.4.2 Typical examples of structure 691
  - 9.4.3 Shipyard layout 691
  - 9.4.4 Ship drawing office, Loftwork and CAD/CAM 692
    - 9.4.4.1 Ship drawing office 692
    - 9.4.4.2 Loftwork following drawing office 696
    - 9.4.4.3 Computer Aided Design (CAD)/Computer Aided Manufacturing (CAM) 698
- 9.5 Ship economics 703
  - 9.5.1 Shipowners and operators 703
    - 9.5.1.1 Types of trade 703
    - 9.5.1.2 Methods of employment 704
  - 9.5.2 Economic criteria 705
    - 9.5.2.1 The basis of these criteria 705
    - 9.5.2.2 Interest 705
    - 9.5.2.3 Present worth 705
    - 9.5.2.4 Repayment of principal 705
    - 9.5.2.5 Sinking fund factor 706
    - 9.5.2.6 Net present value 706
    - 9.5.2.7 Required freight rate 706
    - 9.5.2.8 Yield 706
    - 9.5.2.9 Inflation and exchange rates 706
  - 9.5.3 Operating costs 706
    - 9.5.3.1 Capital charges 706
    - 9.5.3.2 Capital amortization 706
    - 9.5.3.3 Profit and taxes 707
    - 9.5.3.4 Depreciation 707
    - 9.5.3.5 Ship values 707
  - 9.5.4 Daily running costs 707
    - 9.5.4.1 Crew costs 707
    - 9.5.4.2 Provisions and stores 707
    - 9.5.4.3 Maintenance and repair 708
    - 9.5.4.4 Insurance 708
    - 9.5.4.5 Administration and general charges 708
  - 9.5.5 Voyage costs 708
    - 9.5.5.1 Bunkers 708
    - 9.5.5.2 Port and canal dues, pilotage, towage etc. 709
  - 9.5.6 Cargo handling costs 709
- 9.6 Optimization in design and operation 709
  - 9.6.1 Overview 709
  - 9.6.2 Introduction to methodology of optimization 709
  - 9.6.3 Scope of application in ship design 712
  - 9.6.4 Economic basics for optimization 713
    - 9.6.4.1 Discounting 713
    - 9.6.4.2 Initial costs (building costs) 714
    - 9.6.4.3 Annual income and expenditure 715
    - 9.6.4.4 The 'cost-difference' method 716
    - 9.6.4.5 Discontinuities in propulsion unit costs 717
  - 9.6.5 Discussion of some important parameters 717
    - 9.6.5.1 Width 717

9.6.5.2 Length	717
9.6.5.3 Block coefficient	718
9.6.5.4 Speed	719
9.6.6 Special cases of optimization	720
9.6.6.1 Optimization of repeat ships	720
9.6.6.2 Optimizing the dimensions of containerships	721
9.6.7 Developments of the 1980s and 1990s	722
9.6.7.1 Concept exploration models	722
9.6.7.2 Optimization shells	722
References	724

10 Underwater vehicles — 728

- 10.1 Introduction — 730
- 10.2 A bit of history — 730
  - 10.2.1 Introduction — 730
  - 10.2.2 What is an ROV? — 730
  - 10.2.3 In the beginning — 731
  - 10.2.4 Today's observation-class vehicles — 735
- 10.3 ROV design — 735
  - 10.3.1 Underwater vehicles to ROVs — 736
    - 10.3.1.1 Power source for the vehicle — 737
    - 10.3.1.2 Degree of autonomy — 737
    - 10.3.1.3 Communications linkage to the vehicle — 737
    - 10.3.1.4 Special-use ROVs — 738
  - 10.3.2 Autonomy plus: 'Why the tether?' — 738
    - 10.3.2.1 An aircraft analogy — 738
    - 10.3.2.2 Underwater vehicle variations — 739
    - 10.3.2.3 Why the tether? — 739
    - 10.3.2.4 Tele-operation versus remote control — 739
    - 10.3.2.5 Degrees of autonomy — 739
  - 10.3.3 The ROV — 740
    - 10.3.3.1 What is the perfect ROV? — 740
    - 10.3.3.2 ROV classifications — 741
    - 10.3.3.3 Size considerations — 741
    - 10.3.3.4 Buoyancy and stability — 741
    - 10.3.3.5 Dynamic stability — 745
    - 10.3.3.6 Vehicle control — 755
    - 10.3.3.7 Deployment techniques — 757
- 10.4 ROV components — 758
  - 10.4.1 Introduction — 758
  - 10.4.2 Mechanical and electro/mechanical systems — 759
    - 10.4.2.1 Frame — 759
    - 10.4.2.2 Buoyancy — 759
    - 10.4.2.3 Propulsion and thrust — 759
  - 10.4.3 Primary subsystems — 766
    - 10.4.3.1 Lighting — 766
    - 10.4.3.2 Cameras — 768
    - 10.4.3.3 Sensors — 768
    - 10.4.3.4 Manipulator and tool pack — 770
  - 10.4.4 Electrical considerations — 770
    - 10.4.4.1 The tether — 770
    - 10.4.4.2 Power source — 773
    - 10.4.4.3 AC versus DC considerations — 773
    - 10.4.4.4 Data throughput — 773
    - 10.4.4.5 Data transmission and protocol — 774
    - 10.4.4.6 Underwater connectors — 774
  - 10.4.5 Control systems — 775
    - 10.4.5.1 The control station — 777
    - 10.4.5.2 Motor control electronics — 778
- References — 782

11 Marine safety — 784

- 11.1 Background — 786
  - 11.1.1 International trade and shipping — 786
  - 11.1.2 The actors in shipping — 786
  - 11.1.3 The shipowner — 786
  - 11.1.4 Safety and economy — 788
  - 11.1.5 Maritime safety regime — 789
  - 11.1.6 Why safety improvement is difficult — 791
  - 11.1.7 The risk concept — 791
  - 11.1.8 Acceptable risk — 792
  - 11.1.9 Conflict of interest — 792
  - 11.1.10 Expertise and rationality — 793
- 11.2 Regulatory authorities — 794
  - 11.2.1 Introduction — 794
    - 11.2.1.1 The structure of control — 794
  - 11.2.2 International maritime organization (IMO) — 795
    - 11.2.2.1 SOLAS — 795
    - 11.2.2.2 International Convention on Load Lines, 1966 — 796
    - 11.2.2.3 STCW convention — 796
    - 11.2.2.4 MARPOL — 797
    - 11.2.2.5 The ISM Code — 797

xiv   Contents

- 11.2.3 Flag State Control 798
  - 11.2.3.1 The Seaworthiness Act 799
  - 11.2.3.2 Delegation of Flag State Control 799
  - 11.2.3.3 Effectiveness of Flag State Control 799
  - 11.2.3.4 The Flag State Audit Project 799
- 11.2.4 Port state control 800
  - 11.2.4.1 UNCLOS 800
  - 11.2.4.2 MOU PSC 802
- 11.3 Classification societies 802
  - 11.3.1 Background 802
  - 11.3.2 Rules and regulations 805
  - 11.3.3 Lloyds Register 805
  - 11.3.4 Lloyds Register classification symbols 805
  - 11.3.5 Classification of ships operating in ice 806
  - 11.3.6 Structural design programs 806
  - 11.3.7 Periodical Surveys 807
  - 11.3.8 Hull Planned Maintenance Scheme 808
  - 11.3.9 Damage repairs 808
- 11.4 Safety of marine systems 808
  - 11.4.1 Introduction 808
    - 11.4.1.1 Background 808
    - 11.4.1.2 Safety and reliability development in the maritime industry 808
    - 11.4.1.3 Present status 809
    - 11.4.1.4 Databases 809
  - 11.4.2 Ship safety and accident statistics 810
    - 11.4.2.1 Background 810
    - 11.4.2.2 Code of practice for the safety of small fishing vessels 811
    - 11.4.2.3 The Fishing Vessels (Safety Provisions) Safety Rules 1975 812
    - 11.4.2.4 Accident data for fishing vessels 812
    - 11.4.2.5 Data analysis 816
    - 11.4.2.6 Containership accident statistics 818
    - 11.4.2.7 Conclusion 821
  - 11.4.3 Safety analysis techniques 822
    - 11.4.3.1 Background 822
    - 11.4.3.2 Qualitative safety analysis 822
    - 11.4.3.3 Quantitative safety analysis 823
    - 11.4.3.4 Cause and effect relationship 825
    - 11.4.3.5 Preliminary hazard analysis (PHA) 826
    - 11.4.3.6 What-if analysis 827
    - 11.4.3.7 HAZard and OPerability (HAZOP) studies 827
    - 11.4.3.8 Fault tree analysis (FTA) 829
    - 11.4.3.9 Event tree analysis 834
    - 11.4.3.10 Markov chains 835
    - 11.4.3.11 Failure mode, effects and critical analysis (FMECA) 836
    - 11.4.3.12 Other analysis methods 838
    - 11.4.3.13 Conclusion 838
  - 11.4.4 Formal safety assessment of ships and its relation to offshore safety case approach 839
    - 11.4.4.1 Offshore safety assessment 839
    - 11.4.4.2 Formal ship safety assessment 845
    - 11.4.4.3 Risk criteria 847
    - 11.4.4.4 Discussion and conclusion 848
  - 11.4.5 Formal safety assessment (FSA) 849
    - 11.4.5.1 Formal safety assessment 849
- 11.5 Safety management of ship stability 853
  - 11.5.1 Introduction 853
    - 11.5.1.1 Need to introduce a ship stability management system 853
    - 11.5.1.2 Tools of efficient stability management 853
    - 11.5.1.3 The master's range of judgement for operational stability assessment 854
    - 11.5.1.4 Seakeeping guidance and survivability criteria 855
  - 11.5.2 Guidelines on in-service ship stability 859
    - 11.5.2.1 Purpose of guidelines for operational stability 859
    - 11.5.2.2 Loading and stability manual 859
    - 11.5.2.3 Guidelines on the management of ship stability 860

11.5.2.4	Guidance to the master for avoiding dangerous situations in following and quartering seas	861
11.5.2.5	International safety management code (ISM)	866
11.5.3	The human factor. Maritime education and training	867
11.5.4	Operational stability in the future – A wishful forecast	868
References		869

12 Glossary of terms and definitions — 876
12.1 Abbreviations — 878
12.2 Symbols — 879
12.3 Terms and definitions — 881

Author Biographies — 887

Index — 889

Preface

Maritime engineering covers a wide range of scientific, technical and engineering topics, with supporting input on economic, legal and insurance matters. The **Maritime Engineering Reference Book** is designed to serve as a first point of reference for those wishing to know the basics and background information in the field of maritime engineering and to provide accessible references for those who wish to read about particular subjects in more detail.

This book is aimed at a broad readership including practicing naval architects, marine engineers and seagoing officers, together with scientists and engineers not specialized in the maritime field. It should be of use to students of naval architecture, ship science and marine engineering, and other science and engineering students with interests in marine matters. It should also appeal to others involved with maritime operations including financial and legal work, surveying, insurance and marine policy.

The contents have been chosen to provide an overview of maritime engineering, covering the physical features of the marine environment, marine vehicle types, ship and marine vehicle design, construction, operation and safety.

The book has been compiled using extracts from the following twenty books within the range of maritime books in the Elsevier Butterworth-Heinemann collection:

Barrass, C.B. and Derrett, D.R. (2006) *Ship Stability for Masters and Mates*.
Barrass, C.B. (2004) *Ship Design and Performance for Masters and Mates*.
Bertram, V. (1998) *Practical Ship Hydrodynamics*.
Biran, A.B. (2003) *Ship Hydrostatics and Stability*.
Carlton, J.S. (2007) *Marine Propellers and Propulsion*, 2nd Edition.
Christ, R.D. and Wernli, R.L. (2007) *The ROV Manual*.
Eyres, D.J. (2007) *Ship Construction*, 6th Edition.
Jensen, J.J. (2001) *Load and Global Response of Ships*.
Kobylinski, L.K. and Kastner, S. (2003) *Stability and Safety of Ships*.
Kristiansen, S. (2004) *Maritime Transportation: Safety Management and Risk Analysis*.
McGeorge, H.D. (1999) *Marine Auxiliary Machinery*, 7th Edition.
Molland, A.F. and Turnock, S.R. (2007) *Marine Rudders and Control Surfaces*.
Pillay, A. and Wang, J. (2003) *Technology and Safety of Marine Systems*.
Rawson, K.J. and Tupper, E.C. (2001) *Basic Ship Theory*, 5th Edition.
Schneekluth, H. and Bertram, V. (1998) *Ship Design for Efficiency and Economy*.
Shenoi, R.A. and Dodkins, A.R. (2000) *Design of Ships and Marine Structures Made from FRP Composite Materials*, in Kelly. A. and Zweben, C. (eds), Comprehensive Composite Materials, Vol. 6, Elsevier Science Ltd, Oxford, UK.
Taylor, D.A. (1996) *Introduction to Marine Engineering*.
Tupper, E.C. (2004) *Introduction to Naval Architecture*, 4th Edition.
Watson, D.G.M. (1998) *Practical Ship Design*.
Woodyard, D.F. (2004) *Pounder's Marine Diesel Engines and Gas Turbines*, 8th Edition.

The extracts have been taken directly from the above source books, with some small editorial changes and additions. These changes have entailed the re-numbering of Sections and Figures, the linking of Sections within a Chapter, cross-referencing between Chapters and the insertion of additional and more recent references where appropriate. In view of the breadth of content and style of the source books, there is some overlap and repetition of material between Chapters and significant differences in style, but these features have been left in order to retain the flavour and readability of the individual Chapters.

The book is arranged into 12 Chapters:

1. The marine environment
2. Marine vehicle types
3. Flotation and stability
4. Ship structures
5. Powering
6. Marine engines and auxiliary machinery
7. Seakeeping
8. Manoeuvring
9. Ship design, construction and operation
10. Underwater vehicles
11. Marine safety
12. Glossary of terms and definitions

The Chapters can be studied independently, depending on the interests of the reader. References are provided at the end of each Chapter to facilitate access to some of the original sources of information and further depth of study where necessary.

Grateful appreciation is extended to the twenty-five authors of the source books from which this Reference Book has been compiled.

The Editor acknowledges the help and support of Lyndsey Dixon, Associate Editor, and her team for guidance in establishing the structure of the book and in bringing the book to publication.

Tony Molland
September 2008

1 The marine environment

Contents

1.1 The ship in the marine environment
1.2 Wind
1.3 Variations in level of sea surface
1.4 Regular waves
1.5 The sinusoidal wave
1.6 Irregular waves
1.7 Spectrum formulae by Pierson/Moskowitz and Bretschneider
1.8 The JONSWAP sea spectrum
1.9 Maximum wave height in a stationary random sea
1.10 Long-term statistics of irregular seaway
1.11 Wave data from observations
1.12 Wave climate
1.13 Freak waves
1.14 Oceanography
1.15 Ambient air
1.16 Climatic extremes
1.17 Marine pollution
References (Chapter 1)

The various Sections of this Chapter have been taken from the following books, with the permission of the authors:

Christ, R.D and Wernli, S.R. (2007) *The ROV manual*. Butterworth-Heinemann, Oxford, UK. [Section 1.14]

Kobylinski, L.K. and Kastner, S. (2003) *Stability and Safety of Ships*. Elsevier, Oxford, UK. [Sections 1.1–1.12]

Rawson K.J. and Tupper E.C. (2001) *Basic Ship Theory*. 5th Edition, Combined Volume. Butterworth-Heinemann, Oxford, UK. [Sections 1.15, 1.16]

Tupper, E.C. (2004) *Introduction to Naval Architecture*. 4th Edition. Butterworth-Heinemann, Oxford, UK. [Sections 1.13, 1.17]

1.1 The ship in the marine environment

A ship or any ocean vehicle or structure is exposed to the marine environment. It is a complicated and often hostile environment. Environmental forces at sea come from wind, seaway, current, tidal waves, and waves from earthquakes (tsunamis). From the practical point of view, the seafarer has to cope with wind and seaway. Generally, seaway is generated by the wind at the sea surface.

The occurrence and magnitude of wind and seaway depend on the sea area and on the time of the year. Wind and seaway vary randomly and can be described by statistical methods based on probability theory. In detail, we look at the rate of occurrence, the magnitude, and the time variations of wind and seaway.

It is convenient to make a distinction between long-term (in terms of days up to years) and short-term time (in terms of hours) variations of the seaway. While the long-term approach allows for the rate of occurrence and the severity of the seaway, the short-term time variations are important for the dynamic ship response in a particular seaway of constant energy. Seaway is represented by gravity waves of the water at the sea surface. The exciting wave forces vary in time. The ship responds to the oscillating external forces as a dynamic system.

Wind and wave data have been assembled by observation, by measurement, and by mathematical description. Goals of the near future are, for example, to apply the non-linear pattern of extreme irregular seas in ship operation, and to have sea on-line data on the bridge. The literature on the sea environment is abundant. This chapter gives a general insight into the physical features of the marine environment.

1.2 Wind

By tradition, the magnitude of the wind is defined by the Beaufort Scale (Admiral Beaufort, England, 1806). The Beaufort wind scale is based on observation of the sea, by way of a rough grouping from 1 to 12 Bft. The observed wave pattern in deep sea is related to the generating wind force. Storm at Bft. 11 is described as 'Waves are so high that ships within sight are hidden in the troughs; visibility poor'. Beaufort 12 means a hurricane, with the deep sea criterion describing the sea status as follows, de Beurs (1957): 'The sea is white with streaky foam as covered by a dense white curtain; air filled with spray; visibility very poor'.

The Beaufort numbers also correspond to a rating of wind according to ascending wind velocity. Each Bft. number relates to a range of wind velocities. Any wind above 32.5 m/s (63.2 kn.) is Bft. 12. The Beaufort scale is given in Table 1.1 where wind velocity is given at a height of $z_1 = 6$ m. The scale is also depicted in Figure 1.1.

The upper and lower limits of the Beaufort wind regions are approximated by a quadratic polynomial function, with n as Bft. number, v_{w1} upper limit and v_{w2} lower limit of the velocity range:

$$v_{w1} = 0.1424 \cdot n^2 + 1.4127 \cdot n - 0.0434 \quad (1.1)$$

$$v_{w2} = 0.1569 \cdot n^2 + 0.9112 \cdot n - 0.6352 \quad (1.2)$$

The Bft. wind velocities are average values of the horizontal wind at sea. A detailed analysis of the wind profile above the sea surface shows an increase of wind velocity with respect to the vertical distance from the sea surface, see Figure 1.2. The wind velocity at $z_0 = 10$ m above sea level has been used as a reference or characteristic wind speed, van Koten (1976).

Only for detailed analysis and calculation of wind forces the vertical wind distribution must be taken into account. The wind profile is approximated by

$$V_z = V_0 \cdot (z/z_0)^\alpha \quad (1.3)$$

The exponent α is 0.12 for wind at sea surface.

In order to look at the time variation of the wind speed, we can plot the mean energy versus the average occurrence cycle in time. Figure 1.3 gives an example of the so-called spectrum of the ocean wind velocity, data taken from van Koten (1976), see also Price and Bishop (1974). We see four energy peaks, which define four distinctly different ranges of wind energy with respect to their time variation:

(1) In the first peak, the repetition cycle of the wind is only a few minutes and less than a minute (about 0.5 to 3 minutes). This shortest time variation of the wind is of interest for the wind action on the ship and her dynamic response. A wind with its rapid time varation taken into account is called a "gust". In gusts, the maximum wind speed can

Table 1.1 The Beaufort scale.

Wind force $^\circ B$	Description	Limits of speed	
		knots	m/s
0	Calm	1	0.3
1	Light air	1–3	0.3–1.5
2	Light breeze	4–6	1.6–3.3
3	Gentle breeze	7–10	3.4–5.4
4	Moderate breeze	11–16	5.5–7.9
5	Fresh breeze	17–21	8.0–10.7
6	Strong breeze	22–27	10.8–13.8
7	Near gale	28–33	13.9–17.1
8	Gale	34–40	17.2–20.7
9	Strong gale	41–47	20.8–24.4
10	Storm	48–55	24.5–28.4
11	Violent storm	56–63	28.5–32.6
12	Hurricane	64 and over	32.7 and over

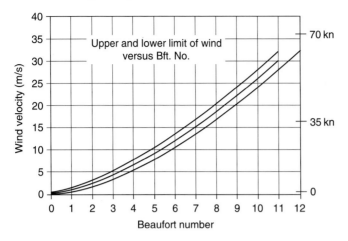

Figure 1.1 Range of wind velocity at Beaufort scale.

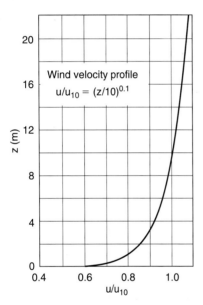

Figure 1.2 Wind velocity profile.

be about 50% more than the mean wind speed. Davenport (1964) developed a gust spectrum.

The sudden occurrence of gusts has caused ship losses. Sailing ships are in particular endangered from extreme wind and gust. The German sailing vessel NIOBE capsized in the Baltic Sea caused by a thunderstorm gust in 1928, see Horn et al. (1953). In 1956, the German training vessel PAMIR, under sails, capsized in a hurricane in the Atlantic. Extreme wind, seaway, and shifting of grain cargo caused the loss, see Wendel and Platzoeder (1958).

Peaks 2, 3 and 4 show, whether the ship meets severe wind in the time of the year (4), and how long these prevailing wind conditions will last (2 and 3), see Figure 1.4. However, these long periods of excitation have no effect on the dynamic ship response:

(2) Peak 2 shows the daily cycles.
(1/2) The mean velocity taken as the average constant wind velocity is small between peak 1 and 2, with periods of 10 to 60 minutes. This again is not critical for a dynamic excitation of ship motion by the wind.
(3) The repetition cycle of the wind is from one day up to one week. This accounts for the well-known time length of storm conditions.
(4) The repetition cycle of the wind is one year, showing the dependence of the wind on the time of the year.

Extensive research has been put into the problem of seaway generation. Hasselmann *et al.* (1973, 1976) showed mathematically the non-linear transfer of energy.

1.3 Variations in level of sea surface

A cyclic rise and fall of the sea level is called a wave. However, we must keep in mind that rise and fall of wind-generated waves are due to the progressive orbital motion of wave particles. Figure 1.4 shows the relative energy of the variations in the level of the sea surface, plotted versus the frequency of the orbital motion. We see pronounced peaks in the energy distribution of the sea level, which correspond to well-known phenomena at sea.

Surge, at small frequencies less than 0.001 Hz (1 Hz = 1 cycle per second): This corresponds to a wave period of >1000 s (>15 min).

Surf beat: peak at about 0.016 Hz. This corresponds to a wave period of 1 min, and wavelength of 6 km.

Swell: peak at about 0.056 Hz, period 18 s, wavelength 500 m.

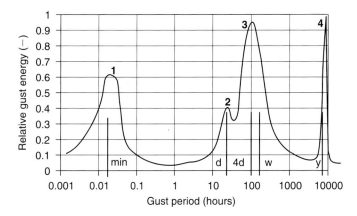

Figure 1.3 Spectrum of ocean wind velocity.

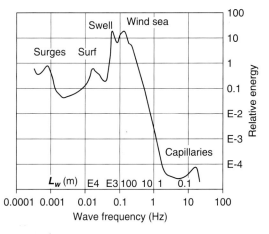

Figure 1.4 Frequency spectrum of sea surface level.

Wind sea: peak at 0.13 Hz, period 7.5 s, wavelength 90 m.

Capillaries: The lowest peak at the large frequency tail of the spectral curve marks the capillary waves.

1.4 Regular waves

1.4.1 The trochoid

The wave contour is given by a trochoid function, based on the kinetic motion of a point inside a wheel at a smaller radius fixed to the rolling motion of the wheel, Figure 1.5. Equations (1.4) and (1.5) give the parametric x and ζ Cartesian co-ordinates of the trochoidal wave.

$$x = \frac{L_w}{2\pi}\left(\hat{\alpha} - \pi \frac{H_w}{L_w} \cdot \sin\alpha\right) \quad (1.4)$$

$$\zeta = -\frac{H_w}{2} \cdot \cos\alpha \quad (1.5)$$

For hydrostatic calculations, the trochoid is applied to estimate the underwater shape of the ship hull in a longitudinal wave, see e.g. the stability regulations of the German Navy, Arndt et al. (1982). However, sine waves can include the orbital motion of the water particles, as discussed in the next Sections.

1.4.2 Higher order waves. Stokes and Airy Theory

The first people to treat surface waves in deep water mathematically were Airy (1845) and Stokes (1849, 1880). In applying potential theory, Stokes developed formulae not only for the exact wave contour, but also for the motion of the water particles in the wave. Parameters in the equations are the wave steepness H_w/L_w and the relative water depth d/L_w. A detailed account of all related formulae can be found in Wiegel (1964). Wehausen and Laitone (1960) gave the fundamental status of potential theory in surface waves. Ever since, many more researchers have contributed to the field of ocean waves.

For a large water depth from d/L_w between 1 and 0.5, the wave contour by Stokes to the third order is:

$$\zeta(x,t) = a_1 \cos(kx - \omega t) + a_2 \cos(2(kx - \omega t)) + a_3 \cos(3(kx - \omega t)) \quad (1.6)$$

Where

$\varepsilon = kx - \omega t$ with $x = 0 \ldots L_w$ and $t = 0 \ldots T_w$

$$a_1 = \frac{1}{2} \cdot H_w$$

$$a_2 = \frac{\pi}{4} \cdot \frac{H_w^2}{L_w}$$

$$a_3 = \frac{3\pi^2}{16} \cdot \frac{H_w^3}{L_w^2}$$

Figure 1.6 shows the profile of a regular Stokes wave by the sum of two sinusoidal components.

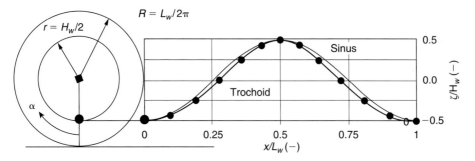

Figure 1.5 Trochoidal wave.

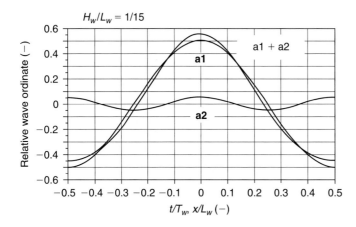

Figure 1.6 Higher order wave contour by Stokes theory.

The second order component increases the wave crest and raises the level in the wave trough. The third component is too small to contribute (here 0.5 mm). The wave crest shows a sharper peak than the sinusoidal wave, and the wave trough is wider and flat. This is closer to reality.

The wave length by StokeTs depending on wave steepness H_w/L_w and relative water depth d/L_w is Wiegel (1964):

$$L_w = L_{w0} \tanh \frac{2\pi d}{L_w} \left\{ 1 + \left(\pi \frac{H_w}{L_w}\right)^2 \frac{14 + 4\cosh(4\pi d/L_w)}{16 \sinh^4 (2\pi d/L_w)} \right\} \quad (1.7)$$

With

$$L_{w0} = \frac{gT_w^2}{2\pi} \quad (1.8)$$

For waves of small wave steepness H_w/L_w Equation (1.7) simplifies to:

$$L_w = L_{w0} \tanh \frac{2\pi d}{L_w} \quad (1.9)$$

With $\tanh x = \dfrac{e^x - e^{-x}}{e^x + e^{-x}}$

The hyperbolic tangent varies between zero and 1. In general, the deeper the water and the less steep the wave, the closer the Stokes wave comes to a sinusoidal wave. This results from Equation (1.7), by d/L_w approaching infinity and H_w/L_w approaching zero. The simplified first order Stokes wave coincides with the sinusoidal wave already derived by Airy (1845).

For decreasing water depth d, the more complicated mathematical description with elliptical functions is applied (so-called cnoidal waves). The solitary wave is the limiting case for very shallow water. Figure 1.7 compares the different wave contours, according to Wiegel (1964). With less relative water depth, the wave crest becomes larger, while the wave trough is widened and raised.

The maximum steepness of a wave according to Michell (1893) is:

$$Max\left(\frac{H_w}{L_w}\right) = 0.142 \cdot \tanh\left(\frac{2\pi d}{L_w}\right) \quad (1.10)$$

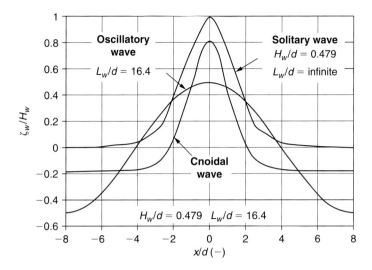

Figure 1.7 Comparison of different wave profiles.

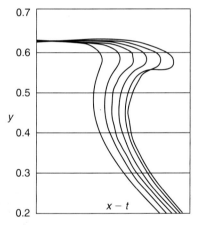

Figure 1.8 Breaking wave.

For deep water waves, this results in a maximum wave steepness of 0.142 or 1/7. In other words, a wave cannot be higher than 1/7 of the wave length. When the wave reaches a steepness of 1/7, the wave breaks. Figure 1.8 shows the profile of a breaking wave as calculated by Longuet-Higgins and Cokelet (1976).

1.5 The sinusoidal wave

1.5.1 Basic relationships to describe regular waves in deep water

The wave pattern we observe at the water surface results from orbital motions of water particles generated by wind energy transfer. We speak of gravity waves, as gravity acceleration acts on the mass of the water particles and tries to smooth the water surface. In the most simplified fashion, we only look at regular sinusoidal waves in a two-dimensional space for unidirectional waves in deep water. Figure 1.9 shows a regular sinusoidal wave with the orbital wave motion.

The basic wave parameters are		Derived basic wave parameters are	
wave height	H_w	wave steepness	H_w/L_w
wave length	L_w	circular wave frequency	ω
wave period	T_w	wave number	k
		wave celerity (phase velocity)	c

The orbital angle is $\varepsilon = kx - \omega t$. Circular frequency ω is defined as the angular velocity of the orbital motion of the water particles in the wave. This is the ratio of the angle of one cycle, 2π, to the time period T_w needed:

$$\omega = \frac{2\pi}{T_w} \text{ (rad/s)} \qquad (1.11)$$

The inverse of the period is the wave frequency f_w:

$$f_w = \frac{1}{T_w} = \frac{\omega}{2\pi} \qquad (1.12)$$

The dimension of the frequency f_w is in Hz (Heinrich Hertz, 1857–1894, who discovered the electromagnetic waves). In fact, 1 Hz is equal to 1/s. In comparison, the dimension of ω is rad/s, which also gives 1/s. The difference in the numerical value is the

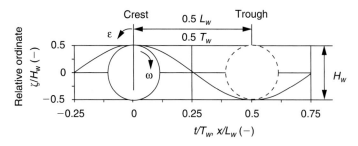

Figure 1.9 Orbital motion, wave length and wave period.

factor 2π, so it is appropriate to make the distinction in using a different notation for f_w and $\omega = 2\pi f_w$.

Accordingly, the wave number k is the ratio of 2π to the wave length L_w:

$$k = \frac{2\pi}{L_w} \text{ (rad/m)} \quad (1.13)$$

Celerity c is also called phase velocity. It is the velocity of the wave contour propagating at the sea surface:

$$c = \frac{L_w}{T_w} \quad (1.14)$$

Wave length is proportional to the wave period squared. The derivation is based on assumptions of continuity and of conservation of energy using the Bernoulli equation (Prandtl 1931, 1952):

$$L_w = \frac{g}{2\pi} T_w^2 \quad (1.15)$$

By Equations (1.14) and (1.15), we find the wave celerity as a function of the wave period or the wave length

$$c = \frac{g}{2\pi} \cdot T_w \quad (1.16)$$

$$c = \sqrt{\frac{gL_w}{2\pi}} \quad (1.17)$$

Based on the above relationships, different formulae can be derived, which contain basically the same information:

$$c = \frac{g}{\omega} \quad (1.18)$$

$$c = \sqrt{\frac{g}{k}} \quad (1.19)$$

$$k = \frac{\omega^2}{g} \quad (1.20)$$

It should be pointed out, that there is a distinction between the velocity c at which the wave crest propagates, and the tangential velocity of the water particles v_t due to the orbital motion. We speak of the phase velocity c, also referred to as wave celerity, as against the orbital (tangential) wave velocity v_t:

$$v_t = 0.5 \cdot H_w \cdot \omega \quad (1.21)$$

The orbital acceleration is:

$$a_t = v_t \cdot \omega = 0.5 \cdot H_w \cdot \omega^2 \quad (1.22)$$

Table 1.2 gives numerical results for regular sinusoidal deep-water waves of different length, plotted in Figure 1.10. The wave height is chosen to a constant 2.50 m.

1.5.2 Normal dispersion of a wave field

Longer waves have a larger celerity than shorter waves (Equation 1.17). This results in a large scatter of waves originating in the same storm field. The phenomenon is called normal dispersion of waves.

By swell, we mean the longest waves generated in a storm. Due to the larger velocity of crest propagation compared with shorter waves, swell approaches areas outside the originating wind field. Swell advances a longer distance in the same time than shorter waves.

Although swell has smaller wave steepness than the shorter waves, it can have considerable wave height due to its large wave length.

The above formulae and values in Table 1.2 demonstrate that longer waves have a larger wave period, increasing with the square root of the wavelength. In the next three equations, the constants are also approximated in the (metric) SI system with $g = 9.81$ m/s^2:

$$T_w = \sqrt{\frac{2\pi L_w}{g}} \cong 0.8\sqrt{L_w} \quad (1.23)$$

Table 1.2 Sinusoidal wave data.

L_w m	k rad/m	ω rad/s	c m/s	c kn	f_w Hz	T_w s	$100\,H_w/L_w$ %	v_t m/s	$100\,v_t/c$ %	a_t m/s²
15	0.419	2.03	4.8	2.5	0.323	3.1	16.67	2.53	52.3	5.14
25	0.251	1.57	6.2	3.2	0.250	4.0	10.00	1.96	31.4	3.08
50	0.126	1.11	8.8	4.5	0.177	5.7	5.00	1.39	15.7	1.54
100	0.063	0.79	12.5	6.4	0.125	8.0	2.50	0.98	7.9	0.77
150	0.042	0.64	15.3	7.9	0.102	9.8	1.67	0.80	5.2	0.51
200	0.031	0.56	17.7	9.1	0.088	11.3	1.25	0.69	3.9	0.39
250	0.025	0.50	19.8	10.2	0.079	12.7	1.00	0.62	3.1	0.31
300	0.021	0.45	21.6	11.1	0.072	13.9	0.83	0.57	2.6	0.26
350	0.018	0.42	23.4	12.0	0.067	15.0	0.71	0.52	2.2	0.22
400	0.016	0.39	25.0	12.8	0.062	16.0	0.63	0.49	2.0	0.19
450	0.014	0.37	26.5	13.6	0.059	17.0	0.56	0.46	1.7	0.17
500	0.013	0.351	27.9	14.4	0.056	17.9	0.50	0.44	1.6	0.15

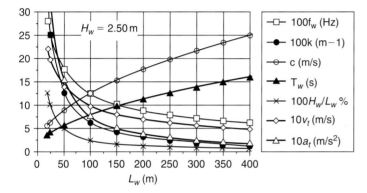

Figure 1.10 Parameters of single sine wave versus wave length.

Longer waves have a smaller wave frequency:

$$\omega_w = \sqrt{\frac{g}{k}} = \sqrt{\frac{2\pi g}{L_w}} \cong \frac{7.85}{\sqrt{L_w}}$$

$$\omega_w = \sqrt{\frac{g}{k}} = \sqrt{\frac{2\pi g}{L_w}} \cong \frac{7.85}{\sqrt{L_w}}$$

$$\Leftrightarrow f_w = \frac{\omega_w}{2\pi} = \frac{1}{T_w} = \sqrt{\frac{g}{2\pi L_w}} \cong \frac{1.25}{\sqrt{L_w}} \quad (1.24)$$

1.5.3 Orbital motion of water particles in a wave

The orbital motion of the water particles has its maximum at the water surface. With increasing vertical distance from the surface, the orbital radius r decreases exponentially.

$$r(z) = 0.5 \cdot H_w \cdot \exp(kz) \quad \text{for} \quad z \leq 0 \quad (1.25)$$

as shown in Table 1.3, Figure 1.11 and Figure 1.12.

At the water surface, the radius is equal to half the wave height $0.5\,H_W$. At half the wave length underneath the surface, the orbital radius of the wave motion has been reduced to 4% of the radius at the surface.

Table 1.3 Decay of orbital radius with distance from surface.

$-z/L_w$	0	0.25	0.5	0.75	1
$r/(0.5H_w) = \exp(kz)$	1	0.21	0.04	0.009	0.002

The wave is progressing from left to right with celerity c. Water particles orbit clockwise with the angular velocity ω. Clockwise is mathematically negative. The angle ε of the orbital position is mathematically positive, counter-clockwise, see also Figure 1.9.

The horizontal and vertical components of the orbital radius are

$$r_x(z) = r(x) \sin \varepsilon \quad (1.26)$$

$$r_z(z) = r(z) \cos \varepsilon \quad (1.27)$$

The wave ordinate ζ is equal to the vertical component of the orbit at the surface, $r_z(z = 0)$:

$$\zeta(x,t) = \frac{H_w}{2} \cos \varepsilon \quad \text{with} \quad d > \frac{L_w}{2} \quad (1.28)$$

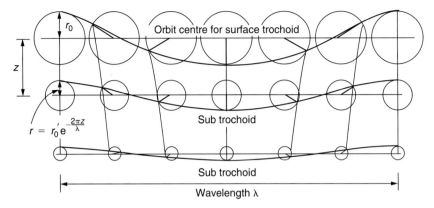

Figure 1.11 Decrease in orbital radius (Tupper, 2004).

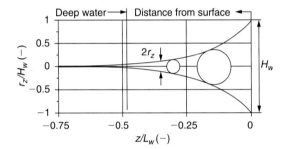

Figure 1.12 Decay of orbital radius in deep water.

The angle ε depends on way of progress x of the wave at the surface and on time t:

$$\varepsilon = kx - \omega t \qquad (1.29)$$

The above derivations are valid for a regular sinusoidal wave. The sinusoidal wave is a good approximation for small steepness ($H_w/L_w \leq 0.05$) in deep water ($d/L_w > 0.05$).

Deep water means a relative water depth d/L_w of more than 0.5. For a 50 m wave, more than 25 m water depth is considered deep, whereas for a 300 m wave, deep water starts at 150 m.

1.6 Irregular waves

The natural seaway on the oceans is irregular. It is also referred to as random sea, or as confused sea. The sea shows rarely a unidirectional, regular sinusoidal wave pattern, but we observe a mixture of waves of different length, height and direction.

The natural seaway can be decomposed to a sum of partial sinusoidal waves, each having a relatively small steepness, even for a severe sea. Therefore, the spectral approach with a sum of partial waves constitutes a valid representation for a random sea.

From careful observation, certain typical or characteristic parameters can be estimated, such as a significant wave height, period, and direction of progress. St. Denis and Pierson (1953) introduced a mathematical description of natural seaways. Their work was a milestone to allow a calculation of random seas and linear ship motion.

The unidirectional, irregular wave pattern ζ is seen as the sum of regular partial waves, as shown in Figure 1.13 (at $x = 0$). From a record of the irregular sea, Fourier analysis can calculate partial waves. An irregular record can be plotted again as a sum of the partial waves, according to Equation (1.30).

$$\zeta(x,t) = \sum_{i=1}^{n} c_i \cos(k_i x - \omega_i t + \varepsilon_i) \qquad (1.30)$$

ζ	wave ordinate, expressing surface elevation	x, t	way of progress, time
i	number of wave component (partial wave)	n	total number of partial waves
c_i	amplitude of the ith partial wave	k_i	wave number (Equation 1.13)
ω_i	circular frequency of partial wave	ε_i	phase angle of partial wave
$S_{\zeta\zeta}$	seaway spectrum (Equation 1.32)	m_0	area of spectrum
H_v, T_v	visually observed height and period	H_w	wave height, general

The phase ε_i is randomly distributed. All wave phases have an equal probability of occurrence, expressed by a constant probability density $1/(2\pi)$ in the range $(0, 2\pi)$.

The amplitudes c_i of the partial waves are calculated by using a seaway spectrum $S_{\zeta\zeta}$:

$$c_i = \sqrt{2 S_{\zeta\zeta i} \delta \omega_i} \qquad (1.31)$$

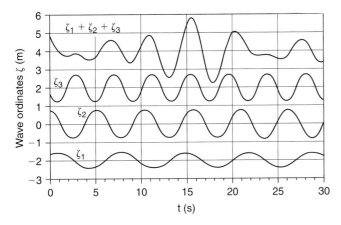

Figure 1.13 Irregular wave from sum of regular waves.

The spectral value of a sine wave component is proportional to one-half the square of the amplitude, divided by the corresponding frequency bandwidth, $\delta\omega_i$:

$$S_{\zeta\zeta_i} = \frac{\delta m_{0i}}{\delta\omega_i} = \frac{(c_i)^2}{2\delta\omega_i} \quad (1.32)$$

The seaway spectrum $S_{\zeta\zeta}$ is defined as the distribution of the total wave energy m_0 with respect to the circular wave frequency ω. From integration of $S_{\zeta\zeta}$ for all positive frequencies ω, we derive the total energy m_0 of the stationary seaway:

$$m_0 = \int_0^\infty S_{\zeta\zeta}(\omega) \cdot d\omega \quad (1.33)$$

The energy, E_w, of a single sine wave is proportional to its height, H_w, squared, per unit area of sea surface, and the sum of kinetic and potential energy:

$$E_w = \frac{1}{8} \cdot \rho \cdot g \cdot H_w^2 \quad \text{With} \quad E_{kin} = E_{pot} = \frac{1}{2} E_w \quad (1.34)$$

For the sake of comparison of wave energies, considering constant values of ρ and g, it is sufficient to use the spectrum $S_{\zeta\zeta}$, without the constant factor ρg, as a measure of energy. The whole area under the spectral curve is the average total energy m_0 of the seaway for stationary conditions, $m_0(m^2) = \sum_i \frac{E_{wi}}{\rho g}$. Without a significant change of the the wave pattern and its energy contents for some time, the seaway is stationary. Practically, we speak of at least 20 minutes constant average energy to consider the sea as stationary. Stationary conditions to apply a short-term statistic can exist in the order of hours.

Half the variance, $0.5\overline{\zeta^2}$ (mean square $\overline{\zeta^2}$ of the surface elevation, z) is equal to the area of the seaway spectrum, m_0 (Parseval Theorem). The bar denotes the statistical average:

$$m_0 = \frac{1}{2}\overline{\zeta^2} = \lim \frac{1}{2T} \int_0^T \zeta^2(t) dt \quad \text{for} \quad T \to \infty \quad (1.35)$$

Inserting $H_{wi} = 2c_i$ in Equation (1.34), the energy of a partial wave is, as used in Equation (1.32):

$$\delta m_{0i} = \frac{E_{wi}}{\rho g} = \frac{1}{8} H_{wi}^2 = \frac{1}{2} c_i^2 = S_{\zeta\zeta_i}\delta\omega_i \quad (1.36)$$

The wave parameters from observation, wave height, H_v, and period, T_v (subscript v for visual), see Section 1.11, can be used to calculate the Bretschneider spectrum Equation (1.42). Comparing wave records and simultaneous observations from weather ships, different relationships to estimate the spectral parameters ($\overline{H}_{1/3}$, $\overline{T}_1$) were recommended in Lewis (1989), Faltinsen (1990). Clauss, et al. (1992) compared formulae and found the simplest approach to be a realistic assumption, see also Price and Bishop (1974):

$$H_v \cong \overline{H}_{1/3} \quad \text{and} \quad T_v = \overline{T}_1 \quad (1.37)$$

$$\overline{T}_1 = 2\pi \frac{m_0}{m_1} = 2\pi \frac{\int_0^\infty S_{\zeta\zeta}(\omega) \cdot d\omega}{\int_0^\infty S_{\zeta\zeta}(\omega)\omega \cdot d\omega} \quad (1.38)$$

$$m_n = \int_0^\infty S_{\zeta\zeta}(\omega) \cdot \omega^n \cdot d\omega$$
$$= n^{th} \text{ order spectral moment} \quad (1.39)$$

H_{wi}	height of ith wave component	H_s	significant height
$\overline{H}_{1/3}$	mean of 1/3 largest heights, Equation (1.52)	H_v	visual height
$\overline{T}_1$	mean period of all sine components	T_v	visual period

The point spectrum $S_{\zeta\zeta}$ can be multiplied by a spreading function $f(\mu)$ to have a directional spectrum. With μ the angular range of wave components on either side of the dominant wave direction, the spreading is often used as:

$$S_{\zeta\zeta}(\omega,\mu) = S_{\zeta\zeta}(\omega) f(\mu) \quad (1.40)$$

$$f(\mu) = \frac{2}{\pi} \cdot \cos^2\mu \quad \text{when} \quad \mu \in \left[-\frac{\pi}{2}; \frac{\pi}{2}\right] \quad (1.41)$$

1.7 Spectrum formulae by Pierson/Moskowitz and Bretschneider

Oceanographers have studied the ocean pattern extensively. Basically, they fitted measurements into general formulae. Most popular is the wave spectrum for fully developed seas in deep water and unlimited fetch by Pierson and Moskowitz (1963). It depends on the wind at 19.5 m height above the surface.

In order to account for the specific sea area and fresh wind sea, a two-parameter version of the Pierson and Moskowitz (P-M) seaway spectrum was developed by Bretschneider (1952, 1961). His formulation uses the significant wave height, $H_{1/3}$, and the mean period, T_1. The modified P-M or Bretschneider spectrum, see a detailed comparison with the P-M spectrum in Principles of Naval Architecture (Lewis 1989), is:

$$S_{\zeta\zeta}(\omega) = A\omega^{-5} \exp(-B\omega^{-4}) \quad \text{for} \quad \omega \geq 0 \quad (1.42)$$

The parameters A and B are as follows:

$$A = 173 \cdot \frac{\overline{H}_{1/3}^2}{\overline{T}_1^4} [m^2 s^{-4}] \quad B = 691 \frac{1}{\overline{T}_1^4} [s^{-4}] \quad (1.43)$$

$$\omega_{\text{mod}} = \left[\frac{4}{5} B\right]^{1/4} \quad \text{modal frequency (of spectral peak)} \quad (1.44)$$

International organizations, such as ISSC (International Ship Structures Congress) and similarly ITTC (International Towing Tank Conference) have adopted the Modified P-M or Bretschneider spectrum, when no other specific seaway spectrum is known. In the Bretschneider formula, Equation (1.42), only H_v–T_v combinations need to be known (replacing $\overline{H}_{1/3}, \overline{T}_1$

using Equation 1.37). Hattendorf (1974) compiled characteristic values of sea areas (NA North Atlantic, NN Northern North Sea, and WB Western Baltic Sea), see Tables 1.4, 1.5, and Hogben et al. (1986).

Table 1.4 Significant wave heights and periods from observation.

Bft. No.	H_v (m) NA	NN	WB	T_v (s) NA	NN	WB
3	1.7	1.0	0.45	6.3	4.6	2.9
6	3.1	3.0	1.2	7.4	6.1	4.4
9	6.45	6.6	2.5	9.1	8.4	5.8
12	9.2	7.7	4.3	10.6	10.5	6.5

Table 1.5 North Atlantic (H_v, T_v), see spectra in Figure 1.15.

Bft	H_v(m)	T_v(s)	Bft	H_v(m)	T_v(s)	Bft	H_v(m)	T_v(s)
4	1.95	6.5	7	4.00	8.0	10	7.45	9.6
5	2.40	6.9	8	5.25	8.5	11	8.40	10.1
6	3.10	7.4	9	6.45	9.1	12	9.20	10.6

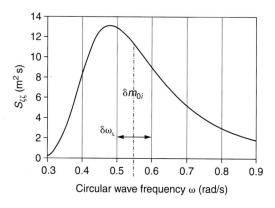

Figure 1.14 Spectral energy for partial wave.

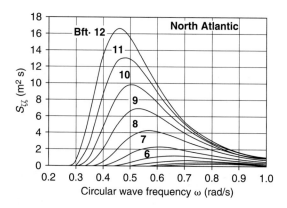

Figure 1.15 Bretschneider Seaway spectra.

Note the shift of the spectrum to smaller frequencies (i.e. longer waves, respectively) at larger energy m_0 due to larger Beaufort number.

Seaway spectra are narrow-banded. This means, the total energy is concentrated at a narrow frequency range, expressed by the spectral breadth parameter ε_b in terms of the frequency moments of the power spectral density function $S_{\zeta\zeta}$, Cartwright and Longuet-Higgins (1956), see also Hutchison (1990):

$$\varepsilon_b^2 = \frac{m_0 m_4 - m_2^2}{m_0 m_4} \quad (1.45)$$

where the m_n moments are defined as in Equation (1.39).

1.8 The JONSWAP sea spectrum

The Joint North Sea Wave Program, a major international study undertaken along a straight line off the island of Sylt in the German Bight with wind offshore, led to the JONSWAP (or J) spectrum, Hasselmann (1973). It is more peaked than the P-M and Bretschneider spectra, Figure 1.16, and has been widely applied in Ocean engineering calculations, in particular for extreme sea conditions. The JONSWAP seaway spectrum needs more parameters, Lewis (1989):

$$S_{\zeta\zeta}(\omega) = \alpha g^2 \omega^{-5} \exp\left[-\frac{5}{4} \cdot \left(\frac{\omega}{\omega_{\text{mod}}}\right)^{-4}\right] \cdot \gamma^p \quad (1.46)$$

$$p = \exp\left[\frac{-(\omega - \omega_{\text{mod}})^2}{2\sigma^2 \omega_{\text{mod}}^2}\right] \quad (1.47)$$

Fetch parameter from statistical fit $\alpha = 0.076 \cdot (gx/v_{w10}^2)^{-0.22}$ with fetch x and wind velocity v_{w10}; width parameter $\sigma_a = 0.07$ for $\omega \leq \omega_{\text{mod}}$ and $\sigma_b = 0.09$ for $\omega > \omega_{\text{mod}}$; enhancement factor γ^p; modal peak frequency:

$$\omega_{\text{mod}} = \frac{2\pi g \cdot 3.5}{v_{w10}} \cdot \left(\frac{gx}{v_{w10}^2}\right)^{-0.33}$$

Faltinsen (1990) and Clauss et al. (1992) give further modifications of the J spectrum. Wind force, wind duration, and fetch, determine the generation of seaway in a certain area. Fetch determines the peak frequency of the seaway spectrum. Seaway builds up with fetch and duration from a fresh wind sea into a fully developed seaway. The fully developed sea is saturated and cannot store any more energy, so further wind transfers energy to longer waves. Figure 1.17 shows the wave generation (parameters height and period) with respect to wind duration, the fetch (length of sea surface in wind direction at which the wind is generating waves), and the velocity of the wind, DHI (1981).

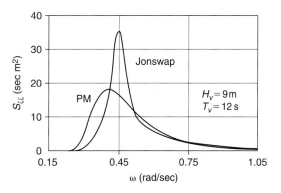

Figure 1.16 Comparison of Bretschneider and higher peaked JONSWAP spectra.

1.9 Maximum wave height in a stationary random sea

The irregular wave height can be described statistically (and other parameters to describe the seaway, such as period, etc).

The Rayleigh function describes the short-term distribution of amplitudes of a narrow band process ($\varepsilon_b = 0$). The probability of a random variable x exceeding a threshold x_i is:

$$P(x > x_i) = \exp\left(-\frac{x_i^2}{2m_0}\right) \quad (1.48)$$

Inserting the wave height as double of the wave amplitude, we get the probability distribution of the wave height H_w exceeding a threshold $H_{1/n}$, as shown in Figure 1.18.

$x = 0.5 H_w$, $x_i = 0.5 H_{1/n}$, $H_v = 4\sqrt{m_0}$ $\Rightarrow$

$$P(H_w > H_{1/n}) = \frac{1}{n} = \int_{H_{1/n}}^{\infty} p(H_w)\, dH_w =$$

$$= \exp\left(-\frac{H_{1/n}^2}{8 m_0}\right) = \exp\left(-\frac{2 H_{1/n}^2}{H_v^2}\right); H_{1/n} > 0 \quad (1.49)$$

The quantities of the Rayleigh distribution are:

P	Probability
p	Probability density
m_0	area of spectrum
H_w	wave height, independent variable
H_v	visual wave height (see Equation 1.38)
$H_{1/n}$	wave height that is exceeded by a probability of $1/n$
$\overline{H}_{1/n}$	Average of the $1/n$th largest wave heights exceeding $H_{1/n}$, i.e. $\overline{H}_{1/n} > H_{1/n}$
$\overline{H}_{1/3}$	Significant height, average of the 1/3 highest waves exceeding $H_{1/3}$, see Figure 1.20.

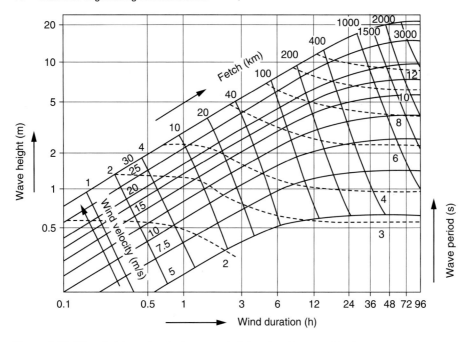

Figure 1.17 Wave height and period versus wind duration, force, and fetch.

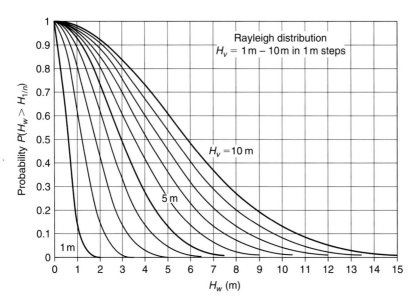

Figure 1.18 Probability of exceeding wave height $H_w = H_{1/n}$ at different sea states (H_v).

The probability of occurrence of the wave height is the probability density p, as the derivative of the probability P with respect to $H_{1/n}$, as shown in Figure 1.19.

$$p(H_w > H_{1/n}) = \frac{H_{1/n}}{4m_0} \exp\left(-\frac{H_{1/n}^2}{8m_0}\right); \quad H_{1/n} > 0$$

(1.50)

Equation (1.49) leads to the threshold, $H_{1/n}$, for any probability P, here chosen as $1/n$:

$$H_{1/n} = \sqrt{8m_0\left(-\ln\frac{1}{n}\right)} = H_v\sqrt{-\frac{1}{2}\ln\frac{1}{n}} \Rightarrow \frac{H_{1/3}}{H_v}$$
$$= \sqrt{-0.5 \cdot \ln(0.33)} \approx 0.74$$

(1.51)

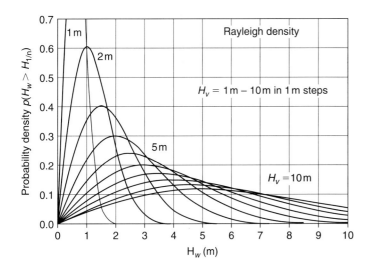

Figure 1.19 Frequency of occurrence of wave height, parameter $H_v = 4\sqrt{m_0}$ defining sea state.

Table 1.6 Average largest wave heights with short-term frequency of occurrence (plotted in Figure 1.21).

$P(H_w > H_{1/n}) = 1/n$	Modal	1/2	1/3	1/10	1/100	10^{-3}	10^{-4}
occurrence rate 100/n (%)	60	50	33.3	10	1	0.1	0.01
$\overline{H}_{1/n}/\sqrt{m_0}$	2.0	2.5	**4.0**	5.1	6.5	7.7	8.9

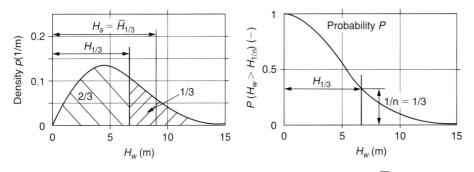

Figure 1.20 Wave height distribution according to Rayleigh, defining $H_{1/n}$ and $\overline{H}_{1/n}$ ($m_0 = 5.06 \text{ m}^2$).

The centre of the area above $H_{1/n}$ defines $\overline{H}_{1/n}$, see Table 1.6 and Figure 1.20, from Equation (1.52):

$$\overline{H}_{1/n} = n \int_{H_{1/n}}^{\infty} H_w \cdot p(H_w) dH_w, \text{ in particular:}$$

$$\overline{H}_{1/3} = 4\sqrt{m_0} \qquad (1.52)$$

The average largest heights from Equation (1.52) are given in Table 1.6 and in Figure 1.21.

Example $n = 3$, Figure 1.20: $m_0 = 5.06 \text{ m}^2$; $\sqrt{m_0} = 2.25 \text{ m}$; $H_v \cong \overline{H}_{1/3} = 4 \cdot 2.25 \text{ m} = 9 \text{ m}$; Equation (1.51):

$$H_{1/3} = \sqrt{8m_0(-\ln 1/n)} = \sqrt{8 \cdot 5.06 \text{ m}^2 \cdot (-\ln(1/3))} = 6.7 \text{ m};$$

$$H_{1/3} \approx 0.74 \overline{H}_{1/3}$$

Inserting into Equation (1.49), we find again our initial assumed probability 1/3:

$$P(H_w > H_{1/3}) = \exp(-6.7 \text{m} \cdot 6.7 \text{m}/(8 \cdot 5.06 \text{ m}^2))$$
$$= \exp(-1.1084) \cong 0.33 \cong 1/n$$

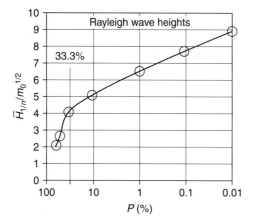

Figure 1.21 Largest wave height in stationary random sea (short-term).

The modal value is the most frequently occurring height (i.e. probability density p is a maximum), in our example:

$$H_{mod} = 2\sqrt{m_0} = 0.5 \cdot \overline{H}_{1/3} = 0.5 \cdot 9\,\text{m} = 4.5\,\text{m},$$

see Figure 1.20.

A larger spectral breadth parameter Equation (1.46), ($0 \leq \varepsilon_b \leq 1$) shifts the density of the heights towards a Gaussian function Cartwright/Longuet-Higgins (1956), Price and Bishop, (1974). For most cases is $\varepsilon_b < 0.6$, so the error with Rayleigh ($\varepsilon_b = 0$) is less than 10%.

Pierson (discussion to Chung-Tu Teng, et al. (1994)) pointed at the height formula derived from the P-M spectrum for the fully developed sea:

$$H_w = 0.0212 \cdot v_w^2 \tag{1.53}$$

Wind velocity, v_w, is measured at 19.5 m above the sea surface. The corresponding wave heights are given in Table 1.7 and plotted in Figure 1.22.

1.10 Long-term statistics of irregular seaway

Long-term statistics gathers data for a longer time span (week, month, year), to include time variations of seaway energy. Extreme value analysis fits long-term statistical data to theoretical probability distributions. Gumbel (1958) developed the theory of the statistics of extremes. Longuet-Higgins (1952) developed the statistical distribution of the wave heights.

One extreme value distribution is the Weibull-function, shown in Figure 1.23. Fitted are measurements at station 46001 in the North Pacific for one month, Chung-Tu Teng et al. (1994).

Table 1.7 Extreme P-M height versus wind velocity.

v_w (m/s)	26.6	27.5	28.3	29.1	29.7	30.7	31.5
v_w (knots)	52	54	55	57	58	60	61
H_w(m)	15	16	17	18	19	20	21

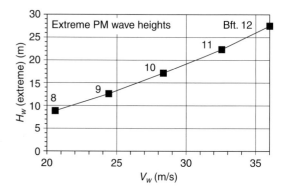

Figure 1.22 Extreme P-M height of fully developed sea.

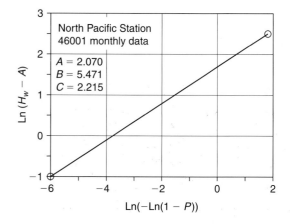

Figure 1.23 Weibull distribution of extreme wave heights.

Weibull distribution:

$$P(H_w \leq x) = 1 - \exp\left[-\left(\frac{x-A}{B}\right)^c\right] \tag{1.54}$$

A, B, C are parameters for location, scale, and shape, selected to fit measured data to the distribution.

An extreme value of a wave height is then associated with a return interval in the time domain, called return period. A common procedure is to assign a probability P to an event X of duration δt within the return period T_r:

$$P(X \leq x) = 1 - \frac{\delta t}{T_r} \tag{1.55}$$

Common return periods are the assumed lifetime of a ship (30 years), for stationary designs (platforms, dikes) 50 or 100 years.

To illustrate a large return period, we calculate:

$$T_r = 30 \text{ years}$$
$$= 30\text{y} \cdot 365 \text{ d/y} \cdot 24 \text{ h/d} \cdot 3600 \text{ s/h} = 9.4 \cdot 10^8 \text{ s}$$

Say some event of 9.4 s duration occurs once in the return period of 30 years:

$$\frac{\delta t}{T_r} = \frac{9.4 \text{ s}}{9.4 \cdot 10^8} = 10^{-8} = P(X > x)$$

Figure 1.24 shows ranges of the long-term probability of exceeding wave heights in the North Atlantic and the North Sea, Roren and Furnes (1976).

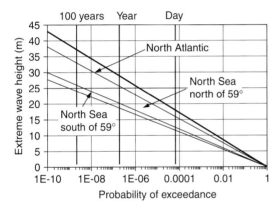

Figure 1.24 Extreme wave height distributions (the straight lines separate 3 areas).

1.11 Wave data from observations

One of the most extensive data collection of waves was set up by the British National Physical Laboratory in collaboration with the Meteorological Office, published as 'Ocean Wave Statistics' by Hogben and Lumb (1967). Ocean data was based on the meteorological logbooks of voluntary observing ships numbering about 500 ships for all shipping routes on the oceans. It contained frequency distributions of visually observed wave heights, periods and directions for 50 sea areas. The wave data reports archived by a number of member states started in 1949 under the auspices of the World Meteorological Organization (WMO).

The succeeding volume is 'Global Wave Statistics', published by British Maritime Technology Limited (BMT), Hogbeṅ, et al. (1986). The book supplies statistics of ocean wave climate for the entire globe for 104 areas of the ocean. This improved data is more reliable because it is based on much larger samples covering longer periods of time.

A huge number of visual observations of both waves and winds reported from ships in normal service all over the world have been evaluated by a quality enhancing analysis and data fitting using a computer program. Although the statistics presented relate only to waves, wind observations have been used to improve the reliability of the wave statistics. Furthermore, instrumental data from nine stations were used to validate results, e.g. Ocean Weather Station India (59° N, 19° W) in the North Atlantic.

For each of the 104 areas, data was given in tables with the categories:

- Annual and four seasons,
- All directions and 8 directional sectors.

The compiled data tables are scatter diagrams representing joint frequency distributions of wave height and period, (H_v, T_v). Each entry in the tables is the probability in parts per thousand that waves in the respective class will lie in the specific ranges of height and period. The visual wave heights, H_v, may be assumed to correspond to *significant heights*, $H_S = \overline{H}_{1/3}$, see Equation (1.37).

The significant wave height $\overline{H}_{1/3}$ is defined as the average of the height from crest to trough, of the highest third of the waves in a sample. When used as input to the Bretschneider spectrum, $\overline{H}_{1/3}$ must result in the same value by the formula Equation (1.52) $\overline{H}_{1/3} = 4 \cdot m_0^{1/2}$. Therefore with the zero order moment m_0, the area of the seaway spectrum, we have a good control on the spectral calculation.

Observed periods are very unreliable, according to a detailed analysis of data by Hogben et al. (1986), so they preferred mathematical modelling of the table data. Their judgement might explain the large discrepancies to be found in the literature on connecting visual periods with different mathematical estimates.

Table 1.8 shows an example of global wave data with a total of N = 111470 observations (corresponding to the 1000 mill in the lower right corner of the table). The vertical scale is wave height, corresponding to the rows, i. The horizontal scale is the wave period, corresponding to the columns, j. Wave heights are grouped into 1 metre sampling intervals from zero to 14 m, with the last range > 14 m, periods into 1 second interval from 4 s to 13 s, with additional ranges < 4 s and > 13 s. Figures 1.25 and 1.26 depict the relative frequency of the joint probability distribution $p(H_v, T_v)$ with the data from Table 1.8.

Figures 1.27 and 1.28 show the marginal distribution of the wave height (from the right most column Table 1.8), and of the wave period (from the last row in Table 1.8).

The new global wave data, Hogben et al. (1986) takes into account both sea and swell heights and therefore

18 Maritime engineering reference book

Table 1.8 Long-term waves, Sea area 8, North West Atlantic, annual, all directions, observed relative number 1000n/N (‰) of heights and periods, N = 111470.

T_v, s	4.5	5.5	6.5	7.5	8.5	9.5	10.5	11.5	12.5	13.5	total
H_v, m											total
0.5	1	8	13	8	2						32
1.5		8	42	66	42	15	3	1			177
2.5		3	27	77	86	48	16	4	1		262
3.5		1	11	45	73	57	26	8	2		223
4.5			3	20	41	41	24	9	3	1	142
5.5			1	7	19	24	17	7	3	1	79
6.5				3	8	12	10	5	2	1	41
7.5				1	4	8	5	3	1	1	23
8.5					2	3	3	2	1		11
9.5	↓ i				1	1	2	1	1		6
10.5		n_{ij}				1	1	1			3
11.5			→ j					1			1
Total	1	20	97	227	278	210	108	41	14	4	**1000**

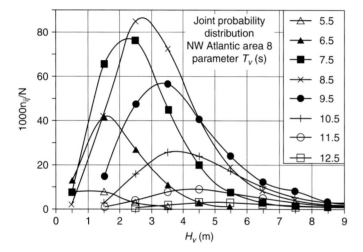

Figure 1.25 Joint relative frequency of significant wave height and period, sampling interval 1 m and 1 s.

result in higher wave heights than the first ocean data, Hogben and Lumb (1967). They also derive the wave periods given by mathematically modelling the joint probability of measured data of the periods with the wave heights observed. Thus the given tables avoid using the visual observations of the wave period, which have been shown to be very unreliable. Generally, quite large differences between the two publications in the statistics of heights and periods exist, and it is advisable to use only the newest data available.

1.12 Wave climate

Information on wind-generated wavTes has been based on:

Visual observation
Instrumental buoy measurements
Remote sensing from satellites
Calculation with spectral models using barometric fields and wind field analysis

Although observation has the lowest quality of data, it has the largest statistical data available. It underestimates the recurrence of heavy storms. Buoy measurement data are smaller in volume and limited to a small number of years. Satellites and calculation models will give better information in the near future, see Woolf et al. (2003) and Wimmer et al. (2006).

We can speak of a 'wave climate' to include the whole wave states at sea. Storms and waves can be classified into typical structures. Boukhanovsky et al. (2000) proposed to apply five storm classes and four wave classes. The storm class depends on the shape of the wind force versus time. The wave class depends on the shape of the seaway spectrum versus frequency.

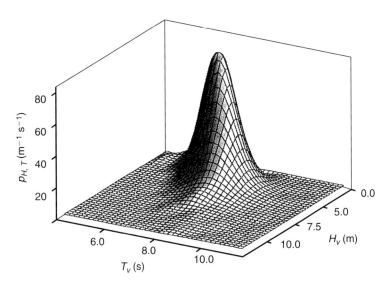

Figure 1.26 Probability density $p(H_v, T_v)$ from Table 1.8 in 3 D, area 8, NWA (sampling range 1m, 1s).

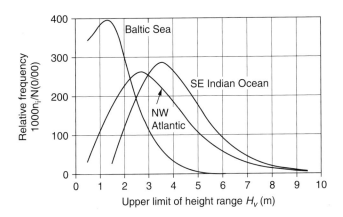

Figure 1.27 Relative frequency of observed wave height (sampling interval 1m).

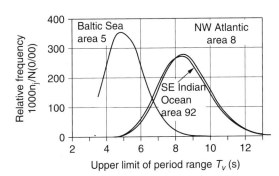

Figure 1.28 Relative frequency of observed wave period (sampling interval 1 s).

Storm class I is generating waves during limited time
Storm class II is generating a fully developed sea.
Storm classes III, IV, V: combinations of class I and II.

The storm class is characterized by storm duration, time of growth and time of decay. Figure 1.29 shows the dimensionless wind force versus time. Table 1.9 gives the frequency of occurrence of the storm shapes and the mean shapes of duration (from measurements in the Black Sea, Boukhanovsky et al. (2000)). While I is triangular, II is more trapezoidal in shape. Storm class III is similar to IV, but the peaks exchanged (not drawn).

Storm class I, with 50% occurrence in the Black Sea, has the largest percentage of the five storm classes, see Table 1.9. Wave class II (wind waves) shows for all classes of wind and storm the largest percentage.

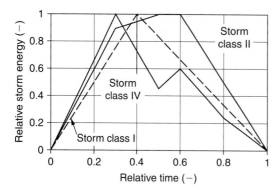

Figure 1.29 Storm shapes classes I, II and IV.

Table 1.9 Percentage of storm and wave classes in the Black Sea.

Storm class	I	II	III	IV	V
Occurrence	**50%**	15%	6%	19%	10%
Wave class	Occurrence (%) of wave class related with storm class				
I (swell)	41	33	31	33	24
II (wind waves)	46	54	53	55	61
III complex	9	9	13	8	11
IV complex	4	4	3	4	4

The sea surface $\zeta(x, y, t)$ is stochastic in the time and space domain. Directional seaway spectra, $S_{\zeta\zeta}$, constitute 'wave weather'. $S_{\zeta\zeta}(\omega, \mu)$ varies in space (x, y) and in time t. It is non-stationary and not homogeneous. Seaway spectra are classified according to:

- S swell   WW wind waves   CS complex sea.

Wind waves and swells are discriminated into four classes with typical spectral shapes according to Table 1.9, see Figure 1.30. It must be kept in mind, that all spectra are average estimates of statistical data, so on-line spectra can show large scatter.

Wind and wave climates allow analysing operational conditions at sea in more detail. Satellites will give direct data on the wave climate distribution, and can even detect rare but feared freak waves. So we can expect to get more direct seaway information on the bridge in the near future. The Atlas of the Oceans by J.T. Holland and I.R. Young (1996) gives information on global meteorology and the GEOSAT satellite mission.

1.13 Freak waves

Between 1993 and 1997 more than 582 ships were lost, totalling some 4.5 million tonnes. Some of these, possibly a third, were lost in bad weather. Each year some 1200 seafarers are lost. As Faulkner (2003) has reported, for centuries mariners have reported meeting 'walls of water', 'holes in the sea' or 'waves from nowhere'. For most of the time they have not been fully believed, being thought guilty of exaggeration. However, investigations following the loss of *MV Derbyshire* and subsequent research show that freak waves do occur and that, although not frequent, they are not rare. See also Bishop et al. (1991) and Faulkner and Williams (1996).

Faulkner discusses four types of abnormal waves apart from Tsunamis which do no harm to ships well out to sea. They are as follows.

Extreme waves in normal stationary seas
These follow the usual laws and occur because waves of different frequencies are superimposed and at times their peaks will coincide. In a sample of 1500 waves a wave k times the significant height of the system will have decreasing probabilities of occurrence as k increases:

k	2.0	2.2	2.4	2.9
% probability	40	9.0	1.5	0.07

Opposing waves and surface currents
That these conditions lead to high waves has been known for a long time. The usually quoted example is that of the Agulhas Current off SE Africa. A 4 knot current opposing a 15 second wave increases its height by about 90%.

Standing waves
These are transient waves which appear and disappear. They occur at the centres of tropical storms, in crossing seas and near steep coastlines.

Progressive abnormal waves
These arise from a number of causes including:

- Accretion of waves within a group.
- Coalescing independent wave groups.
- Energy transfer to the faster bigger waves in a group until they become unstable.
- Non-linear interactions between colliding waves.

Damage due to abnormal waves
Faulkner quotes a number of cases, including:

- Five container ships (4000–4500 twenty-foot equivalent units (TEU)) caught in a North Pacific storm which lost 700 containers overboard. Some reported rolling to 35 to 40 degrees.

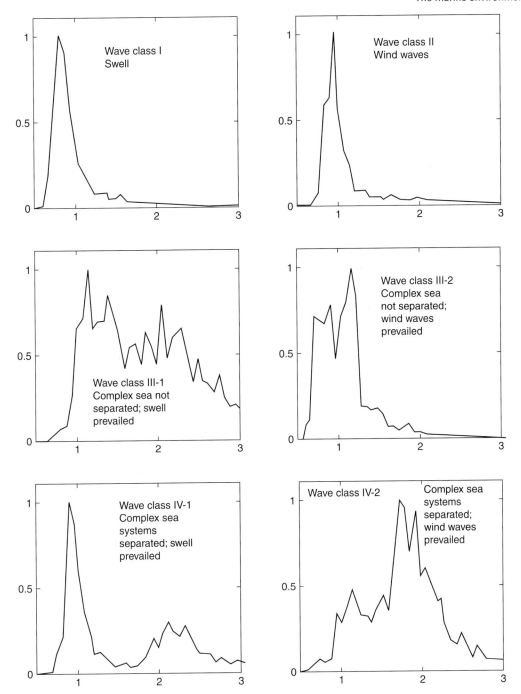

Figure 1.30 Classification of wave spectra.

- A 29.5 m Agulhas wave that smashed a crows nest window 18 m above the safe working load (SWL) on a 256 000 dwt tanker.
- The liner *Michael Angelo* had its bridge, 0.26L from forward, completely smashed, as were windows 24 m above the waterline.

1.14 Oceanography

This Section considers the properties of the oceans such as water properties, circulation, currents and tides. How these properties affect the operation of underwater vehicles, such as submarines and remotely operated

Table 1.10 Earth's water supply.

Water source	Water volume, in cubic miles	Water volume, in cubic kilometers	Percent of freshwater	Percent of total water
Oceans, seas, and bays	321 000 000	1 338 000 000	–	96.5
Ice caps, glaciers, and permanent snow	5 773 000	24 064 000	68.7	1.74
Groundwater				
Fresh	2 526 000	10 530 000	30.1	0.76
Saline	3 088 000	12 870 000	–	0.94
Soil moisture	3959	16 500	0.05	0.001
Ground ice and permafrost	71 970	300 000	0.86	0.022
Lakes				
Fresh	21 830	91 000	0.26	0.007
Saline	20 490	85 400	–	0.006
Atmosphere	3095	12 900	0.04	0.001
Swamp water	2752	11 470	0.03	0.0008
Rivers	509	2120	0.006	0.0002
Biological water	269	1120	0.003	0.0001

Source: US Geological Survey.

vehicles (ROVs) and equipment, is discussed. Special thanks go to Steve Fondriest of Fondriest Environmental Inc. for his contribution to the fundamentals of environmental monitoring and data collection instrumentation.

1.14.1 Distribution of water on earth

Earth is the only planet known to have water resident in all three states (solid, liquid, and gas). It is also the only planet to have known liquid water currently at its surface. Distribution of the earth's water supply is provided in Table 1.10.

As shown in Table 1.10, most (97%) of the world's water supply is in the oceans. Water can dissolve more substances, and in greater quantities, than any other liquid. It is essential to sustain life, a moderator of our planet's temperature, a major contributor to global weather patterns.

The oceans cover 70.8% of the earth's surface, far overreaching earth's land mass. Of the ocean coverage, the Atlantic covers 16.2%, the Pacific 32.4%, the Indian Ocean 14.4%, and the margin and adjacent areas the balance of 7.8%. It is also interesting to note that the Pacific Ocean alone covers 3.2% more surface area on earth than all of the land masses combined.

1.14.2 Properties of water

Water is known as the 'universal solvent'. While pure water is the basis for life on earth, as more impurities are added to that fluid the physical and chemical properties change drastically. The chemical makeup of the water mixture in which an ROV may operate will directly dictate operational procedures and parameters if a successful operation is to be achieved.

Two everyday examples of water's physical properties and their effect on our lives are (1) ice floats in water and (2) we salt our roads in wintertime to 'melt' snow on the road. Clearly, it is important to understand the operating environment and its effect on ship and ROV operations. To accomplish this, the properties and chemical aspects of water and how they're measured will be addressed to determine their overall effect.

The early method of obtaining environmental information was by gathering water samples for later analysis in a laboratory. Today, the basic parameters of water are measured with a common instrument named the 'CTD sonde'. Some of the newest sensors can analyse a host of parameters logged on a single compact sensing unit.

Fresh water is an insulator, with the degree of electrical conductivity increasing as more salts are added to the solution. By measuring the water's degree of electrical conductivity, a highly accurate measure of salinity can be derived. Temperature is measured via electronic methods and depth is measured with a simple water pressure transducer. The CTD probe measures 'conductivity/ temperature/depth', which are the basic parameters in the sonic velocity equation. Newer environmental probes are available for measuring any number of water quality parameters such as pH, dissolved oxygen and CO_2, turbidity, and other parameters.

The measurable parameters of water are needed for various reasons. A discussion of the most common measurement variables the commercial or scientific ROV operator will encounter, the information those parameters provide, and the tools/techniques to measure them follow.

1.14.2.1 *Chlorophyll*

In various forms, chlorophyll is bound within the living cells of algae and other phytoplankton found in surface water. Chlorophyll is a key biochemical component in the molecular apparatus that is responsible for photosynthesis, the critical process in which the energy from sunlight is used to produce life-sustaining oxygen. In the photosynthetic reaction, carbon dioxide is reduced by water and chlorophyll assists this transfer.

1.14.2.2 Circulation

The circulation of the world's water is controlled by a combination of gravity, friction, and inertia. Winds push water, ice, and water vapour around due to friction. Water vapour rises. Fresh and hot water rise. Salt and cold water sink. Ice floats. Water flows downhill. The high-inertia water at the Equator zooms eastward as it travels toward the slower-moving areas near the Poles (coriolis effect – an excellent example of this is the Gulf Stream off the East Coast of the USA). The waters of the world intensify on the Western boundary of oceans due to the earth's rotational mechanics (the so-called 'Western intensification' effect). Add into this mix the gravitational pull of the moon, other planets and the sun, and one has a very complex circulation model for the water flowing around our planet.

In order to break this complex model into its component parts for analysis, oceanographers generally separate these circulation factions into two broad categories, 'currents' and 'tides'. Currents are broadly defined as any horizontal movement of water along the earth's surface. Tides, on the other hand, are water movement in the vertical plane due to periodic rising and falling of the ocean surface (and connecting bodies of water) resulting from unequal gravitational pull from the moon and sun on different parts of the earth. Tides will cause currents, but tides are generally defined as the diurnal and semi-diurnal movement of water from the sun/moon pull.

A basic understanding of these processes will arm ROV operators with the ability to predict conditions at the work site, thus assisting in accomplishing the work task.

Per Bowditch (2002), 'currents may be referred to according to their forcing mechanism as either wind-driven or thermohaline. Alternatively, they may be classified according to their depth (surface, intermediate, deep, or bottom). The surface circulation of the world's oceans is mostly wind driven. Thermohaline currents are driven by differences in heat and salt. The currents driven by thermohaline forces are typically subsurface.' If performing a deep dive with an ROV, count on having a surface current driven by wind action and a subsurface current driven by thermohaline forces—plan for it and it will not ruin the day.

An example of the basic differences between tides and currents follow:

- In the Bay of Fundy's Minas Basin (Nova Scotia, Canada), the highest tides on planet earth occur near Wolfville. The water level at high tide can be as much as 45 feet (16 metres) higher than at low tide. Small Atlantic tides drive the Bay of Fundy/Gulf of Maine system near resonance to produce the huge tides. High tides happen every 12 hours and 25 minutes (or nearly an hour later each day) because of the changing position of the moon in its orbit around the earth. Twice a day at this location, large ships are alternatively grounded and floating. This is an extreme example of tides in action.
- At the Straits of Gibraltar, there is a vertical density current through the Straits. The evaporation of water over the Mediterranean drives the salinity of the water in that sea slightly higher than that of the Atlantic Ocean. The relatively denser high salinity waters in the Mediterranean flow out of the bottom of the Straits while the relatively lower (less dense) salinity waters from the Atlantic flow in on the surface. This is known as a density current. Trying to conduct an ROV operation there will probably result in a very bad day.
- Currents flow from areas of higher elevation to lower elevation. By figuring the elevation change of water over the area, while computing the water distribution in the area, one can find the volume of water that flows in currents past a given point (volume flow) in the stream, river, or body of water. However, the wise operator will find it much easier to just look it up in the local current/tide tables. There are people who are paid to make these computations on a daily basis, which is great as an intellectual exercise, but it is not recommended to 'recreate the wheel'.

(a) Currents

The primary generating forces for ocean currents are wind and differences in water density caused by variations in heat and salinity. These factors are further affected by the depth of the water, underwater topography, shape of the basin in which the current is running, extent and location of land, and the earth's rotational deflection. The effect of the tides on currents is addressed in the next Section.

Each body of water has its peculiar general horizontal circulation and flow patterns based upon a number of factors. Given water flowing in a stream or river, water accelerates at choke points and slows in wider basins per the equations of Bernoulli. Due to the momentum of the water, at a river bend, the higher volume of water (and probably the channel) will be on the outside of the turn. Vertical flow patterns are even more predictable with upwelling and downwelling patterns generally attached to the continental margins.

Just as there are landslides on land, so are there mudslides under the ocean. Mud and sediment detach from a subsea ledge and flow downhill in the oceans, bringing along with it a friction water flow known as a turbidity current. Locked in the turbidity current are suspended sediments. This increase in turbidity can degrade camera optics if operating in an area of turbidity currents; take account of this during project planning.

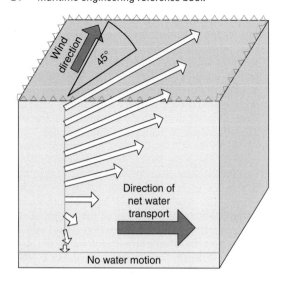

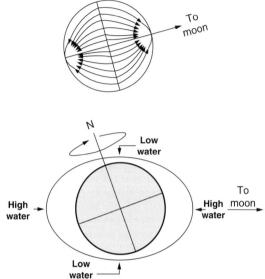

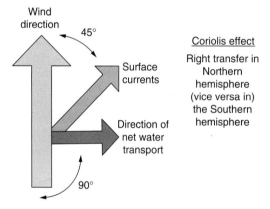

Figure 1.31 Ekman spiral.

Figure 1.32 Tidal movement in conjunction with planets.

Currents remain generally constant over the course of days or weeks, affected mostly by the changes in temperature and salinity profiles caused by the changing seasons.

Of particular interest to underwater vehicle operators is the wind-driven currents culled into the so-called 'Ekman spiral' (Figure 1.31). The model was developed by physicist V Walfrid Ekman from data collected by arctic exploration legend Fridtjof Nansen during the voyage of the *Fram*. From this model, wind drives idealized homogeneous surface currents in a motion 45° from the wind line to the right in the Northern hemisphere and to the left in the Southern hemisphere. Due to the friction of the surface water's movement, the subsurface water moves in an ever-decreasing velocity (and ever-increasing vector) until the momentum imparted by the surface lamina is lost (termed the 'depth of frictional influence'). Although the 'depth of frictional influence' is variable depending upon the latitude and wind velocity, the Ekman frictional transfer generally ceases at approximately 100 m depth. The net water transfer is at a right angle to the wind.

(b) Tides

Tides are generally referred to as the vertical rise and fall of the water due to the gravitational effects of the moon, sun, and planets upon the local body of water (Figure 1.32). Tidal currents are horizontal flows caused by the tides. Tides rise and fall. Tidal currents flood and ebb. The ROV operator is concerned with the amount and time of the tide, as it affects the drag velocity and vector computations on water flow across the work site. Tidal currents are superimposed upon non-tidal currents such as normal river flows, floods, and freshets. All of these factors are put together to find what the actual current will be for the job site (Figure 1.33).

On a vertical profile, the tide may interact with the general flow pattern from a river or estuary; the warm fresh water may flow from a river on top of the cold salt water (a freshet, as mentioned above) as the salt water creeps for a distance up the river. According to Van Dorn (1993), fresh water has been reported over 300 miles at sea off the Amazon. If a brackish water estuary is the operating area, problems to be faced will include variations in water density, water flow vector/speed, and acoustic/turbidity properties.

1.14.2.3 *Compressibility*

For the purposes of observation-class ROV operations, sea water is essentially incompressible. There is a slight

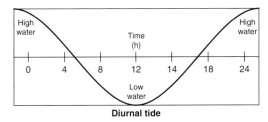
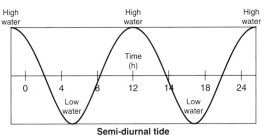
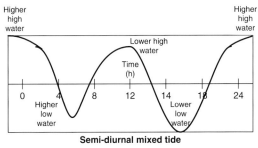

Figure 1.33 Tide periods over a 24-hour span.

compressibility factor, however, that does directly affect the propagation of sound through water. This compressibility factor will affect the sonic velocity computations at varying depths, see 'Sonic velocity and sound channels' later in this Section.

1.14.2.4 *Conductivity*

Conductivity is the measure of electrical current flow potential through a solution. In addition, because conductivity and ion concentration are highly correlated, conductivity measurements are used to calculate ion concentration in solutions.

Commercial and military operators observe conductivity for gauging water density for vehicle ballasting and such and for determining sonic velocity profiles for acoustic positioning and sonar use. Water quality researchers take conductivity readings to determine the purity of water, to watch for sudden changes in natural or waste water, and to determine how the water sample will react in other chemical analyses. In each of these applications, water quality researchers count on conductivity sensors and computer software to sense environmental waters, log and analyse data, and present this data.

1.14.2.5 *Density*

Density is mass per unit volume measured in SI units in kilograms per cubic metre (or, on a smaller scale, grams per cubic centimetre). The density of sea water depends upon salinity, temperature, and pressure. At a constant temperature and pressure, density varies with water salinity. This measure is of particular importance to the operation of an underwater vehicle for determination of the neutral buoyancy for the vehicle.

The density range for sea water is from 1.02200 to $1.030000 \, g/cm^3$ (Thurman, 1994). In an idealized stable system, the higher density water sinks to the bottom while the lower density water floats to the surface. Water under the extreme pressure of depth will naturally be denser than surface water, with the change in pressure (through motion between depths) being realized as heat. Just as the balance of pressure/volume/temperature is prevalent in the atmosphere, so is the temperature/salinity/pressure model in the oceans.

A rapid change in density over a short distance is termed a 'pycnocline' and can trap any number of energy sources from crossing this barrier, including sound (sonar and acoustic positioning systems), current, and neutrally buoyant objects in the water column (underwater vehicles). Changing operational area from lower latitude to higher latitude produces a mean temperature change in the surface layer. As stated previously, the deep ocean is uniformly cold due to the higher density cold water sinking to the bottom of the world's oceans. The temperature change from the warm surface at the tropics to the lower cold water can be extreme, causing a rapid temperature swing within a few metres of the surface. This surface layer remains near the surface, causing a small tight 'surface duct' in the lower latitudes. In the higher latitudes, however, the difference between ambient surface temperature and the temperature of the cold depth is less pronounced. The thermal mixing layer, as a result, is much larger, over a broader range of depth between the surface and the isothermal lower depths, and less pronounced (Figure 1.34). In Figure 1.34(a), density profiles by latitude and depth are examined to display the varying effects of deep-water temperature/density profiles versus ambient surface temperatures. Figures 1.34(b) and (c) look at the same profiles, but focusing on temperature and salinity. Figure 1.34(d) demonstrates a general profile for density at low to midlatitudes. The mixed layer is water of constant temperature due to the effects of wave mechanics/mixing.

A good example of the effect of density on ROV operations comes from a scientific mission conducted in 2003 in conjunction with *National Geographic* magazine. The mission was to the cenotes (sink holes) of the Northern Yucatan in Mexico. Cenotes are a series of pressure holes in a circular arrangement, centred around Chicxulub, the theoretical meteor impact point, purportedly left over from the K-T event from 65 million years ago that killed the dinosaurs.

26 Maritime engineering reference book

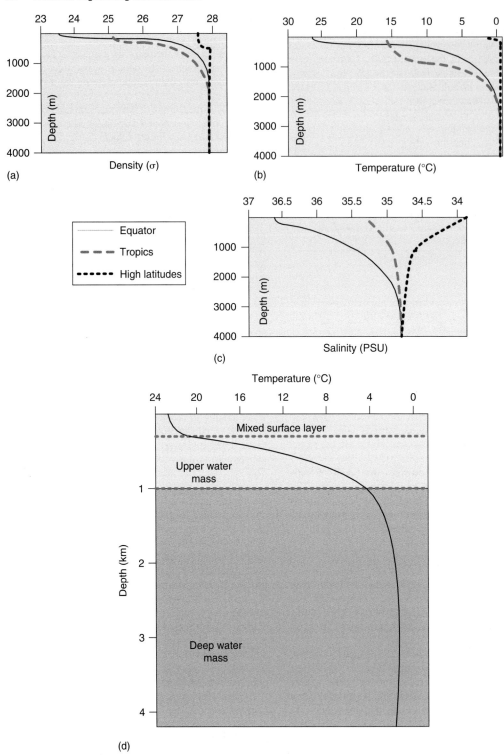

Figure 1.34 Density profiles with varying latitudes along with a general density curve.

The top water in the cenote is fresh water from rain runoff, with the bottom of the cenote becoming salt water due to communication, via an underground cave network, with the open ocean. This column of still water is a near perfect unmixed column of fresh water on top with salt water below. A micro-ROV was being used to examine the bottom of the cenote as well as to sample the salt water/freshwater (halocline) layer. The submersible was ballasted to the fresh water on top layer. When the vehicle came to the salt water layer, the submersible's vertical thruster had insufficient downward thrust to penetrate into the salt water below and kept 'bouncing' off the halocline. The submersible had to be re-ballasted for salt water in order to get into that layer and take the measurements, but the vehicle was useless on the way down due to it being too heavily ballasted to operate in fresh water.

1.14.2.6 Depth

Depth sensors, discussed below, measure the distance from the surface of a body of water to a given point beneath the surface, either the bottom or any intermediate level.

Depth readings are used by researchers and engineers in coastal and ocean profiling, dredging, erosion and flood monitoring, and construction applications.

Bathymetry is the measurement of depth in bodies of water. Further, it is the underwater version of topography in geography. Bottom contour mapping details the shape of the sea floor, showing the features, characteristics, and general outlay. Tools for bathymetry and sea bottom characterization follow.

(a) Echo sounder

An echo sounder measures the round trip time it takes for a pulse of sound to travel from the source at the measuring platform (surface vessel or on the bottom of the submersible) to the sea bottom and return. When mounted on a vessel, this device is generally termed a fathometer and when mounted on a submersible it is termed an altimeter.

According to Bowditch (2002), 'the major difference between various types of echo sounders is in the frequency they use. Transducers can be classified according to their beam width, frequency, and power rating. The sound radiates from the transducer in a cone, with about 50% actually reaching the sea bottom. Beam width is determined by the frequency of the pulse and the size of the transducer. In general, lower frequencies produce a wider beam, and at a given frequency, a smaller transducer will produce a wider beam. Lower frequencies penetrate deeper into the water, but have less resolution in depth. Higher frequencies have a greater resolution in depth, but less range, so the choice is a trade-off. Higher frequencies also require a smaller transducer. A typical low-frequency transducer operates at 12 kHz and a high-frequency one at 200 kHz'. Many smaller ROV systems have altimeters on the same frequency as their imaging system for easier software integration (same software can be used for processing both signals) and reduced cost purposes (Figure 1.35).

Computation of depth as determined by an echo sounder is determined via the following formula:

$$D = (V \times T/2) + K + D_r \quad (1.56)$$

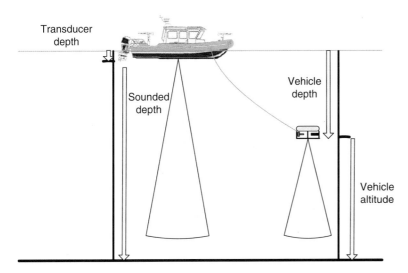

Figure 1.35 Vessel-mounted and sub-mounted sounders.

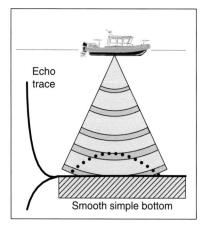

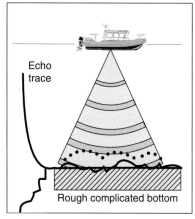

Figure 1.36 Acoustic seabed classification.

where D is depth below the surface (or from the measuring platform), V is the mean velocity of sound in the water column, T is time for the round trip pulse, K is the system index constant, and D_r is the depth of the transducer below the surface.

(b) Optic-acoustic seabed classification
Traditional sea floor classification methods have, until recently, relied upon the use of mechanical sampling, aerial photography, or multi-band sensors (such as Landsat™) for major bottom-type discrimination (e.g. mud, sand, rock, sea grass, and corals). Newer acoustic techniques for collecting hyperspectral imagery are now available through processing of acoustic backscatter.

Acoustic seabed classification analyses the amplitude and shape of acoustic backscatter echoed from the sea bottom for determination of bottom texture and makeup, Figure 1.36. Sea floor roughness causes impedance mismatch between water and the sediment. Further, reverberation within the substrate can be analysed in determining the overall composition of the bottom being insonified. Acoustic data acquisition systems and a set of algorithms that analyse the data allow for determining the seabed acoustic class.

1.14.2.7 Dissolved gases

Just as a soda dissolves CO_2 under the pressure of the soda bottle, so does the entire ocean dissolve varying degrees of gases used to sustain the life and function of the aquatic environment. The soda bottle remains at gas/fluid equilibrium under the higher-than-atmospheric pressure condition until the bottle is opened and the pressure within the canister is lowered. At that point, the gas bubbles out of solution until the gas/air mixture comes back into balance. If, however, that same bottle were opened in the high-pressure condition of a saturation diving bell deep within the ocean, that soda would (instead of bubbling) absorb CO_2 into solution until again saturated with that gas.

The degree of dissolved gases within a given area of ocean is dependent upon the balance of all gases within the area. The exchange of gases between the atmosphere and ocean can only occur at the air/ocean interface, i.e. the surface. Gases are dissolved within the ocean and cross the air/water interface based upon the balance of gases between the two substances (Table 1.11). A certain gas is said to be at local saturation if its distribution within the area is balanced given the local environmental conditions. The degree upon which a gas is able to dissolve within the substance is termed its solubility, and in water is dependent upon the temperature, salinity and pressure of the surrounding fluid, (Figure 1.37). Once a substance is at its maximum gas content (given the local environmental conditions) the substance is saturated with that gas. The degree to which a substance can either accept or reject transfer of gas into substance is deemed its saturation solubility.

The net direction of transfer of gases between the atmosphere and ocean is dependent upon the saturation solubility of the water. If the sea is oversaturated with a certain gas, it will off-gas back

Table 1.11 Distribution of gases in the atmosphere and dissolved in sea water.

Gas	Percentage of gas phase by volume		
	Atmosphere	Surface oceans	Total oceans
Nitrogen (N_2)	78	48	11
Oxygen (O_2)	21	36	6
Carbon dioxide (CO_2)	0.04	15	83

Source: Segar (1998).

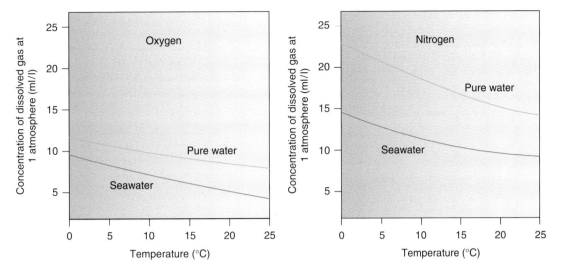

Figure 1.37 Solubility of two sea-level pressure gases based upon temperature.

into the atmosphere, and vice versa. Once equilibrium is again reached, net gas transfer ceases until some environmental variable changes.

Oxygen and CO_2 are dissolved in varying degrees within the world's oceans. Most marine life requires some degree of either of these dissolved gases in order to survive. In shallow water, where photosynthesis takes place, plant life consumes CO_2 and produces O_2. In the deep waters of the oceans, decomposition and animal respiration consumes O_2 while producing CO_2.

Dissolved oxygen, often referred to as DO, is the amount of oxygen that is dissolved in water. It is normally expressed in mg/l (mg/l = ppm) or percent air saturation. Dissolved oxygen can also enter the water by direct absorption from the atmosphere (transfer across the air/water interface).

Aquatic organisms such as plants and animals need dissolved oxygen to live. As water moves past the gills of a fish, microscopic bubbles of oxygen gas, called dissolved oxygen, are transferred from the water to the bloodstream. Dissolved oxygen is consumed by plants and bacteria during respiration and decomposition. A certain level of dissolved oxygen is required to maintain aquatic life. Although dissolved oxygen may be present in the water, it could be at too low a level to maintain life.

The amount of dissolved oxygen that can be held by water depends on the water temperature, salinity, and pressure.

- Gas solubility increases when temperature decreases (colder water holds more O_2).
- Gas solubility increases when salinity decreases (fresh water holds more O_2 than sea water).
- Gas solubility decreases as pressure decreases (less O_2 is absorbed in water at higher altitudes).

There is no mechanism for replenishing the O_2 supply in deep water while the higher pressure of the deep depths allows for greater solubility of gases. As a result, the deep oceans of the world contain huge amounts of dissolved CO_2. The question that remains is to what extent the industrial pollutants containing CO_2 will be controlled before the deep oceans of the world become saturated with this gas. Water devoid of dissolved oxygen (DO) will exhibit lifeless characterization.

Three examples of contrasting DO levels follow:

1. During an internal wreck survey of the USS *Arizona* in Pearl Harbor performed in conjunction with the US National Park Service, it was noted that the upper decks of the wreck were in a fairly high state of metal oxidation. However, as the investigation moved into the lower spaces with less and stable oxygenation due to little or no water circulation, the degree of preservation took a significant turn. On the upper decks, there were vast amounts of marine growth that had decayed the artifacts and encrusted the metal. However, on the lower living quarters, fully preserved uniforms were found where they were left on that morning over 60 years previous, still neatly pressed in closets and on hangars.

2. During an internal wreck survey of a B-29 in Lake Mead, Nevada (again done with the US National Park Service), the wreck area at the bottom of the reservoir was anaerobic, lacking significant levels of dissolved oxygen. The level of preservation of the wreck was amazing, with instrument readings still clearly visible, aluminium skin and structural members still in a fully preserved state and the shiny metal data plate on the engine still readable.

3. During the Cenote project in Northern Yucatan, mentioned earlier in this Section, the fresh water above the halocline was aerobic and alive with all matter of fish, plant, and insect life flourishing. However, once the vehicle descended into the anaerobic salt waters below the halocline, the rocks were bleached, the leaves dropped into the pit were fully preserved and nothing lived.

1.14.2.8 Fresh water

A vast majority of the world's freshwater supply is locked within the ice caps and glaciers of the high Arctic and Antarctic regions. Fresh water is vital to man's survival, thus most human endeavours surround areas of fresh water. Due to the shallow water nature of the freshwater collection points, man has placed various items of machinery, structures, and tooling in and around these locations. The observation-class ROV pilot will, in all likelihood, have plenty of opportunity to operate within the freshwater environment.

The properties of water directly affect the operation of underwater equipment in the form of temperature (affecting components and electronics), chemistry (affecting seals, incurring oxidation, and degrading machinery operation), and specific gravity (buoyancy and performance). These parameters will determine the buoyancy of vehicles, the efficiency of thrusters, the amount and type of biological specimens encountered, as well as the freezing and boiling points of the operational environment. The water density will further affect sound propagation characteristics, directly impacting the operation of sonar and acoustic positioning equipment.

Fresh water has a specific gravity of 1.000 at its maximum density. As water temperature rises, molecular agitation increases the water's volume, thus lowering the density of the fluid per unit volume. In the range between 3.98°C and the freezing point of water, the molecular lattice structure, in the form of ice crystals, again increases the overall volume, thus lowering its density per unit volume (remember, ice floats). The point of maximum density for fresh water occurs at 3.98°C, the point just before formation of ice crystals. At the freezing point of water, the lattice structure rapidly completes, thus significantly expanding the volume per unit mass and lowering the density at that temperature point. A graph describing the temperature/density relationship of pure water is shown in Figure 1.38.

Fresh water has a maximum density at approximately 4°C (see Figure 1.41) yet ocean water has no maximum density above the freezing point. As a result, lakes and rivers behave differently at the freezing point than ocean water. As the weather cools with the approach of winter, the surface water of a freshwater lake is cooled and its density is increased. Surface water sinks and displaces bottom water upward to be cooled in turn.

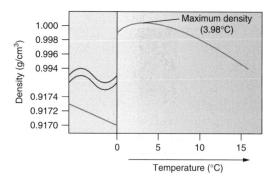

Figure 1.38 Water density versus temperature.

This convection process is called 'overturning'. This overturning process continues until the maximum density is achieved, thus stopping the convection churning process at 4°C. As the lake continues to cool, the crystal structure in the water forms, thus allowing the cooler water at the surface to decrease in density, driving still further the cooler (less dense) surface water upward, allowing ice to form at the surface.

1.14.2.9 Ionic concentration

There are four environmentally important ions: nitrate (NO_3^-), chloride (Cl^-), calcium (Ca^{2+}), and ammonium (NH_4^+). Ion-selective electrodes used for monitoring these parameters are described below.

- *Nitrate* (NO_3^-). Nitrate ion concentration is an important parameter in nearly all water quality studies. Nitrates can be introduced by acidic rainfall, fertilizer runoff from fields, and plant or animal decay or waste.
- *Chloride* (Cl^-). This ion gives a quick measurement of salinity of water samples. It can even measure chloride levels in ocean salt water or salt in food samples.
- *Calcium* (Ca^{2+}). This electrode gives a good indication of hardness of water (as Ca^{2+}). It is also used as an endpoint indicator in EDTA-Ca/Mg hard water titrations.
- *Ammonium* (NH_4^+). This electrode measures levels of ammonium ions introduced from fertilizers. It can also indicate aqueous ammonia levels if sample solutions are acidified to convert NH_3 to NH_4^+.

1.14.2.10 Light and other electro-magnetic transmissions through water

Light and other electro-magnetic transmission through water is affected by the following three factors:

(i) Absorption
(ii) Refraction
(iii) Scattering.

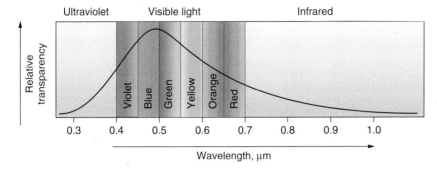

Figure 1.39 Light transparency through water (by wavelength).

All of these factors, which can be measured by an ROV using light sensors, are normally lumped under the general category of attenuation.

(i) Absorption
Electro-magnetic energy transmission capability through water varies with wavelength. The best penetration is gained in the visible light spectrum, Figure 1.39. Other wavelengths of electro-magnetic energy (radar, very low frequency RF, etc.) are able to marginally penetrate the water column (in practically all cases only a few wavelengths), but even with very high intensity transmission, only very limited transmission rates/depths are possible under current technologies. Submerged submarines are able to get RF communications in deep water with very low frequency RF, but at that frequency it may take literally minutes to get through only two alphanumeric characters.

In the ultraviolet range as well as in the infrared wavelengths, electro-magnetic energy is highly attenuated by sea water. Within the visible wavelengths, the blue/green spectrum has the greatest energy transparency, with other wavelengths having differing levels of energy transmission. Disregarding scattering (which will be considered below), within 1 metre of the surface, fully 60% of the visible light energy is absorbed, leaving only 40% of original surface levels available for lighting and photosynthesis. By the 10-metre depth range, only 20% of the total energy remains from that of the surface. By 100 metres, fully 99% of the light energy is absorbed, leaving only 1% visible light penetration – practically all in the blue/green regime, Duxbury and Alison (1997). Beginning with the first metre of depth, artificial lighting becomes increasingly necessary to bring out the true colour of objects of interest below the surface.

Why not put infrared cameras on ROVs? The answer is simple – the visible light spectrum penetration in water favours the use of optical systems, Figure 1.39. IR cameras can certainly be mounted on ROV vehicles, but the effective range of the sensor suffers significantly due to absorption. The sensor may be effective at determining reflective characteristics, but the sensor would be required to be placed at an extremely close range, negating practically all benefits from non-optical IR reflectance.

(ii) Refraction
Light travels at a much slower speed through water, effectively bending (refracting) the light energy as it passes through the medium. This phenomenon is apparent not only with the surface interaction of the sea water, but also with the air/water interface of the ROV's camera system.

(iii) Scattering
Light bounces off water molecules and suspended particles in the water (scattering), further degrading the light transmission capability, in addition to absorption, by blocking the light path. The scattering agents (other than water molecules) are termed suspended solids (e.g. silt, single-cell organisms, salt molecules, etc.) and are measured in mg/l on an absolute scale. Modern electronic instruments have been developed that allow real-time measurement of water turbidity from the ROV submersible or other underwater platform. The traditional physical measure of turbidity, however, is a simple measure of the focal length of a reflective object as it is lost from sight. Termed Secchi depth, a simple reflective Secchi disk (coated with differing colours and textures) is lowered into the water until it just disappears from view.

All of the above issues and parameters will aid in determining the submersible's capability to perform the assigned task within a reasonable timeframe.

1.14.2.11 *Pressure*

The SI unit for pressure is the kilopascal (expressed as kPa). One pascal is equal to one newton per square metre. However, oceanographers normally use ocean pressure with reference to sea level atmospheric pressure. The imperial unit is one atmosphere. The SI unit is the bar. The decibar is a useful measure of water pressure and is equal to 1/10 bar. Sea water generally increases by one atmosphere of pressure for every

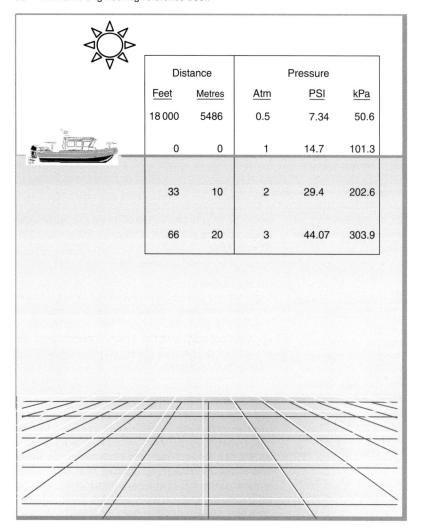

Figure 1.40 Pressure in atmospheres from various levels.

33 feet of depth (approximately equaling 10 metres). Therefore, one decibar is approximately equal to one metre of depth in sea water, Figure 1.40.

The sea-water pressure directly affects all aspects of submersible operation. The design of the submersible's air-filled components must withstand the pressures of depth, the floatation must stand up to the pressure without significant deformation, thus losing buoyancy and sinking the vehicle, and tethers must be sturdy enough to withstand the depth while maintaining their neutral buoyancy. These aspects are discussed further in Chapter 10.

1.14.2.12 *Salt water and salinity*

The world's water supply consists of everything from pure water to water plus any number of dissolved substances due to water's soluble nature. Water quality researchers measure salinity to assess the purity of drinking water, monitor salt water intrusion into freshwater marshes and groundwater aquifers, and to research how the salinity will affect the ecosystem.

The two largest dissolved components of typical sea water are chlorine (56% of total) and sodium (31% of total), with the total of all lumped under the designation of 'salts'. Components of typical ocean water dissolved salts are comprised of major constituents, minor constituents and trace constituents. An analysis of one kilogram of sea water, detailing only the major constituents of dissolved salts, is provided in Table 1.12.

The total quantity of dissolved salts in sea water is expressed as salinity, which can be calculated

Table 1.12 Dissolved salts in water.

Component	Mass in grams
Pure water	965.31
Major constituents	
Chlorine	19.10
Sodium	10.62
Magnesium	1.28
Sulfur	2.66
Calcium	0.40
Potassium	0.38
Minor constituents	0.24
Trace constituents	0.01
Total (in grams)	1000.00

from conductivity and temperature readings. Salinity was historically expressed quantitatively as grams of dissolved salts per kilogram of water (expressed as ‰) or, more commonly, in parts per thousand (PPT). To improve the precision of salinity measurements, salinity is now defined as a ratio of the electrical conductivity of the sea water to the electrical conductivity of a standard concentration of potassium chloride solution. Thus, salinity is now defined in practical salinity units (PSU), although one may still find the older measure of salt concentration in a solution as parts per thousand (PPT) or ‰ used in the field.

Ocean water has a fairly consistent makeup, with 99% having between 33 and 37 PSU in dissolved salts. Generally, rain enters the water cycle as pure water then gains various dissolved minerals as it travels toward the ocean. Water enters the cycle with a salt content of 0 PSU, mixes with various salts to form brackish water (in the range of 0.5–30 PSU) as it blends within rivers and estuaries, then homogenizes with the ocean water (75% of the ocean's waters have between 33 and 34 PSU of dissolved salts) as the cycle ends, then renews with evaporation.

Just as a layer of rapid change in temperature (the thermocline) traps sound and other energy, so does an area of rapid change in salinity, known as a halocline. These haloclines are present both horizontally, see cenote example earlier in this Section, and vertically (e.g. rip tides at river estuaries).

As the salinity of water increases, the freezing point decreases. As an anecdote to salinity, there are brine pools under the Antarctic ice amidst the glaciers in the many lakes of Antarctica's McMurdo Dry Valleys. A team recently found a liquid lake of super-concentrated salt water, seven times saltier than normal sea water, locked beneath 62 feet (19 metres) of lake ice, a record for lake ice cover on earth.

Salts dissolved in water change the density of the resultant sea water for these reasons:

- The ions and molecules of the dissolved substances are of a higher density than water.
- Dissolved substances inhibit the clustering of water molecules (particularly near the freezing point), thus increasing density and lowering the freezing point.

Unlike fresh water, ocean water continues to increase in density up to its freezing point of approximately −2°C. At 0 PSU (i.e. fresh water), maximum density is approximately 4°C with a freezing point of 0°C. At 24.7 PSU and above, ocean water has a freezing point of its maximum density; Therefore, there is no maximum density temperature above the freezing point. The maximum density point scales in a linear fashion between 0 and 24.7 PSU (see Figure 1.41). Thus, ocean water continues to increase in density as it cools and sinks in the open ocean. The deep waters of the world's oceans are uniformly cold as a result.

Comparisons of salinity and temperature effects upon water density yield the following:

- At a constant temperature, variation of the salinity from 0 to 40 PSU changes the density by about 0.035 specific gravity (or about 3.5% of the density for 0 PSU water at 4°C).
- At a constant salinity (i.e. 0 PSU), raising the temperature from 4°C (maximum density) to 30°C (highest temperature generally found in surface water) yields a decrease in density of 0.0043 (1.000 to 0.9957) for a change of 0.4%.

Clearly, salinity has a much higher effect upon water density than does temperature.

As a practical example, suppose a 100-kilogram submersible is ballasted for 4°C fresh water at exactly neutral buoyancy. If that same submersible were to be transferred to salt water, an additional ballasting weight of approximately 3.5 kilograms

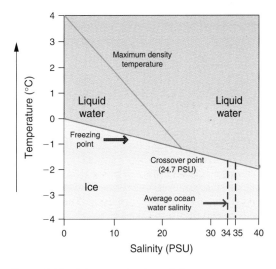

Figure 1.41 Salt-water density, salinity versus temperature.

would be required to maintain that vehicle at neutral buoyancy.

Ice at 0°C has a density of 0.917 g/cm³, which is about 8% less than that of water at the same temperature. Obviously, water expands when it freezes, bursting pipes and breaking apart water-encrusted rocks, thus producing revenues for marine and land plumbing contractors.

1.14.2.13 *Solar radiation*

Solar radiation is the electro-magnetic radiation emitted by the sun and is measured in some underwater scientific applications.

1.14.2.14 *Sonic velocity and sound channels*

Sound propagation (vector and intensity) in water is a function of its velocity, and velocity is a function of water density and compressibility. As such, sound velocity is dependent upon temperature, salinity, and pressure, and is normally derived expressing these three variables (Figure 1.42). The speed of sound in water changes by 3–5 metres per second per °C, by approximately 1.3 metres per second per PSU salinity change, and by about 1.7 metres per second per 100 metres change in depth (compression). The speed of sound in sea water increases with increasing pressure, temperature, and salinity (and vice versa).

The generally accepted underwater sonic velocity model was derived by W.D. Wilson in 1960. A simplified version of Wilson's (1960) formula on the speed of sound in water follows:

$$c = 1449 + 4.6T - 0.055T^2 + 0.0003T^3 + 1.39(S - 35) + 0.017D \quad (1.57)$$

where c is the speed of sound in metres per second, T is the temperature in °C, S is the salinity in PSU, and D is the depth in metres.

Temperature/salinity/density profiles are important measurements for sensor operations in many underwater environments, and they have a dramatic effect on the transmission of sound in the ocean. A change in overall water density over short range due to any of these three variables (or in combination) is termed a pycnocline. Overall variations of pressure and temperature are depicted graphically in Figure 1.43.

This layering within the ocean, due to relatively impervious density barriers, causes the formation of sound channels within bodies of water. These 'channels' trap sound, thus channeling it over possibly long ranges. Sound will also refract based upon its travel across varying density layers, bending toward the more dense water and affecting both range and bearing computations for acoustics (Figure 1.44). Over short ranges (tens or hundreds of metres) this may not be a substantial number

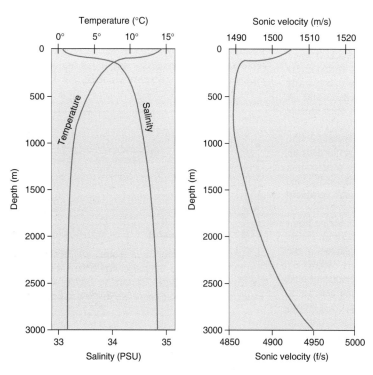

Figure 1.42 Sonic velocity profiles with varying temp/salinity/depth.

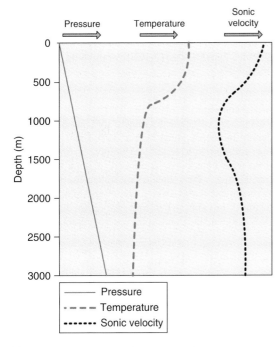

Figure 1.43 Variations of pressure and temperature with depth producing sound velocity changes.

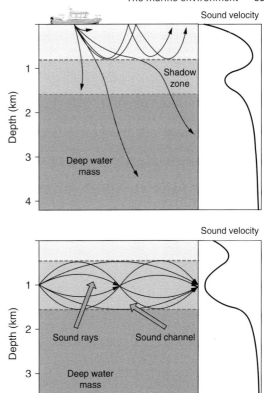

Figure 1.44 Sound velocity profiles and channeling at various depths and layers.

and can possibly be disregarded, but for the longer distances of some larger ROV systems, this becomes increasingly a factor to be considered.

1.14.2.15 *Turbidity*

Turbidity, which causes light scattering – see Section 1.14.2.10 (iii), the measure of the content of suspended solids in water, is also referred to as the 'cloudiness' of the water. Turbidity is measured by shining a beam of light into the solution. The light scattered off the particles suspended in the solution is then measured, and the turbidity reading is given in nephelometric turbidity units (NTU). Water quality researchers take turbidity readings to monitor dredging and construction projects, examine microscopic aquatic plant life, and to monitor surface, storm, and wastewater.

1.14.2.16 *Viscosity*

Viscosity is a liquid's measure of internal resistance to flow, or resistance of objects to movement within the fluid. Viscosity varies with changes in temperature and salinity, as does density. Sea water is more viscous than fresh water, which will slightly affect computations of vehicle drag. Values of kinematic viscosity (v) are given in Table 1.13.

Table 1.13 Viscosity of fresh water, sea water, and air (Molland and Turnock (2007).

Temperature (°C)		10	15	20
Kinematic viscosity (m²/s)	FW × 10⁶	1.30	1.14	1.00
	SW × 10⁶	1.35	1.19	1.05
(Pressure = 1 atm)	Air × 10⁵	1.42	1.46	1.50

Note: FW, fresh water; SW, sea water.

1.14.2.17 *Water quality*

Water quality researchers count on sensors and computer software to sense environmental waters, log and analyse data. Factors to be considered in water quality measurement are discussed below.

(i) Alkalinity and pH
The acidity or alkalinity of water is expressed as pH (potential of hydrogen). This is a measure of the concentration of hydrogen (H^+) ions. Water's pH is

expressed as the logarithm of the reciprocal of the hydrogen ion concentration, which increases as the hydrogen ion concentration decreases, and vice versa. When measured on a logarithmic scale of 0 to 14, a pH of 0 is the highest acidity, a pH of 14 is the highest alkalinity, and a pH of 7 is neutral, Figure 1.45. Pure water is pH neutral, with sea water normally at a pH of 8 (mildly alkaline).

pH measurements help determine the safety of water. The sample must be between a certain pH to be considered drinkable, and a rise or fall in pH may indicate a chemical pollutant. Changes in pH affect all life in the oceans; Therefore, it is most important to aquatic biology to maintain a near neutral pH. As an example, shellfish cannot develop calcium carbonate hard shells in an acidic environment.

(ii) ORP (oxidation reduction potential)
ORP is the measure of the difference in electrical potential between a relatively chemically inert electrode and an electrode placed in a solution. Water quality researchers use ORP to measure the activity and strength of oxidizers (those chemicals that accept electrons) and reducers (those that lose electrons) in order to monitor the reactivity of drinking water and groundwater.

(iii) Rhodamine
Rhodamine, a highly fluorescent dye, has the unique quality to absorb green light and emit red light. Very few substances have this capability, so interference from other compounds is unlikely, making it a highly specific tracer. Water quality researchers use rhodamine to investigate surface water, wastewater, pollutant time of travel, groundwater tracing, dispersion and mixing, circulation in lakes, and storm water retention.

(iv) Specific conductance
Specific conductance is the measure of the ability of a solution to conduct an electrical current. However, unlike the conductivity value, specific conductance readings compensate for temperature. In addition, because specific conductance and ion concentration are highly correlated, specific conductance measurements are used to calculate ion concentration in solutions. Specific conductance readings give the researcher an idea of the amount of dissolved material in the sample. Water quality researchers take specific conductance readings to determine the purity of water, to watch for sudden changes in natural or wastewater, and to determine how the water sample will react in other chemical analyses.

(v) Total dissolved solids
TDS (total dissolved solids) is the measure of the mass of solid material dissolved in a given volume of water, and is measured in grams per litre. The TDS value is calculated based on the specific conductance reading and a user-defined conversion factor. Water quality researchers use TDS measurements to evaluate the purity or hardness of water, to determine how the sample will react in chemical analyses, to watch for sudden changes in natural or wastewater, and to determine how aquatic organisms will react to their environment.

1.14.2.18 *Water temperature*

Water temperature is a measure of the kinetic energy of water and is expressed in degrees Fahrenheit (F) or Celsius (C). Water temperature varies according to season, depth and, in some cases, time of day. Because most aquatic organisms are cold-blooded, they require a certain temperature range to survive. Some organisms prefer colder temperatures and others prefer warmer temperatures. Temperature also affects

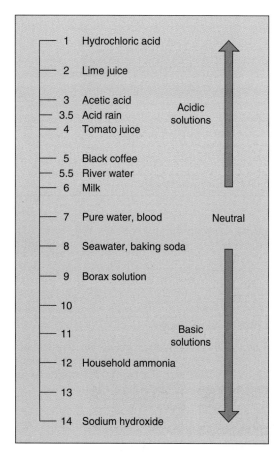

Figure 1.45 Graphical presentation of common items with their accompanying pH.

the water's ability to dissolve gases, including oxygen. The lower the temperature, the higher the solubility. Thermal pollution, the artificial warming of a body of water because of industrial waste or runoff from streets and parking lots, is becoming a common threat to the environment. This artificially heated water decreases the amount of dissolved oxygen and can be harmful to cold-water organisms.

In limnological research, water temperature measurements as a function of depth are often required. Many reservoirs are controlled by selective withdrawal dams, and temperature monitoring at various depths provides operators with information to control gate positions. Power utility and industrial effluents may have significant ecological impact with elevated temperature discharges. Industrial plants often require water temperature data for process, use, and heat transfer calculations.

Pure water freezes at 32°F (0°C) and boils at 212°F (100°C). ROV operations do not normally function in boiling water environments; The focus is, therefore, on the temperature ranges in which most ROV systems operate (0–30°C). The examination of salinity will be in the range from fresh water to the upper limit of sea water.

Temperature in the oceans varies widely both horizontally and vertically. On the high temperature side, the Persian Gulf region during summertime will achieve a maximum of approximately 32°C. The lowest possible value is at the freezing point of −2°C experienced in polar region(s).

The vertical temperature distribution nearly everywhere, except the polar regions, displays a profile of decreasing water temperature with increasing depth. Assuming constant salinity, colder water will be denser and will sink below the warmer water at the surface.

There is usually a mixed layer of isothermal (constant temperature) water from the surface to some near-surface depth due to wind mixing and convective overturning (thermally driven vertical density mixing) that changes with the seasons (Figure 1.46). The layer

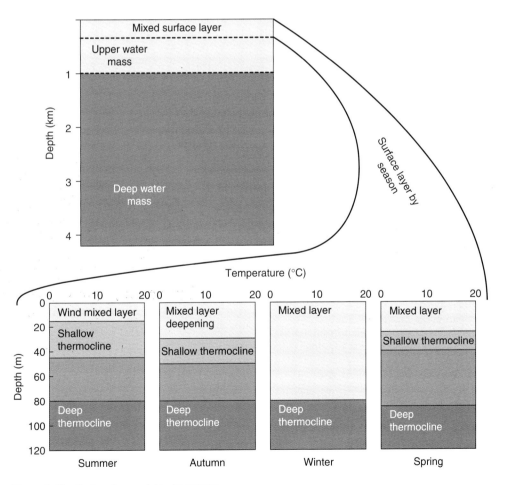

Figure 1.46 Surface layer mixing by season.

is thin at the equator and thick at the poles. The layer where there is a rapid change in temperature over a short distance is termed a thermocline and has some interesting characteristics. Due to the rapid temperature gradient, this thermocline forms a barrier that can trap sound energy, light energy, and any number of suspended particles. The degree of perviousness of the barrier is determined by the relative strength or degree of change over distance. For the ROV operator, a thermocline in the area of operation can hinder the function of acoustic positioning, sonar, and any sounding equipment aboard attempting to burn through the layer. It is especially of concern to those involved in anti-submarine warfare.

1.14.3 Coastal zone classifications and bottom types

General coastal characteristics tend to be similar for thousands of kilometres. Most coasts can be classified as either erosional or depositional depending upon whether their primary features were created by erosion of land or deposition of eroded materials. Erosional coasts have developed where the shore is actively eroded by wave action or where rivers or glaciers caused erosion when the sea level was lower than its present level. Depositional coasts have developed where sediments accumulate either from a local source or after being transported to the area in rivers and glaciers or by ocean currents and waves.

Of primary interest to the ROV operator, with regard to coastal zones, is the general classification of these zones and its affect upon general water turbidity in the operational area. Depositional coasts tend to have a higher quantity of suspended solids in the water column, thus a higher turbidity and degraded camera performance. Erosional coasts tend to possess less suspended particles, thus featuring better camera optics. Further, the depositional source will greatly affect the level of turbidity since mud deposited from a river estuary will have a higher turbidity, than will a rock and sand drainage area.

As stated earlier, oceans cover 70.8% of the earth's surface. Of that composition, the distribution between continental margins and deep-sea basins is provided in Table 1.14.

A substantial amount of scientific and oil exploration/production work is done in the continental margins with ROVs. The continental margins are, in large part, depositional features. Their characteristics are driven by run-off deposited from the adjacent continent.

Sediments are carried from the marine estuaries, then deposited onto the continental shelf. As the sea floor spreads due to tectonic forces, the sediments

Table 1.14 Ocean coverage distribution.

	$10^6 \ km^2$	Percentage of earth's Surface area
Continental margin	93	18.2
Deep basins	268	52.6
Total	361	70.8

fall down the continental slope and come to rest on the abyssal plain. Substantial amounts of oil and gas deposits are locked in these sediments, and are the focus of exploration and production efforts. The general bottom characteristics of this shelf will be mud and sediment.

1.15 Ambient air

A ship is completely immersed in two fluids—air for the upper portion and water for the lower portion. Because the specific heat of water is much greater than that for air, the former suffers a much slower variation of temperature due to climatic changes. In defining the temperature environment of the ship, it is necessary to quote temperatures for both the air and water. Further, it is common experience that the humidity of air is important in determining the degree of comfort in hot climates. This idea is elaborated later in this Section, but for the present, it is enough to know that for air it is necessary to specify the temperature that would be recorded on both a 'wet bulb' and a 'dry bulb' thermometer.

Standard temperatures considered by the UK Ministry of Defence as being appropriate to different regions of the world at various times of the year are given in Table 1.15.

For air conditioning calculations, conditions given in Table 1.16 are recommended.

These temperatures are important to the ship design in a number of ways. The air temperature, for instance, influences:

(a) the amount of heating or cooling required for air entering the ship to maintain a given internal temperature;
(b) the amount of thermal insulation which can be economically justified;
(c) efficiency of some machinery;
(d) the amount of icing likely to occur in cold climates and thus the standard of stability required to combat this, see Chapter 3, and the need to provide special de-icing facilities.

Table 1.15 Standard climatic figures, UK Ministry of Defence.

Region	Average max. Summer temperature (°C)			Average min. Winter temperature (°C)		
	Air		Deep sea	Air		Deep sea
	Dry bulb	Wet bulb	Water temp. near surface	Dry bulb	Wet bulb	Water temp. near surface
Extreme Tropics	34.5	30	33			
Tropics	31	27	30			
Temperature	30	24	29			
Temperate Winter				−4	—	2
Sub-Arctic Winter				−10	—	1
Arctic and Antarctic Winter				−29	—	−2

Table 1.16 Recommended air conditioning temperatures.

	Dry bulb (°C)	Wet bulb (°C)	Relative humidity (%)
Red Sea passenger vessels	32	30	85
Other passenger vessels	31	27	70
Extreme Tropical cargo vessels and tankers	32	29	78

The water temperature influences:

(a) density of water and thus the draught;
(b) the efficiency of, and hence the power output from, machinery, e.g. steam machinery may suffer a loss of about 3% of power in the tropics compared with temperate conditions. The effect on other machinery may be greater, but all types are affected to some degree including auxiliaries such as electrical generators, pumps, and air conditioning plant;
(c) whether ice is likely to be present and if so how thick and, thus, whether special strengthening of the ship is necessary.

Both air and water temperatures can cause expansion or contraction of materials leading to stresses where two dissimilar materials are joined together. Temperature differences between the air and water can lead to relative expansion or contraction of the upper deck relative to the keel causing the ship to hog or sag. This leads to stresses in the main hull girder and can upset the relative alignment of items some distance apart longitudinally. Similarly, if one side of the ship is in strong sunlight the main hull will tend to bend in a horizontal plane.

The above temperatures are used in calculations for assessing the capacity required of ventilation or air conditioning systems or for the degradation in machinery power generation.

Equipment should be tested in accordance with standard specifications:

(a) in locations where the ambient temperature rises and remains at a relatively constant level, e.g. in machinery spaces where the temperature is mainly influenced by the heat generated by machinery;
(b) where the adjacent air temperatures are indirectly affected by solar heating, e.g. equipment in the open but not fully exposed;
(c) fully exposed to solar radiation. In this case not only surface heating is important but also the degradation of materials caused by the ultraviolet radiation;
(d) in low temperature environments.

The tests may have to cater for high or low humidity conditions and may need to allow for diurnal variations.

1.16 Climatic extremes

The designer must allow for extreme climatic conditions although these may only be met on a few occasions. These are usually specified by an owner. British Standards Specifications and the MIL Specifications of the USA are usual. Care is necessary to ensure that the latest specifications are used.

Dust and sand. These cause abrasion of surfaces and penetrate into joints where movements due to vibration,

Table 1.17 Extreme recorded air temperature (°C).

	Harbour	At sea
Tropics	52	38
Arctic	−40	−30

and to changes in temperature, may bring about serious wear. Movements of particles can also set up static electrical charges. Although arising from land masses, dust and sand can be carried considerable distances by high winds. Specifications define test conditions, e.g. grade of dust or sand, concentration, air velocity and duration.

Rain. Equipment should be able to operate without degradation of performance in a maximum rainfall intensity of 0.8 mm/min at 24°C and a wind speed of 8 m/s for periods of 10 min with peaks of 3 mm/min for periods of 2 min. The test condition is 180 mm/h of rain for 1 hour. Driving sea spray can be more corrosive than rain.

Icing. Equipments fitted in the mast head region must remain operational and safe with an ice accretion rate of 6.4 mm/h with a total loading of 24 kg/m^2. Corresponding figures for equipment on exposed upper decks are 25 mm/h and 70 kg/m^2. Such equipments must be able to survive a total loading of 120 kg/m^2. Equipments fitted with de-icing facilities should be able to shed 25 mm ice per hour.

Hail. Equipment should be able to withstand without damage or degradation of performance a hailstorm with stones of 6 to 25 mm, striking velocity 14 to 25 m/s, duration 7 min.

High winds. Exposed equipment should remain secure and capable of their design performance in a relative steady wind to ship speed of 30 m/s with gusts of 1 min duration of 40 m/s. They must remain secure, albeit with some degradation of performance in wind to ship speed of 36 m/s with gusts of 1 min duration of 54 m/s.

Green sea loading. With heavy seas breaking over a ship, exposed equipment and structure are designed to withstand a green sea frontal pulse loading of 70 kPa acting for 350 ms with transients to 140 kPa for 15 ms.

Solar radiation. Direct sunlight in a tropical summer will result in average surface temperatures of up to 50°C for a wooden deck or vertical metal surfaces and 60°C for metal decks. Equipment should be designed to withstand the maximum thermal emission from solar radiation which is equivalent to a heat flux of 1120 W/m^2 acting for 4 hours causing a temperature rise of about 20°C on exposed surfaces.

Mould growth. Mould fungi feed by breaking down and absorbing certain organic compounds. Growth generally occurs where the relative humidity (rh) is greater than 65% and the temperature is in the range 0 to 50°C. Most rapid growth occurs with rh greater than 95% and temperature 20 to 35°C and the atmosphere is stagnant. Equipments should generally be able to withstand exposure to a mould growth environment for 28 days (84 for some critical items) without degradation of performance. Moulds can damage or degrade performance by:

(a) direct attack on the material, breaking it down. Natural products such as wood, cordage, rubber, greases, etc., are most vulnerable;
(b) indirectly by association with other surface deposits which can lead to acid formation;
(c) physical effects, e.g. wet growth may form conducting paths across insulating materials or change impedance characteristics of electronic circuits.

1.17 Marine pollution

As well as the effect of the environment on the ship, it is important to consider the effect of the ship on the environment. Over the years the pollution of the world's oceans has become a cause for increasing international concern. In 1972 the United Nations held a Conference on the Human Environment in Stockholm. This conference recommended that ocean dumping anywhere should be controlled. One outcome was the Convention on the Prevention of Marine Pollution by Dumping of Wastes and Other Matter – *the London Convention* – which came into force in 1975. This convention covers dumping from any source and it must be remembered that wastes dumped from ships amount to only about 10% of the pollutants that enter the sea annually. Land sources account for 44% of pollutants, the atmosphere for 33% (most originating from the land), 12% come from maritime transportation and 1% from offshore production. Thus, the naval architect is concerned directly with rather less than a quarter of the total waste problem. The convention bans the dumping of some materials and limits others. It also controls the location and method of disposal.

In terms of quantity, oil is the most important pollutant arising from shipping operations. The harmful effects of large oil spillages have received wide publicity because the results are so concentrated if the spillage is close to the coast. Most incidents occur during the loading or discharge of oil at a terminal but far greater quantities of oil enter the sea as a result of normal tanker operations such as the cleaning of cargo residues. Important as these are they are only one aspect of marine pollution to be taken into account in designing a ship. Many chemicals carried at sea are a much greater threat to the environment. Some chemicals are so dangerous that their carriage in bulk is banned. In these cases they may be carried in drums. Fortunately, harmful chemicals are carried in smaller ships than tankers, and the ship design is quite complex to enable their cargoes to be carried safely. One ship may carry several chemicals, each posing its own problems.

The detailed provisions of the regulations concerning pollution should be consulted for each case, MARPOL (2002). For sewage, broadly the limitations imposed are that raw sewage may not be discharged at less than 12 nautical miles (NM) from land; macerated and disinfected sewage at not less than 4 NM; only discharge from approved sewage treatment plants is permitted at less than 4 NM. No dunnage may be dumped at less than 25 NM from land and no plastics at all. Levels of pollution from all effluents must be very low.

The rules can have a significant affect upon the layout of, and equipment fitted in, ships. Sources of waste are grouped in vertical blocks to facilitate collection and treatment. Crude oil washing of the heavy oil deposits in bulk carrier oil tanks and segregated water ballast tanks are becoming common. Steam cleaning of tanks is being discontinued. Sewage presents some special problems. It can be heat treated and then burnt. It can be treated by chemicals but the residues have still to be disposed of. The most common system is to use treatment plant in which bacteria are used to break the sewage down. Because the bacteria will die if they are not given enough 'food', action must be taken if the throughput of the system falls below about 25% of capacity, as when, perhaps, the ship is in port. There is usually quite a wide fluctuation in loading over a typical 24 hour day. Some ships, typically ferries, prefer to use holding tanks to hold the sewage until it can be discharged in port.

In warships the average daily arisings from garbage amount to 0.9 kg per person food waste and 1.4 kg per person other garbage. It is dealt with by a combination of incinerators, pulpers, shredders and compactors.

References (Chapter 1)

Airy, G.B. (1845). Tides and waves, *in Encyclopedia Metropolitana*, p. 192.

Arndt, B., Brandl, H. and Vogt, K. (1982). '20 Years of experience – stability regulations of the West-German Navy. *Proc. of STAB'82: 2nd International Conference on Stability of Ships and Ocean Vehicles,* Tokyo, p. 765.

Bishop, R.E.D., Price, W.G. and Temeral, P. (1991). A theory on the loss of the *MV Derbyshire*. *Trans. RINA*, Vol. 133.

Boukhanovsky, A.V., Degtyarev, A.B., Lopatoukhin, L.J. and Rozhkov, Y.V. (2000). 'Stable states of wave climate: Applications for risk estimation'. *Proc. of STAB'2000: 7th International Conference on Stability of Ships and Ocean Vehicles,* Vol. 2, Launceston, Tasmania, pp. 831–846.

Bowditch, N. (2002). *The American Practical Navigator*. National Imagery and Mapping Agency, ISBN 1-57785-271-0.

Bretschneider, C.L. (1952). The generation and decay of wind waves in deep water. *Trans. American Geophys. Union*, Vol. 33, No. 3.

Bretschneider, C.L. (1961). *A one-dimensional gravity wave spectrum*. Ocean Wave Spectra, Prentice Hall, Englewood Cliffs, New Jersey.

Cartwright, D.E. and Longuet-Higgins, M.S. (1956). The statistical distribution of the maxima of a random function. *Proc. of Royal Soc.*, Vol. 237, Series A, London.

Christ, R.D. and Wernli, S.R. (2007). *The ROV manual*. Butterworth-Heinemann, Oxford, UK.

Chung-Tu Teng, Timpe, G. and Palao, I.M. (1994). The development of design waves and wave spectra for use in ocean structure design. *SNAME Transactions*, Vol. 102, pp. 475–499.

Clauss, G., Lehmann, E. and Östergaard, C. (1992). *Offshore structures: conceptual design and hydromechanics*. Springer, Berlin.

Davenport, A.G. (1964). Note on the distribution of the largest value of a random function with application to gust loading. *Proc. of Royal Soc.*

de Beurs (1957). 'Speed and pitching' Holland.

Department of the Army, Corps of Engineers. (2002). *Coastal Engineering Manual*. Publication number EM 1110-2-1100.

DHI (1981). Deutsches Hydrographisches Institut: Handbuch des Atlantischen Ozeans, No. 2057, Hamburg.

Duxbury, A.C. and Alison, B. (1997). *An Introduction to the World's Oceans*, 5th edition. William C. Brown, ISBN 0-697-28273-2.

Faltinson, O.M. (1990). *Sea Loads on Ships and Offshore Structures*. Cambridge University Press.

Faulkner, D. and Williams, R.A. (1996). Design for abnormal ocean waves. *Trans. RINA*, Vol. 138.

Faulkner, D. (2003). Freak waves – what can we do about them? Lecture to *The Honourable Company of Master Mariners and the Royal Institute of Navigation*, London.

Fondriest Environmental. Informational website page on environmental sensors at: http://www.fondriest.comlparameter.htm. Reprinted with permission.

Gumbel, E.J. (1958). *Statistics of extremes*. Columbia University Press.

Hasselmann, K. (1973). Measurements of wind-wave growth and swell decay during the Joint North Sea Wave Project (JONSWAP). *Dt. Hydrogr. Z. Reihe A*, No. 8, p. 12.

Hasselmann, K. *et al.* (1976). A parametric wave prediction model. *J. Phys. Oceanography*, Vol. 6, No. 2.

Hattendorf. (1974). *Handbuch der Werften XII*. Verlag Hansa, Hamburg.

Hogben, N. and Lumb, F.E. (1967). *Ocean wave statistics*. Her Majesty's Stationary Office, London.

Hogben, N., Dacunha, N.M.C. and Oliver, G.F. (1986). *Global wave statistics*. Compiled and edited by British Maritime Technology Ltd, Unwin Brothers.

Horn, F., Süchting, W., Klindwort, E. and Hebecker, O. (1953). *Erkenntnisse und Erfahrungen auf dem Gebiet der Schiffsstabilität*. STG, Hamburg.

Hutchison, B.L. (1990). Seakeeping studies: A status report. *Transactions SNAME*, Vol. 98, pp. 263–317.

Kobylinski, L.K. and Kastner, S. (2003). *Stability and Safety of Ships*. Elsevier, Oxford, UK.

Lewis, E.V. (ed) (1989). *Principles of Naval Architecture*, Second Revision. SNAME.

Longuet-Higgins, M.S. and Cokelet, E.D. (1976). The calculation of steep gravity waves. *Proc. of BOSS'76*, Trondheim.

MARPOL (2002). *MARPOL 73/78* Consolidated edition, 2002. IMO Publication (IMO – 1B 520E). IMO, London.

Michell, J.H. (1893). On the highest waves in water. Phil. Mag., 5th Series 36, pp. 430–437.

Molland, A.F. and Turnock, S.R. (2007). *Marine Rudders and Control Surfaces*. Butterworth-Heinemann, Oxford, UK.

Pierson, W.J. and Moskowitz, L. (1963). A proposed spectral form for fully developed wind seas, Technical report for U.S. Naval Oceanographic Office, New York University (also *J. Geophys. Res.*, Vol. 59, No. 24, 1964).

Price, W.G. and Bishop, R.E.D. (1974). *Probabilistic theory of ship dynamics*. Chapman and Hall, London.

Rawson, K.J. and Tupper, E.C. (2001). *Basic Ship Theory*, 5th edition. Combined Volume. Butterworth-Heinemann, Oxford, UK.

Roren, E.M.Q. and Furnes, O. (1976). State of the art, behaviour of structures and structural design. *Proc. of BOSS'76*, NIT, Trondheim.

Segar, D.A. (1998). *Introduction to Ocean Science*. Wadsworth, ISBN 0-314-09705-8.

St. Denis, M. and Pierson, W.J. (1953). On the motions of ships in confused seas. *Transactions SNAME*, Vol. 61.

Stokes, G.G. (1849). *Transactions Cambridge Phil. Soc.*, p. 441.

Stokes, G.G. (1880). *Mathematical and Physical papers*, Vol. 1. 314. Cambridge University Press.

Teather, R.G. (1994). *Royal Canadian Mounted Police Encyclopedia of Underwater Investigations*. Best, ISBN 0-941332-26-8.

Thurman, H.V. (1994). *Introductory Oceanography*, 7th edition. Macmillan, ISBN 0-02-420811-6.

Tupper, E.C. (2004). *Introduction to Naval Architecture*, 4th edition. Butterworth-Heinemann, Oxford, UK.

US Geological Survey. Educational website page on water science at: http://ga.water.usgs.gov/edu/earthhowmuch.html.

Van Dorn, W.G. (1993). *Oceanography and Seamanship*, 2nd edition. Cornell Maritime Press, ISBN 0-87033-434-4.

van Koten, H. (1976). Fatigue analysis of marine structures. *Proc. of BOSS'76*, Trondheim.

Wehausen, J.V. and Laitone, E.V. (1960). Surface waves *Handbuch der Physik*, Vol. 9, Springer, Berlin, pp. 446–778.

Wendel, K. and Platzoeder, W. (1958). Der Untergang des Segelschulschiffes Pamir. *Hansa*, Vol. 95, Hamburg.

Wiegel, R.L. (1964). *Oceanographical engineering*. Prentice Hall, Englewood Cliffs New Jersey.

Wilson, W.D. (1960). Equation for the speed of sound in sea water. *Journal of the Acoustic Society of America*.

Wimmer, W., Challenor, P.G. and Retzler, C. (2006). Extreme wave heights in the North Atlantic from altimeter data. *Renewable Energy*, Vol. 31, No. 2.

Woolf, D.K., Cotton, P.D. and Challenor, P.G. (2003). Measurements of the offshore wave climate around the British Isles by satellite altimeter. *Philosophical Transactions of the Royal Society of London*, A, Vol. 361, No. 1802.

2 Marine vehicle types

Contents

2.1 Overview
2.2 Merchant ships
2.3 High speed craft
2.4 Yachts
2.5 Warships
References (Chapter 2)

The various Sections of this Chapter have been taken from the following books, with the permission of the authors:

Eyres, D.J. (2007) *Ship Construction*. Butterworth-Heinemann, Oxford, UK. [Section 2.2.4]
Rawson K.J. and Tupper E.C. (2001) *Basic Ship Theory*. 5th Edition, Combined Volume. Butterworth-Heinemann, Oxford, UK. [Sections 2.2.9, 2.4]
Tupper, E.C. (2004) *Introduction to Naval Architecture*. Butterworth-Heinemann, Oxford, UK. [Sections 2.2.1–2.2.3, 2.2.5–2.2.8, 2.3, 2.5]

2.1 Overview

Seagoing marine vehicles may be divided into *transport*, including cargo, container and passenger ships and *non-transport* including fishing vessels, service craft such as tugs and supply vessels, and warships. An overview of the wide range of ship types is given in Figure 2.1 (Eyres, 2007). Each vessel type has a particular role to play and each will have a different set of design and operational conditions. This chapter provides a brief review of the main design and operational features of the principal types of marine vehicle.

2.2 Merchant ships

The development of merchant ship types has been dictated largely by the nature of the cargo and the trade routes. They can be classified accordingly with the major types being:

- general cargo ships
- container ships
- tankers
- dry bulk carriers
- passenger ships
- tugs.

2.2.1 General cargo ships

The industry distinguishes between *break bulk* cargo which is packed, loaded and stowed separately and *bulk* cargo which is carried loose in bulk. The general cargo carrier (Figure 2.2) is a flexible design of vessel which will go anywhere and carry a wide variety of cargo. The cargo may be break bulk or containers. Such vessels have several large clear open cargo-carrying spaces or holds. One or more decks may be present within the holds. These are known as 'tween decks and provide increased flexibility in loading and unloading, permit cargo segregation and improved stability. Access to the holds is by openings in the deck known as hatches.

Hatches are made as large as strength considerations permit in order to reduce the amount of horizontal movement of cargo within the ship. Typically the hatch width is about a third of the ship's beam. Hatch covers are of various types. Pontoon hatches are quite common in ships of up to 10 000 dwt, for the upper deck and 'tween decks, each pontoon weighing up to 25 tonnes. They are opened and closed using a gantry or cranes. In large bulk carriers side rolling hatch covers are often fitted, opening and closing by movement in the transverse direction. Another type of cover is the folding design operated by hydraulics. The coamings of the upper or weather deck hatches are raised above the deck to reduce the risk of flooding in heavy seas. They are liable to distort a little due to movement of the structure during loading and unloading of the ship. This must be allowed for in the design of the securing arrangements. Coamings can provide some compensation for the loss of hull strength due to the deck opening.

A double bottom is fitted along the ship's length, divided into various tanks. These may be used for fuel, lubricating oils, fresh water or ballast water. Fore and aft peak tanks are fitted and may be used to carry ballast and to trim the ship. Deep tanks are often fitted and used to carry liquid cargoes or water ballast. Water ballast tanks can be filled when the ship is only partially loaded in order to provide a sufficient draught for stability, better weight distribution for longitudinal strength and better propeller immersion.

Cranes and derricks are provided for cargo handling. Typically cranes have a lifting capacity of 10–25 tonnes with a reach of 10–20 m, but they can be much larger. General cargo ships can carry cranes or gantries with lifts of up to 150 tonnes. Above this, up to about 500 tonnes lift they are referred to as heavy lift ships.

The machinery spaces are often well aft but there is usually one hold aft of the accommodation and machinery space to improve the trim of the vessel when partially loaded. General cargo ships are generally smaller than the ships devoted to the carriage of bulk cargoes. Typically their speeds range from 12 to 18 knots.

Refrigerated cargo ships (Reefers)
A refrigeration system provides low temperature holds for carrying perishable cargoes. The holds are insulated to reduce heat transfer. The cargo may be carried frozen or chilled and holds are at different temperatures according to requirements. The possible effect of the low temperatures on surrounding structure must be considered. Refrigerated fruit is carried under modified atmosphere conditions. The cargo is maintained in a nitrogen-rich environment in order to slow the ripening process. The costs of keeping the cargo refrigerated, and the nature of the cargo, make a shorter journey time desirable and economic and these vessels are usually faster than general cargo ships with speeds up to 22 knots. Up to 12 passengers are carried on some, this number being the maximum permitted without the need to meet full passenger ship regulations.

2.2.2 Container ships

Container ships (Figure 2.3) are a good example of an integrated approach to the problem of transporting goods. Once goods are placed in the container at a factory or depot, they can be carried by road, rail or sea, being transferred from one to another at road or rail depots or a port. The container need not be opened until it reaches its destination. This makes the operation more secure. The maritime interest is primarily in the ports and ships but any element of the overall system may impose restrictions on what can be done. Height

46 Maritime engineering reference book

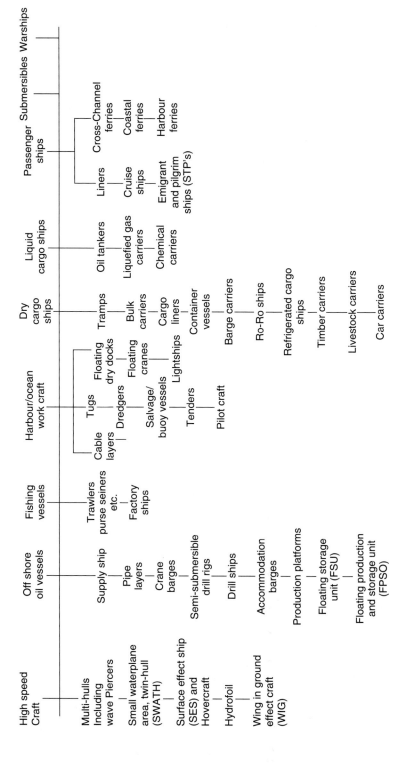

Figure 2.1 Ship types (Eyres, 2007).

Marine vehicle types 47

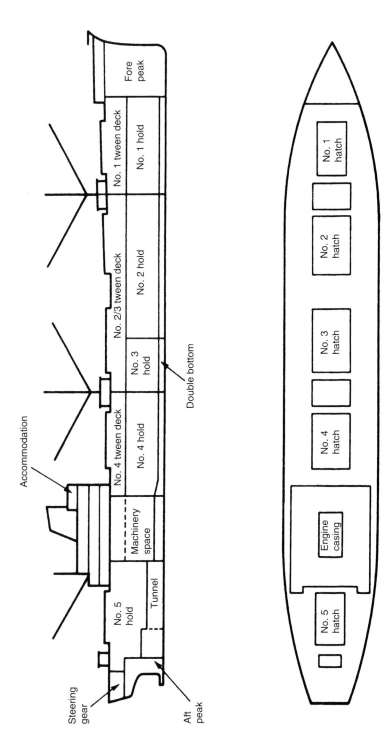

Figure 2.2 General cargo ship.

48 Maritime engineering reference book

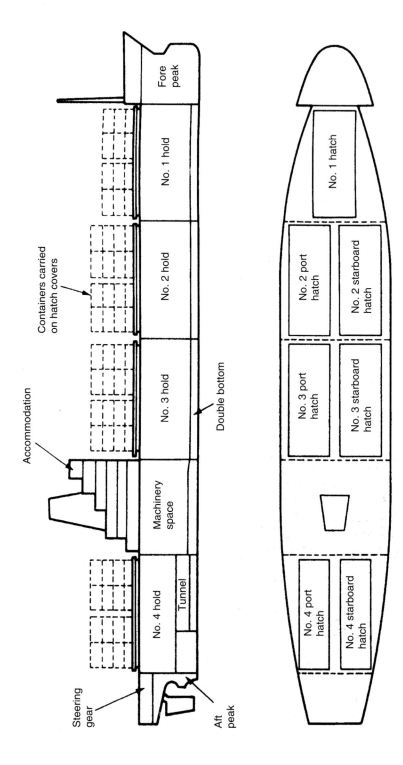

Figure 2.3 Container ship.

of container is likely to be dictated by the tunnels and bridges involved in land transport. Weight is likely to be dictated by the wheel loadings of lorries. The handling arrangements at the main terminals and ports are specially designed to handle the containers quickly and accurately. The larger container ships use dedicated container ports and tend not to have their own cargo handling gantries.

The containers themselves are simply reusable boxes made of steel or aluminium. They come in a range of types and sizes. Details can be obtained from the web site of one of the operators. Nominal dimensions are lengths of 20, 40 and 45 ft, width of 8 ft and height 8.5 or 9.5 ft. Internal volumes and weight of goods that can be carried vary with the material. For a 20 ft general-purpose steel container the internal capacity is about 33 m^3, weight empty is about 2.3 tonnef and the maximum payload is about 21.7 tonnef. Figures for a 40 ft container would be about 68 m^3, 3.8 tonnef and 26.7 tonnef, respectively. Aluminium containers will have about the same volume, weigh less and be able to carry a larger payload. They are used for most general cargoes and liquid carrying.

The cargo-carrying section of the ship is divided into several holds with the containers racked in special frameworks and stacked one upon the other within the hold space. Containers may also be stacked on hatch covers and secured by special lashings. Some modern ships dispense with the hatch covers, pumps dealing with any water that enters the holds. Each container must be of known all up weight and stowage arrangements must ensure the ship's stability is adequate as well as meeting the offloading schedule if more than one port is involved. The ship's deadweight will determine the total number of containers carried.

Cargo holds are separated by a deep web-framed structure to provide the ship with transverse strength. The structure outboard of the container holds is a box-like arrangement of wing tanks providing longitudinal and torsional strength. The wing tanks may be used for water ballast and can be used to counter the heeling of the ship when discharging containers. A double bottom is fitted which adds to the longitudinal strength and provides additional ballast space, see Figure 9.35.

Accommodation and machinery spaces are usually located aft leaving the maximum length of full-bodied ship for container stowage. The overall capacity of a container ship is expressed in terms of the number of standard 20 ft units it can carry, that is, the number of *twenty-foot equivalent units* (TEU). Thus a 40-foot container is classed as 2 TEU. The container ship is one application where the size of ship seems to be ever increasing to take advantage of the economies of scale. By the turn of the century 6000 TEU ships had become the standard for the main trade routes, and some 80 ships of 8000 TEU were on order plus some of 9200 TEU. Concept work was underway for ships of 14 000 TEU size. Container ships tend to be faster than most general cargo ships, with speeds up to 30 knots. The larger ships can use only the largest ports. Since these are fitted out to unload and load containers the ship itself does not need such handling gear. Smaller ships are used on routes for which the large ships would be uneconomic, and to distribute containers from the large ports to smaller ports. That is, they can be used as feeder ships. Since the smaller ports may not have suitable handling gear the ships can load and offload their own cargos.

Some containers are refrigerated. They may have their own independent cooling plant or be supplied with cooled air from the ship's refrigeration system. Because of the insulation required refrigerated containers have less usable volume. Temperatures would be maintained at about $-27°C$ and for a freezer unit about $-60°C$. They may be carried on general cargo ships or on dedicated refrigerated container ships. One such dedicated vessel is a 21 knot, 30 560 dwt ship of 2046 TEU capacity. The ship has six holds of which five are open. The hatchcover-less design enables the cell structure, in which the containers are stowed, to be continued above deck level giving greater security to the upper containers. Another advantage of the open hold is the easier dissipation of heat from the concentration of reefer boxes.

Barge carriers are a variant of the container ship. Standard barges are carried into which the cargo has been previously loaded. The barges, once unloaded, are towed away by tugs and return cargo barges are loaded. Minimal or even no port facilities are required and the system is particularly suited to countries with extensive inland waterways.

2.2.3 Roll-on roll-off ships (Ro-Ro ships)

These vessels (Figure 2.4) are designed for wheeled cargo, often in the form of trailers. The cargo can be rapidly loaded and unloaded through stern or bow doors and sometimes sideports for smaller vehicles. Train ferries were an early example of Ro-Ro ships.

The cargo-carrying section of the ship is a large open deck with a loading ramp usually at the aft end. Internal ramps lead from the loading deck to the other 'tween deck spaces. The cargo may be driven aboard under its own power or loaded by straddle carriers or fork lift trucks. One or more hatches may be provided for containers or general cargo, served by deck cranes. Where cargo, with or without wheels, is loaded and discharged by cranes the term lift-on lift-off (Lo-Lo) is used.

The structure outboard of the cargo decks is a box-like arrangement of wing tanks to provide longitudinal strength and adequate transverse stability. A double bottom is fitted along the complete length. Transverse bulkheads are limited to below the lowest vehicle deck so the side structure must provide adequate transverse and torsional strength. The machinery space

50 Maritime engineering reference book

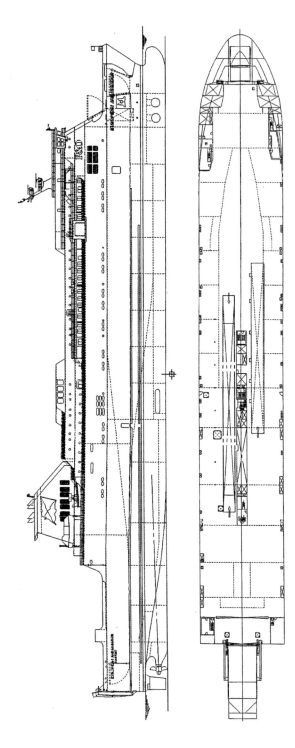

Figure 2.4 Ro-Ro Ship (courtesy RINA).

and accommodation are located aft. Only a narrow machinery casing actually penetrates the loading deck. Sizes range considerably with about 16 000 dwt (28 000 displacement tonne) being quite common and speeds are relatively high and usually in the region of 18–22 knots.

The use of Ro-Ro ships as passenger ferries is discussed later.

2.2.4 Car carriers

The increasing volume of car and truck production in the East (Japan, Korea and China) and a large customer base in the West has seen the introduction and rapid increase in number of ships specifically designed and built to facilitate the delivery of these vehicles globally.

Probably the ugliest ships afloat, car carriers are strictly functional having a very high box-like form above the waterline to accommodate as many vehicles as possible on, in some cases as many as a dozen decks. Whilst most deck spacing is to suit cars some 'tween deck heights may be greater and the deck strengthened to permit loading of higher and heavier vehicles. Within such greater deck spacing liftable car decks may be fitted for flexibility of stowage. The spacing of fixed car decks can vary from 1.85 to 2.3 metres to accommodate varying shape and height of cars. Transfer arrangements for vehicles from the main deck are by means of hoistable ramps which can be lifted and lowered whilst bearing the vehicles. Loading and discharging vehicles onto and off the ship is via a large quarter ramp at the stern and a side shell or stern ramp. The crew accommodation and forward wheelhouse, providing an adequate view forward, sit atop the uppermost continuous weather deck. Propulsion machinery is situated aft with bow thruster/s forward to aid mooring/manoeuvring.

The ship shown in Figure 2.5 has an overall length of 148 metres, a beam of 25 metres and a speed of 19 knots on a 7.2 metre draft. It can carry some 2140 units. A unit being an overall stowage area of 8.5 square metres per car and representing a vehicle 4.125 metres in length and 1.55 metres wide plus an all round stowage margin.

2.2.5 Bulk cargo carriers

The volume of cargoes transported by sea in bulk increased rapidly in the second half of the 20th Century, leading to specialist ships. These were ships carrying cargoes which did not demand packaging and which could benefit from the economies of scale. Most bulk carriers are single deck ships, longitudinally framed with a double bottom, with the cargo-carrying section of the ship divided into holds or tanks. The hold or tank arrangements vary according to the range of cargoes to be carried. Framing is contained within the double bottom and wing tanks to leave the inner surfaces of the holds smooth. They are categorized as:

- *Panamax*. The dimensions of the ship being limited by the need to be able to transit the Panama Canal. The beam must be less than 32.25 m.
- *Suezmax*. The dimensions of the ship being limited by the need to be able to transit the Suez Canal. Draught to be less than 19 m.
- *Capesize*. Without the restrictions of the above types.
- *Handysize*. Generally less than about 50 000 tonnes.
- *Aframax*. This is a term applied to tankers in the range 80 000–120 000 dwt.

Bulk carriers can also be sub-divided into tankers and dry bulk carriers. The requirements, for instance the permitted lengths of cargo holds, vary with the size of ship and the following comments are for general guidance only.

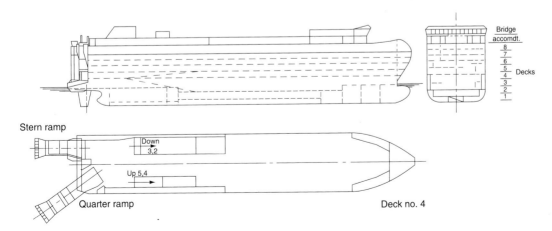

Figure 2.5 Car carrier.

2.2.5.1 Tankers

Tankers are used for the transport of liquids. They include:

- crude oil carriers;
- product tankers;
- gas tankers;
- chemical carriers.

Crude oil carriers

These carry the unrefined crude oil and they have significantly increased in size in order to obtain the economies of scale and to respond to the demands for more and more oil. Designations, such as Ultra Large Crude Carrier (ULCC) and Very Large Crude Carrier (VLCC), have been used for these huge vessels. The ULCC is a ship of 300 000 dwt or more; the VLCC is 200 000–300 000 dwt. Crude oil tankers with deadweight tonnages in excess of half a million have been built although the current trend is for somewhat smaller (130 000–150 000 dwt) vessels.

The cargo-carrying section of the tanker is usually divided into tanks by longitudinal and transverse bulkheads. The size and location of these cargo tanks is dictated by the International Maritime Organization (IMO) Convention MARPOL 1973/1978 which became internationally accepted in 1983. IMO requirements are built into those of the various classification societies. These regulations require the use of segregated ballast tanks and their location such that they provide a barrier against accidental oil spillage. The segregated ballast tanks must be such that the vessel can operate safely in ballast without using any cargo tank for water ballast. See Chapter 11 for more information on IMO and MARPOL.

Tankers ordered after 1993 had to comply with the MARPOL double hull regulation (Figure 2.6). This is opposed to single hull tankers where one or more cargo holds are bounded in part by the ship's shell plating. In the double hull design the cargo tanks are completely surrounded by wing and double bottom tanks which can be used for ballast purposes. The USA, under its 1990 Oil Pollution Act required all newly built tankers trading in US waters to be of the double hull design. There has been debate on whether a double hull is the best way of reducing pollution following grounding or collision. IMO and classification societies are prepared to consider alternatives to the double hull. One alternative favoured by some is the mid-height depth deck design. In such ships a deck is placed at about mid-depth which will be well below the loaded waterline. This divides the cargo tanks into upper and lower tanks. A trunk is taken from the lower tank through the upper tank and vented. The idea is that if the outer bottom is breached the external water pressure will be greater than the pressure of hull from the lower tank and this will force oil up the vent trunk. Thus water enters the ship rather than oil escaping from it. Such tankers would still incorporate segregated ballast tanks outboard of the cargo tanks to safeguard against collision. Subsequent debate within IMO and the EU has led to a speeding up of the timetable for phasing out single hull tankers. For the detailed provisions recourse should be had to the regulations of the authorities concerned.

Segregated ballast tanks would include all the double bottom tanks beneath the cargo tanks, wing tanks and the fore and aft peak tanks. Each cargo tank would be discharged by pumps fitted in the aft pump room, each tank having its own suction arrangement which connects to the pumps, and a

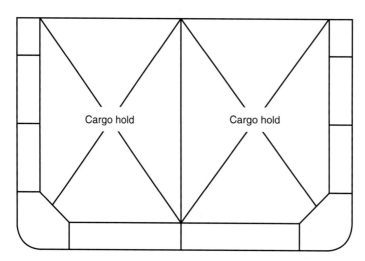

Figure 2.6 Typical section of double hull tanker.

network of piping discharges the cargo to the deck from where it is pumped ashore. The accommodation and machinery spaces would be located aft and separated from the tank region by a cofferdam. Where piping serves several tanks, means must be provided for isolating each tank.

Experience shows that once any initial protective coatings breakdown, permanent ballast tanks suffer corrosion and regular inspection is vital. The builder must provide a high quality coating system and a back-up anode system to give a coverage of 10 mA/m^2 should be included to control corrosion after coating breakdown.

More and more ships are being fitted with equipment to measure actual strains during service. A typical system comprises a number of strain gauges at key points in the structure together with an accelerometer and pressure transducer to monitor bottom impacts. Results are available on the bridge to assist the master in the running of the ship. The information is stored and is invaluable in determining service loadings and long term fatigue data.

Product carriers
After the crude oil is refined the various products are transported in product carriers. The refined products carried include gas oil, aviation fuel and kerosene. Product carriers are smaller than crude oil carriers and, because several different products are carried, they have greater sub-division of tanks. The cargo tank arrangement is again dictated by MARPOL 73/78. Individual 'parcels' of various products may be carried at any one time which results in several separate loading and discharging pipe systems. The tank surfaces are usually coated to prevent contamination and enable a high standard of tank cleanliness to be achieved after discharge. Sizes range from about 18 000 up to 75 000 dwt with speeds of about 14–16 knots.

Liquefied gas carriers
The most commonly carried liquefied gases are *liquefied natural gas* (LNG) and *liquefied petroleum gas* (LPG). They are kept in liquid form by a combination of pressure and low temperature. The combination varies to suit the gas being carried.

The bulk transport of natural gases in liquefied form began in 1959 and has steadily increased since then. Specialist ships are used to carry the different gases in a variety of tank systems, combined with arrangements for pressurizing and refrigerating the gas. Natural gas is released as a result of oil-drilling operations. It is a mixture of methane, ethane, propane, butane and pentane. The heavier gases, propane and butane, are termed 'petroleum gases'. The remainder, largely methane, is known as 'natural gas'. The properties, and behaviour, of these two basic groups vary considerably, requiring different means of containment and storage during transit.

Natural gas carriers
Natural gas is, by proportion, 75–95% methane and has a boiling point of $-162°C$ at atmospheric pressure. Methane has a critical temperature of $-82°C$, which means it cannot be liquefied by the application of pressure above this temperature. A pressure of 47 bar is necessary to liquefy methane at $-82°C$. LNG carriers are designed to carry the gas in its liquid form at atmospheric pressure and a temperature in the region of $-164°C$. The ship design must protect the steel structure from the low temperatures, reduce the loss of gas and avoid its leakage into the occupied regions of the vessel, Ffooks (1993).

Tank designs are either self-supporting, membrane or semi-membrane. The self-supporting tank is constructed to accept any loads imposed by the cargo. A membrane tank requires the insulation between the tank and the hull to be load bearing. Single or double metallic membranes may be used, with insulation separating the two membrane skins. The semi-membrane design has an almost rectangular cross section and the tank is unsupported at the corners.

Figure 2.7 shows a novel design of a small LNG carrier in which boil off gases are totally contained within the tanks leaving easier choice of main propulsion diesels. More typically, an LNG carrier has some five tanks of almost rectangular cross section, each having a central dome. They are supported and separated from the ship's structure by insulation which is a lattice structure of wood and various foam compounds.

The tank and insulation structure is surrounded by a double hull. The double bottom and ship's side regions are used for oil or water ballast tanks whilst the ends provide cofferdams between the cargo tanks. A pipe column is located at the centre of each tank and is used to route the pipes from the submerged cargo pumps out of the tank through the dome. The accommodation and machinery spaces are located aft and separated from the tank region by a cofferdam. LNG carriers have steadily increased in size and ships of around $140 000 \text{ m}^3$ capacity are now on order. Speeds range from 16 to 19 knots.

Petroleum gas carriers
Petroleum gas may be propane, propylene, butane or a mixture. All three have critical temperatures above normal ambient temperatures and can be liquefied at low temperatures at atmospheric pressure, normal temperatures under considerable pressure, or some intermediate combination of pressure and temperature. The design must protect the steel hull where low

54 Maritime engineering reference book

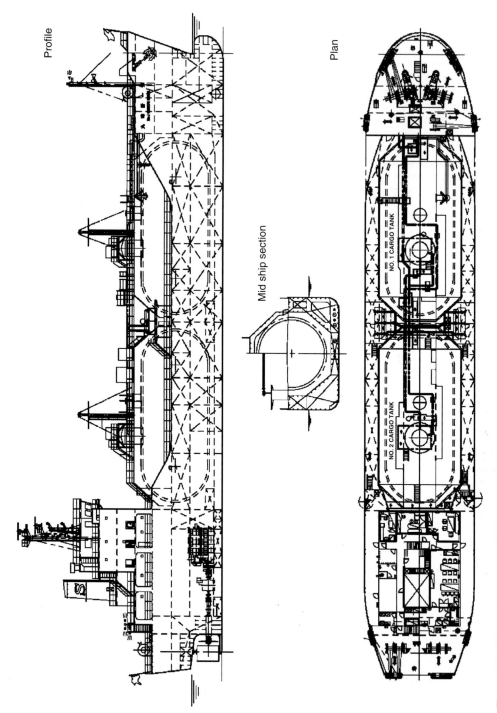

Figure 2.7 LNG carrier (courtesy RINA).

temperatures are used, reduce the gas loss, avoid gas leakage and perhaps incorporate pressurized tanks. The fully pressurized tank operates at about 17 bar and is usually spherical or cylindrical in shape for structural efficiency.

Semi-pressurized tanks operate at a pressure of about 8 bar and temperatures in the region of $-7°C$. Insulation is required and a reliquefaction plant is needed for the cargo boil-off. Cylindrical tanks are usual and may penetrate the deck. Fully refrigerated atmospheric pressure tank designs may be self-supporting, membrane or semi-membrane types as in LNG tankers. The fully refrigerated tank designs operate at temperatures of about $-45°C$. A double hull type of construction is used.

An LPG carrier of about 50 000 dwt is shown in Figure 2.8. It is a flushed deck vessel with four holds. Within the holds there are four independent, insulated, prismatic cargo tanks, supported by a load bearing structure designed to take account of the interaction of movements and forces between the tanks and adjoining hull members. Topside wing, hopper side and double bottom tanks are mainly used for water ballast. Fuel is carried in a cross bunker forward of the engine room. Machinery and accommodation are right aft. It can carry various propane/butane ratios to provide flexibility of operation.

The double hull construction, cargo pumping arrangements, accommodation and machinery location are similar to an LNG carrier. A reliquefaction plant is, however, carried and any cargo boil-off is returned to the tanks. LPG carriers exist in sizes up to around 95 000 m^3. Speeds range from 16 to 19 knots.

Chemical carriers

A wide variety of chemicals is carried by sea. The cargo is often toxic and flammable so the ships are subject to stringent requirements to ensure safety of the ship and the environment. Different cargos are segregated by cofferdams. Spaces are provided between the cargo tanks and the ship's hull, machinery spaces and the forepeak bulkhead. Great care is taken to prevent fumes spreading to manned spaces.

2.2.5.2 Dry bulk carriers

These ships carry bulk cargoes such as grain, coal, iron ore, bauxite, phosphate and nitrate. Towards the end of the 20th Century more than 1000 million tonnes of these cargoes were being shipped annually, including 180 million tonnes of grain. Apart from saving the costs of packaging, loading and offloading times are reduced.

As the volume of cargo carried increased so did the size of ship, taking advantage of improving technology. By the 1970s ships of 200 000 dwt were operating and even larger ships were built later. This growth in size has not been without its problems. In the 28 months from January 1990 there were 43 serious bulk carrier casualties of which half were total losses. Three ships, each of over 120 000 dwt, went missing. Nearly 300 lives were lost as a result of these casualties. To improve the safety of these ships IMO adopted a series of measures during the 1990s. These reflect the lessons learned from the losses of ships in the early 1990s and that of *MV Derbyshire* in 1980 whose wreck was found and explored by remotely controlled vehicles. Among factors being addressed are age, corrosion, fatigue, freeboard, bow height and strength of hatch covers. A formal safety assessment was carried out to guide future decisions on safety matters for bulk carriers.

In a general-purpose bulk carrier (Figure 2.9), only the central section of the hold is used for cargo. The partitioned tanks which surround the hold are used for ballast purposes. This hold shape also results in a self-trimming cargo. During unloading the bulk cargo falls into the space below the hatchway facilitating the use of grabs or other mechanical unloaders. Large hatchways are a particular feature of bulk carriers. They reduce cargo handling time during loading and unloading.

Combination carriers are bulk carriers which have been designed to carry any one of several bulk cargoes on a particular voyage, for instance ore, crude oil or dry bulk cargo.

Stability and loading manuals are supplied to every ship to provide the Master with the information to load, discharge and operate the ship safely, see Chapter 11, Section 11.5.2.2. Loading computer programs are designed to provide, for any condition of loading, a full set of deadweight, trim, stability and longitudinal strength calculations. IMO codes on trimming bulk cargoes require the cargo, with particular attention to cargoes that may liquefy, to be trimmed reasonable level to the boundaries of the compartment to minimize the risk of bulk material shift. The very high loading rates, up to 16 000 tonnes/hour, make the loading task one that needs careful attention.

An ore carrier usually has two longitudinal bulkheads which divide the cargo section into wing tanks and a centre hold which is used for ore. A deep double bottom is fitted. Ore, being a dense cargo, would have a very low centre of gravity if placed in the hold of a normal ship leading to an excess of stability in the fully loaded condition. The deep double bottom raises the centre of gravity and the behaviour of the vessel at sea is improved. The wing tanks and the double bottoms provide ballast capacity. The cross section would be similar to that for an ore/oil carrier shown in Figure 2.10.

An ore/oil carrier uses two longitudinal bulkheads to divide the cargo section in centre and wing tanks which are used for the carriage of oil. When ore is carried, only the centre tank section is used for cargo. A double bottom is fitted but used only for water ballast.

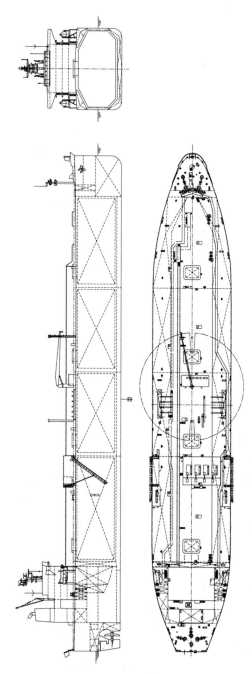

Figure 2.8 LPG carrier (courtesy RINA).

Marine vehicle types 57

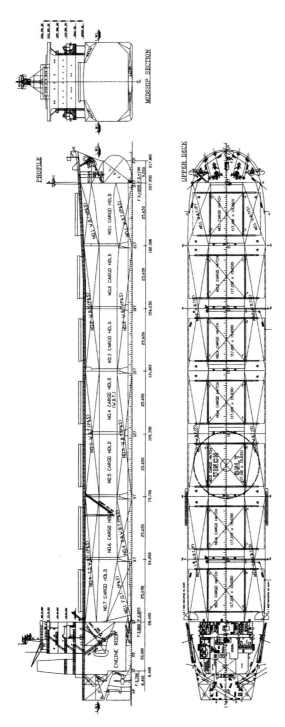

Figure 2.9 General purpose bulk carrier (courtesy RINA).

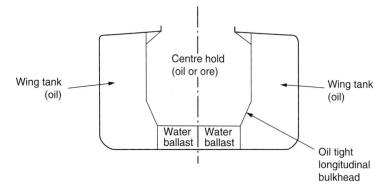

Figure 2.10 Section of oil/ore carrier.

The ore/bulk/oil (OBO) bulk carrier is currently the most popular combination bulk carrier. It has a cargo-carrying cross section similar to the general bulk carrier but the structure is significantly stronger. Large hatches facilitate rapid cargo handling. Many bulk carriers do not carry cargo handling equipment, since they trade between special terminals with special equipment. Combination carriers handling oil cargoes have their own cargo pumps and piping systems for discharging oil. They are required to conform to the requirement of MARPOL. Deadweight capacities range from small to upwards of 200 000 tonnes. Taking a 150 000/160 000 tonne deadweight Capesize bulk carrier as typical, the ship is about 280 m in length, 45 m beam and 24 m in depth. Nine holds hold some 180 000 m^3 grain in total, with ballast tanks of 75 000 m^3 capacity. The speed is about 15.5 knots on 14 MW power. Accommodation about 30.

2.2.6 Passenger ships

Passenger ships can be considered in two categories, the cruise ship and the ferry. The ferry provides a link in a transport system and often has Ro-Ro facilities in addition to its passengers.

Considerable thought has been given to achieving rapid, and safe, evacuation and this is an area where computer simulation has proved very useful. For instance, quicker access is possible to lifeboats stowed lower in the ship's superstructure, chutes or slides can be used for passengers to enter lifeboats already in the water, either directly into the boat or by using a transfer platform. It is important that such systems should be effective in adverse weather conditions and when the ship is heeled. Shipboard arrangements must be designed bearing in mind the land-based rescue organizations covering the areas in which the ship is to operate.

Free fall lifeboats, used for some years on offshore installations, are increasingly being fitted to tankers and bulk carriers. Drop heights of 30 m are now accepted and heights of 45 m have been tested. However, safe usage depends upon the potential users being fit and well trained. These conditions can be met in ships' crews but is problematic for passenger ships. Passengers in, say, a cruise ship may not be fit and may even be partially handicapped.

As might be expected it is passenger ships that are most affected by changes in standards and thinking of society as a whole. In 1997 the Maritime and Coastguard agency issued a guidance note on the needs of disabled people. In 2000 the Ferries Working Group of the Disabled Persons Transport Advisory Committee (DPTAC) issued more detailed guidance, DPTAC (2000).

Cruise ships
Cruise ships (Figure 2.11) have been a growth area. Between 1990 and 2000 the cruise market grew by 60% and the size of ship has also grown with vessels now capable of carrying 3600 passengers at 22 knots. However, the largest cruise ships cannot use some ports and harbours in the more attractive locations. The ship has to anchor well out and ferry passengers ashore by smaller boats. This takes time and there are now a number of small or medium sized ships to cater for passengers who want to visit the smaller islands.

In a cruise ship passengers are provided with a high standard of accommodation and leisure facilities. This results in a large superstructure as a prominent feature of the vessel. The many tiers of decks are fitted with large open lounges, ballrooms, swimming pools and promenade areas. Stabilizers are fitted to reduce rolling and bow thrusters are used to improve manoeuvrability.

Large ferries
Ocean-going ferries are a combination of Ro-Ro and passenger vessel. The vessel has three layers, the lower machinery space, the vehicle decks and the passenger accommodation. A large stern door and sometimes

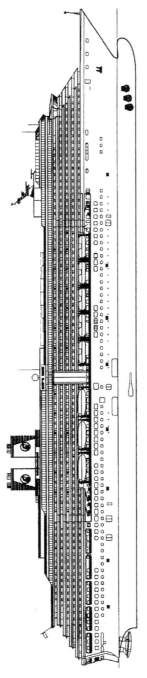

Figure 2.11 Cruise liner (courtesy RINA).

also a bow door provide access for the wheeled cargo to the various decks which are connected by ramps. Great care is needed to ensure these doors are watertight and proof against severe weather. There is usually a secondary closure arrangement in case the main door should leak. The passenger accommodation varies with length of the journey. For short-haul or channel crossings public rooms with aircraft-type seats are provided and for long distance ferries cabins and sleeping berths. Stabilizers and bow thrusters are usually fitted to improve seakeeping and manoeuvring. Size varies according to route requirements and speeds are usually around 20–22 knots.

When used as ferries, vehicles usually enter at one end and leave at the other. This speeds up loading and unloading but requires two sets of doors. There has been considerable debate on the vulnerability of Ro-Ro ships that should water get on to their vehicle decks. Various means of improving stability in the event of collision and to cater for human error in not securing entry doors, have been proposed. Since the loss of the *Herald of Free Enterprise* regulations have been tightened up. The later loss of the *Estonia* gave an additional impetus to a programme of much needed improvements.

2.2.7 Tugs

Tugs perform a variety of tasks and their design varies accordingly. They move dumb barges, help large ships manoeuvre in confined waters, tow vessels on ocean voyages and are used in salvage and firefighting operations. Tugs can be categorized broadly as inland, coastal or ocean going. Put simply, a tug is a means of applying an external force to any vessel it is assisting. That force may be applied in the direct or the indirect mode. In the former the major component of the pull is provided by the tug's propulsion system. In the latter most of the pull is provided by the lift generated by the flow of water around the tug's hull, the tug's own thrusters being mainly employed in maintaining its attitude in the water.

The main features of a tug (Figure 2.12) are an efficient design for free running and a high thrust at zero speed (the *bollard pull*), an ability to get close alongside other vessels, good manoeuvrability and stability.

Another way of classifying tugs is by the type and position of the propulsor units:

(1) **Conventional tugs** have a normal hull, propulsion being by shafts and propellers, which may be open or nozzled, and of fixed or controllable pitch, or by steerable nozzles or vertical axis propellers. They usually tow from the stern and push with the bow.
(2) **Stern drive tugs** have the stern cut away to accommodate twin azimuthing propellers. These propellers, of fixed or controllable pitch, are in nozzles and can be turned independently through 360° for good manoeuvrability. Because the drive is through two right angle drive gears these vessels are sometimes called Z-drive tugs. They usually have their main winch forward and tow over the bow or push with the bow.
(3) **Tractor tugs** are of unconventional hull form, with propulsors sited under the hull about one-third of the length from the bow, and a stabilizing skeg aft. Propulsion is by azimuthing units or vertical axis propellers. They usually tow over the stern or push with the stern.

In most tug assisted operations the ship is moving at low speed. Concern for the environment, following the *Exxon Valdez* disaster, led to the US Oil Pollution Act of 1990. To help tankers, or any ship carrying hazardous cargo, which found themselves unable to steer, escort tugs were proposed.

In this concept the assisted ship may be moving at 10 knots or more. Success depends upon the weather conditions and the proximity of land or underwater hazards, as well as the type and size of tug. Some authorities favour a free-running tug so as not to endanger ship or tug in the majority (incident free) of operations. In this case the tug normally runs ahead of the ship. It has the problem of connecting up to the ship in the event of an incident. For this reason other authorities favour the tug being made fast to the escorted ship either on a slack or taut line.

The direct pull a tug can exert falls off with speed and indirect towing will be more effective at higher speeds. Tugs can be used as part of an integrated tug/barge system. This gives good economy with one propelled unit being used with many dumb units.

The trends in tug design in the last decade of the 20th Century were:

- Tugs with azimuthing propulsion have effectively replaced conventional single and twin screw propelled tugs for harbour work.
- Lengths range up to 45 m but tugs of 30–35 m dominate.
- Powers are generally 2500–3000 kW with a few as high as 5000 kW.
- B/D and B/L ratios have increased to provide greater stability.
- Bollard pull varies with installed power and type of propulsion, being 60–80 tonnef at 5000 kW.
- Free-running speeds range from 10 to 15 knots and tend to increase linearly with the square root of the length.

2.2.8 Icebreakers and ice strengthened ships

The main function of an icebreaker is to clear a passage through ice at sea, in rivers or in ports so that other ships can use the areas which would

Marine vehicle types 61

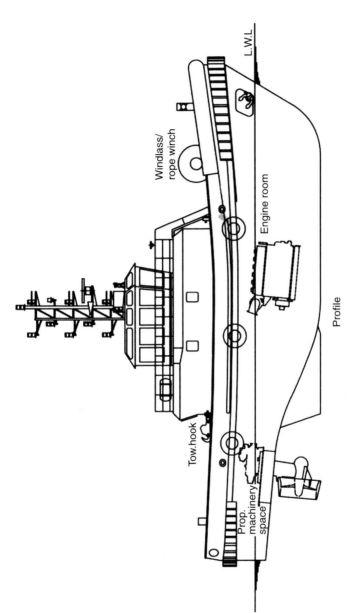

Figure 2.12 Tug (courtesy RINA).

otherwise be denied to them. Icebreakers are vital to the economy of nations such as Russia with ports that are ice bound for long periods of the year and which wish to develop the natural resources within the Arctic. Icebreakers need:

- To be specially strengthened with steels which remain tough at low temperature.
- Extra structure in the bow and along the waterline.
- High power propulsion and manoeuvring devices which are not susceptible to ice damage. The shape of the stern is important here.
- A hull form that enables them to ride up over the ice. This is one way of forcing a way through ice; the ship rides over the ice edge and uses its weight to break the ice. The ship may be 'rocked' by transferring ballast water longitudinally. The hull is well rounded and may roll heavily as protruding stabilizers are unacceptable.
- Good hull sub-division.
- Special hull paints.

Icebreakers are expensive to acquire and operate.

Other ships which need to operate in the vicinity of ice are strengthened to a degree depending upon the perceived risk. Usually they can cope with continuous 1 year old ice of 50–100 cm thickness. Typically they are provided with a double hull, thicker plating forward and in the vicinity of the waterline, with extra framing. They have a flat hull shape and a rounded bow form. Rudders and propellers are protected from ice contact by the hull shape. Inlets for engine cooling water must not be allowed to become blocked.

2.2.9 Fishing vessels

Fishing vessels have evolved over thousands of years to suit local conditions. Fish which live at the bottom of the sea like sole, hake and halibut and those which live near the bottom like cod, haddock and whiting are called demersal species. Those fish which live above the bottom levels, predominantly such as herring and mackerel, are called pelagic species. There are also three fundamental ways of catching fish:

(a) by towing trawls or dredges;
(b) by surrounding the shoals by nets, purse seines;
(c) by static means, lines, nets or pots.

These distinctions enable fishing vessels to be classified in accordance with Figure 2.13.

The commonest type of fishing vessel is the trawler which catches both demersal and pelagic species. The trawl used for the bottom is long and stocking shaped and is dragged at a few knots by cables led to the forward gantry on the ship. When the trawl is brought up it releases its catch in the cod end down the fish hatch in the trawl deck. Operations are similar when trawling for pelagic species but the trawl itself has a wider mouth and is altogether larger.

Trawlers suffer the worst of weather and are the subject of special provision in the freeboard regulations. They must be equipped with machinery of the utmost reliability since failure at a critical moment could endanger the ship. Both diesel and diesel-electric propulsion are now common. Ice accretion in the upperworks is a danger in certain weather, and a minimum value of GM of about 0.75 m is usually required by the owner. Good range of stability is also important and broaching to is an especial hazard.

Despite great improvements in trawler design significant numbers of vessels are lost every year and many of them disappear without any very good explanation. It is probable that such losses are due to the coincidence of two or more circumstances like broaching to, open hatches, choked freeing ports, loss of power, critical stability conditions, etc.

To give adequate directional stability when trawling, experience has shown that considerable stern trim is needed, often as much as 5 degrees. Assistance in finding shoals of fish is given by sonar or echo sounding gear installed in the keel. No modern trawler is properly equipped without adequate radar, communication equipment and navigation aids. A typical stern fishing trawler is shown in Figure 2.14.

2.3 High speed craft

These may have civil or military application so they are considered before going on to consider specific warship types.

A number of hull configurations and propulsion systems are discussed; each designed to overcome problems with other types or to confer some desired advantage. Thus catamarans avoid the loss of stability at high speed suffered by round bilge monohulls. They also provide large deck areas for passenger use or deployment of research or defence equipment. Hydrofoil craft benefit from reduced resistance by lifting the main hull clear of the water. Air cushion vehicles give the possibility of amphibious operation. The effect of waves on performance is minimized in the **S**mall **W**aterplane **A**rea **T**win **H**ull (SWATH) concept. Some craft are designed to reduce wash so that they can operate at higher speeds in restricted waters.

The choice of design depends upon the intended service. In some cases a hybrid is used. Although most applications of these concepts have been initially to small craft some are now appearing in the medium size, especially for high speed ferry services.

In commercial applications one of the special characteristics, such as those mentioned above, may be the deciding factor in the adoption of a particular hull form. In other cases, particularly for ferries, it

Marine vehicle types 63

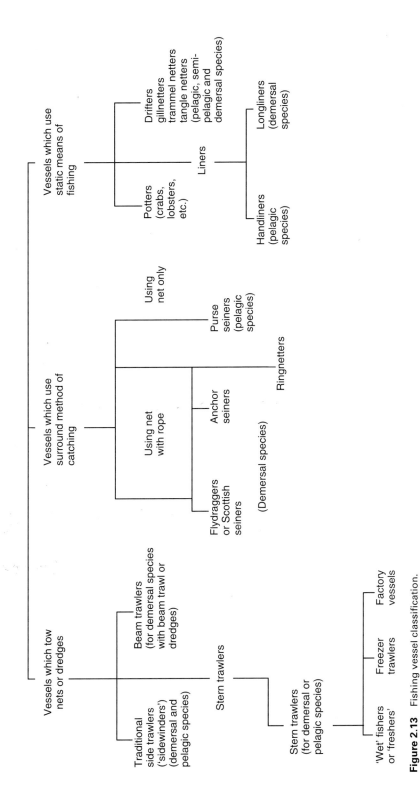

Figure 2.13 Fishing vessel classification.

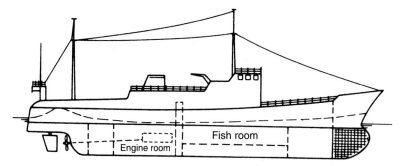

Figure 2.14 Stern trawler.

may be the extra speed which is a feature of several forms. One way of assessing the relative merits of different forms is what is termed the *transport efficiency factor* which is the ratio of the product of payload and speed to the total installed power.

2.3.1 Monohulls

Many high speed small monohulls have had hard chines. Round bilge forms at higher speeds have had stability problems. Hard chine forms with greater beam and reduced length give improved performance in calm water but experience high vertical accelerations in a seaway. Their ride can be improved by using higher deadrise angles leading to a 'deep vee' form. Current practice favours round bilge for its lower power demands at cruising speed and seakindliness, with the adoption of hard chines for Froude numbers above unity for better stability. One advantage of the round bilge form in seakeeping is that it can be fitted more readily with bilge keels to reduce rolling.

2.3.2 Surface effect ships (SESs)

The earliest form of SES was the hovercraft in which the craft was lifted completely clear of the water on an air cushion created by blowing air into a space under the craft and contained by a skirt. For these craft propulsion is by airscrew of jet engines. In some later craft rigid sidewalls remain partially immersed when the craft is raised on its cushion and the skirt is limited to the ends. The sidewalls mean that the craft is not amphibious and cannot negotiate very shallow water. They do, however, improve directional stability and handling characteristics in winds. They also reduce the leakage of air from the cushion and so reduce the lift power needed and they enable more efficient water propellers or waterjets to be used for propulsion. For naval applications it is usual for the sidewalls to provide sufficient buoyancy so that when at rest with zero cushion pressure the cross structure is still clear of the water.

The effect of the air cushion is to reduce the resistance at high speeds, making higher speeds possible for a given power. For ferries, which operate close to their maximum speed for the major part of their passage, it is desirable to operate at high Froude number to get beyond the wavemaking hump in the resistance curve. This, and the wish to reduce the cushion perimeter length for a given plan form area, means that most SESs have a low length to beam ratio. In other applications the craft may be required to operate efficiently over a range of speeds. In this case a somewhat higher length to beam ratio is used to give better fuel consumption rates at the lower cruising speeds.

SESs are employed as ferries on a number of short-haul routes. For the cushion craft shown in Figure 2.15 the passenger seating is located above the central plenum chamber with the control cabin one deck higher. Ducted air propellers and rudders are located aft to provide forward propulsion and lateral control. Centrifugal fans driven by diesel engines create the air cushion. Manoeuvrability is helped by air jet driven bow thrusters.

Early SESs were relatively high cost, noisy craft requiring a lot of maintenance, particularly of the skirts which quickly became worn. As a result of experience with the early, mainly military, craft later versions are considerably improved in all these respects. Naval applications include landing small numbers of covert forces and in mine hunting. In the former, an amphibious craft can cross the exposed beach quickly. The latter use arises from the relative immunity of SESs to underwater explosions.

2.3.3 Hydrofoil craft

Hydrofoil craft make use of hydrodynamic lift generated by hydrofoils attached to the bottom of the craft. When the craft moves through the water a lift force is generated to counteract the craft's weight, the hull is raised clear of the water and the resistance is reduced. High speeds are possible without using unduly large powers. Once the hull is clear of the water and not, therefore, contributing buoyancy, the lift required of the foils is effectively constant. As speed increases either the submerged area of

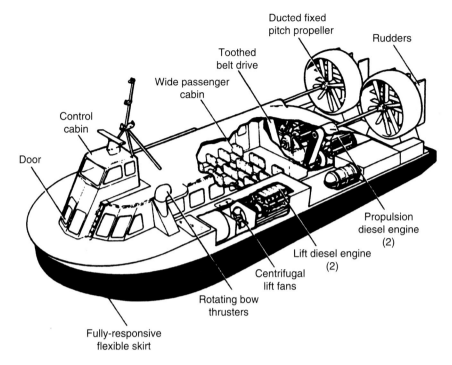

Figure 2.15 Air cushion vehicle.

foil will reduce or their angle of incidence must be reduced. This leads to two types of foil system:

(1) Completely submerged, incidence controlled. The foils remain completely submerged, reducing the risk of cavitation, and lift is varied by controlling the angle of attack of the foils to the water. This is an 'active' system and can be used to control the way the craft responds to oncoming waves.
(2) Fixed surface piercing foils. The foils may be arranged as a ladder either side of the hull or as a large curved foil passing under the hull. As speed increases the craft rises so reducing the area of foil creating lift. This is a 'passive' system.

Foils are provided forward and aft, the balance of area being such as to provide the desired ride characteristics. The net lift must be in line with the centre of gravity of the craft. Like the SES, the hydrofoil has been used for service on relatively short-haul journeys. Both types of craft have stability characteristics which are peculiarly their own.

2.3.4 Multi-hulled vessels

These include sailing catamarans, trimarans, offshore rigs, diving support vessels and ferries. Catamarans are not new as two twin hulled paddle steamers of about 90 m length were built in the 1870s for cross channel service. They were liked by passengers for their seakeeping qualities but were overtaken fairly soon by other developments. The upper decks of catamarans provide large areas for passenger facilities in ferries or for helicopter operations. Their greater wetted hull surface area leads to increased frictional resistance but the relatively slender hulls can have reduced resistance at higher speeds, sometimes assisted by interference effects between the two hulls. A hull separation of about 1.25 times the beam of each hull is reasonable in a catamaran. Manoeuvrability is good.

High transverse stability and relatively short length mean that seakeeping is not always good. This has been improved in the wave piercing catamarans developed to reduce pitching, and in SWATH designs where the waterplane area is very much reduced and a large part of the displaced water volume is well below the waterline. The longitudinal motions can be reduced by using fins or stabilizers.

As a development of twin hull vessels the trimaran form has been proposed. Many design studies indicated many advantages with no significant disadvantages. To prove the concept, and particularly to prove the viability of the structure, a 98 m, 20 knot, demonstrator – *RV Triton* – was completed in 2000. Its structure was designed in accordance with the High Speed and Light Craft Rules of DNV. The main hull is of round bilge form. The side hulls are of multi-chine design on the outboard face with a plane inboard face. The main hull structure is conventional and integrated with a box girder like cross deck from which the side hulls extend. Propulsion is diesel electric with a single propeller, and rudder, behind the main hull with small side hull thrusters. The trials were extensive and in most cases successfully vindicated the theories. The pentamaran forms are developments of the trimaran with a slender main hull and two small hulls each side.

Comparisons of monohulls with multi-hull craft are difficult. Strictly designs of each type should be optimized to meet the stated requirements. Only then can their relative merits and demerits be established. For simpler presentations it is important to establish the basis of comparison be it equal length, displacement, or carrying capacity.

Multi-hull designs have a relatively high structural weight and often use aluminium to preserve payload. Wave impact on the cross structure must be avoided or minimized so high freeboard is needed together with careful shaping of the undersides. Because of their small waterplane areas, SWATH ships are very sensitive to changes in load and its distribution so weight control is vital.

2.3.5 Rigid inflatable boats (RIBs)

Inflatable boats have been in use for many years and, with a small payload, can achieve high speed. The first rigid inflatables came into being in the 1960s with an inflatable tube surrounding a wooden hull. Much research has gone into developing very strong and durable fabrics for the tubes to enable them to withstand the harsh treatment these craft get. Later craft have used reinforced plastic and aluminium hulls. RIBs come in a wide range of sizes and types. Some are open, some have enclosed wheelhouse structures; some have outboard motors, others have inboard engines coupled to propellers or waterjets. Lengths range from about 4–16 m and speeds can be as high as 80 knots.

Uses are also wide in scope, ranging from leisure through commercial to rescue and military. Users include the military, coastguards, customs and excise, the RNLI, oil companies and emergency services. Taking the RNLI use as an example, the rigid lower hull is shaped to make the craft more seakindly and the inflatable collar safeguards against sinking by swamping.

2.3.6 Comparison of high speed types

All the types discussed in this Section have advantages and disadvantages. As stated above for the multi-hulls, a proper comparison requires design studies to be created of each prospective type, to meet the requirement. However, some special requirement, such as the need to operate over land and sea, may suggest one particular form. For instance, a craft capable of running up on to a hard surface points to an air cushion vehicle. Many of these types of craft in use today are passenger carrying. SESs with speeds of over 40 knots are common, and can compete with air transport on some routes. Hydrofoils enjoy considerable popularity for passenger carrying on short ferry routes because of their shorter transit times. Examples are the surface piercing Rodriguez designs and the Boeing Jetfoil with its fully submerged foil system. Catamarans are used for rather larger high speed passenger ferries.

2.4 Yachts

For many years, the design, construction and sailing of yachts has been a fascinating art about which whole books are regularly published. This is because the science is too complex for precise solution – and indeed, few yachtsmen would wish it otherwise. Some tenable theories have, in fact, been evolved to help in explaining certain of the performance characteristics of sailing boats, for example, see Claughton et al. (2006) and Larsson and Eliasson (2000).

A yacht, of course, obeys the fundamental theory described generally in this book for all surface ships. In addition, a yacht is subject to air forces acting on the sails and to water forces due to its peculiar underwater shape – forces which are negligible for ordinary surface ships. Sailing before the wind, a yacht is propelled by:

(a) the vector change of momentum of the wind, deflected by the sails;
(b) lift generated by the sails acting as aerofoils; because an aerofoil requires an angle of incidence, yachtsmen prefer sailing with wind on the quarter rather than dead astern, particularly when flying a spinnaker, to give them more thrust.

When sailing into the wind, the yacht is propelled only by a component of the lift due to the sails, acting as aerofoils. Lift and associated drag depend upon the set of the sails, their sizes, shapes, stretch and material, the angle of heel of the boat, the relative wind velocity and the presentation of the sails to the wind and to each other. Because the yacht does not quite point in the direction in which it is going and is also heeled, the hull too acts as an aerofoil, experiencing hydrodynamic lift and drag which exactly balance out the air forces when the boat is in steady motion.

The transverse couple produced by air and water forces is reacted to by the hydrostatic righting moment to keep the boat in stable equilibrium. Large angle stability and dynamical stability are clearly of great importance.

Longitudinally, the relative position of hydrodynamic lift, the centre of lateral resistance and the lateral component of air lift determine whether the rudder carries weather helm, as shown in Figure 2.16, or lee helm. Lee helm is dangerous because, if the tiller is dropped or the rudder goes free by accident, the boat will not come up into the wind but veer away and increase heel. Ideally, to minimize rudder drag, a yacht should carry slight weather helm in all attitudes and it is towards this 'balance' that yacht designers aim.

The resistance of a yacht calculated or measured in the manner conventional for surface ships, is not a very helpful guide to the yacht designer. Minimum resistance is required at small angles of yaw and small angles of heel, and these are different from the conventional figures. Resistance in waves is also of considerable importance and varies, of course, with the response of the boat to a particular sea – because of augment of resistance due to pitching, a yacht may well sail faster in Force 4 conditions than it does in Force 5, or better in the Portland Reaches than off Rhode Island.

A yacht designer must therefore achieve minimum resistance yawed, heeled and in waves, good longitudinal balance in all conditions and satisfactory stability. The rig must not upset the longitudinal balance and must give maximum performance for sail

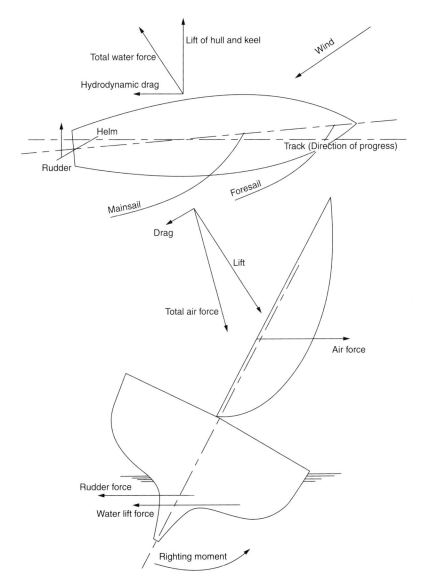

Figure 2.16 Principal forces acting on a boat sailing into the wind.

area permitted, in all conditions and direction of wind, particularly close to the wind. While some theory helps, the process overall remains much an art.

2.5 Warships

Some very interesting problems attach to the design of warships. A fighting ship needs to carry sensors to detect an enemy and weapons to defend itself and attack others. It must be difficult for an enemy to detect and be able to take punishment as well as inflict it. Its ability to survive depends upon its *susceptibility* to being hit and its *vulnerability* to the effects of a striking weapon. Susceptibility depends upon its ability to avoid detection and then, failing that, to prevent the enemy weapon hitting.

2.5.1 Stealth

A warship can betray its presence by a variety of *signatures*. All must be as low as possible to avoid detection by an enemy, to make it more difficult for enemy weapons to home in and to prevent the triggering of sensitive mines. The signatures include:

(1) *Noise* from the propulsor, machinery or the flow of water past the ship. An attacking ship can detect noise by passive sonars without betraying its own presence. Noise levels can be reduced by special propulsor design, by fitting anti-noise mountings to noisy machines and by applying special coatings to the hull. Creating a very smooth hull reduces the risk of turbulence in the water.

(2) *Radar cross section*. When illuminated by a radar a ship causes a return pulse depending upon its size and geometry. By arranging the structural shape so that the returning pulses are scattered over a wide arc the signal picked up by the searching ship is much weaker. Radar absorbent materials can be applied to the outer skin to absorb much of the incident signal.

(3) *Infrared* emissions from areas of heat. The reader will be aware that instruments are used by rescue services to detect the heat from human bodies buried in debris. The principle is the same. The ship is bound to be warmer than its surroundings but the main heat concentrations can be reduced. The funnel exhaust can be cooled and can be pointed in a direction from which the enemy is less likely to approach.

(4) *Magnetic*. Many mines are triggered by the changes in the local magnetic field caused by the passage of a ship. All steel ships have a degree of in-built magnetism which can be countered by creating opposing fields with special coils carrying electrical current. This treatment is known as *degaussing*. In addition the ship distorts the earth's magnetic field. This effect can be reduced in the same way but the ship needs to detect the strength and direction of the earth's field in order to know what correction to apply.

(5) *Pressure*. The ship causes a change in the pressure field as it moves through the water and mines can respond to this. The effect can be reduced by the ship going slowly and this is the usual defensive measure adopted.

It is impossible to remove the signatures completely. Indeed there is no need in some cases as, for instance, the sea has a background noise level which helps to 'hide' the ship. The aim of the designer is to reduce the signatures to levels where the enemy must develop more sophisticated sensors and weapons, must take greater risk of being detected in order to detect, and to make it easier to seduce weapons aimed at the ship by means of countermeasures. Enemy radars can be jammed but acts such as this can themselves betray the presence of a ship. Passive protection methods are to be preferred.

2.5.2 Sensors

Sensors require careful siting to give them a good field of view and to prevent the ship's signatures or motions degrading their performance. Thus search radars must have a complete 360° coverage and are placed high in the ship. Hull mounted sonars are usually fitted below the keel forward where they are remote from major noise sources and where the boundary layer is still relatively thin. Some ships carry sonars that can be towed astern to isolate them from ship noises and to enable them to operate at a depth from which they are more likely to detect a submarine.

Weapon control radars need to be able to match the arcs of fire of the weapons they are associated with. Increasingly this means 360° as many missiles are launched vertically at first and then turn in the direction of the enemy. Often more than one sensor is fitted. Sensors must be sited so as not to interfere with, or be affected by, the ship's weapons or communications.

2.5.3 Own ship weapons

Even a ship's own weapons can present problems for it, apart from the usual ones of weight, space and supplies. They require the best possible arcs and these must allow for the missile trajectory. Missiles create an efflux which can harm protective coatings on structure as well as more sensitive equipment on exposed decks. The weapons carry a lot of explosive material and precautions are needed to reduce the risk of premature detonation. Magazines are protected as much as possible from penetration by enemy light weapons and special firefighting systems are fitted and venting arrangements provided to prevent high pressure build up in the magazine in the event of a detonation. Magazine safety is covered by special regulations and trials.

2.5.4 Enemy weapons

Most warships adopt a policy of layered defence. The aim is to detect an enemy, and the incoming weapon, at the greatest possible range and engage it with a long range defence system. This may be a hard kill system, to take out the enemy vehicle or weapon, or one which causes the incoming weapon to become confused and unable to press home its attack. If the weapon is not detected in time, or penetrates the first line of defence, a medium range system is used and then a short range one. Where an aircraft carrier is present in the task force, its aircraft would usually provide the first line of defence. It is in the later stages that decoys may be released. The incoming weapon's homing system locks on to the decoy and is diverted from the real target although the resulting explosion may still be uncomfortably close. The shortest range systems are the *close in weapon systems*. These essentially are extremely rapid firing guns which put up a veritable curtain of steel in the path of the incoming weapon. At these very short ranges even a damaged weapon may still impact the ship and cause considerable damage.

2.5.5 Sustaining damage

Even very good defence systems can be defeated, if only by becoming saturated. The ship, then, must be able to withstand at least some measure of damage before it is put out of action completely and even more before it is sunk. The variety of conventional attack to which a ship may be subject is shown in Figure 2.17.

The effects on the ship will generally involve a combination of structural damage, fire, flooding, blast, shock and fragment damage. The ship must be designed to contain these effects within as small a space as possible by *zoning*, separating out vital functions so that not all capability is lost as a result of one hit, providing extra equipments (redundancy) and protection of vital spaces. This latter may be by providing splinter proof plating or by siting well below the waterline.

An underwater explosion is perhaps the most serious threat to a ship. This can cause whipping and shock.

2.5.6 Vulnerability studies

General ship vulnerability was discussed earlier. Whilst important for all ships it is especially significant for warships because they can expect to receive damage in action with the enemy. Each new design is the subject of a vulnerability assessment to highlight any weak elements. The designer considers the probability of each of the various methods of attack an enemy might deploy, their chances of success and the likely effect upon the ship. The likelihood of retaining various degrees of *fighting capability*, and finally of simply surviving, is calculated. A fighting capability would be a function such as being able to destroy an incoming

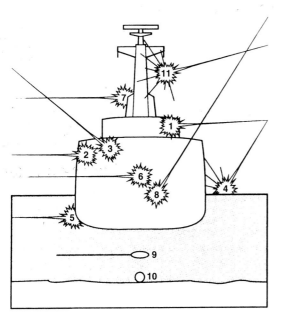

Low capacity, contact
1 cannon shell,
  HE and AP

High capacity, contact
2 HE shell
3 HE bomb
4 HE bomb, near miss
5 contact torpedo
  or mine

Medium capacity, contact
6 missile, sea skimming,
  and SAP shell
7 missile, high level
8 medium case bomb

High capacity, non-contact
9 magnetic-fuzed
  torpedo
10 ground mine
11 proximity-fuzed missile

Figure 2.17 Conventional weapon attack (courtesy RINA).

enemy missile. The contribution of each element of the ship and its systems to each fighting capability is noted, usually in diagrammatic form. For instance, to destroy a missile would require some detection and classification radar, a launcher and weapon, as well as electrics and chilled water services and a command system. Some elements will contribute to more than one capability. For each form of attack the probability of the individual elements being rendered non-operative is assessed using a blend of calculation, modelling and full scale data. If one element is particularly liable to be damaged, or especially important, it can be given extra protection or it can be duplicated to reduce the overall vulnerability. This modelling is similar to that used for reliability assessments. The assessments for each form of attack can be combined, allowing for the probability of each, to give an overall vulnerability for the design. The computations can become quite lengthy. Some judgements are very difficult to make and the results must be interpreted with care. For instance, reduced general services such as electricity may be adequate to support some but not all fighting capabilities. What then happens, in a particular battle, will depend upon which capabilities the command needs to deploy at that moment. For this reason the vulnerability results are set in the context of various engagement scenarios. In many cases, the full consequences of an attack will depend upon the actions taken by the crew in damage limitation. For instance, how effectively they deal with fire and how rapidly they close doors and valves to limit flooding. Recourse must be made to exercise data and statistical allowances made for human performance.

Whilst such analyses may be difficult they can highlight design weaknesses early in the design process when they can be corrected at little cost.

2.5.7 Types of warship

Warships are categorized by their function, for instance the:

- aircraft carrier;
- guided missile cruiser;
- destroyer;
- frigate;
- mine countermeasure vessel;
- submarine;
- support ship;
- amphibious operations ship.

It is not possible to describe all these types but the following notes will provide a flavour of what is involved in warship design. It will be realized that various ship types usually operate as a group or task force. This will be reflected in their attributes. Thus a frigate will often act as an escort and must be sufficiently fast and manoeuvrable to hold and change station as the group changes course. The task force will exercise defence in depth – a layered defence. If a carrier is present its aircraft will provide a long range surveillance and attack capability. Then there will be long and medium range missile systems on the cruisers and destroyers, decoy systems to seduce the incoming weapon and, finally, a close-in weapon system. The range of sensors provided must match the requirements of the weapons fitted.

2.5.7.1 *Frigates and destroyers*

These tend to be maids of all work but with a main function which is usually anti-submarine or anti-aircraft. Their weapon and sensor fits and other characteristics reflect this (Figure 2.18). Usually the main armament is some form of missile system designed to engage the enemy at some distance from the ship, be it an aircraft or submarine. The missile having been fired from a silo or specialist launcher, may be guided all the way to the target by sensors in the ship or may be self-directing and homing. In the latter case, having been directed in the general direction of the target, the weapon's own sensors acquire the target and control the final stages of attack, leaving the ship free to engage other targets. The use of helicopters greatly extends the area of ocean over which the ship can exert an influence.

2.5.7.2 *Mine countermeasures vessels*

Mine countermeasure ships may be either sweepers or hunters of mines, or combine the two functions in one hull. Modern mines can lie on the bottom and only become active when they sense a target with quite specific signature characteristics. They may then explode under the target or release a homing weapon. They may only react after a selected number of ships have passed nearby, or only at selected times. All these features make them difficult to render harmless.

Since sweeping mines depends upon either cutting their securing wires or setting them off by simulated signatures, to which they will react, the latest mines are virtually unsweepable. They need to be hunted, detection being usually by a high-resolution sonar. They can then be destroyed by placing a small charge alongside the mine to trigger it. The charge is usually laid by a remotely operated underwater vehicle. Because mine countermeasure vessels themselves are a target for the mines they are trying to destroy, the ship signatures must be extremely low and the hulls very robust. Nowadays hulls are often made from glass reinforced plastics and much of the equipment is specially made from materials with low magnetic properties, see also Chapter 9, Section 9.3.4.

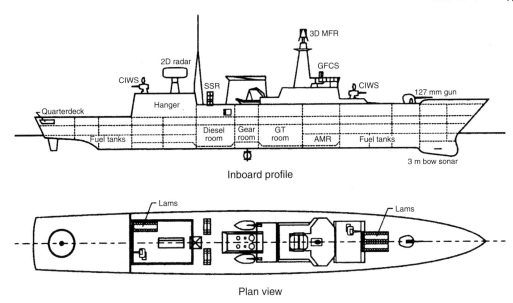

Figure 2.18 Frigate (courtesy RINA).

2.5.7.3 *Submarines*

The submarine with its torpedoes proved a very potent weapon during the two world wars in the first half of the 20th Century in spite of its limited range of military missions. It could fire torpedoes, lay mines and land covert groups on an enemy coast. Since then its fighting capabilities have been extended greatly by fitting long range missile systems which can be deployed against land, sea or air targets. The intercontinental ballistic missile, with a nuclear warhead, enabled the submarine to become the principal deterrent system of the major powers; cruise missiles can be launched against land targets well inland and without the need for the vessel to come to the surface. It still remains a difficult target for an enemy to locate and attack. Thus today the submarine is a versatile, multi-role vessel.

Submarines are dealt with here under warships because to date all large submarines have been warships. They present a number of special challenges to the naval architect:

- Although intended to operate submerged for most of the time they must be safe and manoeuvrable on the surface.
- Stability can be critical in the transition between the submerged and surfaced conditions.
- They are unstable in depth. Going deeper causes the hull to compress reducing the buoyancy force. Depth will go on increasing unless some action is taken.
- Underwater they manoeuvre in three dimensions, often at high speed. They must not betray their presence to an enemy by breaking the surface. Nor can they go too deep or they will implode. Thus manoeuvres are confined to a layer of water only a few ship lengths in depth.
- Machinery must be able to operate independently of the earth's atmosphere.
- The internal atmosphere must be kept fit for the crew to breathe and free of offensive odours.
- Periscopes are provided for the command to see the outside world and other surface piercing masts can carry radar and communications.
- Escape and rescue arrangements must be provided to assist in saving the crew from a stricken submarine.

The layout of a typical conventional submarine is shown in Figure 2.19. Its main feature is a circular pressure hull designed to withstand high hydrostatic pressure. Since it operates in three dimensions the vessel has hydroplanes for controlling depth as well as rudders for movement in the horizontal plane. Large tanks, mainly external to the pressure hull, are needed which can be flooded to cause the ship to submerge or blown, using compressed air, for surfacing. Propulsion systems are needed for both the surfaced and submerged conditions. For nuclear submarines, or those fitted with some other form of *air independent propulsion* (AIP), the same system can be used. 'Conventional' submarines use diesels for surface operations and electric drive, powered by batteries, when submerged. An air intake pipe or 'snort' mast can be fitted to enable air to be drawn into the boat at periscope. Batteries are being constantly

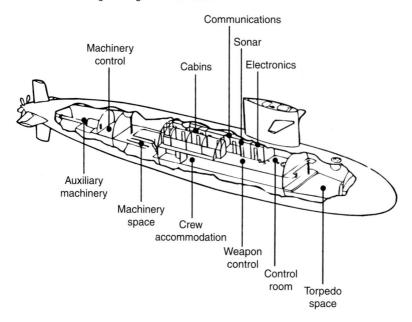

Figure 2.19 Submarine (courtesy RINA).

improved to provide greater endurance underwater and much effort has been devoted to developing AIP systems to provide some of the benefits of nuclear propulsion without the great expense. Closed-cycle diesel engines, fuel cells and Stirling engines are possibilities. The systems still require a source of oxygen such as high-test peroxide or liquid oxygen. Fuel sources for fuel cell application include sulphur free diesel fuel, methanol and hydrogen. Nuclear propulsion is expensive and brings with it problems of disposing of spent reactor fuel. For these reasons increasing interest is being taken in fuel cells.

Having given a submarine a propulsion capability for long periods submerged, it is necessary to make provision for better control of the atmosphere for the crew. The internal atmosphere can contain many pollutants, some becoming important because they can build up to dangerous levels over a long time. A much more comprehensive system of atmosphere monitoring and control is needed than that fitted in earlier conventional submarines.

Clancy (1993) describes in some detail *USS Miami* (SSN-755) and *HMS Triumph* (S-93) plus the ordering and build procedures, roles and missions. The reference includes the weapons and sensors fitted, dimensions, diving depths and speeds. Diagrammatic layouts of these submarines (and sketches of others) are given. Anechoic tiling and radar absorbent materials, to improve stealth are mentioned.

The hydroplanes, fitted for changing and maintaining depth, can only exert limited lift, particularly when the submarine is moving slowly, so the vessel must be close to neutral buoyancy when submerged and the longitudinal centres of buoyancy and weight must be in line. It follows that the weight distribution before diving, and the admission of ballast water when diving, must be carefully controlled. The first task when submerged is to 'catch a trim', that is adjust the weights by the small amounts needed to achieve the balance of weight and buoyancy. Since there is no waterplane, when submerged the metacentre and centre of buoyancy will be coincident and BG will be the same for transverse and longitudinal stability. On the surface the usual stability principles apply but the waterplane area is relatively small. The stability when in transition from the submerged to the surfaced state may be critical and needs to be studied in its own right. The usual principles apply to the powering of submarines except that for deep operations there will be no wavemaking resistance. This is offset to a degree by the greater frictional resistance due to the greater wetted hull surface.

The pressure hull, with its transverse bulkheads, must be able to withstand the crushing pressures at deep diving depth. Design calculations usually assume axial symmetry of structure and loads. This idealization enables approximate and analytical solutions to be applied with some accuracy. Subsequently detailed analyses can be made of non axi-symmetric features such as openings and internal structure. The dome ends at either end of the pressure hull are important features subject usually to finite element analysis and model testing. Buckling of the hull is possible but to be avoided. Assessments are made of *inter-frame collapse* (collapse of the short cylinder of plating between frames under radial compression); inter-bulkhead collapse (collapse of the pressure hull plating with the frames between bulkheads) and frame tripping.

The design is developed so that any buckling is likely to be in the inter-frame mode and by keeping the risk of collapse at 1.5 times the maximum working pressure acceptably small. The effects of shape imperfections and residual stresses are allowed for empirically. Small departures from circularity can lead to a marked loss of strength and the pressure causing yield at 0.25% shape imperfection on radius can be as little as half that required for perfect circularity.

If a stricken submarine is lying on the seabed the crew would await rescue if possible. For rescue at least one hatch is designed to enable a rescue submersible to mate with the submarine. The crew can then be transferred to the surface in small groups without getting wet or being subject to undue pressure. The first such rescue craft, apart from some early diving bells, were the two Deep Submergence Rescue Vessels (DSRVs) of the USN. However, deteriorating conditions inside the damaged submarine may mean that the crew cannot await rescue in this way; the pressure may be rising due to water entry or the atmosphere may become polluted. In such cases the crew can escape from the submarine in depths down to 180 m. One- and two-man escape towers are fitted to allow rapid compression so limiting the body's uptake of gas which would otherwise lead to the 'bends'. A survival suit is worn to protect against hypothermia and a hood holds a bubble of gas for breathing. In Russian submarines emergency escape capsules are provided.

So far commercial applications of submarines have been generally limited to submersibles some of which have been very deep diving. Many are unmanned, remotely operated vehicles. Most of these applications have been associated with deep ocean research, the exploitation of the ocean's resources, rescuing the crews of stricken submarines or for investigations of shipwrecks. A growing use is in the leisure industry for taking people down to view the colourful sub-surface world. In some types of operation the submersible may be the only way of tackling a problem such as the servicing of an oil wellhead in situ which is too deep for divers.

The search for, location and exploration of the wreck of *MV Derbyshire* used the capabilities of the Woods Hole Oceanographic Institution (WHOI). WHOI operates several research ships together with a number of submersibles including:

- The *Alvin*, a three-person submersible capable of diving to 4500 m.
- *Argo*, a deep-towed search and survey vehicle providing optical and acoustic imaging down to 6000 m.
- *Jason/Media*, an ROV system. *Media* serves as a transition point from the armoured cable to the neutrally buoyant umbilical. *Jason* surveys and samples the seabed using a range of equipments – sonar and photographic – and both vehicles can operate at 6000 m. See also Chapter 10.

References (Chapter 2)

Allan, R.G. (1997). From Kori to escort: the evolution of tug design 1947–1997. *Ship and Boat International, RINA*.

Barge Carriers – a Revolution in Marine Transport. *The Naval Architect,* April 1973.

Bhave, and Ghosh Roy. (1975). Special Trade Passenger Ships. *The Naval Architect*, January.

Brown, D.K. and Tupper, E.C. (1989). The naval architecture of surface warships. *TRINA*, Vol. 131.

Brown, D.K. and Moore, G. (2003). *Rebuilding the Royal Navy*. Chatham Publishing, London.

Burrows. (1997). The North Sea Platform Supply Vessel. *ImarE Trans.*, Part 1.

Clancy, T. (1993). *Submarine*. Harper Collins.

Claughton, A., Wellicome, J.F., and Shenoi, R.A. (eds) (2006). *Sailing Yacht. Design*: Vol. 1 Theory, Vol. 2 Practice. Published by The University of Southampton, Southampton, UK.

Dokkum, K. van. (2003). *Ship Knowledge, A Modern Encyclopedia*. Dokmar.

Dorey, A.L. (1990). High speed small craft. The 54th Parsons Memorial Lecture. *TRINA.*

DPTAC (2000). The design of large passenger ships and passenger infrastructure: guidance on meeting the needs of disabled people. *Marine Guidance Note*, Vol. 31, No. M.

Easton, R.W.S. (1983). Modern warships: design and construction. *Trans. I.Mar.E.*, Vol. 93.

Farell, (1975). Chemical Tankers – The Quiet Evolution. *The Naval Architect*, July.

Ferreiro, L.D. and Stonehouse, M.H. (1994). A comparative study of US and UK frigate design. *TRINA*, Vol. 136.

Ffooks, R. (1993). *Natural Gas by Sea – The Development of a New Technology*, 2nd edition. Witherby and Company Limited.

Garzke, W.H., Yoerger, D.R., Harris, S., Dulin, R.O. and Brown, D.K. (1993). Deep underwater exploration vessels – past, present and future, *SNAME* Centennial Meeting.

Gas carrier update. (2005). *The Naval Architect*. RINA, London.

Gillmer, T.C. (1977). *Modern Ship Design*. Naval Institute Press, Annapolis, Maryland.

Greenhorn, J. (1989). The assessment of surface ship vulnerability to underwater attack. *TRINA*, Vol. 131.

Guidelines for the Design and Construction of Offshore Supply Vessels, IMO publication *(IMO-807E)*.

Honnor, A.F. and Andrews, D.J. (1982). HMS invincible: the first of a new genus of aircraft carrying ships. *TRINA*, Vol. 124.

IMO (1992). Comparative study on oil tanker design.

Kuo, C. (1996). Defining the safety case concept. *NA*.

Larsson, L. and Eliasson, R.E. (2000). *Principles of Yacht Design*. Adlard Coles Nautical, London.

Meek, M. (1970). The First OCL Container Ship. *Trans. RINA*, Vol. 112.

Modern Car Ferry Design and Development. *The Naval Architect*, January, 1980.

Murray, J. (1960). Merchant Ships 1860–1960. *Trans. RINA*, Vol. 102.

Pattison, D.R. and Zhang, J.W. (1994). Trimaran ships. *TRINA*, Vol. 137.

Payne, S.M. (1990). The Evolution of the Modern Cruise Liner. *The Naval Architect*.

Payne, S.M. (1992). From Tropicale to Fantasy: a decade of cruise ship development. *TRINA*, Vol. 135.

Payne, S.M. (1994). The return of the true liner, – A design critique of the modern fast cruise ship. *The Naval Architect*, September.

RINA Conference (1992). International Conference on Tankers and Bulk Carriers – the Way Ahead.

RINA Conference (1996). Safety of Passenger Ro-Ro Vessels. 1996. Conference Proceedings – Royal Institution of Naval Architects Publications.

RINA Conference (2003). Design and Operation of Container Ships 2003. Conference Proceedings – Royal Institution of Naval Architects Publications.

RINA Conference (2004). Design and Operation of Double Hull Tankers 2004. Conference Proceedings – Royal Institution of Naval Architects Publications.

RINA Conference (2004). Design and Operation of Gas Carriers 2004. Conference Proceedings – Royal Institution of Naval Architects Publications.

RINA Conference (2004). High Speed Craft 2004 Conference Proceedings – Royal Institution of Naval Architects Publications.

RINA Conference (2005). Design and Operation of Bulk Carriers 2005. Conference Proceedings – Royal Institution of Naval Architects Publications.

RINA International Symposia on Naval Submarines. Held every three years.

RINA Symposium (1985). SWATH ships and advanced multi-hulled vessels.

RINA Symposium (1993). Fast passenger craft – new developments and the Nordic initiative.

RINA Symposium (2000). Hydrodynamics of high speed craft.

RINA Symposium (2000). *RV Triton*, Demonstrator project.

RINA Symposium (2001). Design and operation of bulk carriers – post *MV Derbyshire*.

RINA Symposium (2003). Passenger ship safety.

RINA Symposium (2003). The modern yacht.

The Nautical Institute (1998). Improving ship operational design.

Thomas, T.R. and Easton, M.S. (1992). The Type 23 Duke Class frigate. *TRINA*, Vol. 134.

Thornton, A.T. (1992). Design visualisation of yacht interiors. *TRINA*, Vol. 134.

Tinsley, D. (2000). *Dole Chile*: an innovative container ship design from Germany. *The Noval Architect*.

Tinsley, D. (2000). Preparing for the mammoth container ship. The Naval Architect. *RINA*.

Yun, L. and Bliault, A. (2000). *Theory and design of air cushion craft*. Elsevier Butterworth-Heinemann, Oxford, UK.

3 Flotation and stability

Contents

3.1 Equilibrium
3.2 Stability at small angles
3.3 Hydrostatic curves
3.4 Problems in trim and stability
3.5 Free surfaces
3.6 The inclining experiment
3.7 Stability at large angles
3.8 Weight movements
3.9 Dynamical stability
3.10 Flooding and damaged stability
3.11 Intact stability regulations
3.12 Damage stability regulations
References (Chapter 3)

The various Sections of this Chapter have been taken from the following books, with the permission of the authors:

Biran, A. (2003). *Ship Hydrostatics and Stability*. Butterworth-Heinemann, Oxford, UK.
  [Sections 3.11, 3.12]
Tupper, E.C. (2004). *Introduction to Naval Architecture*. Butterworth-Heinemann, Oxford, UK.
  [Sections 3.1–3.10]

3.1 Equilibrium

3.1.1 Equilibrium of a body floating in still water

Archimede's Principle states that a body immersed in a fluid experiences an upthrust equal to the weight of the fluid displaced, and this is fundamental to the equilibrium of a body floating in still water.

A body floating freely in still water experiences a downward force acting on it due to gravity. If the body has a mass m, this force will be mg and is known as the *weight*. Since the body is in equilibrium there must be a force of the same magnitude and in the same line of action as the weight but opposing it. Otherwise the body would move. This opposing force is generated by the hydrostatic pressures which act on the body, Figure 3.1. These act normal to the body's surface and can be resolved into vertical and horizontal components. The sum of the vertical components must equal the weight. The horizontal components must cancel out otherwise the body would move sideways. The gravitational force mg can be imagined as concentrated at a point G which is the centre of mass, commonly known as the *centre of gravity*. Similarly the opposing force can be imagined to be concentrated at a point B.

Consider now the hydrostatic forces acting on a small element of the surface, da, a depth y below the surface.

Pressure = density × gravitational acceleration × depth = $\rho g y$

The normal force on an element of area $da = \rho g y \, da$

If ϕ is the angle of inclination of the body's surface to the horizontal then the vertical component of force is:

$(\rho g y \, da) \cos \phi = \rho g$ (volume of vertical element)

Integrating over the whole volume the total vertical force is:

$\rho g \nabla$ where ∇ is the immersed volume of the body.

This is also the weight of the displaced water. It is this vertical force which 'buoys up' the body and it is known as the *buoyancy force* or simply *buoyancy*. The point, B, through which it acts is the centroid of volume of the displaced water and is known as the *centre of buoyancy*.

Since the buoyancy force is equal to the weight of the body, $m = \rho \nabla$.

In other words the mass of the body is equal to the mass of the water displaced by the body. This can be visualized in simple physical terms. Consider the underwater portion of the floating body to be replaced by a weightless membrane filled to the level of the free surface with water of the same density as that in which the body is floating. As far as the water is concerned the membrane need not exist, there is a state of equilibrium and the forces on the skin must balance out.

3.1.2 Underwater volume

The shape and dimensions of the ship hull form are represented by a body plan which, together with vertical sections (sheer plan or profile) and horizontal sections (waterlines) make up the lines plan (Figure 3.2). Development of the lines plan and body sections are discussed in more detail in Chapter 9.

Once the ship form is defined the underwater volume can be calculated. If the immersed areas of a number of sections throughout the length of a ship

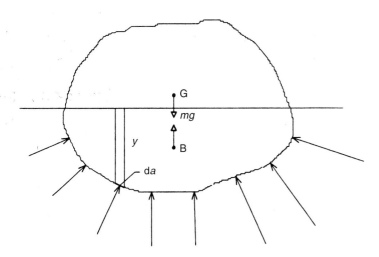

Figure 3.1 Floating body.

78 Maritime engineering reference book

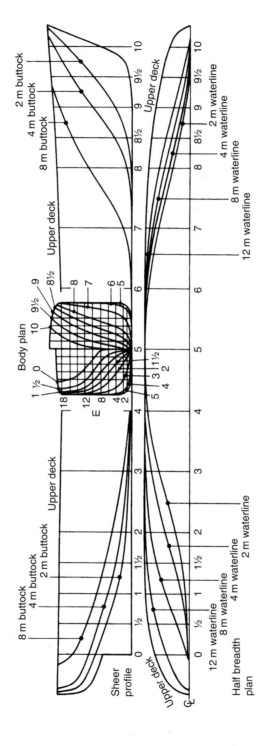

Figure 3.2 Lines plan (Tupper 2004).

are calculated, a sectional area curve can be drawn as in Figure 3.3. The underwater volume is:

$$\nabla = \int A\,dx \qquad (3.1)$$

If immersed cross-sectional areas are calculated to a number of waterlines parallel to the design waterline, then the volume up to each can be determined and plotted against draught as in Figure 3.4. The volume corresponding to any given draught T can be picked off, provided the waterline at T is parallel to those used in deriving the curve.

A more general method of finding the underwater volume, known as the *volume of displacement*, is to make use of *Bonjean* curves. These are curves of immersed cross-sectional areas plotted against draught for each transverse section. They are usually drawn on the ship profile as in Figure 3.5. Suppose the ship is floating at waterline WL. The immersed areas for this waterline are obtained by drawing horizontal lines, shown dotted, from the intercept of the waterline with the middle line of a section to the Bonjean curve for that section. Having the areas for all the sections, the underwater volume and its longitudinal centroid, its centre of buoyancy, can be calculated.

When the displacement of a ship was calculated manually, it was customary to use what was called a *displacement sheet*. A typical layout is shown in Figure 3.6. The displacement from the base up to, in this case, the 5 m waterline was determined by using Simpson's rule applied to half ordinates measured at waterlines 1 m apart and at sections taken at every tenth of the length. The calculations were done in two ways. Firstly the areas of sections were calculated and integrated in the fore and aft direction to give volume. Then areas of waterplanes were calculated and integrated vertically to give volume. The two volume values, A and B in the figure, had to be the same if the arithmetic had been done correctly, providing a check on the calculation. The displacement sheet was also used to calculate the vertical and longitudinal positions of the centre of buoyancy. The calculations are now done by computer. The calculation lends itself very well to the use of Excel spreadsheets.

3.2 Stability at small angles

3.2.1 Concept

The concept of the stability of a floating body can be explained by considering it to be inclined from the upright by an external force which is then removed. In Figure 3.7 a ship floats originally at waterline W_0L_0 and after rotating through a small angle at waterline W_1L_1.

The inclination does not affect the position of G, the ship's centre of gravity, provided no weights are free to move. The inclination does, however, affect the underwater shape and the centre of buoyancy moves from B_0 to B_1. This is because a volume, v, represented by W_0OW_1, has come out of the water and an equal volume, represented by L_0OL_1, has been immersed.

If g_e and g_i are the centroids of the emerged and immersed wedges and $g_e g_i = h$, then:

$$B_0 B_1 = \frac{v \times h}{\nabla} \qquad (3.2)$$

where ∇ is the total volume of the ship.

In general a ship will trim slightly when it is inclined at constant displacement. For the present this is ignored but it means that strictly B_0, B_1, g_e, etc., are the projections of the actual points on to a transverse plane.

The buoyancy acts upwards through B_1 and intersects the original vertical at M. This point is termed the *metacentre* and for small inclinations can be taken as fixed in position. The weight $W = mg$ acting downwards and the buoyancy force, of equal magnitude, acting upwards are not in the same line but form a couple $W \times GZ$, where GZ is the perpendicular on to B_1M drawn from G. As shown this couple will restore the

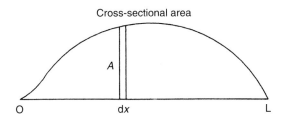

Figure 3.3 Cross-sectional area curve.

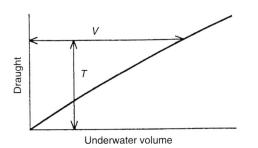

Figure 3.4 Volume curve.

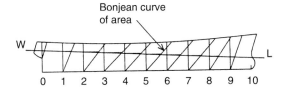

Figure 3.5 Bonjean curves.

80 Maritime engineering reference book

Figure 3.6 Displacement sheet.

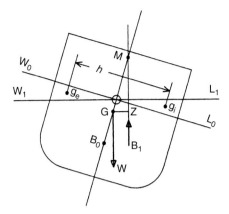

Figure 3.7 Small angle stability.

body to its original position and in this condition the body is said to be in stable equilibrium. $GZ = GM \sin \phi$ and is called the *righting lever* or *lever* and *GM* is called the *metacentric height*. For a given position of G, as M can be taken as fixed for small inclinations, *GM* will be constant for any particular waterline. More importantly, since G can vary with the loading of the ship even for a given displacement, *BM* will be constant for a given waterline. In Figure 3.7 M is above G, giving positive stability, and *GM* is regarded as positive in this case.

If, when inclined, the new position of the centre of buoyancy, B_1, is directly under G, the three points M, G and Z are coincident and there is no moment acting on the ship. When the disturbing force is removed the ship will remain in the inclined position. The ship is said to be in neutral equilibrium and both *GM* and *GZ* are zero.

A third possibility is that, after inclination, the new centre of buoyancy will lie to the left of G. There is then a moment $W \times GZ$ which will take the ship further from the vertical. In this case the ship is said to be unstable and it may heel to a considerable angle or even capsize. For unstable equilibrium M is below G and both *GM* and *GZ* are considered negative.

The above considerations apply to what is called the *initial stability* of the ship, that is when the ship is upright or very nearly so. The criterion of initial stability is the metacentric height. The three conditions can be summarized as:

M above G	GM	and	GZ positive	stable
M at G	GM	and	GZ zero	neutral
M below G	GM	and	GZ negative	unstable

3.2.2 Transverse metacentre

The position of the metacentre is found by considering small inclinations of a ship about its centreline, Figure 3.8.

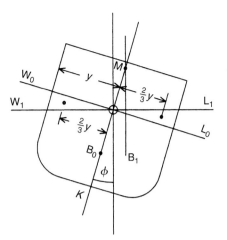

Figure 3.8 Transverse metacentre.

For small angles, say 2 or 3 degrees, the upright and inclined waterlines will intersect at O on the centreline. The volumes of the emerged and immersed wedges must be equal for constant displacement.

For small angles the emerged and immersed wedges at any section, W_0OW_1 and L_0OL_1, are approximately triangular. If y is the half-ordinate of the original waterline at the cross section the emerged or immersed section area is:

$$\frac{1}{2} y \times y \tan \phi = \frac{1}{2} y^2 \phi \qquad (3.3)$$

for small angles, and the total volume of each wedge is:

$$\int \frac{1}{2} y^2 \phi \, dx \qquad (3.4)$$

integrated along the length of the ship.

This volume is effectively moved from one side to the other and for triangular sections the transverse movement will be $4y/3$ giving a total transverse shift of buoyancy of:

$$\int \frac{1}{2} y^2 \phi \, dx \times 4y/3 = \phi \int 2y^3/3 \, dx \qquad (3.5)$$

since ϕ is constant along the length of the ship.

The expression within the integral sign is the second moment of area, or the moment of inertia, of a waterplane about its centreline. It may be denoted by I, whence the transverse movement of buoyancy is:

$$I\phi \quad \text{and} \quad \nabla \times BB_1 = I\phi \qquad (3.6)$$

so that $BB_1 = I\phi/\nabla$ where ∇ is the total volume of displacement.

Referring to Figure 3.8 for the small angles being considered $BB_1 = BM\phi$ and $BM = I/\nabla$. Thus the height of the metacentre above the centre of buoyancy is found by dividing the second moment of area of the waterplane about its centreline by the volume of displacement. The height of the centre of buoyancy above the keel, KB, is the height of the centroid of the underwater volume above the keel, and hence the height of the metacentre above the keel is:

$$KM = KB + BM \tag{3.7}$$

The difference between KM and KG gives the metacentric height, GM.

3.2.3 Transverse metacentre for simple geometrical forms

Vessel of rectangular cross section

Consider the form in Figure 3.9 of breadth B and length L floating at draught T. If the cross section is uniform throughout its length, the volume of displacement $= LBT$.

The second moment of area of waterplane about the centreline $= LB^3/12$. Hence:

$$BM = \frac{LB^3}{12LBT} = B^2/12T \tag{3.8}$$

Height of centre of buoyancy above keel, $KB = T/2$ and the height of metacentre above the keel is:

$$KM = T/2 + B^2/12T \tag{3.9}$$

The height of the metacentre depends upon the draught and beam but not the length. At small draught relative to beam, the second term predominates and at zero draught KM would be infinite.

To put some figures to this, consider the case where B is 15 m for draughts varying from 1 to 6 m. Then:

$$KM = \frac{T}{2} + \frac{15^2}{12T} = 0.5T + \frac{18.75}{T} \tag{3.10}$$

KM values for various draughts are shown in Table 3.1 and KM and KB are plotted against draught in Figure 3.10. Such a diagram is called a *metacentric diagram*. KM is large at small draughts and falls rapidly with increasing draught. If the calculations were extended KM would reach a minimum value and then start to increase. The draught at which KM is minimum can be found by differentiating the equation for KM with respect to T and equating to zero. That is, KM is a minimum at T given by:

$$\frac{dKM}{dT} = \frac{1}{2} - \frac{B^2}{12T^2} = 0,$$

$$\text{giving } T^2 = \frac{B^2}{6} \quad \text{or} \quad T = \frac{B}{\sqrt{6}}$$

In the example KM is minimum when the draught is 6.12 m.

Vessel of constant triangular section

Consider a vessel of triangular cross section floating apex down, the breadth at the top being B and the depth D. The breadth of the waterline at draught T is given by:

$$b = (T/D) \times B$$
$$I = (L/12) \times [(T/D) \times B]^3$$
$$\nabla = L \times (T/D) \times B \times T/2$$
$$BM = I/\nabla = B^2T/6D^2$$
$$KB = 2T/3$$
$$KM = 2T/3 + B^2T/6D^2 \tag{3.11}$$

Table 3.1 KM values.

T	$0.5T$	$18.75/T$	KM
1	0.5	18.75	19.25
2	1.0	9.37	10.37
3	1.5	6.25	7.75
4	2.0	4.69	6.69
5	2.5	3.75	6.25
6	3.0	3.12	6.12

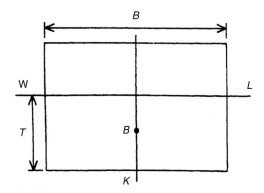

Figure 3.9 Rectangular section vessel.

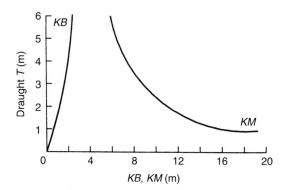

Figure 3.10 Metacentric diagram.

In this case the curves of both *KM* and *KB* against draught are straight lines starting from zero at zero draught.

Vessel of circular cross section
Consider a circular cylinder of radius *R* and centre of section O, floating with its axis horizontal. For any waterline, above or below O, and for any inclination, the buoyancy force always acts through O. That is, *KM* is independent of draught and equal to *R*. The vessel will be stable or unstable depending upon whether *KG* is less than or greater than *R*.

3.2.4 Metacentric diagrams

The positions of B and M have been seen to depend only upon the geometry of the ship and the draughts at which it is floating. They can therefore be determined without knowledge of the loading of the ship that causes it to float at those draughts. A *metacentric diagram,* in which *KB* and *KM* are plotted against draught, is a convenient way of defining the positions of B and M for a range of waterplanes parallel to the design or load waterplane.

Trim
Suppose a ship, floating at waterline W_0L_0 (Figure 3.11), is caused to trim slightly, at constant displacement, to a new waterline W_1L_1 intersecting the original waterplane in a transverse axis through F.
The volumes of the immersed and emerged wedges must be equal so, for small θ:

$$\int 2 y_f(x_f\theta)\,dx = \int 2y_a(x_a\theta)\,dx \qquad (3.12)$$

where y_f and y_a are the waterplane half breadths at distances x_f and x_a from F.
This is the condition that F is the centroid of the waterplane and F is known as the *centre of flotation*. For small trims at constant displacement a ship trims about a transverse axis through the centre of flotation.

If a small weight is added to a ship it will sink and trim until the extra buoyancy generated equals the weight and the centre of buoyancy of the added buoyancy is vertically below the centre of gravity of the added weight. If the weight is added in the vertical line of the centre of flotation then the ship sinks bodily with no trim, as the centre of buoyancy of the added layer will be above the centroid of area of the waterplane. Generalizing this, a small weight placed anywhere along the length can be regarded as being initially placed at F to cause sinkage and then moved to its actual position, causing trim. In other words, it can be regarded as a weight acting at F and a trimming moment about F.

3.2.5 Longitudinal stability

The principles involved are the same as those for transverse stability but for longitudinal inclinations, the stability depends upon the distance between the centre of gravity and the longitudinal metacentre. In this case the distance between the centre of buoyancy and the longitudinal metacentre will be governed by the second moment of area of the waterplane about a transverse axis passing through its centroid. For normal ship forms this quantity is many times the value for the second moment of area about the centreline. Since BM_L is obtained by dividing by the same volume of displacement as for transverse stability, it will be large compared with BM_T and often commensurate with the length of the ship. It is thus virtually impossible for an undamaged conventional ship to be unstable when inclined about a transverse axis.

$$KM_L = KB + BM_L = KB + I_L/\nabla \qquad (3.13)$$

where I_L is the second moment of the waterplane areas about a transverse axis through its centroid, the centre of flotation.

If the ship in Figure 3.11 is trimmed by moving a weight, *w*, from its initial position to a new position *h* forward, the trimming moment will be *wh*. This will cause the centre of gravity of the ship to move from G to G_1 and the ship will trim causing B to move to B_1 such that:

$$GG_1 = wh/W \qquad (3.14)$$

and B_1 is vertically below G_1.
The trim is the difference in draughts forward and aft. The change in trim angle can be taken as the change in that difference divided by the longitudinal distance between the points at which the draughts are measured. From Figure 3.11:

$$\tan\theta = t/L = GG_1/GM_L = wh/WGM_L \qquad (3.15)$$

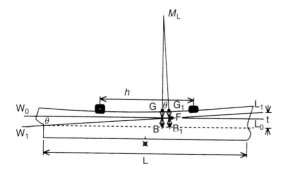

Figure 3.11 Trim changes.

from which:

$$wh = t \times W \times GM_L/L \qquad (3.16)$$

This is the moment that causes a trim t, so the moment to cause unit change of trim is:

$$WGM_L/L \qquad (3.17)$$

The *moment to change trim*, MCT, one metre is a convenient figure to quote to show how easy a ship is to trim.

The value of MCT is very useful in calculating the draughts at which a ship will float for a given condition of loading. Suppose it has been ascertained that the weight of the ship is W and the centre of gravity is x forward of amidships and that at that weight with a waterline parallel to the design waterline it would float at a draught T with the centre of buoyancy y forward of amidships. There will be a moment $W(y - x)/$MCT taking it away from a waterline parallel to the design one. The ship trims about the centre of flotation and the draughts at any point along the length can be found by simple ratios.

3.3 Hydrostatic curves

3.3.1 Surface ships

It has been shown how the displacement, position of B, M and F can be calculated. It is customary to obtain these quantities for a range of waterplanes parallel to the design waterplane and plot them against draught, draught being measured vertically. Such sets of curves are called *hydrostatic curves*, Figure 3.12.

The curves in the figure show moulded and extreme displacement. The former is defined in Chapter 12. It is the latter, normally shown simply as the displacement curve and which allows for displacement to outside of plating and outside the perpendiculars, bossings, bulbous bows, etc., which is relevant to the discussion of flotation and stability. Clearly the additions to the moulded figure can have a measurable effect upon displacement and the position of B.

It will be noted that the curves include one for the increase in displacement for unit increase in draught. If a waterplane has an area A, then the increase in displaced volume for unit increase in draught at that waterplane is $1 \times A$. The increase in displacement will be $\rho g A$. For $\rho = 1025\,\text{kg/m}^3$ and $g = 9.81\,\text{m/s}^2$ increase in displacement per metre increase in draught is:

$$1025 \times 9.81 \times 1 \times A = 10\,055 A \text{ newtons.}$$

The increase in displacement per unit increase in draught is useful in approximate calculations when weights are added to a ship. Since its value varies with draught it should be applied with care.

Hydrostatic curves are useful for working out the draughts and the initial stability, as represented by GM, in various conditions of loading. This is done for all normal working conditions of the ship and the results supplied to the master, see Section 11.5.2.2.

3.3.2 Fully submerged bodies

A fully submerged body presents a special case. Firstly there is no waterplane and therefore no metacentre. The forces of weight and displacement will always act vertically through G and B respectively. Stability then will be the same for inclination about any axis. It will be positive if B is above G. Secondly a submarine or submersible is an elastic body and will compress as the depth of submergence increases. Since water is effectively incompressible, there will be a reducing

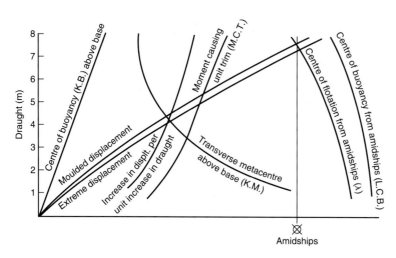

Figure 3.12 Hydrostatic curves.

buoyancy force. Thus the body will experience a net downward force that will cause it to sink further so that the body is unstable in depth variation. In practice the decrease in buoyancy must be compensated for by pumping water out from internal tanks or by forces generated by the control surfaces, the hydroplanes. Care is needed when first submerging to arrange that weight and buoyancy are very nearly the same. If the submersible moves into water of a different density there will again be an imbalance in forces due to the changed buoyancy force. There is no 'automatic' compensation such as a surface vessel experiences when the draught adjusts in response to density changes.

The stability and control of submarines is discussed in Chapter 8, Section 8.18, and more detailed accounts can be found in Burcher and Rydill (1998).

3.4 Problems in trim and stability

3.4.1 Determination of displacement from observed draughts

Suppose draughts at the perpendiculars are T_a and T_f as in Figure 3.13. The mean draught will be $T = (T_a + T_f)/2$ and a first approximation to the displacement could be obtained by reading off the corresponding displacement, Δ, from the hydrostatic curves. In general, W_0L_0 will not be parallel to the waterlines for which the hydrostatics were computed. If waterline W_1L_1, intersecting W_0L_0 at amidships, is parallel to the design waterline then the displacement read from the hydrostatics for draught T is in fact the displacement to W_1L_1. It has been seen that because ships are not symmetrical fore and aft they trim about F. As shown in Figure 3.13, the displacement to W_0L_0 is less than that to W_1L_1, the difference being the layer $W_1L_1L_2W_2$, where W_2L_2 is the waterline parallel to W_1L_1 through F on W_0L_0. If λ is the distance of F forward of amidships then the thickness of layer = $\lambda \times t/L$ where $t = T_a - T_f$.

If i is the increase in displacement per unit increase in draught:

Displacement of layer = $\lambda \times ti/L$ and the actual displacement
$$= \Delta - \lambda \times ti/L \quad (3.18)$$

Whether the correction to the displacement read off from the hydrostatics initially is positive or negative depends upon whether the ship is trimming by the bow or stern and the position of F relative to amidships. It can be determined by making a simple sketch.

If the ship is floating in water of a different density to that for which the hydrostatics were calculated a further correction is needed in proportion to the two density values, increasing the displacement if the water in which ship is floating is greater than the standard.

This calculation for displacement has assumed that the keel is straight. It is likely to be curved, even in still water, so that a draught taken at amidships may not equal $(d_a + d_f)/2$ but have some value d_m giving a deflection of the hull, δ. If the ship sags the above calculation would underestimate the volume of displacement. If it hogs it would overestimate the volume. It is reasonable to assume the deflected profile of the ship is parabolic, so that the deflection at any point distant x from amidships is $\delta[1 - (2x/L)^2]$, and hence:

$$\text{Volume correction} = \int b\delta[1 - (2x/L)^2] dx \quad (3.19)$$

where b is the waterline breadth.

Unless an expression is available for b in terms of x this cannot be integrated mathematically. It can be evaluated by approximate integration using the ordinates for the waterline.

3.4.2 Longitudinal position of the centre of gravity

Suppose a ship is floating in equilibrium at a waterline W_0L_0 as in Figure 3.14 with the centre of gravity distant x

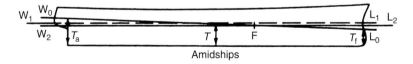

Figure 3.13

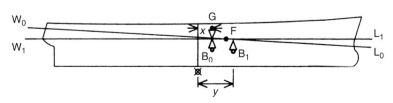

Figure 3.14

from amidships, a distance yet to be determined. The centre of buoyancy B_0 must be directly beneath G. Now assume the ship brought to a waterline W_1L_1 parallel to those used for the hydrostatics, which cuts off the correct displacement. The position of the centre of buoyancy will be at B_1, distant y from amidships, a distance that can be read from the hydrostatics for waterline W_1L_1. It follows that if t was the trim, relative to W_1L_1, when the ship was at W_0L_0:

$$\Delta(y - x) = t \times \text{(moment to cause unit trim) and:}$$

$$x = y - \frac{t \times \text{MCT}}{\Delta} \qquad (3.20)$$

giving the longitudinal centre of gravity.

3.4.3 Direct determination of displacement and position of G

The methods described above for finding the displacement and longitudinal position of G are usually sufficiently accurate when the trim is small. To obtain more accurate results and for larger trims the Bonjean curves can be used. If the end draughts, distance L apart, are observed then the draught at any particular section can be calculated, since:

$$T_x = T_a - (T_a - T_f)\frac{x}{L} \qquad (3.21)$$

where x is the distance from where T_a is measured.

These draughts can be corrected for hog or sag if necessary. The calculated draughts at each section can be set up on the Bonjean curves and the immersed areas read off. The immersed volume and position of the centre of buoyancy can be found by approximate integration. For equilibrium, the centre of gravity and centre of buoyancy must be in the same vertical line and the position of the centre of gravity follows. Using the density of water in which the ship is floating, the displacement can be determined.

3.4.4 Heel due to moving weight

In Figure 3.15 a ship is shown upright and at rest in still water. If a small weight w is shifted transversely through a distance h, the centre of gravity of the ship, originally at G, moves to G_1 such that $GG_1 = wh/W$. The ship will heel through an angle ϕ causing the centre of buoyancy to move to B_1 vertically below G_1 to restore equilibrium. It is seen that:

$$\frac{GG_1}{GM} = \tan\phi \quad \text{and} \quad \tan\phi = \frac{wh}{W \times GM} \qquad (3.22)$$

This applies whilst the angle of inclination remains small enough for M to be regarded as a fixed point.

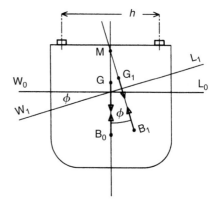

Figure 3.15 Moving weight.

3.4.5 Wall-sided ship

It is interesting to consider a special case when a ship's sides are vertical in way of the waterline over the whole length. It is said to be wall-sided, see Figure 3.16. The vessel can have a turn of bilge provided it is not exposed by the inclination of the ship. Nor must the deck edge be immersed. Because the vessel is wall-sided the emerged and immersed wedges will have sections which are right-angled triangles of equal area. Let the new position of the centre of buoyancy B_1, after inclination through ϕ, be α and β relative to the centre of buoyancy position in the upright condition. Then using the notation shown in the figure:

$$\begin{aligned}
\text{Transverse moment of volume shift} &= \int \frac{y}{2} \times y\tan\phi\, dx \times \frac{4y}{3} \\
&= \int \frac{2}{3} y^3 \tan\phi\, dx \\
&= \tan\phi \int \frac{2}{3} y^3\, dx \\
&= I\tan\phi \qquad (3.23)
\end{aligned}$$

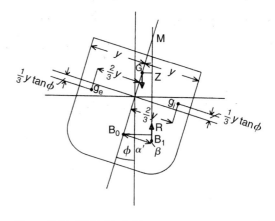

Figure 3.16 Wall-sided ship.

where I is the second moment of area of the waterplane about the centreline. Therefore:

$$\alpha = I \tan \phi / \nabla = B_0 M \tan \phi \quad \text{since } B_0 M = \frac{I}{\nabla}$$

Similarly the vertical moment of volume shift is:

$$\int \frac{1}{2} y^2 \tan \phi \times \frac{2}{3} y \tan \phi \, dx$$
$$= \int \frac{y^3}{3} \tan^2 \phi \, dx = \frac{I}{2} \tan^2 \phi \quad (3.24)$$

and:

$$\beta = \frac{1}{2} I \tan^2 \phi / \nabla = \frac{1}{2} B_0 M \tan^2 \phi \quad (3.25)$$

From the figure it will be seen that:

$$B_0 R = \alpha \cos \phi + \beta \sin \phi$$
$$= B_0 M \tan \phi \cos \phi + \frac{1}{2} B_0 M \tan^2 \phi \sin \phi$$
$$= \sin \phi \left(B_0 M + \frac{1}{2} B_0 M \tan^2 \phi \right) \quad (3.26)$$

Now

$$GZ = B_0 R - B_0 G \sin \phi$$
$$= \sin \phi \left(B_0 M - B_0 G + \frac{1}{2} B_0 M \tan^2 \phi \right)$$
$$= \sin \phi \left(GM + \frac{1}{2} B_0 M \tan^2 \phi \right) \quad (3.27)$$

This is called the *wall-sided formula*. It is often reasonably accurate for full forms up to angles as large as 10°. It will not apply if the deck edge is immersed or the bilge emerges. It can be regarded as a refinement of the simple expression $GZ = GM \sin \phi$.

3.4.6 Suspended weights

Consider a weight w suspended freely from a point h above its centroid. When the ship heels slowly the weight moves transversely and takes up a new position, again vertically below the suspension point. As far as the ship is concerned the weight seems to be located at the suspension point. Compared to the situation with the weight fixed, the ship's centre of gravity will be effectively reduced by GG_1 where:

$$GG_1 = wh/W \quad (3.28)$$

This can be regarded as a loss of metacentric height of GG_1.

Weights free to move in this way should be avoided but this is not always possible. For instance, when a weight is being lifted by a shipboard crane, as soon as the weight is lifted clear of the deck or quayside its effect on stability is as though it were at the crane head. The result is a rise in G which, if the weight is sufficiently large, could cause a stability problem. This is important to the design of heavy lift ships.

3.5 Free surfaces

3.5.1 Effect of liquid free surfaces

A ship in service will usually have tanks which are partially filled with liquids. These may be the fuel and water tanks the ship is using or may be tanks carrying liquid cargoes. When such a ship is inclined slowly through a small angle to the vertical the liquid surface will move so as to remain horizontal. In this discussion a quasi-static condition is considered so that slopping of the liquid is avoided. Different considerations would apply to the dynamic conditions of a ship rolling. For small angles, and assuming the liquid surface does not intersect the top or bottom of the tank, the volume of the wedge that moves is:

$$\int \frac{1}{2} y^2 \phi \, dx, \text{ integrated over the length, } l, \text{ of the tank.}$$

Assuming the wedges can be treated as triangles, the moment of transfer of volume is:

$$\int \frac{1}{2} y^2 \phi \, dx \times \frac{4y}{3} = \phi \int \frac{2}{3} y^3 \, dx = \phi I_1 \quad (3.29)$$

where I_1 is the second moment of area of the liquid, or free surface. The moment of mass moved $= \rho_f \phi I_1$, where ρ_f is the density of the liquid in the tank. The centre of gravity of the ship will move because of this shift of mass to a position G_1 and:

$$GG_1 = \rho_f g \phi I_1 / W = \rho_f g \phi I_1 / \rho g \nabla = \rho_f \phi I_1 / \rho \nabla \quad (3.30)$$

where ρ is the density of the water in which the ship is floating and ∇ is the volume of displacement.

The effect on the transverse movement of the centre of gravity is to reduce GZ by the amount GG_1 as in Figure 3.17(b). That is, there is an effective reduction in stability. Since $GZ = GM \sin \phi$ for small angles, the influence of the shift of G to G_1 is equivalent to raising G to G_2 on the centreline so that $GG_1 = GG_2 \tan \phi$ and the righting moment is given by:

$$W(GM \sin \phi - GG_2 \cos \phi \tan \phi)$$
$$= W(GM - GG_2) \sin \phi \quad (3.31)$$

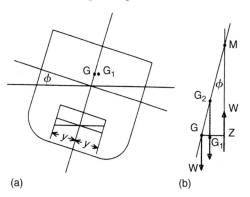

Figure 3.17 Fluid free surface.

Thus the effect of the movement of the liquid due to its free surface, is equivalent to a rise of GG_2 of the centre of gravity, the 'loss' of GM being:

$$\text{Free surface effect } GG_2 = \rho_f I_1 / \rho \nabla \qquad (3.32)$$

Another way of looking at this is to draw an analogy with the loss of stability due to the suspended weight. The water in the tank with a free surface behaves in such a way that its weight force acts through some point above the centre of the tank and height I_1/v above the centroid of the fluid in the tank, where v is the volume of fluid. In effect the tank has its own 'metacentre' through which its fluid weight acts. The fluid weight is $\rho_f v$ and the centre of gravity of the ship will be effectively raised through GG_2 where:

$$W \times GG_2 = \rho \nabla \times GG_2 = (\rho_f v)(I_1/v) = \rho_f I_1 \quad (3.33)$$

and

$$GG_2 = \rho_f I_1 / \rho \nabla \text{ as before.} \qquad (3.34)$$

This loss is the same whatever the height of the tank in the ship or its transverse position. If the loss is sufficiently large, the metacentric height becomes negative and the ship heels over and may even capsize. It is important that the free surfaces of tanks should be kept to a minimum. One way of reducing them is to subdivide wide tanks into two or more narrow ones. In Figure 3.18 a double bottom tank is shown with a central division fitted.

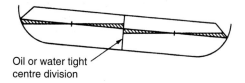

Oil or water tight centre division

Figure 3.18 Tank subdivision.

If the breadth of the tank is originally B, the width of each of the two tanks, created by the central division, is $B/2$. Assuming the tanks have a constant section, and have a length, l, the second moment of area without division is $lB^3/12$. With centre division the sum of the second moments of area of the two tanks is $(l/12)(B/2)^3 \times 2 = lB^3/48$.

That is, the introduction of a centre division has reduced the free surface effect to a quarter of its original value. Using two bulkheads to divide the tank into three equal width sections, reduces the free surface to a ninth of its original value. Thus subdivision is seen to be very effective and it is common practice to subdivide the double bottom of ships. The main tanks of ships carrying liquid cargoes must be designed taking free surface effects into account and their breadths are reduced by providing centreline or wing bulkheads.

Free surface effects should be avoided where possible and where unavoidable must be taken into account in the design. The operators must be aware of their significance and arrange to use the tanks in ways intended by the designer.

3.6 The inclining experiment

As the position of the centre of gravity is so important for initial stability it is necessary to establish it accurately. It is determined initially by calculation by considering all weights making up the ship – steel, outfit, fittings, machinery and systems – and assessing their individual centres of gravity. From these data can be calculated the displacement and centre of gravity of the light ship. For particular conditions of loading the weights of all items to be carried must then be added at their appropriate centres of gravity to give the new displacement and centre of gravity. It is difficult to account for all items accurately in such calculations and it is for this reason that the lightship weight and centre of gravity are measured experimentally.

The experiment is called the *inclining experiment* and involves causing the ship to heel to small angles by moving known weights known distances tranversely across the deck and observing the angles of inclination. The draughts at which the ship floats are noted together with the water density. Ideally the experiment is conducted when the ship is complete but this is not generally possible. There will usually be a number of items both to go on and to come off the ship (e.g. staging, tools etc.). The weights and centres of gravity of these must be assessed and the condition of the ship as inclined corrected.

A typical set up is shown in Figure 3.19. Two sets of weights, each of w, are placed on each side of the ship at about amidships, the port and starboard sets being h apart. Set 1 is moved a distance h to a position alongside

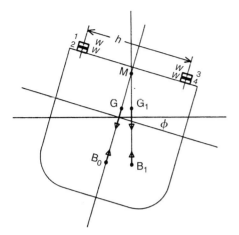

Figure 3.19 Inclining experiment.

sets 3 and 4. G moves to G_1 as the ship inclines to a small angle and B moves to B_1. It follows that:

$$GG_1 = \frac{wh}{W} = GM \tan \phi \quad \text{and}$$
$$GM = wh \cot \phi / W \quad (3.35)$$

ϕ can be obtained in a number of ways. The commonest is to use two long pendulums, one forward and one aft, suspended from the deck into the holds. If d and l are the shift and length of a pendulum respectively, $\tan \phi = d/l$.

To improve the accuracy of the experiment, several shifts of weight are used. Thus, after set 1 has been moved, a typical sequence would be to move successively set 2, replace set 2 in original position followed by set 1. The sequence is repeated for sets 3 and 4. At each stage the angle of heel is noted and the results plotted to give a mean angle for unit applied moment. When the metacentric height has been obtained, the height of the centre of gravity is determined by subtracting GM from the value of KM given by the hydrostatics for the mean draught at which the ship was floating. This KG must be corrected for the weights to go on and come off. The longitudinal position of B, and hence G, can be found using the recorded draughts.

To obtain accurate results a number of precautions have to be observed. First the experiment should be conducted in calm water with little wind. Inside a dock is good as this eliminates the effects of tides and currents. The ship must be floating freely when records are taken so any mooring lines must be slack and the brow must be lifted clear. All weights must be secure and tanks must be empty or pressed full to avoid free surface effects. If the ship does not return to its original position when the inclining weights are restored it is an indication that a weight has moved in the ship, or that fluid has moved from one tank to another, possibly through a leaking valve. The number of people on board must be kept to a minimum, and those present must go to defined positions when readings are taken. The pendulum bobs are damped by immersion in a trough of water.

The draughts must be measured accurately at stem and stern, and must be read at amidships if the ship is suspected of hogging or sagging. The density of water is taken by hydrometer at several positions around the ship and at several depths to give a good average figure. If the ship should have a large trim at the time of inclining it might not be adequate to use the hydrostatics to give the displacement and the longitudinal and vertical positions of B. In this case detailed calculations should be carried out to find these quantities for the inclining waterline.

The Merchant Shipping Acts require every new passenger ship to be inclined upon completion and the elements of its stability determined.

3.7 Stability at large angles

3.7.1 Atwood's formula

So far only a ship's *initial stability* has been considered. That is for small inclinations from the vertical. When the angle of inclination is greater than, say, 4 or 5 degrees, the point, M, at which the vertical through the inclined centre of buoyancy meets the centreline of the ship, can no longer be regarded as a fixed point. Metacentric height is no longer a suitable measure of stability and the value of the *righting arm, GZ*, is used instead.

Assume the ship is in equilibrium under the action of its weight and buoyancy with W_0L_0 and W_1L_1 the waterlines when upright and when inclined through ϕ respectively. These two waterlines will cut off the same volume of buoyancy but will not, in general, intersect on the centreline but at some point S, Figure 3.20.

A volume represented by W_0SW_1 has emerged and an equal volume, represented by L_0SL_1 has been immersed. Let this volume be v. Using the notation in Figure 3.20, the horizontal shift of the centre of buoyancy, is given by:

$$B_0R = v \times h_e h_i / \nabla \quad \text{and}$$
$$GZ = B_0R - B_0G \sin \phi \quad (3.36)$$

This expression for GZ is often called *Atwood's formula*, Atwood and Pengelly (1960).

3.7.2 Curves of statical stability

By evaluating v and $h_e h_i$ for a range of angles of inclination it is possible to plot a curve of GZ against ϕ. A typical example is Figure 3.21.

GZ increases from zero when upright to reach a maximum at A and then decreases becoming zero again at some point B. The ship will capsize if the applied moment is such that its lever is greater than the value of GZ at A. It becomes unstable once the point B has been passed. OB is known as the *range of stability*.

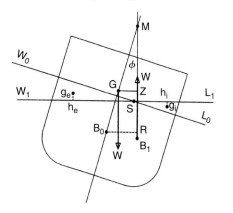

Figure 3.20 Atwood's formula.

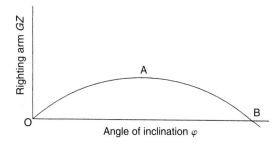

Figure 3.21 Curve of statical stability.

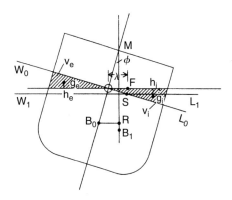

Figure 3.22 Emerged and immersed wedges.

The curve of GZ against ϕ is termed the GZ curve or *curve of statical stability*.

Because ships are not wall-sided, it is not easy to determine the position of S and so find the volume and centroid positions of the emerged and immersed wedges. One method is illustrated in Figure 3.22. The ship is first inclined about a fore and aft axis through O on the centreline. This leads to unequal volumes of emerged and immersed wedges which must be compensated for by a bodily rise or sinkage. In the case illustrated the ship rises. Using subscripts e and i for the emerged and immersed wedges respectively, the geometry of Figure 3.22 gives:

$$B_0 R = \frac{v_e(h_e O) + v_i(h_i O) - \lambda(v_i - v_e)}{\nabla} \quad (3.37)$$

and:

$$\begin{aligned}
GZ &= B_0 R - B_0 G \sin \phi \\
&= \frac{v_e(h_e O) + v_i(h_i O) - \lambda(v_i - v_e)}{\nabla} - B_0 G \sin \phi
\end{aligned} \quad (3.38)$$

For very small angles GZ still equates to $GM\phi$, so the slope of the GZ curve at the origin equals the metacentric height. That is $GM = dGZ/d\phi$ at $\phi = 0$. It is useful in drawing a GZ curve to erect an ordinate at $\phi = 1$ rad, equal to the metacentric height, and joining the top of this ordinate to the origin to give the slope of the GZ curve at the origin.

The wall-sided formula, derived earlier, can be regarded as a special case of Atwood's formula. For the wall-sided ship:

$$GZ = \sin\phi \left(GM + \frac{1}{2} B_0 M \tan^2 \phi \right) \quad (3.39)$$

If the ship has a positive GM it will be in equilibrium when GZ is zero, that is:

$$0 = \sin\phi \left(GM + \frac{1}{2} B_0 M \tan^2 \phi \right) \quad (3.40)$$

This equation is satisfied by two values of ϕ. The first is $\sin \phi = 0$, or $\phi = 0$. This is the case with the ship upright as is to be expected. The second value is given by:

$$\begin{aligned}
GM + \frac{1}{2} B_0 M \tan^2 \phi &= 0 \quad \text{or} \\
\tan^2 \phi &= -2GM/B_0 M
\end{aligned} \quad (3.41)$$

With both GM and $B_0 M$ positive there is no solution to this meaning that the upright position is the only one of equilibrium. This also applies to the case of zero GM, it being noted that in the upright position the ship has stable, not neutral, equilibrium due to the term in $B_0 M$.

When, however, the ship has a negative GM there are two possible solutions for ϕ in addition to that of zero, which in this case would be a position of unstable equilibrium. These other solutions are at ϕ either side of the upright φ being given by:

$$\tan \phi = \left(\frac{2GM}{B_0 M} \right)^{0.5} \quad (3.42)$$

The ship would show no preference for one side or the other. Such an angle is known as an *angle of loll*. The ship does not necessarily capsize although if ϕ is large enough the vessel may take water on board through side openings. The GZ curve for a ship lolling is shown in Figure 3.23.

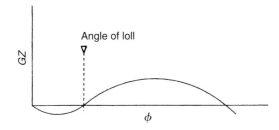

Figure 3.23 Angle of loll.

If the ship has a negative GM of 0.08 m, associated with a B_0M of 5 m, ϕ, which can be positive or negative, is:

$$\phi = \tan^{-1}\left(\frac{2 \times 0.08}{5}\right)^{0.5} = \tan^{-1} 0.179 = 10.1°$$

This shows that small negative GM can lead to significant loll angles. A ship with a negative GM will loll first to one side and then the other in response to wave action. When this happens the master should investigate the reasons, although the ship may still be safe.

3.7.3 Metacentric height in the lolled condition

Continuing with the wall-side assumption, if ϕ_1 is the angle of loll, the value of GM for small inclinations about the loll position, will be given by the slope of the GZ curve at that point. Now:

$$GZ = \sin\phi\left(GM + \frac{1}{2}B_0M \tan^2\phi\right)$$
$$\frac{dGZ}{d\varphi} = \cos\phi\left(GM + \frac{1}{2}B_0M \tan^2\phi\right)$$
$$+ \sin\phi B_0M \tan\phi \sec^2\phi \quad (3.43)$$

substituting ϕ_1 for ϕ gives $dGZ/d\phi = 0 + B_0M \tan^2\phi_1/\cos\phi_1 = -2GM/\cos\phi_1$.

Unless ϕ_1 is large, the metacentric height in the lolled position will be effectively numerically twice that in the upright position although of opposite sign.

3.7.4 Cross curves of stability

Cross curves of stability are drawn to overcome the difficulty in defining waterlines of equal displacement at various angles of heel.

Figure 3.24 shows a ship inclined to some angle ϕ. Note that S is not the same as in Figure 3.22. By calculating, for a range of waterlines, the displacement and perpendicular distances, SZ, of the centroids of these volumes of displacement from the line YY through S, curves such as those in Figure 3.25 can be drawn. These curves are known as *cross curves of stability*

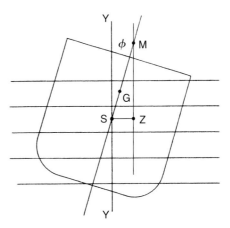

Figure 3.24 Heeled waterlines.

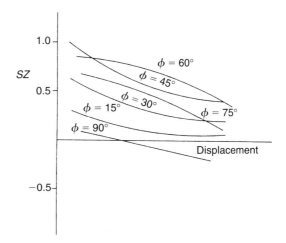

Figure 3.25 Cross curves of stability.

and depend only upon the geometry of the ship and not upon its loading. They therefore apply to all conditions in which the ship may operate.

3.7.5 Curves of statical stability from cross curves

For any desired displacement of the ship, the values of SZ can be read from the cross curves. Knowing the position of G for the desired loading enables SZ to be corrected to GZ by adding or subtracting $SG \sin\phi$, when G is below or above S respectively.

3.7.6 Features of the statical stability curve

There are a number of features of the GZ curve which are useful in describing a ship's stability. It has already been shown that the slope of the curve at the origin is a measure of the initial stability GM. The maximum ordinate of the curve multiplied by the displacement equals the largest steady heeling moment the ship can sustain without capsizing. Its value and the angle at which it occurs are both important. The value at which *GZ* becomes zero, or 'disappears', is the largest angle from which a ship will return once any disturbing moment is removed. This angle is called the *angle of vanishing stability*. The range of angle over which *GZ* is positive is termed the *range of stability*. Important factors in determining the range of stability are freeboard and reserve of buoyancy. The influence of hull shape on stability in discussed by Burcher (1979).

The angle of deck edge immersion varies along the length of the ship. However, often it becomes immersed over a reasonable length within a small angle band. In such cases the *GZ* curve will exhibit a point of inflexion at that angle. It is the product of displacement and *GZ* that is important in most cases, rather than *GZ* on its own. Required features of the stability curve to meet regulations are described in Section 3.11.

3.8 Weight movements

3.8.1 Transverse movement of weight

Sometimes a weight moves permanently across the ship. Perhaps a piece of cargo has not been properly secured and moves when the ship rolls. If the weight of the item is w and it moves horizontally through a distance h, there will be a corresponding horizontal shift of the ship's centre of gravity, $GG_1 = wh/W$, where W is the weight of the ship, Figure 3.26. The value of *GZ* is reduced by $GG_1 \cos \phi$ and the modified righting arm $= GZ - (wh/W) \cos \phi$.

Unlike the case of the suspended weight, the weight will not in general return to its original position when the ship rolls in the opposite direction. If it doesn't the righting lever, and righting moment, are reduced for inclinations to one side and increased for angles on the other side. If $GG_1 \cos \phi$ is plotted on the stability curve, Figure 3.27, for the particular condition of loading of the ship, the two curves intersect at B and C. B gives the new equilibrium position of the ship in still water and C the new angle of vanishing stability. The range of stability and the maximum righting arm are greatly reduced on the side to which the ship lists. For heeling to the opposite side the values are increased but it is the worse case that is of greater concern and must be considered. Clearly every precaution should be taken to avoid shifts of cargo.

Bulk cargoes
A related situation can occur in the carriage of dry bulk cargoes such as grain, ore and coal. Bulk cargoes settle down when the ship goes to sea so that holds which were full initially, have void spaces at the top. All materials of this type have an *angle of repose*. If the ship rolls to a greater angle than this the cargo

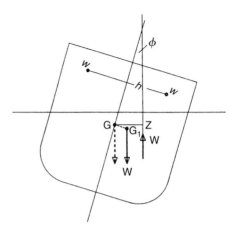

Figure 3.26 Transverse weight shift.

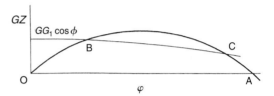

Figure 3.27 Modified *GZ* curve.

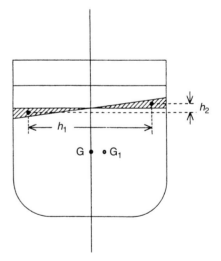

Figure 3.28 Cargo shift.

may move to one side and not move back later. Consequently there can be a permanent transfer of weight to one side resulting in a permanent list with a reduction of stability on that side. In the past many ships have been lost from this cause.

Figure 3.28 shows a section through the hold of a ship carrying a bulk cargo. When the cargo settles down at sea its centre of gravity is at G. If the ship rolls the cargo could take up a new position shown by the inclined line, causing some weight, w, to move horizontally by h_1 and vertically by h_2. As a result the ship's G will move:

$$\frac{wh_1}{\Delta} \text{ to one side and } \frac{wh_2}{\Delta} \text{ higher} \qquad (3.44)$$

The modified righting arm becomes:

$$G_1Z_1 = GZ - \frac{w}{\Delta}[h_1 \cos \phi + h_2 \sin \phi] \qquad (3.45)$$

where GZ is the righting arm before the cargo shifted.

Compared with the stability on initial loading there will have been a slight improvement due to the settling of the cargo.

Preventing shift of bulk cargoes
Regulations have existed for some time to minimize the movement of bulk cargoes and, in particular, grain. First, when a hold is filled with grain in bulk it must be trimmed so as to fill all the spaces between beams and at the ends and sides of holds.

Also centreline bulkheads and shifting boards are fitted in the holds to restrict the movement of grain. They have a similar effect to divisions in liquid carrying tanks in that they reduce the movement of cargo. Centreline bulkheads and shifting boards were at one time required to extend from the tank top to the lowest deck in the holds and from deck to deck in 'tween deck spaces. The present regulations require that the shifting boards or divisions extend downwards from the underside of deck or hatch covers to a depth determined by calculations related to an assumed heeling moment of a filled compartment.

The centreline bulkheads are fitted clear of the hatches, and are usually of steel. Besides restricting cargo movement they can act as a line of pillars supporting the beams if they extend from the tank top to the deck. Shifting boards are of wood and are placed on the centreline in way of hatches. They can be removed when bulk cargoes are not carried.

Even with centreline bulkheads and shifting boards spaces will appear at the top of the cargo as it settles down. To help fill these spaces feeders are fitted to provide a head of grain which will feed into the empty spaces. Hold feeders are usually formed by trunking in part of the hatch in the 'tween decks above. Feeder capacity must be 2% of the volume of the space it feeds. Precautions such as those outlined above permit grain cargoes to be carried with a high degree of safety.

3.9 Dynamical stability

So far stability has been considered as a static problem. In reality it is a dynamic one. One step in the dynamic examination of stability is to study what is known as a ship's *dynamical stability*. The work done in heeling a ship through an angle $\delta\phi$ will be given by the product of the displacement, GZ at the instantaneous angle and $\delta\phi$. Thus the area under the GZ curve, up to a given angle, is proportional to the energy needed to heel it to that angle. It is a measure of the energy it can absorb from wind and waves without heeling too far. This energy is solely potential energy because the ship is assumed to be heeled slowly. In practice a ship can have kinetic energy of roll due to the action of wind and waves. This is considered in the next section.

Influence of wind on stability
In a beam wind the force generated on the above water surface of the ship is resisted by the hydrodynamic force produced by the slow sideways movement of the ship through the water. The wind force may be taken to act through the centroid of the above water area and the hydrodynamic force as acting at half draught, Figure 3.29. For ships with high freeboard the variation of wind speed with height as described in Chapter 1 on the external environment, should be allowed for. For all practical purposes the two forces can be assumed equal.

Let the vertical distance between the lines of action of the two forces be h and the projected area of the above water form be A. To a first order as the ship heels, both h and A will be reduced in proportion to $\cos \phi$.

The wind force will be proportional to the square of the wind velocity, V_w, and can be written as:

$$\text{wind force} = kAV_w^2 \cos \phi \qquad (3.46)$$

where k is an empirical constant. The moment will be:

$$M = kAhV_w^2 \cos^2 \phi \qquad (3.47)$$

The curve of wind moment can be plotted with the ΔGZ curve as in Figure 3.30. If the wind moment builds up or is applied slowly the ship will heel to an angle represented by A and in this condition the range of stability will be from A to B. The problem would then be analogous to that of the shifted weight. On the other hand, if the moment is applied suddenly, say by

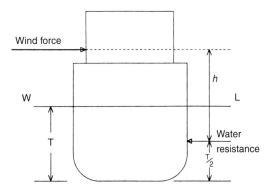

Figure 3.29 Heeling due to wind.

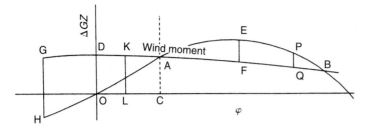

Figure 3.30 Wind moment.

a gust of wind, the amount of energy applied to the ship as it heeled to A would be represented by the area DACO. The ship would only absorb energy represented by area OAC and the remaining energy would carry it beyond A to some angle F such that area AEF = area DAO. Should F be beyond B the ship will capsize, assuming the wind is still acting.

A severe case for a rolling ship is if it is inclined to its maximum angle to windward and about to return to the vertical when the gust hits it. Suppose this position is represented by GH in Figure 3.30. The ship would already have sufficient energy to carry it to some angle past the upright, say KL in the figure. Due to damping this would be somewhat less than the initial windward angle. The energy put into the ship by the wind up to angle L is now represented by the area GDKLOH. The ship will continue to heel until this energy is absorbed, perhaps reaching angle Q.

Angle of heel due to turning

When a ship is turning under the action of its rudder, the rudder holds the hull at an angle of attack relative to the direction of advance. The hydrodynamic force on the hull, due to this angle, acts towards the centre of the turning circle causing the ship to turn. Under the action of the rudder and hull forces the ship will heel to an angle that can be determined in a similar way to the above. Heel due to the rudder is discussed in Chapter 8, Section 8.8.

3.10 Flooding and damaged stability

3.10.1 Background

So far only the stability of an intact ship has been considered. In the event of collision, grounding or just springing a leak, water can enter the ship. If unrestricted, this flooding would eventually cause the ship to founder, that is sink bodily, or capsize, that is turn over. To reduce the probability of this, the hull is divided into a series of watertight compartments by means of bulkheads. In action, warships are expected to take punishment from the enemy so damage stability is clearly an important consideration in their design. However, damage is a possibility for any ship.

Bulkheads cannot ensure complete safety in the event of damage. If the hull is opened up over a sufficient length several compartments can be flooded. This was the case in the tragedy of the *Titanic*. Any flooding can cause a reduction in stability and if this reduction becomes great enough the ship will capsize. Even if the reduction does not cause capsize it may lead to an angle of heel at which it is difficult, or impossible, to launch lifeboats. The losses of buoyancy and stability due to flooding are considered in the following sections.

A major consideration for any ship, but particularly important for one carrying large numbers of people, is the ability to get everyone off safely in the event that the master has to order 'abandon ship'. This means not only boats and rafts capable of taking all on board but the ability to launch them and get the people into them. This may require limiting the angles the ship may take up when damaged, marking escape routes clearly and the provision of chutes to get people from the ship (especially from ships with large freeboard). Time is a major consideration and allowance must be made for the time from initial damage to the acceptance that the ship has to be abandoned – by the passengers as well as the master. Passengers may feel disorientated if woken from deep sleep and some may panic or act irrationally.

3.10.2 Sinkage and trim when a compartment is open to the sea

Suppose a forward compartment is open to the sea, Figure 3.31. The buoyancy of the ship between the

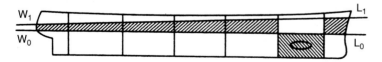

Figure 3.31 Compartment open to the sea.

containing bulkheads is lost and the ship settles in the water until it picks up enough buoyancy from the rest of the ship to restore equilibrium. At the same time the position of the LCB moves and the ship must trim until G and B are again in a vertical line. The ship which was originally floating at waterline W_0L_0 now floats at W_1L_1. Should W_1L_1 be higher at any point than the deck at which the bulkheads stop (the *bulkhead deck*) it is usually assumed that the ship would be lost as a result of the water pressure in the damaged compartment forcing off the hatches and leading to unrestricted flooding fore and aft. In practice the ship might still remain afloat for a considerable time.

Most compartments in a ship contain items which will reduce the volume of water that can enter. Even 'empty' spaces usually have frames or beams in them. At the other extreme some spaces may already be full of ballast water or fuel. The ratio of the volume that is floodable to the total volume is called the *permeability* of the space. Formulae for calculating permeabilities for merchant ships are laid down in the Merchant Ship (Construction) Rules. Typical values are presented in Table 3.2 and are discussed further in Section 3.12.1. Although not strictly accurate, the same values of permeability are usually applied as factors when assessing the area and inertias of the waterplane in way of damage.

Table 3.2 Permeabilities.

Space	Permeability (%)
Watertight compartment	97 (warship)
	95 (merchant ship)
Accommodation spaces	95 (passengers or crew)
Machinery compartments	85
Cargo holds	63
Stores	63

To calculate the damaged waterline successive approximation is needed. The assumptions of small changes do not apply. There are two approaches: the *lost buoyancy* and *added weight methods*. These give different GM values but the same righting moment.

Lost buoyancy method

First the volume of the damaged compartment, Figure 3.32, up to the original waterplane, and the area of waterplane lost, are calculated making allowance for the permeability. Suppose the area of original waterplane is A and the area lost is μa, where μ is the permeability. Let the lost volume of buoyancy be μv. A first approximation to the parallel sinkage suffered is given by:

$$z = \frac{\mu v}{A - \mu a}$$

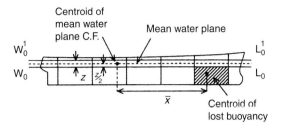

Figure 3.32 Lost buoyancy method.

A second approximation will almost certainly be needed because of the variations in waterplane area with draught. This can be made by taking the characteristics of a waterplane at sinkage $z/2$. The longitudinal centre of flotation and the moment to change trim can be calculated for this intermediate waterplane, again allowing for the permeability. Using subscript m to denote the values for the intermediate waterplane:

$$\text{sinkage} = \frac{\mu v}{A_m} \quad \text{and} \quad \text{trim} = \frac{\mu v \bar{x}}{\text{MCT}_m} \quad (3.48)$$

where $\bar{x}$ is the centroid of the lost volume from the CF.

The new draughts can be calculated from the sinkage and trim. A further approximation can be made if either of these is very large, or the results can be checked from first principles using the Bonjean curves allowing for the flooding and permeability.

In the lost buoyancy method the position of G remains unaltered unless the damage has been so severe as to remove structure or equipment from the ship.

Added weight method

In this method the water entering the damaged compartment is regarded as an added weight. Permeability would have to be allowed for in assessing this weight, and allowance must be made for the free surface of the water that has entered, but all the hydrostatic data used are those for the intact ship. Initially the calculation can proceed as for any added weight, but when the new waterline is established allowance must be made for the extra water that would enter the ship up to that waterplane. Again a second iteration may be needed and the calculation is repeated until a sufficiently accurate answer is obtained.

In the description of both methods it is assumed that the compartment that has been breached extends above the original and the final waterlines. If it does not then the actual floodable volumes must be used, and the assumed waterplane characteristics amended accordingly. It will be clear that it is highly desirable for the ship to have reasonable amounts of potential buoyancy above the intact waterplane as a 'reserve'. This is termed *reserve of buoyancy*.

3.10.3 Stability in the damaged condition

Consider first the lost buoyancy method and the metacentric height. The effect of the loss of buoyancy in the damaged compartment is to remove buoyancy (volume v) from a position below the original waterline to some position above this waterline so that the centre of buoyancy will rise. If the vertical distance between the centroids of the lost and gained buoyancy is bb_1 the rise in centre of buoyancy $= \mu v bb_1/\nabla$. BM will decrease because of the loss of waterplane inertia in way of the damage. If the damaged inertia is I_d, $BM_d = I_d/\nabla$. The value of KG remains unchanged so that the damaged GM, which may be more, but is generally less, than the intact GM is:

$$\text{damaged } GM = GM \text{ (intact)} + \frac{\mu v bb_1}{\nabla} - \frac{I_d}{\nabla} \quad (3.49)$$

If the added weight method is used then the value of KG will change and the height of M can be found from the hydrostatics for the intact ship at the increased draught. The free surface of the water in the damaged compartment must be allowed for.

3.10.4 Asymmetrical flooding

When there are longitudinal bulkheads in the ship there is the possibility of the flooding not extending right across the ship causing the ship to heel. In deciding whether a longitudinal bulkhead will be breached it is usually assumed that damage does not penetrate more than 20% of the breadth of the ship. Taking the case illustrated in Figure 3.33 and using the added weight approach, the ship will heel until:

$$\rho g \nabla GM \sin\phi = \mu \rho g v z \quad \text{or} \quad \sin\phi = \frac{\mu v z}{\nabla GM} \quad (3.50)$$

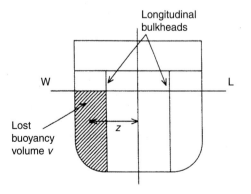

Figure 3.33 Asymmetrical flooding.

As with the calculation for trim, this first angle will need to be corrected for the additional weight of water at the new waterline, and the process repeated if necessary.

Large heels should be avoided and usually means are provided to flood a compartment on the opposite side of the ship. This is termed *counterflooding* or cross-flooding. The ship will sink deeper in the water but this is usually a less dangerous situation than that posed by the heel.

It should be noted that air can be trapped above the flooding water surface. If the top envelope of the compartment is airtight, flooding is stopped. If not, it is only slowed down. Between the position of intact conditions and the final damage position, assuming an equilibrium position is found, the vessel can pass through intermediate positions more dangerous than the final one. It is necessary to check if such positions exist and if the ship can survive them (Barnaby 1963).

3.10.5 Floodable length

So far the consequences of flooding a particular compartment have been studied. The problem can be looked at the other way by asking what length of ship can be flooded without loss of the ship. Loss is generally accepted to occur when the damaged waterline is tangent to the bulkhead deck line at side. The *bulkhead deck* is the uppermost weathertight deck to which transverse watertight bulkheads are carried. A margin is desirable and the limit is taken when the waterline is tangent to a line drawn 76 mm below the bulkhead deck at side. This line is called the *margin line*. The *floodable length* at any point along the length of the ship is the length, with that point as centre, which can be flooded without immersing any part of the margin line when the ship has no list.

Take the ship shown in Figure 3.34 using subscripts 0 and 1 to denote the intact ship data for the intact and damaged waterlines. Loss of buoyancy $= V_1 - V_0$ and this must be at such a position that B_1 moves back to B_0 so that B is again below G. Hence:

$$\bar{x} = \frac{V_1 \times B_0 B_1}{V_1 - V_0} \quad (3.51)$$

This then gives the centroid of the lost buoyancy and, knowing $(V_1 - V_0)$ it is possible to convert this into a length of ship that can be flooded. The calculation would be one of reiteration until reasonable figures are obtained.

The calculations can be repeated for a series of waterlines tangent to the margin line at different positions along the length. This will lead to a curve of floodable length as in Figure 3.35. The ordinate at any point represents the length which can be flooded with the centre at the point concerned. Thus if l is the floodable length at some point the positions of

Flotation and stability 97

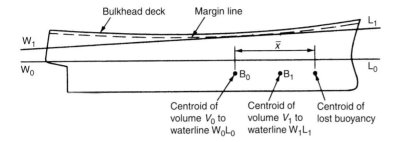

Figure 3.34 Margin line.

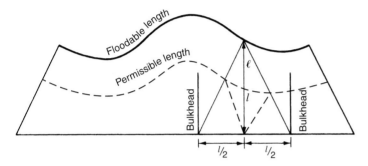

Figure 3.35 Floodable length.

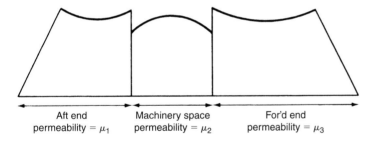

Figure 3.36 Floodable length with permeability.

bulkheads giving the required compartment length are given by setting off distances $l/2$ either side of the point. The lines at the ends of the curves, called the *forward and after terminals* will be at an angle $\tan^{-1} 2$ to the base if the base and ordinate scales are the same.

The permeabilities of compartments will affect the floodable length and it is usual to work out average permeability figures for the machinery spaces and for each of the two regions forward and aft. This leads to three curves for the complete ship as shown in Figure 3.36. The condition that a ship should be able to float with any one compartment open to the sea is a minimum requirement for ocean-going passenger ships. As described in the next section, the Merchant Shipping Regulations set out formulae for calculating permeabilities and a *factor of subdivision* which must be applied to the floodable length curves giving *permissible length*, Figure 3.35. The permissible length is the product of the floodable length and the factor of subdivision. The factor of subdivision depends upon the length of the ship and a *criterion of service numeral* or more simply *criterion numeral*. This numeral represents the criterion of service of the ship and takes account of the number of passengers, the volumes of the machinery and accommodation spaces and the total ship volume. It decreases in a regular and continuous manner with the ship length and factors related to whether the ship carries predominantly cargo or passengers. Broadly, the factor of subdivision ensures that one, two or three compartments can be flooded before the margin line is immersed leading to what are called *one-, two-* or *three-compartment ships*. That is, compartment standard is the inverse of the factor of subdivision. In general terms the factor of subdivision decreases with length of ship and is lower for passenger ships than cargo ships. Regulations relating to floodable length are discussed in Section 3.12.

Practical applications of small and large angle stability and flooding may be found in Barrass and Derrett (2006).

3.11 Intact stability regulations

3.11.1 Introduction

The preceding sections have described the basic physics of stability at small and large angles. At this point we may ask what is satisfactory stability or, in simpler terms, how stable a ship must be. Analysing the data of vessels that behaved well, and especially the data of vessels that did not survive storms or other adverse conditions, various researchers and regulatory bodies prescribe criteria for deciding if the stability is satisfactory. In this section, we present examples of such criteria. Stability regulations prescribe criteria for approving ship designs, accepting new buildings, or allowing ships to sail out of harbour. If a certain ship fulfils the requirements of given regulations, it does not mean that the ship can survive all challenges, but her chances of survival are good because stability regulations are based on considerable experience and reasonable theoretical models. Conversely, if a certain ship does not fulfil certain regulations, she will not necessarily capsize, only the risks are higher and the owner has the right to reject the design or the authority in charge has the right to prevent the ship from sailing out of harbour. Stability regulations are, in fact, **codes of practice** that provide reasonable safety margins. The codes are compulsory not only for designers and builders, but also for ship masters who must check if their vessels meet the requirements in a proposed loading condition.

It is not possible to include all existing stability regulations; we only choose a few representative examples. Neither is it possible to present all the provisions of any single regulation. We only want to draw the attention of the reader to the existence of such codes of practice, and to help the reader in understanding and using the regulations. Technological developments, experience accumulation and especially major marine disasters, can impose revisions of existing stability regulations. For all the reasons mentioned above, before checking the stability of a vessel according to given regulations, the Naval Architect must read in detail their newest, official version.

All stability regulations specify a number of loading conditions for which calculations must be carried out. Some regulations add a sentence like 'and any other condition that may be more dangerous'. It is the duty of the Naval Architect in charge of the project to identify such situations, if they exist, and check if the stability criteria are met for them.

Further descriptions of statutory regulations are given in Chapter 11, and the stability manual in Section 11.5.2.2.

3.11.2 The IMO code on intact stability

The Inter-Governmental Maritime Consultative Organization was established in 1948 and was known as IMCO. That name was changed in 1982 to **IMO – International Maritime Organization**. The purpose of IMO is the inter-governmental cooperation in the development of regulations regarding shipping, **maritime safety**, navigation, and the prevention of marine pollution from ships. IMO is an agency of the United Nations and has 161 members. Further description of IMO is given in Chapter 11. The regulations described in this section were issued by IMO in 1995, and are valid 'for all types of ships covered by IMO instruments' (see IMO 1995). The intact stability criteria of the code apply to 'ships and other marine vehicles of 24 m in length and above'. Countries that adopted these regulations enforce them by issuing corresponding national ordinances. Also, the Council of the European Community published the Council Directive 98/18/EC on 17 March 1998.

3.11.2.1 *Passenger and cargo ships*

The code uses frequently the terms **angle of flooding**, **angle of downflooding**; they refer to the smallest angle of heel at which an opening that cannot be closed weathertight submerges. Passenger and cargo ships covered by the code shall meet the following general criteria:

1. The area under the righting-arm curve should not be less than 0.055 m rad up to 30°, and not less than 0.09 m rad up to 40° or up to the angle of flooding if this angle is smaller than 40°.
2. The area under the righting-arm curve between 30° and 40°, or between 30° and the angle of flooding, if this angle is less than 40°, should not be less than 0.03 m rad.
3. The maximum righting arm should occur at an angle of heel preferably exceeding 30°, but not less than 25°.
4. The initial metacentric height, GM_0, should not be less than 0.15 m.

Passenger ships should meet two further requirements. First, the angle of heel caused by the crowding of passengers to one side should not exceed 10°. The mass of a passenger is assumed equal to 75 kg. The centre of gravity of a standing passenger is assumed to lie 1 m above the deck, while that of a seated passenger is taken as 0.30 m above the seat. The second additional requirement for passenger ships refers to the angle of heel caused by the centrifugal force developed in turning. The heeling moment due to that force is calculated with the formula:

$$M_T = 0.02 \frac{V_0^2}{L_{WL}} \Delta \left(KG - \frac{T_m}{2} \right) \quad (3.52)$$

where V_0 is the service speed in m s^{-1}. Again, the resulting angle shall not exceed 10°. The reason for limiting the angle of heel is that at larger values passengers may panic. The application of this criterion is exemplified in Figure 3.37.

In addition to the general criteria described above, ships covered by the code should meet a **weather criterion** that considers the effect of a beam wind applied when the vessel is heeled windwards. We explain this criterion with the help of Figure 3.38.

The code assumes that the ship is subjected to a constant wind heeling arm calculated as:

$$\ell_{w1} = \frac{PAZ}{1000 g \Delta} \quad (3.53)$$

where $P = 504$ N m^{-2}, A is the projected lateral area of the ship and deck cargo above the waterline, in m^2, Z is the vertical distance from the centroid of A to the centre of the underwater lateral area, or approximately to half-draught, in m, Δ is the displacement mass in tonnes, and $g = 9.81$ m s^{-2}. Unlike the model developed in Section 3.9 (model used by the US Navy), IMO accepts the more severe assumption that the wind heeling arm does not decrease as the heel angle increases. The code uses the notation θ for heel angles; we shall follow our convention and write ϕ. The static angle caused by the wind arm l_{w1} is ϕ_0. Further, the code assumes that a wind gust appears while the ship is heeled to an angle ϕ_1 windward from the static angle, ϕ_0. The angle of roll is given by:

$$\phi_1 = 109 k X_1 X_2 \sqrt{rs} \quad (3.54)$$

where ϕ_1 is measured in degrees, X_1 is a factor given in Table 3.2.2.3-1 of the code, X_2 is a factor given in Table 3.2.2.3-2 of the code, and k is a factor defined as follows:

- $k = 1.0$ for round-bilge ships;
- $k = 0.7$ for a ship with sharp bilges;
- k as given by Table 3.2.2.3-3 of the code for a ship having bilge keels, a bar keel or both.

By using the factor k, the IMO code considers indirectly the effect of damping on stability. More specifically, it acknowledges that sharp bilges, bilge keels and bar keels reduce the roll amplitude. By assuming that the ship is subjected to the wind gust

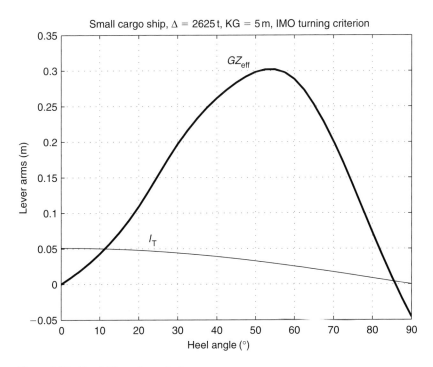

Figure 3.37 The IMO turning criterion.

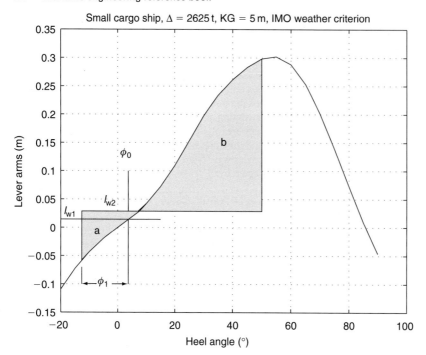

Figure 3.38 The IMO weather criterion.

while heeled windward from the static angle, the dynamical effect appears more severe.

The factor r is calculated from

$$r = 0.73 + 0.6 \frac{OG}{T_m} \quad (3.55)$$

where OG is the distance between the waterline and the centre of gravity, positive upwards. The factor s is given in Table 3.2.2.3-4 of the code, as a function of the roll period, T. The code prescribes the following formula for calculating the roll period, in seconds,

$$T = \frac{2CB}{\sqrt{GM_{eff}}} \quad (3.56)$$

where

$$C = 0.373 + 0.023\left(\frac{B}{T_m}\right) - 0.043\left(\frac{L_{WL}}{100}\right) \quad (3.57)$$

The code assumes that the lever arm of the wind gust is

$$\ell_{w2} = 1.5\,\ell_{w1} \quad (3.58)$$

Plotting the curve of the arm ℓ_{w2} we distinguish the areas a and b. The area b is limited to the right at 50° or at the angle of flooding, whichever is smaller. The area b should be equal to or greater than the area a. This provision refers to dynamical stability, as explained in Section 3.9. When applying the criteria described above, the Naval Architect must use values corrected for the free-surface effect, that is GM_{eff} and GZ_{eff}. The free-surface effect is calculated for the tanks that develop the greatest moment, at a heel of 30°, while half full. The code prescribes the following equation for calculating the free-surface moment:

$$M_F = vb\gamma k \sqrt{\delta} \quad (3.59)$$

where v is the tank capacity in m³, b is the maximum breadth of the tank in m, γ is the density of the liquid in t m⁻³, δ is equal to the block coefficient of the tank, $v/b\ell h$, with h, the maximum height and ℓ, the maximum length, and k, a coefficient given in Table 3.3.3 of the code as function of b/h and heel angle. The contribution of small tanks can be ignored if $M_F/\Delta_{min} < 0.01$ m at 30°. We would like to remind the reader that present computer programs for hydrostatic calculations yield values of the free-surface lever arms for any tank form described in the input, and for any heel angle. It is our opinon that, when available, such values should be preferred to those obtained with Equation (3.59).

The code specifies the loading cases for which stability calculations must be performed. For example, for cargo ships the criteria shall be checked for the following four conditions:

1. Full-load departure, with cargo homogeneously distributed throughout all cargo spaces.

2. Full-load arrival, with 10% stores and fuel.
3. Ballast departure, without cargo.
4. Ballast arrival, with 10% stores and fuel.

3.11.2.2 *Cargo ships carrying timber deck cargoes*

Section 4.1 of the code applies to cargo ships that carry on their deck timber cargo extending longitudinally between superstructures and transversally on the full deck breadth, excepting a reasonable gunwale. Where there is no limiting superstructure at the aft, the cargo should extend at least to the after end of the aftermost hatch. For such ships the area under the righting-arm curve should not be less than 0.08 m rad up to 40° or up to the angle of flooding, whichever is smaller. The effective metacentric height should be positive in all stages of loading, voyage and unloading. The calculations should take into account the absorption of water by the deck cargo, and the water trapped within the cargo.

3.11.2.3 *Fishing vessels*

Section 4.2 of the code applies to decked sea-going vessels; they should fulfil the first three general requirements described in Subsection 3.11.2.1, while the metacentric height should not be less than 0.35 m for single-deck ships. If the vessel has a complete superstructure, or the ship length is equal to or larger than 70 m, the metacentric height can be reduced with the agreement of the government under whose flag the ship sails, but it should not be less than 0.15 m. The weather criterion applies in full to ships of 45 m length and longer. For fishing vessels whose length ranges between 24 and 45 m the code prescribes a wind gradient such that the pressure ranges between 316 and 504 Nm^{-2} for heights of 1–6 m above sea level. Decked vessels shorter than 30 m must have a minimum metacentric height calculated with a formula given in paragraph 4.2.6.1 of the code.

3.11.2.4 *Mobile offshore drilling units*

Section 4.6 of the code applies to mobile drilling units whose keels were laid after 1 March 1991. The wind force is calculated by considering the shape factors of structural members exposed to the wind, and a height coefficient ranging between 1.0 and 1.8 for heights above the waterline varying from 0 to 256 m. The area under the righting-arm curve up to the second static angle, or the downflooding angle, whichever is smaller, should exceed by at least 40% the area under the wind arm. The code also describes an alternative intact-stability criterion for two-pontoon, column-stabilized semi-submersible units.

3.11.2.5 *Dynamically supported craft*

A vessel is a **dynamically supported craft (DSC)** in one of the following cases:

1. If, in one mode of operation, a significant part of the weight is supported by other than buoyancy forces.
2. If the craft is able to operate at Froude numbers, $F_n = V/\sqrt{gL}$, equal or greater than 0.9.

The first category includes air-cushion vehicles and hydrofoil boats. Hydrofoil boats float, or sail, in the **hull-borne** or **displacement mode** if their weight is supported only by the buoyancy force predicted by Archimedes' principle. At higher speeds hydrodynamic forces develop on the foils and they balance an important part of the boat weight. Then, we say that the craft operates in the **foil-borne** mode.

Section 4.8 of the code applies to DSC operating between two ports situated in different countries. The requirements for hydrofoil boats are described in Subsection 4.8.7 of the code. The heeling moment in turning, in the displacement mode, is calculated as:

$$M_R = \frac{0.196 V_0^2 \Delta KG}{L} \qquad (3.60)$$

where V_0 is the speed in turning, in ms^{-1}, and M_R results in kN m. The formula is valid if the radius of the turning circle lies between $2L$ and $4L$. The resulting angles of inclination should not exceed 8°.

The wind heeling moment, in the displacement mode, in kN m, should be calculated as:

$$M_V = 0.001 P_V A_V Z \qquad (3.61)$$

and is considered constant within the whole heeling range. The area subjected to wind pressure, A_V, is called here **windage area**. The wind pressure, P_V, corresponds to force 7 on the Beaufort scale. For boats that sail 100 nautical miles from the land, Table 4.8.7.1.1.4 of the code gives P_V values ranging between 46 and 64 Pa, for heights varying from 1 to 5 m above the waterline. The **windage area lever**, Z, is the distance between the waterline and the centroid of the windage area. A minimum capsizing moment, M_C, is calculated as shown in paragraph 4.8.7.1.1.5.1 of the code and as illustrated in Figure 3.39. The curve of the righting arm is extended to the left to a roll angle ϕ_z averaged from model or sea tests. In the absence of such data, the angle is assumed equal to 15°. Then, a horizontal line is drawn so that the two grey areas shown in the figure are equal. The ordinate of this line defines the value M_C. According to the theory developed in Section 3.9 the ship capsizes if this moment is applied dynamically. The stability is considered sufficient if $M_C/M_V \geq 1$.

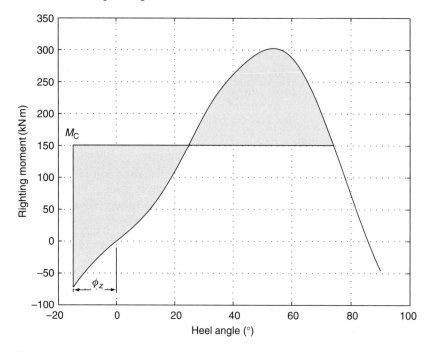

Figure 3.39 Defining the minimum capsizing moment of a dynamically supported craft (DSC).

The code also prescribes criteria for the transient and foil-borne modes. Such criteria consider the forces developed on the foils, a subject that is not discussed in this book.

3.11.2.6 Container ships greater than 100 m

Section 4.9 of the code defines a form factor C depending on the main dimensions of the ship and the configuration of hatches (Figure 4.9-1 in the code). The minimum values of areas under the righting-arm curve are prescribed in the form a/C, where a is specified for several heel intervals.

3.11.2.7 Icing

Chapter 5 of the code bears the title 'Ice considerations'. The following values, prescribed for fishing vessels, illustrate the severity of the problem. Stability calculations should be carried out assuming ice accretion (this is the term used in the code) with the surface densities:

- 30 kg m^{-2} on exposed weather decks and gangways;
- 7.5 kg m^{-2} for projected lateral areas on each side, above the waterplane.

The code specifies the geographical areas in which ice accretion can occur.

3.11.2.8 Inclining and rolling tests

Chapter 7 of the code contains the instructions for carrying on inclining experiments for all ships covered by the regulations, and roll-period tests for ships up to 70 m in length. The relationship between the metacentric height, GM_0, and the roll period, T, is given as:

$$GM_0 = \left(\frac{fB}{T}\right)^2 \quad (3.62)$$

where B is the ship breadth. See also Section 7.2.11.6.

An interesting part of the Annex refers to the plot of heel-angle tangents against heeling moments; it explains the causes of deviations from a straight line, such as free surfaces of liquids, restrictions of movements, steady wind or wind gust.

3.11.2.9 High-speed craft

Regulations relating to high-speed craft stability may be found in IMO (2000). A high-speed craft is defined as a craft capable of a maximum speed V in m/s equal to or exceeding:

$$V = 3.7 \, \nabla^{0.1667}$$

where ∇ is the displacement (m^3) corresponding to the design waterline.

Requirements for initial stability and properties of the GZ curve are given for multihulls (Annex 7) and monohulls (Annex 8).

Multihulls: $GZ_{max} \geq 10°$, and area A under GZ curve up to angle ϕ shall be at least:

$$A = 0.055 \times 30 / \phi \text{ m.rad}$$

where ϕ is the least of (i) the downflooding angle ϕ_f, (ii) the angle at maximum GZ or (iii) 30°.

Requirements are also given for heeling due to wind, heeling due to passenger crowding and high-speed turning, together with criteria for residual stability after damage.

Monohulls: $GM \geq 0.15$ m, $GZ_{max} \geq 15°$, and area under GZ curve shall not be less than 0.07 m.rad up to $\phi = 15°$ where the maximum GZ occurs at $\phi = 15°$, and 0.055 m.rad up to $\phi = 30°$ when the maximum righting lever occurs at 30° or above. For $15° < \phi_{max} < 30°$, the area A under GZ curve shall be at least:

$$A = 0.055 + 0.001 (30° - \phi_{max}) \text{ m.rad}$$

For 30° to 40°, or 30° to ϕ_f, the area under the GZ curve shall not be less than 0.03 m.rad, and GZ at least 0.2 m for $\phi > 30°$.

Further criteria are given for heeling due to wind and for residual stability after damage.

3.11.3 Regulations of the US Navy

In 1944, an American fleet was caught by a tropical storm in the Pacific Ocean. In a short time three destroyers capsized, a fourth one escaped because a funnel broke down under the force of the wind. This disaster influenced the development of stability regulations for the US Navy. They were first published by Sarchin and Goldberg, Sarchin and Goldberg (1962). These regulations were subsequently adopted by other navies.

The intact stability is checked under a wind whose speed depends on the service conditions. Thus, all vessels that must withstand tropical storms should be checked for winds of 100 knots. Ocean-going ships that can avoid the centre of tropical storms should be checked under a wind of 80 knots, while coastal vessels that can avoid the same dangers should be checked for winds of 60 knots. Coastal vessels that can be called to anchorage when expecting winds above Force 8, and all harbour vessels should be checked under the assumption of 60-knots winds.

We explain the weather criterion in Figure 3.40. The righting arm, GZ, is actually the effective righting arm, GZ_{eff}, calculated by taking into account the free-surface effect. The wind arm is obtained from the formula

$$l_V = \frac{0.017 V_w^2 A \ell \cos^2 \phi}{1000 \Delta} \quad (3.63)$$

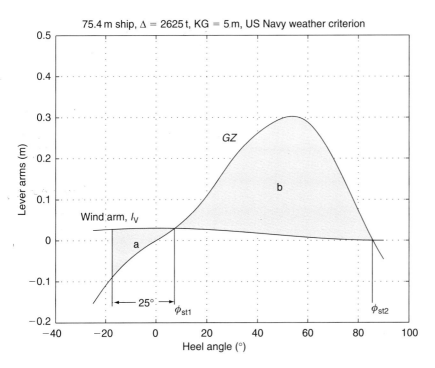

Figure 3.40 The US Navy weather criterion.

where V_w is the wind velocity in knots, A, the projected above water area in m², ℓ, the distance between half-draught and the centroid of the above water area in m, and Δ, the displacement in tonnes. The first angle of static equilibrium is ϕ_{st1}. The criterion for static stability requires that the righting arm at this angle be not larger than 0.6 of the maximum righting arm. To check dynamical stability the regulations assume that the ship is subjected to a gust of wind while heeled 25° to the windward of ϕ_{st1}. We distinguish then the area a between the wind heeling arm and the righting-arm curves up to ϕ_{st1}, and the area b between the two curves, from the first static angle, ϕ_{st1}, up to the second static angle, ϕ_{st2} (see Figure 3.40), or up to the angle of downflooding, whichever is less (see Figure 3.41). The ratio of the area b to the area a should be at least 1.4.

The designer can take into account the wind gradient, that is the variation of the wind speed with height above the waterline. Then, the 'nominal' wind speed defined by the service area is that measured at 10 m (30 ft) above the waterline. Performing a regression about new data presented by Watson (1998) we found the relationship

$$\frac{V_W}{V_0} = 0.73318 h^{0.13149} \qquad (3.64)$$

where V_W is the wind speed at height h, V_0 is the nominal wind velocity, and h is the height above sea level, in m. In Figure 3.42, the points indicated by Watson (1998) appear as asterisks, while the values predicted by Equation (3.64) are represented by the continuous line. An equation found in literature has the form $V_W/V_0 = (h/10)^b$. Regression over the data given by Watson yielded $b = 0.73318$, but the resulting curve fitted less well than the curve corresponding to Equation (3.64).

To apply the wind gradient one has to divide the sail area into horizontal strips and apply in each strip the wind ratio yielded by Equation (3.64). Let R_i be that ratio for the ith strip. The results for the individual strips should be integrated by one of the rules for numerical integration. The coefficient in Equation (3.63) should be modified to 0.0195 and then, the wind arm is given by

$$\ell_V = \frac{0.0195 V_0^2}{1000\Delta} h \left(\sum \alpha_i R_i^2 A_i \ell_i \right) \cos^2 \phi \qquad (3.65)$$

where V_0 is the nominal wind speed, h is the common height of the horizontal strips, α_i is the trapezoidal multiplier, A_i is the area of the ith strip, and ℓ_i the vertical distance from half-draught to the centroid of the ith strip. It can be easily shown that

$$\ell_i = \frac{2i-1}{2} h + \frac{T}{2} \qquad (3.66)$$

To explain the criterion for stability in turning we use Figure 3.43. The heeling arm due to the centrifugal force is calculated from

$$l_{TC} = \frac{V^2(KG - T/2)}{gR} \cos \phi \qquad (3.67)$$

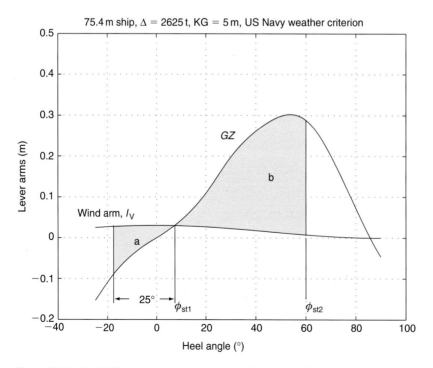

Figure 3.41 The US Navy weather criterion, downflooding angle 60°.

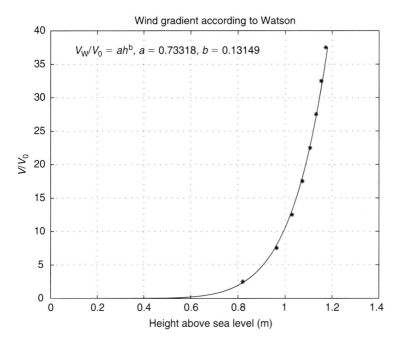

Figure 3.42 Wind gradient.

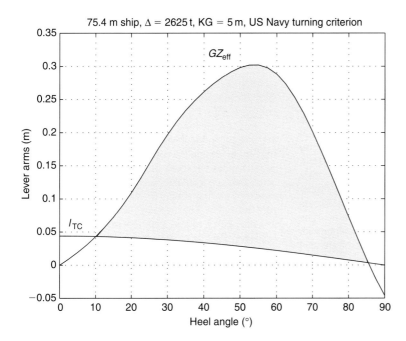

Figure 3.43 The US Navy turning criterion.

where V is the ship speed in ms^{-1} and R is the turning radius in m.

Ideally, R should be taken as one half of the **tactical diameter** measured from model or sea tests at full scale, see Chapter 8, Section 8.6.1.4. Where this quantity is not known, an estimation must be made, as in Section 3.11.4, about the UK Navy, where an approximate relationship is given. The stability is considered satisfactory if:

1. the angle of heel does not exceed 15°;
2. the heeling arm at the angle of static equilibrium is not larger than 0.6, the maximum righting-arm value;

3. the grey area in the figure, called **reserve of dynamical stability** is not less than 0.4 of the whole area under the positive righting-arm curve.

If the downflooding angle is smaller than the second static angle, the area representing the reserve of stability should be limited to the former value.

Another hazard considered in the regulations of the US Navy is the lifting of heavy weights over the side. The corresponding heeling arm is yielded by

$$l_W = \frac{wa}{\Delta} \cos \phi \qquad (3.68)$$

where w is the lifted mass, a is the transverse distance from the centreline to the boom end, and Δ is the displacement mass including w. The criteria of stability are the same as those required for stability in turning.

The crowding of personnel to one side causes an effect similar to that of a heavy weight lifted transversely to one side. The heeling arm is yielded by Equation (3.68), assuming that the personnel moved to one side as far as possible when five men crowd in one square metre. Again, the stability is considered sufficient if the requirements given for stability in turning are met.

3.11.4 Regulations of the UK Navy

The stability standard of the Royal Navy evolved from the criteria published by Sarchin and Goldberg (1962). The first British publication appeared in 1980 as NES 109. The currently valid version is Issue 4 (see MoD, 1999a). The document should be read in conjunction with the publication SSP 42 (MoD, 1999b). The British standard is issued by the Ministry of Defence (MoD) and is applicable to vessels with a military role, to vessels designed to MoD standards but without a military role, and to auxiliary vessels. Vessels with a military role are exposed to enemy action or to similar dangers during peacetime exercises. We shall discuss here only the provisions related to such vessels. The standard NES 109 has two parts, the first dealing with conventional ships, the second with unconventional vessels. The second category includes:

1. monohull vessels of rigid construction having a speed in knots larger than $4\sqrt{L_{WL}}$, where the waterline length is measured in m;
2. multi-hull vessels;
3. dynamically supported vessels.

In this Section, we briefly discuss only the provisions for conventional vessels. According to NES 109 the displacement and KG values used in stability calculations should include growth margins. For warships the weight growth margin should be 0.65% of the lightship displacement, for each year of service. The KG margin should be 0.45% of the lightship KG, for each year of service.

The shape of the righting-arm curve should be such that:

- the area under the curve, up to 30°, is not less than 0.08 m rad;
- the area up to 40° is not less than 0.133 m rad;
- the area between 30° and 40° is not less than 0.048 m rad;
- the maximum GZ is not less than 0.3 m and should occur at an angle not smaller than 30°.

One can immediately see that all these requirements are considerably more severe than those prescribed by IMO 95 for merchant ships.

The stability under beam winds should be checked for the following wind speeds:

- 90 knots for ocean-going vessels;
- 70 knots for ocean-going or coastal vessels that can avoid extreme conditions;
- 50 knots for coastal vessels that can be called to anchorage to avoid winds over Force 8, and for harbour vessels.

These values are lower than those required by the US Navy and partially coincide with those specified by the German Navy. The angle of heel caused by the wind should not exceed 30°. The criterion for statical stability is the same as that of the US Navy, that is, the righting arm at the first static angle should not be greater than 0.6, the maximum righting arm. As in the American regulations, it is assumed that the ship rolls 25° windwards from the first static angle, and it is required that the reserve of stability should not be less than 1.4 times the area representing the wind heeling energy. Figure 1.3 in the UK regulations shows that the area representing the reserve of stability is limited at the right by the downflooding angle. When checking stability in turning the corresponding ship speed should be 0.65 times the speed on a straight-line course. If no better data are available, it should be assumed that the radius of turning equals 2.5 times the length between perpendiculars. The angle of heel in turning should be less than 20°, a requirement less severe than that of the US Navy. The static criterion, regarding the value of the righting arm at the first static angle, and the dynamic criterion, regarding the reserve of stability, are the same as those of the US Navy.

To check stability when lifting a heavy mass over the side, the heeling arm should be calculated from

$$l_W = \frac{w(a \cos \phi + d \sin \phi)}{\Delta} \qquad (3.69)$$

where a is the horizontal distance of the tip of the boom from the centreline, and d is the height of the point of suspension above the deck. Stability is considered sufficient if the following criteria are met:

1. The angle of heel is less than 15°.
2. The righting arm at the first static angle is less than half the maximum righting arm.

3. The reserve of stability is larger than half the total area under the righting-arm curve. The area representing the reserve of stability is limited at the right by the angle of downflooding.

It can be easily seen that criteria 2 and 3 are more stringent than those of the US Navy.

The NES 109 standard also specifies criteria for checking stability under icing. A thickness of 150 mm should be assumed for all horizontal decks, with an ice density equal to 950 kg m^{-3}. Only the effect on displacement and KG should be considered, and not the effect on the sail area.

3.11.5 A criterion for sail vessels

The revival of the interest for large sailing vessels and several accidents justified new researches and the development of codes of stability for this category of ships. Thus, the UK Department of Transport sponsored a research carried out at the Wolfson Unit for Marine Technology and Industrial Aerodynamics at the University of Southampton (Deakin, 1991). The result of the research is the code of stability described in this section. A more recent research is presented by Cleary et al. (1996). The authors compare the stability criteria for sailing ships adopted by the US Coast Guard, the Wolfson Unit, Germanischer Lloyd, Bureau Veritas, Ateliers & Chantiers du Havre, and Dr Ing Alimento of the University of Genoa. These criteria are illustrated by applying them to one ship, the US Coast Guard training barque *Eagle*, formerly *Horst Wessel* built in 1936 in Germany.

In this section, we describe the intact stability criteria of 'The code of practice for safety of large commercial sailing & motor vessels' issued by the UK Maritime and Coastguard Agency (Maritime, 2001). The code 'applies to vessels in commercial use for sport or pleasure ... that are 24 metres in load line length and over ... and that do not carry cargo and do not carry more than 12 passengers.' For shorter sailing vessels, the UK Marine Safety Agency published another code, namely 'The safety of small commercial sailing vessels.'

The research carried out at the Wolfson Unit yielded a number of interesting results:

1. Form coefficients of sail rigs vary considerably and are difficult to predict. We mean here the coefficient c in

$$p = \frac{1}{2} c \rho v^2$$

where p is the pressure, ρ, the air density, and V, the speed of the wind component perpendicular to the sail.

2. The wind-arm curve behaves like $\cos^{1.3} \phi$.

3. Wind gusts do not build up instantly, as conservatively assumed (see Section 3.9). The wind speed of gusts due to atmospheric turbulence are unlikely to exceed 1.4 times the hourly mean, have rise times of 10 to 20 s and durations of less than a minute. Other gusts, due to other atmospheric phenomena, are known as **squalls** and they can be much more dangerous. Because the rise-up times of significant gusts are usually larger than the natural roll periods of sailing vessels, ships do not respond as described in Section 3.9, but have time to find equilibrium positions close to the intersection of the gust-arm curve and the righting-arm curve.

4. Sails considerably increase the damping of the roll motion, limiting the response to a wind gust and enhancing the effect described above. Thus, the heel angle caused by a wind gust is smaller than that predicted by the balance of areas representing wind energy and righting-arm work (Section 3.9).

Based on the above conclusions, the criterion of intact stability adopted by the UK Maritime and Coastguard Agency does not consider the sail rig and the wind moment developed on it. The code simply provides the skipper with a means for appreciating the maximum allowable heel angle under a steady wind, if wind gusts are expected. Sailing at the recommended angle will avoid the submergence under gusts of openings that could lead to ship loss.

The code defines the downflooding angle as the angle at which openings having an 'aggregate area' whose value in metres is greater than $\Delta/1500$, submerge. The displacement, Δ, is measured in tonnes. Deakin (1991) explains that under his assumptions the mass of water flowing through the above openings during 5 minutes equals the ship displacement. No ship is expected to float after a flooding of this extent, and five minutes are considered a maximum reasonable time of survival. For those who wish to understand Deakin's reasoning we remind that the flow through an orifice is proportional to the orifice area multiplied by the fluid speed

$$Q = a c_V \sqrt{2gh} \qquad (3.70)$$

where a is the orifice area, c_V, a discharge coefficient always smaller than 1, g, the acceleration of gravity, and h, the level of water above the orifice. The authors of the code assume $c_V = 1$ and $h = 1$ m. We calculate

$$Q = \frac{\Delta}{1500} \times 1 \times \sqrt{2 \times 9.81 \times 1} = 0.003\Delta \text{ m}^3 \text{ s}^{-1}$$
$$(3.71)$$

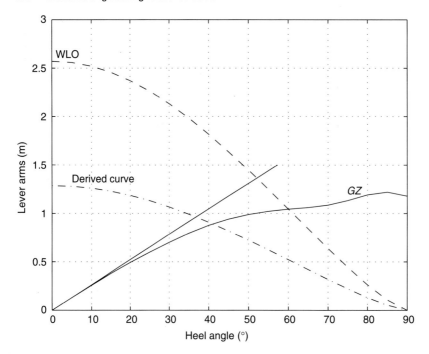

Figure 3.44 Intact stability criterion for sail ships.

It follows that in sea water 5.5 minutes are required for a mass of water equal to the displacement mass.

We use Figure 3.44 to describe the criterion for intact stability. The righting-arm curve is marked GZ; it is based on the data of an actual training yacht. At the downflooding angle we measure the value of the righting arm, GZ_f. We assume here the downflooding angle $\phi_f = 60°$. We calculate a gust-wind lever in upright condition

$$WLO = \frac{GZ_f}{\cos^{1.3} \phi_f} \qquad (3.72)$$

The dashed line curve represents the gust arm. Under the assumptions that the gust speed is 1.4 times the speed of the steady wind, the pressure due to steady wind is one half that of the gust, and so is the corresponding heeling arm. Therefore, we draw the 'derived curve' as the dash-dot line beginning at $W LO/2$ and proportional to $\cos^{1.3} \phi$. This curve intercepts the GZ curve at the angle of **steady heel**, here a bit larger than 40°.

The code requires that:

1. The *GZ* curve should have a positive range not shorter than 90°.
2. If the downflooding angle is larger than 60°, ϕ_f should be taken as 60°.
3. The angle of steady heel should not be less than 15°.

3.11.6 A code of practice for small workboats and pilot boats

The regulations presented in this section (see Maritime, 1998) apply to small UK commercial sea vessels of up to 24 m load line length and that carry cargo and/or not more than 12 passengers. The regulations also apply to service or pilot vessels of the same size. By 'load line length' the code means either 96% of the total waterline length on a waterline at 85% depth, or the length from the fore side of the stem to the axis of the rudder stock on the above waterline.

The lightship displacement to be used in calculations should include a margin for growth equal to 5% of the lightship displacement. The *x*-coordinate of the centre of gravity of this margin shall equal *LCG*, and the *z*-coordinate shall equal either the height of the centre of the weather deck amidships or the lightship *KG*, whichever is the higher. Curves of statical stability shall be calculated for the following loading cases:

- loaded departure, 100% consumables;
- loaded arrival, 10% consumables;
- other anticipated service conditions, including possible lifting appliances.

The stability is considered sufficient if the following two criteria are met in addition to criteria 1–4 in Section 3.11.2.1.

1. The maximum of the righting-arm curve should occur at an angle of heel not smaller than 25°.

2. The effective, initial metacentric height, GM_{eff}, should not be less than 0.35 m.

If a multihull vessel does not meet the above stability criteria, the vessel shall meet the following alternative criteria:

1. If the maximum of the righting-arm curve occurs at 15°, the area under the curve shall not be less than 0.085 m rad. If the maximum occurs at 30°, the area shall not be less than 0.055 m rad.
2. If the maximum of the righting-arm curve occurs at an angle ϕ_{GZmax} situated between 15° and 30°, the area under the curve shall not be less than

$$A = 0.055 + 0.002(30° - \phi_{GZmax}) \quad (3.73)$$

where A is measured in m rad.

3. The area under the righting-arm curve between 30° and 40°, or between 30° and the angle of downflooding, if this angle is less than 40°, shall not be less than 0.03 m rad.
4. The righting arm shall not be less than 0.2 m at 30°.
5. The maximum righting arm shall occur at an angle not smaller than 15°.
6. The initial metacentric height shall not be less than 0.35 m.

The intact stability of new vessels of less than 15 m length that carry a combined load of passengers and cargo of less than 1000 kg is checked in an inclining experiment. The passengers, the crew without the skipper, and the cargo are transferred to one side of the ship, while the skipper may be assumed to stay at the steering position. Under these conditions the angle of heel shall not exceed 7°.

For vessels with a watertight weather deck the freeboard shall be not less than 75 mm at any point. For open boats the freeboard to the top of the gunwale shall not be less than 250 mm at any point.

3.11.7 Regulations for internal water vessels

3.11.7.1 EC regulations

The European prescriptions for internal-navigation ships are contained in directive 82/714/CEE of October 1982. In September 1999, a proposal for modifications was submitted to the European parliament. The proposal details the internal waterways of Europe for which it is valid.

Intact stability is considered sufficient if:

- the heel angle due to the crowding of passengers on one side does not exceed 10°;
- the angle of heel due to the combined effect of crowding, wind pressure and centrifugal force does not exceed 12°.

In calculations it should be assumed that fuel and water tanks are half full. The considered wind pressure is 0.1kN m^{-2}. At the angles of heel detailed above, the minimum freeboard should not be less than 0.2 m. If lateral windows can be opened, a minimum safety distance of 0.1 m should exist.

3.11.7.2 Swiss regulations

The Swiss regulations for internal navigation are contained in an ordinance of 8 October 1978. Some modifications are contained in an ordinance of 9 March 2001 of the Swiss Parliament (Der Schweizerische Bundesrat). According to them cargo ships should be tested under a wind pressure of 0.25kN m^{-2}. The heeling moment in turning, in kN m, should be calculated as

$$M_{TC} = \frac{cV^2\Delta}{L_{WL}}\left(KG - \frac{T}{2}\right) \quad (3.74)$$

where $c \geq 0.4$ is a coefficient to be supplied by the builder or the operator. Stability is considered sufficient if under the above assumptions the heeling angle does not exceed 5° and the deck side does not submerge. The metacentric height should not be less than 1 m. The required wind pressure is definitely lower than that required for sea-going ships. On the other hand, the other requirements are more stringent.

3.11.8 Summary of intact stability regulations

The IMO Code on Intact Stability applies to ships and other marine vehicles of 24 m length and above. The metacentric height of passenger and cargo ships should be at least 0.15 m, and the areas under the righting-arm curve, between certain heel angles, should not be less than the values indicated in the document. Passenger vessels should not heel in turning more than 10°. In addition, passenger and cargo ships should meet a weather criterion in which it is assumed that the vessel is subjected to a wind arm that is constant throughout the heeling range. The heeling arm of wind gusts is assumed equal to 1.5 times the heeling arm of the steady wind. If a wind gust appears while the ship is heeled windwards by an angle prescribed by the code, the area representing the reserve of buoyancy should not be less than the area representing the heel energy. The former area is limited to the right by the angle of downflooding or by 50°, whichever is less.

The IMO code contains special requirements for ships carrying timber on deck, for fishing vessels, for mobile offshore drilling units, for dynamically supported craft, and for containerships larger than 100 m. The code also contains recommendations for inclining and for rolling tests.

The stability regulations of the US Navy prescribe criteria for statical and dynamical stability under wind, in turning, under passenger crowding on one side, and when lifting heavy weights over the side. The static criterion requires that the righting arm at the first static angle should not exceed 60% of the maximum righting arm. When checking dynamical stability under wind, it is assumed that the ship rolled 25° windwards from the first static angle. Then, the area representing the reserve of stability should be at least 1.4 times the area representing the heeling energy. When checking stability in turning, or under crowding or when lifting heavy weights, the angle of heel should not exceed 15° and the reserve of stability should not be less than 40% of the total area under the righting-arm curve.

The stability regulations of the UK Navy are derived from those of the US Navy. In addition to static and dynamic criteria such as those mentioned above, the UK standard includes requirements concerning the areas under the righting-arm curve. The minimum values are higher than those prescribed by IMO for merchant ships. While the wind speeds specified by the UK standard are lower than those in the US regulations, the stability criteria are more severe.

A quite different criterion is prescribed in the code for large sailing vessels issued by the UK Ministry of Transport. As research proved that wind-pressure coefficients of sail rigs cannot be predicted, the code does not take into account the sail configuration and the heeling moments developed on it. The document presents a simple method for finding a heel angle under steady wind, such that the heel angle caused by a gust of wind would be smaller than the angle leading to downflooding and ship loss. The steady heel angle should not exceed 15°, and the range of positive heeling arms should not be less than 90°.

Additional regulations mentioned in this chapter are a code for small workboats issued in the UK, and codes for internal-navigation vessels issued by the European Parliament and by the Swiss Parliament.

3.12 Damage stability regulations

3.12.1 SOLAS

Regulation 5 of the convention, SOLAS (2004) specifies how to calculate the permeabilities to be considered. Thus, the permeability, in percentage, throughout the machinery space shall be

$$85 + 10\left(\frac{a-c}{v}\right)$$

where a is the volume of passenger spaces situated under the margin line, within the limits of the machinery space, c is the volume of between-deck spaces, in the same zone, appropriated to cargo, coal, or store, and v, the whole volume of the machinery space below the margin line.

The percent permeability of spaces forward or abaft of the machinery spaces should be found from

$$63 + 35\frac{a}{v}$$

where a is the volume of passenger spaces under the margin line, in the respective zone, and v, the whole volume, under the margin line, in the same zone.

The **maximum permissible length**, Figure 3.35, of a compartment having its centre at a given point of the ship length is obtained from the floodable length by multiplying the latter by an appropriate number called **factor of subdivision**. For example, a factor of subdivision equal to 1 means that the margin line should not submerge if one compartment is submerged, while a factor of subdivision equal to 0.5 means that the margin line should not submerge when two compartments are flooded, termed a two-compartment standard.

Regulation 6 of the convention shows how to calculate the factor of subdivision as a function of the ship length and the nature of the ship service. First, SOLAS defines a factor, A, applicable to ships primarily engaged in cargo transportation

$$A = \frac{58.2}{L - 60} + 0.18$$

For $L = 131$, $A = 1$. Another factor, B, is applicable for ships primarily engaged in passenger transportation

$$B = \frac{30.3}{L - 42} + 0.18$$

For $L = 79$, $B = 1$.

A **criterion of service numeral**, C_s, is calculated as function of the ship length, L, the volume of machinery and bunker spaces, M, the volume of passenger spaces below the margin line, P, the number of passengers for which the ship is certified, N, and the whole volume of the ship below the margin line, V. There are two formulas for calculating C_s; their choice depends upon the product $P_1 = KN$, where $K = 0.056L$. If P_1 is greater than P,

$$C_s = 72\frac{M + 2P_1}{V + P_1 - P}$$

otherwise

$$C_s = 72\frac{M + 2P}{V}$$

For ships of length 131 m and above, having a criterion numeral $C_s \leq 23$, the subdivision abaft the forepeak is governed by the factor A. If $C_s \geq 123$ the subdivision is governed by the factor B. For $23 < C_s < 123$, the subdivision factor should be interpolated as

$$F = A - \frac{(A-B)(C_s - 23)}{100}$$

If $79 \leq L < 131$, a number S should be calculated from

$$S = \frac{3.754 - 25L}{13}$$

If $C_s = S$, $F = 1$. If $C_s \geq 123$, the subdivision is governed by the factor B. If C_s lies between S and 123, the subdivision factor is interpolated as

$$F = 1 - \frac{(1-B)(C_s - S)}{123 - S}$$

If $79 \leq L < 131$ and $C_s < S$, or if $L < 79$, $F = 1$.

Regulation 7 of the convention contains special requirements for the subdivision of passenger ships. Regulation 8 specifies the criteria of stability in the final condition after damage. The heeling arm to be considered is the one that results from the largest of the following moments:

- crowding of all passengers on one side;
- launching of all fully loaded, davit-operated survival craft on one side;
- due to wind pressure.

We call **residual righting lever arm** the difference

GZ − heeling arm

The range of positive residual arm shall be not less than 15°. The area under the righting-arm curve should be at least 0.015 m rad, between the angle of static equilibrium and the smallest of the following:

- angle of progressive flooding;
- 22° if one compartment is flooded, 27° if two or more adjacent compartments are flooded.

The moment due to the crowding of passengers shall be calculated assuming 4 persons per m² and a mass of 75 kg for each passenger. The moment due to the launching of survival craft shall be calculated assuming all lifeboats and rescue boats fitted on the side that heeled down, while the davits are swung out and fully loaded. The wind heeling moment shall be calculated assuming a pressure of 120 N m^{-2}.

3.12.2 Probabilistic regulations

Wendel (1960a) introduces the notion of **probability of survival after damage**. A year later, a summary in French appears in Anonymous (1961). This paper mentions a translation into French of Wendel's original paper (in Bulletin Technique du Bureau Veritas, February 1961) and calls the method 'une nouvelle voie', that is 'a new way'. Much has been written since then on the probabilistic approach; we mention here only a few publications, such as Rao (1968), Wendel (1970), Abicht and Bakenhus (1970), Abicht, Kastner and Wendel (1977), Wendel (1977). Over the years Wendel used new and better statistics to improve the functions of probability density and probability introduced by him. The general idea is to consider the probability of occurrence of a damage of length y and transverse extent t, with the centre at a position x on the ship length. Statistics of marine accidents should allow the formulation of a function of probability density, $f(x, y, t)$. The probability itself is obtained by triple integration of the density function. The IMO regulation A265 introduces probabilistic regulations for passenger ships, and SOLAS 1974, Part B1, defines probabilistic rules for cargo ships. Concisely, Regulation 25 of the SOLAS convention defines a **degree of subdivision**

$$R = (0.002 + 0.0009L^3)^{1/3} \quad (3.75)$$

where L is measured in metres. An **attained subdivision index** shall be calculated as

$$A = \Sigma p_i s_i \quad (3.76)$$

where p_i represents the probability that the ith compartment or group of compartments may be flooded, and s_i is the probability of survival after flooding the ith compartment or group of compartments. The attained subdivision index, A, should not be less than the required subdivision index, R.

Early details of the standard for subdivision and damage stability of dry cargo ships are given by Gilbert and Card (1990). A critical discussion of the IMO 1992 probabilistic damage criteria for dry cargo ships appears in Sonnenschein and Yang (1993). The probabilistic SOLAS regulations are discussed in some detail by Watson (1998) who also exemplifies them numerically. Ravn et al. (2002) exemplify the application of the rules to Ro–Ro vessels.

Serious criticism of the SOLAS probabilistic approach to damage can be found in Björkman (1995). Quoting from the title of the paper, 'apparent anomalies in SOLAS and MARPOL requirements'. Watson (1998) writes, 'There would seem to be two main objections to the probabilistic rules. The first of these is the extremely large amount of calculations required, which although acceptable in the computer age, is scarcely to be welcomed. The other objection is the lack of guidance that it gives to a designer, who may be even driven to continuing use of the deterministic method in initial design, changing to the probabilistic later – and hoping this does not entail major changes!'

112 Maritime engineering reference book

The 'CORDIS RTD PROJECTS' database of the European Communities, 2000, defines as follows the objective of project HARDER:

'The process of harmonisation of damage stability regulations according to the probabilistic approach is undergoing scrutiny . . . before being proposed for adoption by IMO . . . However, ongoing investigations started revealing serious lack of robustness and consistency and more importantly a worrying lack of rationale in the choice of parameters that are likely to affect the evolution of the overall design and safety of ships.

A recent application of existing tools by a committee of the relevant IMO working group . . . revealed that, before confidence in the whole process is irreversibly affected, concerted effort at European level must address the thorough validation of calculations, the proper choice of parameters and the definition of levels of acceptance . . .'

A report on the progress of the project HARDER is contained in the IMO document SLF 45/3/3 of 19 April 2002. The report covers 'Investigations and proposed formulations for the factor 'S': the probability of survival after flooding'. The approach adopted in the project HARDER is explained by Rusås (2002). As the probabilistic regulations are bound to change, we do not detail them in this book.

3.12.3 The US Navy

The regulations of the US Navy are contained in a document known as DDS-079-1. Part of the regulations are classified, part of those that are not classified can be found in Nickum (1988) or Watson (1998). For a ship shorter than 30.5 m (100 ft) the flooding of any compartment should not submerge her beyond the margin line. Ships longer than 30.5 m and shorter than 91.5 m (300 ft) should meet the same submergence criterion with two flooded compartments. Ships longer than 91.5 m should meet the submergence criterion with a damage extent of $0.15L$ or 21 m, whichever is greater.

When checking stability under wind, the righting arm, GZ, should be reduced by $0.05\cos\phi$ to account for unknown unsymmetrical flooding or transverse shift of loose material. As for intact condition (see Figure 3.40), the standard identifies two areas between the righting-arm and the wind-arm curves. The area A_1 is situated between the angle of static equilibrium and the angle of downflooding or 45°, whichever is smaller. The area A_2 is situated to the left, from the angle of static equilibrium to an angle of roll. The wind velocity and the angle of roll should be taken from DDS-079-1. As in the intact condition, the standard requires that $A_1/A_2 \geq 1.4$.

The US Navy uses the concept of **V lines** to define a zone in which the bulkheads must be completely watertight. We refer to Figure 3.45. Part (a) of the figure shows a longitudinal ship section near a bulkhead. Let us assume that after checking all required combinations of flooded compartments, the highest waterline on the considered bulkhead is WL; it intersects the bulkhead at O. In part (b) of the figure, we show the transverse section AB that contains the bulkhead. The intersection of WL with the bulkhead passes though the point Q. The standard assumes that unsymmetrical flooding can heel the vessel by 15°. The waterline corresponding to this angle is W_1L_1. Rolling and transient motions can

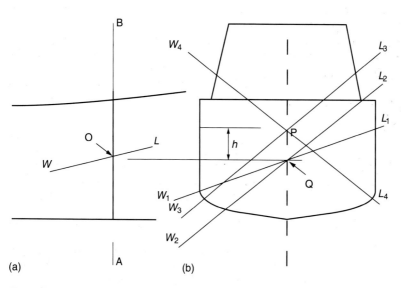

Figure 3.45 V lines.

increase the heel angle by a value that depends on the ship size and should be taken from the standard. We obtain thus the waterline W_2L_2. Finally, to take into account the relative motion in waves (that is the difference between ship motion and wave-surface motion) we draw another waterline translated up by $h = 1.22$ m (4 ft); this is waterline W_3L_3. Obviously, unsymmetrical flooding followed by rolling can occur to the other side too so that we must consider the waterline W_4L_4 symmetrical of W_3L_3 about the centreline. The waterlines W_3L_3 and W_4L_4 intersect at the point P. We identify a V-shaped limit line, W_4PL_3, hence the term 'V lines'. The region below the V lines must be kept watertight; severe restrictions refer to it and they must be read in detail.

3.12.4 The UK Navy

The standard of damage stability of the UK Navy is defined in the same documents NES 109 and SSP 24 that contain the prescriptions for intact stability (see Section 3.11.4). We briefly discuss here only the rules referring to vessels with a military role. The degree of damage to be assumed depends on the ship size, as given in Table 3.3.

Table 3.3

Waterline length	Damage extent
$L_{WL} < 30$ m	any single compartment
$30 \leq L_{WL} \leq 92$	any two adjacent main compartments, that is compartments of minimum 6-m length
>92 m	damage anywhere extending 15% of L_{WL} or 21 m, whichever is greater.

The permeabilities to be used are given in Table 3.4.

Table 3.4

Watertight, void compartment and tanks	0.97
Workshops, offices, operational and accommodation spaces	0.95
Vehicle decks	0.90
Machinery compartments	0.85
Store rooms, cargo holds	0.60

The wind speeds to be considered depend on the ship displacement, Δ, measured in tonnes, Table 3.5.

Table 3.5

Displacement Δ, tonnes	Nominal wind speed, knots
$\Delta \leq 1000$	$V = 20 + 0.005\Delta$
$1000 < \Delta \leq 5000$	$V = 5.06 \ln \Delta - 10$
$5000 < \Delta$	$V = 22.5 + 0.15\sqrt{\Delta}$

The following criteria of stability should be met (see also Figure 3.40):

1. Angle of list or loll not larger than 30°;
2. Righting arm GZ at first static angle not larger than 0.6 maximum righting arm;
3. Area A_1 greater than $A_{\min}$ as given by

$\Delta \leq 5000$ t	$A_{\min} = 2.74 \times 10^{-2} - 1.97 \times 10^{-6} \Delta$ m rad
$5000 < \Delta < 50000$ t	$A_{\min} = 0.164\Delta^{-0.265}$
$\Delta > 50000$ t	consult Sea Technology Group

4. $A_1 > A_2$;
5. Trim does not lead to downflooding;
6. $GM_L > 0$

Like the US Navy, the UK Navy uses the concept of V lines to define a zone in which the bulkheads must be completely watertight; some values, however, may be more severe. We refer again to Figure 3.45. Part (a) of the figure shows a longitudinal ship section near a bulkhead. Let us assume that after checking all required combinations of flooded compartments, the highest waterline on the considered bulkhead is WL; it intersects the bulkhead at O. In part (b) of the figure, we show the transverse section AB that contains the bulkhead. The intersection of WL with the bulkhead passes though the point Q. The standard assumes that unsymmetrical flooding can heel the vessel by 20°. The waterline corresponding to this angle is W_1L_1. Rolling and transient motions can increase the heel angle by 15°, leading to the waterline W_2L_2. Finally, to take into account the relative motion in waves (that is the difference between ship motion and wave-surface motion) we draw another waterline translated up by $h = 1.5$ m; this is waterline W_3L_3. Obviously, unsymmetrical flooding followed by rolling can occur to the other side too so that we must consider the waterline W_4L_4. The waterlines W_3L_3 and W_4L_4 intersect at the point P. Thus, we identify a V-shaped limit line, W_4PL_3, hence the term 'V lines'. The region below the V lines must be kept watertight; severe restrictions refer to it and they must be read in detail.

3.12.5 The German Navy

The BV 1003 regulations are rather laconic about flooding and damage stability. The main requirement refers to the extent of damage. For ships under 30 m length, only one compartment should be assumed flooded. For larger ships a damage length equal to

$$0.18 L_{WL} + 3.6 \text{ m}, \qquad (3.77)$$

but not exceeding 18 m, should be considered. Compartments shorter than 1.8 m should not be taken into account as such, but should be attached to the adjacent compartments. The leak may occur at any place along the ship, and all compartment combinations that can be flooded in the prescribed leak length should be considered. The damage may extend transversely till a longitudinal bulkhead, and vertically from keel up to the bulkhead deck.

Damage stability is considered sufficient if:

- the deck-at-side line does not submerge;
- without beam wind, and if symmetrically flooded, the ship floats in upright condition;
- in intermediate positions the list does not exceed 25° and the residual arm is larger than 0.05 m;
- under a wind pressure of 0.3kN m^{-2} openings of intact compartments do not submerge, the list does not exceed 25° and the residual lever arm is larger than 0.05 m.

If not all criteria can be met, the regulations allow for decisions based on a probabilistic factor of safety.

3.12.6 A code for large commercial sailing or motor vessels

The code published by the UK Maritime and Coastguard Agency, Maritime 2001, specifies that the free flooding of any one compartment should not submerge the vessel beyond the margin line. The damage should be assumed anywhere, but not at the place of a bulkhead. A damage of the latter kind would flood two adjacent compartments, a hypothesis not to be considered for vessels under 85 m. Vessels of 85 m and above should be checked for the flooding of two compartments.

In the damaged condition the angle of equilibrium should not exceed 7° and the range of positive righting arms should not be less than 15° up to the flooding angle. In addition, the maximum righting arm should not be less than 0.1 m and the area under the righting-arm curve not less than 0.015 m rad. The permeabilities to be used in calculations are given in Table 3.6.

The expression 'not a substantial amount of them' is not detailed.

Table 3.6

stores	0.60
stores, but not a substantial amount of them	0.95
accommodation	0.95
machinery	0.85
liquids	0.95 or 0, whichever leads to worse predictions

3.12.7 A code for small workboats and pilot boats

The code published by the UK Maritime and Coastguard Agency contains damage provisions for vessels up to 15 m in length and over, certified to carry 15 or more persons and to operate in an area up to 150 miles from a safe haven. The regulations are the same as those described for sailing vessels in Section 3.12.6, except that there is no mention of the two-compartment standard for lengths of 85 m and over.

3.12.8 EC regulations for internal water vessels

The following prescriptions are taken from a proposal to modify the directive 82/714 CEE, of 4 October 1982, issued by the European Parliament. The intact-stability provisions of the same document are summarized in Section 3.11.

A collision bulkhead should be fitted at a distance of minimum $0.04L_{WL}$ from the forward perpendicular, but not less than 4 m and no more than $0.04L_{WL} + 2$ m. Compartments abaft of the collision bulkhead are considered watertight only if their length is at least $0.10L_{WL}$, but not less than 4 m. Special instructions are given if longitudinal watertight bulkheads are present.

The minimum permeability values to be considered are given in Table 3.7.

Table 3.7

passenger and crew spaces	0.95
machinery spaces, including boilers	0.85
spaces for cargo, luggage, or provisions	0.75
double bottoms, fuel tanks	either 0.95 or 0

Following the flooding of any compartment the margin line should not submerge. The righting moment in damage condition, M_R, should be calculated for the downflooding angle or for the angle at which the bulkhead deck submerges, whichever is the smallest. For all flooding stages, it is required that

$$M_R > 0.2 M_P = 0.2 \times 1.5 bP \qquad (3.78)$$

where M_P is the moment due to passenger crowding on one side, b is the maximum available deck breadth at 0.5 m above the deck, and P is the total mass of the persons aboard. The regulations assume 3.75 persons per m², and a mass of 75 kg per person. The document explains in detail how to calculate the available deck area, that is the deck area that can be occupied by crowding persons.

References

Abicht, W. and Bakenhaus, J. (1970). Berechnung der Wahrscheinlichkeit des Überstehens von Verletzungen quer – und langsunterteilter Schiffe. In *Handbuch der Werften*, Vol. X.

Abicht, W., Kastner, S. and Wendel, K. (1977). *Stability of ships, safety from capsizing, and remarks on subdivision and freeboard*. Rept. 19: Hanover University, Hanover.

Abicht, W., Kastner, S. and Wendel, K. (1977). Stability of ships, safety from capsizing, and remarks on subdivision and freeboard. *Proceedings of the Second West European Conference on Marine Technology*, 23–27 May, Paper No. 9.

Anonymous (1961). Sécurité et compartimentage. Bulletin du Beareau Veritas, 45, No. 5, May, 91.

Attwood, E.L. and Pengelly, H.S. (1960). Theoretical Naval Architecture. New Edition, expanded by A.J. Sims. Longmans, London.

Barnaby, K. (1963). *Basic Naval Architecture*. Hutchinson, London.

Barrass, C.B. and Derrett, D.R. (2006). *Ship Stability for Masters and Mates*, 6th edition. Butterworth-Heinemann, Oxford, UK.

Biran, A. (2003). *Ship Hydrostatics and Stability*. Butterworth-Heinemann, Oxford, UK.

Björkman, A. (1995). On probabilistic damage stability. *The Naval Architect*, October. RINA, London.

Burcher, R.K. (1980). The influence of hull shape on transverse stability. *Trans. RINA*, Vol. 122.

Burcher, R.K. and Rydill, L. (1998). *Concepts in Submarine Design*. Cambridge Ocean Technology Series 2. Cambridge University Press, Cambridge.

Cleary, C., Daidola, J.C. and Reyling, C.J. (1966). Sailing ship intact stability criteria. *Marine Technology*, Vol. 33, No. 3, SNAME.

Deakin, B. (1991). The development of stability standards for UK sailing vessels. *The Naval Architect*, January. RINA, London.

Gilbert, R.R. and Card, J.C. (1990). The new international standard for subdivision and damage stability of dry cargo ships. *Marine Technology*, Vol. 27, No. 2, SNAME.

IMO (1995). *Code on Intact Stability for All Types of Ships Covered by IMO Instruments – Resolution A749(18)*. IMO, London.

IMO (2000). International Code of Safety for High Speed Craft. HSC Code 2000. MSC 97/73.

Maritime and Coastguard Agency (1998). *The Code of Practice for Safety of Small Workboats and Pilot Boats*. The Stationery Office, London.

Maritime and Coastguard Agency (2001). *The Code of Practice for Safety of Large Commercial Sailing and Motor Vessels*, 4th impression. The Stationery Office, London.

MARPOL (2002). 73/78 Consolidated Edition, 2002, IMO Publication (IMO-1B 520E).

MoD (1999a). *Naval Engineering Standard NES 109 – Stability Standard for Surface Ships – Part 1, Conventional Ships*. Issue 4.

MoD (1999b). *SSP24 – Stability of Surface Ships – Part 1 – Conventional Ships*. Issue 2, Abbey Wood, Bristol. Defence Procurement Agency, Unauthorised version circulated for comments.

Nickum, G. (1998). Subdivision and damage stability. In E.V. Lewis (ed), *Principles of Naval Architecture*, Vol. 1, 2nd Revision, SNAME.

Rao, K.A.V. (1968). Einfluss der Lecklänge auf den Sicherheitsgrad von Schiffen. *Schiffbautechnik*, Vol. 18, No. 1.

Ravn, E.S., Jensen, J.J., Baatrup, J. et al. (2002). Robustness of the probabilistic damage stability concept to the degree of details in the subdivision. *Stability of Ships, Lecture notes*. Graduate Course, Department of Mechanical Engineering, Maritime Engineering. Technical University of Denmark, Lyngby.

Rusås, S. (2002). Stability of ships: probability of survival. *Stability of Ships, Lecture notes*. Graduate Course, Department of Mechanical Engineering, Maritime Engineering. Technical University of Denmark, Lyngby.

Sarchin, T.H. and Goldberg, L.L. (1962). Stability and buoyancy criteria for US naval surface warships. *Trans. Vol. 70, SNAME*.

SOLAS (2004). SOLAS Consolidated Edition 2004. Consolidated text of the International Conference for the Safety of Life at Sea, 1974, and its Protocol of 1988, Articles, Annexes and Certificates. Incorporating all amendments in effect from 1 January, 2001. Publication (IMO – 110E). IMO, London.

Sonnenschein, R.J. and Yang, Ch. (1993). One-compartment damage survivability versus 1992 IMO probabilistic damage criteria for dry cargo ships. *Marine Technology*, Vol. 30, No. 1, January. SNAME.

Tupper, E.C. (2004). *Introduction to Naval Architecture*. Butterworth-Heinemann, Oxford, UK.

Watson, D.G.M. (1998). *Practical Ship Design*. Elsevier Science, Oxford, UK.

Wendel, K. (1960). Die Wahrscheinlichkeit des Überstehens von verletzungen. *Schiffstechnik*, Vol. 7, No. 36.

Wendel, K. (1970). Unterteilung von Schiffen. *Handbuch der Werften*, Vol. X.

Wendel, K. (1977). Die Bewertung von Unterteilungen. *Zeitschrift der Technischen Universität Hannover*, Volume published at 25 years of existence of the Department of Ship Technique.

4 Ship structures

Contents

4.1 Main hull strength
4.2 Structural design and analysis
4.3 Ship vibration
References (Chapter 4)

The various Sections of this Chapter have been taken from the following books, with the permission of the authors:

Jensen, J.J. (2001). *Load and Global Response of Ships*, Elsevier, Oxford, UK. [Section 4.3.11]
Rawson, K.J. and Tupper, E.C. (2001). *Basic Ship Theory*, Combined Volume, 5th edition. Butterworth-Heinemann, Oxford, UK. [Sections 4.1, 4.2]
Tupper, E.C. (2004). *Introduction to Naval Architecture*, 4th edition. Elsevier Butterworth-Heinemann, Oxford, UK. [Section 4.3 (excluding 4.3.11)]

4.1 Main hull strength

4.1.1 Introduction

Few who have been to sea in rough weather can doubt that the structure of a ship is subject to strain. Water surges and crashes against the vessel which responds with groans and shudders and creaks; the bow is one moment surging skywards, the next buried beneath green seas; the fat middle of the ship is one moment comfortably supported by a wave and the next moment abandoned to a hollow. The whole constitutes probably the most formidable and complex of all structural engineering problems in both the following aspects:

(a) the determination of the loading
(b) the response of the structure.

As with most complex problems, it is necessary to reduce it to a series of unit problems which can be dealt with individually and superimposed. The smallest units of structure which have to be considered are the panels of plating and single stiffeners which are supported at their extremities by items which are very stiff in comparison; they are subject to normal and edge loads under the action of which their dishing, bowing and buckling behaviour relative to the supports may be assessed. Many of these small units together constitute large flat or curved surfaces of plating and sets of stiffeners called grillages, supported at their edges by bulkheads or deck edges which are very stiff in comparison; they are subject to normal and edge loading and their dishing and buckling behaviour as a unit relative to their supports may be assessed. Finally, many bulkheads, grillages and decks, together constitute a complete hollow box whose behaviour as a box girder may be assessed. It is to this last unit, the whole ship girder, that this section is confined, leaving the smaller units for later consideration.

Excluding inertia loads due to ship motion, the loading on a ship derives from only two sources, gravity and water pressure. It is impossible to conceive a state of the sea whereby the loads due to gravity and water pressure exactly cancel out along the ship's length. Even in still water, this is exceedingly unlikely but in a seaway where the loading is changing continuously, it is inconceivable. There is therefore an uneven loading along the ship and, because it is an elastic structure, it bends. It bends as a whole unit, like a girder on an elastic foundation and is called the *ship girder*. The ship will be examined as a floating beam subject to the laws deduced in other textbooks for the behaviour of beams.

In still water, the loading due to gravity and water pressure are, of course, weight and buoyancy. The distribution of buoyancy along the length follows the curve of areas while the weight is conveniently assessed in unit lengths and might, typically, result in the block diagram of Figure 4.2. (Clearly, the areas representing total weight and total buoyancy must be equal.) This figure would give the resultants dotted which would make the ship bend concave downwards or *hog*. The reverse condition is known as *sagging*. Because it is not difficult to make some of the longer cargo ships break their backs when badly loaded, consideration of the still water hogging or sagging is vital in assessing a suitable cargo disposition. It is the first mate's yardstick of structural strength.

It is not difficult to imagine that the hog or sag of a ship could be much increased by waves. A long wave with a crest amidships would increase the upward force there at the expense of the ends and the hogging of the ship would be increased. If there were a hollow amidships and crests towards the ends sagging would be increased (Figure 4.3). The loads to which the complete hull girder is subject are, in fact:

(a) those due to the differing longitudinal distribution of the downward forces of weight and the upward forces of buoyancy, the ship considered at rest in still water;
(b) the additional loads due to the passage of a train of waves, the ship remaining at rest;
(c) loads due to the superposition on the train of waves of the waves caused by the motion of the ship itself through still water;
(d) the variations of the weight distribution due to the accelerations caused by ship motion.

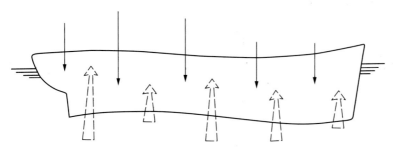

Figure 4.1 Loading on ship hull.

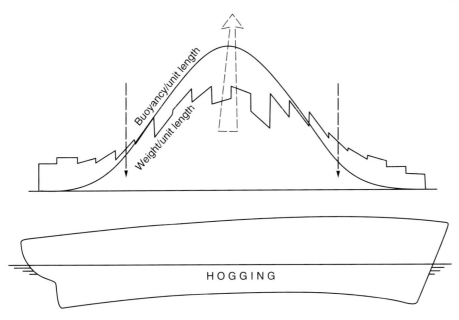

Figure 4.2 Still water hogging.

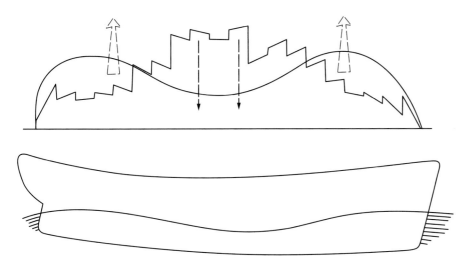

Figure 4.3 Sagging on a wave.

Consideration of the worst likely loading effected by (a) and (b) is the basis of the standard calculation. The effects of (c) and (d) are smaller and are not usually taken into account except partly by the statistical approach outlined later.

4.1.2 The standard calculation

This is a simple approach but one that has stood the test of time. It relies on a comparison of a new design with previous successful design. The calculated stresses are purely notional and based on those caused by a single wave of length equal to the ship's length, crest normal to the middle line plane and with

(a) a crest amidships and a hollow at each end causing maximum hogging, and
(b) a hollow amidships and a crest at each end causing maximum sagging.

The ship is assumed to be momentarily still, balanced on the wave with zero velocity and acceleration and the response of the sea is assumed to be that appropriate to static water. In this condition, the curves of weight and buoyancy are deduced. Subtracted one from the

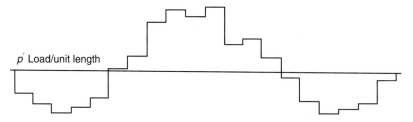

Figure 4.4 Loading curve, p'.

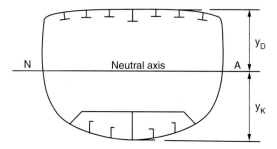

Figure 4.5 Simplified structural section.

other, the curves give a curve of net loading p'. Now, a fundamental relationship at a point in an elastic beam is

$$p' = \frac{dS}{dx} = \frac{d^2M}{dx^2}$$

where p' is the load per unit length, S is the shearing force, M is the bending moment, and x defines the position along the beam.

$$\therefore S = \int p'\, dx \quad \text{and} \quad M = \int S\, dx$$

Thus, the curve of loading p' must be integrated along its length to give a curve of shearing force and the curve of shearing force must be integrated along its length to give the bending moment curve. From the maximum bending moment, a figure of stress can be obtained,

$$\text{stress } \sigma = \frac{M}{I/y}$$

I/y being the modulus of the effective structural section.

A closer look at each of the constituents of this brief summary is now needed.

4.1.2.1 The wave

Whole books have been written about ocean waves and an outline of the properties of ocean waves is given in Chapter 1. Nevertheless, there is no universally accepted standard ocean wave which may be assumed for the standard longitudinal strength calculation. While the shape is agreed to be trochoidal, the observed ratios of length to height are so scattered that many 'standard' lines can be drawn through them. Fortunately, this is not of primary importance; while the calculation is to be regarded as comparative, provided that the same type of wave is assumed throughout for design and type ships, the comparison is valid.

A trochoid is a curve produced by a point at radius r within a circle of radius R rolling on a flat base. The equation to a trochoid with respect to the axes shown in Figure 4.6, is

$$x = R\theta - r\sin\theta$$
$$z = r(1 - \cos\theta)$$

One accepted standard wave is that having a height from trough to crest of one twentieth of its length from crest to crest. In this case, $L = 2\pi R$ and $r = h/2 = L/40$ and the equation to the wave is

$$x = \frac{L}{2\pi}\theta - \frac{L}{40}\sin\theta$$
$$z = \frac{L}{40}(1 - \cos\theta)$$

The co-ordinates of this wave at equal intervals of x are:

$\dfrac{x}{L/20}$	0	1	2	3	4	5	6	7	8	9	10
$\dfrac{z}{L/20}$	0	0.034	0.128	0.266	0.421	0.577	0.720	0.839	0.927	0.982	1.0

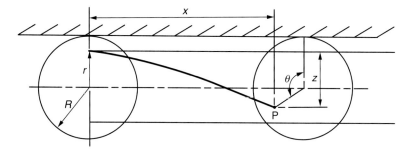

Figure 4.6 Construction of a trochoid.

Research has shown that the $L/20$ wave is somewhat optimistic for wavelengths from 90 m up to about 150 m in length. Above 150 m, the $L/20$ wave becomes progressively more unsatisfactory and at 300 m is probably so exaggerated in height that it is no longer a satisfactory criterion of comparison. This has resulted in the adoption of a trochoidal wave of height $0.607\sqrt{L}$ as a standard wave in the comparative longitudinal strength calculation. This wave has the equation

$$x = \frac{L}{2\pi}\theta - \frac{0.607\sqrt{L}}{2}\sin\theta$$

$$z = \frac{0.607\sqrt{L}}{2}(1-\cos\theta)$$

x, z and L in metres

The $0.607\sqrt{L}$ wave has the slight disadvantage that it is not non-dimensional, and units must be checked with care when using this wave and the formulae derived from it. Co-ordinates for plotting are conveniently calculated from the equations above for equal intervals of θ.

The length of the wave is, strictly, taken to be the length of the ship on the load waterline; in practice, because data is more readily available for the displacement stations, the length is often taken between perpendiculars (if this is different) without making an appreciable difference to the bending moment. Waves of length slightly greater than the ship's length can, in fact, produce theoretical bending moments slightly in excess of those for waves equal to the ship's length, but this has not been an important factor while the calculation continued to be comparative.

Waves steeper than $L/7$ cannot remain stable. Standard waves of size $L/9$ are used not uncommonly for the smaller coastal vessels. It is then a somewhat more realistic basis of comparison.

4.1.2.2 Weight distribution

Consumable weights are assumed removed from those parts of the ship where this aggravates the particular condition under investigation; in the sagging condition, they are removed from positions near the ends and, in the hogging condition, they are removed amidships. The *influence lines* of a similar design should be consulted before deciding where weights should be removed, and the decision should be verified when the influence lines for the design have been calculated (see later), provided that the weights are small enough.

The longitudinal distribution of the weight is assessed by dividing the ship into a large number of intervals. Twenty displacement intervals are usually adequate. The weight falling within each interval is assessed for each item or group in the schedule of weights and tabulated. Totals for each interval divided by the length give mean weights per unit length. It is important that the centre of gravity of the ship divided up in this way should be in the correct position. To ensure this, the centre of gravity of each individual item should be checked after it has been distributed.

One of the major items of weight requiring distribution is the hull itself, and this will sometimes be required before detailed structural design of the hull has been completed. A useful first approximation to the hull weight distribution is obtained by assuming that two-thirds of its weight follows the still water buoyancy curve and the remaining one-third is distributed in the form of a trapezium, so arranged, that the centre of gravity of the whole hull is in its correct position, Figure 4.7.

Having obtained the mean weight per unit length for each interval, it is plotted as the mid-ordinate of

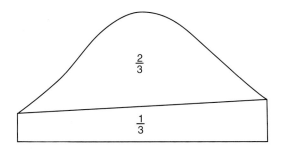

Figure 4.7 Approximate hull weight distribution.

the interval and a straight line is drawn through it, parallel to the chord of the buoyancy curve. This is a device for simplifying later stages of the calculation which introduces usually insignificant inaccuracies. A sawtooth distribution of weight per unit length, as shown in Figure 4.3 results.

4.1.2.3 Buoyancy and balance

If the standard calculation is performed by hand, the wave is drawn on tracing paper and placed over the contracted profile of the ship on which the Bonjean curve has been drawn at each ordinate. Figure 4.8 shows one such ordinate. It is necessary for equilibrium to place the wave at a draught and trim such that

(a) the displacement equals the weight and
(b) the centre of buoyancy lies in the same vertical plane as the centre of gravity.

Simple hydrostatic pressure is assumed; the areas immersed at each ordinate can be read from the Bonjean curves where the wave profile cuts the ordinate, and these areas are subjected to normal approximate integration to give displacement and LCB position. The position of the wave to meet the two conditions can be found by trial and error. It is usual to begin at a draught to the midships point of the trochoid about 85% of still water draught for the hogging condition and 120% for the sagging wave. A more positive way of achieving balance involves calculating certain new tools. Consider a typical transverse section of the ship, x m forward of amidships, at which the Bonjean curve shows the area immersed by the first trial wave surface to be $A\, m^2$.

If $\nabla\, m^3$ and $M\, m^4$ are the volume of displacement and the moment of this volume before amidships for the trial wave surface, then, over the immersed length

$$\nabla = \int A\, dx \quad \text{and} \quad M = \int Ax\, dx$$

Suppose that $\nabla_0\, m^3$ and $M_0\, m^4$ are the volume of displacement and moment figures for equilibrium in still water and that ∇ is less than ∇_0 and M is less than M_0. The adjustment to be made to the trial wave waterline is therefore a parallel sinkage and a trim by the bow in order to make $\nabla = \nabla_0$ and $M = M_0$. Let this parallel sinkage at amidships be z_0 and the change of trim be m radians, then the increased immersion at the typical section is

$$z = (z_0 + mx)\, m$$

Let the slope of the Bonjean curve be $s = dA/dz\, m$. Assuming that the Bonjean curve is straight over this distance, then

$$A + zs = A + (z_0 + mx)s\, m^2$$

In order to satisfy the two conditions, $\nabla = \nabla_0$ and $M = M_0$, therefore,

$$\int [A + (z_0 + mx)s]\, dx = \nabla_0$$

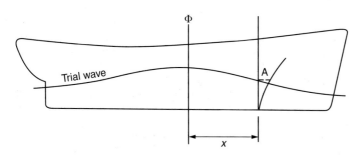

Figure 4.8 Wave on profile.

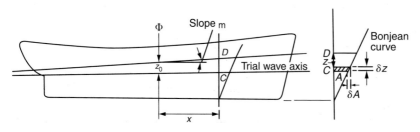

Figure 4.9

and

$$\int [A + (z_0 + mx)s]x\,dx = M_0$$

Now put

$$\int s\,dx = \delta\nabla \text{ m}^2$$

$$\int sx\,dx = \delta M \text{ m}^3$$

$$\int sx^2\,dx = \delta I \text{ m}^4$$

Then

$$z_0 \cdot \delta\nabla + m\delta M = \nabla_0 - \nabla$$

and

$$z_0 \cdot \delta M + m\delta I = M_0 - M$$

$\delta\nabla$, δM and δI can all be calculated by approximate integration from measurement of the Bonjean curves. From the pair of simultaneous equations above the required sinkage z_0 and trim m can be calculated. A check that this wave position in fact gives the required equilibrium should now be made. If the trial waterplane was a poor choice, the process may have to be repeated. Negative values of z_0 and m, mean that a parallel rise and a trim by the stern are required.

Having achieved a balance, the curve of buoyancy per metre can be drawn as a smooth curve. It will have the form shown in Figure 4.2 for the hogging calculation and that of Figure 4.3 for the sagging condition.

4.1.2.4 Loading, shearing force and bending moment

Because the weight curve has been drawn parallel to the buoyancy curve, the difference between the two, which represents the net loading p' for each interval, will comprise a series of rectangular blocks. Using now the relationship $S = \int p'dx$, the loading curve is integrated to obtain the distribution of shearing force along the length of the ship. The integration is a simple cumulative addition starting from one end and the shearing force at the finishing end should, of course, be zero; in practice, due to the small inaccuracies of the preceding steps, it will probably have a small value. This is usually corrected by canting the base line, i.e. applying a correction at each section in proportion to its distance from the starting point.

The curve of shearing force obtained is a series of straight lines. This curve is now integrated in accordance with the relationship $M = \int S\,dx$ to obtain the distribution of bending moment M. Integration is again a cumulative addition of the areas of each trapezium and the inevitable final error, which should be small, is distributed in the same way as is the shearing force error. If the error is large, the calculations must be repeated using smaller intervals for the weight distribution.

The integrations are performed in a methodical, tabular fashion. In plotting the curves, there are several important features which arise from the expression:

$$p' = \frac{dS}{dx} = \frac{d^2M}{dx^2}$$

which will assist and act as checks. These are

(a) when p' is zero, S is a maximum or a minimum and a point of inflexion occurs in the M curve.
(b) when p' is a maximum, a point of inflexion occurs in the S curve,
(c) when S is zero, M is a maximum or minimum.

A typical set of curves is shown in Figure 4.10.

The foregoing calculations will normally be carried out by computer, rather than by hand, using a commercial longitudinal strength software package. The calculation process will, however, broadly follow the same procedures as described.

4.1.2.5 Second moment of area

It is now necessary to calculate the second moment of area, I, of the section of the ship girder. For a simple girder this is a straightforward matter. For a ship

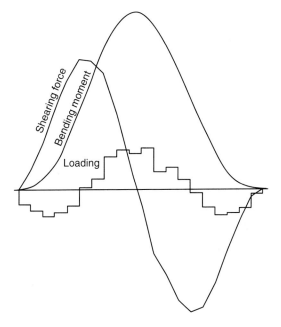

Figure 4.10 Typical loading, shearing force and bending moment curves.

girder composed of plates and sections which may not extend far longitudinally or which might buckle under compressive load or which may be composed of differing materials, a certain amount of adjustment may be judged necessary. Unless there is some good reason to the contrary (such as inadequate jointing in a wood sheathed or armoured deck), it is assumed here that local structural design has been sufficient to prevent buckling or tripping or other forms of load shirking. In fact, load shirking by panels having a width/thickness ratio in excess of seventy is likely and plating contribution should be limited to seventy times the thickness (see Section 4.2). No material is assumed to contribute to the section modulus unless it is structurally continuous for at least half the length of the ship amidships. Differing materials are allowed for in the manner described later. In deciding finally on which parts of the cross section to include, reference should be made to the corresponding assumptions made for the designs with which final stress comparisons are made. Similar comparisons are made also about members which shirk their load.

Having decided which material to include, the section modulus is calculated in a methodical, tabular form. An *assumed neutral axis* (ANA) is first taken near the mid-depth, Figure 4.11. Positions and dimensions of each item forming the structural mid-section are then measured and inserted into a table of the type shown in Table 4.1.

The strength section of some ships is composed of different materials, steel, light alloy, wood or plastic. How is the second moment of area calculated for these composite sections? Consider a simple beam composed of two materials, suffixes a and b. From the theory of beams, it is known that the stress is directly proportional to the distance from the neutral axis and that, if R is the radius of curvature of the neutral axis and E is the elastic modulus,

$$\text{stress } \sigma = \frac{E}{R} h$$

Consider the typical element of area A of Figure 4.12. For equilibrium of the cross-section, the net force must be zero, therefore

$$\Sigma(\sigma_a A_a + \sigma_b A_b) = 0$$

$$\Sigma\left(\frac{E_a}{R} A_a h_a + \frac{E_b}{R} A_b h_b\right) = 0$$

i.e.

$$\Sigma\left(A_a h_a + \frac{E_b}{E_a} A_b h_b\right) = 0$$

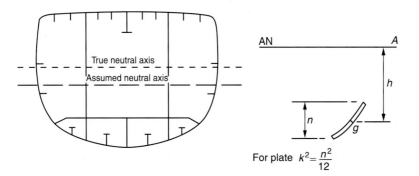

Figure 4.11 Midship section.

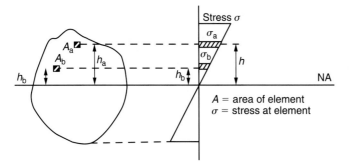

Figure 4.12 Composite section.

Now the contribution to the bending moment by the element A is the force σA multiplied by the distance from the neutral axis, h. For the whole section,

$$M = \Sigma(\sigma_a A_a h_a + \sigma_b A_b h_b)$$

$$= \frac{E_a}{R} \Sigma \left(A_a h_a^2 + \frac{E_b}{E_a} A_b h_b^2 \right)$$

$$= \frac{E_a}{R} \cdot I_{\text{eff}}.$$

where I_{eff} is the effective second moment of area.

It follows from the equations above that the composite section may be assumed composed wholly of material a, provided that an effective area of material b, $(E_b/E_a)A_b$, be used instead of its actual area. The areas of material b used in columns 2, 4, 5 and 7 of Table 4.1 must then be the actual areas multiplied by the ratio of the elastic moduli. The ratio E_b/E_a for different steels is approximately unity; for wood/steel it is about $\frac{1}{16}$ in compression and $\frac{1}{25}$ in tension. For aluminium alloy/steel it is about $\frac{1}{3}$ and for glass reinforced plastic/steel it is between $\frac{13}{15}$ and $\frac{1}{30}$. The figures vary with the precise alloys or mixtures and should be checked.

4.1.2.6 Bending stresses

Each of the constituents of the equation, $\sigma = M/(I/y)$, has now been calculated, maximum bending moment M, second moment of area, I, and the levers y for two separate conditions, hogging and sagging. Values of the maximum direct stress at deck and keel arising from the standard comparative longitudinal strength calculation can thus be found.

Direct stresses occurring in composite sections are, following the work of the previous section

in material a, $\quad \sigma_a = \dfrac{E_a y}{R} = \dfrac{M}{I_{\text{eff}}} y$

in material b, $\quad \sigma_b = \dfrac{E_b y}{R} = \dfrac{My}{I_{\text{eff}}} \dfrac{E_b}{E_a}$

It is at this stage that the comparative nature of the calculation is apparent, since it is now necessary to decide whether the stresses obtained are acceptable or whether the strength section needs to be modified. It would, in fact, be rare good fortune if this was unnecessary, although it is not often that the balance has to be repeated in consequence. The judgement on the acceptable level of stress is based on a comparison with similar ships in similar service for which a similar calculation has previously been performed. This last point needs careful checking to ensure that the same wave has been used and that the same assumptions regarding inclusion of material in weight and modulus calculations are made. For example, a device was adopted many years ago to allow for the presence of rivet holes, either by decreasing the effective area of section on the tension side by $\frac{2}{11}$ or by increasing the tensile stress in the ratio $\frac{11}{9}$. Although this had long been known to be an erroneous procedure, some authorities continued to use it in order to maintain the basis for comparison for many years.

Stresses different from those found acceptable for the type ships may be considered on the following bases:

(a) Length of ship. An increase of acceptable stress with length is customary on the grounds that standard waves at greater lengths are less likely to be met. This is more necessary with the $L/20$ wave

Table 4.1 Modulus calculation.

1	2	3	4	5	6	7
Item	A (cm^2)	h (m)	Ah (cm^2 m)	Ah^2 (cm^2 m^2)	k^2 (m^2)	Ak^2 (cm^2 m^2)
Each item above ANA						
Totals above ANA	ΣA_1		$\Sigma A_1 h_1$	$\Sigma A_1 h_1^2$		$\Sigma A_1 k_1^2$
Each item below ANA						
Totals below ANA	ΣA_2		$\Sigma A_2 h_2$	$\Sigma A_2 h_2^2$		$\Sigma A_2 k_2^2$

where A = cross-sectional area of item, h = distance from ANA, k = radius of gyration of the structural element about its own NA.

Subscript 1 is used to denote material above the ANA and subscript 2 for material below.

$$\text{Distance of true NA above ANA} = \frac{\Sigma A_1 h_1 - \Sigma A_2 h_2}{\Sigma A_1 + \Sigma A_2} = d$$

Second moment of area about true NA $= \Sigma A_1 h_1^2 + \Sigma A_2 h_2^2 + \Sigma A_1 k_1^2 + \Sigma A_2 k_2^2 - (\Sigma A_1 + \Sigma A_2)d^2 = I$

Lever above true neutral axis from NA to deck at centre $= y_D$

Lever below true neutral axis from NA to keel $= y_K$

than the $0.607\sqrt{L}$ wave for which the probability varies less with length. If a standard thickness is allowed for corrosion, it will constitute a smaller proportion of the modulus for larger ships.

(b) Life of ship. Corrosion allowance is an important and hidden factor. Classification societies demand extensive renewals when survey shows the plating thicknesses and modulus of section to be appreciably reduced. It could well be economical to accept low initial stresses to postpone the likely time of renewal. Comparison with type ship ought to be made before corrosion allowance is added and the latter assessed by examining the performance of modern paints and anti-corrosive systems (see Chapter 9).

(c) Conditions of service. Classification societies permit a reduced modulus of section for service in the Great Lakes or in coastal waters. Warship authorities must consider the likelihood of action damage by future weapons and the allowance to be made in consequence. Warships are not, of course, restricted to an owner's route.

(d) Local structural design. An improvement in the buckling behaviour or design at discontinuities may enable higher overall stress to be accepted.

(e) Material. Modern high grade steels permit higher working stresses and the classification societies encourage their use, subject to certain provisos.

(f) Progress. A designer is never satisfied; a structural design which is entirely successful suggests that it was not entirely efficient in the use of materials and he is tempted to permit higher stresses next time. Such progress must necessarily be cautiously slow.

A fuller discussion of the nature of failure and the aim of the designer of the future occurs later in the Chapter. Approximate values of total stress which have been found satisfactory in the past are given below:

Ships	Wave	Design stresses Deck	Keel
		N/mm^2	N/mm^2
100 m frigate	L/20	110	90
150 m destroyer	L/20	125	110
200 m general cargo vessel	$0.607\sqrt{L}$	110	90
250 m aircraft carrier	L/20	140	125
300 m oil tanker	$0.607\sqrt{L}$	140	125

4.1.2.7 Shear stresses

The shearing force at any position of the ship's length is that force which tends to move one part of the ship vertically relative to the adjacent portion, Figure 4.13. It tends to distort square areas of the sides into rhomboids. The force is distributed over the section, each piece of material contributing to the total. It is convenient to consider shear stress, the force divided by the area and this is divided over the cross-section of a simple beam according to the expression

$$\tau = \frac{SA\bar{y}}{Ib}$$

$A\bar{y}$ is the moment about the NA of that part of the cross-section above section PP where the shear stress τ is required. I is the second moment of area of the whole cross-section and b is the total width of material at section P (Figure 4.14). The distribution of shear stress over a typical cross-section of a ship is shown in Figure 4.13. The maximum shear stress occurs at the neutral axis at those points along the length where the shearing force is a maximum. A more accurate distribution would be given by shear flow theory applied to a hollow box girder, discussed later.

Acceptable values of shear stress depend on the particular type of side construction. Failure under the action of shear stress would normally comprise wrinkling of panels of plating diagonally. The stress at which this occurs depends on the panel dimensions, so while the shear stress arising from the standard longitudinal strength calculation clearly affects side plating thickness and stiffener spacings, it does not have a profound effect on the structural cross-section of the ship.

Already, there is a tendency to assume that the stresses obtained from the standard comparative calculation actually occur in practice, so that the design of local structure may be effected using these stresses. Because local structure affected by the ship girder stresses cannot otherwise be designed, there is no choice. Strictly, this makes the local structure comparative. In fact, the stresses obtained by the comparative calculation for a given wave profile have been shown by full scale measurements to err on the safe side so that their use involves a small safety factor. Often, other local loading is more critical. This is discussed more fully in Section 4.2.1.2.

4.1.2.8 Influence lines

The ship will not often, even approximately, be in the condition assumed for the standard calculation. It is important for designers and operators to know at a glance, the effect of the addition or removal of weight on the longitudinal strength. Having completed the standard calculation, the effects of small additions of weight are plotted as influence lines in much the same way as for bridges and buildings (see Figure 4.16).

An influence line shows the effect on the *maximum* bending moment of the addition of a unit weight at any position along the length. The height of the line at P represents the effect on the maximum bending moment at X of the addition of a unit weight at P. Two influence lines are normally drawn, one for the

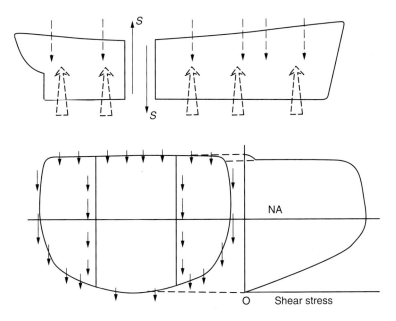

Figure 4.13 Shear loading on a ship guider.

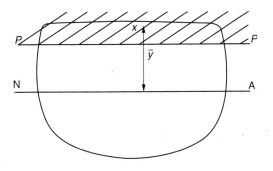

Figure 4.14

hogging and one for the sagging condition. Influence lines could, of course, be drawn to show the effect of additions on sections other than that of the maximum BM but are not generally of much interest.

Consider the addition of a weight P, x aft of amidships, to a ship for which the hogging calculation has been performed as shown in Figure 4.17. It will cause a parallel sinkage s and a change of trim t over the total length L. If A and I are the area and least longitudinal second moment of area of the curved waterplane and the ship is in salt water of reciprocal weight density u, then, approximately,

$$\text{sinkage } s = \frac{uP}{A} \quad \text{and} \quad \text{trim } t = \frac{P(x-f)uL}{I}$$

The increase in bending moment at OZ is due to the moment of added weight less the moment of buoyancy of the wedges aft of OZ less the moment of buoyancy of the parallel sinkage aft of OZ.

Moment of buoyancy of parallel sinkage about

$$\text{OZ} = \frac{sM_a}{u} = \frac{PM_a}{A}$$

Moment of buoyancy of wedges

$$= \frac{1}{u}\int 2yx_1 \times \frac{(x_1 - \overline{f-e})}{L} t\,dx_1$$

$$= \frac{I_a t}{uL} - \frac{M_a(f-e)t}{uL} = \{I_a - M_a(f-e)\}\frac{P(x-f)}{I}$$

Moment of weight which is included only if the weight is aft of OZ and therefore only if positive

$= P[x - e]$, positive values only.

Increase in BM for the addition of a unit weight,

$$\frac{\delta M}{P} = -\{I_a - M_a(f-e)\}\frac{(x-f)}{I} - \frac{M_a}{A} + [x-e]$$

Note that this expression is suitable for negative values of x (i.e. for P forward of amidships), provided that the expression in square brackets, [ ], is discarded if negative. A discontinuity occurs at OZ, the ordinate of maximum bending moment. The influence lines are straight lines which cut the axis at points about 0.2–0.25 of the length from amidships. It is within this length, therefore, that weights should be removed to aggravate the hogging condition and outside this length that they should be removed to aggravate the sagging condition.

128   Maritime engineering reference book

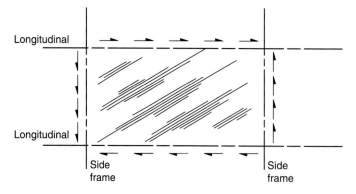

Figure 4.15

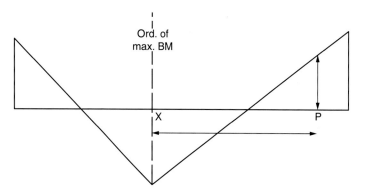

Figure 4.16 Influence lines.

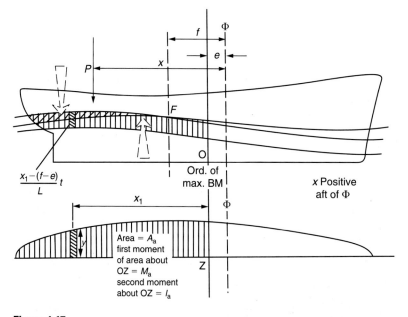

Figure 4.17

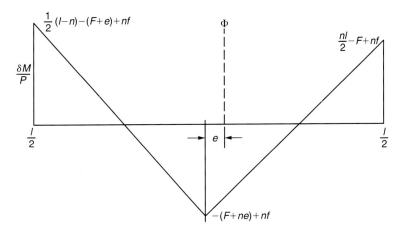

Figure 4.18

Certain simplifying assumptions are sometimes made to produce a simpler form of this equation. There is little point in doing this because no less work results. However, if $(f - e)$ is assumed negligible, n put equal to I_a/I and F put equal to M_a/A

$$\frac{\delta M}{P} = -n(x - f) - F + [x - e]$$

This gives the results shown in Figure 4.18.

A note of caution is necessary in the use of influence lines. They are intended for small weight changes and are quite unsuitable for large changes such as might occur with cargo, for example. In general, weight changes large enough to make a substantial change in stress are too large for the accuracy of influence lines.

4.1.2.9 Changes to section modulus

Being a trial and error process, the standard calculation will rarely yield a suitable solution first time. It will almost always be necessary to return to the structural section of the ship to add or subtract material in order to adjust the resulting stress. The effect of such an addition, on the modulus I/y, is by no means obvious. Moreover, it is not obvious in the early stages, whether the superstructure should be included in the strength section or whether the construction should be such as to discourage a contribution from the superstructure. Some measure of the effect of adding material to the strength section is needed if we are to avoid calculating a new modulus for each trial addition.

In order to provide such a measure, consider (Figure 4.19) the addition of an area a at a height z above the neutral axis of a structural section whose second moment of area has been calculated to be

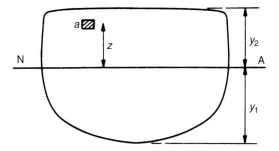

Figure 4.19

Ak^2, area A, radius of gyration k and levers y_1 to the keel and y_2 to the deck. Now

$$\text{stress } \sigma = \frac{M}{I/y}$$

By considering changes to I and y it can be shown that for a given bending moment, M, stress σ will obviously be reduced at the deck and will be reduced at the keel if $z > k^2/y_1$, when material is added within the section, $z < y_2$.

If the material is added above the deck, $z > y_2$ then the maximum stress occurs in the new material and there will be a reduction in stress above the neutral axis (Figure 4.20) if

$$a > \frac{A(z/y_2 - 1)}{(z^2/k^2 + 1)}$$

The position is not quite so simple if the added material is a superstructure. See later section on superstructures.

4.1.2.10 Slopes and deflections

The bending moment on the ship girder has been found by integrating first the loading p' with

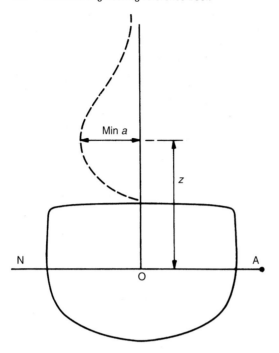

Figure 4.20

respect to length to give shearing force S, and then by integrating the shearing force S with respect to length to give M. There is a further relationship:

$$M = EI \frac{d^2 y}{dx^2}$$

or

$$\frac{dy}{dx} = \frac{1}{E} \int \frac{M}{I} dx \quad \text{and} \quad y = \int \frac{dy}{dx} dx$$

Thus, the slope at all points along the hull can be found by integrating the M/EI curve and the bent shape of the hull profile can be found by integrating the slope curve. It is not surprising that the errors involved in integrating approximate data four times can be quite large. Moreover, the calculation of the second moment of area of all sections throughout the length is laborious and the standard calculation is extended this far only rarely—it might be thus extended, for example, to give a first estimate of the distortion of the hull between a master datum level and a radar aerial some distance away. It may also be done to obtain a deflected profile in vibration studies.

4.1.2.11 Horizontal flexure

Flexure perpendicular to the plane with which we are normally concerned may be caused by vibration, flutter, uneven lateral forces or bending while rolling. Vibrational modes are discussed in Section 4.3. Unsymmetrical bending can be resolved into bending about the two principal axes of the cross-section, Figure 4.21. The only point in the ship at which the maxima of the two effects combine is the deck edge and if the ship were to be balanced on a standard wave which gave a bending moment M, the stress at the deck edge would be

$$\sigma = \frac{Mz}{I_{yy}} \cos\phi + \frac{My}{I_{zz}} \sin\phi$$

However, it is not quite so easy as this; if the ship were heading directly into a wave train, it would not be rolling. Only in quartering or bow seas will bending and rolling be combined and, in such a case, the maximum bending moments in the two planes would not be in phase, so that the deck edge effect will be less than that obtained by superposition of the two maxima. Some limited research into this problem indicates that horizontal bending moment maxima are likely to be of the order of 40% of the vertical bending moment maxima and the ratio of the stresses is likely to be about 35% in common ship shapes. The increase in deck edge stress over that obtained with head seas is thought to be of the order of 20–25%. This is an excellent reason to avoid stress concentrations in this area.

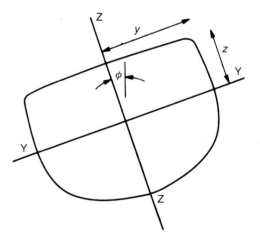

Figure 4.21

The horizontal bending moment is addressed by Jensen (2001) and forces due to rolling are discussed by Kobylinski and Kastner (2003).

4.1.2.12 Behaviour of a hollow box girder

The simple theory of bending of beams assumes that plane sections of the beam remain plane and that the direct stress is directly proportional to the distance from the neutral axis. A deck parallel to the neutral axis would therefore be expected to exhibit constant stress across its width. In fact, the upper flange of a hollow box girder like a ship's hull can receive its load only by shear at the deck edge, Figure 4.22. The diffusion of this shear into a plane deck to create the direct stresses is a difficult problem mathematically; the diffusion from the edge elements across the deck is such that the plane sections do not remain plane, and the mathematics shows that the direct stresses reduce towards the middle of the deck. This has been borne out by observations in practice. The effect is known as *shear lag* and its magnitude depends on the type of loading and dimensions of the ship; it is more pronounced, for example, under concentrated loads. In common ship shapes, it accounts for a difference in stress level at mid-deck of only a few per cent. It is more important in the consideration of the effects of superstructures and of the effective breadth of plating in local strength problems which are discussed in Section 4.2.

4.1.2.13 *Wave pressure correction*

This is usually known as the *Smith Correction*, Jensen (2001). In calculating the buoyancy per unit length at a section of the ship, the area of the section given by the Bonjean curve cut by the wave surface was taken. This assumes that the pressure at a point P on the section is proportional to the depth h of the point below the wave surface and the buoyancy $= w \int h \, dB = w \times$ area immersed (Figure 4.23).

This simple hydrostatic law is not true in a wave because the wave is caused by the orbital motion of water particles. It can be shown that the pressure at a point P in a wave at a depth h below the wave surface is the same as the hydrostatic pressure at a depth h', where h' is the distance between the mean (or still water) axis of the surface trochoid and the sub-surface trochoid through P (Figure 4.24). Thus, the buoyancy at a section of a ship is:

$$w \int h' \, dB = w \times \text{(effective area immersed)}.$$

Setting up h' from the sub-surface trochoid through P and similar heights from a series of sub-surface trochoids drawn on the wave profile, the effective immersed area up to W'L is obtained. New Bonjean curves can thus be plotted (Figure 4.25). The formula for h' is

$$h' = H - \frac{r_0^2}{2R}\left\{1 - \exp\left(-\frac{2H}{R}\right)\right\}$$

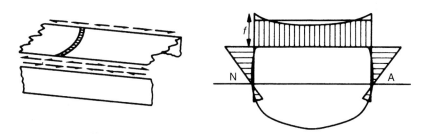

Figure 4.22 Direct stresses in practice.

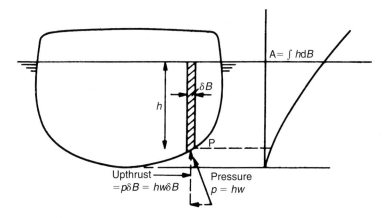

Figure 4.23

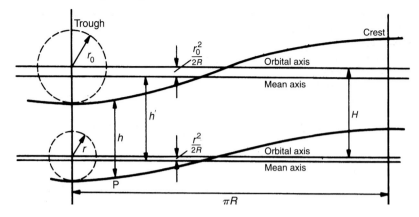

Figure 4.24

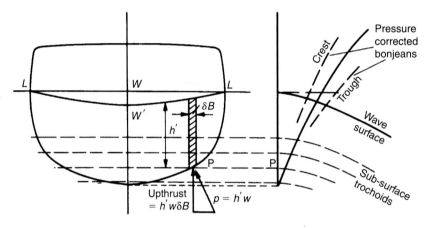

Figure 4.25

An approximation to the Smith correction is:

$$\text{corrected wave BM} = \text{wave BM} \times \exp\left(-\frac{nT}{L}\right)$$

where T is the draught, L the length and n a constant given by:

Block coeff. C_B	n, sagging	n, hogging
0.80	5.5	6.0
0.60	5.0	5.3

4.1.2.14 Longitudinal strength standards by rule

Until 1960 the Classification Societies prescribed the structure of merchant ships through tables of dimensions. They then changed to the definition of applied load and structural resistance by formulae. In the 1990s the major Classification Societies under the auspices of the International Association of Classification Societies (IACS) agreed a common minimum standard for the longitudinal strength of ships supported by the statistics of structural failure.

There is today widespread acceptance of the principle that there is a very remote probability that load will exceed strength during the whole lifetime of a ship. This probability may be as low as 10^{-8} at which level the IACS requirement is slightly more conservative than almost every Classification Society standard.

Loading on a merchant ship is separated into two parts:

(a) the bending moment and shear force due to the weight of the ship and the buoyancy in still water,
(b) the additional effects induced by waves.

Still water loading is calculated by the simple methods described at the beginning of this chapter without, of course, the wave which is replaced by the straight waterline of interest. Several such waterlines will usually be of concern. Stresses caused by such loading may be as much as 40% of the total stresses allowed and incorrect acceptance of cargo or unloading of cargo has caused spectacular failures of the ship girder. With such ships as bulk carriers in particular,

Operators must follow the sequences of loading and unloading recommended by their Classification Societies with scrupulous care. Not only may the ship break its back by bending but failure could be caused by the high shear forces that occur between full and empty holds. Indeed, a simple desktop computer to assess changes as they are contemplated has become common since high capacity cargo handling has evolved.

Wave induced bending moment (WIBM) may be represented by the formulae:

Hogging WIBM $= 0.19\, MCL^2 BC_b$ kN m

Sagging WIBM $= -0.11\, MCL^2 B(C_b + 0.7)$ kN m

where L and B are in metres and $C_b \geq 0.6$

and $C = 10.75 - \left(\dfrac{300 - L}{100}\right)^{1.5}$ for $90 \leq L \leq 300$ m

$= 10.75$ for $300 < L < 350$ m

$= 10.75 - \left(\dfrac{L - 350}{150}\right)^{1.5}$ for $350 \leq L$

M is a distribution factor along the length of the ship.

$M = 1.0$ between $0.4L$ and $0.65L$ from the stern

$= 2.5x/L$ at x metres from the stern up to $0.4L$

$= 1.0 - \dfrac{x - 0.65L}{0.35L}$ at x metres from the stern between $0.65L$ and L.

Wave induced shear force is given by IACS as

Hogging condition $S = 0.3\, F_1 CLB(C_b + 0.7)$ kN

Sagging condition $S = -0.3\, F_2 CLB(C_b + 0.7)$ kN

where L, B and C_b are as given above and F_1 and F_2 by Figure 4.27.

The modulus of cross section amidships must be such that bending stress caused by combining still

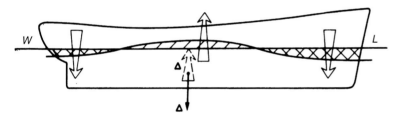

Figure 4.26 Wave induced bending forces.

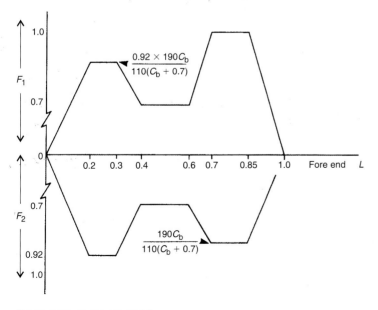

Figure 4.27 Factors F_1 and F_2.

water and wave induced BM is less than 175 N/mm². There is the further proviso that, despite the waters in which the vessel may sail, this midship section modulus shall not be less than:

$$CL^2 B(C_b + 0.7) \text{ cm}^3$$

which represents a stress of 110 N/mm².

The maximum permissible total shear stress, wave induced plus still water is also 110 N/mm².

The formulae given above embrace most common ship types. Special consideration must be given to vessels with large deck openings, large flare, heated cargoes or which fall outside limits when:

$$L \geq 500 \text{ m} \quad L/B \leq 5$$
$$C_b < 0.60 \quad B/D \geq 2.5$$

The formulae for wave induced bending moment and shear force have been revised over the years by the Classification Societies, and latest recommendations should be derived from sources such as Lloyds Register (2007) or other Classification Societies.

Classification Societies now rely upon computer programs based on the methods presently to be described in this Chapter. Fundamental to this is the representation of the wave loading by wave spectra and by strip theory, Chapter 7. Strip theory is linear and, in essence, assumes the vessel to be wallsided. Thus, wave induced loading would be the same in both hogging and sagging conditions which is known in practice not to be so. Sagging stresses are greater by about 10% than hogging stresses due in part to the support at the end of the ship provided by flare. This accounts for the different coefficients given in the formulae above, providing some degree of non-linearity in the treatment.

This gratifying conformity among Classification Societies is, of course, reviewed periodically. It may be reviewed if statistics show that standards have become inadequate or if oceanographers demonstrate that world wave patterns are changing with climate. Variations remain among the Classification Societies in the more detailed aspects of structural design and in allowances made for high quality steels. Variations also occur in the stresses permitted among different ship types and with restricted service conditions. Under the auspices of IACS, member societies have been engaged in the development of common structural rules for ships. The first two of these common structural rules, for bulk carriers of 90 metres or more in length and for oil tankers of 150 metres or more in length, came into force on 1st April, 2006. These common rules will be incorporated into each member society's rule book (Eyres, 2007).

4.1.2.15 *Full scale trials*

Full scale trials can be carried out for a variety of purposes.

(a) To load a hull, possibly to destruction, noting the (ultimate) load and the failure mechanisms.
(b) To measure the strains in a ship at sea, over a short period, in a seaway which is itself measured.
(c) To measure strains in ships at sea over a prolonged period of service with the crew noting any signs of failure. This is the only trial which is likely to throw light on fatigue failure.

There have been several attempts to measure the behaviour of the ship girder by loading ships and recording their behaviour at various positions by strain gauges and measurements of deflection. *Neverita* and *Newcombia*, *Clan Alpine* and *Ocean Vulcan* were subject to variations in loading by filling different tanks while they were afloat. *Wolf, Preston, Bruce* and *Albuera* were supported at the middle or the ends in dry dock and weight added until major structural failure occurred.

Short duration sea trials pose difficulty in finding rough weather and in measuring the wave system accurately. In one successful trial two frigates of significantly different design were operated in close company in waves with significant heights up to 8 metres. Electrical resistance gauges were used to record stress variations with time and the seas measured using wave buoys. Of particular interest was the slamming which excites the hull girder whipping modes. It was found that the slam transient increased the sagging bending moment much more than the hogging. In frigates whipping oscillations are damped out quite quickly but they can persist longer in some vessels. Such trials, using two ships, are reported by Andrew and Lloyd (1981).

Long term trials are of great value. They provide data that can be used in the more representative calculations discussed later. For many years now a number of RN frigates have been fitted with automatic mechanical or electrical strain gauges recording the maximum compressive and tensile deck stresses in each four hour period. Ship's position, speed and sea conditions are taken from the log. Maximum bending stresses have exceeded the $L/20$ standard values by a factor of almost two in some cases. This is not significant while the standard calculation is treated as purely comparative. It does show the difficulties of trying to extrapolate from past experience to designs using other constructional materials (e.g. reinforced plastic or concrete) or with a distinctly different operating pattern.

4.1.2.16 *The nature of failure*

Stress has never caused any material to fail. Stress is simply a convenient measure of the material behaviour which may 'fail' in many different ways. 'Failure' of a structure might mean permanent strain, cracking, unacceptable deflection, instability, a short life or even a resonant vibration. Some of these criteria are conveniently measured in terms of stress. In defining an acceptable level of stress for a ship, what 'failure' do we have in mind?

Structural failure of the ship girder may be due to one or a combination of (a) Cracking, (b) Fatigue failure, (c) Instability.

It is a fact that acceptable stress levels are at present determined entirely by experience of previous ships in which there have been a large number of cracks in service. The fact that these might have been due to poor local design is at present largely disregarded, suggesting that some poor local design or workmanship somewhere in the important parts of the hull girder is inevitable.

However good the design of local structure and details might be, there is one important influence on the determination of acceptable levels of stress. The material built into the ship will have been rolled and, finally, welded. These processes necessarily distort the material so that high stresses are already built in to the structure before the cargo or sea impose any loads at all. Very little is known about these built-in stresses. Some may yield out, i.e. local, perhaps molecular, straining may take place which relieves the area at the expense of other areas. The built-in stress clearly affects fatigue life to an unknown degree. It may also cause premature buckling. Thus, with many unknowns still remaining, changes to the present practice of a stress level determined by a proliferation of cracks in previous ships, must be slow and cautious. Let us now examine this progress.

4.1.2.17 Realistic assessment of longitudinal strength

Study of the simple standard longitudinal strength calculations so far described has been necessary for several good reasons. First, it has conveyed an initial look at the problem and the many assumptions which have had to be made to derive a solution which was within the capabilities of the tools available to naval architects for almost a hundred years. Moreover, the standard has been adequate on the whole for the production of safe ship designs, provided that it was coupled with conventions and experience with similar structure which was known to have been safe. As a comparative calculation it has had a long record of success. Second, it remains a successful method for those without ready access to modern tools or, for those who do enjoy such access, it remains an extremely useful starting point for a process which like any design activity is iterative. That is to say, the structural arrangement is guessed, analysed, tested against standards of adequacy, refined, reanalysed and so on until it is found to be adequate. Analysis by the standard calculation is a useful start to a process which might be prolonged. Third, much of the argument which has been considered up to now remains valid for the new concepts which must now be presented.

Of course, the standard calculation can be performed more readily now with the help of computers. The computer, however, has permitted application of mathematics and concepts of behaviour which have not been possible to apply before. It has permitted an entirely new set of standards to replace the static wave balance and to eradicate many of the dubious assumptions on which it was based. Indeed, it is no longer necessary even to assume that the ship is statically balanced. The basis of the new methods is one of realism; of a moving ship in a seaway which is continuously changing.

During a day at sea, a ship will suffer as many as 10,000 reversals of strain and the waves causing them will be of all shapes and sizes. At any moment, the ship will be subject not to a single $L/20$ wave but a composite of many different waves. Their distribution by size can be represented by a histogram of the numbers occurring within each range of wave lengths. In other words, the waves can be represented statistically. Now the statistical distribution of waves is unlikely to remain constant for more than an hour or two by which time wind, weather and sea state will modify the statistics. Over the life of a ship such sea state changes will have evened out in some way and there will be a lifetime statistical description of the sea which will be rather different from the short-term expectation.

This provides the clue to the new approaches to longitudinal strength. What is now sought as a measure is the likelihood that particular bending moments that the sea can impose upon the ship will be exceeded. This is called the probability of exceedence and it will be different if it is assessed over one hour, four hours, one day or 25 years. What the new standard does is to ensure that there is a comfortably small probability of exceedence of that bending moment which would cause the ship to fail during its lifetime. With some 30 million or so strain reversals during a lifetime, the probability of exceedence of the ship's strength needs to be very small indeed. If the frequency distribution of applied bending moment is that given in Figure 4.28 and the ship's strength is S, then the probability of failure is

$$p(\text{failure}) = \int_{S}^{\infty} f(\text{BM})\, d\,\text{BM}$$

The procedure should be read in conjunction with the behaviour of elements of the structure in Section 4.2, the statistics of waves in Chapter 1 and the motions of ships in a seaway in Chapter 7. Thus, the more realistic approach to longitudinal strength will consider:

(a) loading imposed upon the ship;
(b) response of the longitudinal structure;
(c) assessment of structural safety.

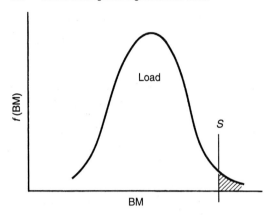

Figure 4.28

4.1.2.18 Realistic assessment of loading longitudinally

It is necessary first to return to the effect of a single wave, not a trochoid which is too inconvenient mathematically but to a sinewave of a stated height and length or frequency. We need to examine the effect of this wave, not upon a rock-like ship presumed to be static, but upon a ship which will move, in particular to heave and to pitch. Movement $q(t)$ of any system with one degree of freedom subject to an excitation $Q(t)$ is governed by the differential equation.

$$\ddot{q}(t) + 2k\omega_0 \dot{q}(t) + \omega_0 q = Q(t)$$

where k is the damping factor and ω_0 is the natural frequency. If now the input to the system $Q(t)$ is sinusoidal $= x \cos \omega t$, the solution to the equation, or the output, is

$$q(t) = HQ(t - \varepsilon)$$

where ε is a phase angle and H is called the response amplitude operator (RAO)

$$\text{RAO} = \frac{1}{\{(\omega_0^2 - \omega^2)^2 + 4k^2\omega_0^2\}^{1/2}}$$
$$= \frac{\text{amplitude of output}}{\text{amplitude of input}}$$

Both H and ε are functions of the damping factor k and the tuning factor ω/ω_0.

This form of solution is a fairly general one and applies when the input is expressed in terms of wave height and the output is the bending moment amidships. The multiplier H is of course more complex but depends upon damping, wave frequency and ship shape, heave position and pitch angle. It is called the bending moment response amplitude operator and may be calculated for a range of wave frequencies, Figure 4.29.

The calculation of the RAOs—and indeed the heave and pitch of the ship subject to a particular wave—is performed by standard computer programs using, say, strip theory, see Chapter 7.

An array of sources and sinks can be devised to represent the flow past a prismatic body even when there is an air/water interface. Flow past the body, the forces on the body, the consequent movement of the

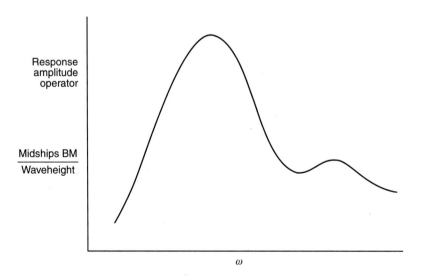

Figure 4.29

body and the elevation or depression of the interface can all be calculated. A ship can be represented by a series of short prismatic bodies or strips, all joined together and the total forces and bending moments upon such a hull calculated, the different elevations representing the self-generated waves. Waves imposed upon the ship may also be represented in the same way. The potential functions needed for this representation are pulsating ones and interference occurs between waves created and imposed. These are affected by the beam of the ship relative to the wave length of the self-generated waves and this has resulted in several different approaches to strip theory using slightly different but important assumptions. There are other important assumptions concerned with the boundaries. The general term strip theory embraces all such studies, including slender body theory and systematic perturbation analysis, see Chapter 7.

The mathematics is based upon several assumptions, the most important of which has been that the relationship of bending moment and wave height is linear, i.e. bending moment proportional to waveheight. Another effect of the assumptions is that the mathematics makes the ship wallsided. Despite these obviously incorrect assumptions, the results from strip theory are remarkably accurate and with steady refinements to the theory will improve further.

It is now necessary to combine the effect of all of those waves which constitute the sea as oceanographers have statistically described. For a short time ahead they have found that each maximum hog or sag bending moment can be well represented by a Rayleigh distribution

$$p(\mathrm{BM}_{max}) = \frac{\mathrm{BM}_{max}}{m_0} \exp\left(-\frac{\mathrm{BM}_{max}^2}{2m_0}\right)$$

where m_0 is the total energy in the ship response which is the mean square of the response

$$m_0 = \int_0^\infty \int_{-\pi}^{\pi} \mathrm{RAO}^2\, S(\omega, \theta)\, d\theta\, d\omega$$

S being the total energy of the waves in the direction θ. The probability that BM_{max} will exceed a value B in any one cycle is given by

$$p(\mathrm{BM}_{max} > B) = \exp\left(-\frac{B^2}{2m_0}\right)$$

or the probability of exceedence in n cycles

$$p(\mathrm{BM}_{max} > B, n) = 1 - \{1 - \exp(-B^2/2m_0)\}^n$$

This illustrates the general approach to the problem. It has been necessary to define $S(\omega, \theta)$, the spectrum of wave energy which the ship may meet in the succeeding hour or two. Over the entire life of the ship, it may be expected to meet every possible combination of wave height and frequency coming from every direction. Such long-term statistics are described by a two-parameter spectrum agreed by the ISSC which varies slightly for different regions of the world. The procedure for determining the probability of exceedence is a little more complicated than described above but the presentation of the results is similar.

Figure 4.30 shows such a result. At any given probability of exceedence the short time ahead is predicted by a Rayleigh or a Gaussian distribution based on the mean value of the sea characteristics pertaining.

Other approaches to the problem are possible. Some authorities, for example, use as a standard the likelihood that a particular bending moment will be exceeded during any period of one hour (or four hours) during the life of the ship. This has the advantage of being conveniently compared with measurement of statistical strain gauges installed in ships which record the maximum experienced every hour (or four hours).

Figure 4.31 is such a plot on a log logscale which shows linear results that can be easily extrapolated.

We may now capitalize upon the study we made of the simple trochoidal wave. The effective wave height H_e is defined as that trochoidal wave of ship length which by the static standard wave calculation (without the Smith correction) gives the same wave bending moment as the worst that the ship would experience during its lifetime. That which appears to fit frigates very well is:

$$H_e = 2.2\, L^{0.3} \text{ metres}$$

Figure 4.32 shows various effective wave height formulae compared with frigates at probability of once in a lifetime of 3×10^7 reversals and with merchant ships at 10^8.

Comparisons show that the $L/20$ wave gives 50–70% of the remote expectation for frigates while $0.607\sqrt{L}$ runs through the 10^8 spots, overestimating the load for very large ships. As a first estimate of required longitudinal strength therefore a designer may safely use the static standard strength calculation associated with a wave height of $L/9$ up to 50 m, $2.2\, L^{0.3}$ around 100 m and $0.607\sqrt{L}$ m up to about 300 m.

The next section considers the ability of a ship to withstand longitudinal bending.

4.1.2.19 Realistic structural response

The totally elastic response of the ship treated as if it were a simple beam has already been considered. Small variations to simple beam theory to account for shear lag were also touched upon. There are some

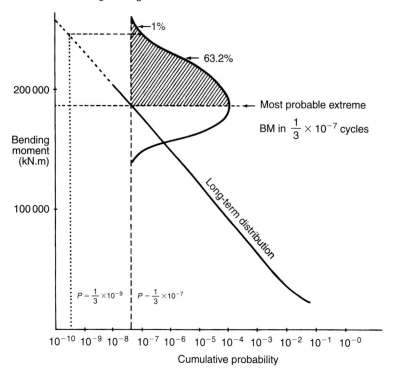

Figure 4.30 Cumulative probability.

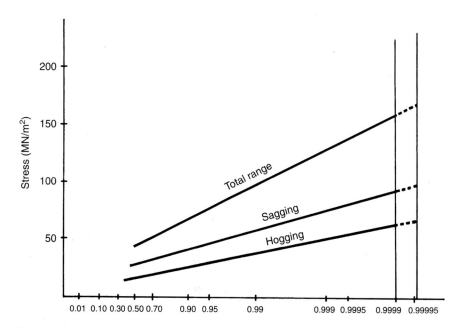

Figure 4.31 Probability of non-exceedence in a 4-hour period.

other reasons why a hollow box girder like a ship may not behave like a simple beam. Strain locked in during manufacture, local yielding and local buckling will cause redistribution of the load-bearing capability of each part and its contribution to the overall cross section. If extreme loads are to be just contained, it is necessary to know the ultimate load-bearing ability of the cross section.

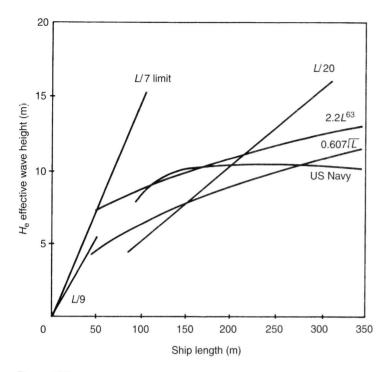

Figure 4.32

Let us first go to one extreme, and it is again necessary to draw upon some results of Section 4.2. It is there explained that the ultimate resistance to bending of a beam after the load has been increased to the point when the beam becomes totally plastic is

$$M_p = \sigma_Y S$$

where S is the sum of the first moments of area of cross section each side of the neutral axis. This surely is the ultimate strength of a perfect beam which is turned into a plastic hinge. Common sense tells us that such a state of affairs is most unlikely in the large hollow box girder which is the ship. While the tension side might conceivably become totally plastic, the compression side is likely to buckle long before that. This is taken into account in the concept of the ultimate longitudinal resistance to load which forms the basis on which ultimate strength is now assessed, Caldwell (1965), Paik (2004). The essence of the method is to build up the total resistance to bending from a summation of the contribution of each element. It is not difficult to imagine that as compressive load in a deck is increased there comes a time when the panels will buckle, shirk their load and throw an increased burden upon their adjacent longitudinal stiffeners. These in turn may buckle as the load is further increased, shirk their contribution and throw an extra burden on 'hard' areas like corners of decks and junctions with longitudinal bulkheads.

This is how a realistic assessment of cross-sectional resistance should be made, a method which is now in general use. Every element which constitutes the effective cross section is first examined and a stress-strain curve is plotted. When the buckling behaviour of the element is embraced, the curves are called load shortening curves and are usually plotted in non-dimensional form as a family of curves with a range of initial assumed imperfections, Figure 4.33.

The cross section of the ship girder is then assumed to bend with plane sections remaining plane to take up a radius of curvature R, the cross section rotating through an angle θ (Figure 4.34). At any distance h above the neutral axis an element n of area A_n will be strained by an amount ε_n. Its load-bearing capacity can be picked off the relevant load shortening curve, say, at Q (Figure 4.33). All such elements, so calculated, will lead to a bending resistance of the cross section

$$M = \sum_n \sigma_n A_n h_n$$

Carried out for a range of values of θ or R, a load shortening curve for the whole cross section can be calculated to provide a good indication of the collapse bending moment, Figure 4.35. With the heavy structure of the bottom in compression when the ship hogs, the ultimate BM approaches the plastic moment M_p but a ratio of 0.6 in the sagging condition is not unusual.

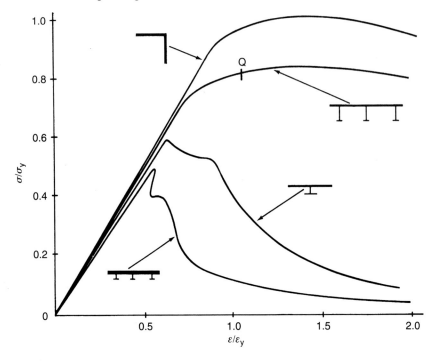

Figure 4.33

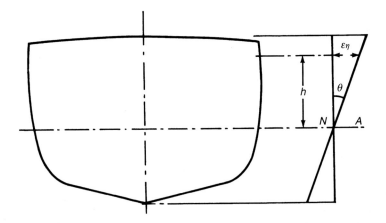

Figure 4.34

In the calculation of M, the neutral axis does not stay still except when the behaviour is wholly elastic so that the additional condition for equilibrium of cross section must be imposed

$$\sum_n \sigma_n A_n = 0$$

For this reason, it is convenient to consider changes to θ in increments which are successively summed and the process is called incremental analysis. Load shortening curves for the elements may embrace any of the various forms of buckling and indeed plastic behaviour whether it be assisted by locked-in manufacturing stresses or not.

While there are some obvious approximations to this method and some simplifications, it is undoubtedly the most realistic assessment of the collapse strength of the ship girder yet. The near horizontal part of Figure 4.35 clearly represents the failure load, or collapse strength or ultimate strength.

4.1.2.20 Assessment of structural safety

At its simplest, failure will occur when the applied load exceeds the collapse load. It is first necessary to decide what probability of failure is acceptable,

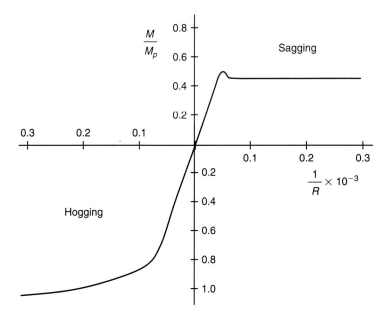

Figure 4.35

morally, socially or economically. History has shown that a comfortable level is a probability of 0.01 or 1% that a single wave encounter will load the ship beyond its strength. With 30 million such encounters during a lifetime—every 10 seconds or so for 25 years—this represents a probability of 0.33×10^{-9}. Some argue that this is altogether too remote and the ship could be weaker. In fact, noting the logarithmic scale, there is not a great deal of difference in the applied load at these very remote occurrences (Figure 4.30). Because the statistical method is still essentially comparative, all that is necessary is to decide upon one standard and use it to compare new designs with known successful practice or, rarely, failure. This is what all major authorities now do. It is worth taking a glimpse at further developments.

From the previous section, we have been able to declare a single value of the strength S which the loading must exceed for failure to occur. S may also be assumed to vary statistically because it may itself possess a variability. The strength, for example, will be affected by the thickness of plating rolled to within specified tolerances because that is what happens in practice. It will be dependent upon slightly different properties of the steel around the ship, upon ship production variations, upon defects, upon welding procedures and initial distortion which differs slightly. All such small variations cannot be precisely quantified at the design stage but the range over which they are likely can be. Instead of a single figure S for the strength, a probability density function can be constructed. Figure 4.36 more properly shows the realistic shapes of loading and strength. For bending moment M, the shaded area

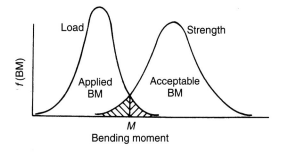

Figure 4.36

under the load curve to the right of M represents the probability that the load will exceed M. The shaded area under the strength curve to the left of M represents the probability that the ship will not be strong enough to withstand M.

We are thus concerned with the shaded overlap at the bottom. If we are studying extreme loads whose probability of occurrence is remote we have to define with some precision the shapes of the very tiny tails of the distributions.

Instead, most authorities match the two distributions to known mathematical shapes and allow the mathematics to take care of the tails. The most usual shapes to be assumed are Rayleigh or Gaussian which are defined by two properties only, the mean and the root mean square (or variance). This leads to a relationship between load P and strength S

$$S = v_P v_S \, P$$

where v_P is the partial safety factor deduced from the mean and variance of the applied loading distribution and v_S is the partial safety factor assessed for the mean and variance of the strength variability. A further partial safety factor is often added, v_c, which assesses subjectively (i.e. as a matter of judgement) the gravity of failure, Jensen (2001).

4.1.2.21 *Hydroelastic analysis*

The ship in a seaway is an elastic body which enjoys bodily movements in all six directions and distorts also about and along all three axes. Its distortions affect the load applied by the sea and both structural and hydrodynamic damping affect the problem. The sea itself is a random process.

Until recently a reasonably complete mathematical description and solution of this formidable problem had never been achieved. It has now been possible to describe the dynamic behaviour of the ship in terms of modes of distortion which also embrace such solid body movements as pitch and heave, thus bringing together seakeeping and structural theories. Superposition of each element of behaviour in accordance with the excitation characteristics of the sea enables the total behaviour to be predicted, including even slamming and twisting of the hull, Bishop and Price (1979).

This powerful analytical approach has been used to examine the overall stress distribution in large ships, showing that important problems emerge at sections of the ship other than amidships. Combination of shear force and bending moment cause principal stresses much higher than had been suspected previously. Areas of particular concern are those about 20% of the ship's length from the stern or from the bow, where slamming may further exacerbate matters. A lack of vigilance in the detail design or the production of the structure in these areas could, it has been suggested, have been responsible for some bulk carrier and VLCC fractures, Bishop *et al.* (1991).

4.1.2.22 *Slamming (see also Chapter 7)*

One hydroelastic phenomenon which has been known for many years as slamming has now succumbed to theoretical treatment. When flat areas of plating, usually forward, are brought into violent contact with the water at a very acute angle, there is a loud bang and the ship shudders. The momentum of the ship receives a check and energy is imparted to the ship girder to make it vibrate. Strain records show that vibration occurs in the first mode of flexural vibration imposing a higher frequency variation upon the strain fluctuations due to wave motion. Amplitudes of strain are readily augmented by at least 30% and sometimes much more, so that the phenomenon is an important one.

The designer can do a certain amount to avoid excessive slamming simply by looking at the lines 30–40% of the length from the bow and also right aft to imagine where acute impact might occur. The seaman can also minimize slamming by changes of speed and direction relative to the wave fronts. In severe seas the ship must slow down.

Extreme values of bending moment acceptable by the methods described in this chapter already embrace the augmentation due to slamming. This is because the relationships established between full scale measurements and the theory adopted make such allowance.

4.1.3 Material considerations

A nail can be broken easily by notching it at the desired fracture point and bending it. The notch introduces a stress concentration which, if severe enough, will lead to a bending stress greater than the ultimate and the nail breaks on first bending. If several bends are needed failure is by fatigue, albeit, low cycle fatigue.

A stress concentration is a localized area in a structure at which the stress is significantly higher than in the surrounding material. It can conveniently be conceived as a disturbance or a discontinuity in the smooth flow of the lines of stress such as a stick placed in a fast flowing stream would cause in the water flow. There are two types of discontinuity causing stress concentrations in ships:

(a) discontinuities built into the ship unintentionally by the methods of construction, e.g. rolling, welding, casting, etc.;
(b) discontinuities deliberately introduced into the structural design for reasons of architecture, use, access, e.g. hatch openings, superstructures, door openings, etc.

Stress concentrations cannot be totally avoided either by good design or high standards of workmanship. Their effects, however, can be minimized by attention to both and it is important to recognize the effects of stress concentrations on the ship girder. Many ships and men were lost because these effects were not recognized in the early Liberty ships of the 1940s.

In general, stress concentrations may cause yield, brittle fracture or buckling. There is a certain amount of theory which can guide the designer, but a general understanding of how they arise is more important in their recognition and treatment because it is at the detail design stage that many can be avoided or minimized.

4.1.3.1 *Geometrical discontinuties*

The classical mathematical theory of elasticity has produced certain results for holes and notches in

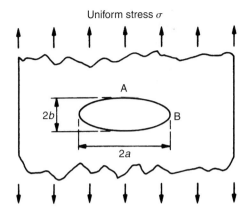

Figure 4.37

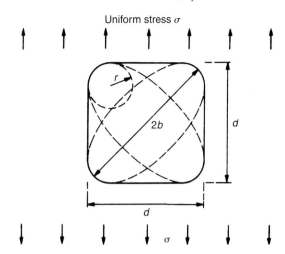

Figure 4.38

laminae. The stress concentration factors at A and B of Figure 4.37 of an elliptical hole in an infinite plate under uniform tension in the direction of the b-axis are given by:

$$j_A = \frac{\text{stress}_A}{\sigma} = -1$$

$$j_B = \frac{\text{stress}_B}{\sigma} = 1 + \frac{2a}{b}$$

For the particular case of a circular hole, $a = b$ and $j_B = 3$, i.e. the stress at the sides of a circular hole is three times the general tensile stress level in the plate, while at top and bottom there is a compressive stress equal to the general stress level. If a crack is thought of, ideally, as a long thin ellipse, the equation above gives some idea of the level of stress concentration at the ends; a crack twenty times as long as its width, for example, lying across the direction of loading would cause a stress, at the ends, forty-one times the general stress level and yielding or propagation of the crack is likely for very modest values of σ.

A square hole with radiused corners might be represented for this examination by two ellipses at right angles to each other and at 45 degrees to the direction of load. For the dimensions given in Figure 4.38 the maximum stress concentration factor at the corners is given approximately by:

$$j = \frac{1}{2}\sqrt{\frac{b}{r}}\left\{1 + \frac{\sqrt{(2b + 2r)}}{\sqrt{b} - \sqrt{r}}\right\}$$

Figure 4.39 shows the effect of the variation of corner radius r on length of side b for a square hole with a side parallel to the direction of stress and for a square hole at 45 degrees. The figure shows:

(a) that there may be a penalty of up to 25% in stress in failing to align a square hole with rounded corners with the direction of stress;

(b) that there is not much advantage in giving a corner radius greater than about one-sixth of the side;

(c) that the penalty of corner radii of less than about one-twentieth of the side, is severe. Rim reinforcement to the hole can alleviate the situation.

These results are suitable for large hatches. With the dimensions as given in Figure 4.40, the maximum stress concentration factor can be found with good accuracy from the expression:

$$j_{\max} = \frac{2 - 0.4^{B/b}}{2 - 0.4^{l/B}} \left\{1 + \frac{0.926}{1.348 - 0.826^{20r/B}}\left[0.577 - \left(\frac{b}{B} - 0.24\right)^2\right]\right\}$$

It is of importance that the maximum stress occurs always about 5–10 degrees around the corner and the zero stress 50–70 degrees round. Butts in plating should be made at this latter point. Figure 4.41 shows the results for a hole with $l = B$. Note that the concentration factor is referred not to the stress in the clear plate but to the stress at the reduced section.

4.1.3.2 Built-in stress concentrations

The violent treatment afforded a plate of mild steel during its manufacture, prevents the formation of a totally unstrained plate. Uneven rolling or contraction, especially if cooling is rapid, may cause areas of strain, even before the plate is selected for working into a ship. These are often called 'built in' or 'locked up' stresses (more accurately, strains). Furthermore, during the processes of moulding and welding, more strains are built in by uneven cooling. However careful the

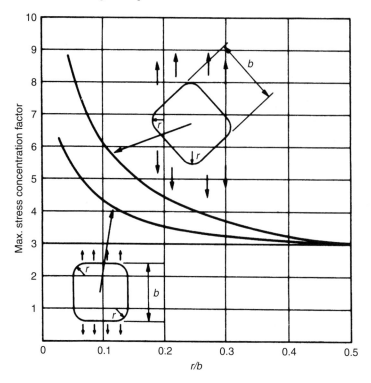

Figure 4.39

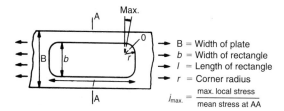

Figure 4.40 Rectangular hole in finite plate.

welders, there will be some, perhaps minute, holes, cracks and slag inclusions, lack of penetration and undercutting in a weld deposit.

Inspection of important parts of the structure will minimize the number of visual defects occurring, and radiography can show up those below the surface, but lack of homogeneity cannot be observed by normal inspection procedures, even though the 'built-in' stress may exceed yield. What then happens to them when the ship is subjected to strain?

In ductile materials, most of the concentrations 'yield out', i.e. the concentration reaches the yield point, shirks further load and causes a re-distribution of stress in the surrounding material. If the concentration is a crack, it may propagate to an area of reduced stress level and stop. If it becomes visible, a temporary repair is often made at sea by drilling a circular hole at its end, reducing the concentration factor. This is a common first aid treatment. There is considerably more anxiety if the material is not so ductile as mild steel, since it does not have so much capacity for 'yielding out'; furthermore, a high yield steel is often employed in places where a general high stress level is expected so that cracks are less able to propagate to areas of reduced stress level and stop. A further anxiety in all materials is the possibility of fatigue, since concentrations at which there is, locally, a high stress level will be able to withstand few reversals.

4.1.3.3 Crack extension, brittle fracture

Cracks then, cannot be prevented but can be minimized. It is important that they are observed and rectified before they cause catastrophic failure. They can extend under the action of fatigue or due to brittle fracture. Even in heavy storms fatigue cracks are only likely to increase in length at a rate measured in mm/s. A brittle fracture, however, can propogate at around 500 m/s. Thus brittle fracture is of much greater concern. The loss of Liberty ships has already been mentioned. More recent examples have been the MV *Kurdistan* which broke in two in 1979 and the MV *Tyne Bridge* which experienced a 4 m long crack in 1982, Corlett *et al.*

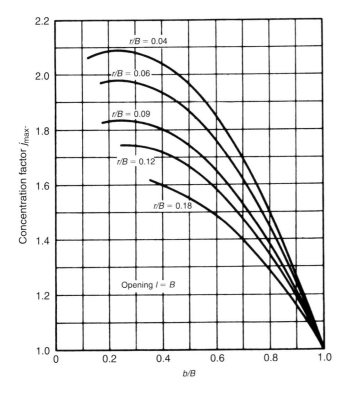

Figure 4.41 Stress concentration factor for a hole with $l = B$.

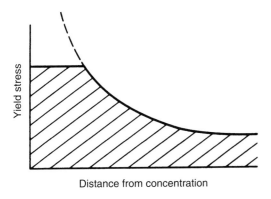

Figure 4.42 Yielding out.

(1988) and Department of Transport (1988). Some RN frigates damaged in collision in the 'Cod War' in the 1970s exhibited brittle fracture showing that thin plates are not necessarily exempt from this type of failure as had been generally thought up to that time.

The critical factors in determining whether brittle fracture will occur are stress level, length of crack and material toughness, this last being dependent upon temperature and strain rate. The stress level includes the effects of stress concentrations and residual stresses due to the fabrication processes. The latter are difficult to establish but, as an illustration, in frigates a compressive stress of about 50 MPa is introduced in hull plating by welding the longitudinals, balanced by local regions in the vicinity of the weld where tensile stresses are at yield point.

At low temperatures fracture of structural steels and welds is by cleavage. Once the threshold toughness for crack initiation is exceeded, the energy required for crack extension is so low that it can be provided by the release of stored elastic energy in the system. Unless fracture initiation is avoided structural failure is catastrophic. At higher temperatures fracture initiation is by growth and coalescence of voids. Subsequent crack extension is sustained only by increased load or displacement. The temperature marking the transition in fracture mode is termed the transition temperature. This temperature is a function of loading rate, structural thickness, notch acuity and material microstructure.

Ideally one would like a simple test that would show whether a steel would behave in a 'notch ductile' manner at a given temperature and stress level. This does not exist because the behaviour of the steel depends upon the geometry and method of loading. For instance, cleavage fracture is favoured by high triaxial stresses and these are promoted by increasing plate thickness. The choice then, is between a simple test like the Charpy test (used extensively in quality control) or a more expensive test which attempts to create more representative conditions (e.g. the Wells Wide Plate or Robertson Crack Arrest tests). More recently the development of linear elastic fracture mechanics based on stress intensity factor, K, has been followed by usable elastic-plastic methodologies based on crack tip opening displacement. CTOD or δ, and the J contour integral, has in principle made it possible to combine the virtues of both types of test in one procedure.

For a through thickness crack of length $2a$ subject to an area of uniform stress, σ, remote from stress concentration the elastic stress intensity factor is given by:

$$K = \sigma(\pi a)^{1/2}$$

The value at which fracture occurs is K_c and it has been proposed that $K_c = 125\,\text{MPa}(\text{m})^{1/2}$ would provide a high assurance that brittle fracture initiation could be avoided. A fracture parameter, J_c, can be viewed as extending K_c into the elastic–plastic regime, with results presented in terms of KJ_c which has the same units as K_c. Approximate equivalents are:

$$KJ_c = [J_c E]^{1/2} = [2\delta_c \sigma_Y E]^{1/2}$$

It would be unwise to assume that cracks will never be initiated in a steel structure. For example, a running crack may emerge from a weld or heat affected zone unless the crack initiation toughness of the weld procedures meets that of the parent plate. It is prudent, therefore, to use steels which have the ability to arrest cracks. It is recommended that a crack arrest toughness of the material of between 150 and 200 MPa $(\text{m})^{1/2}$ provides a level of crack arrest performance to cover most situations of interest in ship structures.

Recommendations are:

(a) To provide a high level of assurance that brittle fracture will not initiate, and a steel with a Charpy crystallinity less than 70% at 0 °C be chosen.
(b) To provide a high level of crack arrest capability together with virtually guaranteed fracture initiation avoidance, a steel with Charpy crystallinity less than 50% at 0°C be chosen.
(c) For crack arrest strakes a steel with 100% fibrous Charpy fracture appearance at 0°C be chosen.

If the ship is to operate in ice, or must be capable of withstanding shock or collision without excessive damage, steels with higher toughness would be appropriate.

4.1.3.4 Fatigue

Provided ships are inspected regularly for cracking, the relatively slow rate of fatigue crack growth means that fatigue is not a cause for major concern in relation to ship safety. If, however, cracks go undetected their rate of growth will increase as they become larger and they may reach a size that triggers brittle fracture. Also water entering, or oil leaking, through cracks can cause problems and repair can be costly. Fatigue, then, is of concern particularly as most cracks occurring in ship structures are likely to be fatigue related. It is also important to remember that fatigue behaviour is not significantly affected by the yield strength of the steel. The introduction of higher strength steels and acceptance of higher nominal stress levels (besides the greater difficulty of welding these steels) means that fatigue may become more prevalent. Thus it is important that fatigue is taken into account in design as far as is possible.

Design for fatigue is not easy—some would say impossible. However, there are certain steps a designer can, and should, take. Experience, and considerable testing, show that incorrect design of detail is the main cause of cracking. The situation may be summarised by saying that design for fatigue is a matter of detail design and especially a matter of design of welded connections, Dover et al. (2001). Methods used rely very heavily on experimental data. The most common to date has been one using the concept of a nominal stress. Typically for steel the fatigue characteristics are given by a log/log plot of stress range against number of cycles to failure. This S–N curve as it is termed takes the form of a straight line with life increasing with decreasing stress range until a value below which the metal does not fatigue. As a complication there is some evidence that in a corrosive atmosphere there is no lower limit. However, in laboratories, tests of welded joints lead to a series of S–N lines of common slope. The various joints are classified by number, the number being the stress range (N/mm^2) at 20 million cycles based on a mean test value less two standard deviations which corresponds to a survival probability of 97.7%. As an example, a cruciform joint, K butt weld with fillet welded ends is in Class 71, Figure 4.43.

These data relate to constant amplitude loading and they are not too sensitive to mean stress level. However, a ship at sea experiences a varying load depending upon the conditions of sea and loading under which it operates. This is usually thought of in terms of a spectrum of loading and a transfer factor must be used to relate the stress range under spectrum loading to the data for constant amplitude. Testing at Hamburg suggests that a transfer factor of 4 is appropriate for the range of notch cases existing in ship structures, assuming 20 million cycles as typical of the average merchant ship life. It also recommends a safety factor of 4/3. For a Class 71 detail this gives a permissible stress range of $71 \times 4 \times 3/4 = 213$. This must be checked against the figures derived from the longitudinal strength calculation.

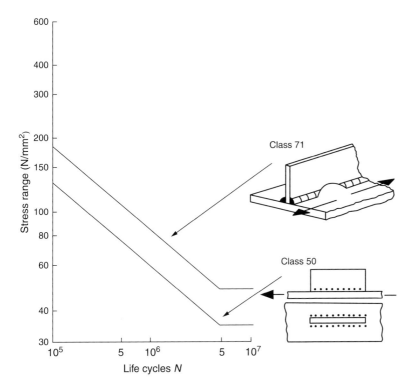

Figure 4.43 S–N curves (based on Hamburg work).

As is discussed in the next section, many structures are now analysed by finite element methods. These are capable of analysing local detail on which fatigue strength depends, Violette and Shenoi (1999), but interpretation of the results is made difficult by the influence of the mesh size used. The smaller the mesh, and the closer one approaches a discontinuity, the higher the stress calculated. The usual 'engineering' solution is to use a relatively coarse mesh and compare the results with the figures accepted in the nominal stress approach described above. The best idea of acceptable mesh size is obtained by testing details which have been analysed by finite element methods and comparing the data for varying mesh size.

4.1.3.5 *Discontinuities in structural design*

In the second category of stress concentrations, are those deliberately introduced by the designer. Theory may assist where practical cases approximate to the assumptions made, as, for example, in the case of a side light or port hole. Here, in a large plate, stress concentration factors of about three may be expected. At hatches too, the effects of corner radii may be judged from Figures 4.39 and 4.41. Finite element techniques outlined in Section 4.2 have extended the degree to which theory may be relied upon to predict the effect of large openings and other major stress concentrations.

Often, the concentrations will be judged unacceptably high and reinforcement must be fitted to reduce the values. One of the most effective ways to do this is by fitting a rim to the hole or curved edge. A thicker insert plate may also help, but the fitting of a doubling plate is unlikely to be effective unless means of creating a good connection are devised.

Designers and practitioners at all levels must be constantly on the watch for the discontinuity, the rapid change of structural pattern and other forms of stress raiser. It is so often bad local design which starts the major failure. Above all, the superposition of one concentration on another must be avoided—but the reinforcement rim where the concentration is lowest, avoid stud welds and fittings at structural discontinuities, give adequate room between holes, grind the profile smooth at the concentration.

4.1.3.6 *Superstructures and deckhouses*

These can constitute severe discontinuities in the ship girder. They can contribute to the longitudinal strength but are unlikely to be fully effective. Their effectiveness can be improved by making them long,

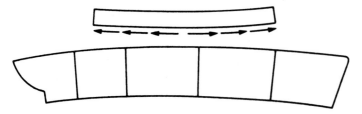

Figure 4.44

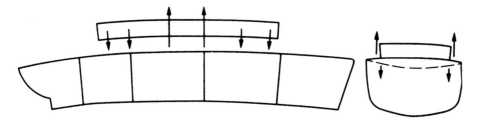

Figure 4.45

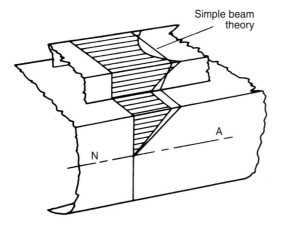

Figure 4.46 Direct stress in a superstructure.

minimizing changes in plan and profile, extending them the full width of the hull and paying careful attention to their connections to the hull.

The only contact between the upper deck and the superstructure is along the bottom of the superstructure sides through which the strains and forces must be transmitted. Because the upper deck stretches, so also must the lower edge of the superstructure sides thus causing shear forces which tend to distort the superstructure into a shape opposite to that of the hull, Figure 4.44. The two are held together, however, and there must be normal forces on the superstructure having the opposite effect. The degree of these normal forces must depend upon the flexibility of the deck beams and main transverse bulkheads if the superstructure is set in from the ship's sides. The effect of the shear forces will depend on the manner in which the shear is diffused into the superstructure, and the shear lag effects are likely to be more appreciable there than in the main hull girder.

If the effects of shear lag are ignored, the results are suitable for the middle portions of long superstructures. This approach embraces the effects of shear diffusion but ignores the concentrated forces from main transverse bulkheads, and it is suitable for short superstructures or for those which extend out to the ship's side. An efficiency of superstructure is defined as:

$$\eta = \frac{\sigma_0 - \sigma}{\sigma_0 - \sigma_1}$$

where σ_0 is the upper deck stress which would occur if there were no superstructure present, σ is the upper deck stress calculated and σ_1 is the upper deck stress with a fully effective superstructure. Curves are supplied from which the factors leading to the efficiency may be calculated. As might be expected, the efficiency depends much on the ratio of superstructure length to its transverse dimensions.

The square ends of the superstructure constitute major discontinuities, and may be expected to cause large stress concentrations. They must not be superimposed on other stress concentrations and should be avoided amidships or, if unavoidable, carefully reinforced. For this reason, expansion joints are to be used with caution; while they relieve the superstructure of some of its stress by reducing its efficiency, they introduce stress concentrations which may more than restore the stress level locally. To assist the normal forces, superstructure ends should coincide with main transverse bulkheads.

Unwanted hull–superstructure interactions can be avoided by using low modulus material in the superstructure such as reinforced plastic which offers tensile and compressive strengths comparable to the yield strength of mild steel with an E-value less than a tenth that of steel. In this case the superstructure will not make any significant contribution to longitudinal strength.

4.2 Structural design and analysis

4.2.1 Introduction

4.2.1.1 *Overview*

The object of this section is to provide an understanding of the structural behaviour of the ship and a recognition of the unit problems. Applied mechanics and mathematics will have provided the tools; now they must be applied to specific problems. The scope and the limitations of different theories must be known if they are to be used with success and the engineer needs to be aware of the various works of reference. Recognition of the problem and knowledge of the existence of a theory suitable for its solution are exceedingly valuable to the practising engineer. There is rarely time to indulge an advanced and elegant theory when a simple approach provides an answer giving an accuracy compatible with the loading or the need. On the other hand, if a simple approach is inadequate, it must be discarded. It is therefore understanding and recognition which this section seeks to provide.

Optimum design is often assumed to mean the minimum weight structure capable of performing the required service. While weight is always significant, cost, ease of fabrication and ease of maintenance are also important, Kuo *et al.* (1984). Cost can increase rapidly if non-standard sections or special quality materials are used; fabrication is more difficult with some materials and, again, machining is expensive. This section discusses methods for assessing the minimum requirements to provide against failure. The actual structure decided upon must reflect all aspects of the problem. Comprehensive reviews of developments in the design and analysis of ship structures are contained in the Proceedings of the meetings of the International Ship and Offshore Structures Congress, such as ISSC (2006) and earlier meetings in 2003 and 2000.

The whole ship girder provides the background and the boundaries for local structural design. The needs of the hull girder for areas at deck, keel and sides must be met. Its breakdown into plating and stiffener must be determined; there is also much structure required which is not associated with longitudinal strength; finally, there are many particular fittings which require individual design. An essential preliminary to the analysis of any structure or fitting, however, is the assessment of the loading and criterion of failure.

4.2.1.2 *Loading and failure*

Because it is partly the sea which causes the loading, some of the difficulties arising in Section 4.1 in defining the loading apply here also. The sea imposes on areas of the ship impact loads which have not yet been extensively measured, although the compilation of a statistical distribution of such loading continues. While more realistic information is steadily coming to hand, a loading which is likely to provide a suitable basis for comparison must be decided upon and used to compare the behaviour of previous successful and unsuccessful elements. For example, in designing a panel of plating in the outer bottom, hydrostatic pressure due to draught might provide a suitable basis of comparison, and examination of previous ships might indicate that when the ratio of permanent set to thickness exceeded a certain percentage of the breadth to thickness ratio, extensive cracking occurred.

Some loads to which parts of the structure are subject are known with some accuracy. Test water pressure applied during building (to tanks for example), often provides maximum loads to which the structure is subject during its life; bulkheads cannot be subjected to a head greater than that to the uppermost watertight deck, unless surging of liquid or shift of cargo is allowed for; forces applied by machinery are generally known with some accuracy; acceleration loads due to ship motion may be known statistically. More precise analytical methods are warranted in these cases.

Having decided on the loading, the next step is to decide on the ultimate behaviour, which, for brevity, will be called failure. From the point of view of structural analysis, there are four possible ways of failing, by (a) direct fracture, (b) fatigue fracture, (c) instability, and (d) unacceptable deformation.

(a) *Direct fracture* may be caused by a part of the structure reaching the ultimate tensile, compressive, shear or crushing strengths. If metallurgical or geometrical factors inhibit an otherwise ductile material, it may fail in a brittle manner before the normally expected ultimate strength. It should be noted, that yield does not by itself cause fracture and cannot therefore be classed as failure in this context;
(b) *Fatigue fracture*. The elastic fatigue lives of test specimens of materials are fairly well documented.

The elastic fatigue lives of complex structures are not divinable except by test, although published works of tests on similar structural items may give a lead. The history of reversals in practical structures at sea also requires definition. Corrosion fatigue is a special case of accelerated failure under fatigue when the material is in a corrosive element such as sea water. The 'bent nail' fatigue is another special case, in which yield is exceeded with each reversal and the material withstands very few reversals;

(c) *Instability.* In a strut, buckling causes an excessive lateral deflection; in a plating panel, it may cause load shirking by the panel or wrinkling; in a cylinder under radial pressure, instability may cause the circumference to corrugate; in a plating stiffener it may cause torsional tripping. Most of these types of instability failure are characterized by a relatively rapid increase in deflection for a small increase in load and would generally be regarded as failure if related to the whole structure; where only part of the structure shirks its load, as sometimes happens, for example, with panels of plating in the hull girder section, overall 'failure' does not necessarily occur.

(d) *Deformation.* A particular deflection may cause a physical foul with machinery or may merely cause alarm to passengers, even though there is no danger. Alignment of machinery may be upset by excessive deflection. Such deformations may be in the elastic or the elasto-plastic range. The stiffness of a structure may cause an excessive amplitude of vibration at a well used frequency. Any of these could also constitute failure.

For each unit of structure in a ship, first the loading must be decided and then the various ways in which it would be judged to have failed must be listed and examined. What are these units of structure?

4.2.1.3 *Structural units of a ship*

There are four basic types of structure with which the ship designer must deal: (a) plating-stiffener combinations, (b) panels of plating, (c) frameworks, (d) fittings.

(a) *Plating-stiffener combinations.* The simplest form of this is a single simple beam attached to a plate. Many parallel beams supporting plating constitute a grillage with unidirectional stiffening. Beams intersecting at right angles constitute an orthogonally stiffened grillage. These various units may be initially flat or curved, loaded in any plane and possess a variety of shapes and boundaries.

(b) *Panels of plating.* These are normally rectangular and supported at the four edges, subject to normal or in-plane loads. Initially, they may be nominally flat or dished.

(c) *Frameworks.* These may be portals of one or more storeys. Frameworks may be constituted by the transverse rings of side frames and deck beams or the longitudinal ring of deck girder, bulkhead stiffeners and longitudinal. They may be circular as in a submarine. Loading may be distributed or concentrated in their planes or normal to their planes.

(d) *Fittings.* There is a great variety of fittings in ships the adequacy of whose strength must be checked. Particular ones include control surfaces such as rudders and hydroplanes, shaft brackets and spectacle plates, masts, derricks, davits and machinery seatings.

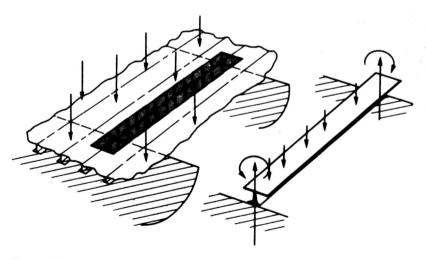

Figure 4.47

4.2.2 Stiffened plating

4.2.2.1 *Simple beams*

Very many of the local strength problems in a ship can be solved adequately by the application of simple beam theory to a single stiffener-plating combination. This is permissible if the boundaries of the unit so isolated are truly represented by forces and moments that adjacent units apply to it. Frequently, when there is a series of similarly loaded units, the influence of adjacent units on the edges will be zero; similarity longitudinally might also indicate that the end slopes are zero. These edge constraints have a large influence and in many cases will not be so easily determined. It is important that the deflection of the supporting structure is negligible compared with the deflection of the isolated beam, if the unit is to be correctly isolated; this is likely to be true if the supports are bulkheads but not if they are orthogonal beams. A summary of results for common problems in simple beams is given in Figure 4.48.

According to the Bernoulli–Euler hypothesis from which the simple theory of bending is deduced, sections plane before bending remain plane afterwards and

$$\sigma = \frac{M}{I/y}$$

For many joists and girders this is very closely accurate. Wide flanged beams and box girders, however, do not obey this law precisely because of the manner in which shear is diffused from the webs into the flanges and across the flanges. The direct stresses resulting from this diffusion do not quite follow this law but vary from these values because sections do not remain plane. Distribution of stress across stiffened plating under bending load is as shown by the wavy lines in Figure 4.49, and this effect is known as the shear lag effect. While the wavy distribution of stress cannot be found without some advanced mathematics, the maximum stress can still be found by simple beam theory if, instead of assuming that all of the plating is partially effective, it is assumed that part of the plating is wholly effective. This effective breadth of flange, λ (Figure 4.49), is used to calculate the effective second moment of area of cross-section. It is dependent on the type of loading and the geometry of the structure. Because it is quite close to the neutral axis, the effective breadth of plating is not very influential and a figure of thirty thicknesses of plating is commonly used and sufficiently accurate; otherwise $\lambda = B/2$ is used.

There remains, in the investigation of the single stiffener-plating combination, the problem of behaviour under end load. Classical Euler theories assume perfect struts and axial loads which never occur in practice. Many designers use these or the Rankine–Gordon formula which embraces the overriding case of yield, together with a factor of safety often as high as twenty or thirty. This is not a satisfactory approach, since it disguises the actual behaviour of the member. Very often, in ship structures there will be a lateral load, which transforms the problem from one of elastic instability into one of bending with end load. Typical solutions appear in Figure 4.48. The lateral load might be sea pressure, wind, concentrated weights, personnel load, cargo or flooding pressure. However, there will, occasionally, arise problems where there is no lateral load in the worst design case. How should the designer proceed then?

Practical structures are always, unintentionally, manufactured with an initial bow, due to welding distortion, their own weight, rough handling or processes of manufacture. It can be shown that the deflection of a strut with an initial simple bow y_0 is given by:

$$y = \frac{P_E}{P_E - P} y_0$$

(y is the total deflection, including y_0).

$P_E/(P_E - P)$ is called the *exaggeration factor*. P_E is the classical Euler collapse load, $P_E = \pi^2 EI/l^2$, l being the effective length, Figure 4.50.

The designer must therefore decide first what initial bow is likely; while some measurements have been taken of these in practical ship structures, the designer will frequently have to make a common-sense estimate. Having decided the value, and calculated the Euler load, the designer can find the maximum deflection from the equation above.

The maximum bending moment for a member in which end rotation is not constrained is, of course,

$$\text{Max BM} = P y_{max}$$

In general, so far as end loading is concerned, the assumption of simple support is safe.

4.2.2.2 *Grillages*

Consider the effect of a concentrated weight W on two simply-supported beams at right angles to each other as shown in Figure 4.51. This is the simplest form of grillage. Assume that the beams, defined by suffixes 1 and 2, intersect each other in the middle and that each is simply supported. What is not immediately obvious is how much the flexure of beam 1 contributes to supporting W and how much is contributed by the flexure of beam 2. Let the division of W at the middles be R_1 and R_2, then

$$R_1 + R_2 = W$$

Problem	Bending moments	Deflections
 (cantilever with end load W)	$BM_A = Wl$	$\delta_B = \dfrac{Wl^3}{3EI}$
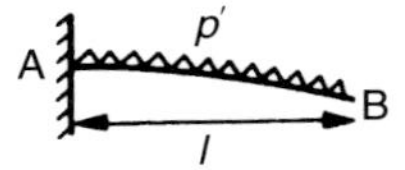 (cantilever with distributed load p')	$BM_A = \dfrac{p'l^2}{2}$	$\delta_B = \dfrac{p'l^4}{8EI}$
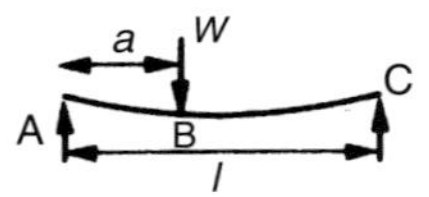 (simply supported beam with point load W at a)	$BM_B = Wa\left(\dfrac{l-a}{l}\right)$	$\delta = \dfrac{Wx}{6EI}\left(\dfrac{l-a}{l}\right)(2al - a^2 - x^2) + \dfrac{W}{6EI}[x-a]^{3*}$ if $a = \dfrac{l}{2}$, $\delta_B = \dfrac{Wl^3}{48EI}$ * Disregard last term if $x < a$.
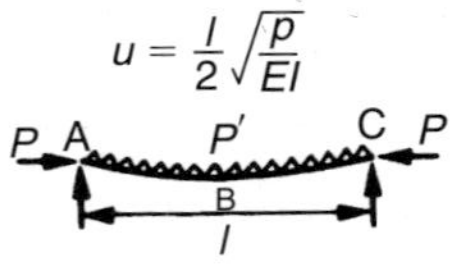 $u = \dfrac{l}{2}\sqrt{\dfrac{P}{EI}}$ (beam with axial load P and distributed load p')	$BM_B = \dfrac{p'l^2}{8} \cdot \dfrac{2(1-\cos u)}{u^2 \cos u}$	$\delta = \dfrac{p'l^2}{4Pu^2}\left\{\dfrac{\cos u(1-2x/l)}{\cos u} - 1\right\} - \dfrac{p'x}{2P}(l-x)$ if $P = 0$, $\delta_B = \dfrac{5p'l^4}{384EI}$

Figure 4.48 Properties of simple beams.

Problem	Bending moments	Deflections
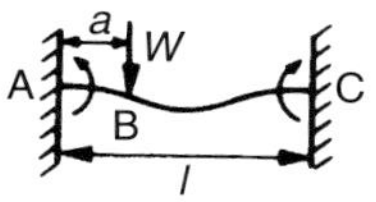 (fixed-pinned beam with point load W at distance a from A, length l)	$BM_C = Wa\left(\dfrac{l-a}{l}\right)^2$ $BM_A = W(l-a)\left(\dfrac{a}{l}\right)^2$ if $a = \dfrac{l}{2}$, $BM_A = BM_C = \dfrac{Wl}{8}$	$\delta_{\max} = \dfrac{2Wa^3}{3EI}\left(\dfrac{l-a}{l+2a}\right)^2$ if $a = \dfrac{l}{2}$, $\delta_B = \dfrac{Wl^3}{192EI}$
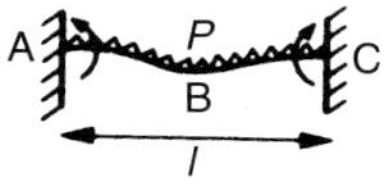 (fixed-fixed beam with distributed load p')	$BM_A = BM_C = \dfrac{p'l^2}{12}$	$\delta = \dfrac{p'x^2(l-x)^2}{24EI}$ $\delta_B = \dfrac{p'l^4}{384EI}$
(beam-column with axial P and transverse W at a; $u = \dfrac{l}{2}\sqrt{\dfrac{P}{EI}}$) 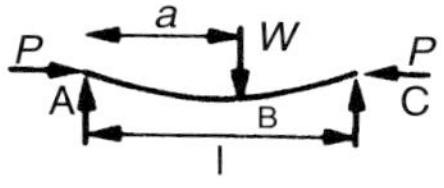	For AB $BM = \dfrac{Wl}{2u}\cdot\dfrac{\sin\dfrac{2ux}{l}\sin 2u\left(1-\dfrac{a}{l}\right)}{\sin 2u}$ max at $x = \dfrac{\pi l}{4u}$ or a For BC $BM = \dfrac{Wl}{2u}\cdot\dfrac{\sin\dfrac{2au}{l}\sin 2u\left(1-\dfrac{x}{l}\right)}{\sin 2u}$ max at $x = l\left(1 - \dfrac{\pi}{4u}\right)$ or a	For AB $\delta = \dfrac{Wl}{2uP}\left\{\dfrac{\sin 2u\left(1-\dfrac{a}{l}\right)\sin\dfrac{2ux}{l}}{\sin 2u} - \dfrac{2xu}{l}\left(1-\dfrac{a}{l}\right)\right\}$ For BC $\delta = \dfrac{Wl}{2uP}\left\{\dfrac{\sin\dfrac{2au}{l}\sin 2u\left(1-\dfrac{x}{l}\right)}{\sin 2u} - \dfrac{2au}{l}\left(1-\dfrac{x}{l}\right)\right\}$

Figure 4.48 (Continued)

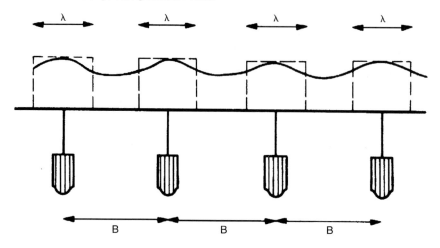

Figure 4.49 Shear lag effects.

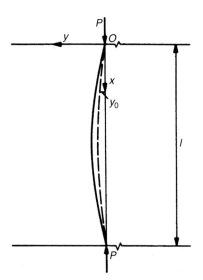

Figure 4.50

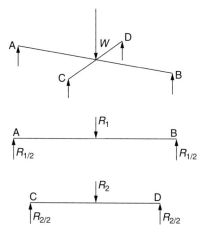

Figure 4.51

Examining each beam separately, the central deflections are given by

$$\delta_1 = \frac{R_1 l_1^3}{48 E I_1}$$

and

$$\delta_2 = \frac{R_2 l_2^3}{48 E I_2}$$

But, if the beams do not part company, $\delta_1 = \delta_2$

$$\therefore \frac{R_1 l_1^3}{I_1} = \frac{R_2 l_2^3}{I_2}$$

Since $R_1 + R_2 = W$,

$$\delta = \frac{W l_1^3}{48 E I_1} \times \frac{1}{\left(1 + \dfrac{I_2 l_1^3}{I_1 l_2^3}\right)}$$

and the maximum bending moments in the two beams are

$$\text{Max BM}_1 = \frac{R_1 l_1}{4} = \frac{W l_1}{4\left(1 + \dfrac{I_2 l_1^3}{I_1 l_2^3}\right)}$$

$$\text{Max BM}_2 = \frac{R_2 l_2}{4} = \frac{W l_2}{4\left(1 + \dfrac{I_1 l_2^3}{I_2 l_1^3}\right)}$$

This has been an exceedingly simple problem to solve. It is not difficult to see, however, that computation of this sort could very quickly become laborious. Three beams in each direction, unaided by symmetry could give rise to nine unknowns solved by nine simultaneous equations. Moreover, a degree of fixity at the edges introduces twelve unknown moments while moments at the intersections cause twist in the orthogonal beams. Edge restraint, uneven spacing of stiffeners, differing stiffeners, contribution from plating, shear deflection and other factors all further contribute to making the problem very difficult indeed.

Mathematical theories have evolved to solve a great range of such problems whose solutions are available through computer programs and, in some cases, by data sheets.

4.2.2.3 Swedged plating

Fabrication costs can be reduced in the construction of surfaces which need not be plane, by omitting stiffeners altogether and creating the necessary flexural rigidity by corrugated or swedged plating. Main transverse bulkheads in a large oil tanker, for example, may have a depth of swedge of 25 cm. Such plating is incorporated into the ship to accept end load in the direction of the swedges as well as lateral loading. It tends to create difficulties of structural discontinuity where the swedge meets conventional stiffeners, for example where the main transverse bulkhead swedges meet longitudinal deck girders.

Buckling of some faces of the plating is possible if the swedges are not properly proportioned and it is this difficulty, together with the shear diffusion in the plating, which makes the application of simple beam theory inadequate. Properly designed, corrugated plating is highly efficient.

4.2.2.4 Comprehensive treatment of stiffened plating

As will be discussed presently, panels or stiffened plating may shirk their duty by buckling so that they do not make their expected contribution to the overall ship's sectional modulus. This shirking, or load shortening, is illustrated in Figure 4.52. This shows that the load shortening depends upon:

(a) the imperfections of the stiffeners in the form of a bow
(b) plating panel slenderness ratio $\beta = (b/t)\sqrt{\sigma_Y/E}$
(c) the ratio of stiffener cross section A_s to the overall cross section $A (A_S/A = 0.2$ average imperfections of stiffener)
(d) the stiffener slenderness ratio

$$\lambda = \frac{l}{\pi k}\sqrt{\frac{\sigma_Y}{E}}$$

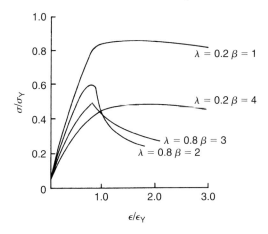

Figure 4.52 Load shortening curves for panels with tee bar stiffeners.

where k is the radius of gyration of the plating/stiffener combination and l is the stiffener length.

Note from Figure 4.52 the sudden nature of the change, occurring especially for high values of λ. A_s/A is also found to be very significant as might be expected. Readers are referred to published papers for a more complete appraisal such as Smith *et al.* (1992) and also for load shortening curves for plating panels alone like that shown in Figure 4.78.

4.2.3 Panels of plating

Knowledge of the behaviour of panels of plating under lateral pressure (sometimes called plate elements) has advanced rapidly. It is important for the designer to understand the actual behaviour of plates under this loading, so that the theory or results most suitable for the application can be selected. To do this, consider how a panel behaves as the pressure is increased.

4.2.3.1 Behaviour of panels under lateral loading

Consider, at first, the behaviour of a rectangular panel with its four edges clamped, and unable to move towards each other. As soon as pressure is applied to one side, elements in the plate develop a flexural resistance much like the elements of a simple beam but in orthogonal directions. Theories relating the pressure to the elastic flexural resistance of the plate alone are called *small deflection theories*. As the pressure is increased and deflection of the same order as the plate thickness results, the resistance of the plate to the pressure stiffens because of the influence of membrane tension. This influence is dependent upon the deflection, since the resistance is due to the resolute of the tension

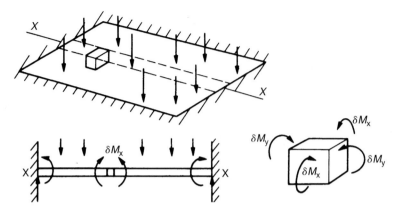

Figure 4.53 Flexural resistance of a plate.

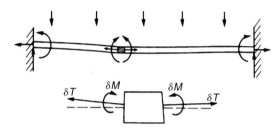

Figure 4.54 Membrane tension effects in a plate.

against the direction of the pressure. Clearly, it has but a small influence when the deflection is small. Figure 4.54 shows a typical section through the plate; the orthogonal section would be similar. Elastic theories which take into account both flexural rigidity and membrane tension are called *large deflection theories*.

With a further increase in pressure, yield sets in, deflection increases more rapidly and the effect of membrane tension becomes predominant. The plate is partly elastic and partly plastic and theories relating to the behaviour following the onset of yield are called *elasto-plastic theories*, Hansen (1996), Jensen (2001). Yield is first reached at the middles of the two longer sides on the pressure side of the plate; soon afterwards, yield is reached on the other side of the plate and this area of plasticity spreads towards the corners. Plasticity spreads with increase in pressure in the stages illustrated in Figure 4.55. The plastic areas, once they have developed through the thickness, are called hinges since they offer constant resistance to rotation. Finally, once the hinges have joined to form a figure of eight, distortion is rapid and the plate distends like a football bladder until the ultimate tensile strength is reached.

The precise pattern of behaviour depends on the dimensions of the panel but this description is typical.

There can be no doubt that after the first onset of yield, a great deal of strength remains. Unless there are good reasons not to do so, the designer would be foolish not to take advantage of this strength to effect an economical design. Once again, this brings us to an examination of 'failure'. As far as a panel under lateral pressure is concerned, failure is likely to be either fatigue fracture or unacceptable deflection. Deflection considered unacceptable for reasons of appearance, to avoid the 'starved horse' look, might nowadays be thought an uneconomical criterion. In considering fatigue fracture, it must be remembered that any yield will cause some permanent set; removal of the load and reapplication of any lesser load will not increase the permanent set and the plate will behave elastically. A plate designed to yield under a load met very rarely, will behave elastically for all of its life save for the one loading which causes the maximum permanent set. Indeed, initial permanent set caused by welding distortion will permit the plate to behave elastically, thereafter, if this is taken as the maximum acceptable permanent set.

In ship's structures, small deflection elastic theories would be used for plates with high fluctuating loading such as those opposite the propeller blades in the outer bottom and where no permanent set can be tolerated in the flat keel, or around sonar domes, for example, in the outer bottom. Large deflection elastic theory is applicable where deflection exceeds about a half the thickness, as is likely in thinner panels. Elasto-plastic theories are appropriate for large areas of the shell, for decks, bulkheads and tanks.

4.2.3.2 *Available results for flat plates under lateral pressure*

The three ranges covered by the different theories are illustrated in Figure 4.56. It is clear that unless the correct theory is chosen, large errors can result.

Ship structures 157

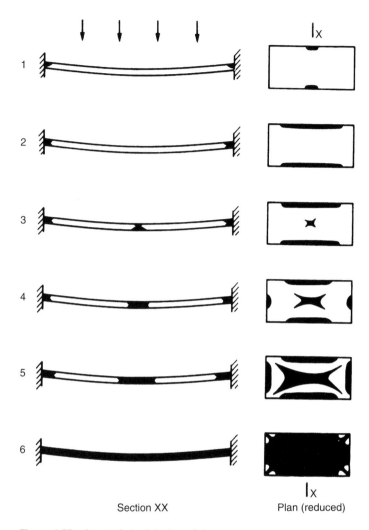

Figure 4.55 Onset of plasticity in a plate.

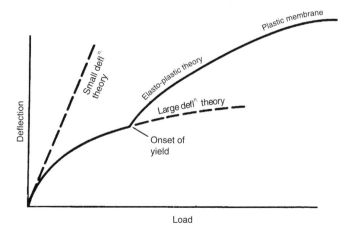

Figure 4.56 Comparison of plate theories.

For plates with totally clamped edges (which are rare):

$$\text{Central deflection} = k_1 \cdot \frac{pb^4}{384D}$$

$$\text{Maximum stress} = k_2 \cdot \frac{p}{2}\left(\frac{b}{t}\right)^2$$

where

$$D = \frac{Et^3}{12(1-\nu^2)}$$

k_1 and k_2 are non-dimensional and the units should be consistent.

n	1	1.25	1.50	1.75	2	∞
k_1	0.486	0.701	0.845	0.928	0.976	1.0
k_2	0.616	0.796	0.906	0.968	0.994	1.0

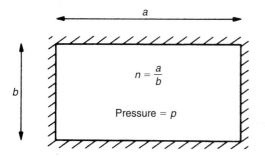

Figure 4.57

Large deflection elastic theory gives results as shown in Figure 4.58.

Elasto-plastic results are based upon the maximum allowable pressure defined, arbitrarily, as the lesser pressure which will cause either

(a) a central plastic hinge in a long plate or 1.25 times the pressure which causes yield at the centre of a square plate; or
(b) the membrane tension to be two-thirds the yield stress.

The first criterion applies to thick plates and the latter to thin plates. Design curves giving maximum permissible pressure, deflection and permanent set based on these criteria are shown in Figure 4.59. Results from an important extension to plate theory, in which pressures have been calculated which will not permit any increase in an initial permanent set, i.e. the plate, after an initial permanent set (caused perhaps by welding) behaves elastically, are presented in Figure 4.60 for long plates.

In considering the real behaviour of panels forming part of a grillage it has been shown that the pull-in at the edges has an appreciable effect on panel behaviour, and that a panel in a grillage does not have the edge constraint necessary to ensure behaviour in the manner of Figure 4.59. The edge constraint which can give rise to membrane tension arises from the hoop effects in the plane of the boundary. Figure 4.61 gives design curves assuming that the edges of the panel are free to move inwards.

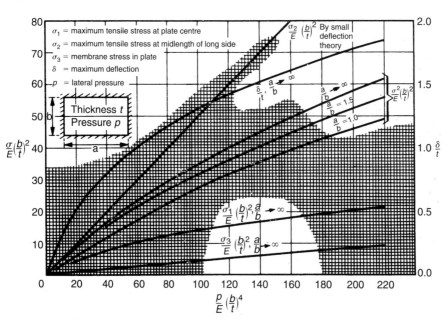

Figure 4.58 Large deflection results.

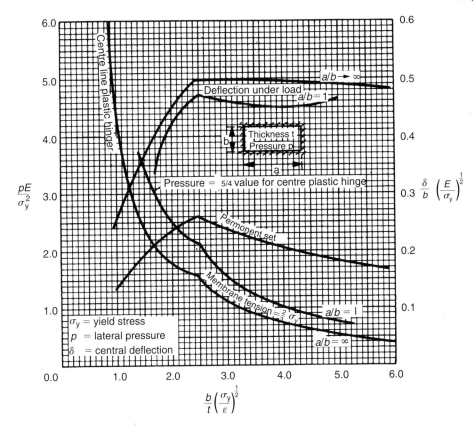

Figure 4.59 Elasto-plastic results.

An example illustrating the effects of different end fixity is given in Rawson and Tupper (2001).

4.2.3.3 Buckling of panels

Buckling of panels in the direction of the applied lateral pressure is known as snap through buckling. This is likely where the initial permanent set is $1\frac{1}{2} - 3$ times the thickness.

Buckling due to edge loading has been dealt with on a theoretical basis. For a panel simply supported at its edges, the critical buckling stress is given by

$$\sigma_c = \frac{k\pi^2 E}{12(1-\nu^2)} \left(\frac{t}{b}\right)^2$$

where k is given by Figure 4.62. When a plate does buckle in this way, it may not be obvious that is has occurred; in fact, the middle part of the panel shirks its load which is thrown on to the edge stiffeners. It is common practice to examine only the buckling behaviour of these stiffeners associated with a width of plating equal to thirty times its thickness.

Buckling due to shear in the plane of the plate causes wrinkling in the plate at about 45 degrees. Such a failure has been observed in the side plating of small ships at the sections of maximum shear. The critical shear stress is given by

$$\tau_c = kE \left(\frac{t}{b}\right)^2$$

where $k = 4.8 + 3.6\,(b/a)^2$ for edges simply supported or $k = 8.1 + 5.1(b/a)^2$ for edges clamped. A more truly representative examination of panel behaviour under biaxial compression and lateral pressure is given in the comprehensive work by Smith *et al.* (1992). The results of this research enables a designer to determine optimum panel shapes and to take into account initial strains and imperfections in the plating.

4.2.4 Frameworks

4.2.4.1 Overview

Analysis of the three-dimensional curved shape of the hull between main bulkheads is the correct approach to the determination of its strength. Finite element techniques can be used for this, but the reduction of the problem to two-dimensional strips or frames

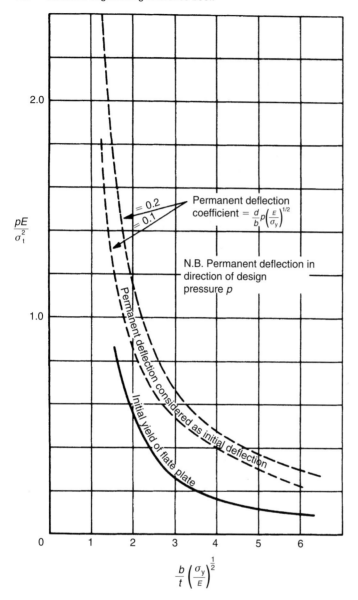

Figure 4.60

remains important. In reducing the problem to two-dimensional frameworks, it is necessary to be aware that approximations are being made.

There are, in general, three types of plane framework with which the ship designer is concerned:

(a) orthogonal portals
(b) ship-shape rings
(c) circular rings.

Portals arise in the consideration of deckhouses, superstructures and similar structures and may have one or more storeys. If the loading in the plane of the portal is concentrated, the effect of the structure perpendicular to the plane of the portal is likely to be one of assistance to its strength. If the loading is spread over many portals, one of which is being isolated for analysis, the effects of the structure perpendicular to its plane will be small, from considerations of symmetry unless there is sideway when the in-plane stiffness of the plating will be appreciable. Thus, the reduction of the problems to two-dimensional frameworks is, in general, pessimistic and safe, although each problem should be examined on its merits.

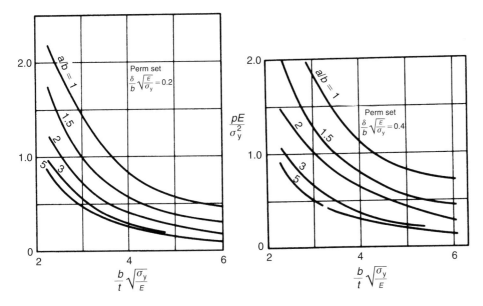

Figure 4.61 Design curves for panels with edges free to move inwards.

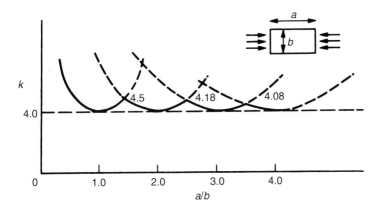

Figure 4.62

Ship-shape rings arise by isolating a transverse slice of the ship, comprising bottom structure, side frames and deck beams, together with their associated plating. Treatment of the complete curved shell in this manner is likely to be highly pessimistic because the effect of longitudinal structure in keeping this ring to shape must be considerable. These longitudinals connect the ring to transverse bulkheads with enormous rigidity in their own planes. The calculations are, nevertheless, performed to detect the likely bending moment distribution around the ring so that material may be distributed to meet it. Transverse strength calculations are, therefore, generally comparative in nature except, perhaps, in ships framed predominantly transversely.

Circular rings occur in submarines and other pressure vessels, such as those containing nuclear reactors.

4.2.4.2 Methods of analysis

Standard textbooks on structural analysis treat the more common framed structures met in ship design such as Southwell (1940) and Gere and Timoshenko (1991). A brief summary of four methods of particular use to the naval architect may, however, be worthwhile. These methods are: (a) moment distribution, (b) slope-deflection, (c) energy methods, and (d) limit design methods.

(a) The *moment distribution* of Hardy Cross is particularly suitable for portal problems where members are straight and perpendicular to each other. Bending moment distribution is obtained very readily by this method but slopes and deflection are not obtained without the somewhat more general relaxation methods of R. V. Southwell.

The moment distribution process is one of iteration in methodical sequence as follows:

(i) all joints of the framework are assumed frozen in space, the loading affecting each beam as if it were totally encastre, with end fixing moments;
(ii) one joint is allowed to rotate, the total moment at the joint being distributed amongst all the members forming the joint according to the formula $(I/l)/(\Sigma I/l)$; the application of such a moment to a beam causes a carry-over of one-half this value (if encastre at the other end) to the far end, which is part of another joint;
(iii) this joint is then frozen and a half of the applied moment is carried over to each remote end (sometimes, the carry-over factor is less than one-half—indeed, when the remote end is pinned, it is zero);
(iv) the process is repeated at successive joints throughout the framework until the total moments at each joint are in balance.

This process prevents sidesway of the framework which occurs unless there is complete symmetry. This is detected by an out-of-balance moment on the overall framework. As a second cycle of operations therefore, sufficient side force is applied to liquidate this out-of-balance without allowing joint rotation, thus causing new fixing moments at the joints. These are then relaxed by repeating the first cycle of operations and so on. For a more comprehensive treatment, the reader is referred to standard textbooks on the analysis of frameworks.

(b) *Slope deflection analysis* is based upon the fundamental equation

$$M = EI \frac{d^2 y}{dx^2}$$

Thence, the area of the M/EI-curve, $= \int \frac{M}{EI} dx$, gives the change of slope, dy/dx. Integrated between two points in a beam, $\int \frac{M}{EI} dx$ gives the difference in the slopes of the tangents at the two points. Further, $\int \frac{Mx}{EI} dx$ between A and B, i.e. the moment of the M/EI-diagram about a point A gives the distance AD between the tangent to B and the deflected shape as shown in Figure 4.63. These two properties of the M/EI-diagram are used to determine the distribution of bending moment round a framework.

Consider the application of the second principle to a single beam AB subjected to an external loading and end fixing moments M_{AB} and M_{BA} at which the slopes are θ_{AB} and θ_{BA}, positive in the direction shown in Figure 4.63.

Let the first moment of the free bending moment diagram (i.e. that due to the external loading assuming the ends to be pinned) about the ends A and B be respectively m_A and m_B. Then, taking moments of the M/EI-diagram about A,

$$\frac{M_{AB}}{EI} \frac{l^2}{2} - \frac{(M_{AB} + M_{BA})}{2EI} \frac{2l^2}{3} + \frac{m_A}{EI} = -l\theta_{BA} + \delta$$

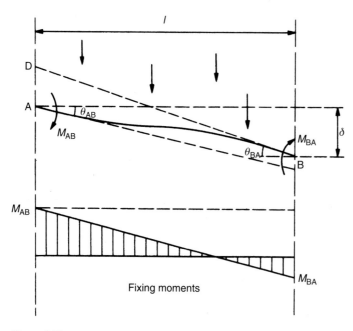

Figure 4.63

i.e.

$$\theta_{BA} = -\frac{l}{6EI}\left(M_{AB} - 2M_{BA} + \frac{6m_A}{l^2}\right) + \frac{\delta}{l}$$

Taking moments about B,

$$\frac{M_{AB}}{EI}\frac{l^2}{2} - \frac{(M_{AB} + M_{BA})}{2EI}\frac{l^2}{3} + \frac{m_B}{EI} = l\theta_{AB} - \delta$$

i.e.

$$\theta_{AB} = \frac{l}{6EI}\left(2M_{AB} - M_{BA} + \frac{6m_B}{l^2}\right) + \frac{\delta}{l}$$

These expressions are fundamental to the slope deflection analysis of frameworks. They may be applied, for example, to the simple portal ABCD of Figure 4.64, expressions being obtained for the six slopes, two of which will be zero and two pairs of which (if C and B are rigid joints) will be equated. On eliminating all these slopes there will remain five equations from which the five unknowns, M_{AB}, M_{BA}, M_{CD}, M_{DC} and δ can be found. This method has the advantage over moment distribution of supplying distortions, but it becomes arithmetically difficult when there are several bays. It is suitable for computation by computer where repetitive calculations render a program worth writing. Sign conventions are important.

(c) The *energy method* most useful in the context of this chapter is based on a theorem of Castigliano. This states that the partial derivative of the total strain energy U with respect to each applied load is equal to the displacement of the structure at the point of application in the direction of the load:

$$\frac{\partial U}{\partial P} = \delta_p$$

The expressions for strain energies U due to direct load, pure bending, torsion and shear are:

Direct load P, member of cross-sectional area A,

$$U = \int \frac{P^2}{2AE}\,dx$$

Bending moment M, curved beam of second moment I,

$$U = \int \frac{M^2}{2EI}\,ds$$

Torque T, member of polar second moment J,

$$U = \int \frac{T^2}{2CJ}\,dx$$

Shearing force S, element of cross-sectional area A (Figure 4.65).

$$U = \int \frac{S^2}{2AC}\,dx = \int \frac{\tau^2 A\,dx}{2C}$$

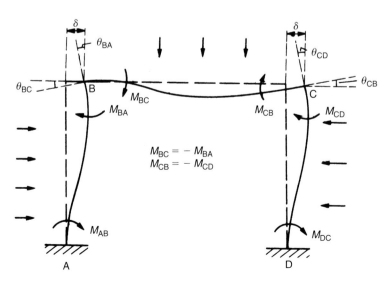

Figure 4.64

The strain energy in a cantilever of rectangular cross-section $a \times b$, for example, with an end load W is given by

$$U = \int_0^l \frac{M^2}{2EI} dx + \int_0^{ab} \int_0^l \frac{\tau^2}{2C} dA dx$$

Now shear stress τ varies over a cross-section according to the expression $\tau = (SA/Ib)\bar{y}$. For the rectangular cantilever then,

$$\tau = \frac{6W}{a^3 b}\left(\frac{a^2}{4} - y^2\right) \text{ and } dA = b dy.$$

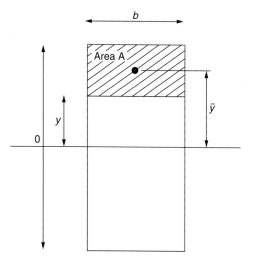

Figure 4.65

Then

$$U = \int_0^l \frac{W^2 x^2 dx}{2EI} + 2\int_0^{a/2} \int_0^l \frac{18W^2}{a^6 bC}\left(\frac{a^4}{16} - \frac{a^2 y^2}{2} + y^4\right) dy dx$$
$$= \frac{W^2 l^3}{6EI} + \frac{3W^2 l}{5abC}$$

End deflection

$$\delta = \frac{\partial U}{\partial W} = \frac{Wl^3}{3EI} + \frac{6Wl}{5abC}$$

The first expression is the bending deflection as given in Figure 4.48, and the second expression is the shear deflection.

In applying strain energy theorems to the ring frameworks found in ship structures, it is common to ignore the effects of shear and direct load which are small in comparison with those due to bending. Confining attention to bending effects, the generalized expression becomes:

$$\delta = \frac{\partial U}{\partial P} = \int \frac{M}{EI} \frac{\partial M}{\partial P} ds$$

Applying this, by example, to a simple ship-shape ring with a rigid centre line bulkhead, Figure 4.66, which can be replaced by three unknown forces and moments, these can be found from the three expressions, since all displacements at B are zero:

$$0 = \frac{\partial U}{\partial H} = \int \frac{M}{EI} \frac{\partial M}{\partial H} ds$$
$$0 = \frac{\partial U}{\partial V} = \int \frac{M}{EI} \frac{\partial M}{\partial V} ds$$
$$0 = \frac{\partial U}{\partial M} = \int \frac{M}{EI} \frac{\partial M}{\partial M_B} ds,$$

summed for members BC and CDE.

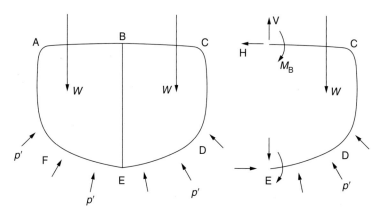

Figure 4.66

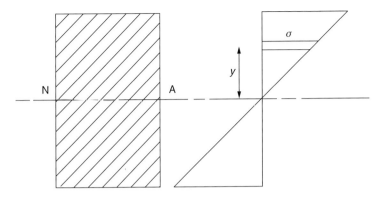

Figure 4.67

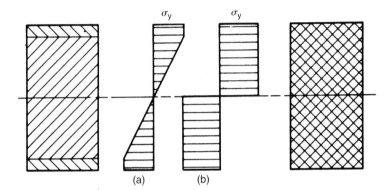

Figure 4.68

Expressions for the bending moment M in terms of V, H, M_B and the applied load can be written down for members BC and CDE, added, and the above expressions determined. Hence V, H and M_B can be found and the bending moment diagram drawn. This method has the advantage that it deals readily with frames of varying inertia, the integrations being carried out by Simpson's rule. Like the slope deflection method, the arithmetic can become formidable.

(d) The *limit design method* is called also plastic design or collapse design. This method uses knowledge of the behaviour of a ductile material in bending beyond the yield point. A beam bent by end couples M within the elastic limit has a cross-section in which the stress is proportional to the distance from the neutral axis,

$$M = \sigma Z \text{ where } Z = \frac{I}{y}, \text{ the section modulus.}$$

Bent further, yield is reached first in the outer fibres and spreads until the whole cross-section has yielded, when the plastic moment, $M_p = a_y S$ where S is the addition of the first moments of area of each side about the neutral axis and is called the plastic modulus. The ratio S/Z is called the *shape factor* which has a value 1.5 for a rectangle, about 1.2 for a rolled steel section and about 1.3 for a plating-stiffener combination. When a beam has yielded across its section, its moment of resistance is constant and the beam is said to have formed a plastic hinge.

The least load which forms sufficient plastic hinges in a framework to transform it into a mechanism is called the collapse load. A portal, for example, which has formed plastic hinges as shown in Figure 4.69 has become a mechanism and has 'collapsed'. Let us apply this principle to an encastre beam (Figure 4.70).

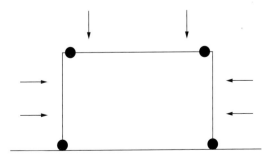

Figure 4.69

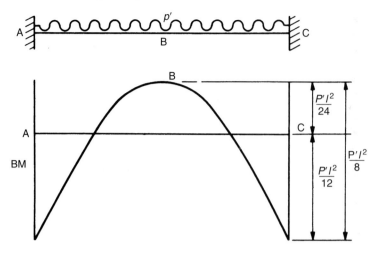

Figure 4.70

If yield is assumed to represent 'failure' by an elastic method of design, the maximum uniformly distributed load that the beam can withstand is (from Figure 4.48).

$$p' = \frac{12Z\sigma_y}{l^2}$$

Using the definition of failure for the plastic method of design, however, collapse occurs when hinges occur at A, B and C, and they will all be equal to $\frac{1}{2}p'(l^2/8)$, i.e.

$$p' = \frac{16S\sigma_y}{l^2}$$

If the shape factor for the beam section is taken as 1.2, the ratio of these two maximum carrying loads is 1.6, i.e. 60% more load can be carried after the onset of yield before the beam collapses.

The plastic design method is often more conveniently applied through the principle of virtual work whence the distance through which an applied load moves is equated to the work done in rotating a plastic hinge. In the simple case illustrated in Figure 4.71, for example,

$$W\frac{l\theta}{2} = M_p\theta + M_p\theta + M_p(2\theta) = 4M_p\theta$$

i.e. collapse load, $W = 8M_p/l$. This method can be used for finding the collapse loads of grillages under concentrated load; in this case, various patterns of plastic hinges may have to be tried in order to find the least load which would cause a mechanism (Figure 4.72).

Knowledge of the collapse loads of superstructures under nuclear blast or extreme winds is necessary, and the limit design method is the only way of calculating them. The method is suitable for the determination of behaviour of other parts of a ship structure subject to

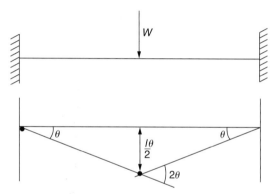

Figure 4.71

once-in-a-lifetime extreme loads such as bulkheads. Use of the method with a known factor of safety (or load factor), can ensure normal behaviour in the elastic range and exceptional behaviour in the plastic range—it is indeed, the only method which illustrates the real load factor over working load.

4.2.4.3 Elastic stability of a frame

The type of elastic instability of major concern to the designer of plated framed structures in a ship is that causing tripping, i.e. the torsional collapse of a stiffener sideways when the plating is under lateral load. Tripping is more likely,

(a) with unsymmetrical stiffener sections,
(b) with increasing curvature,
(c) with the free flange of the stiffener in compression, rather than in tension,
(d) at positions of maximum bending moment, especially in way of concentrated loads.

Recommended spacing, l, for tripping brackets is summarized in Figure 4.73 for straight tee stiffeners and for curved tee stiffeners for which R/W is greater than 70.

For curved stiffeners for which R/W is less than 70, Figure 4.73 may be used by putting $l' = Rl/70W$. For straight unsymmetrical stiffeners l should not exceed $8W$. For curved unsymmetrical stiffeners for which R is more than $4f^2h/t^2$ (unsymmetrical stiffeners should not otherwise be used), $l = (RW/5)^{\frac{1}{2}}$ if the flange is in compression and $(2RW/5)^{\frac{1}{2}}$ if in tension.

The elastic stability of a circular ring frame under radial and end compression is the basis of important investigations in the design of submarines. Theoretically a perfectly circular ring under radial compression will collapse in a number of circumferential corrugations or lobes. Under additional end loading, a ring stiffened cylinder may collapse by longitudinal corrugation; this load, too, alters the number of circumferential lobes due to radial pressure. Elastic instability of the whole ring stiffened cylinder between bulkheads is also possible. Finally, built-in distortions in a practical structure have an important effect on the type of collapse and the magnitude of the collapse loads. These problems involve lengthy mathematics and will not be pursued. The government of the diving depth of the submarine by consideration of such elastic stability should, nevertheless, be understood.

4.2.4.4 End constraint

The degree of rotational end constraint has more effect on deflections than it has on stresses. End

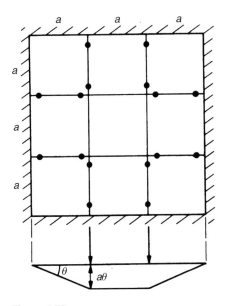

Figure 4.72

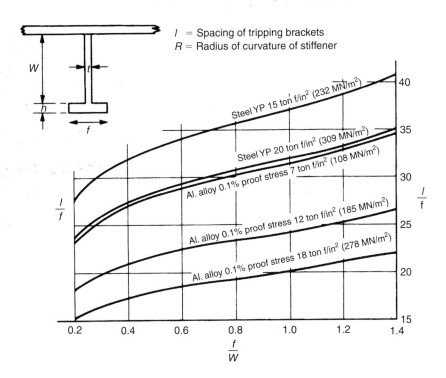

Figure 4.73 Spacing of tripping brackets.

constraints in practical ship structures approach the clamped condition for flexural considerations, provided that the stiffeners are properly continuous at the joint. The degree of rotational end constraint of a member is due to

(a) the stiffness of the joint itself. It is relatively simple to produce a joint which can develop the full plastic moment of the strongest of the members entering the joint; no more is necessary;
(b) the effects of the other members entering the joint. These can be calculated to give the actual rotational stiffness pertaining to the member.

A square joint can provide entirely adequate stiffness. Brackets may be introduced to cheapen fabrication and they also reduce, slightly, the effective span of the member. The reduction in span (Figure 4.74) is

$$b' = \frac{b}{1 + d/B}$$

An example illustrating different analysis methods for frameworks is given in Rawson and Tupper (2001).

4.2.5 Finite element analysis (FEA)

The displacement δ of a simple spring subject to a pull p at one end is given by $p = k\delta$ where k is the stiffness. Alternatively, $\delta = fp$ where f is the flexibility and $f = k^{-1}$.

If the forces and displacements are not in the line of the spring or structural member but are related to a set of Cartesian co-ordinates, the stiffness will differ in the three directions and, in general

$$\mathbf{p}_1 = \mathbf{k}_{11}\delta_1 + \mathbf{k}_{12}\delta_2$$
and $$\mathbf{p}_2 = \mathbf{k}_{21}\delta_1 + \mathbf{k}_{22}\delta_2$$

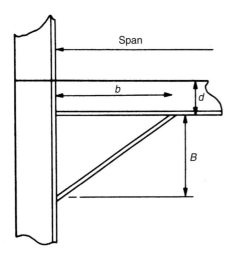

Figure 4.74

This pair of equations is written in the language of matrix algebra

$$\mathbf{P} = \mathbf{Kd}$$

$\mathbf{P}$ is the complete set of applied loads and $\mathbf{d}$ the resulting displacements. $\mathbf{K}$ is called the stiffness matrix and is formed of such factors $\mathbf{k}_{11}$ which are called member stiffness matrices (or sub-matrices). For example, examine a simple member subject to loads p_X and p_Y and moments m at each end causing displacements δ_X, δ_Y and θ.

For equilibrium,

$$m_1 + m_2 + p_{Y_2}l = 0 = m_1 + m_2 - p_{Y_1}l$$

also $p_{X_1} + p_{X_2} = 0$

For elasticity,

$$p_{X_1} = -p_{X_2} = \frac{EA}{l}\left(\delta_{X_1} - \delta_{X_2}\right)$$

From the slope deflection analysis discussed earlier can be obtained:

$$m_1 = \frac{6EI}{l^2}\delta_{Y_1} + \frac{4EI}{l}\theta_1 - \frac{6EI}{l^2}\delta_{Y_2} + \frac{2EI}{l}\theta_2$$

These equations may be arranged

$$\begin{bmatrix} p_{X_1} \\ p_{Y_1} \\ m_1 \end{bmatrix} = \begin{bmatrix} \frac{EA}{l} & 0 & 0 \\ 0 & \frac{12EI}{l^3} & \frac{6EI}{l^2} \\ 0 & \frac{6EI}{l^2} & \frac{4EI}{l} \end{bmatrix} \begin{bmatrix} \delta_{X_1} \\ \delta_{Y_1} \\ \theta_1 \end{bmatrix}$$

$$+ \begin{bmatrix} -\frac{EA}{l} & 0 & 0 \\ 0 & -\frac{12EI}{l^3} & \frac{6EI}{l^2} \\ 0 & -\frac{6EI}{l^2} & \frac{2EI}{l} \end{bmatrix} \begin{bmatrix} \delta_{X_2} \\ \delta_{Y_2} \\ \theta_2 \end{bmatrix}$$

i.e. $\quad \mathbf{p} = \mathbf{k}_{11}\delta_1 + \mathbf{k}_{12}\delta_2$

This very simple example is sufficient to show that a unit problem can be expressed in matrix form. It also suggests that we are able to adopt the very powerful mathematics of matrix algebra to solve structural problems which would otherwise become impossibly complex to handle. Furthermore, computers can be quite readily programmed to deal with matrices. A fundamental problem is concerned with the inversion of the matrix to discover the displacements arising from applied loads, viz.

$$\mathbf{d} = \mathbf{K}^{-1}\mathbf{P}$$

Now strains are related to displacements,

ε = Bd

For plane strain for example,

$$\boldsymbol{\epsilon} = \begin{bmatrix} \varepsilon_x \\ \varepsilon_y \\ \gamma_{zy} \end{bmatrix} = \begin{bmatrix} \dfrac{\partial u}{\partial x} \\ \dfrac{\partial v}{\partial y} \\ \dfrac{\partial u}{\partial y} + \dfrac{\partial v}{\partial x} \end{bmatrix} = \mathbf{Bd}$$

and stress is related to strain,

σ = Dε

D is a matrix of elastic constants which, for plane stress in an isotropic material is

$$D = \dfrac{E}{1-v^2}\begin{bmatrix} 1 & v & 0 \\ v & 1 & 0 \\ 0 & 0 & \dfrac{1}{2}(1-v) \end{bmatrix}$$

There are other relationships which are valuable such as the transformation matrix which changes reference axes. These together form the set of tools required for the solution of structural problems using finite element techniques. This brief description can do little more than explain the concept and the reader should examine standard textbooks. Finite element analysis is approached broadly as follows:

(a) The structure is divided up by imaginary lines meeting at nodes, forming finite elements which are often triangular or rectangular and plane (but may be irregular and three-dimensional).
(b) For each element, a displacement function is derived which relates the displacements at any point within the element to the displacements at the nodes. From the displacements strains are found and from the strains, stresses are derived.
(c) Forces at each node are determined equivalent to the forces along the boundaries of the element.
(d) Displacements of elements are rendered compatible with their neighbour's (this is not often totally possible).
(e) The whole array of applied loads and internal forces are arranged to be in equilibrium.

It is not within the scope of this book to describe how this analysis is carried out. It requires a good knowledge of the shorthand of matrix algebra and draws upon the work described in this chapter concerning various unit problems of beams, panels, grillages and frameworks and also the concepts of relaxation techniques and minimum strain energy.

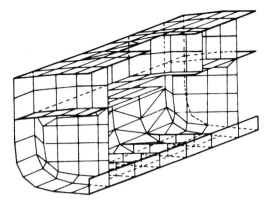

Figure 4.75

It places in the hands of the structural analyst, a tool of enormous power and flexibility. There is no longer any need to make the assumption of simple beam theory for the longitudinal strength of the ship; the ship girder may now be built up from finite elements (Figure 4.75) and the effects of the loads applied by sea and gravity determined. Indeed, this is now the basis for the massive suites of computer programs available for the analysis of total ship structure. The effects of the sea spectra are translated by strip theory into loads of varying probability and the effects of those loadings upon a defined structure are determined by finite element analysis. It is not yet a perfect tool. Moreover, it is a tool of analysis and not of design which is often best initiated by approximate and cheaper methods before embarking upon the expense of these programs. Detailed descriptions of FEA can be found in textbooks such as Rao (2005) and Zienkiewicz et al. (2005).

4.2.6 Realistic assessment of structural elements

The division of the ship into small elements which are amenable to the types of analysis presented earlier in this chapter remains useful as a rough check upon more advanced methods. There is now becoming available a large stock of data and analytical methods which do not have to adopt some of the simplifying assumptions that have been necessary up to now. This is due in large measure to the widespread use of finite element techniques and the computer programs written for them, e.g. ANSYS (2008). Experimental work has carefully sifted the important parameters and relevant assumptions from the unimportant, so that the data sheets may present the solutions in realistic forms most useful to the structural designer. Once again the designer has to rely on information derived from computer analysis which cannot be checked readily, so that the wise will need to fall back from time to time upon simple analysis of sample elements to give confidence in them.

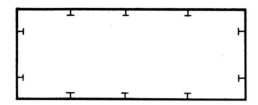

Figure 4.76

As explained in Section 4.1, the elemental behaviour of the whole ship cross section may be integrated to provide a knowledge of the total strength. Judgement remains necessary in deciding what elements should be isolated for individual analysis. This judgement has been assisted by extensive experimentation into box girders of various configurations under end load. Figure 4.76 shows the cross section of a typical specimen.

The wisdom of generations of ship designers has steadily evolved a structure which:

(a) has more cross-sectional area in the stiffener than in the plating;
(b) has longitudinals more closely spaced than transverses;
(c) favours quite deep longitudinals, preferably of symmetrical cross section.

Such structures tend to provide high collapse loads in compression and an efficient use of material. What happens as the load is steadily increased is first a buckling of the centres of panels midway between stiffeners. Shirking of load by the panel centres throws additional load upon each longitudinal which, as load is further increased, will finally buckle in conjunction with the strip of plate to which it is attached. This throws all of the load upon the 'hard' corners which are usually so stiff in compression that they remain straight even after plasticity has set in. It has been found that these hard corners behave like that in conjunction with about a half the panel of plating in each direction. Thus the elements into which the box girder should be divided are plating panels, longitudinals with a strip of plating and hard corners.

Finite element analysis of these elements is able to take account of two factors which were previously the subject of simplifying assumptions. It can account for built-in manufacturing strains and for initial distortion. Data sheets or standard programs are available from software houses for a very wide range of geometry and manufacturing assumptions to give the stress-strain (or more correctly the load shortening curves) for many elements of structure. These can be integrated into total cross-sectional behaviour in the manner described in Section 4.1.

By following the general principles of structural design above—i.e. close deep longitudinals, heavy transverse frames and panels longer than they are wide—two forms of buckling behaviour can usually be avoided:

(a) Overall grillage buckling of all plating and stiffeners together; this is likely only when plating includes most of the material and stiffeners are small.
(b) Tripping of longitudinals by sideways buckling or twisting; this is likely with stiffeners, like flat bars which have low torsional stiffness.

The remaining single stiffener/plating behaviour between transverse frames may now be examined with varying material geometry and imperfections.

Data sheets on the behaviour of panels of plating shown in Figures 4.77 and 4.78 demonstrate firstly the wisdom of the general principles enunciated above and secondly how sensitive to imperfections they are. With moderate initial distortion and average built-in strains some 50–80% of the theoretical yield of a square panel can be achieved (Figure 4.77). Long panels as in transversely framed ships (Figure 4.78) achieve only 10% or so.

Data sheets for stiffened plating combinations are generally of the form shown in Figure 4.79. On the tension side the stress/strain relationship is taken to be of the idealized form for ductile mild steel. In compression, the effects of progressive buckling of flanges is clearly seen. Sometimes the element is able to hold its load-bearing capacity as the strain is increased; in other cases the load-bearing capacity drops off from a peak in a form called catastrophic buckling.

Because the finite element analysis is able to account for the separate panel buckling the width of the associated plating may be taken to the mid panel so that the 30 t assumption is no longer needed. Moreover, many of the elemental data sheets now available are for several longitudinals and plating acting together.

4.2.7 Composite materials

Composite materials are used extensively in many marine applications. An outline of the applications and fabrication methods of composites in the marine field is given in Chapter 9. In the marine field, attention is focused mainly on fibre reinforced plastics (FRP). Details of the methodologies used in structural analysis and design using FRP can be found in Smith (1990), Shenoi and Wellicome (1993) and Kelly and Zweben (2000). Various investigations into the behaviour of FRP can be found in references such as Smith and Chalmers (1987), Clarke *et al.* (1998) and Jeong and Shenoi (2001).

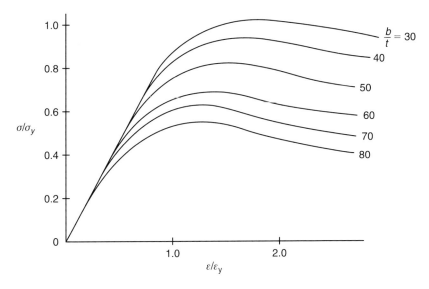

Figure 4.77

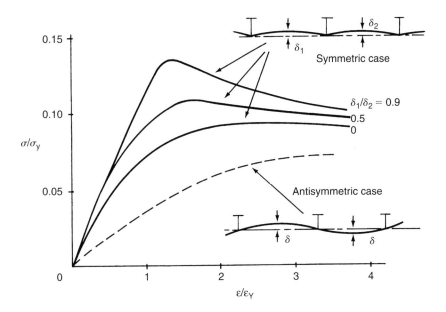

Figure 4.78

4.3 Ship vibration

4.3.1 Overview

Vibrations are dealt with as either *local vibrations* or *main hull vibrations*. The former are concerned with a small part of the structure, perhaps an area of deck. The frequencies are usually higher, and the amplitudes lower, than the main hull vibrations. Because there are so many possibilities and the calculations can be complex they are not usually studied directly during design except where large excitation forces are anticipated. Generally the designer avoids machinery which generate disturbing frequencies close to those of typical ship type structures. Any faults are corrected as a result of trials experience. This is often more economic than carrying out extensive design calculations as the remedy is usually a matter of adding a small amount of additional stiffening.

Main hull vibrations are a different matter. If they do occur the remedial action may be very expensive. They

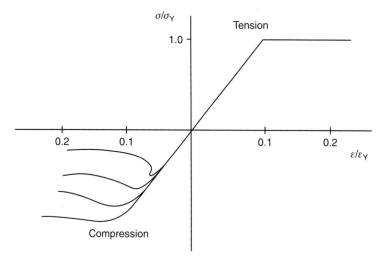

Figure 4.79

must therefore be looked at in design. The hull may bend as a beam or twist like a rod about its longitudinal axis. These two modes of vibration are called *flexural* and *torsional* respectively. Flexing may occur in a vertical or horizontal plane but the vertical flexing is usually the more worrying. Except in lightly structured ships and container ships the torsional mode is not usually too important, see Jensen (2001) and Pedersen (1983).

4.3.2 Flexural vibrations

When flexing in the vertical or horizontal planes the structure has an infinite number of degrees of freedom and the mode of vibration is described by the number of *nodes* which exist in the length. The fundamental mode is the two-node as shown in Figure 4.80.

This yields a displacement at the ends of the ship since there is no rigid support there. This is often referred to as a *free-free* mode and differs from that which would be taken up by a structural beam where there would be zero displacement at one end at least. The next two higher modes have three and four nodes. All are free-free and can occur in both planes. Associated with each mode is a natural frequency of free vibration, the frequency being higher for the higher modes. If the ship were of uniform rigidity and uniform mass distribution along its length and was supported at its ends, the frequencies of the higher modes would be simple multiples of the fundamental. In practice ships differ from this although perhaps not as much as might be expected, as is shown in Table 4.2 (Dieudonne 1959). It will be noted that the greater mass of a loaded ship leads to a reduction in frequency.

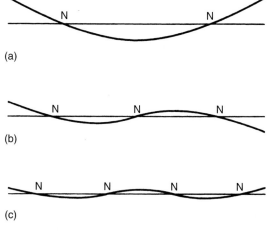

Figure 4.80 (a) Two-node; (b) Three-node; (c) Four-node.

4.3.3 Torsional vibration

In this case the displacement is angular and a one-node mode of vibration is possible. Figure 4.81 shows the first three modes.

4.3.4 Coupling

It is commonly assumed for analysis purposes that the various modes of vibration are independent and can be treated separately. In some circumstances, however, vibrations in one mode can generate

Table 4.2 Typical ship vibration frequencies (cpm).

Ship type	Length (m)	Condition of loading	Frequency of vibration						
			Vertical				Horizontal		
			2 node	3 node	4 node	5 node	2 node	3 node	4 node
Tanker	227	Light	59	121	188	248	103	198	297
		Loaded	52	108	166	220	83	159	238
Passenger ship	136		104	177			155	341	
Cargo ship	85	Light	150	290			230		
		Loaded	135	283			200		
Cargo ship	130	Light	106	210			180	353	
		Loaded	85	168			135	262	
Destroyer	160	Average action	85	180	240		120	200	

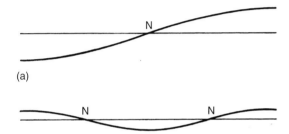

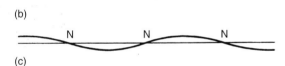

Figure 4.81 (a) One-node; (b) Two-node; (c) Three-node.

vibration in another. In this case the motions are said to be *coupled*. For instance in a ship a horizontal vibration will often excite torsional vibration because of the non-uniform distribution of mass in the vertical plane.

4.3.5 Formulae for ship vibration

The formulae for uniform beams suggests that for the ship an approximation will be given by a formula of the type:

$$\text{Frequency} = \text{Const.} \left(\frac{EI}{Ml^3} \right)^{0.5} \quad (4.1)$$

Suggestions for the value of the constant for different ship types have been made but these can only be very approximate because of the many variables involved in ships. The most important are:

(1) Mass and stiffness distribution along the length.
(2) Departure from ordinary simple theory due to shear deflection and structural discontinuities.
(3) Added mass.
(4) Rotary inertia.

4.3.6 Direct calculation of vibration

Empirical formulae enable a first estimate to be made at the frequency of vibration. The accuracy will depend upon the amount of data available from ships on which to base the coefficients. It is desirable to be able to calculate values directly taking account of the specific ship characteristics and loading. These days a full finite element analysis could be carried out to give the vibration frequencies, including the higher order modes, see Section 4.2.5. Before such methods became available there were two methods used for calculating the two-node frequency:

(1) The *deflection method* or *full integral method*.
(2) The *energy method*.

(1) The deflection method
In this method the ship is represented as a beam vibrating in simple harmonic motion in which, at any moment, the deflection at any position along the length is $y = f(x)\sin pt$. The function $f(x)$ for non-uniform mass and stiffness distribution is unknown but it can be approximated by the curve for the free-free vibration of a uniform beam.

Differentiating y twice with respect to time gives the acceleration at any point as proportional to y and the square of the frequency. This leads to the dynamic loading. Integrating again gives the shear force and another integration gives the bending moment. A double integration of the bending moment curve gives the deflection curve. At each stage the constants of integration can be evaluated from the end conditions. The deflection curve now obtained can be compared with that originally assumed for *f(x)*. If they differ significantly a second approximation can be obtained by using the derived curve as the new input to the calculation.

In using the deflection profile of a uniform beam it must be remembered that the ship's mass is not uniformly distributed, nor is it generally symmetrically distributed about amidships. This means that in carrying out the integrations for shear force and bending moment the curves produced will not close at the ends of the ship. In practice there can be no force or moment at the ends so corrections are needed. A bodily shift of the base line for the shear force curve and a tilt of the bending moment curve are used. See also Section 4.1.2.4.

In the calculation the mass per unit length must allow for the mass of the entrained water using one of the methods described for dealing with added virtual mass, see Section 7.2.11, and Landweber and Macagno (1957), Lewis (1929) and Townsin (1969). The bending theory used ignores shear deflection and rotary inertia effects. Corrections for these are made at the end by applying factors to the calculated frequency.

(2) The energy method
This method uses the principle that, in the absence of damping, the total energy of a vibrating system is constant. Damping exists in any real system but for ships it is acceptable to ignore it for the present purpose. Hence the sum of the kinetic and potential energies is constant.

In a vibrating beam the kinetic energy is that of the moving masses and initially this is assumed to be due to linear motion only. Assuming simple harmonic motion and a mass distribution, the kinetic energy is obtained from the accelerations deduced from an assumed deflection profile and frequency. The potential energy is the strain energy of bending.

When the beam is passing through its equilibrium position the velocity will be a maximum and there will be no bending moment at that instant. All the energy is kinetic. Similarly when at its maximum deflection the energy is entirely potential. Since the total energy is constant the kinetic energy in the one case can be equated to the potential energy in the other.

As in the deflection method the initial deflection profile is taken as that of a uniform bar. As before allowance is made for shear deflection and for rotary inertia. Applying this energy method to the case of the simply supported, uniform section, beam with a concentrated mass M at mid-span and assuming a sinusoidal deflection curve, yields a frequency of:

$$\frac{1}{2\pi}\left(\frac{\pi^4 EI}{2Ml^3}\right)^{0.5} \text{ compared with}$$

$$\frac{1}{2\pi}\left(\frac{48EI}{Ml^3}\right)^{0.5} \text{ for the exact solution.}$$

Since $\pi^4/2$ is 48.7 the two results are in good agreement. This simple example suggests that as long as the correct end conditions are satisfied there is considerable latitude in the choice of the form of the deflection profile.

Calculation of higher modes
It might be expected that the frequencies of higher modes could be obtained by the above methods by assuming the appropriate deflection profile to match the mode needed. Unfortunately, instead of the assumed deflection curve converging to the correct one it tends to diverge with successive iterations. This is due to the profile containing a component of the two-node profile which becomes dominant. Whilst ways have been developed to deal with this, one would today choose to carry out a finite element analysis.

4.3.7 Approximate formulae

It has been seen that the mass and stiffness distributions in the ship are important in deriving vibration frequencies. Such data is not available in the early design stages when the designer needs some idea of the frequencies for the ship. Hence there has always been a need for simple empirical formulae. Schlick (1884) suggested that:

$$\text{Frequency} = \text{Const.}\left(\frac{EI_a}{ML^3}\right)^{0.5} \quad (4.2)$$

where I_a is the moment of inertia of the midship section.

This formula has severe limitations and various authorities have proposed modifications to it.

Burrill (1934–1935) suggested one allowing for added mass and shear deflection.

The frequency was given as:

$$\frac{\text{Const.} \times \left(\dfrac{I}{\Delta L^3}\right)^{0.5}}{\left(1 + \dfrac{B}{2T}\right)^{0.5}(1 + r_s)^{0.5}} \quad (4.3)$$

where r_s is the deflection correction factor.

Todd (1961) adapted Schlick to allow for added mass, the total virtual displacement being given by:

$$\Delta_v = \Delta\left(\frac{B}{3T} + 1.2\right) \quad (4.4)$$

He concluded that I should allow for superstructures in excess of 40% of the ship length. For ships with and without superstructure the results for the two-node vibration generally obeyed the rule:

$$\text{Frequency} = 238\,660\left(\frac{I}{\Delta_v L^3}\right)^{0.5} + 29 \quad (4.5)$$

if I is in m^4, dimensions in m and Δ_v is in MN.

By approximating the value of I, Todd proposed:

$$\text{Frequency} = \text{Const.} \times \left(\frac{BD^3}{\Delta_v L^3}\right)^{0.5} \quad (4.6)$$

where B is breadth and D is depth.

Typical values of the constant in SI units, were found to be

Large tankers (full load)	11 000
Small tankers (full load)	8 150
Cargo ships (60% load)	9 200

Many other approximate formulae have been suggested. The simpler forms are acceptable for comparing ships which are closely similar. The designer must use the data available to obtain the best estimate of frequency allowing for the basic parameters which control the physical phenomenon.

4.3.8 Amplitudes of vibration

The amplitude of oscillation of a simple mass spring combination depends upon the damping and magnification factor. The situation for a ship is more complex. Allowance must be made for at least the first three or four modes, superimposing the results for each. This can be done by finite element analysis and once the amplitude has been obtained the corresponding hull stress can be evaluated.

The question then arises as to whether the amplitude of vibration is acceptable. Limitations may be imposed by the reactions of humans, equipment or by strength considerations. Sensitive equipment can be protected by placing them on special mounts and this is done quite extensively in warships in particular. Human beings respond mainly to the vertical acceleration they experience. Curves are published (BS 6634; ISO 6954) indicating the combinations of frequency and displacement that are likely to be acceptable.

4.3.9 Checking vibration levels

It will be appreciated by now that accurate calculation of vibration levels is difficult. It is possible to put a check upon the levels likely to be achieved as the ship nears structural completion by using a vibration exciter. The exciter is simply a device for generating large vibratory forces by rotating an out of balance weight. Placed at appropriate positions in the ship it can be activated and the structural response to known forces measured.

4.3.10 Reducing vibration

Ideally vibration would be eliminated completely but this is not a realistic goal. In practice a designer aims to:

(1) Balance all forces in reciprocating and rotary machinery and in the propeller.
(2) Provide good flow into the propeller and site it clear of the hull.
(3) Avoid resonance by changing the stiffness of components or varying the exciting frequencies.
(4) Use special mounts to shield sensitive equipment from the vibration.
(5) Fit a form of vibration damper, either active or passive.

The two main sources of vibration are the machinery and propellers. With improvements in engine balancing and the increased use of resilient engine mountings, engine excited hull vibration should not be significant, see also Thomas *et al.* (2000). The propeller operates in a non-uniform wake field (see Chapter 5, Section 5.2) and, as a result, the propeller-induced forces remain the principal source of excitation. See also Ward and Willshare (1975) who examined full-scale measurements of propeller excited vibration. Adequate propeller-hull clearances (e.g. see recommendations of classification societies), the use of propeller skew and the use of hull shapes to produce a more uniform circumferential wake distribution, can minimize the effects. An outline of the origins of the propeller-induced forces, as presented by Jensen (2001), is given in the next section.

4.3.11 Propeller-induced forces

When the propeller of the ship rotates in the inhomogeneous wake field, periodic pressure forces will arise in the stern. These hydrodynamic forces will act partly on the propeller and be transferred to the hull girder via the bearings of the propeller axis and partly on the plating of the stern in the form of a pulsating water pressure, see Figure 4.82.

It is common to both types of loads that it is very difficult to calculate them by theoretical methods because of the complicated hydrodynamic flow conditions around the propeller. Therefore, it is often necessary to use model experiments and empirical

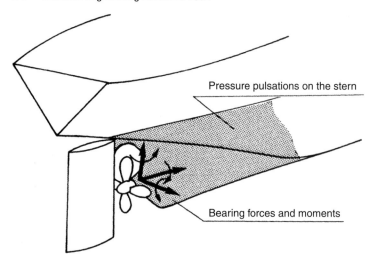

Figure 4.82 Propeller-induced periodic forces.

formulas. Reference may be made to Breslin and Andersen (1994).

The magnitude of the periodic forces and moments transferred from the propeller to the propeller axis can in principle be determined by calculating the hydrodynamic lift L on each propeller blade. To do this, it is required that the inhomogeneous wake field around the propeller is known, which is difficult to do theoretically. The lift is determined by 'lifting surface' analyses, For example, see Section 5.4. To get an idea of the propeller-induced forces and moments, it is assumed in the following that the resulting lift L_j on a propeller blade No. j is known. The lift is a function of the position of the blade, given by the angle θ relative to a vertical position of the propeller blade, see Figure 4.83.

For each blade, the lift L_j can be divided into two force components: The blade thrust $T_j(\theta)$ and the resistance $P_j(\theta)$, having effect in respectively the direction of the propeller axis and perpendicularly to the axis of the propeller blade. The distance r from the propeller axis to the point of action of these forces must, as the lift, be determined by hydrodynamic calculations. This distance is here assumed to be independent of the blade position θ, but variation of r with θ can easily be included. The lift L_j and thus also T_j and P_j are periodic with the period 2π. Hence, T_j and P_j can be expanded in Fourier series

$$T_j(\theta) = \frac{1}{2} a_0 + \sum_{n=1}^{\infty} a_n \cos n\theta$$

$$p_j(\theta) = \frac{1}{2} b_0 + \sum_{n=1}^{\infty} b_n \cos n\theta \qquad (4.7)$$

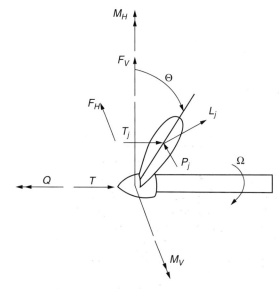

Figure 4.83 Resulting forces and moments on the propeller.

It can be proved from hydrodynamics that the Fourier coefficients a_n and b_n only depend on the corresponding nth component in the wake field. In Equation (4.7) it is assumed that the propeller axis lies in the centre line plane of the ship. If not as for ships with two propellers, there will also be sine components in the Fourier expansions due to the asymmetric wake field.

Subsequently, the resulting load components on the propeller axis at the propeller can be determined by adding up the loads T_j and P_j from the total of Z similar propeller blades:

Propeller thrust $\quad T(\theta) = \sum_{j=1}^{Z} T_j(\theta_j)$

Propeller moment $\quad Q(\theta) = r \sum_{j=1}^{Z} P_j(\theta_j)$

Vertical Force $\quad F_V(\theta) = \sum_{j=1}^{Z} P_j(\theta_j) \sin \theta_j$

Vertical Bending Moment $M_V(\theta) = r \sum_{j=1}^{Z} T_j(\theta_j) \cos \theta_j$

Horizontal Force $\quad F_H(\theta) = \sum_{j=1}^{Z} P_j(\theta_j) \cos \theta_j$

Horizontal Bending Moment

$$M_H(\theta) = r \sum_{j=1}^{Z} T_j(\theta_j) \sin \theta_j \quad (4.8)$$

where $\quad \theta_j = \theta + \dfrac{2\pi}{Z}(j-1) \quad (4.9)$

Then Equation (4.7) is inserted in Equation (4.8), so that the load components T, Q, F_V, M_V, F_H, and M_H are expressed in the coefficients a_n and b_n. The expressions can be reduced considerably by application of the formulas:

$$\sum_{j=1}^{Z} \cos\left[n\left(\theta + \frac{2\pi}{Z}(j-1)\right)\right]$$
$$= \begin{cases} Z \cos kZ\theta & \text{for } n = kZ, k \text{ integer} \\ 0 & \text{otherwise} \end{cases}$$

$$\sum_{j=1}^{Z} \sin\left[n\left(\theta + \frac{2\pi}{Z}(j-1)\right)\right]$$
$$= \begin{cases} Z \sin kZ\theta & \text{for } n = kZ, k \text{ integer} \\ 0 & \text{otherwise} \end{cases} \quad (4.10)$$

The validity of Equation (4.10) follows from (with $i = \sqrt{-1}$):

$$\sum_{j=1}^{Z} e^{in(\theta + \frac{2\pi}{Z}(j-1))} = e^{in\theta} \frac{1 - e^{i2\pi n}}{1 - e^{i2\pi n/Z}}$$
$$= \begin{cases} e^{ikZ\theta} Z & \text{for } n = kZ, k \text{ integer} \\ 0 & \text{otherwise} \end{cases}$$

The results become

$$T = \sum_{j=1}^{Z}\left[\frac{1}{2}a_0 + \sum_{n=1}^{\infty} a_n \cos n\theta_j\right]$$
$$= Z\left[\frac{1}{2}a_0 + \sum_{k=1}^{\infty} a_{kZ} \cos kZ\theta\right]$$

$$Q = rZ\left[\frac{1}{2}b_0 + \sum_{k=1}^{\infty} b_{kZ} \cos kZ\theta\right]$$

$$F_V = \sum_{j=1}^{Z}\left[\frac{1}{2}b_0 + \sum_{n=1}^{\infty} b_n \cos\theta_j\right]\sin\theta_j$$
$$= \frac{1}{2}b_0 \sum_{j=1}^{Z} \sin\theta_j + \frac{1}{2}\sum_{n=1}^{\infty} b_n$$
$$\sum_{j=1}^{Z}(\sin[(n+1)\theta_j] - \sin[(n-1)\theta_j])$$
$$= \frac{1}{2}Z\sum_{k=1}^{\infty}(b_{kZ-1} - b_{kZ+1})\sin kZ\theta \quad (4.11)$$

$$M_V = \frac{r}{2}Z\left[a_1 + \sum_{k=1}^{\infty}(a_{kZ-1} + a_{kZ+1})\cos kZ\theta\right]$$

$$F_H = \frac{1}{2}Z\left[b_1 + \sum_{k=1}^{\infty}(b_{kZ-1} + b_{kZ+1})\cos kZ\theta\right]$$

$$M_H = \frac{r}{2}Z\sum_{k=1}^{\infty}(a_{kZ-1} - a_{kZ+1})\sin kZ\theta$$

It is seen from the results, Equation (4.11), that all load components are periodic with the period $2\pi/Z$, because the same propeller configuration occurs each time a new blade gets in the same position as the preceding blade. If the propeller axis rotates with the constant frequency Ω then

$$\theta = \Omega t$$

and the load components, Equation (4.11), will thus only contain periodic components with frequencies which are multiples of the *blade frequency* $Z\Omega$.

In addition to the propeller thrust T and the moment Q, also the vertical bending moment M_V and the horizontal force F_H have a time-independent component. These mean values may be of importance in the determination of the lay-up of the propeller axis. Moreover, it is seen that only the harmonic components of the wake field corresponding to multiples of the blade frequency $\pm \Omega$ enter into the expressions for F_V, M_V, F_H and M_H, while only components which are multiples of the blade frequency form part of the propeller thrust and moment.

As a rule, the most important components in Equation (4.11) in relation to generation of hull vibrations are the terms which vary with the blade frequency. If only these terms are kept, the result is as follows:

$$T_1 = Za_Z \cos Z\Omega t$$
$$Q_1 = rZb_z \cos Z\Omega t$$
$$F_{V1} = \frac{1}{2}Z(b_{Z-1} - b_{Z+1})\sin Z\Omega t$$
$$M_{V1} = \frac{r}{2}Z(a_{Z-1} + a_{Z+1})\cos Z\Omega t$$
$$F_{H1} = \frac{1}{2}Z(b_{Z-1} + b_{Z+1})\cos Z\Omega t$$
$$M_{H1} = \frac{r}{2}Z(a_{Z-1} - a_{Z+1})\sin Z\Omega t \quad (4.12)$$

Table 4.3 Vibration response and endurance test levels for surface warships.

Ship type	Region	Standard test level Peak values and frequency range	Endurance tests
Minesweeper size and above	Masthead	1.25 mm, 5 to 14 Hz 0.3 mm, 14 to 23 Hz 0.125 mm, 23 to 33 Hz	1.25 mm, 14 Hz 0.3 mm, 23 Hz 0.125 mm, 33 Hz Each 1 hour
	Main	0.125 mm, 5 to 33 Hz	0.125 mm, 33 Hz For 3 hours
Smaller than minesweeper	Masthead and main	0.2 mm or a velocity of 63 mm/s whichever is less. 7 to 300 Hz	0.2 mm, 50 Hz For 3 hours
	Aftermost $\frac{1}{8}$ of ship length	0.4 mm or a velocity of 60 mm/s whichever is less. 7 to 300 Hz	0.4 mm, 24 Hz For 3 hours

The odd harmonic components in the wake field are usually much smaller than the even components in the wake field. It follows then from Equation (4.12) that, for a propeller with an even number of blades, the most important periodic loads will be T_1 and Q_1 while, for a propeller with an odd number of blades, F_{V1}, M_{V1}, F_{H1} and M_{H1} will be the dominant vibratory loads.

For conventional ships, the size of the time-varying loads T_1, Q_1, ... is of the order of magnitude of 5–20% of respectively the mean propeller thrust and moment.

The significance of the time-varying loads on the propeller is mainly that they may cause too large vibrations of the propeller axis. Their contribution to the generation of hull girder vibrations is normally considerably smaller than the contribution from the pulsating hydrodynamic forces induced on the stern as a consequence of the inhomogeneous wake field around the rotating propeller. This is especially so, if the propeller cavitates, as this effect strongly enhances the latter load but does not increase substantially the forces on the propeller.

There is no reliable theoretical method for determination of the hydrodynamic pressure induced by the rotating propeller on the stern, especially not if the propeller cavitates. However, as those loads may often lead to vibration problems, it is of great importance to be able to estimate their size and their variation with characteristic geometric quantities for the propeller and the stern. Halden (1980) gives an attempt to obtain such bases of estimation through correlation with extensive measurements.

4.3.12 Vibration testing of equipment

Most equipments are fitted in a range of ships and in different positions in a ship. Thus their design cannot be tailored to too specific a vibration specification. Instead they are designed to standard criteria and then samples are tested to confirm that the requirements have been met. These tests include endurance testing for several hours in the vibration environment. Table 4.3 gives test conditions for naval equipments to be fitted to a number of warship types.

In Table 4.3 the masthead region is that part of the ship above the main hull and superstructure. The main hull includes the upper deck, internal compartments and the hull.

References (Chapter 4)

Andrew, R.N. and Lloyd, A.R.J.M. (1981). Full scale comparative measurements of the behaviour of two frigates in severe head seas. *Trans. RINA*, Vol. 123.

ANSYS (2008). *ANSYS Structural*. ANSYS Inc, Canonsburg, USA.

Bishop, R.E.D. and Price, W.G. (1979). *Hydroelasticity of Ships*. Cambridge University Press, London.

Bishop, R.E.D., Price, W.G. and Temeral, P. (1991). A theory on the loss of the *MV Derbyshire*. *Trans. RINA*, Vol. 133.

Breslin, J. and Andersen, P. (1994). *Hydrodynamics of Ship Propellers*. Cambridge University Press, London, Cambridge Ocean Technology Series 3.

Burrill, L.C. (1934–1935). Ship vibration: Simple methods of estimating critical frequencies. *Trans. NECIES*.

BS (1985). British Standard 6634, 1985; Overall evaluation of vibration in merchant ships.

BS (1993). British Standard 7608, 1993 Code of practice for fatigue design and assessment of steel structures.

Caldwell, J.B. (1965). Ultimate Longitudinal Strength. *Trans. RINA*, Vol. 107.

Chalmers, D.W. (1993). *Design of ship's structures*. HMSO.

Clarke, S.D., Shenoi, R.A., Hicks, I.A. and Cripps, R.M. (1998). Fatigue characteristics for FRP sandwich structures of RNLI lifeboats. *Trans. RINA*, Vol. 140.

Clarkson, J. (1965). *The Elastic Analysis of Flat Grillages*. Cambridge University Press, Cambridge.

Corlett, E.C.B., Coleman, J.C. and Hendy, N.R. (1988). Kurdistan – the anatomy of a marine disaster. *Trans. RINA*, Vol. 130.

Department of transport (1988). A Report into the Circumstances Attending the Loss of the *MV Derbyshire*. Appendix 7: Examination of Fractured Deck Plate of *MV Tyne Bridge*.

Dieudonne, J. (1959). Vibration in ships. *Trans RINA*, Vol. 101.

Dover, W.D., Collins, R. and Etude, L.S. (2001). Fatigue strength of welded connections in offshore structures and ships. *Trans. RINA*, Vol. 143.

Eyres, D.J. (2007). *Ship Construction*, 6th edition. Butterworth-Heinemann, Oxford, UK.

Faulkner, D. and Sadden, J.A. (1979). Towards a unified approach to ship structural Safety. *Trans. RINA*, Vol. 121.

Faulkner, D. and Willams, R.A. (1996). Design for abnormal ocean waves. *Trans. RINA*, Vol. 138.

Gere, J.M. and Timoshenko, S.P. (1991). *Mechanics of Materials*, 3rd edition. Chapman and Hall, London.

Gibbons, G. (2003). Fatigue in ship structures. *The Naval Architect*, January. RINA, London.

Gibbons, G. (2003). Fracture in ship structure. *The Naval Architect*, January. RINA, London.

Halden, K.O. (1980). Early design-stage approach to reducing hull surface forces due to propeller cavitation. *Trans*. Vol. 88. SNAME.

Hansen, A.M. (1996). Strength of midship sections. *Marine Structures*, Vol. 9.

ISO (1984). ISO 6954 Guidelines for the overall evaluation of vibration in merchant ships.

ISSC (2006). *Proceedings of 16th International Ship and Offshore Structures Congress*, P.A. Frieze, R.A. Shenoi (eds), 2006. University of Southampton, Southampton, UK.

Jensen, J.J. (2001). *Load and Global Response of Ships*. Elsevier, Oxford, UK.

Jeong, H.K. and Shenoi, R.A. (2001). Structural reliability of fibre reinforced composite plates. *Trans. RINA*, Vol. 143.

Johnson, A.J. (1950–1951). Vibration tests on all welded and riveted 10 000 ton dry cargo ships. *Trans. NECIES*.

Kelly, A. and Zweben, C. (Editors in Chief) (2000). *Comprehensive Composite Materials*, 6 Volumes. Elsevier, Oxford, UK.

Kobylinski, L.K. and Kastner, S. (2003). *Stability and Safety of Ships*. Elsevier, Oxford, UK.

Kuo, C., MacCallum, K.J. and Shenoi, R.A. (1984). An effective approach to structural design for production. *Trans. RINA*, Vol. 126.

Landweber, L. and Macagno, M.C. (1957). Added mass of two-dimensional forms oscillating in a free surface. *Journal of Ship Research*, SNAME.

Lewis, F.M. (1929). The inertia of the water surrounding a vibrating ship. *Trans. SNAME*.

Lloyds Register (2007). Lloyds Register – *Rules and Regulations for the Classification of Ships,* Part 3, Ship structures. Lloyds Register, London.

McCallum, J. (1974). The strength of fast cargo ships. *Trans. RINA*.

Meek, M. *et al.* (1972). The structural design of the OCL container ships. *Trans RINA*, Vol.114.

Morel, P., Beghin, D. and Baudin, M. (1995). Assessment of the vibratory behaviour of ships. *RINA Conference on Noise and Vibration,* RINA, London.

Muckle, W. (1954). The buoyancy curve in longitudinal strength calculations. *Shipbuilder and Marine Engine Builder*, February.

Murray, J.M. (1965). Notes on the longitudinal strength of tankers. *Trans. NECIES*.

Nishida, S. (1994). *Failure Analysis in Engineering Applications*. Butterworth-Heinemann, Oxford, UK.

Paik, J.K. (2004). Principles and criteria for ultimate limit state design and strength assessment of ship hulls. *Trans. RINA*, Vol. 146.

Pedersen, P.T. (1983). Beam model for tosional-bending response of ship hulls. *Trans. RINA*, Vol. 125, pp. 171–182.

Petershagen, H. (1986). Fatigue problems in ship structures. *Advances in Marine Structure*. Elsevier Applied Science Publishers, London.

Rao, S.S. (2005). *The Finite Method in Engineering*, 4th edition. Butterworth-Heinemann, Oxford, UK.

Rawson, K.J. and Tupper, E.C. (2001). *Basic Ship Theory*, Combined Volume, 5th edition. Butterworth-Heinemann, Oxford, UK.

Schlick, O. (1884). Vibration of steam vessels. *Trans. RINA*.

Shenoi, R.A. and Wellicome, J.F. (1993). *Composite Materials in Marine Structures*, Vols. 1 and 2. Cambridge University Press, Cambridge, UK.

Smith, C.S. (1990). *Design of Marine Structures in Composite Materials*. Elsevier Applied Science, London.

Smith, C.S., Anderson, N., Chapman, J.C., Davidson, P.C. and Dowling, P.J. (1992). Strength of stiffened plating under combined compression and lateral pressure. *Trans. RINA*, Vol. 134.

Smith, C.S. and Chalmers, D.W. (1987). Design of ship superstructures in fibre reinforced plastic. *The Naval Architect,* RINA, London.

Somerville, W.L., Swan, J.W. and Clarke, J.D. (1977). Measurement of residual stresses and distortions in stiffened panels. *Journal of Strain Analysis*, Vol. 12, No. 2.

Southwell, R. (1940). *Relaxation Methods in Engineering Sciences*. Oxford University Press, Oxford, UK.

Sumpter, J.D.G. (1986). Design against fracture in welded structures. *Advances in Marine Structure*. Elsevier Applied Science Publishers, London.

Sumpter, J.D.G., Bird, J., Clarke, J.D. and Caudrey, A.J. (1989). Fracture toughness of ship steels. *Trans. RINA*, Vol. 131.

Taylor, J.L. (1924–1925). The theory of longitudinal bending of ships. *Trans. NECIES*.

Taylor, J.L. (1927–1928). Ship vibration periods. *Trans. NECIES*.

Taylor, J.L. (1930). Vibration of ships. *Trans. RINA*.

Thomas, B.H., Harper, H.W., Hargreaves, R.E. and McClenahan, R.H. (2000). Practical experience of machinery induced vibration and lessons to be learned in ship design and procurement. *Trans. RINA*, Vol. 142.

Todd, F.H. (1961). *Ship Hull Vibration*. Arnold.

Townsin, R.L. (1969). Virtual mass reduction factors: J values for ship vibration calculations derived from tests with beams including ellipsoids and ship models. *Trans. RINA*, Vol. 111.

Tupper, E.C. (2004). *Introduction to Naval Architecture*, 4th edition. Elsevier Butterworth-Heinemann, Oxford, UK.

Violette, F.L.M. (1994). The effect of corrosion on structural detail design. *RINA International Conference on Marine Corrosion Prevention*. RINA, London.

Violette, F.L.M. and Shenoi, R.A. (1999). On the fatigue performance prediction of ship structural details. *Trans. RINA*, Vol. 141.

Ward, G. and Willshare, G.T. (1975). Propeller excited vibration with particular reference to full scale measurements. *Trans. RINA*, Vol. 117.

Yuille, I.M. and Watson, L.B. (1960). Transverse strength of single hulled ships. *Trans. RINA*, Vol. 102.

Zienkiewicz, O.C., Taylor, R.L. and Zhu, L.Z. (2005). *The Finite Element Method: Its Basis and Fundamentals*, 6th edition. Butterworth-Heinemann, Oxford, UK.

5 Powering

Contents

5.1 Resistance and propulsion
5.2 Wake
5.3 Propeller performance characteristics
5.4 Propeller theories
5.5 Cavitation
5.6 Propeller design
5.7 Service performance and analysis
References (Chapter 5)

The various Sections of this Chapter have been taken from the following books, with the permission of the authors:

Bertram, V. (2000). *Practical Ship Hydrodynamics*. Butterworth-Heinemann, Oxford, UK.
 [Sections 5.1.3.4, 5.1.7.3]
Carlton, J.S. (2007). *Marine Propellers and Propulsion*. 2nd edn. Butterworth-Heinemann, Oxford, UK.
 [All Sections except 5.1.3.4 and 5.1.7.3]

5.1 Resistance and propulsion

Prior to the mid-nineteenth century comparatively little was known about the laws governing the resistance of ships and the power that was required to give a particular speed. Brown (1983) gives an account of the problems of that time and depicts the role of William Froude, who can be justly considered as the father of ship resistance studies.

5.1.1 Froude's analysis procedure

William Froude (1955) recognized that ship models of geometrically similar form would create similar wave systems, albeit at different speeds. Furthermore, he showed that the smaller models had to be run at slower speeds than the larger models in order to obtain the same wave pattern. His work showed that for a similarity of wave pattern between two geometrically similar models of different size the ratio of the speeds of the models was governed by the relationship

$$\frac{V_1}{V_2} = \sqrt{\frac{L_1}{L_2}} \tag{5.1}$$

By studying the comparison of the specific resistance curves of models and ships Froude noted that they exhibited a similarity of form although the model curve was always greater than that for the ship (Figure 5.1). This led Froude to the conclusion that two components of resistance were influencing the performance of the vessel and that one of these, the wave-making component R_w, scaled with $V/\sqrt{L}$ and the other did not. This second component, which is due to viscous effects, derives principally from the flow of the water around the hull but also is influenced by the air flow and weather acting on the above-water surfaces. This second component was termed the frictional resistance R_F.

Froude's major contribution to the ship resistance problem, which has remained useful to the present day, was his conclusion that the two sources of resistance might be separated and treated independently. In this approach, Froude suggested that the viscous resistance could be calculated from frictional data whilst that wave-making resistance R_w could be deduced from the measured total resistance R_T and the calculated frictional resistance R_F as follows:

$$R_W = R_T - R_F \tag{5.2}$$

In order to provide the data for calculating the value of the frictional component Froude performed his famous experiments at the Admiralty owned model tank at Torquay. These experiments entailed towing a series of planks ranging from 10 to 50 ft in length, having a series of surface finishes of shellac varnish, paraffin wax, tin foil, graduation of sand roughness and other textures. Each of the planks was 19 in. deep and $\frac{3}{16}$ in. thick and was ballasted to float on its edge. Although the results of these experiments suffered from errors due to temperature differences, slight bending of the longer planks and laminar flow on some of the shorter planks, Froude was able to derive an empirical formula which would act as a basis for the calculation of the frictional resistance component R_F in Equation (5.2). The relationship Froude derived took the form

$$R_F = fSV^n \tag{5.3}$$

in which the index n had the constant value of 1.825 for normal ship surfaces of the time and the coefficient f varied with both length and roughness, decreasing with length but increasing with roughness. In Equation (5.3), S is the wetted surface area.

As a consequence of this work Froude's basic procedure for calculating the resistance of a ship is as follows:

1. Measure the total resistance of the geometrically similar model R_{TM} in the towing tank at a series of speeds embracing the design $V/\sqrt{L}$ of the full-size vessel.
2. From this measured total resistance subtract the calculated frictional resistance values for the model R_{FM} in order to derive the model wave making resistance R_{WM}.
3. Calculate the full-size frictional resistance R_{FS} and add these to the full-size wave making

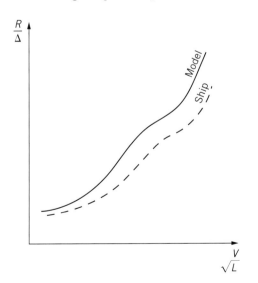

Figure 5.1 Comparison of a ship and its model's specific resistance curves.

resistance R_{WS}, scaled from the model value, to obtain the total full-size resistance R_{TS}.

$$R_{TS} = R_{WM}\left(\frac{\Delta_S}{\Delta_M}\right) + R_{FS} \quad (5.4)$$

In equation (5.4) the suffixes M and S denote model and full scale, respectively and Δ is the displacement.

The scaling law of the ratio of displacements derives from Froude's observations that when models of various sizes, or a ship and its model, were run at corresponding speeds dictated by Equation (5.1), their resistances would be proportional to the cubes of their linear dimensions or, alternatively, their displacements. This was, however, an extension of a law of comparison which was known at that time.

Froude's law, Equation (5.1), states that the wave making resistance coefficients of two geometrically similar hulls of different lengths are the same when moving at the same $V/\sqrt{L}$ value, V being the ship or model speed and L being the waterline length. The ratio $V/\sqrt{L}$ is termed the speed length ratio and is of course dimensional; however, the dimensionless Froude number can be derived from it to give

$$F_n = \frac{V}{\sqrt{(gL)}} \quad (5.5)$$

in which g is the acceleration due to gravity (9.81 m/s²). Care needs to be exercised in converting between the speed length ratio and the Froude number:

$F_n = 0.3193 \dfrac{V}{\sqrt{L}}$ where V is in m/s; L is in metres

$F_n = 0.1643 \dfrac{V}{\sqrt{L}}$ where V is in knots; L is in metres

Froude's work with his plank experiments was carried out prior to the formulation of the Reynolds number criteria and this undoubtedly led to errors in his results: for example, the laminar flow on the shorter planks. Using dimensional analysis, after the manner shown in Section 5.3, it can readily be shown today that the resistance of a body moving on the surface, or at an interface of a medium, can be given by

$$\frac{R}{\rho V^2 L^2} = \phi\left\{\frac{VL\rho}{\mu}, \frac{V}{\sqrt{gL}}, \frac{V}{a}, \frac{\sigma}{g\rho L^2}, \frac{p_0 - p_v}{\rho V^2}\right\} \quad (5.6)$$

In this equation the left-hand side term is the resistance coefficient C_R whilst on the right-hand side of the equation:

The 1st term is the Reynolds number R_n.
The 2nd term is the Froude number F_n (Equation (5.5))
The 3rd term is the Mach number M_a.
The 4th term is the Weber number W_e.
The 5th term is the Cavitation number σ_0.

For the purposes of ship propulsion the 3rd and 4th terms are not generally significant and can, therefore, be neglected. Hence Equation (5.6) reduces to the following for all practical ship purposes:

$$C_R = \phi\{R_n, F_n, \sigma_0\} \quad (5.7)$$

in which

ρ is the density of the water
μ is the dynamic viscosity of the water
p_0 is the free stream undisturbed pressure
p_v is the water vapour pressure.

5.1.2 Components of calm water resistance

In the case of a vessel which is undergoing steady motion at slow speeds, that is where the ship's weight balances the displacement upthrust without the significant contribution of hydrodynamic lift forces, the components of calm water resistance can be broken down into the contributions shown in Figure 5.2. From this figure it is seen that the total resistance can be decomposed into two primary components, pressure and skin friction resistance, and these can then be broken down further into more discrete components. In addition to these components there is of course the air resistance and added resistance due to rough weather: these are, however, dealt with separately in Sections 5.1.8 and 5.1.5, respectively.

Each of the components shown in Figure 5.2 can be studied separately provided that it is remembered that each will have an interaction on the others and, therefore, as far as the ship is concerned, need to be considered in an integrated way.

5.1.2.1 *Wave making resistance R_W*

Lord Kelvin (Kelvin (1904 a–c)) in 1904 studied the problem of the wave pattern caused by a moving pressure point. He showed that the resulting system of waves comprises a divergent set of waves together with a transverse system which are approximately normal to the direction of motion of the moving point. Figure 5.3 shows the system of waves so formed. The pattern of waves is bounded by two straight lines which in deep water are at an angle ϕ to the direction of motion of the point, where ϕ is given by

$$\phi = \sin^{-1}\left(\tfrac{1}{3}\right) = 19.471°$$

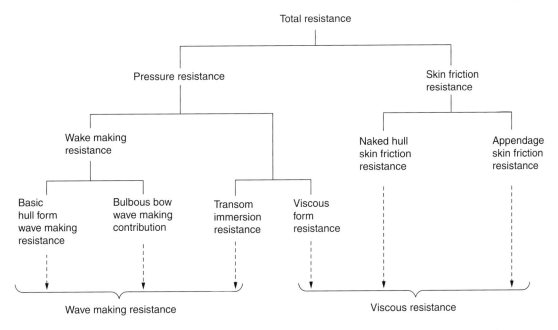

Figure 5.2 Components of ship resistance.

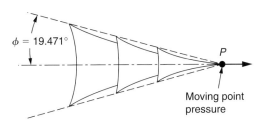

Figure 5.3 Wave pattern induced by a moving-point pressure in calm water.

The interference between the divergent and transverse systems gives the observed waves their characteristic shape, and since both systems move at the same speed, the speed of the vessel, the wavelength λ between successive crests is

$$\lambda = \frac{2\pi}{g} V^2 \qquad (5.8)$$

The height of the wave systems formed decreases fairly rapidly as they spread out laterally because the energy contained in the wave is constant and it has to be spread out over an increasingly greater length. More energy is absorbed by the transverse system than by the divergent system, and this disparity increases with increasing speed.

A real ship form, however, cannot be represented adequately by a single-moving pressure point as analysed by Kelvin. The simplest representation of a ship, Figure 5.4, is to place a moving pressure field near the bow in order to simulate the bow wave system, together with a moving suction field near the stern to represent the stern wave system. In this model the bow pressure field will create a crest near the bow, observation showing that this occurs at about $\lambda/4$ from the bow, whilst the suction field will introduce a wave trough at the stern: both of these wave systems have a wavelength $\lambda = 2\pi V^2/g$.

The divergent component of the wave system derived from the bow and the stern generally do not exhibit any strong interference characteristics. This is not the case, however, with the transverse wave systems created by the vessel, since these can show a strong interference behaviour. Consequently, if the bow and stern wave systems interact such that they are in phase a reinforcement of the transverse wave patterns occurs at the stern and large waves are formed in that region. For such a reinforcement to take place, Figure 5.5(a), the distance between the first crest at the bow and the stern must be an odd number of half-wavelengths as follows:

$$L - \frac{\lambda}{4} = k\frac{\lambda}{2} \quad \text{where } k = 1, 3, 5, \ldots, (2j+1) \\ \text{with } j = 0, 1, 2, 3, \ldots$$

From which

$$\frac{4}{2k+1} = \frac{\lambda}{L} = \frac{2\pi V^2}{gL} = 2\pi (F_n)^2$$

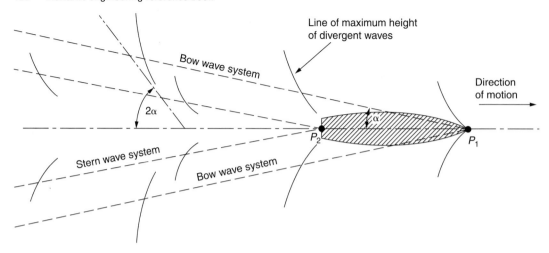

Figure 5.4 Simple ship wave pattern representation by two pressure points.

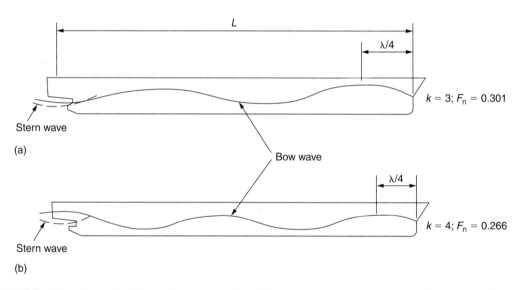

Figure 5.5 Wave reinforcement and cancellation at stern: (a) wave reinforcement at stern and (b) wave cancellation at stern.

that is,

$$F_n = \sqrt{\frac{2}{\pi(2k+1)}} \quad (5.9)$$

For the converse case when the bow and stern wave systems cancel each other, and hence produce a minimum wave making resistance condition, the distance $L - \lambda/4$ must be an even number of half wave lengths (Figure 5.5(b)):

$$L - \frac{\lambda}{4} = k\frac{\lambda}{2} \quad \text{where } k = 2, 4, 6, \ldots, 2j \\ \text{with } j = 1, 2, 3, \ldots$$

Hence

$$F_n = \sqrt{\frac{2}{\pi(2k+1)}}$$

as before, but with k even in this case.

Consequently from Equation (5.9), Table 5.1 can be derived, which for this particular model of wave action identifies the Froude numbers at which reinforcement (humps) and cancellation (hollows) occur in the wave making resistance.

Each of the conditions shown in Table 5.1 relates sequentially to maximum and minimum conditions

Table 5.1 Froude numbers corresponding to maxima and minima in the wave making resistance component.

K	F_n	Description
1	0.461	1st hump in R_w curve
2	0.357	1st hollow in R_w curve
3	0.301	2nd hump in R_w curve
4	0.266	2nd hollow in R_w curve
5	0.241	3rd hump in R_w curve
⋮	⋮	⋮

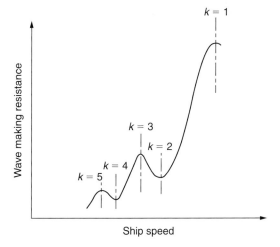

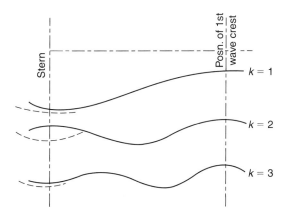

Figure 5.6 Form of wave making resistance curve.

in the wave making resistance curves. The 'humps' occur because the wave profiles and hence the wave making resistance are at their greatest in these conditions whilst the converse is true in the case of the 'hollows'. Figure 5.6 shows the general form of the wave making resistance curve together with the schematic wave profiles associated with the various values of k.

The hump associated with $k = 1$ is normally termed the 'main hump' since this is the most pronounced hump and occurs at the highest speed. The second hump, $k = 3$, is called the 'prismatic hump' since it is influenced considerably by the prismatic coefficient of the particular hull form.

The derivation of Figure 5.6 and Table 5.1 relies on the assumptions made in its formulation; for example, a single pressure and suction field, bow wave crest at $\lambda/4$; stern trough exactly at the stern, etc. Clearly, there is some latitude in all of these assumptions, and therefore the values of F_n at which the humps and hollows occur vary. In the case of warships the distance between the first crest of the bow wave and the trough of the stern wave has been shown to approximate well to $0.9L$, and therefore this could be used to rederive Equation (5.9), and thereby derive slightly differing values of Froude numbers corresponding to the 'humps' and 'hollows'. Table 5.2 shows these differences, and it is clear that the greatest effect is formed at low values of k. Figure 5.6 for this and the other reasons cited is not unique but is shown here to provide awareness and guidance on wave making resistance variations.

Table 5.2 Effect of difference in calculation basis on prediction of hump and hollow Froude numbers.

k	1	2	3	4	5
$L - \lambda/4$ basis	0.46	0.36	0.30	0.27	0.24
$0.9L$ basis	0.54	0.38	0.31	0.27	0.24

A better approximation to the wave form of a vessel can be made by considering the ship as a solid body rather than two point sources. Wigley initially used a simple parallel body with two pointed ends and showed that the resulting wave pattern along the body could be approximated by the sum of five separate disturbances of the surface (Figure 5.7). From this figure it is seen that a symmetrical disturbance corresponds to the application of Bernoulli's theorem with peaks at the bow and stern and a hollow, albeit with cusps at the start and finish of the parallel middle body, between them. Two wave forms starting with a crest are formed by the action of the bow and stern whilst a further two wave forms commencing with a trough originates from the shoulders of the parallel middle body. The sum of these five wave profiles is shown in the bottom of Figure 5.7 and compared with a measured profile which shows good general agreement. Since the wavelength λ varies with speed and the points at which the waves originate are fixed, it is easy to understand that the whole profile of the resultant wave form will change with speed length ratio.

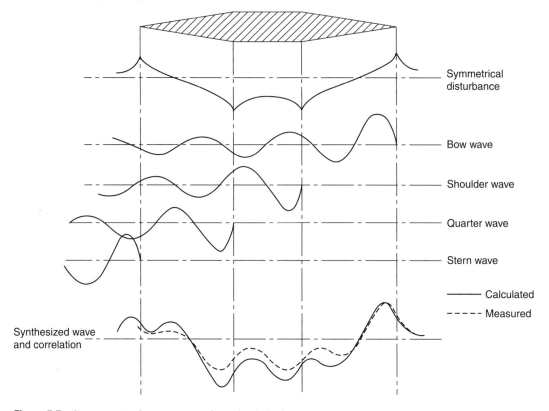

Figure 5.7 Components of wave systems for a simple body.

This analysis procedure was extended by Wigley for a more realistic hull form comprising a parallel middle body and two convex extremities. Figure 5.8 shows the results in terms of the same five components and the agreement with the observed wave form.

Considerations of this type lead to endeavouring to design a hull form to produce a minimum wave making resistance using theoretical methods. The basis of these theories is developed from Kelvin's work on a travelling pressure source; however, the mathematical boundary conditions are difficult to satisfy with any degree of precision. Results of work based on these theories have been mixed in terms of their ability to represent the observed wave forms, and consequently there is still considerable work to undertake in this field.

5.1.2.2 The contribution of the bulbous bow

Bulbous bows are today commonplace in the design of ships. Their origin is to be found before the turn of the century, but the first application appears to have been in 1912 by the US Navy. The general use in merchant applications appears to have waited until the late 1950s and early 1960s.

The basic theoretical work on their effectiveness was carried out by Wigley (1936) in which he showed that if the bulb was nearly spherical in form, then the acceleration of the flow over the surface induces a low-pressure region which can extend towards the water surface. This low-pressure region then reacts with the bow pressure wave to cancel or reduce the effect of the bow wave. The effect of the bulbous bow, therefore, is to cause a reduction, in the majority of cases, of the effective power required to propel the vessel, the effective power P_E being defined as the product of the ship resistance and the ship speed at a particular condition. Figure 5.9 shows a typical example of the effect of a bulbous bow from which it can be seen that a bulb is, in general, beneficial above a certain speed and gives a penalty at low speeds. This is because of the balance between the bow pressure wave reduction effect and increase in frictional resistance caused by the presence of the bulb on the hull.

The effects of the bulbous bow in changing the resistance and delivered power characteristics can be

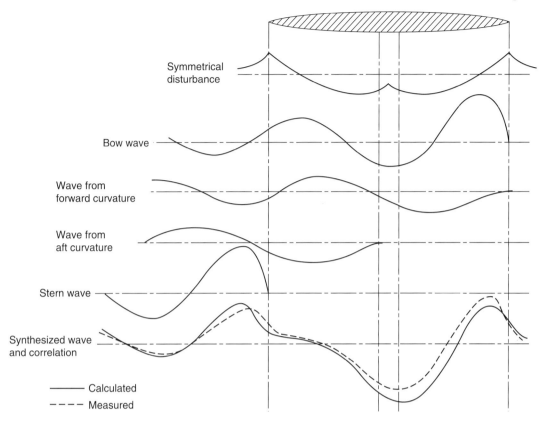

Figure 5.8 Wave components for a body with convex ends and a parallel middle body.

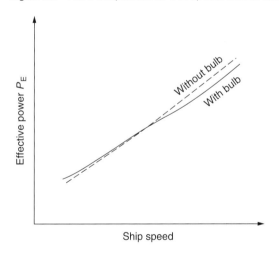

Figure 5.9 Influence of a bulbous bow of the effective power requirement.

attributed to several causes. The principal of these are as follows:

1. The reduction of bow pressure wave due to the pressure field created by the bulb and the consequent reduction in wave making resistance.
2. The influence of the upper part of the bulb and its intersection with the hull to introduce a downward flow component in the vicinity of the bow.
3. An increase in the frictional resistance caused by the surface area of the bulb.
4. A change in the propulsion efficiency induced by the effect of the bulb on the global hull flow field.
5. The change induced in the wave breaking resistance.

The shape of the bulb is particularly important in determining its beneficial effect. The optimum shape for a particular hull depends on the Froude number associated with its operating regime, and bulbous bows tend to give good performance over a narrow range of ship speeds. Consequently, they are most commonly found on vessels which operate at clearly defined speeds for much of their time. The actual bulb form, Figure 5.10, is defined in relation to a series of form characteristics as follows:

1. length of projection beyond the forward perpendicular;
2. cross-sectional area at the forward perpendicular (A_{BT});

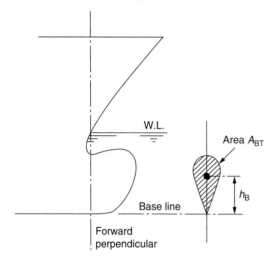

Figure 5.10 Bulbous bow definition.

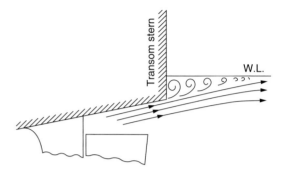

Figure 5.11 Flow around an immersed transom stern.

3. height of the centroid of cross section A_{BT} from the base line (h_B);
4. bulb section form and profile;
5. transition of the bulb into the hull.

With regard to section form many bulbs today are designed with non-circular forms so as to minimize the effects of slamming in poor weather. There is, however, still considerable work to be done in relating bulb form to power saving and much contemporary work is proceeding. For current design purposes reference can be made to the work of Inui (1962), Todd (1967), Yim (1974) and Schneekluth (1987).

In addition to its hydrodynamic behaviour the bulb also introduces a further complication into resistance calculations. Traditionally the length along the waterline has formed the basis of many resistance calculation procedures because it is basically the fundamental hydrodynamic dimension of the vessel. The bulbous bow, however, normally projects forward of the forward point of the definition of the waterline length, and since the bulb has a fundamental influence on some of the resistance components, there is a case for redefining the basic hydrodynamic length parameter for resistance calculations.

5.1.2.3 *Transom immersion resistance*

In modern ships a transom stern is now normal practice. If at the design powering condition a portion of the transom is immersed, this leads to separation taking place as the flow form under the transom passes out beyond the hull (Figure 5.11). The resulting vorticity that takes place in the separated flow behind the transom leads to a pressure loss behind the hull which is taken into account in some analysis procedures.

The magnitude of this resistance is generally small and, of course, vanishes when the lower part of the transom is dry. Transom immersion resistance is largely a pressure resistance that is scale independent.

5.1.2.4 *Viscous form resistance*

The total drag on a body immersed in a fluid and travelling at a particular speed is the sum of the skin friction components, which is equal to the integral of the shearing stresses taken over the surface of the body, and the form drag, which is in the integral of the normal forces acting on the body.

In an inviscid fluid the flow along any streamline is governed by Bernoulli's equation and the flow around an arbitrary body is predictable in terms of the changes between pressure and velocity over the surface. In the case of Figure 5.12(a) this leads to the net axial force in the direction of motion being equal to zero since in the two-dimensional case shown in Figure 5.12(a),

$$\oint p \cos \theta \, ds = 0 \qquad (5.10)$$

When moving in a real fluid, a boundary layer is created over the surface of the body which, in the case of a ship, will be turbulent and is also likely to separate at some point in the after body. The presence of the boundary layer and its growth along the surface of the hull modifies the pressure distribution acting on the body from that of the potential or inviscid case. As a consequence, the left-hand side of Equation (5.10) can no longer equal zero and the viscous form drag R_{VF} is defined for the three-dimensional case of a ship hull as

$$R_{VF} = \sum_{k=1}^{n} p_k \cos \theta_k \delta S_k \qquad (5.11)$$

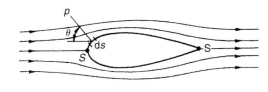

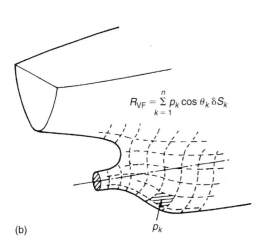

Figure 5.12 Viscous form resistance calculation: (a) inviscid flow case on an arbitrary body and (b) pressures acting on shell plate of a ship.

in which the hull has been split into n elemental areas δS_k and the contribution of each normal pressure p_k acting on the area is summed in the direction of motion (Figure 5.12(b)).

Equation (5.11) is an extremely complex equation to solve since it relies on the solution of the boundary layer over the vessel and this is a solution which at the present time can only be approached using considerable computational resources for comparatively simple hull forms. As a consequence, the viscous form resistance is normally accounted for using empirical or pseudo-empirical methods at this time.

5.1.2.5 *Naked hull skin friction resistance*

The original data upon which to calculate the skin friction component of resistance was that provided by Froude in his plank experiments at Torquay. This data, as discussed in the previous section, was subject to error and in 1932 Schoenherr re-evaluated Froude's original data in association with other work in the light of the Prandtl–von Karman theory. This analysis resulted in an expression of the friction coefficient C_F as a function of Reynolds number R_n

and the formulation of a skin friction line, applicable to smooth surfaces, of the following form:

$$\frac{0.242}{\sqrt{C_F}} = \log(R_n \cdot C_F) \quad (5.12)$$

This equation, known as the Schoenherr line, was adopted by the American Towing Tank Conference (ATTC) in 1947 and in order to make the relationship applicable to the hull surfaces of new ships an additional allowance of 0.0004 was added to the smooth surface values of C_F given by Equation (5.12). By 1950 there was a variety of friction lines for smooth turbulent flows in existence and all, with the exception of Froude's work, were based on Reynolds number. Phillips-Birt (1970) provides an interesting comparison of these friction formulations for a Reynolds number of 3.87×10^9 which is applicable to ships of the length of the former trans-Atlantic liner *Queen Mary* and is rather less than that for the large supertankers: in either case lying way beyond the range of direct experimental results. The comparison is shown in Table 5.3 from which it is seen that close agreement is seen to exist between most of the results except for the Froude and Schoenherr modified line. These last two, whilst giving comparable results, include a correlation allowance in their formulation. Indeed the magnitude of the correlation allowance is striking between the two Schoenherr formulations: the allowance is some 30 per cent of the basic value.

Table 5.3 Comparison of C_F values for different friction lines for a Reynolds number $R_n = 3.87 \times 10^9$ (taken from Phillips-Birt (1970).

Friction line	C_F
Gerbers	0.00134
Prandtl–Schlichting	0.00137
Kemph–Karham	0.00103
Telfer	0.00143
Lackenby	0.00140
Froude	0.00168
Schoenherr	0.00133
Schoenherr +0.0004	0.00173

In the general application of the Schoenherr line some difficulty was experienced in the correlation of large and small model test data and wide disparities in the correlation factor C_A were found to exist upon the introduction of all welded hulls. These shortcomings were recognized by the 1957 International Towing Tank Conference (ITTC) and a modified line was accepted. The 1957 ITTC line is expressed as

$$C_F = \frac{0.075}{(\log_{10} R_n - 2.0)^2} \quad (5.13)$$

and this formulation, which is in use with most ship model basins, is shown together with the Schoenherr line in Figure 5.13. It can be seen that the present ITTC line gives slightly higher values of C_F at the lower Reynolds numbers than the Schoenherr line whilst both lines merge towards the higher values of R_n.

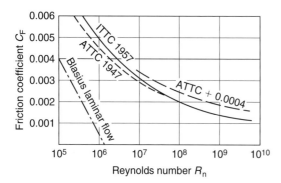

Figure 5.13 Comparison of ITTC (1957) and ATTC (1947) friction lines.

The frictional resistance R_F derived from the use of either the ITTC or ATTC lines should be viewed as an instrument of the calculation process rather than producing a definitive magnitude of the skin friction associated with a particular ship. As a consequence when using a Froude analysis based on these, or indeed any friction line data, it is necessary to introduce a correlation allowance into the calculation procedure. This allowance is denoted by C_A and is defined as

$$C_A = C_{T(\text{measured})} - C_{T(\text{estimated})} \quad (5.14)$$

In this equation, as in the previous equation, the resistance coefficients C_T, C_F, C_W and C_A are non-dimensional forms of the total, frictional, wave making and correlation resistances, and are derived from the basic resistance summation

$$R_T = R_W + R_V$$

by dividing this equation throughout by $\frac{1}{2}\rho V_s^2 S$, $\frac{1}{2}\rho V_s^2 L^2$ or $\frac{1}{2}\rho V_s^2 \nabla^{2/3}$ according to convenience.

5.1.2.6 Appendage skin friction

The appendages of a ship such as the rudder, bilge keels, stabilizers, transverse thruster openings and so on introduce a skin friction resistance above that of the naked hull resistance.

At the ship scale the flow over the appendages is turbulent, whereas at model scale it would normally be laminar unless artificially stimulated, which in itself may introduce a flow modelling problem. In addition, many of the hull appendages are working wholly within the boundary layer of the hull, and since the model is run at Froude identity and not Reynolds identity this again presents a problem. As a consequence the prediction of appendage resistance needs care if significant errors are to be avoided. The calculation of this aspect is further discussed in Section 5.1.3.

In addition to the skin friction component of appendage resistance, if the appendages are located on the vessel close to the surface then they will also contribute to the wave making component since a lifting body close to a free surface, due to the pressure distribution around the body, will create a disturbance on the free surface. As a consequence, the total appendage resistance can be expressed as the sum of the skin friction and surface disturbance effects as follows:

$$R_{\text{APP}} = R_{\text{APP(F)}} + R_{\text{APP(W)}} \quad (5.15)$$

where $R_{\text{APP(F)}}$ and $R_{\text{APP(W)}}$ are the frictional and wave making components, respectively, of the appendages. In most cases of practical interest to the merchant marine $R_{\text{APP(W)}} \simeq 0$ and can be neglected: this is not the case, however, for some naval applications.

5.1.2.7 Viscous resistance

Figure 5.2 defines the viscous resistance as being principally the sum of the form resistance, the naked hull skin friction and the appendage resistance. In the discussion on the viscous form resistance it was said that its calculation by analytical means was an extremely complex matter and for many hulls of a complex shape was not possible with any degree of accuracy at the present time.

Hughes (1954) attempted to provide a better empirical foundation for the viscous resistance calculation by devising an approach which incorporated the viscous form resistance and the naked hull skin friction. To form a basis for this approach Hughes undertook a series of resistance tests using planks and pontoons for a range of Reynolds numbers up to a value of 3×10^8. From the results of this experimental study Hughes established that the frictional resistance coefficient C_F could be expressed as a unique inverse function of aspect ratio $\mathcal{R}$ and, fucrthermore, that this function was independent of Reynolds number. The function derived from this work had the form:

$$C_F = C_F\big|_{\mathcal{R}=\infty} \cdot f\left(\frac{1}{\mathcal{R}}\right)$$

in which the term $C_F\big|_{\mathcal{R}=\infty}$ is the frictional coefficient relating to a two-dimensional surface; that is, one having an infinite aspect ratio.

This function permitted Hughes to construct a two-dimensional friction line defining the frictional resistance of turbulent flow over a plane smooth surface. This took the form

$$C_F|_{AR=\infty} = \frac{0.066}{[\log_{10} R_n - 2.03]^2} \qquad (5.16)$$

Equation (5.16) quite naturally bears a close similarity to the ITTC 1957 line expressed by Equation (5.13). The difference, however, is that the ITTC and ATTC lines contain some three-dimensional effects, whereas Equation (5.16) is defined as a two-dimensional line. If it is plotted on the same curve as the ITTC line, it will be found that it lies just below the ITTC line for the full range of R_n and in the case of the ATTC line it also lies below it except for the very low Reynolds numbers.

Hughes proposed the calculation of the total resistance of a ship using the basic relationship

$$C_T = C_V + C_W$$

in which $C_V = C_F|_{AR=\infty} + C_{FORM}$, thereby giving the total resistance as

$$C_T = C_F|_{AR=\infty} + C_{FORM} + C_W \qquad (5.17)$$

in which C_{FORM} is a 'form' resistance coefficient which takes into account the viscous pressure resistance of the ship. In this approach the basic skin friction resistance coefficient can be determined from Equation (5.16). To determine the form resistance the ship model can be run at a very slow speed when the wave making component is very small and can be neglected; when this occurs, that is to the left of point A in Figure 5.14, then the resistance curve defines the sum of the skin friction and form resistance components. At the point A, when the wave making resistance is negligible, the ratio

$$\frac{AC}{BC} = \frac{\text{viscous resistance}}{\text{skin friction resistance}}$$

$$= \frac{\text{skin friction resistance} + \text{viscous form resistance}}{\text{skin friction resistance}}$$

$$= 1 + \frac{\text{viscous form resistance}}{\text{skin friction resistance}}$$

and if $k = \frac{\text{viscous form resistance}}{\text{skin friction resistance}}$

then $\frac{AC}{BC} = (1+k) \qquad (5.18)$

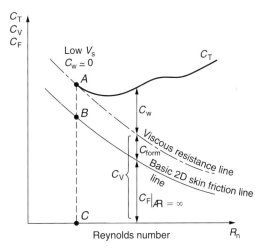

Figure 5.14 Hughes model of ship resistance.

In Equation (5.18), $(1 + k)$ is termed the form factor and is assumed constant for both the ship and its model. Indeed the form factor is generally supposed to be independent of speed and scale in the resistance extrapolation method. In practical cases the determination of $(1 + k)$ is normally carried out using a variant of the Prohashka method by a plot of C_T against F_n^4 and extrapolating the curve to $F_n = 0$ (Figure 5.15). From this figure the form factor $(1 + k)$ is deduced from the relationship

$$1 + k = \lim_{F_n \to 0} \left(\frac{R}{R_F} \right)$$

This derivation of the form factor can be used in the resistance extrapolation only if scale-independent pressure resistance is absent; for example, there must be no immersion of the transom and slender appendages which are oriented to the direction of flow.

Although traditionally the form factor $(1 + k)$ is treated as a constant with varying Froude number the fundamental question remains as to whether it is valid to assume that the $(1 + k)$ value, determined at vanishing Froude number, is valid at high speed. This is of particular concern at speeds beyond the main resistance hump where the flow configuration around the hull is likely to be very different from that when $F_n = 0$, and therefore a Froude number dependency can be expected for $(1 + k)$. In addition a Reynolds dependency may also be expected since viscous effects are the basis of the $(1 + k)$ formulation. The Froude and Reynolds effects are, however, likely to effect most the high-speed performance and have a lesser influence on general craft.

The extrapolation from model to full scale using Hughes' method is shown in Figure 5.16(a), from which it is seen that the two-dimensional skin

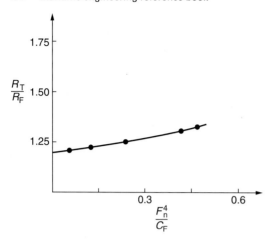

Figure 5.15 Determination of $(1 + k)$ using Prohaska method.

friction line, Equation (5.16), is used as a basis and the viscous resistance is estimated by scaling the basic friction line by the form factor $(1 + k)$. This then acts as a basis for calculating the wave making resistance from the measured total resistance on the model which is then equated to the ship condition along with the recalculated viscous resistance for the ship Reynolds number. The Froude approach (Figure 5.16(b)), is essentially the same, except that the frictional resistance is based on one of the Froude, ATTC (Equation (5.12)) or ITTC (Equation (5.13)) friction lines without a $(1 + k)$ factor. Clearly the magnitude of the calculated wave making resistance, since it is measured total resistance minus calculated frictional resistance, will vary according to the friction formulation used. This is also true of the correlation allowances as defined in Equation (5.15), and therefore the magnitudes of these parameters should always be considered in the context of the approach and experimental facility used.

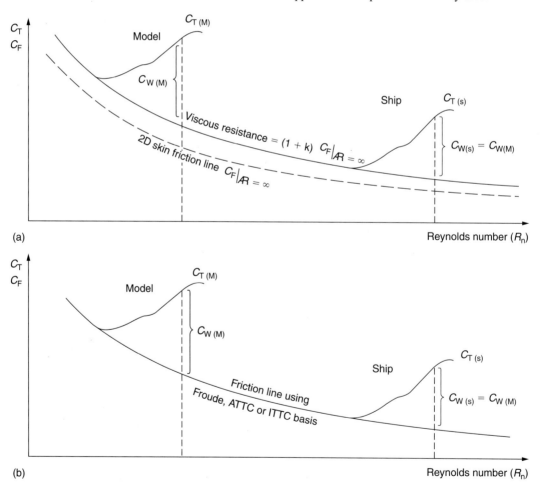

Figure 5.16 Comparison of extrapolation approaches: (a) extrapolation using Hughes approach and (b) extrapolation using Froude approach.

In practice both the Froude and Hughes approaches are used in model testing; the latter, however, is most frequently used in association with the ITTC 1957 friction formulation rather than Equation (5.16).

5.1.3 Methods of resistance evaluation

To evaluate the resistance of a ship the designer has several options available. These range, as shown in Figure 5.17, from what may be termed the traditional methods through to advanced Computational Fluid Dynamics (CFD) methods. The choice of method depends not only on the capability available but also on the accuracy desired, the funds available and the degree to which the approach has been developed. Figure 5.17 identifies four basic classes of approach to the problem; the traditional and standard series, the regression-based procedures; the direct model test and the CFD approach. Clearly these are somewhat artificial distinctions, and consequently break down on close scrutiny; they are, however, convenient classes for discussion purposes.

Unlike the CFD and direct model test approaches, the other methods are based on the traditional naval architectural parameters of hull form; for example, block coefficient, longitudinal centre of buoyancy, prismatic coefficient, etc. These form parameters have served the industry well in the past for resistance calculation purposes; however, as requirements become more exacting and hull forms become more complex these traditional parameters are less able to reflect the growth of the boundary layer and wave making components. As a consequence much current research is being expended in the development of form parameters which will reflect the hull surface contours in a more equable way.

5.1.3.1 Traditional and standard series analysis methods

A comprehensive treatment of these methods would require a book in itself and would also lie to one side of the main theme of this text. As a consequence an outline of four of the traditional methods starting with that of Taylor and proceeding through Ayer's analysis to the later methods of Auf'm Keller and Harvald are presented in order to illustrate the development of these methods.

Taylor's method (1910–1943)
Admiral Taylor in 1910 published the results of model tests on a series of hull forms. This work has since been extended (Taylor (1943)) to embrace a range of $V/\sqrt{L}$ from 0.3 to 2.0. The series comprised some 80 models in which results are published for beam draught ratios of 2.25, 3.0 and 3.75 with five displacement length ratios. Eight prismatic coefficients were used spanning the range 0.48 to 0.80, which tends to make the series useful for the faster and less full vessels.

The procedure is centred on the calculation of the residual resistance coefficients based on the data for each B/T value corresponding to the prismatic and $V/\sqrt{L}$ values of interest. The residual resistance component C_R is found by interpolation from the

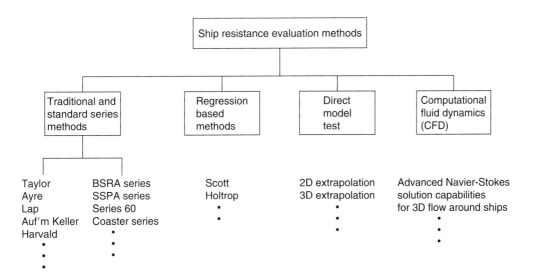

Figure 5.17 Ship resistance evaluation methods and examples.

three B/T values corresponding to the point of interest. The frictional resistance component is calculated on a basis of Reynolds number and wetted surface area together with a hull roughness allowance. The result of this calculation is added to the interpolated residuary resistance coefficient to form the total resistance coefficient C_T from which the naked effective horsepower is derived for each of the chosen $V/\sqrt{L}$ values from the relation

$$\text{EHP}_n = AC_T V_S^3 \qquad (5.19)$$

where A is the wetted surface area.

Ayre's method (1942)
Ayre (1948) developed method in 1927, again based on model test data, using a series of hull forms relating to colliers. In his approach, which in former years achieved widespread use, the method centres on the calculation of a constant coefficient C_2 which is defined by Equation (5.20)

$$\text{EHP} = \frac{\Delta^{0.64} V_S^3}{C_2} \qquad (5.20)$$

This relationship implies that in the case of full-sized vessels of identical forms and proportions, the EHP at corresponding speeds varies as $(\Delta^{0.64} V_S^3)$ and that C_2 is a constant at given values of $V/\sqrt{L}$. In this case the use of $\Delta^{0.64}$ avoids the necessity to treat the frictional and residual resistances separately for vessels of around 30 m.

The value of C_2 is estimated for a standard block coefficient. Corrections are then made to adjust the standard block coefficient to the actual value and corrections applied to cater for variations in the beam–draught ratio, position of the LCB and variations in length from the standard value used in the method's derivation.

Auf'm Keller method
Auf'm Keller (1973) extended the earlier work of Lap (1954) in order to allow the derivation of resistance characteristics of large block coefficient, single-screw vessels. The method is based on the collated results from some 107 model test results for large single-screw vessels and the measurements were converted into five sets of residuary resistance values. Each of the sets is defined by a linear relationship between the longitudinal centre of buoyancy and the prismatic coefficient. Figure 5.18 defines these sets, denoted by the letters A to E, and Figure 5.19 shows the residuary resistance coefficient for set A. As a consequence it is possible to interpolate between the sets for a particular LCB versus C_P relationship.

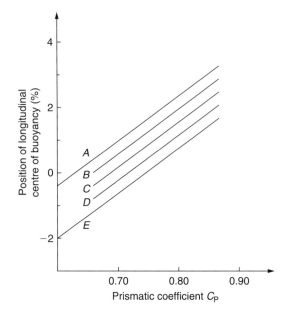

Figure 5.18 Definition of ship class.

The procedure adopted is shown in outline form by Figure 5.20 in which the correction for ζ_r and the ship model correlation C_A are given by Equation (5.21) and Table 5.4, respectively:

$$\% \text{ change in } \zeta_r = 10.357[e^{1.129(6.5-L/B)} - 1] \qquad (5.21)$$

As in the case of the previous two methods the influence of the bulbous bow is not taken into account but good experience can be achieved with the method within its area of application.

Harvald method
The method proposed by Harvald (1978) is essentially a preliminary power prediction method designed to obtain an estimate of the power required to drive a vessel. The approach used is to define four principal parameters upon which to base the estimate; the four selected are:

1. the ship displacement (Δ),
2. the ship speed (V_s),
3. the block coefficient (C_b),
4. the length displacement ratio ($L/\nabla^{1/3}$).

By making such a choice all the other parameters that may influence the resistance characteristics need to be standardized, such as hull form, B/T ratio, LCB, propeller diameter, etc. The method used by Harvald is to calculate the resistance of a standard form for a range of the four parameters cited above and then evaluate the shaft power using a Quasi-Propulsive

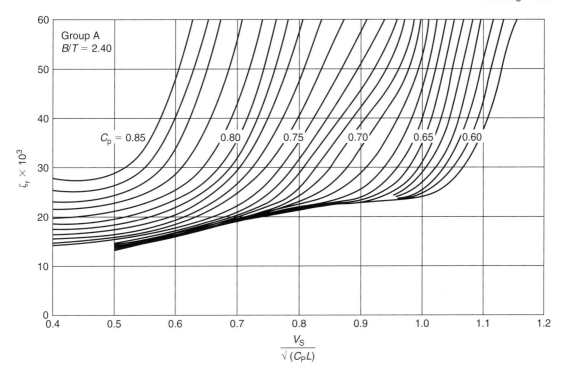

Figure 5.19 Diagram for determining the specific residuary resistance as a function of $V_s/\sqrt{(C_p L)}$ and C_p (Reproduced with permission from Auf'm Keller (1973)).

Coefficient (QPC) based on the wake and thrust deduction method discussed in Section 5.2 and a propeller open water efficiency taken from the Wageningen B Series propellers. The result of this analysis led to the production of seven diagrams for a range of block coefficient from 0.55 to 0.85 in 0.05 intervals of the form shown in Figure 5.21. From these diagrams an estimate of the required power under trial conditions can be derived readily with the minimum of effort. However, with such a method it is important to make allowance for deviations of the actual form from those upon which the diagrams are based.

Standard series data

In addition to the more formalized methods of analysis there is a great wealth of data available to the designer and analyst in the form of model data and more particularly in model data relating to standard series hull forms. That is, those in which the geometric hull form variables have been varied in a systematic way. Much data has been collected over the years and Bowden (1970) gives a very useful guide to the extent of the data available for single-screw ocean-going ships between the years 1900 and 1969. Some of the more recent and important series and data are given in Pattullo (1974), Moor (1960a and b; 1965/66; 1973, 1974), Moor et al. (1961), Moor and Pattullo (1968), Lackenby et al. (1966), SSPA (1969), van Manan et al. (1961). Unfortunately, there is little uniformity of presentation in the work as the results have been derived over a long period of time in many countries of the world. The designer therefore has to accept this state of affairs and account for this in his calculations. In addition hull form design has progressed considerably in recent years and little of these changes is reflected in the data cited in these references. Therefore, unless extreme care is exercised in the application of such data, significant errors can be introduced into the resistance estimation procedure.

In more recent times the Propulsion Committee of the ITTC have been conducting a cooperative experimental programme between tanks around the world, ITTC (1987). The data so far reported relates to the Wigley parabolic hull and the Series 60, $C_b = 0.60$ hull forms.

5.1.3.2 Regression-based methods

Ship resistance prediction based on statistical regression methods has been a subject of some interest for a number of years. Early work by Scott in the 1970s (Scott (1972, 1973)), resulted in methods

198 Maritime engineering reference book

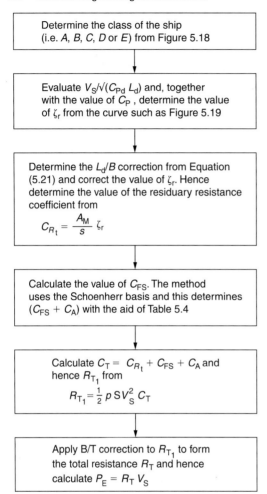

Figure 5.20 Auf'm Keller resistance calculation.

Table 5.4 Values of C_A used in Auf'm Keller method, Auf'm Keller (1973).

Length of vessel (m)	Ship model correlation allowance
50–150	$0.0004 \rightarrow 0.00035$
150–210	0.0002
210–260	0.0001
260–300	0
300–350	-0.0001
350–450	-0.00025

for predicting the trial performance of single- and twin-screw merchant ships.

The theme of statistical prediction was then taken up by Holtrop in a series of papers (Holtrop (1977, 1978, 1988) and Holtrop and Mennen (1978, 1982)). These papers trace the development of a power prediction method based on the regression analysis of random model and full-scale test data together with, in the latest version of the method, the published results of the Series 64 high-speed displacement hull terms. In this latest version the regression analysis is now based on the results of some 334 model tests. The results are analysed on the basis of the ship resistance equation.

$$R_T = R_F(1 + k_1) + R_{APP} + R_W + R_B + R_{TR} + R_A \quad (5.22)$$

In this equation the frictional resistance R_F is calculated according to the 1957 ITTC friction formulation, Equation (5.13), and the hull form factor $(1 + k_1)$ is based on a regression equation and is expressed as a function of afterbody form, breadth, draught, length along the waterline, length of run, displacement, prismatic coefficient:

$$(1 + k_1) = 0.93 + 0.487118(1 + 0.011 C_{\text{stern}})$$
$$\times (B/L)^{1.06806}(T/L)^{0.46106}$$
$$\times (L_{WL}/L_R)^{0.121563}(L_{WL}^3/\nabla)^{0.36486}$$
$$\times (1 - C_P)^{-0.604247} \quad (5.23)$$

in which the length of run L_R is defined by a separate relationship, if unknown, as follows:

$$L_R = L_{WL}\left(1 - C_P + \frac{0.06 C_P \, \text{LCB}}{(4C_P - 1)}\right)$$

The sternshape parameter C_{stern} in Equation (5.23) is defined in relatively discrete and coarse steps for different hull forms, as shown in Table 5.5:

The appendage resistance according to the Holtrop approach is evaluated from the equation

$$R_{APP} = \tfrac{1}{2}\rho V_S^2 C_F(1 + k_2)_{\text{equv}} \sum S_{APP} + R_{BT} \quad (5.24)$$

in which the frictional coefficient C_F of the ship is again determined by the ITTC 1957 line and S_{APP} is the wetted area of the particular appendages of the vessel. To determine the equivalent $(1 + k_2)$ value for the appendages, denoted by $(1 + k_2)_{\text{equv}}$, appeal is made to the relationship

$$(1 + k_2)_{\text{equv}} = \frac{\sum(1 + k_2)S_{APP}}{\sum S_{APP}} \quad (5.25)$$

The values of the appendage form factors are tentatively defined by Holtrop as shown in Table 5.6.

If bow thrusters are fitted to the vessel their influence can be taken into account by the term R_{BT} in Equation (5.24) as follows:

$$R_{BT} = \pi \rho V_S^2 d_T C_{BTO}$$

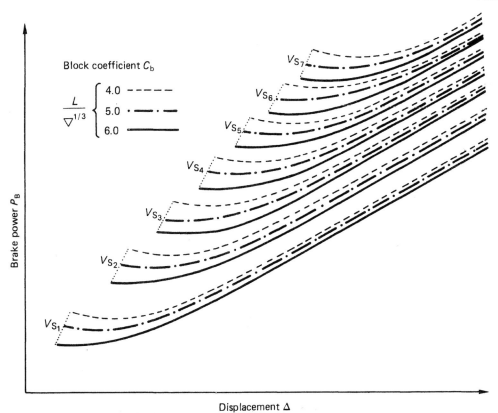

Figure 5.21 Harvald estimation diagram for ship power.

Table 5.5 C_{stern} parameters according to Holtrop.

Afterbody form	C_{stern}
Pram with gondola	−25
V-shaped sections	−10
Normal section ship	0
U-shaped sections with Hogner stern	10

Table 5.6 Tentative appendage form factors $(1 + k_2)$.

Appendage type	$(1 + k_2)$
Rudder behind skeg	1.5–2.0
Rudder behind stern	1.3–1.5
Twin-screw balanced rudders	2.8
Shaft brackets	3.0
Skeg	1.5–2.0
Strut bossings	3.0
Hull bossings	2.0
Shafts	2.0–4.0
Stabilizer fins	2.8
Dome	2.7
Bilge keels	1.4

in which d_T is the diameter of the bow thruster and the coefficient C_{BTO} lies in the range 0.003 to 0.012. When the thruster lies in the cylindrical part of the bulbous bow, $C_{BTO} \to 0.003$.

The prediction of the wave making component of resistance has proved difficult and in the last version of Holtrop's method (Holtrop (1988)) a three-banded approach is proposed to overcome the difficulty of finding a general regression formula. The ranges proposed are based on the Froude number F_n and are as follows:

Range 1: $F_n > 0.55$
Range 2: $F_n < 0.4$
Range 3: $0.4 < F_n < 0.55$

within which the general form of the regression equations for wave making resistance in ranges 1 and 2 is

$$R_W = K_1 K_2 K_3 \nabla \rho g \, \exp[K_4 F_n^{K_6} + K_5 \cos(K_7/F_n^2)] \tag{5.26}$$

The coefficients K_1, K_2, K_3, K_4, K_5, K_6 and K_7 are defined by Holtrop (1988) and it is of interest to note that the coefficient K_2 determines the influence of the bulbous bow on the wave resistance. Furthermore, the difference in the coefficients of Equation (5.26) between ranges 1 and 2 above lie in the coefficients K_1 and K_4. To accommodate the intermediate range, range 3, a more or less arbitrary interpolation formula is used of the form

$$R_W = R_W|_{F_n = 0.4} + \frac{(10F_n - 4)}{1.5}$$

$$\times [R_W|_{F_n = 0.55} - R_W|_{F_n = 0.4}] \quad (5.27)$$

The remaining terms in Equation (5.22) relate to the additional pressure resistance of the bulbous bow near the surface R_B and the immersed part of the transom R_{TR} and are defined by relatively simple regression formulae. With regard to the model–ship correlation resistance the most recent analysis has shown the formulation in Holtrop and Menen (1982) to predict a value some 9 to 10 per cent high; however, for practical purposes that formulation is still recommended by Holtrop:

$$R_A = \tfrac{1}{2} \rho V_s^2 S C_A$$

where

$$C_A = 0.006(L_{WL} + 100)^{-0.16} - 0.00205$$

$$+ 0.003 \sqrt{(L_{WL}/7.5)} C_B^4 K_2 (0.04 - c_4) \quad (5.28)$$

in which $c_4 = T_F/L_{WL}$ when $T_F/L_{WL} \leq 0.04$
and $c_4 = 0.04$ when $T_F/L_{WL} > 0.04$

where T_F is the forward draught of the vessel and S is the wetted surface area of the vessel.

K_2 which also appears in Equation (5.26) and determines the influence of the bulbous bow on the wave resistance is given by

$$K_2 = \exp[-1.89 \sqrt{c_3}]$$

where

$$c_3 = \frac{0.56(A_{BT})^{1.5}}{BT(0.31\sqrt{A_{BT}} + T_F - h_B)}$$

in which A_{BT} is the transverse area of the bulbous bow and h_B is the position of the centre of the transverse area A_{BT} above the keel line with an upper limit of $0.6T_F$ (see Figure 5.10).

Equation (5.28) is based on a mean apparent amplitude hull roughness $k_S = 150 \mu m$. In cases where the roughness may be larger than this use can be made of the ITTC-1978 formulation, which gives the increase in roughness as

$$\Delta C_A = (0.105 k_S^{1/3} - 0.005579)/L^{1/3} \quad (5.29)$$

The Holtrop method provides a most useful estimation tool for the designer. However, like many analysis procedures it relies to a very large extent on traditional naval architectural parameters. As these parameters cannot fully act as a basis for representing the hull curvature and its effect on the flow around the vessel there is a natural limitation on the accuracy of the approach without using more complex hull definition parameters. At the present time considerable research is proceeding in this direction to extend the viability of the resistance prediction method.

5.1.3.3 Direct model tests

Model testing of a ship in the design stage is an important part of the design process and one that, in a great many instances, is either not explored fully or is not undertaken. In the author's view this is a false economy, bearing in mind the relatively small cost of model testing as compared to the cost of the ship and the potential costs that can be incurred in design modification to rectify a problem or the through life costs of a poor performance optimization.

General procedure for model tests
Whilst the detailed procedures for model testing differ from one establishment to another the underlying general procedure is similar. Here the general concepts are discussed, but for a more detailed account reference can be made to Phillips-Birt (1970). With regard to resistance and propulsion testing there are fewer kinds of experiment that are of interest: the resistance test, the open water propeller test, the propulsion test and the flow visualization test. The measurement of the flow field is discussed in Carlton (2007).

Resistance tests
In the resistance test the ship model is towed by the carriage and the total longitudinal force acting on the model is measured for various speeds (Figure 5.22). The breadth and depth of the towing tank essentially governs the size of the model that can be used. Todd's original criterion that the immersed cross-section of the vessel should not exceed 1 per cent of the tank's cross-sectional area was placed in doubt after the famous *Lucy Ashton* experiment. This showed that to avoid boundary interference from the tank walls and bottom this proportion should be reduced to the order of 0.4 per cent.

The model, constructed from paraffin wax, wood or glass-reinforced plastic, requires to be manufactured to a high degree of finish and turbulence simulators placed at the bow of the model in order to stimulate the transition from a laminar into a turbulent boundary layer over the hull. The model is positioned under the carriage and towed in such a way that it is free to heave and pitch, and ballasted to the required draught and trim.

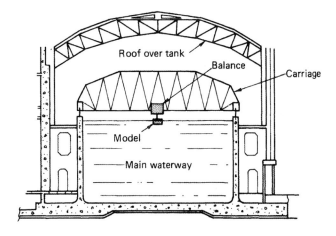
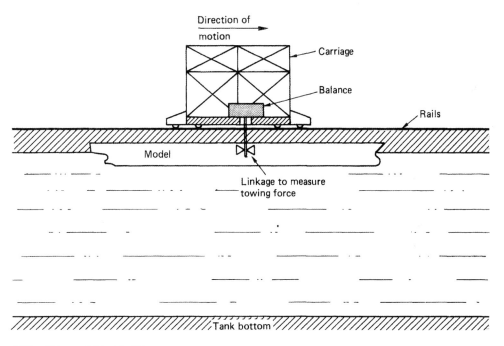

Figure 5.22 Ship model test facility.

In general there are two kinds of resistance tests: the naked hull and the appended resistance test. If appendages are present local turbulence tripping is applied in order to prevent the occurrence of uncontrolled laminar flow over the appendages. Also the propeller should be replaced by a streamlined cone to prevent flow separation in this area.

The resistance extrapolation process follows Froude's hypothesis and the similarity law is followed. As such the scaling of the residual, or wave making component, follows the similarity law

$$R_{W\,ship} = R_{W_{model}} \lambda^3 (\rho_s/\rho_M)$$

provided that $V_S = V_M\sqrt{\lambda}$, where $\lambda = L_S/L_M$.

In general, the resistance is scaled according to the relationship

$$R_s = [R_M - R_{F_M}(1+k)]\lambda^3 \left(\frac{\rho_S}{\rho_M}\right) + R_{F_S}(1+k) + R_A$$

$$= [R_M - F_D]\lambda^3 \left(\frac{\rho_S}{\rho_M}\right) \qquad (5.30)$$

in which

$$F_D = \tfrac{1}{2}\rho_M V_M^2 S_M (1+k)(C_{F_M} - C_{F_S}) - \frac{\rho_M}{\rho_S} R_A/\lambda^3$$

that is,

$$F_D = \tfrac{1}{2}\rho_M V_M^2 S_M[(1+k)](C_{F_M} - C_{F_S}) - C_A] \quad (5.31)$$

The term F_D is known as both the scale effect correction on resistance and the friction correction force. The term R_A in Equation (5.30) is the resistance component, which is supposed to allow for the following factors: hull roughness; appendages on the ship but not present during the model experiment; still air drag of the ship and any other additional resistance component acting on the ship but not on the model. As such its non-dimensional form C_A is the incremental resistance coefficient for ship–model correlation.

When $(1 + k)$ in Equation (5.30) is put to unity, the extrapolation process is referred to as a two-dimensional approach since the frictional resistance is then taken as that given by the appropriate line, Froude flat plate data, ATTC or ITTC 1957, etc.

The effective power (P_E) is derived from the resistance test by the relationship

$$P_E = R_S V_S \quad (5.32)$$

Open water tests

The open water test is carried out on either a stock or actual model of the propeller to derive its open water characteristics in order to derive the propulsion coefficients. The propeller model is fitted on a horizontal driveway shaft and is moved through the water at an immersion of the shaft axis frequently equal to the diameter of the propeller (Figure 5.23).

The loading of the propeller is normally carried out by adjusting the speed of advance and keeping the model revolutions constant. However, when limitations in the measuring range, such as a J-value close to zero or a high carriage speed needed for a high J-value, are reached the rate of revolutions is also varied. The measured thrust values are corrected for the resistance of the hub and streamlined cap, this correction being determined experimentally in a test using a hub only without the propeller.

The measured torque and corrected thrust are expressed as non-dimensional coefficients K_{TO} and K_{QO} in the normal way (see Section 5.3); the suffix O being used in this case to denote the open rather than the behind condition. The open water efficiency and the advance coefficient are then expressed as

$$\eta_0 = \frac{J}{2\pi}\frac{K_{TO}}{K_{QO}}$$

and

$$J = \frac{V_c}{nD}$$

where V_c is the carriage speed.

Unless explicitly stated it should not be assumed that the propeller open water characteristics have been corrected for scale effects. The data from these

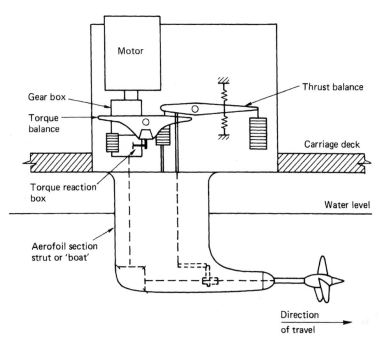

Figure 5.23 Propeller open water test using towing tank carriage.

Propulsion tests

In the propulsion test the model is prepared in much the same way as for the resistance test and turbulence stimulation on the hull and appendages is again applied. For this test, however, the model is fitted with the propeller used in the open water test together with an appropriate drive motor and dynamometer. During the test the model is free to heave and pitch as in the case of the resistance test.

In the propulsion test the propeller thrust T_M, the propeller torque Q_M and the longitudinal towing force F acting on the model are recorded for each tested combination of model speed V_M and propeller revolutions n_M.

Propulsion tests are carried out in two parts. The first comprises a load variation test at one or sometimes more than one constant speed whilst the other comprises a speed variation test at constant apparent advance coefficient or at the self-propulsion point of the ship. The ship self-propulsion point being defined when the towing force (F) on the carriage is equal to the scale effect correction on viscous resistance (F_D), Equation (5.31).

The required thrust T_S and self-propulsion point of the ship is determined from the model test using the equation:

$$T_S = \left[T_M + (F_D - F) \frac{\partial T_M}{\partial F} \right] \lambda^3 \frac{\rho_S}{\rho_M} \quad (5.33)$$

In Equation (5.33) the derivative $\partial T_M / \partial F$ is determined from the load variation tests which form the first part of the propulsion test. In a similar way the local variation test can be interpolated to establish the required torque and propeller rotational speed at self-propulsion for the ship.

In the extrapolation of the propulsion test to full scale the scale effects on resistance (F_D), on the wake field and on the propeller characteristics need to be taken into account. At some very high speeds the effects of cavitation also need to be taken into account. This can be done by analysis or through the use of specialized facilities.

Flow visualization tests

Various methods exist to study the flow around the hull of a ship. One such method is to apply stripes of an especially formulated paint to the model surface, the stripes being applied vertical to the base line. The model is then towed at Froude identity and the paint will smear into streaks along the hull surface in the direction of the flow lines.

In cases where the wall shear stresses are insufficient tufts are used to visualize the flow over the hull. In general, woollen threads of about 5 cm in length will be fitted onto small needles driven into the hull surface. The tufts will be at a distance of between 1 and 2 cm from the hull surface and the observation made using an underwater television camera. The interaction phenomenon between the propeller and ship's hull can also be studied in this way by observing the behaviour of the tufts with and without the running propeller.

Model test facilities

Many model test facilities exist around the world almost all of which possess a ship model towing tank. Some of the model facilities available are listed in Table 5.7; this, however, is by no means an exhaustive list of facilities but is included here to give an idea of the range of facilities available. A current list of test facilities is held by ITTC.

Two-dimensional extrapolation method

This as discussed previously is based on Froude's original method without the use of a form factor. Hence the full-scale resistance is determined from

$$R_S = (R_M - F_D)\lambda^3 \left(\frac{\rho_S}{\rho_M} \right)$$

where

$$F_D = \tfrac{1}{2} \rho_M V_M^2 S_M (C_{F_M} - C_{F_S} - C_A)$$

and when Froude's friction data is used C_A is set to zero, but this is not the case if the ATTC-1947 or ITTC-1957 line is used.

When the results of the propulsion test are either interpolated for the condition when the towing force (F) is equal to F_D or when F_D is actually applied in the self-propulsion test the corresponding model condition is termed the 'self-propulsion point of the ship'. The direct scaling of the model data at this condition gives the condition generally termed the 'tank condition'. This is as follows:

$$\left.\begin{aligned}
P_{DS} &= P_{DM} \lambda^{3.5} \left(\frac{\rho_S}{\rho_M} \right) \\
T_S &= T_M \lambda^3 \left(\frac{\rho_S}{\rho_M} \right) \\
n_S &= n_M / \sqrt{\lambda} \\
V_S &= V_M \sqrt{\lambda} \\
R_S &= (R_M - F_D) \lambda^3 \left(\frac{\rho_S}{\rho_M} \right)
\end{aligned}\right\} \quad (5.34)$$

Table 5.7 Examples of towing tank facilities around the world (Reproduced with permission from Clayton and Bishop (1982)).

Facilities	Length (m)	Width (m)	Depth (m)	Maximum carriage speed (m/s)
European facilities				
Qinetiq Haslar (UK)	164	6.1	2.4	7.5
	270	12.0	5.5	12.0
Experimental and	76	3.7	1.7	9.1
Electronic Lab.	188	2.4	1.3	13.1
B.H.C. Cowes (UK)	197	4.6	1.7	15.2
MARIN Wageningen (NL)	100	24.5	2.5	4.5
	216	15.7	1.25	5
	220	4.0	4.0	15/30
	252	10.5	5.5	9
MARIN Depressurized Facility, Ede (NL)	240	18.0	8.0	4
Danish Ship Research Laboratories	240	12.0	6.0	14
Ship Research Institute of Norway (NSFI)	27	2.5	1.0	2.6
	175	10.5	5.5	8.0
SSPA. Göteborg, Sweden	260	10.0	5.0	14.0
Bassin d'Essais de Carènes, Paris	155	8.0	2.0	5
	220	13.0	4.0	10
VWS West Germany	120	8.0	1.1	4.2
	250	8.0	4.8	20
H.S.V. Hamburg West Germany	30	6.0	1.2	0.0023–1.9
	80	4.0	0.7	3.6
	80	5.0	3.0	3.6
	300	18.0	6.0	8.0
B.I.Z. Yugoslavia	37.5	3.0	2.5	3
	23	12.5	6.2	8
	293	5.0	3.5	12
North American Facilities				
NSRDC Bethesda USA	845	15.6	6.7	10
	905	6.4	3.0–4.8	30
NRC, Marine Dynamics and Ship Laboratory, Canada	137	7.6	3.0	8
Far East Facilities				
Meguro Model Basin, Japan	98	3.5	2.25	7
	235	12.5	7.25	10
	340	6.0	3.0	20
Ship Research Institute, Mitaka Japan	20	8.0	0–1.5	2
	50	8.0	4.5	2.5
	140	7.5	0–3.5	6
	375	18.0	8.5	15
KIMM – Korea	223	16.0	7.0	
Hyundai – Korea	232	14.0	6.0	

The power and propeller revolutions determined from the tank condition as given by Equation (5.34) require to be converted into trial prediction figures for the vessel. In the case of the power trial prediction this needs to be based on an allowance factor for the results of trials of comparable ships of the same size or alternatively on the results of statistical surveys. The power trial allowance factor is normally defined as the ratio of the shaft power measured on trial to the power delivered to the propeller in the tank condition.

The full-scale propeller revolutions prediction is based on the relationship between the delivered power and the propeller revolutions derived from the tank condition. The power predicted for the trial condition is then used in this relationship to devise the corresponding propeller revolutions. This propeller speed is corrected for the over- or underloading effect and often corresponds to around $\frac{1}{2}$ per cent decrease of rpm for a 10 per cent increase of power. The final stage in the propeller revolutions prediction is to account for the scale effects in the wake and propeller blade friction. For the trial condition these scale effects are of the order of

$\frac{1}{2}\sqrt{\lambda}$% for single-screw vessels
$1 - 2$% for twin-screw vessels

The allowance for the service condition on rotational speed is of the order one per cent.

Three-dimensional extrapolation method
The three-dimensional extrapolation method is based on the form factor concept. Accordingly the resistance is scaled under the assumption that the viscous resistance of the ship and its model is proportional to the frictional resistance of a flat plate of the same length and wetted surface area when towed at the same speed, the proportionality factor being $(1 + k)$ as discussed in Section 5.1.2. In addition it is assumed that the pressure resistance due to wave generation, stable separation and induced drag from non-streamlined or misaligned appendages follow the Froude similarity law.

The form factor $(1 + k)$ is determined for each hull from low-speed resistance or propulsion measurements when the wave resistance components are negligible. In the case of the resistance measurement of form factor then this is based on the relationship:

$$(1 + k) = \lim_{F_n \to 0} \left(\frac{R}{R_F} \right)$$

In the case of the propulsion test acting as a basis for the $(1 + k)$ determination then this relationship takes the form

$$(1 + k) = \lim_{F_n \to 0} \left(\frac{F - T/(\partial T/\partial F)}{(F|_{T=0}/R)R_F} \right)$$

The low-speed measurement of the $(1 + k)$ factor can only be validly accomplished if scale-independent pressure resistance is absent, which means, for example, that there is no immersed transom. In this way the form factor is maintained independent of speed and scale in the extrapolation method.

In the three-dimensional method the scale effect on the resistance is taken as

$$F_D = \tfrac{1}{2}\rho_M V_M^2 S_M[(1 + k)(C_{F_M} - C_{F_S}) - C_A]$$

in which the form factor is normally taken relative to the ITTC-1957 line and C_A is the ship-model correlation coefficient. The value of C_A is generally based on an empirically based relationship and additional allowances are applied to this factor to account for extreme hull forms at partial draughts, appendages not present on the model, 'contract' conditions, hull roughness different from the standard of 150 μm, extreme superstructures or specific experience with previous ships.

In the three-dimensional procedure the measured relationship between the thrust coefficient K_T and the apparent advance coefficient is corrected for wake scale effects and for the scale effects on propeller blade friction. At model scale the model thrust coefficient is defined as

$$K_{TM} = f(F_n, J)_M$$

whereas at ship scale this is

$$K_{TS} = f\left(F_n, J\left(\frac{1 - w_{TS}}{1 - w_{TM}}\right) + \Delta K_T\right)$$

According to the ITTC 1987 version of the manual for the use of the 1978 performance reduction method, the relationship between the ship and model Taylor wake fractions can be defined as

$$w_{TS} = (t + 0.04) + (w_{TM} - t - 0.04)$$
$$\times \frac{(1 + k)C_{FS} + \Delta C_F}{(1 + k)C_{FM}}$$

where 0.04 is included to take account of the rudder effect and ΔC_F is the roughness allowance given by

$$\Delta C_F = \left[105\left(\frac{k_s}{L_{WL}}\right)^{1/3} - 0.64\right] \times 10^{-3}$$

The measured relationship between the thrust and torque coefficient is corrected for the effects of friction over the blades such that

$$K_{TS} = K_{TM} + \Delta K_T \quad \text{and} \quad K_{QS} = K_{QM} + \Delta K_Q$$

where the factors ΔK_T and ΔK_Q are determined from the ITTC procedure as discussed in Section 5.3.

The load of the full-scale propeller is obtained from the relationship

$$\frac{K_T}{J^2} = \frac{S}{2D^2} \frac{C_{TS}}{(1 - t)(1 - w_{TS})^2}$$

and with K_T/J^2 as the input value the full-scale advance coefficient J_{TS} and torque coefficient K_{QTS} are read off from the full-scale propeller characteristics and the following parameters calculated:

$$n_S = \frac{(1 - w_{TS})V_S}{J_{TS}D}$$

$$P_{DS} = 2\pi \rho D^5 n_S^3 \frac{K_{QTS}}{\eta_R} \times 10^{-3}$$

$$T_S = \frac{K_T}{J^2} J_{TS}^2 \rho D^4 n_S^2$$

$$Q_S = \frac{K_{QTS}}{\eta_R} \rho D^5 n_S^2 \qquad (5.35)$$

The required shaft power P_S is found from the delivered power P_{DS} using the shafting mechanical efficiency η_S as

$$P_S = P_{DS}/\eta_S$$

5.1.3.4 *Computational fluid dynamics*

(a) Wave resistance computations
The wave resistance problem considers the steady motion of a ship in initially smooth water assuming an ideal fluid, i.e. especially neglecting all viscous effects. The ship will create waves at the freely deformable water surface. The computations involve far more information than the mere resistance which is of minor importance in many applications and usually computed quite inaccurately. But the expression 'wave resistance problem' is easier than 'steady, inviscid straight-ahead course problem', and thus more popular.

The work of the Australian mathematician J. H. Michell in 1898 is seen often as the birth of modern theoretical methods for ship wave resistance predictions. While Michell's theory cannot be classified as computational fluid dynamics in the modern sense, it was a milestone at its time and is still inspiring mathematicians today. Michell expressed the wave resistance of a thin wall-sided ship as:

$$R_w = \frac{4}{\pi} \rho V^2 v^2 \int_1^\infty \frac{\lambda^2}{\sqrt{\lambda^2 - 1}} |A(\lambda)|^2 \, d\lambda \qquad (5.36)$$

with:

$$A(\lambda) = -iv\lambda \int_S e^{v\lambda^2 z + iv\lambda x} f(x, z) dz \, dx \qquad (5.37)$$

V is the ship speed, ρ water density, $v = g/V^2$, g gravity acceleration, $f(x, z)$ halfwidth of ship, x longitudinal coordinate (positive forward), z vertical coordinate (from calm waterline, positive upwards), S ship surface below the calm waterline. The expression gives realistic results for very thin bodies (width/length ratio very small) for arbitrary Froude number, and for slender ships (width/length ratio and depth/length ratio very small) for high Froude numbers. Michell's theory (including all subsequent refinements) is in essence unacceptable for real ship geometries and ship speeds. However, on occasion it is still useful. An example may be the prediction of the wave resistance of a submarine near the free surface with a streamlined snorkel piercing the free surface. While CFD can discretize the main submarine, it will neglect all appendages of much smaller scale. Then Michell's theory can be applied to analyse the additional influence of the snorkel which will have a very large Froude number based on the chord length of its profile cross-section. Söding (1995) gives a Fortran routine to compute Michell's integral.

The classical methods (thin ship theories, slender-body theories) introduce simplifications which imply limitations regarding the ship's geometry. Such methods can be successful for vessels with high length-displacement ratios, such as high-speed semi-displacement catamarans, Couser *et al.* (1998). Many ship geometries are generally not thin or slender enough. The differences between computational and experimental results are consequently unacceptable. Practical applications in industry are based almost exclusively on boundary element methods. These remain the most important tools for naval architects despite the recently increased application of viscous flow tools.

These boundary element methods represent the flow as a superposition of Rankine sources and sometimes also dipoles or vortices. Classical methods using so-called Kelvin or Havelock sources fulfil automatically a crude approximation of the dynamical and kinematical free surface conditions. Kelvin sources are complicated and require great care in their numerical evaluation. Rankine sources on the other hand are quite simple. The potential of a Rankine point source is a factor divided by the distance between the point source and the considered point in the fluid domain. The factor is called the source strength. The derivative of the potential in arbitrary spatial direction gives the velocity in this direction. This mathematical operation is simple to perform for Rankine sources.

Boundary element methods discretize surfaces into a finite number of elements and corresponding number of collocation points. A desired (linear) condition is fulfilled exactly at these collocation points by proper adjustment of the initially unknown source strengths. One hopes/claims that between these points the boundary condition is fulfilled at least in good approximation. Laplace's equation and the decay condition (far away the ship does not disturb the flow) are automatically fulfilled. Mirror images of the panels at the bottom of the fluid domain walls may enforce a no-penetration condition there for shallow-water cases. Repeated use of mirror images at vertical canal walls can enforce in similar fashion the side-wall condition. For numerical reasons, this is preferable to a treatment of the side walls as collocation points similar as for the ship hull.

In the wave resistance problem, we consider a ship moving with constant speed V in water of constant depth and width. For inviscid and irrotational flow, this problem is equivalent to a ship being fixed in an inflow of constant speed. The following simplifications are generally assumed:

- Water is incompressible, irrotational, and inviscid.
- Surface tension is negligible.
- There are no breaking waves.
- The hull has no knuckles which cross streamlines.
- Appendages and propellers are not included in the model. (The inclusion of a propeller makes little sense as long as viscous effects are not also included.)

The governing field equation is Laplace's equation. A unique description of the problem requires further conditions on all boundaries of the modelled fluid domain:

- Hull condition: water does not penetrate the ship's surface.
- Transom stern condition: for ships with a transom stern, we generally assume that the flow separates and the transom stern is dry. Atmospheric pressure is then enforced at the edge of the transom stern. The condition is usually linearized assuming that the water flows only in the longitudinal direction. This can only approximately reflect the real conditions at the stern, but apparently works well as long as the transom stern is moderately small as for most container ships. For fast ships which have a very large transom stern, several researchers report problems. For submerged transom sterns at low speed, the potential flow model is inapplicable and only field methods are capable of an appropriate analysis.
- Kinematic condition: water does not penetrate the water surface.
- Dynamic condition: there is atmospheric pressure at the water surface. Beneath an air cushion, this conditions modifies to the air cushion pressure. The inclusion of an air cushion in wave resistance computations has been reported in various applications. However, these computations require the user to specify the distribution of the pressure, especially the gradual decline of the pressure at the ends of the cushion. In reality, this is a difficult task as the dynamics of the air cushion and the flexible skirts make the problem more complicated. Subsequently, the computations must be expected to be less accurate than for conventional displacement hulls.
- Radiation condition: waves created by the ship do not propagate ahead. (This condition is not valid for shallow water cases when the flow becomes unsteady and soliton waves are pulsed ahead. For subcritical speeds with depth Froude number $F_{nh} < 1$, this poses no problem.)

- Decay condition: the flow is undisturbed far away from the ship.
- Open-boundary condition: waves generated by the ship pass unreflected any artificial boundary of the computational domain.
- Equilibrium: the ship is in equilibrium, i.e. trim and sinkage are changed in such a way that the dynamical vertical force and the trim moment are counteracted.
- Bottom condition (shallow-water case): no water flows through the sea bottom.
- Side-wall condition (canal case): no water flows through the side walls.
- Kutta condition (for catamaran/SWATH): at the stern/end of the strut the flow separates. The Kutta condition describes a phenomenon associated with viscous effects. Potential flow methods use special techniques to ensure that the flow separates. However, the point of separation has to be determined externally 'by higher insight'. For geometries with sharp aftbodies (foils), this is quite simple. For twin-hull ships, the disturbance of the flow by one demi-hull induces a slightly non-uniform inflow at the other demi-hull. This resembles the flow around a foil at a very small angle of incident. A simplified Kutta condition suffices usually to ensure a realistic flow pattern at the stern: Zero transverse flow is enforced. This is sometimes called the 'Joukowski condition'.

The decay condition substitutes the open-boundary condition if the boundary of the computational domain lies at infinity. The decay condition also substitutes the bottom and side wall condition if bottom and side wall are at infinity, which is the usual case.

Hull, transom stern, and Kutta condition are usually enforced numerically at collocation points. Also a combination of kinematic and dynamic condition is numerically fulfilled at collocation points. Combining dynamic and kinematic boundary conditions eliminates the unknown wave elevation, but yields a non-linear equation to be fulfilled at the *a priori* unknown free surface elevation.

Classical methods linearize the differences between the actual flow and uniform flow to simplify the non-linear boundary condition to a linear condition fulfilled at the calm-water surface. This condition is called the Kelvin condition. For practical purposes this crude approximation is nowadays no longer accepted.

Dawson (1977) proposed to use the potential of a double-body flow and the undisturbed water surface as a better approximation. Double-body linearizations were popular until the early 1990s. The original boundary condition of Dawson was inconsistent. This inconsistency was copied by most subsequent publications following Dawson's approach. Sometimes this inconsistency is accepted deliberately to avoid evaluation of higher derivatives, but in most cases and possibly also in the original it was simply an oversight. Dawson's approach requires the evaluation of terms on the free surface along streamlines of the double-body flow. This required either more or less elaborate schemes for streamline tracking or some 'courage' in simply applying Dawson's approach on smooth grid lines on the free surface which were algebraically generated.

Research groups in the UK and Japan proposed in the 1980s non-linear approximations for the correct boundary condition. These non-linear methods should not be confused with 'fully non-linear' methods that fulfil the correct non-linear boundary condition iteratively. They are not much better than Dawson's approach and no longer considered state of the art.

The first consistently linearized free surface condition for arbitrary approximations of the base flow and the free surface elevation was developed in Hamburg by Söding. This condition is rather complicated involving up to third derivatives of the potential, but it can be simply repeated in an iterative process which is usually started with uniform flow and no waves.

Fully non-linear methods were first developed in Sweden and Germany in the late 1980s. The success of these methods quickly motivated various other research groups to copy the techniques and apply the methods commercially. The most well-known codes used in commercial applications include SHIPFLOW-XPAN, SHALLO, RAPID, SWIFT, and FSWAVE/VSAERO. The development is very near the limit of what potential flow codes can achieve. The state of the art is well documented in two PhD theses, Raven (1996) and Janson (1996). Despite occasional other claims, all 'fully non-linear' codes have similar capabilities when used by their designers or somebody well trained in using the specific code. Everybody loves his own child best, but objectively the differences are small. All 'fully non-linear' codes in commercial use share similar shortcomings when it comes to handling breaking waves, semi-planing or planing boats or extreme non-linearities. It is debatable if these topics should be researched following an inviscid approach in view of the progress that viscous free-surface CFD codes have made.

Once the unknown velocity potential is determined, Bernoulli's equation determines the wave elevation. In principle, a linearized version of Bernoulli's equation might be used. However, it is computationally simpler to use the non-linear equation. Once the potential is determined, the forces can also be determined by direct pressure integration on the wetted hull. The wave resistance may also be determined by an analysis of the wave pattern (wave cut analysis) which is reported to be often more accurate. The z-force and y-moment are used to adjust the position of the ship in fully non-linear methods.

Waves propagate only downstream (except for rare shallow-water cases). This radiation condition has to be enforced by numerical techniques. Most methods employ special finite difference (FD) operators to compute second derivatives of the potential in the free surface condition. Dawson proposed a four-point FD operator for second derivatives along streamlines. Besides the considered collocation point, the FD operator uses the next three points upstream. Dawson's method automatically requires grids oriented along streamlines of the double-body flow approximate solution. Dawson determined his operator by trial and error for a two-dimensional flow with a simple Kelvin condition. His criteria were that the wave length should correspond to the analytically predicted wave length and the wave amplitude should remain constant some distance behind the disturbance causing the waves.

Dawson approximated the derivative of any function H with respect to ℓ at the point i numerically by:

$$H_{\ell i} \approx CA_i H_i + CB_i H_{i-1} + CC_i H_{i-2} + CD_i H_{i-3} \quad (5.38)$$

$H_{\ell i}$ is the derivative with respect to ℓ at point P_i. H_i to H_{i-3} are the values of the function H at points P_i to P_{i-3}, all lying on the same streamline of the double-body flow upstream of P_i. The coefficients CA_i to CD_i are determined from the arc lengths L_j ($j = 1$ to $i - 3$) of the streamline between point P_i and point P_j:

$$L_j = \int_{P_i}^{P_j} d\ell \quad \text{on the streamline}$$
$$CA_i = -(CB_i + CC_i + CD_i)$$
$$CB_i = L_{i-1}^2 L_{i-3}^2 (L_{i-3} - L_{i-2})(L_{i-3} + L_{i-2})/D_i$$
$$CC_i = -L_{i-1}^2 L_{i-3}^2 (L_{i-3} - L_{i-1})(L_{i-3} + L_{i-1})/D_i$$
$$CD_i = L_{i-1}^2 L_{i-2}^2 (L_{i-2} - L_{i-1})(L_{i-2} + L_{i-1})/D_i$$
$$D_i = L_{i-1} L_{i-2} L_{i-3} (L_{i-3} - L_{i-1})$$
$$(L_{i-2} - L_{i-1})(L_{i-3} - L_{i-2})$$
$$\times (L_{i-3} + L_{i-2} + L_{i-1})$$

This four-point FD operator dampens the waves to some extent and gives for usual discretizations (about 10 elements per wave length) wave lengths which are about 5% too short. Strong point-to-point oscillations of the source strength occur for very fine grids. Various FD operators have been subsequently investigated to overcome these disadvantages. Of all these, only the spline interpolation developed at MIT was really convincing as it overcomes all the problems of Dawson Nakos (1990), Nakos and Sclavounos (1990).

An alternative approach to FD operators involves 'staggered grids' as developed in Hamburg. This technique adds an extra row of source points (or panels) at the downstream end of the computational domain and an extra row of collocation points at the upstream end (Figure 5.24). For equidistant grids this can also

Figure 5.24 'Shifting' technique (in 2d).

be interpreted as shifting or staggering the grid of collocation points vs. the grid of source elements. This technique shows absolutely no numerical damping or distortion of the wave length, but requires all derivatives in the formulation to be evaluated numerically.

Only part of the water surface can be discretized. This introduces an artificial boundary of the computational domain. Disturbances created at this artificial boundary can destroy the whole solution. Methods based on FD operators use simple two-point operators at the downstream end of the grid which strongly dampen waves. At the upstream end of the grid, where waves should not appear, various conditions can be used, e.g. the longitudinal component of the disturbance velocity is zero. Nakos (1990) has to ensure in his MIT method (SWAN code) based on spline interpolation that waves do not reach the side boundary. This leads to relatively broad computational domains. Time-domain versions of the SWAN code use a 'numerical beach'. For the wave resistance problem, the time-domain approach seems unnecessarily expensive and is rarely used in practice. Norwegian researchers tried to reduce the computational domain by matching the panel solution for the near-field to a thin-ship-theory solution in the far-field. However, this approach saved only little computational time at the expense of a considerably more complicated code and was subsequently abandoned. The 'staggered grid' technique is again an elegant alternative. Without further special treatment, waves leave the computational domain without reflection.

Most methods integrate the pressure on the ship's surface to determine the forces (especially the resistance) and moments. 'Fully non-linear' methods integrate over the actually wetted surface while older methods often take the CWL as the upper boundary for the integration. An alternative to pressure integration is the analysis of the wave energy behind the ship (wave cut analysis). The wave resistance coefficients should theoretically tend to zero for low speeds. Pressure integration gives usually resistance coefficients which remain finite for small Froude numbers. However, wave cut analysis requires larger grids behind the ship leading to increased computational time and storage. Most developers of wave resistance codes have at some point tried to incorporate wave cut analysis to determine the wave resistance more accurately. So far the evidence has not yet been compelling enough to abandon the direct pressure integration.

Most panel methods give as a direct result the source strengths of the panels. A subsequent computation determines the velocities at the individual points. Bernoulli's equation then gives pressures and wave elevations (again at individual points). Integration of pressures and wave heights finally yields the desired forces and moments which in turn are used to determine dynamical trim and sinkage ('squat').

Fully non-linear state of the art codes fulfil iteratively an equilibrium condition (dynamical trim and sinkage) and both kinematic and dynamic conditions on the actually deformed free surface. The differences in results between 'fully non-linear' and linear or 'somewhat non-linear' computations are considerable (typically 25%), but the agreement of computed and measured resistances is not better in 'fully non-linear' methods. This may in part be due to the computational procedure or inherent assumptions in computing a wave resistance from experimental data (usually using a form factor method), but also due to computational errors in determining the resistance which are of similar magnitude as the actual resistance. One reason for the unsatisfactory accuracy of the numerical procedures lies in the numerical sensitivity of the pressure integration. The pressure integration involves basically subtracting forces of same magnitude which largely cancel. The relative error is strongly propagated in such a case. Initial errors stem from the discretization. For example, integration of the hydrostatic pressure for the ship at rest should give zero longitudinal force, but usual discretizations show forces that may lie within the same order of magnitude as the wave resistance. Still, there is consensus that panel methods capture the pressure distribution at the bow quite accurately. The vertical force is not affected by the numerical sensitivity. Predictions for the dynamical sinkage differ usually by less than 5% for a large bandwidth of Froude numbers. Trim moment is not predicted as well due to viscous effects and numerical sensitivity. This tendency is amplified by shallow water.

Panel methods are still the most important CFD instrument for form improvement of ships. Worldwide research aims at faster methods, wider applications, and higher accuracy:

- *Faster methods*
  Cluster or multigrid techniques could make existing methods faster by one order of magnitude (Söding

(1996)). The price is a more complicated program code. At present, the actual computation in practice accounts for less than 10% of the total response time (from receiving the hull description to delivering the CFD report) and even less of the cost. So the incentive to introduce these techniques is low in practice.

- *Wider applications*
  Non-Linear solutions are limited today to moderate non-linearities. Ships with strong section flare close to the waterline and fast ships still defy most attempts to obtain non-linear solutions. Planing boats feature complex physics including spray. Panel methods as described here will not be applied successfully to these ships for some years, although first attempts in this direction appeared in the early 1990s. State of the art computations for planing boats are based on special non-CFD methods, e.g. slender-body approaches.

- *Higher accuracy*
  The absolute accuracy of the predicted resistance is unsatisfactory. Patch methods as proposed by Söding, may overcome this problem to some extent. But the intersection between water surface and ship will remain a problem zone, because the problem is ill-posed here within a potential flow model. The immediate vicinity of the bow of a ship always features to some extent breaking waves and spray not included by the currently used methods. The exact simulation of plunging waves is impossible for panel methods. Ad-hoc solutions are subject to research, but no convincing solution has been published yet. One approach of overcoming these limitations lies in methods discretizing the fluid volume rather than boundary element methods. Such methods can simulate flows with complicated free surface geometries (breaking waves, splashes) allowing the analyses of problems beyond the realm of BEM applications.

(b) Viscous flow computations

RANS solvers are state of the art for viscous ship flows. A computational prediction of the total calm-water resistance using RANS solvers to replace model tests would be desirable, but so far the accuracy of the RANS predictions is largely perceived as still insufficient. Nevertheless, RANS solvers are widely applied to analyse:

- the flow around aftbodies of ships
- the flow around appendages

The first research applications for RANS solutions with wavemaking for ships appeared in the late 1980s. By the late 1990s various research groups also presented results for ships free to trim and sink. However, most computations for actual ship design projects in practice still neglected all free-surface effects (double-body flow). Most computations, especially those for practical design applications, were limited to Reynolds numbers corresponding to model tests. Sometimes, potential flow computations were used as preprocessors to determine trim and sinkage and the wave elevation, before RANS computations started with fixed boundaries.

Various applications to ship design and research applications are found in the literature. Representative for the state of the art for ship design applications are surveys by leading companies in the field such as Flowtech (Larsson (1997, 1998), or HSVA (Bertram and Jensen (1994)), Bertram (1998)). The state of the art in research is documented in validation workshops like the Tokyo 1994 workshop and the Gothenborg 2000 workshop. RANS computations require considerable skill and experience in grid generation and should therefore as a rule be executed by experts usually found in special consulting companies or modern towing tanks. Comprehensive reviews of developments in theoretical and numerical modelling of resistance are given in ITTC (2002a, 2005a).

5.1.4 Propulsive coefficients

The propulsive coefficients of the ship performance form the essential link between the effective power required to drive the vessel, obtained from the product of resistance and ship speed, and the power delivered from the engine to the propeller.

The power absorbed by and delivered to the propeller P_D in order to drive the ship at a given speed V_S is

$$P_D = 2\pi n Q \quad (5.39)$$

where n and Q are the rotational speed and torque at the propeller. Now the torque required to drive the propeller Q can be expressed for a propeller working behind the vessel as

$$Q = K_{Qb} \rho n^2 D^5 \quad (5.40)$$

where K_{Qb} is the torque coefficient of the propeller when working in the wake field behind the vessel at a mean advance coefficient J. By combining Equations (5.39) and (5.40) the delivered power can be expressed as

$$P_D = 2\pi K_{Qb} \rho n^3 D^5 \quad (5.41)$$

If the propeller were operating in open water at the same mean advance coefficient J the open water torque coefficient K_{Qo} would be found to vary slightly from that measured behind the ship model. As such the ratio K_{Qo}/K_{Qb} is known as the relative rotative efficiency η_r

$$\eta_r = \frac{K_{Qo}}{K_{Qb}} \quad (5.42)$$

this being the definition stated in Section 5.3.

Hence, Equation (5.41) can then be expressed in terms of the relative rotative efficiency as

$$P_D = 2\pi \frac{K_{Qo}}{\eta_r} \rho n^3 D^5 \qquad (5.43)$$

Now the effective power P_E is defined as

$$P_E = RV_s$$
$$= P_D QPC$$

where the QPC is termed the quasi-propulsive coefficient.

Hence, from the above, in association with Equation (5.43).

$$RV_s = P_D QPC$$
$$= 2\pi \frac{K_{Qo}}{\eta_r} \rho n^3 D^5 QPC$$

which implies that

$$QPC = \frac{RV_s \eta_r}{2\pi K_{Qo} \rho n^3 D^5}$$

Now the resistance of the vessel R can be expressed in terms of the propeller thrust T as $R = T(1-t)$, where t is the thrust deduction factor as explained later. Also from Section 5.2 the ship speed V_s can be defined in terms of the mean speed of advance V_a as $V_a = V_s(1-w_t)$, where w_t is the mean Taylor wake fraction. Furthermore, since the open water thrust coefficient K_{To} is expressed as $T_o = K_{To}\rho n^2 D^4$, with T_o being the open water propeller thrust at the mean advance coefficient J,

$$\frac{T_o}{K_{To}} = \rho n^2 D^4$$

and the QPC can be expressed from the above as

$$QPC = \frac{T_o(1-t)V_a K_{To}\eta_r}{(1-w_t)2\pi K_{Qo} n D T_o}$$

which reduces to

$$QPC = \left(\frac{1-t}{1-w_t}\right)\eta_0 \eta_r$$

since, from Equation (5.69),

$$\eta_0 = \frac{J}{2\pi}\frac{K_{To}}{K_{Qo}}$$

The quantity $(1-t)/(1-w_t)$ is termed the hull efficiency η_h and hence the QPC is defined as

$$QPC = \eta_h \eta_0 \eta_r \qquad (5.44)$$

or, in terms of the effective and delivered powers,

$$P_E = P_D QPC$$

that is,

$$P_E = P_D \eta_h \eta_0 \eta_r \qquad (5.45)$$

5.1.4.1 Relative rotative efficiency

The relative rotative efficiency (η_r), as defined by Equation (5.42), accounts for the differences in torque absorption characteristics of a propeller when operating in mixed wake and open water flows. In many cases the value of η_r lies close to unity and is generally within the range

$$0.95 \leq \eta_r \leq 1.05$$

In a relatively few cases it lies outside this range. Holtrop (1988) gives the following statistical relationships for its estimation:

$$\left.\begin{array}{l} \text{For conventional stern single-screw ships:} \\ \quad \eta_r = 0.9922 - 0.05908(A_E/A_0) \\ \qquad\quad + 0.07424(C_P - 0.0225\text{ LCB}) \\ \text{For twin-screw ships} \\ \quad \eta_r = 0.9737 + 0.111(C_P - 0.0225\text{ LCB}) \\ \qquad\quad - 0.06325\ P/D \end{array}\right\} \quad (5.46)$$

If resistance and propulsion model tests are performed, then the relative rotative efficiency is determined at model scale from the measurements of thrust T_m and torque Q_m with the propeller operating behind the model. Using the non-dimensional thrust coefficient K_{Tm} as input data the values J and K_{Qo} are read off from the open water curve of the model propeller used in the propulsion test. The torque coefficient of the propeller working behind the model is derived from

$$K_{Qb} = \frac{Q_M}{\rho n^2 D^5}$$

Hence the relative rotative efficiency is calculated as

$$\eta_r = \frac{K_{Qo}}{K_{Qb}}$$

The relative rotative efficiency is assumed to be scale independent.

5.1.4.2 Thrust deduction factor

When water flows around the hull of a ship which is being towed and does not have a propeller fitted a certain pressure field is set up which is dependent on the hull form. If the same ship is now fitted with a propeller and is propelled at the same speed the pressure field around the hull changes due to the action of the propeller. The propeller increases the velocities of the flow over the hull surface and hence reduces the local pressure field over the after part of the hull surface. This has the effect of increasing, or augmenting, the resistance of the vessel from that which was measured in the towed resistance case and this change can be expressed as

$$T = R(1 + a_r) \qquad (5.47)$$

where T is the required propeller thrust and a_r is the resistance augmentation factor. An alternative way of expressing Equation (5.47) is to consider the deduction in propeller effective thrust which is caused by the change in pressure field around the hull. In this case the relationship

$$R = T(1 - t) \qquad (5.48)$$

applies, in which t is the thrust deduction factor. The correspondence between the thrust deduction factor and the resistance augmentation factor can be derived from Equations (5.47) and (5.48) as being

$$a_r \left(\frac{t}{1-t} \right)$$

If a resistance and propulsion model test has been performed, then the thrust deduction factor can be readily calculated from the relationship defined in the 1987 ITTC proceedings

$$t = \frac{T_M + F_D - R_c}{T_M}$$

in which T_M and F_D are defined previously and R_c is the resistance corrected for differences in temperature between the resistance and propulsion tests:

$$R_c = \frac{(1+k)C_{FMC} + C_R}{(1+k)C_{FM} + C_R} R_{TM}$$

where C_{FMC} is the frictional resistance coefficient at the temperature of the self-propulsion test.

In the absence of model tests an estimate of the thrust deduction factor can be obtained from the work of Holtrop (1988) and Harvald (1978). In the Holtrop approach the following regression-based formulas are given:

For single-screw ships:

$$\left. \begin{aligned} t &= \frac{0.25014(B/L)^{0.28956}(\sqrt{(B/T)}/D)^{0.2624}}{(1 - C_p + 0.0225 \, \text{LCB})^{0.01762}} \\ &\quad + 0.0015 C_{stern} \end{aligned} \right\} \qquad (5.49)$$

For twin-screw ships:

$$t = 0.325 C_B - 0.18885 D / \sqrt{(BT)}$$

In Equation (5.49) the value of the parameter C_{stern} is found from Table 5.5.

The alternative approach of Harvald to the calculation of the thrust deduction factor is to assume that it comprises three separate components as follows:

$$t = t_1 + t_2 + t_3 \qquad (5.50)$$

in which t_1, t_2 and t_3 are basic values derived from hull from parameters, a hull form correction and a propeller diameter correction, respectively. The values of these parameters for single-screw ships are reproduced in Figure 5.25.

5.1.4.3 Hull efficiency

The hull efficiency can readily be determined once the thrust deduction and mean wake fraction are known. However, because of the pronounced scale effect of

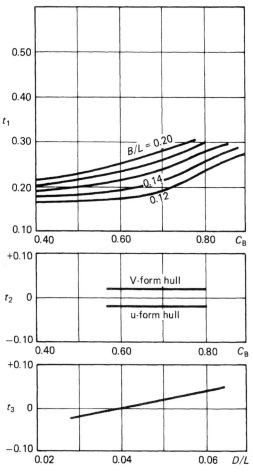

Figure 5.25 Thrust deduction estimation of Harvald for single-screw ships (Reproduced with permission from Harvald (1978)).

the wake fraction there is a difference between the full-scale ship and model values. In general, because the ship wake fraction is smaller than the corresponding model value, due to Reynolds effects, the full-scale efficiency will also be smaller.

5.1.4.4 Quasi-propulsive coefficient

It can be deduced from Equation (5.44) that the value of the QPC is dependent upon the ship speed, pressure field around the hull, the wake field presented to the propeller and the intimate details of the propeller design such as diameter, rate of rotation, radial load distribution, amount of cavitation on the blade surfaces, etc. As a consequence, the QPC should be calculated from the three component efficiencies given in Equation (5.44) and not globally estimated.

Of particular interest when considering general trends is the effect that propeller diameter can have on the QPC; as the diameter increases, assuming the rotational speed is permitted to fall to its optimum value, the propeller efficiency will increase and hence for a given hull from the QPC will tend to rise. In this instance the effect of propeller efficiency dominates over the hull and relative rotative efficiency effects.

5.1.5 Influence of rough water

The discussion so far has centred on the resistance and propulsion of vessels in calm water or ideal conditions. Clearly the effect of bad weather is either to slow the vessel down for a given power absorption or, conversely, an additional input of power to the propeller in order to maintain the same ship speed.

In order to gain some general idea of the effect of weather on ship performance appeal can be made to the NSMB Trial Allowances 1976, Jong (1976). These allowances were based on the trial results of 378 vessels and formed an extension to the 1965 and 1969 diagrams. Figure 5.26 shows the allowances for ships with a trial displacement between 1000 and 320 000 tonnes based on the Froude extrapolation method and coefficients. Analysis of the data upon which this diagram was based showed that the most significant variables were the displacement, Beaufort wind force, model scale and the length between perpendiculars. As a consequence a regression formula was suggested as follows:

$$\text{trial allowance} = 5.75 - 0.793\Delta^{1/3} + 12.3B_n$$
$$+ (0.0129L_{pp} - 1.864B_n)\lambda^{1/3} \quad (5.51)$$

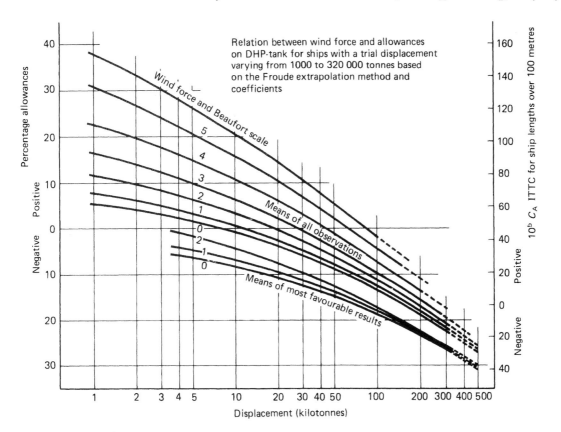

Figure 5.26 NSMB 1976 trial allowances (Reproduced with permission from Jong (1976)).

where B_n and λ are the Beaufort number and the model scale, respectively.

Apart from global indicators and correction factors such as Figure 5.26 or Equation (5.51) considerable work has been undertaken in recent years to establish methods by which the added resistance due to weather can be calculated for a particular hull form. Latterly particular attention has been paid to the effects of diffraction in short waves which is a particularly difficult area.

In general estimation methods range from those which work on data bases for standard series hull forms whose main parameter have been systematically varied to those where the calculation is approached from fundamental considerations. In its most simplified form the added resistance calculation is of the form

$$R_{TW} = R_{TC}(1 + \Delta_R) \tag{5.52}$$

where R_{TW} and R_{TC} are the resistances of the vessel in waves and calm water, respectively, and Δ_R is the added resistance coefficient based on the ship form parameters, speed and irregular sea state. Typical of results of calculation procedures of this type are the results shown in Figure 5.27 for a container ship operating in different significant wave heights H_S and a range of heading angles from directly ahead ($\theta = 0°$) to directly astern ($\theta = 180°$).

Shintani and Inoue (1984) have established charts for estimating the added resistance in waves of ships based on a study of the Series 60 models. This data takes into account various values of C_B, B/T, L/B and LCB position and allows interpolation to the required value for a particular design. In this work the compiled results have been empirically corrected by comparison with model test data in order to enhance the prediction process.

In general the majority of the practical estimation methods are based in some way on model test data: either for deriving regression equations or empirical correction factors.

In the case of theoretical methods to estimate the added resistance and power requirements in waves, methods based on linear potential theory tend to under-predict the added resistance when compared to equivalent model tests. In recent years some non-linear analysis methods have appeared which indicate that if the water surface due to the complete non-linear flow is used as the steady wave surface profile then the accuracy of the added resistance calculation can be improved significantly (Raven (1996), Hermans (2004)). Although CFD analyses are relatively limited, those published so far show encouraging results when compared to measured results, for example Orihara and Miyata (2003).

In the context of added resistance numerical computations have suggested that the form of the bow above the calm water surface can have a significant influence on the added resistance in waves. Such findings have also been confirmed experimentally and have shown that a blunt-bow ship could have its added resistance reduced by as much as 20 to 30 per cent while having minimal influence on the calm water resistance.

5.1.6 Restricted water effects

Restricted water effects derive essentially from two sources. These are first a limited amount of water under the keel and secondly, a limitation in the width of water each side of the vessel which may or may not be in association with a depth restriction.

In order to assess the effects of restricted water operation, these being particularly complex to define mathematically, the ITTC (ITTC (1987)) have expressed typical influencing parameters. These are as follows:

1. An influence exists on the wave resistance for values of the Froude depth number F_{nh} in excess of 0.7. The Froude depth number is given by

$$F_{nh} = \frac{V}{\sqrt{(gh)}}$$

where h is the water depth of the channel.

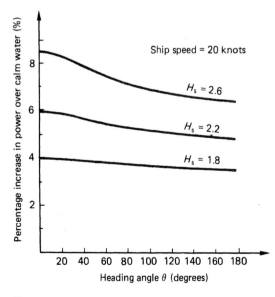

Figure 5.27 Estimated power increase to maintain ship speed in different sea states for a container ship.

2. The flow around the hull is influenced by the channel boundaries if the water depth to draught ratio (h/T) is less than 4. This effect is independent of the Froude depth number effect.
3. There is an influence of the bow wave reflection from the lateral boundary on the stern flow if either the water width to beam ratio (W/B) is less than 4 or the water width to length ratio (W/L) is less than unity.
4. If the ratio of the area of the channel cross-section to that of the mid-ship section (A_c/A_M) is less than 15, then a general restriction of the waterway will start to occur.

In the case of the last ratio it is necessary to specify at least two of the following parameters: width of water, water depth or the shape of the canal section because a single parameter cannot identify unconditionally a restriction on the water flow.

The most obvious sign of a ship entering into shallow water is an increase in the height of the wave system in addition to a change in the ship's vibration characteristics. As a consequence of the increase in the height of the wave system the assumption of small wave height, and consequently small wave slopes, cannot be used for restricted water analysis. This, therefore, implies a limitation to the use of linearized wave theory for this purpose; as a consequence higher-order theoretical methods need to be sought. Currently several researchers are working in this field and endeavouring to enhance the correlation between theory and experiment.

The influence of shallow water on the resistance of high-speed displacement monohull and catamaran forms is discussed by Molland *et al.* (2004), and test data presented for a series of models.

Barrass (1979) suggests the depth/draught ratio at which shallow water just begins to have an effect is given by the equation

$$h/T = 4.96 + 52.68(1 - C_w)^2$$

in which the C_w is the water-plane coefficient. Alternatively, Schneekluth (1987) provides a set of curves based on Lackenby's work (Figure 5.28) to enable the estimation of the speed loss of a vessel from deep to shallow water. The curves are plotted on a basis of the square of Froude depth number to the ratio $\sqrt{A_M}/h$. Beyond data of this type there is little else currently available with which to readily estimate the added resistance in shallow water.

One further effect of shallow water is the phenomenon of ship squat. This is caused by a venturi effect between the bottom of the vessel and the bottom of the seaway which causes a reduction of pressure to occur. This reduction of pressure then induces the ship to increase its draught in order to maintain equilibrium. Barras developed a relationship for ship

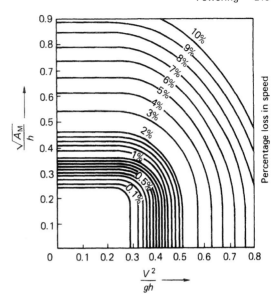

Figure 5.28 Loss of speed in transfer from deep to shallow water (Reproduced from Schneekluth (1987)).

squat by analysing the results from different ships and model tests with block coefficients in the range 0.5 to 0.9 for both open water and in restricted channel conditions. In his analysis the restricted channel conditions were defined in terms of h/T ratios in the range 1.1 to 1.5. For the conditions of unrestricted water in the lateral direction such that the effective width of the waterway in which the ship is travelling must be greater than $[7.7 + 45(1 - C_w)^2]B$, the squat is given by

$$S_{max} = (C_b(A_M/A_C)^{2/3}V_s^{2.08})/30 \text{ for } F_{nh} \leq 0.7$$

5.1.7 High-speed hull form resistance

In the case of a conventional displacement ship the coefficient of wave making resistance increases with the Froude number based on waterline length until a value of $F_n \simeq 0.5$ is reached. After this point it tends to reduce in value such that at high Froude numbers, in excess of 1.5, the wave making resistance becomes a small component of the total resistance. The viscous resistance, however, increases due to its dependence on the square of the ship speed; this is despite the value of C_F reducing with Froude number. As a consequence of this rise in the viscous resistance a conventional displacement hull requires excessive power at high speed and other hull forms and modes of support require to be introduced. Such forms are the planing hull form, the hydrofoil and the hovercraft.

The underlying principle of high-speed planing craft resistance and propulsion have been treated by

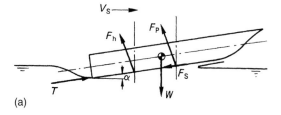

Figure 5.29(a) Forces experienced by a planing craft.

several authors: for example, DuCane and Clayton and Bishop (1982). These authors not only examine high-speed displacement and planing craft but also hydrofoils and hovercraft. As a consequence for the detailed principles of their motion reference can be made to these works.

The forces acting on a planing hull are shown by Figure 5.29(a) in which the forces shown as W, F_p, F_h, F_s and T are defined as follows:

W is the weight of the craft;
F_p is the net force resulting from the variation of pressure over the wetted surface of the hull;
F_h is the hydrostatic force acting at the centre of pressure on the hull;
F_s is the net skin friction force acting on the hull;
T is the thrust of the propulsor.

By the suitable resolution of these forces and noting that for efficient planing, the planing angle should be small it can be shown that the total resistance comprises three components:

$$R_T = R_I + R_{WV} + R_{FS} \qquad (5.53)$$

where R_I is the induced resistance or drag derived from the inclination of F_p from the vertical due to the trim angle of the craft;
R_{WV} is the derives from the wave making and viscous pressure resistance;
R_{FS} is the skin friction resistance.

At high speed the wave making resistance becomes small but the vessel encounters an induced drag component which is in contrast to the case for conventional displacement hulls operating at normal speeds.

To estimate the resistance properties of high-speed displacement and planing craft use can be made of either standard series data or specific model test results.

5.1.7.1 Standard series data

A considerable amount of data is available by which an estimate of the resistance and propulsion characteristics can be made. Table 5.8 identifies some of the data published in the open literature for this purpose.

Table 5.8 Published data for displacement and planing craft.

Standard series data	
Displacement data	Planing data
Norstrom Series (1936)	Series 50 (1949)
de Groot Series (1955)	
Marwood and Silverleaf (1960)	Series 62 (1963)
Series 63 (1963)	Series 65 (1974)
Series 64 (1965)	
SSPA Series (1968)	
NPL Series (1984)	
NSMB Series (1984)	
Robson Naval Combatants (1988)	
Southampton Catamaran Series (1996)	

In addition to basic test data of this type various regression-based analysis are available to help the designer in predicting the resistance characteristics of these craft; for example, van Oortmerssen (1971) and Mercier and Savitsky (1973). In addition Savitsky and Ward Brown (1976) offer procedures for the rough water evaluation of planing hulls.

5.1.7.2 Model test data

In specific cases model test data is derived for a particular hull form. In these cases the principles for model testing outlined in Status (1981) and the various ITTC proceedings should be adhered to in order to achieve valid test results.

Multi-hull resistance

The wave resistance of a multi-hull vessel is commonly approximated by considering the waves generated by each hull of the vessel acting in isolation to be superimposed on each other (Tuck and Lazauskas (1998) and Day et al. (2003)). If this approach is followed through then an expression for the wave resistance for a pair of non-staggered identical hulls takes the form

$$R_W = 0.5\pi\rho V_s^2 \int |A(\theta)|_{SH}^2 \cdot F(\theta) \cdot \cos^3\theta \, d\theta$$

where $\int |A(\theta)|_{SH}^2$ refers to the amplitude function for the side hull and $F(\theta)$ is a hull interference function and is dependent on the hull separation, ship length and Froude number. However, it is important to phase the waves generated by each hull correctly if their transverse components are to be cancelled. This cancellation effect is a function of the Froude number and the longitudinal relative positions of the hulls. Moreover, the cancellation effect of the transverse waves will be beneficial for a range of Froude numbers around that for which the cancellation is designed to occur.

An approximation of the type discussed above does not, however, take into account that the waves generated by one hull will be incident upon another hull whereupon they will be diffracted by that hull. These diffracted waves comprise a reflected and transmitted wave which implies that the total wave system of the multi-hull ship is not a superposition of the waves generated by each hull in isolation. In this context it is the divergent waves at the Kelvin angle that are responsible for the major part of the interaction. Three-dimensional Rankine panel methods are helpful for calculating the wave patterns around multi-hull ships and when this is done for catamarans, it is seen that in some cases relatively large wave elevations occur between the catamaran hulls in the after regions of the ship.

A regression-based procedure was developed (Pham *et al.* (2001)) to assess the wave resistance of hard chine catamarans within the range:

$$10 \leq L/B \leq 20$$
$$1.5 \leq B/T \leq 2.5$$
$$0.4 \leq C_b \leq 0.6$$
$$6.6 \leq L/\nabla^{1/3} \leq 12.6$$

Within this procedure the coefficient of wave making resistance C_w is given by

$$C_w = \exp(\alpha)(L/B)^{\beta 1}(B/T)^{\beta 2} C_b^{\beta 3} (s/L)^{\beta 4}$$

where the coefficient α, $\beta 1$, $\beta 2$, $\beta 3$ and $\beta 4$ are functions of Froude number and s is the spacing between the two demi-hulls.

In this procedure two interference factors are introduced following the formulation of Day *et al.* (1997), one relating to the wave resistance term (τ) and the other a body interference effect expressed as a modified factor $(1 + \beta_k) = 1.42$ as established by Insel and Molland (1991). This permits the total resistance coefficient to be expressed as

$$C_T = 2(1 + \beta_k)C_F + \tau C_w$$

Subsequently, an optimization scheme has been developed, Anantha *et al.* (2006), for hard chine catamaran hull form basic design based on the earlier work of Pham *et al.* (2001).

Further useful experimental data for a systematic series of high-speed semi-displacement catamaran forms is given by Molland *et al.* (1996).

5.1.7.3 *Summary of problems for fast and unconventional ships*

Model testing has a long tradition for the prediction and optimization of ship performance of conventional ships. The scaling laws are well established and the procedures correlate model and ship with a high level of accuracy.

The same scaling laws generally apply to high-speed craft, but two fundamental problems may arise:

1. Physical quantities may have major effects on the results which cannot be deduced from classical model tests. The physical quantities in this context are: surface tension (spray), viscous forces and moments, aerodynamic forces, cavitation.
2. Limitations of the test facilities do not allow an optimum scale. The most important limitations are generally water depth and carriage speed.

Fast and unconventional ships are often 'hybrid' ships, i.e. they produce the necessary buoyancy by more than one of the three possible options: buoyancy, dynamic lift (foils or planing), aerostatic lift (air cushion). For the propulsion of fast ships, subcavitating, cavitating, and ventilated propellers as well as waterjets with flush or pitot-type inlets are used. Due to viscous effects and cavitation, correlation to full-scale ships causes additional problems.

Generally we cannot expect the same level of accuracy for a power prediction as for conventional ships. The towing tank should provide an error estimate for each individual case. Another problem arises from the fact that the resistance curves for fast ships are often quite flat near the design point as are the curves of available thrust for many propulsors. For example, errors in predicted resistance or available thrust of 1% would result in an error of the attainable speed of also about 1%, while for conventional cargo ships the error in speed would often be only 1/3%, i.e. the speed prediction is more accurate than the power prediction.

The main problems for model testing are discussed individually:

- *Model tank restrictions*
  The physics of high-speed ships are usually highly non-linear. The positions of the ship in resistance (without propeller) and propulsion (with propeller) conditions differ strongly. Viscosity and free surface effects, including spray and overturning waves, play significant roles making both experimental and numerical predictions difficult.

  Valid predictions from tank tests for the resistance of the full-scale ship in unrestricted water are only possible if the tank is sufficiently large as compared to the model to allow similarity in flow. Blockage, i.e. the ratio of the submerged cross-section of the model to the tank cross-section, will generally be very low for models of high-speed ships. However, shallow-water effects depend mainly on the model speed and the tank water depth. The depth Froude number F_{nh} should not be greater than 0.8 to be free of significant shallow-water effects.

  The frictional resistance is usually computed from the frictional resistance of a flat plate of similar length as the length of the wetted underwater body

of the model. This wetted length at test speed differs considerably from the wetted length at zero speed for planing or semi-planing hull forms. In addition the correlation requires that the boundary layer is fully turbulent. Even when turbulence stimulators are used, a minimum Reynolds number has to be reached. We can be sure to have a turbulent boundary layer for $R_n > 5 \cdot 10^6$. This gives a lower limit to the speeds that can be investigated depending on the used model length.

Figure 5.29(b) illustrates, using a towing tank with water depth $H = 6$ m and a water temperature 15°, how an envelope of possible test speeds evolve from these two restrictions. A practical limitation may be the maximum carriage speed. However, at HSVA the usable maximum carriage speed exceeds the maximum speed to avoid shallow-water effects.

- *Planing hulls*
  In the planing condition a significant share of the resistance is frictional and there is some aerodynamic resistance. At the design speed, the residual resistance, i.e. the resistance component determined from model tests, may only be 25% to 30% of the total resistance. In model scale, this part is even smaller. Therefore the measurements of the model resistance must be very accurate. Resistance of planing hulls strongly depends on the trim of the vessel. Therefore a careful test set-up is needed to ensure that the model is towed in the correct direction. The most important problem, however, is the accurate determination of the wetted surface and the wetted length which is needed to compute the frictional resistance for both the model and the ship. The popular use of side photographs are not adequate. Preferably underwater photographs should be used. In many cases, the accurate measurement of trim and sinkage may be adequate in combination with hydrostatic computation of wetted surface and length. As the flotation line of such vessels strongly depends on speed, proper arrangement of turbulence stimulation is needed as well.

  Depending on the propulsion system, planing vessels will have appendages like rudders and shafts. For typical twin-screw ships with shafts, one pair of I-brackets and one pair of V-brackets, the appendage resistance could account for 10% of the total resistance of the ship. As viscous resistance is a major part in the appendage resistance and as the Reynolds number of the appendages will be small for the model in any case or the appendage may be within the boundary layer of the vessel, only a crude correlation of the appendage resistance is possible: the resistance of the appendage is determined in model scale by comparing the resistance of the model with and without appendages. Then an empirical correction for transferring the appendage resistance to the full-scale ship is applied. In many cases, it may be sufficient to perform accurate measurements without any appendages on the model and then use an empirical estimate for the appendage resistance.

- *Craft with hydrofoils*
  Hydrofoils may be used to lift the hull out of the water to reduce resistance. Besides classical hydrofoils which are lifted completely out of the water and are fully supported by foil lift, hybrid hydrofoils may be used which are partially supported by buoyancy and partially by foil lift, e.g. catamarans with foils between the two hulls. When performing and evaluating resistance and propulsion tests for such vessels, the following problems have to be kept in mind:
  – The Reynolds number of the foils and struts will always be very low. Therefore the boundary

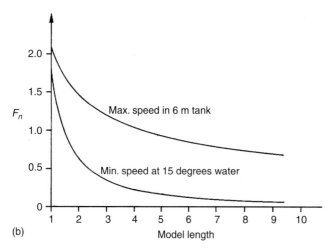

Figure 5.29(b) Possible speed range to be safely investigated in a 6 m deep towing tank at 15° water temperature.

layer on the foil may become partially laminar. This will influence the lift and the frictional resistance of the foils in a way requiring special correlation procedures to compensate at least partially these scaling errors. The uncertainty level is still estimated as high as 5% which is definitely higher than for conventional craft.
– Cavitation may occur on the full-scale hydrofoil. This may not only cause material erosion, but it will also influence the lift and drag of the foils. Significant cavitation will certainly occur if the foil loading exceeds 10^5 N/m^2. With configurations not fully optimized for cavitation avoidance, significant cavitation is expected for foil loadings in excess of $6 \cdot 10^4$ N/m^2 already. Another important parameter is the vessel's speed. Beyond 40 knots, cavitation has to be expected on joints to struts, flaps, foil tips and other critical parts. At speeds beyond 60 knots, cavitation on the largest part of the foil has to be expected. When model testing these configurations in model tanks, no cavitation will occur. Therefore similarity of forces cannot be expected. To overcome this problem, resistance and propulsion tests could be performed in a free surface cavitation tunnel. However, due to the usually small cross-sections of these tunnels, shallow-water effects may be then unavoidable. Therefore HSVA recommends the following procedure:

1. Perform tests in the towing tank using non-cavitating foils from stock, varying angle of attack, and measure the total resistance and the resistance of the foils.
2. Test the foils (including struts) in a cavitation tunnel varying angle of attack, observe cavitation and measure forces.
3. Combine the results of both tests by determining the angle of attack for similar lift of foils and summing the resistance components.

In the preliminary design phase, the tests in the cavitation tunnel may be substituted by corresponding flow computations.

- *Surface effect ships (SES)*
SESs combine aerostatic lift and buoyancy. The wave resistance curve of an SES exhibits humps and hollows as in conventional ships. The magnitude of the humps and hollows in wave resistance depends strongly on the cushion L/B ratio. Wavemaking of the submerged hulls and the cushion can simply be scaled according to Froude similarity as long as the tank depth is sufficient to avoid shallow-water effects. Otherwise a correction based on the potential flow due to a moving pressure patch is applied. Due to the significant influence of trim, this method has some disadvantages. To determine the wetted surface, observations inside the cushion are required with a video camera. The frictional resistance of the seals cannot be separated out of the total resistance. The pressure distribution between seals and cushion has to be controlled and the air flow must be determined. Also the model aerodynamic resistance in the condition under the carriage has to be determined and used for separating the wave resistance. Generally separate wind tunnel tests are recommended to determine the significant aerodynamic resistance of such ships.

- *Propulsion with propellers*
    – *Conventional propellers*
    Most of the problems concerning the scaling of resistance also appear in the propulsion test, as they determine the propeller loading. The use of a thrust deduction fraction is formally correct, but the change in resistance is partially due to a change of trim with operating propellers. For hydrofoils, the problem is that cavitation is not present at model scale. Therefore, for cases with propeller loading where significant cavitation is expected, additional cavitation tests are used to determine the thrust loss due to cavitation. Z-drives which may even be equipped with contra-rotating propellers are expensive to model and to equip with accurate measuring devices. Therefore propulsion tests with such units are rarely performed. Instead the results of resistance and open-water tests of such units in a proper scale are numerically combined.

    – *Cavitating propellers*
    Certain high-speed propellers are designed to operate with a controlled extent of cavitation on the suction side of the blades. They are called super-cavitating or partially cavitating (Newton-Rader) propellers. These propulsors cannot be tested in a normal towing tank. Here again either resistance tests or propulsion tests with non-cavitating stock propellers are performed and combined with open-water tests in a cavitation tunnel.

    – *Surface-piercing propellers*
    Surface-piercing or ventilated propellers operate directly at the free surface. Thus the suction side is ventilated and therefore the collapse of cavitation bubbles on the blade surface is avoided. Due to the operation at the free surface, Froude similarity has to be maintained in model tests. On the other hand, thrust and torque, but more important also the side and vertical forces, strongly depend on the cavitation number. The vertical force may amount up to 40% of the thrust and therefore will strongly influence the resistance of planing vessels or SES, ships where this type of propeller is typically employed. Model tests on surface piercing

propellers are reported by Ferrando *et al.* (2002), and a comprehensive review of such propellers is given in ITTC (2002b).

- *Waterjet propulsion*
A common means of propulsion for high-speed ships is the waterjet. Through an inlet in the bottom of the craft water enters into a bent duct to the pump, where the pressure level is raised. Finally the water is accelerated and discharged in a nozzle through the transom. Power measurements on a model of the complete system cannot be properly correlated to full scale. Only the inlet and the nozzle are built to scale and an arbitrary model pump with sufficient capacity is used. The evaluation of waterjet experiments is difficult and involves usually several special procedures involving a combination of computations, e.g. the velocity profile on the inlet by boundary layer or RANS computations, and measured properties, e.g. pressures in the nozzle. The properties of the pump are determined either in separate tests of a larger pump model, taken from experience with other pumps, or supplied by the pump manufacturer. A special committee of the ITTC was formed to cover waterjet propulsion and latest recommendations and literature references may be found in the ITTC proceedings, ITTC (2005c). A more detailed account of the design and operation of waterjets is given by Carlton (2007). Many of the problems associated with the hydrodynamics of fast craft are addressed by Faltinsen (2006).

5.1.8 Air resistance

The prediction of the air resistance of a ship can be evaluated in a variety of ways ranging from the extremely simple to undertaking a complex series of model tests in a wind tunnel.

At its simplest the still air resistance can be estimated as proposed by Holtrop (1988) who followed the simple approach incorporated in the ITCC-1978 method as follows:

$$R_{AIR} = \frac{1}{2}\rho_a V_S^2 A_T C_{air} \qquad (5.54)$$

in which V_S is the ship speed, A_T is the transverse area of the ship and C_{air} is the air resistance coefficient, taken as 0.8 for normal ships and superstructures. The density of air ρ_a is normally taken as 1.23 kg/m³.

For more advanced analytical studies appeal can be made to the works of van Berlekom (1981) and Gould (1982). The approach favoured by Gould is to determine the natural wind profile on a power law basis and select a reference height for the wind speed. The yawing moment centre is then defined relative to the bow and the lateral and frontal elevations of the hull and superstructure are subdivided into so-called 'universal elements'. In addition the effective wind speed and directions are determined from which the Cartesian forces together with the yawing moment can be evaluated.

Regression equations for air drag coefficients for merchant ships are given by Isherwood (1973) and, following wind tunnel tests, Molland and Barbeau (2003) present air drag coefficients for the superstructures of fast displacement catamarans.

The determination of the air resistance from wind tunnel measurement would only be undertaken in exceptional cases and would most probably be associated with flow visualization studies to, for example, design suitable locations for helicopter landing and take-off platforms. For more commercial applications the cost of undertaking wind tunnel tests cannot be justified since air resistance is by far the smallest of the resistance components.

5.2 Wake

A body, by virtue of its motion through the water, causes a wake field in the sense of an uneven flow velocity distribution to occur behind it; this is true whether the body is a ship, a submarine, a remotely operated vehicle or a torpedo. The wake field at the propulsor plane arises from three principal causes: the streamline flow around the body, the growth of the boundary layer over the body and the influence of any wave-making components. The latter effect naturally is dependent upon the depth of immersion of the body below the water surface. Additionally, and equally important, is the effect that the propulsor has on modifying the wake produced by the propelled body.

5.2.1 General wake field characteristics

The wake field is strongly dependent on ship type and so each vessel can be considered to have a unique wake field. Figure 5.30 shows three wake fields for different ships. Figure 5.30(a) relates to a single-screw bulk carrier form in which a bilge vortex can be seen to be present and dominates the flow in the thwart-ship plane of the propeller disc. The flow field demonstrated by Figure 5.30(b) relates again to a single-screw vessel, but in this case to a fairly fast and fine lined vessel having a 'V'-formed afterbody unlike the 'U'-form of the bulk carrier shown in Figure 5.30(a). In Figure 5.30(b) it is seen, in contrast to the wake field produced by the 'U'-form hull, that a high-speed axial flow field exists for much of the propeller disc except for the sector embracing the top dead centre location, where the flow is relatively slow and in some cases may even reverse in direction. Definitions of 'U'- and 'V'-form hulls are shown in Figure 5.31; however,

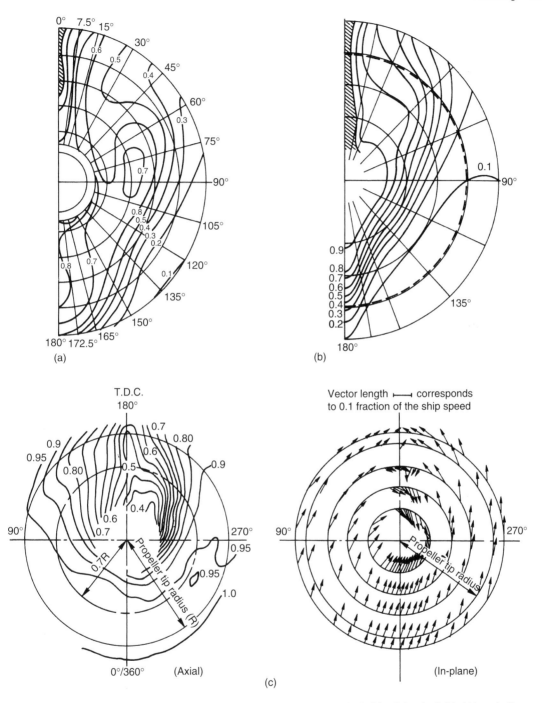

Figure 5.30 Typical wake field distributions: (a) axial wake field – U-form hull; (b) axial wake field – V-form hull; (c) axial and in-plane wake field – twin-screw hull (parts (a) and (b) Reproduced with permission from Huse (1974).

there is no 'clear-cut' transition from one form to another, and Figure 5.30(a) and (b) represent extremes of both hull form types. Both of the flow fields discussed so far relate to single-screw hull forms and, therefore, might be expected to exhibit a reflective symmetry about the vertical centre plane of the vessel. For a twin-screw vessel, however, no such symmetry naturally exists, as seen by Figure 5.30(c), which shows the wake field for a twin-screw ferry. In this figure the location of the shaft

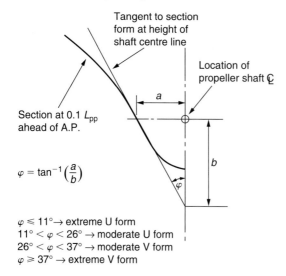

$\varphi \leqslant 11° \rightarrow$ extreme U form
$11° < \varphi < 26° \rightarrow$ moderate U form
$26° < \varphi < 37° \rightarrow$ moderate V form
$\varphi \geqslant 37° \rightarrow$ extreme V form

Figure 5.31 Definition of U- and V-form hulls.

supports, in this case 'A' brackets, is clearly seen, but due to the position of the shaft lines relative to the hull form, symmetry of the wake field across the 'A' bracket centre line cannot be maintained. Indeed, considerable attention needs to be paid to the design of the shaft supports, whether these are 'A' brackets, bossings or gondolas, in order that the flow does not become too disturbed or retarded in these locations, otherwise vibration and noise may arise and become difficult problems to solve satisfactorily. This is also of equal importance for single-, twin- or triple-screw ships.

It is of interest to note how the parameter φ, Figure 5.31, tends to influence the resulting wake field at the propeller disc of a single-screw ship. For the V-form hull (Figure 5.30(b)), one immediately notes the very high wake peak at the top dead centre position of the propeller disc and the comparatively rapid transition from the 'dead-water' region to the near free stream conditions in the lower part of the disc. This is caused by the water coming from under the bottom of the ship and flowing around the curvature of the hull, so that the fluid elements which were close to the hull, and thus within its boundary layer, also remain close to the hull around the bilge and flow into the propeller close to the centre plane. Consequently, a high wake peak is formed in the centre plane of the propeller disc.

The alternative case of a wake field associated with an extreme U-form hull is shown in Figure 5.30(a); here the flow pattern is completely different. The water flowing from under the hull is in this case unable to follow the rapid change of curvature around the bilge and, therefore, separates from the hull surface. These fluid elements then flow upwards into the outer part of the propeller disc and the region above this separated zone is then filled with water flowing from above: this creates a downward flow close to the hull surface. The resultant downward flow close to the hull and upward flow distant from the hull give rise to a rotational motion of the flow into the propeller disc which is termed the bilge vortex. The bilge vortex, therefore, is a motion which allows water particles in the boundary layer to be transported away from the hull and replaced with water from outside the boundary layer; the effect of this is to reduce the wake peak at the centre plane of the propeller disc.

Over the years, in order to help designers produce acceptable wake fields for single-screw ships, several hull form criteria have been proposed, as outlined, for example, in Huse (1974) and Carlton and Bantham (1978). Criteria of these types basically reduce to a series of guidelines such as:

1. The angle of run of the waterlines should be kept to below 27–30° over the entire length of run. Clearly it is useless to reduce the angle of run towards the stern post if further forward the angles increase to an extent which induces flow separation.
2. The stern post width should not exceed 3 per cent of the propeller diameter in the ranges 0.2 to 0.6R above the shaft centre line.
3. The angle of the tangent to the hull surface in the plane of the shaft centre line (see Figure 5.31) should lie within the range 11 to 37°.

The detailed flow velocity fields of the type shown in Figure 5.30 and used in propeller design are almost without exception derived from model tests. Today it is still the case that some 80 to 85 per cent of all ships that are built do not have the benefit of a model wake field test.

5.2.2 Wake field definition

In order to make use of the wake field data it needs to be defined in a suitable form. There are three principal methods: the velocity ratio, Taylor and Froude methods, although today the method based on Froude's wake fraction is rarely, if ever, used. The definitions of these methods are as follows.

Velocity ratio method. Here the iso-velocity contours are expressed as a proportion of the ship speed (V_s) relative to the far-field water speed. Accordingly, water velocity at a point in the propeller disc is expressed in terms of its axial, tangential and radial components, v_a, v_t and v_r, respectively:

$$\frac{v_a}{V_s}, \quad \frac{v_t}{V_s} \quad \text{and} \quad \frac{v_r}{V_s}$$

Figure 5.30(c) is expressed using above velocity component definitions. The velocity ratio method has today become perhaps the most commonly used method of wake field representation, due first to the relative conceptual complexities the other, and older, representations have in dealing with the in-plane propeller components, and second, the velocity ratios are more convenient for data input into analytical procedures.

Taylor's method. In this characterization the concept of 'wake fraction' is used. For axial velocities the Taylor wake fraction is defined as:

$$w_T = \frac{V_s - v_A}{V_s} = 1 - \left(\frac{v_A}{V_s}\right) \quad (5.55)$$

that is, one minus the axial velocity ratio or, alternatively, it can be considered as the loss of axial velocity at the point of interest when compared to the ship speed and expressed as a proportion of the ship speed. For the other in-plane velocity components we have the following relationships:

$$w_{T_t} = 1 - \left(\frac{v_t}{V_s}\right) \quad \text{and} \quad w_{T_r} = 1 - \left(\frac{v_r}{V_s}\right)$$

However, these forms are rarely used today, and preference is generally given to expressing the tangential and radial components in terms of their velocity ratios v_t/V_s and v_r/V_s.

Notice that in the case of the axial components the subscript 'a' is omitted from w_T.

Froude method. This is similar to the Taylor characterization, but instead of using the vehicle speed as the reference velocity the Froude notation uses the local velocity at the point of interest. For example, in the axial direction we have:

$$w_F = \frac{V_s - v_a}{v_a} = \left(\frac{V_s}{v_a}\right) - 1$$

For the sake of completeness it is worth noting that the Froude and Taylor wake fractions can be transformed as follows:

$$w_F = \frac{w_T}{1 - w_T} \quad \text{and} \quad w_T = \frac{w_F}{1 + w_F}$$

Mean velocity or wake fraction. The mean axial velocity within the propeller disc is found by integrating the wake field on a volumetric basis of the form:

$$\left.\begin{array}{l}\overline{W}_T = \dfrac{\int_{r_h}^{R} r \int_0^{2\pi} w_T \, d\phi \, dr}{\pi(R^2 - r_h^2)} \\[2ex] \left(\dfrac{\overline{v}_a}{V_s}\right) = \dfrac{\int_{r_h}^{R} r \int_0^{2\pi} \left(\dfrac{v_a}{V_s}\right) d\phi \, dr}{\pi(R^2 - r_h^2)}\end{array}\right\} \quad (5.56)$$

Much debate has centred on the use of the volumetric or impulsive integral form for the determination of mean wake fraction, for example Prohaska and van Lammeren (1937) and van Lammeren (1938); however, modern analysis techniques generally use the volumetric basis as a standard.

Fourier analysis of wake field. Current propeller analysis techniques rely on being able to describe the wake field encountered by the propeller at each radial location in a reasonably precise mathematical way. Figure 5.32 shows a typical transformation of the wake field velocities at a particular radial location of a polar wake field plot, similar to those shown in Figure 5.30, into a mean and fluctuating component. Figure 5.32 then shows diagrammatically how the total fluctuating component can then be decomposed into an infinite set of sinusoidal components of various harmonic orders. This follows from Fourier's theorem, which states that any periodic function can be represented by an infinite set of sinusoidal functions. In practice, however, only a limited set of harmonic components are used, since these are sufficient to define the wake field within both the bounds of calculation and experimental accuracy: typically the first eight to ten harmonics are those which might be used, the exact number depending on the propeller blade number. A convenient way, therefore, of describing the velocity variations at a particular radius in the propeller disc is to use Fourier analysis techniques and to define the problem using the global reference frame. Using this basis, the general approximation of the velocity distribution at a particular radius becomes:

$$\frac{v_a}{V_s} = \sum_{k=0}^{n} \left[a_k \cos\left(\frac{k\phi}{2\pi}\right) + b_k \sin\left(\frac{k\phi}{2\pi}\right) \right] \quad (5.57)$$

Equation (5.57) relates to the axial velocity ratio; similar equations can be defined for the tangential and radial components of velocity.

5.2.3 The nominal wake field

The nominal wake field is the wake field that would be measured at the propeller plane without the presence or influence of the propeller modifying the flow at the stern of the ship. The nominal wake field $\{w_n\}$ of a ship can be considered to effectively comprise three components: the potential wake, the frictional wake and the wave-induced wake, so that the total nominal wake field $\{w_n\}$ is given by

$$\{w_n\} = \{w_p\} + \{w_v\} + \{w_w\} + \{\Delta w\} \quad (5.58)$$

where the suffixes denote the above components, respectively, and the curly brackets denote the total wake field rather than values at a particular point. The component $\{\Delta w\}$ is the correlation or relative

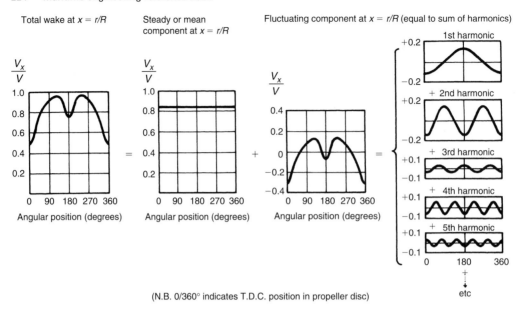

Figure 5.32 Decomposition of wake field into mean and fluctuating components.

interaction component representing the non-linear part of the wake field composition.

The potential wake field $\{w_p\}$ is the wake field that would arise if the vessel were working in an ideal fluid, that is one without viscous effects. As such the potential wake field at a particular transverse plane on the body is directly calculable using analytical methods, and it matters not whether the body is moving ahead or astern. Clearly, for underwater bodies, and particularly for bodies of revolution, the calculation procedures are comparatively simpler to use than for surface ship forms. For calculations on ship forms use is made of panel methods which today form the basis of three-dimensional, inviscid, incompressible flow calculations. The general idea behind these methods is to cover the surface with three-dimensional body panels over which there is an unknown distribution of singularities; for example, point sources, doublets or vortices. The unknowns are then solved through a system of simultaneous linear algebraic equations generated by calculating the induced velocity at control points on the panels and applying the flow tangency condition. In recent years many such programmes have been developed by various institutes and software houses around the world. For axisymmetric bodies in axial flow a distribution of sources and sinks along the axis will prove sufficient for the calculation of the potential wake.

In contrast to calculation methods an approximation to the potential wake at the propeller plane can be found by making a model of the vehicle and towing it backwards in a towing tank, since in this case the viscous effects at the propeller plane are minimal.

In general, the potential wake field can be expected to be a small component of the total wake field, as shown by Harvald (1972). Furthermore, since the effects of viscosity do not have any influence on the potential wake, the shape of the forebody does not have any influence on this wake component at the stern.

The frictional wake field $\{w_v\}$ arises from the viscous nature of the water passing over the hull surface. This wake field component derives from the growth of the boundary layer over the hull surface, which, for all practical purposes, can be considered as being predominantly turbulent in nature at full scale. To define the velocity distribution within the boundary layer it is normal, in the absence of separation, to use a power law relationship of the form:

$$\frac{v}{V} = \left(\frac{y}{\delta}\right)^n$$

where v is the local velocity at a distance y from the boundary surface, V is the free stream velocity and δ is the boundary layer thickness, which is normally defined as the distance from the surface to where the local velocity attains a value of 99 per cent of the free stream velocity. The exponent n for turbulent boundary layers normally lies with in the range 1/5 to 1/9.

A further complication within the ship boundary layer problem is the onset of separation which will occur if the correct conditions prevail in an adverse pressure gradient; that is a pressure field in which the pressure increases in the direction of flow. Consider, for example, Figure 5.33(a), which shows the flow around some parts of the hull. At station 1 the normal viscous boundary layer has developed; further along the hull at station 2 the velocity of fluid elements close to the surface is less than at station 1, due to the steadily increasing pressure gradient. As the elements continue further downstream they may come to a stop under the action of the adverse pressure gradient, and actually reverse in direction and start moving back upstream as seen at station 3. The point of separation occurs when the velocity gradient $\partial v/\partial n = 0$ at the surface, and the consequence of this is that the flow separates from the surface leaving a region of reversed flow on the surface of the body. Re-attachment of the flow to the surface can subsequently occur if the body geometry and the pressure gradient become favourable.

The full prediction by analytical means of the viscous boundary layer for a ship form is a very complex procedure, and at the present time only partial success has been achieved using large computational codes. A typical calculation procedure for a ship form divides the hull into three primary areas for computation: the potential flow zone, the boundary layer zone and the stern flow and wake zone (Figure 5.33(b)). Whilst considerable effort has been expended on RANS codes to give accurate predictions for all ship forms, at present the most common procedure for determining the total wake field is by model tests in a towing tank.

The wake component due to wave action $\{w_w\}$ is due to the movement of water particles in the system of gravity waves set up by the ship on the surface of the water. Such conditions can also be induced by a vehicle operating just below the surface of the water. Consequently, the wave wake field depends largely on Froude number, and is generally presumed to be of a small order. Harvald (1950) has undertaken experiments from which it would appear that the magnitudes of $\{w_w\}$ are generally less than about 0.02 for a ship form.

5.2.4 Estimation of wake field parameters

From the propeller design viewpoint the determination of the wake field in which the propeller operates is of fundamental importance. The mean wake field determines, along with other parameters of power, revolutions and ship speed, the overall design dimensions of the propeller, and the variability of the wake field about the mean wake influences the propeller blade section design and local pitch. Clearly, the most effective way at present of determining the detailed characteristics of the wake field is from model tests; this, however, is not without problems in the areas of wake scaling and propeller interaction. In the absence of model wake field data the designer must resort to other methods of prediction; these can be in the form of regression equations, the plotting of historical analysis data derived from model or full-scale trials, or from his own intuition and experience, which in the case of an experienced designer must never be underestimated. In

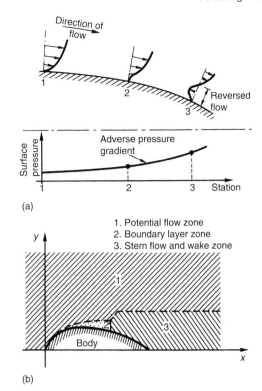

Figure 5.33 Flow boundary layer considerations: (a) origin of separated flow and (b) typical flow computational zones.

the early stages of design the methods cited above are likely to be the ones used.

The determination of the mean wake has received much attention over the years. Harvald (1950) discusses the merits of some two dozen methods developed in the period from 1896 through to the late 1940s for single-screw vessels. From this analysis he concluded that the most reliable, on the basis of calculated value versus value from model experiment, was due to Schoenherr (1939):

$$\bar{w}_a = 0.10 + 4.5 \frac{C_{pv} C_{ph}(B/L)}{(7 - C_{pv})(2.8 - 1.8 C_{ph})} + \frac{1}{2}(E/T - D/B - k\eta)$$

where L is the length of the ship,
 B is the breadth of the ship,
 T is the draught of the ship,
 D is the propeller diameter,
 E is the height of the propeller shaft above the keel,
 C_{pv} is the vertical prismatic coefficient of the vessel,
 C_{ph} is the horizontal prismatic coefficient of the vessel,

226   Maritime engineering reference book

η is the angle of rake of the propeller in radians and

k is the coefficient (0.3 for normal sterns and 0.5–0.6 for sterns having the deadwood cut way).

In contrast, the more simple formula of Taylor (1933) was also found to give acceptable values as a first approximation; this was

$$\bar{w}_a = 0.5C_b - 0.05$$

where C_b is the block coefficient of the vessel.

The danger with using formulae of this type and vintage today is that hull form design has progressed to a considerable extent in the intervening years. Consequently, whilst they may be adequate for some simple hull forms their use should be undertaken with great caution and is, therefore, not to be recommended as a general design tool.

Amongst others the more modern methods that were proposed by Harvald (1977) and illustrated in Figure 5.34 are useful. This method approximates the mean axial wake fraction and thrust deduction by the following relationships:

$$\left.\begin{array}{l} \bar{w}_a = w_1 + w_2 + w_3 \\ t = t_1 + t_2 + t_3 \end{array}\right\} \quad (5.59)$$

where t_1, w_1 are functions of B/L and block coefficient,
t_2, w_2 are functions of the hull forms and
t_3, w_3 are propeller diameter corrections.

Alternatively, the later work by Holtrop and Mennen (1988) and developed over a series of papers resulted in the following regression formulae for single- and twin-screw vessels, Holtrop (1988).

Single screw:

$$\begin{aligned} \bar{w}_a = {} & C_9(1 + 0.015C_{\text{stern}})[(1+k)C_F + C_A]\frac{L}{T_A} \\ & \times \left\{ 0.050776 + 0.93405C_{11} \right. \\ & \times \frac{[(1+k)C_F + C_A]}{(1.315 - 1.45C_p + 0.0225\text{ LCB})} \right\} \\ & + 0.27915(1 + 0.015C_{\text{stern}}) \\ & \times \sqrt{\frac{B}{L(1.315 - 1.45C_p + 0.0225\text{ LCB})}} \\ & + C_{19}(1 + 0.015C_{\text{stern}}) \end{aligned}$$

Twin screw:

$$\bar{w}_a = 0.3095C_b + 10C_b[(1+k)C_F + C_A] \\ - 0.23\frac{D}{\sqrt{BT}}$$

(5.60)

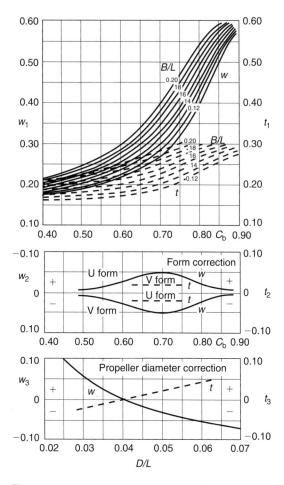

Figure 5.34 The wake and thrust deduction coefficient for single-screw ships (Reproduced with permission from Harvald (1977)).

where:

$$\begin{aligned} C_9 &= C_8 \quad (C_8 < 28) \\ &= 32 - 16/(C_8 - 24) \quad (C_8 > 28) \end{aligned}$$

and

$$\begin{aligned} C_8 &= BS/(LDT_A) \quad (B/T_A < 5) \\ &= S(7B/T_A - 25)/(LD(B/T_A - 3)) \quad (B/T_A > 5) \\ C_{11} &= T_A/D \quad (T_A/D < 2) \\ &= 0.0833333(T_A/D)^2 + 1.3333 \quad (T_A/D > 2) \\ C_{19} &= 0.12997/(0.95 - C_B) \\ &\quad - 0.11056/(0.95 - C_p) \quad (C_p < 0.7) \\ &= 0.18567/(1.3571 - C_M) - 0.71276 \\ &\quad + 0.38648C_p \quad (C_p > 0.7) \end{aligned}$$

and

Single-screw afterbody form	C_{stern}
Pram with gondola	−25
V-shaped sections	−10
Normal section shape	0
U-shaped sections with Hogner stern	10

These latter formulae were developed from the results of single- and twin-screw model tests over a comparatively wide range of hull forms. The limits of applicability are referred to in the papers and should be carefully studied before using the formulae.

In the absence of model tests the radial distribution of the mean wake field, that is the average wake value at each radial location, is difficult to assess. Traditionally, this has been approximated by the use of van Lammeren's diagrams (van Lammeren et al. (1942)), which are reproduced in Figure 5.35. Van Lammeren's data is based on the single parameter of vertical prismatic coefficient, and is therefore unlikely to be truly representative for all but first approximations to the radial distribution of mean wake. Harvald (1950) re-evaluated the data in which he corrected all the data to a common value of D/L of 0.004 and then arranged the data according to block coefficient and breadth-to-length ratio as shown in Figure 5.36 for single-screw models together with a correction for frame shape. In this study Harvald drew attention to the considerable scale effects that occurred between model and full scale. He extended his work to twin-screw vessels, shown in Figure 5.37, for a diameter-to-length ratio of 0.03, in which certain corrections were made to the model test data

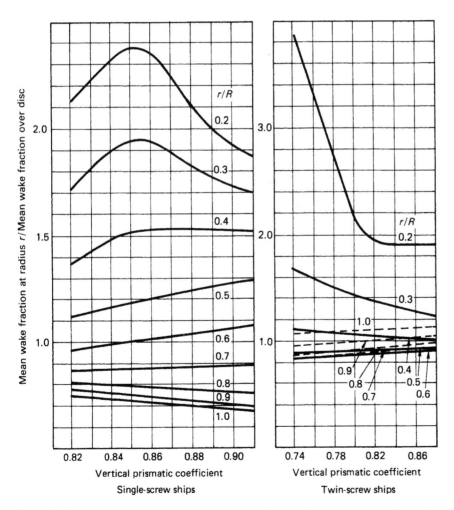

Figure 5.35 van Lammeren's curves for determining the radial wake distribution (Reproduced with permission from Harvald (1950)).

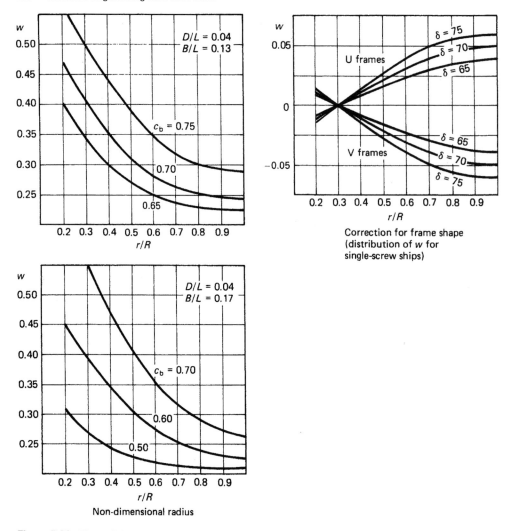

Figure 5.36 The radial variation of the wake coefficient of single-screw ships ($D/L = 0.04$) (Reproduced with permission from Harvald (1950)).

partly to correct for the boundary layer of the shaft supports. The twin-screw data shown in the diagram refers to the use of bossings to support the shaft lines rather than the modern practice of 'A' and 'P' brackets.

It must be emphasized that all of these methods for the estimation of the wake field and its various parameters are at best approximations to the real situation and not a substitute for properly conducted model tests.

5.2.5 Effective wake field

Classical propeller theories assume the flow field to be irrotational and unbounded; however, because the propeller normally operates behind the body which is being propelled these assumptions are rarely satisfied. When the propeller is operating behind a ship the flow field in which the propeller is operating at the stern of the ship is not simply the sum of the flow field in the absence of the propeller together with the propeller-induced velocities calculated on the basis of the nominal wake. In practice a very complicated interaction takes place which gives rise to noticeable effects on propeller performance. Figure 5.38 shows the composition of the velocities that make up the total velocity at any point in the propeller disc. From the propeller design viewpoint it is the effective velocity field that is important since this is the velocity field that should be input into propeller design and analysis procedures. The

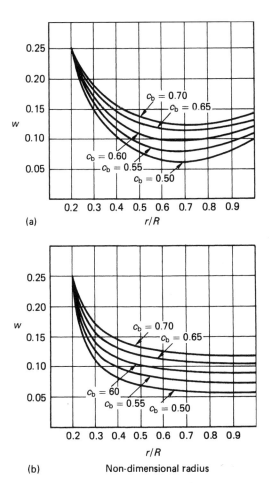

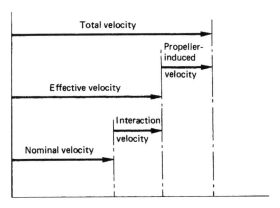

Figure 5.38 Composition of the wake field.

Figure 5.37 (a) The radial variation of the wake coefficient for models having twin screws ($D/L = 0.03$) and (b) the radial variation of the wake coefficient for twin-screw ships ($D/L = 0.03$) (Reproduced with permission from Harvald (1950)).

effective velocity field can be seen from the figure to be defined in one of two ways:

$$\left. \begin{array}{l} \text{effective velocity} = \text{nominal velocity} \\ \qquad\qquad\qquad + \text{interaction velocity} \\ \text{or} \\ \text{effective velocity} = \text{total velocity} \\ \qquad\qquad\qquad - \text{propeller induced} \\ \qquad\qquad\qquad\quad \text{velocity} \end{array} \right\} \quad (5.61)$$

If the latter of the two relationships is used, an iterative procedure can be employed to determine the effective wake field if the total velocity field is known from measurements just ahead of the propeller. The procedure used for this estimation is shown in Figure 5.39 and has been shown to converge. However,

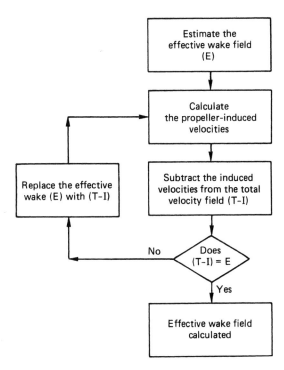

Figure 5.39 (T–I) approach to effective wake field estimation.

this procedure has the disadvantage of including within it all the shortcomings of the particular propeller theory used for the calculation of the induced velocities. As a consequence this may lead to an incorrect assessment of the interaction effects arising, for example, from the differences in the theoretical treatment of the trailing vortex system of the propeller.

An alternative procedure is to use the former of the two formulations of effective velocity defined in Equation (5.61). This approach makes use of the nominal wake field measured in the towing tank, this being a considerably easier measurement than that of measuring the total velocity, since for the nominal velocity measurement the propeller is absent. Several approaches to this problem have been proposed, including those known as the V-shaped segment and force-field approaches. The V-shaped segment method finds its origins in the work of Huang and Groves (1980), which was based on investigations of propeller–wake interaction for axisymmetric bodies. This approach is perhaps the simplest of all effective wake estimation procedures since it uses only the nominal wake field and principal propeller dimensions as input without undertaking detailed hydrodynamic computations. In the general case of a ship wake field, which contrasts with the axisymmetric basis upon which the method was first derived by being essentially non-uniform, the velocity field is divided into a number of V-shaped segments over which the general non-uniformity is replaced with an equivalent uniform flow. The basis of a V-shaped segment procedure is actuator disc theory, and the computations normally commence with an estimate of the average thrust loading coefficient based on a mean effective wake fraction; typically such an estimate comes from standard series open water data. From this estimate an iterative algorithm commences in which an induced velocity distribution is calculated, which then allows the associated effective velocities and their radial locations to be computed. Procedures of this type do not take into account any changes of flow structure caused by the operating propeller since they are based on the approximate interaction between a propeller and a thick stern boundary layer.

An alternative, and somewhat more complex, effective wake estimation procedure is the force-field method. Such approaches usually rely for input on the nominal wake field and the propeller thrust together with an estimate of the thrust deduction factor. These methods calculate the total velocity field by solving the Euler and continuity equations describing the flow in the vicinity of the propeller. The propeller action is modelled by an actuator disc having only an axial force component and a radial thrust distribution which is assumed constant circumferentially at each radial station. The induced velocities, which are identified within the Euler equations, can then, upon convergence, be subtracted from the total velocity estimates at each point of interest to give the effective wake distribution.

Clearly methods of effective wake field estimation such as the V-shaped segment, force-field and the (T–I) approaches are an essential part of the propeller design and analysis procedure. However, all of these methods lack the wider justification from being subjected to correlation, in open literature, between model and full-scale measurements. Indeed the number of vessels upon which appropriate wake field measurements have been undertaken is minimal for a variety of reasons; typically cost, availability and difficulty of measurement. The latter reason has at least been partially removed with the advent of laser-Doppler techniques which allow effective wake field measurement; nevertheless, this is still a complex procedure.

The analytical treatment of effective wake prediction has gained pace during recent years. A coupled viscous and potential flow procedure was developed by Kerwin et al. (1994, 1997) for the design of an integrated propulsor driving an axisymmetric body. In this method the flow around the body was computed with the aid of a RANS code with the propulsor being represented by body forces whose magnitudes were estimated using a lifting surface method. As such, in this iterative procedure the RANS solver estimated the total velocity field from which the propeller-induced velocities were subtracted to derive the effective propulsor inflow. Warren et al. (2000) used a similar philosophy in order to predict propulsor-induced manoeuvring forces in which a RANS code was used for flow calculations over a hull, the appendages and a duct. The time averaged flow field was then input into a three-dimensional lifting surface code which estimated the time varying forces and pressures which were then re-input into the RANS solver in an iterative fashion until convergence was achieved. Hsin et al. (2002) developed Kerwin's ideas to a podded propulsor system in order to predict hull–propeller interaction.

Choi and Kinnas (2001, 2003) developed an unsteady effective wake prediction methodology by coupling an unsteady lifting surface cavitating propeller procedure with a three-dimensional Euler code. In this arrangement the propeller effect is represented by unsteady body forces in the Euler solver such that the unsteady effective wake both spatially and temporally can be estimated. Using this method it was found that the predicted total velocity distribution in front of the propeller was in good agreement with measured data. Lee et al. (2003) studied rudder sheet cavitation with some success when comparing theoretical predictions with experimental observation. In this procedure a vortex lattice method was coupled to a three-dimensional Euler solver and boundary element method; the latter being used to calculate the cavitating flow around the rudder.

Considerable progress is being made in the estimation of effective wake using advanced computational analyses and this trend looks set to continue for the foreseeable future.

5.2.6 Wake field scaling

Since the model of the ship which is run in the towing tank is tested at Froude identity, that is equal

Froude numbers between the ship and model, a disparity in Reynolds number exists which leads to a relative difference in the boundary layer thickness between the model and the full-scale ship; the model having the relatively thicker boundary layer. Consequently, for the purposes of propeller design it is necessary to scale, or contract as it is frequently termed, the wake measured on the model so that is becomes representative of that on the full-size vessel. Figure 5.40 illustrates the changes that can typically occur between the wake fields measured at model and full scale and with and without a propeller. The results shown in Figure 5.40 relate to trials conducted on the research vessel *Meteor* in 1967 and show respectively pitot measurements made with a 1/14th scale model; the full-scale vessel being towed without a propeller and measurements, again at full scale, made in the presence of the working propeller.

In order to contract nominal wake fields in order to estimate full-scale characteristics two principal methods have been proposed in the literature and are in comparatively wide use. The first method is due to Sasajima et al. (1966) and is applicable to single-screw ships. In this method it is assumed that the displacement wake is purely potential in origin and as such is independent of scale effects, and the frictional wake varies linearly with the skin friction coefficient. Consequently, the total wake at a point is considered to comprise the sum of the frictional and potential components. The total contraction of the wake field is given by

$$c = \frac{C_{fs} + \Delta C_{fs}}{C_{fm}}$$

where C_{fs} and C_{fm} are the ship and model ITTC 1957 friction coefficients expressed by

$$C_f = \frac{0.075}{(\log_{10} R_n - 2)^2}$$

and ΔC_{fs} is the ship correlation allowance.

The contraction in Sasajima's method is applied with respect to the centre plane in the absence of any potential wake data, this being the normal case. However, for the general case the contraction procedure is shown in Figure 5.41 in which the ship frictional wake (w_{fs}) is given by

$$w_{fs} = w_{fm} \frac{(1 - w_{ps})}{(1 - w_{pm})}$$

The method was originally intended for full-form ships having block coefficients in the order of 0.8 and L/B values of around 5.7. Numerous attempts by a number

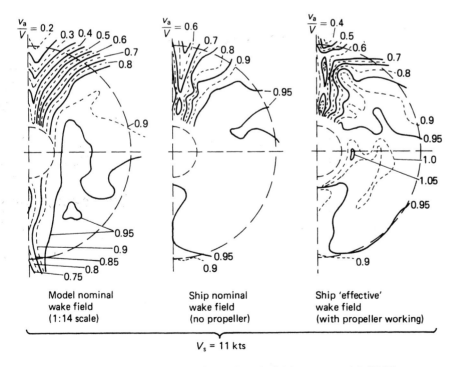

Figure 5.40 Comparison of model and full-scale wake fields – meteor trials (1967).

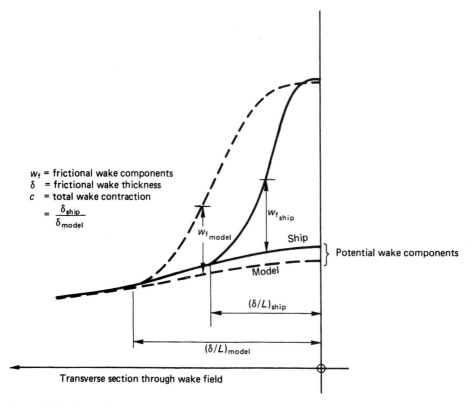

Figure 5.41 Basic of Sasajima wake scaling method.

of researchers have been made to generalize and improve the method. The basic idea behind Sasajima's method is to some extent based on the flat plate wake idealization; however, to account for the full range of ship forms encountered in practice, that is those with bulbous sterns, flat afterbodies above the propeller and so on, a more complete three-dimensional contraction process needs to be adopted. Hoekstra (1975) developed such a procedure in the mid-1970s in which he introduced, in addition to the centre-plane contraction, a concentric contraction and a contraction to a horizontal plane above the propeller.

In this procedure the overall contraction factor (c) is the same as that used in the Sasajima approach. However, this total contraction is split into three component parts:

$$c = ic + jc + kc \quad (i + j + |k| = 1)$$

where i is the concentric contraction, j is the centre-plane contraction and k is the contraction to a horizontal surface above the propeller.

In Hoekstra's method the component contractions are determined from the harmonic content of the wake field; as such the method makes use of the first six Fourier coefficients of the circumferential wake field at each radius. The contraction factors are determined from the following relationships:

$$i = \frac{F_i}{|F_i|+|F_j|+|F_k|} \qquad j = \frac{F_j}{|F_i|+|F_j|+|F_k|}$$

$$\text{and } k = \frac{F_k}{|F_i|+|F_j|+|F_k|}$$

in which

$$F_i = \int_{r_{hub}}^{2R} S_i(r)\mathrm{d}r, \quad F_j = \int_{r_{hub}}^{2R} S_j(r)\mathrm{d}r$$

$$\text{and } F_k = \int_{r_{hub}}^{2R} S_k(r)\mathrm{d}r$$

with

$$S_i = 1 - A_0$$
$$\quad + \begin{cases} A_2 + A_4 + A_6 - \tfrac{1}{2}S_k & \text{if } S_k \geq S_j \\ -S_j + (A_2 + A_4 + A_6) & \text{if } S_k < S_j \end{cases}$$
$$S_j = -[A_2 + A_4 + A_6$$
$$\quad + |\max(A_2 \cos 2\phi + A_4 \cos 4\phi$$
$$\quad + A_6 \cos \phi|] \quad (\phi \neq 0, \pi, 2\pi)$$
$$S_k = 2(A_1 + A_3 + A_5)$$

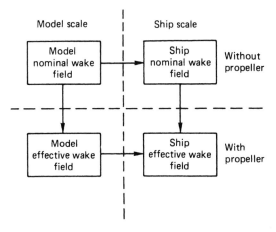

Figure 5.42 Relationship between model and ship wake field.

where A_n ($n = 0, 1, ..., 6$) are the Fourier coefficients and at the hub S_i is taken as unity with $S_j = S_k = 0$.

The method as proposed by Hoekstra also makes an estimation of the scale effect on the wake peak velocity in the centre plane and for the scale effect on any bilge vortices that may be present. The method has been shown to give reasonable agreement in a limited number of cases of full scale to model correlation. However, there have been very few sets of trail results available upon which to base any firm conclusions of this or any other wake field scaling procedure.

Figure 5.42 essentially draws the discussions of effective wake and wake scaling together. In most design or analysis situations the engineer is in possession of the model nominal wake field and wishes to derive the ship or full-scale effective wake field characteristics. There are essentially two routes to achieve this. The most common is to scale the derived nominal wake field from model to full scale and then to derive the effective wake field at ship scale from the derived nominal full-scale wake.

Carlton (2007) goes on to discuss wake quality assessment and measurement of the wake field.

5.3 Propeller performance characteristics

For discussion purposes the performance characteristics of a propeller can conveniently be divided into open water and behind-hull properties. In the case of open water characteristics, these relate to the description of the forces and moments acting on the propeller when operating in a uniform fluid stream; hence the open water characteristics, with the exception of inclined flow problems, are steady loadings by definition. The behind-hull characteristics are those generated by the propeller when operating in a mixed wake field behind a body. Clearly these latter characteristics have both a steady and unsteady component by the very nature of the environment in which the propeller operates. In this section the discussion will centre on the open water characteristics since these form the basic performance parameters about which the behind-hull characteristics are generated when the propeller is working behind a body.

5.3.1 General open water characteristics

The forces and moments produced by the propeller are expressed in their most fundamental form in terms of a series of non-dimensional characteristics: these are completely general for a specific geometric configuration. The non-dimensional terms used to express the general performance characteristics are as follows:

$$\text{thrust coefficient } K_T = \frac{T}{\rho n^2 D^4}$$

$$\text{torque coefficient } K_Q = \frac{Q}{\rho n^2 D^5}$$

$$\text{advance coefficient } J = \frac{V_a}{nD}$$

$$\text{cavitation number } \sigma = \frac{p_0 - e}{\frac{1}{2}\rho V^2} \quad (5.62)$$

where in the definition of cavitation number, V is a representative velocity which can either be based on free stream advance velocity or propeller rotational speed. Whilst for generalized open water studies the former is more likely to be encountered there are exceptions when this is not the case, notably at the bollard pull condition when $V_a = 0$ and hence $\sigma_0 \to \infty$. Consequently, care should be exercised to ascertain the velocity term being employed when using design charts or propeller characteristics for analysis purposes.

To establish the non-dimensional groups involved in the above expressions (Equation (5.62)), the principle of dimensional similarity can be applied to geometrically similar propellers. The thrust of a marine propeller when working sufficiently far away from the free surface so as to not cause surface waves may be expected to depend upon the following parameters:

(a) The diameter (D).
(b) The speed of advance (V_a).
(c) The rotational speed (n).
(d) The density of the fluid (ρ).
(e) The viscosity of the fluid (μ).
(f) The static pressure of the fluid at the propeller station ($p_0 - e$).

Hence the thrust (T) can be assumed to be proportional to ρ, D, V_a, n, μ and $(p_0 - e)$:

$$T \propto \rho^a D^b V_a^c n^d \mu^f (p_0 - e)^g$$

Since the above equation must be dimensionally correct it follows that

$$MLT^{-2} = (ML^{-3})^a L^b (LT^{-1})^c (T^{-1})^d \\ \times (ML^{-1}T^{-1})^f (ML^{-1}T^{-2})^g$$

and by equating indices for M, L and T we have

for mass M: $\quad 1 = a + f + g$
for length L: $\quad 1 = -3a + b + c - f - g$
for time T: $\quad -2 = -c - d - f - 2g$

from which it can be shown that

$$a = 1 - f - g \\ b = 4 - c - 2f - g \\ d = 2 - c - f - 2g$$

Hence from the above we have

$$T \propto \rho^{(1-f-g)} D^{(4-c-2f-g)} V_a^c n^{(2-c-f-2g)} \mu^f (p_0 - e)^g$$

from which

$$T = \rho n^2 D^4 \left(\frac{V_a}{nD}\right)^c \cdot \left(\frac{\mu}{\rho nD^2}\right)^f \cdot \left(\frac{p_0 - e}{\rho n^2 D^2}\right)^g$$

These non-dimensional groups are known by the following:

thrust coefficient $\quad K_T = \dfrac{T}{\rho n^2 D^4}$

advance coefficient $\quad J = \dfrac{V_a}{nD}$

Reynolds number $\quad R_n = \dfrac{\rho n D^2}{\mu}$

cavitation number $\quad \sigma_0 = \dfrac{p_0 - e}{\frac{1}{2}\rho n^2 D^2}$

$\therefore K_T \propto \{J, R_n, \sigma_0\}$

that is

$$K_T = f(J, R_n, \sigma_0) \qquad (5.63)$$

The derivation for propeller torque K_Q is an analogous problem to that of the thrust coefficient just discussed. The same dependencies in this case can be considered to apply, and hence the torque (Q) of the propeller can be considered by writing it as a function of the following terms:

$$Q = \rho^a D^b V_a^c n^d \mu^f (p_0 - e)^g$$

and hence by equating indices we arrive at

$$Q = \rho n^2 D^5 \left(\frac{V_a}{nD}\right)^c \left(\frac{\mu}{\rho nD^2}\right)^f \cdot \left(\frac{p_0 - e}{\rho n^2 D^2}\right)^g$$

which reduces to

$$K_Q = g(J, R_n, \sigma_0) \qquad (5.64)$$

where the torque coefficient

$$K_Q = \frac{Q}{\rho n^2 D^5}$$

With the form of the analysis chosen the cavitation number and Reynolds number have been non-dimensionalized by the rotational speed. These numbers could equally well be based on advance velocity, so that

$$\sigma_0 = \frac{p_0 - e}{\frac{1}{2}\rho V^2} \quad \text{and} \quad R_n = \frac{\rho V D}{\mu}$$

Furthermore, by selecting different groupings of indices in the dimensional analysis it would be possible to arrive at an alternative form for the thrust loading:

$$T = \rho V_a^2 D^2 \phi(J, R_n, \sigma_0)$$

which gives rise to the alternative form of thrust coefficient C_T defined as

$$C_T = \frac{T}{\frac{1}{2}\rho V_a^2 (\pi D^2/4)} = \frac{8T}{\pi \rho V_a^2 D^2} \qquad (5.65)$$

$$C_T = \Phi(J, R_n, \sigma_0)$$

Similarly it can be shown that the power coefficient C_P can also be given by

$$C_P = \phi(J, R_n, \sigma_0) \qquad (5.66)$$

In cases where the propeller is sufficiently close to the surface, so as to disturb the free surface or to draw air, other dimensionless groups will become important. These will principally be the Froude and Weber numbers, and these can readily be shown to apply by introducing gravity and surface tension into the foregoing dimensional analysis equations for thrust and torque.

A typical open water diagram for a set of fixed pitch propellers working in a non-cavitating environment at forward, or positive, advance coefficient is shown in Figure 5.43. This figure defines, for the particular propeller, the complete set of operating conditions at positive advance and rotational speed, since the propeller under steady conditions can only operate along the characteristic line defined by its pitch ratio P/D. The diagram is general in the sense that, subject to scale effects, it is applicable to any propeller having the same geometric form as the one for which the

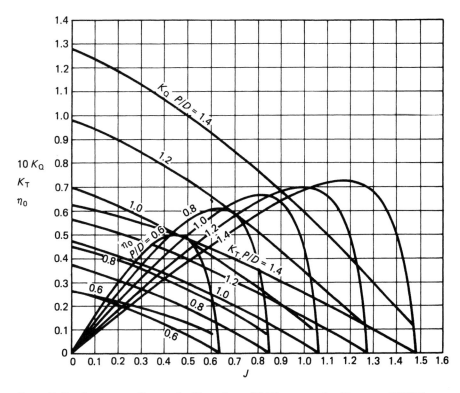

Figure 5.43 Open water diagram for Wageningen B5-75 screw series (Courtesy: MARIN).

characteristic curves were derived, but the subject propeller may have a different diameter or scale ratio and can work in any other fluid, subject to certain Reynolds number effects. When, however, the K_T, K_Q versus J diagram is used for a particular propeller of a given geometric size and working in a particular fluid medium, the diagram, since the density of the fluid and the diameter then become constants, effectively reduces from general definitions of K_T, K_Q and J to a particular set of relationships defining torque, thrust, revolutions and speed of advance as follows:

$$\left\{\frac{Q}{n^2}, \frac{T}{n^2}\right\} \text{ versus } \left(\frac{V_a}{n}\right)$$

The alternative form of the thrust and torque coefficient which stems from Equations (5.65) and (5.66) and which is based on the advance velocity rather than the rotational speed, is defined as follows:

$$C_T = \frac{T}{\frac{1}{2}\rho A_0 V_a^2}$$
$$C_P = \frac{P_D}{\frac{1}{2}\rho A_0 V_a^3}$$ (5.67)

From Equation (5.67) it can be readily deduced that these thrust loading and power loading coefficients can be expressed in terms of the conventional thrust and torque coefficient as follows:

$$C_T = \frac{8}{\pi}\frac{K_T}{J^2}$$

and

$$C_P = \frac{8}{\pi}\frac{K_Q}{J^3}$$ (5.68)

The open water efficiency of a propeller (η_o) is defined as the ratio of the thrust horsepower to delivered horsepower:

$$\eta_o = \frac{\text{THP}}{\text{DHP}}$$

Now since $\text{THP} = TV_a$
and $\text{DHP} = 2\pi n Q$

where T is the propeller thrust, V_a, the speed of advance, n, the rotational speed of the propeller and Q, the torque. Consequently, with a little mathematical manipulation we may write

$$\eta_o = \frac{TV_a}{2\pi n Q}$$

that is

$$\eta_o = \frac{K_T}{K_Q} \frac{J}{2\pi} \quad (5.69)$$

The K_Q, K_T versus J characteristic curves contain all of the information necessary to define the propeller performance at a particular operating condition. Indeed, the curves can be used for design purposes for a particular basic geometry when the model characteristics are known for a series of pitch ratios. This, however, is a cumbersome process and to overcome these problems Admiral Taylor derived a set of design coefficients termed B_p and δ; these coefficients, unlike the K_T, K_Q and J characteristics, are dimensional parameters and so considerable care needs to be exercised in their use. The terms B_p and δ are defined as follows:

$$B_p = \frac{(\text{DHP})^{1/2} N}{V_a^{2.5}} \quad (5.70)$$

$$\delta = \frac{ND}{V_a}$$

where DHP is the delivered horsepower in British or metric units depending on the, diagram used
N is the propeller rpm
V_a is the speed of advance (knots)
D is the propeller diameter (ft).

From Figure 5.44, which shows a typical propeller design diagram, it can be seen that it essentially comprises a plotting of B_p, as abscissa, against pitch ratio as ordinate with lines of constant δ and open water efficiency superimposed. This diagram is the basis of many design procedures for marine propellers, since the term B_p is usually known from the engine and ship characteristics. From the figure a line of optimum propeller open water efficiency can be seen as being the locus of the points on the diagram which have the highest efficiency for a give value of B_p. Consequently, it is possible with this diagram to select values of δ and P/D to maximize the open water efficiency η_o for a given powering condition as defined by the B_p parameter. Hence a basic propeller geometry can be derived in terms of diameter D, since $D = \delta V_a / N$, and P/D. Additionally, this diagram can be used for a variety of other design purposes, such as, for example, rpm selection; however, these aspects of the design process will be discussed later in Section 5.6.

It will be seen that the B_p versus δ diagram is limited to the representation of forward speeds of advance only, that is, where $V_a > 0$, since $B_p \to \infty$ when $V_a = 0$. This limitation is of particular importance when considering the design of tugs and other similar craft, which can be expected to spend an important part of their service duty at zero ship speed, termed bollard pull, whilst at the same time developing full power. To overcome this problem, a different sort of design diagram was developed from the fundamental K_T, K_Q versus J characteristics, so that design and analysis problems at or close to zero speed of advance can be considered. This diagram is termed the $\mu - \sigma$ diagram, and a typical example of one is shown in Figure 5.45. In this diagram the following relationships apply:

$$\mu = n \sqrt{\frac{\rho D^5}{Q}}$$

$$\phi = V_a \sqrt{\frac{\rho D^3}{Q}}$$

$$\sigma = \frac{TD}{2\pi Q} \quad (5.71)$$

where D is the propeller diameter (m)
Q is the delivered torque (kgf m)
ρ is the mass density of water (kg/m^4 s^2)
T is the propeller thrust (kgf)
n is the propeller rotational speed (rev/s)
V_a is the ship speed of advance (m/s).

Diagrams of the type shown in Figure 5.45 are non-dimensional in the same sense as those of the fundamental K_T, K_Q characteristics and it will be seen that the problem of zero ship speed, that is when $V_a = 0$, has been removed, since the function $\phi \to 0$ as $V_a \to 0$. Consequently, the line on the diagram defined by $\phi = 0$ represents the bollard pull condition for the propeller. It is important, however, not to confuse propeller thrust with bollard pull, as these terms are quite distinct and mean different things. Propeller thrust and bollard pull are exactly what the terms imply; the former relates to the hydrodynamic thrust produced by the propeller, whereas the latter is the pull the vessel can exert through a towline on some other stationary object. Bollard pull is always less than the propeller thrust by a complex ratio, which is dependent on the underwater hull form of the vessel, the depth of water, the distance of the vessel from other objects, and so on.

In the design process it is frequently necessary to change between coefficients, and to facilitate this process. Tables 5.9 and 5.10 are produced in order to show some of the more common relationships between the parameters.

Note the term σ in Tables 5.9 and 5.10 and in Equation (5.71) should not be confused with cavitation number, which is an entirely different concept. The term σ in the above tables and Equation (5.71) relates to the $\mu - \sigma$ diagram, which is a non-cavitating diagram.

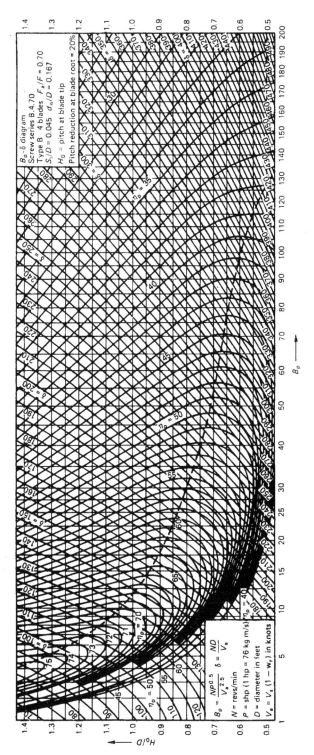

Figure 5.44 Original B4-70 $B_p - \delta$ diagram (Courtesy: MARIN).

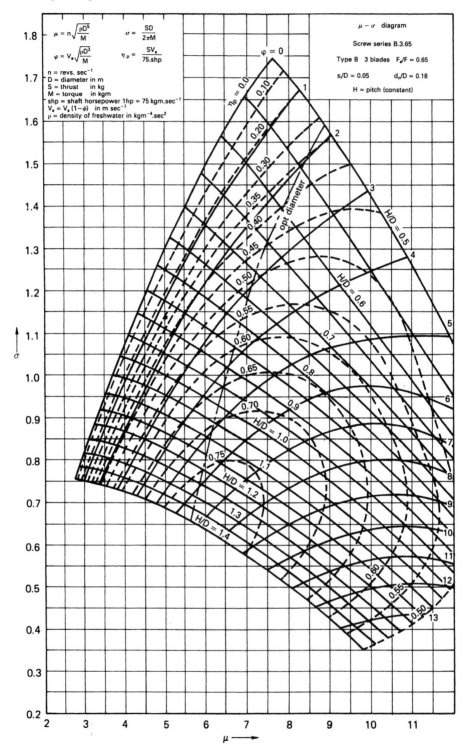

Figure 5.45 Original B3.65 $\mu - \sigma$ diagram (Courtesy: MARIN).

Table 5.9 Common functional relationships (British units).

$$K_Q = 9.5013 \times 10^6 \left(\frac{P_D}{N^3 D^5}\right) \text{(salt water)}$$

$$B_P = 23.77\sqrt{\frac{\rho K_Q}{J^5}}$$

$$J = \frac{101.33 V_a}{ND} = \frac{101.33}{\delta}$$

$$\mu = \frac{1}{\sqrt{K_Q}} = 3.2442 \times 10^{-4} \sqrt{\frac{N^3 D^5}{P_D}} \text{(salt water)}$$

$$\phi = \frac{J}{\sqrt{K_Q}} = J\mu$$

$$\sigma = \frac{\eta_o}{J} = \frac{\eta_o \mu}{\phi} = \frac{K_T}{2\pi K_Q}$$

where:
P_D is the delivered horsepower in Imperial units
Q is the delivered torque at propeller in (lbf ft)
T is the propeller thrust (lbf)
N is the propeller rotational speed in (rpm)
n is the propeller rotational speed in (rev/s)
D is the propeller diameter in (ft)
V_a is the propeller speed of advance in (knots)
v_a is the propeller speed of advance in (ft/s)
ρ is the mass density of water (1.99 slug/ft³ sea water; 1.94 slug/ft³ for fresh water).

Table 5.10 Common functional relationships (Metric units).

$$K_Q = 2.4669 \times 10^4 \left(\frac{P_D}{N^3 D^5}\right) \text{(salt water)}$$

$$B_P = 23.77\sqrt{\frac{\rho K_Q}{J^5}}$$

$$J = \frac{30.896 V_a}{ND} = \frac{101.33}{\delta}$$

$$\mu = \frac{1}{\sqrt{K_Q}} = 6.3668 \times 10^{-3} \sqrt{\frac{N^3 D^5}{P_D}} \text{(salt water)}$$

$$\phi = \frac{J}{\sqrt{K_Q}} = J\mu$$

$$\sigma = \frac{\eta_o}{J} = \frac{\eta_o \mu}{\phi} = \frac{K_T}{2\pi K_Q}$$

where:
P_D is the delivered horsepower (metric units)
Q is the delivered propeller torque (kp m)
T is the propeller thrust (kp)
N is the propeller rotational speed (rpm)
n is the propeller rotational speed (rev/s)
D is the propeller diameter (m)
V_a is the propeller speed of advance (knots)
v_a is the propeller speed of advance (m/s)
ρ is the mass density of water (104.48 sea water) (101.94 fresh water)

5.3.2 Effect of cavitation on open water characteristics

Cavitation, which is a two-phase flow phenomenon, is discussed more fully in Section 5.5; however, it is pertinent here to recognize the effect that cavitation development can have on the propeller open water characteristics.

Cavitation for the purposes of generalized analysis is defined by a free stream cavitation number σ_0 which is the ratio of the static to dynamic head of the flow. For our purposes in this Section we will consider a cavitation number based on the static pressure at the shaft centre line and the dynamic head of the free stream flow ahead of the propeller:

$$\sigma_0 = \frac{\text{static head}}{\text{dynamic head}} = \frac{p_0 - e}{\frac{1}{2}\rho V_a^2}$$

where p_0 is the absolute static pressure at the shaft centre line and e is the vapour pressure at ambient temperature. Consequently, a non-cavitating flow is one where $(p_0 - e) \gg \frac{1}{2}\rho V_a^2$, that is one where σ_0 is large. As σ_0 decreases in value cavitation takes more effect as demonstrated in Figure 5.46. This figure illustrates the effect that cavitation has on the K_T and K_Q curves and, for guidance purposes only, shows a typical percentage of cavitation on the blades experienced at various cavitation numbers in uniform flow. It is immediately apparent from the figure that moderate levels of cavitation do not affect the propulsion performance of the propeller and significant cavitation activity is necessary in order to get thrust and torque breakdown. Furthermore, it will frequently be noted that the K_T and K_Q curves rise marginally above the non-cavitating line just prior to their rapid decline after thrust or torque breakdown.

It is, however, important not necessarily to associate the other problems of cavitation, for example, hull-induced vibration and erosion of the blade material, with the extent of cavitation necessary to cause thrust and torque performance breakdown. Relatively small levels, in terms of the extent, of cavitation, given the correct conditions, are sufficient to give rise to these problems.

5.3.3 Propeller scale effects

Open water characteristics are frequently determined from model experiments on propellers run at high speed and having diameters of the order of 200 to 300 mm. It is, therefore, reasonable to pose the question of how the reduction in propeller speed and increase in diameter at full scale will affect the propeller performance characteristics. Figure 5.47 shows the principal features of scale effect, from which it can be seen that whilst

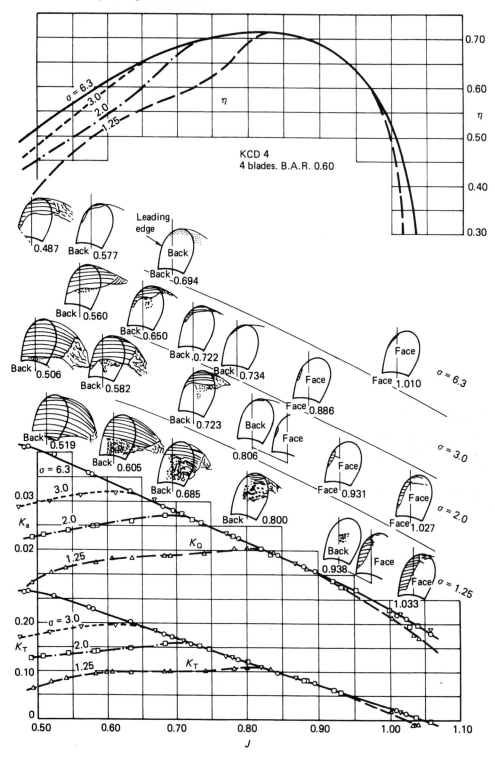

Figure 5.46 Curves of K_T, K_Q and η and cavitation sketches for KCD 4 (Reproduced from Burrill and Emerson (1963)).

which is based on a simplification of Lerbs' equivalent profile procedure. Lerbs showed that a propeller can be represented by the characteristics of an equivalent section at a non-dimensional radius of around $0.70R$ or $0.75R$, these being the two sections normally chosen. The method calculates the change in propeller performance characteristics as follows.

The revised thrust and torque characteristics are given by

$$\left.\begin{array}{l} K_{T_s} = K_{T_m} - \Delta K_T \\ K_{Q_s} = K_{Q_m} - \Delta K_Q \end{array}\right\} \quad (5.72)$$

where the scale corrections ΔK_T and ΔK_Q are given by

$$\Delta K_T = -0.3 \Delta C_D \left(\frac{P}{D}\right)\left(\frac{cZ}{D}\right)$$

$$\Delta K_Q = 0.25 \Delta C_D \left(\frac{cZ}{D}\right)$$

and in Equation (5.72) the suffixes s and m denote the full-scale ship and model test values respectively. The term ΔC_D relates to the change in drag coefficient introduced by the differing flow regimes at model and full scale, and is formally written as

$$\Delta C_D = C_{DM} - C_{DS}$$

where

$$C_{DM} = 2\left(1 + \frac{2t}{C}\right)\left[\frac{0.044}{(R_{nx})^{1/6}} - \frac{5}{(R_{nx})^{2/3}}\right]$$

and

$$C_{DS} = 2\left(1 + \frac{2t}{c}\right)\left(1.89 + 1.62 \log_{10}\left(\frac{c}{K_p}\right)\right)^{-2.5}$$

In these relationships t/c is the section thickness to chord ratio; P/D is the pitch ratio; c is the section chord length and R_{nx} is the local Reynolds number, all relating to the section located $0.75R$. The blade roughness K_p is taken as 30×10^{-6} m.

In this method it is assumed that the full-scale propeller blade surface is hydrodynamically rough and the scaling procedure considers only the effect of Reynolds number on the drag coefficient.

An alternative approach to the use of Equation (5.72) has been proposed by Vasamarov (1983) in which the correction for the Reynolds effect on propeller open water efficiency is given by

$$\eta_{os} = \eta_{om} - F(J)\left[\left(\frac{1}{R_{nm}}\right)^{0.2} - \left(\frac{1}{R_{ns}}\right)^{0.2}\right] \quad (5.73)$$

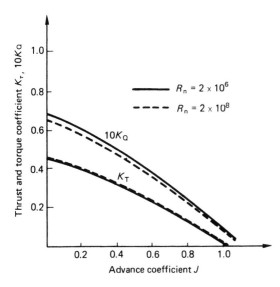

Figure 5.47 Principal features of scale effect.

the thrust characteristic is largely unaffected the torque coefficient is somewhat reduced for a given advance coefficient.

The scale effects affecting performance characteristics are essentially viscous in nature, and as such are primarily due to boundary layer phenomena dependent on Reynolds number. Due to the methods of testing model propellers and the consequent changes in Reynolds number between model and full scale, or indeed a smaller model and a larger model, there can arise a different boundary layer structure to the flow over the blades. Whilst it is generally recognized that most full-scale propellers will have a primarily turbulent flow over the blade surface this need not be the case for the model where characteristics related to laminar flow can prevail over significant parts of the blade.

In order to quantify the effect of scale on the performance characteristics of a propeller an analytical procedure is clearly required. There is, however, no common agreement as to which is the best procedure. In a survey conducted by the 1987 ITTC it was shown that from a sample of 22 organizations, 41 per cent used the ITTC 1978 procedure; 23 per cent made corrections based on correlation factors developed from experience; 13 per cent, who dealt with vessels having open shafts and struts, made no correction at all; a further 13 per cent endeavoured to scale each propulsion coefficient whilst the final 10 per cent scaled the open water test data and then used the estimated full-scale advance coefficient. It is clear, therefore, that research is needed in this area in order to bring a measure of unification between organizations.

At present the principal analytical tool available is the 1978 ITTC performance prediction method,

where

$$F(J) = \left(\frac{J}{J_0}\right)^\alpha$$

From the analysis of the function $F(J)$ from open water propeller data, it has been shown that J_0 can be taken as the zero thrust advance coefficient for the propeller. Consequently, if model tests are undertaken at two Reynolds numbers and the results analysed according to Equation (5.73); then the function $F(J)$ can be uniquely determined.

Yet another approach has recently been proposed (Voitkounski (1985)) in which the scale effect is estimated using open water performance calculations for propellers having similar geometric characteristics to the Wageningen B-series.

The results of the analysis are presented in such a way as

$$\left.\begin{array}{l} 1 - \dfrac{K_T}{K_{T_I}} = f(R_n, K_T) \\ 1 - \dfrac{\eta_0}{\eta_{0_I}} = g(R_n, K_T) \end{array}\right\} \quad (5.74)$$

where the suffix I represents the values of K_T and η_0 for an ideal fluid. Consequently, if model values of the thrust and torque at the appropriate advance coefficient are known, that is K_{T_m}, K_{Q_m}, together with the model Reynolds number, then from Equation (5.74) we have

$$\frac{K_{T_m}}{K_{T_I}} = 1 - f(R_{n_m}, K_{T_m})$$

$$\Rightarrow K_{T_I} = \frac{K_{T_m}}{(1 - f(R_{n_m}, K_{T_m}))} = \left.\frac{K_{T_m}}{1 - \left(1 - \dfrac{K_{T_m}}{K_{T_I}}\right)}\right|_{R_{n_m}}$$

Similarly

$$\eta_{0_I} = \left.\frac{\eta_{0_m}}{1 - \left(1 - \dfrac{\eta_{0_m}}{\eta_{0_I}}\right)}\right|_{R_{n_m}}$$

From which the ideal values of K_{T_I} and η_{0_I} can be determined for the propeller in the ideal fluid. Since the effect of scale on the thrust coefficient is usually small and the full-scale thrust coefficient will lie between the model and ideal values the assumption is made that

$$K_{T_s} \simeq \left(\frac{K_{T_M} + K_{T_I}}{2}\right)$$

that is the mean value, and since the full-scale Reynolds number R_{n_s} is known, the functions

$$f(R_{n_s}, K_{T_s}) \text{ and } g(R_{n_s}, K_{T_s})$$

can be determined from which the full-scale values of K_{T_s} and η_{0_s} can be determined from Equation (5.74):

$$K_{T_s} = K_{T_I}[1 - f(R_{n_s}, K_{T_s})]$$
$$\eta_{0_s} = \eta_{0_I}[1 - g(R_{n_s}, K_{T_s})]$$

from which the full-scale torque coefficient can be derived as follows:

$$K_{Q_s} = \frac{J}{2\pi} \frac{K_{T_s}}{\eta_{0_s}}$$

The essential difference between these latter two approaches is that the scale effect is assumed to be a function of both Reynolds number and propeller loading rather than just Reynolds number alone as in the case of the present ITTC procedure. It has been shown that significant differences can arise between the results of the various procedures. Scale effect correction of model propeller characteristics is not a simple procedure and much attention needs to be paid to the effects of the flow structure in the boundary layer and the variations of the lift and drag characteristics within the flow regime. With regard to the general question of scaling, the above methods were primarily intended for non-ducted propellers operating on their own. Nevertheless, the subject of scaling is still not fully understood. Although the problem is complicated by the differences in friction and lift coefficient, the scale effect is less predictable due to the quantity of both laminar flow in the boundary layer and the separation over the blade surfaces. Consequently, there is the potential for the extrapolation process from model to full scale to become unreliable since only averaged amounts of laminar flow are taken into account in the present estimation procedures.

To try and overcome this difficulty a number of techniques have been proposed, particularly those involving leading edge roughness and the use of trip wires, but these procedures still lack rigour in their application to extrapolation. Bazilevski (2001), in a range of experimental conditions using trip wires of 0.1 mm diameter located at 10 per cent of the chord length from the leading edge showed that the experimental scatter on the measured efficiency could be reduced from 13.6 to 1.5 per cent with the use of turbulence stimulation. It was found that trip wires placed on the suction surface of the blades were more effective than those placed on the pressure face and that the effectiveness of the trip wire was dependent upon the ratio of wire diameter to the boundary layer thickness. Boorsma (2000) considered an alternative method of turbulence tripping by the use of sand grain roughness on the leading edge based on the correlation of a sample of five propellers. In his work he showed that the rotation rate correlation factor at constant power could be reduced from 2.4 to 1.7 per cent and, furthermore, concluded that turbulence tripping was not always effective at the inner blade radii.

It is often considered convenient in model experiments to perform model tests at a higher rotational speed than would be required by strictly adhering to the Froude identity. If this is done this then tends to minimize any flow separation on the trailing edge or laminar flow on the suction side of the blade. Such a procedure is particularly important when the propeller is operating in situations where relatively low turbulence levels are encountered in the inflow and where stable laminar flow is likely to be present. Such a situation may be found in cases where tractor thrusters or podded propulsors are being investigated. Ball and Carlton (2006) show examples of this type of behaviour relating to model experiments with podded propulsors.

Clearly compound propellers such as contra-rotating screws and ducted propellers will present particular problems in scaling. In the case of the ducted propeller the interactions between the propeller, the duct and the hull are of particular concern and importance. In addition there is also some evidence to suggest that vane wheels are particularly sensitive to Reynolds number effects since both the section chord lengths and the wheel rotational speed are low, which can cause difficulty in interpreting model test data.

Holtrop (2001) proposed that the scaling of structures like ducts can be addressed by considering the interior of the duct as a curved plate. In this analysis an assumed axial velocity of nP_{tip} is used to determine a correction to the longitudinal towing force ΔF given by

$$\Delta F = 0.5 \rho_m \Pi (nP_{tip})^2 (C_{Fm} - C_{Fs}) c_m D_m$$

where n, P_{tip}, c and D are the rotational speed, pitch at the blade tip, duct chord and diameter, respectively.

In the case of podded propulsor housings the problem is rather more complicated in that there is a dependence upon a number of factors. For example, the shape of the housing and its orientation with respect to the local flow, the interaction with the propeller wake and the scale effects of the incident flow all have an influence on the scaling problem.

5.3.4 Specific propeller open water characteristics

Before proceeding to outline the various standard series available to the propeller designer or analyst, it is helpful to briefly consider the types of characteristic associated with each of the principal propeller types, since there are important variants between, say, fixed pitch and controllable pitch propellers or non-ducted and ducted propellers.

5.3.4.1 *Fixed pitch propellers*

The preceding discussions in this section have used as examples the characteristics relating to fixed pitch propellers since these are the simplest form of propeller characteristic. Figure 5.43 is typical of this type of propeller in that the propeller, in the absence of significant amounts of cavitation, as already discussed, is constrained to operate along a single set of characteristic thrust and torque lines.

5.3.4.2 *Controllable pitch propellers*

With the controllable pitch propeller the additional variable of pitch angle introduces a three-dimensional nature to the propeller characteristics, since the total characteristics comprise sets of K_T and K_Q versus J curves for each pitch angle as seen in Figure 5.48. Indeed, for analysis purposes the performance characteristics can be considered as forming a surface, in contrast to the single line for the fixed pitch propeller.

When analysing the performance of a controllable pitch propeller at off-design conditions use should not be made of fixed pitch characteristics beyond say 5° or 10° from design pitch since the effects of section distortion, can affect the performance characteristics considerably.

A further set of parameters arises with controllable pitch propellers and these are the blade spindle torques, a knowledge of which is of considerable importance when designing the blade actuating mechanism. The total spindle torque, which is the torque acting about the spindle axis of the blade and which requires either to be balanced by the hub mechanism in order to hold the blades in the required pitch setting or, alternatively, to be overcome when a pitch change is required, comprises three components as follows:

$$Q_s(J, \Delta\theta) = Q_{SH}(J, \Delta\theta) + Q_{SC}(n, \Delta\theta) + Q_{SF}(J, \Delta\theta) \tag{5.75}$$

where Q_s is the total spindle torque at a given value of J and $\Delta\theta$;

Q_{SH} is the hydrodynamic component of spindle torque due to the pressure field acting on the blade surfaces;

Q_{SC} is the centrifugal component resulting from the blade mass distribution;

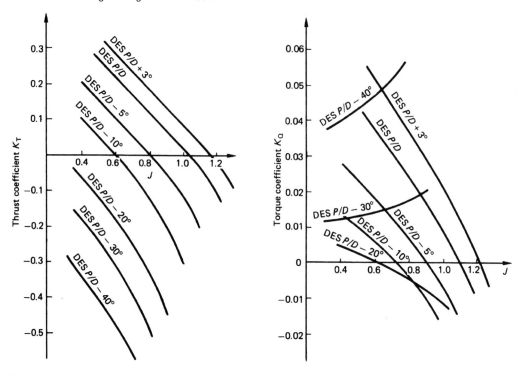

Figure 5.48 Typical controllable pitch propeller characteristic curves.

Q_{SF} is the frictional component of spindle torque resulting from the relative motion of the surfaces within the blade hub.

The latter component due to friction is only partly in the domain of the hydrodynamicist, since it depends both on the geometry of the hub mechanism and the system of forces and moments generated by the blade pressure field and mass distribution acting on the blade palm.

Figure 5.49 shows typical hydrodynamic and centrifugal blade spindle torque characteristics for a controllable pitch propeller. In Figure 5.49 the spindle torques are expressed in the coefficient form of K_{QSH} and K_{QSC}. These coefficients are similar in form to the conventional propeller torque coefficient in so far as they relate to the respective spindle torques as follows:

$$K_{QSH} = \frac{Q_{SH}}{\rho n^2 D^5}$$

$$K_{QSC} = \frac{Q_{SC}}{\rho_m n^2 D^5} \qquad (5.76)$$

where ρ is the mass density of water and ρ_m is the mass density of the blade material. Clearly, since the centrifugal component is a mechanical property of the blade only, it is independent of advance coefficient. Hence K_{QSC} is a function of $\Delta\theta$ only.

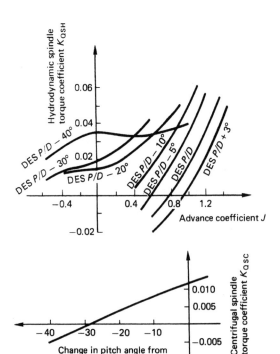

Figure 5.49 Typical controllable pitch propeller spindle torque characteristic curves.

5.3.4.3 Ducted propellers

Whilst the general aspects of the discussion relating to non-ducted, fixed and controllable pitch propellers apply to ducted propellers, the total ducted propulsor thrust is split into two components: the algebraic sum of the propeller and duct thrusts and any second-order interaction effects. To a first approximation, therefore, the total propulsor thrust T can be written as

$$T = T_p + T_n$$

where T_p is the propeller thrust and T_n is the duct thrust.

In non-dimensional form this becomes

$$K_T = K_{TP} + K_{TN} \qquad (5.77)$$

where the non-dimensionalization factor is $\rho n^2 D^4$ as before.

The results of model tests normally present values of K_T and K_{TN} plotted as a function of advance coefficient J as shown in Figure 5.50 for a fixed pitch ducted propulsor. The torque characteristic is, of course, not split into components since the propeller itself absorbs all of the torque of the engine. In general the proportion of thrust generated by the duct to that of the total propulsor thrust is a variable over the range of advance coefficient. In merchant practice by far the greater majority of ducted propellers are designed with accelerating ducts, as discussed in Chapter 6, Section 6.2.2. For these duct forms the ratio of K_{TN}/K_T is of the order of 0.5 at the bollard pull, or zero advance coefficient, condition, but this

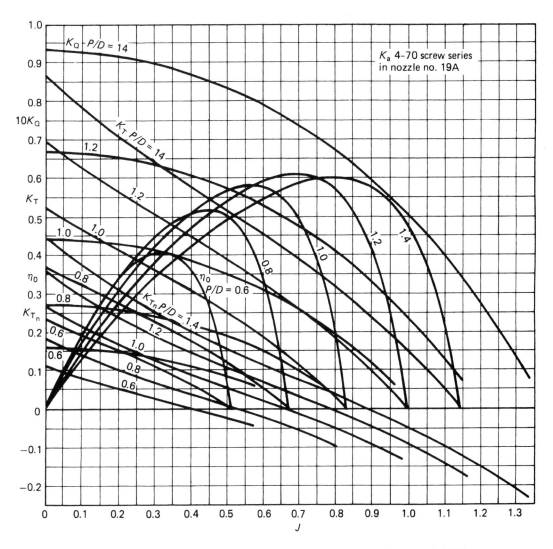

Figure 5.50 Open water test results of K_a 4–70 screw series with nozzle no. 19A (Courtesy: MARIN).

usually falls to around 0.05 or 0.10 at the design free running condition. Indeed, if the advance coefficient is increased to a sufficiently high level, then the duct thrust will change sign, as seen in Figure 5.50, and act as a drag; however, this situation is unlikely to arise in normal practice. When decelerating ducts are used, analogous conditions arise, but the use of these ducts is confined to certain specialist cases, normally those having a low radiated noise requirement.

5.3.4.4 *High-speed propellers*

With high-speed propellers much of what has been said previously will apply depending upon the application. However, the high-speed propeller will be susceptible to two other factors. The first is that cavitation is more likely to occur, and consequently the propeller type and section blade form must be carefully considered in so far as any supercavitating blade section requirements need to be met. The second factor is that many highspeed propellers are fitted to shafts with considerable rake angles. This rake angle, when combined with the flow directions, gives rise to two flow components acting at the propeller plane as seen in Figure 5.51. The first of these is parallel to the shaft and has a magnitude $V_a \cos(\lambda)$ and the second is perpendicular to the shaft with a magnitude $V_a \sin(\lambda)$ where λ is the relative shaft angle as shown in the figure. It will be appreciated that the second, or perpendicular, component immediately presents an asymmetry when viewed in terms of propeller relative velocities, since on one side of the propeller disc the perpendicular velocity component is additive to the propeller rotational velocity whilst on the other side it is subtractive (see Figure 5.51). This gives rise to a differential loading of the blades as they rotate around the propeller disc, which causes a thrust eccentricity and side force components. Figure 5.52 demonstrates these features which of course will apply generally to all propellers working in non-uniform flow but are more noticeable with high-speed propellers due to the speeds and inclinations involved. The magnitude of these eccentricities can be quite large; for example, in the case of unity pitch ratio with a shaft rake of 20°, the transverse thrust eccentricity indicated by Figure 5.52 may well reach $0.40R$. Naturally due to the non-uniform tangential wake field the resulting cavitation pattern will also be anti-symmetric.

5.3.5 Standard series data

Over the years there have been a considerable number of standard series propellers tested in many different establishments around the world. To discuss them all in any detail would clearly be a large undertaking requiring considerable space; consequently, those most commonly used today by propeller designers and analysts are referenced here.

The principal aim in carrying out systematic propeller tests is to provide a data base to help the designer understand the factors which influence propeller performance and the inception and form of cavitation on the blades under various operating conditions.

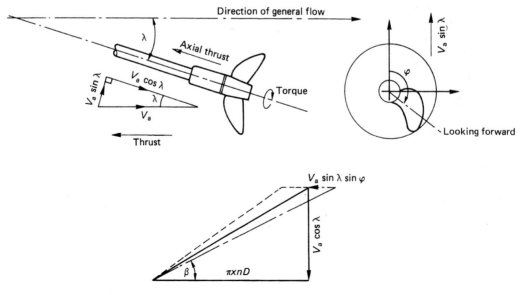

Figure 5.51 Inclined flow velocity diagram.

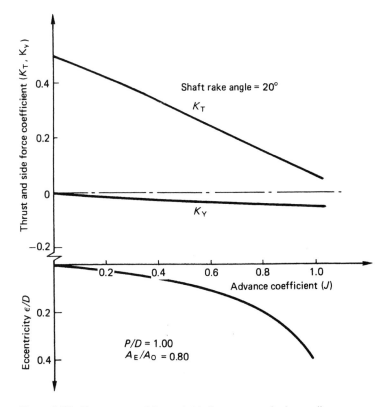

Figure 5.52 Thrust eccentricity and side forces on a raked propeller.

Table 5.11 Fixed pitch, non-ducted propeller series summary.

Series	Number of propellers in series	Range of parameters			D(mm)	r_h/R	Cavitation data available	Notes
		Z	A_E/A_O	P/D				
Wageningen B-series	≈120	2–7	0.3–1.05	0.6–1.4	250	0.169	No	Four-bladed propeller has non-constant pitch dist
Au-series	34	4–7	0.4–0.758	0.5–1.2	250	0.180	No	
Gawn-series	37	3	0.2–1.1	0.4–2.0	508	0.200	No	
KCA-series	≈30	3	0.5–1.25	0.6–2.0	406	0.200	Yes	
Ma-series	32	3 and 5	0.75–1.20	1.0–1.45	250	0.190	Yes	
Newton–Rader series	12	3	0.5–1.0	1.05–2.08	254	0.167	Yes	
KCD-series	24	3–6 (mainly 4)	0.587 Principal 0.44–0.8	0.6–1.6	406	0.200	Yes	Propellers not geosyms
Meridian series	20	6	0.45–1.05	0.4–1.2	305	0.185	Yes	Propellers not geosyms

A second purpose is to provide design diagrams, or charts, which will assist in selecting the most appropriate dimensions of actual propellers to suit full-size ship applications.

Table 5.11 summarizes the principal fixed pitch non-ducted propeller series. Carlton (2007) describes these series in some detail, using the sources listed in the References.

5.4 Propeller theories

5.4.1 Early theories

Carlton (2007) describes the development of the early propeller theories including the momentum theory due to Rankine (1865) and Froude (1878, 1889), and the blade element theory of Froude (1878). The work

248 Maritime engineering reference book

of Prandtl, Betz (1919, 1920) and Goldstein (1929) is reviewed, together with the work of Burrill (1944), Lerbs (1952) and Eckhardt and Morgan's design method, Eckhardt and Morgan (1955). Carlton goes on to describe lifting surface models through to boundary element and CFD methods, and these are discussed in the following Sections.

5.4.2 Lifting surface models

Figure 5.53 shows in a conceptual way the basis of the lifting surface model. Essentially the blade is replaced by an infinitely thin surface which takes the form of the blade camber line and upon which a distribution of vorticity is placed in both the spanwise and chordal directions. Early models of this type used this basis for their formulations and the solution of the flow problem was in many ways analogous to the thin aerofoil approach. Later lifting surface models then introduced a distribution of sources and sinks in the chordal directions so that, in conjunction with the incident flow field, the section thickness distribution could be simulated and hence the associated blade surface pressure field approximated. The use of lifting surface models, as indeed for other models of propeller action, is for both the solution of the design and analysis problems. In the design problem the geometry of the blade is only partially known in so far as the radial distributions of chord, rake skew and section thickness distributions are known. The radial distribution of pitch and the chordwise and radial distribution of camber remain to be determined. In order to solve the design problem the source and vortex distributions representing the blades and their wake need to be placed on suitable reference surfaces to enable the induced velocity field to be calculated. Linear theories assume that the perturbation velocities due to the propeller are small compared with the inflow velocities. In this way the blades and their wake can simply be projected onto stream surfaces formed by the undisturbed flow. However, in the majority of practical design cases the resulting blade geometry deviates substantially from this assumption, and as a consequence the linear theory is generally not sufficiently accurate.

The alternative problem, the analysis problem, differs from the design solution in that the propeller geometry is completely known and we are required to determine the flow field generated under known conditions of advance and rotational speed. The analysis exercise divides into two comparatively well-defined types: the steady flow and the unsteady flow solutions. In the former case the governing equations are the same as in the design problem, with the exception that the unknowns are reversed. The circulation distribution over the blades is now the unknown. As a consequence, the singular integral which gives the velocity induced

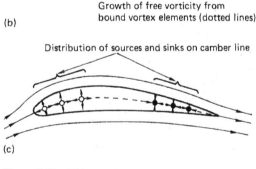

Figure 5.53 Lifting surface concept: (a) lifting surface model for a propeller blade; (b) lifting surface concept at section A–A to simulate blade loading and (c) source–sink distribution to simulate section thickness.

by a known distribution of circulation in the design problem becomes an integral equation in the analysis problem, which is solved numerically by replacing it with a system of linear algebraic equations. In the case of unsteady propeller flows their solution is complicated by the presence of shed vorticity in the blade wake that depends on the past history of the circulation around the blades. Accordingly, in unsteady theory the propeller blades are assumed to generate lift in gusts, for which an extensive literature exists for the general problem: for example, McCroskey (1982), Crighton (1985) and the widely used Sear's function. The unsteadiness of the incident flow is characterized by the non-dimensional parameter k, termed the reduced frequency parameter. This parameter is defined as the product of the local semichord, and the frequency of encounter divided by the relative inflow speed. For the purposes of unsteady flow calculations the wake or inflow velocity field is characterized at each radial station by the harmonic components of the circumferential velocity distribution (Figure 5.32), and with the assumption that the propeller responds linearly to changes in inflow, the unsteady flow problem reduces to one of estimating the response of the propeller to each harmonic. In the case of a typical marine propeller the reduced frequency k corresponding to the first harmonic is of the order of 0.5, whilst the value corresponding to the blade rate harmonic will be around two or three. From classical two-dimensional theory of an aerofoil encountering sinusoidal gusts, it is known that the effects of flow unsteadiness become significant for values of k greater than 0.1. As a consequence the response of a propeller to all circumferential harmonics of the wake field is unsteady in the sense that the lift generated from the sections is considerably smaller than that predicted from the equivalent quasi-steady value and is shifted in phase relative to the inflow.

In the early 1960s many lifting surface procedures made their appearance due mainly to the various computational capabilities that became available generally at that time. Prior to this, the work of Strscheletsky (1950), Guilloton (1957) and Sparenberg (1959) laid the foundations for the development of the method. Pien (1961) is generally credited with producing the first of the lifting surface theories subsequent to 1960. The basis of this method is that the bound circulation can be assumed to be distributed over the chord of the mean line, the direction of the chord being given by the hydrodynamic pitch angle derived from a separate lifting line calculation. This lifting line calculation was also used to establish the radial distribution of bound circulation. In Pien's method the free vortices are considered to start at the leading edge of the surface and are then continued into the slipstream in the form of helical vortex sheets. Using this theoretical model the required distortion of the chord into the required mean line can be determined by solving the system of integral equations defining the velocities along the chord induced by the system of bound and free vortices. The theory is linearized in the sense that a second approximation is not made using the vortex distribution along the induced mean line.

Pien's work was followed by that of Kerwin (1961), van Manen and Bakker (1962), Yamazaki (1962) English (1962), Cheng (1965), Murray (1967), Hanaoka (1969), van Gent (1975, 1977) and a succession of papers by Breslin, Tsakonas and Jacobs spanning something over thirty years' continuous development of the method. Indeed, much of this latter development is captured in the book by Breslin and Anderson (1992). Typical of modern lifting surface theories is that by Brockett (1981). In this method the solid boundary effects of the hub are ignored; this is consistent with the generally small magnitude of the forces being produced by the inner regions of the blade. Furthermore, in Brockett's approach it is assumed that the blades are thin, which then permits the singularities which are distributed on both sides of the blades to collapse into a single sheet. The source strengths, located on this single sheet, are directly proportional to the derivative of the thickness function in the direction of flow, conversely the vortex strengths are defined. In the method a helicoidal blade reference surface is defined together with an arbitrary specified radial distribution of pitch. The trailing vortex sheet comprises a set of constant radius helical lines whose pitch is to be chosen to correspond either to that of the undisturbed inflow or to the pitch of the blade reference surface. Brockett uses a direct numerical integration procedure for evaluating the induced velocities. However, due to the non-singular nature of the integrals over the other blades and the trailing vortex sheets the integrands are approximated over prescribed sets of chordwise and radial intervals by trigonometric polynomials. The integrations necessary for both the induced velocities and the camber line form are undertaken using predetermined weight functions. Unfortunately the integral for the induced velocity at a point on the reference blade contains a Cauchy principle-value singularity. This is solved by initially carrying out the integration in the radial direction and then factoring the singularity out in the chordwise integrand. A cosine series is then fitted to the real part of the integrand, the Cauchy principal value of which was derived by Glauert in 1948.

5.4.3 Lifting–line–lifting-surface hybrid models

The use of lifting surface procedures for propeller design purposes clearly requires the use of computers having a reasonably large capacity. Such capabilities are not always available to designers and as a consequence there has developed a generation of hybrid models essentially based on lifting line procedures and incorporating lifting surface corrections together with various cavitation prediction methods.

It could be argued that the very early methods of analysis fell, to some extent, into this category by relying on empirical section and cascade data to correct basic high aspect ratio calculations. However, the real evolution of these methods can be considered to have commenced subsequent to the development of the correction factors by Morgan et al. (1968). The model of propeller action proposed by van Oossanen (1974) typifies an advanced method of this type by providing a very practical approach to the problem of propeller analysis. The method is based on the Lerbs induction factor approach (Lerbs (1952)), but because this was a design procedure the Lerbs method has to be used in the inverse sense, which is notoriously unstable. To overcome this instability in order to determine the induced velocities and circulation distribution for a given propeller geometry, van Oossanen introduced an additional iteration for the hydrodynamic pitch angle. In order to account for the effects of non-uniform flow, the average of the undisturbed inflow velocities over the blade sections is used to determine the advance angle at each blade position in the propeller disc, and the effect of the variation of the undisturbed inflow velocities is accounted for by effectively distorting the geometric camber distribution. The effect of the bound vortices is also included because of their non-zero contribution to the induced velocity in a non-uniform flow. The calculation of the pressure distribution over the blades at each position in the propeller disc is conducted using the Theodorsen approach after first distorting the blade section camber and by defining an effective angle of attack such that a three-dimensional approximation is derived by use of a two-dimensional method.

So as to predict propeller performance correctly, particularly in off-design conditions, van Oossanen calculates the effect of viscosity on the lift and drag properties of the blade sections. The viscous effects on lift are accounted for by boundary layer theory, in which the lift curve slope is expressed in terms of the boundary layer separation and the zero lift angle is calculated as a function of the relative wake thickness on the suction and pressure sides. In contrast, the section drag coefficient is based on an equivalent profile analysis of the experimental characteristics of the Wageningen B series propellers.

The cavitation assessment is calculated from a boundary layer analysis and is based on the observation that cavitation inception occurs in the laminar-turbulent transition region of the boundary layer. The extent of the cavitation is derived by calculating the value of Knapp's dynamic similarity parameter for spherical cavities for growth and decline, based on the results of cavitation measurements on profiles.

This method has proved a particularly effective analysis tool for general design purposes and Figure 5.54 underlines the value of the method in estimating the extent of cavitation and its comparison with observations.

5.4.4 Vortex lattice methods

The vortex lattice method of analysis is in effect a subclass of the lifting surface method. In the case of propeller design and analysis it owes its origins largely to Kerwin, working at the Massachussetts Institute of Technology, although in recent years others have taken up the development of the method: for example, Szantyr (1984).

In the vortex lattice approach the continuous distributions of vortices and sources are replaced by a finite set of straight line elements of constant strength whose end points lie on the blade camber surface. From this system of line vortices the velocities are computed at a number of suitably located control points between the elements. In the analysis problem the vortex distributions (Figure 5.55) are unknown functions of time and space, and as a consequence have to be determined from the boundary conditions of the flow condition being analysed. The source distributions, however, can be considered to be independent of time, and their distribution over the blade is established using a stripwise application of thin aerofoil theory at each of the radial positions. As such, the source distribution is effectively known, leaving the vortex distribution as the principal unknown. Kerwin and Lee (1978) consider the vortex strength at any point as a vector lying in the blade or vortex sheet which can be resolved into spanwise and chordwise components on the blades, with the corresponding components termed shed and trailing vorticity in the vortex sheets emanating from the blades (Figure 5.55). Based on this approach the various components of the vortex system can be defined with respect to time and position by applying Kelvin's theorem in association with the pressure continuity condition over the vortex wake. Hence the distributed spanwise vorticity can be determined from the boundary conditions of the problem.

In essence there are four principal characteristics of the vortex lattice model which need careful consideration in order to define a valid model. These are as follows:

1. The element's orientation.
2. The spanwise distribution of elements and control points.
3. Chordwise distribution of elements and control points.
4. The definition of the Kutta condition in numerical terms.

With regard to element distribution Greeley and Kerwin (1982) proposed for steady flow analysis that the radial interval from the hub r_h to the tip R be divided into M equal intervals with the extremities of the lattice inset one-quarter interval from the ends of the blade. The end points of the discrete vortices are located at radii r_m given by

$$r_m = \frac{(R - r_h)(4m - 3)}{4M + 2} + r_h (m = 1, 2, 3, \ldots, M + 1)$$

(5.78)

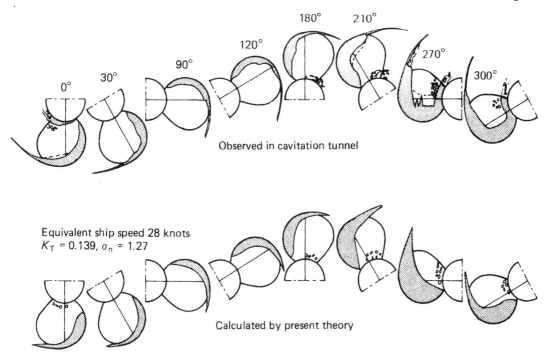

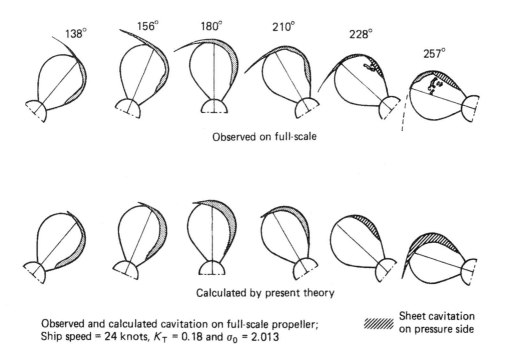

Figure 5.54 Comparison of observed and predicted cavitation by van Oossanen's hybrid method of propeller analysis (Courtesy: MARIN).

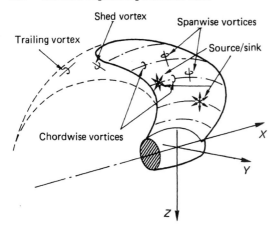

Figure 5.55 Basic components of lifting surface models.

In the case of the chordwise distribution of singularities they chose a cosine distribution in which the vortices and control points are located at equal intervals of $\tilde{s}$, where the chordwise variable s is given by:

$$S = 0.5(1 - \cos \tilde{s}) \quad (0 \leq \tilde{s} \leq \pi)$$

If there are N vortices over the chord, the positions of the vortices, $S_v(n)$, and the control points, $S_c(i)$, are given by

$$\left. \begin{array}{l} S_v(n) = 0.5\left\{1 - \cos\left[\dfrac{(n - \frac{1}{2})\pi}{N}\right]\right\} \quad n = 1, 2, \ldots, N \\ \text{and} \\ S_c(i) = 0.5\left\{1 - \cos\left[\dfrac{i\pi}{N}\right]\right\} \quad i = 1, 2, \ldots, N \end{array} \right\}$$

(5.79)

With this arrangement the last control point is at the trailing edge and two-dimensional calculations show that this forces the distribution of vorticity over the chord to have the proper behaviour near the trailing edge; that is, conformity with the Kutta condition. In the earlier work Kerwin and Lee (1978) showed that for the solution of both steady and unsteady problems the best compromise was to use a uniform chordwise distribution of singularities together with an explicit Kutta condition:

$$S_n(n) = \dfrac{n - 0.75}{N} \quad (n = 1, 2, \ldots, N) \quad (5.80)$$

Lan (1974) showed that chordwise spacing of singularity and control points proposed by Equation (5.79) gave exact results for the total lift of a flat plate or parabolic camber line and was more accurate than the constant spacing arrangement, Equation (5.80), in determining the local pressure near the leading edge. This choice, as defined by Equation (5.79), commonly referred to as cosine spacing, can be seen as being related to the conformal transformation of a circle into a flat or parabolically cambered plate by a Joukowski transformation.

The geometry of the trailing vortex system has an important influence on the accuracy of the calculation of induced velocities on the blade. The normal approach in lifting surface theories is to represent the vortex sheet emanating from each blade as a pure helical surface with a prescribed pitch angle. Cummings (1968), Loukakis (1971) and Kerwin (1976) developed conceptually more advanced wake models in which the roll-up of the vortex sheet and the contraction of the slipstream were taken into account. Current practice with these methods is to consider the slipstream to comprise two distinct portions: a transition zone and an ultimate zone as shown in Figure 5.56. The transition zone of the slipstream is the one where the roll-up of the trailing vortex sheet and the contraction of the slipstream are considered to occur and the ultimate zone comprises a set of Z helical tip vortices together with either a single rolled-up hub vortex or Z helical hub vortices. Hence the slipstream model is defined by some five parameters as follows (see Figure 5.56):

1. Radius of the rolled-up tip vortices (r_w).
2. Angle between the trailing edge of the blade tip and the roll-up point (θ_w).
3. Pitch angle of the outer extremity of the transition slipstream (β_T).
4. Pitch angle of the ultimate zone tip vortex helix (β_w).
5. Radius of the rolled-up hub vortices (r_{wh}) in the ultimate zone if this is not considered to be zero.

In using vortex lattice approaches it had been found that whilst a carefully designed lattice arrangement should be employed for the particular blade which is being analysed, the other $Z - 1$ blades can be represented by significantly coarser lattice without causing any important changes in the computed results. This, therefore, permits economies of computing time to be made without loss of accuracy. Kerwin (1976) shows a comparison of the radial distributions of pitch and camber obtained by the vortex lattice approach and by traditional lifting surface methods, Brockett (1981), (Figure 5.57). Although the results are very similar, some small differences are seen to occur particularly with respect to the camber at the inner radii.

The problem of vortex sheet separation and the theoretical prediction of its effects at off-design conditions are currently occupying the attention of many hydro-dynamicists around the world. At these conditions the vortex sheet tends to form from the leading edge at some radius inboard from

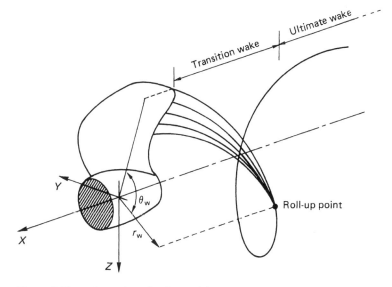

Figure 5.56 Deformation of wake model.

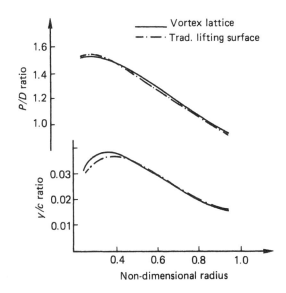

Figure 5.57 Comparison of results obtained between traditional lifting surface and vortex lattice methods (Kerwin).

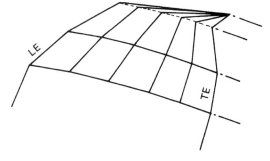

Figure 5.58 Simplified leading-edge vortex separation model (Kerwin and Lee).

the tip rather than at the tip. Kerwin and Lee (1978) developed a somewhat simplified representation of the problem which led to a substantial improvement in the correlation of theoretical predictions with experimental results. In essence their approach is shown in Figure 5.58, in which for a conventional vortex lattice arrangement the actual blade tip is replaced by a vortex lattice having a finite tip chord. The modification is to extend the spanwise vortex lines in the tip panel as free vortex lines which join at a 'collection point', this then becomes the origin of the outermost element of the discretized vortex sheet. The position of the collection point is established by setting the pitch angle of the leading-edge free vortex equal to the mean of the undisturbed inflow angle and the pitch angle of the tip vortex as it leaves the collection point. Greeley and Kerwin (1982) developed the approach further by establishing a semi-empirical method for predicting the point of leading-edge separation. The basis of this method was the collapsing of data for swept wings in a non-dimensional plotting of critical leading-edge suction force, as determined from inviscid theory as a function of a local leading-edge Reynolds number, as shown in Figure 5.59. This then allowed the development of an approximate model in which the free vortex sheet was placed at

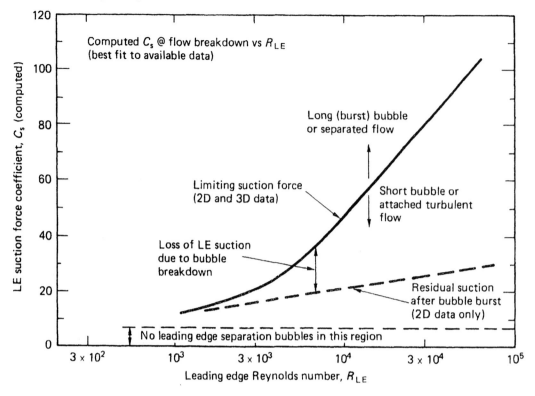

Figure 5.59 Empirical relationship between the value of the leading-edge suction force coefficient at the point of flow breakdown as a function of leading-edge Reynolds number (Reproduced with permission from Greeley and Kerwin (1982)).

a height equal to 16-blade boundary-layer thickness and the resulting change in the calculated chordwise pressure distribution found.

Lee and Kerwin et al. developed the vortex lattice code PUF-3 in its original form. However, in recent years the code was extended to include a number of further features amongst which were wake alignment and shaft inclination, mid-chord cavitation, thickness-load coupling, the influence of the propeller boss, duct effects and right- and left-handed rotational options. This extended form of the code is known as MPUF-3A which is also coupled to a boundary element method to solve for the diffraction potential on the ship's hull. This latter process is done once the propeller problem has been solved and then determines the unsteady pressure fluctuations acting on the hull due to the action of the propeller.

More recently a three dimensional Euler equation solver based on a finite volume method has been developed at the University of Texas. This capability, assuming that the inflow velocity field is known sufficiently far upstream of the propeller from model tests or computations, estimates the effective wake field at the propeller. Currently, the propeller is represented by time-averaged body forces over the volume that the propeller forces are covering while they are rotating. This procedure is coupled to the MPUF-3A code in an iterative manner such that the Euler solver defines the global flow and effective wake field while the MPUF-3A code solves for the flow around the blades, including cavitation, and provides the propeller body forces to be used in the Euler solver.

5.4.5 Boundary element methods

Boundary element methods for propeller analysis have been developed in recent years in an attempt to overcome two difficulties of lifting surface analyses. The first is the occurrence of local errors near the leading edge and the second in the more widespread errors which occur near the hub where the blades are closely spaced and relatively thick. Although the first problem can to some extent be overcome by introducing a local correction derived by Lighthill (1951), in which the flow around the leading edge of a two-dimensional, parabolic half-body is matched to the three-dimensional flow near the leading edge derived from lifting surface theory, the second problem remains.

Boundary element methods are essentially panel methods, and their application to propeller technology began in the 1980s. Prior to this the methods were pioneered in the aircraft industry, notably by Hess and Smith, Maskew and Belotserkovski. Hess and Valarezo (1985) introduced a method of analysis based on the earlier work of Hess and Smith (1967) in 1985. Subsequently, Hoshino (1989) has produced a surface panel method for the hydrodynamic analysis of propellers operating in steady flow. In this method the surfaces of the propeller blades and hub are approximated by a number of small hyperboloidal quadrilateral panels having constant source and doublet distributions. The trailing vortex sheet is also represented by similar quadrilateral panels having constant doublet distributions. Figure 5.60, taken from Hoshino (1989), shows a typical representation of the propeller and vortex sheet combination using this approach. The strengths of the source and doublet distributions are determined by solving the boundary value problems at each of the control points which are located on each panel. Within this solution the Kutta condition is obviously obeyed at the trailing edge.

Using methods of this type good agreement between theoretical and experimental results for blade pressure distributions and open water characteristics has been achieved. Indeed a better agreement of the surface pressure distributions near the blade–hub interface has been found to exist than was the case with conventional lifting surface methods.

Kinnas and his colleagues at the University of Texas, Austin, have in recent years done a considerable amount of development on boundary element codes. The initial development of the PROPCAV code in 1992 developed the boundary element method to solve for an unsteady cavitating flow around propellers which were subject to non-axisymmetric inflow conditions (Kinnas and Fine (1992)). Subsequently this approach has been extended to include the effects of non-cylindrical propeller bosses, mid-chord cavitation on the back and face of the propeller (Young and Kinnas (2001)), the modelling of unsteady developed tip vortex cavitation (Lee and Kinnas (2004)) and the influence of fully unsteady trailing wake alignment (Lee and Kinnas (2005)). Good correlation has been shown to exist between the results of this computational method and the measured performance of the DTMB 4383, 72° skew propeller at model scale for both non-cavitating and cavitating flows. Currently the effects of viscosity are estimated by using uniform values of friction coefficient applied to the wetted parts of the propeller blades; however, the code is being coupled to an integral boundary layer solver in order to better account for the effects of viscosity. This solver will both determine the friction acting on the propeller blades and estimate the influence of the viscous effects on the blade pressure distributions. Such a capability may also permit the influence of viscosity on the location of the cavity detachment in the case of mid-chord cavitation as well as on the location of the leading vortex.

A method proposed by Greco et al. (2004) aimed at enhancing the slip stream flow prediction when using a boundary element method showed that the estimated position of the tip vortex was in good agreement with experimental data. In essence the propeller induced trailing wake was determined as part of the flow field solution using an iterative method in which the wake surface is aligned to the local flow. The numerical predictions from this method were then correlated with the vorticity field derived from laser Doppler velocity measurements made in a cavitation tunnel.

Within the framework of the MARIN based Cooperative Research Ships organization Vaz and Bosschers (2006) have been developing a three-dimensional sheet cavitation model using a boundary element model of the marine propeller. This developing approach has been tested against the results from two model propellers under steady flow conditions: the propellers being the MARIN S and the INSEAN E779A propellers. In the case of the former propeller which was designed to exhibit only sheet cavitation, two conditions were examined. At low loading the cavity extent was underpredicted but at moderate loadings the correlation was acceptable. In the second case, the INSEAN propeller had a higher tip loading than the S propeller with the cavitation having partial and super-cavitation in the tip region together with a cavitating tip vortex. For this propeller the cavity extents were predicted reasonably well. This method is currently still in its development phase and it is planned to extend the validation to behind conditions and also cavity volume variations, the latter being done through the low-frequency hull pressure pulses which are, in addition to the contribution from the non-cavitating propeller, mainly influenced by the cavity volume accelerations.

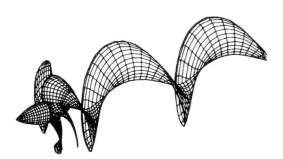

Figure 5.60 Panel arrangement on propeller and trailing vortex wake for boundary element representation (Reproduced with permission from Hoshino (1989)).

5.4.6 Methods for specialist propulsors

The discussion in this Chapter has so far concentrated on methods of design and analysis for conventional propellers. It is also pertinent to comment on the application of these methods to specialist propulsor types: particularly controllable pitch propellers, ducted propellers, contra-rotating propellers and supercavitating propellers.

The controllable pitch propeller, in its design pitch conditions, is in most respects identical to the conventional fixed pitch propeller. It is in its off-design conditions that special analysis procedures are required to determine the blade loads, particularly the blade spindle torque and hence the magnitude of the actuating forces required. Klaassen and Arnoldus (1964) made an early attempt at describing the character of these forces and the methods of translating these into actuating forces. This work was followed by that of Gutsche (1965) in which he considered the philosophical aspects of loading assumptions for controllable pitch propellers Rusetskiy (1968), however, developed hydrodynamic models based on lifting line principles to calculate the forces acting on the blades during the braking, ring vortex and contra-flow stages of controlled pitch propeller off-design performance. This procedure whilst taking into account section distortion by means of the effect on the mean line is a straightforward procedure which lends itself to hand calculation. The fundamental problem with the calculation of a controllable pitch propeller at off-design conditions is not that of resolving the loadings acting on the blades into their respective actuating force components, but of the calculating the blade loadings on surface pressure distributions under various, and in some cases extreme, flow regimes and with the effects of blade section distortion. The basic principles of Rusetskiy's method were considered and various features enhanced by Hawdon et al. (1975), particularly in terms of section deformation and the flow over these deformed sections. Lifting-line-based procedures continued to the main method of approaching the calculation of the hydrodynamic loading components until the 1980s: the centrifugal spindle torque is a matter of propeller geometry and the frictional spindle torque, dependent on mechanics and the on magnitude of the resultant hydrodynamic and centrifugal components. Pronk (1980) considered the calculation of the hydrodynamic loading by the use of a vortex lattice approach based on the general principles of Kerwin's work. In this way the computation of the blade hydrodynamics lost many of the restrictions that the earlier methods required to be placed on the calculation procedure.

As early as 1879 Parsons fitted and tested a screw propeller having a complete fixed shrouding and guide vanes. However, the theoretical development of the ducted propeller essentially started with the work of Kort (1934). In its early form the duct took the form of a long channel through the hull of the ship but soon gave way to the forerunners of the ducted propellers we know today, comprising an annular aerofoil placed around the outside of a fixed or controllable pitch propeller.

Following Kort's original work, Steiss (1936) produced a one-dimensional actuator disc theory for ducted propeller action; however, development of ducted propeller theory did not really start until the 1950s. Horn and Amtsberg (1961) developed an earlier approach by Horn, in which the duct was replaced by a distribution of vortex rings of varying circulation along the length of the duct, and in 1955 Dickmann and Weissinger (1955) considered the duct and propeller to be a single unit replaced by a vortex system. In this system the propeller is assumed to have an infinite number of blades and constant bound vortex along the span of the blade. The slipstream is assumed to be a cylinder of constant radius and no tangential induced velocities are present in the slipstream. Despite the theoretical work the early design methods, several of which are used today, were essentially pseudo-empirical methods. Typical of these are those presented by van Manen (1959, 1962, 1966) and they represent a continuous development of the subject which was based on the development of theoretical ideas supported by the results of model tests, chiefly the K_a ducted propeller series. Theoretical development, however, continued with contributions by Morgan (1961), Dyne (1967) and Oosterveld (1970) for ducted propellers working in uniform and wake adapted flow conditions.

Chaplin developed a non-linear approach to the ducted propeller problem and subsequently Ryan and Glover (1972) presented a theoretical design method which avoided the use of a linearized theory for the duct by using surface vorticity distribution techniques to represent both the duct and the propeller boss. The representation of the propeller was by means of an extension of the Strscheletsky approach and developed by Glover (1970) in earlier studies on heavily loaded propellers with slipstream contraction. The treatment of the induced velocities, however, was modified in order to take proper account of the induced velocities of the duct to achieve good correlation with experimental results. In this way the local hydrodynamic pitch angle at the lifting line was defined as

$$\beta_i = \tan^{-1}\left[\frac{V_a + u_{ap} + u_{ad}}{\omega r - u_{tp}}\right]$$

where u_{ap} is the axial induced velocity of the propeller,

u_{ad} is the axial induced velocity of the duct and
u_{tp} is the tangential induced velocity of the propeller.

Subsequently, Caracostas (1978) extended the Ryan and Glover work to off-design operation conditions using some of the much earlier Burrill (1944) philosophies of propeller analysis.

Tsakonas and Jacobs (1978) extended the theoretical approach to ducted propeller operation by utilizing unsteady lifting surface theory to examine the interaction of the propeller and duct when operating in a non-uniform wake field. In this work they modelled the duct and propeller geometry in the context of their camber and thickness distributions. In addition to the problem of the interactions between the duct and propeller there is also the problem of the interaction between the ducted propulsor and the body which is being propelled. Falcao de Campos (1983) has recently studied this problem in the context of axisymmetric flows. The basic approach pursued in this study assumes the interaction flow between the ducted propulsor and the hull, which ultimately determines the performance of the duct and propeller, is inviscid in nature and can, therefore, be treated using Euler's equations of motions. Whilst this approach is valid for the global aspects of the flow, viscous effects in the boundary layers on the various components of the ducted propulsor system can be of primary importance in determining the overall forces acting on the system. As a consequence Falcao de Campos considers these aspects in some detail and then moves on to consider the operation of a ducted propeller in axisymmetric shear flow. The results of his studies led to the conclusion that inviscid flow models can give satisfactory predictions of the flow field and duct performance over a wide range of propeller loadings, provided that the circulation around the duct profile can be accurately determined and a detailed account of the viscous effects on the duct can be made in the establishment of the criteria for the determination of the duct circulation. Additionally, Kerwin et al. (1987) in their extension of the MPUF-3A code to ducted propellers developed an orifice equation model in order to take account of the viscous flow through the propeller tip to duct gap.

The main thrust of ducted propeller research has been in the context of the conventional accelerating or decelerating duct forms, including azimuthing systems, although this latter aspect has been treated largely empirically. The pumpjet is a closely related member of the ducted propeller family and has received close attention for naval applications. Clearly, as a result much of the research is classified but certain aspects of the work are published in open literature. Two particularly good treatments of the subject are to be found in McCormick and Eisenhuth (1962) and Henderson et al. (1964), and a more recent exposition of the subject is given by Wald (1970). In this latter work equations have been derived to describe the operation of a pumpjet which is closely integrated into the hull design and ingests a portion of the hull boundary layer. From this work it is shown that maximum advantage of this system is attained only if full advantage is taken of the separation inhibiting effect of the propulsor on the boundary layer of the afterbody, a fact not to be underestimated in other propulsor configurations. Another closely related member of the ducted propeller family is the ring propeller, which comprises a fixed pitch propeller with an integrally mounted or cast annular aerofoil, with low length to diameter ratio, at the blade tips. In addition to the tip-mounted annular aerofoil designs, some of which can be found in small tugs or coasters, there have been design proposed where the ring has been sited at some intermediate radial location on the propeller: one such example was the English Channel packet steamer *Cote d'Azur,* built in 1950. Work on ring propellers has mainly been confined to model test studies and reported by van Gunsteren (1969) and Keller (1966). From these studies the ring propeller is shown to have advantages when operating in off-design conditions, with restricted diameter, or by giving added protection to the blades in ice, but has the disadvantage of giving a relatively low efficiency.

Contra-rotating propellers, as discussed in Section 6.2.4, have been the subject of interest which has waxed and waned periodically over the years. The establishment of theoretical methods to support contra-rotating propeller development has a long history starting with the work of Greenhill (1988) who produced an analysis method for the Whitehead torpedo; however, the first major advances in the study was made by Rota (1909) who carried out comparative tests with single and contra-rotating propellers on a steam launch. In a subsequent paper, Rota (1914), he further developed this work by comparing and contrasting the results of the work contained in his earlier paper with some propulsion experiments conducted by Luke (1914). Little more appears to have been published on the subject until Lerbs (1955) introduced a theoretical treatment of the problem, and a year later van Manen and Sentic (1956) produced a method based on vortex theory supported by empirical factors derived from open water experiments. Morgan (1960) subsequently produced a step-by-step design method based on Lerbs' theory, and in addition he showed that the optimum diameter can be obtained in the usual way for a single-screw propeller but assuming the absorption of half the required thrust of power and that the effect on efficiency of axial spacing between the propellers was negligible. Whilst Lerbs' work was based on lifting line principles, Murray (1967) considered the application of lifting surface to the theory of contra-rotating propellers.

Van Gunsteren (1970) developed a method for the design of contra-rotating propellers in which the interaction effects of the two propellers are largely determined with the aid of momentum theory. Such an approach allows the slipstream contraction effects and an allowance for the mutually induced pressures in the cavitation calculation to be taken into account in a relatively simple manner. The radial distributions of the mutually induced velocities are calculated by lifting line theory; however, the mutually induced effects are separated from self-induced effects in such a way that each propeller of the pair can be designed using a procedure for simple propellers. Agreement between this method and experimental results indicated a reasonable level of correlation.

Tsakonas et al. (1983) has extended the development of lifting surface theory to the contra-rotating propeller problem by applying linearized unsteady lifting surface theory to propeller systems operating in uniform or non-uniform wake fields. In this latter approach the propeller blades lie on helicoidal surfaces of varying pitch, and have finite thickness distributions, together with arbitrary definitions of blade outline, camber and skew. Furthermore, the inflow field of the after propeller is modified by accounting for the influence of the forward propeller so that the potential and viscous effects of the forward propeller are incorporated in the flow field of the after propeller. It has been shown that these latter effects play an important role in determining the unsteady loading on the after propeller and as a consequence cannot be ignored. Subsequently, work at the Massachussetts Institute of Technology has extended panel methods to rotor–stator combinations.

High-speed and more particularly supercavitating propellers have been the subject of considerable research effort. Two problems present themselves: the first is the propeller inflow and the second is the blade design problem. In the first case of the oblique flow characteristics these have to some extent been dealt with empirically, as discussed in Section 5.3. In the case of calculating the performance characteristics, the oblique flow characteristics manifest themselves as an in-plane flow over the propeller disc, whose effect needs to be taken into account. Theoretical work on what was eventually to become a design method started in the 1950s with the work of Tulin on steady two-dimensional flows over slender symmetrical bodies, Tulin (1953), although supercavitating propellers had been introduced by Posdunine as early as 1943. This work was followed by other studies on the linearized theory for supercavitating flow past lifting foils and struts at zero cavitation number (Tulin (1955, 1956)), in which Tulin used the two-term Fourier series for the basic section vorticity distribution. Subsequently Johnson (1957) in developing a theoretical analysis for low drag supercavitating sections used three- and five-term expressions. Tachmindji and Morgan (1957) developed a practical design method based on a good deal of preceding research work which was extended with additional design information (Tachmindji and Morgan (1969)). The general outline of the method essentially followed a similar form to the earlier design procedure set down by Eckhardt and Morgan (1955).

A series of theoretical design charts for two-, three- and four-bladed supercavitating propellers was developed by Caster (1959). This work was based on the two-term blade sections and was aimed at providing a method for the determination of optimum diameter and revolutions. Anderson (1974) developed a lifting line theory which made use of induction factors and was applicable to normal supercavitating geometry and for non-zero cavitation numbers. However, it was stressed that there was a need to develop correction factors in order to get satisfactory agreement between the lifting line theory and experimental results.

Supercavitating propeller design generally requires an appeal to theoretical and experimental results – not unlike many other branches of propeller technology. In the case of theoretical methods Young and Kinnas (2003, 2004) have extended their boundary element code to the modelling of supercavitating and surface piercing propellers analysis. With regard to the experimental data to support the design of supercavitating propellers the designer can make appeal to the works of Newton and Radar (1961), van den Voorde and Esveldt (1962) and Taniguchi and Tanibayashi (1962).

5.4.7 Computational fluid dynamics methods

During the last ten years considerable advances have been made in the application of computational fluid dynamics to the analysis and design of marine propellers. This has now reached a point where in the analysis case useful insights into the viscous and cavitating behaviour or propellers can be obtained from these methods. While progress has been made with the codes in the design case, these have not yet reached a level where these methods have gained wide acceptance but, no doubt, this will happen in the coming years.

A number of approaches for modelling the flow physics have been developed. Typically for the analysis of the flow around cavitating non-cavitating propellers these approaches are the Reynolds Averaged Navier–Stokes (RANS) method, Large Eddy Simulation (LES) techniques, Detached Eddy Simulations (DES) and Direct Numerical Simulations (DNS). However, in terms of practical propeller computations, as distinct from research exercises, the application of many of these methods is limited by the amount of computational effort required to derive a solution. As such, the RANS codes appear to have found most favour because the computational times are rather lower than for the other methods. Most of the approaches have a number of common basic features in that they employ multi-grid acceleration and finite volume approximations. There are, nevertheless, a number of differences to be found between various practitioners in that a variety of approaches are used for the grid topology, cavitating flow modelling and turbulence modelling. In this latter context there is a range of turbulence models in use, for example $k - \varepsilon$, $k - \omega$, and Reynolds stress models are frequently seen being deployed, with results from the latter two methods yielding good correlations.

Computational grid formation has proved a difficult area in marine propeller analysis, particularly in terms of achieving a smooth distribution of grid cells. Moreover, important structures in the flow field such as shaft lines and A-bracket structures require careful modelling with localized grid refinements. These considerations also apply to flow structures such as propeller blade tip vortices. Notwithstanding these issues, when considering propulsion test simulations these are characterized by widely different spatial and time scales for the hull and propeller.

If structured curvilinear grids are used in the modelling process this may result in a large number of cells which, in turn, may produce a complicated and time-consuming grid generation process. This has led to unstructured grids being favoured since these can easily handle complex geometries and the clustering of grid cells in regions of the flow where large parameter gradients occur. Rhee and Joshi (2003) analysed a five-bladed c.p.p. propeller in open water conditions using hybrid unstructured meshes in which they used prismatic cells in the boundary layer with a system of tetrahedral cells filling in the remainder of the computational domain far from the solid boundaries. This approach allowed them to have a detailed model of the boundary layer flows while retaining many of the advantages of an unstructured mesh. In this formulation of the problem they used a $k - \omega$ turbulence model. When correlating their computed results the K_T and K_Q values were 8 per cent and 11 per cent different from the measured model test values, and while good agreement was found between the circumferential averaged axial and tangential velocities the predicted radial velocities were less accurate. Additionally, the turbulent velocity fluctuations in the wake region were also found to be underpredicted. An alternative grid generation approach for complex geometries, the Chimera technique, is becoming relatively popular, Muscari and Di Mascio (2005). In this approach simple structured grids, called sub-grids, are used for limited parts of the fluid domain and these sub-grids may overlap each other. All of these sub-grids are then embedded into a parent grid that extends across the whole fluid domain. This method has been used to address tip vortex and propeller flows as described in Hsiao and Chahine (2001) and Kim et al. (2003).

Notwithstanding the present difficulties in the computational fluid dynamics method at this relatively early stage in its development, one of the underlying values of these types of study is in giving insights into phenomenological behaviour where classical extrapolation techniques are not applicable. For example, Abdel-Maksoud and his colleagues have examined the scale effects on ducted propellers and also the influence of the hub cap shape on propeller performance (Abdel-Maksoud et al. (2002, 2004)). Similarly, Wang et al. (2003) have examined the three-dimensional viscous flow field around an axisymmetric body with an integrated ducted propulsor and other work has been done on podded propulsors.

In developing the method further in order to reach its full potential research is required in a number of areas. In particular, it is necessary to achieve a robust and reliable modelling of the boundary layer and similarly with wakes and two-phase flow behaviour. In addition, as discussed by Kim and Rhee (2004) who analysed the interaction between turbulence modelling and local mesh refinements it is apparent that an adequate grid resolution of the flow field regions where vertical flow dominates is particularly important.

Comprehensive reviews of developments in theoretical and numerical modelling of propellers and propulsion is given in ITTC (2002b, 2005b).

5.5 Cavitation

Cavitation is a general fluid mechanics phenomenon that can occur whenever a liquid is used in a machine which induces pressure and velocity fluctuations in the fluid. Consequently, pumps, turbines, propellers, bearings and even the human body, in for example the heart and knee-joints, are all examples of machines where the destructive consequences of cavitation may occur.

The history of cavitation has been traced back to the middle of the eighteenth century, when some attention was paid to the subject by the Swiss mathematician Euler (1756) in a paper read to the Berlin Academy of Science and Arts in 1754. In that paper Euler discussed the possibility that a phenomenon that we would today call cavitation occurs on a particular design of water wheel and the influence this might have on its performance.

However, little reference to cavitation pertaining directly to the marine industry has been found until the mid-nineteenth century, when Reynolds (1873) wrote a series of papers concerned with the causes of engine racing in screw propelled steamers. These papers introduced the subject of cavitation as we know it today by discussing the effect it had on the performance of the propeller: when extreme cases of cavitation occur, the shaft rotational speed is found to increase considerably from that expected from the normal power absorption relationships.

The trial reports of *HMS Daring* in 1894 noted this overspeeding characteristic, as did Sir Charles Parsons, shortly afterwards, during the trials of his experimental steam turbine ship *Turbinia*. The results of the various full-scale experiments carried out in these early investigations showed that an improvement in propeller performance could be brought about by the increase in blade surface area. In the case of the *Turbinia*, which originally had a single propeller on each shaft and initially only achieved just under twenty knots on trials, Parsons found that to absorb the full power required on each shaft it was necessary to adopt a triple propeller arrangement to increase the surface area to the required proportions. Consequently, he used three propellers mounted in tandem on each shaft, thereby deploying a total of nine propellers distributed about the three propeller shafts. This arrangement not only allowed the vessel to absorb the full power at the correct shaft speeds, but also permitted the quite remarkable trial speed of 32.75 knots to be attained.

In an attempt to appreciate fully the reasons for the success of these decisions, Parsons embarked on a series of model experiments designed to investigate the nature of cavitation. To accomplish this task, Parsons constructed in 1895 an enclosed circulating channel. This apparatus allowed the testing of 2 in. diameter propellers and was a forerunner of cavitation tunnels as we know them today. However, recognizing the limitations of this tunnel, Parson constructed a much larger tunnel fifteen years later in which he could test 12 in. diameter propeller models. Subsequently, other larger tunnels were constructed in Europe and America during the 1920s and 1930s, each incorporating the lessons learned from its predecessors. More recently a series of very large cavitation facilities have been constructed in various locations around the world. Typical of these are the depressurized towing tank at MARIN in Ede; the large cavitation tunnel at SSPA in Gothenburg, the HYCAT at HSVA in Hamburg; the Grande Tunnel Hydrodynamique at Val de Reuil in France and the Large Cavitation Channel (LCC) in Memphis, Tennessee.

5.5.1 The basic physics of cavitation

The underlying physical process which governs the action of cavitation can, at a generalized level, be considered as an extension of the well-known situation in which a kettle of water will boil at a lower temperature when taken to the top of a high mountain. In the case of cavitation development the pressure is allowed to fall to a low level while the ambient temperature is kept constant, which in the case of a propeller is that of the surrounding sea water. Parsons had an early appreciation of this concept and he, therefore, allowed the atmospheric pressure above the water level in his tunnels to be reduced by means of a vacuum pump, which enabled cavitation to appear at much lower shaft speeds, making its observation easier.

If cavitation inception were to occur when the local pressure reaches the vapour pressure of the fluid then the inception cavitation number σ_i would equal the minimum pressure coefficient Cp_{min}. However, a number of other influencing factors prevent this simple relationship from being valid. For example, the ability of the fluid to withstand tensions; nuclei requiring a finite residence time in which to grow to an observable size and measurement and calculation procedures normally produce time averaged values of pressure coefficients. Consequently, the explanation of cavitation as being simply a water boiling phenomenon, although true, is an oversimplification of the actual physics that occur. To initially appreciate this, consider first the phase diagram for water shown in Figure 5.61. If it is assumed that the temperature is sufficiently high for the water not to enter its solid phase, then at either point B or C one would expect the water to be both in its liquid state and have an enthalpy equivalent to that state. For example, in the case of fresh water at standard pressure and at a temperature of 10°C this would be of the order of 42 kJ/kg. However, at points A, which lies in the vapour phase, the fluid would be expected to have an enthalpy equivalent to a superheated vapour, which in the example quoted above, when the pressure was dropped to say 1.52 kPa, would be in excess of 2510 kJ/kg. The differences in these figures is primarily because the fluid gains a latent

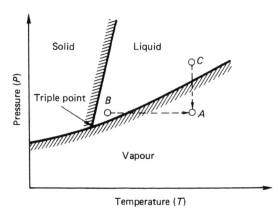

Figure 5.61 Phase diagram for water.

Table 5.12 Saturation temperature of fresh water.

Pressure (kPa)	0.689	6.894	13.79	27.58	55.15	101.3	110.3
Saturation temperature (°C)	1.6	38.72	52.58	67.22	83.83	100.0	102.4

enthalpy change as the liquid–vapour line is crossed, so that at points B and C the enthalpies are

$$h_{B,C} = h_{\text{fluid}}(p,t)$$

and at the point A the fluid enthalpy becomes

$$h_A = h_{\text{fluid}} + h_{\text{latent}} + h_{\text{superheat}}$$

Typically for fresh water the liquid–vapour line is defined by Table 5.12.

Secondly, it is important to distinguish between two types of vaporization. The first is the well-known process of vaporization across a flat surface separating the liquid and its vapour. The corresponding variation in vapour pressure varies with temperature as shown in Table 5.12, and along this curve the vapour can coexist with its liquid in equilibrium. The second way in which vaporization can occur is by cavitation, which requires the creation of cavities within the liquid itself. In this case the process of creating a cavity within the liquid requires work to be done in order to form the new interface. Consequently, the liquid can be subjected to pressures below the normal vapour pressure, as defined by the liquid–vapour line in Figure 5.61, or Table 5.12, without vaporization taking place. As such, it is possible to start at a point such as C, shown in Figure 5.61 which is in the liquid phase, and reduce the pressure slowly to a value well below the vapour pressure, to reach the points A with the fluid still in the liquid phase. Indeed, in cases of very pure water, this can be extended further, so that the pressure becomes negative; when a liquid is in these over-expanded states it is said to be in a metastable phase. Alternatively it is possible to bring about the same effect at constant pressure by starting at a point B and gradually heating the fluid to a metastable phase at the point A. If either of these paths, constant pressure or temperature, or indeed some intermediate path, is followed, then eventually the liquid reaches a limiting condition at some point below the liquid–vapour line in Figure 5.61, and either cavitates or vaporizes.

The extent to which a liquid can be induced metastably to a lower pressure than the vapour pressure depends on the purity of the water. If water contains a significant amount of dissolved air, then as the pressure decreases the air comes out of the solution and forms cavities in which the pressure will be greater than the vapour pressure. This effect applies also when there are no visible bubbles; submicroscopic gas bubbles can provide suitable nuclei for cavitation purposes.

Hence cavitation can either be vaporous or gaseous or, perhaps, a combination of both. Consequently, the point at which cavitation occurs can be either above or below the vapour pressure corresponding to the ambient temperatures.

In the absence of nuclei a liquid can withstand considerable negative tensions without undergoing cavitation. For example, in the case of a fluid, such as water, which obeys van der Waals' equation:

$$\left(p + \frac{a}{V^2}\right)(V - b) = RT \quad (5.81)$$

a typical isotherm is shown in Figure 5.62, together with the phase boundary for the particular temperature. In addition, the definition of the tensile strength of the liquid is also shown on this figure. The resulting limiting values of the tensions that can be withstood form a wideband; for example, at room temperature, by using suitable values for a and b in Equation (5.81), the tensile strength can be shown to be about 500 bars. However, some researchers have suggested that the tensile strength of the liquid is the same as the intrinsic pressure a/V^2 in Equation (5.81); this yields a value of around 10 000 bars. In practice, water subjected to rigorous filtration and pre-pressurization seems to rupture at tensions of the order of 300 bars. However, when solid, non-wetted nuclei having a diameter of about 10^{-6} cm are present in the water it will withstand tensions of only the order of tens of bars. Even when local pressure conditions are known accurately it is far from easy to predict when cavitation

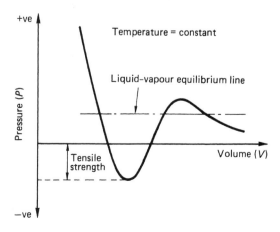

Figure 5.62 Van der Waals' isotherm and definition of tensile strength of liquid.

will occur because of the necessity to estimate the size and distribution of the nuclei present.

Despite the extensive literature on the subject, both the understanding and predictability of bubble nucleations is a major problem for cavitation studies. There are in general two principal models of nucleation; these are the stationary crevice model and the entrained nuclei models. Nuclei in this sense refers to clusters of gas or vapour molecules of sufficient size to allow subsequent growth in the presence of reduced pressure. The stationary nuclei are normally assumed to be harboured in small crevices of adjacent walls whilst, in contrast, the travelling nuclei are assumed to be entrained within the mainstream of the fluid. Consequently, entrained nuclei are considered the primary source of cavitation, although of course cavitation can also be initiated from stationary nuclei located in the blade surface at the minimum pressure region. Of the nucleation models proposed those of Harvey et al. (1944(a), 1944(b), 1945, 1947) and subsequently by others are probably the most important, Strasberg (1959), Flynn (1964), Winterton (1977) and Cram (1980). These models propose that entrained microparticles in the liquid, containing in themselves unwetted acute angled microcrevices, are a source of nucleation. This suggests that if a pocket of gas is trapped in a crevice then, if the conditions are correct, it can exist in stable equilibrium rather than dissolve into the fluid. Consider first a small spherical gas bubble of radius R in water. For equilibrium, the pressure difference between the inside and outside of the bubble must balance the surface tension force:

$$p_v - p_1 = \frac{2S}{R} \quad (5.82)$$

where p_v is the vapour and/or gas pressure (internal pressure)
p_1 is the pressure of the liquid (external pressure)
S is the surface tension.

Now the smaller the bubble becomes, according to Equation (5.82) the greater must be the pressure difference across the bubble. Since, according to Henry's law, the solubility of a gas in a liquid is proportional to gas pressure, it is reasonable to assume that in a small bubble the gas should dissolve quickly into the liquid. Harvey et al., however, showed that within a crevice, provided the surface is hydrophobic, or imperfectly wetted, then a gas pocket can continue to exist. Figure 5.63 shows in schematic form the various stages in the nucleation process on a microparticle. In this figure the pressure reduces from left to right, from which it is seen that the liquid–gas interface changes from a convex to concave form and eventually the bubble in the crevice of the microparticle grows to a sufficient size so that a part breaks away to form a bubble entrained in the body of the fluid.

Other models of nucleation have been proposed – for example those of Fox and Herzfeld (1954) and Plesset (1963) – and no doubt these also play a part in the overall nucleation process, which is still far from well understood. Fox and Herzfeld suggested that a skin of organic impurity, for example fatty acids, accumulates on the surface of a spherical gas bubble in order to inhibit the dissolving of the gas into the fluid as the bubble decreases in size; this reduction in size causes the pressure differential to increase, as seen by Equation (5.82). In this way, it is postulated that the nuclei can stabilize against the time when the bubble passes through a low-pressure region, at which point the skin would be torn apart and a cavity initiated. The 'skin' model has in latter years been refined and improved by Yount (1979, 1982). Plesset's unwetted mote model suggested that such motes can provide bubble nucleation without the presence of gases other than the inevitably present vapour of the liquid. The motes, it is suggested, would provide weak spots in the fluid about which tensile failure of the liquid would occur at pressures much less than the theoretical strength of the pure liquid.

An additional complicating factor arises from the flow over propeller blades being turbulent in nature, consequently any nuclei in the centre of the turbulent eddies may experience localized pressures which are rather lower than the mean or time averaged pressure that has been either calculated or measured. As a result the local pressure within the eddy formations may fall below the vapour pressure of the fluid while the average pressure remains above that level.

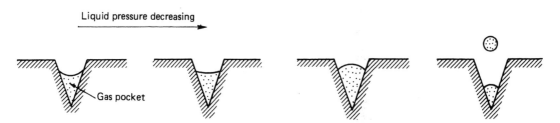

Figure 5.63 Nucleation model for a crevice in an entrained microparticle (Harvey).

Cavitation gives rise to a series of other physical effects which, although of minor importance to ship propulsion, are interesting from the physical viewpoint and deserve passing mention especially with regard to material erosion. The first is sonoluminescence, which is a weak emission of light from the cavitation bubble in the final stage of its collapse. This is generally ascribed to the very high temperatures resulting from the essentially adiabatic compression of the permanent gas trapped within the collapsing cavitation bubbles. Schlieren and interferometric pictures have succeeded in showing the strong density gradients or shock waves in the liquid around collapsing bubbles. When bubbles collapse surrounding fluid temperatures as high as 100 000 K have been suggested and Wheeler (1960) has concluded that temperature rises of the order of 500 to 800°C can occur in the material adjacent to the collapsing bubble. The collapse of the bubbles is completed in a very short space of time (milli- or even microseconds) and it has been shown that the resulting shock waves radiated through the liquid adjacent to the bubble may have a pressure difference as high as 4000 atm.

The earliest attempt to analyse the growth and collapse of a vapour or gas bubble in a continuous liquid medium from a theoretical viewpoint appears to have been made by Besant (1859). This work was to some extent ahead of its time, since bubble dynamics was not an important engineering problem in the mid-1800s and it was not until 1917 that Lord Rayleigh laid the foundations for much of the analytical work that continues to the present time (Rayleigh (1917)). His model considered the problem of a vapour-filled cavity collapsing under the influence of a steady external pressure in the liquid, and although based on an over-simplified set of assumptions, Rayleigh's work provides a good model of bubble collapse and despite the existence of more modern and advanced theories is worthy of discussion in outline form.

In the Rayleigh model the pressure p_v within the cavity and the pressure at infinity p_0 are both considered to be constant. The bubble is defined using a spherical coordinate system whose origin is at the centre of the bubble whose initial steady state radius is R_0 at time $t = 0$. At some later time t, under the influence of the external pressure p_0 which is introduced at time $t = 0$, the motion of the bubble wall is given by

$$\frac{d^2 R}{dt^2} + \frac{3}{2R}\left(\frac{dR}{dt}\right)^2 = \frac{1}{\rho R}(p_v - p_0) \qquad (5.83)$$

where ρ is the density of the fluid. By direct integration of Equation (5.83), assuming that both p_v and p_0 are constant, Rayleigh described the collapse of the cavity in terms of its radius R at a time t as being

$$\left(\frac{dR}{dt}\right)^2 = \frac{2}{3}\frac{(p_0 - p_v)}{\rho}\left(\left(\frac{R_0}{R}\right)^3 - 1\right) \qquad (5.84)$$

By integrating Equation (5.84) numerically it is found that the time to collapse of the cavity t_0, known as the 'Rayleigh collapse time', is

$$t_0 = 0.91468 R_0 \left(\frac{\rho}{p_0 - p_v}\right)^{1/2} \qquad (5.85)$$

This time t presupposes that at the time $t = 0$, the bubble is in static equilibrium with a radius R_0. The relationship between bubble radius and time in non-dimensional terms is derived from the above as being

$$\frac{t}{t_0} = 1.34 \int_{R/R_0}^{1} \frac{dx}{(1/x^2 - 1)^{1/2}} \qquad (5.86)$$

and the results of this equation, shown in Figure 5.64, have been shown to correspond to experimental observations of a collapsing cavity.

The Rayleigh model of bubble collapse leads to a series of significant results from the viewpoint of cavitation damage; however, because of simplifications involved it cannot address the detailed mechanism of cavitation erosion. The model shows that infinite velocities and pressures occur at the point when the bubble vanishes and in this way points towards the basis of the erosion mechanism. The search for the detail of this mechanism has led to considerable effort on the part of many researchers in recent years. Such work has introduced not only the effects surface tension, internal gas properties and viscosity effects, but also

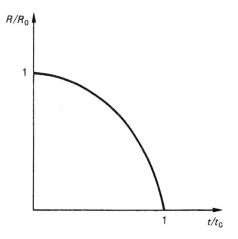

Figure 5.64 Collapse of a Rayleigh cavity.

those of bubble asymmetries which predominate during the collapse process. Typical of these advanced studies is the work of Mitchell and Hammitt (1973) who also included the effects pressure gradient and relative velocity as well as wall proximity. An alternative approach by Plesset and Chapman (1971) used potential flow assumptions which precluded the effects of viscosity, which in the case of water is unlikely to be of major importance. Plesset and Chapman focused on the bubble collapse mechanism under the influence of wall proximity, which is of major significance in the study of cavitation damage, as is discussed in Carlton (2007). Their approach was based on the use of cylindrical coordinates as distinct from Mitchell's spherical coordinate approach, and this allowed them to study the microjet formation during collapse to a much deeper level because the spherical coordinates required the numerical analysis to be terminated as the microjet approaches the initial bubble centre. Figure 5.65 shows the results of a computation of an initially spherical bubble collapsing close to a solid boundary, together with the formation of the microjet directed towards the wall.

Subsequently, Chahine has studied cloud cavity dynamics by modelling the interaction between bubbles. In his model he was able to predict the occurrence of high pressures during collapse principally by considering the coupling between bubbles in an idealized way through symmetric distributions of identically sized bubbles.

Bark in his researches at Chalmers University has demonstrated the effects of cavity rebound following the initial collapse of a cavity. Figure 5.66 shows this effect over a sequence of four-blade passages in terms of a propeller radiated hull pressure signature.

From the figure it is immediately obvious that the hull pressure signature is very variable, particularly in terms of amplitude, from blade passage to blade passage. Moreover, the influence of the cavity rebound in comparison to the cavity growth and initial decay parts of the signature is significant and, can if the physical conditions are correct, one or more rebound events can take place. This variability in the signature which comprises spatial, temporal and phenomenological attributes underlines the importance

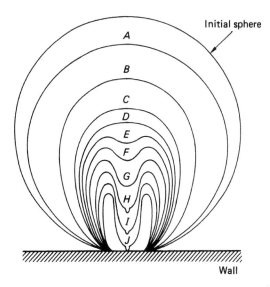

Figure 5.65 Computed bubble collapse (Plesset–Chapman).

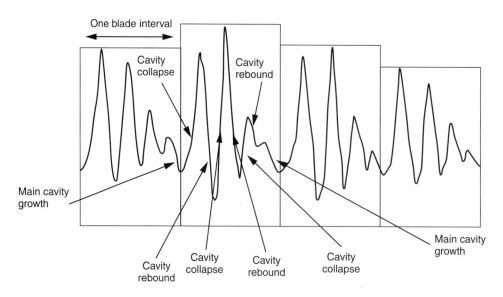

Figure 5.66 Influence of cavity rebound on a hull radiated pressure signature.

of analysing these signatures correctly in order that information is not inadvertently lost when analysing hull surface pressure signatures.

In extending the study of the physics of cavitation and, in particular, its aggressiveness towards material erosion, Fortes-Patella and her colleagues have been developing a model of cavitation action, Fortes-Patella (1998a, 1998b, 1999, 2000, 2001), Challier *et al.* (2000), Coutier-Delgosha (2001, 2003(a), 2003(b)), Lohrberg *et al.* (2001(a), 2001(b)). In essence the method is based on the study of the pressure wave characteristics emitted during bubble collapse: in particular, focusing on the relationship between the initial and collapsing states. Within this work a better agreement was found between experiment and calculation for pressure wave models of erosion. They also showed that there was no influence of material on the flow pressure pulse histogram and that the number of pits normalized by surface area and time was found by experiment to be proportional to $\lambda^{2.7}$, where λ is the geometric scaling factor. This result is close to the cubic law which was noted by Lecoffre (1995) and it was found that volume damage rate does not appear to significantly change with scale. It was shown that the flow speed (V) does, however, have a significant influence on the erosion in that the number pits per unit area and time is proportional to V^5 and the pit volume, normalized on the same basis, is proportional to V^7. In this context pit depth did not seem to significantly vary with the flow speed.

The computational model developed is based on a series of energy transformations within an overall energy balance scenario as outlined in Figure 5.67.

Within this model the terms P_{pot}, P_{pot}^{mat}, P_{waves}^{mat} represent the potential power of the vapour structure, the flow aggressiveness potential and the pressure wave power, respectively, while the η^{**} and η^* represent transmission efficiencies and β is the transmission factor for fluid–material interaction.

5.5.2 Types of cavitation experienced by propellers

Cavitating flows are by definition multi-phase flow regions. The two phases that are most important are the water and its own vapour; however, in almost all cases there is a quantity of gas, such as air, which has significant effects in both bubble collapse and inception – most importantly in the inception mechanism. As a consequence cavitation is generally considered to be a two-phase, three-component flow regime. Knapp *et al.* (1970) classified cavitation into fixed, travelling or vibratory forms, the first two being of greatest interest in the context of propeller technology.

A fixed cavity is one in which the flow detaches from the solid boundary of the immersed object to form a cavity or envelope which is fixed relative to the object upon which it forms, and in general such cavities have a smooth glassy appearance. In contrast, as their name implies, travelling cavities move with the fluid flowing past the body of interest. Travelling cavities originate either by breaking away from the surface of a fixed cavity, from which they can then enter the flow stream, or from nuclei entrained within the fluid medium. Figure 5.68 differentiates between these two basic types of cavitation.

The flow conditions at the trailing edge of a cavity are not dissimilar, but rather more complicated, to those of water passing over a weir. The cavity shedding mechanism is initiated by the re-entrant jet in that it forms between the underside of the cavity and the propeller blade surface. The behaviour of the re-entrant jet, therefore, is of importance in the phenomenological behaviour of the cavitation on the blades. In studies on twisted aerofoils undertaken by Foeth and van Terwisga (2006) they have concluded that the cavity topology principally determines the direction of the re-entrant flow and that convex cavity

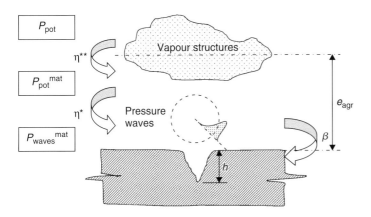

Figure 5.67 Basis of the Fortes-Patella *et al.* model.

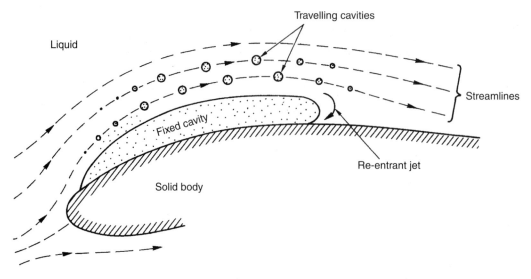

Figure 5.68 Fixed and travelling cavities.

shapes appear to be intrinsically unstable. Moreover, the condition when the re-entrant jet reaches the leading edge of the aerofoil is not the only determinant in shedding because the side entrant jets of convex cavities have both a chordal and spanwise motion. These motions focus in the closure region of the sheet cavity where they tend to disturb the flow which then initiates a break-off from the main sheet cavity structure. Interestingly, during the collapse of the sheet cavity structures it was noted that they degenerated into vortical structures which leads to the conclusion that a mixing layer exists with its characteristic spanwise and streamwise vortices.

Uhlman (2006) has studied the fully non-linear axisymmetric potential flow past a body of revolution using the boundary integral method and devised a model for the exact formulation of the re-entrant jet cavity closure condition. The results of this modelling approach were shown to be in good agreement with experimental results and were consistent with momentum flux requirements. This re-entrant jet model represents an enhancement over the earlier Riabouchinsky-type cavity closure model since the boundary conditions entail the physical conditions of constant pressure and no flux.

The cavitation patterns which occur on marine propellers are usually referred to as comprising one or more of the following types: sheet, bubble, cloud, tip vortex or hub vortex cavitation.

Sheet cavitation initially becomes apparent at the leading edges of the propeller blades on the back or suction surface of the blades if the sections are working at positive incidence angles. Conversely, if the sections are operating at negative incidence this type of cavitation may initially appear on the face of the blades. Sheet cavitation appears because when the sections are working at non-shock-free angles of incidence, large suction pressures build up near the leading edge of the blades of the 'flat plate' type of distribution. If the angles of incidence increase in magnitude, or the cavitation number decreases, then the extent of the cavitation over the blade will grow both chordally and radially. As a consequence the cavitation forms a sheet over the blade surface whose extent depends upon the design and ambient conditions. Figure 5.69(a) shows an example of sheet cavitation on a model propeller, albeit with tip vortex cavitation also visible. Sheet cavitation is generally stable in character, although there are cases in which a measure of instability can be observed. In these cases the reason for the instability should be sought, and if it is considered that the instability will translate to full scale, then a cure should be sought, as this may lead to blade erosion or unwanted pressure fluctuations.

Bubble cavitation (Figure 5.69(b)), is primarily influenced by those components of the pressure distribution which cause high suction pressures in the mid-chord region of the blade sections. Thus the combination of camber line and section thickness pressure distributions have a considerable influence on the susceptibility of a propeller towards bubble cavitation. Since bubble cavitation normally occurs first in the mid-chord region of the blade, it tends to occur in non-separated flows. This type of cavitation, as its name implies, appears as individual bubbles growing, sometimes quite large in character, and contracting rapidly over the blade surface.

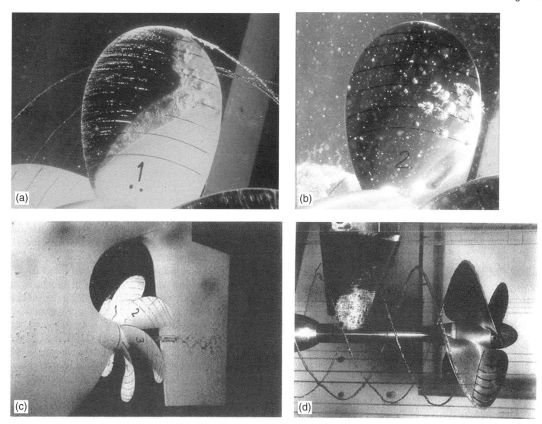

Figure 5.69 Types of cavitation on propellers (MARIN): (a) sheet and cloud cavitation together with a tip vortex; (b) mid-chord bubble cavitation together with a tip vortex and some leading edge streak cavitation; (c) hub vortex cavitation with traces of LE and tip vortex in top of propeller disc (Courtesy: MARIN) and (d) tip vortex cavitation.

Cloud cavitation is frequently to be found behind strongly developed stable sheet cavities and generally in moderately separated flow in which small vortices form the origins for small cavities. This type of cavitation (Figure 9.69(a) with traces on Figure 9.69(b)) appears as a mist or 'cloud' of very small bubbles and its presence should always be taken seriously.

The vortex types of cavitation, with few exceptions, occur at the blade tips, the leading edge and hub of the propeller and they are generated from the low-pressure core of the shed vortices. The hub vortex is formed by the combination of the individual vortices shed from each blade root, and although individually these vortices are unlikely to cavitate, under the influence of a converging propeller cone the combination of the blade root vortices has a high susceptibility to cavitate. When this occurs the resulting cavitation is normally very stable and appears to the observer as a rope with strands corresponding to the number of blades of the propeller. Tip vortex cavitation is normally first observed some distance behind the tips of the propeller blades. At this time the tip vortex is said to be 'unattached', but as the vortex becomes stronger, either through higher blade loading or decreasing cavitation number, it moves towards the blade tip and ultimately becomes attached. Figures 9.69(c) and (d) shows typical examples of the hub and tip vortices, respectively.

In addition to the principal classes of cavitation, there is also a type of cavitation that is sometimes referred to in model test reports as 'streak' cavitation. This type of cavitation, again as its name implies, forms relatively thin streaks extending from the leading edge region of the blade chordally across the blades.

Propulsor–Hull Vortex (PHV) cavitation was reported by Huse (1971) in the early 1970s. This type of cavitation may loosely be described as the 'arcing' of a cavitating vortex between a propeller tip and the ship's hull. Experimental work with flat, horizontal plates above the propeller in a cavitation tunnel shows that PHV cavitation is most pronounced for small tip clearances. In addition, it has been observed that the advance coefficient also has a significant influence on its occurrence; the lower the advance coefficient the more likely PHV cavitation is to occur. Figure 5.70 shows a probable mechanism for PHV cavitation formation. In the figure it is postulated that at high

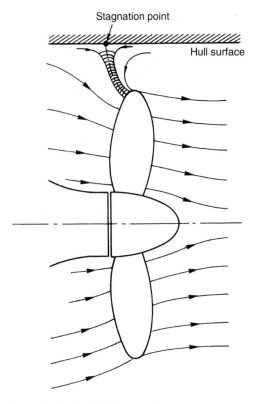

Figure 5.70 Basis for PHV cavitation.

loading the propeller becomes starved of water due to the presence of the hull surface above and possibly the hull in the upper part of the aperture ahead of the propeller. To overcome this water starvation the propeller endeavours to draw water from astern, which leads to the formation of a stagnation streamline from the hull to the propeller disc, as shown. The PHV vortex is considered to form due to turbulence and other flow disturbances close to the hull, causing a rotation about the stagnation point, which is accentuated away from the hull by the small radius of the control volume forming the vortex. This theory of PHV action is known as the 'pirouette effect' and is considered to be the most likely of all the theories proposed. Thus the factors leading to the likelihood of the formation of PHV cavitation are thought to be:

1. low advance coefficients,
2. low tip clearance,
3. flat hull surfaces above the propeller.

Van der Kooij and Gent studied the problem of propeller–hull vortex cavitation for the ducted propeller case and concluded that the occurrence of PHV cavitation depended strongly on hull–duct clearance and propeller blade position.

Methods of overcoming the effects of PHV cavitation are discussed in Carlton (2007), Chapter 23.

The foregoing observations relate principally to indications gained from undertaking model tests. In recent years, however, considerably more full-scale observations have been made using both the conventional hull window penetrations and more recently using the boroscope technique. This has increased the understanding of the full-scale behaviour of cavitation and its correlation to model-scale testing. Figure 5.71, taken from Carlton and Fitzsimmons (2006), shows a consecutive sequence of boroscope images taken under natural daylight conditions of the tip vortex development emanating from the propeller blades of an 8500 TEU container ship. This continuous sequence, comprising eight images, was taken at a time interval of 1/25 seconds. In the figure the rising propeller blade can be clearly seen on the right-hand side of the images and the behaviour of the vortices emanating from the two blades immediately preceding the rising blade can be observed on the left. The observation was made from the hull above the propeller over a period of 0.28 seconds with the ship proceeding on a steady course at constant speed. At the arbitrary time $t = 0$ the vortex from the blade immediately leading the rising blade exhibits a well-formed structure having some circumferential surface texture and small variations in radius with slight tendency towards expansion near the top of the picture. By 0.04 seconds later the cavitating structural expansion has started to grow with the expansion showing a distinct asymmetric behaviour towards the propeller station. In the subsequent frames this asymmetric expansion progressively increases and exhibits a tendency for the principal area of asymmetry to become increasingly distinct from the main vortex structure. By the time $t = 0.16$ seconds a new vortex is clearly following the tendency of its predecessor as indeed the described vortex followed the behaviour of its own predecessor shown at time $t = 0$.

The complexity of the tip vortex mechanisms was discussed by Carlton and Fitzsimmons (2004) in relation to observations made on a number of ships. In that paper a mechanism derived from full-scale observation of LNG ship propeller cavitation was described to explain an expansive mechanism for the tip vortex structure. This was in effect an interaction between the tip vortex and the supercavitating parts of the sheet cavity at the blade tip region where the supercavitating part of the blade sheet cavity was rapidly expanded under the action of the tip vortex. It is, therefore, interesting to note that Lücke (2006) has identified from model tests two mechanisms for tip vortex bursting: one following the conventional aerodynamic treatment of vortex bursting and the other very similar to that described at full scale

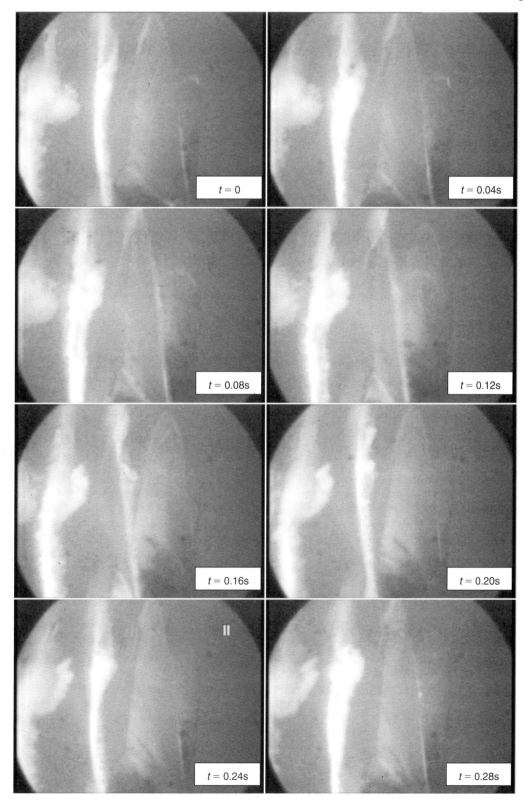

Figure 5.71 A sequence of images of the tip vortex emanating from the propeller of an 8500 TEU container ship.

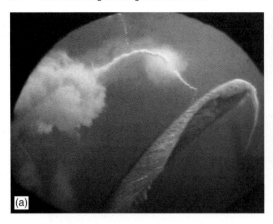 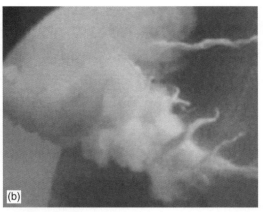

Figure 5.72 Full-scale cavitating sheet and vortex cavitation on an LNG ship: (a) cavitation on a straight course and (b) tip vortex behaviour during turning.

above, thereby, suggesting a possible model to full scale similarity. While the earlier descriptions of this phenomenon centred on steady course ship operation at constant speed, the complexity of the tip vortex development was found to increase significantly when the ship began to undertake open water turning manoeuvres. An example of this behaviour is shown in Figure 5.72 in which the expansive cloud seen in Figure 5.72(a) and developed during the cavity collapse phase under uniform straight course conditions has, in the turning manoeuvre, extended its trailing volume region and developed a system of ring-like vortex structures circumscribing this trailing part of the cavitating volume (Figure 5.72(b)). However, in interpreting these structures it must be recalled that only the cavitating part of the vortex structure is visible in these images and the complete vortex structure, including the cavitating and non-cavitating parts, is considerably larger.

Manoeuvres have been found to generate extremely complex interactions between cavitation structures on a propeller blade and also between the propeller and the hull as well as between propellers in multi-screw ships. In the case of a high speed, twin-screw passenger ship when undertaking berthing manoeuvring in port strong cavitation interaction was observed between the propellers. This interaction took the form of vorticity shed from one propeller blade and directing itself transversely across the ship's afterbody to interact with the cavity structures on the adjacent propeller blades. Figure 5.73 captured this interaction taking place by means of a digital camera viewing through a conventional hull window arrangement. The complexity of the cavity structure and the locus of its travel is immediately apparent.

In the case of propeller–hull interaction Figure 5.74 shows a cavitating propeller–hull vortex captured by a boroscope observation in a steep buttock flow field

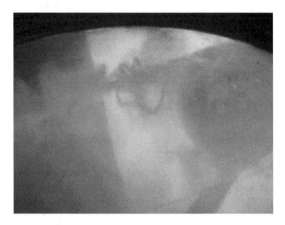

Figure 5.73 Cavitation interaction between propellers.

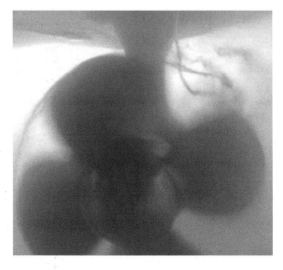

Figure 5.74 Example of a propeller–hull vortex emanating from a podded propulsion unit.

which then entered the propeller disc of a podded propulsor. In this image the relatively strong tip vortices can be seen emanating from the propeller blades while the tip vortex rises vertically towards the hull.

Vortex interaction, particularly at off-design conditions may cause troublesome excitation of the ship structure by generating a combined harmonic and broadband signature. Figure 5.75 shows a series of images demonstrating the interaction of vortex cavitation emanating from one of the propellers of a twin-screw ship when operating at full shaft speed and reduced blade pitch at 8 knots: this is discussed further in terms of its effects and consequences in Section 5.6. From the images it can be seen that the propeller is emitting both a cavitating tip and a leading edge vortex from the blades. At first these vortices travel back in the flow field largely independently, certainly as far as perturbation to the cavitating part of the vortex structures are concerned (3:39:28): however, even at this early stage some small influence on the tip vortex can be seen.

As the vortices coexist the mutual interference builds up (3:39:06) with the cavitating part of the vortices thickening and becoming less directionally stable as well as inducing some ring vortices which encircle both of the vortex structures. A short time later (3:39:07) ring vortex structures are being developed to a far greater degree with the vortices thickening and being surrounded by much greater coaxial cloudiness; although some cloudiness, as can be seen, was present one second earlier.

Finally, due to the interaction of the two vortices they eventually destroy their basic continuous helical form and break up into intermittent ring formations following each other along a helical track (3:22:02). However, even at this late stage some of the earlier encircling ring structures are still present together with coaxial cloudy regions around the main core of the vortex. As these new ring structures pass downstream (3:39:28) their cavitating cores lose thickness and gain a strong cloudy appearance.

In the case of a patrol boat which was powered by a triple screw, fixed pitch propeller arrangement the underlying problem was a poor design basis for the propeller: principally a large slow-running propeller with a high P/D ratio. The propellers experienced

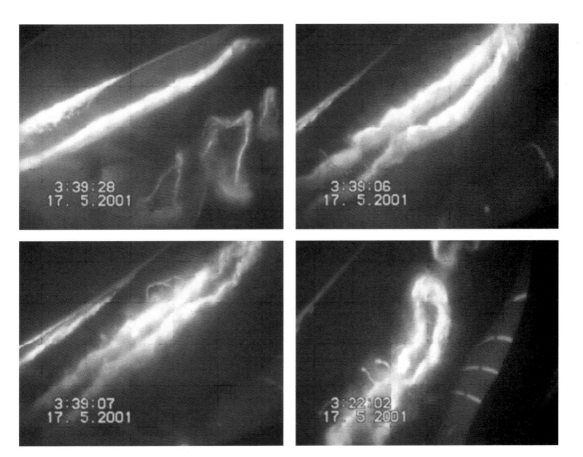

Figure 5.75 Vortex interaction mechanisms from a use ship propeller.

high angles of attack due to the variations in the tangential component of the velocity field induced by the shaft angle, ten degrees, which gave rise to a set of face and root cavitation erosion problems which could not be reconciled without recourse to artificial means. The blades were designed and manufactured to ISO 484 Class S, however, the cavitation problem was exacerbated by a lack of consistent definition of the blade root section geometry, which is outside the ISO standard, and which permitted arbitrary section forms to result in the root regions. The blade roots originally were also very close to the leading edge of the boss which caused problems in blending the blade leading edge onto the hub. This initial hub was designed with a small leading edge radius.

Figure 5.76(a) shows a typical cavitation pattern observed on the propeller blades when near the top dead centre position with the propeller operating at close to its sprint condition. From the figure a large, but relatively benign, back sheet cavity can be seen which spreads over a significant portion of the blade chord length. In the root region this sheet cavity transforms itself into a complex thick structure having a much more cloudy nature together with embedded vortex structures. Indeed, whenever the ring type structures as seen in the centre of the picture have been observed, at either model or full scale, these have often indicated a strong erosion potential. Root cavitation structures of this type are extremely aggressive in terms of cavitation erosion and being shed from the downstream end of a root cavity may indicate a mechanism which, in association with ship speed, can produce two or more isolated erosion sites along the root chord.

Kennedy *et al.* (1993) studied the cavitation performance of propeller blade root fillets. They found that the critical region of the fillet as far as cavitation inception was concerned was from its leading edge to around 20 per cent of the chord length. Moreover fillet forms, for example those with small radii, which give rise to vortex structures should be avoided. However, the relatively low Reynolds number at which typical propeller model tests are conducted may not permit the observation of vortex cavitation structures in the blade roots and this presents a further difficulty for the designer.

5.5.3 Cavitation considerations in design

The basic cavitation parameter used in propeller design is the cavitation number which was introduced in Section 5.3.2. In its most fundamental form the cavitation number is defined as

$$\text{cavitation number} = \frac{\text{static pressure head}}{\text{dynamic pressure head}} \quad (5.87)$$

The relationship has, however, many forms in which the static head may relate to the shaft centre line immersion to give a mean value over the propeller disc, or may relate to a local section immersion either at the top dead centre position or some other instantaneous position in the disc. Alternatively, the dynamic head may be based upon either single velocity components such as the undisturbed free stream advance velocity and the propeller rotational speed or the vectorial combination of these velocities in either the mean or local sense. Table 5.13 defines some of the more common cavitation number formulations used in propeller technology: the precise one chosen depends upon the information known or the intended purpose of the data.

The cavitating environment in which a propeller operates has a very large influence not only on the detail of the propeller design but also upon the type of propeller that is used. For example, whether it is better to use for a given application a conventional, supercavitating or surface piercing propeller. A useful

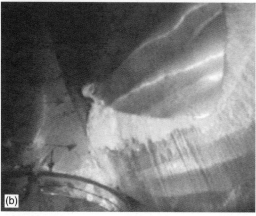

Figure 5.76 Root cavitation on a high-speed propeller: (a) in the sprint condition and (b) below the sprint condition.

Table 5.13 Common formulations of cavitation numbers.

Definition	Symbol	Formulation
Free stream-based cavitation no.	σ_0	$\dfrac{p_0 - p_v}{\frac{1}{2}\rho v_A^2}$
Rotational speed-based cavitation no.	σ_n	$\dfrac{p_0 - p_v}{\frac{1}{2}\rho(\pi xnD)^2}$
Mean cavitation no.	σ	$\dfrac{p_0 - p_v}{\frac{1}{2}\rho[v_A^2 + (\pi xnD)^2]}$
Local cavitation no.*	σ_L	$\dfrac{p_0 - p_v + xRg\cos\theta}{\frac{1}{2}\rho[\{v_A(x,\theta) + u_A(x,\theta)\}^2 + \{\pi xnD - v_T(x,\theta) - u_T(x,\theta)\}^2]}$

u_A and u_T are the propeller-induced velocities
v_A and v_T are the axial and tangential wake velocities

*Sometimes the local cavitation number is calculated without the influence of u_A and u_T, and also v_T when this is not known.

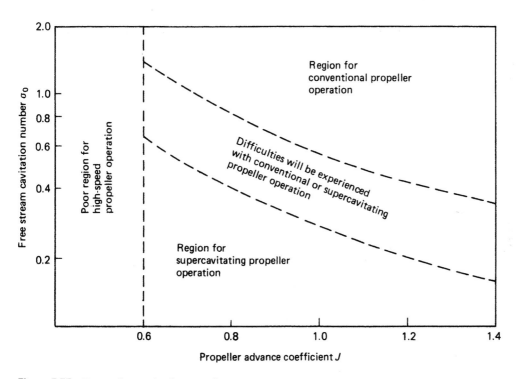

Figure 5.77 Zones of operation for propellers.

initial guide to determining the type of propeller most suited to a particular application is afforded by the diagram shown in Figure 5.77, which was derived from the work of Tachmindji and Morgan. The diagram is essentially concerned with the influence of inflow velocities, propeller geometric size and static head and attempts from these parameters, grouped into advance coefficient and cavitation number, to give guidance on the best regions in which to adopt conventional and supercavitating propellers. Clearly the 'grey' area in the middle of the diagram is dependent amongst other variables on both the wake field fluctuations and also shaft inclination angle. Should neither the conventional nor supercavitating propeller option give a reasonable answer to the particular design problem, then the further options of waterjet or surface piercing propulsors need to be explored, since these extend the range of propulsion alternatives.

From the early works of Parsons and Barnaby and Thornycroft on models and at full scale it was correctly concluded that extreme back or suction side cavitation of the type causing thrust breakdown could be avoided by increasing the blade surface area. Criteria were subsequently developed by relating the mean thrust to the required blade surface area in the form of a limiting thrust loading coefficient. The first such criterion of 77.57 kPa (11.25 lbf/in.^{2}) was derived in the latter part of the last century. Much development work was undertaken in the first half of the century in deriving refined forms of these thrust loading criteria for design purposes; two of the best known are those derived by Burrill (1978) and Keller (1966).

Burrill's method, which was proposed for fixed pitch, conventional propellers, centres around the use of the diagram shown in Figure 5.78. The mean cavitation number is calculated based on the static head relative to the shaft centre line, and the dynamic head is referred to the 0.7R blade section. Using this cavitation number $\sigma_{0.7R}$, the thrust loading coefficient τ_c is read off from Figure 5.78 corresponding to the permissible level of back cavitation. It should, however, be remembered that the percentage back cavitation allowances shown in the figure are based on cavitation tunnel estimates in uniform axial flow. From the value of τ_c read off from the diagram the projected area for the propeller can be calculated from the following:

$$A_P = \frac{T}{\frac{1}{2}\tau_c\rho[V_A^2 + (0.7\pi nD)^2]} \quad (5.88)$$

To derive the expanded area from the projected area, Burrill provides the empirical relationship which is valid for conventional propeller forms only:

$$A_E \simeq \frac{A_P}{(1.067 - 0.229 P/D)} \quad (5.89)$$

The alternative blade area estimate is due to Keller and is based on the relationship for the expanded area ratio:

$$\frac{A_E}{A_O} = \frac{(1.3 + 0.3Z)T}{(p_0 - p_v)D^2} + K \quad (5.90)$$

where p_0 is the static pressure at the shaft centre line (kgf/m^2)

p_v is the vapour pressure (kgf/m^2)
T is the propeller thrust (kgf)
Z is the blade number and
D is the propeller diameter (m)

The value of K in Equation (5.90) varies with the number of propellers and ship type as follows: for single-screw ships $K = 0.20$, but for twin-screw ships it varies within the range $K = 0$ for fast vessels through to $K = 0.1$ for the slower twin-screw ships.

Both the Burrill and Keller methods have been used with considerable success by propeller designers as a means of estimating the basic blade area ratio associated with a propeller design. In many cases, particularly for small ships and boats, these methods and even more approximate ones perhaps form the major part of the cavitation analysis; however, for larger vessels and those for which measured model wake field data is available, the cavitation analysis should proceed considerably further to the evaluation of the pressure distributions around the sections and their tendency towards cavitation inception and extent.

The nature of the pressure distribution around an aerofoil is highly dependent on the angle of attack of the section. Figure 5.79 shows typical velocity distributions for an aerofoil in a non-cavitating flow at positive, ideal and negative angles of incidence. This figure clearly shows how the areas of suction on the blade surface change to promote back, mid-chord or face cavitation in the positive, ideal or negative incidence conditions, respectively. When cavitation occurs on the blade section the non-cavitating pressure distribution is modified with increasing significance as the cavitation number decreases. Balhan (1951) showed, by means of a set of two-dimensional aerofoil experiments in a cavitation tunnel, how the pressure distribution changes. Figure 5.80 shows a typical set of results at an incidence of 5° for a Karman–Trefftz profile with thickness and camber chord ratios of 0.0294 and 0.0220, respectively. From the figure the change in form of the pressure distribution for cavitation numbers ranging from 4.0 down to 0.3 can be compared with the results from potential theory; the Reynolds number for these tests was within the range 3×10^6 to 4×10^6. The influence that these pressure distribution changes have on the lift coefficient can be deduced from Figure 5.81, which is also taken from Balhan and shows how the lift coefficient varies with cavitation number and incidence angle of the aerofoil. From this figure it is seen that at moderate to low incidence the effects are limited to the extreme low cavitation numbers, but as incidence increases to high values, 5° in propeller terms, this influence spreads across the cavitation number range significantly.

For propeller blade section design purposes the use of 'cavitation bucket diagrams' is valuable, since they capture in a two-dimensional sense the cavitation behaviour of a blade section. Figure 5.82(a) outlines the basic features of a cavitation bucket diagram. This diagram is plotted as a function of the section angle of attack against the section cavitation number, however, several versions of the diagrams have been produced: typically, angle of attack may be replaced by lift coefficient and cavitation number by minimum pressure coefficient. From the diagram, no matter what its basis, four primary areas are identified: the cavitation-free area and the areas where back sheet, bubble and face cavitation can be expected. Such diagrams are produced from systemic calculations on

Powering 275

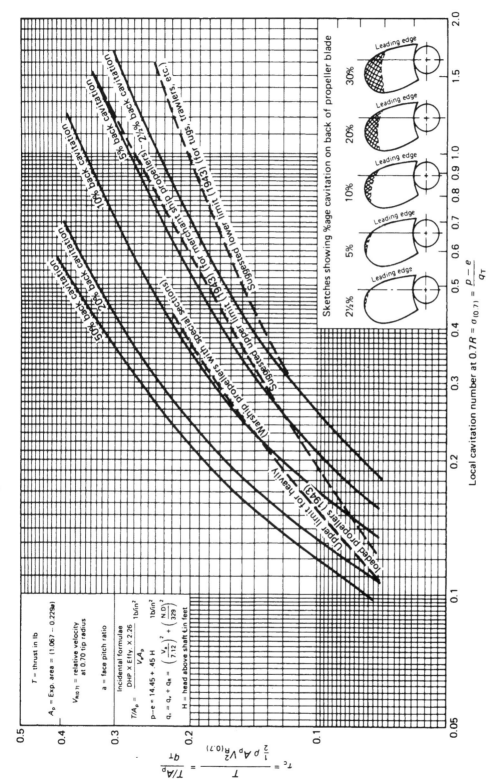

Figure 5.78 Burrill cavitation diagram for uniform flow (Reproduced from Burrill (1978)).

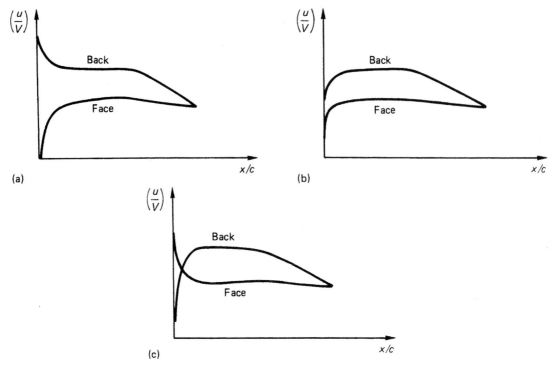

Figure 5.79 Typical section velocity distributions: (a) positive incidence; (b) ideal incidence and (c) negative incidence.

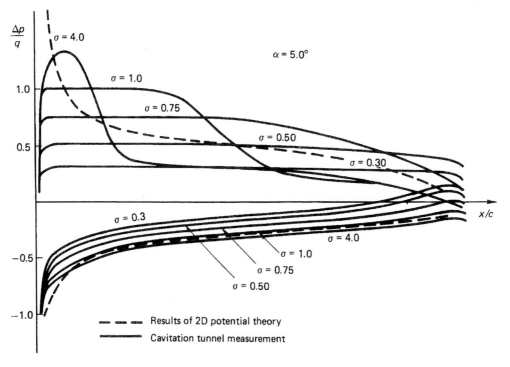

Figure 5.80 The effect of cavitation on an aerofoil section pressure distribution (Balhan (1951)).

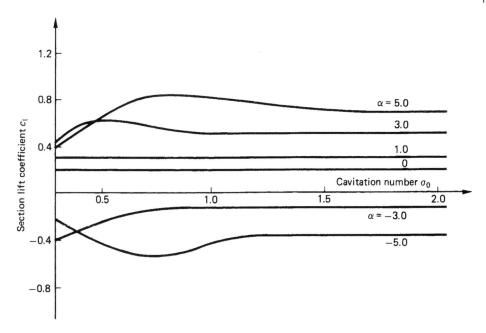

Figure 5.81 The effect of cavitation on the section lift coefficient (Balhan (1951)).

a parent section form and several cases are supported by experimental measurement (see e.g. Shen (1985)). The width of the bucket defined by the parameter α_d is a measure of the tolerance of the section to cavitation-free operation. Figure 5.82(b) shows an example of a cavitation bucket diagram based on experimental results using flat-faced sections. This work, conducted by Walchner and published in 1947, clearly shows the effect of the leading edge form on the section cavitation inception characteristics. Furthermore, the correlation with the theoretical limiting line can be seen for shockless entry conditions.

Whilst useful for design purposes the bucket diagram is based on two-dimensional flow characteristics, and can therefore give misleading results in areas of strong three-dimensional flow; for example, near the blade tip and root.

Propeller design is based on the mean inflow conditions that have either been measured at model scale or estimated empirically using procedures as discussed in Section 5.2. When the actual wake field is known, the cavitation analysis needs to be considered as the propeller passes around the propeller disc. This can be done either in a quasi-steady sense using procedures based on lifting line methods with lifting surface collections, or by means of unsteady lifting surface and boundary element methods. The choice of method depends in essence on the facilities available to the analyst and both approaches are commonly used. Figure 5.54 shows the results of a typical analysis carried out for a twin-screw vessel, Oossanen (1974).

Figure 5.54 also gives an appreciation of the variability that exists in cavitation extent and type on a typical propeller when operating in its design condition.

The calculation of the cavitation characteristics can be done either using the pseudo-two-dimensional aerofoil pressure distribution approach in association with cavitation criteria or using a cavitation modelling technique; the latter method is particularly important in translating propeller cavitation growth and decay into hull-induced pressures. The use of the section pressure distributions calculated from either a Theodoressen or Weber basis to determine the cavitation inception and extent has been traditionally carried out by equating the cavitation number to the section suction pressure contour as seen in Figure 5.83(a). Such analysis, however, does not take account of the time taken for a nucleus to grow from its size in the free stream to a visible cavity and also for its subsequent decline as well as the other factors discussed in Section 5.5.1. Although these parameters of growth and decline are far from fully understood, attempts have been made to derive engineering approximations for calculation purposes. Typical of these is that by van Oossanen (1974) in which the growth and decay is based on Knapp's similarity parameter (Knapp (1952)). In van Oossanen's approach (Figure 5.83(b)), at a given value of cavitation number σ the nuclei are expected to grow at a position x_{c1}/C on the aerofoil and reach a maximum size at x_{c2}/C, whence the cavity starts to decline in size until it vanishes at a position x_{c3}/C. Knapp's similarity parameter, which is based on Rayleigh's equation

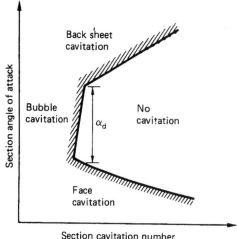

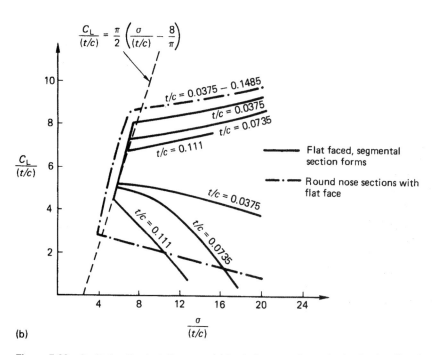

Figure 5.82 Cavitation 'bucket' diagrams: (a) basic features of a cavitation bucket diagram and (b) Walchner's foil experiments with flat-faced sections.

for bubble growth and collapse, defines a ratio K_n as follows:

$$K_n = \frac{t_D \sqrt{(\Delta p)_D}}{t_G \sqrt{(\Delta p)_G}} \quad (5.91)$$

where t and Δp are the total change times and effective liquid tension producing a change in size, respectively, and the suffixes D and G refer to decline and growth. Van Oossanen undertook a correlation exercise on the coefficient K_n for the National Advisory Committee for Aeronautics (NACA) 4412 profile, which resulted in a multiple regression-based formula for K_n as follows:

$$\log_{10} K_n = 9.407 - 84.88(\sigma/\sigma_i)^2 + 75.99(\sigma/\sigma_i)^3 \\ - \frac{0.5607}{(\sigma/\sigma_i)} + \log_{10}(\theta_{inc}/c) \\ \times \left[1.671 + 4.565(\sigma/\sigma_i) - 32(\sigma/\sigma_i)^2 \right. \\ \left. + 25.87(\sigma/\sigma_i)^3 - \frac{0.1384}{(\sigma/\sigma_i)} \right] \quad (5.92)$$

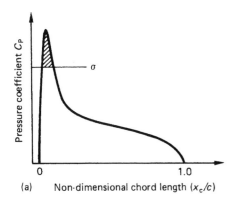

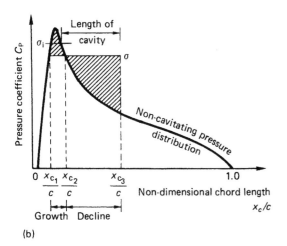

Figure 5.83 Determination of cavitation extent: (a) traditional approach to cavitation inception and (b) van Oossanen's approach to cavitation inception.

in which θ_{inc} is the momentum thickness of the laminar boundary layer at the cavitation inception location. For calculation purposes it is suggested that if the ratio (θ_{inc}/c) is greater than 0.0003 bubble cavitation occurs and for smaller values sheet cavitation results. As a consequence of Equation (5.92) it becomes possible to solve Equation (5.91) iteratively in order to determine the value of x_{c3} since Equation (5.91) can be rewritten as

$$K_n = \frac{\int_{x_{c2}/c}^{x_{c3}/c} \frac{\mathrm{d}(x_c/c)}{V_{x_c}(x_c/c)} \sqrt{\int_{x_{c2}/c}^{x_{c3}/c} [\sigma + C_P(x_c/c)]\mathrm{d}(x_c/c)}}{\int_{x_{c1}/c}^{x_{c2}/c} \frac{\mathrm{d}(x_c/c)}{V_{x_c}(x_c/c)} \sqrt{\int_{x_{c1}/c}^{x_{c2}/c} -[\sigma + C_P(x_c/c)]\mathrm{d}(x_c/c)}}$$

where V_{x_c} is the local velocity at x_c.

The starting point x_{c1} for the cavity can, for high Reynolds numbers in the range $1 \times 10^5 < R_{x_{\text{tr}}} < 6 \times 10^7$, be determined from the relationship derived by Cebeci (1972) as follows:

$$R_{\theta_{\text{tr}}} = 1.174 \left[1 + \frac{22\,400}{R_{x_{\text{tr}}}}\right] R_{x_{\text{tr}}}^{0.46} \quad (5.93)$$

where $R_{\theta_{\text{tr}}}$ is the Reynolds number based on momentum thickness and local velocity at the position of transition, and $R_{x_{\text{tr}}}$ is the Reynolds number based on free stream velocity and the distance of the point of transition from the leading edge. For values of $R_{x_{\text{tr}}}$ below this range, the relationship

$$R_{\theta_{\text{ci}}} = 4.048 R_{x_{\text{ci}}}^{0.368} \quad (5.94)$$

holds in the region $1 \times 10^4 < R_{x_{\text{ci}}} < 7 \times 10^5$. In this case $R_{\theta_{\text{ci}}}$ is the Reynolds number based on local velocity and momentum thickness at the point of cavitation inception, and $R_{x_{\text{ci}}}$ is the Reynolds number based on distance along the surface from the leading edge and free stream velocity at the position of cavitation inception.

Having determined the length of the cavity, van Oossanen extended this approach to try and approximate the form of the pressure distribution on a cavitating section, and for these purposes assumed that the cavity length is less than half the chord length of the section. From work on the pressure distribution over cavitating sections it is known that the flat part of the pressure distribution, Figure 5.80 for example, corresponds to the location of the actual cavity. Outside this region, together with a suitable transition zone, the pressure returns approximately to that of a non-cavitating flow over the aerofoil. Van Oossanen conjectured that the length of the transition zone is approximately equal to the length of the cavity and the resulting pressure distribution approximation is shown in Figure 5.84.

Ligtelijn and Kuiper (1983) conducted a study to investigate the importance of the higher harmonics in the wake distribution on the type and extent of cavitation, and as a consequence give guidance on how accurately the wake should be modelled. Their study compared the results of lifting surface calculations with the results of model tests in a cavitation tunnel where the main feature of the wake field was a sharp wake peak. It was concluded that the lower harmonics of the wake field principally influence the cavity length prediction and that the difference between two separate calculations based on four and ten harmonics was negligible.

Considerable work has been done in attempting to model cavitation mathematically. The problem is essentially a free streamline problem, since there is a flow boundary whose location requires determination as an integral part of the solution. Helmholtz and

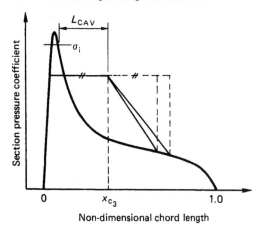

Figure 5.84 Van Oossanen's approximate construction of a cavitation pressure distribution on an aerofoil section.

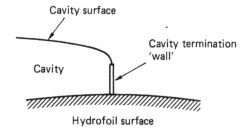

Figure 5.85 Riabouchinsky-type cavity termination 'wall'.

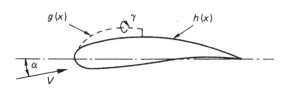

Figure 5.86 Uhlman's non-linear model of a two-dimensional partially cavitating flow.

Kirchoff in the latter part of the nineteenth century attempted a solution of the flow past a supercavitating flat plate at zero cavitation number using complex variable theory. Subsequently, Levi-Civita extended this work to include the flow past curved bodies. The zero cavitation number essentially implied an infinite cavity, and the next step in the solution process was to introduce finite cavitation numbers which, as a consequence, introduce finite sized cavities. The finite cavity, however, requires the cavity to be terminated in an acceptable mathematical and physical manner. Several models have been proposed, amongst which the Riabouchinsky cavity termination model, which employs a 'wall' to provide closure of the cavity (Figure 5.85), and the more physically realistic re-entrant jet model (Figure 5.68), are examples. These models, most of which were developed in the late 1940s, are non-linear models which satisfy the precise kinematic and dynamic boundary conditions over the cavity surface. As a consequence considerable analytical complexity is met in their use. Tulin (1953) developed a linearized theory for zero cavitation number and this was extensively applied and extended such that Geurst (1959) and Geurst and Verbrugh (1959) introduced the linearized theory for partially cavitating hydrofoils operating at finite cavitation numbers, and extended this work with a corresponding theory for supercavitating hydrofoils, Geurst (1960).

Three-dimensional aspects of the problem were considered by Leehey (1971) who proposed a theory for supercavitating hydrofoils of finite span. This procedure was analogous to the earlier work of Geurst on two-dimensional cavitation problems in that it uses the method of matched asymptotic expansions from which a comparison can be made with the earlier work. Uhlman (1978), using a similar procedure, developed a method of analysis for partially cavitating hydrofoils of finite span. With the advent of large computational facilities significantly more complex solutions could be attempted. Typical of these is the work of Jiang (1977) who examined the three-dimensional problem using an unsteady numerical lifting surface theory for supercavitating hydrofoils of finite span using a vortex source lattice technique.

Much of the recent work is based on analytical models which incorporate some form of linearizing assumptions. However, techniques now exist, such as boundary integral or surface singularity methods, which permit the solution of a Neumann, Dirichlet or mixed boundary conditions to be expressed as an integral of appropriate singularities distributed over the boundary of the flow field. Uhlman (1983), taking advantage of these facilities, has presented an exact non-linear numerical model for the partially cavitating flow about a two-dimensional hydrofoil (Figure 5.86). His approach uses a surface vorticity technique in conjunction with an iterative procedure to generate the cavity shape and a modified Riabouchinsky cavity termination wall to close the cavity. Comparison with Tulin and Hsu's earlier thin cavity theory (Tulin and Hsu (1977)) shows some significant deviations between the calculated results of the non-linear and linear approaches to the problem.

Stern and Vorus (1983) developed a non-linear method for predicting unsteady sheet cavitation on propeller blades by using a method which separates the velocity potential boundary value problem into a static and dynamic part. A sequential solution technique is adopted in which the static potential problem relates to the cavity fixed instantaneously relative to the blade whilst the dynamic potential solution addresses the instantaneous reaction of the cavity to the static potential and predicts the cavity deformation and motion relative to the blades. In this approach, because the non-linear character of the unsteady cavitation is preserved, the predictions from the method contain many of the observed characteristics of both steady and unsteady cavitation behaviour. Based on this work two modes of cavity collapse were identified, one being a high-frequency mode where the cavity collapsed towards the trailing edge whilst the second was a low-frequency mode where the collapse was towards the leading edge.

Isay (1981), in association with earlier work by Chao, produced a simplified bubble grid model in order to account for the compressibility of the fluid, surrounding a single bubble. From this work the Rayleigh–Plesset Equation (5.83) was corrected to take account of the compressibility effects of the ambient fluid as follows:

$$\frac{d^2 R}{dt^2} + \frac{3}{2R}\left(\frac{dR}{dt}\right)^2 = \frac{1}{\rho R}\left[p_G - \frac{2S}{R} - p_\infty e^{-\alpha/\alpha_{**}} + p_v e^{-\alpha/\alpha_{**}}\right] \quad (5.95)$$

where p_v and p_∞ are vapour pressure and local pressure in the absence of bubbles, α is the local gas volume ratio during bubble growth, α_{**} is an empirical parameter and S is the surface tension. Furthermore, Isay showed that bubbles growing in an unstable regime reach the same diameter in a time-dependent pressure field after a short distance. This allows an expression to be derived for the bubble radius just prior to collapse. Mills (1991) extended the above theory, which was based on homogeneous flow, to inhomogeneous flow conditions met within propeller technology and where local pressure is a function of time and position on the blade. Following this theoretical approach Equation (5.83) then becomes

$$\frac{3\omega^2}{2}\left(\frac{\partial R_{\phi 0}}{\partial \chi}\right)^2 + R_{\phi 0}\omega^2\left(\frac{\partial^2 R_{\phi 0}}{\partial \chi^2}\right) = \frac{1}{\rho}[p(R_{\phi 0}) - p(\chi, \phi_0 - \chi)] \quad (5.96)$$

from which computation for each class of bubble radii can be undertaken.

In Equation (5.96) χ is the chordwise coordinate, ω is the rate of revolution and ϕ_0 is the instantaneous blade position.

For computation purposes the gas volume $\alpha_{\phi 0}$ at a position on the section channel can be derived from

$$\frac{\alpha_{\phi 0}(\chi)}{1 + \alpha_{\phi 0}(\chi)} = \frac{4\pi}{3}\sum_{j=1}^{J}\xi_{0j}\cdot R_{0j}^3(x, \phi_0 - \chi) \quad (5.97)$$

in which ξ_{0j} is the bubble density for each class and R_0 is the initial bubble size. Using Equations (5.96) and (5.97) in association with a blade undisturbed pressure distribution calculation procedure (Section 5.4), the cavity extent can be estimated over the blade surface.

Vaz and Bosschers (2006) have adapted a partially non-linear model in which the kinematic and dynamic boundary conditions are applied in the partially cavitating flow case on the surface of the blade below the cavity surface. In contrast for the supercavitating case the conditions are applied at the cavity surface. While the method can, in the case of the two-dimensional partial cavitation, be improved by a Taylor expansion of the velocities on the cavity surface based on the cavity thickness, this has yet to be implemented in their present three-dimensional formulation of the problem. Nevertheless, their analysis cavitation modelling procedure when applied to the prediction of sheet cavitation in steady flow for the MARIN S propeller, design such that sheet cavitation is only present, has shown good correlation with the experimental observations although this correlation appeared to be load dependent: showing underprediction at lower loadings. In an alternative correlation exercise with the INSEAN E779A propeller which has a higher tip loading leading to both partial and supercavitation in the outer regions of the blade in addition to a cavitating tip vortex the blade cavitation extent was reasonably well predicted. This method is being extended to the prediction of unsteady cavitation prediction.

In an alternative approach to propeller sheet cavitation prediction, Sun and Kinnas (2006) have used a viscous–inviscid interactive method of analysis. In their approach the inviscid wetted and cavitating flows are analysed using a low-order potential boundary element analysis based on a thin cavity modelling approach. Then by making the assumption of a two-dimensional boundary layer acting in strips along the blade, the effects of viscosity on the wetted and cavitating flows are taken into account by coupling the inviscid model with a two-dimensional integral boundary layer analysis procedure. Comparison of the results from this procedure with the first iteration of a fully non-linear cavity approach Kinnas et al. (1994) and Brewer and Kinnas (1997), which itself had been validated from a FLUENT Computational Fluid Dynamics (CFD) modelling, has shown good correlation with the differences being negligible when the cavities are thin.

The importance of CFD in cavitation prediction has been increasing. The current multi-phase flow capabilities of some of the more advanced Reynolds Averaged Navier–Strokes (RANS) solvers are being found to be helpful in gaining insights into the cavitation performance of marine propellers. Moreover, the current pace of development of this branch of the technology is permitting more quantitative evaluations to be undertaken with confidence. For example, Bulten and Oprea (2006) consider the use of CFD methods for propeller tip cavitation inception. They have found that provided that local mesh refinement is utilized in the tip region which enables a detailed analysis of the flow in the tip vortex, then at model-scale cavitation inception can be predicted reasonably in the case of the DTRC 4119 propeller. Furthermore they have extended their studies to the consideration of McCormick's scaling law for cavitation inception but suggest that further work is necessary before definitive conclusions can be drawn on the correlation. In the corresponding case of rudder cavitation prediction the multi-phase capabilities of the more advanced commercial packages have been shown to give good agreement with observation given that the inflow from the propeller is modelled with some accuracy. Carlton (2007) goes on to discuss cavitation inception, cavitation induced damage, (e.g. Moulijn et al. (2006)), cavitation tests on propellers, cavitation pressure data and cavitation on rudders (e.g. Carlton et al. (2006)).

5.6 Propeller design

Other Sections in this Chapter have considered different aspects of the propeller. This Section attempts to provide a basis for drawing together the various threads of the subject, so that the propeller and its design process can be considered as an integrated entity. The completed propeller depends for its success on the satisfactory integration of several scientific disciplines. These disciplines are principally hydrodynamics, stress analysis, metallurgy and manufacturing technology with supportive inputs from mathematics, dynamics and thermodynamics. It is not uncommon to find that several of the requirements of the principal disciplines for a particular design are in partial conflict, to a greater or lesser extent, in their aim to satisfy a particular set of requirements and constraints. The test of the designer is in how satisfactorily he resolves these conflicts to produce an optimal design: optimal in the sense of satisfying the various constraints. It may therefore be inferred that in propeller technology, as in all other forms of engineering design, there is no single correct solution to a particular propulsion problem.

5.6.1 The design and analysis loop

The phases of the design process, given that there is a requirement for a particular ship to be propelled, can be summarized in the somewhat abstract terms of design textbooks as shown in Figure 5.87. From the figure, it is seen that the creation of the artefact commences with the definition of the problem, which requires the complete specification for the design. This specification must include a complete definition of the inputs and the required outputs together with the limitations on these quantities and the constraints on the design. Following the design definition phase, the process passes on to the synthesis phase, in which the basic design is formulated from the various 'building blocks' that the designer has at his disposal. In order to provide the optimal solution, the synthesis phase cannot exist in isolation and has to be conducted with the analysis and optimization phase in an interactive loop in order to refine the design to that required – that is, a design that complies with the original specification and also has the optimal property. The design loop must be flexible enough, should an unresolvable conflict arise with the original definition of the problem, to allow an appeal for change to the definition of the design problem from

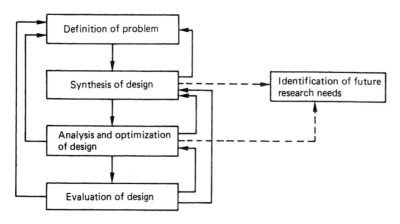

Figure 5.87 Phases of engineering design.

either of the synthesis or analysis and optimization phases. Indeed it is also likely that either of these phases of design will lead to the identification of areas for longer-term research to aid future design problems. As was noted, design is an interactive process in which one passes through several steps, evaluates the results, and then returns to an earlier phase of the procedure. Consequently, we may synthesize many components of the design, analyse and optimize them, and return to the synthesis to see what effect this has on the remaining parts of the system. Analysis may also include model testing in either a towing tank or cavitation tunnel. When the design loop of synthesis, analysis and optimization is complete, the process then passes on the evaluation phase. This phase is the final proof of the design, from which its success is determined, since it usually involves the testing of a prototype. In propeller design, the luxury of a prototype is rare, since the propeller is a unit volume item under normal circumstances. Hence, the evaluation stage is normally the sea trial phase of the ship-building programme. Nevertheless, when design does not perform as expected, then it is normal, as in the generalized design process, to return to an earlier phase to explore the reasons for failure and propose modification.

Theses general design ideas, although abstract, are nevertheless useful and directly applicable to the propeller design process. How then are they applied? In the first instance it must be remembered that in general a propeller can only be designed for a single design point which involves a unique specification of a power, rotational speed, ship speed and a mean radial wake field. The controllable pitch propeller is the partial exception to this rule when it would be normal to consider two or more design points. Although there is a unique design point in general the propeller operates in a variable circumferential wake field and may be required to work at off-design points: in some instances the sea trial condition is an off-design point. Therefore, there is in addition to the synthesis phase of Figure 5.87 also an analysis and optimization phase, as also shown in the figure.

In the case of propeller design, the conceptual design approach shown in Figure 5.87 can be translated in the following way. The definition of the problem is principally the specification of the design point, or points in the case of controllable pitch propellers, for the propeller together with the constraints which are applicable to that particular design or the vessel to which it is to be fitted. In both of these activities the resulting specification should be a jointly agreed document into which the owner, shipbuilder, engine builder and propeller designer have contributed: to do otherwise can lead to a grossly inadequate or unreasonable specification being developed. Following the production of the design specification the synthesis of the design can commence. This design will be for the propeller type agreed during the specification stage, because it is very likely that some preliminary propeller design studies will have been conducted during the design specification phase. At that time propeller type, blade number and so on are most likely to have been chosen. As a consequence, during the synthesis phase the basic design concept will be worked up into a detailed design proposal typically using, for advanced designs, a wake adapted lifting line with lifting surface correction capability. The choice of method, however, will depend on the designer's own capability and the data available, and may, for small vessels, be an adaptation of a standard series propeller which may work in a perfectly satisfactory manner from the cost-effectiveness point of view.

The design that results from the synthesis phase, assuming the former of the two synthesis approaches have been adopted, will then pass into the analysis and optimization phase. This phase may contain elements of both theoretical analysis and model testing. The theoretical analysis will vary, depending upon the designer's capabilities and the perceived cost benefit of this stage, from adaptations of Burrill's vortex analysis procedure through to unsteady lifting surface, vortex lattice or boundary element capabilities (Section 5.4.5). With regard to model testing in this phase, this can embrace a range of towing tank studies for resistance and propulsion purposes through to cavitational tunnel studies for determination of cavitation characteristics and noise prediction. The important lesson in propeller technology is to appreciate that each of the analysis techniques, theoretical and model testing, gives a partial answer, since although today our understanding of the various phenomena has progressed considerably from that of say twenty or thirty years ago, there are still many areas where our understanding is far from complete. As a consequence, the secret of undertaking a good analysis and optimization phase is not simply to take the results of the various analysis at face value, but to examine them in the light of previous experience and a knowledge of their various strengths and weaknesses to form a balanced view of the likely performance of the proposed propellers; this is only the essence of good engineering practice.

Figure 5.88 translates the more abstract concept of the phases of engineering design, shown in the previous figure, into a propeller related design concept in the light of the foregoing discussion.

5.6.2 Design constraints

The constraints on propeller design may take many forms: each places a restriction on the designer and in most cases if more than one constraint is placed then this places a restriction on the upper bound of

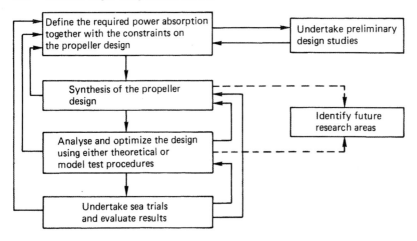

Figure 5.88 The phases of propeller design.

performance that can be achieved in any one area. For example, if a single constraint is imposed, requiring the most efficient propeller for a given rotational speed, then the designer will most likely choose the optimum propeller with the smallest blade area ratio, consistent with any blade cavitation erosion criteria, in order to maximize efficiency. If then a second constraint is imposed, requiring the radiated pressures on the hull surface not to exceed a certain value, then the designer will start to increase the blade chord lengths and adjust other design parameters in order to control the cavitation. Therefore, since the blade area is no longer minimized, this will cause a reduction in efficiency but enhance the hull pressure situation.

Although this is a somewhat simplified example, it adequately illustrates the point and as a consequence, it is important that all concerned with the ship design consider the various constraints in the full knowledge of their implications and the realization that the setting of unnecessary or over-strict constraints will most likely lead to a degradation in the propeller's overall performance.

5.6.3 Choice of propeller type

The choice of propeller type for a particular propulsion application can be a result of the consideration of any number of factors. These factors may, for example, be the pursuit of maximum efficiency, noise reduction, ease of manoeuvrability, cost of installation and so on. Each vessel and its application has to be considered on its application has to be considered on its own merits taking into account the items listed in Table 5.14.

In terms of optimum open water efficiency van Manen (1966) developed a comparison for a variety of propeller forms based on the results of systematic series data. In addition to the propeller data from experiments

Table 5.14 Factors affecting choice of propulsor.

Role of vessel
Special requirements
Initial installation costs
Running costs
Maintenance requirements
Service availability
Legislative requirements

at MARIN he also included data relating to fully cavitating and vertical axis propellers (Tachmindji and Morgan (1958) and van Manen (1963)) and the resulting comparison is shown in Figure 5.89. The figure shows the highest obtainable open water efficiency for the different types of propeller as a function of the power coefficient B_p. As can be seen from the legend at the top of the figure the lightly loaded propellers of fast ships lie to the left-hand side whilst the more heavily loaded propellers of the large tankers and bulk carriers and also the towing vessels lie to the right-hand side of the figure. Such a diagram is able to give a quick indication of the type of propeller that will give the best efficiency for a given type of ship. As is seen from the diagram the accelerating duct becomes a more attractive proposition at high values of B_p whereas the contra-rotating and conventional propellers are most efficient at the lower values of B_p.

In cases where cavitation is a dominant factor in the propeller design such as in high-speed craft, Tachmindji *et al.* (1957) developed a useful basic design diagram to determine the applicability of different propeller types with respect to cavitating conditions of these types of craft. This diagram is reproduced in Figure 5.90. From the figure it is seen that it comprises a series of regions which define the applicability of different types of propeller. In the top

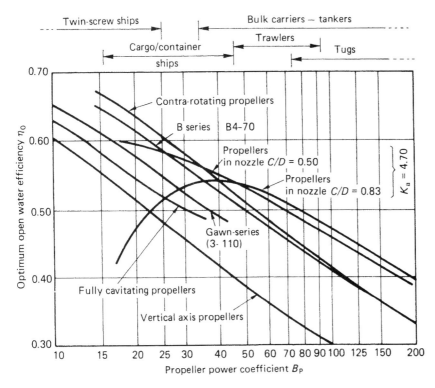

Figure 5.89 Typical optimum open water efficiencies for different propeller types (Reproduced with permission from van Manen (1966)).

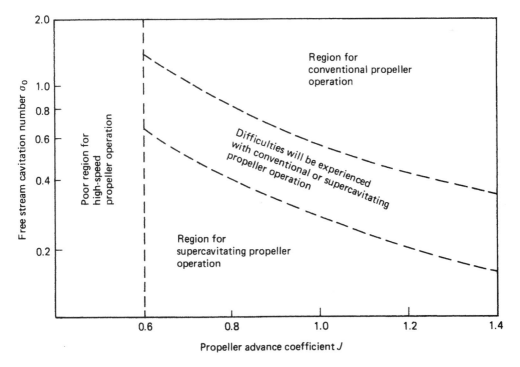

Figure 5.90 The effect of cavitation number on propeller type for high-speed propellers.

Table 5.15 Change in the propeller defect incidence with time for propellers in the range 5000 < BHP < 10 000(1960–1989).

	1960–64	1965–69	1970–74	1975–79	1980–84	1985–89
Fixed pitch propeller	0.018	0.044	0.067	0.066	0.065	0.044
Controllable pitch propellers	0.080	0.161	0.128	0.157	0.106	0.079

Defects recorded in defect incidence per year per unit.

right-hand region are to be found the conventional propellers fitted to most merchant vessels, whilst in the bottom right-hand region are the conditions where supercavitating propellers will give the best efficiencies. Propellers that fall towards the left-hand side of the diagram are seen to give low efficiency for any type of propeller and since low advance coefficient implies high B_p the correspondence between these Figures 5.90 and 5.89 can be seen by comparison.

The choice between fixed pitch propellers and controllable pitch propellers has been a long contested debate between the proponents of the various systems. In Chapter 6, Section 6.2 it is shown that the controllable pitch propellers have gained a significant share of the Ro/Ro, ferry, fishing, offshore and tug markets with vessels of over 2000 BHP. This is clearly because there is a demand for either high levels of manoeuvrability or a duality of operation that can best be satisfied with a controllable pitch propeller rather than a two-speed reduction gearbox for these types of vessel. For the classes of vessel which do not have these specialized requirements, then the simpler fixed pitch propeller appears to provide a satisfactory propulsion solution. With regard to reliability of operation, as might be expected the controllable pitch propeller has a higher failure rate due to its increased mechanical complexity. Table 5.15 details the failure rates for both fixed pitch propellers and controllable pitch propellers over a period of about a quarter of a century (Carlton (1989). In either case, however, it is seen that the propeller has achieved the status of being a very reliable marine component.

The controllable pitch propeller does have the advantage of permitting constant shaft speed operation of the propeller. Although this generally establishes a more onerous set of cavitation conditions, it does readily allow the use of shaft-driven generators should the economics of the ship operation dictate that this is advantageous. In addition, in this present age of environmental concern, there is some evidence to suggest (Carlton (1990)) that the NO_x exhaust emissions can be reduced on a volumetric basis at intermediate engine powers when working at constant shaft speed. Figure 5.91 shows this trend from which a reduction in the

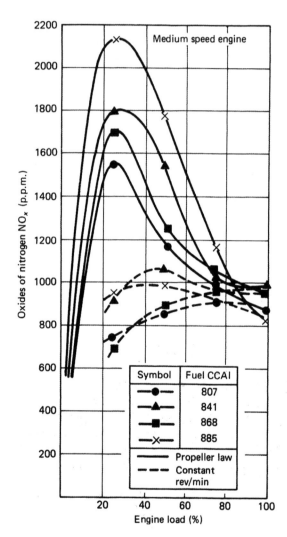

Figure 5.91 Influence of engine operating conditions and fuel CCAI number.

NO emissions, these forming about 90 per cent of the total NO_x component, can be seen at constant speed operation for a range of fuel qualities. Such data, however, needs to be interpreted in the context of mass emission for particular ship applications.

In cases where manoeuvrability or directional control is important, the controllable pitch propeller, steerable duct, azimuthing propeller and the cycloidal propeller can offer various solutions to the problem, depending on the specific requirements.

By way of summary, Table 5.16 lists some of the important features of the principal propeller types.

5.6.4 The propeller design basis

The term 'propeller design basis' refers to the power, rotational speed and ship speed that are chosen to act as the basis for the design of the principal propeller geometric features. This is an extremely important matter even for the controllable pitch propellers, since in this latter case the design helical sections will only be absolutely correct for one pitch setting. This discussion, however, will largely concentrate on the fixed pitch propeller since for this type of propeller the correct choice of the design basis is absolutely critical to the performance of ship.

The selection of the design basis starts with a consideration of the mission profile for the vessel. Each vessel has a characteristic mission profile which is determined by the owner to meet the commercial needs of the particular service under the economic conditions prevailing. It must be recognized that the mission profile of a particular ship may change throughout its life, depending on a variety of circumstances. When this occurs it may then be necessary to change the propeller design, as witnessed by the slow steaming of the large tankers after the oil crisis of the early 1970s and the consequent change of propellers by many owners in order to enhance the ship's efficiency at the new operating conditions. The mission profile is determined by several factors, but is governed chiefly by the vessel type and its intended trade pattern; Figure 5.92 shows three examples relating to a container ship, a Ro/Ro ferry and a warship. The wide divergence in the form of these curves amply illustrates that the design basis for a particular vessel must be chosen with care such that the propeller will give the best overall performance in the areas of operation required. This may well require several preliminary design studies in order to establish the best combination of diameter, pitch ratio and blade area to satisfy the operational constraints of the ships.

In addition to satisfying the mission profile requirements it is also necessary that the propeller and engine characteristics match, not only when the vessel is new but also after the vessel has been in service for some years. Since the diesel engine at the present time is used for the greater majority of propulsion pants, we will use this as the primary basis for the discussion. The diesel engine has a general characteristic of the type shown in Figure 5.93 with a propeller demand curve superimposed on it which is shown in this instance to pass through the Maximum Continuous Rating (MCR) of the engine. It should not, however, be assumed that in

Table 5.16 Some important characteristics of propeller types.

Propeller type	Characteristics
Fixed pitch propellers	Ease of manufacture Design for a single condition (i.e. design point) Blade root dictates boss length No restriction on blade area or shape Rotational speed varies with power absorbed Relatively small hub size
Controllable pitch propellers	Can accommodate multiple operating conditions Constant or variable shaft speed operation Restriction on blade area to maintain blade reversibility Blade root is restricted by palm dimensions Increased mechanical complexity Larger hub size, governed by spindle torque requirements
Ducted propellers	Can accommodate fixed and controllable pitch propellers Duct form should be simple to facilitate manufacture Enhanced thrust at low ship speed Duct form can be either accelerating or decelerating Accelerating ducts tend to distribute thrust equally between duct and propeller at bollard pull Ducts can be made steerable
Azimuthing units	Good directional control of thrust Increased mechanical complexity Can employ either ducted or non-ducted propellers of either fixed or controllable pitch type
Cycloidal propellers	Good directional control of thrust Avoids need for rudder on vessel Increased mechanical complexity
Contra-rotating propellers	Provides ability to cancel torque reaction Enhanced propulsive efficiency in appropriate conditions Increased mechanical complexity Can be used with fixed shaft lines or azimuthing units

the general case the propeller demand curve must pass through the MCR point of the engine. The propeller demand curve is frequently represented by the so-called 'propeller law', which is a cubic curve. This, however, is an approximation, since the propeller demand is dependent on all of the various hull resistance and

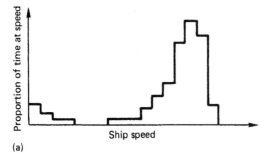

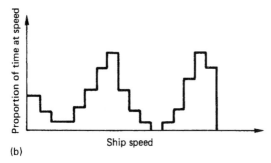

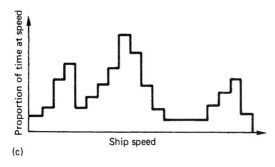

Figure 5.92 Examples of ship mission profiles: (a) container ship; (b) Ro/Ro passenger ferry and (c) warship.

propulsion components, and therefore has a more complex functional relationship. See also Chapter 6, Section 6.3, regarding diesel engine performance. In practice, however, the cubic approximation is generally valid over limited power ranges. If the pitch of the propeller has been selected incorrectly, then the propeller will be either over-pitched (stiff), curve A, or under-pitched (easy), curve B. In either case, the maximum power of the engine will not be realized, since in the case of over-pitching the maximum power attainable will be X at a reduced rpm, this being governed by the engine torque limit. In the alternative under-pitching case, the maximum power attainable will be Y at 100 per cent rpm, since the engine speed limit will be the governing factor.

In addition to purely geometric propeller features, a number of other factors influence the power absorption characteristics. Typical of these factors are sea conditions, wind strength, hull condition in terms of roughness and fouling, and, of course, displacement. It is generally true that increased severity of any of these conditions requires an increase in power to drive the ship at the same speed. This has the effect of moving the power demand curve of the propeller (Figure 5.93) to the left in the direction of curve A. As a consequence, if the propeller is designed to operate at the MCR condition when the ship is clean and in a light displacement with favourable weather, such as might be found on a trial condition, then the ship will not be able to develop full power in subsequent service when the draughts are deeper and the hull fouls or when the weather deteriorates. Under these conditions the engine torque limit will restrict the brake horsepower developed by the engine.

Clearly this is not a desirable situation and a method of overcoming this needs to be sought. This is most commonly achieved by designing the propeller to operate at a few revolutions fast when the vessel is new, so that by mid-docking cycle the revolutions will have fallen to the desired value. In addition, when significant changes of draught occur between the trial and the operating conditions, appropriate allowances need to be made for this effect. Figure 5.94 illustrates one such scenario, in which the propeller has been selected so that in the most favourable circumstances, such as the trial condition, the engine is effectively working at a derated condition and hence the ship will not attain its maximum speed this is because the engine will reach its maximum speed before reaching its maximum power. As a consequence in poorer weather or when the vessel fouls or works at a deeper draught, the propeller characteristic moves to the left so that the maximum power becomes available. Should it be required on trials to demonstrate the vessel's full-speed capability, then engine manufacturers often allow an overspeed margin with a restriction on the time the engine can operate at this condition. This concept of the difference in performance of the vessel on trial and in service introduces the term of a 'sea margin', which is imposed by the prudent owner in order to ensure the vessel has sufficient power available in service and throughout the docking cycle.

In practice the propeller designer will use a derated engine power as the basis for the propeller design. This is to prevent excessive maintenance costs in keeping the engine at peak performance throughout its life. Hence the propeller is normally based on a Normal Continuous Rating (NCR) of between 85 per cent and 90 per cent of the MCR conditions; Figure 5.95 shows a typical propeller design point for a vessel working with a shaft generator. For this ship an NCR of 85 per cent of the MCR was chosen and the power of the shaft generator P_G deducted from the NCR.

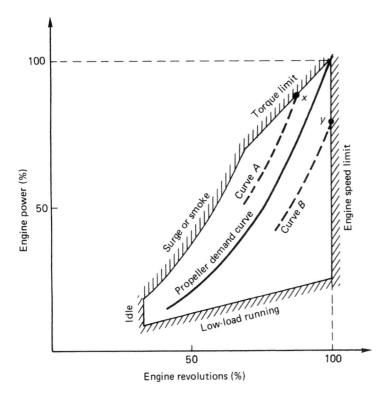

Figure 5.93 Engine characteristic curve.

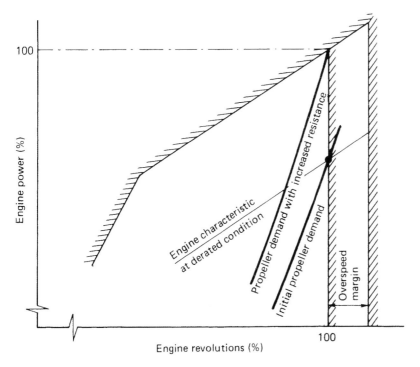

Figure 5.94 Change in propeller demand due to weather, draught changes and fouling.

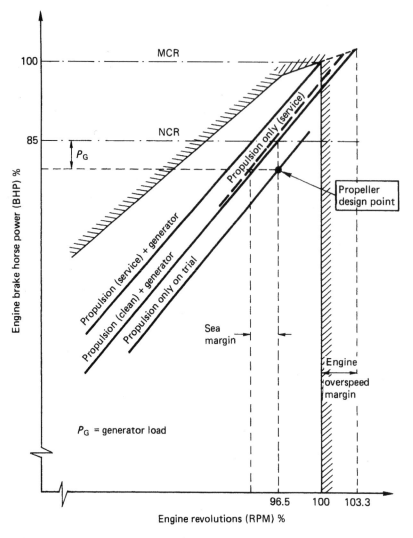

Figure 5.95 Typical propeller design point.

This formed the propeller design power and the rotational speed for the propeller design power. The rotational speed for the propeller design was then fixed such that the power absorbed by the propeller in service, together with the generator power when in operation, could absorb the MCR of the engine at 100 per cent rpm. This was done by deducting the power required by the generator from the combined service propeller and generator demand curve to arrive at the service propulsion only curve and then applying the sea margin which enables the propeller to run fast on trial. In this way the design power and revolutions basis became fixed.

In the particular case of a propeller intended for a towing duty, the superimposition of the propeller and engine characteristics presents an extreme example of the relationship between curve A and the propeller demand curve shown in Figure 5.93. In this case, however, curve A is moved far to the left because of the added resistance to the vessel caused by the tow. Such situations normally require correction by the use of a two-speed gearbox in the case of the fixed pitch propeller, or by the use of a controllable pitch propeller.

The controllable pitch propeller presents an interesting extension to the fixed pitch performance maps shown in Figures 5.93 to 5.95. A typical example is shown in Figure 5.96, in which the controllable pitch propeller characteristic is superimposed on an engine characteristic. The propeller demand curve

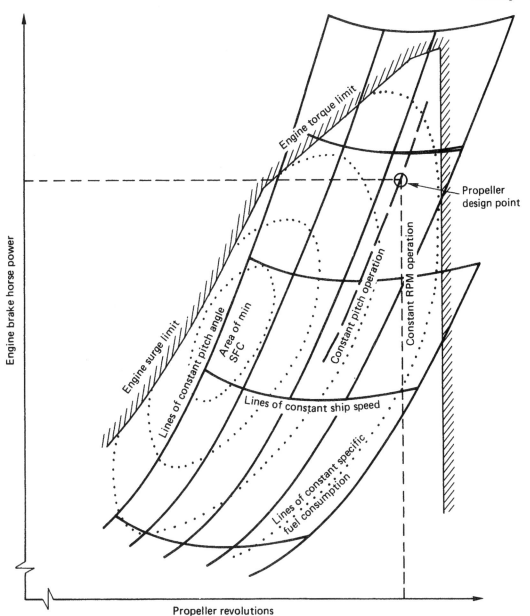

Figure 5.96 Controllable pitch propeller characteristic curve superimposed on a typical engine mapping.

through the design point clearly does not pass through the minimum specific fuel consumption region of the engine maps: this is much the same as for the fixed pitch propeller. However, with the controllable pitch propeller it is possible to adjust the pitch at partial load condition to move towards this region. In doing so it can be seen that the propeller mapping may come very close to the engine surge limit which is not a desirable feature. Nevertheless, the controllable pitch propeller pitch–rpm relationship, frequently termed the 'combinator diagram' can be programmed to give an optimal overall efficiency for the vessel.

In general, in any shaft line three power definitions are assumed to exist, these being the brake horsepower, the shaft horsepower and the delivered horsepower. The following definitions generally apply:

Brake power (P_B)   The power delivered at the engine coupling or flywheel.

Shaft power (P_S)   The power available at the output coupling of the gearbox, if fitted. If no gearbox is fitted then $P_S = P_B$. Also, if a shaft-driven generator is fitted on the line shaft, then two shaft powers exist; one before the generator P_{SI} and the one aft of the generator $P_{SA} = P_{SI} - P_G$. In this latter case some bearing losses may also be taken into account.

Delivered power (P_D)   The power available at the propeller after the bearing losses have been deducted.

In design terms, where no shaft generators exist to absorb power it is normally assumed that P_D is between 98 and 99 per cent of the value of P_S depending on the length of the line shafting and the number of bearings. When a gearbox is installed, then P_S usually lies between 96 and 98 per cent of the value of P_B, depending on the gearbox type.

5.6.5 Use of standard series data in design

Standard series data is one of the most valuable tools the designer has at his disposal for preliminary design and feasibility study purposes. Design charts, or in many cases today regression formulae, based on standard series data can be used to explore the principal dimensions of a propeller and their effect on performance and cavitation prior to the employment of more detailed design or analysis techniques. In many cases, however, propellers are designed solely on the basis of standard series data, the only modification being to the section thickness distribution for strength purposes. This practice not only commonly occurs for small propellers but is also seen to a limited extent on the larger merchant propellers.

When using design charts, however, the user should be careful of the unfairness that exists between some of the early charts, and therefore should always, where possible, use a cross-plotting technique with these earlier charts between the charts for different blade area ratios. These unfairnesses arose in earlier times when scale effects were less well understood than they are today, and in several of the series this has now been eradicated by recalculating the measured results to a common Reynolds number base.

Some examples of the use of standard series data are given below. In each of these cases, which are aimed to illustrate the use of the various design charts, the hand calculation procedure has been adopted. This is quite deliberate, since if the basis of the procedure is understood, then the computer-based calculations will be more readily accepted and be able to be critically reviewed. The examples shown are clearly not exhaustive, but serve to demonstrate the underlying use of standard series data.

5.6.5.1 Determination of diameter

To determine the propeller diameter D for a propeller when absorbing a certain delivered power P_D and a rotational speed N and in association with a ship speed V_s, it is first necessary to determine a mean design Taylor wake fraction (w_T) from either experience, published data or model test results. From this the mean speed of advance V_a can be determined as $V_a = (1 - w_T)V_s$. This then enables the power coefficient B_p to be determined as follows:

$$B_p = \frac{P_D^{1/2} N}{V_a^{2.5}}$$

which is then entered into the appropriate design chart as seen in Figure 5.97(a). The value of δ_{opt} is then read off from the appropriate 'constant δ line' at the point of intersection of this line and the maximum efficiency line for required B_p value. From this the optimum diameter D_{opt} can be calculated from the equation

$$D_{opt} = \frac{\delta_{opt} V_a}{N} \qquad (5.98)$$

If undertaking this process manually this should be repeated for a range of blade area ratios in order to interpolate for the required blade area ratio in general optimum diameter will decrease for increasing blade area ratio, see insert to Figure 5.97(a).

Several designers have produced regression equations for calculating the optimum diameter. One such example produced by van Gunsteren and based on the Wageningen B series, is particularly useful and is given here as

$$\delta_{opt} = 100 \left[\frac{B_p^3}{(155.3 + 75.11 B_p^{0.5} + 36.76 B_p)} \right]^{0.2}$$

$$\times \left[0.9365 + \frac{1.49}{Z} - \left(\frac{2.101}{Z} - 0.1478 \right)^2 \right]$$

$$\times \frac{A_E}{A_O} \qquad (5.99)$$

where B_p is calculated in British units of British horsepower, rpm and knots; Z is the number of blades and A_E/A_O is the expanded area ratio.

Having calculated the optimum diameter in either of these ways, it then needs to be translated to a behind hull diameter D_b in order to establish the diameter for the propeller when working under the influence of the ship rather than in open water. Section 5.6.6 discusses this aspect of design.

5.6.5.2 Determination of mean pitch ratio

Assuming that the propeller B_p value together with its constituent quantities and the behind hull diameter D_b axe known, then to evaluate the mean pitch ratio of the standard series equivalent propeller is an easy matter. First the behind hull δ value is calculated as

$$\delta_b = \frac{ND_b}{V_a} \quad (5.100)$$

from which this value together with the power coefficient B_p is entered onto the $B_p - \delta$ chart, as shown in Figure 5.97(b). From this chart the equivalent pitch ratio (P/D) can be read off directly. As in the case of propeller diameter this process should be repeated for a range of blade area ratios in order to interpolate for the required blade area ratio. It will be found, however, that P/D is relatively insensitive to blade area ratio under normal circumstances.

In the case of the Wageningen B series all of the propellers have constant pitch with the exception of the four-blade series, where there is a reduction of pitch towards the root. In this latter case, the P/D value derived from the chart needs to be reduced by 1.5 per cent in order to arrive at the mean pitch.

5.6.5.3 Determination of open water efficiency

This is derived at the time of the mean pitch determination when the appropriate value of η_o can be read off from the appropriate constant efficiency curve corresponding to the value of B_p and δ_b derived from Equation (5.100).

5.6.5.4 Required propeller rpm to give required P_D or P_E

In this case, which is valuable in power absorption studies, a propeller would be defined in terms of its diameter, pitch ratio and blade area ratio and the problem is to define the rpm to give a particular delivered power P_D or, by implication, P_E. In addition it is necessary to specify the speed of advance V_a either

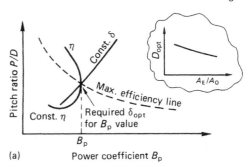

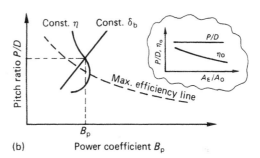

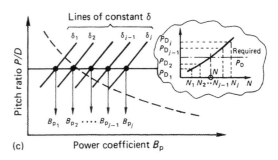

Figure 5.97 Examples of use of standard series data in the $B_p \sim \delta$ from: (a) diameter determination; (b) pitch ratio and open water efficiency determination and (c) power absorption analysis of a propeller.

as a known value or as an initial value to converge in an iterative loop.

The procedure is to form a series of rpm values, N_j (where $j = 1,\ldots, k$) from which a corresponding set of δ_j can be produced. Then by using the $B_p - \delta$ chart in association with the P/D values, a set of B_{pj} values can be produced, as seen in Figure 5.97(c). From these values the delivered powers P_{Dj} be calculated, corresponding to the initial set of N_j, and the required rpm can be deduced by interpolation to correspond to the particular value of P_D required. The value of P_D is, however, associated with the blade area ratio of the chart, and consequently this procedure needs to be repeated for a range of A_E/A_O values to allow the unique value of P_D to be determined for the actual A_E/A_O of the propeller.

By implication this can be extended to the production of the effective power to correlate with the initial value of V_a chosen. To accomplish this the open water efficiency needs to be read off at the same time as the range of B_{pj} values to form a set of η_{oj} values. Then the efficiency η_o can be calculated to correspond with the required value of P_D in order to calculate the effective power P_E as

$$P_E = \eta_o \eta_H \eta_r P_D$$

Figure 5.98 demonstrates the algorithm for this calculation, which is typical of many similar

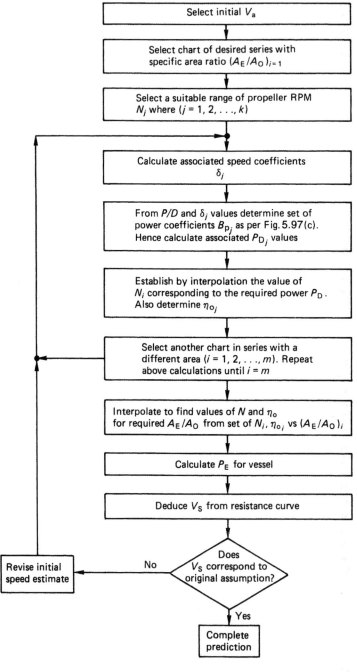

Figure 5.98 Calculation algorithm for power absorption calculations by hand calculation.

procedures that can be based on standard series analysis to solve particular problems.

5.6.5.5 Determination of propeller thrust at given conditions

The estimation of propeller thrust for a general free running condition is a trivial matter once the open water efficiency η_o has been determined from a $B_p - \delta$ diagram and the delivered power and speed of advance are known. In this case the thrust becomes

$$T = \frac{P_D \eta_o}{V_a} \quad (5.101)$$

However, at many operating conditions such as towing or the extreme example of zero ship speed, the determination of η_o is difficult or impossible since, when V_a is small, then $B_p \to \infty$ and, therefore, the $B_p - \delta$ chart cannot be used. In the case when $V_a = 0$ the open water efficiency η_o loses significance because it is the ratio of thrust power to the delivered power and the thrust power is zero because $V_a = 0$; in addition, Equation (5.101) is meaningless since V_a is zero. As a consequence, a new method has to be sought.

Use can be made either of the standard $K_T - J$ propeller characteristics or alternatively of the $\mu - \sigma$ diagram. In the case of the $K_T - J$ curve, if the pitch ratio and the rpm and V_a are known, then the advance coefficient J can be determined and the appropriate value of K_T read off directly, and from this the thrust can be determined. Alternatively, the $\mu - \sigma$ approach can be adopted as shown in Figure 5.99.

5.6.5.6 Effects of cavitation

In all propellers the effects of cavitation are important. In the case of general merchant propellers some standard series give guidance on cavitation in the global sense; see for example, the KCD series of propellers where generalized face and back cavitation limits are given. The problem of cavitation for merchant ship propellers, whilst addressed early in the design process, is nevertheless generally given more detail assessment, in terms of pressure distribution, etc., in later stages of the design. In the case of high-speed propellers, however, the effects of cavitation need particular consideration at the earliest stage in the design process along with pitch ratio, diameter and, by implications, the choice of propeller rpm.

Many of the high-speed propeller series include the effects of cavitation by effectively repeating the model tests at a range of free stream cavitation numbers based on advance velocity. Typical in this respect is the KCA series. From propeller series of this kind the influence of cavitation on the propeller

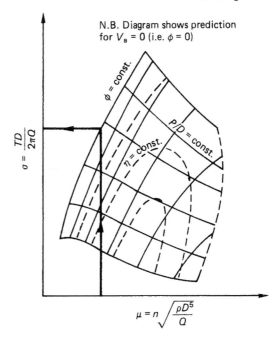

Figure 5.99 The use of the $\mu - \sigma$ chart in thrust prediction.

design can be explored, for example, by taking a series of charts for different blade area ratios and plotting for a given advance coefficient K_Q against the values of σ tested to show the effect of blade area against thrust or torque breakdown for a given value of cavitation number. Figure 5.100 demonstrates this approach. In the design process for high-speed propellers several analogous design studies need to be undertaken to explore the effects different diameters, pitch ratios and blade areas on the cavitation properties of the propeller.

5.6.6 Design considerations

The design process of a propeller should not simply be a mechanical process of going through a series of steps such as those defined in the previous section. Like any design it is a creative process of resolving the various constraints to produce an optimal solution. An eminent propeller designer once said 'It is very difficult to produce a bad propeller design but it is equally difficult to produce a first class design.' These words are very true and should be engraved on any designer's heart.

5.6.6.1 Direction of rotation

The direction of rotation of the propeller has important consequences for manoeuvring and also

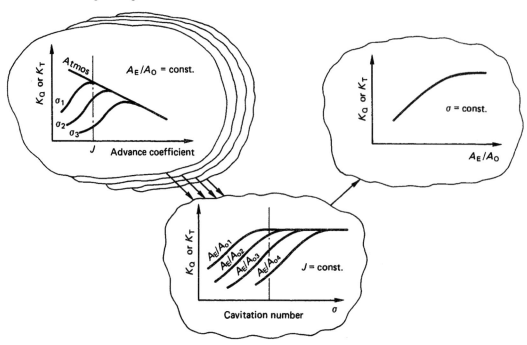

Figure 5.100 The use of high-speed standard series data to explore the effects of cavitation.

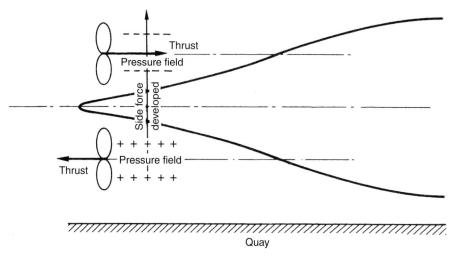

Figure 5.101 Side force developed by reversing thrusts of propellers on a twin-screw vessel due to pressure field in hull.

for cavitation and efficiency considerations with twin-screw vessels. In terms of manoeuvring, for a single-screw vessel the influence on manoeuvring is entirely determined by the 'paddle wheel effect'. When the vessel is stationary and the propeller started, the propeller will move the after-body of the ship in the direction of rotation: that is in the sense of a paddle or road wheel moving relative to the ground. Thus with a fixed pitch propeller, this direction of initial movement will change with the direction of rotation, that is ahead or astern thrust, whilst in the case of a controllable pitch propeller the movement will tend to be unidirectional. In the case of twin-screw vessels, certain differences become apparent. In addition to the paddle wheel effect other forces due to the pressure differential on the hull and shaft eccentricity come into effect. The pressure differential, due to reverse thrusts of the propellers, on either side of the hull gives a lateral force and turning moment, Figure 5.101, which remains

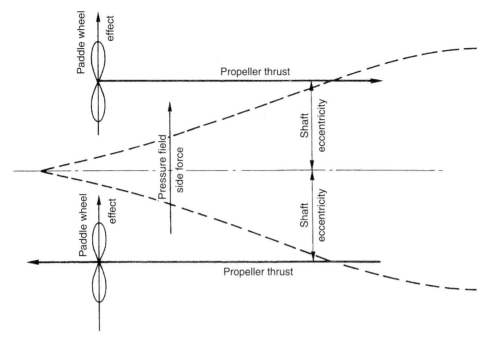

Figure 5.102 Induced turning moment components.

largely unchanged for fixed and controllable pitch propellers and direction of rotation. The magnitude of this thrust is of course a variable depending on the underwater hull form: in the case of some gondola hull forms, it is practically non-existent. However, in the general case of manoeuvring van Gunsteren undertook an analysis between rotation direction and fixed and controllable pitch propellers to produce a ranking of the magnitude of the turning moment produced. This analysis took into account shaft eccentricity, the axial pressure field and the paddle wheel effect (Figure 5.102) based on full-scale measurements, Voorde (1968), for frigates. The results of his analysis are shown in Table 5.17 for manoeuvring with two propellers giving equal thrusts and in Table 5.18 for manoeuvring on a single propeller.

Whilst the magnitudes in Tables 5.17 and 5.18 relate to particular trials, they do give guidance on the effect of propeller rotation on manoeuvrability. The negative signs were introduced to indicate a turning moment contrary to nautical intuition. From the manoeuvrability point of view it can be deduced that fixed pitch propellers are best when outward turning; however, no such clear-cut conclusion exists for the controllable pitch propeller.

From the propeller efficiency point of view, it has been found that the rotation present in the wake field, due to the flow around the ship, at the propeller disc can lead to a gain in propeller efficiency when the direction of rotation of the propeller is opposite to the direction of rotation in the wake field. However, if concern over cavitation extent is present, then this can to some extent be helped by considering the propeller rotation in relation to the wake rotation. If the problem exists for a twin-screw ship at the tip, then the blades should turn in the opposite sense to the rotation in the wake, whilst if the concern is at root then the propellers should rotate in the same sense as the wake rotation. As a consequence the dangers of blade tip and tip vortex cavitation need to be carefully considered against the possibility of root cavitation.

5.6.6.2 *Blade number*

The number of blades is primarily determined by the need to avoid harmful resonant frequencies of the ship structure and the machinery. However, as blade number increases for a given design the extent of the suction side sheet cavity generally tends to decrease.

Table 5.17 Turning moment ranking of two propellers producing equal thrusts (complied from Gunsteren).

Twin-screw installation (reverse thrusts)	Turning moment ranking
F.p.p.; inward turning	−2.1
F.p.p.; outward turning	10.1
C.p.p.; inward turning	3.3
C.p.p.; outward turning	4.6

C.p.p.: controllable pitch propeller.
F.p.p.: fixed pitch propeller.

Table 5.18 Turning moment ranking of one propeller operating on a twin-screw installation (complied from Gunsteren).

Twin-screw installation (single propeller operation)	Direction of thrust	Turning moment ranking
F.p.p.; inward turning	Forward	−1.2
F.p.p.; inward turning	Astern	−1.1
F.p.p.; outward turning	Forward	5.6
F.p.p.; outward turning	Astern	4.5
C.p.p.; inward turning	Forward	−1.2
C.p.p.; inward turning	Astern	4.5
C.p.p.; outward turning	Forward	5.6
C.p.p.; outward turning	Astern	−1.1

At the root, the cavitation problems can be enhanced by choosing a high blade number, since the blade clearances become less in this case.

In addition to resonant excitation and cavitation considerations, it is also found that both propeller efficiency and optimum propeller diameter increase as blade number reduces. As a consequence of this latter effect, it will be found, in cases where a limiting propeller diameter is selected, that propeller rotational speed will be dependent on blade number to some extent.

The cyclical variations in thrust and torque forces generated by the propeller are also dependent on blade number.

5.6.6.3 *Diameter, pitch–diameter ratio and rotational speed*

The choice of these parameters is generally made on the basis of optimum efficiency. However, efficiency is only moderately influenced by small deviations in the diameter, P/D and revolutions when the delivered horsepower is held constant. The effect of these parameters on the cavitation behaviour of the propeller is extremely important, and so needs careful exploration at the preliminary design stage.

For example, it is likely the propellers of high-powered or fast ships should have an effective pitch diameter ratio larger than the optimum value determined on the basis of optimum efficiency. Furthermore, it is generally true that a low rotational speed of the propeller is a particularly effective means of retarding the development of cavitation over the suction faces of the blades.

In Section 5.6.5.1 the optimum diameter calculation was discussed. For an actual propeller working behind a ship the diameter needs to be reduced from the optimum value predicted from the standard series data, and traditionally this was done by reducing the optimum diameter by 5 per cent and 3 per cent, for single- and twin-screw vessels, respectively. This correction is necessary because the resultant propulsion efficiency of the vessel is a function of both the open water propeller efficiency, to which the chart optimum diameter refers, and the propeller–hull interaction effects. Hawdon *et al.* (1984) conducted a study into the effect of the character of the wake field on the optimum diameter. From this study they derived a relationship of the form shown in Figure 5.103; however, the authors note that, in addition to the mean effective wake, it is necessary to take into account the radial wake distribution as implied by the distinction between different hull forms.

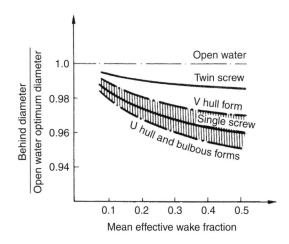

Figure 5.103 Correction to optimum open water diameter, Hawdon *et al.* (1984).

5.6.6.4 *Blade area ratio*

In general, the required expanded area ratio when the propeller is operating in a wake field is larger than that required to simply avoid cavitation at shock-free angles of attack. Furthermore, a larger variation in the section angle of attack can to some extent be supported by increasing the expanded area ratio of the propeller. Nevertheless, in the case of a controllable pitch propeller there is a limit to the extent of the blade area due to the requirement of blade–blade passing in order to obtain reversibility of the blades. Notwithstanding the advantages of increasing blade area, it must be remembered that this leads to an increase section drag and hence a loss in efficiency of the propeller (see Figure 5.97(b)).

5.6.6.5 *Section form*

In terms of section form, the most desirable thickness distribution from the cavitation viewpoint is an elliptic form. This, however, is not very practical in section drag terms and in practice the National Advisory Committee for Aeronautics (NACA), 16, 65 and 66 (modified) forms are the most utilized. With regard to mean lines, the NACA $a = 1.0$ is not generally considered a good form since the effect of viscosity on life for this camber line is large and there is doubt as to whether the load distribution can be achieved in practice. The most favoured form would seem to be the NACA $a = 0.8$ or 0.8 (modified) although a number of organizations use proprietory section forms.

5.6.6.6 *Cavitation*

Sheet cavitation is generally caused by the suction peaks in the way of the leading edge being too high whilst bubble cavitation tends to be induced by too high cambers being used in the mid-chord region of the blade.

The choice of section pitch and the associated camber line should aim to minimize or eradicate the possibility of face cavitation. Hence the section form and its associated angle of attack requires to be designed so that it can accommodate the full range of negative incidence.

There are few propellers in service today which do not cavitate at some point around the propeller disc. The secret of design is to accept that cavitation will occur but to minimize its effects, both in terms of the erosive and pressure impulse effects.

The initial blade design can be undertaken using one of the basic estimation procedures, notably the Burrill cavitation chart or the Keller formula (see Section 5.5). These methods usually give a reasonable first approximation to the blade area ratio required for a particular application. The full propeller design process needs to incorporate within it procedures to design the radial distribution of chord length and camber in association with cavitation criteria rather than through the use of standard outlines.

If the blade area, or more specifically the section chord length, is unduly restricted, then in order to generate the same lift from section, this being a function of the product cc_1, the lift coefficient must increase. This generally implies a larger angle of attack or camber, which in turn leads to higher suction pressures, and hence greater susceptibility to cavitation. Hence, in order to minimize the extent of cavitation, the variations in the angle of attack around the propeller disc should only give rise to lift coefficients in the region of shock-free flow entry for the section if this is possible.

In general terms the extent of sheet cavitation, particularly with high-powered fast ships, tends to be minimized when the blade section thickness is chosen to be sufficiently high to fall just below the inception of bubble cavitation on the blades. With respect to the other section parameters, the selection of the blade camber and pitch should normally be such that the attitude of the resultant section can accommodate the negative incidence range that the section has to meet in practice whilst the radial distribution of chord length needs to be selected in association with the variations in in-flow angle.

Tip vortex cavitation is best controlled by adjustment of the radial distribution of blade loading near the tip. The radial distribution of bound circulation at the blade tip lies within the range:

$$0 \leq \frac{d\Gamma}{dr} < \infty \quad (5.102)$$

Hence, the closer $d\Gamma/dr$ is to zero, the greater will be the control of the tip vortex strength. In addition to the control exerted by Equation (5.102) further control can be exerted by choosing the highest number of propeller blades, since this means that the total load is distributed over a greater number of blades.

5.6.6.7 *Skew*

The use of skew has been shown to be effective in reducing both shaft vibratory forces and hull pressure-induced vibration (see Carlton (2007), Chapter 23). The effectiveness of a blade skew distribution for retarding cavitation development depends to a very large extent on the matching of the propeller skew with the skew of the maximum or minimum in-flow angles in the radial sense.

5.6.6.8 *Hub form*

It is clearly advantageous for the propeller hub to be as small as possible consistent with its strength and the flexibility it gives to the blade root section design.

In addition to the hub diameter consideration, the form of the hub is of considerable importance.

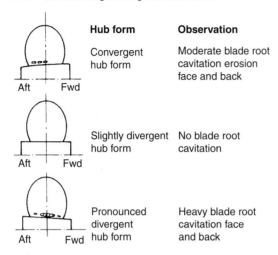

Figure 5.104 Observed blade root cavitation erosion on a fast patrol craft propeller.

A convergent hub form is normally quite satisfactory for slow merchant vessels; however, for higher-speed ships and fast patrol vessels or warships experience indicates that a slightly divergent hub form is best from the point of view of avoiding root erosion problems. In the case of a fast patrol craft van Gunsteren and Pronk (1973) experimented with different hub profiles, and the results are shown in Figure 5.104. The convergent hub enlarges the flow disc area between the hub and the edge of the slipstream, which has only minimal contraction, from forward to aft, and therefore decelerates the flow which results in positive pressure gradient. This may introduce flow separation that promotes cavitation. The strongly divergent hub accelerates the flow, and therefore reduced the pressure, which again promotes cavitation.

In addition to the use of a slightly divergent hub form, where appropriate, the use of a parallel or divergent cone (Figure 5.105) can assist greatly in reducing the strength of the root vortices and their erosive effects on the rudder.

5.6.6.9 Shaft inclination

If the propeller shafts are inclined in any significant way, this gives rise to a cyclic variation in the advance angle of the flow entering the propeller. The amplitude of this variation is given by

$$\Delta\beta = \frac{\sin\phi}{1 + \left(\dfrac{\pi x}{J}\right)} \qquad (5.103)$$

where ϕ is the inclination of the shaft relative to the flow and β is the advance angle at the particular radius. It should be noted that ϕ varies between a

Figure 5.105 Truncated fairwater cone fitted to a high-speed patrol vessel.

static and dynamic trim condition, and this can in some cases be quite significant. In addition to the consequences for cavitation at the root sections, since Equation (5.103) gives a larger value for $\Delta\beta$ at the root than at the tip, shaft inclination can give rise to significant lateral and shaft eccentricity forces and moments.

5.6.6.10 Duct form

When a ducted propeller is selected a choice of duct form is required. In choosing a duct form for normal commercial purposes it is necessary to ensure that the form is both hydrodynamically reasonable and also practical and easy to manufacture. For many commercial purposes a duct form of the Wageningen 19a type will suffice when a predominantly unidirectional accelerating duct form is required. When an improved astern performance is required, then a duct based on the Wageningen No. 37 form usually provides an acceptable compromise between ahead and astern operation.

The use of decelerating duct forms are comparatively rare outside of naval practice and generally operate at rather higher B_p values than the conventional accelerating duct form.

5.6.6.11 The balance between propulsion efficiency and cavitation effects

The importance of attaining a balance between the achievements of maximum propulsion efficiency and attaining an acceptable cavitation performance has been noted on a number of occasions during this Section. These references have mostly been in the context of the design point for the propeller and, by implication, for ships with a relatively narrow operational spectrum. Important as this is, for ships which operate under very variable conditions this balance has then to be maintained across a wide spectrum of operating conditions.

A typical example of such a situation might, for example, be a cruise ship and Figure 5.106 illustrates the problems that can occur if this balance is not maintained and the design specification is incorrectly developed. In this case a high maximum ship speed was required and the builder offered a premium for achieving a maximum speed above the contract speed with the given power installation. As a consequence the ship was designed to achieve as high a speed as possible since no mention in the contract had been made of the importance of acceptable vibratory performance at lower ship speeds. The result was a most pleasing performance from the controllable pitch propellers at the ship's maximum contract speed from both an efficiency and cavitation viewpoint. However, when the ship operated on legs of the cruise schedule which called for lower speeds complex cavitating tip vortex structures were developed by the propeller which gave rise to broadband excitation of the hull structure at these lower operational speeds. In Figure 5.106 it can be seen that the resulting excitation levels in the restaurant, when evaluated in accordance with the ISO 6954 (2000) Code, were rather higher at eight knots than was the case with the higher design speed of 27 knots.

The example, therefore, underlines the importance of attaining the correct balance of performance characteristics across the operating spectrum of the ship and, moreover, of defining the design specification having due regard of the way that it is intended to operate the ship.

5.6.6.12 *Propeller tip considerations*

There are many factors which can be deployed in the design of the propeller tip in order to influence the behavioural characteristics of the propeller, particularly in relation to noise and cavitation. Apart from increasing the strength of the blade tips for ships such as dredgers or which regularly take the ground, one of the primary aims in designing the blade tip is to influence the path and diffusivity of the tip vortex as well as minimizing any unwelcome interactions between supercavitating tip sheet cavities and the tip vortex. Moreover, in some designs there is the desire to increase the tip loading by the use of end-plates. However, among the more important influencing parameters are the chordal and radial profile of thickness, camber and section length, the use of tip plates or winglets and tip skew and rake.

Vonk *et al.* (2006) examined the influence on tip rake on propeller efficiency and cavitation behaviour through a series of computational fluid dynamic studies. They suggested that the cavitation characteristics in the mid-chord areas, where bubble cavitation can arise, and in the tip region can be enhanced by the use of aft tip rake. Conversely, they concluded that forward tip rake, although not generally helpful to the cavitation characteristics, has a greater potential for improving the

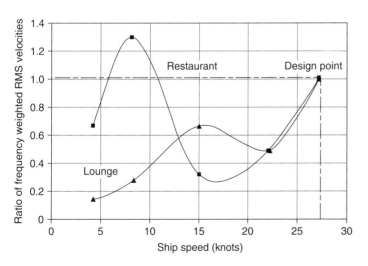

Figure 5.106 Vibratory behaviour of cruise ship whose propellers had not been designed for use across the operating spectrum.

propeller efficiency. Notwithstanding this by utilizing the cavitation benefits of aft tip rake, then for the same set of cavitation criteria the design can be adjusted to yield a greater efficiency in the design balancing process. Dang (2004) examined the behaviour of a forward raked propeller. In these types of propeller there is a tendency to generate a pre-swirl which is in the opposite direction to the rotation of the tip vortex and this tends to disperse the vortices from the tip region. Indeed, there is a large measure of similarity between the conclusions derived from this work and that of Vonk *et al.* Friesch *et al.* (2003) described a series of model- and full-scale trial measurements on a Kappel propeller. The Kappel propeller being one where the tip end-plate is integrated into the blade providing a smooth curved transition towards the suction side of the blade, Andersen and Andersen (1986) and Andersen *et al.* (2005). In the research programme described by Friesch it was demonstrated that for a product tanker the propulsive efficiency was higher in the case of the Kappel propeller than for a conventional propeller. Moreover, it was shown that the frictional component and scale effect of the Kappel propeller were larger than for the conventional propeller and a new surface strip method was produced in order to scale the frictional forces over the blade.

5.6.6.13 *Propellers operating in partial hull tunnels*

Where a ship's draught may be restricted for operational reasons there is sometimes benefit in designing the hull form so as to have partial tunnels. Figure 5.107 shows typical configurations for both a single- and twin-screw ship.

As has been previously discussed, in general, the slowest turning, largest diameter propeller is likely to return the highest propulsive efficiency: moreover, slow rotational speed can also have cavitation benefits. However, where operating draughts are restricted propeller immersion can be a dominating factor in the propeller design, not only from a reduced cavitation number perspective but also from the ever attendant possibility of air-drawing into the propeller disc. To counteract these effects the designing of partial ducts into the hull-form permits the largest propeller diameter, slowest turning propeller to be installed in a flow field which also frequently can have attenuated ship boundary layer influences and minimal risk of air-drawing taking place. Such arrangements have been fitted to single- and multi-screw ships and if correctly designed may enhance not only the propeller efficiency but also the hull efficiency to a limited extent.

An alternative reason for the employment of partial tunnels is to be found in the case of lifeboats where for reasons of giving a measure of a protection to people in the water, the propeller is located within a tunnel. However, in these cases the tunnel is normally rather more encasing than that shown in Figure 5.107.

5.6.6.14 *Composite propeller blades*

Although all propellers are subject to hydroelastic effects the isotropic behaviour of the conventional propeller metals tends generally to reduce these effects to negligible proportions except for highly biased skew or particularly specialized designs. The anisotropic behaviour of carbon fibre composites allows the designer extra degrees of freedom in exploiting the potential advantages of hydroelasticity.

Because the carbon fibre material is normally supplied in tapes with the fibre having specific orientations, typically $0°$, $\pm 45°$, 0 to $90°$, etc., the primary strength of the fibres also corresponds to these directions. Consequently, the lay-up of the fibres and the way they are combined in the matrix will give different deflection properties in each of the radial and chordal directions of the blade. Therefore, the blade can be designed to deflect in ways which are beneficial

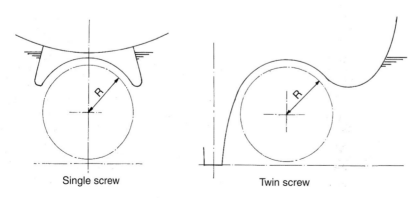

Figure 5.107 Examples of partial tunnels.

from a power absorption or cavitation inception and control perspective as the rotational speed is increased. This implies that the design process must be fully hydroelastic in the sense that a finite element procedure, capable of accommodating composite material lay-up mechanics, and a hydrodynamic analysis code are integrated into a convergent solution capability. Research into composite propellers is reported by Searle (1998), Kane and Smith (2003), AIR (2003) and Young et al. (2006).

Currently, with certain naval exceptions, composite propellers have only been produced in relatively small sizes, amongst the biggest to date probably being for the experimental trimaran *Triton*. Nevertheless, in addition to their potential hydrodynamic advantages there is also a weight advantage since composite blades are much lighter than those made from conventional materials. In this context, composite propellers, in their larger sizes, commonly have a metallic boss with the composite blades keyed into the boss using a number of proprietary configurations. An additional feature with composite propellers is that it is likely that radiated noise emissions can be reduced significantly, perhaps up to around 5 dB in certain cases.

5.6.6.15 *The propeller basic design process*

In order to outline the overall basic design process for a propeller an example has been chosen, in this case for the design of a small coastal ferry, and the resulting EXCEL spreadsheet for one operating condition is shown in Table 5.19. Within the overall design process many such spreadsheets are developed and cross-plotted in order to arrive at the final basic design. Moreover, such processes are integrated into other similar capabilities relating to hull resistance and propulsion analysis in order to achieve an integrated design.

5.6.7 The design process

The level of detail to which a propeller design process is taken is almost as variable as the number of propeller designers in existence. The principal manufacturers all have detailed design capabilities, albeit based on different methods. Whilst computational capability of the designer plays a large part in the detail of the design process, the information available upon which to base the design is also an important factor: there is little value in using advanced and high-level computational techniques requiring detailed input when gross assumptions have to be made concerning the basis of the design. Figures 5.108 and 5.109 show two extreme examples of the design processes used in propeller technology.

Leaving to one side the design of propellers which are standard 'off the shelf' designs such as may be found on outboard motor boats, the design process shown in Figure 5.108 represents the most basic form of propeller design that could be considered acceptable by any competent designer. Such a design process might be expected to be applied to, say, a small fishing boat or large workboat, where little is known of the inflow into the propeller. It is not unknown, however, to see standard series propellers applied to much larger vessels of a 100 000 tonnes deadweight and above; such occurrences are, however, comparatively rare and more advanced design processes normally need to be used for these vessels. The design of high-speed propellers can also present a complex design problem. In the calculation of such propellers the second box, which identifies the calculation of the blade dimensions, may involve a considerable amount of chart work with standard series data: this is particularly true if unfavourable cavitation conditions are encountered.

Blade stresses should always, in the author's view, be calculated as a separate entity by the designer, using as a minimum the cantilever beam technique followed by a fatigue estimate based on the material's properties. The use of classification society minimum thicknesses should always be used as a check to see that the design satisfies these conditions, since they are generalized minimum standards of strength.

Since many standard series propellers are of the flat face type an increase in thickness gives an implied increase in camber which will increase the propeller blade effective pitch. After the propeller has been adjusted for strength the design needs to be analysed for power absorption using the methods of Sections 5.6.5 and 5.6.6 in order to derive the appropriate blade pitch distribution. During the design process the question of design tolerances need to be addressed whatever level of design is used, otherwise significant departures between design and practice will occur.

Whilst Figure 5.108 shows the simplest form of design method and such processes and used to design perfectly satisfactory propellers for many vessels, more complex design procedures become necessary when increasing constraints are placed on the design and increasing amounts of basic information are available upon which to base the design. Variants of the design process shown in Figure 5.108 normally increase in complexity when a mean circumferential wake distribution is substituted for the mean wake fraction. This then enables the propeller to become wake adapted through the use of lifting line or higher-order design methods and the analysis phase may then embody a blade element, lifting line, lifting surface or boundary element based analysis procedure for different angular positions in the propeller disc: this presupposes that model wake data is available rather than the mean radial wake distribution being estimated from the procedures discussed in Section 5.2. As the complexity of the design procedure increases, the process outlined in Figure 5.109 is approached, which embodies most of the advanced design and analysis techniques available today.

Table 5.19 Typical basic propeller design calculation.

PROPELLER BASIC DESIGN
Program E/BD1
Twin Screw Passenger Ferry
T = 2.88 m
10th July 2005

Ship speed (Vs)	**15.5**	kts			
Delivered power	**836**	kW	Immersion to CL	**2.2** m	
	1137	hp(m)	Height of stern wave	**1.2** m	
Revolutions	**300**	rpm	Total immersion	**3.4** m	
Wake fraction	**0.112**				
Speed of Advance	**13.764** kts		Static head (p0-e)	19.47 lbf/in^2	
Diameter	**1.980** m		Dynamic head (qt)	38.82 lbf/in^2	
Blade number	**4**		Cavitation number	0.501	
$W_{max} - W_{min}$	**0.3**		Tc (Burrill)	**0.183**	

W_{max}	**0.35**	d/R	1.110	
Tip clearance	**1.0** m	Non-cavitating po	1.3 kPa (N.B. for (d/r) ≤ 2	
Ship displacement	**1118** m^3	Cavitating pc	2.5 kPa (d/r) ≤ 1	1.0 kPa (d/r) > 1
		Blade rate hull pressure	**2.8 kPa (d/r) ≤ 1**	**1.6 kPa (d/r) > 1**

Wageningen B4 analysis

BAR	0.40	0.55	0.70	0.85	
P/D	**0.841**	**0.859**	**0.869**	**0.875**	
$\eta 0$	**0.679**	**0.693**	**0.694**	**0.683**	
Thrust	8.15	8.32	8.33	8.20	tonnes
Ap	1.7	1.7	1.7	1.7	m^2
Ad	1.9	1.9	2.0	1.9	m^2
Ae/A0	0.616	0.632	0.634	0.625	

Basic propeller				
Propeller Ae/A0		0.627	Kq	0.0192
P/D mean		**0.869**	Kt	**0.123**
$\eta 0$		**0.675**	J	**0.678**
Thrust		4.9 tonnes		

Expanded blade area (0.1667 R)   1.93 m^2

r/R	wc	Chord (mm)	pc	Pitch (mm)	tc	Thick (mm)	Ca	Area (mm^2)	t/c
1.0000	0.000	0	1.026	**1765**	0.055	**4.0**	1.000	0	
0.9375	0.702	410	1.037	**1784**	0.107	**7.7**	0.769	2432	0.019
0.8750	0.920	538	1.043	**1795**	0.167	**12.0**	0.732	4735	0.022
0.7500	1.143	668	1.040	**1789**	0.293	**21.1**	0.714	10067	0.032
0.6250	1.206	705	1.030	**1772**	0.433	**31.2**	0.712	15653	0.044
0.5000	1.176	688	1.008	**1734**	0.588	**42.3**	0.709	20640	0.062
0.3750	1.072	627	0.971	**1671**	0.770	**55.4**	0.704	24464	0.088
0.2500	0.886	518	0.909	**1564**	1.000	**72.0**	0.695	25923	0.139
0.1667	0.712	416	0.832	**1432**	1.190	**85.7**	0.688	24541	0.206

CSR/MCR powering ratio	0.850		
Developed power (stressing)	984 kW		
Revolutions (stressing)	316.7 rpm		
Ship speed (stressing)	16.4 kts		
Density of material	**0.271** lb/in^3	Area coefficient	**0.690**

Table 5.19 Continued.

Blade thickness @ 0.25 R	**72** mm	←
Mean blade thickness	31 mm	
Mass of blade	83 kg	
Position of blade centre of gravity	**0.51** R	
Centrifugal force	**46** kN	
Centrifugal lever	**100** mm	
Position of centre of thrust	**0.7** R	
Position of centre of torque	**0.66** R	
Blade number	4	
Bending stress due to thrust	23.6 MPa	
Bending stress due to torque	11.1 MPa	
Centrifugal bending stress	15.5 MPa	
Direct centrifugal stress	1.8 MPa	
Tensile stress @ 0.25 R	51.9 MPa	
Allowable tensile stress	49 MPa	
New thickness @ 0.25 R	**74.1** mm	

Modulus coefficient	**0.110**	
Section area	25737 mm^2	
Section modulus	295412.517 mm^3	
Pitch angle @ 0.25 R	45.16°	

r/R	Area	SM	Area × SM	Lever	A.SM.L	Lever	A.SM.L^2
1.0000	0	0.5	0	0	0	0	0
0.8750	4735	2	9469	1	9469	1	9469
0.7500	10067	1	10067	2	20133	2	40266
0.6250	15653	2	31305	3	93915	3	281746
0.5000	20640	1	20640	4	82558	4	330233
0.3750	24464	2	48928	5	244640	5	1223202
0.2500	25923	0.5	12962	6	77770	6	466617

Crown of boss radius	0.23	R
Volume of blade to 0.25 R	0.0110 m^3	
Volume 0.25 R to cob	0.0006 m^3	
Volume of fillets	0.0003 m^3	
Blade volume	0.0119 m^3	
Blade weight	90	kgf
Weight of blade to 0.25 R	83	kgf

Shaft power basis	1004 kW	
Shaft RPM basis	300 rpm	
Blade chord @ cob	494 mm	
Blade pitch @ cob	1532 mm	
Boss length	368 mm	
Shaft tensile strength	600 N/mm^2	
Fitting factor (k)	**1.22**	
Tail shaft diameter	198 mm	
cob diameter	455.4 mm	
Shaft taper ratio	1 in **30**	

Boss posn	Fwd	Cob	Aft
r/R	0.24	0.23	0.19
Ext. radius	239.1	227.7	191.3 mm
Int. radius	98.9	92.7	86.6 mm

cob/tailshaft dia. ratio	2.30
Fwd boss dia/tail shaft dia	2.42
Volume of boss	0.094 m^2
MI of boss	11 kg.m^2 (Wk^2)

Propeller weight	**1062 kgf**
	1.1 tonnes

Centroid of blade beyond 0.25 R	0.505 R
MI blade beyond 0.25 R about tip	0.0030 m^5
MI blade to cob about tip	0.0035 m^5
MI of blade about tip	0.0035 m^5
MI of blade about shaft centre-line	27 kg.m^2 (Wk^2)
Radius of gyration of blade	0.557 R

Dry moment of inertia of propeller about shaft CL	**120 kg.m^2 (Wk2)**
	0.1 tonne.m^2 (Wk2)
Radius of gyration of propeller	**0.340 R**

Each designer, however, will use different theoretical methods and his correlation with full-scale experience will be dependent on the methods used. This underlines the reason why it may be dangerous and unjust to criticize a designer for not using the most up-to-date theoretical methods, since the extent of his theoretical to full-scale correlation database may outweigh the advantages given by use of more up-to-date methods.

Theoretical design methods, and analysis methods too for that matter, will only take the designer so far.

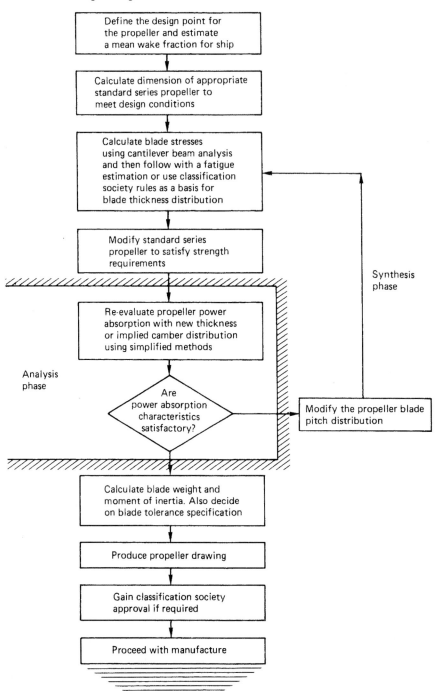

Figure 5.108 Example of a simplified design procedure.

Current knowledge is lacking in many detailed aspects of propeller design; nowhere this is more true than in defining the flow at the blade–boss interface of all propellers. In such case careful assumptions regarding the assumed blade loading at the root have to be made in the context of the anticipated severity of the in-flow conditions – this may dictate that a zero circulation or some other condition, determined

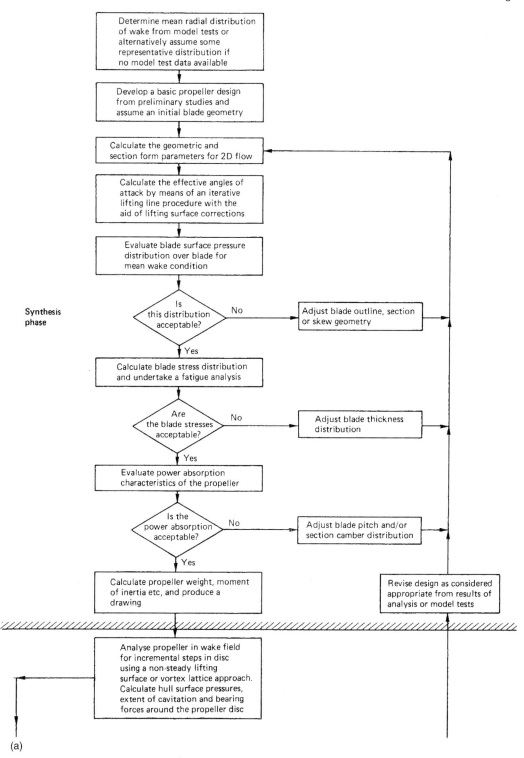

Figure 5.109 Example of a fully integrated synthesis and analysis procedure.

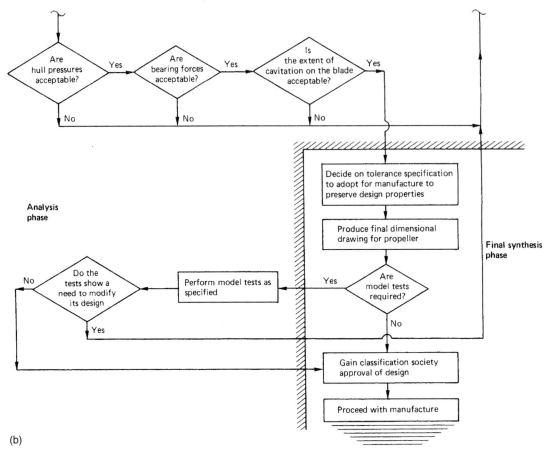

(b)

Figure 5.109 Continued

from experience, is an appropriate assumption. In either case the actual circulation which occurs on the blade will not be known due to the nature of the complex three-dimensional flow regime in this region of the blade. Another classic example is the definition of the geometric and flow conditions that cause singing, although in this case the remedy is well known from normal propeller types.

It will, however, be noted that each of the design processes shown in Figure 5.108 and 5.109 contain the elements of synthesis and analysis phase shown in Figure 5.108. Much has been written on the subject of propeller design and analysis by many practitioners of the subject. The references in this Section contain a considerable amount of this information, however, other references that contain work specifically related to design and analysis which is not referred to elsewhere in this Section include Burrill (1955, 1964), Hannan (1971), Cummings et al. (1972), Sinclair and Emerson (1973), Boswell and Cox (1974), Parsons (1975), Oossenen (1976), Dyne (1980), Hawdon and Patience (1981), Handler (1982), Holden and Kvinge (1983), Beek and Verbeek (1983), Norton and Elliot (1988) and Klintrop (1989).

The traditional approach to the detailed design of propeller blades has been that the propeller blade sections are designed for the mean inflow conditions around the circumference at a set of specific radii in the propeller disc. During this process the design is then balanced against the various constraints and velocity excursions relating to that particular design. Kinnas et al. (1997, 1998) have explored an alternative approach of optimizing the design for the actual flow conditions without the necessity of employing circumferential flow averaging processes. Their method uses a B-spline representation of the blade and determines the blade performance characteristics via second order Taylor expansions of the thrust, torque, cavitation extent and volume in the region of the solution using the MPUF-3A code. However, to converge to an optimum solution using this procedure a considerable amount of computer time is required which tends to limit the method's general applicability. To overcome these problems they developed an approach in which the optimum blade geometry that was being sought is found from a set of geometries which have been scaled from a basic geometry. In this alternative procedure the blade performance is computed from the MPUF-3A code for selected geometries within the set of geometries derived from the basic propeller and then interpolation curves are used to establish continuous analytical functions of performance. These functions are then used within an optimization procedure to establish the required final optimum blade geometry. Deng (2005) presents the detail of the optimization method used in this procedure.

5.7 Service performance and analysis

In general the performance of a ship in service is different from that obtained on trial. Apart from any differences due to loading conditions, and for which due correction should be made, these differences arise principally from the weather, fouling and surface deterioration of the hull and propeller.

The subject of service performance quite naturally, therefore, can be divided into four component parts for discussion purposes as follows:

1. Effects of weather – both sea and wind.
2. Hull roughness and fouling.
3. Propeller roughness and fouling.
4. The monitoring of ship performance.

As such the discussion in this Section will essentially fall into these four categories.

5.7.1 Effects of weather

The influence of the weather, both in terms of wind and sea conditions, is an extremely important factor in ship performance analysis. The analytical aspects of the prediction of the effects of wind and sea state was discussed in Section 5.1.5, and therefore need not be reiterated here. In the case of the service data returned from the ship for analysis purposes it is insufficient to simply record wind speed and sea state according to the Beaufort scale. In the case of wind, it is important to record both its speed and direction, since both of these parameters clearly influence the drag forces experienced by the vessel. With regard to sea conditions, this is somewhat more complex since in many instances the actual sea state will contain both a swell component and a local surface disturbance which are not related. For example, if a sea is not fully developed, then the apparent Beaufort number will not be representative of the conditions actually prevailing at the time. Consequently, both the swell and surface disturbance effects and their direction relative to the ship's heading need to be taken into account if a realistic evaluation is to be made of the weather effects in the analysis procedures. In making these comments it is fully recognized, in the absence of instrumented data as opposed to subjective judgment, that the resulting data will contain an observational error band on the part of the deck officer. Nevertheless, an experienced estimate of the conditions is essential to good analysis practice.

5.7.2 Hull roughness and fouling

The surface texture or hull roughness of a vessel is a continuously changing parameter which has a comparatively significant effect on the ship performance. This effect derives from the way in which the roughness of the hull surface influences the boundary layer and its growth over the hull. Hence, the effect of hull roughness can be considered as an addition to the frictional component of resistance of the hull. Table 5.20 shows typical comparative proportions of frictional resistance (C_F) to total resistance (C_T) at design speed for a series of ship types.

From this table it is clearly seen that the frictional components play a large role for almost all types of vessel. Naturally the larger full-form vessels have the largest frictional components.

The roughness of a hull can be considered to be the sum of two separate components as follows:

hull surface roughness = permanent roughness
+ temporary roughness

in which the permanent roughness refers to the amount of unevenness in the steel plates and the temporary roughness is that caused by the amount and composition of marine fouling.

Permanent roughness derives from the initial condition of the hull plates and the condition of the painted surface directly due to either the application or the drying of the paint on the hull. The condition of the hull plates embraces the bowing of the ships plates, weld seams and the condition of the steel surface. The bowing of the plates or 'hungry horse' appearance has a comparatively small effect on resistance, generally not greater than about one per cent. Similarly, the welded seams also have a small

Table 5.20 Typical proportions of frictional to total resistance for a range of ship types.

Ship type		C_F/C_T
ULCC – 516 893 dwt	(loaded)	0.85
Crude carrier – 140 803 dwt	(loaded)	0.78
	(ballast)	0.63
Product tanker – 50 801 dwt	(loaded)	0.67
Refrigerated cargo ship – 8500 dwt		0.53
Container ship – 37000 dwt		0.62
Ro/Ro ferry		0.55
Cruise liner		0.66
Offshore tug supply vessel		0.38

contribution: for example, a VLCC or container ships might incur a penalty of the order $\frac{3}{4}$ per cent and so it may be cost effective to remove these by grinding the surface of the weld. By far the greatest influence on resistance is to be found in the local surface topography of the steel plates. This topography is governed by a wide range of variables: corrosion, mechanical damage, deterioration of the paint film, a build-up of old coatings, rough coating caused by poor application, cold flow resulting from too short a drying time prior to immersion, scoring of the paint film resulting from scrubbing to remove fouling, poor cleaning prior to repainting, etc. Consequently, it can be seen that the permanent roughness, which is permanent in the sense of providing the base surface after building or dry-docking during service, cannot be eliminated by subsequent coating, and therefore, to improve it in terms of local surface topography, complete removal of the old coatings is necessary to restore the hull surface.

In contrast, temporary roughness can be removed or reduced by the removal of the fouling organisms or subsequent coating treatment. It is caused in a variety of different ways: for example, the porosity of leached-out anti-fouling, the flaking of the current coating caused by internal stresses, and corrosion caused by the complete breakdown of the coating system and by marine fouling. Whilst permanent roughness can be responsible for an annual increment of, say, 30 to 60 μm in roughness perhaps, the effects of marine fouling can be considerably more dramatic and can be responsible, given the right circumstances, for 30 to 40 per cent increases in fuel consumption in a relatively short time.

The sequence of marine fouling commences with slime, comprising bacteria and diatoms, which then progresses to algae and in turn on to animal foulers such as barnacles, culminating in the climax community. Within this cycle Christie (1981) describes the colonization by marine bacteria of a non-toxic surface as being immediate, their numbers reaching several hundreds in a few minutes, several thousands within a few hours and several millions within two to three days. Diatoms tend to appear within the first two or three days and grow rapidly, reaching peak numbers within the first fortnight. Depending on local conditions this early diatom growth may be overtaken by fouling algae.

The mixture of bacteria, diatoms and algae in this early stage of surface colonization is recognized as the primary slime film. The particular fouling community which will eventually establish itself on the surface is known as the climax community and is particularly dependent on the localized environment. In conditions of good illumination this community may be dominated by green algae, or by barnacles or mussels, as is often observed on static structures such as pier piles or drilling rigs.

The vast numbers and diversity of organism comprising the primary slime film results in the inevitable formation of 'slime' on every submerged marine surface, whether it is 'toxic' or 'non-toxic'. The adaptability of the bacteria is such that these organisms are found in nature colonizing habitats varying in temperature from below 0 to 75°C. The adaptability of diatoms is similarly impressive; they can be found in all aquatic environments from fresh water to hyper-saline conditions and are even found growing on the undersides of ice floes. These life cycles and the adaptability of the various organisms combine to produce a particularly difficult control problem.

Severe difficulty of fouling control is not, however, restricted to microfouling; recent years have seen the emergence of oceanic, stalked barnacles as a serious problem fouling VLCCs working between the Persian Gulf and Northern Europe. This group of barnacles is distinguished from the more familiar 'acorn' barnacles in both habitat and structure.

Whereas acorn barnacles are found in coastal waters, characteristically attached directly to fixed objects such as rocks, buoys, ships, pilings and sometimes to other organisms, such as crabs, lobsters and shellfish; stalked barnacles are usually found far from land attached to flotsam or to larger animals such as whales, turtles and sea snakes by means of a long, fleshy stalk. The species, the most important of which is the Conchoderma, is recognized as a problem for large slow-moving vessels, and much research dealing with their life cycle and habits has been undertaken. The conclusions of this work indicate that VLCCs become fouled with Conchoderma while under way in open ocean. The results of the shipboard studies suggest that vessels travelling between the Gulf and Northern Europe are most likely to become fouled in the Atlantic Ocean between the Canary Islands and South Africa and particularly in an area between 17°S and 34°S. Adult Conchoderma, however, have been reported to be in every ocean in the world, and so there are no areas of warm ocean where vessels can be considered immune from attack.

The fouling of underwater surfaces is clearly dependent on a variety of parameters such as ship type, speed, trading pattern, fouling pattern, dry-dock interval, basic roughness and so on. To assist in quantifying some of these characteristics Evans and Svensen (1987) produced a general classification of ports with respect to their fouling or cleaning characteristics; Table 5.21 reproduces this classification.

Paint systems have developed from traditional anti-fouling coatings to self-polishing anti-fouling (SPA) and reactivatable anti-foulings (RA) in order to provide greater protection against fouling problems. SPAs are based on components which dissolve slowly in sea water and due to the friction of the sea water passing over the hull, toxins are continuously released. Thus,

Table 5.21 Port classification according to Evans and Svensen (1987).

Clean ports	Fouling ports		Cleaning ports	
	Light	Heavy	Non-scouring	Scouring
Most UK ports	Alexandria	Freetown	Bremen	Calcutta
Auckland	Bombay	Macassar	Bribane	Shanghai
Cape Town	Colombo	Mauritius	Buenos Aires	Yangtze Ports
Chittagong	Madras	Rio de Janeiro	E. London	
Halifax	Mombasa	Scurabaya	Hamburg	
Melbourne	Negapatam	Lagos	Hudson Ports	
Valparaiso	Karadii		La Plata	
Wellington	Pernambuco		St Lawrence Ports	
Sydney*	Santos		Manchester	
	Singapore			
	Suez			
	Tuticorin			
	Yokohama			

*Variable conditions.

this overcomes the weakness of traditional anti-fouling where only part of the anti-fouling is water soluble, and where an inactive layer slowly develops through which the toxins have to migrate. Reactivatable coatings depend on a mechanical polishing with special brushes in order to remove the inactive layer formed at the surface of the anti-fouling. Both SPA and RA systems depend upon high-quality anti-corrosive systems to act as a basis, and the service life is proportional to the thickness of the film at application. Figure 5.110 shows in schematic form the action of an SPA type of coating.

The life of an SPA coating, which if correctly applied can extend hull protection considerably beyond that afforded by traditional anti-foulings. Typically in the order of five years, dependent on the ship speed, hull permanent roughness, distance travelled and the thickness and polishing rate of the coats applied. The wear-off rate or polishing rate of anti-foulings is not always completely uniform, since it depends on both the turbulence structure of the flow and the local friction coefficient. The flow structure and turbulence intensities and distribution within the boundary layer change with increasing ship speed, which gives a thinner lamina sublayer, and consequently a hydrodynamically rougher surface, since more of the roughness peaks penetrate the sublayer at higher ship speeds. A further consequence of the reduced lamina sublayer at high speed is that the diffusion length for the chemically active ingredients is shorter, which leads to a faster chemical reaction, and therefore faster renewal, at the surface. In addition to the ship speed considerations, the hull permanent roughness is also of considerable importance. Whilst this will not in general affect the polishing rate of the coating, one will find that in the region of the peaks the anti-fouling will polish

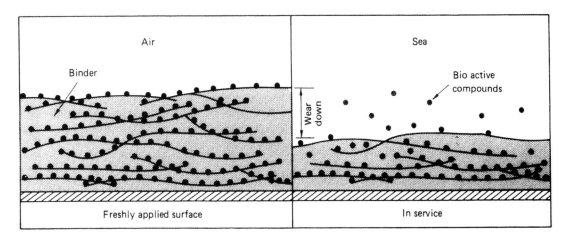

Figure 5.110 Principle of self-polishing process.

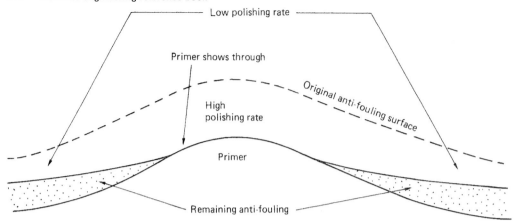

Figure 5.111 Influence of surface roughness on polishing anti-fouling paints.

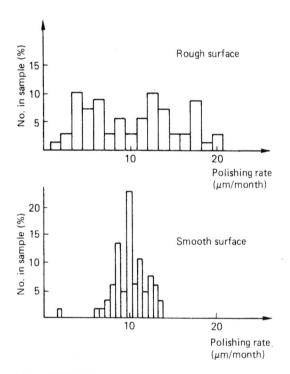

Figure 5.112 Influence of roughness on polishing rate (Bidstrup et al. (1981)).

through more quickly since the coating surface will be worked harder by the increased shear stresses and turbulent vortices. Figure 5.111 shows this effect in schematic form. Whilst the average polishing rate for the coating will be the same for a rough or smooth hull, the standard deviation on the distribution curve for polishing rate will give a much bigger spread for rough hulls. Figure 5.112 demonstrates this effect by showing the results of model experiments (Bidstrup et al. (1981)) for both a smooth and rough surface, 50 and 500 μm, respectively; similar effects are noted on vessels at sea. Consequently, it will be seen that the paint coating needs to be matched carefully to the operating and general conditions of the vessel.

Although particularly successful in minimizing the hull resistance over a docking cycle, hull coatings containing toxins have been the subject of a progressive banning regime by the International Maritime Organization. This occurred first with pleasure craft based in coastal and estuarial marinas where it was noted that mutations in marine life were occurring and then latterly the ban was extended to large commercial ships. This led to intensive research efforts into maintaining hull performance during the service life of ships and has represented a significant challenge given that the fouling sequence normally commences with slime which then progresses to algae, following which the animal foulers tend to take up residence and finally culminate in a climax community. Prior to the introduction of bioactive compounds hull fouling led, in ship performance terms, to regular and frequent increases in hull resistance over a docking cycle. With the introduction of anti-fouling paints this sawtooth characteristic largely disappeared and the necessity to dry-dock for fouling reasons reduced as discussed in Section 5.7.5.

A number of coating solutions are being evolved for use in the post biocide era with various benefits claimed. In the case of the silicone-based elastomeric coatings, which have no known toxic effects, these prevent marine life from adhering to the hull surface by virtue of the coating's properties provided that the ship speed is maintained above a critical value, typically in the region of 17 to 18 knots and does not spend long periods stationary in port. Moreover, some advantage in terms of a reduced turbulent flow wall shear stress

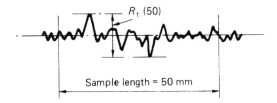

Figure 5.113 Definition of $R_t(50)$ roughness measure.

is also likely and a number of sea trials are currently in progress to demonstrate whether this is the case.

The standard measure of hull roughness that has been adopted within the marine industry is $R_{t(50)}$. This is a measure of the maximum peak-to-valley height over 50 mm lengths of the hull surface, as shown in Figure 5.113. When undertaking a survey of a hull, several values of $R_{t(50)}$ will be determined at a particular location on the hull and these are combined to give a mean hull roughness (MHR) at that location defined by

$$\text{MHR} = \frac{1}{n}\sum_{i=1}^{n} h_i \qquad (5.104)$$

where h_i are the individual $R_{t(50)}$ values measured at that location.

The Average Hull Roughness (AHR) is an attempt to combine the individual MHR values into a single parameter defining the hull conditions at a particular time. Typically the vessel may have been divided up into a number of equal areas, perhaps 100, and a value of MHR determined for each area. These MHR values are then combined in the same way as Equation (5.104) to give the AHR for the vessel:

$$\text{AHR for vessel} = \frac{\sum_{j=1}^{m} w_j (\text{MHR})_j}{\sum_{j=1}^{m} w_j} \qquad (5.105)$$

where w_j is a weight function depending on the location of the patch on the hull surface. For many purposes w_j is put equal to unity for all j values; however, by defining the relation in the general way some flexibility is given to providing a means for weighting important areas of the hull with respect to hull roughness. Most notable here are the regions in the fore part of the vessel.

Townsin *et al.* (1981) suggest that if a full hull roughness survey is made, the AHR will be statistically correct using $w_j = 1$ in Equation (5.105). However, should some stations be left out for reasons of access, etc., then the AHR can be obtained in the following way:

$$\begin{aligned}\text{AHR for vessel} = &(\text{MHR of sides})\\ &\times \text{fraction of the sides covered}\\ &+(\text{MHR of flats})\\ &\times \text{fraction of the flats covered}\\ &+(\text{MHR of boot topping})\\ &\times \text{fraction of the boot}\\ &\quad \text{topping covered} \qquad (5.106)\end{aligned}$$

Much debate has centred on the use of a simple parameter such as $R_{t(50)}$ in representing non-homogenous surfaces. The arguments against this parameter suggest that the lack of data defining the surface in terms of its texture is serious and has led to the development of replica-based criteria for predicting power loss resulting from hull roughness, Anon. (1987). With this method, the surface of the actual ship is compared to those reproduced on replica cards, which themselves have been cast from other ships in service and the surfaces tested in a water tunnel to determine their drag. When a particular card has been chosen as being representative of a particular hull surface, a calculation of power penalty is made by use of diagrams relating the principal ship particulars; these diagrams having been constructed from a theoretical analysis procedure.

There is unfortunately limited data to be found that gives a statistical analysis and correlation with measured roughness functions for typical hull surfaces. Amongst the tests carried out, Musker (1977), Johannson (1985) and Walderhaug (1986), feature as well-known examples. In the case of Musker, for example, he found that the measured roughness function for a set of five surfaces did not show a good correlation with $R_{t(50)}$ and used a combination of statistical parameters to improve the correlation.

The parameters used in his study were:

1. the standard deviation (σ_r);
2. the average slope (S_p);
3. the skewness of the height distribution (S_k);
4. the Kurtosis of the distribution (K_u);

and he combined them into an 'equivalent height' (h') which correlated with the measured roughness function using a filtered profile with a 2 mm long wavelength cut off. The relationship used was

$$h' = \sigma_r(1 + aS_p)(1 + bS_k K_u) \qquad (5.107)$$

With regard to $R_{t(50)}$ as a parameter, Townsin (1987) concludes that for rough surfaces, including surface-damaged and deteriorated anti-fouling coatings – in excess of around 250 μm AHR, it is an unreliable parameter to correlate with added drag. However, for new and relatively smooth hulls it appears to correlate

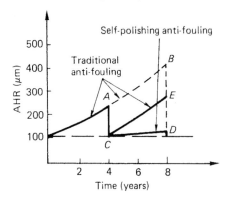

Figure 5.114 Effect of different coatings on hull roughness (Reproduced from Naess (1980)).

Table 5.22 Typical annual hull roughness increments.

Coating type	Annual increase in roughness (μm/year)
Self-polishing paints	10–30
Traditional coating	40–60

well with other available measures of roughness function, and so can form a basis to assess power penalties for ships.

It is found that the majority of new vessels have AHR of the order of 90 to 130 μm provided that they have been finished in a careful and proper manner. McKelvie (1981) notes, however, that values for new vessels of 200 to 250 μm have not been uncommon in the period preceding 1981. The way in which this value increases with time is a variable depending on the type of coating used. To illustrate this Figure 5.114 shows a typical scenario, Naess (1980), for a vessel in the first eight years of its life. In the figure it will be seen that the initial roughness AHR increased after four years to a value of around 250 μm using traditional anti-fouling coatings (Point A on the diagram). If the vessel is shot blasted, it can be assumed that the initial hull roughness could be reinstated since an insignificant amount of corrosion should have taken place. If, after cleaning, the vessel is treated with a reactivatable or SPA, after a further period of four years in service the increase in roughness would be small. Alternatively, if the vessel had been treated with traditional anti-fouling, as in the previous four-year period, then a similar increase in roughness would be noted. As illustrated in the diagram, the rate of increasing roughness depends on the coating system employed and the figures shown in Table 5.22 will give some general indication of the probable increases.

Clearly, significant deviations can occur in these roughening rates in individual circumstances for a wide variety of reasons. Figure 5.115, which is taken from Townsin (1986) shows the scatter that can be obtained over a sample of some 86 surveys conducted over the two-year period 1984/85.

Assuming that the AHR can be evaluated, this value then has to be converted into a power penalty if it is to be of any practical significance beyond being purely an arbitrary measure of paint quality. Lackenby (1962) proposed an early approximation that for every 25 μm increase in roughness an increase in fuel consumption of around 2.5 per cent could be anticipated.

Bowden and Davison (1974) proposed the relationship

$$\frac{\Delta P_1 - \Delta P_2}{P} \times 100\% = 5.8[(k_1)^{1/3} - (k_2)^{1/3}] \quad (5.108)$$

where k_1 and k_2 are the AHR for the rough and smooth ship, respectively, and ΔP_1 and ΔP_2 are the power increments associated with these conditions, P is the maximum continuous power rating of the vessel.

The relationship was adopted by the 1978 International Towing Tank Conference (ITTC) as the basis for the formulation of power penalties and appeared in those proceedings in the form:

$$\Delta C_F \times 10^3 = 105\left(\frac{k_s}{L}\right)^{1/3} - 0.64 \quad (5.109)$$

in which k_s is the mean apparent amplitude of the surface roughness over a 50 mm wavelength and L is the ship length. With Equation (5.109) a restriction in length of 400 m was applied, and it is suitable for resistance extrapolation using a form factor method and the 1957 ITTC friction line. It assumes a standard roughness of 150 μm.

Townsin (1987) has produced a modified expression for the calculation of ΔC_F based on the AHR parameter and applicable to new and relatively smooth vessels:

$$\Delta C_F \times 10^3 = 44\left[\left(\frac{\mathrm{AHR}}{L}\right)^{1/3} - 10(R_n)^{-1/3}\right] + 0.125 \quad (5.110)$$

The effects of the distribution of roughness on the skin friction of ships have been explored by Kauczynski and Walderhaug (1987). They showed the most important part of the hull with respect to the increase in resistance due to roughness is the bow region. However, the length of the significant part of this portion of the hull decreases as the block coefficient increases. In the case of vessels with higher block coefficients, of the order of 0.7 to 0.8, the afterbody also plays a significant role. Figure 5.116, based on Kauczynski and Walderhaug (1987), illustrates this

Powering 315

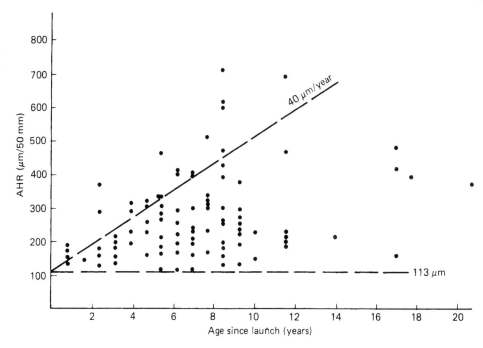

Figure 5.115 Survey of hull roughness conducted during period 1984–85 (Reproduced with permission from Townsin et al. (1986)).

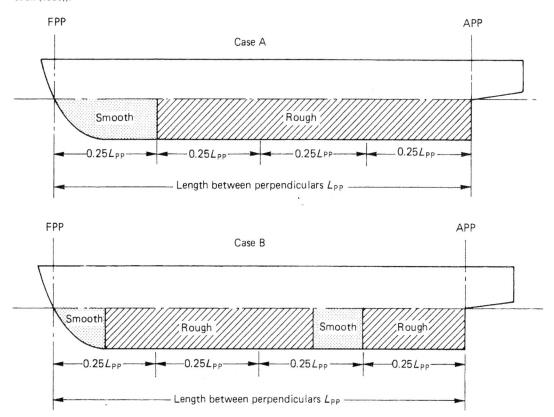

Figure 5.116 Hull smoothing regimes considered by Kauczynski and Walderhaug (Reproduced from Kauczynski and Walderhaug (1987)).

point by considering two smoothing regimes for a vessel. In case A, a smooth strip equal to 25 per cent of L_{WL} was fixed to the bow, whereas in case B the smooth area was divided into two equal portions, both with a length equal to 12.5 per cent L_{WL}. In both cases the smoothed areas were equal. Calculations showed that the reduction in C_F compared to the whole rough surface were 0.105×10^{-3} and 0.119×10^{-3} for cases A and B, respectively, thus showing an advantage for the smoothing regimes of case B. In order to compute the value of C_F corresponding to paint roughness, Kauczynski and Walderhaug based their calculations on a conformal mapping technique for describing the hull form and used a momentum integral method for the calculation of the three-dimensional turbulent boundary layer characteristics. The results of these calculations for five hull forms of the Series 60 models with block coefficients between 0.60 and 0.80 have shown that the increase of frictional resistance due to roughness ΔC_F is a function of block coefficient, Reynolds number, $R_{t(50)}$ and $R_{t(1)}$. A regression procedure was applied by the authors to these results in order to give a readily applicable approximation of the form

$$\Delta C_F = a_0 + a_i \bar{k}_B^{*1/i} + b_j \Delta C_B^{*j} \frac{\bar{k}_B^*}{L_{WL}} \bar{k}_1^* R_n^* \quad (5.111)$$

where $i, j = 1, 2, 3$ and

$$\bar{k}_B^* = \frac{\bar{k}_B / L_{WL}}{3.32 \times 10^9}; \quad R_n^* = \frac{R_n}{2.7 \times 10^6}; \quad \bar{k}_1^* = \frac{k_1}{105}$$

with

$$\Delta C_B^* = \frac{C_B - 0.6}{0.2}$$

In order to derive the coefficients a_0 and a_i in Equation (5.111) a further polynomial expression has been derived as follows:

$$a_i = \sum_{n=1}^{4} \sum_{m=1}^{5} f_{i,p} (\bar{k}_1^*)^{n-1} (R_n^*)^{m-1}$$

where $i = 0, 1, 2, 3$
$p = m + 5(n - 1)$

The coefficients $f_{i,p}$ are given by Table 5.23 for all values of $p = 1, 2, 3, \ldots, 20$. The coefficients b_j are given by Table 5.24.

The calculation procedure is subject to the constraints imposed by the model series and the conditions examined. Thus $\bar{k}_{1_{max}}$, $(\bar{k}_B/L_{WL})_{max}$, $R_{n_{max}}$ and $C_{B_{min}}$ are defined as $105 \mu m$, 3.32×10^9, 2.7×10^6 and 0.6, respectively. The method described has been examined in comparison with others, notably those by Hohansson, Townsin and Bowden, for a 16 knot, 350 m tanker, and the results are shown in Figure 5.117. The

Table 5.23 Values of coefficient $f_{i,p}$ (taken from Kauczynski and Walderhaug (1987)).

P	$f_{0,p} \times 10^3$	$f_{1,p} \times 10^3$	$f_{2,p} \times 10^3$	$f_{3,p} \times 10^3$
1	−0.05695	−0.08235	−0.48093	0.43460
2	−0.25473	−0.73105	1.01946	−1.37640
3	−0.18337	−2.01563	1.31724	−0.11176
4	0.38401	0.79786	2.02432	−2.30461
5	−0.27985	0.27460	−2.56908	2.26801
6	0.12397	0.47117	1.30053	−1.43575
7	1.95506	−10.87320	35.18020	−24.04790
8	−4.89111	17.57430	−63.66010	49.25690
9	1.70315	−5.44915	19.77400	−15.92990
10	0.72533	−2.50564	9.33041	−7.31122
11	−0.07676	−0.74104	0.62533	−0.06440
12	−2.93232	9.38549	−36.49980	29.61980
13	1.88597	−0.39504	6.04098	−11.67790
14	6.04607	−23.66800	88.78930	−64.38880
15	−5.02286	17.10700	−64.93670	49.93150
16	0.07829	0.00438	0.56607	−0.56425
17	0.04596	0.25232	−0.09525	−0.39173
18	3.04651	−10.75950	42.36420	−31.51610
19	−7.47250	23.29540	−93.06020	72.79150
20	4.26166	−13.00580	51.38600	−40.80970

Table 5.24 Coefficients b_j (taken from Kauczynski and Walderhaug (1987)).

j	$b_j \times 10^3$
1	−0.09440
2	0.01126
3	0.13756

range of values predicted for $(\bar{k}_B/\bar{k}_1)$ in the range 4 to 8, typical values for painted surfaces, embrace the result from Bowden's formula. Nevertheless, Bowden's formula does not consider the effects of R_n and C_B, and consequently in other examples differences may occur. With regard to Townsin and Johansson's formulae, close agreement is also seen in the region where $\bar{K}_1$ is of the order of $30 \mu m$.

Walderhaug (1986) suggests an approximation to the procedure outlined above, which has the form

$$\Delta C_F \times 10^3 \simeq 0.5 \left(\frac{k_E \times 10^6}{L} \right)^{0.2}$$
$$\times \left[1 + \left(\frac{C_B - 0.75}{0.7} \right)^2 \right]$$
$$\times \left[\ln \left(1 + \frac{u_\tau k_E}{v} \right) \right]^{0.7} \quad (5.112)$$

where the effective roughness k_E is given by

$$k_E = \left(\frac{\bar{k}}{\lambda} \right)_1 (R_{t(50)} - K_A)$$

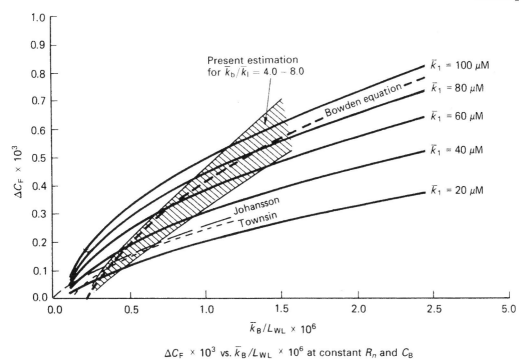

Figure 5.117 Comparison of roughness ΔC_F values (Reproduced with permission from Kauczynski and Walderhaug (1987)).

with the roughness to wavelength ratio $(\bar{k}/\lambda)_1$ at $= 1$ mm and the admissible roughness k_A given by

$$k_A = \frac{fv}{V}(\ln R_n)^{1.2}$$

with $f = 2.5$ for painted surfaces and the friction velocity

$$u_\tau = \frac{V}{(\ln R_n)^{1.2}}$$

5.7.3 Hull drag reduction

Methods involving the injection of small quantities of long-chain polymers into the turbulent boundary layer surrounding a hull form, such as polyethylene oxide, were shown in the 1960s to significantly reduce resistance, provided the molecular weight and concentration were chosen correctly. Experiments conducted at that time suggested that the reduction in drag was linked to changes in the structure of the turbulence by the addition of the long-chain polymers. Frenkiel et al. (1976) and Berman (1978) discuss these effects in detail.

These methods which rely on the injection of such substances into the sea, however, are unlikely to be environmentally acceptable today. Nevertheless, current research is focusing on a range of methods involving boundary layer fluid injection and manipulation. These methods embrace the injection of low-pressure air, either in the formation of air bubble interfaces between the hull and the sea water or through the provision of an air cushion trapped by an especially developed hull form. Some attention is also devoted to the injection of non-toxic or environmentally friendly fluids into the hull boundary layer.

5.7.4 Propeller roughness and fouling

Propeller roughness is a complementary problem to that of hull roughness and one which is no less important. As in the hull roughness case, propeller roughness arises from a variety of causes, chief of which are marine growth, impingement attack, corrosion, cavitation erosion, poor maintenance and contact damage.

The marine growth found on propellers is similar to that observed on hulls except that the longer weed strands tend to get worn off. Notwithstanding this, weed having a length of the order of 10 to 20 mm is not uncommon on the minor regions of the blade, as

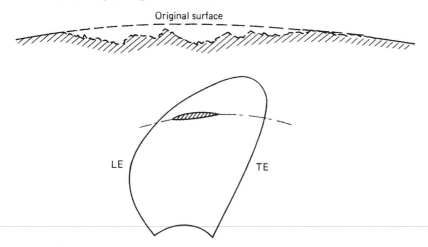

Figure 5.118 Typical cavitation damage profile.

indeed are stalked barnacles which are frequently found alive on the blades after a vessel has docked subsequent to a considerable journey. Marine fouling of these types increase the power absorption of the propeller considerably, which for a fixed pitch propeller will result in a reduction of service rotational speed.

Impingement attack resulting from the passage of the water and the abrasive particles held in suspension over the blade surfaces normally affects the blades in the leading edge region and particularly in the outer radii of the blade where the velocities are highest. This results in a comparatively widespread area of fairly shallow depth surfaces roughness. Similarly with corrosion of either the chemical or electrochemical kind. Furthermore, with both corrosive and impingement roughness the severity of the attack tends to be increased with the turbulence levels in the boundary layer of the section. Consequently, subsequent to an initial attack, increased rates of surface degradation could be expected with time.

Cavitation erosion is normally, but not always, confined to localized areas of the blade. It can vary from a comparatively slight and relatively stable surface deterioration of a few millimetres in depth to a very rapid deterioration of the surface reaching depths of the order of the section thickness in a few days. Fortunately, the later scenario is comparatively rare. Cavitation damage, however, presents a highly irregular surface, as seen in Figure 5.118, which will have an influence on the drag characteristics of the blade sections. Blade-to-blade differences are likely to occur in the erosion patterns caused by cavitation, and also to some extent with the forms of roughness. This will of course influence the individual drag characteristics of the sections.

Finally, poor maintenance and contact damage influence the surface roughness; in the former case perhaps by the use of too coarse grinding discs and incorrect attention to the edge forms of the blade, and in the latter case, by gross deformation leading both to a propeller drag increase and also to other secondary problems; for example, cavitation damage. With regard to the frequency of propeller polishing there is a consensus of opinion between many authorities that it should be undertaken in accordance with the saying 'little and often' by experienced and specialized personnel. Furthermore, the pursuit of super-fine finishes to blades is generally not worth the expenditure, since these high polishes are often degraded significantly during transport or in contact with ambient conditions.

The effects of surface roughness on aerofoil characteristics have been known for a considerable period of time. These effects are principally confined to the drag coefficient and a typical example taken from Abbot and von Doenhoff (1958) is seen in Figure 5.119 for a National Advisory Committee for Aeronautics (NACA) 65-209 profile.

The effect on section lift is small since the lift coefficient is some 20 to 30 times greater than the drag coefficient and studies conducted by the ITTC showed that the influence of roughness on the lift coefficient can be characterized by the relationship

$$\Delta C_L = -1.1 \Delta C_D \qquad (5.113)$$

Results such as those shown in Figure 5.119 are based on a uniform distribution of sand grain roughness over the section surface. In practice, however, this is far from the case, and this implies that a multi-parameter statistical representation of the propeller surface embracing both profile and texture might be more appropriate than a single parameter such as the maximum peak-to-valley height. Grigson (1981) shows two surfaces to illustrate this point (Figure 5.120) which have approximately the same

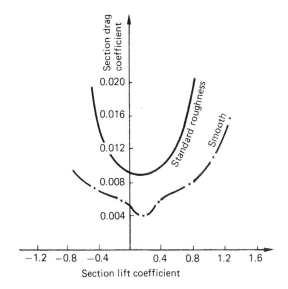

Figure 5.119 Effect of roughness on NACA 65-209 profile.

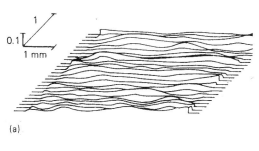

(a)

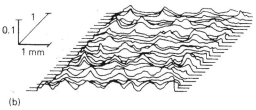

(b)

Figure 5.120 Example of two different textures having approximately the same roughness amplitude (Reproduced with permission from Grigson (1981)).

roughness amplitudes but quite different textures. In general propeller surface roughness is of the Colebrook–White type and can be characterized in terms of the mean apparent amplitude and a surface texture parameter.

The topography of a surface can be reduced into three component terms: roughness, waviness and form errors as shown in Figure 5.121. Clearly, the definition of which category any particular characteristic lies is related to the wavelength of

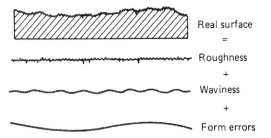

Figure 5.121 Reduction of surface profile into components.

the characteristic. The International Standards Organization (ISO) has used two standards in the past; these are the peak-to-valley average (PVA) and the centre-line average (CLA or R_a) and, the definition of these terms are as follows:

Peak-to-valley average (PVA) — This is the sum of the average height of the peaks and the average depth of the valleys. It does not equate to the R_t parameter, since this latter term implies the maximum rather than the average value.

Centre-line average (CLA or R_a) — This is the average deviation of the profile about the mean line and is given by the relation

$$R_a = \frac{1}{l}\int_{x=0}^{l} |y(x)|\,dx \quad (5.114)$$

where l is the length of the line over which the roughness distribution $y(x)$ is measured.

There is unfortunately very little correspondence between the values derived from a PVA or CLA analysis. Some idea of the range of correspondence can be deduced from Figure 5.122, taken from SMM(18) for mathematically defined forms. The authors of SMM(18) suggest a value of the order of 3.5 when converting from CLA to PVA for propeller surfaces. The difference between these two measurement parameters is important when comparing the 1966 and 1981 ISO surface finish requirements for propellers, since the former was expressed in terms of PVA whilst the latter was in CLA. Sherrington and Smith (1987) discuss the wider aspects of characterizing the surface topography of engineering surfaces.

Table 5.25 itemizes these requirements for Class 'S' and Class '1' propellers.

Several methods of surface roughness assessment exist and these range from stylus-based instruments through to the 'Rubert' comparator gauge. For

Type	Shape	PVA (μm)	PVA/CLA
Sinusoidal		8	$\pi = 3.142$
Triangular		10	4
Occasional triangular protrusions	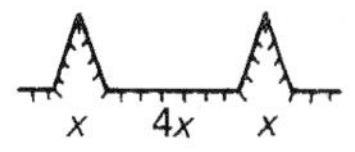	15.5	6.2
Occasional triangular indentations	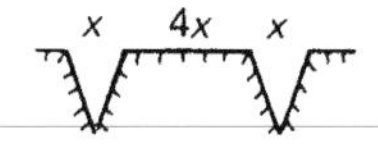	15.5	6.2
Occasional triangular protrusions and indentations	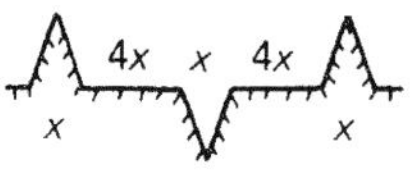	50	20
Parabolic indentations (mathematical scratches)		9.7	3.9
Occasional parabolic indentations (scratches)	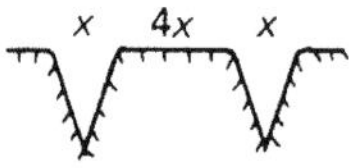	22.5	9
Parabolic cusps		9.7	3.9
Occasional parabolic cusps	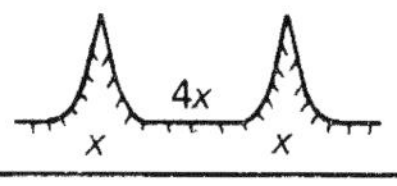	22.5	9

CLA = centre-line average (the average deviation from the mean line)
PVA = peak-to-valley average (the average height of the peaks plus the average depth of the valleys)

Figure 5.122 Comparison between CLA and PVA measurements of roughness for constant CLA value of 2.5 μm R_a (Reproduced with permission from SMM(18)).

Table 5.25 ISO surface finish requirements.

Specification		Class 'S'	Class 'I'	Units
ISO R484	1966	3	9	μm (PVA)
ISO R484/1 ISO R484/2	1981	3	6	μm (Ra)

the stylus-based instruments it has been generally found that a wavelength cut-off value of the order of 2.5 mm gives satisfactory values for the whole range of propellers. The stylus-based instrument will give a direct measure of the surface profile, which is in contrast to the comparator gauge method in which the surface of the blades at particular points are 'matched' to the nearest surface on the reference gauge. The 'Rubert' gauge which is perhaps the most commonly used comprises six individual surfaces tabulated A through to F as seen in Figure 5.123. These surfaces have been the subject of extensive measurement exercises by a number of authorities. Townsin et al. (1985) undertook a series of studies to determine the value of Muskers' apparent height h' from both his original definition and a series of approximations. The values derived for the apparent roughness together with the maximum peak-to-valley

Powering 321

observation on a blade. This is because the roughness will vary over a blade and different parts of the blade will be more significant than others, chiefly the outer sections since the flow velocities are higher. Furthermore, differences will exist from blade to blade. To overcome this problem a matrix of elements should be superimposed on the suction and pressure surfaces, as shown in Figure 5.124. In each of the twelve regions defined by the matrix on each surface of the blade several roughness measurements should be taken in the direction of the flow and widely spaced apart. A minimum of three measurements is recommended in each patch from which a mean value can be taken, Townsin *et al.* (1985).

5.7.5 Generalized equations for the roughness-induced power penalties in ship operation

Townsin *et al.* (1985) established a valuable and practical basis upon which to analyse the effects of roughness on the hull and propeller of a ship. In this analysis they established a set of generalized equations, the derivations of which form the basis of this section. The starting point for their analysis is to consider the power delivered to the propeller in order to propel a ship at a given speed V_s through the water:

$$P_D = \frac{RV_s}{\text{QPC}}$$

where R is the resistance of the ship at the speed V_s and the QPC is the quasi-propulsive coefficient given by

$$\text{QPC} = \eta_H \eta_r \eta_0$$
$$= \eta_H \eta_r \frac{K_T}{K_Q} \frac{J}{2\pi}$$

Consequently, the basic relationship for the delivered power P_D can be re-expressed as follows:

$$P_D = \frac{\pi \rho S V_s^3 C_T K_Q}{K_T J \eta_H \eta_r} \quad (5.116)$$

by writing the ship resistance R as $\frac{1}{2} S V_s^2 C_T$. Equation (5.116) can be linearized by taking logarithms and differentiating the resulting equation to give

$$\frac{dP_D}{P_D} = \frac{d\rho}{\rho} + \frac{dS}{S} + \frac{3dV_s}{V_s} + \frac{dC_T}{C_T} + \frac{dK_Q}{K_Q} - \frac{dK_T}{K_T}$$
$$- \frac{dJ}{J} - \frac{d\eta_H}{\eta_H} - \frac{d\eta_r}{\eta_r}$$

In this equation it can be assumed for all practical purposes that the density (ρ), the wetted surface area (S) and the relative rotative efficiency (η_r) are

Figure 5.123 The Rubert gauge.

Table 5.26 Rubert gauge surface parameters.

Rubert surface	h' equation (5.107) (μm)	h' (approximation) equation (5.115) (μm)	R_t (2.5) (μm)	R_a (2.5) (μm)
A	1.32	1.1	6.7	0.65
B	3.4	5.4	14.2	1.92
C	14.8	17.3	31.7	4.70
D	49.2	61	50.8	8.24
E	160	133	97.2	16.6
F	252	311	153.6	29.9

Note: a and b in equation (5.107) taken as 0.5 and 0.2, respectively.

amplitude $R_t(2.5)$ quoted by the manufacturers of the Rubert gauge is given in Table 5.26. Also in this table is shown the approximation to h' derived from the relation

$$h' \simeq 0.0147 R_a^2(2.5) P_c \quad (5.115)$$

where P_c is the peak count per unit length and is used as a texture parameter.

When measuring the roughness of a propeller surface it is not sufficient to take a single measurement or

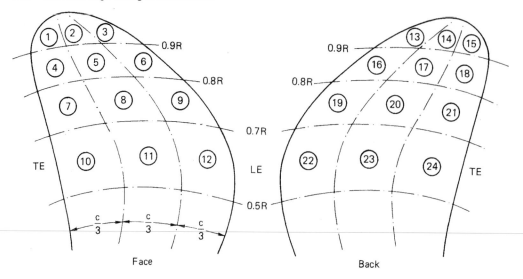

Figure 5.124 Definition of patches for recording propeller roughness.

unaffected by increases in roughness of the order normally expected in ships in service. As consequence these terms can be neglected in the above equation to give

$$\frac{dP_D}{P_D} = \frac{3dV_s}{V_s} + \frac{dC_T}{C_T} + \frac{dK_Q}{K_Q} - \frac{dK_T}{K_T} - \frac{dJ}{J} - \frac{d\eta_H}{\eta_H}$$

In addition, since roughness, as distinct from biological fouling, is likely to cause only relatively small changes in the power curve, these can then be approximated to linear functions. Consequently, the differentials can be considered in terms of finite differences:

$$\frac{\Delta P_D}{P_D} = \frac{3\Delta V_s}{V_s} + \frac{\Delta C_T}{C_T} + \frac{\Delta K_Q}{K_Q} - \frac{\Delta K_T}{K_T} - \frac{\Delta J}{J} - \frac{\Delta \eta_H}{\eta_H}$$

(5.117)

This equation clearly has elements relating to both the propeller and the hull, and can be used to determine the power penalty for propulsion at constant ship speed V_s:

$$\frac{\Delta P_D}{P_D} = \frac{\Delta C_T}{C_T} + \frac{\Delta K_Q}{K_Q} - \frac{\Delta J}{J} - \frac{\Delta K_T}{K_T} - \frac{\Delta \eta_H}{\eta_H}$$

(5.118)

Clearly, it will simplify matters considerably if Equation (5.118) can be decoupled into hull and propeller components, and therefore treated separately. This can be done subject to certain simplifications in the following way.

The terms $\Delta K_T/K_T$ and $\Delta K_Q/K_Q$ can be divided into two components; one due to propeller roughness and one due to the change in operating point assuming the propeller remained smooth:

$$\left.\begin{array}{l}\dfrac{\Delta K_Q}{K_Q} = \left(\dfrac{\Delta K_Q}{K_Q}\right)_R + \left(\dfrac{\Delta K_Q}{K_Q}\right)_J \\[2mm] \dfrac{\Delta K_T}{K_T} = \left(\dfrac{\Delta K_T}{K_T}\right)_R + \left(\dfrac{\Delta K_T}{K_T}\right)_J\end{array}\right\}$$

(5.119)

where the suffixes R and J denote propeller roughness and operating point, respectively. This distinction is shown in Figure 5.125 for the torque coefficient characteristics. The relative changes to the propeller characteristic due to roughness alone can be estimated from Lerb's theory of equivalent profiles.

Considering the second term in each of Equations (5.119), since for a smooth propeller

$$\Delta K_Q = \frac{dK_Q}{dJ} \Delta J$$

and similarly for ΔK_T, we write for the change in operating point terms in Equations (5.119):

$$\left.\begin{array}{l}\left(\dfrac{\Delta K_Q}{K_Q}\right)_J = \dfrac{J}{K_Q}\left(\dfrac{dK_Q}{dJ}\right)\left(\dfrac{\Delta J}{J}\right) \\[3mm] \left(\dfrac{\Delta K_T}{K_T}\right)_J = \dfrac{J}{K_T}\left(\dfrac{dK_T}{dJ}\right)\left(\dfrac{\Delta J}{J}\right)\end{array}\right\}$$

(5.120)

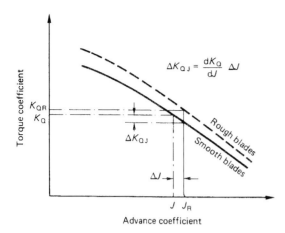

Figure 5.125 Effect of change of operating advance on propeller torque characteristics with rough and smooth blades.

Equations (5.121) and (5.122) and noting that C_T is wholly viscous so that $\Delta C_T = \Delta C_V$, we obtain

$$\frac{\Delta J}{J} = \left(\frac{1 - w_{TR}}{1 - w_T}\right)\left[\frac{1 + (\Delta K_T/K_T)}{1 + (\Delta C_V/C_T)}\right]^{1/2} - 1 \quad (5.123)$$

By applying the binomial theorem to Equation (5.123) and since $\Delta K_T/K_T$ and $\Delta C_V/C_T$ are small,

$$\frac{\Delta J}{J} = \left(\frac{1 - w_{TR}}{1 - w_T}\right)\left[1 + \frac{1}{2}\left(\frac{\Delta K_T}{K_T}\right) - \left(\frac{\Delta C_V}{C_T}\right)\right]$$

Hence from Equations (5.119) and (5.120) and substituting these into the above an explicit relationship can be found for the term $\Delta J/J$ as follows:

$$\frac{\Delta J}{J} = \frac{\left(\dfrac{1 - w_{TR}}{1 - w_T}\right)\left[1 + \dfrac{1}{2}\left(\dfrac{\Delta K_T}{K_T}\right)_R - \dfrac{\Delta C_V}{C_T}\right] - 1}{1 - \dfrac{1}{2}\left(\dfrac{1 - w_{TR}}{1 - w_T}\right)\dfrac{J}{K_T}\left(\dfrac{dK_T}{dJ}\right)}$$

In this equation the terms C_V and w_T relate to the hull roughness, excluding any propeller-induced wake considerations, and the term $(\Delta K_T/K_T)_R$ relates to the propeller roughness. Separating these terms out, we have

$$\frac{\Delta J}{J} = \frac{\left(\dfrac{1 - w_{TR}}{1 - w_T}\right)\left[1 - \dfrac{\Delta C_V}{C_T}\right] - 1}{1 - \dfrac{1}{2}\left(\dfrac{1 - w_{TR}}{1 - w_T}\right)\dfrac{J}{K_T}\dfrac{dK_T}{dJ}}$$

$$+ \frac{\dfrac{1}{2}\left(\dfrac{\Delta K_T}{K_T}\right)_R}{\left(\dfrac{1 - w_T}{1 - w_{TR}}\right) - \dfrac{1}{2}\dfrac{J}{K_T}\dfrac{dK_T}{dJ}} \quad (5.124)$$

The first term in Equation (5.124) is a function of hull roughness only and is the relative change in advance coefficient due to hull roughness only $(\Delta J/J)_H$. The second term is a function of both propeller and hull roughness; this can, however, be reduced to a propeller roughness function by assuming that

$$\left(\frac{1 - w_T}{1 - w_{TR}}\right) \simeq 1$$

when the change in propeller roughness can be approximated by the function

$$\left(\frac{\Delta J}{J}\right)_R \simeq \frac{\dfrac{1}{2}\left(\dfrac{\Delta K_T}{K_T}\right)_R}{1 - \dfrac{1}{2}\left(\dfrac{J}{K_T}\right)\left(\dfrac{dK_T}{dJ}\right)}$$

Now the term ΔJ is the difference between the rough and smooth or original operating points, as seen in Figure 5.125:

$$\Delta J = J_R - J$$

that is,

$$\Delta J = \frac{V_s}{D}\left[\frac{(1 - w_{TR})}{N_R} - \frac{(1 - w_T)}{N}\right]$$

since V_s is assumed constant from Equation (5.118). Hence by referring to the original operating point

$$\frac{\Delta J}{J} = \left(\frac{1 - w_{TR}}{1 - w_T}\right)\frac{N}{N_R} - 1 \quad (5.121)$$

Furthermore, since $\Delta K_T = (K_{TR} - K_T)$,

$$\frac{\Delta K_T}{K_T} = \left(\frac{K_{TR}}{K_T} - 1\right)$$

that is,

$$\frac{\Delta K_T}{K_T} = \frac{T_R}{T}\left(\frac{N}{N_R}\right)^2 - 1 \quad (5.122)$$

and by assuming an identity of thrust deduction between the rough and original smooth condition for a given ship speed, this implies

$$\frac{T_R}{T} = \frac{R_R}{R} = \frac{C_{TR}}{C_T} = \frac{C_T + \Delta C_T}{C_T} = \left[1 + \frac{\Delta C_T}{C_T}\right]$$

Hence, substituting this relationship into Equation (5.122), eliminating propeller revolutions between

Consequently, the total change in advance coefficient, Equation (5.124), can be decoupled into the sum of independent changes in hull and propeller roughness:

$$\frac{\Delta J}{J} \simeq \left(\frac{\Delta J}{J}\right)_{\text{Hull rough}} + \left(\frac{\Delta J}{J}\right)_{\text{Prop. rough}}$$

This immediately allows the power penalty $\Delta P_D/P_D$, expressed by Equation (5.118), to be decoupled into the following:

$$\left(\frac{\Delta P_D}{P_D}\right)_{\text{Prop. rough}} = \left(\frac{\Delta K_Q}{K_Q}\right)_{\text{Prop. rough}} - \left(\frac{\Delta J}{J}\right)_{\text{Prop. rough}}$$
$$- \left(\frac{\Delta K_T}{K_T}\right)_{\text{Prop. rough}} \quad (5.125)$$

and

$$\left(\frac{\Delta P_D}{P_D}\right)_{\text{Hull rough}} = \frac{\Delta C_V}{C_T} - \frac{\Delta \eta_H}{\eta_H} + \left(\frac{\Delta K_Q}{K_Q}\right)_{\text{Hull rough}} - \left(\frac{\Delta J}{J}\right)_{\text{Hull rough}}$$

but since propellers generally work in a region of the propeller curve, where the ratio, over small changes, of K_T/K_Q is relatively constant, this latter equation reduces to

$$\left(\frac{\Delta P_D}{P_D}\right)_{\text{Hull rough}} = \frac{\Delta C_V}{C_T} - \frac{\Delta \eta_H}{\eta_H} - \left(\frac{\Delta J}{J}\right)_{\text{Hull rough}}$$

$$(5.126)$$

Equations (5.118), (5.125) and (5.126) form the generalized equations of roughness-induced power penalties in ship operation. These latter two equations can, however, be expanded to give more explicit relationship for the hull and propeller penalties.

In the case of Equation (5.125) for the propeller penalty, the propeller roughness effects $(\Delta K_Q/K_Q)$ and $(\Delta K_T/K_T)$ can be estimated from Lerb's equivalent profile method, from which the following relationship can be derived in association with Burrill's analysis:

$$\left(\frac{\Delta P_D}{P_D}\right)_{\text{Prop. rough}} = \left[\frac{2.2}{(P/D)} - 1.1 + \left\{\frac{3.3(P/D) - 2J}{2.2(P/D) - J}\right\}\right.$$
$$\left. \times (0.45(P/D) + 1.1)\right] \left(\frac{\Delta c_D}{c_L}\right)_{0.7}$$

$$(5.127)$$

For the hull roughness contribution, Townsin et al. (1985) shows that by assuming a constant thrust deduction factor and employing the ITTC 1978 formula for wake scaling such that

$$w_{TR} = t + (w_T - t)\left[\frac{\Delta C_{FS} - \Delta C_{FT}}{(1+k)C_{FS} + \Delta C_{FT}} + 1\right]$$

Then the full roughness power penalty becomes

$$\left(\frac{\Delta P_D}{P_D}\right)_{\text{Hull rough}} = \left[1 - \left(\frac{w_T - t}{1 - w_{TR}}\right)\frac{C_T}{(1+k)C_{FS} + \Delta C_{FT}}\right]$$
$$\times \left(\frac{\Delta C_F}{C_T}\right) - \left(\frac{\Delta J}{J}\right)_{\text{Hull rough}} \quad (5.128)$$

where P_D is the delivered power at the propeller,
C_T is the ship thrust coefficient,
C_F is the ship frictional coefficient,
C_{FS} is the smooth ship frictional coefficient,
J is the advance coefficient,
P is the propeller pitch,
D is the propeller diameter,
Δc_d is the change in reference section drag, coefficient,
c_l is the reference section lift coefficient,
ΔC_{FT} is the increment in ship skin friction, coefficient in trial condition.

5.7.6 Monitoring ship performance

The role of the ship service analysis is summarized in Figure 5.126. Without for the moment considering the means of transmitting the data from the vessel, this information should have two primary roles for the ship operator. The first is to develop a data bank of information from which standards of performance under varying operational and environmental conditions can be derived. The resulting standards of performance, derived from this data, then become the basis of operational and chartering decision by providing a reference for a vessel's performance in various weather conditions and a reliable comparator against which the performance of sister or similar vessels can be measured. The role for the data records is to enable the analysis of trends of either the hull or machinery to be undertaken, from which the identification of potential failure scenarios and maintenance decisions can be derived.

Table 5.27 identifies the most common set of parameters that are traditionally recorded to a greater or lesser extent by seagoing personnel in the ship's engine and bridge logbooks. It is this information which currently forms the database from which analysis can proceed. In the table the measurement of shaft power

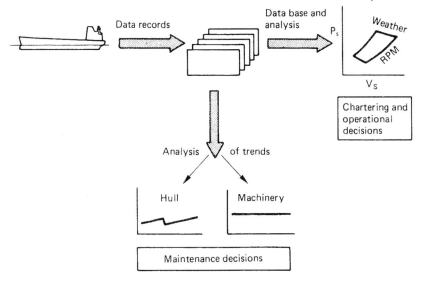

Figure 5.126 Role of ship service analysis.

Table 5.27 Traditionally recorded parameters in ship log books.

Deck log
Ship draughts (fore and aft)
Time and distance travelled (over the ground)
Subjective description of the weather (wind, sea state, etc.)
Ambient air and sea water temperature
Ambient air pressure
General passage information

Engine log
Cooling sea water temperature at inlet and outlet
Circulating fresh water cooling temperature and pressures for all engine components
Lubricating oil temperature and pressures
Fuel lever, load indicator and fuel pump settings
Engine/shaft revolution count
Turbocharger speed
Scavenge and injection pressures
Exhaust gas temperatures (before and after turbocharger)
Main engine fuel and lubricating oil temperatures
Bunker data
Generator and boiler performance data
Evaporator and boiler performance data
Torsion meter reading*

*denotes if fitted

has been noted with an asterisk, this is to draw attention to the fact that this extremely important parameter is only recorded in relatively few cases, due to the lack of a torsion meter having been fitted, and, as such this cannot be considered to be a commonly available parameter.

Traditionally, Admiralty coefficient (A_c) based methods have formed the basis of many practical service performance analysis procedures used by shipowners and managers. Current practice with some ship operators today is to simply plot a curve of Admiralty coefficient against time. Figure 5.127(a) shows a typical example of such a plot for a 140 000 tonnes dwt bulk carrier, from which it can be seen that it is difficult to interpret in any meaningful way due to the inherent scatter in this type of plot. One can, nevertheless, move a stage further with this type of study by analysing the relationship between the Admiralty coefficient and the apparent slip (S_a) as seen in Figure 5.127(b) which shows a convergence in the data and invites the drawing of a trend line through the data. The form of these coefficients is given by the well-known relationships

$$A_c = \frac{\Delta^{2/3} V_s^3}{P_s}$$

and

$$S_a = 1 - 30.86 \left(\frac{V_s}{PN}\right) \text{ metric}$$

The data for this type of analysis is extracted from the ship's deck and engine room log abstracts and the resulting curves of A_c plotted against S_a would normally be approximated by a linear relationship over the range of interest. Furthermore, apparent slip can be correlated to the weather encountered by the vessel by converting the description of the sea state,

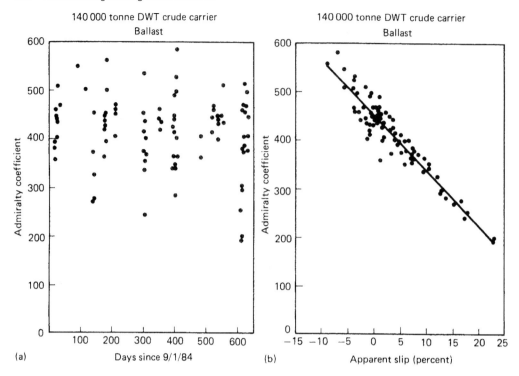

Figure 5.127 Common ship service procedures in use by the shipping community: (a) Admiralty coefficient versus time and (b) Admiralty coefficient versus apparent slip.

as recorded by the ship's navigating officers, to wave height according to an approved scale for that purpose. The wave heights derived in this way can be modified to take account of their direction relative to the ship and, having established the wave height versus apparent slip lines for the propeller, the Admiralty coefficient or other similar variable can be plotted against the appropriate line using the recorded apparent slip from the log book. Methods such as these, whilst providing a basis for analysis, can lead in some circumstances to misinterpretation. Furthermore, the Admiralty coefficient, although a useful criterion, is a somewhat 'blunt instrument' when used as a performance criterion since it fails to effectively distinguish between the engine and hull-related parameters. The same is also true for the alternative version of this equation, termed the fuel coefficient, in which the shaft horsepower (P_s) is replaced with the fuel consumption. This latter derivative of the Admiralty coefficient serves where the vessel is not fitted with a torsion meter, see Section 6.3.8.

Several coefficients of performance have been proposed based on various combinations of the parameters listed in Table 5.27. Whipps (1985), for example, attempts to split the overall performance of the vessel into two components – the responsibility of the engine room and the responsibility of the bridge watch-keepers.

Accordingly, three coefficients of performance are proposed:

1. K_1 – nautical miles/tonne of fuel (overall performance).
2. K_2 – metres travelled/shp/h (navigational performance).
3. K_3 – grams of fuel/shp/h (engine performance).

Clearly, these coefficients require the continuous or frequent monitoring of the parameters concerned and the presentation of the coefficients of performance to the ship's staff on a continuous or regular basis. Experience with these and other similar monitoring techniques suggests that they do aid the ship's staff to enhance the performance of the vessel by making them aware of the economic consequences of their decisions at the time of their actions in terms that are readily understandable, this latter aspect being particularly important.

More recently Bazari (2006) has considered the application of energy auditing to ship operation and design. This process is designed to undertake energy audits during a ship's operation either singly or across a fleet, particularly where there are a number of ships of the same design. From the results of these audits it then becomes possible to assess the potential for improvement in propulsion efficiency. This procedure

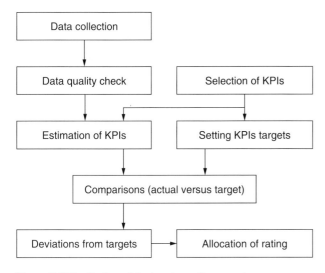

Figure 5.128 Outline of the benchmarking or rating process.

involves three principal activities in the benchmarking or rating process shown in Figure 5.128. The three main activities include:

1. Selection of Key Performance Indicators (KPIs) and specifying their reference target values.
2. Data collection and assuring the data quality.
3. Estimating the KPIs, comparing these to the reference targets, estimating deviations and allocating a rating to the ship.

This analysis procedure is applicable to many ship types; for example, passenger ships, tankers and container ships. However, to carry out the process effectively it is essential to give consideration of all aspects of ship design, machinery procurement, ship operation, alternative technologies and fuels within the analysis process and to take a holistic view of the ship operation.

In the case of new ships significant reductions in the ship's overall fuel consumption are considered feasible using these auditing processes to make improvements to the ship design and use of energy-efficient machinery. While procuring a more energy-efficient ship may be slightly more expensive in the first instance, when fuel prices are high or show a general upward trend, the extra initial investment may well be recovered in the ship's operational account. Indeed it has been found that the majority of the effort within the auditing process, given that the hydrodynamic design process has been satisfactorily undertaken, is concentrated on the engineering systems; the use of energy-efficient machinery; optimization of hotel, HVAC and refrigeration systems and the wider use of shore services. These considerations need to be input at the conceptual design phase of the ship and reviewed at a pre-contract specification stage to ensure that energy efficiency is fully considered as part of the ship design process.

When applied to ships that are in service the primary focus of the auditing processes should be on the reduction of fuel consumption. This can be achieved as outlined in Figure 5.128 using a combination of benchmarking, energy audits and performance monitoring. Within this process a systematic and holistic investigation needs to be undertaken which considers both technical and non-technical aspects of the operation. Furthermore, to obtain optimum results from the process it is often better if this is done by both an independent auditing practitioner and the ship's operator so that at least the two viewpoints are fully considered and agreed in the auditing process by which joint ownership of the result can then be achieved. This position has to be attained from a comprehensive level of data gathering and analysis combined with a shipboard energy survey. Moreover, in addition to the technical systems, the process should also take into account the operational profile of the ship and its main machinery together with any reference data from other similar ships.

In order to progress beyond the basic stages of performance monitoring it is necessary to attempt to address the steady-state ship powering equations:

$$P_s = \frac{RV_s}{\eta_0 \eta_r \eta_m} \left[\frac{1 - w_T}{1 - t} \right]$$

$$R = (1 - t)T$$

These equations clearly require a knowledge of the measured shaft power and thrust together with the ship and shaft speeds in association with the appropriate weather data. All of these parameters are potentially available, with the possible exception of shaft axial thrust. Thrust measurement has, in the past, proved notoriously difficult. In many instances this is due to the relative order of the magnitudes of the axial and torsional strain in the shaft, and has generally only been attempted for specific measurement exercises under carefully controlled circumstances, using techniques such as the eight gauge Hylarides bridge, (Carlton (2007), Chapter 17). When this measurement has been attempted on a continuous service basis the long-term stability of the measurement has frequently been a problem.

Consequently, it is generally possible to attempt only a partial solution to the steady-state powering equations defined above. To undertake this partial solution, the first essential is to construct a propeller analysis model so as to determine the thrust, torque and hence efficiency characteristics with advanced coefficient. The method of constructing these characteristics can vary depending on the circumstances and the data available and, as such, can range from standard series open water curves to more detailed lifting line, vortex lattice, techniques or boundary element. The resulting model of propeller action should, however, have the capability to accommodate allowances for propeller roughness and fouling, since this can, and does, influence the power absorption and efficiency characteristics to a marked extent. For analysis purposes it is clearly desirable to have as accurate a representation of the propeller characteristics as possible, especially if quantitative cost penalties are the required outcome of the exercise. However, if it is only performance trends that are required, then the absolute accuracy requirements can be relaxed somewhat since the rates of change of thrust and torque coefficients, dK_q/dJ and dK_t/dJ, are generally similar for similar types of propeller.

Figure 5.129 demonstrates an analysis algorithm. It can be seen that the initial objective, prior to developing standards of actual performance, is to develop two time series – one expressing the variation of effective wake fraction and the other expressing the specific fuel consumption with time. In the case of the effective wake fraction analysis this is a measure, over a period of time, of the change in the condition of the underwater surfaces of the vessel, since if either the hull or propeller surfaces deteriorate, the effective analysis wake fraction can be expected to reflect this change in particular ways. The second series, relating the specific fuel consumption to time, provides a global measure of engine performance. Should this latter parameter tend to deteriorate, and it is shown by analysis that it is not a false trend, for example an instrument failure, then the search for the cause of the fault can be carried on using the other parameters listed in Table 5.27. Typically, these other parameters might be exhaust temperatures, turbocharger performance, bearing temperatures and so on.

The capabilities of this type of analysis can be seen in Figure 5.130, which relates to the voyage performance of a bulk carrier of some thirty years ago. The upper time series relates to the specific fuel consumption, from which it is apparent that, apart from the usual scatter, little deterioration takes place in this global engine characteristic over the time interval shown. The second series is that of the analysis effective wake fraction, from which it can be seen that a marked increase in the wake fraction occurs during each docking cycle and coincident with the dry-docking periods, when cleaning and repainting takes place, the wake fraction falls to a lower level. It is of interest to note that after each dry-docking the wake fraction never actually regains its former value, and consequently underlines the fact that hull deterioration has at least two principal components. The first is an irreversible increase with age of the vessel and is the general deterioration of the hull condition with time, while the second is on a shorter-time cycle and related to repairable hull deterioration and biological fouling (see Section 5.7.2).

Comparison of this analysis with that of a recent 140 000 dwt crude oil carrier (Figure 5.131) shows how the deterioration between docking cycles has been reduced despite the docking cycle having increased from the order of a year in Figure 5.130 to around three and a half years in this latter example. As might be expected, in Figure 5.131 there is still an upward trend in the wake fraction with time, but nothing so dramatic as in the earlier case. The improvement in this case is almost entirely due to the use of modern paints and good propeller maintenance. Indeed, modern paint technology has advanced to such an extent that providing good application practice is adhered to then it can be expected that the hull condition will deteriorate relatively slowly.

Instrumentation errors are always a potential source of concern in performance analysis methods. Such errors are generally in the form of instrument drift, leading to a progressive distortion of the reading and, these can generally be detected by the use of trend analysis techniques. Alternatively, they are in the form of a gross distortion of the reading in which case the principles of deductive logic can be applied.

On many ships today a complete record in the ship's logs of all of the engine measured parameters and ship's operational entries are made relatively few times a day. Typically, one entry per day for the comprehensive set of data on many deep sea vessels assuming an automatic data logging system is not installed. Provided that the vessel is working on deep sea passages lasting a number of days, then

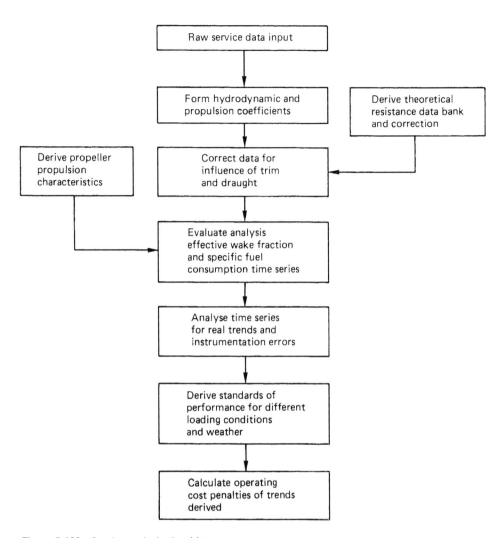

Figure 5.129 Service analysis algorithm.

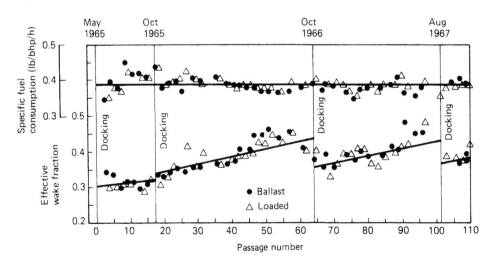

Figure 5.130 Service analysis for a bulk carrier (mid-1960s).

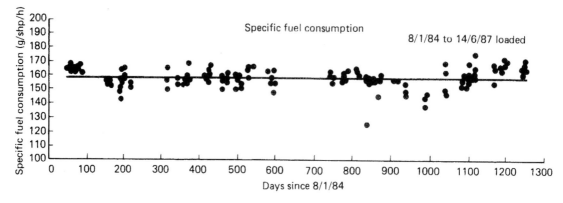

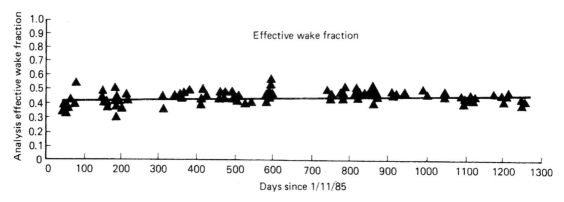

Figure 5.131 Service analysis for a 140 000 tonnes dwt crude carrier.

this single entry practice, although not ideal from the analysis viewpoint, will probably be satisfactory for the building up of a profile of the vessel's operating characteristics over a period of time. Data logging by automatic or semi-automatic means clearly enhances this situation and leads to a much more accurate profile of the ship operation in a much shorter-time frame. This is to some extent only an extension of the present procedures for alarm monitoring. In the alternative case, of a short sea route ferry for example, this once or twice per day level of recording is not appropriate since the vessel may make many passages in a day lasting of the order of one or two hours.

Over a suitable period of time a data bank of information can be accumulated for a particular ship or group of vessels. This data bank enables the average criterion of performance for the ship to be derived. A typical example of such a criterion is shown in Figure 5.132 for a medium-sized container ship. This diagram, which is based on the actual ship measurements and corrected for trim, draught and fouling, relates the principal operational parameters of power, ship and shaft speed, and weather. Consequently, such data, when complied for different trim draught and hull conditions can provide a reliable guide to performance for chartering purposes on any particular class of trade route.

Trim and draught have important influences on the performance of the vessel. Draught is clearly a variable determined by the cargo that is being carried. Trim, however, is a variable over which, for a great many vessels, some control can be exercised by the ship's crew. If this is done effectively and with due regard to weather conditions, then this can result in considerable savings in the transport efficiency of the vessel.

The traditional method of data collection is via the deck and engine room log, and this is the most commonly used method today. In terms of current data processing, capabilities, which involve both significant statistical trend analysis and detailed hydrodynamic analysis components, this method of data collection is far from ideal since, of necessity, it involves the translation of the data from one medium to another.

The immediate solution to this problem is to be found in the use of the desktop or personal computers working in either the on-line or off-line mode. Such computers are ideal for many shipboard applications since, in addition to having large amounts of memory,

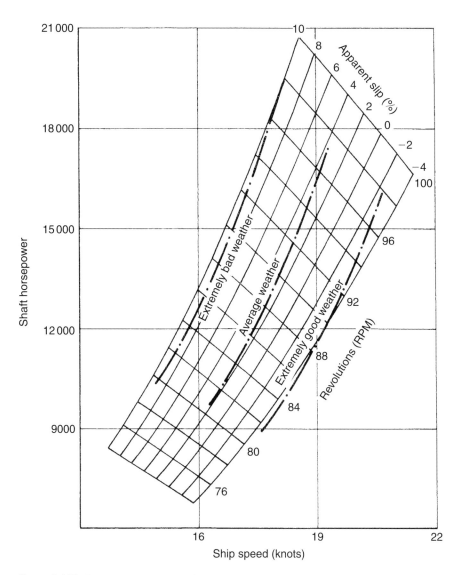

Figure 5.132 Typical power diagram for a container ship.

they are small and user friendly, can readily be provided with custom-built software and are easily obtainable in most parts of the world. When used in the off-line mode, they are to some extent an extension of the traditional method of log entries, where instead of being written by hand in the book the data is typed directly into the computer for both storage on disc media and also for the production of the normal log sheet. This permits both easy transfer of the data to shore-based establishments for analysis and the undertaking of simple trend analysis studies on board. Such 'on board' analyses methods produce tangible benefits if conducted advisedly.

When the small computer is used in the 'on-line' mode the measured parameters are input directly from the transducers via a data acquisition system to the computer and its disc storage. In this way, a continuous or periodic scanning of the transducers can take place and the data, or representative samples of it, can be stored on disc as well as providing data for a continuous statistical analysis. Such methods readily raise alarms when a data parameter moves outside a predetermined bound in a similar way to conventional alarm handling.

This clearly has advantage in terms of man hours but does tend to add to the degree of remoteness between operator and machine unless considerable attention has been paid to the 'user-friendliness' of the system. This ergonomic aspect of data presentation is particularly important if the system is to be accepted and used to its full potential by the ship's operating personnel. All too often poorly designed computer-based monitoring equipment is seen lying largely discarded in the engine control room or on the bridge either because it has been insufficiently ruggedly designed for marine use and is, thus, prone to failure and regular breakdown or, more frequently, because the engineering design has been adequately undertaken, but the data and information it presents is not in an easily assimilative form for the crew and the manuals describing its operation contain too much specialized jargon which is unfamiliar to the operator. This underlines the importance of choosing a monitoring system which satisfies the company's commercial objectives as well as being compatible with the operator's actual and perceived requirements.

These small on-line systems are the first step towards an integrated ship management system embracing the activities of the deck, engine room and catering departments. Such systems are beginning to make their appearance. With these systems, the vessel's operating staff and shore-based managers are still required to assimilate this data, albeit presented in a much more generally comprehensible form than has previously been the case, and use it in the context of the commercial constraints, classification society requirements and statutory regulations. The next generation of computer-based systems is likely to involve the use of expert systems and neural network technologies. By introducing this type of technology, the ship operator is provided with an interpretive back-up of accumulated knowledge and expertise which has been introduced into the computer-based system and will be available in the form of supplementary information and suggested courses of action at the time the particular problem occurs. Such capabilities will, if carefully constructed, introduce a level of consistency of decision making, hitherto unprecedented, and also prevent knowledge and experience being lost when staff members leave or retire from a company. In the specific case of diagnostic information, such as might occur from the data transmitted from a diesel engine, the artificial intelligence aspects and neural network will encompass pattern recognition techniques or determine the probability of a particular fault or failure scenario. Katsonlakos *et al.* (1989) consider some machinery and ship management-based approaches in this area.

Figure 5.133 schematically illustrates a system where a shipboard-based monitoring system provides a level of intelligent support to the ship operators for local operational and maintenance decision making. The data collected by the various sensory systems can then be transmitted as data streams via narrow and broad band communication links to a shore-based station at some convenient location which undertakes more detailed and long-term analysis of the data and also supports a database for the vessel. Such a shore-based analysis facility is then in a position to provide data for the various interested communities, since the analysis will contain not only commercial data but also technical data on both ambient conditions and any impending short- and long-term failures.

By way of example of the advanced models available or in the process of development, for a given trade route or operating pattern, the operational economics look to establish the most efficient routing and voyage planning for a ship so as to avoid the penalties of added resistance when encountering poor weather.

The assessment of the added resistance of a ship can be conveniently made from model tests or, alternatively, estimated from non-linear computational methods, Section 5.1, and such an assessment is made for a variety of sea conditions. Given, therefore, the knowledge of the ship's powering behaviour in a variety of sea conditions and where the propeller design point was fixed with respect to the slow-speed diesel engine operating diagram, use can be made of weather forecast information to optimize the voyage plan. Such planing processes have been put to good use in passenger liner trades for voyage time-keeping purposes, but can also be used to minimize voyage costs in the sense of optimizing the voyage plan with respect to any number of voyage attributes. These attributes might be the ship performance characteristics, engine performance parameters, ship loading and so on which can then be relaxed with respect to the constraints acting on the voyage, for example anticipated poor weather, port slots and the wider issues surrounding the transport chain of which the container ship voyage is but one part. As such, it is possible formulate a mathematical problem:

$$\text{Min } V_c = f(A_1, A_2, A_3, \ldots; C_1, C_2, C_3)$$

to minimize the voyage overall cost V_c against a set of voyage attributes A_n and constraints C_n. If the voyage attributes and constraints can be linearized then the solution to this exercise is relatively trivial in mathematical terms, however, for most practical situations the solution will exhibit at least some non-linear characteristics which then makes the cost minimization function more complex problem to solve but, nevertheless, soluble in many cases using available numerical methods.

Powering 333

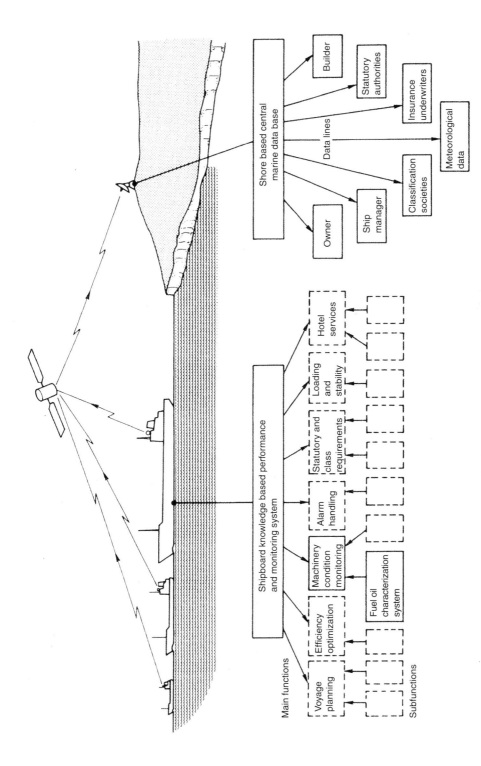

Figure 5.133 Computer-based integrated ship management structure.

References (Chapter 5)

Abbott, Ira H., and Doenhoff A.E. von. (1958). *Theory of Wing Sections*. Dover, New York.

Abdel-Maksoud, M. and Heinke, H.J. (2002). Scale effects on ducted propellers. *Proc. 24th Symp. on Naval Hydrodyn.*, Fukuoka, Japan.

Abdel-Maksoud, M., Hellwig, K., and Blaurock, J. (2004). Numerical and experimental investigation of the hub vortex flow of a marine propeller. *Proc. 25th Symp. on Naval Hydrodyn.*, St. John's, Newfoundland.

AIR (2003). *Fertigung-Technologie GmbH*. Rostock, Germany.

Anantha Subramanian, V., Dhinesh, G. and Deepti, J.M. (2006). Resistance optimisation of hard chine high speed catamarans. *J. Ocean Technol.*, Vol. I, No. I, Summer.

Anderson, P. (1974). Lifting-line theory and calculations for supercavitating propellers. *ISP*.

Andersen, S.V. and Andersen, P. (1986). Hydrodynamical design of propellers with unconventional geometry, *Trans. RINA*.

Andersen, P., Friesch, J., Kappel, J.J., Lundegaard, L. and Patience, G. (2005). Development of a marine propeller with nonplanar lifting surfaces. *Marine Technol.*, Vol. 42, No. 3, pp. 144–158, July.

Anon. (1987). A hull roughness evaluation project. *Marine Propulsion*, February.

Auf'm Keller, W.H. (1973). Extended diagrams for determining the resistance and required power for single-crew ships. *ISP*, Vol. 20.

Ayre, Sir Amos L. (1948). Approximating EHP – revision of data given in papers of 1927 *Trans. NECIES*.

Balhan, J. (1951). Metingen aan Enige bij Scheepsschroeven Gebruikelijke Profielen in Vlakke Stroming met en Zonder Cavitatie. *NSMB Publ.*, No. 99.

Ball, W., and Carlton, J.S. (2006). Podded Propulsor shaft loads from free-running model experiments in calm-water and waves. *Int. J. Maritime Eng.*, Vol. 148, No. Part A4, *Trans. RINA*.

Barrass, C.B. (1979). Ship-handling problems in shallow water. *MER*, November.

Bazari, Z. (2006). Ship Energy Performance Benchmarking/Rating: Methodology and Application. *Trans. WMTCConf. (ICMES 2006)*, London, March.

Bazilevski, Y.S. (2001). On the Propeller Blade Turbulization in Model Tests. *Laventiev Lectures*, Vol. 16, St Petersburg, pp. 201–206.

Beek, T. van. and Verbeek, R. (1983). Design aspects of efficient marine propellers. *Jahrbuch der Schiffbautechnis-chen Gesellschaft*.

Berlekom, van. W.B. (1981). Winds forces on modern ship forms – effects on performance. *Transactions of the North East Institute of Engineers and Shipbuilders*, Vol. 97, No. 4.

Berman, N.S. (1978). Drag reduction by polymers. *Ann. Review Fluid Mech.*, Vol. 10, pp. 27–64.

Bertram, V. and Jensen, G. (1994). Recent applications of computational fluid dynamics. *Ship Technology Research.*, 41/3, pp. 131–334.

Bertram, V. (1998). Marching towards the numerical ship model basin. *Euromech Conf. 374*, Poitiers, pp. 3–17.

Bertram, V. (2000). *Practical Ship Hydrodynamics*. Butterworth-Heinemann, Oxford, UK.

Besant, W.H. (1859). *Hydrostatics and Hydrodynamics*. Art ISE Cambridge University Press, Cambridge.

Betz, A. (1919). Ergebn. aerodyn. Vers. Anst. Gottingen.

Betz, A. (1920). Eine Erweiterung der Schrauben Theorie. *Zreits. Flugtech*.

Bidstrup, K. et al. (1981). Self activating module systems, a new approach to polishing antifouling paints. *21st Annual Marine and Offshore Coatings Conference*, California.

Bjarne, E. (1973). Systematic studies of control-rotating propellers for merchant ships, *IMAS Conference*, *Trans. I. Mar. E.*

Blount, D.L. and Hubble, E.N. (1981). Sizing for segmental section commercially available propellers for small craft. *Propellers 1981 Conference, Trans. SNAME.*

Boorsma, A. (2000). Improving Full Scale Ship Powering Predictions by Application of Propeller Leading Edge Roughness. *M.Sc. Thesis*, Delft, December.

Boswell, R.J. (1971). Design, Cavitation Performance, and Open Water Performance of a Series of Research Skewed Propellers. NSRDC Report No. 3339, March.

Boswell, R.J. and Cox, G.G. (1974). Design and model evaluation of a highly skewed propeller for a cargo ship. *Mar. Tech.*, January.

Bowden, B.S. (1970). A Survey of Published Data for Single-Screw Ocean Going Ships. NPL Ship Report 139, May.

Bowden, B.S. and Davison, N.J. (1974). Resistance increments due to hull roughness associated with form factor extrapolation methods. *NMI* Ship TM 3800, January.

Breslin, J.P. and Anderson, P. (1992). *Hydrodynamics of Ship Propellers*. Cambridge.

Brewer, W.H. and Kinnas, S.A. (1997). Experimental and viscous flow analysis on a partially cavitating hydrofoil. *J. Ship Res.*, Vol. 41, pp. 161–171.

Brockett, T.E. (1981). Lifting surface hydrodynamics for design of rotating blades. *Propellers '81 Symp. SNAME.*

Brown, D.K. (1983). *A Century of Naval Construction*. Conway.

Bulten, N. and Oprea A.L. (2006). Evaluation of McCormick's rule propeller tip cavitation inception based on CFD results. *6th Int. Symp. on Cavitation CAV2006*, Wageningen, September.

Burrill, L.C. (1944). Calculation of marine propeller performance characteristics. *Trans. NECIES*, Vol. 60.

Burrill, L.C. (1947). On propeller theory. *Trans. IESS*.

Burrill, L.C. (1955). *Considérations sur le Diamètre Optimum des Hélices*. Assoc. Tech. Maritime et Aéro-nautique, Paris.

Burrill, L.C. and Emerson, A. (1963). Propeller cavitation: some observations from the 16 in. propeller tests in the New King's College cavitation tunnel. *Trans. NECIES*, Vol. 79.

Burrill, L.C. (1964). Progress in marine theory, design and research in Great Britain. *Appl. Mech. Convention, Trans. I. Mech. E.*

Burrill, L.C. and Emerson, A. (1978). Propeller cavitation: further tests on 16 in. propeller models in the King's College Cavitation Tunnel. *Trans. NECIES*, Vol. 195.

Cane, P.du. *High Speed Craft*. Temple Press, London, 1951.

Caracostas, N. (1978). Off-design performance analysis of ducted propellers. *Propellers '78 Symp., Trans. SNAME*.

Carlton, J.S. (1989). Propeller service experience. *7th Lips Symposium*.

Carlton, J.S. (2007). *Marine Propellers and Propulsion*. Butterworth-Heinemann, Oxford, UK.

Carlton, J.S. and Bantham, I. (1978). Full scale experience relating to the propeller and its environment. *Propellers 78 Symposium, Trans. SNAME*, 1978.

Carlton, J.S. (1990). (Marine) Diesel engine exhaust emissions when operating with variable quality fuel and under service conditions. *Trans. I. Mar. E., IMAS 90 Conf.*, May.

Carlton, J.S. and Fitzsimmons, P.A. (2004). Cavitation: some full scale experience of complex structures and methods of analysis and observation. *Proc. 27th ATTC*, St John's Newfoundland, August.

Carlton, J.S. and Fitzsimmons, P.A. (2006). Full scale observations relating to propellers. *6th Int. Symp. on Cavitation CAV2006*, Wageningen, September.

Carlton, J.S., Fitzsimmons, P.A., Radosavljevic, D. and Boorsma, A. (2006). Factors influencing rudder erosion: experience of computational methods, model tests and full scale observations. *NAV2006 Conf.*, Genoa, June.

Caster, E.B. (1959). TMB 3-Bladed Supercavitating Propeller Series. DTMB Report No. 1245.

Caster, E.B. TMB 2, 3 and 4 Bladed Supercavitating Propeller Series. DTMB Report No. 1637.

Cebeci, T., Mosinskis, G.J. and Smith, A.M.O. (1972). Calculation of Viscous Drag on Two-Dimensional and Axi-Symmetric Bodies in Incompressible Flows. American Institute of Aeronautics and Astronautics. Paper No. 72-1.

Challier, G., Fortes-Patella, R. and Reboud, J.L. (2000). Interaction between pressure waves and spherical bubbles: discussion about cavitation erosion mechanism. *2000 ASME Fluids Eng. Summer Conf.*, Boston, June.

Cheng, H.M. (1965). Hydrodynamic Aspect of Propeller Design Based on Lifting Surface Theory Parts I and II. DTMB Report No. 1802, 1964, 1803, 1965.

Choi, J.K. and Kinnas, S.A. (2001). Prediction of Non-Axisymmetric Effective Wake by a Three-Dimensional Euler Solver. *J. Ship Res.*, Vol. 45, No. 1.

Choi, J.K. and Kinnas, S.A. (2003). Prediction of Unsteady Effective Wake by a Euler Solver/Vortex-Lattice Coupled method. *J. Ship Res.*, Vol. 47, No. 2.

Christie, A.O. (1981). Hull roughness and its control. *Marintech Conference*, Shanghai, China.

Chu, C., Chan, Z.L., She, Y.S. and Yuan, V.Z. (1979). The 3-bladed JD–CPP series. *4th Lips Propeller Symp.*, October.

Clayton, B.R. and Bishop, R.E.D. (1982). *Mechanics of Marine Vehicles*. E. and F.N. Spon, London.

Couser, P.R., Wellicome, J.F. and Molland, A.F. (1998). An improved method for the theoretical prediction of the wave resistance of transom-stern hulls using a slender body approach. *International Shipbuilding Progress*, Vol. 45, No. 444, December.

Coutier-Delgosha, O., Fortes-Patella, R. and Reboud, J.L. (2003a). Evaluation of the turbulence model influence on the numerical simulations of unsteady cavitation. *Trans. ASME.*, Vol. 125, January.

Coutier-Delgosha, O., Perin, J., Fortes-Patella, R. and Reboud, J.L. (2003b). A numerical model to predict unsteady cavitating flow behaviour in inducer blade cascades. *5th Int. Symp. on Cavitation (CAV2003)*, Osaka, November.

Coutier-Delgosha, O., Reboud, J.L. and Fortes-Patella, R. (2001). Numerical study of the effect of the leading edge shape on cavitation around inducer blade sections. *4th Int. Symp. on Cavitation (CAV2001)*.

Cram, L.A. (1980). *Cavitation and Inhomogeneities in Underwater Acoustics*. Springer, Berlin.

Crighton, D.G. (1985). The Kutta conditions in unsteady flows. *Ann. Rev. FluidMech.*, Vol. 17.

Cummings, D.E. (1968). Vortex Interaction in a Propeller Wake. Department of Naval Architecture Report No. 68–12, MIT.

Cummings, R.A. *et al.* (1972). Highly skewed propellers. *Trans. SNAME*, November.

Dang, J. (2004). Improving cavitation performance with new blade sections for marine propellers. *Int. Shipbuilding Progress*, 51.

Dawson, C.W. (1977). A practical computer method for solving ship–wave problems. *2nd Int. Conf. on Numerical Ship Hydrodynamics*, Berkeley.

Dawson, J. (1953). Resistance and propulsion of single-screw coasters. I, II, III and IV. *Trans. IESS*, Paper Nos. 1168, 1187, 1207 and 1247 in 1953, 1954, 1956, 1959, respectively.

Day, A.H., Doctors, L.J. and Armstrong, N. (1997). Concept evaluation for large, very high speed vessels. *5th Int. Conf. on Fast Sea Transportation. Fast.*

Day, S., Clelland, D. and Nixon, E. (2003). Experimental and numerical investigation of 'Arrow' Trimarans. *Proc. FAST2003*, Vol. III, Session D2.

Deng, Y. (2005). Performance database interpolation and constrained nonlinear optimisation applied to propulsor blade design. *Master's Thesis*, Department of Civil Engineering, The University of Texas at Austin.

Dickmann, H.E., and Weissinger, J. (1955). Beitrag zur. Theories Optimalar Dusenscharauben (Kortdusen). *Jahrbuch der Schiffbautechnischon Gessellschaft*, Vol. 49.

Dyne, G. (1980). A note on the design of wake-adapted propellers. *J. Ship. Res.*, Vol. 24, No. 4, December.

Dyne, G. A. (1967). Method for the Design of Ducted Propellers in a Uniform Flow. SSPA Report No. 62.

Eckhardt, M.K. and Morgan, W.B. (1955). A propeller design method. *Trans. SNAME*, Vol. 63.

Emerson, A. and Sinclair, L. (1978). Propeller design and model experiments. *Trans. NECIES*.

English, J.W. (1962). The Application of a Simplified Lifting Surface Technique to the Design of Marine Propellers. Ship Div. Report No. SH R30/62 NPL.

Euler, M. (1756). Théorie plus complète des machines qui sont mises en movement par la réaction de l'eau. L'Académie Royale des Sciences et Belles Lettres, Berlin.

Evans, J.P. and Svensen, T.E. (1987). Voyage simulation model and its application to the design and operation of ships. *Trans. RINA*.

Falcao de Campos, J.A.C. (1983). On the Calculation of Ducted Propeller Performance in Axisymmetric Flows. NSMB Report No. 696.

Faltinsen, O.M. (2005). *Hydrodynamics of High-Speed Marine Vehicles*. Cambridge University Press, Cambridge.

Ferrando, M., Scamardella, A., Bose, N., Liu, P. and Veitch, B. (2002). Performance of a family of surface piercing propellers. *Trans. RINA*, Vol. 144.

Flynn, H.G. (1964). *Physical Acoustics*, 1. Academic Press, New York.

Foeth, E-J. and Terwisga, J. van. (2006) The structure of unsteady cavitation. I. Observations of an attached cavity on a three dimensional hydrofoil. Part II: Applying time-resolve PIV to attached cavitation. *6th Int. Symp. on Cavitation CAV2006*, Wageningen, September.

Fortes-Patella, R., Challier, G. and Reboud, J.L. (1999). Study of pressure wave emitted during spherical bubble collapse. *Proc. 3rd ASME/JSME Joint Fluids Eng. Conf.*, San Francisco, July.

Fortes-Patella, R., Challier, G. and Reboud, J.L. (2001). Cavitation erosion mechanism: numerical simulation of the interaction between pressure waves and solid boundaries. *4th Int. Symp. on Cavitation (CAV2001)*.

Fortes-Patella, R. and Reboud, J.L. (1998a). A new approach to evaluate the cavitation erosion power. *3rd Int. Symp. on Cavitation (CAV1998)*.

Fortes-Patella, R. and Reboud, J.L. (1998b). Energetical approach and impact efficiency in cavitation erosion. *3rd Int. Symp. on Cavitation (CAV1998)*.

Fortes-Patella, R., Reboud, J.L. and Archer, A. (2000). Cavitation damage measurement by 3D laser profilometry. *Wear*, Vol. 246, pp. 59–67.

Fox, F.E. and Herzfeld, K.F. (1954). Gas bubbles with organic skin as cavitation nuclei. *J. Acoustic. Soc. Amer.*, Vol. 26.

Frenkiel, F.N., Landahl, M.T. and Lumley, J.L. (1976). Structure of turbulence and drag reduction. *IUTAM Symp.*, Washington, DC, June.

Friesch, J., Anderson, P. and Kappel, J.J. (2003). Model/full scale correlation investigation for a new marine propeller. *NAV2003 Conf.*, Palermo, Italy.

Froude, R.E. (1889). On the part played in the operation of propulsion differences in fluid pressure. *Trans. RINA*, Vol. 30.

Froude, W. (1878). On the elementary relation between pitch, slip and propulsive efficiency. *Trans. RINA*, Vol. 19.

Froude, W. (1955). The papers of William Froude. *Trans. RINA*.

Gawn, R.W.L. (1952). Effect of pitch and blade width on propeller performance. *Trans. RINA*.

Gawn, R.W.L. and Burrill, L.C. (1957). Effect of cavitation on the performance of a series of 16 in. model propellers. *Trans. RINA*.

Gent, W. van. (1975). Unsteady lifting surface theory of ship screws: derivation and numerical treatment of integral equation. *J. Ship Res.*, Vol. 19, No. 4.

Gent, W. van. (1977). On the Use of Lifting Surface Theory for Moderately and Heavily Loaded Ship Propellers NSMB Report No. 536.

Geurst, J.A. (1959). Linearised theory for partially cavitated hydrofoils. *ISP*, Vol. 6.

Geurst, J.A. (1960). Linearised theory for fully cavitated hydrofoils. *ISP*, Vol. 7.

Geurst, J.A. and Verbrugh, P.J. (1959). A note on cambered effects of a partially cavitated hydrofoil. *ISP*, Vol. 6.

Glover, E.J. (1970). Slipstream deformation and its influence on marine propeller design. Ph.D. Thesis, Newcastle upon Tyne.

Goldstein, S. (1929). On the vortex theory of screw propellers. *Proc. Royal Soc.*, Ser. A, Vol. 123.

Gould, R.W.F. (1982). The Estimation of Wind Loadings on Ship Superstructures. *RINA. Marine Technology Monograph*, No. 8.

Greco, L., Salvatore, F. and Di Felice, F. (2004) Validation of a quasi-potential flow model for the analysis of marine propellers wake. *Proc. 25th Symp. on Naval Hydrodyn.*, St. John's, Newfoundland.

Greeley, D.A. and Kerwin, J.E. (1982). Numerical methods for propeller design and analysis in steady flow. *Trans. SNAME*, Vol. 90.

Greenhill, A.G. (1988). A theory of the screw propeller. *Trans. RINA*.

Grigson, C.W.B. (1981). Propeller roughness, its nature and its effect upon the drag coefficients of blades and ship power. *Trans. RINA*.

Guilloton, R. (1957). Calcul des vitesses induites en vue du trace des helices. *Schiffstechnik*, Vol. 4.

Gunsteren, L.A. van (1969). Ringpropellers. *SNAME & I. Mar. E. Meeting*, Canada.

Gunsteren, L.A. van. (1970). Application of momentum theory in counter-rotating propeller design. *ISP*.

Gunsteren, L.A. van and Pronk, C. (1973). Propeller design concepts. *ISP*, Vol. 20 July, No. 227.

Gunsteren, L.A. van. Design and performance of controllable pitch propellers. *Trans. SNAME*, New York Sect.

Gutsche, F. (1965). Loading assumptions for controllable pitch propellers. *Schiffbauforschung*, Vol. 4.

Gutsche, F.A. and Schroeder, G. (1963). Freifahruersuche an Propellern mit festen und verstellbaren Flügeln 'voraus' und zuruck'. *Schiffbauforschung*, Vol. 2, No. 4.

Hanaoka, T. (1969). Numerical Lifting Surface Theory of a Screw Propeller in Non-Uniform Flow (Part 1 – Fundamental Theory). *Rep. Ship Res. Inst.*, Vol. 6, No. 5, Tokyo.

Handler, J.B. *et al.* (1982). Large diameter propellers of reduced weight. *Trans. SNAME*, Vol. 90.

Hannan, T.E. (1971). *Strategy for Propeller Design*. Thos. Reed Publ.

Hannan, T.E. Principles and Design of the Marine Screw Propeller.

Hansen, E.O. (1967). Thrust and Blade Spindle Torque Measurements of Five Controllable Pitch Propeller Designs for MS0421. NSRDC Report No. 2325, April.

Harvald, Sv.Aa. (1978). Estimation of power of ships. *ISP.*, Vol 25, (No. 283).

Harvald, Sv.Aa. (1950). *Wake of Merchant Ships*. Danish Technical Press.

Harvald, Sv.Aa. (1972). Potential and frictional wake of ships. *Trans. RINA*.

Harvald, Sv.Aa. (1977). *Estimation of Power of Ships*. ISP.

Harvey, E.N. *et al.* (1944a). Bubble formulation in animals. I. Physical factors. *J. Cell. Comp. Physiol.*, Vol. 24, No. 1.

Harvey, E.N. *et al.* (1944b). Bubble formation in animals. II. Gas nuclei and their distribution in blood and tissues. *J. Cell. Comp. Physiol.*, Vol. 24, No. 1.

Harvey, E.N. *et al.* (1945). Removal of gas nuclei from liquids and surfaces. *J. Amer. Chem. Soc.*, Vol. 67.

Harvey, E.N. *et al.* (1947). On cavity formation in water. *J. Appl. Phys.*, Vol. 18, No. 2.

Hawdon, L., Carlton, J.S. and Leathard, F.I. (1975). The analysis of controllable pitch propeller characteristics at off design conditions. *Trans. I. Mar. E.*

Hawdon, L. and Patience, G. (1981). Propeller design for economy. *3rd Int. Marine Propulsion Conf.*, London.

Hawdon, L., Patience, G. and Clayton, J.A. (1984). The effect of the wake distribution on the optimum diameter of marine propellers. *Trans. NECIES, NEC100 Conf.*

Henderson, R.E., McMahon, J.F. and Wislicenus, G.F. (1964). A Method for the Design of Pumpjets. Ordnance Research Laboratory, Pennsylvania State University.

Hermans, A.J. (2004). Added resistance by means of time-domain models in seakeeping. *19th WWWFB*, Cortona, Italy.

Hess, J.L. and Smith, A.M.O. (1967). Calculation of potential flow about arbitrary bodies. *Prog. Aeronaut. Sci.*, Vol. 8.

Hess, J.L. and Valarezo, W.O. (1985). Calculation of Steady Flow about Propellers by Means of a Surface Panel Method. *AIAA*, Paper No. 85.

Hoekstra, M. (1975). Prediction of full scale wake characteristics based on model wake survey. *Int. Shipbuilding*, Prog. No. 250, June.

Holden, K. and Kvinge, T. (1983). On application of skew propellers to increase propulsion efficiency. *5th Lips Propeller Symposium*.

Holtrop, J. (1977). A statistical analysis of performance test results. *ISP*, Vol. 24, February.

Holtrop, J. (1978). Statistical data for the extrapolation of model performance tests. *ISP*.

Holtrop, J. and Mennen, G.G.J. (1978). A statistical power prediction method. *ISP*, Vol. 25, October.

Holtrop, J. and Mennen, G.G.J. (1982). An approximate power prediction method. *ISP*, Vol. 29, July.

Holtrop, J. (1988). A statistical resistance prediction method with a speed dependent form factor. *SMSSH '88*, Varna, October.

Holtrop, J. (1988). A statistical re-analysis of resistance and propulsion data. *ISP*, Vol. 31, November.

Holtrop, J. (2001). Extrapolation of Propulsion Tests for Ships with Appendages and Complex Propulsors. *Mar. Technol.*, Vol. 38.

Horn, F. and Amtsberg, H. (1961). *Entwurf von Schiffdusen-systemen (Kortdusen). Jahrbuch der Schiffbautech- nishen*, Gesellschaft, Vol. 44, 1950.

Hoshino, T. (1989). Hydrodynamic analysis of propellers in steady flow using a surface panel method. *Trans. Soc. Nav. Arch Jap.*

Hsiao, C.-T. and Chahine, G.L. (2001). Numerical simulation of bubble dynamics in a vortex flow using Navier–Stokes computations and moving Chimera grid scheme. *CAV2001*, Pasadena.

Hsin, C.Y., Chou, S.K. and Chen, W.C. (2002). A New Propeller Design Method for the POD Propulsion System. *Proc. 24th Symp. on Naval Hydrodynam.*, Fukuoka, Japan.

Huang, T.T. and Groves, N.C. (1980). Effective wake: theory and experiment. *13th ONR Symposium*.

Hughes, G. (1954). Friction and form resistance in turbulent flow and a proposed formulation for use in model and ship correlation. *Trans. RINA*, Vol. 96.

Huse, E. (1971). Propeller–Hull Vortex Cavitation. Norw. Ship Model Exp. Tank Publ. No. 106, May.

Huse, E. (1974). Effect of Afterbody Forms and Afterbody Fins on the Wake Distribution of Single Screw Ships. NSFI Report No. R31–74.

Insel, M. and Molland, A.F. (1992). An investigation into the resistance components of high speed displacement catamarans. *Trans. RINA*, Vol. 134.

Inui, T. (1962). *Wavemaking resistance of ships. Trans. SNAME*, Vol. 70.

Isay, W.H. (1981). *Kavitation*. Schiffahrts-Verlag 'Hansa', Hamburg.

Isherwood, R.M. (1973). Wind Resistance of Merchant Ships. *Trans. of the Royal Institution of Naval Architects*, Vol. 115.

ITTC (1987). *18th ITTC Conference*, Japan.

ITTC (2002a). Report of the Resistance Committee. *Proceedings of 23rd International Towing Tank Conference*, Vol.1. Venice, Italy. Published by INSEAN, Rome.

ITTC (2002b). Report of the Propulsion Committee. *Proceedings of 23rd International Towing Tank Conference*, Vol. 1. Venice, Italy. Published by INSEAN, Rome.

ITTC (2005a). Report of the Resistance Committee. *Proceedings of 24th International Towing Tank Conference*, Vol. 1. Edinburgh, UK. Published by The University of Newcastle upon Tyne, UK.

ITTC (2005b). Report of the Propulsion Committee. *Proceedings of 24th International Towing Tank Conference*, Vol. 1. Edinburgh, UK. Published by The University of Newcastle upon Tyne, UK.

ITTC (2005c). Report of the Specialist Committee on Validation of Waterjet Test Procedures. *Proceedings of 24th International Towing Tank Conference*, Vol. 1l. Edinburgh, UK. Published by The University of Newcastle upon Tyne, UK.

Janson, C.E. (1996). *Potential flow panel methods for the calculation of free surface flows with lift* PhD Thesis, Chalmers University of Technology.

Jiang, C.W. (1977). Experimental and Theoretical Investigations of Unsteady Supercavitating Hydrofoils of Finite Span. Ph.D. Thesis, MIT.

Johansson, L.E. (1985). The local effect of hull roughness on skin friction. Calculations based on floating element data and three dimensional boundary layer theory. *Trans. RINA*, Vol. 127.

Johnson, V.E. Theoretical Determination of Low-Drag Supercavitating Hydrofoils and Their Two-Dimensional Characteristics at Zero Cavitation Number. NACA Report No. 1957.

Jong, H.J. de, and Fransen, H.P. (1976). N.S.M.B. Trial allowances 1976. *ISP*.

Kane, C. and Smith, J. (2003). Composite blades in marine propulsors. *Proceedings of International Conference on Advanced Marine Materials: Techniques and Applications*. RINA, London.

Katsoulakos, P.S., Hornsby, C.P.W. and Rowe, R. (1989). Advanced ship performance information systems. *Trans. I. Mar. E. (C)*, Vol. 101, Conf. 3.

Kauczynski, W. and Walderhaug, H. (1987). Effects of distributed roughness on the skin friction of ships. *ISP*, Vol. 34, No. 389, January.

KcKelvie, A.N. (1981). Hull preparation and paint application: their effect on smoothness. *Marine Eng. Review*, July.

Keller, J. (1966). auf'm. Enige Aspecten bij het Ontwerpen van Scheepsschroeven. *Schip en Werf*, No. 24.

Keller, W.H. Auf'm (1966). Comparative Tests with B-Series Screws and Ringpropellers. NSMB Report No. 66–047, DWT.

Kelvin, L. (1904a). On deep water two-dimensional waves produced by any given initiating disturbance. *Proc. Roy. Soc.* (Edin.), Vol. 25, pp. 185–196.

Kelvin, L. (1904b). On the front and rear of a free procession of waves in deep water. *Proc. Roy. Soc.* (Edin.) Vol. 25, pp. 311–327.

Kelvin, L. (1904c). Deep water ship waves. *Proc. Roy. Soc.* (Edin.) Vol. 25, pp. 562–587.

Kennedy, J.L., Walker, D.L., Doucet, J.M. and Randell, T. (1993). Cavitation performance of propeller blade root fillets. *Trans. SNAME*.

Kerwin, J.E. (1961). *The solution of Propeller Lifting Surface Problems by Vortex Lattice Methods*. Report Department of Ocean Engineering MIT.

Kerwin, J.E. (1976). *A Deformed Wake Model for Marine Propellers*. Department of Ocean Engineering Rep, 76–6, MIT.

Kerwin, J.E., Chang-Sup Lee. (1978). Prediction of steady and unsteady marine propeller performance by numerical lifting-surface theory. *Trans. SNAME*, Paper No. 8, Annual Meeting.

Kerwin, J.E., Kinnas, S.A., Lee, J.-T. and Shih, W.Z. (1987). A surface panel method for the hydrodynamic analysis of ducted propellers. *Trans. SNAME*, Vol. 95.

Kerwin, J.E., Keenan, D.P., Black, S.D. and Diggs, J.G. (1994). A Coupled Viscous/Potential Flow Design Method for Wake Adapted Multi-Stage Ducted Propulsors. *Trans. SNAME*, Vol. 102.

Kerwin, J.E., Taylor, T.E., Black, S.D. and McHugh, G.P. (1997). A Coupled Lifting Surface Analysis Technique for Marine Propulsors in Steady Flow. *Proc. Propeller/Shafting Symp.*, Virginia.

Kim, J., Paterson, E. and Stern, F. (2003). Verification and validation and sub-visual cavitation and acoustic modelling for ducted marine propulsor. *Proc. 8th Symp. on Numerical Ship Hydrodyn.*, Seoul.

Kim, S.E. and Rhee, S.H. (2004). Toward high-fidelity prediction of tip-vortex around lifting surfaces. *Proc. 25th Symp. on Naval Hydrodyn.*, St John's, Newfoundland.

Kinnas, S.A. and Fine, N.E. (1992). A nonlinear boundary element method for the analysis of unsteady propeller sheet cavitation. *Proc. 19th Symp. on Naval Hydrodyn.*, Seoul.

Kinnas, S.A., Griffin, P.E., Choi, J-K. and Kosal, E.M. (1998). Automated design of propulsor blades for high-speed ocean vehicle applications. *Trans. SNAME*, Vol. 106, pp. 213–240.

Kinnas, S.A., Mishima, S. and Brewer, W.H. (1994). Nonlinear analysis of viscous flow around cavitating hydrofoils. *20th Symp. on Naval Hydrodynamics*, University of California.

Klaassen, H. and Arnoldus, W. (1964). Actuating forces in controllable pitch propellers. *Trans. I. Mar.E.*, Vol. 1.

Klintorp, H. (1989). Integrated design for efficient operation. *Motor Ship*, Vol. January.

Knapp, R.T. (1952). Cavitation Mechanics and its Relation to the Design of Hydraulic Equipment. *Proc. I. Mech. E. Pt.*, No. 166.

Knapp, R.T., Daily, J.W. and Hammitt, F.G. (1970). *Cavitation*. McGraw-Hill, New York.

Kort, L. (1934). Der neue Dusen schraubenartrieb. *Werft Reederei Hafen*, Vol. 15.

Koushan, K. (2006). Dynamics of ventilated propeller blade loading on thrusters. *WTMC Conf. Trans. I. Mar. EST*, London.

Kuiper, G. (1992). *The Wageningen Propeller Series*. MARIN, May.

Lackenby, H. (1962). The resistance of ships with special reference to skin friction and hull surface conditions. 34th Thomas Lowe Grey Lecture. *Trans. I. Mech. E.*

Lackenby, H. and Parker, M.N. (1966). The BSRA methodical series – an overall presentation – variation of resistance with breadth–draught ratio and length–displacement ratio. *Trans. RINA*, Vol. 108.

Lammeren, W.P.A. van. (1938). Analysis der voort-stuwingscomponenten in verband met het schaleffect bij scheepsmodelproven, blz.58.

Lammeren, W.P.A. van, Manen, J.D. van. and Oosterveld, M.W.C. (1969). The Wageningen B-screw series. *Trans. SNAME*.

Lammeren, W.P.A. van, Troost, L. and Koning, J.G. (1942). *Weerstand en Voortsluwing van Schepen*.

Lan, C.E. (1974). A quasi-vortex lattice method in thin wing theory. *J. Airer*, Vol. 11.

Lap, A.J.W. (1954). Diagrams for determining the resistance of single screw ships. *ISP*, Vol. 1, No. 4.

Larsson, L. (1997). CFD in ship design–Prospects and limitations. *Ship Technology Research*, Vol. 44, pp. 133–154.

Larsson, L., Regnström, B., Broberg, L., Li, D. Q. and Janson, C. E. (1998). Failures, fantasies, and feats in the theoretical/numerical prediction of ship performance. *22. Symp. Naval Shiphydrodyn.*, Washington.

Lecoffre, Y. (1995) Cavitation erosion, hydrodynamic scaling laws, practical methods of long term damage prediction. *CAV1995 Conf.*, Deauville.

Lee, H., Kinnas, S.A., Gu, H. and Naterajan, S. (2003). Numerical Modelling of Rudder Sheet Cavitation Including Propeller/Rudder Interaction and the Effects of a Tunnel. *CAV 2003*, Osaka, Japan.

Lee, H.S. and Kinnas, S.A. (2004). Application of boundary element method in the prediction of unsteady blade sheet and developed tip vortex cavitation on marine propellers. *J. Ship Res.*, Vol. 48, No. 1, pp. 15–30.

Lee, H.S. and Kinnas, S.A. (2005). Unsteady wake alignment for propellers in nonaxisymmetric flows. *J. Ship Res.*, Vol. 49, No. 3.

Leehey, P. (1971). Supercavitating hydrofoil of finite span. *Proc. IUTAM Symp. on Non-Steady Flow of Water at High Speeds*, Leningrad.

Lerbs, H.W. (1952). Moderately loaded propellers with a finite number of blades and an arbitrary distribution of circulation. *Trans. SNAME*, Vol. 60.

Lerbs, H.W. (1955). Contra-Rotating Optimum Propellers Operating in a Radially Non-Uniform Wake. DTMB Report No. 941.

Lighthill, M.J. (1951). A new approach to thin aerofoil theory. *Aeronaut. Quart.*, Vol. 3.

Ligtelijn, J. Th. and Kuiper, G. (1983). Intentional cavitation as a design parameter. *2nd PRADS Symp.*

Lingren, H. (1961). Model tests with a family of three and five bladed propellers. *SSPA* Paper No. 47, Göteborg.

Lohrberg, H., Stoffel, B., Fortes-Patella, R. and Reboud, J.L. (2001a). Numerical and experimental investigations on the cavitating flow in a cascade of hydrofoils. *4th Int. Symp. on Cavitation (CAV2001)*.

Lohrberg, H., Stoffel, B., Coutier-Delgosha, O., Fortes-Patella, R. and Reboud, J.L. (2001b). Experimental and numerical studies on a centrifugal pump with curved blades in cavitating condition. *4th Int. Symp. on Cavitation (CAV2001)*.

Loukakis, T.A. (1971). A New Theory for the Wake of Marine Propellers. Department of Naval Architecture Report No. 71–7, MIT.

Lücke, T. (2006). Investigations of propeller tip vortex bursting. *Proc. Nav 2006 Conf.*, Genoa, June.

Luke, W.J. (1914). Further experiments upon wake and thrust deduction. *Trans. RINA*.

Manen, J.D. van and Sentic, A. (1956). Contra-rotating propellers. *ISP*, Vol. 3.

Manen, J.D. and van Superina, A. (1959). The design of screw propellers in nozzles. *ISP*, Vol. 6, March.

Manen, J.D. van, et al. (1961). Scale effect experiments on the victory ships and models. I, II, III and IV in three papers. *Trans. RINA*, 1955, 1958 and 1961.

Manen, J.D. van (1962). Effect of radial load distribution on the performance of shrouded propellers. *Int. Shipbuilding Prog.*, Vol. 9, p. 93.

Manen, J.D. van and Bakker, A.R. (1962). Numerical results of Sparenberg's lifting surface theory for ship screws. *Proc. 4th Symp. Nav. Hydro.*, Washington.

Manen, J.D. van. (1963). Ergebnisse systematischer ver-suche mit propellern mit annahernd senkrecht slehender achse. *Jahrbuch*, STG.

Manen, J.D. van. and Oosterveld, M.W.C. (1966). Analysis of ducted-propeller design. *Trans. SNAME*, Vol. 74.

Manen, J.D. van. (1966). The choice of the propeller. *Marine Technol.*, Vol. 3, No. 2, April.

Manen, J.E. van and Oosterveld, M.W.C. (1969). Model Tests on Contra-Rotating Propellers. Publication No. 317, NSMB Wageningen.

MARIN Report No. 26, September, 1986.

McCormick, B.W. and Eisenhuth, J.J. (1962). The design and performance of propellers and pumpjets for underwater propulsion. *American Rocket Soc., 17th Annual Meeting*.

McCroskey, W.J. (1982). Unsteady aerofoils. *Ann. Rev. FluidMech.*, Vol. 14.

Mercier, J.A., Savitsky, D. (1973). Resistance of Transom-Stern Craft in the Pre-Planing Regime. Davidson Laboratory, Stevens Institute of Tech., Rep. No. 1667.

Mills, L. (1991). Die Anwendung der Blasendy namik auf die theoretische mit experimentellen Daten. Ph.D. Thesis, Hamburg.

Mishima, S. and Kinnas, S.A. (1997). Application of a numerical optimisation technique to the design of cavitating hydrofoil sections. *J. Ship Res.*, Vol. 40, No. 2, pp. 93–107.

Mitchell, T.M. and Hammitt, F.G. (1973). Asymmetric cavitation bubble collapse. *Trans. ASME J., Fluids Eng.*, Vol. 95, No. 1.

Molland, A.F., Wellicome, J.F. and Couser, P.R. (1996). Resistance experiments on a systematic series of high speed displacement catamaran forms: Variation of length-displacement ratio and breadth-draught ratio. *Transactions of the Royal Institution of Naval Architects*, Vol. 138, pp. 59–71.

Molland, A.F. and Barbeau, T-E. (2003). An investigation into the aerodynamic drag on the superstructures of fast catamarans. *Trans. of the Royal Institution of Naval Architects*, Vol. 145.

Molland, A.F., Wilson, P.A., Taunton, D.J., Chandraprabha, S. and Ghani, P.A. (2004). Resistance and wash measurements on a series of high speed displacement monohull and catamaran forms in shallow water. *Trans. of the Royal Institution of Naval Architects*, Vol. 146.

Moor, D.I. (1960a). The © of some $0.80C_b$ forms. *Trans. RINA*.

Moor, D.I. (1960b). The effective horse power of single screw ships – average modern attainment with particular reference to variations of C_b and LCB. *Trans. RINA*.

Moor, D.I. (1965/1966). Resistance, propulsion and motions of high speed single screw cargo liners. *Trans. NECIES*, Vol. 82.

Moor, D.I. (1974). Resistance and propulsion qualities of some modern single screw trawler and bulk carrier forms. *Trans. RINA*.

Moor, D.I., Parker, M.N. and Pattullo, R.N. (1961). The BSRA methodical series – an overall presentation – geometry of forms and variations of resistance with block coefficient and longitudinal centre of buoyancy. *Trans. RINA*, Vol. 103.

Moor, D.I. and Pattullo, R.N.M. (1968). The Effective Horsepower of Twin-Screw Ships – Best Modern Attainment for Ferries and Passenger Liners with Particular Reference to Variations of Block Coefficient. BSRA Report No. 192.

Moor, D.I. (1973). Standards of ship performance. *Trans. IESS*, Vol. 117.

Morgan, W.B. (1960). The design of contra-rotating propellers using Lerbs' theory. *Trans. SNAME*.

Morgan, W.B. (1961). Theory of Ducted Propeller with Finite Number of Blades. University of California, Institute of Engineering Research, May.

Morgan, W.B. and Wrench, J.W. (1965). Some computational aspects of propeller design. *Meth. Comp. Phys.*, Vol. 4.

Morgan, W.B., Silovic, V. and Denny, S.B. (1968). Propeller lifting-surface corrections. *Trans. SNAME*, Vol. 76.

Moulijn, J.C., Friesch, J., Genderen, M. van, Junglewitz, A. and Kuiper, G. (2006). A criterion for the erosiveness of face cavitation. *6th Int. Symp. on Cavitation CAV2006*, Wageningen, September.

Murray, M.T. (1967). *Propeller Design and Analysis by Lifting Surface Theory.* ARL.

Muscari, R., Di Mascio, A. (2005). Simulation of the flow around complex hull geometries by an overlapping grid approach. *Proc 5th Osaka Colloquium*, Japan.

Musker, A.J. (1977). Turbulent shear-flow near irregularly rough surfaces with particular reference to ship's hulls. Ph.D. Thesis, University of Liverpool, December.

Naess, E. (1980). Reduction of drag resistance by surface roughness and marine fouling. *Norwegian Maritime Research*, No. 4.

Nakos, D. (1990). Ship wave patterns and motions by a three-dimensional Rarakine panel method. Ph.D. thesis, MIT.

Nakos, D. and Sclovunos, P. (1990). Steady and unsteady wave patterns. *J. Fluid Mechanics*, Vol. 215, pp. 256–288.

Newton, R.N. and Rader, H.P. (1961). Performance data for propellers for high speed craft. *Trans. RINA*, Vol. 103.

Norton, J.A. and Elliot, K.W. (1988). Current practices and future trends in marine propeller design and manufacture. *Mar. Tech.*, Vol. 25, No. 2, April.

Oortmerssen, G.van. (1971). A power prediction method and its application to small ships. *ISP*, Vol. 18, No. 207.

Oossanen, P. van. (1974). Calculation of Performance and Cavitation Characteristics of Propellers Including Effects of Non-Uniform Flow and Viscosity. NSMB Publ. No. 457.

Oossanen, P. van. (1976). Trade Offs in the Design of Sub-Cavitating Propellers. NSMB Publ. No. 491.

Oosterveld, M.W.C. (1970). Wake Adapted Ducted Propellers. NSMB Report No. 345.

Oosterveld, M.W.C. (1973). Ducted propeller characteristics. *RINA Symp. on Ducted Propellers*, London.

Oosterveld, M.W.C. and Ossannen, P. van (1975). Further computer-analysed data of the Wageningen B-screw series. *ISP*, Vol. 22.

Orihara, H. and Miyata, H. (2003). Evaluation of added resistance in regular incident waves by computational fluid dynamics motion simulation using an overlapping grid system. *J. Marine Sci. and Technol.*, Vol. 8, No. 2, pp. 47–60.

Parsons, C.G. (1975). Tuning the screw. *Marine Week*, September 12.

Pattullo, R.N.M. (1974). The resistance and propulsion qualities of a series of stern trawlers – variations of longitudinal position of centre of buoyancy, breadth, draught and block coefficient. *Trans. RINA*.

Peck, J.G. and Moore, D.H. (1973). Inclined-shaft propeller performance characteristics. SNAME, Spring Meeting.

Pham, X.P., Kantimahanthi, K. and Sahoo, P.K. (2001). Wave resistance prediction of hard chine catamarans using regression analysis. *2nd Int. Euro-Conference on High Performance Marine Vehicles*, Hamburg.

Phillips-Birt, D. (1970). *Ship Model Testing*. Leonard-Hill.

Pien, P.C. (1961). The calculation of marine propellers based on lifting surface theory. *J. Ship Res.*, Vol. 5, No. 2.

Plesset, M.S. (1963). Bubble Dynamics. CIT Rep. 5.23. Calif. Inst. of Tech., February.

Plesset, M.S. and Chapman, R.B. (1971). Collapse of an initially spherical vapour cavity in neighbourhood of a solid boundary. *J. Fluid Mech.*, Vol. 47, No. 2.

Prohaski, C.W. and Lammeren, W.P.A. van (1937). *Mitstrommessung on Schiffsmodellen*, Heft 16, S.257, Schiffbau.

Pronk, C. (1980). *Blade Spindle Torque and Off-Design Behaviour of Controllable Pitch Propellers*. Delft University.

Rankine, W.J. (1865). On the mechanical principles of the action of propellers. *Trans. RINA*, Vol. 6.

Raven, H.C. (1996). A solution method for the non-linear ship wave resistance problem. PhD Thesis, TU Delft.

Rayleigh, L. (1917). On the pressure developed in a liquid during the collapse of a spherical cavity. *Phil. Mag.*, Vol. 34.

Reynolds, O. (1873). The causes of the racing of the engines of screw steamers investigated theoretically and by experiment. *Trans. INA*.

Rhee, S.H. and Joshi, S. (2003). CFD validation for a marine propeller using an unstructured mesh based RANS method. *Proc. FEDSM'03*, Honolulu.

Rota, G. (1909). The propulsion of ships by means of contrary turning screws on a common axis. *Trans. RINA*.

Rota, G. (1992). Further experiments on contrary turning co-axial screw propellers. *Trans. RINA*.

Rusetskiy, A.A. (1968). *Hydrodynamics of Controllable Pitch Propellers*. Idatelstvo Sudostroyeniye, Leningrad.

Ryan, P.G. and Glover, E.J. (1972). A ducted propeller design method: a new approach using surface vorticity distribution techniques and lifting line theory. *Trans. RINA*, Vol. 114.

Sasajima, H., Tanaka, I. and Suzuki, T. (1966). Wake distribution of full ships. *J. Soc. Nav. Arch.*, Vol. 120, Japan.

Savitsky, D. and Ward Brown, P. (1976). Procedures for hydrodynamic evaluation of planing hulls in smooth and rough waters. *Maritime Technol.*, Vol. 13, No. 4, October.

Schneekluth, H. (1987). *Ship Design for Efficiency and Economy*. Butterworths, London.

Schoenherr, K.E. (1939). Propulsion and propellers. *Principles of Naval Architecture*, Vol. 2, p. 149.

Scott, J.R. (1972). A method of predicting trial performance of single screw merchant ships. *Trans. RINA*.

Scott, J.R. (1973). A method of predicting trial performance of twin screw merchant ships. *Trans. RINA*.

Searle, T. (1998). Composite propellers, the final frontier. *Design Engineering*.

Series 60 (1960). The DTMB Series 60 – Presented in a set of papers by various authors 1951, 1953, 1954, 1956, 1957 and 1960. *Trans. SNAME*.

Shen, Y.T. (1985). Wing section for hydrofoils. 3. experimental verifications. *J.ShipRes.*, Vol. 29.

Sherrington, I. and Smith, E.H. (1987). Parameters for characterising the surface topography of engineering components. *Trans. I. Mech. E.*

Shintani, A. and Inoue, R. (1984). Influence of hull form characteristics on propulsive performance in waves. *SMWP*, December.

Sinclair, L. and Emerson, A. (1973). Calculation, experiment and experience in merchant ship propeller design. *IMAS 73 Conf. Trans. I. Mar. E.*

SMM (18). Propeller Surface Roughness and Fuel Economy. SMM Technical Brief No. 18.

Söding, H. (1995). Wave resistance by Michell integral. *Ship Technology Research*, Vol. 42, pp. 163–164.

Söding, H. (1996). Advances in panel methods. *21. Symp. Naval Hydrodyn.,* Trondheim, pp. 997–1006.

Sparenberg, J.A. (1959). Application of lifting surface theory to ship screws. *Proc. K. Ned. Akad. Wet.*, Vol. 62, No. S, Ser B.

SSPA (1969). The SSPA Standard Series published in a set of papers from SSPA 1948–1959 and summarized in 1969.

Status (1981). Status of Hydrodynamic Technology as Related to Model Tests of High Speed Marine Vehicles. DTNSRDC Rep. No. 81/026.

Steiss, W. (1936). Erweiterte Stahltheorie für Dusenschraubon mit und ne Leitapparat. *Werft Reediri und Hafen*, Vol. 17.

Stern, F. and Vorus, W.G. (1983). A non-linear method for predicting unsteady sheet cavitation on marine propellers. *J. Ship. Res.*, Vol. 17.

Strasberg, M.J. (1959). *Acoust. Soc. Am.*, Vol. 31.

Strom-Tejsen, J. and Porter, R.R. (1972). Prediction of controllable-pitch propeller performance in off-design conditions. *Third Ship Control System Symp.*, Paper VII B-1, Bath, UK.

Strscheletsky, M. (1950). *Hydrodynamische Grundlagen zur Berechnung der Schiffschranben.* G. Braun, Karlsruhe.

Sun, H. and Kinnas, S.A. (2006). Simulation of sheet cavitation on propulsor blades using a viscous/inviscid interactive method. *6th Int. Symp. on Cavitation CAV2006*, Wageningen, September.

Szantyr, J.A. (1984). A new method for the analysis of unsteady propeller cavitation and hull surface pressures. *Trans. RINA*.

Tachmindji, A.J. and Morgan, W.B. (1957). The Design and Performance of Supercavitating Propellers. DTMB Report No. 807.

Tachmindji, A.J. and Morgan, W.B. (1958). The design and estimated performance of a series of supercavitating propellers. *2nd Symp. on Naval Hydrodynamics*, Washington.

Tachmindji, A.J. and Morgan, W.B. (1969). The design and estimated performance of a series of supercavitating propellers. *Proc. 2nd Symp. on Nav. Hydro.*, ONR, Washington.

Taniguchi, K. and Tanibayashi, H. (1962). Cavitation tests on a series of supercavitating propellers. *IAHR Symp. on Cav. Hydraulich Mach.*, Japan.

Taniguchi, K., Tanibayashi, H. and Chiba, N. (1969). Investigation into the Propeller Cavitation in Oblique Flow. Mitsubishi Technical Bulletin No. 45, March.

Taschmindji, A.J., Morgan, W.B., Miller, M.L. and Hecher, R. (1957). The Design and Performance of Supercavitating Propellers. DTMB Rep. C-807, Bethesda, Washington, DC, February.

Taylor, D.W. (1933). *Speed and Power of Ships*.

Taylor, D.W. (1943). *The Speed and Power of Ships*. US Govt. Printing Office, Washington.

Todd, F.H. (1967). Resistance and propulsion, Chapter 7, *Principles of Naval Architecture*. SNAME.

Townsin, R.L. (1987). Developments in the circulation of rough underwater surface power penalties. *Celena 25th Anniversary Symp.*, Genoa.

Townsin, R.L., Byrne, D., Svensen, T.E. and Milne, A. (1981). Estimating the technical and economic penalties of hull and propeller roughness. *Trans. SNAME*.

Townsin, R.L., Byrne, D., Svensen, T.E. and Milne, A. (1986). Fuel economy due to improvements in ship hull surface conditions 1976–1986. *ISP*, July.

Townsin, R.L., Spencer, D.S., Mosaad, M. and Patience, G. (1985). Rough propeller penalties. *Trans. SNAME*, Vol. 16.

Troost, L. (1938). Open water test series with modern propeller forms. *Trans. NECIES*, Vol. 54, p. 1938.

Troost, L. (1940). Open water test series with modern propeller forms. II. Three bladed propellers. *Trans. NECIES*.

Troost, L. (1951). Open water test series with modern propeller forms. III. Two bladed and five bladed propellers – extension of the three and four bladed B-series. *Trans. NECIES*, Vol. 67.

Tsakonas, S., Jacobs, W.R. (1978) Propeller-duct interaction due to loading and thickness effects. *Propellers 78 Symp., Trans. SNAME*.

Tsakonas, S., Jacobs, W.R. and Liao, P. (1983). Prediction of steady and unsteady loads and

hydrodynamic forces on counter-rotating propellers. *J. Ship Res.*, Vol. 27.
Tuck, E.O. and Lazauskas, L. (1998). Optimum spacing of a family of multi-hulls. *Ship Technol. Res.*, Vol. 45, pp. 180–195.
Tulin, M.P. (1953). Steady Two Dimensional Cavity Flow about Slender Bodies. DTMB Report No. 834.
Tulin, M.P. (1956). Supercavitating flow past foils and struts. *Proc. Symp. on Cavitation in Hydrodyn.*, NPL.
Tulin, M.P. and Burkart, M.F. (1955). Linearized. Theory for Flows about Lifting Foils at Zero Cavitation Number. DTMB Report No. 638.
Tulin, M.P. and Hsu, C.C. (1977). The theory of leading-edge cavitation on lifting surfaces with thickness. *Symp. on Hydro. of Ship and Offshore Prop. Syst.*, March.
Uhlman, J.S. (1978). A partially cavitated hydrofoil of finite span. *J.F.E.*, Vol. 100.
Uhlman, J.S. (1983). The surface singularity method applied to partially cavitating hydrofoils. *5th Lips Symp.*, May.
Uhlman, J.S. (2006). A note on the development of a nonlinear axisymmetric re-entrant jet cavitation model. *J. Ship Res.*, Vol. 50, No. 3, September.
Vasamarov, K.G. and Minchev, A.D. (1983). A propeller hydrodynamic scale effect. BSHC Report No. PD-83-108.
Vaz, G. and Bosschers, J. (2006). Modelling three dimensional sheet cavitation on marine propellers using a boundary element method. *6th Int. Symp. on Cavitation CAV2006*, Wageningen, September.
Voitkounski, Y.I. (ed.) (1985). *Ship Theory Handbook*, Vol. 1. Sudostroeme, Leningrad.
Vonk, K.P., Terwisga, T. van and Ligtelijn, J. Th. (2006). Tip rake: improved propeller efficiency and cavitation behaviour. *WMTC Conf.*, London.
Voorde, C.B. van den and Esveldt, J. (1962). Tunnel tests on supercavitating propellers. *ISP*, Vol. 9.
Voorde, C.B. van den (1968). Effect of Inward and Outward Turning Propellers on Manoeuvring Alongside for Ships of the Frigate Class. NSMB Report No. 68-015-BT.
Wald, Q.E. (1970). Analysis of the integral pumpjet. *J. Ship Res.*, December.
Walderhaug, H. (1986). Paint roughness effects on skin friction. *ISP*, Vol. 33, No. 382, June.
Wang, T., Zhou, L.D. and Zhang, X. (2003). Numerical simulation of 3-D viscous flow field around axisymmetric body with integrated ducted propulsion. *J. Ship Mech.*, Vol. 7, No. 2.
Warren, C.L., Taylor, T.E. and Kerwin, J.E. (2000). Coupled Viscous/Potential Flow Method for the Prediction of Propulsor-Induced Manoeuvring Forces. *Proc. Propeller/Shafting Symp*, Virginia. Beach.
Wheeler, W.H. (1960). Indentation of metals by cavitation. *ASME Trans.*, Vol. 82, Series D.
Whipps, S.L. (1985) On-line ship performance monitoring system: operational experience and design requirements. *Trans. I. Mar. E.* (TM), Vol. 98, Paper 8.
Wigley, W.C.S. (1936). The theory of the bulbous bow and its practical application. *Trans. NECIES*, Vol. 52.
Winterton, R.H.S. (1977). *J Phys. D: Appl. Phys.*, Vol. 10.
Yamazaki, R. (1962). On the theory of screw propellers. *Proc. 4th Symp. Nav. Hydro.*, Washington.
Yazaki, A. (1962). Design diagrams of modern four, five, six and seven-bladed propellers developed in Japan. *4th Naval Hydrodynamics Symp.*, National Academy of Sciences, Washington.
Yim, B. (1974). A simple design theory and method for bulbous bows of ships. *J. Ship Res.*, Vol. 18.
Young, Y.L. and Michael, T.J. et al. (2006). Numerical and experimental investigations of composite marine propellers. *Proceedings of 26th Symposium on Naval Hydrodynamics*, Sept. 2006, Rome, Italy. Publ. by ONR/INSEAN.
Young, Y.L. and Kinnas, S.A. (2001). A BEM for the prediction of unsteady midchord face and/or back propeller cavitation. *J. Fluid, Eng.*, Vol. 123, No. 2, pp. 311–319.
Young, Y.L. and Kinnas, S.A. (2003). Numerical modelling of supercavitating propeller flows. *J. Ship Res.*, Vol. 47, No. 1, pp. 48–62.
Young, Y.L. and Kinnas, S.A. (2004). Performance prediction of surface-piercing propellers. *J. Ship Res.*, Vol. 48, No. 4, pp. 288–304.
Yount, D.E. (1979). *J Acoust. Soc. Am.*, Vol. 65.
Yount, D.E. (1982). *J Acoust. Soc. Am.*, Vol. 71.

6 Marine engines and auxiliary machinery

Contents

6.1 Introduction
6.2 Propulsion systems
6.3 Diesel engine performance
6.4 Engine and plant selection
6.5 Propulsion engines
6.6 Auxiliary machinery and equipment
6.7 Instrumentation and control
References (Chapter 6)

The various Sections of this Chapter have been taken from the following books, with the permission of the authors:

Carlton, J.S. (2007) *Marine Propellers and Propulsion*. 2nd Edition. Butterworth-Heinemann, Oxford, UK. [Section 6.2]

McGeorge, H.D. (1995) *Marine Auxiliary Machinery*. 7th Edition. Butterworth-Heinemann, Oxford, UK. [Section 6.6.1 (excluding 6.6.1.6)]

Taylor, D. A. (1996) *Introduction to Marine Engineering*. Revised 2nd Edition. Butterworth-Heinemann, Oxford, UK. [Sections 6.5.3, 6.6.1.6, 6.6.2, 6.6.3, 6.7]

Woodyard, D.F. (2004) *Pounder's Marine Diesel Engines and Gas Turbines*. 8th Edition. Butterworth-Heinemann, Oxford, UK. [Sections 6.3–6.5 (excluding 6.5.3)]

6.1 Introduction

This Chapter provides an overview and typical examples of the main and auxiliary machinery and equipment found on ships. Machinery is often divided into the main or propulsion engines, electrical generation, systems such as electrical, piping, refrigeration and air conditioning, fire fighting and protection, deck machinery and cargo handling equipment, bow thrusters and stabilizers, instrumentation and control, safety equipment and other auxiliary machinery and equipment. The auxiliary machinery may be in support of the main propulsion engines and include heat exchangers and compressed air, or in support of ship and cargo handling such as propellers and shafting, steering gear and deck cranes, or in support of ship services such as ballast water arrangements and sewage systems.

6.2 Propulsion systems

The range of propulsion systems that are either currently in use or have been under development are reviewed. The principal propulsion devices are briefly reviewed by outlining their major features and characteristics together with their general areas of application.

6.2.1 Fixed pitch propellers

The fixed pitch propeller has traditionally formed the basis of propeller production over the years in either its mono-block or built-up forms. Carlton (2007) reviews the early development of the screw propeller. Whilst the mono-block propeller is commonly used today the built-up propeller, whose blades are cast separately from the boss and then bolted to it after machining, is now rarely used. This was not always the case since in the early years of the last century built-up propellers were very common, partly due to the inability to achieve good quality large castings at that time and partly to difficulties in defining the correct blade pitch. In both these respects the built-up propeller has obvious advantages. Nevertheless, built-up propellers generally have a larger boss radius than its fixed pitch counterpart and this can cause difficulty with cavitation problems in the blade root section regions in some cases.

Mono-block propellers cover a broad spectrum of design types and sizes, ranging from those weighing only a few kilograms for use on small power-boats to those, for example, destined for large container ships which can weigh around 130 tonnes and require the simultaneous casting of significantly more metal in order to produce the casting. Figure 6.1 shows a collage of various types of fixed pitch propeller in use today. These types range from a large four-bladed propeller fitted to a bulk carrier and is seen in the figure in contrast to a man standing on the dock bottom, through highly skewed propellers for merchant and naval applications, to small high-speed patrol craft and surface piercing propellers.

As might be expected, the materials of manufacture vary considerably over such a wide range of designs and sizes. For the larger propellers, over 300 mm in diameter, the non-ferrous materials predominate: high-tensile brass together with the manganese and nickel–aluminium bronzes are the most favoured types of materials. However, stainless steel has also gained limited use. Cast iron, once a favourite material for the production of spare propellers, has now virtually disappeared from use. Alternatively, for small propellers, use is frequently made of materials such as the polymers, aluminium, nylon and more recently carbon fibre composites.

For fixed pitch propellers the choice of blade number, notwithstanding considerations of blade-to-blade clearances at the blade root to boss interface, is largely an independent variable and is normally chosen to give a mismatch to the range of hull, superstructure and machinery vibration frequencies which are considered likely to cause concern. Additionally, blade number is also a useful parameter in controlling unwelcome cavitation characteristics. Blade numbers generally range from two to seven, although in some naval applications, where considerations of radiated noise become important, blade numbers greater than these have been researched and used to solve a variety of propulsion problems. For merchant vessels, however, four, five and six blades are generally favoured, although many tugs and fishing vessels frequently use three-blade designs. In the case of small work or pleasure power-boats two and three-bladed propellers tend to predominate.

The early propeller design philosophies centred on the optimization of the efficiency from the propeller. Whilst today this aspect is no less important, and, in some respects associated with energy conservation, has assumed a greater importance, other constraints on design have emerged. These are in response to calls for the reduction of vibration excitation and radiated noise from the propeller. This latter aspect has of course been a prime concern of naval ship and torpedo propeller designers for many years; however, pressure to introduce these constraints, albeit in a generally less stringent form, into merchant ship design practice has grown in recent years. This has been brought about by the 'increases in power transmitted per shaft; the use of after deckhouses; the maximization of the cargo carrying capacity, which imposes constraints on the hull lines; ship structural failure and international legislation.

Marine engines and auxiliary machinery 347

Figure 6.1 Typical fixed pitch propellers: (a) large four-bladed propeller for a bulk carrier; (b) high-speed patrol craft propeller; (c) seven-bladed balanced high-screw design; (d) surface piercing propeller and (e) biased high-skew, low-blade-area ratio propeller.

For the majority of vessels of over 100 tonnes displacement it is possible to design propellers on whose blades it is possible to control, although not eliminate, the effects of cavitation in terms of its erosive effect on the material, its ability to impair hydrodynamic performance and it being the source of vibration excitation. In this latter context it must be remembered that there are very few propellers which are free from cavitation since the greater majority experience cavitation at some position in the propeller disc: submarine propellers when operating at depth, the propellers of towed array frigates and research vessels when operating under part load conditions are notable exceptions, since these propellers are normally designed to be subcavitating to meet stringent noise emission requirements to minimize either detection or interference with their own instruments. Additionally, in the case of propellers operating at significant water depths such as in the case of a submarine, due account must be taken of the additional hydrostatic pressure-induced thrust which will have to be reacted by the ship's thrust block.

For some small, high-speed vessels where both the propeller advance and rotational speeds are high and the immersion low, a point is reached where it is not possible to control the effects of cavitation acceptably within the other constraints of the propeller design. To overcome this problem, all or some of the blade sections are permitted to fully cavitate, so that the cavity developed on the back of the blade extends beyond the trailing edge and collapses into the wake of the blades in the slipstream. Such propellers are termed supercavitating propellers and frequently find application on high-speed naval and pleasure craft. Figure 6.2(c) illustrates schematically this design philosophy in contrast to non-cavitating and partially cavitating propeller sections, shown in Figure 6.2(a) and (b), respectively.

When design conditions dictate a specific hydrodynamic loading together with a very susceptible cavitation environment, typified by a low cavitation number, there comes a point when even the supercavitating propeller will not perform satisfactorily: for example, if the propeller tip immersion becomes so small that the propeller tends to draw air from the surface, termed ventilation, along some convenient path such as along the hull surface or down a shaft bracket. Eventually, if the immersion is reduced sufficiently by either the design or operational constraints the propeller tips will break surface. Although this condition is well known on cargo vessels when operating in ballast conditions and may, in these cases, lead to certain disadvantages from the point of view of material fatigue and induced vibration, the surface breaking concept can be an effective means of propelling relatively small high-speed craft. Such propellers are termed surface piercing propellers and their design immersion, measured from the free surface to the shaft centre line, can be reduced to zero; that is, the propeller operates half in and half out of the water. In these partially immersed conditions the propeller blades are commonly designed to operate such that the pressure face of the blade remains fully wetted and the suction side is fully ventilated or dry. This is an analogous operating regime to the supercavitating propeller, but in this case the blade surface suction pressure is at atmospheric conditions and not the vapour pressure of water.

6.2.2 Ducted propellers

Ducted propellers, as their name implies, generally comprise two principal components: the first is an annular duct having an aerofoil cross section which may be either of uniform shape around the duct and, therefore, symmetric with respect to the shaft centre line, or have certain asymmetric features to accommodate the wake field flow variations. The second component, the propeller, is a special case of a non-ducted propeller in which the design of the blades has been modified to take account of the flow interactions caused by the presence of the duct in its flow field. The propeller for these units can be either of the fixed or controllable pitch type and in some special applications, such as torpedo propulsion, may be a contra-rotating pair. Ducted propellers, sometimes referred to as Kort nozzles by way of recognition of the Kort Propulsion Company's initial patents and long association with this type of propeller, have found application for many years where high thrust at low speed is required; typically in towing and trawling situations. In such cases, the duct generally contributes some 50% of the propulsor's total thrust at zero ship speed, termed the bollard pull condition. However, this relative contribution of the duct falls to more modest amounts with increasing ship speed and it is also possible for a duct to give a negative contribution to the propulsor thrust at high advance speeds, see

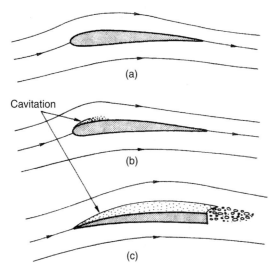

Figure 6.2 Propeller operating regimes: (a) non-cavitating; (b) partially cavitating and (c) supercavitating.

Section 5.3.4.3. This latter situation would nevertheless be a most unusual design condition to encounter.

There are nominally two principal types of duct form, the accelerating and decelerating duct, and these are shown in Figure 6.3(a), (b), (c) and (d), respectively. The underlying reason for this somewhat artificial designation can be appreciated, in global terms by considering their general form in relation to the continuity equation of fluid mechanics. This can be expressed for incompressible flow in a closed conduit between two stations a-a and b-b as,

$$\rho A_a v_a = \rho A_b v_b \qquad (6.1)$$

where v_a is the velocity at station a-a;
v_b is the velocity at station b-b;
A_a is the cross section area at station a-a;
A_b is the cross section area at station b-b and
ρ is the density of the fluid.

Within this context station b-b can be chosen in way of the propeller disc whilst a-a is some way forward although not necessarily at the leading edge. In the case of Figure 6.3(a), which shows the accelerating duct, it can be seen that A_a is greater than A_b since the internal diameter of the duct is greater at station a-a. Hence, from Equation (6.1) and since water is incompressible, v_a must be less than v_b which implies an acceleration of the water between stations a-a and b-b; that is, up to the propeller location. The converse situation is true in the case of the decelerating duct shown in Figure 6.3(d). To determine precisely which form the duct actually is, if indeed this is important, the induced velocities of the propeller also need to be taken into account in the velocity distribution throughout the duct.

By undertaking a detailed hydrodynamic analysis it is possible to design complex duct forms intended for specific application and duties. Indeed, attempts at producing non-symmetric duct forms to suit varying wake field conditions have been made which result in a duct with both varying aerofoil section shape and incidence, relative to the shaft centre line, around its circumference. However, with duct forms it must be appreciated that the hydrodynamic desirability for a particular form must be balanced against the practical manufacturing problem of producing the desired shape if an economic, structurally sound and competitive duct is to result. This tenet is firmly underlined by appreciating that ducts have been produced for a range of propeller diameters from 0.5 m or less up to around 8.0 m. For these larger sizes, fabrication problems can be difficult, not least in maintaining the circularity of the duct and providing reasonable engineering clearances between the blade tips and the duct: recognizing that from the hydrodynamic viewpoint that the clearance should be as small as possible.

Many standard duct forms are in use today but those most commonly used are shown in Figure 6.3. While the duct shown in Figure 6.3(a), the

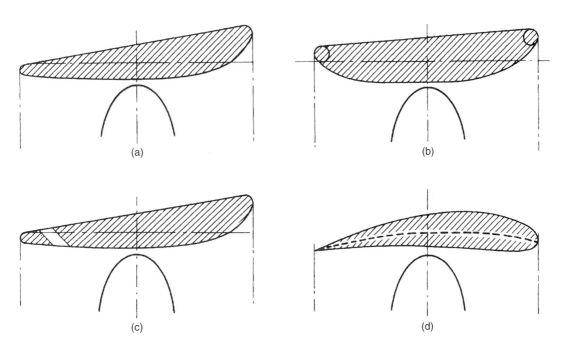

Figure 6.3 Duct types: (a) accelerating duct; (b) 'pull–push' duct; (c) Hannan slotted duct and (d) decelerating duct.

Wageningen 19A form, is probably the most widely used and has a good ahead performance, its astern performance is less good due to the aerofoil form of the duct having to work in reverse: that is, the trailing edge effectively becomes the leading edge in astern operations. This is of relatively minor importance in, say, a trawler or tanker, since for the majority of their operating lives they are essentially unidirectional ships. However, this is not true for all vessels since some, such as tugs, are expected to have broadly equal capabilities in both directions. In cases where a bidirectional capability is required a duct form of the type illustrated in Figure 6.3(b), the Wageningen No. 37 form, might be selected since its trailing edge represents a compromise between a conventional trailing and leading edge of, for example, the 19A form. For this type of duct the astern performance is improved but at the expense of the ahead performance, thereby introducing an element of compromise in the design process. Several other methods of overcoming the disadvantages of the classical accelerating duct form in astern operations have been patented over the years. One such method is the 'Hannan slot', shown in Figure 6.3(c). This approach, whilst attempting to preserve the aerodynamic form of the duct in the ahead condition allows water when backing to enter the duct both in the conventional manner and also through the slots at the trailing edge in an attempt to improve the astern efficiency of the unit.

When the control of cavitation and more particularly the noise resulting from cavitation is of importance, use can be made of the decelerating duct form. A duct form of this type, Figure 6.3(d), effectively improves the local cavitation conditions by slowing the water before passing through the propeller. Most applications of this duct form are found in naval situations, for example, with submarines and torpedoes. Nevertheless, some specialist research ships also have needs which can be partially satisfied by the use of this type of duct in the appropriate circumstances.

An interesting development of the classical ducted propeller form is found in the pump jet, Figure 6.4. The pump jet sometimes comprises a row of inlet guide vanes, which double as duct supports, followed by a row of rotor blades which are finally followed by a stator blade row. Typically, rotor and stator blade numbers might lay between 15 and 20, respectively, each row having a different blade number. Naturally there are variants of this basic design in which the blade numbers may be reduced or the inlet guide vanes dispensed with. The efficiency achievable from the unit is dependent upon the design of the rotor, the rotor–stator interaction, the final stator row in converting the swirl component of the flow generated by the rotor into useful thrust and the reduction of the guide vane size in order to limit skin

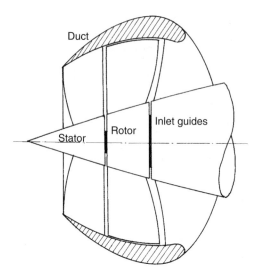

Figure 6.4 Outline of a pump jet.

friction losses: hence, the desirability of not using them if possible. The pump jet in this form is largely restricted to military applications and should not be confused with a type of directional thruster.

The ducts of ducted propellers, in addition to being fixed structures rigidly attached to the hull, are in some cases found to be steerable. The steerable duct, which obviates the need for a rudder, is mounted on pintles whose axes lie on the vertical diameter of the propeller disc. This then allows the duct to be rotated about the pintle axes by an inboard steering motor and consequently the thrust of the propeller can be directed towards a desired direction for navigation purposes. Clearly, however, the arc through which the thrust can be directed is limited by geometric constraints. Applications of this type can range from small craft, such as harbour tugs, to comparatively large commercial vessels as shown by Figure 6.5. A further application of the steerable ducted propeller which has gained considerable popularity in recent years, particularly in the offshore field, is the azimuthing thruster where in many cases these units can be trained around a full 360°.

6.2.3 Podded and azimuthing propulsors

Azimuthing thrusters have been in common use for many years and can have either non-ducted or ducted propeller arrangements. They can be further classified into pusher or tractor units as seen in Figure 6.6. The essential difference between the azimuthing and podded propellers lies in where the engine or motor driving the propeller is sited. If the motor is sited in the ship's hull then the system

end of the shaft line. Nevertheless, variants of this arrangement do exist. The propellers associated with these propulsors have been of the fixed pitch type and are commonly built-up although their size is not particularly large. Currently, the largest size of unit is around the 23 MW capacity and the use of podded propulsors has been mainly in the context of cruise ships and ice breakers where their manoeuvring potential have been fully realized. Clearly, however, there are a number of other ship types which might benefit from their application. Figure 6.7 shows a typical example of a large podded propulsor unit.

Tractor arrangements of podded and azimuthing propulsors generally have an improved inflow velocity field since they do not have a shafting and A-bracket system ahead of them to cause a disturbance to the inflow. This tends to help suppress the blade rate harmonic pressures since the relatively undisturbed wake field close to zero azimuthing angles is more conducive to maintaining low rates of growth and collapse of cavities. However, there is a tendency for these propellers to exhibit broadband excitation characteristics and during the design process care has to be exercised to minimize these effects. At high azimuthing angles then the flow field is more disturbed.

Each of these systems posses significant manoeuvrability advantages, however, when used in combinations of two or more care has to be exercised in preventing the existence of sets of azimuthing angle where the propulsors can mutually interfere with each other. If this occurs large fluctuating forces and moments can be induced on the shaft system and significant vibration can be encountered.

6.2.4 Contra-rotating propellers

The contra-rotating propeller principle, comprising two coaxial propellers sited one behind the other and rotating in opposite directions, has traditionally been associated with the propulsion of aircraft, although Ericsson's original proposal of 1836 used this method as did de Bay's design for the *Iolair*.

Contra-rotating propulsion systems have the hydrodynamic advantage of recovering part of the slipstream rotational energy which would otherwise be lost to a conventional single screw system. Furthermore, because of the two propeller configuration, contra-rotating propellers possess a capability for balancing the torque reaction from the propulsor which is an important matter for torpedo and other similar propulsion problems. In marine applications of contra-rotating propulsion it is normal for the aftermost propeller to have a smaller diameter than the forward propeller and, in this way, accommodate the slipstream contraction effects. Similarly, the blade numbers of

Figure 6.5 Steerable ducted propeller.

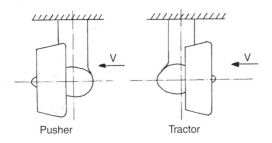

Figure 6.6 Pusher and tractor thruster units.

would be termed an azimuthing propulsor and most commonly the mechanical drive would be of a Z or L type to the propeller shaft. Frequently, the drive between the vertical and horizontal shafts is via spiral bevel gears.

In the case of a podded propulsor the drive system normally comprises an electric motor directly coupled to a propeller shaft which is supported on two rolling element bearing systems: one frequently being a radial bearing closest to the propeller while the other is spherical roller bearing at the opposite

Figure 6.7 Typical podded propulsor unit.

the forward and aft propellers are usually different; typically, four and five for the forward and aft propellers, respectively.

Contra-rotating propeller systems have been the subject of considerable theoretical and experimental research as well as some practical development exercises. Whilst they have found a significant number of applications, particularly in small high-speed outboard units, operating for example at around 1500 to 2000 rpm, the mechanical problems associated with the longer line shafting systems of larger vessels have generally precluded them from use on merchant ships. Interest in the concept has had a cyclic nature: interest growing and then waning. A recent upsurge in interest in 1988, however, has resulted in a system being fitted to a 37 000 dwt bulk carrier, IHI (1993), and subsequently to a 258 000 dwt VLCC in 1993.

6.2.5 Overlapping propellers

This again is a two-propeller concept. In this case the propellers are not mounted coaxially but are each located on separate shaft systems with the distance between the shaft centre lines being less than the diameter of the propellers. Figure 6.8 shows a typical arrangement of such a system; again this is not a recent idea and references may be found dating back over a hundred years: for example, Taylor's design of 1830.

As in the case of the contra-rotating propeller principle, recent work on this concept has been largely confined to research and development, and the system has rarely been used in practice. Research

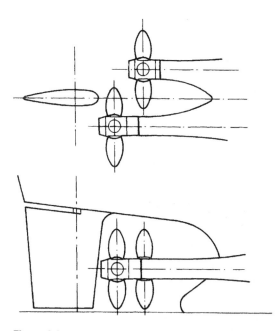

Figure 6.8 Overlapping propellers.

has largely centred on the effects of the shaft spacing to propeller diameter ratio on the overall propulsion efficiency in the context of particular hull forms, Kerlen *et al.* (1970) and Restad *et al.* (1973). The principal aim of this type of propulsion arrangement is to gain as much benefit as possible from the low-velocity portion of the wake field and thereby, increase propulsion efficiency. Consequently, the benefits derived from this propulsion concept are intimately related to the propeller and hull propulsion coefficients.

Despite one propeller working partially in the wake of the other, cavitation problems are not currently thought to pose insurmountable design problems. However, significant increases in the levels of fluctuating thrust and torque have been identified when compared to single-screw applications. In comparison to the twin-screw alternative, research has indicated that the overlapping arrangement may be associated with lower building costs, and this is portrayed as one further advantage for the concept.

When designing this type of propulsion system several additional variables are presented to the designer. These are the direction of propeller rotation, the distance between the shafts, the longitudinal clearance between the propellers and the stern shape. At the present time there are only partial answers to these questions. Research tends to suggest that the best direction of rotation is outward, relative to the top-dead-centre position and that the optimum distance between the shafts lies below 0.8 D. In addition there are indications that the principal effect of the longitudinal spacing of the propellers is to be found in vibration excitation and that propulsion efficiency is comparatively insensitive to this variable.

6.2.6 Tandem propellers

Tandem propeller arrangements are again not a new propulsion concept. Perhaps the best-known example is that of Parson's *Turbinia* where three propellers were mounted on each of the three propellers in order to overcome the effects of cavitation induced thrust breakdown, Figure 6.9. Indeed, the principal reason for the employment of tandem propellers has been to ease difficult propeller loading situations; however, these occasions have been relatively few. The disadvantage of the tandem propeller arrangement when applied to conventional single and twin-screw ships is that the weights and axial distribution of the propellers create large bending moments which have to be reacted principally by the stern tube bearings.

Some azimuthing and podded propulsor arrangements, however, employ this arrangement by having a propeller located at each end of the propulsion shaft, either side of the pod body. In this way the load is shared by the tractor and pusher propellers and the weight induced shaft moments controlled.

6.2.7 Controllable pitch propellers

Unlike fixed pitch propellers whose only operational variable is rotational speed, the controllable pitch propeller provides an extra degree of freedom in its ability to change blade pitch. However, for some propulsion applications, particularly those involving shaft-driven generators, the shaft speed is held constant, thus reducing the number of operating variables again to one. While this latter arrangement is very convenient for electrical power generation it can cause difficulties in terms of the cavitation characteristics of the propeller by inducing back and face cavitation at different propulsion conditions.

The controllable pitch propeller has found application in the majority of the propeller types and applications so far discussed in this Section with the

Figure 6.9 Tandem propeller arrangement on a shaft line of *Turbinia*.

possible exception of the contra-rotating and tandem propellers, although even in this extreme example of mechanical complexity some development work has been undertaken for certain specialist propulsion problems. In the last fifty years the controllable pitch propeller has grown in popularity from representing a small proportion of the propellers produced to its current position of having a very substantial market share. This growth is illustrated by Figure 6.10 which shows the proportion of controllable pitch propeller systems when compared to the total number of propulsion systems classed with Lloyd's Register during the period 1960 to 2004, taken at five-year intervals. From this figure it can be seen that currently the controllable pitch propeller has about a 35% market share when compared to fixed pitch propulsion systems, whilst Table 6.1 shows the relative distribution of controllable pitch propellers within certain classes of ship type. From the table it is seen that the controllable pitch propeller is currently most favoured in the passenger ship and ferry, general cargo, tug and trawling markets, noting of course that Table 6.1 relates to vessels with installed powers of greater than 2000 bhp.

The controllable pitch propeller, although of necessity possessing a greater degree of complexity than the fixed pitch alternative, does possess a number of important advantages. Clearly, manoeuvring is one such advantage in that fine thrust control can be achieved without necessarily the need to accelerate and decelerate the propulsion machinery. Furthermore, fine control of thrust is particularly important in certain cases: for example, in dynamic positioning situations or where frequent berthing manoeuvres are required such as in short sea route ferry operations. Moreover, the basic controllable pitch propeller hub design can in many instances be modified to accommodate the feathering of the propeller blades. The feathering position is the position where the blades are aligned approximately fore and aft and in the position in which they present least resistance to forward motion when not rotating. Such arrangements find applications on double-ended ferries or in small warships. In this latter application,

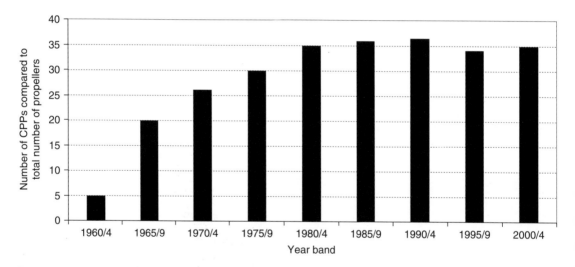

Figure 6.10 Market share of controllable pitch propellers.

Table 6.1 Percentage relative distribution of controllable pitch propellers to the total number of propellers by ship type classed with Lloyd's Register and having installed powers greater than 2000 bhp.

Ship type	1960–1964	1965–1969	1970–1974	1975–1979	1980–1984	1985–1989	1990–1994	1995–1999	2000–2004
Tankers	1	7	15	14	23	13	21	17	10
Bulk carriers	1	9	10	5	5	12	0	1	1
Container ships	0	13	24	3	1	13	18	10	9
General cargo	2	12	20	29	42	43	45	55	80
Passenger ships and ferries	24	64	82	100	94	100	88	78	63
Tugs and offshore vessels	29	50	44	76	85	100	77	73	78
Fishing vessels	48	54	87	90	93	92	100	90	89

the vessel could, typically, have three propellers; the two wing screws being used when cruising with centre screw not rotating implying, therefore, that it would benefit from being feathered in order to produce minimum resistance to forward motion in this condition. Then when the sprint condition is required all three propellers could be used at their appropriate pitch settings to develop maximum speed.

The details and design of controllable pitch propeller hub mechanisms are outside the scope of this Section since it is primarily concerned with the hydrodynamic aspects of ship propulsion. It will suffice to say, therefore, that each manufacturer has an individual design of pitch actuating mechanism, but that these designs can be broadly grouped into two principal types; those with inboard and those with outboard hydraulic actuation. Figure 6.11 shows these principal types in schematic form. For further discussion and development of these matters reference can be made to the works of Plumb (1987), Smith (1983) and Brownlie (1998), respectively, which provide introductions to this subject. Alternatively, propeller manufacturers' catalogues frequently provide a source of outline information on this aspect of controllable pitch propeller design.

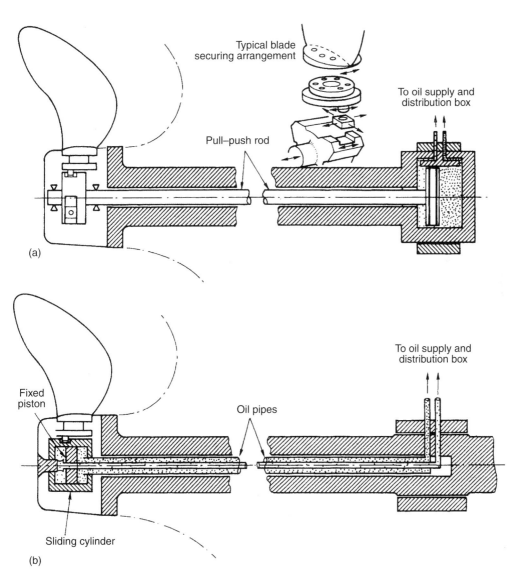

Figure 6.11 Controllable pitch propeller schematic operating systems: (a) pull–push rod system and (b) hub piston system.

The hub boss, in addition to providing housing for the blade actuation mechanism, must also be sufficiently strong to withstand the propulsive forces supplied to and transmitted from the propeller blades to the shaft. In general, therefore, controllable pitch propellers tend to have larger hub diameters than those for equivalent fixed pitch propellers. Typically the controllable pitch propeller hub has a diameter in the range 0.24 to 0.32 D, but for some applications this may rise to as high as 0.4 or even 0.5 D. In contrast, fixed pitch propeller boss diameters are generally within the range 0.16 to 0.25 D. The large boss diameters may give rise to complex hydrodynamic problems, often cavitation related, but for the majority of normal applications the larger diameter of the controllable pitch propeller hub does not generally pose problems that cannot be either directly or indirectly solved by known design practices. See also Section 6.6.2.9 and Figure 6.98.

Certain specialist types of controllable pitch propeller have been designed and patented in the past. Two examples are the self-pitching propeller and the Pinnate propeller, both of which are modern versions of much earlier designs. Self-pitching propellers are a modern development of Griffiths' work in 1849. The blades are sited on an external crank which is pinned to the hub and they are free to take up any pitch position, Miles et al. (1993). The actual blade pitch position taken up in service depends on a balance of the blade loading and spindle torque components which are variables depending on, amongst other parameters, rotational speed: at zero shaft speed but with a finite ship speed the blades are designed to feather. At the present time these propellers have only been used on relatively small craft.

The Pinnate design is to some extent a controllable pitch–fixed pitch propeller hybrid. It has a blade activation mechanism which allows the blades to change pitch about a mean position by varying angular amounts during one revolution of the propeller. The purpose of the concept is to reduce both the magnitude of the blade cyclical forces and cavitation by attempting to adjust the blades for the varying inflow velocity conditions around the propeller disc. Trials of these types of propeller have been undertaken on small naval craft and Simonsson (1983) describes these applications.

6.2.8 Waterjet propulsion

The origin of the waterjet principle can be traced back to 1661, when Toogood and Hayes produced a description of a ship having a central water channel in which either a plunger or centrifugal pump was installed to provide the motive power. In more recent times waterjet propulsion has found considerable application on a wide range of small high-speed craft while its application to larger craft is growing with tunnel diameters of upwards of 2 m being considered.

The principle of operation of the present-day waterjet is that in which water is drawn through a ducting system by an internal pump which adds energy after which the water is expelled aft at high velocity. The unit's thrust is primarily generated as a result of the momentum increase imparted to the water. Figure 6.12 shows, in outline form, the main features of the waterjet system.

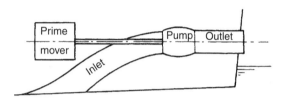

Figure 6.12 Waterjet configuration.

The pump configuration adopted for use with a waterjet system depends on the specific speed of the pump; specific speed N_s being defined in normal hydraulic terms as

$$N_s = \frac{(N)Q^{1/2}}{H^{3/4}} \quad (6.2)$$

where Q is the quantity of fluid discharged, N is the rotational speed and H is the head.

For low values of specific speed centrifugal pumps are usually adopted whereas for intermediate and high values of N_s axial pumps and inducers are normally used, respectively. The prime movers usually associated with these various pumps are either gas turbines or highspeed diesel engines.

Waterjet propulsion offers a further dimension to the range of propulsion alternatives and tends to be used where other propulsion forms are rejected for some reason: typically for reasons of efficiency, cavitation extent, noise or immersion and draught. For example, in the case of a small vessel travelling at say 45 knots one might expect that a conventional propeller would be fully cavitating, whereas in the corresponding waterjet unit the pump should not cavitate. Consequently, the potential for waterjet application, neglecting any small special purpose craft with particular requirements, is where conventional, transcavitating and supercavitating propeller performance is beginning to fall off. Indeed surface piercing propellers and waterjet systems are to some extent competitors for some similar applications. Waterjet units, however, tend to be heavier than conventional propeller-based systems and, therefore, might be expected to find favour with larger craft; for example, large wave-piercing ferries.

In terms of manoeuvrability the waterjet system is potentially very good, since deflector units are normally fitted to the jet outlet pipe which then direct the water flow and hence introduce turning forces by changing the direction of the jet momentum. Similarly for stopping manoeuvres, flaps or a 'bucket' can be introduced over the jet outlet to redirect the flow forward and hence apply an effective reactive retarding force to the vessel.

6.2.9 Cycloidal propellers

Cycloidal propeller development started in the 1920s, initially with the Kirsten–Boeing and subsequently the Voith–Schneider designs. It is interesting to note that the Kirsten–Boeing design was very similar in its hydrodynamic action to the horizontal waterwheel developed by Robert Hooke some two and half centuries earlier in 1681.

The cycloidal or vertical axis propellers basically comprise a set of vertically mounted vanes, six or eight in number, which rotate on a disc mounted in a horizontal or near horizontal plane. The vanes are constrained to move about their spindle axis relative to the rotating disc in a predetermined way by a governing mechanical linkage. Figure 6.13(a) illustrates schematically the Kirsten–Boeing principal. It can be seen from the figure that the vanes' relative attitude to the circumference of the circle, which governs their tracking path, is determined by referring the motion of the vanes to a particular point on that circumference. As such, it can be deduced that each vane makes half a revolution about its own pintle axis during one revolution of the entire propeller disc. The thrust magnitude developed by this propeller design is governed by rotational speed alone and the direction of the resulting thrust by the position of the reference point on the circumference of the vane-tracking circle.

The design of the Voith–Schneider propeller is rather more complex since it comprises a series of linkages which enable the individual vane motions to be controlled from points other than on the circumference of the vane-tracking circle. Figure 6.13(b) demonstrates this for a particular value of the eccentricity (e) of the vane-control centre point from the centre of the disc. By controlling the eccentricity, which in turn governs the vane-pitch angles, both the thrust magnitude and direction can be controlled independently of rotational speed. In the case of the Voith–Schneider design, in contrast to the Kirsten–Boeing propeller, the individual vanes make one complete revolution about their pintle axes for each complete revolution of the propeller disc. In many cases the units are provided with guards to help protect the propulsor blades from damage from external sources.

Vertical axis propellers do have considerable advantages when manoeuvrability or station keeping and this is an important factor in the ship design, since the resultant thrust can be readily directed along any navigational bearing and have variable magnitude. Indeed, this type of propeller avoids the necessity for a separate rudder installation on the vessel. Despite the relative mechanical complexity, these propellers have shown themselves to be reliable in operation over many years of service.

6.2.10 Paddle wheels

Paddle propulsion, as is well known, predates screw propulsion. However, this form of propulsion has almost completely disappeared except for a very few specialized applications. These are to be found largely on lakes and river services either as tourist or nostalgic attractions, or alternatively, where limited draughts are encountered. Nevertheless, the Royal Navy, until a few years ago, also favoured their use on certain classed of harbour tug where they were found to be exceptionally manoeuvrable. The last example of a seagoing paddle steamer, the *Waverley*, is seen in Figure 6.14.

The principal reason for the demise of the paddle wheel was its intolerance of large changes of draught and the complementary problem of variable immersion in seaways. Once having been superseded by screw propulsion for ocean-going vessels their use was largely confined through the first half of the last century to river steamers and tugs. Paddle wheels,

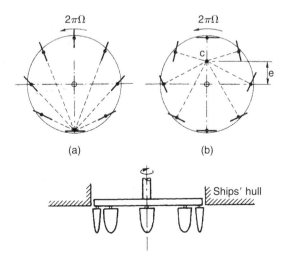

Figure 6.13 Vertical axis propeller principle: (a) Kirsten–Boeing propeller and (b) Voith-Schneider propeller.

Figure 6.14 *P.S. Waverley:* Example of a side wheel paddle steamer.

however, also suffered from damage caused by flotsam in rivers and were relatively expensive to produce when compared to the equivalent fixed pitch propeller.

Paddle design progressed over the years from the original simple fixed float designs to the feathering float system which then featured throughout much of its life. Figure 6.15 shows a typical feathering float paddle wheel design from which it can be seen that the float attitude is governed from a point just slightly off-centre of the wheel axis. Feathering floats are essential to good efficiency on relatively small diameter and deeply immersed wheels. However, on the larger wheels, which are not so deeply immersed, feathering floats are not essential and fixed float designs were normally adopted. This led to the practice of adopting feathered wheels in side-mounted wheel applications, such as were found on the Clyde or Thames excursion steamers, because of the consequent wheel diameter restriction imposed by the draught of the vessel. In contrast, on the stern wheel propelled vessels, such as those designed for the Mississippi services, the use of fixed floats was preferred since the wheel diameter restriction did not apply.

The design of paddle wheels is considerably more empirical than that of screw propellers today, nevertheless, high propulsion efficiencies were achieved and these were of similar orders to equivalent screw-propelled steamers. Ideally, each float of the paddle wheel should enter the water 'edgeways' and without shock having taken due account of the relative velocity of the float to the water. Relative velocity in still water has two components: the angular speed due to the rotation of the wheel and the speed of the vessel V_a. From Figure 6.16 it can be seen that at the point of entry A, a resultant vector $\overline{a}$ is produced from the combination of advance speed V_a and the rotational vector ωR. This resultant vector represents the absolute velocity at the point of entry and to avoid shock at entry, that is a vertical thrusting action of the float, the float should be aligned parallel to this vector along the line YY. However, this is not possible practically and the best that can be achieved is to align the floats to the point B and this is achieved by a linkage EFG which is introduced into the system. Furthermore, from Figure 6.16 it is obvious that the less the immersion of the wheel (h), the less is the advantage to be gained from adopting a feathering float system. This explains why the fixed float principal is adopted for large, lightly immersed wheels.

With regard to the overall design parameters, based on experience it was found that the number of fixed floats on a wheel should be about one for every foot of diameter of the wheel and for feathering designs this number was reduced to around 60 or 70% of the fixed float 'rule'. The width of the floats used in a particular design was of the order of 25 to 40% of the float length for feathering designs, but this figure was reduced for the fixed float paddle wheel to between 20 and 25%. A further constraint on the immersion of the floats was that the peripheral speed at the top of the floats should not exceed the ship speed and, in general, feathering floats were immersed in the water up to about half a float width whilst with sternwheelers, the tops of the floats were never far from the water surface.

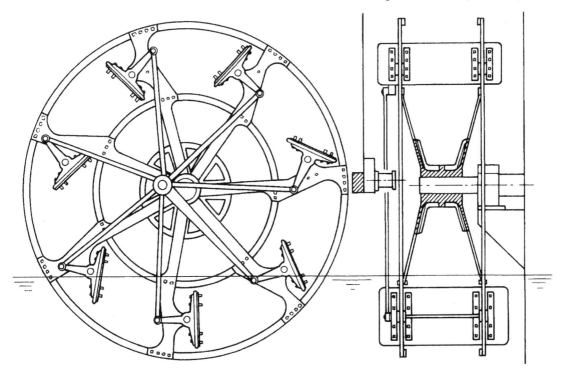

Figure 6.15 Paddle wheel (Reproduced from Hamilton (1948)).

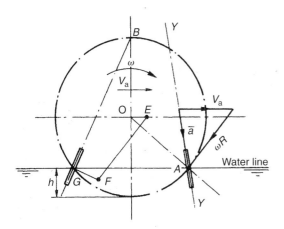

Figure 6.16 Paddle wheel float relative velocities.

The empirical nature of paddle design was recognized as being unsatisfactory and in the mid-1950s Volpich and Bridge (1954, 1955, 1956), conducted systematic experiments on paddle wheel performance at the Denny tank in Dumbarton. Unfortunately, this work came at the end of the time when paddle wheels were in use as a common form of propulsion and, therefore, never achieved its full potential.

6.2.11 Magnetohydrodynamic propulsion

Magnetohydrodynamic propulsion potentially provides a means of ship propulsion without the aid of either propellers or paddles. The laws governing magnetohydrodynamic propulsion were known in the nineteenth century and apart from a few isolated experiments such as that by Faraday when he attempted to measure the voltage across the Thames induced by its motion through the earth's magnetic field and the work of Hartmann on electromagnetic pumps in 1918, the subject had largely to wait for engineering development until the 1960s.

The idea of electromagnetic thrusters was first patented in the USA by Rice (1961) during 1961. Following this patent the USA took a leading role in both theoretical and experimental studies culminating in a report from the Westinghouse Research Laboratory in 1966. This report showed that greater magnetic field densities were required before the idea could become practicable in terms of providing a realistic alternative for ship propulsion. In the 1970s superconducting coils enabled further progress to be made with this concept.

The fundamental principal of electromagnetic propulsion is based upon the interaction of a magnetic field B produced by a fixed coil placed inside the ship and an electric current passed through the sea water from electrodes in the bottom of the ship or across a duct, as shown diagrammatically in Figure 6.17. Since the magnetic field and the current are in mutually orthogonal directions, then the resulting Lorentz force provides the necessary pumping action. The Lorentz force is $J \times B$ where J is the induced current density. Iwata et al. (1983) and subsequently in (1990) present an interesting description of the state of the art of superconducting propulsion.

In theory the electrical field can be generated either internally or externally, in the latter case by positioning a system of electrodes in the bottom of the ship. This, however, is a relatively inefficient method for ship propulsion. The environmental impact of the internal system is considerably reduced due to the containment of the electromagnetic fields. Most work, therefore, has concentrated on systems using internal magnetic fields and the principle of this type of system is shown in Figure 6.18(a) in which a duct, through which sea water flows, is surrounded by superconducting magnetic coils which are immersed in a cryostat. Inside the duct are placed two electrodes, which create the electric field necessary to interact with the magnetic field in order to create the Lorentz forces necessary for propulsion. Nevertheless, the efficiency of a unit is low due to the losses caused by the low conductivity of sea water. The efficiency, however, is proportional to the square of the magnetic flux intensity and to the flow speed, which is a function of ship speed. Consequently, in order to arrive at a reasonable efficiency it is necessary to create a strong magnetic flux intensity by the use of powerful magnets. In order to investigate the full potential of these systems at prototype scale a small craft, Yamato 7, was built for trial purposes by the Japanese and Figure 6.18(b) shows a cross section through one of the prototype propulsion units, indicating the arrangement of the six dipole propulsion ducts within the unit. Figure 6.19 shows the experimental craft, Yamato 1.

Electromagnetic propulsion does have certain potential advantages in terms of providing a basis for noise and vibration-free hydrodynamic propulsion.

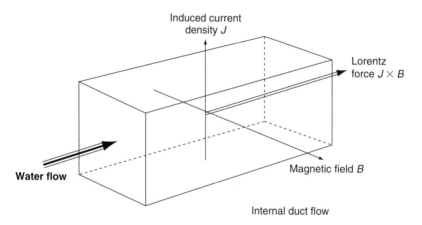

Figure 6.17 Magnetohydrodynamic propulsion principle.

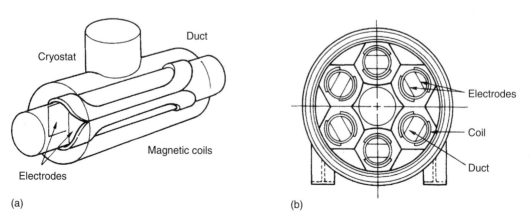

Figure 6.18 Internal magnetic field electromagnetic propulsion unit: (a) the dipole propulsion unit with internal magnetic field and (b) a cross section through a prototype propulsion unit.

Figure 6.19 *Yamato 1*: Experimental magnetohydrodynamic propulsion craft.

However, a major obstacle to the development of electromagnetic propulsion until relatively recently was that the superconducting coil, in order to maintain its zero-resistance property, required to be kept at the temperature of liquid helium, 4.2 K ($-268°$). This clearly requires the use of thermally well-insulated vessels in which the superconducting coil could be placed in order to maintain these conditions. The criticality of this thermal condition can be seen from Figure 6.20 which indicates how the resistance of a superconductor behaves with temperature and eventually reaches a critical temperature when the resistance falls rapidly to zero. Superconductors are also sensitive to current and magnetic fields; if either become too high then the superconductor will fail in the manner shown in Figure 6.21.

Superconductivity began with the work of Kamerlingh Onnes at Leiden University in 1911 when he established the superconducting property for mercury in liquid helium; for this work he won a Nobel Prize. Work continued on superconductivity, however, progress was slow in finding metals which would perform at temperatures as high as that of liquid nitrogen, $-196°C$. By 1973 the best achievable temperature was 23 K. However, in 1986 Muller and Bednorz in Zurich turned their attention to ceramic oxides which had hitherto been considered as insulators. The result of this shift of emphasis was to immediately increase the critical temperature to 35 K by the use of a lanthanum, barium, copper oxide compound: this discovery led to Muller and Bednorz also being awarded a Nobel Prize for their work. Consequent on this discovery, work in the USA, China, India and Japan intensified, leading to the series of rapid developments depicted in Table 6.2.

Whilst these advances are clearly encouraging since they make the use of superconducting coils easier

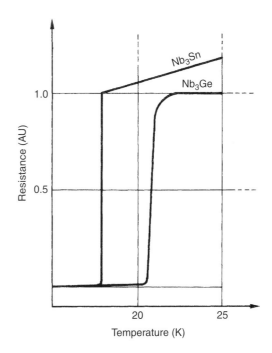

Figure 6.20 Superconducting effect.

from the thermal insulation viewpoint, many ceramic oxides are comparatively difficult to produce. First, the process by which the superconductor is made is very important if the correct molecular structure is to be obtained and second, ceramics are brittle. Consequently, whilst this form of propulsion clearly has potential and significant advances have been made, both in the basic research and application, much work still has to be done before this type of propulsion can become a reality on a commercial scale or even the concept fully tested.

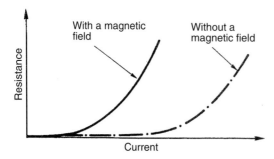

Figure 6.21 Effect of a magnetic field on a superconductor.

Table 6.2 Development of superconducting ceramic oxides.

Date	Ceramic oxide	Superconducting temperatures (K)
September 1986	La—Ba—Cu—O	35
January 1987	Y—Ba—Cu—O	93
January 1988	Bi—Sr—Ca—Cu—O	118
February 1988	Tl—Ba—Ca—Cu—O	125

6.2.12 Superconducting motors for marine propulsion

Notwithstanding the problems for magnetohydrodynamic propulsion, superconductivity has in the last few years shown its potential for the production of marine propulsion motors using the high-temperature superconductors of Bi-2223 material $[(Bi,Pb)_2Sr_2Ca_2Cu_3O_x]$ which have a T_c of 110 K but operate at a temperature of 35 to 40 K. This material has, at the present time, been demonstrated to be the most technically viable material for propulsion motors. In the USA a 5 MW demonstrator machine has proved satisfactory and a 25 MW demonstrator is being constructed to demonstrate the potential for marine propulsion purposes. In addition to other marine propulsion applications the relatively small diameter of these machines, if finally proved satisfactory, may have implications for podded propulsors since the hub diameter may be then reduced given that this diameter is principally governed by the electric motor size.

6.3 Diesel engine performance

6.3.1 Rating

An important parameter for a marine diesel engine is the rating figure, usually stated as bhp or kW per cylinder at a given rev/min, Figure 6.22.

Although enginebuilders talk of continuous service rating (csr) and maximum continuous rating (mcr), as well as overload ratings, the rating which concerns a shipowner most is the maximum output guaranteed by the enginebuilder at which the engine will operate continuously day in and day out. It is most important that an engine be sold for operation at its true maximum rating and that a correctly sized engine be installed in the ship in the first place; an under-rated main engine, or more particularly an auxiliary, will inevitably be operated at its limits most of the time. It is wrong for a ship to be at the mercy of two or three undersized and thus overrated auxiliary engines, or a main engine that needs to operate at its maximum continuous output to maintain the desired service speed.

Prudent shipowners usually insist that the engines be capable of maintaining the desired service speed fully loaded, when developing not more than 80% (or some other percentage) of their rated brake power. However, such a stipulation can leave the full-rated power undefined and therefore does not necessarily ensure a satisfactory moderate continuous rating, hence the appearance of continuous service rating and maximum continuous rating. The former is the moderate in-service figure, the latter is the enginebuilder's set point of mean pressures and revolutions which the engines can carry continuously, Figure 6.23.

Normally a ship will run sea trials to meet the contract trials speed (at a sufficient margin above the required service speed) and the continuous service rating should be applied when the vessel is in service. It is not unknown for shipowners to then stipulate that the upper power level of the engines in service should be somewhere between 85–100% of the service speed output, which could be as much as 20% less than the engine maker's guaranteed maximum continuous rating.

6.3.2 Maximum rating

The practical maximum output of a diesel engine may be said to have been reached when one or more of the following factors operate:

1. The maximum percentage of fuel possible is being burned effectively in the cylinder volume available (combustion must be completed fully at the earliest possible moment during the working stroke).
2. The stresses in the component parts of the engine generally, for the mechanical and thermal conditions prevailing, have attained the highest safe level for continuous working.
3. The piston speed and thus revolutions per minute cannot safely be increased.

For a given cylinder volume, it is possible for one design of engine effectively to burn considerably more fuel than one of another design. This may be the result of more effective scavenging, higher

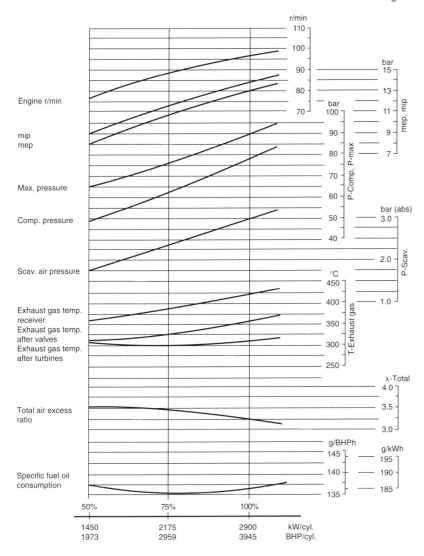

Figure 6.22 Typical performance curves for a two-stroke engine.

pressure turbocharging, by a more suitable combustion chamber space and design, and by a more satisfactory method of fuel injection. Similarly, the endurance limit of the materials of cylinders, pistons and other parts may be much higher for one engine than for another; this may be achieved by the adoption of more suitable materials, by better design of shapes, thicknesses, etc., more satisfactory cooling and so on. A good example of the latter is the bore cooling arrangements now commonly adopted for piston crowns, cylinder liner collars and cylinder covers in way of the combustion chamber.

The piston speed is limited by the acceleration stresses in the materials, the speed of combustion and the scavenging efficiency: that is, the ability of the cylinder to become completely free of its exhaust gases in the short time of one part cycle. Within limits, so far as combustion is concerned, it is possible sometimes to increase the speed of an engine if the mean pressure is reduced. This may be of importance for auxiliary engines built to match a given alternator speed.

For each type of engine, therefore, there is a top limit beyond which the engine should not be run continuously. It is not easy to determine this maximum continuous rating; in fact, it can only be satisfactorily established by exhaustive tests for each size and type of engine, depending on the state of development of the engine at the time.

If a cylinder is overloaded by attempting to burn too much fuel, combustion may continue to the end of the working stroke and perhaps also until after

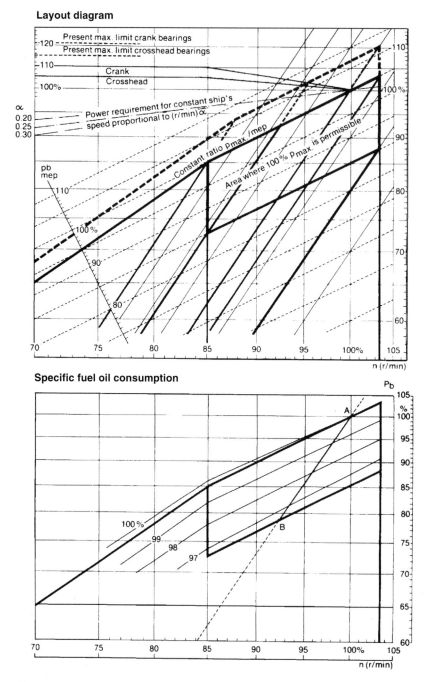

Figure 6.23 Layout diagrams showing maximum and economy ratings and corresponding fuel consumptions.

exhaust has begun. Besides suffering an efficiency loss, the engine will become overheated and piston seizures or cracking of engine parts may result; or, at least, sticking piston rings will be experienced, as well as dirty and sticking fuel valves.

6.3.3 Exhaust temperatures

The temperature of the engine exhaust gases can be a limiting factor for the maximum output of an engine. An exhaust-temperature graph plotted with mean indicated pressures as abscissae and exhaust temperatures as ordinates will generally indicate when the economical combustion limit, and sometimes when the safe working limit, of an engine has been attained. The economical limit is reached shortly after the exhaust temperature begins to curve upwards from what was, previously, almost a straight line.

Very often the safe continuous working load is also reached at the same time, as the designer naturally strives to make all the parts of an engine equally suitable for withstanding the respective thermal and mechanical stresses to which they are subjected.

When comparing different engine types, however, exhaust temperature cannot be taken as proportionate to mean indicated pressure. Sometimes it is said and generally thought that engine power is limited by exhaust temperature. What is really meant is that torque is so limited and exhaust temperature is a function of torque and not of power. The exhaust temperature is influenced by the lead and dimensions of the exhaust piping. The more easily the exhaust gases can flow away, the lower their temperature, and vice versa.

6.3.4 Derating

An option available to reduce the specific fuel consumption of diesel engines is derated or so-called 'economy' ratings. This means operation of an engine at its normal maximum cylinder pressure for the design continuous service rating, but at lower mean effective pressure and shaft speed.

By altering the fuel injection timing to adjust the mean pressure/maximum pressure relationship the result is a worthwhile saving in fuel consumption. The power required for a particular speed by a given ship is calculated by the naval architect and, once the chosen engine is coupled to a fixed pitch propeller, the relationship between engine power, propeller revolutions and ship speed is set according to the fixed propeller curve. A move from one point on the curve to another is simply a matter of giving more or less fuel to the engine.

Derating is the setting of engine performance to maximum cylinder pressures at lower than normal shaft speeds, at a point lower down the propeller curve. For an existing ship and without changing the propeller this will result in a lower ship speed, but in practice when it is applied to newbuildings, the derated engine power is that which will drive the ship at a given speed with the propeller optimized to absorb this power at a lower than normal shaft speed.

Savings in specific fuel consumption by fitting a derated engine can be as much as 5 g/bhph. However, should it be required at some later date to operate the engine at its full output potential (normally about 15–20% above the derated value) the ship would require a new propeller to suit both higher revolutions per minute and greater absorbed power. The injection timing would also have to be reset.

6.3.5 Mean effective pressures

The term brake mean effective pressure (bmep) is widely quoted by enginebuilders, and is useful for industrial and marine auxiliary diesel engines that are not fitted with a mechanical indicator gear. However, the term has no useful meaning for shipboard propulsion engines. It is artificial and superfluous as it is derived from measurements taken by a dynamometer (or brake), which are then used in the calculation of mechanical efficiency. Aboard ship, where formerly the indicator and now pressure transducers producing PV diagrams on an oscilloscope, are the means of recording cylinder pressures, mean indicated pressure (mip) is the term used, particularly in the calculation of indicated power.

Many ships now have permanently mounted torsionmeters. By using the indicator to calculate mean indicated pressure and thus indicated power, and the torsionmeter to calculate shaft power from torque readings and shaft revolutions, the performance of the engine both mechanically and thermally in the cylinders can be readily determined.

Instruments such as pressure transducers, indicators, tachometers and pressure gauges (many of which are of the electronic digital or analogue type of high reliability) allow the ship's engineer to assess accurately the performance of the engine at any time.

The values of brake power, mean indicated pressure and revolutions per minute are, of course, capable of mutual variation within reasonable limits, the power developed per cylinder being the product of mean indicated (or effective) pressure, the revolutions per minute and the cylinder constant (based on bore and stroke). The actual maximum values for power and revolutions to be used in practice are those quoted by the enginebuilder for the given continuous service rating.

6.3.6 Propeller slip

The slip of the propeller is normally recorded aboard ship as a useful pointer to overall results. While it may be correct to state that the amount of apparent slip is no indication of propulsive efficiency in a new ship design, particularly as a good design may have a relatively high propeller slip, the daily variation in slip (based on ship distance travelled compared with the product of propeller pitch and revolutions turned by the engine over a given period of time) can be symptomatic of changes in the relationship of propulsive power and ship speed; and slip, therefore, as an entity, is a useful parameter. The effects on ship speed 'over the ground' by ocean currents is sometimes considerable.

For example, a following current may be as much as 2.5% and heavy weather ahead may have an effect of more than twice this amount.

6.3.7 Propeller law

An enginebuilder is at liberty to make the engine mean pressure and revolutions what he will, within the practical and experimental limits of the engine design. It is only after the maximum power and revolutions are decided and the engine has been coupled to a propeller that the propeller law operates in its effect upon power, mean pressure and revolutions.

shp varies as V^3
shp varies as N^3
Q varies as N^2
P varies as N^2

where shp = aggregate shaft horsepower of engine, metric or imperial;
V = speed of ship in knots
N = revolutions per minute
Q = torque, in kg metres or lbft = Pr
P = brake mean pressure kgf/cm^2 or lbf/in^2
r = radius of crank, metres or feet

If propeller slip is assumed to be constant:

$shp = KV^3$

where K = constant from shp and N for a set of conditions.

But N is proportional to V, for constant slip,

$\therefore \quad shp = K_1 N^3$

where K_1 = constant from shp and N for a set of conditions.

But $\quad shp = \dfrac{p \times A \times c \times r \times 2\pi \times N}{33\,000}$

$= K_1 N^3$ (imperial)

when A = aggregate area of pistons, cm^2 or in^2
c = 0.5 for two-stroke, 0.25 for four-stroke engines:

or $\quad Q = PAcr = \dfrac{33\,000}{2\pi \times N} \times K_1 N^3 = K_2 N^2$ (imperial)

or

$Q = PAcr = \dfrac{4500}{2\pi \times N} \times K_1 N^3 = K_2 N^2$ (metric)

i.e. $Q = K_2 N^2$

where K_2 = constant, determinable from Q and N for a set of conditions.

$PAcr = Q$ or $\dfrac{Q}{Acr} = \dfrac{K_2}{Acr} \times N^2$

i.e. $\quad P = K_3 N^2$

where K_3 = constant determinable from P and N for a set of conditions.

The propeller law index is not always 3, nor is it always constant over the full range of speeds for a ship. It could be as much as 4 for short high speed vessels but 3 is normally satisfactory for all ordinary calculations. The index for N, when related to the mean pressure P, is one number less than that of the index for V.

Propeller law is most useful for enginebuilders at the testbeds where engine loads can be applied with the dynamometer according to the load and revolutions calculated from the law, thus matching conditions to be found on board the ship when actually driving a propeller.

6.3.8 Fuel coefficient

An easy yardstick to apply when measuring machinery performance is the fuel coefficient:

$$C = \dfrac{\Delta^{2/3} \times V^3}{F} \quad (6.3)$$

where C = fuel coefficient
Δ = displacement of ship in tons
V = speed in knots
F = fuel burnt per 24 hours in tons

This method of comparison is applicable only if ships are similar, are run at approximately corresponding speeds, operate under the same conditions, and burn the same quality of fuel. The ship's displacement in relation to draught is obtained from a scale provided by the shipbuilders.

6.3.9 Admiralty coefficient

$$A_c = \dfrac{\Delta^{2/3} \times V^3}{p} \quad (6.4)$$

where A_c = Admiralty coefficient, dependent on ship form, hull finish and other factors, and p is power.

If A_c is known for a ship the approximate power can be calculated for given ship conditions of speed and displacement. See also Section 5.7.6 for use of Admiralty coefficient when monitoring ship performance.

6.3.10 Apparent propeller slip

$$\text{Apparent slip, per cent} = \left(\frac{P \times N - 101.33 \times V}{P \times N}\right) \times 100 \tag{6.5}$$

where P = propeller pitch in ft
V = speed of ship in knots
101.33 is one knot in ft/min

The true propeller slip is the slip relative to the wake stream, which is something very different. The engineer, however, is normally interested in the apparent slip.

6.3.11 Propeller performance

Many variables affect the performance of a ship's machinery at sea so the only practical basis for a contract to build to a specification and acceptance by the owner is a sea trial where everything is under the builder's control. The margin between the trial trip power and sea service requirements of speed and loading must ensure that the machinery is of ample capacity. One important variable on the ship's performance is that of the propeller efficiency.

Propellers are designed for the best combinations of blade area, diameter, pitch, number of blades, etc, and are matched to a given power and speed of propulsion engine; and in fact each propeller is specifically designed for the particular ship, see Chapter 5, Section 5.6. It is important that the engine should be able to provide heavy torque when required, which implies an ample number of cylinders with ability to carry high mean pressures. However, when a propeller reaches its limit of thrust capacity under head winds an increase in revolutions can be to no avail.

In tank tests with models for powering experiments the following particulars are given:

Quasi-propulsive coefficient (η_D)

$$\eta_D = \frac{\text{model resistance} \times \text{speed}}{2\pi \times \text{torque} \times \text{rev/min}} = \frac{\text{work got out per min}}{\text{work put in per min}}$$

Total shaft (delivered) power at propeller (P_D)

$$P_D = \frac{P_E \times a}{\eta_D} \tag{6.6}$$

where P_E = effective power for model as determined by tank testing;
a = increase for appendages and air resistance equal to 10–12% of the naked model P_E, for smooth water conditions.

The shaft power at the propeller for smooth sea trials can be about 10% more than in tank tests. The additional power, compared with sea trials, for sea service is about 11–12% more for the South Atlantic and 20–25% more for the North Atlantic. This is due to the normal weather conditions in these areas. The size of the ship affects these allowances: a small ship needs a greater margin. By way of example: 15% margin over trial conditions equals 26.5% over tank tests. See also Sections 5.15 and 5.7.

The shaft power measured by torsionmeter abaft the thrust block exceeds the P_D by the power lost in friction at the sterntube and plummer blocks and can be as much as 5–6%.

The brake power (P_B) exceeds the torsionmeter measured power by the frictional power lost at the thrust block. P_B can only be calculated onboard ship by multiplying the recorded indicated power by the mechanical efficiency stated by the enginebuilder.

Required P_B for engine = P_D + sea margin + power lost at sterntube and plummer blocks + power loast at thrust block.

Typical values for the quasi-propulsive coefficient (n_D) are: tanker 0.67–0.72; slow cargo vessel 0.72–0.75; fast cargo liner 0.70–0.73; ferry 0.58–0.62; passenger ship 0.65–0.70.

6.3.12 Power build-up

Figures 6.24 and 6.25 are typical diagrams showing the propulsion power data for a twin-screw vessel. In Figure 6.24 curve A is the ehp (P_E) at the trial draught; B is the ehp (P_E) corrected to the contract draught; C is the power at trial draught on the Firth of Clyde; D is the *shp* corrected to the contract draught; E is the power service curve from voyage results; F shows the relation between speed of the ship and engine revolutions on trials; G is the service result. The full rated power of the propelling engines is 18 000 bhp; the continuous service rating is 15 000 bhp.

In Figure 6.24 curves A to D show *shp* as ordinates and the speed of the ship as abscissae. In Figure 6.25 powers are shown as ordinates, revolutions as abscissae.

Figure 6.26 shows the relationship between revolutions, power and brake mean pressure for the conditions summarized in Figures 6.24 and 6.25. A fair line drawn through the observed points for the whole range shows the *shp* to increase approximately as the cube of the revolutions and the square of the bmep. For the range 95–109 rev/min, the index increases to 3.5 for the power and 2.5 for the bmep. Between 120 and 109 rev/min a more closely drawn curve shows the index to rise to 3.8 and the bmep to 2.8. In Figure 6.26 ordinates and abscissae are plotted to a logarithmic base, thus reducing the power/revolution and the pressure/revolution curves to straight lines, for simplicity.

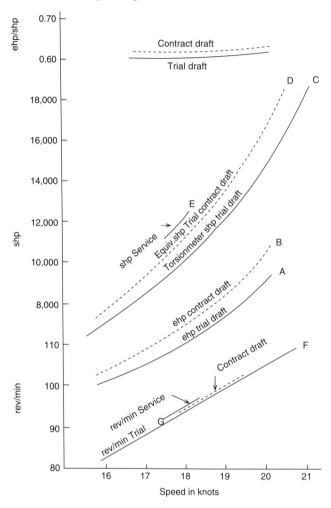

Figure 6.24 Propulsion data.

6.3.13 Trailing and locking of propeller

In Figure 6.27 there are shown the normal speed/power curves for a twin-screw motor vessel on the measured mile and in service.

The effect upon the speed and power of the ship when one of the propellers is trailed, by 'Freewheeling', is indicated in the diagram. The effect of one of the propellers being locked is also shown.

Figure 6.28 shows the speed/power curves for a four-screw motorship:

1. When all propellers are working.
2. When the vessel is propelled only by the two centre screws, the outer screws being locked.
3. When the ship is being propelled only by the two wing screws, the two inner screws being locked.

6.3.14 Astern running

Figure 6.29 summarizes a series of tests made on the trials of a twin-screw passenger vessel, 716 ft long, 83 ft 6 in beam, trial draught 21 ft forward, 26 ft aft, 26 000 tons displacement.

As plotted in Figure 6.29, tests I to VI show distances and times, the speed of approach being as stated at column 2 in Table 6.3. The dotted curves show reductions of speed and times.

The dotted curves A to F respectively correspond to curves I to VI. In test I, after the ship had travelled, over the ground, a distance of two nautical miles (1 nm = 6080 ft) the test was terminated and the next test begun.

In Table 6.4 a typical assortment of observed facts related to engine stopping and astern running is

Marine engines and auxiliary machinery 369

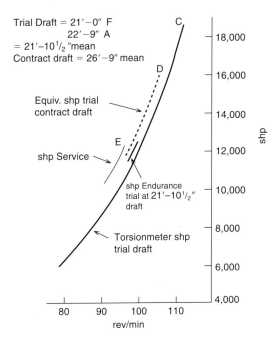

Figure 6.25 Propulsion data.

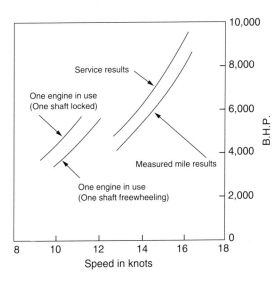

Figure 6.27 Speed/power curves, twin-screw vessel.

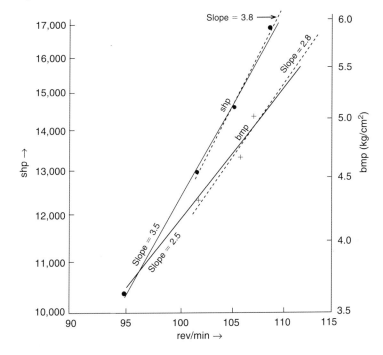

Figure 6.26 Engine trials: power, revolutions and mean pressure.

given. Where two or three sets of readings are given, these are for different vessels and/or different engine sizes.

Trials made with a cargo liner showed that the ship was brought to rest from 20 knots in 65 seconds. Another cargo liner, travelling at 16 knots, was brought to a stop in a similar period. The engine, running full power ahead, was brought to 80 rev/min astern in 32 seconds, and had settled down steadily at full astern revolutions in 50 seconds.

A shipbuilder will think of ship speed in terms of the trial performance in fair weather but to the shipowner ship speed is inevitably related to scheduled performance on a particular trade route. Sea trials are invariably run with the deadweight limited to fuel, fresh water and ballast. Because of the difference between loaded and trials draught, the hull resistance may be 25–30% greater for the same speed. This has a consequential effect upon the relation between engine torque and power, and in the reaction on propeller efficiency.

Adverse weather, marine growth and machinery deterioration necessitate a further power allowance, if the service speed is to be maintained. The mean wear and tear of the engine may result in a reduction of output by 10–15% or a loss of speed by up to one knot may be experienced.

When selecting a propulsion engine for a given ship a suitable power allowance for all factors such as weather, fouling, wear and tear, as well as the need to maintain the service speed at around 85% of the maximum continuous rating, should all be taken into consideration. See also Section 5.7.

6.4 Engine and plant selection

6.4.1 Introduction

Choosing a propulsion engine or engines and the most suitable plant configuration for a given newbuilding or retrofit project is not a simple decision. It dictates careful study of the machinery options available and the operating profile of the ship.

In the past the shipowner or designer had the straight choice of a direct-coupled low speed two-stroke

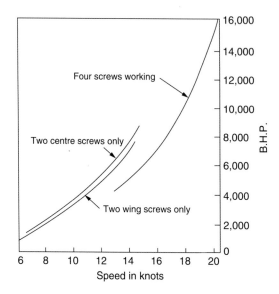

Figure 6.28 Speed/power curves, quadruple-screw ship.

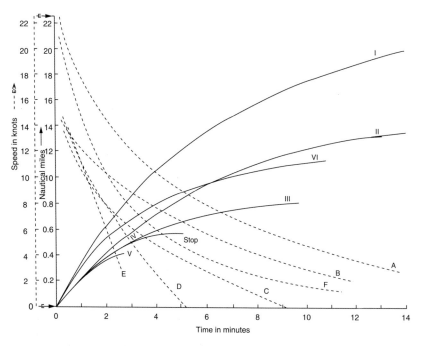

Figure 6.29 Ship stopping trials.

Table 6.3 Ship stopping trials.

1 Test No.	2 Ahead speed of approach knots (rev/min)	3 Propellers	4 Propellers stopped min)		5 Ship stopped (min)	6 Distance travelled (nautical miles)
			P.	S.		
I	23.0 (119)	Trailing; unlocked	16.4	14.9	—	2.0
II	14.5 (75)	Trailing; unlocked	12.5	12.1	17.0	1.4
III	13.5 (75)	Trailing; locked	1.5	1.5	13.0	0.8
IV	15.0 (75)	Ahead running checked; no additional astern power	1.3	1.3	5.2	0.5
V	14.8 (75)	Engine stopped; astern as quickly as possible	0.7	0.8	3.1	0.4
VI	22.4 (116)	Trailing; unlocked	1.5	1.6	15.0	1.1

Table 6.4 Engine reversing and ship stopping.

Ship	Engine type	Ahead (rev/min)	Time for engine stopping (sec)	Engine moving astern (sec)	Astern running		Ship stopped	
					rev/min	sec	min	sec
Large passenger	D.A. 2C. (twin)	65	—	30	—	—	4	2
Small fast passenger	S.A. 2C. tr. (twin)	217	53	63.5	160	72.5	2	15
		195	35	45	160	60	1	54
		220	45	59	170	81	2	5
Passenger	Diesel-electric (twin)	92	110	—	80	225	5	0
Cargo	S.A. 2C. (twin)	112	127.5	136	100	141	—	
Cargo	D.A. 2C. (single)	90	24	26	90	115	2	26
		88	12	14	82	35	3	25
		116	35	40	95	50	—	
Cargo	S.A. 2C. (single)	95	31	33	75	45	—	
		110	12	21	110	53	3	21
		116	10	55	110	110	4	6

engine driving a fixed pitch propeller or a geared medium speed four-stroke engine driving either a fixed or controllable pitch (CP) propeller. Today, ships are entering service with direct-coupled (and sometimes geared) two-stroke engines driving fixed or CP propellers, geared four-stroke engines or high/medium speed diesel–electric propulsion plants. Diverse diesel–mechanical and diesel–electric configurations can be considered (Figures 6.30 and 6.31).

Low speed engines are dominant in the mainstream deepsea tanker, bulk carrier and containership sectors while medium speed engines are favoured for smaller cargo ships, ferries, cruise liners, RoRo freight carriers and diverse specialist tonnage such as icebreakers, offshore support and research vessels. The overlap territory has nevertheless become more blurred in recent years: mini-bore low speed engines can target shortsea and even river/coastal vessels while new generations of high powered large bore medium speed designs contest the traditional deepsea arenas of low speed engines.

Many shipowners may remain loyal to a particular type or make and model of engine for various reasons, such as reliable past operating experience, crew familiarity, spares inventories and good service support. But new or refined engine designs continue to proliferate as enginebuilders seek to maintain or increase market share or to target a new sector. In some cases an owner who has not invested in new buildings for several years may have to consider a number of models and plant configurations which are unfamiliar.

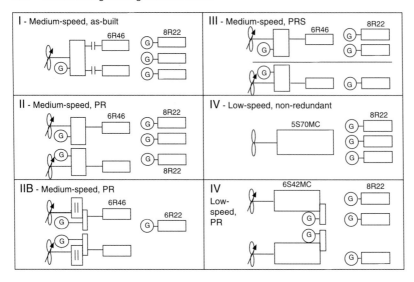

Figure 6.30 Low and medium speed diesel-based propulsion machinery options for a 91 000 dwt tanker and associated auxiliary power generating (G) source.

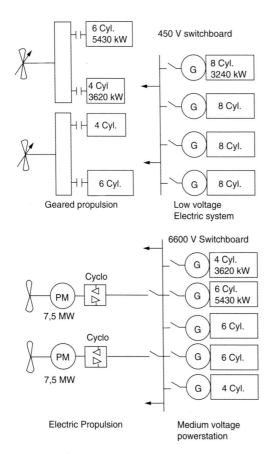

Figure 6.31 Schematics of geared medium speed diesel propulsion plant with gensets (top) and diesel–electric plant serving propulsion and auxiliary power demands (below).

Much decision-making on the engine focuses on cost considerations—not just the initial cost but the type of fuel which can be reliably burned, maintenance costs, desirable manning levels and availability/price of spares. The tendency now is to assess the total life cycle costs rather than the purchase price of the main engine. Operating costs over, say, 20 years may vary significantly between different types and makes of engine, and the selected plant configuration in which the engine has to function. Key factors influencing the choice of engine may be summarized as:

- Capability to burn heavy fuel of poor quality without detrimental impact on the engine components and hence maintenance/spares costs.
- The maintenance workload: the number of cylinders, valves, liners, rings and bearings requiring periodic attention in relation to the number of crew carried (bearing in mind that lower manning levels and less experienced personnel are now more common).
- Suitability for unattended operation by exploiting automated controls and monitoring systems.
- Propulsive efficiency: the ability of the engine or propeller shaft to be turned at a low enough speed to drive the largest diameter (and hence most efficient) propeller.
- Size and weight of the propulsion machinery.
- Cost of the engine.

The size of the machinery space is largely governed by the size of the main engine which may undermine the cargo-carrying capacity of the ship. The available headroom is also important in some ships, notably ferries with vehicle decks, and insufficient headroom and surrounding free space may make it difficult or impossible for some engines to be installed or overhauled.

6.4.2 Diesel–mechanical drives

6.4.2.1 *Overview*

The direct drive of a fixed pitch propeller by a low speed two-stroke engine remains the most popular propulsion mode for deepsea cargo ships. At one time a slight loss of propulsive efficiency was accepted for the sake of simplicity but the introduction of long stroke and, more recently, super- and ultra-long stroke crosshead engines has reduced such losses. For a large ship a direct-coupled speed of, say, 110 rev/min is not necessarily the most suitable since a larger diameter propeller turning at speeds as low as 60 rev/min is more efficient than one of a smaller diameter absorbing the same power at 110 rev/min. The longer stroke engines now available develop their rated outputs at speeds ranging from as low as 55 rev/min (very large bore models) up to around 250 rev/min for the smallest bore models. It is now possible to specify a direct-drive engine/propeller combination which will yield close to the optimum propulsive efficiency for a given ship design.

Large bore low speed engines develop high specific outputs, allowing the power level required by many ship types to be delivered from a small number of cylinders. Operators prefer an engine with the fewest possible cylinders, as long as problems with vibration and balance are not suffered. Fewer cylinders obviously influence the size of the engine and the machinery space, the maintenance workload, and the amount of spares which need to be held in stock. In most deepsea ships the height restriction on machinery is less of a problem than length, a larger bore engine with fewer cylinders therefore underwriting a shorter engineroom and more space for cargo. Larger bore engines also generally return a better specific fuel consumption than smaller engines and offer a greater tolerance to heavy fuels of poor quality.

A direct-coupled propulsion engine cannot operate unaided since it requires service pumps for cooling and lubrication, and fuel/lube oil handling and treatment systems. These ancillaries need electrical power which is usually provided by generators driven by medium or high speed diesel engines. Many genset enginebuilders can now offer designs capable of burning the same heavy fuel grade as the main engine as well as marine diesel oil or blended fuel (heavy fuel and distillate fuel mixed in various proportions, usually 70:30) either bunkered as an intermediate fuel or blended onboard. 'Unifuel' installations—featuring main and auxiliary engines arranged to burn the same bunkers—are now common.

6.4.2.2 *Auxiliary power generation*

The cost of auxiliary power generation can weigh heavily in the choice of main machinery. Developments have sought to maximize the exploitation of waste heat recovery to supplement electricity supplies at sea, to facilitate the use of alternators driven by the main engine via speed-increasing gearing or mounted directly in the shaftline, and to power other machinery from the main engine.

Gear-based constant frequency generator drives allow a shaft alternator to be driven by a low speed engine in a fixed pitch propeller installation, with full alternator output available between 70% and 104% of propeller speed. A variety of space-saving arrangements are possible with the alternator located alongside or at either end of a main engine equipped with compact integral power take-off gear. Alternatively, a thyristor frequency converter system can be specified to serve an alternator with a variable main engine shaft speed input in a fixed pitch or CP propeller installation.

The economic attraction of the main engine-driven generator for electrical power supplies at sea is that it exploits the high thermal efficiency, low specific fuel consumption and low grade fuel-burning capability of the ship's diesel prime mover. Other advantages are that the auxiliary diesel gensets can be shut down, yielding benefits from reduced running hours in terms of lower fuel and lubricating oil consumptions, maintenance demands and spares costs.

System options for electricity generation have been extended by the arrival of power turbines which, fed with exhaust gas surplus to the needs of modern high efficiency turbochargers, can be arranged to drive alternators in conjunction with the main engine or independently.

These small gas turbines are also in service in integrated systems linking steam turbo-alternators, shaft alternators and diesel gensets; the various power sources—applied singly or in combination—promise optimum economic electricity production for any ship operating mode. Some surplus electrical output can also be tapped to support the propulsive effort via a shaft alternator switched to function as a propulsion motor.

Such a plant is exploited in a class of large low speed engine-powered containerships with significant reefer capacity whose overall electrical load profile is substantial and variable. Crucial to its effectiveness is a computer-controlled energy management system which co-ordinates the respective contributions of the various power sources to achieve the most economical mode for a given load demand.

Integrated energy-saving generating plants have been developed over the years by the major Japanese shipbuilding groups for application to large tankers and bulk carriers. The systems typically exploit waste heat (from low speed main engine exhaust gas, scavenge air and cooling water) to serve a steam turbo-alternator, air conditioning plant, heaters and distillers. System refinements were stimulated by the diminishing amount of energy available from the exhaust gas of low speed engines, in terms of both temperature and volumes, with the progressive rise in thermal efficiencies. The ability of the conventional waste heat boiler/turbo-alternator set to meet electrical demands at sea was compromised, any shortfall having to be plugged by supplementary oil firing of an auxiliary boiler or by running a diesel genset and/or shaft alternator. The new integrated systems, some also incorporating power gas turbines, maximize the exploitation of the waste heat available in ships whose operating profiles and revenues can justify the added expense and complexity.

6.4.2.3 Geared drives

The most common form of indirect drive of a propeller features one or more medium speed four-stroke engines connected through clutches and couplings to a reduction gearbox to drive either a fixed pitch or CP propeller (Figures 6.32 and 6.33). The CP propeller eliminates the need for a direct-reversing engine while the gearing allows a suitable propeller speed to be selected. There is inevitably a loss of efficiency in the transmission but in most cases this would be cancelled out by the improvement in propulsive efficiency when making a comparison of direct-coupled and indirect drive engines of the same power. The additional cost of the transmission can also be offset by the lower cost of the four-stroke engine since two-stroke designs, larger and heavier, cost more. The following generic merits are cited for geared multi-medium or high speed engine installations:

- Ships with more than one main engine benefit from enhanced availability through redundancy: in the event of one engine breaking down another can maintain navigation. The number of engines engaged can also be varied to secure the most economical mode for a given speed or deployment profile. Thus, when a ship is running light, partially loaded or slow steaming, one engine can be deployed at its normal (high efficiency) rating and some or all of the others shut down. In contrast, in similar operational circumstances, a single direct-coupled engine might have to be run for long periods at reduced output with lower efficiency.
- The ability to vary the number of engines deployed allows an engine to be serviced at sea, easing maintenance planning. This flexibility is particularly valued in an era when port turnround times are minimized. Engines can also be overhauled in port without the worry of authorities demanding an unscheduled shift of berth or sudden departure.
- By modifying the number of engines per ship and the cylinder numbers per engine to suit individual power requirements the propulsion plant for a fleet can be standardized on a single engine model, with consequent savings in spares costs and inventories, and benefits in crew familiarity. The concept can also be extended to the auxiliary power plant through 'uniform machinery' installations embracing main and genset engines of the same model.
- Compact machinery spaces with low headrooms can be created, characteristics particularly valued for RoRo ferries.

Designers of marine gearing, clutches and couplings have to satisfy varying and sometimes conflicting demands for operational flexibility, reliability, low noise and compactness from transmission systems.

Advances in design, materials and controls have contributed to innovative solutions for versatile single- and multi-engine propulsion installations featuring power take-offs for alternator drives and power take-ins to boost propulsive effort. In many propulsion installations a gearbox is expected to: determine the propeller speed and direction of rotation, and provide a reversing capability; provide a geometric coupling that can connect and separate the flow of power between the engine and propeller shaft or waterjet drive; and absorb the thrust from the propeller.

An impressive flexibility of operating modes can be arranged from geared multi-engine propulsion plants which are, in practice, overwhelmingly based on four-stroke machinery (a number of geared two-stroke engine installations have been specified, however, for special purpose tonnage such as offshore shuttle tankers).

6.4.2.4 Father-and-son layouts

Flexibility is enhanced by the adoption of a so-called 'father-and-son' (or 'mother-and-daughter') configuration: a partnership of similar four-stroke engine models, but with different cylinder numbers, coupled to a common reduction gearbox to drive a propeller shaft. Father-and-son pairs have been specified to power large twin-screw cruise vessels. An example is provided by the 1995-built P&O liner *Oriana* whose 40 000 kW propulsion plant is based on two nine-cylinder and two six-cylinder MAN B&W L58/64 medium speed engines. Each father-and-son pair drives a highly skewed Lips CP propeller via a Renk Tacke low noise gearbox which

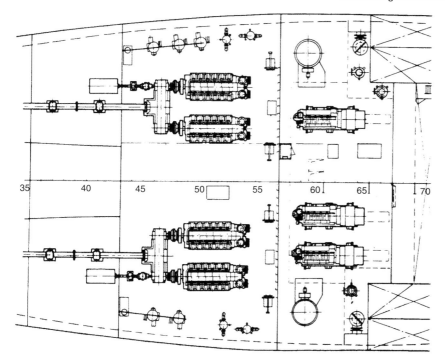

Figure 6.32 A typical multi-engine geared twin-screw installation.

also serves a 4200 kW shaft alternator/propulsion motor.

Propulsion can be effected either by: the father-and-son engines together; the father engines alone or the son engines alone; and with or without the shaft alternators operating as propulsion motors (fed with electrical power by the diesel gensets).

6.4.3 Diesel–electric drive

6.4.3.1 Overview

An increasingly popular form of indirect drive is diesel–electric propulsion based on multi-medium speed main gensets. New generations of AC/AC drive systems exploiting cycloconverter or synchroconverter technology have widened the potential of electric propulsion after years of confinement to specialist niches, such as icebreakers, research vessels and cablelayers.

The diesel–electric mode is now firmly entrenched in large cruise ships and North Sea shuttle tanker propulsion, references have been established in shortsea and deepsea chemical carriers, as well as in Baltic RoRo passenger/freight ferries, and project proposals argue the merits for certain classes of containership. Dual-fuel diesel–electric propulsion has also penetrated the offshore supply vessel and LNG carrier sectors.

A number of specialist groups contest the high powered electric drive market, including ABB Marine, Alstom, STN Atlas and Siemens which highlight system capability to cope with sudden load changes, deliver smooth and accurate ship speed control, and foster low noise and vibration. Applications have been stimulated by continuing advances in power generation systems, AC drive technology and power electronics.

Two key AC technologies have emerged to supersede traditional DC electric drives: the Cycloconverter system and the Load Commutated Inverter (LCI) or Synchrodrive solution. Both are widely used in electric propulsion but, although exploiting the same basic Graetz bridge prevalent in DC drive systems, they have different electrical characteristics. Development efforts have also focused on a further AC system variant, the voltage source Pulse Width Modulation (PWM) drive.

A variable-speed AC drive system comprises a propulsion motor (Figure 6.34) and a frequency converter, with the motor speed controlled by the converter altering the input frequency to the motor.

Both Cyclo and PWM drives are based on advanced AC variable speed technology that matches or exceeds the performance characteristics of conventional DC drives. Each features motor speed control with maximum torque available from zero to full speed in either direction, facilitating

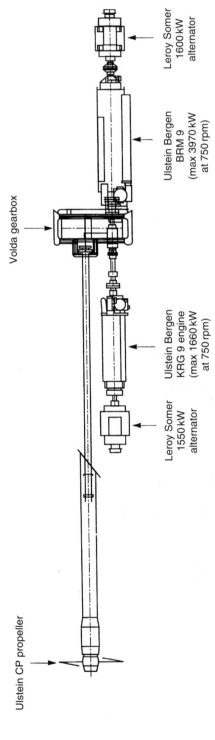

Figure 6.33 One of the twin medium speed geared propulsion/auxiliary power lines of a Norwegian coastal ferry.

Figure 6.34 One of two 15 000 kW STN Atlas propulsion motors installed on the cruise liner *Costa Victoria*.

operation with simple and robust fixed pitch (rather than controllable pitch) propellers. Smooth control underwrites operation at very low speeds, and vector control yields a rapid response to enhance plant and ship safety. Integrated full electric propulsion (IFEP) is now standard for cruise liners and specified for a widening range of other ship types including worships, in which the economic and operational benefits of the central power station concept can be exploited.

Electric propulsion requires motors to drive the propellers and gensets to supply the power. It seems somewhat illogical to use electric generators, switchgear and motors between prime movers and propeller when direct-coupling or geared transmission to the shaft may be all that is necessary. There are obviously sound reasons for some installations to justify the complication and extra cost of electric propulsion, Lloyd's Register citing:

6.4.3.2 *Flexibility of layout*

The advantage of electric transmission is that the prime movers and their associated generators are not constrained to have any particular relationship with the load—a cable run is a very versatile medium. In a propulsion system it is therefore possible to install the main diesel gensets and their support' services in locations remote from the propeller shaft (Figure 6.35). Diesel gensets have even been installed in containers on deck to provide propulsive power, and other vessels equipped with a 10 000 kW generator mounted in a block at the stern above the RoRo deck.

Another example of the flexibility facilitated by an electric propulsion system is a semi-submersible rig installation whose gensets are mounted on the main deck and propulsion motors in the pontoons. In cruise ships, ferries and tankers the layout flexibility allows the naval architect to create compact machinery installations releasing extra revenue-earning space for passenger accommodation/amenities and cargo.

Opportunities to design and build tankers more cost-effectively are also offered. The optimized location of the main machinery elements allows the ship's overall length to be reduced and steel costs correspondingly lowered for a given cargo capacity; alternatively, the length of the cargo tank section can be extended within given hull dimensions.

Diesel–electric solutions facilitate the modularization and delivery of factory-tested turnkey packages to the building yard, with the complete main gensets mounted on common bedplates ready for coupling to their support systems. The compactness of the machinery outfit fosters shorter runs for the cabling and ancillary systems, and the engine casing and exhaust gas piping are also shortened.

It is difficult for a diesel–electric plant to match the fuel economy of a single direct-drive low speed main engine which is allowed to operate at its optimum load for the long transoceanic leg of a tanker's voyage. But some types of tanker (oil products and chemical tonnage, as well as North Sea shuttle carriers) are deployed in varied service profiles and spend a considerable time at part loads: for example, in restricted waters, during transit in ballast and manoeuvring.

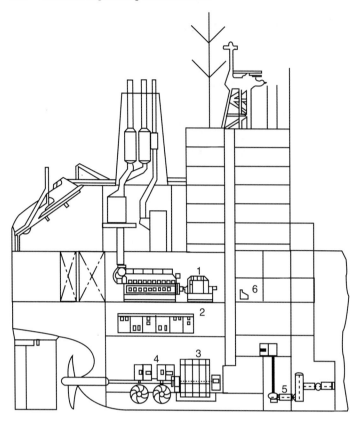

Figure 6.35 Machinery arrangement in a diesel–electric tanker.
1. Diesel gensets
2. Main switchboard and cycloconverters
3. Propulsion motor
4. Stern thrusters
5. Cargo pumps
6. Engine control room

A diesel–electric tanker can exploit the abundance of electrical power to drive a low noise cargo pumping outfit, perhaps reducing time in port; and (particularly in the case of a dynamically positioned shuttle carrier) to serve powerful bow and stern thrusters during sensitive manoeuvring at the loading buoy (Figure 6.36).

6.4.3.3 Load diversity

Certain types of tonnage have a requirement for substantial amounts of power for ship services when the demands of the propulsion system are low: for example, tankers and any other ship with a significant cargo discharging load. A large auxiliary power generating plant for cruise ships and passenger ferries is dictated by the hotel services load (air conditioning, heating and lighting) and the heavy demands of transverse thrusters during manoeuvring. The overall electrical demand may be 30–40% of the installed propulsion power and considerable standby capacity is also called for to secure system redundancy for safety. These factors have helped to promote the popularity of the medium speed diesel–electric 'power station' concept for meeting all propulsion, manoeuvring and hotel energy demands in large passenger ships.

6.4.3.4 *Economical part load running*

This is best achieved when there is a central power station feeding propulsion and ship services. A typical medium speed diesel–electric installation features four main gensets (although plants with up to nine sets are in service) and, with parallel operation of all the sets, it is easy to match the available generating capacity to the load demand. In a four-engine plant, for example, increasing the number of sets in operation from two that are fully loaded to three partially loaded will result in three sets operating at 67% load: not ideal but not a serious operating condition.

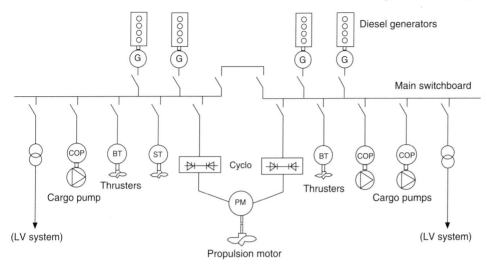

Figure 6.36 Diesel–electric 'power station' meeting the load demands of propulsion, cargo pumping, thrusters and hotel services in a tanker.

It is not necessary to operate gensets at part load to provide the spare capacity for the sudden loss of a set: propulsion load reduction may be available instantaneously and in most ships a short-term reduction in propulsive power does not represent a hazard. The propulsion plant controls continuously monitor generating capability, any generator overload immediately resulting in the controls adjusting the input to the propulsion motors. During manoeuvring, propulsion plant requirements are below system capacity and the failure of one generator is unlikely to present a hazardous situation.

6.4.3.5 *Ease of control*

The widespread use of CP propellers means that control facilities once readily associated with electric drives are no longer able to command the same premium. It is worth noting, however, that electric drives are capable of meeting the most exacting demands with regard to dynamic performance, exceeding by a very wide margin anything that is required of a propulsion system.

6.4.3.6 *Low noise*

An electric motor can provide a drive with very low vibration characteristics, a quality valued for warships, research vessels and cruise ships where, for different reasons, a low noise signature is required. In the case of warships and research vessels it is noise into the water which is the critical factor; in cruise ships it is structure-borne noise and vibration to passenger spaces that are desirably minimized. Electric transmission allows the prime movers to be mounted in such a way that the vibration transmitted to the seatings, and hence to the water or passenger accommodation, is minimized. The use of double resilient mounting may be necessary for very low noise installations.

6.4.3.7 *Environmental protection and ship safety*

Controls to curb noxious exhaust gas emissions—tightening nationally, regionally and globally—also favour the specification of electric transmission since the constant speed running of diesel prime movers optimized to the load is conducive to lower NOx emission levels. An increasing focus on higher ship safety through redundancy of propulsion plant elements is another positive factor. A twin-propeller installation is not the only method of achieving redundancy. Improving the redundancy of a single-screw ship can also be secured by separating two or more propulsion motors in different compartments and coupling them to a reduction gear located in a separate compartment. The input shafts are led through a watertight bulkhead.

6.4.3.8 *Podded propulsors (See also Section 6.2.4)*

Significant economic and technical benefits in ship design, construction and operation are promised by electric podded propulsors, whose appeal has extended since the early 1990s from icebreakers and cruise liners to offshore vessels, ferries and tankers.

A podded propulsor (or pod) incorporates its electric drive motor in a hydrodynamically-optimized submerged housing which can be fully rotated with the propeller(s) to secure 360-degree azimuthing and thrusting capability (Figures 6.7 and 6.37).

Figure 6.37 Three 14 000 kW Azipod propulsors power Royal Caribbean Cruises' Voyager-class liners.

The motor is directly coupled to the fixed pitch propeller (s) mounted at either or both ends of the pod. Pusher and tractor or tandem versions can be specified, with input powers from 400 kW to 30 000 kW supplied by a shipboard generating plant. The merits cited for pods over traditional diesel-electric installations with shaftlines highlight:

- Space within the hull otherwise reserved for conventional propulsion motors can be released and exploited for other purposes.
- More creative freedom for ship designers since the propulsors and the prime movers require no direct physical connection.
- Steering capability is significantly better than with any conventional rudder system; stern thruster(s) can be eliminated, along with rudder(s) and shaftline(s).
- Excellent reversing capability and steering during astern navigation, and enhanced crash stop performance.
- Low noise and vibration characteristics associated with electric drive systems are enhanced by the motor's underwater location; and hull excitations induced by the propeller are very low thanks to operation in an excellent wake field.
- Propulsor unit deliveries can be late in the shipbuilding process, reducing 'dead time' investment costs; savings in overall weight and ship construction hours are also promised.

6.4.3.9 Combined systems

A combination of geared diesel and diesel–electric drive is exploited in some offshore support vessels whose deployment profile embraces two main roles: to transport materials from shore bases to rigs and platforms, and to standby at the offshore structures for cargo and anchor handling, rescue and other support operations. In the first role the propulsive power required is that necessary to maintain the free running service speed. In the second role little power is required for main propulsion but sufficient electrical power is vital to serve winches, thrusters and cargo handling gear.

The solution for these dual roles is to arrange for the main propellers to be driven by medium speed engines through a reduction gear which is also configured to drive powerful shaft alternators. By using one shaft alternator as a motor and the other to generate electrical power it is possible to secure a diesel–electric main propulsion system of low power and at the same time provide sufficient power for the thrusters. Such a twin-screw plant could simultaneously have one main engine driving the alternator and the propeller while the other main engine is shut down and its associated propeller driven by the shaft alternator in propulsion motor mode.

A further option is to fit dedicated propulsion motors as well as the shaft alternators, yielding greater plant flexibility and fuel saving potential but at the expense of extra cost and complexity. For the above power plants to operate effectively CP main and thruster propellers are essential.

Combined medium or high speed diesel engine and gas turbine (CODAG) systems are now common for large fast ferry propulsion, while some large diesel–electric cruise liners feature medium speed engines and gas turbines driving generators in a CODLAG or CODEG arrangement.

6.5 Propulsion engines

6.5.1 Diesel engines

6.5.1.1 *Low speed engines*

Low speed two-stroke engine designers have invested heavily to maintain their dominance of the mainstream deepsea propulsion sector formed by tankers, bulk carriers and containerships. The long-established supremacy reflects the perceived overall operational economy, simplicity and reliability of single, direct-coupled crosshead engine plants. Other factors are the continual evolution of engine programmes by the designer/licensors in response to or anticipation of changing market requirements, and the extensive network of enginebuilding licensees in key shipbuilding regions. Many of the standard ship designs of the leading yards, particularly in Asia, are based on low speed engines.

The necessary investment in R&D, production and overseas infrastructure dictated to stay competitive, however, took its toll over the decades. Only three low speed engine designer/licensors—MAN B&W Diesel, Mitsubishi and Sulzer (now part of the Wärtsilä Corporation)—survived into the 1990s to contest the international arena.

The roll call of past contenders include names either long forgotten or living on only in other engineering sectors: AEG-Hesselman, Deutsche Werft, Fullagar, Krupp, McIntosh and Seymour, Neptune, Nobel, North British, Polar, Richardsons Westgarth, Still, Tosi, Vickers, Werkspoor and Worthington. The last casualties were Doxford, Götaverken and Stork whose distinctive engines remain at sea in diminishing numbers. The pioneering designs displayed individual flair within generic classifications which offered two- or four-stroke, single- or double-acting, and single- or opposed-piston configurations. The Still concept even combined the Diesel principle with a steam engine: heat in the exhaust gases and cooling water was used to raise steam which was then supplied to the underside of the working piston.

Evolution decreed that the surviving trio of low speed crosshead engine designers should pursue a common basic configuration: two-stroke engines with constant pressure turbocharging and uniflow scavenging via a single hydraulically-operated exhaust valve in the cylinder head. Current programmes embrace mini-to-large bore models with short, long and ultra-long stroke variations to match the propulsive power demands and characteristics of most deepsea (and even some coastal/shortsea) cargo tonnage. Installations can be near-optimized for a given duty from a permutation involving the engine bore size, number of cylinders, selected output rating and running speed. Bore sizes range from 260 mm to 1080 mm, stroke/bore ratios up to 4.2:1, in-line cylinder numbers from four to 14, and rated speeds from around 55 to 250 rev/min. Specific fuel consumptions as low as 154 g/kW h are quoted for the larger bore models whose economy can be enhanced by optional Turbo Compound Systems in which power gas turbines exploit exhaust energy surplus to the requirements of modern high efficiency turbochargers.

Progress in the performance development of low speed engines in the popular circa-600 mm bore class is illustrated in Figure 6.38.

Recent years have seen the addition of intermediate bore sizes to enhance coverage of the power/speed spectrum and further optimize engine selection. Both MAN B&W Diesel and Sulzer also extended their upper power limits in the mid-1990s with the introduction of super-large bore models—respectively of 980 mm and 960 mm bore sizes—dedicated to the propulsion of new generations of 6000 TEU-plus containerships with service speeds of 25 knots or more. The 12-cylinder version of MAN B&W's current K98MC design, delivering 68 640 kW, highlights the advance in specific output achieved since the 1970s when the equivalent 12-cylinder B&W K98GF model yielded just under 36 800 kW. Large bore models tailored to the demands of new generation VLCC and ULCC propulsion have also been introduced.

Successively larger and faster generations of post-panamax containerships have driven the development of specific output and upper power limits by low speed engine designers. To ensure that containership designs with capacities up to and exceeding 10 000 TEU can continue to be specified with single engines, both MAN B&W and Wärtsilä extended their respective MC and Sulzer RTA programmes. Wärtsilä raised the specific rating of the Sulzer RTA96C design by 4% to 5720 kW/cylinder at 102 rev/min, and introduced an in-line 14-cylinder model delivering up to 80 080 kW. (The previous power ceiling had been 65 880 kW from a 12-cylinder model.)

MAN B&W responded to the challenge with new 13- and 14-cylinder variants of the K98MC and MC-C series, offering outputs from 74 230 kW to 80 080 kW at 94 or 104 rev/min. These series can also be extended to embrace 15- to 18-cylinder models, if called for, taking the power threshold to just under 103 000 kW. Such an output would reportedly satisfy the propulsive power demands of containerships with capacities up to 18 000 TEU and service speeds of 25–26 knots. In 2003 MAN B&W Diesel opened another route to higher powers: a 1080 mm bore version of the MC engine was announced with a rating of 6950 kW/cylinder at 94 rev/min, the 14-cylinder model thus offering 97 300 kW.

V-cylinder configurations of existing low speed engine designs have also been proposed by MAN B&W Diesel to propel mega-containerships, promising significant savings in weight and length per unit power over traditional in-line cylinder models. These

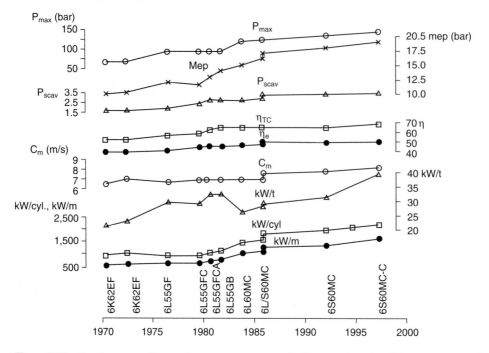

Figure 6.38 Development of key performance parameters for low speed engines (circa 600 mm bore) over a 30-year period (MAN B&W Diesel).

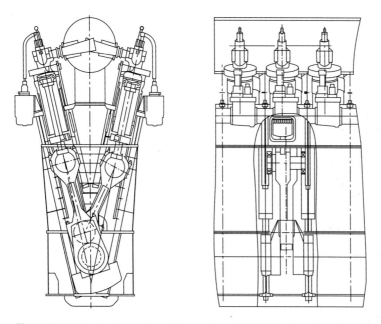

Figure 6.39 V-cylinder versions of larger bore MC/MC-C engines have been proposed by MAN B&W Diesel.

engines would allow the higher number of cylinders to be accommodated within existing machinery room designs (Figure 6.39).

Parallel development by the designer/licensors seeks to refine existing models and lay the groundwork for the creation of new generations of low speed engine. Emphasis in the past has been on optimizing fuel economy and raising specific outputs but reliability, durability and overall economy are now priorities in R&D programmes, operators valuing longer component lifetimes, extended periods between overhauls and easier servicing.

Lower production costs through more simple manufacture and easier installation procedures are also targeted, reflecting the concerns of enginebuilder/licensees and shipyards. More compact and lighter weight engines are appreciated by naval architects seeking to maximize cargo space and deadweight capacity within given overall ship dimensions.

In addition, new regulatory challenges—such as noxious exhaust emission and noise controls—must be anticipated and niche market trends addressed if the low speed engine is to retain its traditional territory (for example, the propulsion demands of increasingly larger and faster containerships which might otherwise have to be met by multiple medium speed engines or gas turbines).

A number of features have further improved the cylinder condition and extended the time between overhauls through refinements in piston design and piston ring configurations. A piston cleaning ring incorporated in the top of the cylinder liner now controls ash and carbon deposits on the piston topland, preventing contact between the liner and these deposits which would otherwise remove part of the cylinder lube oil from the liner wall.

Computer software has smoothed the design, development and testing of engine refinements and new concepts but the low speed engine groups also exploit full-scale advanced hardware to evaluate innovations in components and systems.

Sulzer began operating its first Technology Demonstrator in 1990, an advanced two-stroke development and test engine designated the 4RTX54 whose operating parameters well exceeded those of any production engine (Figure 6.40). Until then, the group had used computer-based predictions to try to calculate the next development stage. Extrapolations were applied, sometimes with less than desirable results. The 4RTX54 engine, installed at the Swiss designer's Winterthur headquarters, allowed practical tests with new parameters, components and systems to be carried out instead of just theory and calculations. Operating data gathered in the field could be assessed alongside results derived from the test engine.

The four-cylinder 540 mm bore/2150 mm stroke engine had a stroke bore ratio approaching 4:1 and could operate with mean effective pressures of up to

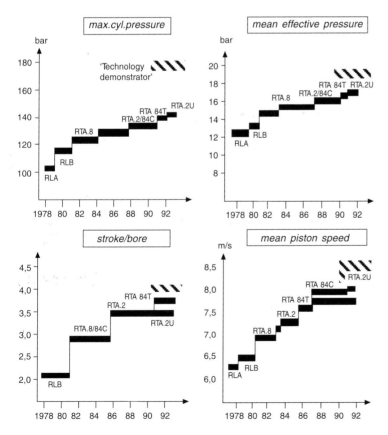

Figure 6.40 Evolution of Sulzer low speed engine parameters from the late 1970s in comparison with those of its RTX54 Two-stroke Technology Demonstrator engine.

20 bar, maximum cylinder pressures up to 180 bar and mean piston speeds up to 8.5 m/sec. Operating without a camshaft—reportedly the first large two-stroke engine to do so—the RTX54 was equipped with combined mechanical, hydraulic, electronic (mechatronics) systems for fuel injection, exhaust valve lift, cylinder lubrication and starting, as well as controllable cooling water flow. The systems underwrote full flexibility in engine settings during test runs.

Sulzer's main objectives from the Technology Demonstrator engine were to explore the potential of thermal efficiency and power concentration; to increase the lifetime and improve the reliability of components; to investigate the merits of microprocessor technology; and to explore improvements in propulsion efficiency. A number of concepts first tested and confirmed on the 4RTX54 engine were subsequently applied to production designs. The upgraded RTA-2U series and RTA84T, RTA84C and RTA96C engines, for example, benefit from a triple-fuel injection valve system in place of two valves. This configuration fosters a more uniform temperature distribution around the main combustion chamber components and lower overall temperatures despite higher loads. Significantly lower exhaust valve and valve seat temperatures are also yielded.

An enhanced piston ring package for the RTA-2U series was also proven under severe running conditions on the 4RTX54 engine. Four rings are now used instead of five, the plasma-coated top ring being thicker than the others and featuring a pre-profiled running face. Excellent wear results are reported. The merits of variable exhaust valve closing (VEC) were also investigated on the research engine whose fully electronic systems offered complete flexibility. Significant fuel savings in the part-load range were realized from the RTA84T 'Tanker' engine which further exploits load-dependent cylinder liner cooling and cylinder lubrication systems refined on the 4RTX54. The 4RTX54 was replaced as a research and testing tool in 1995 by the prototype 4RTA58T engine adapted to serve as Sulzer's next Two-stroke Technology Demonstrator (Figure 6.41).

The widest flexibility in operating modes and the highest degree of reliability are cited by Copenhagen-based MAN B&W Diesel as prime R&D goals underwriting future engine generations, along with:

- Ease of maintenance.
- Production cost reductions.
- Low specific fuel consumption and high plant efficiency over a wide load spectrum.
- High tolerance towards varied heavy fuel qualities.
- Easy installation.
- Continual adjustments to the engine programme in line with the evolving power and speed requirements of the market.
- Compliance with emission controls.
- Integrated intelligent electronic systems.

Continuing refinement of MAN B&W Diesel's MC low speed engine programme and the development of intelligent engines (see section below) are supported by an R&D centre adjacent to the group's Teglholmen factory in Copenhagen.

At the heart of the centre is the 4T50MX research engine, an advanced testing facility which exploits an unprecedented 4.4:1 stroke/bore ratio. Although based on the current MC series, the four-cylinder 500 mm bore/2200 mm stroke engine is designed to operate at substantially higher ratings and firing pressures than any production two-stroke engine available today. An output of 7500 kW at 123 rev/min was selected as an initial reference level for carrying out extensive measurements of performance, component temperatures and stresses, combustion and exhaust emission characteristics, and noise and vibration. The key operating parameters at this output equate to 180 bar firing pressure, 21 bar mean effective pressure and 9 m/sec mean piston speed. Considerable potential was reserved for higher ratings in later test running programmes.

A conventional camshaft system was used during the initial testing period of the 4T50MX engine. After reference test-running, however, this was replaced by electronically controlled fuel injection pumps and exhaust valve actuators driven by a hydraulic servo-system (Figure 6.42). The engine is prepared to facilitate extensive tests on primary methods of exhaust emission reduction, anticipating increasingly tougher regional and international controls in the future. Space was allocated in the R&D centre for the installation of a large NOx-reducing selective catalytic reduction (SCR) facility for assessing the dynamics of SCR-equipped engines and catalyst investigations.

The research engine, with its electronically controlled exhaust valve and injection system, has fully lived up to expectations as a development tool for components and systems, MAN B&W Diesel reports. A vast number of possible combinations of injection pattern, valve opening characteristics and other parameters can be permutated. The results from testing intelligent engine concepts are being tapped for adoption as single mechanical units as well as stand-alone systems for application on current engine types. To verify the layout of the present standard mechanical camshaft system, the 4T50MX engine was rebuilt with a conventional mechanical camshaft unit on one cylinder. The results showed that the continuous development of the conventional system seems to have brought it close to the optimum, and the comparison gave no reason for modifying the basic design.

An example of the degrees of freedom available is shown by a comparison between the general engine

Marine engines and auxiliary machinery 385

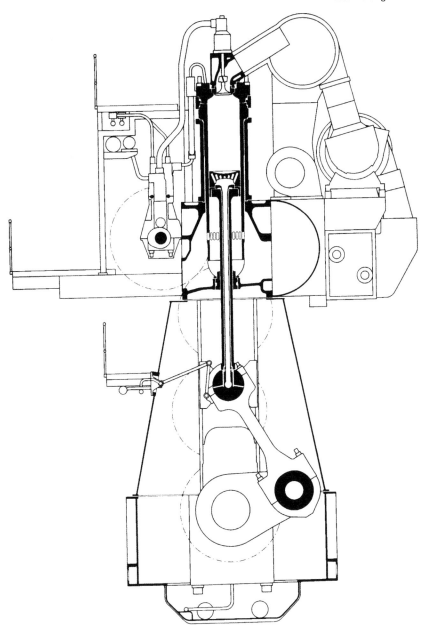

Figure 6.41 The current Sulzer Two-stroke Technology Demonstrator research engine is based on an RTA58T model.

performance with the firing pressure kept constant in the upper load range by means of variable injection timing (VIT) and by variable compression ratio (VCR). The latter is obtained by varying the exhaust valve closing time. This functional principle has been transferred to the present exhaust valve operation with the patented system illustrated in the diagram (Figure 6.43). The uppermost figure shows the design of the hydraulic part of the exhaust valve; below is the valve opening diagram. The fully drawn line represents control by the cam while the dotted line shows the delay in closing, thus reducing the compression ratio at high loads so as to maintain a constant compression pressure in the upper load range. The delay is simply obtained by the oil being trapped in the lower chamber; and the valve closing is determined by the opening of the throttle valve which is controlled by the engine load.

Traditionally, the liner cooling system has been arranged to match the maximum continuous rating

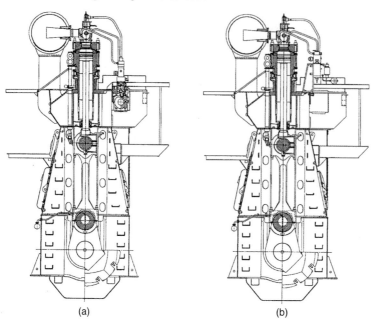

Figure 6.42 MAN B&W Diesel's 4T50MX low speed research engine arranged with a conventional camshaft (a) and with electronically controlled fuel injection pumps and exhaust value actuating pumps (b).

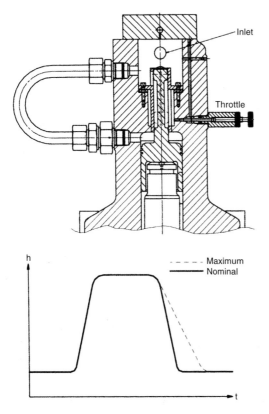

Figure 6.43 Mechanical/hydraulic variable compression ratio (MAN B&W Diesel).

load. Today, however, it seems advantageous to control the inside liner surface temperature in relation to the load. Various possibilities for securing load-dependent cylinder liner cooling have therefore been investigated. One system exploits different sets of cooling ducts in the bore-cooled liner, the water supplied to the different sets depending on the engine load. Tests with the system have shown that the optimum liner temperature can be maintained over a very wide load range. The system is considered perfectly feasible but the added complexity has to be carefully weighed against the service advantages.

The fuel valve used on MC engines operates without any external control of its function. The design has worked well for many years but could be challenged by the desire for maintaining an effective performance at very low loads. MAN B&W Diesel has therefore investigated a number of new designs with the basic aim of retaining a simple and reliable fuel valve without external controls. Various solutions were tested on the 4T50MX engine, among them a design whose opening pressure is controlled by the fuel oil injection pressure level (which is a function of the engine load). At low load the opening pressure is controlled by the spring alone. When the injection pressure increases at higher load, this higher pressure adds to the spring force and the opening pressure increases.

Another example of fuel valve development is aimed at reduced emissions. This type incorporates

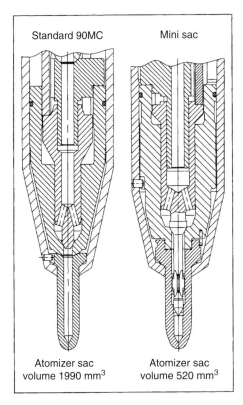

Figure 6.44 Reduced emissions result from a fuel valve with a smaller sac volume (MAN B&W Diesel).

a conventional conical spindle seat as well as a slide valve inside the fuel nozzle, minimizing the sac volume and thus the risk of after-dripping. Significantly lower NOx emissions are reported, as well as reduced smoke and even carbon monoxide, but at the expense of a slightly higher fuel consumption. This type of fuel valve is now included in the options for special low NOx applications of MC engines (Figure 6.44).

The 4T50MX engine was used to test a triple fuel valve-per-cylinder configuration, the measurements mirroring Sulzer's results in yielding reduced temperature levels and a more even temperature distribution than with a two-valve arrangement. The K80MC-C, K90MC/MC-C, S90MC-T and K98MC-C engines were subsequently specified with triple fuel valves to enhance reliability.

Intelligent engines
Both MAN B&W Diesel and New Sulzer Diesel demonstrated 'camshaftless' operation with their research engines, applying electronically controlled fuel injection and exhaust valve actuation systems. Continuing R&D will pave the way for a future generation of highly reliable 'intelligent engines':

those which monitor their own condition and adjust parameters for optimum performance in all operating regimes, including fuel-optimized and emissions-optimized modes. An 'intelligent engine-management system' will effectively close the feedback loop by built-in expert knowledge.

Engine performance data will be constantly monitored and compared with defined values in the expert system; if deviations are detected corrective action is automatically taken to restore the situation to normal. A further step would incorporate not only engine optimizing functions but management responsibilities, such as maintenance planning and spare parts control.

To meet the operational flexibility target, MAN B&W Diesel explains, it is necessary to be able to change the timing of the fuel injection and exhaust valve systems while the engine is running. To achieve this objective with cam-driven units would involve a substantial mechanical complexity which would undermine engine reliability. An engine without a traditional camshaft is therefore dictated.

The concept is illustrated in Figure 6.45 whose upper part shows the operating modes which may be selected from the bridge control system or by the intelligent engine's own control system. The centre part shows the brain of the system: the electronic control system which analyses the general engine condition and controls the operation of the engine systems shown in the lower part of the diagram (the fuel injection, exhaust valve, cylinder lube oil and turbocharging systems).

To meet the reliability target it is necessary to have a system which can actively protect the engine from damage due to overload, lack of maintenance and maladjustments. A condition monitoring system must be used to evaluate the general condition of the engine, thus maintaining its performance and keeping its operating parameters within prescribed limits. The condition monitoring and evaluation system is an on-line system with automatic sampling of all 'normal' engine performance data, supplemented by cylinder pressure measurements. The system will report and actively intervene when performance parameters show unsatisfactory deviations. The cylinder pressure data delivered by the measuring system are used for various calculations:

- The mean indicated pressure is determined as a check on cylinder load distribution as well as total engine output.
- The compression pressure is determined as an indicator of excessive leakage caused by, for example, a burnt exhaust valve or collapsed piston rings (the former condition is usually accompanied by an increased exhaust gas temperature in the cylinder in question).

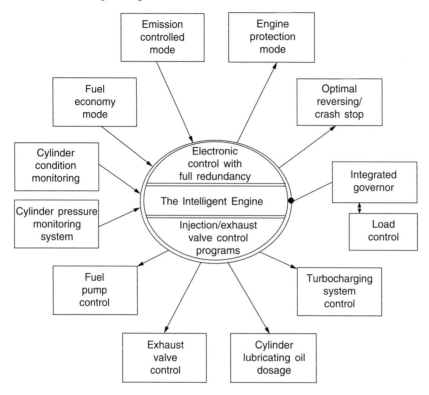

Figure 6.45 Schematic of the Intelligent Engine (MAN B&W Diesel).

- The cylinder wall temperature is monitored as an additional indicator of the piston ring condition.
- The firing pressure is determined for injection timing control and for control of mechanical loads.
- The rate of pressure rise (dp/dt) and rate of heat release are determined for combustion quality evaluation as a warning in the event of 'bad fuels' and to indicate any risk of piston ring problems in the event of high dp/dt values.

The cylinder condition monitoring system is intended to detect faults such as blow-by past the piston rings, cylinder liner scuffing and abnormal combustion. The detection of severe anomalies by the integrated systems triggers a changeover to a special operating mode for the engine, the 'engine protection mode'. The control system will contain data for optimum operation in a number of different modes, such as 'fuel economy mode', 'emission controlled mode', 'reversing/crash stop mode' and various engine protection modes. The load limiter system (load diagram compliance system) aims to prevent any overloading of the engine in conditions such as heavy weather, fouled hull, shallow water, too heavy propeller layout or excessive shaft alternator output. This function will appear as a natural part of future governor specifications.

The fuel injection system is operated without a conventional camshaft, using high pressure hydraulic oil from an engine-driven pump as a power source and an electronically controlled servo system to drive the injection pump plunger. The general concept of the InFI (intelligent fuel injection) system and the InVA (intelligent valve actuation) system for operating the exhaust valves is shown in Figure 6.46. Both systems, when operated in the electronic mode, receive the electronic signals to the control units. In the event of failure of the electronic control system the engine is controlled by a mechanical input supplied by a diminutive camshaft giving full redundancy.

Unlike a conventional, cam-driven pump the InFI pump has a variable stroke and will only pressurize the amount of fuel to be injected at the relevant load. In the electronic mode (that is, operating without a camshaft) the system can perform as a single injection system as well as a pre-injection system with a high degree of freedom to modulate the process in terms of injection rate, timing, duration, pressure, single/double injection, cam profile and so on. Several optimized injection patterns can be stored in the computer and

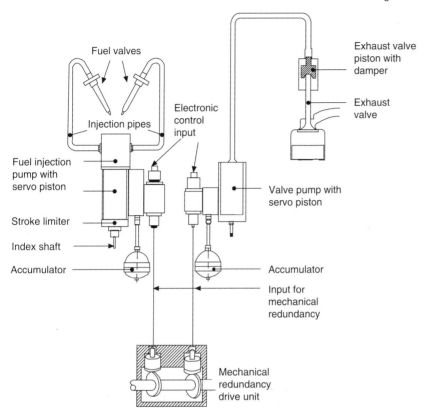

Figure 6.46 Electronically controlled hydraulic systems for fuel injection and exhaust valve operation on MAN B&W Diesel's 4T50MX research engine.

chosen by the control system in order to operate the engine with optimum injection characteristics at several loads: from dead slow to overload as well as for starting, astern running and crash stop. Changeover from one to another of the stored injection characteristics is effected from one injection to the next. The system is able to adjust the injection amount and injection timing for each cylinder individually in order to achieve the same load (mean indicated pressure) and the same firing pressure (Pmax) in all cylinders; or, in protection mode, to reduce the load and Pmax on a given single cylinder if the need arises.

The exhaust valve system (InVA) is driven on the same principles as the fuel injection system, exploiting the same high pressure hydraulic oil supply and a similar facility for mechanical redundancy. The need for controlling exhaust valve operation is basically limited to timing the opening and closing of the valve. The control system is thus simpler than that for fuel injection.

Cylinder lubrication is controllable from the condition evaluation system so that the lubricating oil amount can be adjusted to match the engine load. Dosage is increased in line with load changes and if the need is indicated by the cylinder condition monitoring system (in the event of liner scuffing and ring blow-by, for example). Such systems are already available for existing engines.

The turbocharging system control will incorporate control of the scavenge air pressure if a turbocharger with variable turbine nozzle geometry is used, and control of bypass valves, turbocompound system valves and turbocharger cut-off valves if such valves are incorporated in the system. Valves for any selective catalytic reduction (SCR) exhaust gas cleaning system installed will also be controlled.

Operating modes may be selected from the bridge control system or by the system's own control system. The former case applies to the fuel economy modes and the emission-controlled modes (some of which may incorporate the use of an SCR system). The optimum reversing/crash stop modes are selected by the system itself when the bridge control system requests the engine to carry out the corresponding operation. Engine protection mode, in contrast, will be selected by the condition monitoring and evaluation system independently of actual operating modes (when this is not considered to threaten ship safety).

The fruit of MAN B&W Diesel's and Sulzer's R&D is now available commercially, their respective electronically-controlled ME and RT-flex engines being offered alongside the conventional models and increasingly specified for a wide range of tonnage. The designs are detailed in Woodyard (2004).

Research and development by Mitsubishi, the third force in low speed engines, has successfully sought weight reduction and enhanced compactness while retaining the performance and reliability demanded by the market. The Japanese designer's current UEC-LS type engines yield a specific power output of around three times that of the original UE series of the mid-1950s. The specific engine weight has been reduced by around 30% over that period and the engine length in relation to power output has been shortened by one-third.

Mitsubishi strengthened its long relationship as a Sulzer licensee in 2002 by forging a joint venture with the Wärtsilä Corporation to develop a new 500 mm bore design to be offered in two versions: a 'mechanical' RTA50C and an 'electronic' RT-flex 50C. Detailed descriptions of low speed engines produced by the main manufacturers are given by Woodyard (2004).

6.5.1.2 Medium speed engines

New designs and upgraded versions of established models have maintained the dominance of medium speed four-stroke diesel engines in the propulsion of smaller ships as well as larger specialist tonnage such as cruise vessels, car/passenger ferries and RO-RO freight carriers. The larger bore designs can also target the mainstream cargo ship propulsion market formed by bulk carriers, containerships and tankers, competing against low speed two-stroke machinery. The growth of the fast ferry sector has benefited those medium speed enginebuilders (notably Caterpillar and Ruston) who can offer designs with sufficiently high power/weight and volume ratios, an ability to function reliably at full load for sustained periods, and attractive through-life operating costs. Medium speed engines further enjoy supremacy in the deepsea genset drive sector, challenged only in lower power installations by high speed four-stroke engines.

Significant strides have been made in improving the reliability and durability of medium speed engines in the past decade, both at the design stage and through the in-service support of advanced monitoring and diagnostic systems. Former weak points in earlier generations of medium speed engines have been eradicated in new models which have benefited from finite element method calculations in designing heavily loaded components. Designers now argue the merits of new generations of longer stroke medium speed engines with higher specific outputs allowing a smaller number of cylinders to satisfy a given power demand and foster compactness, reliability, reduced maintenance and easier servicing. Progress in fuel and lubricating oil economy is also cited, along with enhanced pier-to-pier heavy fuel burning capability and better performance flexibility throughout the load range.

Completely bore-cooled cylinder units and combustion spaces formed by liner, head and piston combine good strength and stiffness with good temperature control which are important factors in burning low quality fuel oils. Low noise and vibration levels achieved by modern medium speed engines can be reduced further by resilient mounting systems, a technology which has advanced considerably in recent years.

IMO limits on nitrogen oxides emissions in the exhaust gas can generally be met comfortably by medium speed engines using primary measures to influence the combustion process (in some cases, it is claimed, without compromising specific fuel consumption). Wärtsilä's low NOx combustion technology, for example, embraces high fuel injection pressures (up to 2000 bar) to reduce the duration of injection; a high compression ratio (16:1); a maximum cylinder pressure of up to 210 bar; and a stroke/bore ratio greater than 1.2:1. Concern over smoke emissions, particularly by cruise ship operators in sensitive environmental areas, has called for special measures from engine designers targeting that market, notably electronically-controlled common rail fuel injection and fuel-water emulsification.

Ease of inspection and overhaul—an important consideration in an era of low manning levels and faster turnarounds in port—was addressed in the latest designs by a reduced overall number of components (in some cases, 40% fewer than in the preceding engine generation) achieved by integrated and modular assemblies using multi-functional components. Simplified (often plug-in or clamped) connecting and quick-acting sealing arrangements also smooth maintenance procedures. Channels for lubricating oil, cooling water, fuel and air may be incorporated in the engine block or other component castings, leaving minimal external piping in evidence. Compact and more accessible installations are achieved by integrating ancillary support equipment (such as pumps, filters, coolers and thermostats) on the engine. Lower production costs are also sought from design refinements and the wider exploitation of flexible manufacturing systems to produce components.

The cylinder unit concept is a feature of the latest four-stroke designs, allowing the head, piston, liner and connecting rod to be removed together as a complete assembly for repair, overhaul or replacement by a renovated unit onboard or ashore. This modular approach was adopted by MAN B&W for its L16/24, L21/31 and L27/38 designs, by MTU for its Series

8000 engine, by Rolls-Royce for its Bergen C-series, and by Hyundai for its H21/32 design.

Compactness and reduced weight remain the key attractions of the medium speed engine, offering ship designers the opportunity to increase the cargo capacity and lower the cost of a given newbuilding project, and the ability to achieve via reduction gearing the most efficient propeller speed. Medium speed enginebuilders can offer solutions ranging from single-engine plants for small cargo vessels to multi-engine/twin-screw installations for the most powerful passenger ships, based on mechanical (geared) or electrical transmissions (see Section 6.4). Multi-engine configurations promote plant availability and operational flexibility, allowing the number of prime movers engaged at any time to match the service schedule. The convenient direct drive of alternators and other engineroom auxiliary plant (hydraulic power packs, for example) is also facilitated via power take-off gearing.

Among the design innovations in recent years must be noted Sulzer's use of hydraulic actuation of the gas exchange valves for its ZA50S engine, the first time that this concept (standard on low speed two-stroke engines for many years) had been applied to medium speed four-stroke engines (Figure 6.51). In conjunction with pneumatically controlled load-dependent timing to secure variable inlet closing, hydraulic actuation on the ZA50S engine allows flexibility in valve timing, fostering lower exhaust gas emissions and improved fuel economy.

Variable inlet closing, combined with optimized turbocharging, contributes to a very flat fuel consumption characteristic across the load range of the engine as well as a considerable reduction in smoke levels in part-load operation. The ZA50S engine, like the smaller bore ZA40S design it was derived from, features Sulzer's rotating piston which was also exploited in GMT's upgraded 550 mm bore medium speed engine.

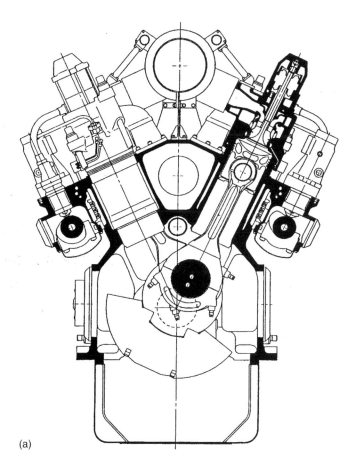

(a)

Figure 6.47 The Japanese ADD 30V engine is distinguished by a single-valve gas exchange system comprising a main valve located at the centre of the cylinder and a control valve placed co-axially against the main valve. The control valve switches the air intake and exhaust channels in the cylinder cover. Both valves are driven hydraulically. Side-mounted fuel injectors are arranged around the cylinder periphery.

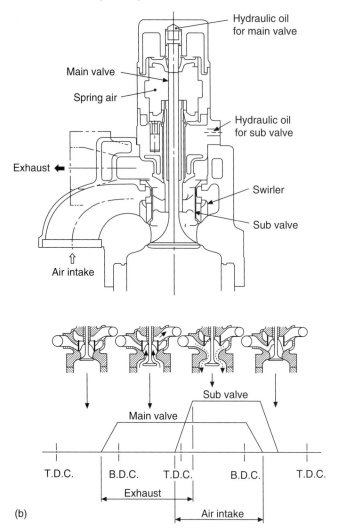

(b)

Figure 6.47 (Continued)

Wärtsilä's 46 engine exploited a number of innovations in medium speed technology, originally including a twin fuel injection system (featuring pilot and main injection valves), thick-pad bearings (large bearings with thick oil films) and pressure-lubricated piston skirts. The twin injection system was later superseded as advances in fuel injection technology allowed a single-valve system to be applied.

The operating flexibility of MAN B&W's L32/40 design benefits from the provision of separate camshafts, arranged on either side of the engine, for the fuel injection and valve actuating gear. One camshaft is dedicated to drive the fuel injection pumps and to operate the starting air pilot valves; the other serves the inlet and exhaust valves. Such an arrangement allows fuel injection and air charge renewal to be controlled independently, and thus engine operation to be more conveniently optimized for either high fuel economy or low exhaust emissions mode. Injection timing can be adjusted by turning the camshaft relative to the camshaft driving gear (an optional facility (Figure 6.48). The valve-actuating camshaft can be provided with different cams for full-load and part-load operations, allowing valve timing to be tailored to the conditions. A valve camshaft-shifting facility is optional, the standard engine version featuring just one cam contour (Figure 6.49).

A carbon cutting ring is now a common feature of medium speed engines specified to eliminate the phenomenon of cylinder bore polishing caused by carbon deposits and hence significantly reduce liner wear. It also fosters a cleaner piston ring area, low and very stable lubricating oil consumption, and reduced blow-by.

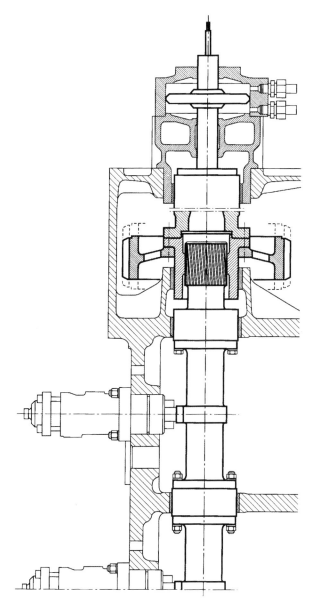

Figure 6.48 Optimization of fuel injection timing on MAN B&W's L32/40 engine is facilitated by turning the dedicated camshaft relative to its driving gear.

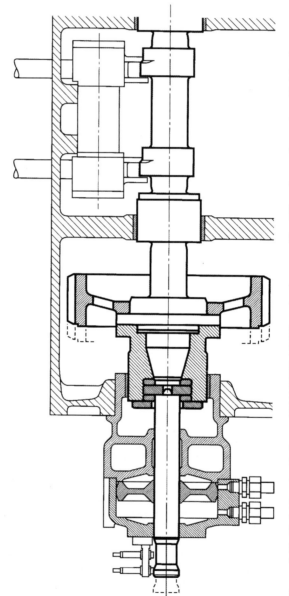

Figure 6.49 Variable valve timing on MAN B&W's L32/40 engine is secured by different cams for part- and full-load operations.

Also termed an anti-polishing or fire ring, a carbon cutting ring comprises a sleeve insert which sits between the top piston ring turning point and the top of the cylinder liner. It has a slightly smaller diameter than the bore of the liner, this reduction being accommodated by a reduced diameter for the top land of the piston. The main effect of the ring is to prevent the build-up of carbon around the edges of the piston crown which causes liner polishing and wear, with an associated rise in lubricating oil consumption. A secondary function is a sudden compressive effect on the ring belt as the piston and carbon cutting ring momentarily interface. Lubricating oil is consequently forced away from the combustion area, again helping to reduce consumption: so effectively, in fact, that Bergen Diesel found it necessary to redesign the ring pack to allow a desirable amount of oil consumption. Lubricating oil consumption, the Norwegian engine designer reports, is cut by more than half and insoluble deposits in the oil reduced dramatically, significantly extending oil filter life. Carbon cutting rings can be retrofitted to deliver their benefits to engines in service. Removal prior to piston withdrawal is simply effected with a special tool.

Designers now also favour a 'hot box' arrangement for the fuel injection system to secure cleaner engine lines and improve die working environment in the machinery room thanks to reduced temperatures; additionally, any fuel leakage from the injection system components is retained within the box.

The major medium speed enginebuilders have long offered 500 mm-bore-plus designs in their portfolios. MAN B&W still fields the L58/64 series and SEMT-Pielstick its 570 mm bore PC4.2 and PC40L series, but MaK's 580 mm bore M601 and the Sulzer ZA50S engines have been phased out, as was Stork-Wärtsilä's TM620 engine in the mid-1990s.

In the 1970s MAN and Sulzer jointly developed a V-cylinder 650 mm bore/stroke design (developing 1325 kW/cylinder at 400 rev/min) that did not proceed beyond prototype testing.

Wärtsilä's 64 series, launched in 1996, took the medium speed engine into a higher power and efficiency territory, the 640 mm bore/900 mm stroke design now offering an output of 2010 kW/cylinder at 333 rev/min. A V12-cylinder model delivers 23 280 kW at 400 rev/min. The range can therefore meet the propulsive power demands of virtually all merchant ship tonnage types with either single- or multi-engine installations. The key introductory parameters were: 10 m/s mean piston speed; 25 bar mean effective pressure; and 190 bar maximum cylinder pressure. The Finnish designers claimed the 64 series to be the first medium speed engine to exceed the 50% thermal efficiency barrier, and suggested that overall plant efficiencies of 57–58% are possible from a combi-cycle exploiting waste heat to generate steam for a turbo-alternator.

At the other end of the medium speed engine power spectrum, the early 1990s saw the introduction of a number of 200 mm bore long stroke designs from leading builders, such as Daihatsu, MaK and Wärtsilä Diesel, contesting a sector already targeted by Sulzer's S20 model. These heavy fuel-burning engines (typically with a 1.5:1 stroke/bore ratio) were evolved for small-ship propulsion and genset drive duties, the development goals addressing overall operating economy, reliability, component durability, simplicity of maintenance and reduced production costs. Low and short overall configurations gave more freedom to naval architects in planning machinery room layouts and eased installation procedures (Figure 6.50).

The *circa*-320 mm bore sector is fiercely contested by designers serving a high volume market created by propulsion and genset drive demands. A number of new designs—including Caterpillar/MaK's M32 and MAN B&W Diesel's L32/40—emerged to challenge upgraded established models such as the Wärtsilä 32.

A Japanese challenger in a medium speed arena traditionally dominated by European designer/licensors arrived in the mid-1990s after several years' R&D by the Tokyo-based Advanced Diesel Engine Development Company. The joint venture embraced Hitachi Zosen, Kawasaki Heavy Industries and Mitsui Engineering and Shipbuilding. The 300 mm bore/480 mm stroke ADD30V design, in a V50-degree configuration, developed up to 735 kW/cylinder at 750 rev/min, significantly more powerful than contemporary medium speed engines of equivalent bore size. A mean effective pressure of around 25 bar and a mean piston speed of 11.5 m/s were exploited in tests with a six-cylinder prototype, although an mep approaching 35 bar and a mean piston speed of 12 m/s are reportedly possible. In addition to a high specific output, the developers sought a design which was also over 30% lighter in weight, 10–15% more fuel economical and with a better part-load performance than established engines. Underwriting these advances in mep and mean piston speed ratings are an anti-wear ceramic coating for the sliding surfaces of the cylinder liners and piston rings, applied by a plasma coating method. A porous ceramic heat shield was also developed for the combustion chamber to reduce heat transfer to the base metal of the piston crown.

A key feature is the single-valve air intake and exhaust gas exchange (Figure 6.47), contrasting with the four-valve (two inlet and two exhaust) heads of other medium speed engines. The greatly enlarged overall dynamic valve area and the reduction in pressure losses during the gas exchange period promote a higher thermal efficiency. The system is based on a heat-resistant alloy main poppet valve located over the centre of the cylinder and a control valve placed co-axially against this main valve. The control valve switches the air intake and exhaust

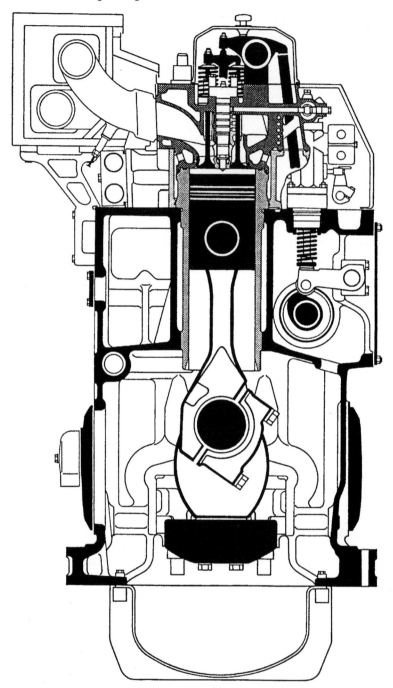

Figure 6.50 MaK's M20 engine, representing a new breed of 200 mm bore designs.

channels in the cylinder cover. Both main and control valves are driven hydraulically.

Fuel injection is executed from the side through multiple injectors arranged around the cylinder periphery instead of a conventional top-mounted central injection system. Combustion characteristics were optimized by raising the fuel injection pressure to around 2000 bar, thus enhancing the fuel-air mixture formation and fostering low NOx emissions without sacrificing fuel economy. A computer-based mechatronics system automatically controls the timing of fuel injection and valve opening/closing to match the operating conditions. The first 6ADD30V production marine engines, built by Mitsui, were

Figure 6.51 The first example of the Sulzer ZA50S engine, a nine-cylinder model, on test. The design (no longer produced) was distinguished by hydraulic actuation of the gas exchange valves.

specified as the prime movers for the diesel–electric propulsion plant of a large Japanese survey vessel.

Offshore industry market opportunities—and the potential of mainstream shipping interest—have encouraged a number of medium speed enginebuilders to develop dual-fuel and gas-diesel designs offering true multi-fuel capabilities with high efficiency and reliability, and low carbon dioxide emissions. The engines can run on gas (with a small percentage of liquid pilot fuel) or entirely on liquid fuel (marine diesel oil, heavy fuel or even crude oil). Switching from one fuel to another is possible without interrupting power generation.

The high cost of R&D to maintain a competitive programme and continuing investment in production resources and global support services have stimulated a number of joint ventures and takeovers in the four-stroke engine sector in recent years. Most notable have been Wärtsilä's acquisition of the former New Sulzer Diesel and Caterpillar's takeover of MaK. Earlier, Wärtsilä had acquired another leading medium speed enginebuilder, the Netherlands-based Stork-Werkspoor Diesel. This trend towards an industry comprising a small number of major multi-national players contesting the world market has continued with the absorption of the British companies Mirrlees Blackstone, Paxman and Ruston (formerly part of Alstom Engines) into the MAN B&W Diesel group. Rolls-Royce inherited the Bergen Diesel interests in Norway through the takeover of Vickers-Ulstein, adding to its Allen programme.

Considerable potential remains for further developing the power ratings of medium speed engines, whose cylinder technology has benefited in recent years from an anti-polishing ring at the top of the liner, water distribution rings, chrome-ceramic piston rings, pressurized skirt lubrication and nodular cast iron/low friction skirt designs.

The pressure-lubricated skirt elevated the scuffing limit originally obtaining by more than 50 bar, reduced piston slap force by 75% and doubled the lifetime of piston rings and grooves. Furthermore, it facilitated a reduction in lube oil consumption and, along with the simultaneous introduction of the nodular cast iron skirt, practically eliminated the risk of piston seizure. The anti-polishing ring dramatically improved cylinder liner lifetime beyond 100 000 hours, and lube oil consumption became controllable and stable over time, most engines today running at rates between 0.1 and 0.5 g/kWh. A further reduction in piston ring and groove wear was also achieved, and the time-between-overhauls extended to 18 000–20 000 hours. The ring itself is a wear part but is turnable in four positions in a four-stroke engine, fostering a lengthy lifetime for the component.

Such elements underwrite a capability to support a maximum cylinder pressure of 250 bar, of which 210–230 bar is already exploited in some engines today. Leading designers such as Wärtsilä suggest it may be possible to work up towards 300 bar with the same basic technology for the cylinder unit, although some areas need to be developed: bearing technology, for example, where there is potential in both geometry and materials. A steel piston skirt may become the most cost effective, and cooling of the piston top will probably change from direct oil cooling to indirect. The higher maximum cylinder pressure can be exploited for increasing the maximum effective pressure or improving the thermal efficiency of the process. Continuing to mould the development of the medium speed engine will be: NOx emissions; CO_2 emissions; fuel flexibility; mean time-between-failures; and reduced maintenance.

Designing modern medium speed engines

Investment in the development of a new medium speed engine may be committed for a number of reasons. The enginebuilder may need to extend its portfolio with larger or smaller designs to complete the available range, to exploit recently-developed equipment and systems, or to enhance the reliability, availability and economy of the programme As an alternative to a new design, it may be possible to upgrade an existing engine having the potential for improved power ratings, lower operational costs, reduced emissions and weight. A recent example is the upgrading by MAN B&W Diesel of its 48/60 engine, whose new B-version offers a 14% higher specific output with considerably reduced fuel consumption and exhaust emission rates, a lower weight, and the same length and height but a narrower overall width than its predecessor (Figure 6.52).

Having set the output range of the new engine, a decision must be made on whether it should be built as an in-line cylinder version only or as a V-type engine as well. In the case of the largest and heaviest medium speed engines, the efforts and costs involved in adding a V-type to the existing portfolio might not be justified if there is a limited potential market for engines above 20 000 kW. Other restraining factors could be the capacity of the foundry for casting very large crankcases or difficulties in transporting engines weighing 400 tonnes or more over land.

Once the power per cylinder of the proposed new engine is known, the first indication of its bore size is determined by the piston load, which relates the specific output to the circular piston surface. Piston loads have increased with the development of better materials: the original 48/60 engine has a piston load of $58 \, kW/cm^2$, while the 48/60B engine achieves $66.4 \, kW/cm^2$. The bore and stroke of an engine are interrelated by the stroke/bore ratio, which in turn is based on the designer's experiences with earlier engines. Some 25 years ago it was not uncommon to design medium speed engines with very similar bore and stroke dimensions (so-called 'square' engines); more recently, the trend has been towards longer stroke designs, which offer clear advantages in optimizing the combustion space geometry to achieve lower NOx and soot emission rates. A longer stroke can reduce NOx emissions with almost no fuel economy penalty and without changing the maximum combustion pressure. The compression

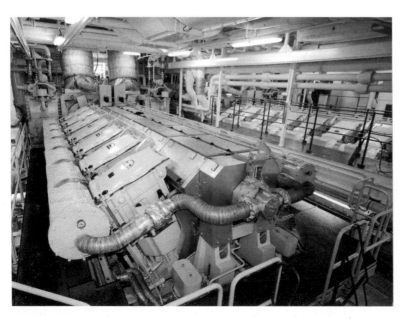

Figure 6.52 MAN B&W Diesel's 48/60 engine, shown here in V14-cylinder form, benefited from a redesign that raised specific output by 14% and reduced fuel consumption and emissions.

ratio can also be increased more easily and, together with a higher firing pressure, fuel consumption rates will decrease. Finally, long stroke engines yield an improved combustion quality, a better charge renewal process inside the cylinder and higher mechanical efficiency.

A good compromise between an optimum stroke/bore ratio and the costs involved, however, is an important factor: it is a general rule that the longer the engine stroke, the higher the costs for an engine per kW of output. The trend towards longer stroke medium speed engine designs is indicated by the table showing the current MAN B&W Diesel family; the first four models were launched between 1984 and 1995, and the remaining smaller models after 1996.

Model	Stroke/bore ratio
58/64	1.1
48/60	1.25
40/54	1.35
32/40	1.25
27/38	1.41
21/31	1.47
16/24	1.5

The above is a simplified outline of the initial design process and the reality is much more complex. The final configuration results from considering a combination of choices, which may have different effects on fuel consumption rates and emissions. Nevertheless, the usual trade-off between fuel economy and NOx emissions can be eliminated. The use of high efficiency turbochargers is also essential.

A decision on engine speed is the next step. The mean piston speed in metres per second can be calculated from the bore and speed of an engine using the following formula: mean piston speed (m/s) = bore (m) × speed (rev/min)/30. The upper limit of the mean piston speed is primarily given by the size and mass of the piston and the high forces acting on the connecting rod and crankshaft during engine running. A mean speed between 9.5 m/s and slightly above 10 m/s is quite common for modern large bore medium speed engines. Any substantial increase above 10 m/s will reduce operational safety and hence reliability. Since medium speed engines may be specified to drive propellers and/or alternators, the selection of engine speed has to satisfy the interrelation between the frequency of an alternator (50 Hz or 60 Hz) and the number of pole pairs.

In designing the individual engine components, two main criteria are addressed: reducing manufacturing costs and reducing the number of overall engine parts to ease maintenance.

An example of the design process, by Rolls-Royce for the new Allen 5000 medium speed engine, is described by Woodyard (2004). Woodyard also gives detailed descriptions of medium speed engines produced by the main manufacturers.

6.5.1.3 *High speed engines*

High speed four-stroke trunk piston engines are widely specified for propelling small, generally specialized, commercial vessels and as main and emergency genset drives on all types of tonnage. The crossover point between high and medium speed diesel designs is not sharply defined but for the purposes of this Section engines running at 1000 rev/min and over are reviewed.

Marine high speed engines traditionally tended to fall into one of two design categories: high performance or heavy duty types. High performance models were initially aimed at the military sector, and their often complex designs negatively affected manufacturing and maintenance costs. Applications in the commercial arena sometimes disappointed operators, the engines dictating frequent overhauls and key component replacement.

Heavy duty high speed engines in many cases were originally designed for off-road vehicles and machines but have also found niches in stationary power generation and locomotive traction fields. A more simple and robust design with modest mean effective pressure ratings compared with the high performance contenders yields a comparatively high weight/power ratio. But the necessary time-between-overhauls and component lifetimes are more acceptable to civilian operators.

In developing new models, high speed engine designers have pursued essentially the same goals as their counterparts in the low and medium speed sectors: reliability and durability, underwriting extended overhaul intervals and component longevity and hence low maintenance costs; easier installation and servicing; compactness and lower weight; and enhanced performance across the power range with higher fuel economy and reduced noxious emissions.

Performance development progress over the decades is highlighted by considering the cylinder dimension and speed of an engine required to deliver 200 kW/cylinder (Figure 6.53). In 1945 a bore of 400 mm-plus and a speed of around 400 rev/min were necessary; in 1970 typical medium speed engine parameters resulted in a bore of 300 mm and a speed of 600 rev/min, while typical high speed engine parameters were 250 mm and 1000 rev/min to yield 200 kW/cylinder. Today, that specific output can be achieved by a 200 mm bore high speed design running at 1500 rev/min.

Flexible manufacturing systems (FMS) have allowed a different approach to engine design. The reduced cost of machining has made possible integrated

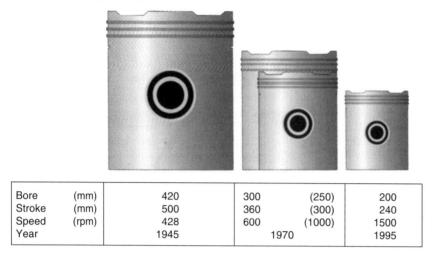

Bore	(mm)	420	300	(250)	200
Stroke	(mm)	500	360	(300)	240
Speed	(rpm)	428	600	(1000)	1500
Year		1945	1970		1995

Figure 6.53 Cylinder dimensions and speeds for medium and high speed engines delivering 200 kW/cylinder (1945, 1970 and 1995). (Reference Wärtsilä).

structural configurations, with more functions assigned to the same piece of metal. The overall number of parts can thus be reduced significantly over earlier engines (by up to 40% in some designs), fostering improved reliability, lower weight and increased compactness without compromising on ease of maintenance. FMS also facilitates the offering of market-adapted solutions without raising cost: individual engines can be optimized at the factory for the proposed application.

A widening market potential for small high speed engines in propulsion and auxiliary roles encouraged the development in the 1990s of advanced new designs for volume production. The *circa*-170 mm bore sector proved a particularly attractive target for leading European and US groups which formed alliances to share R&D, manufacture and marketing—notably Cummins with Wärtsilä Diesel, and MTU with Detroit Diesel Corporation.

High speed engine designs have benefited from such innovations as modular assembly, electronically controlled fuel injection systems, common rail fuel systems and sophisticated electronic control/monitoring systems. Some of the latest small bore designs are even released for genset duty burning the same low grade fuel (up to 700 cSt viscosity) as low speed crosshead main engines.

Evolving a new design
An insight into the evolution of a high speed engine design for powering fast commercial vessels is provided by MTU of Germany with reference to its creation of the successful 130 mm bore Series 2000 and 165 mm bore Series 4000 engines, which together cover an output band from 400 kW to 2720 kW.

MTU notes first that operators of fast tonnage place high value on service life and reliability, with fuel economy and maximized freight capacity also important. In the fast vessel market, conflicting objectives arise between key parameters such as low specific fuel consumption, low weight/power ratio and extended engine service life. If one parameter is improved, at least one of the others is undermined. The engine designer's aim is therefore to optimize co-ordination of the parameters to suit the application.

Knowledge of the anticipated service load profile is vital for determining the specific loads that must be addressed during the engine design stage so that the required maintenance and major overhaul intervals can be established. Load acceptance characteristics and performance map requirements have a strong influence on turbocharging and the maximum possible mean pressures.

Specifying performance map requirements is simultaneously connected with the selection of the lead application, in this case high speed tonnage. The maximum possible mean pressures are determined on the basis of the power-speed map requirements of various vessel types (for example, air cushion, hydrofoil and planing hull types) and the form of turbocharging (sequential or non-sequential, single or two-stage). With increasing mean pressures (higher power concentration), the weight/power ratio of the engine can be improved.

The maximum mean piston speed is derived from a service life requirement (time-between-overhaul) and the target for the weight/power ratio. With increasing

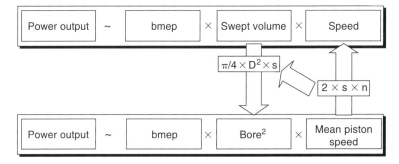

Figure 6.54 Calculation of power output.

mean piston speed, a greater power concentration in a given volume is achieved, thus improving the weight/power ratio. For fast vessel engines, mean piston speeds of 11–12 m/s and mean effective pressures up to 22 bar (single-stage turbocharging) or 30 bar (two-stage turbocharging) are typical. Figure 6.54 shows the correlations of four-stroke engines for determining bore diameter, stroke and speed. The output per cylinder (Pe) is known from the power positioning of the proposed new engine, and the maximum mean pressure and maximum mean piston speed have already been established. The required minimum bore diameter (D) can thus be determined.

The appropriate stroke (s) is determined using the specified stroke/bore ratio (s/D). If large s/D ratios are selected, a large engine height and width results (V-engine); small s/D ratios are associated with somewhat reduced fuel efficiency. For relevant MTU engines with good weight/power ratios, the stroke/bore ratio lies within a range of 1.1 to 1.25. The engine speed appropriate to the established stroke is determined via the resulting mean piston speed. Engine speed is an important factor for the customer as the size of the gearbox required is based on the speed and torque.

The peak firing pressure or peak firing pressure/mean pressure ratio is the most important factor influencing the specific effective fuel consumption. For high efficiency, MTU suggests, peak firing pressure/mean pressure ratios of around 8 should be targeted. With the mean pressure already specified, the peak firing pressure can be established. If, for example, a mean pressure of 22 bar (single-stage turbocharging) is selected, the peak firing pressure should lie in the 160–180 bar range. For two-stage turbocharged engines with correspondingly high mean pressures, peak firing pressures of above 200 bar must be targeted. With two-stage turbocharging, the potential for fuel consumption reduction can be increased using charge air cooling.

Caterpillar

A wide programme of high speed engines from the US designer Caterpillar embraces models with bore sizes ranging from 105 mm to 170 mm. The largest and most relevant to this review is the 170 mm bore/190 mm stroke 3500 series which is produced in V8-, V12- and V16-cylinder versions with standard and higher B-ratings to offer outputs up to around 2200 kW. The engines, with minimum/maximum running speeds of 1200/1925 rev/min, are suitable for propelling workboats, fishing vessels, fast commercial craft and patrol boats. Genset applications can be covered with ratings from 1000 kVA to 2281 kVA.

The series B engines (Figure 6.55) benefited from a number of mechanical refinements introduced to take full advantage of the combustion efficiency improvement delivered by an electronic control system. Electronically controlled unit fuel injectors combine high injection pressures with an advanced injector design to improve atomization and timing. Outputs were raised by 17 to 30% above earlier 3500 series models.

A special high performance variant of the V16-cylinder 3500 series model was introduced to target niche markets, the refinements seeking increased power, enhanced reliability and lower fuel and lubricating oil consumptions without undermining durability. This Phase II high performance version of the 3516 has an upper rating of 2237 kW at 1925 rev/min. It was released for fast passenger vessels with low-load factors with a standard maximum continuous rating of 1939 kW at 1835 rev/min and a 'two hours out of 12' rating of 2088 kW at 1880 rev/min. Optional higher ratings up to 2205 kW at 1915 rev/min can be specified for cooler climate deployment, with revised turbocharger, fuel injector and timing specifications.

Key contributions to higher performance came from high efficiency ABB turbochargers, a seawater aftercooler to supply colder air to the combustion

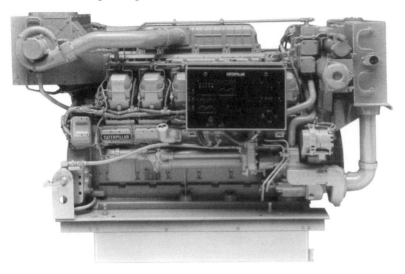

Figure 6.55 Caterpillar 3512B engine with electronic control system.

chambers, larger and more aggressive camshafts, and a new deep crater piston design. The fuel is delivered through strengthened unit injectors designed and manufactured by Caterpillar to secure injection pressures of 1380 bar.

An optimum air-fuel mixture which can be burned extremely efficiently is fostered by the combination of a denser air intake and the high injection pressure. The reported result is a specific fuel consumption range at full load of 198–206 g/kWh with all fuel, oil and water pumps driven by the engine. Modifications to the steel crown/aluminium skirt pistons and rings lowered lubricating oil consumption to 0.55 g/kWh.

A particularly desirable feature for fast ferry propulsion is underwritten by the high efficiency combustion and low crevice volume pistons which help to eliminate visible exhaust smoke at all steady points along the propeller demand curve. The rear gears were widened and hardened to serve the higher pressure unit injectors. New gas-tight exhaust manifolds with bellow expansion joints and stainless steel O-rings improved engineroom air quality by eliminating exhaust gas leaks.

A longer-stroke variant of the Cat 3500 series B engine was introduced after marine field tests undertaken from early 1998, these 3512B and 3516B models offering as much as 13 per cent higher powers than their standard counterparts, with respective maximum commercial ratings of 1380 kW and 1864 kW at 1600 rev/min. Seven per cent improvements in power-to-weight ratio and fuel economy were reported, along with lower emission levels than the standard engines. The higher output was achieved by enlarging the cylinder displacement (increasing the stroke by 25 mm) and without raising cylinder pressure or undermining bearing life or the durability of other key components.

A new single-piece forged crankshaft has more mass and is made from a stronger steel alloy than before to handle the higher loads. The connecting rods are longer and feature stronger shaft geometry; and a more robust rod pin end enhances the durability required for the increased piston speeds and higher inertia loads. The pistons are of the same two-piece design proven in standard Cat 3500 series B engines, a steel crown and aluminium skirt securing high strength and reduced weight. The engine footprint of these more powerful variants remained unchanged; only the dimensions of the higher capacity aftercooler and turbochargers were increased.

All Caterpillar 3500 series-B engines are controlled by a microprocessor-based electronic control module (ECM). Information is collected from engine sensors by the ECM which then analyses the data and adjusts injection timing and duration to optimize fuel efficiency and reduce noxious exhaust emissions. Electronic control also supports onboard and remote monitoring capabilities, the ECM reporting all information through a two-wire Cat Data Link to the instrument panel. The panel records and displays faults as well as operating conditions. An optional Customer Communications Module translates engine data to standard ASCII code for transmission to a PC or via satellite to remote locations.

Caterpillar's Engine Vision System (EVS) is compatible with the high performance 3500 series-B engines and the company's other electronically controlled engines. The EVS displays engine and transmission data, vessel speed, trip data, historical data, maintenance intervals, diagnostics and troubleshooting information. Up to three engines can be monitored simultaneously, the system transferring between the vision display and individual ECMs via the two-wire data link.

An upgrade announced in 2002 introduced the 3500B series II engines with enhancements to their electronic control, monitoring, display and cooling systems as well as new derating and operating speed options designed for specific applications. New electronics included the latest Caterpillar ADEM III control system, allowing more engine parameters to be controlled and monitored, with more accuracy and fault-reporting capability. A new 'programmable droop' capability allows precise governor control for load-sharing applications. A combined cooling system, rather than two separate circuits, became an option. A higher maximum continuous rating of 2000 kW from the 3516B series II engine was offered to yield more power and bollard pull capacity for larger harbour tugs; the higher rating also addressed some types of ferries and offshore service vessels.

Cummins

The most powerful own-design engine in Cummins' high speed programme, the KTA50-M2 model, became available from early 1996 (Figure 6.56). The 159 mm bore/159 mm stroke design is produced by the US group's Daventry factory in the UK in V16-cylinder form with ratings of 1250 kW and 1340 kW for medium continuous duty and 1030 kW and 1180 kW for continuous duty applications. The running speeds range from 1600 rev/min to 1900 rev/min, depending on the duty; typical applications include fishing vessels, tugs, crewboats and small ferries.

The KTA50-M2 engine benefited from a new Holset turbocharger, low temperature after-cooling and gallery-cooled pistons. Cummins' Centry electronics system contributes to enhanced overall performance and fuel economy, providing adjustable all-speed governing, intermediate speed controls, dual power curves, a built-in hour metre and improved transient response. Diagnostic capabilities are also incorporated.

Woodyard (2004) also includes outline descriptions of the high speed engines developed and manufactured by Deutz, GMT, Isotta Fraschini, MAN B&W Holeby, Mitsubishi, MTU, Nigata, Paxman, SEMT-Pielstick, Wärtsilä, Zvezda, Scania and Volvo-Penta.

6.5.2 Gas turbines

6.5.2.1 *Overview*

Gas turbines have dominated warship propulsion for many years but their potential remains to be fully realized in the commercial shipping sector. Breakthroughs in containerships, a small gas carrier and the Baltic ferry *Finnjet* during the 1970s promised a deeper penetration that was thwarted by the rise in bunker prices and the success of diesel engine designers in raising specific power outputs and enhancing heavy fuel burning capability.

In recent years, however, gas turbine suppliers with suitable designs have secured propulsion plant contracts from operators of large cruise ships and

Figure 6.56 Cummins' largest own-design engine, the KTA50-M2 model.

high speed ferries, reflecting the demand for compact, high output machinery in those tonnage sectors, rises in cycle efficiency and tightening controls on exhaust emissions. A new generation of marine gas turbine—superseding designs with roots in the 1960s—will benefit from the massive investment in aero engine R&D over the past decade, strengthening competitiveness in commercial vessel propulsion.

The main candidates for gas turbine propulsion in commercial shipping are:

Cruise ships: the compactness of gas turbine machinery can be exploited to create extra accommodation or public spaces, and the waste heat can be tapped for onboard services. In a large cruise ship project some 20–100 more cabins can be incorporated within the same hull dimensions, compared with a diesel-electric solution, depending on the arrangement philosophy. Compactness is fostered by the smaller number of prime movers and minimal ancillary systems (roughly around 50% fewer than a diesel-based plant).

Large fast passenger ferries and freight carriers: the extremely high power levels demanded for such vessels is difficult to satisfy with diesel machinery alone; coastal water deployment favours the use of marine diesel oil in meeting emission controls.

LNG carriers: an ability to burn both cargo boil-off gas and liquid fuel at higher efficiency than traditional steam turbine propulsion plant should be appreciated.

Fast containerships: the compact machinery allows space for additional cargo capacity.

Among the merits cited for gas turbine propulsion plant in commercial tonnage are:

- High power-to-weight and power-to-volume ratios; aero-derived gas turbines typically exhibit power-to-weight ratios at least four times those of medium speed diesel engines; compactness and weight saving releases machinery space for extra revenue-earning activities; a General Electric LM2500 aero-derived gas turbine unit delivering 25 000 kW, for example, measures 4.75 m long × 1.6 m diameter and weighs 3.5 tonnes.
- Low noise and vibration.
- Ease of installation and servicing fostered by modular packages integrated with support systems and controls.
- Modest maintenance costs, low spare part requirements and ease of replacement.
- Environmental friendliness (lower NOx and SOx emissions than diesel engines).
- Reduced manning levels facilitated by full automation and unmanned machinery space capability.
- Operational flexibility: swift start-up: no warm-up or idling period required; idle can typically be reached within 30 seconds, followed by acceleration to full power; deceleration can be equally as rapid, after which the turbine can be immediately shut down; subsequent restarts, even after high power shutdown, can be instant with no 'cool down' restriction.
- High availability, underwritten by high reliability and rapid repair and/or turbine change options.

6.5.2.2 *Plant configurations*

Optimizing the combination of low power manoeuvring and high power operation is accomplished in many naval applications by using a combined diesel or gas turbine propulsion system (CODOG). Other arrangements can be configured to suit the power demands and/or operational flexibility required for a project: combined diesel and gas turbine (CODAG—Figure 6.57); combined gas turbine and gas turbine (COGAG); combined gas turbine or gas turbine (COGOG); and combined diesel and gas turbine electric propulsion (CODLAG or CODEG).

A notable COGAG plant powers the Stena HSS 1500-class high speed passenger/vehicle ferries, whose service speed of 40 knots is secured by twin General Electric LM2500 and twin LMI600 gas turbines arranged in father-and-son configurations with a total output of 68 000 kW. All four turbines are deployed for the maximum speed mode, with the larger or smaller pairs engaged alone for intermediate speeds; this enables the turbines to operate close to their optimum efficiency at different vessel speeds, with consequent benefits in fuel economy (Figure 6.58).

An example of a CODEG plant is provided by *Queen Mary 2*, the world's largest passenger ship, whose 117 200 kW power station combines twin General Electric LM2500+ sets with four Wärtsilä 16V46 medium speed diesel engines, all driving generators.

Combined-cycle gas turbine and steam turbine electric (COGES) plants embrace gensets driven by gas and steam turbines. Waste heat recovery boilers exploit the gas turbine exhaust and produce superheated steam (at around 30 bar) for the steam turbine genset. Such an arrangement completely changes the properties of the simple-cycle gas turbine: while gas turbine efficiency decreases at low load the steam turbine recovers the lost power and feeds it back into the system. The result is a fairly constant fuel consumption over a wide operating range. Heat for ship services is taken directly from steam turbine extraction (condensing-type turbine) or from the steam turbine exhaust (back pressure turbine), and there is thus normally no need to fire auxiliary boilers.

Installations supplied by General Electric Marine Engines for large Royal Caribbean Cruises' liners pioneered the COGES plant at sea, the first entering

Marine engines and auxiliary machinery 405

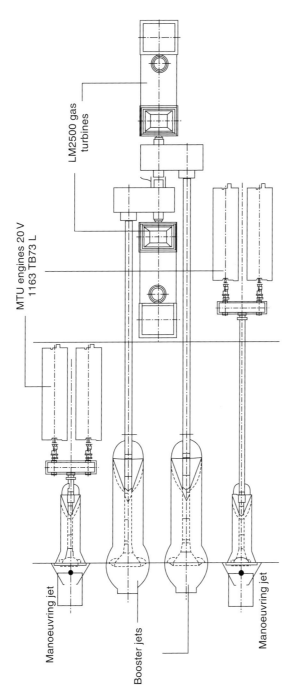

Figure 6.57 CODAG propulsion plant configuration for a large monohull fast ferry.

Figure 6.58 Final preparation of a General Electric LM2500 propulsion module for a large high speed ferry.

service in mid-2000 (Figures 6.59 and 6.60). Each 59 000 kWe outfit comprises a pair of GE LM2500+ gas turbine-generator sets, rated at 25 000 kWe apiece, and a 9000 kWe non-condensing steam turbine-generator. Heat recovery steam generators located in the exhaust ducts of the gas turbines produce the steam to drive the steam turbine and feed auxiliaries such as evaporators and heating systems. Since no additional fuel is consumed to drive the steam turbine, the additional power it generates represents a 15–18% increase in efficiency with the gas turbines operating at rated power. The auxiliary steam represents a further efficiency improvement.

Lightweight gas turbine-generator sets (weighing approximately 100 tons) have allowed naval architects to locate them in the base of a cruise ship's funnel. Such an arrangement replaces the gas turbine inlet and exhaust ducts normally running to the enginerooms with a smaller service trunk housing power lines, fuel and water supplies to the gas turbine package. A significant area on every deck between the funnel and the machinery spaces is thus released for other purposes.

6.5.2.3 Cycles and efficiency

Significant progress has been made in enhancing the thermal efficiency of simple-cycle gas turbines for ship propulsion over the years, R&D seeking to improve part-load economy and reduce the fuel cost penalty compared with diesel engines. In 1960 marine gas turbines had an efficiency of around 25% at their rated power, while second generation aero-derivatives were introduced in the 1970s with efficiencies of around 35%. Subsequent advances—design refinements, new materials and cooling techniques, and the appropriate matching of higher compressor pressure ratios—have resulted in some large simple-cycle turbines achieving efficiencies of over 40%, Figure 6.61.

More complex gas turbine cycles can deliver specific fuel consumptions closely approaching the very flat curve characteristics of larger diesel engines. Part-load efficiency can be improved in a number of ways, notably through the intercooled recuperated (ICR) cycle which uses the exhaust gas to heat the combustor inlet (Figure 6.62). This method was chosen for the WR-21 gas turbine which has achieved a 42% thermal efficiency across 80% of the operating range. (The first advanced-cycle gas turbine, the Rolls-Royce RM60, was operated in *HMS Grey Goose* for four years from 1953. Unfortunately, the technical complexity of the intercooling and recuperation was far ahead of the contemporary production techniques, thwarting commercial success, but the turbine proved reliable and efficient during its comparatively brief service.)

The ICR cycle and other performance/efficiency-enhancing solutions were examined by Mitsubishi Heavy Industries, taking as an example its own MFT-8 aero-derived gas turbine developed for

Marine engines and auxiliary machinery 407

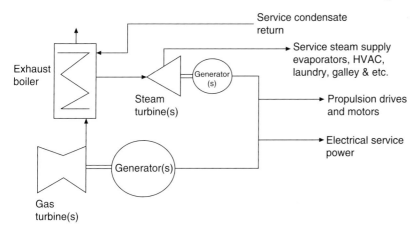

Figure 6.59 Schematic layout of COGES plant for a large cruise ship (GE Marine Engines).

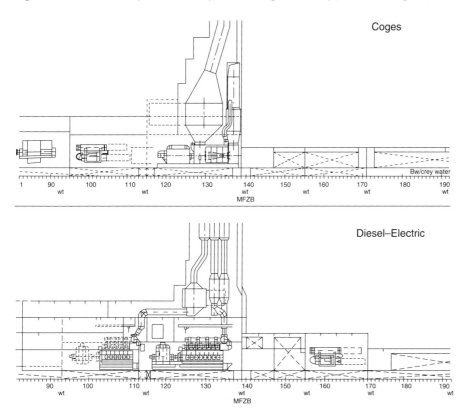

Figure 6.60 Space saving of a COGES plant compared with a diesel-electric installation for a cruise ship (Deltamarin).

marine use. The Japanese designer investigated various methods of improving the thermal efficiency and power output to similar levels as those of the diesel engine, focusing on six gas turbine cycle configurations (Figure 6.63):

- *Simple cycle:* the standard layout comprising compressor, turbine and combustor.

- *Intercooled cycle:* as for the simple cycle but with an intercooler added to increase performance through a reduction of high pressure compressor power.
- *Intercooled recuperated cycle:* as for the intercooled cycle but with a recuperator installed to recover heat from the gas turbine exhaust to the combustor inlet.

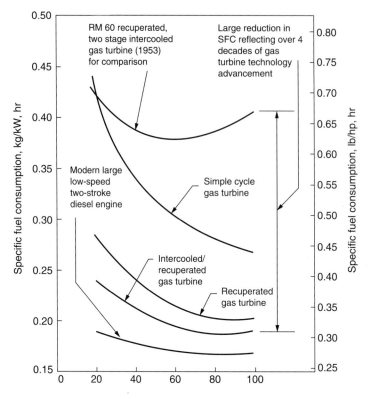

Figure 6.61 Comparison of specific fuel consumption curves against load for various gas turbine cycles and a low speed two-stroke diesel engine.

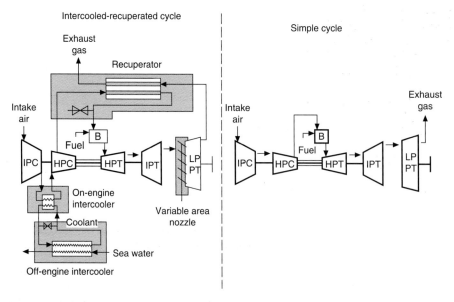

Figure 6.62 Schematic layout of intercooled-recuperated (ICR) cycle and simple-cycle gas turbines (Rolls-Royce).

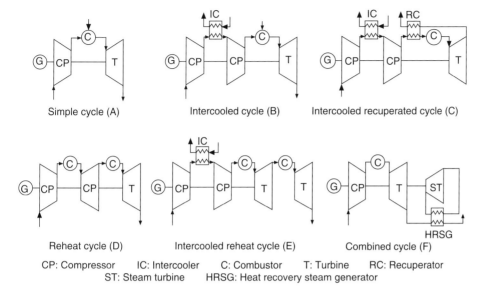

Figure 6.63 Various gas turbine cycle configurations (Mitsubishi).

- *Reheat cycle:* as for the simple cycle but with a reheat combustor added downstream of the high pressure turbine.
- *Intercooled reheat cycle:* a combination of intercooled and reheat cycle configurations.
- *Combined cycle:* a bottoming steam turbine cycle added to a simple-cycle gas turbine.

The main conditions that are changed by varying the cycle, MHI explains, are the turbine inlet temperature and the pressure ratio (1,150°C and 20:1 respectively in the case of the MFT-8 simple-cycle gas turbine). Parametric studies were carried out to establish the optimum performance from each cycle, within the bounds of what is practically achievable in terms of temperatures and pressures. With this in mind, a maximum turbine exhaust temperature of 600°C was specified, and pressure ratios for maximum efficiency and maximum power were determined for each cycle.

The highest efficiency and power was obtained with the combined cycle option, which had the further advantage of a pressure ratio lower than even the simple cycle. The major disadvantage of the combined cycle is the fact that the heat recovery steam generator is physically large and contains very heavy components. Although this is not necessarily a problem in industrial applications, where the combined cycle is commonly used, it is a considerable drawback in marine installations.

MHI's investigations favoured the intercooled recuperated cycle, which showed high efficiency and in power output was second only to the combined cycle. Moreover, maximum efficiency and maximum power were identical, at the achievable pressure ratio figures of 20 in each case. In the other non-simple cycles the two pressure ratio values were far apart, with one or other of the values being higher than 20, a ratio that incurs many practical disadvantages. The only problem with the ICR cycle, according to MHI, is that both the cost and size of the recuperator need to be reduced to make it more applicable to a marine gas turbine.

Although atmospheric pressure variation, an important consideration for aero gas turbines, is relatively small for marine units operating at sea level both temperature and humidity can vary significantly. Both of these parameters influence the specific heat of the air as a working fluid within the gas turbine. Atmospheric variations can have a significant effect on thermal efficiency and specific fuel consumption. Increased air temperature, in particular, reduces efficiency and lowers fuel economy; relative humidity has a less significant influence, though at high inlet temperatures an increase in humidity will have an adverse effect on specific fuel consumption.

6.5.2.4 Emissions

A lower operating temperature and more controlled combustion process enable gas turbines to deliver exhaust emissions with significantly lower concentrations of nitrogen oxides (NOx) and sulphur oxides (SOx) than diesel engines. Gas turbines typically take in three times the amount of air required to combust the fuel and the exhaust is highly diluted with fresh air. Along with a continuous combustion process, this yields very low levels of particulate emissions and a cleaner exhaust.

Combustion in gas turbines is a continuous process, with average temperatures and pressures lower than the peak levels in diesel engines that foster NOx emissions. The fundamental characteristic of continuous combustion in a gas turbine is that residence time at high flame temperatures (a key cause of NOx formation) is capable of being controlled. In a gas turbine a balance between smoke production and NOx generation can be easily secured.

Staged pre-mixed combustion allows NOx emission levels to be reduced without the need for expensive selective catalytic reduction systems typically required by diesel engines to meet the most stringent controls (Figure 6.64). Proven technologies are available to reduce NOx emissions from gas turbines even further:

- Wet technology based on steam or water injection into a standard combustor can lower NOx emissions to less than 1 g/kWh, exploiting a concept proven in land-based industrial applications. The availability of clean water can restrict operation at sea but the technology may be suitable for limited use in coastal areas.
- Dry Low Emissions (DLE) technology yields much lower emissions than current marine engine requirements, well below 1 g/kWh. Lean pre-mix DLE combustion technology maintains a near-optimum fuel-air distribution throughout the combustion zone and the flame temperature in a narrow band favourable both to low NOx and low carbon monoxide production.

SOx emissions are a function of fuel sulphur content. The heavy fuel typically burned by diesel engines may have a sulphur content up to 5% (3–3.5% on average). In contrast, gas turbines fired by marine distillate fuels have a maximum 1–2% sulphur content and the average level in gas turbine fuels is less than 0.5%.

6.5.2.5 *Lubrication*

The quantity of lubricating oil in circulation in a gas turbine is much smaller than in a diesel engine of equivalent output, and similarly the oil consumption is significantly less. (The Rolls-Royce Marine Spey, for example, has a recorded in-service average lube oil consumption of 0.1 litres/h.) The smaller charge, however, means the oil is subject to far greater stress than in a diesel engine, and the reduced consumption means it is not refreshed as often.

Apart from lubricating the bearings and other key components of a gas turbine, the bulk of the oil performs an intensive cooling function. Unlike its diesel engine counterpart, the lubricant does not come into contact with the combustion process and does not have to remove products of combustion or neutralize acids formed by burning sulphur-containing fuels.

In aero-derived gas turbines the concentration of power and heat, coupled with the comparatively small quantity of lubrication oil in circulation, results in peak oil temperatures typically over 200°C. Mineral oils are not suitable for this type of gas turbine because they would quickly oxidize at such temperatures: gums, acids and coke deposits would form and the viscosity of the oil would rapidly increase. Synthetic lubricating oils are therefore favoured because of their intrinsic ability to withstand much higher temperatures than mineral oils.

There are two types of lubricating oil generally available for gas turbine applications: 'standard'

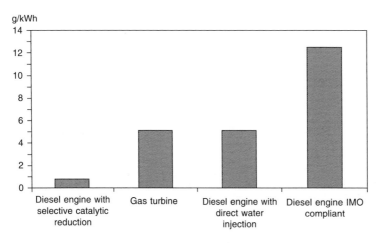

Figure 6.64 Specific NOx emissions for prime movers operating on marine gas oil.

or 'corrosion inhibiting'. Most turbine operators, according to lube oil supplier BP Marine, have moved towards premium corrosion inhibiting (C/I in US military specifications) oils to take advantage of the higher protection afforded to the bearings and other key components.

A low coking propensity, good swell characteristics and compatibility with seals are cited by BP Marine as desirable properties of a synthetic lube oil for aero-derived gas turbines. Coking is the formation of hard, solid particles of carbon due to high temperatures and can result in the blocking of oil ways. The tendency of an oil to break down and produce coke may be exacerbated by turbine operating procedures commonly encountered in fast ferry service, such as rapid acceleration and sudden shut-downs, resulting in the oil being subjected to excessive temperature rises.

Swell is caused by synthetic seals coming into contact with synthetic oils and absorbing the oil. Some swelling is desirable to ensure good sealing but too much can damage the seal and result in leakage. In addition to swell characteristics, the oil and the elastomers which come into contact must be compatible in all other respects so that degradation is avoided.

BP Marine's Enersyn MGT synthetic lubricant for marine gas turbines, a corrosion inhibiting product, is claimed to offer the desirable properties outlined above, along with good thermal, oxidative and hydrolytic stability, and corrosion resistance. It also conforms to the commonly required viscosity of most synthetic oils for gas turbines, nominally 5 cSt at 100°C.

A 5 cSt synthetic lube oil from Castrol Marine, Castrol 5000, is also approved for use in a range of aero-derived marine turbines, promising excellent high temperature and oxidation stability as well as superior load-carrying capabilities. Another synthetic lubricant, Castrol 778, is claimed to exhibit excellent anti-wear and rust protection properties supported by superior oxidation stability. The turbine is protected during extreme cold weather starting and during extended high temperature operation, and deposit or sludge formation is prevented over prolonged drain interval periods.

6.5.2.6 *Air filtration*

The use of air filtration systems tailored to the individual application can significantly enhance the reliability and efficiency of a marine gas turbine installation. A typical filtration system comprises three stages:

- *Vane separator:* a static mechanical device that exploits inertia to remove liquid droplets from an air flow passing through it. The intended purpose of this first stage is to remove the bulk of any large quantities of water and coarse spray that may otherwise overload the remaining components of the filtration system. This type of device can remove droplets down to around 12 microns in size, beyond which inertia has little effect. Unfortunately, the damaging salt aerosol experienced within this environment is predominantly below this size and so will pass to the next filtration stage.
- *Coalescer:* this filter-type device—usually of the one-inch depth pleated variety—is specifically designed to coalesce small water (and particularly salt aerosol) droplets; in other words, to capture small liquid droplets and make them form larger droplets. This comparatively easy task for a filter can be achieved with a relatively open filter material which, although a good coalescer, only has a very limited efficiency against dust particulate.

A low efficiency open media filter has a low pressure loss and thus for this application can be run at much higher airflow face velocities than traditional filters in other applications before it has an impact on gas turbine performance. Velocities are typically between 6 m/s and 8 m/s, compared with 2.5 m/s–3.5 m/s for other applications. This enables a much smaller filter system to be designed for naval vessels, where space is at a premium.

High velocity does not come without a price, however. A coalescer operating at a low velocity is likely to be able to drain all of the liquid it collects—through its media—with carry-over into the duct beyond. But a high velocity coalescer will re-entrain the liquid captured as large droplets, and so a third stage is required, particularly as this liquid is heavily laden with salt.

- *Vane separator:* this third stage device is very similar to the first stage and in many systems is exactly the same. Its purpose is to remove droplets being re-entrained into the airflow by the second-stage coalescer; as such, it must be totally reliable since this liquid has the potential to cause significant damage to the gas turbine because of its salt content.

A successful filtration system serving the gas turbine plant of a fast ferry should have the following features and characteristics, suggests UK-based specialist Altair Filters International:

- It should be a three-stage (vane, coalescer/dust filter, vane) system.
- The filter element should have an appropriate degree of dust filtration performance.
- The filter should have a suitable lifetime.
- The system should be of the high velocity, compact type.

- The system should have a low pressure loss that is not significantly affected in wet conditions.
- The aerodynamics of the intake should be carefully considered.

6.5.2.7 *Marine gas turbine designs*

Most marine gas turbines are derived from aircraft jet engines, whose development from scratch can cost over US$1 billion. Such an investment cannot be justified for the maritime industry but the extensive testing and service experience accumulated by aero engines provides an ideal basis for marine derivatives. Marine gas turbines are available only in specific sizes and ratings, unlike a given diesel engine design which can cover a wide power range with different inline and V-cylinder configurations. The economical application of gas turbines to marine propulsion therefore dictates matching the unit to the project, and hence calls for a technical and economic analysis of the vessel's proposed deployment.

Operational costs include availability for service and maintenance requirements, cost of consumables (fuel and lube oil) and manning costs. All these factors have to be considered for military ship designs but not to the extent that is called for in commercial shipping. Trade-offs among the various parameters can be conducted to determine the right fit of propulsion system to the application.

All marine gas turbines incorporate the same fundamental components (Figure 6.65):

- A compressor to draw in and compress atmospheric air.
- A combustion system into which fuel is injected, mixed with the compressed air, and burned.
- A compressor turbine which absorbs sufficient power from the hot gases to drive the compressor.
- A power turbine which absorbs the remaining energy in the hot gas stream and converts it into shaft power.

Lightweight aero-derived, rather than heavyweight industrial-derived, gas turbine designs are favoured for marine propulsion applications. The aero jet engine extracts only sufficient energy from the gas stream to drive the compressor and accessories, and releases the surplus gas at high velocity through a convergent nozzle to propel the aircraft by reaction.

When converting the jet engine into a shaft drive machine it is necessary to provide an additional free power turbine to absorb the energy left in the gas stream and to transmit that energy in the form of shaft power. The original jet engine is then termed the *Gas Generator* and the whole assembly becomes a *Gas Turbine*. Using a free power turbine reduces the starting power requirement as only the gas generator rotating assemblies have to be turned during the starting cycle. The resulting prime mover benefits from the intensive development and refinement of the original aero engine, both before and during service.

Certain aero engines are already fitted with power turbines and termed turbo-prop or turbo-shaft

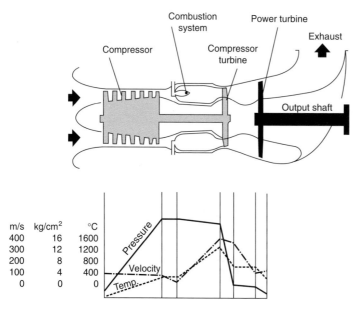

Figure 6.65 Main elements of an aero-derived gas turbine, and the pressure, temperature and air/gas speeds during the cycle.

machines; these are readily adaptable to industrial and marine use. Special technical and production techniques are applied to the re-designed gas turbine to ensure its suitability for sea level operation at high powers, and for the marine environment.

A basic gas generator has one rotating assembly: the compressor and its turbine coupled together. The characteristics of axial-flow compressors vary considerably, however, over the operating range from starting to full power. On some high compression ratio compressors it is necessary to fit automatic blow-off valves and to alter the angle of the inlet guide vanes and first stages of stator blades to ensure efficient operation.

To achieve the necessary stability in larger gas generators, the compressor is divided into two separate units: the low pressure and high pressure compressors, each driven by its own turbine through co-axial shafts. Each compressor is able to operate at its own optimum speed, giving flexibility of operation and efficient compression throughout the running range. Only the high pressure rotor needs to be turned during starting, and therefore even the largest gas generator can be started by battery.

Although such a two-spool compressor is very flexible, it is still necessary in some cases to adjust the inlet guide vanes to deal with the changing flow of air entering the compressor. This is effected automatically by pressure sensors acting upon rams which alter the angle of the guide vanes.

An axial-flow compressor consisting of alternate rows of fixed and rotating blades draws in atmospheric air through an air intake and forces it through a convergent duct formed by the compressor casing into an intermediate casing, where the compressed air is divided into separate flows for combustion and cooling purposes. A typical annular combustion chamber receives about 20% of the air flow for combustion into which fuel is injected and burned.

Initial ignition is executed by electrical igniters, which are switched off when combustion becomes self-sustaining. The resulting expanded gas is cooled by the remainder of the air flow which enters the combustion chamber via slots and holes to reduce the temperature to an acceptable level for entry into the one or more axial-flow stages of the turbine. The turbine drives the compressor, to which it is directly coupled. The remaining high velocity gases are exhausted and are available for use in the power turbine.

A power turbine of the correct 'swallowing' capacity is required to convert the gas flow into shaft power (Figure 6.66). Typically of one or more axial-flow stages, the power turbine may be arranged separately on a baseframe designed to mount the gas generator, to which the turbine is linked by a bellows joint to avoid the need for very accurate alignment and to allow for differential expansion. In some cases, however, expansion is allowed for in the design of the power turbine and the gas generator is mounted directly on to the inter-turbine duct. The power turbine, usually designed to last the life of the plant, is surrounded by an exhaust volute which passes the final exhaust to atmosphere through a stack.

Most marine gas turbines incorporate a free power turbine, which is free to rotate at whatever speed is dictated by the combination of the power output of the Gas Turbine Change Unit (GTCU) and the drive train behind it. The gas turbine is designed so that there are no critical speeds in the running range, and it can operate continuously with the power turbine at any speed from stopped to the full power rev/min of the power turbine. This facility is especially important under manoeuvring conditions where the

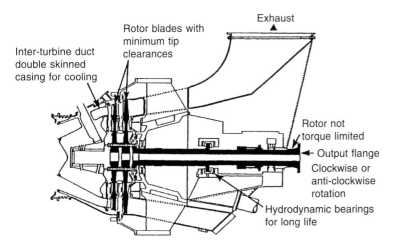

Figure 6.66 Power turbine of a Rolls-Royce Spey SMIC engine.

combination of a free power turbine and reversing waterjets provides a flexible and effective propulsion system.

Aero- and industrial-derived marine gas turbines are available with unit outputs ranging from around 2000 kW to over 50 000 kW. The larger designs (with ratings of 17 000 kW upwards) are reviewed by Woodyard (2004).

6.5.3 Steam turbines

6.5.3.1 Introduction

The steam turbine used to be the first choice for very large power main propulsion units. Its advantages of little or no vibration, low weight, minimal space requirements and low maintenance costs are considerable. Furthermore a turbine can be provided for any power rating likely to be required for marine propulsion. However, the higher specific fuel consumption when compared with a diesel engine, together with the weight and space of the boiler(s) and gearing, offsets these advantages, although refinements such as reheat have narrowed the gap. Steam turbines are of course still used for land-based electrical power generation and, in the marine field, for limited applications such as nuclear powered submarines, LNG carriers burning the boil-off gas and coal-fired ships.

The steam turbine is a device for obtaining mechanical work from the energy stored in steam. Steam enters the turbine with a high energy content and leaves after giving up most of it. The high pressure steam from the boiler is expanded in nozzles to create a high velocity jet of steam. The nozzle acts to convert heat energy in the steam into kinetic energy. This jet is directed into blades mounted on the periphery of a wheel or disc (Figure 6.67). The steam does not 'blow the wheel around'. The shaping of the blades causes a change in direction and hence velocity of the steam jet. Now a change in velocity for a given mass flow of steam will produce a force which acts to turn the turbine wheel, i.e. mass flow of steam (kg/s) × change in velocity (m/s) = force (kgm/s^2).

This is the operating principle of all steam turbines, although the arrangements may vary considerably. The steam from the first set of blades then passes to another set of nozzles and then blades and so on along the rotor shaft until it is finally exhausted. Each set comprising nozzle and blades is called a stage.

6.5.3.2 Turbine types

There are two main types of turbine, the 'impulse' and the 'reaction'. The names refer to the type of force which acts on the blades to turn the turbine wheel.

Impulse
The impulse arrangement is made up of a ring of nozzles followed by a ring of blades. The high pressure, high energy steam is expanded in the nozzle to a lower pressure, high velocity jet of steam. This jet of steam is directed into the impulse blades and leaves in a different direction (Figure 6.68). The changing direction and therefore velocity produces an impulsive force which mainly acts in the direction of rotation of the turbine blades. There is only a very small end thrust on the turbine shaft.

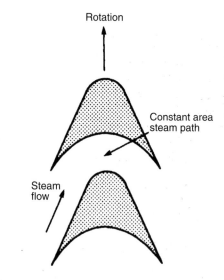

Figure 6.68 Impulse blading.

Reaction
The reaction arrangement is made up of a ring of fixed blades attached to the casing, and a row of similar blades mounted on the rotor, i.e. moving

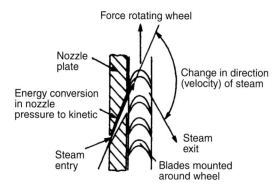

Figure 6.67 Energy conversion in a steam turbine.

blades (Figure 6.69). The blades are mounted and shaped to produce a narrowing passage which, like a nozzle, increases the steam velocity. This increase in velocity over the blade produces a reaction force which has components in the direction of blade rotation and also along the turbine axis. There is also a change in velocity of the steam as a result of a change in direction and an impulsive force is also produced with this type of blading. The more correct term for this blade arrangement is 'impulse-reaction'.

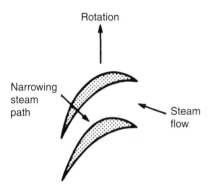

Figure 6.69 Reaction blading.

Compounding

Compounding is the splitting up, into two or more stages, of the steam pressure or velocity change through a turbine.

Pressure compounding of an impulse turbine is the use of a number of stages of nozzle and blade to reduce progressively the steam pressure. This results in lower or more acceptable steam flow speeds and a better turbine efficiency.

Velocity compounding of an impulse turbine is the use of a single nozzle with an arrangement of several moving blades on a single disc. Between the moving blades are fitted guide blades which are connected to the turbine casing. This arrangement produces a short lightweight turbine with a poorer efficiency which would be acceptable in, for example, an astern turbine.

The two arrangements may be combined to give what is called 'pressure-velocity compounding'.

The reaction turbine as a result of its blade arrangement changes the steam velocity in both fixed and moving blades with consequent gradual steam pressure reduction. Its basic arrangement therefore provides compounding.

The term 'cross-compound' is used to describe a steam turbine unit made up of a high pressure and a low pressure turbine (Figure 6.70). This is the

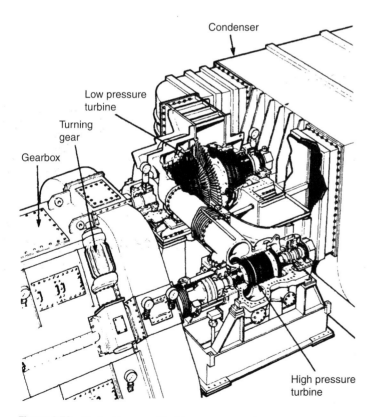

Figure 6.70 Cross-compound turbine arrangement.

usual main propulsion turbine arrangement. The alternative is a single cylinder unit which would be usual for turbo-generator sets, although some have been fitted for main propulsion service.

Reheat

Reheating is a means of improving the thermal efficiency of the complete turbine plant. Steam, after expansion in the high pressure turbine, is returned to the boiler to be reheated to the original superheat temperature. It is then returned to the turbine and further expanded through any remaining stages of the high pressure turbine and then the low pressure turbine.

Named turbine types

A number of famous names are associated with certain turbine types.

Parsons. A reaction turbine where steam expansion takes place in the fixed and moving blades. A stage is made up of one of each blade type. Half of the stage heat drop occurs in each blade type, therefore providing 50% reaction per stage.

Curtis. An impulse turbine with more than one row of blades to each row of nozzles, i.e. velocity compounded.

De Laval. A high speed impulse turbine which has only one row of nozzles and one row of blades.

Rateau. An impulse turbine with several stages, each stage being a row of nozzles and a row of blades, i.e. pressure compounded.

6.5.3.3 Astern arrangements

Marine steam turbines are required to be reversible. This is normally achieved by the use of several rows of astern blading fitted to the high pressure and low pressure turbine shafts to produce astern turbines. About 50% of full power is achieved using these astern turbines. When the turbine is operating ahead the astern blading acts as an air compressor, resulting in windage and friction losses.

6.5.3.4 Turbine construction

The construction of an impulse turbine is shown in Figure 6.71. The turbine rotor carries the various wheels around which are mounted the blades. The steam decreases in pressure as it passes along the shaft and increases in volume requiring progressively larger blades on the wheels. The astern turbine is mounted on one end of the rotor and is much shorter than the ahead turbine. The turbine rotor is supported by bearings at either end; one bearing incorporates a thrust collar to resist any axial loading.

The turbine casing completely surrounds the rotor and provides the inlet and exhaust passages for the

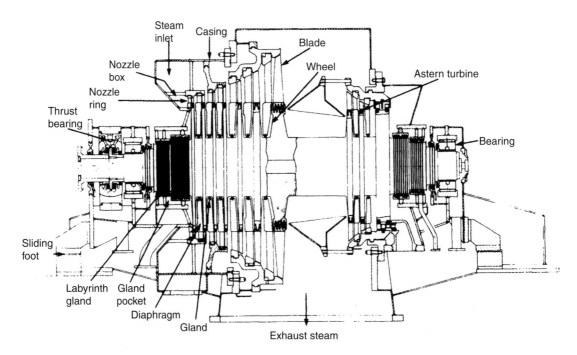

Figure 6.71 Impulse turbine.

steam. At the inlet point a nozzle box is provided which by use of a number of nozzle valves admits varying amounts of steam to the nozzles in order to control the power developed by the turbine. The first set of nozzles is mounted in a nozzle ring fitted into the casing. Diaphragms are circular plates fastened to the casing which are fitted between the turbine wheels. They have a central circular hole through which the rotor shaft passes. The diaphragms contain the nozzles for steam expansion and a gland is fitted between the rotor and the diaphragm.

The construction of a reaction turbine differs somewhat in that there are no diaphragms fitted and instead fixed blades are located between the moving blades.

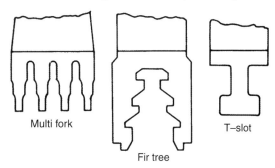

Figure 6.72 Blade fastening.

Rotor

The turbine rotor acts as the shaft which transmits the mechanical power produced to the propeller shaft via the gearing. It may be a single piece with the wheels integral with the shaft or built up from a shaft and separate wheels where the dimensions are large.

The rotor ends adjacent to the turbine wheels have an arrangement of raised rings which form part of the labyrinth gland sealing system. Journal bearings are fitted at each end of the rotor. These have rings arranged to stop oil travelling along the shaft which would mix with the steam. One end of the rotor has a small thrust collar for correct longitudinal alignment. The other end has an appropriate flange or fitting arranged for the flexible coupling which joins the rotor to the gearbox pinion.

The blades are fitted into grooves of various designs cut into the wheels.

Blades

The shaping and types of turbine blades have already been discussed. When the turbine rotor is rotating at high speed the blades will be subjected to considerable centrifugal force and variations in steam velocity across the blades will result in blade vibration.

Expansion and contraction will also occur during turbine operation, therefore a means of firmly securing the blades to the wheel is essential. A number of different designs have been employed (Figure 6.72).

Fitting the blades involves placing the blade root into the wheel through a gate or entrance slot and sliding it into position. Successive blades are fitted in turn and the gate finally closed with a packing piece which is pinned into place. Shrouding is then fitted over tenons on the upper edge of the blades. Alternatively, lacing wires may be passed through and brazed to all the blades.

End thrust

In a reaction turbine a considerable axial thrust is developed. The closeness of moving parts in a high speed turbine does not permit any axial movement to take place: the axial force or end thrust must therefore be balanced out.

One method of achieving this balance is the use of a dummy piston and cylinder. A pipe from some stage in the turbine provides steam to act on the dummy piston which is mounted on the turbine rotor (Figure 6.73). The rotor casing provides the cylinder to enable the steam pressure to create an axial force on the turbine shaft. The dummy piston annular area and the steam pressure are chosen to produce a force which exactly balances the end thrust from the reaction blading. A turbine with ahead and astern blading will have a dummy piston at either end to ensure balance in either direction of rotation.

Another method often used in low pressure turbines is to make the turbine double flow. With this arrangement steam enters at the centre of the shaft and flows along in opposite directions. With an equal division of steam the two reaction effects balance and cancel one another.

Taylor (1996) goes on to describe turbine glands, nozzles, bearings, lubrication, turbine control and gearing. He also includes a chapter on boilers.

6.6 Auxiliary machinery and equipment

Machinery, other than the main propulsion unit, is usually called 'auxiliary', even though it may be in direct support of the main machinery. Typical auxiliary machinery and equipment includes feed systems, pumps and pumping, air compressors, heat exchangers, distillation equipment, oil/water separators, sewage treatment plants and incinerators. All of these items of auxiliary machinery are covered in some detail by McGeorge (1995) and Taylor (1996). This Section provides, as examples, descriptions of ship service systems, shafting and propellers and steering gears.

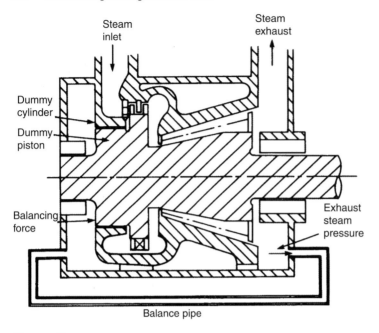

Figure 6.73 Dummy piston balance arrangement.

6.6.1 Ship service systems

Some of the equipment in the machinery space is dedicated to servicing the ship in general and providing amenities for personnel or passengers. Thus the bilge system is available to clear oil/water leakage and residues from machinery and other spaces as well as to provide an emergency pumping capability. The domestic water and sewage systems provide amenities for personnel.

6.6.1.1 Bilge systems

The essential purpose of a bilge system is to clear water from the ship's 'dry' compartments, in emergency. The major uses of the system are for clearing water and oil which accumulates in machinery space bilges as the result of leakage or draining, and when washing down dry cargo holds. The bilge main in the engine room has connections from dry cargo holds, tunnel and machinery spaces. Tanks for liquid cargo and ballast are served by cargo discharge systems and ballast systems respectively. They are not connected to the bilge system unless they have a double function, as for example with deep tanks that are used for dry cargo or ballast. Spectacle blanks or change over chests are fitted to connect/isolate spaces of this kind, as necessary. Accommodation spaces are served by scuppers with non-return valves which are fitted at the ship's side.

Bilge system regulations
Regulations prescribe the requirements for bilge systems and the details of a proposed arrangement must be submitted for approval to the appropriate government department or Classification Society. The number of power operated bilge pumps (usually three or four) that are required in the machinery spaces is governed by the size and type of ship. For smaller vessels one of the pumps may be main engine driven but the other must be independently driven. A bilge ejector is acceptable as a substitute provided that, like the pumps, it is capable of giving an adequate flow rate. At least 120 m/min (400 ft/min) through the pipe is a figure that has been required. Pipe cross section is also governed by the rules, which means that this, combined with linear flow, dictates a discharge rate. Bilge ejectors are supplied with high pressure sea water from an associated pump.

The diameters of bilge main and branch pipes are found as stated above from formulae based on ship size and the Classification Societies generally prescribe the bore of the main bilge line and branch bilge lines and relate the bilge pump capacity of each pump to that required to maintain a minimum water speed in the line. Fire pump capacity is related to the capacity of the bilge pump thus defined:

Bilge main dia. $d_1 = 1.68 \sqrt{L(B+D)} + 25$ mm
Branch dia. $d_2 = 2.16 \sqrt{C(B+D)} + 25$ mm
d_2 not to be less than 50 mm and need not exceed 100 mm.
d_1 must never be less than d_2

where

L = length of ship in m;
B = breadth of ship in m;
D = moulded depth at bulkhead deck in m;
C = length of compartment in m.

Each pump should have sufficient capacity to give a water speed of 122 m/min through the Rule size mains of this bore. Furthermore each bilge pump should have a capacity of not less than:

$$\frac{0.565}{10^3} d_1^2 \text{ m}^3/\text{h}$$

The fire pumps, excluding any emergency fire pump fitted, must be capable of delivering a total quantity of water at a defined head not less than two-thirds of the total bilge pumping capacity. The defined head ranges from 3.2 bar in the case of passenger ships of 4000 tons gross or more to 2.4 bar for cargo ships of less than 1000 tons gross.

Pumps installed for bilge pumping duties must be self-priming or able to be primed. The centrifugal type with an air pump is suitable and there are a number of rotary self-priming pumps available. Engine driven pumps are usually of the reciprocating type and there are still in use many pumps of this kind driven by electric motors through cranks.

The bilge pumps may be used for other duties such as general service, ballast and fire-fighting, which are intermittent. The statutory bilge pumps may not be used for continuous operation on other services such as cooling, although bilge injections can be fitted on such pumps and are a requirement on main or stand-by circulating pumps.

Common suction and discharge chests permit one pump to be used for bilge and ballast duties. The pipe systems for these services must, however, be separate and distinct. The ballast piping has screw lift valves so as to be able to both fill and empty purpose-constructed tanks with sea water. The bilge system is designed to remove water or oily water from 'dry' spaces throughout the vessel and is fitted with screw-down non-return valves to prevent any flooding back to the compartment served. The two could not be connected because they are incompatible. At the pump suction chest, the bilge valve must be of the screw-down non-return type to prevent water from entering the bilge line from sea water or ballast suctions.

Materials which can be used are also given in the construction rules. When steel is used, it requires protection inside and out and both surfaces should be galvanized. The preparation of the surfaces for galvanizing is important as is the continuity of the coating. The external painting of steel pipes may be the only protection used to prevent rust arising from contact with water in the bilges. Flanged joints are made between sections of pipe and support must be adequate. Branch, direct and emergency bilge suctions are provided to conform with the regulations and as made necessary by the machinery space arrangement.

Bilge and ballast system layout
In the system shown (Figure 6.74), the bilge main has suctions from the port and starboard sides of the engine room, from the tunnel well and from the different cargo holds. There are three pumps shown connected to the bilge main. These are the fire and bilge pump, the general service pump and the auxiliary bilge pump. These pumps also have direct bilge suctions to the engine room port side, starboard side and tunnel well respectively. The ballast pump (port side for'd) could be connected to the bilge main but is shown with an emergency bilge suction only. The main sea-water circulating pump at the starboard side of the machinery space also has an emergency suction. This emergency suction or the one on the ballast pump is required by the regulations. The ballast pump is self-priming and can serve as one of the required bilge pumps as well as being the stand-by sea-water circulating pump.

The auxiliary bilge pump is the workhorse of the system and need not be one of the statutorily required bilge pumps. For this installation, it is a low capacity, smooth flow pump which is suited for use in conjunction with the oily/water separator. All bilge suctions have screw-down non-return valves with strainers or mud boxes at the bilge wells. Oily bilges and purifier sludge tanks have suitable connections for discharge to the oily water separator or ashore.

The system is tailored to suit the particular ship. Vessels with open floors in the machinery space may have bilge suctions near the centre line and in such cases, wing suctions would not be necessary provided the rise of floor was sharp enough.

The essential safety role of the bilge system means that bilge pumps must be capable of discharging directly overboard. This system is also used when washing down dry cargo spaces.

When clearing the water and oil which accumulates in machinery space bilges, the discharge overboard must be via the oily/water separator and usually with the use of the special bilge pump, i.e. the auxiliary bilge pump of the system shown.

The following paragraphs are extracted from the International Convention for the Safety of Life at Sea 1974 Chapter 11–1 Regulation 18 which relates to passenger ships:

The arrangement of the bilge and ballast pumping system shall be such as to prevent the possibility of water passing from the sea and from water ballast spaces into the cargo and machinery spaces, or from one compartment to another. Special provision shall be made to prevent any deep tank having bilge and ballast connections being inadvertently run up from the sea when containing cargo, or pumped out through a bilge pipe when containing water ballast.

Provision shall be made to prevent the compartment served by any bilge suction pipe being flooded in the event of the pipe being severed, or otherwise damaged by collision or grounding in any other compartment. For this purpose, where the pipe is at any part situated nearer the side of the ship than one-fifth the breadth of the ship (measured at right angles to the centre line at the level of the deepest subdivision load line), or in a duct keel, a non-return valve shall be fitted to the pipe in the compartment containing the open end.

All the distribution boxes, cocks and valves in connection with the bilge pumping arrangements shall be in positions which are accessible at all times under ordinary circumstances. They shall be so arranged that, in the event of flooding, one of the bilge pumps may be operative on any compartment; in addition, damage to a pump or its pipe connecting to the bilge main outboard of a line drawn at one-fifth of the breadth of the ship shall not put the bilge system out of action. If there is only one system of pipes common to all the pumps, the necessary cocks or valves for controlling the bilge suctions must be capable of being operated from above the bulkhead deck. Where in addition to the main bilge pumping system an emergency bilge pumping system is provided, it shall be independent of the main system and so arranged that a pump is capable of operating on any compartment under flooding condition; in that case only the cocks and valves necessary for the operation of the emergency system need be capable of being operated from above the bulkhead deck.

All cocks and valves mentioned in the above paragraph of this Regulation which can be operated from above the bulkhead deck shall have their controls at their place of operation clearly marked and provided with means to indicate whether they are open or closed.

6.6.1.2 Oil/water separators

Oil/water separators are necessary aboard vessels to prevent the discharge of oil overboard mainly when pumping out bilges. They also find service when deballasting or when cleaning oil tanks. The requirement to fit such devices is the result of international legislation. Legislation was needed because free oil and oily emulsions discharged in a waterway can interfere with natural processes such as photosynthesis and re-aeration, and induce the destruction of the algae and plankton so essential to fish life. Inshore discharge of oil can cause damage to bird life and mass pollution of beaches. Ships found discharging water containing more than 100 mg/litre of oil or discharging more than 60 litres of oil per nautical mile can be heavily fined, as also can the ship's Master.

In consequence it is important that an oil/water separator is correctly installed, used and maintained. It is generally accepted that oil is less dense than water and this is the basis of the design of devices to separate the two liquids. Some of the modern heavy fuels however, have a density at 15°C which approaches, is the same as or is even higher than that of water and this has added to the problems of separation in oil/water separators and in centrifuges. The operation of oil/water separators relies heavily on gravity and a conventional difference in densities. Centrifuges by their speed of rotation, exert a force many times that of gravitational effect and the heater reduces density in comparison with that of water.

Oil/water separators and centrifuges are both employed for the purpose of separating oil and water but there are major differences. Oil/water separators are required to handle large quantities of water from which usually, small amounts of oil must be removed. Various features are necessary to aid removal of the oil from the large bulk of water particularly when the difference in densities is small.

Centrifuges are required to remove (again usually) small quantities of water from a much larger amount of oil. Additionally the centrifuge must separate solids and it must, with respect to fuel, handle large quantities at the rate at which the fuel is consumed.

Principle of operation

The main principle of separation by which commercially available oil/water separators function, is the gravity differential between oil and water.

In oily water mixtures, the oil exists as a collection of globules of various sizes. The force acting on such a globule, causing it to move in the water is proportional to the difference in weight between the oil particle and a particle of water of equal volume. This can be expressed as:

$$F_s = \frac{\pi}{6} D^3 (\rho_w - \rho_o) g \quad (6.7)$$

where:

F_s = separating force
ρ_w = density of water
ρ_o = density of oil
D = diameter of oil globule
g = acceleration due to gravity

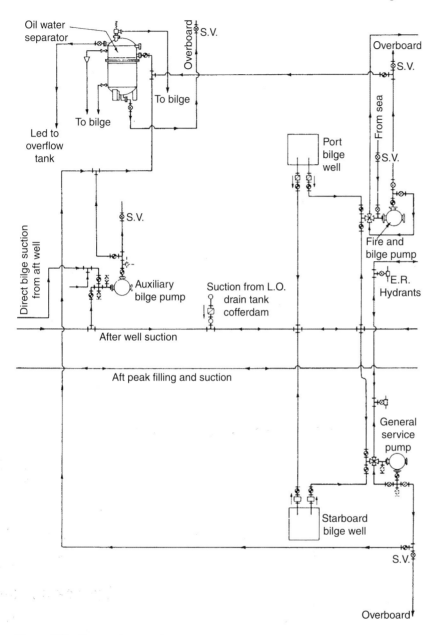

Figure 6.74 Bilge, ballast and fuel main.

The resistance to the movement of the globule depends on its size and the viscosity of the fluids. For small particles moving under streamline flow conditions, the relationship between these properties can be expressed by Stoke's Law:

$$F_r = 3 \pi v \mu d \qquad (6.7)$$

where:

F_r = resistance to movement
μ = viscosity of fluid
v = terminal velocity of particle
d = diameter of particle.

When separation of an oil globule in water is taking place F_s will equal F_r and the above equations can be worked to express the relationship of the terminal (or in this case rising) velocity of the globule with viscosity, relative density and particle size:

$$v = \left(\frac{g}{18\mu}\right)(\rho_w - \rho_o) d^2 \qquad (6.8)$$

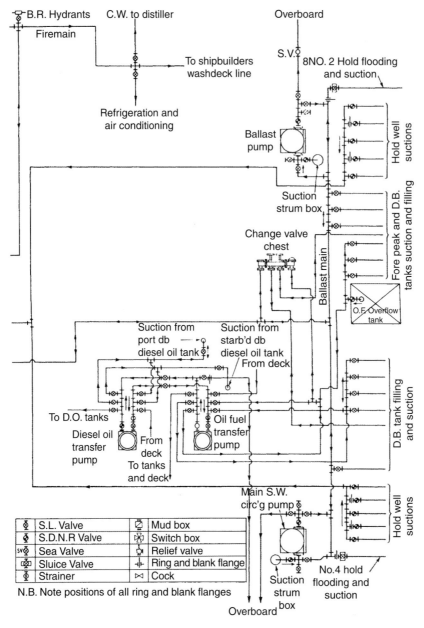

Figure 6.74 (Continued)

In general, a high rate of separation is encouraged by a large size of oil globule, elevated temperature of the system (which increases the specific gravity differential of the oil and water and reduces the viscosity of the oil) and the use of sea water. Turbulence or agitation should be avoided since it causes mixing and re-entrainment of the oil. Laminar or streamlined flow is beneficial.

In addition to the heating coils provided to optimize separation, there are various other means used to improve and speed up operation. The entrance area in oil/water separators is made large so that flow is slow and large slugs of oil can move to the surface quickly. (The low capacity pump encourages slow and laminar flow.) Alternation of flow path in a vertical direction continually brings oil near to the surface, where separation is enhanced by weirs which reduce liquid depth. Angled surfaces provide areas on which oil can accumulate and form globules, which then float upwards. Fine gauze screens are also used as coalescing or coagulating surfaces.

Pumping considerations

A faster rate of separation is obtained with large size oil globules or slugs and any break up of oil globules in the oily feed to the separator should be avoided. This factor can be seriously affected by the type and rating of the pump used. Tests were carried out by a British government research establishment some years ago on the suitability of various pumps for separator feed duties and the results are shown in Table 6.5.

Table 6.5 Pump suitability for oil/water separator duty.

Type	Remarks
Double vane Triple screw Single vane Rotary gear	Satisfactory at 50% derating
Reciprocating Hypocycloidal	Not satisfactory: modification may improve efficiencies to 'satisfactory' level
Diaphragm Disc and shoe Centrifugal Flexible vane	Unsatisfactory

It follows that equal care must be taken with pipe design and installation to avoid turbulence due to sharp bends or constrictions and to calculate correctly liquid flow and pipe size to guarantee laminar flow.

The Simplex-Turbulo oil/water separator

The Simplex-Turbulo oil/water separator (Figure 6.75) consists of a vertical cylindrical pressure vessel containing a number of inverted conical plates. The oily water enters the separator in the upper half of the unit and is directed downwards to the conical plates. Large globules of oil separate out in the upper part of the separator. The smaller globules are carried by the water into the spaces between the plates. The rising velocity of the globules carries them upwards where they become trapped by the under-surfaces of the plates and coalesce until the enlarged globules have sufficient rising velocity to travel along the plate surface and break away at the periphery. The oil rises, is caught underneath an annular baffle and is then led up through the turbulent inlet area by risers to collect in the dome of the separator. The water

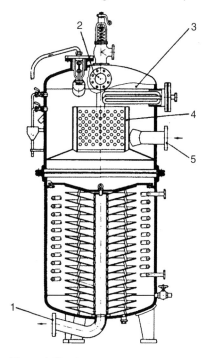

Figure 6.75 Simplex-Turbulo oil/water separator.
1. Clean water run-off connection
2. Outlet
3. Oil accumulation space
4. Riser pipes
5. Inlet connection

leaves the conical plate pack via a central pipe which is connected to a flange at the base of the separator.

Two test cocks are provided to observe the depth of oil collected in the separator dome. When oil is seen at the lower test cock, the oil drain valve must be opened. An automatic air release valve is located in the separator dome. An electronically operated oil drainage valve is also frequently fitted. This works on an electric signal given be liquid level probes in the separator. Visual and audible oil overload indicators may also be fitted. To assist separation steam coils or electric heaters are fitted in the upper part of the separator. Where high viscosity oils are to be separated additional heating coils are installed in the lower part.

Before initial operation, the separator must be filled with clean water. To a large extent the conical plates are self-cleaning but periodically the top of the vessel should be removed and the plates examined for sludge build-up and corrosion. It is important that neither this separator nor any other type is run at over capacity. When a separator is overloaded the flow becomes turbulent, causing re-entrainment of the oil and consequent deterioration of the effluent quality.

To meet the requirement of legislation which came into force in October 1983 and which requires that the oil content of bilge discharges be reduced in

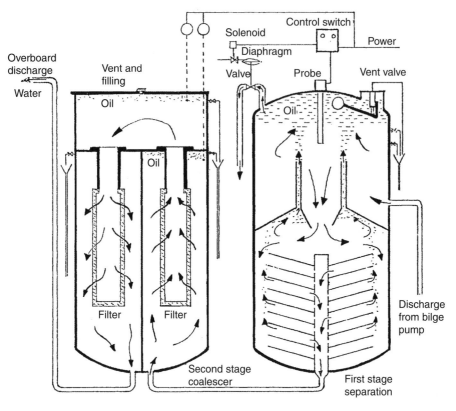

Figure 6.76 Simplex-Turbulo oil/water separator with coalescer.

general to 100 ppm and to 15 ppm in special areas and within 12 nautical miles of land, a second stage coalescer (Figure 6.76) was added in some designs. Filter elements in the second stage remove any small droplets of oil in the discharge and cause them to be held until they form larger droplets (coalesce). As the larger globules form, they rise to the oil collecting space.

Oil content monitoring

In the past, an inspection glass, fitted in the overboard discharge pipe of the oil/water separator permitted sighting of the flow. The discharge was illuminated by a light bulb fitted on the outside of the glass port opposite the viewer. The separator was shut down if there was any evidence of oil carry over, but problems with observation occurred due to poor light and accumulation of oily deposits on the inside of the glasses.

Present-day monitors are based on the same principle. However, whilst the eye can register anything from an emulsion to globules of oil a light-sensitive photo-cell detector cannot. Makers may therefore use a sampling and mixing pump to draw

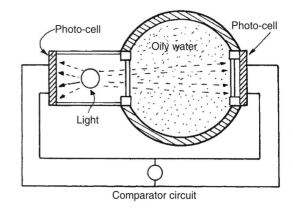

Figure 6.77 Monitor for oily water using direct light.

a representative sample with a general opaqueness more easily registered by the simple photo-cell monitor. Flow through the sampling chamber is made rapid to reduce deposit on glass lenses. They are easily removed for cleaning.

Bilge or ballast water passing through a sample chamber can be monitored by a strong light shining directly through it and on to a photo-cell (Figure 6.77).

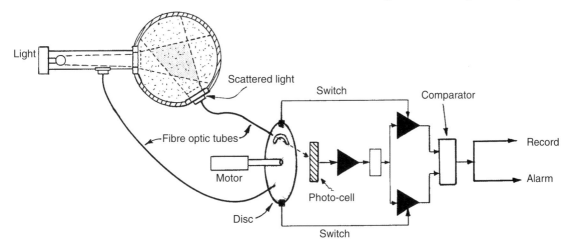

Figure 6.78 Monitor based on scattered light (courtesy Sofrance).

Light reaching the cell decreases with increasing oil content of the water. The effect of this light on the photo-cell compared with that of direct light on the reference cell to the left of the bulb, can be registered on a metre calibrated to show oil content.

Another approach is to register light scattered by oil particles dispersed in the water by the sampling pumps (Figure 6.78). Light reflected or scattered by any oil particles in the flow, illuminates the scattered light window. This light when compared with the source light increases to a maximum and then decreases with increasing oil content of the flow. Fibre optic tubes are used in the device shown to convey light from the source and from the scattered light window to the photo-cell. The motor-driven rotating disc with its slot, lets each light shine alternately on the photo-cell and also, by means of switches at the periphery, causes the signals to be passed independently to a comparator device.

These two methods briefly described, could be used together to improve accuracy, but they will not distinguish between oil and other particles in the flow. Methods of checking for oil by chemical test would give better results but take too long in a situation where excess amounts require immediate shut down of the oily water separator.

Tanker ballast

Sampling and monitoring equipment fitted in the pump room of a tanker can be made safe by using fibre optics to transmit light to and from the sampling chamber (Figure 6.79). The light source and photo-cell can be situated in the cargo control room together with the control, recording and alarm console. The sampling pump can be fitted in the pumproom to keep the sampling pipe short and so minimize time delay. For safety the drive motor is fitted in the machinery space, with the shaft passing through a gas-tight seal in the bulkhead.

Oil content reading of the discharge is fed into the control computer together with discharge rate and ship's speed to give a permanent record. Alarms, automatic shutdown, back-flushing and recalibration are incorporated.

6.6.1.3 *Ballast arrangements*

The ballasting of a vessel which is to proceed without cargo to the loading port is necessary for a safe voyage, sometimes in heavy weather conditions. On arrival at the port the large amount of ballast must be discharged rapidly in readiness for loading. Ballast pump capacity is governed by the volume of water that has to be discharged in a given time. The ballast pump is often also the stand-by sea-water circulating pump (Figure 6.74) but very large ballast discharge capacity is necessary for some ships. Vessels with tanks available for either ballast or oil fuel are fitted with a change-over chest or cock designed to prevent mistakes. An oily water separator on the ballast pump discharge would prevent discharge of oil with the ballast from a tank that had been used for fuel or oil cargo.

Ballast carried in the empty cargo tanks of crude oil carriers has potential for pollution when discharged, particularly if cargo pumps are used for the purpose. Only very large oil/water separators have the capacity to reduce this pollution. Segregated ballast tanks with dedicated ballast pumps prevent the problem.

Fore and aft peak tanks, double bottom and deep tanks used for ballast in dry cargo vessels as well

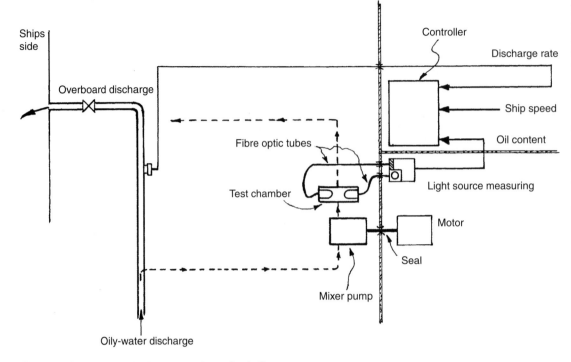

Figure 6.79 Seres monitoring system for tanker ballast.

as ballast spaces in bulk liquid carriers, can be dangerous due to lack of oxygen or the presence of harmful gases. Oxygen may be depleted by corrosion and harmful gases may be produced by organisms or pollutants in the water. The ballast water from some areas has been found to carry dangerous bacteria.

Ballast tank air and overflow pipes must be of the required size relative to the filling lines, that is, 25% greater in area and, in any case, not less than 50 mm bore. They are fitted at the highest part of the tank or at the opposite end to the filling connection. Tanks used for fuel storage also have to fulfil the requirements for fuel tanks. Nameplates are attached to the tops of all air pipes and sounding pipes must have means of identification. The latter are to be of steel with a striker plate at the bottom and must conform to the various rulings. The pipelines for ballasting must be of adequate strength and if of steel, protected by galvanizing or other means. The ballasting of some tanks, such as those in the double bottoms, is carried out by running up by opening appropriate valves, rather than by pumping. Remotely operated valves are installed with modern ballast systems. Pump and valve controls are then centrally located.

Centrifugal pumps with water ring primers, used for ballast pumping, are suitable for use as statutory bilge pumps.

6.6.1.4 Domestic water systems

Systems using gravity tanks to provide a head for domestic fresh and sanitary water, have long been superseded by schemes where supply pressure is maintained by a cushion of compressed air in the service tanks (Figure 6.80). The trade name Pneupress is commonly used to describe the tanks and system.

Fresh water

The fresh water is supplied to the system by one of two pumps which are self-priming or situated at a lower level than the storage tanks. The pump starters are controlled by pressure switches which operate when pressure in the service tank varies within pre-determined limits as water is used. The pump discharges through filters to a rising main, branched to give cold and hot supplies, the latter through a calorifier. A circulating pump may be fitted in circuit with the steam or electrically heated calorifier. An ultra-violet light sterilizer is fitted adjacent to the Pneupress tank of some systems. Ultra-violet light acts in such an arrangement, as a point of use biocide. Although effective as a means of killing bacteria, it does not apparently provide protection in the long term. The Department of Transport requirement for protection of fresh water in storage

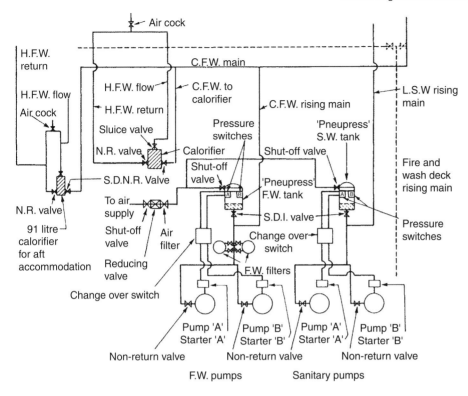

Figure 6.80 Domestic fresh and sanitary water system.

tanks, is that chlorine dosing or the Electro-Katadyn method, be used. Guidance on the procedures to ensure that fresh water is safe for consumption is provided by M notices listed in the References.

Sanitary water
The sanitary system operates on the same Pneupress principle as that described for fresh water. Pumps, if supplying sea water, are protected by filters on the suction side which require regular cleaning. A few sanitary systems use fresh or distilled water to reduce corrosion in pipes and flushing valves, particularly in vacuum systems where water consumption is minimal. Treated liquid effluent is recirculated in the chemical sewage treatment system described later in the next Section; this also operates with a Pneupress system.

Water production
A considerable amount of fresh water is consumed in a ship. The crew uses on average about 70 litre/person/day and in a passenger ship, consumption can be as high as 225 litre/person/day. Water used in the machinery spaces as make up for cooling system losses may be fresh or distilled but distilled water is essential for steam plant where there is a water tube boiler. Steamship consumption for the propulsion plant and hotel services can be as high as 50 tonnes/day.

It is now common practice to take on only a minimal supply of potable water in port and to make up the rest by distillation of sea water. The saved storage capacity for water is available for cargo and increases the earning power of the ship. A vessel which carries sufficient potable water for normal requirements is required, if ocean-going, to carry distillation plant for emergency use.

Modern low pressure evaporators and reverse osmosis systems give relatively trouble-free operation particularly in comparison with the types that were fitted in older ships. They are sufficiently reliable to provide, during continuous and unattended operation, the water needed for the engine room and domestic comsumption. An advantage of low pressure evaporators is that they enable otherwise wasted heat from diesel engine jacket cooling water to be put to good use.

Reverse osmosis systems were installed to give instant water production capacity without extensive modifications (as with vessels commandeered for hostilities in the Falklands War). They are used to advantage on some passenger cruise vessels and are fitted in ships which may remain stopped at sea for various reasons (tankers awaiting orders – outside 20 mile limit).

Warning is given in M Notice M620 that evaporators must not be operated within 20 miles of a coastline and that this distance should be greater in some circumstances. Pollution is present in inshore waters from sewage outfalls, disposal of chemical wastes from industry, drainage of fertilizers from the land and isolated cases of pollution from grounding or collision of ships and spillage of cargo.

Low pressure evaporators

The main object of distillation is to produce water essentially free of salts. Potable water should contain less than 500 mg/litre of suspended solids. Good quality boiler feed will contain less than 2.5 mg/litre. Sea water has a total dissolved solids content in the range 30 000–42 000 mg/litre, depending on its origin but the figure is usually given as 32 000 mg/litre.

Low pressure evaporators for the production of water can be adapted for steamships but operate to greatest advantage with engine cooling water on motorships. The relatively low temperature jacket water entering at about 65°C and leaving at about 60°C will produce evaporation because vacuum conditions reduce the boiling temperature of sea water from 100°C to less than 45°C.

The single effect, high vacuum, submerged tube evaporator shown in Figure 6.81 is supplied with diesel engine cooling water as the heating medium. Vapour evolved at a very rapid rate by boiling of the sea-water feed, tends to carry with it small droplets of salt water which must be removed to avoid contamination of the product. The demister of knitted monel metal wire or polypropylene collects the salt-filled water droplets as they are carried through by the air. These coalesce forming drops large enough to fall back against the vapour flow.

Evaporation of part of the sea water leaves a brine the density of which must be controlled by continual removal through a brine ejector or pump. Air and other gases released by heating of the sea water, but which will not condense, are removed by the air ejector. The evaporator shown has a single combined ejector for extraction of both brine and air.

One of the gases liberated is CO_2 from calcium bi-carbonate in the sea water. Loss of carbon dioxide from calcium bi-carbonate leaves plain calcium carbonate which has poor solubility and a tendency to form soft, white scale. Other potential scale-forming salts are calcium sulphate and magnesium compounds.

Scale is not a major problem where submerged heating coils reach a temperature of only 60°C. This heat is too low for formation of magnesium scales and provided brine density is controlled, calcium sulphate will not cause problems. Continuous removal of the brine by the brine pump or ejector limits density. Approximately half of the sea-water feed is converted into distilled water, the quantity of brine extracted is equivalent to the remainder of the feed delivered. The level of water in the evaporator is maintained constant by means of a brine weir over which excess passes to the ejector.

The small quantity of soft calcium carbonate scale can be removed by periodic cleaning with a commercially available agent or the evaporator can be continually dosed with synthetic polymer to bind the scale-forming salts into a 'flocc' which mostly discharges with the brine. Use of continuous treatment will defer acid cleaning to make it an annual exercise. Without continuous treatment, cleaning may be necessary after perhaps two months. Steam heated evaporators with their higher heating surface temperature, benefit more from chemical dosing, because magnesium scales form when surfaces are at 80°C or more.

Salinometer

The condensate or product, if of acceptable quality, is delivered to the appropriate tanks by the distilled water pump. Quality is continuously tested by the salinometer both at start up and during operation. If the device registers an excess of salinity it will dump the product and activate the alarm using its solenoid valves. The product is recirculated in some installations.

The electric salinometer

Pure distilled water may be considered a non-conductor of electricity. The addition of impurities such as salts in solution increases the conductivity of the water, and this can be measured. Since the conductivity of the water is, for low concentrations, related to the impurity content, a conductivity meter can be used to monitor the salinity of the water. The instrument can be calibrated in units of conductivity (micromhos) or directly in salinity units (older instruments in grains/gall., newer instruments in ppm or mg/litre) and it is on this basis that electric salinometers (Figure 6.82) operate. The probe type electrode cell (Figure 6.83) is fitted into the pipeline from the evaporator, co-axially through a retractable valve which permits it to be withdrawn for examination and cleaning. The cell cannot be removed while the valve is open and consists of two stainless steel concentric electrodes having a temperature compensator located within the hollow inner electrode. It operates within the limits of water

Marine engines and auxiliary machinery 429

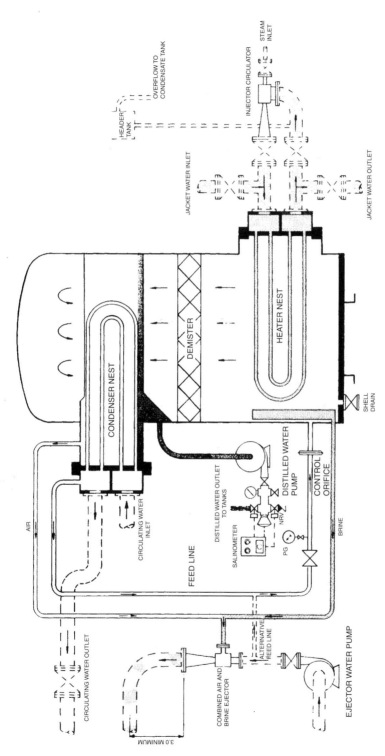

Figure 6.81 High vacuum, submerged tube evaporatory (movac Mk2 – Caird & Rayner).

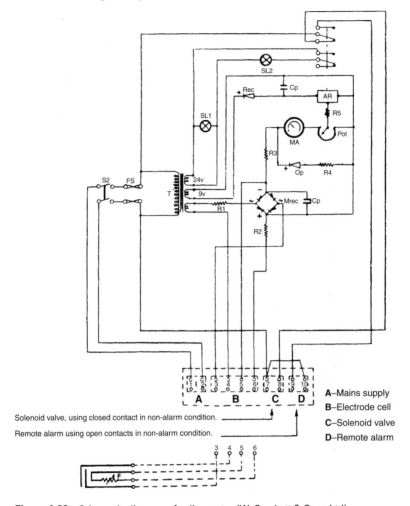

Figure 6.82 Schematic diagram of salinometer (W. Crockatt & Sons Ltd).

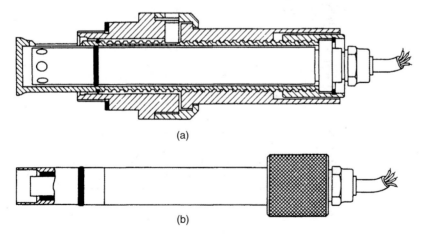

Figure 6.83 Probe type electrode cell.

pressure up to 10.5 bar and water temperatures between 15° and 110°C.

The incoming a.c. mains from control switch S2 through fuses FS feed transformer T. A pilot lamp SL1 on the 24 V secondary winding indicates the circuit is live.

The indicating circuit comprises an applied voltage across the electrode cell and the indicator. The indicator shows the salinity by measuring the current which at a preset value actuates the alarm circuit warning relay. The transformer cell tapped voltage is applied across a series circuit comprising the bridge rectifier Mrec, the current limiting resistor R1 and the electrode cell.

The current from rectifier Mrec divides into two paths, one through the temperature compensator F via resistor R2 and the other through the alarm relay potentiometer (Pot) indicator MA and resistor R3, the two paths joining in a common return to the low potential side of the rectifier.

The indicator is protected from overload by a semi-conductor in shunt across the indicator and potentiometer. When the water temperature is at the lower limit of the compensated range the total resistance of the compensator is in circuit and the two paths are as described above. As the temperature of the water rises, the resistance of the compensator device drops progressively, the electrical path through the compensator now has a lower resistance than the other and a large proportion of the cell current. The compensator therefore ensures that the alteration in the balance of the resistances of the two paths corresponds to the increased water conductivity due to the rise in temperature and a correct reading is thus obtained over the compensated range.

The alarm setting is adjustable and the contacts of the warning relay close to light a lamp or sound a horn when salinity exceeds the acceptable level.

The salinometer is also arranged to control a solenoid operated valve which dumps unacceptable feed water to the bilge or recirculates. The salinometer and valve reset automatically when the alarm condition clears.

Corrosion

The shell of the evaporator may be of cupro-nickel or other corrosion resistant material but more commonly is of steel. The steel shell of evaporators is prone to corrosion. Protection is provided in the form of natural rubber, rolled and bonded to the previously shot-blasted steel. The adhesive is heat cured and the integrity of the rubber checked by spark test.

Reason for distillate treatment

The low operating temperature of the evaporator described is not sufficient to sterilize the product. Despite precautions near the coast, harmful organisms may enter with the sea water and pass through to the domestic water tank and system. Additionally there is a likelihood that while in the domestic tank, water may become infested with bacteria, due to a build-up of a colony of organisms from some initial contamination. Sterilization by the addition of chlorine is recommended in Merchant Shipping Notice M1214. A later notice, M1401, states that the Electro-Katadyn process in use since the 1960s, has also been approved.

Another problem with distilled water is that having none of the dissolved solids common in fresh water it tastes flat. It also tends to be slightly acidic due to its ready absorption of carbon dioxide (CO_2). This condition makes it corrosive to pipe systems and less than beneficial to the human digestive tract.

Chlorine sterilization and conditioning

Initial treatment (Figure 6.84) involves passing the distillate through a neutralite unit containing magnesium and calcium carbonate. Some absorption of carbon dioxide from the water and the neutralizing effect of these compounds removes

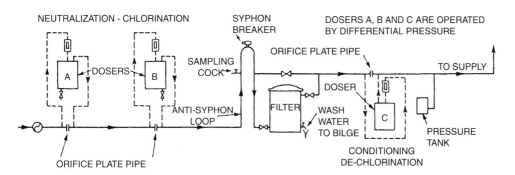

Figure 6.84 Chlorine sterilization and conditioning.

acidity. The addition of hardness salts also gives the water a better taste. The sterilizing agent chlorine, being a gas, is carried into the water as a constituent of sodium hypochlorite (a liquid) or in granules of calcium chloride dissolved in water. The addition is set to bring chlorine content to 0.2 ppm. While the water resides in the domestic tank, chlorine should preserve sterility. In the long term, it will evaporate so that further additions of chlorine may be needed.

The passage of water from storage tanks to the domestic system is by way of a carbon filter which removes the chlorine taste.

Electro-Katadyn method of sterilization
The Electro-Katadyn process (Figure 6.85) accepted as an alternative to chlorination (see Merchant Shipping Notice M1401) involves the use of a driven silver anode to inject silver ions (Ag^+) into the distilled water product of the low temperature evaporator. Silver is toxic to the various risk organisms. Unlike the gas chlorine, it will not evaporate but remains suspended in the water.

The sterilizer is placed close to the production equipment with the conditioning unit being installed after the sterilizer and before the storage tank.

The amount of metal released to water passing through the unit is controlled by the current setting.

If a large volume has to be treated, only part is bypassed through and a high current setting is used to inject a large amount of silver. The bypassed water is then added to the rest in the pipeline. With low water flow, all of the water is delivered through the device and the current setting is such as to give a concentration of 0.1 ppm of silver. The silver content of water in the domestic system should be 0.08 ppm maximum.

Ultra-violet sterilizer
A means for sterilizing potable water at the point of use is provided on many offshore installations and ships by an ultra-violet radiation unit which is positioned after the hydrophore tank and as close as possible to the tap supply points. The stainless steel irradiation chamber contains low pressure mercury vapour tubes, housed in a quartz jacket. Tubes are wired in series with a transformer for safety. A wiper is fitted within the chamber to clean the jackets and lamp observation window. Units of a similar type are used for pretreatment disinfection in some reverse osmosis plant.

Flash evaporators
The evaporator, described above, boils sea water at the saturation temperature corresponding to the uniform pressure through the evaporation and condensing chambers. With flash evaporators (Figure 6.86) the water is heated in one compartment before being released into a second chamber in which the pressure is substantially lower. The drop in pressure changes the saturation temperature below the actual temperature, so that some of the water instantly flashes off as vapour.

Steam in the chamber at sub-atmospheric pressure is condensed by contact with tubes circulated with the salt feed and is removed by a distillate pump. Suitably placed baffles and demisters, similar to those already described, prevent carry-over of saline droplets. The arrangements for continuous monitoring of distillate purity are similar to those described above.

If two or more vessels in series are maintained at progressively lower absolute pressures, the process can be repeated. Incoming salt feed absorbs the latent heat of the steam in each stage, with a resultant gain in economy of heat and fuel. This is known as cascade evaporation, a term which is self-explanatory. Figure 6.86 shows a two stage flash evaporator distiller. The flash chambers are maintained at a very low absolute pressure by ejectors, steam or water operated; the salt feed is heated initially by the condensing vapour in the flash chambers, subsequently in its passage

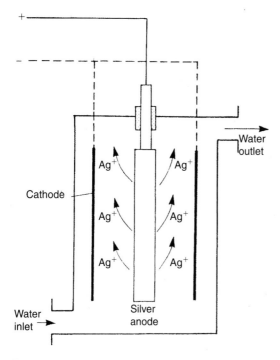

Figure 6.85 Electro-Katadyn sterilization.

Marine engines and auxiliary machinery 433

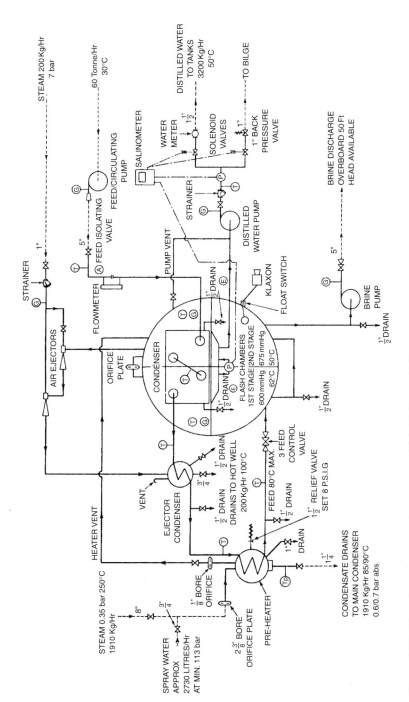

Figure 6.86 Flow diagrams – cascade evaporator (Caird & Rayner Ltd).

through the ejector condenser (when steam-operated ejectors are used) and is raised to its final temperature in a heater supplied with low pressure exhaust steam. Brine density is maintained, as in the case of the evaporator-distillers described previously, by an excess of feed over evaporation and the removal of the excess by a pump. The re-circulation of brine may be provided for in plant.

It should be noted that when distillate is used for drinking it may require subsequent treatment to make it potable.

Reverse osmosis
Osmosis is the term used to describe the natural migration of water from one side of a semi-permeable membrane into a solution on the other side. The phenomenon occurs when moisture from the soil passes through the membrane covering of the roots of plants, with no loss of nutrient liquid from the plant. The membrane acts as a one way barrier, allowing the passage of water but not of the nutrients dissolved in the liquid within the root. Osmosis can be demonstrated in a laboratory with a parchment-covered, inverted thistle funnel partly filled with solution and immersed in a container of pure water. The liquid level in the funnel rises as pure water passes through the parchment and into the solution. The action will continue despite the rise of the head of the salt solution relative to that of the pure water. Osmotic pressure can be obtained by measuring the head of the solution when the action ceases.

The semi-permeable membrane and the parchment are like filters. They allow the water molecules through but not the larger molecules of dissolved substances. The phenomenon is important not only for the absorption of water through the roots of plants but in animal and plant systems generally.

Reverse osmosis is a water filtration process which makes use of semi-permeable membrane-like materials. Salt (sea) water on one side of the membrane (Figure 6.87) is pressurized by a pump and forced against the material. Pure water passes through but the membrane is able to prevent passage of the salts. For production of large amounts of pure water, the membrane area must be large and it must be arranged in a configuration which makes it strong enough to withstand the very high pump pressure needed.

The man-made membrane material used for sea-water purification is produced in the form of flimsy polyamide or polysulphonate sheets, which without backing would not be strong enough. The difficulty of combining the requirements of very large area with adequate reinforcement of the thin sheets is dealt with by making up spirally wound cartridges (Figure 6.88b). The core of the cartridge is a porous tube to which are attached the open edges of a large

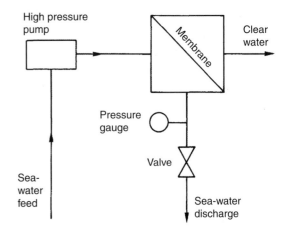

Figure 6.87 Reverse osmosis principle.

number of envelopes each made of two sheets of the membrane material. The envelopes, sealed together on three sides, contain a sheet of porous substance which acts as the path to the central porous tube for water which is squeezed through the membranes. The envelopes are separated by coarse gauze sheets. Assembled envelopes and separators initially have the appearance of a book opened so that the covers are in contact, the spine or binding forming a central tube. The finished cartridge is produced by rotating the actual central tube, so that envelopes and separators are wrapped around it in a spiral, to form a cylindrical shape. Cartridges with end spacers are housed in tubes of stainless steel (Figure 6.88a) or other material. Output of the reverse osmosis plant is governed by the number of cartridge tubes in parallel. Quality is improved by installing sets of tubes in series.

One problem with any filtration system is that deposit accumulates and gradually blocks the filter. Design of the cartridges is therefore such that the sea-water feed passes through the spiral windings and over the membrane sheets with a washing action that assists in keeping the surfaces clear of deposit. A dosing chemical, sodium hexametaphosphate, is also added to assist the action.

The pump delivery pressure for a reverse osmosis system of 60 bar (900 lb/in^2) calls for a robust reciprocating or gear pump. The system must be protected by a relief arrangement.

Pre-treatment and post-treatment
Sea-water feed for reverse osmosis plant is pretreated before being passed through. The chemical sodium hexametaphosphate is added to assist the wash through of salt deposit on the surface of the elements and the sea water is sterilized to remove bacteria

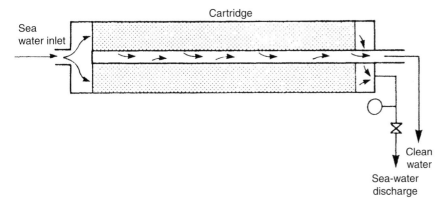

Figure 6.88(a) Cartridge for reverse osmosis.

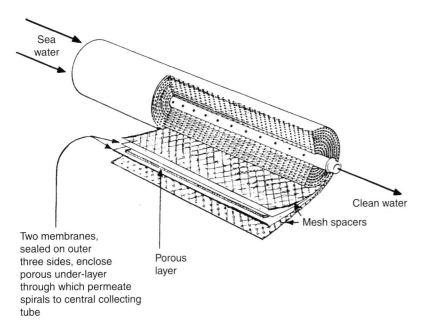

Figure 6.88(b) Spirally wound cartridge for reverse osmosis.

which would otherwise become resident in the filter. Chlorine is reduced by the compressed carbon filter while solids are removed by the other filters.

Treatment is also necessary to make the water product of reverse osmosis potable. The method is much the same as for water produced in low temperature evaporators.

Treatment of water from shore sources

There is a risk that water supplied from ashore may contain harmful organisms which can multiply and infect drinking or washing water storage tanks. All water from ashore, whether for drinking or washing purposes, is to be sterilized. When chlorine is used, the dose must be such as to give a concentration of 0.2 ppm. The Department of Transport recommends in Merchant Shipping Notice number M1214 that because of the risk from legionella bacteria entering the respiratory system by way of fine mist from a shower spray, all water including that for washing only, should be treated by sterilization.

The transfer hose for fresh water is to be marked and kept exclusively for that purpose. The ends must be capped after use and the hose must be stored clear of the deck to reduce the risk of contamination.

Domestic water tanks

Harmful organisms in drinking water storage tanks have caused major health problems on passenger vessels and in general to ship's crews and personnel working on oil platforms. To eliminate this problem, water storage tanks should be pumped out at six-month intervals and, if necessary, the surfaces should be hosed down to clean them. At the 12-month inspection, recoating may be needed in addition to the cleaning. Washing with a 50 ppm solution of chlorine is suggested. Super-chlorinating when the vessel is drydocked consists of leaving a 50 ppm chlorine solution in the tank over a four hour period, followed by flushing with clean water.

The steel tank surfaces may be prepared for coating by wire brushing and priming. Subsequently a cement wash is applied or an epoxy or other coating suitable for use in fresh water tanks.

6.6.1.5 *Sewage systems*

The exact amount of sewage and waste water flow generated on board ship is difficult to quantify. European designers tend to work on the basis of 70 litres/person/day of toilet waste (including flushing water) and about 130–150 litres/person/day of washing water (including baths, laundries, etc.). US authorities suggest that the flow from toilet discharges is as high as 114 litres/person/day with twice this amount of washing water.

The breakdown of raw sewage in water is effected by aerobic bacteria if there is a relatively ample presence of oxygen, but by anaerobic bacteria if the oxygen has been depleted. When the amount of sewage relative to water is small, dissolved oxygen in the water will assist a bio-chemical (aerobic) action which breaks down the sewage into simple, clean components and carbon dioxide. This type of action is produced in biological sewage treatment plant in which air (containing 21% oxygen) is bubbled through to sustain the aerobic bacteria. The final discharge from an aerobic treatment plant has a clean and clear appearance.

The discharge of large quantities of raw sewage into restricted waters such as those of inland waterways and enclosed docks, will cause rapid depletion of any oxygen in the water so that aerobic bacteria are unable to survive. When the self-purification ability of the limited quantity of water is overwhelmed in this way, breakdown by putrefaction occurs. Anaerobic bacteria, not reliant on oxygen for survival, are associated with this action which results in the production of black, turgid water and gases which are toxic and flammable. The process is used deliberately in some shore sewage treatment works to produce gas which is then used as fuel for internal combustion engines on the site.

The very obvious effects of sewage discharge in waterways and enclosed docks prompted the Port of London Authority and others to establish regulations concerning sewage discharge and to provide facilities ashore for ships' crews. The lavatories were vandalized and the scheme was found to be impractical. Legislation imposed nationally by the USA (through the Coast Guard) and the Canadian Government was more effective and together with the anticipation of the ratification of Annexe IV of the 1973 IMCO Conference on Marine Pollution was probably more responsible for the development of holding tanks and on board sewage treatment plant.

Some plants are designed so that the effluent is retained in the vessel for discharge well away from land, or to a receiving facility ashore; others are designed to produce an effluent which is acceptable to port authorities for discharge inshore. In the former type, the plant consists of holding tanks which receive all lavatory and urinal emptyings, including flushing water, while wash-basins, showers and baths are permitted to discharge overboard. Some are designed to minimize the amount of liquid retained by flushing with recycled effluent. It is claimed that such a system only requires about 1% of the retaining capacity of a conventional retention system.

Effluent quality standards

To discharge sewage in territorial waters the effluent quality may have to be within certain standards laid down by the local or national authorities. These will usually be based on one or more of three factors, namely the bio-chemical oxygen demand (BOD), suspended solids content and e-coliform count of the discharge.

Bio-chemical oxygen demand

The bio-chemical oxygen demand (BOD) is determined by incubating at 20°C, a sample of sewage effluent which has been well-oxygenated. The amount of oxygen absorbed over a five-day period is then measured. The test is used in this context to evaluate the effectiveness of treatment as it measures the total amount of oxygen taken up as final and complete breakdown of organic matter by aerobic bacteria in the effluent occurs. The quantity of oxygen used equates to the amount of further breakdown required.

Suspended solids

Suspended solids are unsightly and over a period of time can give rise to silting problems. They are usually a sign of a malfunctioning sewage plant and when very high will be accompanied by a high BOD. Suspended solids are measured by filtering a sample through a pre-weighed pad which is then dried and re-weighed.

Coliform count

The e-coliform is a family of bacteria which live in the human intestine. They can be quantified easily in a laboratory test the result of which is indicative of the amount of human waste present in a particular sewage sample. The result of this test is called the e-coli. count and is expressed per 100 ml.

Holding tanks

Simple holding tanks may be acceptable for ships which are in port for only a very brief period. The capacity would need to be excessively large for long stays because of the amount of flushing water. They require a vent, with the outlet suitably and safely positioned because of gas emissions. A flame trap reduces risk. Inhibiting internal corrosion implies some form of coating and, for washing through of the tank and pump after discharge of the contents at sea, a fresh water connection is required.

Elsan holding and recirculation (zero discharge) system

A retention or holding tank is required where no discharge of treated or untreated sewage is allowed in a port area. The sewage is pumped out to shore reception facilities or overboard when the vessel is proceeding on passage at sea, usually beyond the 12 nautical mile limit.

Straight holding tanks for retention of sewage during the period of a ship's stay in port were of a size large enough to contain not only the actual sewage but also the flushing water. Each flush delivered perhaps 5 litres of sea water. Passenger vessels or ferries with automatic flushing for urinals required very large holding tanks.

Problems resulting from the retention of untreated wastes relate to its breakdown by anaerobic bacteria. Clean breakdown by aerobic organisms occurs where there is ample oxygen, as described previously. In the conditions of a plain retention tank where there is no oxygen, anaerobic bacteria and other organisms thrive. These cause putrefaction, probably with corrosion in the tank and production of toxic and flammable gases.

The Elsan type plant (Figure 6.89) has an initial reception chamber in which separation of liquid and solid sewage takes place. Wastes drop on to a moving perforated rubber belt (driven by an electric motor) which the liquid passes through but solids travel with the belt to fall into a caustic treatment tank. Solids are then transferred by a grinder pump to the sullage or holding tank. The liquid passes via the perforated belt to treatment tanks which contain chlorine and caustic based compounds. These chemicals make the liquid effluent acceptable for use as a flushing fluid. The Pneupress arrangement which supplies liquid for flushing the toilets can deliver recirculated fluid or, when the vessel is on passage, sea water.

Capacity of the holding tank is 2 litres per/person/day. The tank is pumped out at sea, or to shore if the ship is in port for a long period. Tank size is small because liquid effluent passes mainly to the flushing system (excess overflows to the sullage tanks).

Biological sewage treatment

A number of biological sewage treatment plant types are in use at sea but nearly all work on what is called the extended aeration process. Basically this consists of oxygenating by bubbling air through or by agitating the surface. By so doing a family of bacteria is propagated which thrives on the oxygen content and digests the sewage to produce an innocuous sludge. In order to exist, the bacteria need a continuing supply of oxygen from the air and sewage wastes. If plant is shut down or bypassed or if the air supply fails, the bacteria die and the plant cannot function correctly until a new bacteria colony is generated. Change of flushing liquid—as when a ship moves from a sea-water environment to fresh water—drastic change of temperature or excess use of lavatory cleaning agents can also affect the bacteria colony. The process of regeneration can take several days depending on the level of harm caused.

Bacteria which thrive in the presence of oxygen are said to be aerobic. When oxygen is not present, the aerobic bacteria cannot live but a different family of bacteria is generated. These bacteria are said to be anaerobic. Whilst they are equally capable of breaking down sludge, in so doing they generate gases such as hydrogen sulphide and methane. Continuing use of a biological sewage system after a failure of the air supply, could result in propagation of anaerobic bacteria and processes. The gases produced by anaerobic activity are dangerous, being flammable and toxic.

Extended aeration plants used at sea are package plants consisting basically of three inter-connected tanks (Figure 6.90). The effluent may be comminuted (i.e. passed through a device which consists of a rotating knife-edge drum which acts both as a filter and a cutter) or simply passed through a bar screen from where it passes into the first chamber. Air is supplied to this chamber via a diffuser which breaks the air up into fine bubbles. The air is forced

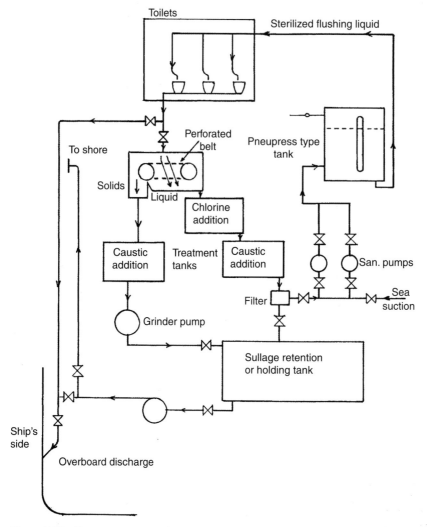

Figure 6.89 Elsan type sewage plant.

through the diffuser by a compressor. After a while a biological sludge is formed and this is dispersed throughout the tank by the agitation caused by the rising air bubbles.

The liquid from the aeration tank passes to a settling tank where under quiescent conditions, the activated sludge, as it is known, settles and leaves a clear effluent. The activated sludge cannot be allowed to remain in the settling tank since there is no oxygen supplied to this area and in a very short time the collected sludge would become anaerobic and give off offensive odours. The sludge is therefore continuously recycled to the aeration tank where it mixes with the incoming waste to assist in the treatment process.

Over a period of time the quantity of sludge in an aeration tank increases due to the collection of inert residues resulting from the digestion process, this build up in sludge is measured in ppm or mg/litre, the rate of increase being a function of the tank size. Most marine biological waste treatment plants are designed to be desludged at intervals of about three months. The desludging operation entails pumping out about three quarters of the aeration tank contents and refilling with clean water.

The clear effluent discharged from a settling tank must be disinfected to reduce the number of coliforms to an acceptable level. Disinfection is achieved by treating the clean effluent with a solution of calcium or sodium hypochlorite, this is usually carried out in a tank or compartment on the end of the sewage treatment unit. The chlorinator shown in Figure 6.90 uses tablets of calcium hypochlorite retained in perforated plastic tubes around which

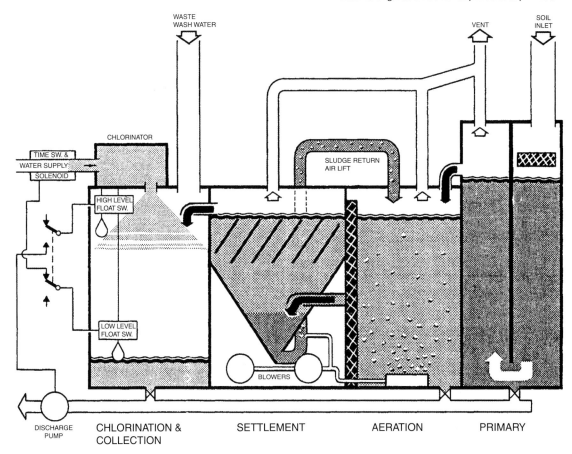

Figure 6.90 Biological sewage treatment plant (Hamworthy).

the clean effluent flows dissolving some of the tablet material as it does so. The treated effluent is then held in the collection tank for 60 minutes to enable the process of disinfection to be completed. In some plants the disinfection is carried out by ultra-violet radiation.

6.6.1.6 *Incinerators*

Stricter legislation with regard to pollution of the sea, limits and, in some instances, completely bans the discharge of untreated waste water, sewage, waste oil and sludge. The ultimate situation of no discharge can be achieved by the use of a suitable incinerator. When used in conjunction with a sewage plant and with facilities for burning oil sludges, the incinerator forms a complete waste disposal package.

One type of incinerator for shipboard use is shown in Figure 6.91. The combustion chamber is a vertical cylinder lined with refractory material. An auxiliary oil-fired burner is used to ignite the refuse and oil sludge and is thermostatically controlled to minimize fuel consumption. A sludge burner is used to dispose of oil sludge, water and sewage sludge and works in conjunction with the auxiliary burner. Combustion air is provided by a forced draught fan and swirls upwards from tangential ports in the base. A rotating-arm device accelerates combustion and also clears ash and non-combustible matter into an ash hopper. The loading door is interlocked to stop the fan and burner when opened.

Solid material, usually in sacks, is burnt by an automatic cycle of operation. Liquid waste is stored in a tank, heated and then pumped to the sludge burner where it is burnt in an automatic cycle. After use the ash box can be emptied overboard.

6.6.2 Shafting and propellers

6.6.2.1 *Overview*

The transmission system on a ship transmits power from the engine to the propeller. It is made up of

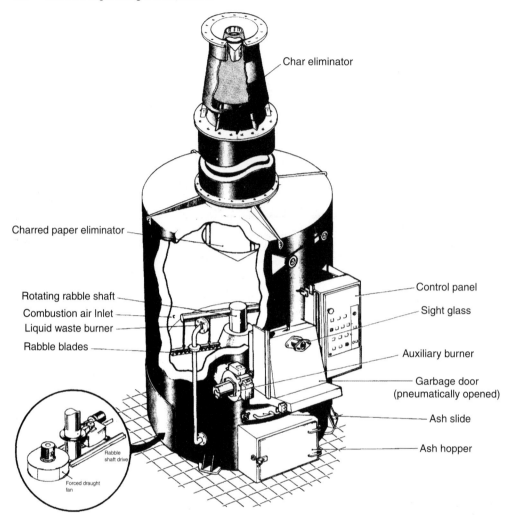

Figure 6.91 Incinerator.

shafts, bearings, and finally the propeller itself. The thrust from the propeller is transferred to the ship through the transmission system.

The different items in the system include the thrust shaft, one or more intermediate shafts and the tailshaft. These shafts are supported by the thrust block, intermediate bearings and the sterntube bearing. A sealing arrangement is provided at either end of the tailshaft with the propeller and cone completing the arrangement. These parts, their location and purpose are shown in Figure 6.92.

6.6.2.2 *Thrust block*

The thrust block transfers the thrust from the propeller to the hull of the ship. It must therefore be solidly constructed and mounted onto a rigid seating or framework to perform its task. It may be an independent unit or an integral part of the main propulsion engine. Both ahead and astern thrusts must be catered for and the construction must be strong enough to withstand normal and shock loads.

The casing of the independent thrust block is in two halves which are joined by fitted bolts (Figure 6.93). The thrust loading is carried by bearing pads which are arranged to pivot or tilt. The pads are mounted in holders or carriers and faced with white metal. In the arrangement shown the thrust pads extend three quarters of the distance around the collar and transmit all thrust to the lower half of the casing. Other designs employ a complete ring of pads. An oil scraper deflects the oil lifted by the thrust collar and directs it onto the pad stops. From here it cascades

Marine engines and auxiliary machinery 441

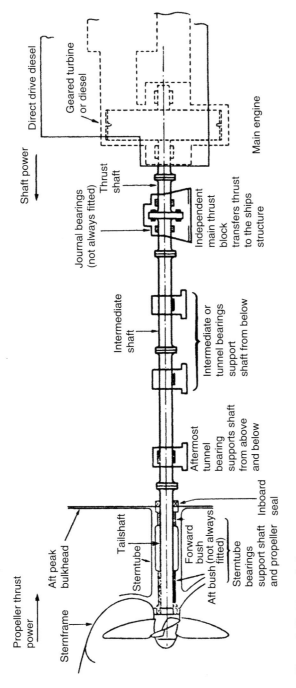

Figure 6.92 Transmission system.

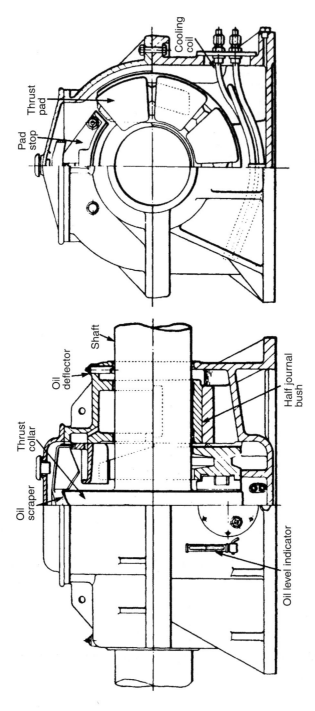

Figure 6.93 Thrust block.

over the thrust pads and bearings. The thrust shaft is manufactured with integral flanges for bolting to the engine or gearbox shaft and the intermediate shafting, and a thrust collar for absorbing the thrust.

Where the thrust shaft is an integral part of the engine, the casing is usually fabricated in a similar manner to the engine bedplate to which it is bolted. Pressurized lubrication from the engine lubricating oil system is provided and most other details of construction are similar to the independent type of thrust block.

6.6.2.3 *Shaft bearings*

Shaft bearings are of two types, the aftermost tunnel bearing and all others. The aftermost tunnel bearing has a top and bottom bearing shell because it must counteract the propeller mass and take a vertical upward thrust at the forward end of the tailshaft. The other shaft bearings only support the shaft weight and thus have only lower half bearing shells.

An intermediate tunnel bearing is shown in Figure 6.94. The usual journal bush is here replaced by pivoting pads. The tilting pad is better able to carry high overloads and retain a thick oil lubrication film. Lubrication is from a bath in the lower half of the casing, and an oil thrower ring dips into the oil and carries it round the shaft as it rotates. Cooling of the bearing is by water circulating through a tube cooler in the bottom of the casing.

6.6.2.4 *Sterntube bearing*

The sterntube bearing serves two important purposes. It supports the tailshaft and a considerable proportion of the propeller weight. It also acts as a gland to prevent the entry of sea water to the machinery space.

Early arrangements used bearing materials such as lignum vitae (a very dense form of timber) which were lubricated by sea water. Most modern designs use an oil lubrication arrangement for a white metal lined sterntube bearing. One arrangement is shown in Figure 6.95.

Oil is pumped to the bush through external axial grooves and passes through holes on each side into internal axial passages. The oil leaves from the ends of the bush and circulates back to the pump and the cooler. One of two header tanks will provide a back pressure in the system and a period of oil supply in the event of pump failure. A low-level alarm will be fitted to each header tank.

Oil pressure in the lubrication system is higher than the static sea-water head to ensure that sea water cannot enter the sterntube in the event of seal failure.

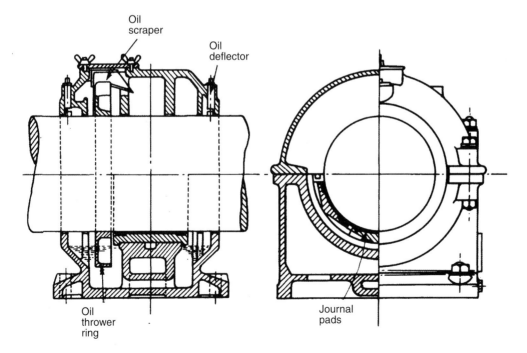

Figure 6.94 Tunnel bearing.

6.6.2.5 *Sterntube seals*

Special seals are fitted at the outboard and inboard ends of the tailshaft. They are arranged to prevent the entry of sea water and also the loss of lubricating oil from the stern bearing.

Older designs, usually associated with sea-water lubricated stern bearings, made use of a conventional stuffing box and gland at the after bulkhead. Oil-lubricated stern bearings use either lip or radial face seals or a combination of the two.

Lip seals are shaped rings of material with a projecting lip or edge which is held in contact with a shaft to prevent oil leakage or water entry. A number of lip seals are usually fitted depending upon the particular application.

Face seals use a pair of mating radial faces to seal against leakage. One face is stationary and the other rotates. The rotating face of the after seal is usually secured to the propeller boss. The stationary face of the forward or inboard seal is the after bulkhead. A spring arrangement forces the stationary and rotating faces together.

6.6.2.6 *Shafting*

There may be one or more sections of intermediate shafting between the thrust shaft and the tailshaft, depending upon the machinery space location. All shafting is manufactured from solid forged ingot steel with integral flanged couplings. The shafting sections are joined by solid forged steel fitted bolts.

The intermediate shafting has flanges at each end and may be increased in diameter where it is supported by bearings.

The propeller shaft or tailshaft has a flanged face where it joins the intermediate shafting. The other end is tapered to suit a similar taper on the propeller boss. The tapered end will also be threaded to take a nut which holds the propeller in place.

6.6.2.7 *Propeller*

The propeller consists of a boss with several blades of helicoidal form attached to it. When rotated it 'screws' or thrusts its way through the water by giving momentum to the column of water passing through it. The thrust is transmitted along the shafting to the thrust block and finally to the ship's structure.

A solid fixed-pitch propeller is shown in Figure 6.96. Although usually described as fixed, the pitch does vary with increasing radius from the boss. The pitch at any point is fixed, however, and for

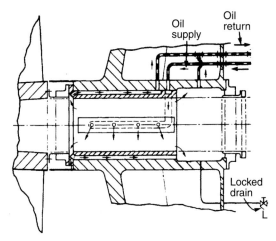

Figure 6.95 Oil lubricated sterntube bearing.

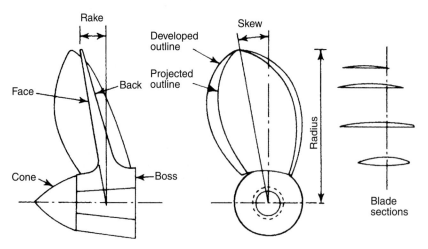

Figure 6.96 Solid propeller.

calculation purposes a mean or average value is used.

A propeller which turns clockwise when viewed from aft is considered right-handed and most single-screw ships have right-handed propellers. A twin-screw ship will usually have a right-handed starboard propeller and a left-handed port propeller.

6.6.2.8 *Propeller mounting*

The propeller is fitted onto a taper on the tailshaft and a key may be inserted between the two: alternatively a keyless arrangement may be used. A large nut is fastened and locked in place on the end of the tailshaft: a cone is then bolted over the end of the tailshaft to provide a smooth flow of water from the propeller.

One method of keyless propeller fitting is the oil injection system. The propeller bore has a series of axial and circumferential grooves machined into it. High pressure oil is injected between the tapered section of the tailshaft and the propeller. This reduces the friction between the two parts and the propeller is pushed up the shaft taper by a hydraulic jacking ring. Once the propeller is positioned the oil pressure is released and the oil runs back, leaving the shaft and propeller securely fastened together.

The Pilgrim Nut is a patented device which provides a pre-determined frictional grip between the propeller and its shaft. With this arrangement the engine torque may be transmitted without loading the key, where it is fitted. The Pilgrim Nut is, in effect, a threaded hydraulic jack which is screwed onto the tailshaft (Figure 6.97). A steel ring receives thrust from a hydraulically pressurized nitrile rubber tyre. This thrust is applied to the propeller to force it onto the tapered tailshaft. Propeller removal is achieved by reversing the Pilgrim Nut and using a withdrawal plate which is fastened to the propeller boss by studs. When the tyre is pressurized the propeller is drawn off the taper. Assembly and withdrawal are shown in Figure 6.97.

6.6.2.9 *Controllable-pitch propeller*

A controllable-pitch propeller is made up of a boss with separate blades mounted into it. An internal mechanism enables the blades to be moved simultaneously through an arc to change the pitch angle and therefore the pitch. A typical arrangement is shown in Figure 6.98.

When a pitch demand signal is received a spool valve is operated which controls the supply of low pressure oil to the auxiliary servo motor. The auxiliary servo motor moves the sliding thrust block assembly to position the valve rod which extends into the propeller hub. The valve rod admits high pressure oil into one side or the other of the main servo motor cylinder. The cylinder movement is transferred by a crank pin and ring to the propeller blades. The propeller blades all rotate together until the feedback signal balances the demand signal and the low pressure oil to the auxiliary servo motor is cut off. To enable emergency control of propeller pitch in the event of loss of power the spool valves can be operated by hand. The oil pumps are shaft driven.

The control mechanism, which is usually hydraulic, passes through the tailshaft and operation is usually from the bridge. Varying the pitch will vary the thrust provided, and since a zero pitch position exists the engine shaft may turn continuously. The blades may rotate to provide astern thrust and therefore the engine does not require to be reversed. See also Section 6.2.7 and Figure 6.11.

6.6.2.10 *Cavitation*

Cavitation, the forming and bursting of vapour-filled cavities or bubbles, can occur as a result of pressure variations on the back of a propeller blade. The results are a loss of thrust, erosion of the blade surface, vibrations in the afterbody of the ship and noise. It is usually limited to high speed heavily loaded propellers and is not a problem under normal operating conditions with a well designed propeller. Cavitation is discussed in more detail in Chapter 5, Section 5.5.

6.6.2.11 *Propeller maintenance*

When a ship is in dry dock the opportunity should be taken to thoroughly examine the propeller, and any repairs necessary should be carried out by skilled dockyard staff.

A careful examination should be made around the blade edges for signs of cracks. Even the smallest of cracks should not be ignored as they act to increase stresses locally and can result in the loss of a blade if the propeller receives a sharp blow. Edge cracks should be welded up with suitable electrodes.

Bent blades, particularly at the tips, should receive attention as soon as possible. Except for slight deformation the application of heat will be required. This must be followed by more general heating in order to stress relieve the area around the repair.

Surface roughness caused by slight pitting can be lightly ground out and the area polished. More serious damage should be made good by welding and subsequent heat treatment. A temporary repair

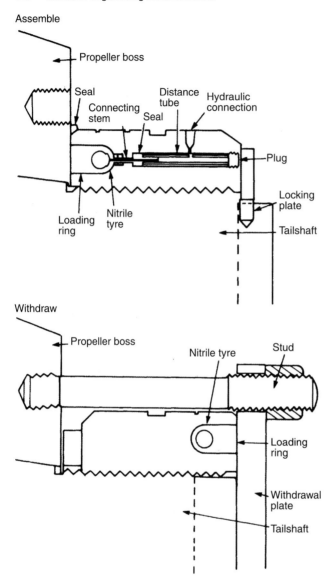

Figure 6.97 Pilgrim Nut operation.

for deep pits or holes could be done with a suitable resin filler.

Routine removed of marine growth on propellers may be carried out, as described in Chapter 5, Section 5.7.4.

6.6.3 Steering gear

6.6.3.1 *Overview*

The steering gear provides a movement of the rudder in response to a signal from the bridge. The total system may be considered made up of three parts, *control equipment, a power unit* and a *transmission to the rudder stock*. The control equipment conveys a signal of desired rudder angle from the bridge and activates the power unit and transmission system until the desired angle is reached. The power unit provides the force, when required and with immediate effect, to move the rudder to the desired angle. The transmission system, the steering gear, is the means by which the movement of the rudder is accomplished.

Certain requirements must currently be met by a ship's steering system. There must be two independent

Marine engines and auxiliary machinery 447

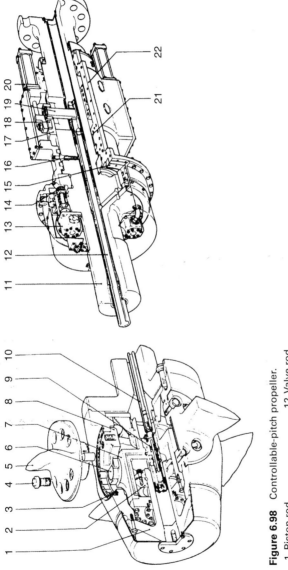

Figure 6.98 Controllable-pitch propeller.

1 Piston rod
2 Piston
3 Blade seal
4 Blade bolt
5 Blade
6 Crank pin
7 Servo motor cylinder
8 Crank ring
9 Control valve
10 Valve rod
11 Mainshaft
12 Valve rod
13 Main pump
14 Pinion
15 Internally toothed gear ring
16 Non-return valve
17 Sliding ring
18 Sliding thrust block
19 Corner pin
20 Auxiliary servo motor
21 Pressure seal
22 Casing

means of steering, although where two identical power units are provided an auxiliary unit is not required. The power and torque capability must be such that the rudder can be swung from 35° one side to 35° the other side with the ship at maximum speed, and also the time to swing from 35° one side to 30° the other side must not exceed 28 seconds. The system must be protected from shock loading and have pipework which is exclusive to it as well as be constructed from approved materials. Control of the steering gear must be provided in the steering gear compartment. Rudder types and estimation of rudder torque are described in Chapter 8, Section 8.19.

Tankers of 10 000 ton gross tonnage and upwards must have two independent steering gear control systems which are operated from the bridge. Where one fails, changeover to the other must be immediate and achieved from the bridge position. The steering gear itself must comprise two independent systems where a failure of one results in an automatic changeover to the other within 45 seconds. Any of these failures should result in audible and visual alarms on the bridge.

Steering gears can be arranged with hydraulic control equipment known as a 'telemotor', or with electrical control equipment. The power unit may in turn be hydraulic or electrically operated. Each of these units will be considered in turn, with the hydraulic unit pump being considered first. A pump is required in the hydraulic system which can immediately pump fluid in order to provide a hydraulic force that will move the rudder. Instant response does not allow time for the pump to be switched on and therefore a constantly running pump is required which pumps fluid only when required. A variable delivery pump provides this facility.

6.6.3.2 Variable delivery pumps

A number of different designs of variable delivery pump exist. Each has a means of altering the pump stroke so that the amount of oil displaced will vary from zero to some designed maximum value. This is achieved by use of a floating ring, a swash plate or a slipper pad.

The radial cylinder (Hele-Shaw) pump is shown in Figure 6.99. Within the casing a short length of shaft drives the cylinder body which rotates around a central valve or tube arrangement and is supported at the ends by ball bearings. The cylinder body is connected to the central valve arrangement by ports which lead to connections at the outer casing for the supply and delivery of oil. A number of pistons fit in the radial cylinders and are fastened to slippers by a gudgeon pin. The slippers fit into a track in the circular floating ring. This ring may rotate, being supported by ball bearings, and can also move from side to side since the bearings are mounted in guide blocks. Two spindles which pass out of the pump casing control the movement of the ring.

The operating principle will now be described by reference to Figure 6.100. When the circular floating ring is concentric with the central valve arrangement the pistons have no relative reciprocating motion in their cylinders (Figure 6.100(a)). As a result no oil is pumped and the pump, although rotating, is not delivering any fluid. If however the circular floating ring is pulled to the right then a relative reciprocating motion of the pistons in their cylinders does occur (Figure 6.100(b)). The lower piston, for instance, as it moves inwards will discharge fluid out through the lower port in the central valve arrangement. As it continues past the horizontal position the piston moves outwards, drawing in fluid from the upper port. Once past the horizontal position on the opposite side, it begins to discharge the fluid. If the circular floating ring were pushed to the left then the suction and discharge ports would be reversed (Figure 6.100(c)).

This pump arrangement therefore provides, for a constantly rotating unit, a no-flow condition and infinitely variable delivery in either direction. The pump is also a positive displacement unit. Where two pumps are fitted in a system and only one is operating, reverse operation might occur. Non-reversing locking gear is provided as part of the flexible coupling and is automatic in operation. When a pump is stopped the locking gear comes into action; as the pump is started the locking gear releases.

The swash plate and slipper pad designs are both axial cylinder pumps. The slipper pad is an improvement on the swash plate which provides higher pressure. An arrangement of a swash plate pump is shown in Figure 6.101. The driving shaft rotates the cylinder barrel, swash plate and pistons. An external trunnion (short shaft) enables the swash plate to be moved about its axis. The cylinders in the barrel are connected to ports which extend in an arc around the fixed port plate.

When the swash plate is vertical no pumping action takes place. When the swash plate is tilted pumping occurs, the length of stroke depending upon the angle of tilt. Depending upon the direction of tilt the ports will be either suction or discharge. This pump arrangement therefore offers the same flexibility in operation as the radial piston type.

6.6.3.3 Telemotor control

Telemotor control is a hydraulic control system employing a transmitter, a receiver, pipes and a

Marine engines and auxiliary machinery 449

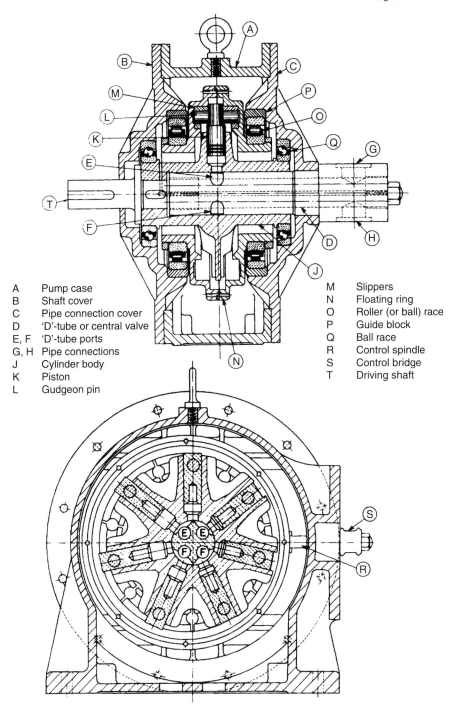

A	Pump case		M	Slippers
B	Shaft cover		N	Floating ring
C	Pipe connection cover		O	Roller (or ball) race
D	'D'-tube or central valve		P	Guide block
E, F	'D'-tube ports		Q	Ball race
G, H	Pipe connections		R	Control spindle
J	Cylinder body		S	Control bridge
K	Piston		T	Driving shaft
L	Gudgeon pin			

Figure 6.99 Hele-Shaw pump.

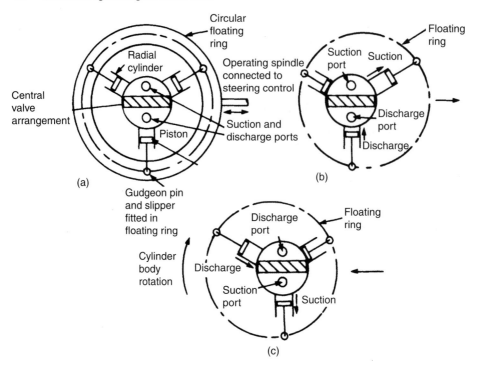

Figure 6.100 Hele-Shaw pump—operating principle.

charging unit. The transmitter, which is built into the steering wheel console, is located on the bridge and the receiver is mounted on the steering gear. The charging unit is located near to the receiver and the system is charged with a non-freezing fluid.

The telemotor system is shown in Figure 6.102. Two rams are present in the transmitter which move in opposite directions as the steering wheel is turned. The fluid is therefore pumped down one pipeline and drawn in from the other. The pumped fluid passes through piping to the receiver and forces the telemotor cylinder unit to move. The suction of fluid from the opposite cylinder enables this movement to take place. The cylinder unit has a control spindle connected to it by a pin. This control spindle operates the slipper ring or swash plate of the variable delivery pump. If the change-over pin is removed from the cylinder unit and inserted in the local handwheel drive then manual control of the steering gear is possible. Stops are fitted on the receiver to limit movement to the maximum rudder angle required. The charging unit consists of a tank, a pump, and shut-off cocks for each and is fitted in the main piping between the transmitter and receiver.

In the transmitter a replenishing tank surrounds the rams, ensuring that air cannot enter the system. A bypass between the two cylinders opens as the wheel passes midships. Also at mid position the supercharging unit provides a pressure in the system which ensures rapid response of the system to a movement of the wheel. This supercharging unit also draws in replenishing fluid if required in the system, and provides a relief valve arrangement if the pressure is too high. Pressure gauges are connected to each main pipeline and air vent cocks are also provided.

In normal operation the working pressure of about 20 to 30 bar, or the manufacturer's given figure, should not be exceeded. The wheel should not be forced beyond the 'hard over' position as this will strain the gear. The replenishing tank should be checked regularly and any lubrication points should receive attention. Any leaking or damaged equipment must be repaired or replaced as soon as possible. The system should be regularly checked for pressure tightness. The rudder response to wheel movement should be checked and if sluggish or slow then air venting undertaken. If, after long service, air venting does not remove sluggishness, it may be necessary to recharge the system with new fluid.

6.6.3.4 *Electrical control*

The electrical remote control system is commonly used in modern installations since it uses a small

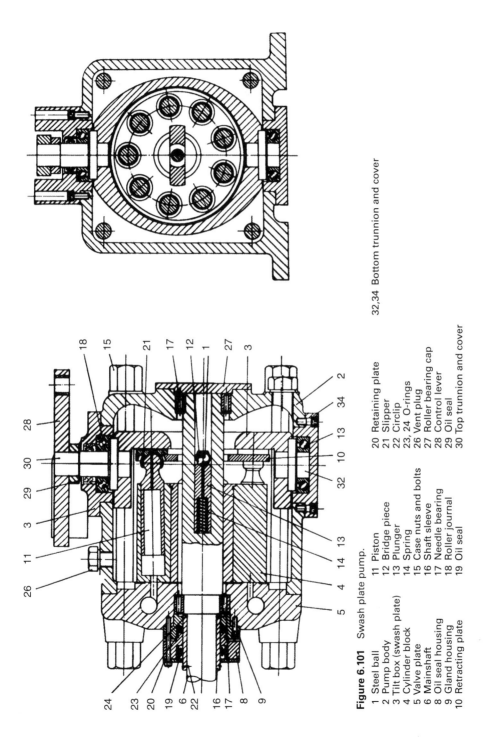

Figure 6.101 Swash plate pump.

1 Steel ball
2 Pump body
3 Tilt box (swash plate)
4 Cylinder block
5 Valve plate
6 Mainshaft
8 Oil seal housing
9 Gland housing
10 Retracting plate
11 Piston
12 Bridge piece
13 Plunger
14 Spring
15 Case nuts and bolts
16 Shaft sleeve
17 Needle bearing
18 Roller journal
19 Oil seal
20 Retaining plate
21 Slipper
22 Circlip
23, 24 O-rings
26 Vent plug
27 Roller bearing cap
28 Control lever
29 Oil seal
30 Top trunnion and cover
32, 34 Bottom trunnion and cover

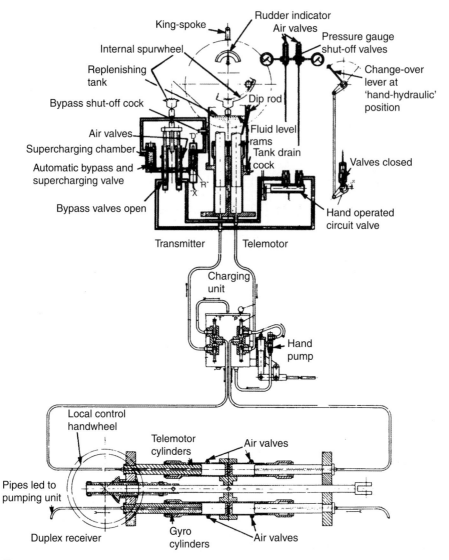

Figure 6.102 Telemotor control system.

control unit as transmitter on the bridge and is simple and reliable in operation.

The control box assembly, which is mounted on the steering gear, is shown in Figure 6.103(a) and (b). Movement of the bridge transmitter results in electrical imbalance and current flow to the motor. The motor drives, through a flexible coupling, a screw shaft, causing it to turn. A screw block on the shaft is moved and this in turn moves the floating lever to which a control rod is attached. The control rod operates the slipper ring or swash plate of the variable delivery pump. A cut-off lever connected to the moving tiller will bring the floating lever pivot and the lever into line at right angles to the screw shaft axis. At this point the rudder angle will match the bridge lever angle and the pumping action will stop. The rotating screw shaft will have corrected the electrical imbalance and the motor will stop. For local manual control, the electrical control is switched off and a small handwheel is connected to the screw shaft. A detent pin holds the handwheel assembly clear when not in use. Rotation of the handwheel will move the floating lever and bring about rudder movement as already described.

6.6.3.5 *Power units*

Two types of hydraulically powered transmission units or steering gear are in common use, the *ram* and the *rotary vane*.

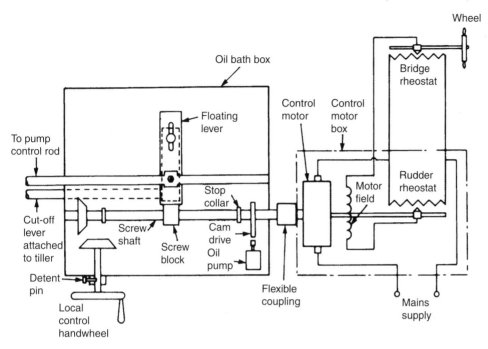

Figure 6.103(a) Control box.

Ram type
Two particular variations, depending upon torque requirements, are possible: the *two-ram* and the *four-ram*. A two-ram steering gear is shown in Figure 6.104.

The rams acting in hydraulic cylinders operate the tiller by means of a swivel crosshead carried in a fork of the rams. A variable delivery pump is mounted on each cylinder and the slipper ring is linked by rods to the control spindle of the telemotor receiver. The variable delivery pump is piped to each cylinder to enable suction or discharge from either. A replenishing tank is mounted nearby and arranged with non-return suction valves which automatically provide make-up fluid to the pumps. A bypass valve is combined with spring-loaded shock valves which open in the event of a very heavy sea forcing the rudder over. In moving over, the pump is actuated and the steering gear will return the rudder to its original position once the heavy sea has passed. A spring-loaded return linkage on the tiller will prevent damage to the control gear during a shock movement.

During normal operation one pump will be running. If a faster response is required, for instance in confined waters, both pumps may be in use. The pumps will be in the no-delivery state until a rudder movement is required by a signal from the bridge telemotor transmitter. The telemotor receiver cylinder will then move: this will result in a movement of the floating lever which will move the floating ring or slipper pad of the pump, causing a pumping action. Fluid will be drawn from one cylinder and pumped to the other, thus turning the tiller and the rudder. A return linkage or hunting gear mounted on the tiller will reposition the floating lever so that no pumping occurs when the required rudder angle is reached.

A four-ram steering gear is shown in Figure 6.105. The basic principles of operation are similar to the two-ram gear except that the pump will draw from two diagonally opposite cylinders and discharge to the other two. The four-ram arrangement provides greater torque and the flexibility of different arrangements in the event of component failure. Either pump can be used with all cylinders or with either the two port or two starboard cylinders. Various valves must be open or closed to provide these arrangements.

The use of a control valve block incorporating rudder shock relief valves, pump isolating valves, ram isolating and bypass valves, offers greater flexibility with a four-ram steering gear. In normal operation one pump can operate all cylinders. In an emergency situation the motor or a pair of hand pumps could be used to operate two port rams, two starboard rams, two forward rams or two after rams.

The crosshead arrangement on the four-ram type steering gear described incorporates what is known as the 'Rapson Slide'. This provides a mechanical advantage which increases with the angle turned

454 Maritime engineering reference book

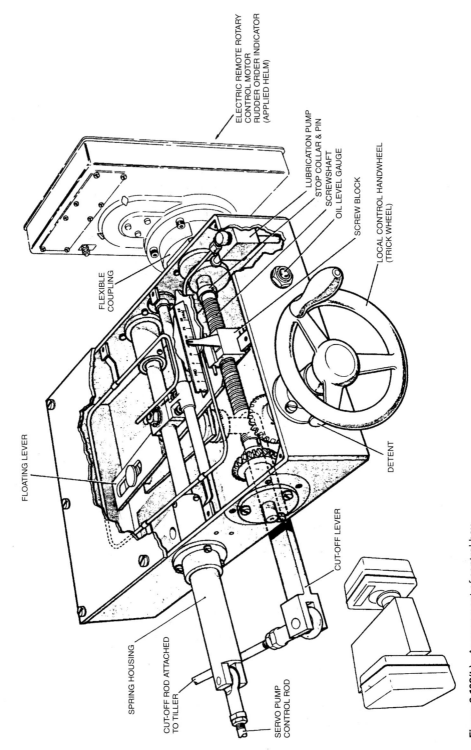

Figure 6.103(b) Arrangement of control box.

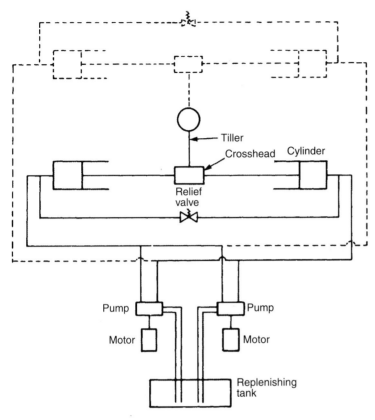

Figure 6.104(a) Diagrammatic arrangement of two-ram steering gear (additional items for four-ram system shown dotted).

through. The crosshead arrangement may use either a forked tiller or a round arm tiller (Figure 6.106). The round arm tiller has a centre crosshead which is free to slide along the tiller. Each pair of rams is joined so as to form a double bearing in which the trunnion arms of the crosshead are mounted. The straight line movement of the rams is thus converted into an angular tiller movement. In the forked tiller arrangement the ram movement is transferred to the tiller through swivel blocks.

To charge the system with fluid it is first necessary to fill each cylinder then replace the filling plugs and close the air cocks. The cylinder bypass valves should be opened and the replenishing tanks filled. The air vents on the pumps should be opened until oil discharges free of air, the pumps set to pump and then turned by hand, releasing air at the appropriate pair of cylinders and pumping into each pair of cylinders in turn using the hand control mechanism. The motor should then be started up and, using the local hand control, operation of the steering gear checked. Air should again be released from the pressurized cylinders and the pumps through the appropriate vents.

During normal operation the steering gear should be made to move at least once every two hours to ensure self lubrication of the moving parts. No valves in the system, except bypass and air vent, should be closed. The replenishing tank level should be regularly checked and, if low, refilled and the source of leakage found. When not in use, that is, in port, the steering motors should be switched off. Also the couplings of the motors should be turned by hand to check that the pump is moving freely. If there is any stiffness the pump should be overhauled. As with any hydraulic system cleanliness is essential when overhauling equipment and only linen cleaning cloths should be used.

Rotary vane type
With this type of steering gear a vaned rotor is securely fastened onto the rudder stock (Figure 6.107). The rotor is able to move in a housing which is solidly attached to the ship's structure. Chambers are formed between the vanes on the rotor and the vanes in the housing. These chambers will vary

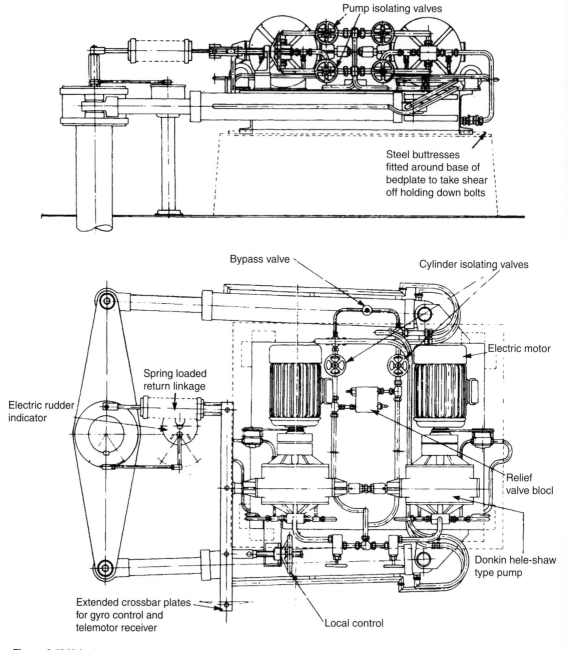

Figure 6.104(b) Actual arrangement of two-ram steering gear.

in size as the rotor moves and can be pressurized since sealing strips are fitted on the moving faces. The chambers either side of the moving vane are connected to separate pipe systems or manifolds. Thus by supplying hydraulic fluid to all the chambers to the left of the moving vane and drawing fluid from all the chambers or the right, the rudder stock can be made to turn anti-clockwise. Clockwise movement will occur if pressure and suction supplies are reversed. Three vanes are usual and permit an angular movement of 70°: the vanes also act as stops limiting rudder movement. The hydraulic fluid is supplied by a variable delivery pump and control will be electrical, as described earlier. A relief valve is fitted in the system to prevent overpressure and allow for shock loading of the rudder.

Marine engines and auxiliary machinery 457

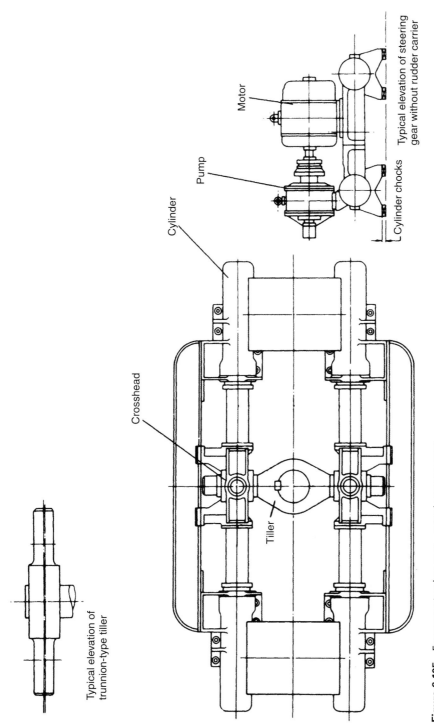

Figure 6.105 Four-ram steering gear—actual arrangement.

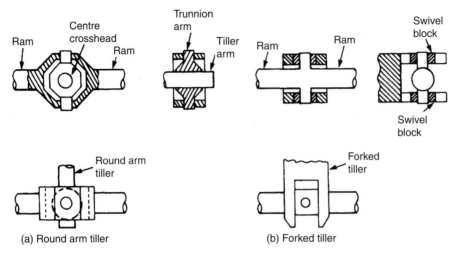

Figure 6.106 Crosshead arrangements.

6.6.3.6 *All-electric steering*

Steering gears which comprise electric control, electric power unit and electrical transmission, are of two types, the Ward–Leonard system and the Direct Single Motor system. Both types have a geared-down motor drive via a pinion to a toothed quadrant.

A Ward–Leonard arrangement is shown diagrammatically in Figure 6.108. A continuously running motor-generator set has a directly coupled exciter to provide the field current of the generator. The exciter field is part of a control circuit, although in some circuits control is directly to the field current of the generator with the exciter omitted. When the control system is balanced there is no exciter field, no exciter output and no generator output, although it is continuously running. The main motor which drives the rudder has no input and thus is stationary. When the wheel on the bridge is turned, and the rheostat contact moved, the control system is unbalanced and a voltage occurs in the exciter field, the exciter, and the generator field. The generator then produces power which turns the rudder motor and hence the rudder. As the rudder moves it returns the rudder rheostat contact to the same position as the bridge rheostat, bringing the system into balance and stopping all current flow.

In the single motor system the motor which drives the rudder is supplied directly from the ship's mains through a contactor type starter. Reversing contacts are also fitted to enable port or starboard movements. The motor runs at full speed until stopped by the control system, so a braking system is necessary to bring the rudder to a stop quickly and at the desired position.

The usual electrical maintenance work will be necessary on this equipment in order to ensure satisfactory operation.

6.6.3.7 *Twin-system steering gears*

To meet the automatic changeover within the 45 seconds required for tankers of 10 000 ton gross tonnage and above, a number of designs are available. Two will be described, one for a ram type steering gear and one for a rotary vane type steering gear. In each case two independent systems provide the power source to move the tiller, the failure of one resulting in a changeover to the other. The changeover is automatic and is achieved within 45 seconds.

The ram type steering gear arrangement is shown diagrammatically in Figure 6.109. A simple automatic device monitors the quantity of oil in the circuit. Where a failure occurs in one of the systems it is located and that circuit is isolated. The other system provides uninterrupted steering and alarms are sounded and displayed.

Consider pump 1 in operation and pump 2 placed on automatic reserve by the selector switch. If a leak develops in circuit 2 the float chamber oil level will fall and proximity switch A on the monitor will be activated to close the solenoid valve 2 which isolates circuit 2 and bypasses the cylinders in that circuit. An alarm will also be given. If the leak is in circuit 1 however, the float chamber oil level will fall further until proximity switch B is activated. This will cut off the power supply to motor 1 and solenoid valve 1

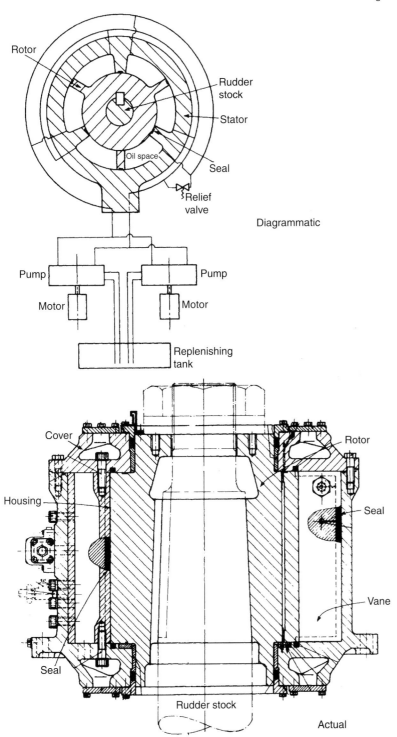

Figure 6.107(a) Rotary vane steering gear.

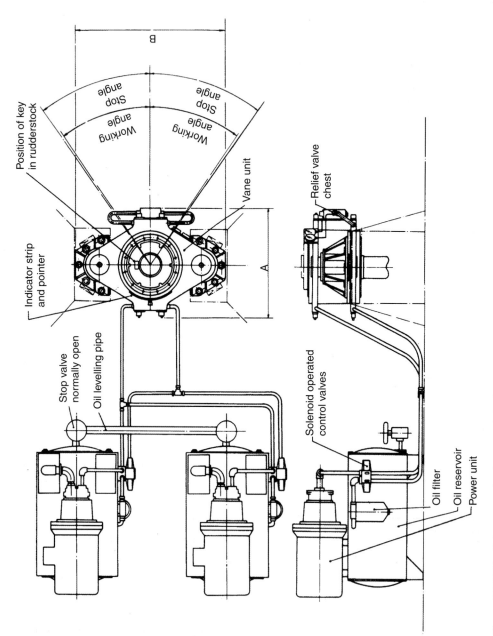

Figure 6.107(b) Rotary vane steering gear. Actual arrangement.

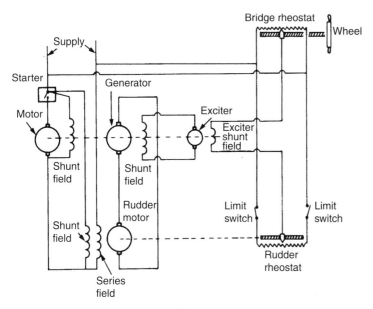

Figure 6.108 Ward–Leonard steering gear.

and connect the supply to motor 2 and solenoid valve 2, thus isolating circuit 1. If pump 2 were running and pump 1 in reserve, a similar changeover would occur. While a two cylinder system has been described this system will operate equally well with four double acting cylinders.

An arrangement based on a rotary vane type steering gear is shown in Figure 6.110. This system involves the use of only one actuator but it is directly fitted to a single tiller and rudderstock and therefore complete duplication of the system does not occur anyway. Self closing lock valves are provided in the two independent hydraulic circuits which operate the actuator. The self closing valves are fitted on the inlet and outlet ports of the actuator and open under oil pressure against the action of a spring. Where an oil pressure loss occurs in one circuit the valves will immediately close under the action of their springs. A low tank level alarm will sound and the other pump can be started. This pump will build up pressure, open the valves on its circuit and the steering gear can immediately operate.

6.6.3.8 *Steering gear testing*

Prior to a ship's departure from any port the steering gear should be tested to ensure satisfactory operation. These tests should include:

1. Operation of the main steering gear.
2. Operation of the auxiliary steering gear or use of the second pump which acts as the auxiliary.
3. Operation of the remote control (telemotor) system or systems from the main bridge steering positions.
4. Operation of the steering gear using the emergency power supply.
5. The rudder angle indicator reading with respect to the actual rudder angle should be checked.
6. The alarms fitted to the remote control system and the steering gear power units should be checked for correct operation.

During these tests the rudder should be moved through its full travel in both directions and the various equipment items, linkages, etc., visually inspected for damage or wear. The communication system between the bridge and the steering gear compartment should also be operated.

6.7 Instrumentation and control

6.7.1 Instrumentation

All machinery must operate within certain desired parameters. Instrumentation enables the parameters, such as temperature and pressure, to be measured or displayed against a scale. A means of control is also required in order to change or alter the displayed readings to meet particular requirements.

The main parameters include the measurement of pressure, temperature, liquid level, flow rate, torque, salinity and oxygen content. The various devices used for measurement of system parameters are

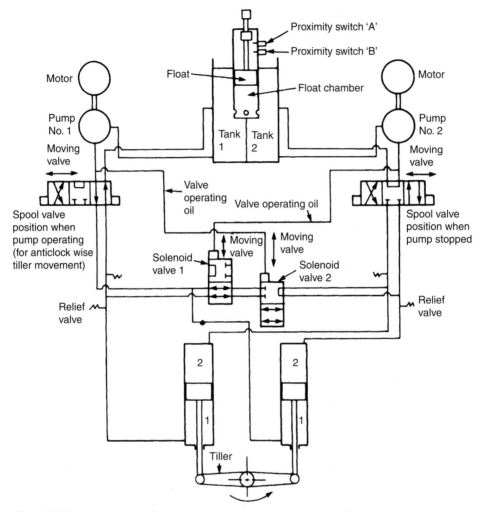

Figure 6.109 Ram type twin circuit system–pump 1 running circuit 2 leaking.

described by McGeorge (1995) and Taylor (1996). The following Sections describe briefly the theory and application of automatic control.

6.7.2 Control

6.7.2.1 *Control theory*

To control a device or system is to be able to adjust or vary the parameters which affect it. This can be achieved manually or automatically, depending upon the arrangements made in the system. All forms of control can be considered to act in a loop. The basic elements present in the loop are a detector, a comparator/controller and a correcting unit, all of which surround the process and form the loop (Figure 6.111). This arrangement is an automatic closed loop if the elements are directly connected to one another and the control action takes place without human involvement. A manual closed loop would exist if one element were replaced by a human operator.

It can be seen therefore that in a closed loop control system the control action is dependent on the output. A detecting or measuring element will obtain a signal related to this output which is fed to the transmitter. From the transmitter the signal is then passed to a comparator. The comparator will contain some set or desired value of the controlled condition which is compared to the measured value signal. Any deviation or difference between the two values will result in an output signal to the controller. The controller will then take action in a manner related to the deviation and provide a signal to a correcting

Marine engines and auxiliary machinery 463

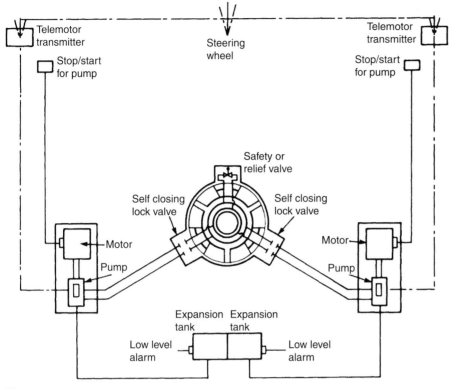

Figure 6.110 Rotary-vane twin circuit system.

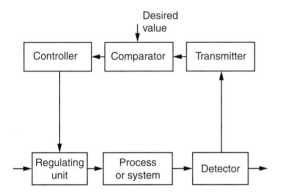

Figure 6.111 Automatic closed loop control.

unit. The correcting unit will then increase or decrease its effect on the system to achieve the desired value of the system variable. The comparator is usually built in to the controller unit.

The transmitter, controller and regulating unit are supplied with an operating medium in order to function. The operating medium may be compressed air, hydraulic oil or electricity. For each medium various types of transmitting devices, controllers and regulating units are used.

Detailed accounts of control theory and marine control systems can be found in Fossen (2002) and Perez (2005).

6.7.2.2 Transmitters

Pneumatic
Many pneumatic devices use a nozzle and flapper system to give a variation in the compressed air signal. A pneumatic transmitter is shown in Figure 6.112. If the flapper moves *away* from the nozzle then the transmitted or output pressure will fall to a low value. If the flapper moves *towards* the nozzle then the transmitted pressure will rise to almost the supply pressure. The transmitted pressure is approximately proportional to the movement of the flapper and thus the change in the measured variable. The flapper movement will be very minute and where measurement of a reasonable movement is necessary a system of levers and linkages must be introduced. This in turn leads to errors in the system and little more than on-off control.

Improved accuracy is obtained when a feedback bellows is added to assist in flapper positioning (Figure 6.113). The measured value acts on one end

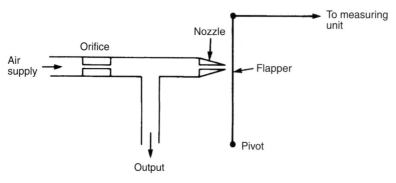

Figure 6.112 Position balance transmitter.

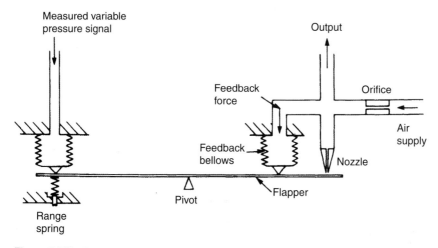

Figure 6.113 Force balance transmitter with feedback.

of the pivoted flapper against an adjustable spring which enables the measuring range to be changed. The opposite end of the flapper is acted upon by the feedback bellows and the nozzle. In operation a change in the measured variable may cause the flapper to approach the nozzle and thus build up the output signal pressure. The pressure in the feedback bellows also builds up, tending to push the flapper away from the nozzle, i.e. a negative feedback. An equilibrium position will be set up giving an output signal corresponding to the measured variable.

Most pneumatic transmitters will have relays fitted which magnify or amplify the output signals to reduce time lags in the system and permit signal transmission over considerable distances. Relays can also be used for mathematical operations, such as adding, subtracting, multiplying or dividing of signals. Such devices are known as 'summing' or 'computing relays'.

Electrical

Simple electrical circuits may be used where the measured variable causes a change in resistance which is read as a voltage or current and displayed in its appropriate units.

Another method is where the measured variable in changing creates a potential difference which, after amplification, drives a reversible motor to provide a display and in moving also reduces the potential difference to zero.

Alternating current positioning motors can be used as transmitters when arranged as shown in Figure 6.114. Both rotors are supplied from the same supply source. The stators are star wound and when the two rotor positions coincide there is no current flow since the e.m.f.s of both are equal and opposite. When the measured variable causes a change in the transmitter rotor position, the two e.m.f.s will be out of balance. A current will flow and the receiver

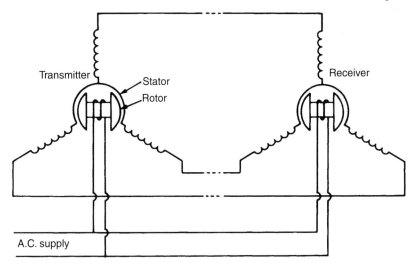

Figure 6.114 A.C. positioning motors.

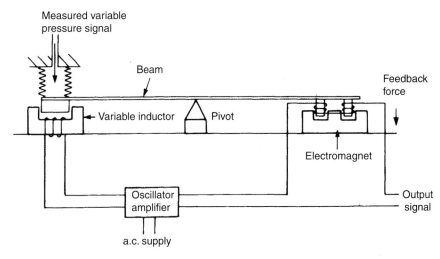

Figure 6.115 Force balance electronic transmitter.

rotor will turn until it aligns with the transmitter. The receiver rotor movement will provide a display of the measured variable.

An electrical device can also be used as a transmitter (Figure 6.115). The measured variable acts on one end of a pivoted beam causing a change in a magnetic circuit. The change in the magnetic circuit results in a change in output current from the oscillator amplifier, and the oscillator output current operates an electromagnet so that it produces a negative feedback force which opposes the measured variable change. An equilibrium position results and provides an output signal.

Hydraulic
The telemotor of a hydraulically actuated steering gear is one example of a hydraulic transmitter. A complete description of the unit and its operation is given in Section 6.6.3.

6.7.2.3 Controller action

The transmitted output signal is received by the controller which must then undertake some corrective action. There will however be various time lags or delays occurring during first the measuring and then the transmission of a signal indicating a change. A delay will also occur in the action of the controller. These delays produce what is known as the transfer function of the unit or item, that is, the relationship between the output and input signals.

The control system is designed to maintain some output value at a constant desired value, and a knowledge of the various lags or delays in the system is necessary in order to achieve the desired control. The controller must therefore rapidly compensate for these system variations and ensure a steady output as near to the desired value as practicable.

Two-step or on-off
In this, the simplest of controller actions, two extreme positions of the controller are possible, either on or off. If the controller were, for example, a valve it would be either open or closed. A heating system is considered with the control valve regulating the supply of heating steam. The controller action and system response is shown in Figure 6.116. As the measured value rises above its desired value the valve will close. System lags will result in a continuing temperature rise which eventually peaks and then falls below the desired value. The valve will then open again and the temperature will cease to fall and will rise again. This form of control is acceptable where a considerable deviation from the desired value is allowed.

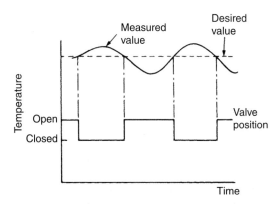

Figure 6.116 Two-step or on-off control.

Continuous action
Proportional action. This is a form of continuous control where any change in controller output is proportional to the deviation between the controlled condition and the desired value. The *proportional band* is the amount by which the input signal value must change to move the correcting unit between its extreme positions. The desired value is usually located at the centre of the proportional band. *Offset* is a sustained deviation as a result of a load change in the process. It is an inherent characteristic of proportional control action. Consider, for example, a proportional controller operating a feedwater valve supplying a boiler drum. If the steam demand, i.e. load, increases then the drum level will fall. When the level has dropped the feedwater valve will open. An equilibrium position will be reached when the feedwater valve has opened enough to match the new steam demand. The drum level, however, will

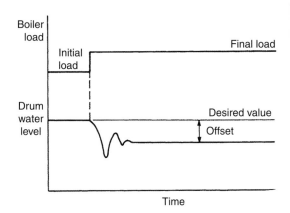

Figure 6.117 System response to proportional controller action.

have fallen to a new value below the desired value, i.e. offset. See Figure 6.117.

Integral action. This type of controller action is used in conjunction with proportional control in order to remove offset. Integral or reset action occurs when the controller output varies at a rate proportional to the deviation between the desired value and the measured value. The integral action of a controller can usually be varied to achieve the required response in a particular system.

Derivative action. Where a plant or system has long time delays between changes in the measured value and their correction, derivative action may be applied. This will be in addition to proportional and integral action. Derivative or rate action is where the output signal change is proportional to the rate of change of deviation. A considerable corrective action can therefore take place for a small deviation which occurs suddenly. Derivative action can also be adjusted within the controller.

Multiple-term controllers. The various controller actions in response to a process change are shown in Figure 6.118. The improvement in response associated with the addition of integral and derivative action can clearly be seen. Reference is often made to the number of terms of a controller. This means the various actions: proportional (P), integral (I), and derivative (D). A three-term controller would therefore mean P + I + D, and two-term usually P + I. A controller may be arranged to provide either split range or cascade control, depending upon the arrangements in the control system. These two types of control are described in the section dealing with control systems.

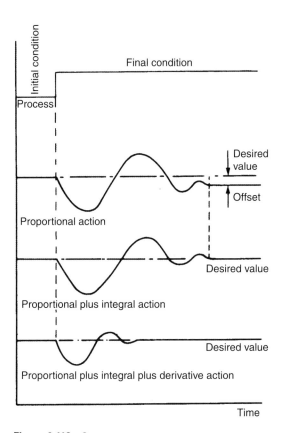

Figure 6.118 System response.

6.7.2.4 Controllers

The controller may be located close to the variable measuring point and thus operate without the use of a transmitter. It may however be located in a remote control room and receive a signal from a transmitter and relay, as mentioned earlier. The controller is required to maintain some system variable at a desired value regardless of load changes. It may also indicate the system variable and enable the desired value to be changed. Over a short range about the desired value the controller will generate a signal to operate the actuating mechanism of the correcting unit. This control signal may include proportional, integral and derivative actions, as already described. Where all three are used it may be known as a 'three-term' controller.

A pneumatic three-term controller is shown in Figure 6.119. Any variation between desired and measured values will result in a movement of the flapper and change in the output pressure. If the derivative action valve is open and the integral action valve closed, then only proportional control occurs. It can be seen that as the flapper moves towards the nozzle a pressure build-up will occur which will increase the output pressure signal and also move the bellows so that the flapper is moved away from the nozzle. This is then a negative feedback which is proportional to the flapper or measured value movement. When the integral action valve is opened, any change in output signal pressure will affect the integral action bellows which will oppose the feedback bellows movement. Varying the opening of the integral action valve will alter the amount of integral action of the controller. Closing the derivative action valve any amount would introduce derivative action. This is because of the delay that would be introduced in the provision of negative feedback for a sudden variable change which would enable the output signal pressure to build up. If the measured variable were to change slowly then the proportional action would have time to build up and thus exert its effect.

An electronic three-term controller is shown in Figure 6.120. The controller output signal is subjected to the various control actions, in this case by electronic components and suitable circuits. Any change in the measured value will move the input potentiometer and upset the balance of the control bridge. A voltage will then be applied to the amplifier and the amplifier will provide an output signal which will result in a movement of the output potentiometer. The balancing bridge will then provide a voltage to the amplifier which equals that from the control bridge. The amplifier output signal will then cease.

The output potentiometer movement is proportional to the deviation between potentiometer positions, and movement will continue while the deviation remains. Integral and derivative actions are obtained by the resistances and capacitors in the circuit. With the integral capacitor fitted while a deviation exists there will be a current through resistance R as a result of the voltage across it. This current flow will charge the integral capacitor and thus reduce the voltage across R. The output

468 Maritime engineering reference book

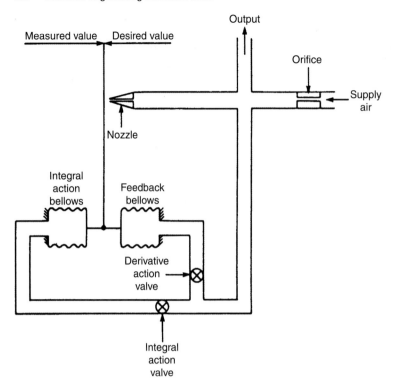

Figure 6.119 Pneumatic three-term controller.

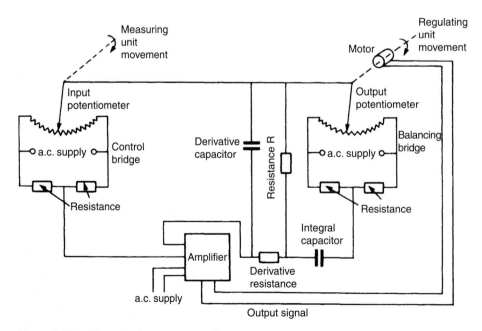

Figure 6.120 Electronic three-term controller.

potentiometer must therefore continue moving until no deviation exists. No offset can occur as it would if only proportional action took place. Derivative action occurs as a result of current flow through the derivative resistor which also charges the derivative capacitor. This current flow occurs only while the balancing bridge voltage is changing, but a larger voltage is required because of the derivative capacitor. The derivative action thus results in a faster return to the equilibrium position, as would be expected. The output potentiometer is moved by a motor which also provides movement for a valve or other correcting unit in the controlled process.

6.7.2.5 Correcting unit

The controller output signal is fed to the correcting unit which then alters some variable in order to return the system to its desired value. This correcting unit may be a valve, a motor, a damper or louvre for a fan or an electric contactor. Most marine control applications will involve the actuation or operation of valves in order to regulate liquid flow.

Pneumatic control valve
A typical pneumatic control valve is shown in Figure 6.121. It can be considered as made up of two parts—the actuator and the valve. In the arrangement shown a flexible diaphragm forms a pressure tight chamber in the upper half of the actuator and the controller signal is fed in. Movement of the diaphragm results in a movement of the valve spindle and the valve. The diaphragm movement is opposed by a spring and is usually arranged so that the variation of controller output corresponds to full travel of the valve.

The valve body is arranged to fit into the particular pipeline and houses the valve and seat assembly. Valve operation may be direct acting where increasing pressure on the diaphragm closes the valve. A reverse acting valve opens as pressure on the diaphragm increases. The diaphragm movement is opposed by a spring which will close or open the valve in the event of air supply failure depending upon the action of the valve.

The valve disc or plug may be single or double seated and have any of a variety of shapes. The various shapes and types are chosen according to the type of control required and the relationship between valve lift and liquid flow.

A non-adjustable gland arrangement is usual. Inverted V-ring packing is used to minimize the friction against the moving spindle.

In order to achieve accurate valve disc positioning and overcome the effects of friction and unbalanced

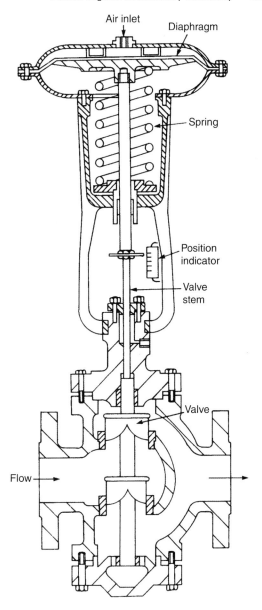

Figure 6.121 Pneumatically controlled valve.

forces a valve positioner may be used. The operating principle is shown in Figure 6.122. The controller signal acts on a bellows which will move the flapper in relation to the nozzle. This movement will alter the air pressure on the diaphragm which is supplied via an orifice from a constant pressure supply. The diaphragm movement will move the valve spindle and also the flapper. An equilibrium position will be set up when the valve disc is correctly positioned. This arrangement enables the use of a separate power source to actuate the valve.

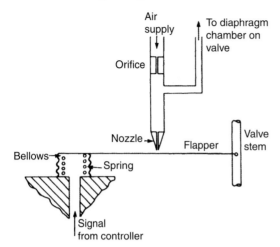

Figure 6.122 Valve positioner.

Actuator operation

The control signal to a correcting unit may be pneumatic, electric or hydraulic. The actuating power may also be any one of these three and not necessarily the same as the control medium.

Electrical control signals are usually of small voltage or current values which are unable to effect actuator movement. Pneumatic or hydraulic power would then be used for actuator operation.

A separate pneumatic power supply may be used even when the control signal is pneumatic, as described in the previous section.

Hydraulic actuator power is used where large or out of balance forces occur or when the correcting unit is of large dimensions itself. Hydraulic control with separate hydraulic actuation is a feature of some types of steering gear, as mentioned in Section 6.6.3.

6.7.2.6 Control systems

Boiler water level

A modern high pressure, high temperature watertube boiler holds a small quantity of water and produces large quantities of steam. Very careful control of the drum water level is therefore necessary. The reactions of steam and water in the drum are complicated and require a control system based on a number of measured elements.

When a boiler is operating the water level in the gauge glass reads higher than when the boiler is shut down. This is because of the presence of steam bubbles in the water, a situation which is accepted in normal practice. If however there occurs a sudden increase in steam demand from the boiler the pressure in the drum will fall. Some of the water present in the drum at the higher pressure will now 'flash off' and become steam. These bubbles of steam will cause the drum level to rise. The reduced mass of water in the drum will also result in more steam being produced, which will further raise the water level. This effect is known as 'swell'. A level control system which used only level as a measuring element would close in the feed control valve—when it should be opening it.

When the boiler load returns to normal the drum pressure will rise and steam bubble formation will reduce, causing a fall in water level. Incoming colder feed water will further reduce steam bubble formation and what is known as 'shrinkage' of the drum level will occur.

The problems associated with swell and shrinkage are removed by the use of a second measuring element, 'steam flow'. A third element, 'feed water flow', is added to avoid problems that would occur if the feed water pressure were to vary.

A three element control system is shown in Figure 6.123. The measured variables or elements are 'steam flow', 'drum level' and 'feed water flow'. Since in a balanced situation steam flow must equal feed flow, these two signals are compared in a differential relay. The relay output is fed to a two-term controller and comparator into which the measured drum level signal is also fed. Any deviation between the desired and actual drum level and any deviation between feed and steam flow will result in controller action to adjust the feed water control valve. The drum level will then be returned to its correct position.

A sudden increase in steam demand would result in a deviation signal from the differential relay and an output signal to open the feed water control valve. The swell effect would therefore not influence the correct operation of the control system. For a reduction in steam demand, an output signal to close the feed water control valve would result, thus avoiding shrinkage effects. Any change in feed water pressure would result in feed water control valve movement to correct the change before the drum level was affected.

Exhaust steam pressure control

Exhaust steam for various auxiliary services may be controlled at constant pressure by appropriate operation of a surplus steam (dump) valve or a make-up steam valve. A single controller can be used to operate one valve or the other in what is known as 'split range control'.

The control arrangement is shown in Figure 6.124. The steam pressure in the auxiliary range is measured by a pressure transmitter. This signal is fed to the controller where it is compared with the desired value. The two-term controller will

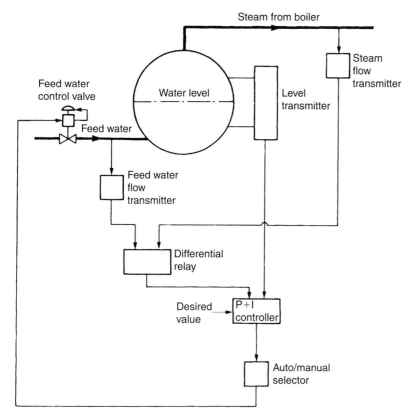

Figure 6.123 Boiler water level control.

provide an output signal which is fed to both control valves. Each valve is operated by a different range of pressure with a 'dead band' between the ranges so that only one valve is ever open at a time. The arrangement is shown in Figure 6.124. Thus if the auxiliary range pressure is high the dump valve opens to release steam. If the pressure is low the make-up valve opens to admit steam.

This split range control principle can be applied to a number of valves if the controller output range is split appropriately.

Steam temperature control
Steam temperature control of high pressure superheated steam is necessary to avoid damage to the metals used in a steam turbine.

One method of control is shown in Figure 6.125. Steam from the primary superheater may be directed to a boiler drum attemperator where its temperature will be reduced. This steam will then be further heated in the secondary superheater. The steam temperature leaving the secondary superheater is measured and transmitted to a three-term controller which also acts as a comparator. Any deviation from the desired value will result in a signal to a summing relay. The other signal to the relay is from a steam flow measuring element. The relay output signal provides control of the coupled attemperator inlet and bypass valves. As a result the steam flow is proportioned between the attemperator and the straight through line. This two-element control system can adequately deal with changing conditions. If, for example, the steam demand suddenly increased a fall in steam temperature might occur. The steam flow element will however detect the load change and adjust the amount of steam attemperated to maintain the correct steam temperature.

Boiler combustion control
The essential requirement for a combustion control system is to correctly proportion the quantities of air and fuel being burnt. This will ensure complete combustion, a minimum of excess air and acceptable exhaust gases. The control system must therefore measure the flow rates of fuel oil and air in order to correctly regulate their proportions.

A combustion control system capable of accepting rapid load changes is shown in Figure 6.126. Two

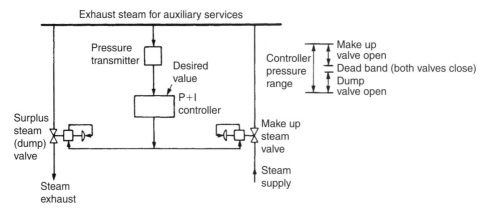

Figure 6.124 Exhaust steam pressure control.

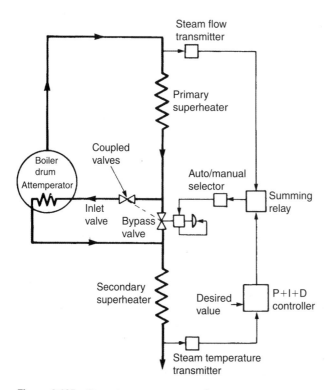

Figure 6.125 Steam temperature control.

control elements are used, 'steam flow' and 'steam pressure'. The steam pressure signal is fed to a two-term controller and is compared with the desired value. Any deviation results in a signal to the summing relay.

The steam flow signal is also fed into the summing relay. The summing relay which may add or subtract the input signals provides an output which represents the fuel input requirements of the boiler. This output becomes a variable desired value signal to the two-term controllers in the fuel control and combustion air control loops. A high or low signal selector is present to ensure that when a load change occurs the combustion air flow is always in excess of the fuel requirements. This prevents poor combustion and black smokey exhaust gases. If the master signal is for an increase in steam flow, then when it is fed to the low signal selector it is blocked since

Marine engines and auxiliary machinery 473

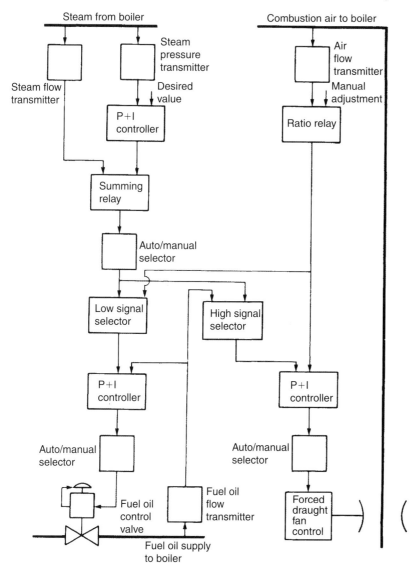

Figure 6.126 Boiler combustion control.

it is the higher input value. When the master signal is input to the high signal selector it passes through as the higher input. This master signal now acts as a variable desired value for the combustion air sub-loop and brings about an increased air flow. When the increased air flow is established its measured value is now the higher input to the low signal selector. The master signal will now pass through to bring about the increased fuel supply to the boiler via the fuel supply sub-loop. The air supply for an increase in load is therefore established before the increase in fuel supply occurs. The required air to fuel ratio is set in the ratio relay in the air flow signal lines.

Cooling water temperature control
Accurate control of diesel engine cooling water temperature is a requirement for efficient operation. This can be achieved by a single controller under steady load conditions, but because of the fluctuating situation during manœuvring a more complex system is required.

The control system shown in Figure 6.127 uses a combination of cascade and split range control. Cascade control is where the output from a master controller is used to automatically adjust the desired value of a slave controller. The master controller obtains an outlet temperature reading from the engine which is compared with a desired value. Any

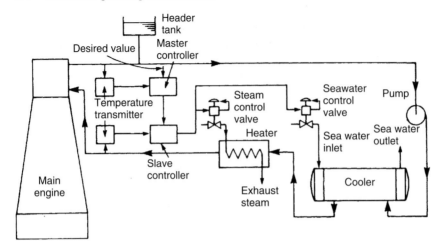

Figure 6.127 Cooling water temperature control.

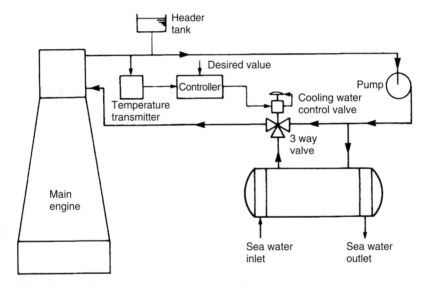

Figure 6.128 Cooling water temperature control.

deviation acts to adjust the desired value of the slave controller. The slave controller also receives a signal from the water inlet temperature sensor which it compares with its latest desired value. Any deviation results in a signal to two control valves arranged for split range control. If the cooling water temperature is high, the sea water valve is opened to admit more cooling water to the cooler. If the cooling water temperature is low, then the sea water valve will be closed in. If the sea water valve is fully closed, then the steam inlet valve to the water heater will be opened to heat the water. Both master and slave controllers will be identical instruments and will be two-term (P + I) in action.

Another method of temperature control involves the use of only a single measuring element (Figure 6.128). A three-way valve is provided in the cooling water line to enable bypassing of the cooler. The cooler is provided with a full flow of sea water which is not controlled by the system. A temperature sensing element on the water outlet provides a signal to a two-term controller (P + I). The controller is provided with a desired value and any deviation between it and the signal will result in an output to the three-way control valve. If the measured temperature is low, more water will be bypassed and its temperature will therefore increase. If the measured temperature is high, then less water will be bypassed, more will be cooled and the temperature will fall. A simple system such as this can be used only after careful analysis of the plant conditions and the correct sizing of equipment fitted.

6.7.2.7 Centralized control

The automatic control concept, correctly developed, results in the centralizing of control and supervisory functions. All ships have some degree of automation and instrumentation which is centred around a console. Modern installations have machinery control rooms where the monitoring of control functions takes place. The use of a separate room in the machinery space enables careful climate control of the space for the dual benefit of the instruments and the engineer.

Control consoles are usually arranged with the more important controls and instrumentation located centrally and within easy reach. The display panels often make use of mimic diagrams. These are line diagrams of pipe systems or items of equipment which include miniature alarm lights or operating buttons for the relevant point or item in the system. A high temperature alarm at, for instance, a particular cylinder exhaust would display at the appropriate place on the mimic diagram of the engine. Valves shown on mimic diagrams would be provided with an indication of their open or closed position, pumps would have a running light lit if operating, etc. The grouping of the controllers and instrumentation for the various systems previously described enables them to become part of the complete control system for the ship.

The ultimate goal in the centralized control room concept will be to perform and monitor every possible operation remotely from this location. This will inevitably result in a vast amount of information reaching the control room, more than the engineer supervisor might reasonably be expected to continuously observe. It is therefore usual to incorporate data recording and alarm systems in control rooms. The alarm system enables the monitoring of certain measured variables over a set period and the readings obtained are compared with some reference or desired value. Where a fault condition is located, i.e. a measured value different from the desired value, audible and visual alarms are given and a print-out of the fault and the time of occurrence is produced. Data recording or data logging is the production of measured variable information either automatically at set intervals or on demand. A diagrammatic layout of a data logging and alarm monitoring system is shown in Figure 6.129.

6.7.2.8 Unattended machinery spaces

The sophistication of modern control systems and the reliability of the equipment used have resulted in machinery spaces remaining unattended for long periods. In order to ensure the safety of the ship and its equipment during UMS operation certain essential requirements must be met:

1. *Bridge control.* A control system to operate the main machinery must be provided on the bridge. Instrumentation providing certain basic information must be provided.
2. *Machinery control room.* A centralized control room must be provided with the equipment to operate all main and auxiliary machinery easily accessible.
3. *Alarm and fire protection.* An alarm system is required which must be comprehensive in coverage of the equipment and able to provide warnings in the control room, the machinery space, the accommodation and on the bridge. A fire detection and alarm system which operates rapidly must also be provided throughout the machinery space, and a fire control point must be provided outside the machinery space with facilities for control of emergency equipment.
4. *Emergency power.* Automatic provision of electrical power to meet the varying load requirements. A means of providing emergency electrical power and essential lighting must be provided. This is usually met by the automatic start up of a stand-by generator.

6.7.2.9 Bridge control

Equipment operation from the machinery control room will be by a trained engineer. The various preparatory steps and logical timed sequence of events which an engineer will undertake cannot be expected to occur when equipment is operated from the bridge. Bridge control must therefore have built into the system appropriate circuits to provide the correct timing, logic and sequence. There must also be protection devices and safety interlocks built into the system.

A bridge control system for a steam turbine main propulsion engine is shown in Figure 6.130. Control of the main engine may be from the bridge control unit or the machinery control room. The programming and timing unit ensures that the correct logical sequence of events occurs over the appropriate period. Typical operations would include the raising of steam in the boiler, the circulating of lubricating oil through the turbine and the opening of steam drains from the turbine.

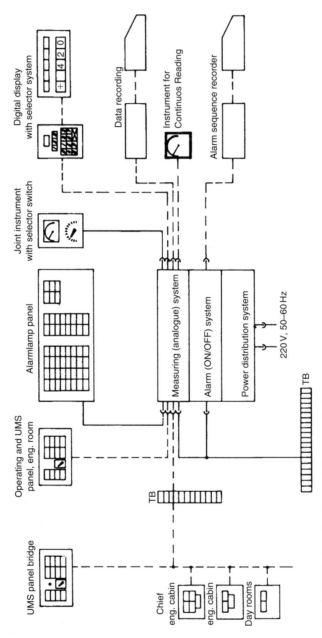

Figure 6.129 Data logging and alarm monitoring system.

Marine engines and auxiliary machinery 477

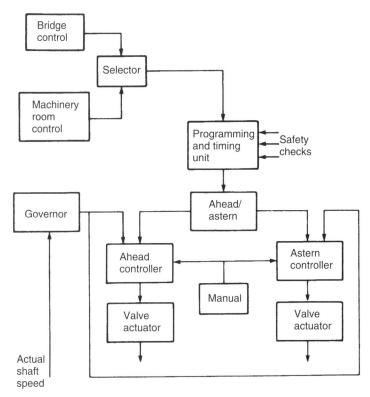

Figure 6.130 Bridge control of steam turbine plant.

The timing of certain events, such as the opening and closing of steam valves, must be carefully controlled to avoid dangerous conditions occurring or to allow other system adjustments to occur. Protection and safety circuits or interlocks would be input to the programming and timing unit to stop its action if, for example, the turning gear was still engaged or the lubricating oil pressure was low. The ahead/astern selector would direct signals to the appropriate valve controller resulting in valve actuation and steam supply. When manœuvring some switching arrangement would ensure that the astern guardian valve was open, bled steam was shut off, etc. If the turbine were stopped it would automatically receive blasts of steam at timed intervals to prevent rotor distortion. A feedback signal of shaft speed would ensure correct speed without action from the main control station.

A bridge control system for a slow-speed diesel main engine is shown in Figure 6.131. Control may be from either station with the operating signal passing to a programming and timing unit. Various safety interlocks will be input signals to prevent engine starting or to shut down the engine if a fault occurred. The programming unit signal would then pass to the camshaft positioner to ensure the correct directional location. A logic device would receive the signal next and arrange for the supply of starting air to turn the engine. A signal passing through the governor would supply fuel to the engine to start and continue operation. A feedback signal of engine speed would shut off the starting air and also enable the governor to control engine speed. Engine speed would also be provided as an instrument reading at both control stations.

A bridge control system for a controllable-pitch propeller is shown in Figure 6.132. The propeller pitch and engine speed are usually controlled by a single lever (combinator). The control lever signal passes via the selector to the engine governor and the pitch-operating actuator. Pitch and engine speed signals will be fed back and displayed at both control stations. The load control unit ensures a constant load on the engine by varying propeller pitch as external conditions change. The input signals are from the fuel pump setting and actual engine speed. The output signal is supplied as a feedback to the pitch controller.

The steering gear is, of course, bridge controlled and is arranged for automatic or manual control. A typical automatic or auto pilot system is shown in Figure 6.133. A three-term controller provides the

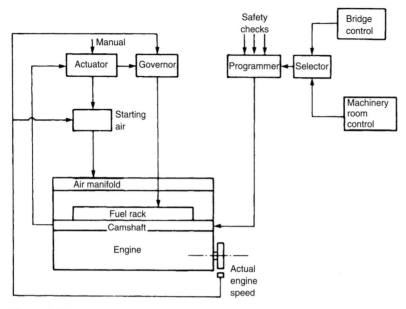

Figure 6.131 Bridge control of slow-speed diesel engine.

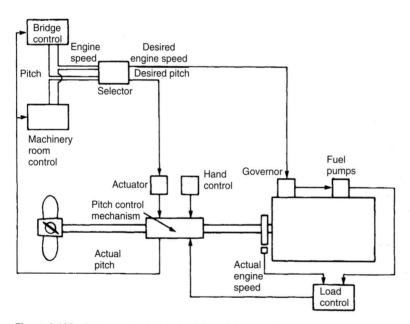

Figure 6.132 Bridge control of controllable-pitch propeller.

output signal where a course deviation exists and will bring about a rudder movement. The various system parts are shown in terms of their system functions and the particular item of equipment involved. The feedback loop between the rudder and the amplifier (variable delivery pump) results in no pumping action when equilibrium exists in the system. External forces can act on the ship or the rudder to cause a change in the ship's actual course resulting in a feedback to the controller and subsequent corrective action. The controller action must be correctly adjusted for the particular external conditions to ensure that excessive rudder movement does not occur.

Marine engines and auxiliary machinery 479

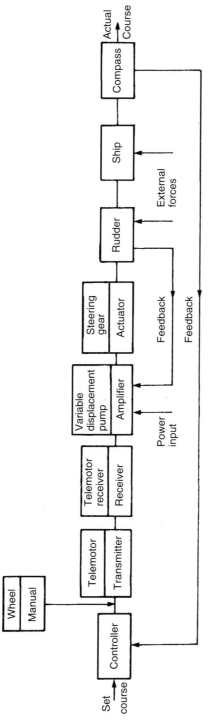

Figure 6.133 Automatic steering system.

Electrical supply control

The automatic provision of electrical power to meet varying load demands can be achieved by performing the following functions automatically:

1. Prime mover start up.
2. Synchronizing of incoming machine with bus-bars.
3. Load sharing between alternators.
4. Safety and operational checks on power supply and equipment in operation.
5. Unloading, stopping and returning to stand-by of surplus machines.
6. Preferential tripping of non-essential loads under emergency conditions and their reinstating when acceptable.

A logic flow diagram for such a system is given in Figure 6.134. Each of three machines is considered able to supply 250 kW. A loading in excess of this will result in the start up and synchronizing of another machine. Should the load fall to a value where a running machine is unnecessary it will be unloaded, stopped and returned to the stand-by condition. If the system should overload through some fault, such as a machine not starting, an alarm will be given and preferential tripping will occur of non-essential loads. Should the system totally fail, the emergency alternator will start up and supply essential services and lighting through its switchboard.

6.7.2.10 *Integrated control*

The various control and monitoring systems described so far may be integrated in order to enable more efficient ship operation and reduce manning. Machinery control systems are being combined with navigation and cargo control systems to bring about 'Efficient Ship' integrated control systems. Combining previously separate sources of data regarding, for example, ship speed and fuel consumption, enables optimizing of ship or engine operating parameters.

An Integrated Control System would be made up of a Bridge System, a Cargo Control System, a Machinery Control System and possibly a Ship Management System.

The Bridge System would include an automatic radar plotting aid display, an electronic chart table, an autopilot, a gyro, log, and echo sounder. The Cargo Control System will vary according to the type of vessel, but will enable loading calculations, cargo management, ballast control and data logging. The Machinery Control System will combine various control systems to enable surveillance to UMS requirements, performance and condition monitoring, generator control and automatic data logging. Ship Management would involve administrative record keeping, word processing, stock control and maintenance planning.

Workstations with computers, monitors and keyboards would be provided in the appropriate locations, such as the machinery control room, on the bridge, in the cargo control room and various ship's offices. A network would connect the various workstations and enable the exchange of information between them.

Inputs from the various monitored items of equipment would be fed to Local Scanner and Control Units (LSCU), which would contain a microprocessor and be effectively a microcomputer. The LSCU is part of a local control loop which can function independently, if necessary. The LSCUs are connected up to a central computer which can interface with them and would act as the workstation for the particular system.

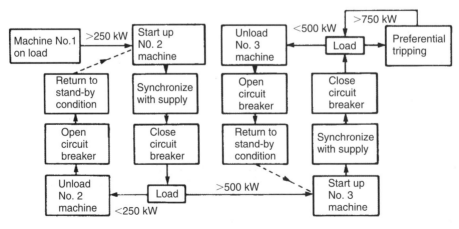

Figure 6.134 Automatic load control of alternators.

Integrating the various systems enables optimal control of a ship and improved efficiency. Fuel consumption figures could be monitored, for example and used to predict an appropriate time to drydock the vessel as hull resistance increased due to fouling. Condition monitoring of machinery would enable maintenance schedules to be planned in order to minimise breakdowns and repair costs. Satellite communications will also enable data to be relayed from ship to shore for analysis by office-based technical staff. See also Section 5.7.6.

References (Chapter 6)

Allanson, J.T. and Charnley, R. (1987). Drinking water from the sea: reverse osmosis, the modern alternative. *Trans I Mar E*, Vol. 88.

Bille, T. (1970). Experiences with controllable pitch propellers. *Trans I Mar E*, Vol. 80, p. 8.

Brownlie, K. (1998). Controllable Pitch Propellers, *I. Mar. EST*. ISBN 1-902536-01-X.

Carlton, J.S. (2007). *Marine Propellers and Propulsion*, 2nd edition. Butterworth-Heinemann, Oxford, UK.

Code for the Construction and Equipment of Ships Carrying Dangerous Chemicals in Bulk (BCH Code), IMO publication *(IMO-772E)*.

Cowley, J. (1982). Steering Gear: New Concepts and Requirements. *Trans I Mar E*, Vol. 94, paper 23.

Crombie, G. and Clay, C.F. (1972). Design feature of and operating experience with turnbull split stern bearings. *Trans I Mar E*, Vol. 84, p. 1.

Developments in water-lubricated bearing technology (2004). *The Naval Architect* February.

Fossen, T.I. (2000). *Marine Control Systems*. Marine Cybernetics AS, Trondheim, Norway.

Gilchrist, A. (1976). Sea Water Distillers. *Trans I Mar E*, Vol. 88.

Hamilton, F.C. (1948). *Famous Paddle Steamers*. Marshall, London.

Hensel, W. (1984). Energy saving in ships power supplies. *Trans I Mar E*, Vol. 96, paper 49.

Herbert, C.W. and Hill, A. (1972). Sterngear design for maximum reliability – the Glacier-Herbert system. *Trans I Mar E*, Vol. 84, p. 11.

Hill, E.C. (1987). *Legionella and Ships Water Systems*. MER, I Mar EST.

IHI (l993). CRP System for Large Merchant Ships, Ship Technology International '93, SPG.

International Code for the Construction and Equipment of Ships Carrying Dangerous Chemicals in Bulk (IBC Code), IMO publication *(IMO-100E)*.

Iwata, A., Tada, E. and Saji, Y. (1983) Experimental and theoretical study of superconducting electromagnetic ship propulsion. Paper No. 2, *5th Lips Propeller Symposium*, May.

Iwata, A. (1990). Superconducting electromagnetic propulsion system. *Bull. of Mar. Eng. Soc.*, Vol. 18, No. 1.

Kerlen, H., Esveldt, J. and Wereldsman, R. (1970). *Propulsion, Cavitation and Vibration Characteristics of Overlapping Propellers for a Container Ship*. Schiffbautechnische Gesellschaft, Berlin, November.

Marine Power and Propulsion – solutions for naval architects (2005). The Royal Institution of Naval Architects.

MARPOL 73/78 Consolidated Edition 2002, IMO publication *(IMO-1B 520E)* (*see* Annex 1 – Chapter I – Regulation 18 – *Pumping, piping and discharge arrangements for oil tankers*).

MARPOL 2005 Amendments 2005, IMO publication *(IMO-1525E)*.

McGeorge, H.D. (1993). *Marine Electrical Equipment and Practice*, 2nd edition. Butterworth-Heinemann, Newnes.

McGeorge, H.D. (1995). *Marine Auxiliary Machinery*. 7th edition. Butterworth-Heinemann, Oxford, UK.

Merchant Shipping Notice No. M1214 Recommendations to Prevent Contamination of Ships Fresh Water Storage and Distribution Systems.

Merchant Shipping Notice No. M1401 Disinfection of Ships Domestic Fresh Water.

Mikkelsen, G. (1984). Auxiliary power generation in today's ships. *Trans 1 Mar E*, Vol. 96, paper 52.

Miles, A., Wellicome, J.F. and Molland, A.F. (1993). The technical and commercial development of self-pitching propellers. *Trans. RINA*, Vol. 135.

Mitchell, R.W.S. and Kievits, F.J. (1974). Gas Turbine Corrosion in the Marine Environment. Proceedings of a Joint Conference: Corrosion in the Marine Environment, *Trans I Mar E*, series B.

Murrell, P.W. and Barclay, L. (1984). Shaft driven generators for marine application. *Trans I Mar E*, Vol. 96, paper 50.

Perez, T. (2005). *Ship Motion Control. Course Keeping and Roll Stabilisation using Rudder and Fins*. Springer.

Plumb, C.M. (1987). Warship Propulsion System Selection. *I. Mar.EST*.

Pressicaud J. P. Correlation between theory and reality in alignment of line shafting, Bureau Veritas.

Pringle, G.G. (1982). Economic power generation at sea: the constant speed shaft driven generator. *Trans I Mar E*, Vol. 94, paper 30.

Queen Mary 2: Genesis of a Queen (2004). The Royal Institution of Naval Architects.

Restad, K., Volcy, G.G., Ganier, H. and Masson, J.C. (1973). Investigation on free and forced vibrations of an LNG tanker with overlapping/propeller arrangement. *Trans. SNAME*.

Rice, W.A. (1961). *US Patent 2997013*, 22, August.

Rose A. (1974). Hydrostatic stem gear, N.E. Coast 1. of Engineers and Shipbuilders. Sterntube Bearings. The Glacier Metal Co. Ltd.

Schneider, P. (1984). Production of auxiliary energy by the main engine. *Trans I Mar E*, Vol. 96, paper 51.

Simonsson, P. (1983). *The Pinnate Propeller*, Department of Mechanics. Royal Institute of Technology, Stockholm.

Sinclair, L. and Emerson, A. (1968). The design and development of propellers for high powered merchant vessels. *Trans I Mar E*, Vol. 80, p. 5.

Smith, D.W. (1983). *Marine Auxiliary Machinery*, 6th edition. Butterworth-Heinemann, UK.

SOLAS Consolidated Edition (2004). IMO publication *(IMO-1D 110E)* (*see* Chapter 11–2 – Part D – *Fire safety measures for tankers)*.

Steering a Safer Course (1988). *MER,* October. p. 31.

Taylor, D.A. (1996). *Introduction to Marine Engineering*. Revised 2nd Edition. Butterworth-Heinemann, Oxford, UK.

The International Convention for the Safety of Life at Sea (1974). as amended in November 1981, Chapter 11–1: Regulation 29.

The Merchant Shipping (Crew Accommodation) Regulations 1978, HMSO.

The Merchant Shipping (Passenger Ship Construction and Survey) Regulations 1984, HMSO.

The Merchant Shipping (Cargo Ships Construction and Survey) Regulations 1984, HMSO.

Volpich and Bridge (1954–55). Paddle wheels: Preliminary model experiments. *Trans. IESS*, Vol. 98.

Volpich and Bridge (1955–56). Systemic model experiments. *Trans. IESS*, Vol. 99.

Volpich and Bridge (1956–57). Further model experiments and ship model correlation. *Trans. IESS*, Vol. 100.

Wilkin. T. A. and Strassheim, W. (1973). Some theoretical and practical aspects of shaft alignment, *I Mar E* IMAS 73 Conference.

Woodyard. D.F. (2004). *Pounder's Marine Diesel Engines and Gas Turbines*, 8th edition. Butterworth-Heinemann, Oxford, UK.

7 Seakeeping

Contents

7.1 Seakeeping qualities
7.2 Ship motions
7.3 Limiting seakeeping criteria
7.4 Overall seakeeping performance
7.5 Data for seakeeping assessments
7.6 Non-linear effects
7.7 Numerical prediction of seakeeping
7.8 Experiments and trials
7.9 Improving seakeeping performance
7.10 Ship motion control
References (Chapter 7)

The various Sections of this Chapter have been taken from the following books, with the permission of the authors:

Bertram, V. (2000) *Practical Ship Hydrodynamics*, Butterworth-Heinemann, Oxford, UK. [Sections 7.3.1.2, 7.7]

Kobylinski, L.K. and Kastner, S. (2003) *Stability and Safety of Ships*. Elsevier, Oxford, UK. [Section 7.2.11]

Molland, A.F. and Turnock, S.R. (2007) *Marine Rudders and Control Surfaces*. Butterworth-Heinemann, Oxford, UK. [Sections 7.10.2.4, 7.10.3]

Rawson K.J. and Tupper E.C. (2001) *Basic Ship Theory*. 5th Edition, Combined Volume. Butterworth-Heinemann, Oxford, UK. [Sections 7.1–7.2.10, 7.3 (excluding 7.3.1.2), 7.4–7.6, 7.8–7.10.2.3]

7.1 Seakeeping qualities

The general term sea worthiness must embrace all those aspects of a ship design which affect its ability to remain at sea in all conditions and to carry out its specified duty. It should, therefore, include consideration of strength, stability and endurance, besides those factors more directly influenced by waves. In this Chapter, the term seakeeping is used to cover these more limited features, i.e. motions, speed and power in waves, wetness and slamming.

The relative importance of these various aspects of performance in waves varies from design to design depending upon what the operators require of the ship, but the following general comments are applicable to most ships.

7.1.1 Motions

Excessive amplitudes of motion are undesirable. They can make shipboard tasks hazardous or even impossible, and reduce crew efficiency and passenger comfort. In warships, most weapon systems require their line of sight to remain fixed in space and to this end each system is provided with its own stabilizing system. Large motion amplitudes increase the power demands of such systems and may restrict the safe arcs of fire.

The phase relationships between various motions are also important. Generally, the phasing between motions is such as to lead to a point of minimum vertical movement about two-thirds of the length of the ship from the bow. In a passenger liner, this area would be used for the more important accommodation spaces. If it is desirable to reduce the vertical movement at a given point, then this can be achieved if the phasing can be changed, for example in a frigate motion at the flight deck can be the limiting factor in helicopter operations. Such actions must inevitably lead to increased movement at some other point. In the frigate, increased movement of the bow would result and wetness or slamming might then limit operations.

7.1.2 Speed and power in waves

When moving through waves the resistance experienced by a ship is increased and, in general, high winds mean increased air resistance. These factors cause the ship speed to be reduced for a given power output, the reduction being aggravated by the less favourable conditions in which the propeller is working. These factors are discussed further in Chapter 5. Other unpleasant features of operating in waves such as motions, slamming and wetness are generally eased by a reduction in speed so that an additional speed reduction may be made voluntarily.

7.1.3 Wetness

When the relative movement of the bow and local wave surface becomes too great, water is shipped over the forecastle. At an earlier stage, spray is driven over the forward portion of the ship by the wind. Both conditions are undesirable and can be lessened by increasing freeboard. The importance of this will depend upon the positioning of upper deck equipment and its sensitivity to salt spray. Spray rails, flare angles and knuckles may all influence the troublesome nature of spray which, in cold climates, causes ice accretion. Wetness is discussed in more detail in Section 7.3.1.3.

7.1.4 Slamming

Under some conditions, the pressures exerted by the water on a ship's hull become very large and slamming occurs. Slamming is characterized by a sudden change in the vertical acceleration of the ship followed by a vibration of the ship girder in its natural frequencies. The conditions leading to slamming are high relative velocity between ship and water, shallow draught and small rise of floor. The area between 10 and 25% of the length from the bow is the area most likely to suffer high pressures and to sustain damage. Slamming is discussed in more detail in Section 7.3.1.2.

7.1.5 Ship routing

Since the ship behaviour depends upon the wave conditions it meets, it is reasonable to question whether overall performance can be improved by avoiding the more severe waves. This possibility has been successfully pursued by some authorities, Satchwell (1989). Data from weather ships are used to predict the speed loss in various ocean areas and to compute the optimum route. In this way, significant savings have been made in voyage times, e.g. of the order of 10–15 hours for the Atlantic crossing.

7.1.6 Importance of good seakeeping

No single parameter can be used to define the seakeeping performance of a design. In a competitive world, a comfortable ship will attract more passengers than a ship with a bad reputation. A ship with less power augment in waves will be able to maintain tighter schedules or will have a lower fuel bill. In extreme cases, the seakeeping qualities of a ship may determine its ability to make a given voyage at all.

Good seakeeping is clearly desirable, but the difficulty lies in determining how far other design features must, or should, be compromised to improve seakeeping. This will depend upon each particular design, but it is essential that the designer has some means of judging the expected performance and the effect on the ship's overall effectiveness. Theory, model experiment and ship trial all have a part to play. Because of the random nature of the sea surface in which the ship operates, considerable use is made of the principles of statistical analysis.

Having improved the physical response characteristics of a ship in waves the overall effectiveness of a design may be further enhanced by judicious siting of critical activities and by fitting control devices such as anti-roll stabilizers.

As with so many other aspects of ship design a rigorous treatment of seakeeping is very complex and a number of simplifying assumptions are usually made. For instance, the ship is usually regarded as responding to the waves as a rigid body when assessing motions and wetness although its true nature as an elastic body must be taken into account in a study of structure, Bishop and Price (1979). In the same way it is instructive, although not correct, to study initially the response of a ship to regular long-crested waves ignoring the interactions between motions, for example when the ship is heaving the disturbing forces will generate a pitching motion. This very simple approach is now dealt with before considering coupled motions.

7.2 Ship motions

7.2.1 Degrees of freedom

A floating body has six degrees of freedom. In order to completely define the ship motion it is necessary to consider movements in all these modes as illustrated in Figure 7.1. The motions are defined as movements of the centre of gravity of the ship and rotations about a set of orthogonal axes through the centre of gravity, G. These are space axes moving with the mean forward speed of the ship but otherwise fixed in space.

It will be noted that roll and pitch are the dynamic equivalents of heel and trim. Translations along the x- and y-axis and rotation about the z-axis lead to no residual force or moment, provided displacement remains constant, as the ship is in neutral equilibrium. For the other translation and rotations, movement is

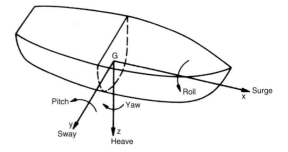

Translation or rotation	Axis	Description	Positive sense
Translation	Along x	Surge	Forwards
	Along y	Sway	To starboard
	Along z	Heave	Downwards
Rotation	About x	Roll	Starboard side down
	About y	Pitch	Bow up
	About z	Yaw	Bow to starboard

Figure 7.1 Ship motions.

opposed by a force or moment provided the ship is stable in that mode. The magnitude of the opposition increases with increasing displacement from the equilibrium position, the variation being linear for small disturbances.

This is the characteristic of a simple spring system. Thus, it is to be expected that the equation governing the motion of a ship in still water, which is subject to a disturbance in the roll, pitch or heave modes, will be similar to that governing the motion of a mass on a spring. This is indeed the case, and for the undamped case the ship is said to move with simple harmonic motion.

Disturbances in the yaw, surge and sway modes will not lead to such an oscillatory motion and these motions, when the ship is in a seaway, exhibit a different character to roll, pitch and heave. These are considered separately and it is the oscillatory motions which are dealt with in the next few sections. It is convenient to consider the motion which would follow a disturbance in still water, both without and with damping, before proceeding to the more realistic case of motions in waves.

7.2.2 Undamped motion in still water

It is assumed that the ship is floating freely in still water when it is suddenly disturbed. The motion following the removal of the disturbing force or moment is now studied for the three oscillatory motions.

7.2.2.1 Rolling

Let ϕ be the inclination of the ship to the vertical at any instant. The moment, acting on a stable ship, will be in a sense such as to decrease ϕ. For small values of ϕ,

moment $= -\Delta \text{GM} \phi$

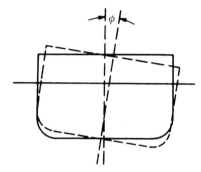

Figure 7.2   Rolling.

Applying Newton's laws of motion

moment = (moment of inertia about $0x$)
(angular acceleration)

i.e.

$$-\Delta \text{GM} \phi = +\frac{\Delta}{g} k_{xx}^2 \frac{d^2\phi}{dt^2} \quad (7.1)$$

i.e.

$$\frac{d^2\phi}{dt^2} + \left(g \frac{\text{GM}}{k_{xx}^2}\right)\phi = 0 \quad (7.2)$$

This is the differential equation denoting simple harmonic motion with period T_ϕ where

$$T_\phi = 2\pi \left(\frac{k_{xx}^2}{g\text{GM}}\right)^{\frac{1}{2}} = \frac{2\pi k_{xx}}{(g\text{GM})^{\frac{1}{2}}} \quad (7.3)$$

It will be noted that the period of roll is independent of ϕ and that this will hold as long as the approximation GZ = GMϕ applies, i.e. typically up to ± 10 degrees. Such rolling is termed *isochronous*.

In practice k_{xx} must be increased to allow for what are usually termed 'added mass' effects due to motion induced in the water although this does not mean that a specific body of water actually moves with the ship, see Section 7.2.11.2. Added mass values vary with frequency but this variation can often be ignored to a first order. Typically the effect increases k_{xx} by about 5%.

Hence

$$T_\phi \propto \frac{1}{(\text{GM})^{\frac{1}{2}}} \quad (7.4)$$

Thus the greater is GM, i.e. the more stable the ship, the shorter the period and the more rapid the motion. A ship with a short period is said to be 'stiff'— compare the stiff spring—and one with a long period is said to be 'tender'. Most people find a long period roll less unpleasant than a short period roll.

Pitching
This is analogous to roll and the motion is governed by the equation

$$\frac{d^2\theta}{dt^2} + \left(\frac{g\text{GM}_L}{k_{yy}^2}\right)\theta = 0 \quad (7.5)$$

and the period of the motion is

$$T_\theta = \frac{2\pi k_{yy}}{(g\text{GM}_L)^{\frac{1}{2}}} \quad \text{for very small angles of pitch.} \quad (7.6)$$

7.2.2.2 Heaving

Let z be the downward displacement of the ship at any instant. The force acting on the ship tends to reduce z and has a magnitude F_z given by

$$F_z = -\frac{A_W z}{u} \quad (7.7)$$

where u is the reciprocal weight density of the water.

Hence, the heaving motion is governed by the equation

$$\frac{\Delta}{g} \frac{d^2 z}{dt^2} = -\frac{A_W z}{u}$$

or

$$\frac{d^2 z}{dt^2} + \frac{gA_W}{u\Delta} z = 0 \quad (7.8)$$

from which

$$\text{period} = 2\pi \left(\frac{u\Delta}{gA_W}\right)^{\frac{1}{2}} \quad (7.9)$$

Δ may be effectively increased by a significant amount (perhaps doubled) by the 'added mass' effect.

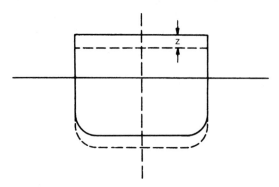

Figure 7.3 Heaving.

7.2.3 Damped motion in still water

Now consider what happens when the motion is damped. It is adequate to illustrate the effect of damping on the rolling motion.

Only the simplest case of damping is considered here namely, that in which the damping moment varies linearly with the angular velocity. It opposes the motion since energy is always absorbed.

Allowing for the entrained water the equation for rolling in still water becomes

$$\frac{\Delta}{g} k_{xx}^2 (1 + \sigma_{xx}) \ddot{\phi} + B \dot{\phi} + \Delta \mathrm{GM} \phi = 0 \quad (7.10)$$

where

$$\frac{\Delta k_{xx}^2}{g} \sigma_{xx} = \text{augment of rolling inertia of ship due to entrained water}$$
$$B = \text{damping constant.}$$

This can be likened to the standard differential equation

$$\ddot{\phi} + 2k\omega_0 \dot{\phi} + \omega_0^2 \phi = 0$$

where

$$\omega_0^2 = \frac{g\mathrm{GM_T}}{k_{xx}^2 (1 + \sigma_{xx})} \quad \text{and}$$
$$k = \frac{Bg}{2\omega_0 \Delta k_{xx}^2 (1 + \sigma_{xx})}$$

which in turn defines the effective period T_ϕ of the motion as

$$T_\phi = \frac{2\pi}{\omega_0}(1 - k^2)^{-\frac{1}{2}} = 2\pi k_{xx} \left(\frac{1 + \sigma_{xx}}{g\mathrm{GM}} \right)^{\frac{1}{2}} (1 - k^2)^{-\frac{1}{2}}$$

(7.11)

When the damping is not proportional to the angular velocity the differential equation is no longer capable of ready solution.

7.2.4 Approximate period of roll

Of the various ship motions the roll period is likely to vary most from design to design and, because of the much greater amplitudes possible, it is often the most significant. Various approximate formulae have been suggested for calculating the period of roll, using Equation (7.3) including:

$$T_\phi = 2\pi \frac{K}{(g\mathrm{GM})^{\frac{1}{2}}} \quad (7.12)$$

Suggested values of K for merchant ships and warships are given by the respective expressions:

Merchant ships

$$\left(\frac{K}{B} \right)^2 = F \left[C_\mathrm{B} C_\mathrm{u} + 1.10 C_\mathrm{u} (1 - C_\mathrm{B}) \right.$$
$$\left. \left(\frac{H}{T} - 2.20 \right) + \frac{H^2}{B^2} \right] \quad (7.13)$$

where
C_u = upper deck area coeff. = $\frac{1}{LB}$ (deck area)

H = effective depth of ship = $D + A/L_\mathrm{pp}$
A = projected lateral area of erections and deck
L_pp = L.B.P.
T = mean moulded draught
F = constant = 0.125 for passenger and cargo ship,
= 0.133 for oil tankers,
= 0.177 for whalers.

Warships

$$\left(\frac{K}{B} \right)^2 = F \left[C_\mathrm{B} C_\mathrm{u} + 1.10 C_\mathrm{e} (1 - C_\mathrm{B}) \right.$$
$$\left. \left(\frac{H_n}{T} - 2.20 \right) + \frac{H_n^2}{B_\mathrm{u}^2} \right] \quad (7.14)$$

where
B_u = max. breadth under water;
C_e = exposed deck area coeff.;
$H_n = D + A_n/L_\mathrm{pp}$;
D = depth from top of keel to upper deck;
A_n = sum of the projected lateral areas of forecastle, under bridge and gun;
F = constant ranging from 0.172 for small warships to 0.177 for large warships.

7.2.5 Motion in regular waves

7.2.5.1 *Assumptions*

In Chapter 1, it is explained that the irregular wave systems met at sea can be regarded as made up of a large number of regular components. A ship's motion record will exhibit a similar irregularity and it can be regarded as the summation of the ship responses to all the individual wave components. Theoretically, this super-position procedure is valid only for those sea states for which the linear theory of motions is applicable, i.e. for moderate sea states. It has been demonstrated by several authorities, however, that provided the basic data is derived from relatively mild regular components, the technique can be applied, with sufficient accuracy for most engineering purposes, to more extreme conditions. Thus, the basic element in ship motions is the response of the ship to a regular train of waves. For mathematical convenience, the wave is assumed to have a sinusoidal profile. The characteristics of such a system are dealt with in Chapter 1.

In the simple approach, it is necessary to assume that the pressure distribution within the wave system is unaffected by the presence of the ship. This is one of the assumptions made by William Froude in his study of ship rolling and is commonly known as 'Froude's Hypothesis'.

7.2.5.2 *Rolling in a beam sea*

The equation for rolling in still water is modified by introducing a forcing function on the right-hand side of the equation. This could be obtained by calculating the hydrodynamic pressure acting on each element of the hull and integrating over the complete wetted surface.

The resultant force acting on a particle in the surface of a wave must be normal to the wave surface. Provided the wave length is long compared with the beam of the ship, it is reasonable to assume that the ship is acted on by a resultant force normal to an 'effective wave surface' which takes into account all the sub-surfaces interacting with the ship. Froude used this idea and further assumed that the 'effective wave slope' was that of the sub-surface passing through the centre of buoyancy of the ship.

With this assumption it can be shown that, approximately, the equation of motion for undamped rolling motion in beam seas becomes

$$\frac{\Delta}{g} k_{xx}^2 (1 + \sigma_{xx}) \ddot{\phi} + \Delta \mathrm{GM}(\phi - \phi') = 0 \quad (7.15)$$

where $\phi' = \alpha \sin \omega t$; α = maximum slope of the surface wave; ω = frequency of the surface wave.

If ϕ_0 and ω_0 are the amplitude and frequency of unresisted rolling in still water, the solution to this equation takes the form

$$\phi = \phi_0 \sin(\omega_0 t + \beta) + \frac{\omega_0^2 \alpha}{\omega_0^2 - \omega^2} \sin \omega t \quad (7.16)$$

The first term is the free oscillation in still water and the second is a forced oscillation in the period of the wave train.

The amplitude of the forced oscillation is

$$\frac{\omega_0^2 \alpha}{\omega_0^2 - \omega^2}$$

When the period of the wave system is less than the natural period of the ship ($\omega > \omega_0$), the amplitude is negative which means that the ship rolls into the wave (Figure 7.4(a)). When the period of the wave is greater than the natural period of the ship, the amplitude is positive and the ship rolls with the wave (Figure 7.4(b)). For very long waves, i.e. ω very small, the amplitude tends to α and the ship remains approximately normal to the wave surface. When the frequencies of the wave and ship are close the amplitude of the forced oscillation becomes very large.

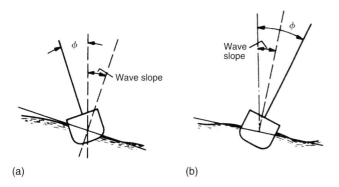

(a)   (b)

Figure 7.4 Rolling in a beam sea.

The general equation for rolling in waves can be written as:

$$\ddot{\phi} + 2k\omega_0\dot{\phi} + \omega_0^2\phi = \omega_0^2\alpha\cos\omega t \qquad (7.17)$$

The solution to this differential equation is

$$\phi = \mu\alpha\cos(\omega t - \varepsilon) \qquad (7.18)$$

where

$$\tan\varepsilon = \frac{2k\Lambda}{1-\Lambda^2}$$

$\Lambda = $ *tuning factor* $= \omega/\omega_0$;
$\mu = $ *magnification factor*
$= 1/\{(1-\Lambda^2)^2 + 4k^2\Lambda^2\}^{\frac{1}{2}}$

Plots of the phase angle ε and magnification factor are presented in Figure 7.5. It will be appreciated that these expressions are similar to those met with in the study of vibrations.

The effect of damping is to cause the free oscillation to die out in time and to modify the amplitude of the forced oscillation. In an ideal regular sea, the ship would oscillate after a while only in the period of the waves. In practice, the maximum forced roll amplitudes occur close to the natural frequency of the ship, leading to a ship at sea rolling predominantly at frequencies close to its natural frequency.

A more detailed discussion of rolling is given in Section 7.2.11.

7.2.5.3 *Pitching and heaving*

In this case, attention is focused on head seas. In view of the relative lengths of ship and wave, it is not reasonable to assume, as was done in rolling, that the wave surface can be represented by a straight line. The principle, however, remains unchanged in that there is a forcing function on the right-hand side of the equation and the motions theoretically exhibit a natural and forced oscillation. Because the response curve is less peaked than that for roll the pitch and heave motions are mainly in the frequency of encounter, i.e. the frequency with which the ship meets successive wave crests.

Another way of viewing the pitching and heaving motion is to regard the ship/sea system as a mass/spring system. Consider pitching. If the ship moved extremely slowly relative to the wave surface it would, at each point, take up an equilibrium position on the wave. This may be regarded as the static response of the ship to the wave and it will exhibit a maximum angle of trim which will approach the maximum wave slope as the length of the wave becomes very large relative to the ship length. In practice, the ship does not have time to respond in this way, and the resultant pitch amplitude will be the 'static' angle multiplied by a magnification factor depending upon the ratio of the frequencies of the wave and the ship and the amount of damping present. This is the standard magnification curve used in the study of vibrations. Provided the damping and natural ship period are known, the pitching amplitude can be obtained from a drawing board study in which the ship is balanced at various points along the wave profile.

Having discussed the basic theory of ship motions, it is necessary to consider in what form the information is presented to the naval architect before proceeding to discuss motions in an irregular wave system.

7.2.6 Presentation of motion data

It is desirable that the form of presentation should permit ready application to ships of differing sizes and to waves of varying magnitude. The following assumptions are made:

(a) Linear motion amplitudes experienced by geometrically similar ships are proportional to the ratio of the linear dimensions in waves which are geometrically similar and in the same linear ratio. That is, the heave amplitude of a 200 m ship in waves 150 m long and 6 m high will be double that of a 100 m ship in waves 75 m × 3 m; $V/\sqrt{L}$ constant;
(b) Angular motion amplitudes are the same for geometrically similar ship and wave combinations, i.e. if the pitch amplitude of the 200 m ship is 2 degrees, then the pitch amplitude for the 100 m ship is also 2 degrees;
(c) For a ship in a given wave system all motion amplitudes vary linearly with wave height;
(d) Natural periods of motions for geometrically similar ships vary with the square root of the linear dimension, i.e. the rolling period of a ship will be three times that of a one-ninth scale model.

These assumptions follow from the mathematical analysis already outlined.

A quite common plot for motions in regular waves, is the amplitude, expressed non-dimensionally, to a base of wave-length to ship length ratio for a series of $V/\sqrt{L}$ values. The ordinates of the curve are referred to as *response amplitude operators,* Figure 7.6.

This system of plotting is non-dimensional, but a slight complication arises with angular motions when using wave spectra which are in terms of wave height. Since wave height is proportional to wave slope, the data can be presented as in Figure 7.7 with no need to differentiate between linear and

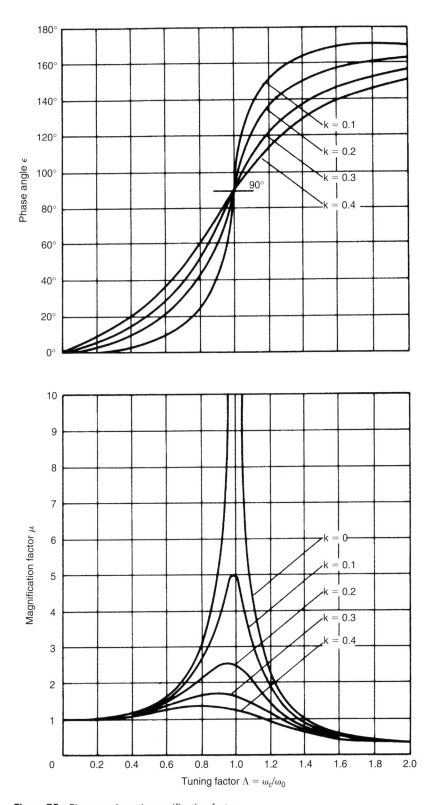

Figure 7.5 Phase angle and magnification factor.

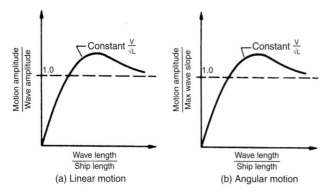

Figure 7.6 Non-dimensional plotting.

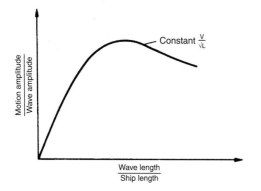

Figure 7.7 Presentation of data for spectral analysis.

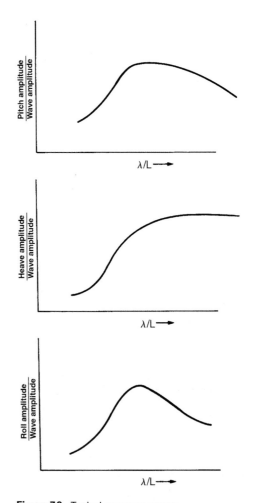

Figure 7.8 Typical response curves.

angular motions, although the curves are no longer non-dimensional.

Some typical response curves are reproduced in Figure 7.8, and more information on such presentations is given by Lloyd (1998).

7.2.7 Motion in irregular seas

The foundations for the study of ship motions in irregular seas were laid in 1905 by R. E. Froude when he wrote in the context of regular wave experiments:

> 'Irregular waves such as those commonly met with at sea—are only a compound of a number of regular systems (individually of a comparatively small magnitude) of various periods, ranging through the whole gamut (so to speak) represented by our diagrams, and more. And the effect of such a compound wave series on the models would be more or less a compound of the effects proper to the individual units composing it.'

It has been seen that for regular waves the motion data can be presented in the form of response amplitude operators (RAO), Figure 7.8 for various ship speeds in waves of varying dimension relative to the ship length. Generally, a designer is concerned with a comparison of two or more designs so that, if one design showed consistently lower RAOs in all waves and at all speeds, the conclusion to be reached would be clear cut. This is not usually the case, and one design will be superior to the other in some conditions and inferior in other conditions. If it is known, using data such as that presented in Chapter 1, that on the intended route, certain waves are most likely to be met then the design which behaves better in these particular waves would be chosen.

Of more general application is the use of the concept of wave spectra. It is shown in Chapter 1 that, provided phase relationships are not critical, the apparently irregular sea surface can be represented mathematically by a spectrum of the type

$$S(\omega) = \frac{A}{\omega^5} \exp\left(-\frac{B}{\omega^4}\right) \qquad (7.19)$$

where ω = circular frequency in radians per second.

A and B are constants which can be expressed in terms of the characteristic wave period and/or the significant wave height.

Since

$$\lambda = \frac{2\pi g}{\omega^2}, \quad \frac{d\lambda}{d\omega} = -\frac{4\pi g}{\omega^3}$$

If λ is to be used as the base for the spectrum instead of ω then the requirement that the total spectral energy is constant leads to

$$S(\lambda) = S(\omega)\frac{d\omega}{d\lambda} = -\frac{A}{4\pi g \omega^2} \exp\left(-\frac{B}{\omega^4}\right) \qquad (7.20)$$

For a known ship length the wave spectrum can be replotted to a base of λ/L to correspond to the base used earlier, Figure 7.8, for the motion response amplitude operators. Then the wave spectrum and motion data such as that presented in Figure 7.9 for heave can be combined to provide the energy spectrum of the motion.

Various motion parameters can then be derived from the spectral characteristics as for the waves themselves. See Chapter 1.

For example average heave amplitude = $1.25\sqrt{m_0}$

where m_0 is the area under the heave spectrum.

In some cases motion data is presented to a base of frequency of encounter of the ship with the wave. The same process can be followed to arrive at the motion spectrum but noting that the wave spectrum is derived from an analysis of the variation of the surface elevation at a fixed point. In this case then it must be modified to allow for the effective or encounter spectrum as experienced by the ship.

If the ship is moving at velocity V at an angle ψ to the direction of advance of the wave system the wave

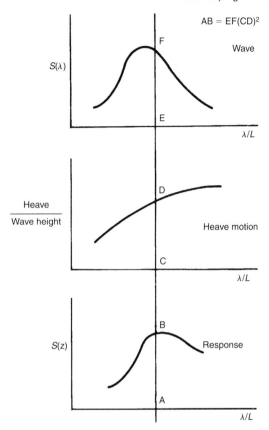

Figure 7.9 Wave spectrum, heave data and response spectrum.

spectrum as experienced by the ship is obtained by multiplying:

(a) abscissae by $\left(1 - \dfrac{\omega V}{g}\cos\psi\right)$

(b) ordinates by $\left(1 - \dfrac{2\omega V}{g}\cos\psi\right)^{-1}$

When the ship is moving directly into the wave system $\cos\psi = -1$.

The effect of ship speed on the shape of the wave spectrum is illustrated in Figure 7.10 which shows a spectrum appropriate to a wind speed of 30 knots and ship speeds of 0, 10, 20 and 30 knots.

To illustrate the procedure for obtaining the motion spectra, consider one speed for the ship and assume that the encounter spectrum for that speed is as shown in Figure 7.11(a). Also, assume that the amplitude response operators for heave of the ship, at that same speed, are as shown in Figure 7.11(b). The ordinate of the wave energy spectrum is proportional to the square of the amplitude of the component

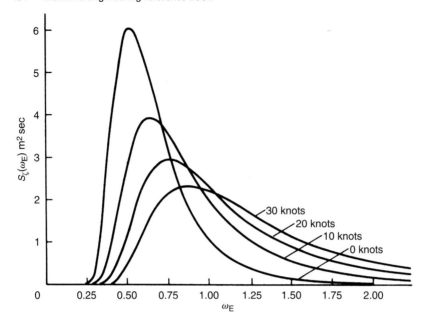

Figure 7.10 Effect of ship speed on encounter spectrum.

waves. Hence, to derive the energy spectrum for the heave motion as shown in Figure 7.11(c), the following relationship is used

$$S_z(\omega_E) = [Y_{z\zeta}(\omega_E)]^2 S_\zeta(\omega_E) \qquad (7.21)$$

i.e.

$$RC = (RB)^2(RA)$$

If the area under the motion energy spectrum is obtained by integration, the significant heave amplitude, etc., can be deduced by using the same multiplying factors as those given in Chapter 1 for waves.

For example, if m_0 is the area under the roll spectrum

average roll amplitude = $1.25\sqrt{m_0}$
significant roll amplitude = $2\sqrt{m_0}$
average amplitude of $\dfrac{1}{10}$ highest rolls = $2.55\sqrt{m_0}$

Any of these quantities, or the area under the spectrum, can be used to compare designs at the chosen speed.

The lower the figure the better the design and the single numeral represents the overall response of the ship at that speed in that wave system. The process can be repeated for other speeds and other spectra. The actual wave spectrum chosen is not critical provided the comparison is made at constant significant wave height and not constant wind speed.

Example. A sea spectrum for the North Atlantic is defined in Table 7.1, $S_\zeta(\omega)$ being in m^{2}s.

It is required to calculate the encounter spectra for a ship heading directly into the wave system at speeds of 10, 20 and 28 knots.

Assuming that the heave response of a ship, 175 m in length, is defined by Figure 7.12 deduce the heave spectra for the three speeds and hence the probability curves for the motion.

It has been shown for the wave spectra, that

$$\omega_E = \omega\left(1 + \dfrac{\omega V}{g}\right)$$

Table 7.1 Sea spectrum.

ω	0.3	0.4	0.5	0.6	0.7	0.8	0.9	1.0	1.1	1.2
$S_\zeta(\omega)$	0.20	2.00	4.05	4.30	3.40	2.30	1.50	1.00	0.70	0.50

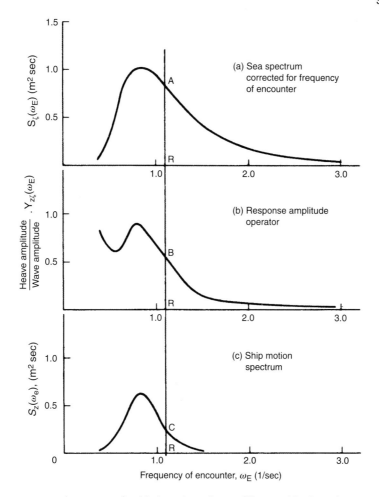

Figure 7.11 Energy spectra and response of a ship in an irregular sea (illustrated for heave).

For 10 knots;

$$V = 10 \times \frac{1852}{3600} = 5.14 \text{ m/s}$$

$$g = 9.807 \text{ m/s}^2 \therefore \omega_E = \omega(1 + 0.525\omega)$$

similarly for 20 and 28 knots ω_E is equal to $\omega(1 + 1.05\omega)$ and $\omega(1 + 1.47\omega)$ respectively.

Figure 7.12 is used by calculating the wave-length appropriate to each ω value. λ and λ/L are tabulated in Tables 7.2 and 7.3 with the response amplitude operators from Figure 7.12. Since curves show response at each speed the RAOs apply to the appropriate ω_E.

It has also been shown that ordinates of the spectrum must be multiplied by

$$\left(1 + \frac{2\omega V}{g}\right)^{-1} = (1 + 1.05\omega)^{-1} \text{ for 10 knots}$$

$$(1 + 2.10\omega)^{-1} \text{ for 20 knots}$$
$$(1 + 2.94\omega)^{-1} \text{ for 28 knots}$$

The calculations can be carried out in tabular fashion as below for 10 knots and repeated for 20 knots and 28 knots.

The ordinates of the heave motion spectrum at each speed are obtained by multiplying the wave spectrum ordinate by the square of the RAO as in Table 7.4.

The heave spectra can now be plotted and the areas under each obtained to give m_0. Values of m_0 so deduced are

10 knots: $m_0 = 0.37$ and $\sqrt{(2 m_0)} = 0.86$
20 knots: $m_0 = 0.62$ and $\sqrt{(2 m_0)} = 1.11$
28 knots: $m_0 = 0.72$ and $\sqrt{(2 m_0)} = 1.20$

The probability that at a random instant of time the heave exceeds some value z is given by $P(z) = 1 - \text{erf}(z/\sqrt{(2 m_0)})$. The error function, erf, is obtained from standard mathematical tables.

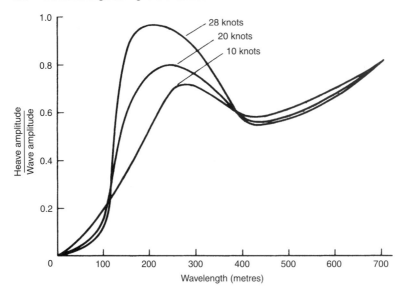

Figure 7.12 Heave response.

Table 7.2

ω	$1 + 0.525\omega$	ω_E	$S_\zeta(\omega)$	$1 + 1.05\omega$	$S_\zeta(\omega_E)$
0.3	1.158	0.347	0.20	1.315	0.15
0.4	1.210	0.484	2.00	1.420	1.41
0.5	1.263	0.632	4.05	1.525	2.65
0.6	1.315	0.789	4.30	1.630	2.64
0.7	1.368	0.958	3.40	1.735	1.96
0.8	1.420	1.136	2.30	1.840	1.25
0.9	1.473	1.326	1.50	1.945	0.77
1.0	1.525	1.525	1.00	2.050	0.49
1.1	1.578	1.736	0.70	2.155	0.32
1.2	1.630	1.956	0.50	2.260	0.22

Table 7.3

ω	λ(m)	λ/L	RAO		
			10 knots	20 knots	28 knots
0.3	689	3.97	0.80	0.80	0.80
0.4	387	2.23	0.60	0.60	0.60
0.5	247	1.42	0.69	0.80	0.95
0.6	171	0.985	0.44	0.69	0.93
0.7	126	0.730	0.28	0.40	0.29
0.8	96.6	0.556	0.18	0.15	0.10
0.9	76.5	0.440	0.12	0.08	0.05
1.0	61.6	0.354	0.10	0.06	0.04
1.1	51.2	0.295	0.08	0.05	0.03
1.2	43.0	0.250	0.07	0.04	0.03

7.2.8 Motion in oblique seas

The procedure outlined above for finding the motion spectra can be applied for the ship at any heading provided the appropriate encounter spectrum is used and the response amplitude operators are available for that heading.

In a regular wave system, as the ship's course is changed from directly into the waves, two effects are introduced, viz.:

(a) the effective length of the wave is increased and the effective steepness is decreased;
(b) the frequency of encounter with the waves is decreased as already illustrated.

An approximation to motions in an oblique wave system can be obtained by testing in head seas with the height kept constant but length increased to $\lambda/\cos\psi$ and with the model speed adjusted to give the correct frequency of encounter. This is a reasonable procedure for vertical motions but it is only an approximation. A further discussion of encounter frequency and motion in oblique seas is given in Section 7.2.11.

7.2.9 Surge, sway and yaw

As already explained, these motions exhibit a different character from that of roll, pitch and

Table 7.4

ω	10 knots			20 knots			28 knots		
	$S_\zeta(\omega_E)$	RAO	$S_Z(\omega_E)$	$S_\zeta(\omega_E)$	RAO	$S_Z(\omega_E)$	$S_\zeta(\omega_E)$	RAO	$S_Z(\omega_E)$
0.3	0.15	0.80	0.098	0.12	0.80	0.079	0.11	0.80	0.068
0.4	1.41	0.60	0.508	1.09	0.60	0.392	0.92	0.60	0.330
0.5	2.65	0.69	1.262	1.98	0.80	1.270	1.64	0.95	1.480
0.6	2.64	0.44	0.511	1.91	0.69	0.910	1.56	0.93	1.350
0.7	1.96	0.28	0.153	1.38	0.40	0.207	1.11	0.29	0.093
0.8	1.25	0.18	0.041	0.86	0.15	0.019	0.69	0.10	0.007
0.9	0.77	0.12	0.011	0.52	0.08	0.003	0.41	0.05	0.001
1.0	0.49	0.10	0.005	0.32	0.06	0.001	0.25	0.04	—
1.1	0.32	0.08	—	0.21	0.05	—	0.17	0.03	—
1.2	0.22	0.07	—	0.14	0.04	—	0.11	0.03	—

heave. They are not subject to the same theoretical treatment as these oscillatory motions but a few general comments are appropriate.

7.2.9.1 *Surge*

At constant power in still water a ship will move at constant speed. When it meets waves there will be a mean reduction in speed due to the added resistance and changed operating conditions for the propeller. The speed is no longer constant and the term surge or surge velocity is used to define the variation in speed about the new mean value. Several effects are present. There is the orbital motion of the wave particles which tends to increase the speed of the ship in the direction of the waves at a crest and decrease it in a trough. In a regular wave system, this speed variation would be cyclic in the period of encounter with the waves. In an irregular sea, the height and hence the resistance of successive waves varies giving rise to a more irregular speed variation. This is superimposed upon the orbital effect which is itself irregular in this case. The propellers will also experience changing inflow conditions due to the waves and the ship's responses. The thrust will vary, partly depending upon the dynamic characteristics of the propulsion machinery and transmission system. The resulting surge is likely to be highly non-linear.

The surge experienced by a vessel of length 146.15 m is shown in Figure 7.13.

The maximum response occurs in waves approximately equal in length to the ship. In waves of this length and 5 m high, the speed oscillation is about ±0.25 knots. The effect varies approximately linearly with speed.

7.2.9.2 *Sway*

When the wave system is other than immediately ahead or astern of the ship, there will be transverse forces arising from similar sources to those causing the surging motion. In a regular sea, these would lead to a regular motion in the period of encounter with the waves but, in general, they lead to an irregular athwartships motion about a mean sideways drift. This variation about the mean is termed sway. It is also influenced by the transverse forces acting on the rudder and hull due to actions to counteract yaw which is next considered.

7.2.9.3 *Yaw*

When the wave system is at an angle to the line of advance of the ship the transverse forces acting will introduce moments tending to yaw the ship. Corrective action by the rudder introduces additional moments and the resultant moments cause an irregular variation in ship's heading about its mean heading. This variation is termed yawing. In a regular sea with an automatic rudder control system, the motion would exhibit a regular period depending on the period of encounter and the characteristics of the control equation, see Section 8.12. In general, however, the motion is quite irregular.

Some of the difficulty of maintaining course in rough weather is indicated in Figure 7.14 which is for a ship of 146 m.

7.2.10 Large amplitude rolling

Linear theory shows that large angles of roll can occur when the wave encounter frequency of a beam

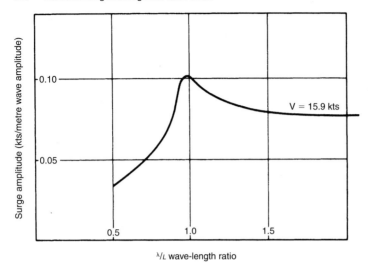

Figure 7.13 Surging in head sea.

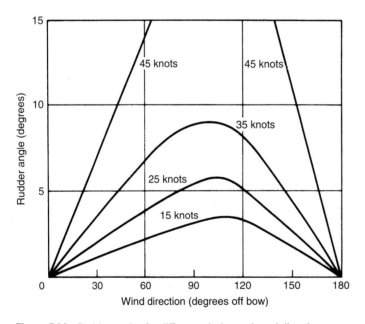

Figure 7.14 Rudder angles for different wind speeds and directions.

sea is close to the ship's natural frequency of roll. The amplitude reached will depend upon the degree of damping and whether any stabilising devices, such as active fins, are employed (see Section 7.10).

Linear theory asumes a steady metacentric height but when a ship is moving through waves this height is a dynamic quantity not a static one. As the wave surface moves along the length of the ship the shape of the underwater form changes, particularly at the bow and stern, an effect accentuated by heave and pitching motions. These changes lead to variations in the effective metacentric height. When a ship is in a following sea metacentric height variations are long period. Particularly in ships with flat transom sterns there may be a loss of stability and the resulting roll amplitudes can be very large.

Another non-linear effect which causes rolling occurs when the dominant encounter period approximates half the natural period of roll in head or following seas. If associated with fairly large stability variations, large roll angles can result. This phenomenon is often called *half cycle* or *parametric* rolling. It starts quite unexpectedly and quickly reaches very large amplitudes. See Biran (2003) and Kobylinski and Kastuer (2003) for detailed discussions of parametric rolling.

Roll motions are considered further in the next Section (7.2.11).

7.2.11 Roll excitation and influence of speed and heading

7.2.11.1 *Motion directions of rigid body*

The oscillatory ship motion is described by three translatory and by three rotational degrees of freedom (DOF). Figure 7.15 is a more detailed version of Figure 7.1 and shows the three translatory motions of the rigid ship body: surge, sway, and heave, and the three rotational motions: roll, pitch, and yaw. The general definition is compiled in Table 7.5.

Looking at the motion directions in more detail, we realise it is practical to define the particular axis of any of the six DOF as follows. The detailed Cartesian co-ordinate systems (axes perpendicular to each other) are shown in Figure 7.15.

$O^* x^* y^* z^*$	system of space fixed axis
$C \bar{x} \bar{y} \bar{z}$	system of direction-fixed axis, when origin C moves with ship speed V
$C x_b y_b z_b$	system of body fixed axis for ship motion
$z_{b1} = \bar{z}$	body axis, for angular yaw motion as well as heave direction. This axis is independent of pitch and roll
y_{b2}	body axis for angular pitch motion. It deflects from direction fixed axis $\bar{y}$ by yaw angle ψ
X_{b3}	body axis for angular roll motion ϕ. It deflects from direction fixed axis $\bar{x}$ by yaw angle ψ and by pitch angle θ (θ perpendicular to ψ). In other words, only for roll we refer to the real ship body, no matter what other motions the ship is in

Mathematically, the ship motion can be expressed by the so-called motion equations. They result from the equilibrium of all forces acting on the rigid ship in the three translatory directions x_1, x_2, x_3, and from the equilibrium of all moments in the three rotational directions x_4, x_5, x_6. For a floating body in equilibrium, the sum of all forces and the sum of all moments acting must be equal to zero,

$$\sum_i F_i = 0 \quad i = 1, 2, 3 \tag{7.22}$$

$$\sum_i M_i = 0 \quad i = 4, 5, 6 \tag{7.23}$$

The common way of expressing the motion equations is to place the reaction forces or moments on the left hand side of the equation, while the excitation forces or moments are on the right hand side,

ship reaction = external excitation

In a seaway, the external excitation depends on the seaway action. The particular forces and moments

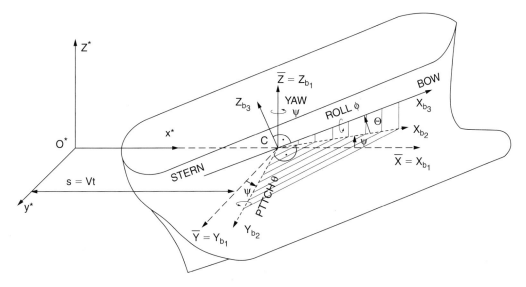

Figure 7.15 Definition of six motion degrees of freedom.

Table 7.5 Degrees of freedom of rigid ship body.

j	oscillatory motion	mode	axis	Positive Sense	Either x_j or specific symbols	
1	SURGE		along $\bar{x}$	Forward	x_1	X
2	SWAY	translation	along $\bar{y}$	to starboard	x_2	y
3	HEAVE		along $\bar{z}$	Upward	x_3	z
4	**ROLL**		about x_{b3}	**Starboard down**	x_4	ρ
5	PITCH	rotation	about y_{b2}	bow up	x_5	ϕ
6	YAW		about z_{b1}	bow to port	x_6	Ψ

of the ship reaction depend on the ship motions and their derivatives in the corresponding degree of freedom. For a ship as a rigid free floating body, the resulting set of motion equations for all six degrees of freedom is as follows,

$$\sum_{j=1}^{6}(a_{ij}\cdot\ddot{x}_j + b_{ij}\cdot\dot{x}_j + c_{ij}\cdot x_j) = d_i$$
$$\text{with}\quad i = 1,\ldots,6 \qquad (7.24)$$

X oscillatory ship motion
$\dot{x}$ velocity of oscillatory motion
$\ddot{x}$ acceleration of oscillatory motion
i direction of force/moment
j motion direction
$i\neq j$ in subscript ij give the coupling coefficients

The coefficients a, b, c are the parameters of the motion equations. They are linear coefficients of force or moment, with respect to the particular motion oscillation and its derivatives, see Equation (7.24). The terms of the forces/moments are defined as follows:

$a\ddot{x}$ inertia force/moment depending on the oscillatory motion acceleration of the ship body
$b\dot{x}$ damping force/moment depending on the motion velocity
Cx restoring force/moment depending on the particular oscillatory motion x
d external excitation force/moment due to the seaway

Equation (7.24) is a set of six coupled differential equations, which includes the combined ship motion in all six degrees of freedom. However, even this set of equations relies on simplified assumptions, mainly, the reaction forces are linearly dependent on (proportional to) the ship motion. This is valid for small motion amplitudes only.

Even with this simplification, one will rarely try to solve the set of six coupled equations. This would require calculating (or measuring) all the partial forces and moments, or all the reaction coefficients a_{ij}, b_{ij}, c_{ij}, and the excitation d_i. Even then, for certain ship motions such as roll, the linearisation does not suffice to give answers for large roll motions including capsize. In order to calculate large roll motion, the Non-linearity must be taken into account.

It is convenient to reduce the effort and to combine only the equations of special interest, such as:

- *Vertical longitudinal plane* with three DOF:
  Three equations describe the coupled motions of *surge, heave, and pitch* (x_1, x_3, x_5). Even more simplified: heave and pitch alone (x_3, x_5).
- *Horizontal plane* with three DOF:
  Three equations describe the coupled motions of *sway, surge and yaw* (x_1, x_2, x_6).

However, for the problems of ship stability and safety from capsize, the roll motion is most important. We have the following combinations with roll as the rotation in the vertical transverse plane:

- *Vertical transverse plane* with two DOF coupled with one DOF for the *horizontal plane*:
  Three equations describe the coupled motions of **roll** with sway and yaw (x_2, x_4, x_6)
- Roll coupled with 2 DOF in the longitudinal vertical plane:
  Three equations describe the coupled motions of **roll** with heave and pitch (x_4, x_5, x_6)
- Greatest simplification using the *roll* motion alone (x_4).

For the sake of simplicity, studying the uncoupled roll motion alone can give valuable insight into the typical roll motion pattern. Based on the solutions of the one DOF roll motion equation, the influence of coupling with other degrees of freedom can be further analysed by extending the number of DOF. In coupling, we have resulting forces/moments in direction i not only from the motion direction j equal to i, but from other directions of motion j as well. An example is heave and pitch influencing the restoring moments in the roll motion equation ($i = 4$) because of the time dependent variation of the underwater shape of the oscillating ship body.

According to Equation (7.24), we want to write only one (for roll) of the set of 3 coupled motion equations for heave, roll, and pitch ($x_j = x_3, x_4, x_5$). The motions are coupled, if forces or moments result from motion directions j other than i. Based on Equation (7.24), the resulting ship motion equation for the moments in the roll direction $\left(\sum M_{i=4} = 0\right)$, depending on the roll ($j = 4$) coupled with heave ($j = 3$) and pitch ($j = 5$) is written as

$$
\begin{array}{ll}
\text{Heave} \quad j = 3 & a_{43} \cdot \ddot{x}_3 + b_{43} \cdot \dot{x}_3 + c_{43} \cdot x_3 + \\
\text{Roll} \quad j = i = 4 & a_{44} \cdot \ddot{x}_4 + b_{44} \cdot \dot{x}_4 + c_{44} \cdot x_4 + \\
\text{Pitch} \quad j = 5 & a_{45} \cdot \ddot{x}_5 + b_{45} \cdot \dot{x}_5 + c_{45} \cdot x_5 = d_4
\end{array}
$$
(7.25)

$i = 4$ moment equation for roll

j direction of motion (see Figure 7.15).

Writing down the other two equations for heave and pitch is a good exercise.

In order to solve this coupled Equation (7.25), i.e. to estimate the roll angle x_4, we must additionally solve the equations for heave ($i = 3$) and for pitch ($i = 5$), accordingly (not given here).

The second row in Equation (7.25) together with the sum of all coupling effects put into M_{4c}, results in the shortened moment equation as follows.

$$a_{44} \cdot \ddot{x}_4 + b_{44} \cdot \dot{x}_4 + c_{44} \cdot x_4 + M_{4c} = d_4 \quad (7.26)$$

M_{4c} sum of all coupling moments for $i = 4$ from the motion directions j other than 4. By disregarding the coupling moments ($M_{4c} = 0$), the linear roll motion equation is then written in the simplest form, replacing x_4 by ϕ and disregarding the subscripts:

$$a\ddot{\phi} + b\dot{\phi} + c\phi = d \quad (7.27)$$

a, b, c proportionality factors

d external roll excitation

7.2.11.2 Mass moment of inertia

The coefficient in the inertia term in Equation (7.27) of the rolling ship is defined as follows,

$$a = I'_T = \Delta \cdot i'^2_T \quad (7.28)$$

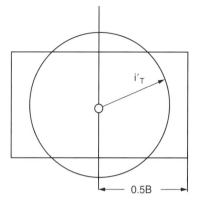

Figure 7.16 Explanation of roll radius of gyration.

where

$$\begin{array}{l} I'_T = I_T + I''_T \\ \Delta = \rho \cdot \nabla \end{array} \quad (7.29)$$

I'_T mass moment of inertia of the rolling ship including the hydrodynamic mass moment of the surrounding water

I_T inertia mass moment of the rolling ship

I''_T hydrodynamic mass moment (also called added mass moment)

Δ displacement mass

ρ water density

∇ displacement volume

$i_{T'}$ roll radius of gyration

We can think of the radius of gyration i'_T being the radius of a solid ring, which replaces the total mass of the ship as shown in Figure 7.16. This radius is enlarged by the inertia effect of the surrounding water with respect to roll acceleration, the so-called hydrodynamic mass moment or added mass moment I''_T, Equation (7.29).

7.2.11.3 Linear restoring moment

For large heel, the static restoring moment (Section 3.7) is:

$$M_{st} = g \cdot \Delta \cdot GZ \quad (7.30)$$

The distance GZ is the righting lever of the ship. The GZ-curve for large heel angles is highly non-linear.

With larger heel, GZ increases to a maximum. At a heel angle larger than the maximum, there is a strong decrease of GZ to zero. Figure 7.17 shows a typical GZ-curve for a ship in still water.

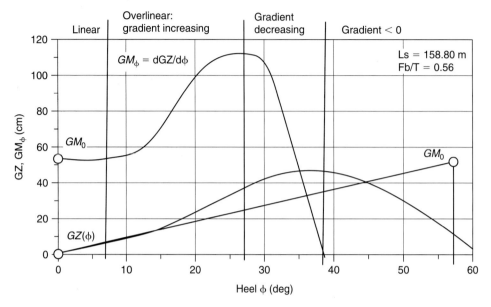

Figure 7.17 Typical still water GZ curve.

The different heel regions of the GZ-curve are characterised by the gradient $dGZ/d\phi$, referred to as the metacentric height GM_ϕ (see Figure 7.17). For most ships at small heel up to about 5 degrees the gradient GM_ϕ is constant, and the righting lever $GZ(\phi)$ can be approximated by a linear function of heel angle ϕ, with GM being the initial metacentric height for $\phi = 0$:

$$GM = \frac{dGZ}{d\phi}(\phi = 0)$$
$$GZ = GM \cdot \phi \quad \text{for } \phi \leq 5 \text{ deg} \quad (7.31)$$

The metacentric height GM_0 is also referred to as "initial stability", see Chapter 3. The linear range depends on the hull flare at the waterline (generally about 5 degrees).

The parameter c in Equation (7.27) is the so-called spring constant. The spring constant is the ratio of the restoring hydrostatic moment versus the inclination. For small heel, it results in the displacement weight (equal to the buoyancy force F_B for the floating ship) multiplied by the metacentric height GM:

$$c = \frac{M_{\text{st}}}{\phi} = \frac{g \cdot \Delta \cdot GM \cdot \phi}{\phi}$$
$$= g \cdot \Delta \cdot GM = F_B \cdot GM \quad (7.32)$$

As discussed in Section 7.2.2.1, for the case when the initial stability is large, the ship is called "stiff", i.e. she is not sensitive to small heeling moments.

On the contrary, for small initial metacentric height, the ship is "tender", i.e. the ship is sensitive to small heeling moments. However, it must be kept in mind that the full GZ-curve at large heel must be taken into account at large exciting moments and in severe seas.

It is a common procedure to calculate both GZ and GM and to draw the calculated GM onto the graph of GZ.

7.2.11.4 *Natural roll period*

The ratio c/a is equal to the natural circular roll frequency squared:

$$\omega_0^2 = \frac{c}{a} = \frac{g \cdot \Delta \cdot GM}{\Delta \cdot i_T'^2} = \frac{g \cdot GM}{i_T'^2} \quad (7.33)$$

In the nautical field, it is practical to refer to the natural roll period, T_0, of the ship. The period is the inverse of the frequency. With $\omega = 2\pi f$ and $f = \frac{1}{T}$, we have:

$$T_0 = \frac{1}{f_0} = \frac{2\pi}{\omega_0} \quad (7.34)$$

The natural roll period, T_0, can be estimated with the ship at free roll in still water conditions, see Figure 7.18 Using a stopwatch is accurate enough. However, IMO requires the average of about 5 cycles be taken, see Section 7.2.11.6.

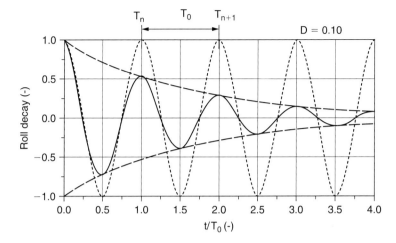

Figure 7.18 Example of roll decay.

7.2.11.5 Roll damping

The oscillating free rolling motion eventually dies out, when no further excitation adds to the energy of the motion. The free roll transfers the roll energy to the surrounding water by potential and friction forces. The decay of the roll is due to damping.

The differential equation of the ship at free roll as a damped oscillator is from Equation (7.27) with roll excitation d equal to zero:

$$a \cdot \ddot{\phi} + b \cdot \dot{\phi} + c \cdot \phi = 0 \quad (7.35)$$

With the initial conditions $\phi(t = 0) = \phi_0$ and $d\phi/dt(t = 0) = 0$.

We introduce the generalised constants:

$$2\delta = \frac{b}{a} \quad \text{and} \quad \omega_0^2 = \frac{c}{a}$$

The differential equation of the free roll becomes:

$$\ddot{\phi} + \frac{b}{a} \cdot \dot{\phi} + \frac{c}{a} \cdot \phi = 0$$
$$\ddot{\phi} + 2\delta \cdot \dot{\phi} + \omega_0^2 \cdot \phi = 0 \quad (7.36)$$

The solution, i.e. the free roll motion is:

$$\phi = \phi_0 \cdot \exp(-\delta\, t) \cdot \cos \omega_0 t \quad (7.37)$$

For small damping, the frequency ω_ϕ of the free roll can be approximated by the natural frequency ω_0 from $\omega_\phi^2 = \omega_0^2(1 - D^2) \approx \omega_0^2$ as $D \ll 1$.

Figure 7.18 shows free roll decay. The envelope of the one-sided amplitudes is, from Equation (7.37), expressed by:

$$\phi/\phi_0 = \exp(-\delta\, t) \quad (7.38)$$

The ratio of two successive (taken only from either port or starboard) roll amplitudes ϕ_n and ϕ_{n+1} at a distance of the natural period T_0 is:

$$\phi(t = T_n) = \phi_0 \cdot \exp(-\delta \cdot T_n) = \phi_n$$
$$\phi(t = T_{n+1}) = \phi_0 \cdot \exp(-\delta \cdot T_{n+1}) = \phi_{n+1}$$
$$\frac{\phi_{n+1}}{\phi_n} = \frac{\phi_0 \cdot \exp(-\delta \cdot T_{n+1})}{\phi_0 \cdot \exp(-\delta \cdot T_n)}$$
$$= \exp(-\delta(T_{n+1} - T_n)) = \exp(-\delta \cdot T_0)$$
$$\frac{\phi_n}{\phi_{n+1}} = \exp(-\delta \cdot T_0)$$
$$\delta = \frac{1}{T_0} \cdot \ln\left(\frac{\phi_n}{\phi_{n+1}}\right) \quad (7.39)$$

The term $b \cdot \dot{\phi}$ in the roll motion Equation (7.35) is the linear damping moment. We have introduced the linear damping coefficient as the logarithmic decrement $\delta = b/2a$ While the dimension of b is Nm/s^{-1}, (damping moment in Nm divided by the roll velocity rad/s see Equation (7.26)), the dimension of δ is s^{-1} (see Equation (7.39)). In order to define a dimensionless damping parameter, we set δ in relation to the natural frequency ω_0, and we get the damping constant D:

$$D = \frac{\delta}{\omega_0} \left[\frac{s^{-1}}{s^{-1}} = 1\right] \quad (7.40)$$

The dimensionless damping $D = b/2\sqrt{a \cdot c}$ includes all three factors a, b, and c of the motion equation. D can be evaluated from the time record of the free roll decay, with δ from Equation (7.39) put into Equation (7.40), and ω_0 from Equation (7.34), we have:

$$D = \frac{\delta}{\omega_0} = \frac{\delta \cdot T_0}{2\pi} = \frac{1}{2\pi} \ln\left(\frac{\phi_n}{\phi_{n+1}}\right) \quad (7.41)$$

To estimate the damping parameter D, successive roll amplitudes at one side are to be measured and put into Equation (7.41). For most ships, the dimensionless damping D is a small value not larger than 0.10. The roll decay also allows measuring the natural rolling period T_0. Basically, a stopwatch recording and reading of the inclinometer are sufficient. A record for documentation and to allow for error detection of the measurement is advisable.

7.2.11.6 $GM - T_0$ Relationship and rolling period test

The radius of gyration is made dimensionless. We divide by half the beam of the ship as shown in Figure 7.16. This ratio is called the rolling coefficient,

$$C_r = \frac{i'_T}{0.5 \cdot B} \quad (7.42)$$

When introducing the rolling coefficient into Equation (7.33) for ω_0 and (7.34) for T_0, we obtain a relationship of the natural roll period T_0 with C_r and GM. (It is common to write without the subscript, but we still mean the upright position: $GM = GM_0$). In the metric system, $\pi \approx \sqrt{g}$, and the simplified form we usually find in regulations and textbooks is as follows:

$$T_0 = \frac{2\pi}{\omega_0} = \frac{2\pi \cdot i'_T}{\sqrt{g \cdot GM}} = \frac{2\pi \cdot C_r \cdot 0.5 B}{\sqrt{g \cdot GM}} = \frac{C_r \cdot B}{\sqrt{GM}}$$

$$T_0 = \frac{C_r \cdot B}{\sqrt{GM}} \quad (7.43)$$

Equation (7.43) can be used to calculate T_0, when both C_r and GM are known for the same loading condition. See also Equations (7.3) and (7.12).

The practical importance of the above relationship lies in estimating the metacentric height GM by conducting the rolling period test. As soon as we know the rolling coefficient, C_r, we can determine the initial metacentric height GM by measuring the natural roll period, T_0, and put into the relationship for GM:

$$GM = \left(\frac{C_r \cdot B}{T_0}\right)^2 \quad (7.44)$$

This equation is called the Weiss formula, Weiss, (1953). He introduced the rolling period test to determine the GM values for a large number of small ships with reduced effort.

The rolling period test should be conducted with the ship in harbour in smooth water with the minimum interference from the wind and tide. The ship can be made to roll by rhythmically lifting up and putting down a weight as far off the centre line as possible, or by people running athwartships in unison. As soon as the roll amplitude is large enough, no more excitation is allowed. The initial roll amplitude for the measured roll decay should not exceed five degrees.

It is advisable to take the time measurements in the region of large roll velocity to have clear-cut points for begin and end of the period. This does not mean to measure when the ship passes exactly the upright position, as this is rarely possible to be estimated correctly. A fixed landmark for reference when the ship is closer to the upright position is suggested. When the ship passes the upright position, roll velocity is at the maximum, and the mark is passed quickly, Figure 7.19. This method is of particular advantage for large roll periods.

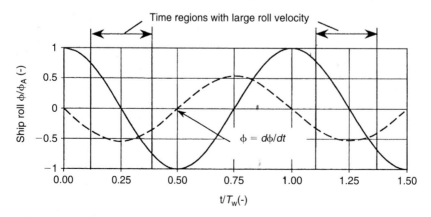

Figure 7.19 Range to measure the period at large roll velocity (at same direction).

IMO (1993) gives detailed instructions for the test procedure. IMO requires the time to be taken for not less than about five complete oscillations. The counting should begin, contrary to the above proposal, when the ship is at the extreme end of a roll. After allowing the roll to completely fade away, this operation should be repeated at least twice more.

Jens (1964) evaluated the accuracy of rolling tests with coasters in the Baltic Sea. He concluded:

- The accuracy of measured rolling periods decreased with larger roll periods. Even very small disturbances (from wind, waves, and rudder action) are affecting the natural roll significantly. The C_r value, (Equation 7.42), varied for one ship condition between 0.7 and 0.83, so the statistical deviation of the measurements was too large. The rolling test should not be used at very small GM values and corresponding large natural roll periods (in the order of ≤ 10 cm and ≥ 20 s).
- The smaller the GM of the ship is, the more accurate is the application of the inclination test to measure GM.
- Distribution of cargo, ballast, and height of deck cargo have a dominant influence on the C_r value.

IMO allows estimating the stability by means of rolling period tests for small ships of up to 70 m in length. IMO Resolution A.749(18) was adopted on 4 November 1993 (succeeding Appendix 2.1 of IMO Resolution 167, Recommendation on Intact Stability for Passenger and Cargo Ships under 100 Metres in Length, London 1981). However, the rolling period test must be seen as a very simplified method, when no other stability information is available. We want to stress again that T_0 is the period of the free roll and must not be confused with the rolling period at excitation T_r.

IMO Res. 749(18), London 1993, gives several reasons to disregard determining stability by means of the rolling period test, or at least the numerical results should be reduced. In such conditions of low stability, the test could give dangerously large values for GM, which are not reliable:

- For a long period of roll, corresponding to a GM of 0.20 m or below,
- Inducing the roll by putting over the helm,
- Rolling test in disturbed waters. Quote: *'Forced oscillations corresponding to the sea period and differing from the natural period at which the ship seems to move should be disregarded.'*

The Weiss formula Equation (7.44) gives GM as a function of:

- Natural roll period, T_o
- beam of the vessel, B
- rolling coefficient, C_r (Equation 7.42).

Application of the Weiss formula requires knowledge of the rolling coefficient C_r. The C_r value depends on the type of vessel, her loading condition, i.e. in particular on the distribution of the cargo in the transverse plane. Resolution A.749(18) gives some statistical values for coasters and fishing boats.

For coasters of normal size, the observed metric C_r–values are:

- (i) empty ship or carrying ballast   0.88
- (ii) ship fully loaded with liquids in tanks
- comprising 20% of total load   0.78
- comprising 5% of total load   0.73

This data on coasters is based on extensive measurements in the Baltic Sea, after serious problems with carrying timber deck cargo had been encountered, Jens, (1964), Thode, (1965).

IMO Resolution A.749(18) (1993) and IMO Circular 707 (1995) present an approximation formula from statistics *(L* ship length, *B* breadth, *d* draught of the ship hull, in meters):

$$0.5 C_r = 0.373 + 0.023 \cdot \left(\frac{B}{d}\right) - 0.043 \cdot \left(\frac{L}{100}\right) \quad (7.45)$$

However, it must be stressed that for any vessel the C_r-value should rather be determined directly. In order to estimate the rolling coefficient of the ship, both the inclining experiment for GM and the rolling period test for T_0 must be carried out at the same loading condition of the ship. This allows gathering experimental data on the actual C_r-values of the particular vessel at various loading conditions.

$$C_r = \frac{T_0 \sqrt{GM}}{B} \quad (7.46)$$

Generally, the application of the Weiss formula is not restricted to small ships. It can be applied to vessels of any size, as soon as both tests (ship inclination test and rolling period test) have been carried out. In fact, to carry out an inclination test for the ship in service, the "Operational Ship Inclining" (OSI) to measure GM directly is highly recommended, see Kobylinski and Kastner (2003), Chapter 13.

7.2.11.7 *Modes of roll excitation in a seaway*

Oscillatory motions at sea mainly result from time varying forces of the seaway acting on the vessel. Large roll motion is dangerous because of large accelerations on the cargo and the danger from capsizing.

We refer to the linear roll motion Equation (7.27) again, $a\ddot{\phi} + b\dot{\phi} + c\phi = \text{d}$.

There are two very distinct ways of roll excitation for a ship in a seaway. The external excitation is expressed by a time variation of the term in the right-hand side of the roll motion equation. The parametric excitation is expressed by a time variation of terms in the left-hand side of the equation.

The external moment acting on the hull in beam seas is expressed by the moment d in Equation (7.27). A ship in beam seas can experience large roll with large inertia forces acting on the cargo. However, following and stern quartering seas at the same stability can be more dangerous with respect to capsizing and total loss of the ship.

Longitudinal and quartering seas can result in a time dependent variation of the ship reaction moments, mainly the righting (restoring) moment.

An excitation due to time variation of ship reaction (on the left-hand side of the motion equation) is called parametric. At parametric resonance, the ship is in danger from capsizing. This is most pronounced in certain conditions in longitudinal and stern quartering seas. Both external and parametric excitations exist simultaneously in quartering seas, see Kobylinski and Kastner (2003), Chapter 11.

7.2.11.8 *Ship roll in beam seas*

In beam seas we have only external excitation, commonly written on the right hand side of the equation of motion. The summation of all moments acting along the longitudinal ship axis gives: dynamic reaction + static reaction = external excitation. For small amplitudes, the roll motion equation is a linear second order differential Equation (7.27).

For small amplitudes we derive the roll excitation in regular beam seas, see Figure 7.20. First, we assume the ship upright in the position between crest and trough. In this intermediate position of the passing wave the wave slope is at the maximum. Assuming a hydrostatic floating condition, the water surface is tilted by the angle ϑ_A. The additionally submerged volume triangle on one side is compensated by the triangle on the other side coming out of the water.

Both the additional and the reducing volume triangles result in a heeling moment, as depicted by the two opposite forces of the triangles.

The vessel tries to reach hydrostatic equilibrium again, by exerting a righting moment with respect to the tilted water surface. For small amplitudes, the righting moment is proportional to the buoyancy force F_B, the GM, and the heel angle $\phi = \vartheta_A$. The amplitude of the beam sea excitation is:

$$d_A = c \cdot \vartheta_A = F_B \cdot GM \cdot \vartheta_A = g \cdot \Delta \cdot GM \cdot \vartheta_A \quad (7.47)$$

The ship experiences the same moment with opposite sign at the other intermediate position between trough and crest.

Now the passing wave varies the excitation periodically with the wave frequency ω. Starting at the wave trough, we have the exciting moment versus time as follows:

$$d = d_A \cdot \sin(\omega \cdot t)$$
$$d = g \cdot \Delta \cdot GM \cdot \vartheta_A \sin(\omega \cdot t)$$

The wave slope, $\vartheta(x)$, is the first derivative of the wave ordinate with respect to the distance x from the wave crest in the travelling direction of the wave, see Figure 7.21.

$$\zeta(x) = 0.5 \cdot H_w \cos(kx)$$
$$\vartheta(x) = \frac{\partial \zeta}{\partial x} = -0.5 H_w \cdot k \cdot \sin kx$$

The wave slope amplitude ϑ_A is then:

$$\vartheta_A = 0.5 H_w \cdot k = \frac{0.5 H_w \cdot 2\pi}{L_w} = \pi \frac{H_w}{L_w} (rad) \quad (7.48)$$

Finally, the exciting moment in beam seas becomes:

$$d = g \cdot \Delta \cdot GM \cdot \pi \frac{H_w}{L_w} \sin(\omega \cdot t) \quad (7.49)$$

We insert this excitation into Equation (7.27). The solution of the differential Equation (7.27) is given

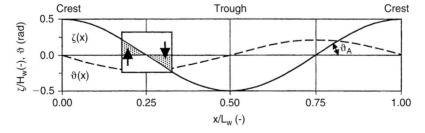

Figure 7.20 Roll excitation of ship in beam seas.

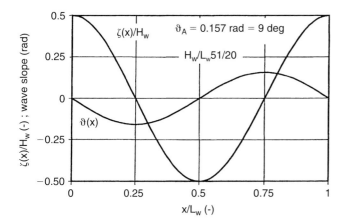

Figure 7.21 Wave slope versus distance from crest.

by the transfer function V_3, defined as the amplitude ratio of roll and wave slope:

$$V_3 = \frac{\phi_A}{\vartheta_A} \quad (7.50)$$

The transfer function V_3 is the dynamic amplification factor,

$$V_3(\eta) = \frac{1}{\sqrt{(1-\eta^2)^2 + 4 \cdot D^2 \cdot \eta^2}} \quad (7.51)$$

The dimensionless wave frequency with respect to the natural roll frequency is the tuning factor η:

$$\eta = \frac{T_0}{T_w} = \frac{\omega}{\omega_0} \quad (7.52)$$

The dimensionless damping D is (from Equation 7.41):

$$D = \frac{\delta}{\omega_0} = \frac{b}{2\sqrt{(c \cdot a)}} = \frac{1}{2\pi} \ln \frac{\phi_n}{\phi_{n+1}} \quad (7.53)$$

The transfer function V_3 of the linear roll in beam seas as given in Equation (7.51) is shown in Figure 7.22, depicted versus the dimensionless wave frequency η. This roll calculation is valid only for comparatively small roll amplitudes, i.e. in the range where the GZ-curve can be approximated by $GM \cdot \phi$. The roll peak is limited by the damping.

The roll amplitude in beam seas shows a distinct dependence on the exciting wave frequency ω due to the difference with ω_0 in the denominator of V_3. We write from Equation (7.51)

$$V_3(\omega) = \frac{\omega_0^2}{\sqrt{(\omega_0^2 - \omega^2)^2 + 4 \cdot \delta^2 \cdot \omega^2}} \quad (7.54)$$

The resulting linear roll motion in beam seas is:

$$\phi = \vartheta_A \cdot V_3 \sin(\omega \cdot t + \gamma_3) \quad (7.55)$$

The phase angle γ_3 between the exciting moment d and the roll is:

$$\gamma_3 = \arctan\left(\frac{2 \cdot D \cdot \eta}{\eta^2 - 1}\right) \quad (7.56)$$

Figure 7.22 shows many important details on roll response, and it is discussed further:

1. The static heel at $\omega = 0$ results from constant excitation independent of time:

$$V_3(\omega = 0) = 1 \Rightarrow \phi_{stat} = \vartheta_A \quad (7.57)$$

2. With the exciting wave frequency, ω, increasing, there is a steady increase of the roll response:

$$\eta < 1 \Rightarrow \omega < \omega_0 \Rightarrow T_w > T_0 \quad (7.58)$$

The left region in Figure 7.22 defines the range between the static heel and the resonance peak. The dynamic response is always larger than the static heel, $V_3 > 1$.

Assuming a constant seaway energy, the corresponding large GM values in this region determine the behaviour of a stiff ship.

3. There is a dominant amplification in the region around $\eta = 1$:

Whereas left and right of the peak the curves for different damping coincide, the peak region increases considerably with smaller damping.

The frequency of the peak response is:

$$\omega_r = \sqrt{\omega_0^2 - 2\delta^2} < \omega_0 \quad (7.59)$$

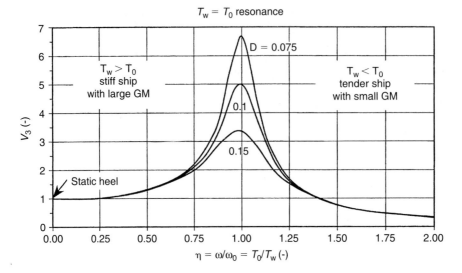

Figure 7.22 Transfer function of roll in beam seas.

The resonant roll amplitude at the peak is:

$$\phi_r = \frac{\omega_0}{2\delta} \cdot \vartheta_A = \frac{1}{2D} \cdot \vartheta_A \qquad (7.60)$$

4. In the frequency range above resonance, we observe a rapid decrease of the roll response:

$$\eta > 1 \Rightarrow \omega > \omega_0 \Rightarrow T_w < T_0 \qquad (7.61)$$

Assuming again a constant seaway, but the ship at various GM, the dynamic roll response is decreasing sharply. The large natural period T_0 of the ship corresponds to a small GM, and we have a tender ship.

5. At very large wave frequency ω, the dynamic roll response is less than the static heel angle, and it approaches zero, $V_3 \to 0$. Therefore, at little or very small GM, when $T_0 >> T_w$, the ship experiences little or almost no roll amplification in purely beam seas.

7.2.11.9 Roll in beam seas at large amplitudes

Figure 7.17 characterises the GZ curve in relation with the gradient $GM_\phi = dGZ/d\phi$. We subdivide the roll amplitudes ϕ_A into regions according to the shape of the GZ curve. Naturally, the specific ranges depend on the particular hull form. The example in Figure 7.17 shows an over-linear GZ-curve.

Small roll, $\phi_A \leq 5^0$ $GM_\phi = GM_0 = const$
Larger roll, $\phi_A > 5^0$ GM_ϕ increasing from GM_0 to max(GM_ϕ)
Large roll GM_ϕ decreasing from max(GM_ϕ) to $GM_\phi = 0$
Extreme roll $GM_\phi < 0$, i.e. above max(GZ)

This is an approximate definition, but it is practically sufficient. The GZ regions depend on the hull form and on ship proportions. Increased breadth-draught ratio B/T increases BM and thus GM. A large freeboard-draught ratio Fb/T results in a large range of stability.

For larger roll amplitudes, more mathematical effort is needed in order to solve the equation of roll motion. The linear solution (Equations 7.51 and 7.54) give a good insight into the typical behaviour including resonance in beam seas. For extreme roll including capsize, numerical timze domain simulations with step by step integration of the roll motion equation have been developed with good results.

At larger roll in beam seas, the deviation of the GZ-curve from the linear approximation results in a different solution of the roll motion equation, Bhattacharyya (1978):

$$a \cdot \ddot{\phi} + b \cdot \dot{\phi} + c \cdot \phi = d \cos(\omega \cdot t) \qquad (7.62)$$

The restoring term was approximated by a cubic function, as shown in Figure 7.23(a) and 7.23(b):

$$c\phi = c_1 \phi + c_2 \phi^3 \qquad (7.63)$$

Figure 7.23(a) and (b) give the calculated non-linear transfer function for two different GZ curves. The non-linear solution shows also a resonance peak as in the linear case of Figure 7.22 but now it is bent to one side:

(a) $c_2 > 0$ GZ over-linear: curve bends to larger ω (to the right).

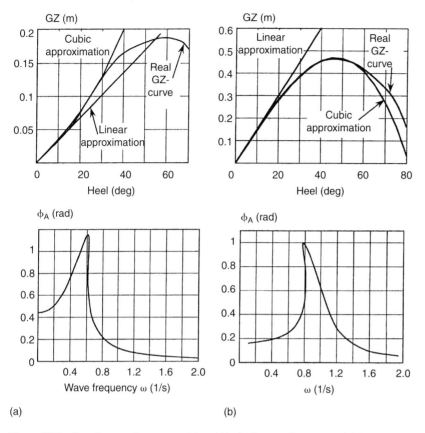

Figure 7.23 Over-linear roll response (a) and Under-linear roll response (b) in beam seas.

In case the *GZ*-curve at large heel deviates to smaller values than given by the linear approximation, we call it under-linear, and the roll response peak is bent towards smaller tuning factor values T_0/T_w proportional to ω for $\omega_0 = $ const, see Figure 7.23(a).

(b) $c_2 < 0$ *GZ* under-linear: curve bends towards smaller ω (to the left).

In case the *GZ*-curve at large heel deviates to larger values than given by the linear approximation, we call it over-linear, and the roll response peak is bent towards larger values of the tuning factor T_0/T_w, see Figure 7.23(b).

7.2.11.10 *GZ-Variation in longitudinal waves*

A ship in longitudinal seas experiences a completely different shape of the underwater volume as compared with the ship in still water and in beam seas. The righting moment of the vessel varies in time with the passing wave. This results in a dynamic excitation of roll motion.

A first approach is to calculate the *GZ*-curves in different wave positions. The hydrostatic approach assumes the water particles in the wave not in motion, although the water surface is elevated according to the wave. Figure 7.24 shows four different positions of the wave crest along the ship (drawn with vertical shift to separate lines).

Figure 7.25 shows typical *GZ*-curves in wave crest and trough compared with the *GZ*-curve in still water. For most hull forms there is a considerable decrease in the crest, and an increase in the trough. The decrease in the crest is larger than the increase in the trough.

The change of *GZ* results from the change in the location of the centre of buoyancy *B* of the heeled ship hull in the longitudinal wave. Weight force, *W*, and the centre of gravity, *G*, remain constant.

The *GZ*-variations can be demonstrated by three typical cross-sections to define the ship

510   Maritime engineering reference book

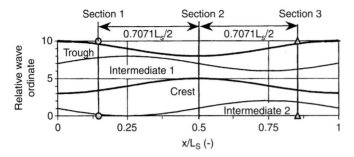

Figure 7.24 Ship in longitudinal wave at different positions relative to the crest.

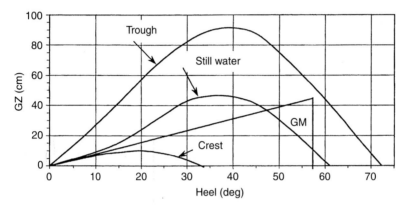

Figure 7.25 *GZ*-curves in wave crest and trough compared with still water.

hull. A simplified hydrostatic procedure allows an estimate of the *GZ*-variations in longitudinal waves. The displacement volume versus ship length is approximated by a cubic function. Three cross sections according to the rule of Tchebychev represent the longitudinal buoyancy distribution of the ship in the wave. The longitudinal position of section 1 is 0.7071 multiplied by half the ship length behind the main section. Section 2 is the main cross section, and section 3 is located 0.7071 multiplied by half the ship length before the main section. The three sections represent the after-body, mid-ship, and the fore-body of the underwater hull, Figure 7.26. The draught should be chosen according to the fully loaded condition because the freeboard is smallest. For ships with a small block coefficient, the ballast condition might be looked at too.

This approach yields sufficient accuracy and serves to understand the influence of the hull form in the particular wave position. The hydrostatic approach is also fundamental for an understanding of any parametric roll excitation and resonance. The wave length is taken as the length on the waterline of the particular loading condition, Kastner (1982).

The wave height-length ratio must decrease with increasing ship length (L in metre) as expressed in Equation (7.64). The formula corresponds to the 0.1% occurrence in the North Atlantic. Although wave steepness decreases, wave height versus length is still increasing. The regular wave has a trochoidal shape (Section 1.1). Figure 7.26 demonstrates the three sections at a constant heel of 30 degrees.

In the wave crest, the freeboard amidships reduces considerably. The freeboard even becomes negative, see Figure 7.26. Due to the lack of buoyancy above the deck side at large heel in the wave crest, the centre of buoyancy in heeled condition B_ϕ shifts towards the centre of gravity G. This shift of B reduces the *GZ*.

At the same time, sections 1 and 3 show an increase in effective freeboard for the ship in the wave crest. This positive effect cannot counteract the large *GZ*-reduction amidships. Thus, from summing up along the ship length, the wave crest results in a reduction of the *GZ*-curve. This reduction is larger with finer ship shapes designed for large speed at small block coefficient C_B. On the contrary, wave trough amidships results in an increase of

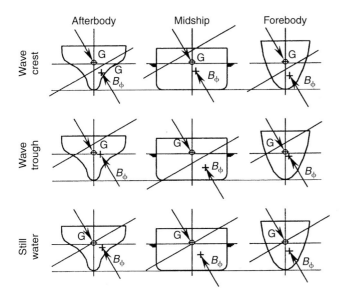

Figure 7.26 Contribution of characteristic ship sections on *GZ*-changes in longitudinal wave.

the righting lever *GZ*. The effective freeboard of the midship section 2 is considerably increased. The overall *GZ* reduction in the crest is larger than the gain in the trough. The mean *GZ* at crest and trough is therefore below the *GZ*-curve in still water. The overall *GZ* reduction in the crest is larger than the gain in the trough.

A detailed calculation for wave lengths from $0.6L_S$ to $1.75L_S$ shows the reduction (*S-C*) of *GZ* in the crest *C* compared with still water (*S*), see Figure 7.27. The maximum of the *GZ* reduction in a crest is not at a wave length equal to ship length, but at a smaller wave length between $0.75L_S$ to $0.9L_S$ for the hull form used, Kastner (1975). For easy comparison of different ships, a wave of ship length is still recommended for the method.

The wave height ought to be chosen according to a formula derived from wave statistics in the North Atlantic at about 1%, see Figure 7.28.

$$\frac{H_w}{L_w} = \frac{1}{10 + 0.05 \cdot L_w} \quad (L_w \text{ in metre}) \quad (7.64)$$

The calculated *GZ*-curves in a longitudinal regular wave can only substitute reality, but they give a reasonably good estimate of the variation of the restoring moment. Blume and Hattendorff (1982) compared the hydrostatic results with measurements

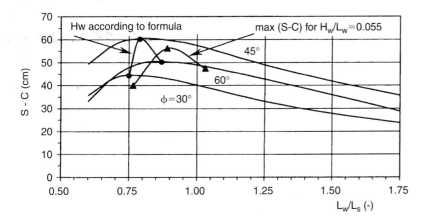

Figure 7.27 Influence of wave-length on hydrostatic *GZ* in a wave crest.

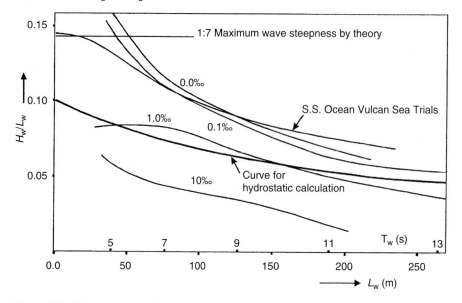

Figure 7.28 Wave steepness for calculating *GZ*-variation in a longitudinal wave.

on models of container ships in following seas. The underwater hull was determined by the wave contour along the hull at different ship speeds. For a Froude number, F_n, up to 0.28, there was almost no difference to the hydrostatic *GZ* at Froude number zero. Only at the large Froude number of 0.36, the reduction in the wave crest was about half the value of the hydrostatic calculation. In the wave trough at $F_n = 0.36$, the increase was about 10% less than the hydrostatic result, Figure 7.29.

Only crest and trough were used above to reflect the largest deviations of the *GZ*-curve. The two intermediate positions, crest at mid after-body (1) and crest at mid fore-body (2), show some deflection from the *GZ*-curve in still water too, see Figure 7.30, but are mainly disregarded for simplified estimates. Generally, *GZ*-variations in longitudinal sea depend on the hull form and hull proportions. With a small block coefficient, and with station flare, the variations will be larger. A large freeboard is of advantage to have less *GZ* reduction in a wave crest.

Kastner (1969, 1970, 1970a) compared the hydrostatic calculation in a regular wave with the wave contour of an irregular seaway. The *GZ*-variations are most pronounced in a wave group, where the passing wave closely resembles a single wave. In following seas at the Froude number 0.20 the seaway energy concentrates in a narrow frequency band, Figure 7.31. The narrow frequency band is discussed further in Section 7.2.11.17.

Authorities like to stay with the *GZ*-curve in still water and implement wave effects into the required *GZ* values. Only in the German Navy rules for stability, the *GZ* crest and *GZ* mean of crest and trough was included as a simple way to account for the wave impact, Wendel (1965), Arndt *et al.* (1982).

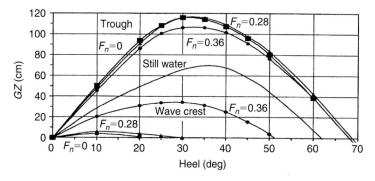

Figure 7.29 Hydrostatic *GZ*-curves in crest and trough with dynamic corrections.

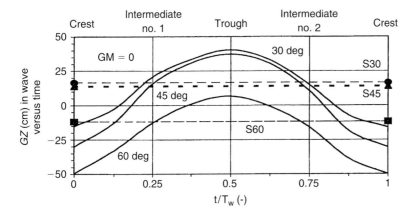

Figure 7.30 GZ at constant heel in longitudinal wave.

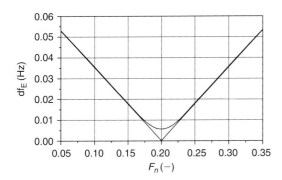

Figure 7.31 Frequency bandwidth of seaway in following seas versus Froude number.

Blume and Hattendorff (1982, 1984) developed a so-called C-factor for usual merchant ship hull forms, which allows including the GZ reduction in waves by a formula based on capsizing model experiments. IMO implemented the C-factor for container ships and fast ships with a small block coefficient (tested C_B was 0.554 through 0.675) into the IMO stability criteria, IMO (1993). Calculation of GZ curves in longitudinal seas is not required. Blume invented a C-factor, depending on hull proportions, form coefficients, ship length and height of centre of gravity above keel:

$$C = \frac{T \cdot D'}{B^2} \cdot \sqrt{\frac{T}{KG} \cdot \left(\frac{C_B}{C_w}\right)^2 \cdot \sqrt{\frac{100}{L_{pp}}}} \quad (7.65)$$

T Mean draught (m)
B Moulded breadth of the ship (m)
KG Height of the centre of gravity above the keel (m); not to be taken less than T
C_B Block coefficient
C_w Waterplane coefficient

Table 7.6 Intact stability criteria based on C factor.

No.	Criterion ≥	Required	Dimension
1a	e_{30}	0.009/C	metre · rad
1b	e_{40}	0.016/C	metre · rad
2	e_{30-40}	0.006/C	metre · rad
3	GZ at ≥ 30°	0.033/C	metre
4	GZ_{max}	0.042/C	metre
5	$e_{GZtotal}$	0.029/C	metre · rad

The effective freeboard D' accounts for the volume of the hatches above deck amidships (from plus and minus L/4 of the main section). Ship length is to be ≥ 100 m, and KG is to be larger than draught T.

The smaller the C-factor, the larger are the GZ values required. For container ships or ships with considerable flare or large water plane areas, IMO asks for hydrostatic values in the form of a required constant divided by C, as given in Table 7.6. Figure 7.32 shows an example GZ curve according to the regulation.

7.2.11.11 Encounter period of ship and waves

The encounter of a ship with the waves governs the time dependent excitation of the ship in waves. The encounter period, T_E, is the time elapsed from wave crest to the next wave crest passing the ship.

Figure 7.33 shows the general co-ordinate system of the ship travelling on the sea surface, i.e. in the horizontal plane. The heading angle is the difference of the wave direction, μ, and the ship course, ξ, in the global co-ordinate system.

$$\chi = \mu - \xi$$

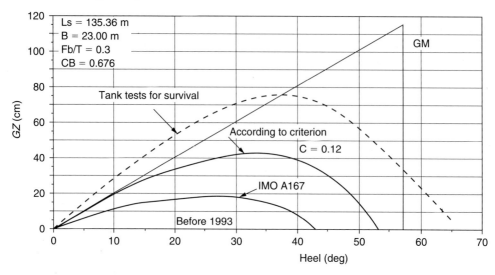

Figure 7.32 GZ-curve required by IMO based on C-factor.

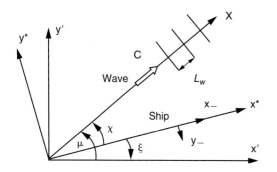

Figure 7.33 Co-ordinate system of surface waves and advancing ship.

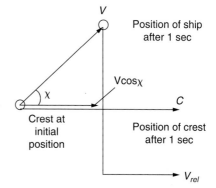

Figure 7.34 Wave celerity and ship speed at heading angle.

T_E	encounter period of ship and wave
c	wave celerity (velocity of wave crest at sea surface)
V	ship speed
V_{rel}	relative velocity wave–ship
$x'y'$	sea surface in geodetic co-ordinates
x^*y^*	sea surface with respect to the ship course
$x\_y\_$	ship equilibrium axis
ξ	angle of ship direction
μ	angle of wave direction
χ	heading angle of ship and waves (zero for following seas)

T_E can be derived by a simple transformation of the waves travelling across the sea surface against the ship, see the vector diagram for the relative velocity ship and waves in Figure 7.34. The vector diagram of velocities can be seen as distances travelled per second.

V_{rel} is the relative distance the wave has travelled in the x-direction with respect to the ship. A shorter distance in the same time of 1 sec means a smaller relative velocity of the wave with respect to the ship. We derive the relative velocity wave–ship,

$$V_{rel} = c - V\cos\chi \qquad (7.66)$$

and the encounter period,

$$T_E = \frac{L_w}{V_{rel}} = \frac{L_w}{c - V\cos\chi} \qquad (7.67)$$

Celerity, length and period (or frequency) of a regular wave correspond to each other according to the definition

$$c = \frac{L_w}{T_w} = L_w f_w \qquad (7.68)$$

The celerity of gravity waves in deep water is proportional to the wave period, but independent of the wave height (derived by applying the Bernoulli equation of hydrodynamics on energy):

$$c = \frac{g}{2\pi} T_w \quad (7.69)$$

With $g/2\pi \cong 1.56 \, \text{m/s}^2$. From (7.68) and (7.69) we get:

$$L_w = \frac{g}{2\pi} T_w^2 \quad (7.70)$$

Substituting (7.69) and (7.70) into Equation (7.67), we find T_E as a function of T_w, V, χ:

$$T_E = \frac{T_w^2}{T_w - \frac{2\pi}{g} V \cos \chi} \quad (7.71)$$

This equation is basically very simple, but it shows important peculiarities depicted in Figures 7.35 (parameter period), 7.36 (parameter length), and 7.37, 7.38 with $f_E = 1/T_E$. These detailed graphs can make aware of the particular encounter situation, and on the influence of changing ship speed and heading. The steeper the wave the faster the celerity c of a single harmonic wave, Pierson (1993). On the contrary, the phase speed of wave crests in random seas is reduced, Pierson (1954). Mathematical progress on non-linear superposition of components of high random waves may lead to modifications of the single harmonic waves as used here. A single straight line alternative to Equation (7.71), as finally used in IMO guidelines, is discussed in Chapter 11.

From Equation (7.69), using Equations (7.68) and (7.70), we find the formula for the wave celerity $c(L_w)$,

$$c = \sqrt{\frac{gL_w}{2\pi}} \quad (7.72)$$

From inserting the celerity c from Equation (7.72) into Equation (7.67), we can replace the wave period T_w by the wave length L_w. The result is T_E as a function of the wave length instead of the period, T_E (χ, V, T_W), as shown in Figure 7.36:

$$T_E = \frac{L_w}{\sqrt{\frac{g}{2\pi} L_w} - V \cos \chi} \quad (7.73)$$

7.2.11.12 Encounter frequency

Let us choose the encounter frequency, f_E, rather than the encounter period, T_E, defined as:

$$f_E = \frac{1}{T_E} \quad \text{(Dimension Hz} = \text{Hertz)} \quad (7.74)$$

Or, the corresponding circular encounter frequency:

$$\omega_E = 2\pi \cdot f_E \quad (\text{Radians s}^{-1}) \quad (7.75)$$

The reason for using frequencies instead of periods is that in following waves we find conditions where encounter periods become very large, and correspondingly the encounter frequencies of waves are very small. The information is the same. It is easier to look at figures such as f_E close to or even equal to zero, than at the corresponding T_E with values approaching infinity. In stern quartering seas at large ship speed, the encounter period T_E for shorter waves goes to infinity. This means practically, the ship runs with these waves for a long time. The corresponding f_E curves show zero encounter frequency.

In the diagrams, zero heading means following seas. The heading of 45 degrees means stern quartering seas; 90 degrees, starboard beam seas; 180 degrees, head seas.

From Figure 7.37 and 7.38 it is obvious that for large ship speed, a number of regular waves coincide at a very narrow frequency band in following and stern quartering seas.

7.2.11.13 Wave group of two regular waves

The linear superposition of partial wave components of a random sea compressed to a small frequency band by the transformation to the ship can be demonstrated by the simplest case with two waves of the same amplitude, but a very small difference in frequency, see Figure 7.39. The background can be found in any book on engineering mechanics, such as 'Hütte', Czichos (1991).

The linear superposition of two neighbouring partial waves:

$$\zeta = \zeta_1 + \zeta_2 = a_1 \cdot \sin(2\pi \cdot f_1 \cdot t) + a_2 \cdot \sin(2\pi \cdot f_2 \cdot t) \quad (7.76)$$

With

$$df = f_1 - f_2 << f_{1,2} \quad (7.77)$$

And

$$f_m = (f_1 + f_2)/2, \quad T_m = \frac{1}{f_m} = \frac{2}{f_1 + f_2} \quad (7.78)$$

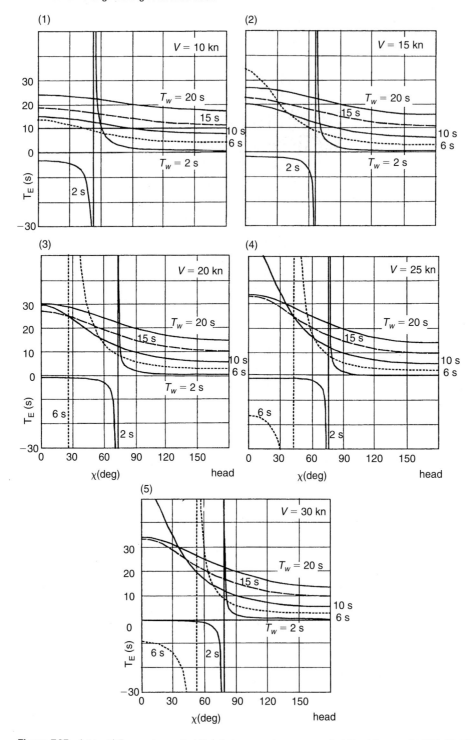

Figure 7.35 (.1 to .5) Encounter period $T_E(\chi)$ at parameter wave period T_w, ship speed V (10, 15, 20, 25, 30 kn).

Seakeeping 517

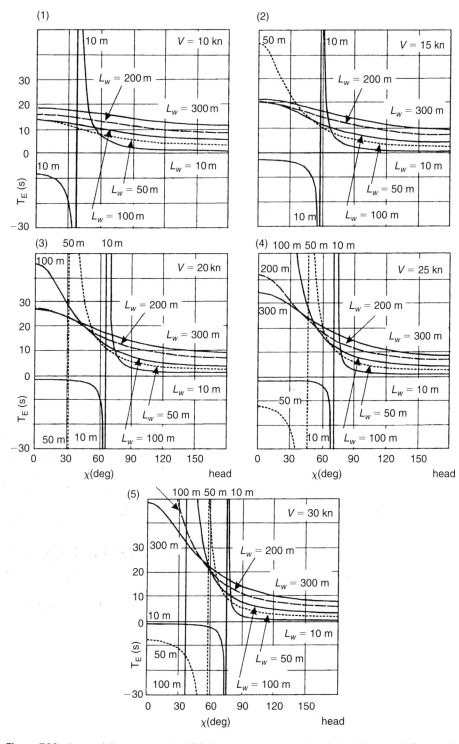

Figure 7.36 (.1 to .5) Encounter period $T_E(\chi)$ at parameter wave length L_w, ship speed V (10, 15, 20, 25, 30 kn).

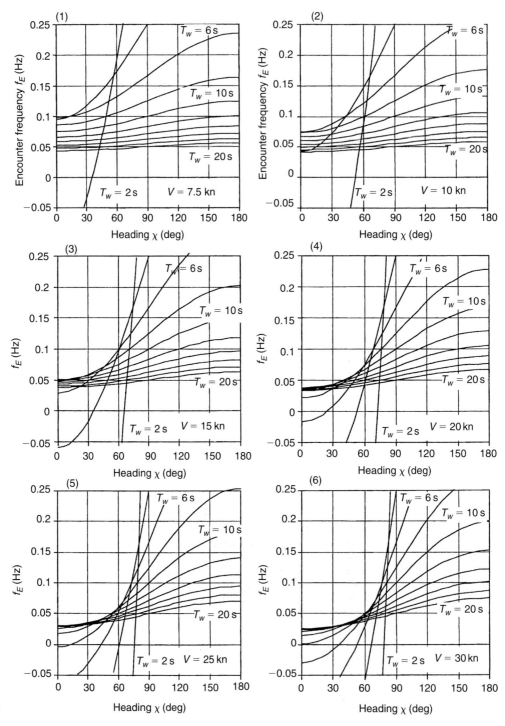

Figure 7.37 (.1 to .6) Encounter frequency $f_E(\chi)$, parameter T_w (from 2 s through 20 s in 2 s steps) for different ship speeds V (7.5 kn to 30 kn), head seas at 180 deg.

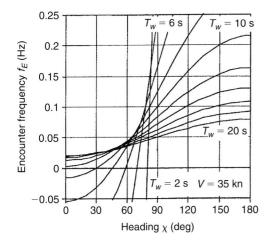

Figure 7.37 (.7) Encounter frequency $f_E(\chi)$, parameter T_w, for ship speed V (35 km).

Results by straightforward evaluation in:

$$\zeta = (a_1 + a_2) \cdot \cos\left(2\pi \frac{df}{2} \cdot t\right) \cdot \sin(2\pi f_m \cdot t) \quad (7.79)$$

$$T_{\mathrm{mod}} = \frac{2}{df} = \frac{2}{f_1 - f_2} \quad (7.80)$$

The resulting wave superposition shows an oscillation with the mean frequency, $f_m = 1/T_m$, while the amplitude varies periodically between $a_1 + a_2$ and zero. The sum of the two frequencies is modulated by the frequency $f_{\mathrm{mod}} = df/2 = 1/T_{\mathrm{mod}}$. The slowly varying time function $\cos(2\pi t/T_{\mathrm{mod}}) = \cos(2\pi t df/2)$ is the envelope of the resulting wave amplitudes.

Figure 7.39 shows a numerical result (encounter periods from Figure 7.37, 20 kn speed, 45 deg heading): $a_1 = a_2 = 1$ m, $f_{E1} = 0.04$ Hz ($T_{E1} = 25$ s),

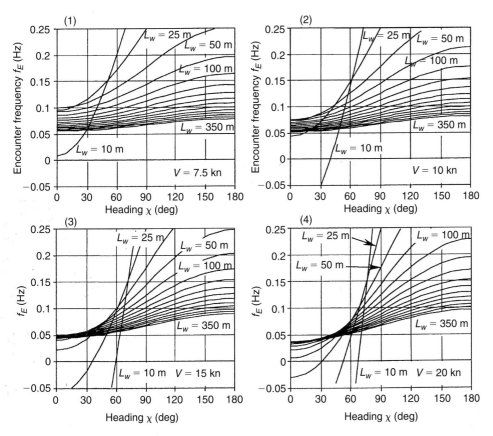

Figure 7.38 (.1 to .4) Encounter frequency $f_E(\chi)$, parameter L_w (in steps of 25 m) for different ship speeds (7.5 kn to 20 kn).

520   Maritime engineering reference book

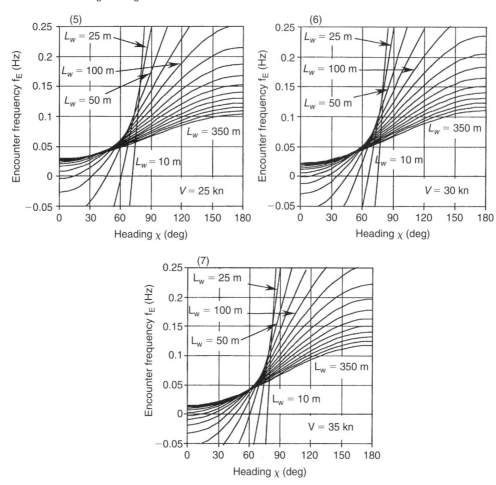

Figure 7.38 (.5 to .7) Encounter frequency $f_E(\chi)$, parameter L_w (in steps of 25 m) for different ship speed (20 kn to 35 kn).

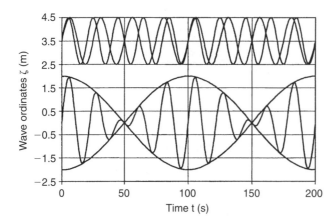

Figure 7.39 Wave group of two regular waves.

$f_{E2} = 0.05\,\text{Hz}$ ($T_{E2} = 20\,\text{s}$). Then $df = 0.01\,\text{Hz}$, $T_{Emod} = 2/0.01\,\text{Hz} = 200\,\text{s}$, $a_1 + a_2 = 2\,m$, $f_{Em} = 0.045\,\text{Hz}$, $T_{Em} = 1/f_{Em} = 22.2\,\text{s}$.

The resulting sum shows the same phenomenon as seafarers have always observed concerning the occurrence of so-called wave groups. A wave group is characterised by a number of waves at large amplitude and almost the same frequency. Between the wave groups, we always find a time sequence where the waves reduce considerably. In the next section, the grouping phenomenon is extended to irregular seas.

7.2.11.14 Wave encounter of a ship in irregular seas

Basically, the same results as above with the single waves can be obtained by transforming the seaway spectra onto the running ship. Figure 7.40 shows an example of an encounter seaway spectrum in following seas. As we recall, see Section 1.4, a seaway spectrum represents the energy of the seaway, distributed over the frequency.

We consider the irregular seaway as a summation of partial regular waves with different partial wave heights, frequencies, and phases. To transform the irregular seaway pattern from the ocean onto the ship advancing with the speed V at the heading χ, we must perform the same transformation as shown above for a single regular wave, but now with all the partial waves which are members of the irregular seas.

St. Denis and Pierson (1953) were the first to present the spectra formulation of the seaway. It is included in many books on ship hydrodynamics (e.g. Principles of Naval Architecture, Lewis,

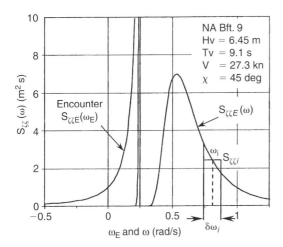

Figure 7.40 Encounter spectrum in stern quartering seas.

(1989)). Kastner (1969, 1970) applied the irregular wave encounter to the simulation of nonlinear roll resonance in irregular seas. Here, we compile the main formulae for the spectral transformation.

From Equation (7.67) we can write for the encounter frequency of a partial wave:

$$f_E = \frac{1}{T_E} = \frac{c - V \cdot \cos \chi}{L_w} \quad (7.81)$$

From Equation (7.68) we set $c/L_w = f_w$, and we have:

$$f_E = f_w \cdot \left(1 - \frac{V \cdot \cos \chi}{c}\right) \quad (7.82)$$

The dimensionless ratio of the ship speed component in the wave direction and the wave celerity is named α,

$$\alpha = \frac{V \cdot \cos \chi}{c} \quad (7.83)$$

And we write Equation (7.82) as:

$$f_E = f_w (1 - \alpha) \quad (7.84)$$

The wave ordinate for a regular wave is given in Chapter 1 as (from Equations (1.28) and (1.29)):

$$\zeta(x,t) = \frac{H_w}{2} \cos \varepsilon = \frac{H_w}{2} \cos(kx - \omega t) \quad (7.85)$$

According to the co-ordinate systems as shown in Figure 7.33, x is the global co-ordinate for the wave direction, and x^* the co-ordinate for the ship. For a ship travelling at speed V in the x^* direction, the relative distance travelled in time t is then

$$x_E = x - t \cdot V \cdot \cos \chi \rightarrow x = x_E + t \cdot V \cdot \cos \chi \quad (7.86)$$

Introducing x from Equation (7.86) into Equation (7.85), we get:

$$\varepsilon_E = kx - \omega t = k(x_E + t \cdot V \cdot \cos \chi) - \omega t = kx_E - \omega\left(1 - V \cos \chi \cdot \frac{k}{\omega}\right) \cdot t$$

We replace the wave number k according to Equation (1.20), $k = \omega^2/g$, and we have:

$$\varepsilon_E = \frac{\omega^2}{g} x_E - \omega\left(1 - V \cos \chi \cdot \frac{\omega^2}{g} \cdot \frac{1}{\omega}\right) \cdot t$$
$$= kx_E - \omega_E t$$

The factor in front of time t is the encounter frequency ω_E:

$$\omega_E = \omega - V \cos\chi \cdot \frac{\omega^2}{g} \tag{7.87}$$

With $c = g/\omega$, we find the version:

$$\omega_E = \omega\left(1 - \frac{V \cdot \cos\chi}{c}\right) = \omega(1 - \alpha) \tag{7.88}$$

This corresponds to Equations (7.82) and (7.84), because $\omega_E = 2\pi f_E$.

It is convenient to replace the ship speed V by the dimensionless speed F_n, the Froude number:

$$F_n = \frac{V}{\sqrt{g \cdot L_s}}$$

We can write the encounter frequency as a function of the Froude number F_n and ship length L_s, from Equation (7.87):

$$\omega_E = \omega - \left(\frac{\omega^2}{g}\right) \cdot V \cos\chi$$

$$= \omega - \left(\frac{\omega^2}{g}\right) \cdot F_n \cdot \cos\chi \cdot \sqrt{g \cdot L_s} \tag{7.89}$$

The corresponding frequency f_E is (dimension Hz = Hertz, with $\frac{2\pi}{\sqrt{g}} \approx 2\ (s \cdot m^{-0.5})$):

$$f_E = f_w - f_w^2 \cdot \left(\frac{2\pi}{\sqrt{g}}\right) \cdot V \cos\chi \cdot \sqrt{L_s}$$

$$\cong f_w - 2 \cdot \sqrt{L_s} \cdot (F_n \cos\chi) \cdot f_w^2 \tag{7.90}$$

Figure 7.41 depicts the quadratic function f_E versus $f_w\ (=\omega/2\pi)$ in stern quartering sea with the Froude number F_n as parameter. Figure 7.43 shows the encounter frequency $f_E = /1/T_E$ for a constant ship speed but different headings from stern to head seas.

In Equation (7.87), we can also express the ship speed by the dimensionless Froude number F_n and ship length Ls, with $V = F_n\sqrt{gL_s}$. We further introduce the dimensionless form of the frequencies:

$$\varpi_E = \omega_E \sqrt{\frac{L_s}{g}} \quad \text{and} \quad \varpi = \omega\sqrt{\frac{L_s}{g}} \tag{7.91}$$

The relationship Equation (7.87) results in the dimensionless form of the encounter frequency:

$$\varpi_E = \varpi(1 - \varpi \cdot F_n \cos\chi) \tag{7.92}$$

What we want to know is the irregular wave pattern of the seaway as the partial waves are passing the advancing ship. The irregular wave pattern at the sea surface can be calculated as the linear sum of all partial regular waves. At the moment, we restrict ourselves to unidirectional waves, i.e. they all run in the same direction.

Let us refer again to the wave ordinate in Equation (7.85), but now writing it for the wave encounter:

$$\zeta_E(x,t) = \frac{H_w}{2}\cos\varepsilon_E = \frac{H_w}{2}\cos(kx_E - \omega_E t)$$

We are going to use the regular wave for any partial wave i of the irregular seaway, and transform it to the ship according to the encounter frequency. Each partial wave is sinusoidal in shape. It depends on partial wave height h_{wi} (equal to the double amplitude $2c_i$), on wave frequency ω_i, (or with respect to the ship on the encounter frequency ω_{Ei}),

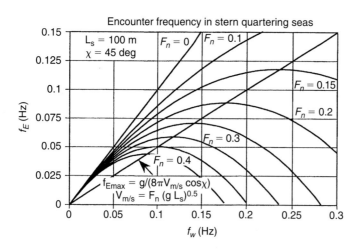

Figure 7.41 Frequency of encounter versus wave frequency, parameter F_n.

and on a randomly varied phase relation ϕ_i. We are going to set $x_E = 0$. This means we are not interested in the wave contour along the ship, but in the time sequence of the irregular waves passing the ship. Then the irregular wave ordinates before and after the transformation are:

$$\zeta(t) = \sum_{i=1}^{n} \zeta_i(t) = \sum_{i=1}^{n} c_i \cos(\omega_i t + \beta_i)$$

$$\zeta_E(t) = \sum_{i=1}^{n} \zeta_{Ei}(t)$$

$$= \sum_{i=1}^{n} c_{Ei} \cos(\omega_{Ei} t + \beta_{Ei}) \qquad (7.93)$$

The phase angle β for all partial waves is randomly distributed. The probability density is constant, this means any particular phase value between zero and 2π has the same probability of occurrence:

$$p(\beta) = \frac{1}{2\pi} = \text{const} \quad \text{for} \quad 0 \leq \beta \leq 2\pi \qquad (7.94)$$

with

$$P = \int_{-\infty}^{\infty} p(\beta) d\beta = 1$$

In order to calculate Equation (7.93), we must also know the amplitudes, c_i, of the partial waves of the irregular seas. The wave heights correspond to the energy contents of the particular components constituting the irregular sea. The amplitude (half wave height) results from the square root of double the spectral energy $S_{\zeta\zeta i}$ multiplied by the frequency width $\delta\omega_i$, see Figure 7.40.

$$c_i = \sqrt{2 S_{\zeta\zeta i} \cdot \delta \omega_i} \qquad (7.95)$$

7.2.11.15 Wave energy and encounter spectra

The sea spectrum $S_{\zeta\zeta}$ as defined in Chapter 1, and discussed in Section 7.2.7, represents the wave energy with respect to the wave frequency ω. When the wave frequency transforms to the encounter frequency, the spectral value $S_{\zeta\zeta}$ transforms accordingly.

With preservation of energy, the total seaway energy at transformation to the travelling ship stays the same:

$$m_{0\zeta\zeta E} = m_{0\zeta\zeta}$$

Preservation of energy applies also to the energy differentials:

$$dm_{0\zeta\zeta E} = dm_{0\zeta\zeta}$$
$$dm_{0\zeta\zeta E} = S_{\zeta\zeta E}(\omega_E) d\omega_E$$
$$dm_{0\zeta\zeta} = S_{\zeta\zeta}(\omega) d\omega$$

So, we have for the encounter spectrum the following expression:

$$S_{\zeta\zeta E}(\omega_E) = S_{\zeta\zeta}(\omega) \frac{d\omega}{d\omega_E} = \frac{S_{\zeta\zeta}(\omega)}{d\omega_E/d\omega} \qquad (7.96)$$

The derivative of Equation (7.87) is

$$\frac{d\omega_E}{d\omega} = 1 - 2\omega \frac{V \cos \chi}{g} \qquad (7.97)$$

We insert Equation (7.97) into Equation (7.96) and get the final expression for the encounter spectrum:

$$S_{\zeta\zeta E}(\omega_E) = \frac{S_{\zeta\zeta}(\omega)}{1 - 2\omega \dfrac{V \cos \chi}{g}} \qquad (7.98)$$

Figure 7.42 depicts an example of encounter seaway spectra at constant ship speed of 25 knots for different heading χ. In beam seas, the encounter spectrum is not influenced by the ship speed. In head seas ($\chi = 180$ deg) the encounter frequency increases, but the value of the encounter spectrum decreases as the energy spreads over a larger frequency range. However, in stern seas ($\chi = 0$) the encounter spectrum becomes narrower and concentrates into a small frequency band. The smallest frequency band can be seen in stern quartering seas ($\chi = 45$ deg and 315 deg).

7.2.11.16 Relevant frequencies of the spectrum of encounter

Let us look at the spectral transformation of the sea spectrum on the advancing ship in more detail. There are four dominant frequencies, ω_{Ev}, ω_v, $\omega_{E\infty}$, and ω_∞, defined as follows.

When the ship is running with the wave at the same velocity, with $V \cos \chi = c$, according to Equation (7.88), the frequency of encounter is zero:

$$\omega_{Ev} = 0$$

From Equation (7.87), we find the corresponding frequency ω_v of the wave running at the same celerity c as the ship advances in the wave direction (see also Section 7.2.7):

$$\omega_{Ev} = 0 = \omega_v - \frac{\omega_v^2}{g} \cdot V \cos \chi = \omega_v \left(1 - \frac{\omega_v}{g} \cdot V \cos \chi \right)$$

$$\omega_v = \frac{g}{V \cdot \cos \chi} \qquad (7.99)$$

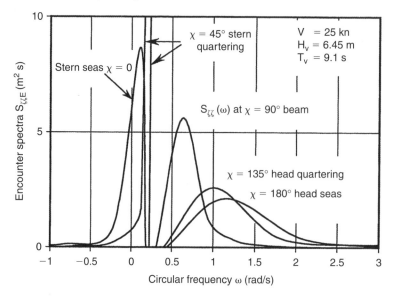

Figure 7.42 Encounter spectra, parameter χ, speed 25 kn.

The maximum of the encounter frequency ω_E corresponds to the wave frequency ω_∞ before transformation to the ship, where the tangent is equal to zero, from Equation (7.97):

$$\frac{d\omega_E}{d\omega} = 1 - 2\omega_\infty \frac{V \cos \chi}{g} = 0$$

$$\omega_\infty = \frac{g}{2 \cdot V \cos \chi} \quad (7.100)$$

The maximum of ω_E results from putting ω_∞ into Equation (7.87), and we have

$$\omega_{E\infty}(\omega_\infty) = \frac{g}{2V \cos \chi} - \frac{V \cos \chi}{g} \cdot \frac{g}{2V \cos \chi} \cdot \frac{g}{2V \cos \chi}$$

$$= \frac{g}{V \cos \chi} \left(\frac{1}{2} - \frac{1}{4} \right)$$

$$\max(\omega_E) = \omega_{E\infty} = \frac{g}{4 \cdot V \cos \chi} \quad (7.101)$$

Inserting ω_∞ from Equation (7.100) in Equation (7.98) for the encounter spectrum, the denominator becomes zero, and $S_{\zeta\zeta E}$ goes towards infinity:

$$S_{\zeta\zeta E}(\omega_{E\infty}) = \frac{S_{\zeta\zeta}(\omega_\infty)}{1 - 2 \cdot \frac{g}{2V \cos \chi} \cdot \frac{V \cos \chi}{g}} = \frac{S_{\zeta\zeta}}{1 - 1}$$

$$S_{\zeta\zeta E\infty} \to \infty \quad (7.102)$$

From the above equations, we see the very simple relationships between the derived relevant frequencies, such as

$$\omega_{E \max} = \omega_{E\infty}$$
$$\omega_v = 2 \cdot \omega_\infty = 4 \cdot \omega_{E\infty}$$
$$\omega_\infty = 2 \cdot \omega_{E\infty}$$
$$\omega_{Ev} = 0$$

Table 7.7 compiles the relevant frequencies of encounter ship and wave, using circular frequency ω, frequency f_w, or period T_w, according to

$$\omega = 2\pi \cdot f_w = \frac{2\pi}{T_w}.$$

The main derivations above have been given with the circular wave frequency ω (dimension s^{-1}). Practically, the wave period T_w is easier to discuss in the world of shipping. When we refer to frequency, it is convenient to use f_w (and f_E respectively) with the dimension Hertz (Hz) as the inverse of the period. Gravity acceleration g in the metric system is 9.81 m/s². The constant $g/(2\pi)$ is $g/(2\pi) = 9.81/6.28 = 1.56$ m/(rad s²) and $2\pi/g = 0.64$ rad s²/m.

Figure 7.43 depicts the encounter frequency f_E for a ship length of 100 m at the Froude number 0.20 and five headings: 0 deg following, 45° stern quartering, 90° beam, 135° head quartering, 180° head seas. The widest spread of f_E appears in head seas, see also Figure 7.37. Figure 7.44 gives a typical result with a narrow encounter spectrum in following seas. Figure 7.45 depicts the corresponding

Table 7.7 Relevant frequencies at spectral seaway transformation onto the ship.

No.	Circular frequency rad $s^{-1} \cong 1/s$	Frequency Hertz (Hz) = 1/s	Period second (s)
1	$\omega_v = \dfrac{g}{V \cos \chi}$	$f_{wV} = \dfrac{g}{2\pi \cdot V \cos \chi}$ $= \dfrac{1}{2\sqrt{L_s} \cdot F_n \cos \chi}$	$T_{wV} = \dfrac{2\pi}{g} \cdot V \cos \chi$ $= 2\sqrt{L_s} \cdot F_n \cos \chi$
2	$\omega_{Ev} = 0$	$f_{Ev} = 0$	$T_{Ev} = \infty$ (infinity)
3	$\omega_\infty = \dfrac{g}{2 \cdot V \cos \chi}$	$f_{w\infty} = \dfrac{g}{2\pi \cdot 2 \cdot V \cos \chi}$ $= \dfrac{1}{4\sqrt{L_s} \cdot F_n \cos \chi}$	$T_{W\infty} = \dfrac{2\pi}{g} \cdot 2V \cos \chi$ $= 4\sqrt{L_s} \cdot F_n \cos \chi$
4	$\omega_{E\infty} = \dfrac{g}{4 \cdot V \cos \chi}$ $= \max(\omega_E)$	$f_{E\infty} = \dfrac{g}{2\pi \cdot 4 \cdot V \cos \chi}$ $= \dfrac{1}{8\sqrt{L_s} \cdot F_n \cos \chi}$	$T_{E\infty} = \dfrac{2\pi}{g} \cdot 4V \cos \chi$ $= 8\sqrt{L_s} \cdot F_n \cos \chi$

Legend:
1. Wave frequency ω_v which transforms to zero encounter frequency ω_{Ev};
2. Zero encounter frequency ω_{Ev} at $V \cos \chi = c$;
3. Wave frequency ω_∞ which transforms to $\omega_{E\infty}$;
4. Maximum encounter frequency $\max(\omega_E) = \omega_{E\infty}$

spectrum before the transformation. By comparing both graphs, we realise the meaning of the relevant frequencies.

The seaway spectrum in Figure 7.45 divides into three regions I, II, and III. Region I at the lower frequency tail and region II are separated by ω_∞. The middle region II and the upper region III are separated by ω_V.

The encounter spectrum folds at the maximum $\omega_{E\infty}$, and the two different regions I and II and up below that folding frequency, Figure 7.44.

Region III of the encounter spectrum has been defined for the negative frequencies, i.e. where the ship is overtaking the waves. Relative to the ship, it appears as if the waves are head seas. This is only due to the ship speed in the wave direction being larger than the velocity of the partial waves, therefore overtaking the stern waves. Region III corresponds to the large frequency tail in the original seaway spectrum $S_{\zeta\zeta}$.

The spectral representation has been normalised, i.e. the area under the curve is set equal to unity.

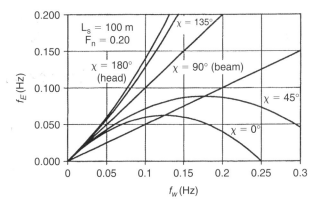

Figure 7.43 Encounter frequency at constant ship speed and different heading.

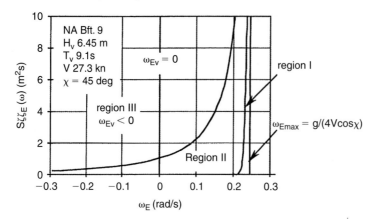

Figure 7.44 Encounter spectrum in stern quartering seas showing 3 regions.

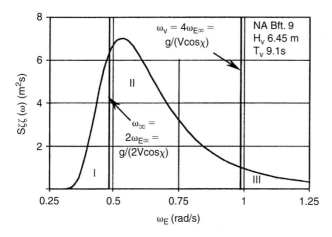

Figure 7.45 Seaway spectrum to be transformed in 3 regions I, II, III.

This allows a quick comparison of different seaway spectra and the energy distribution versus the frequency, independent of the total sea energy.

7.2.11.17 Bandwidth of the transformed sea spectrum

In Figures 7.44 and 7.45, we have seen the energy of the irregular seaway transformed onto the ship with speed and heading becoming more concentrated on a smaller frequency band in following and stern quartering seas. In case the total energy of the sea spectrum concentrates in a narrower frequency band due to the transformation onto the advancing ship, the seaway energy within this band must be larger than with the not yet transformed seaway spectrum, because of the preservation of energy,

$$m_0(S_{\zeta\zeta E}) = m_0(S_{\zeta\zeta}) \qquad (7.103)$$

with

$$m_0(S_{\zeta\zeta E}) = \int_0^\infty S_{\zeta\zeta E} d\omega_E$$

and $\quad m_0(S_{\zeta\zeta}) = \int_0^\infty S_{\zeta\zeta} d\omega$

In other words, the narrower the transformed seaway spectrum is, the closer the ship can experience a nearly single frequency excitation even in an irregular natural seaway.

However, the seaway energy within the narrow peaked region of the transformed seaway spectrum is not evenly distributed, but varies rapidly, while at $f_{E\infty}$

it is going towards infinity. To account for the shape of the transformed seaway spectrum $S_{\zeta\zeta E}$ we can use the first order moment

$$m_1 = \int_0^\infty (f_E - f_{E\infty}) S_{\zeta\zeta E}(f_E) df_E \qquad (7.104)$$

Figure 7.46 shows the result of a numerical evaluation of m_1 (ship length L_s 100 m). The parameter Λ is length ratio L_s/L_w of ship and wave. The wave spectrum used for the calculation has the maximum at the modal frequency of 0.125 Hz, which corresponds to a wave length of 100 m. The minimum bandwidth in stern seas from Figure 7.46 appears at the Froude number F_{n1} of 0.19. For the assumed range of L_s/L_w between 0.75 and 1.5, the smallest value of m_1 covers a Froude number from 0.15 to 0.22.

$$0.15 \leq F_{n1}(m_1 = Min) \leq 0.22 \qquad (7.105)$$

This F_{n1} range can be seen as dangerous in following seas. When the modal value of the spectrum coincides with longer waves than the ship length, the minimal bandwidth shifts to larger ship speed (in our 100 m example at $L_w = L_s/0.15$ to $F_{n3} = 0.22$). For shorter waves the shift is to a smaller Froude number ($F_{n1} = 0.15$ at $L_w = L_s/1.5$). Table 7.8 compiles the corresponding ship speed in knots according to the ship length.

In stern quartering seas, the ship speed must be even larger to have more waves to coincide with the encounter frequency (or vice versa, the encounter periods). This is demonstrated by the crossing of the f_E curves in a narrow region (Figure 7.37 and 7.38).

The resulting F_n range of 0.15 to 0.22 is about half the speed of the ship running with the wave crest.

The Froude number in following sea of the ship speed at wave celerity, is:
With

$$F_n = \frac{V}{\sqrt{gL_s}} \text{ and } c = \sqrt{\frac{gL_w}{2\pi}}$$

$$F_{nw}(V = c) = \frac{1}{\sqrt{2\pi}} \cdot \sqrt{\frac{L_w}{L_s}} \qquad (7.106)$$

For the wave length equal to the ship length,

$$F_{nw}(V = c) = \frac{1}{\sqrt{2\pi}} \cong 0.40 \qquad (7.107)$$

At a narrow band of the encounter frequency, all the energy of the particular waves is going to act upon the ship in the same time sequence. In other words, running a ship in following or stern quartering seas, closely resembles regular wave excitation, no matter how irregular or confused the sea is. This appears to be a very strange result. However, the above equations and the graphical representation show this surprising aspect very clearly. Following and stern quartering seas are most serious with respect to danger from capsizing due to the narrow frequency excitation. In case either the natural period or the half of the natural period coincides with the narrow excitation, a large roll due to resonance builds up. This is discussed further by Kobylinski and Kastner (2003), Chapter 11.

7.2.11.18 Irregular time series of wave encounter

The wave encounter in irregular seas is shown in Figure 7.47, calculated with an Excel program by Skalicky (1998) modified by the author, using the seaway spectrum according to Bretschneider (Chapter 1). Figure 7.47a shows the encounter spectrum in stern quartering seas at V/T = 3 kn/s.

Table 7.8 Speed versus ship length for energy concentration in stern seas.

$A = L_s/L_w$	1.5 $L_w < L_s$		1 $L_w = L_s$		0.75 $L_w > L_s$	
$F_n = V/(gL_s)^{1/2}$	0.15		0.19		0.22	
L_s (m)	L_w (m)	V(kn)	L_w(m)	V(kn)	L_w(m)	V(kn)
25	17	4.6	25	5.8	33	6.7
50	33	6.5	50	8.2	67	9.5
100	67	9.1	100	11.6	133	13.4
200	133	12.9	200	16.4	267	19.0
300	200	15.8	300	20.1	400	23.3

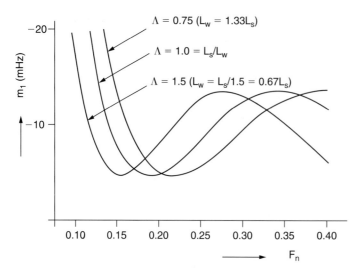

Figure 7.46 First order moment m_1 of encounter spectrum.

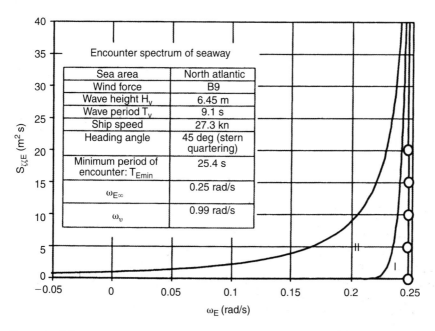

Figure 7.47(a) Encounter spectrum with overlapping of region I and II.

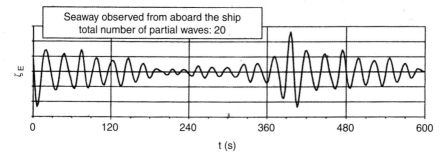

Figure 7.47(b) Irregular time series from calculation showing wave groups, $\chi = 45°$, V/T = 3 (kn/s).

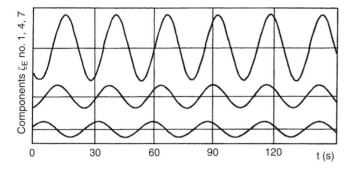

Figure 7.47(c) Time series of 3 large wave components according to Figure 7.4(a). The periods are very close to each other due to the narrow encounter spectrum in stern quartering seas.

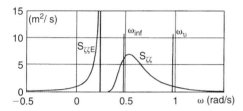

Figure 7.47(d) Comparison of encounter spectrum with original seaway spectrum.

A random choice of an irregular wave sequence ζ_E passing the ship is plotted (Figure 7.47 (b)). Equation (7.93) sums the partial waves by linear superposition, (7.94) gives the random choice of phase and (7.95) the amplitude of the components by the seaway spectrum. The time series, ζ_E, shows the typical pattern of the way the ship experiences irregular seas. Figure 7.47(b) demonstrates the build-up of wave groups and beats at certain encounter conditions due to the small bandwidth of the encounter spectrum in stern quartering and following seas. The wave groups show large waves alternating with long spells of less severe waves.

The Spectrum $S_{\zeta\zeta}$ has been split up into 20 partial wave components ζ_{wi}, the largest being shown in Figure 7.47(c). Figure 7.47(d) shows the transformed spectrum $S_{\zeta\zeta E}$ (Equation (7.98)) together with the original Bretschneider seaway spectrum, $S_{\zeta\zeta}$.

The above shown superposition of sine waves is linear. However, steep waves are of higher order as expressed by Stokes, Section 1.4.2. In steep waves, crests become more pronounced. Figure 7.48 depicts a numerical summation of steep and long crested Stokes waves by third order perturbation expansion, Pierson (1993). Some of the wave crests are nearly double above the mean, compared with the troughs below. So wave crests in steep random waves are higher than a linear model predicts, and the velocity field below the crests does not agree with measurements. This is a well-known property of dangerous waves at sea.

7.3 Limiting seakeeping criteria

7.3.1 Limiting critera

The ability of a ship to carry out its intended mission efficiently may be curtailed by a number of factors. There is a correspondingly wide range of limiting seakeeping criteria. The limit may be set by the ability of the ship itself, or its systems, to operate effectively and safely, or by the comfort or proficiency of passengers or crew. In so far as equipment or personnel performance is degraded when motions (e.g. vertical acceleration) exceed a certain level, careful siting of the related activity within the ship in an area of lesser motions may

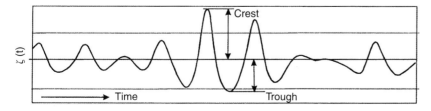

Figure 7.48 Non-linear irregular waves showing increase in crest larger than decrease in trough.

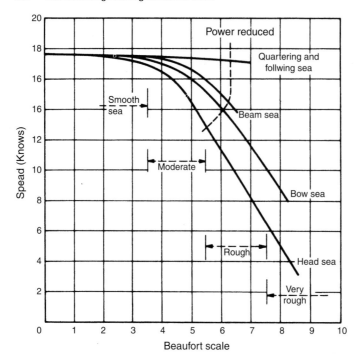

Figure 7.49 Speed Loss in waves.

extend the range of sea conditions in which operation is acceptable. Other features such as slamming or propeller emergence are dependent on overall ship geometry and loading although here again the design of the ship (e.g. its inherent strength in the case of slamming) can determine the acceptable level before damage occurs or conditions become unsafe.

There is a potential danger in applying 'standard' acceptance levels of any criterion to a new design. There must be a judicious choice, both of criteria and acceptance levels, to reflect the particular design, its function and its similarity to previous designs for which operating experience is available. Thus a new design may have been specifically strengthened forward to enable it safely to withstand high slamming loads. Nevertheless guideline figures applicable to general ship types are useful in preliminary design development. Some performance parameters can be assessed in different ways. This may lead to different absolute values of criteria. Hence in using criteria values it is important they be computed for a new design using the same method as that adopted in establishing the acceptable levels.

It is now appropriate to review briefly the seakeeping parameters most frequently used as potential limiting criteria. They are speed and power in waves, slamming, wetness, propeller emergence and impairment of human performance.

7.3.1.1 *Speed and power in waves*

As a wave system becomes more severe, the power needed to drive the ship through it at a given speed increases. The difference arises mainly from the increased resistance experienced by the hull and appendages, but the overall propulsive efficiency also changes due to the changed conditions in which the propeller operates. If the propulsion machinery is already producing full power, it fallows that there must be an enforced reduction in speed. Past a certain severity of waves, the motions of the ship or slamming may become so violent that the captain may decide to reduce speed below that possible with the power available. This is a voluntary speed reduction and might be expected to be made in merchant ships of fairly full form at Beaufort numbers of 6 or more. The speed reduction lessens as the predominant wave direction changes from directly ahead to the beam, Figure 7.49.

Figure 7.50 shows how the power required for various speeds increases with increasing sea state as represented by the Beaufort number. The figure applies to a wave system 10 degrees off the bow and to a ship 150 m long with a longitudinal radius of gyration equal to 22% of the length. Decreasing the longitudinal moment of inertia decreases the additional power required and also results in drier decks forward.

Figure 7.51 shows the reduction in speed which occurs at constant power (5.83 MW) for the ship in the same conditions and shows the significance of varying the longitudinal radius of gyration. The effect of the variation is less significant in large ships than in small. It is associated with a reduction in natural pitching period. Speed loss in waves is discussed in Chapter 5, and in references such as Townsin and Kwon (1983), Townsin *et al.* (1993) and Kwon (2000).

Other ship design features conducive to maintaining higher speed in rough weather are a high length-displacement ratio, $L/\nabla^{1/3}$, and fine form forward. Increased damping by form changes or the deliberate introduction of a large bulbous bow can also help. When it is realized that the passage times of ships in rough weather may be nearly doubled, it is clearly of considerable importance to design the ship, both above and below water, so that it can maintain as high a speed as possible. Wetness is a significant factor influencing the need to reduce speed, and this is dealt with in Section 7.3.1.3.

7.3.1.2 *Slamming*

In rough seas with large relative ship motion, slamming may occur with large water impact loads. Usually, slamming loads are much larger than other wave loads. Sometimes ships suffer local damage from the impact load or large-scale buckling on the deck. For high-speed ships, even if each impact load is small, frequent impact loads accelerate fatigue failures of hulls. Thus, slamming loads may threaten the safety of ships. The expansion of ship size and new concepts in fast ships have decreased relative rigidity causing in some cases serious wrecks.

A rational and practical estimation method of wave impact loads is thus one of the most important prerequisites for safety design of ships and ocean structures. Wave impact has challenged many researchers since von Karman's work in 1929. Today, mechanisms of wave impacts are correctly understood for the 2-d case, and accurate impact load estimation is possible for the deterministic case. The long-term prediction of wave impact loads can be also given in the framework of linear stochastic

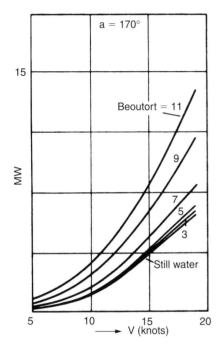

Figure 7.50 Power in waves for a 150 m long ship.

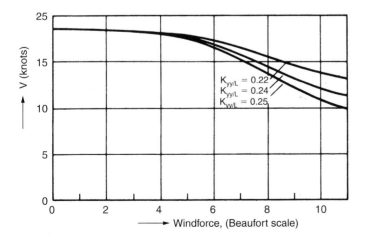

Figure 7.51 Variation in speed at constant power.

theories. However, our knowledge on wave impact is still far from sufficient.

A fully satisfactory theoretical treatment has been prevented so far by the complexity of the problem:

- Slamming is a strongly non-linear phenomenon which is very sensitive to relative motion and contact angle between body and free surface.
- Predictions in natural seaways are inherently stochastic; slamming is a random process in reality.
- Since the duration of wave impact loads is very short, hydro-elastic effects are large.
- Air trapping may lead to compressible, partially supersonic flows where the flow in the water interacts with the flow in the air.

Most theories and numerical applications are for two-dimensional rigid bodies (infinite cylinders or bodies of rotational symmetry), but slamming in reality is a strongly three-dimensional phenomenon. We will here briefly review the most relevant theories. Further recommended literature includes:

- Tanizawa and Bertram (1998) for practical recommendations translated from the Kansai Society of Naval Architects, Japan.
- Mizoguchi and Tanizawa (1996) for stochastical slamming theories.
- Korobkin (1996) for theories with strong mathematical focus.
- ISSC (1995) for a comprehensive compilation (more than 1000 references) of slamming literature.
- ITTC (2005) for a review of slamming.

The wave impact caused by slamming can be roughly classified into four types, Figure 7.52.

1. Bottom slamming occurs when emerged bottoms re-enter the water surface.
2. Bow-flare slamming occurs for high relative speed of bow-flare to the water surface.
3. Breaking wave impacts are generated by the superposition of incident wave and bow wave hitting the bow of a blunt ship even for small ship motion.
4. Wet-deck slamming occurs when the relative heaving amplitude is larger than the height of a catamaran's wet-deck.

Both bottom and bow-flare slamming occur typically in head seas with large pitching and heaving motions. All four water impacts are 3-d phenomena, but have been treated as 2-d for simplicity. For example, types 1 and 2 were idealized as 2-d wedge entry to the calm-water surface. Type 3 was also studied as 2-d phenomenon similar to wave impact on breakwaters. We will therefore review 2-d theories first.

- *Linear slamming theories based on expanding thin plate approximation* Classical theories approximate the fluid as inviscid, irrotational, incompressible, free of surface tension. In addition, it is assumed that gravity effects are negligible. This allows a (predominantly) analytical treatment of the problem in the framework of potential theory.

For bodies with small deadrise angle, the problem can be linearized. Von Karman (1929) was the first to study theoretically water impact (slamming). He idealized the impact as a 2-d wedge entry problem on the calm-water surface to estimate the water impact load on a seaplane during landing, Figure 7.53. Mass, deadrise angle, and initial penetrating velocity of the wedge are denoted as m, β and V_0. Since the impact is so rapid, von Karman assumed very small water surface elevation during impact and negligible gravity effects. Then the added mass is approximately $m_v = \left(\dfrac{1}{2}\right)\pi\rho c^2$. ρ is the water density and c the half width of the wet area implicitly computed from $dc/dt = V \cot \beta$. The momentum before the impact mV_0 must be equal to the sum of the wedge momentum mV and added mass momentum $m_v V$, yielding the impact load as:

$$P = \frac{V_0^2/\tan \beta}{\left(1 + \dfrac{\rho\pi c^2}{2m}\right)^3} \cdot \rho\pi c \qquad (7.108)$$

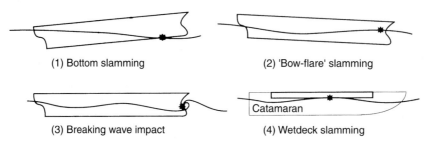

(1) Bottom slamming

(2) 'Bow-flare' slamming

(3) Breaking wave impact

(4) Wetdeck slamming

Figure 7.52 Types of slamming impact of a ship.

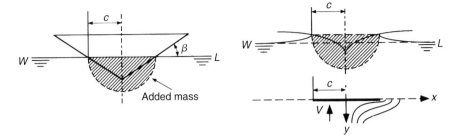

Figure 7.53 Water impact models of von Karman (left) and Wagner (right).

Since von Karman's impact model is based on momentum conservation, it is usually referred to as momentum impact, and because it neglects the water surface elevation, the added mass and impact load are underestimated, particularly for small deadrise angle.

Wagner derived a more realistic water impact theory in 1932. Although he assumed small deadrise angles β in his derivation, the theory was found to be not suitable for $\beta < 3°$, since then air trapping and compressibility of water play an increasingly important role. If β is assumed small and gravity neglected, the flow under the wedge can be approximated by the flow around an expanding flat plate in uniform flow with velocity V, Figure 7.53. Using this model, the velocity potential ϕ and its derivative with respect to y on the plate $y = 0^+$ is analytically given as:

$$\phi = \begin{cases} V\sqrt{c^2 - x^2} & \text{for } x < c \\ 0 & \text{for } x > c \end{cases}$$

$$\partial \phi / \partial y = \begin{cases} 0 & \text{for } x < c \\ V/\sqrt{1 - c^2/x^2} & \text{for } x > c \end{cases}$$

The time integral of the last equation gives the water surface elevation and the half width of the wetted area c. The impact pressure on the wedge is determined from Bernoulli's equation as:

$$\frac{p(x)}{\rho} = \frac{\partial \phi}{\partial t} - \frac{1}{2}(\nabla \phi)^2 = \sqrt{c^2 - x^2}\frac{dV}{dt}$$
$$+ V\frac{c}{\sqrt{c^2 - x^2}}\frac{dc}{dt} - \frac{1}{2}\frac{V^2 x^2}{c^2 - x^2} \quad (7.109)$$

Wagner's theory can be applied to arbitrarily shaped bodies as long as the deadrise angle is small enough not to trap air, but not so small that air trapping plays a significant role. Wagner's theory is simple and useful, even if the linearization is sometimes criticized for its inconsistency as it retains a quadratic term in the pressure equation. This term is indispensable for the prediction of the peak impact pressure, but it introduces a singularity at the edge of the expanding plate ($x = \pm c$) giving negative infinite pressure there. Many experimental studies have checked the accuracy of Wagner's theory. Measured peak impact pressures are typically a little lower than estimated. This suggested that Wagner's theory gives conservative estimates for practical use. However, a correction is needed on the peak pressure measured by pressure gauges with finite gauge area. Special numerical FEM analyses of the local pressure in a pressure gauge can be used to correct measured data. The corrected peak pressures agree well with estimated values by Wagner's theory. Today, Wagner's theory is believed to give accurate peak impact pressure for practical use.

The singularity of Wagner's theory can be removed taking spray into account. An 'inner' solution for the plate is asymptotically matched to an 'outer' solution of the spray region, as, e.g., proposed by Watanabe in Japan in the mid-1980s, (Figure 7.54). The resulting equation for constant falling velocity is consistent and free from singularities. Despite this theoretical improvement, Watanabe's and Wagner's theories predict basically the same peak impact pressure, Figure 7.55.

- *Simple non-linear slamming theories based on self-similar flow*
  We consider the flow near the vertex of a 2-d body immediately after water penetration. We can assume:

  – Near the vertex, the shape of the 2-d body can be approximated by a simple wedge.
  – Gravity accelerations are negligible compared to fluid accelerations due to the impact.
  – The velocity of the body v_0 is constant in the initial stage of the impact. Then the flow can be considered as self-similar depending only on $x/v_0 t$ and $y/v_0 t$, where x, y are Cartesian coordinates and t is time. Russian scientists have converted the problem to a 1-d integral equation for $f(t)$. The resulting integral equation is so complicated that it cannot be solved analytically. However, numerically it has been solved by Faltinsen in Norway up to deadrise

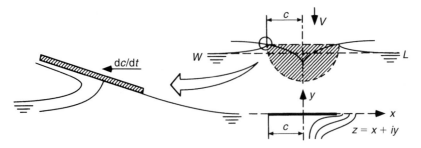

Figure 7.54 Water impact model of Watanabe.

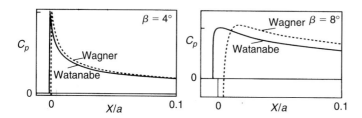

Figure 7.55 Spatial impact pressure distribution.

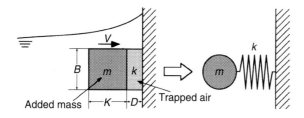

Figure 7.56 Bagnold's model.

angles $\beta \geq 4°$. The peak impact pressure for $\beta = 4°$ was almost identical (0.31% difference) to the value given by Wagner's theory.

- *Slamming theories including air trapping*
  So far, slamming theories have neglected the density of air, i.e. if a deformation of the free surface was considered at all, it occurred only after the body penetrated the water surface. The reality is different. The body is preceded by an air cushion that displaces water already before the actual body entry. The density of air plays an even bigger role if air trapping occurs. This is especially the case for breaking wave impacts. In the 1930s, Bagnold performed pioneering work in the development of theories that consider this effect. Bagnold's impact model is simply constructed from added mass, a rigid wall, and a non-linear air cushion between them, (Figure 7.56). This model allows qualitative predictions of the relation between impact velocity, air cushion thickness, and peak impact pressure. For example, the peak impact pressure is proportional to V and $\sqrt{H}$ for slight impact and weak non-linearity of the air cushion; but for severe impact, the peak impact pressure is proportional to V^2 and H. These scaling laws were validated by subsequent experiments.

  Trapped air bottom slamming is another typical impact with air cushion effect. For two-dimensional bodies, air trapping occurs for deadrise angles $\beta \leq 3°$. Chuang's (1967) experiment for 2-d wedges gave peak impact pressures as in Table 7.9. The impact velocity V is given in m/s.

  For $\beta = 0°$ air trapping is significant and the peak impact pressure is proportional to V. Increasing the deadrise angle reduces the amount of air trapping and thus the non-linearity. For practical use, the peak impact pressure is usually assumed to be proportional to V^2 for all β. This results in a conservative estimate.

Table 7.9 Chuang's (1967) relation for peak impact pressures.

β	$0°$	$1°$	$3°$	$6°$	$10°$	$15°$	$\geq 18°$
P_{peak} (kPa)	$102\,V$	$115\,V^{1.4}$	$189\,V^{1.6}$	$64.5\,V^2$	$31\,V^2$	$17.8\,V^2$	Wagner's theory

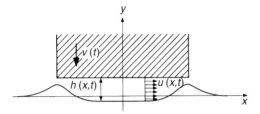

Figure 7.57 One-dimensional air flow model of Verhagen.

Johnson and Verhagen developed 2-d theories for bottom impact with air trapping considering 1-d air flow between water surface and bottom to estimate the water surface distortion and the trapped air volume, Figure 7.57.

The peak impact pressure thus estimated was much higher than measured. This disagreement results from the boundary condition at the edge of the flat bottom, where a jet emits to the open air. The theory assumes that the pressure at the edge is atmospheric pressure. This lets the air between water surface and bottom escape too easily, causing an underestimated trapped air volume. Experiments showed that the pressure is higher than atmospheric. Yamamoto has therefore proposed a modified model using a different boundary condition.

Experiments at the Japanese Ship Research Institute have observed the trapped air impact with high-speed cameras and measured the initial thickness of air trapping. It was much thicker than the estimates of both Verhagen and Yamamoto. The reason is that a mixed area of air and water is formed by the high-speed air flow near the edge. Since the density of this mixed area is much higher than that of air, this area effectively chokes the air flow increasing air trapping.

The mechanism of wave impact with air trapping is in reality much more complicated. Viscosity of air, the effect of air leakage during compression, shock waves inside the air flow, and the complicated deformation of the free surface are all effects that may play an important role. Computational fluid dynamics may be the key to significant success here, but has not yet progressed sufficiently yet as discussed below.

- *Effect of water compressibility*

When a blunt body drops on calm water or a flat bottom drops on a smooth wave crest, usually no air trapping occurs. Nevertheless, one cannot simply use Wagner's theory, because at the top of such a blunt body or wave crest the relative angle between body and free surface becomes zero. Then both Wagner's and Watanabe's theories give infinite impact pressure. In reality, compressibility of liquid is important for a very short time at the initial stage of impact, when the expansion velocity of the wet surface dc/dt exceeds the speed of sound for water $c_w \approx 1500\,\text{m/s}$ producing a finite impact pressure. Korobkin (1996) developed two-dimensional theories which consider compressibility and free-surface deformation. For parabolic bodies dropping on the calm-water surface, he derived the impact pressure simply to $P = \rho c_w V$. Korobkin's theory is far more sophisticated yielding also the time history of the pressure decay, but will not be treated here.

- *3-d slamming theories*

All slamming theories treated so far were two dimensional, i.e. they were limited to cross-sections (of infinite cylinders). In reality slamming for ships is a strongly three-dimensional phenomenon due, for example, to pitch motion and cross-sections in the foreship changing rapidly in the longitudinal direction.

For practical purposes, one tries to obtain quasi three-dimensional solutions based on strip methods or high-speed strip methods. At the University of Michigan, Troesch developed a three-dimensional boundary element method for slamming. However, the method needs to simplify the physics of the process and the geometry of body and free surface and failed to show significant improvement over simpler strip-method approaches when compared to experiments.

Limiting oneself to axisymmetric bodies dropping vertically into the water makes the problem *de facto* two dimensional. The study of 3-d water impact started from the simple extension of Wagner's theory to such cases. The water impact of a cone with small deadrise angle can then be treated in analogy to Wagner's theory as an expanding circular disk. A straightforward extension of Wagner's theory by Chuang (1967) overpredicts the peak impact pressure. Subsequent refinements of the theory resulted in a better estimate of the peak impact pressure:

$$p(r) = \frac{1}{2}\rho V^2 \left(\frac{2}{\pi}\right)^2 \left[\frac{4\cot\beta}{\sqrt{1-r^2/c^2}} - \frac{r^2/c^2}{1-r^2/c^2}\right]$$
(7.110)

r and c correspond to x and c in Figure 7.53. This equation gives about 14% lower peak impact pressures than a straightforward extension of Wagner's theory. Experiments confirmed that the impact pressure on a cone is lower than that on a 2-d wedge of same deadrise angle. So the 3-d effect reduces the impact pressure at least for convex bodies. This indicates that Wagner's theory gives conservative estimates for practical purpose. Since the impact on a ship hull is usually a very local phenomenon, Wagner's equation has been used also for 3-d surfaces using local relative velocity and angle between ship hull and water surface.

Watanabe (1986) extended his two-dimensional slamming theory to three-dimensional oblique impact of flat-bottomed ships. This theory was validated in experiments observing three-dimensional bottom slamming with a high-speed video camera and transparent models. Watanabe classified the slamming of flat-bottomed ships into three types:

1. Slamming due to inclined re-entry of the bottom. The impact pressure runs from stern to bow. No air trapping occurs.
2. Slamming due to vertical (orthogonal) re-entry of the bottom to a wave trough with large-scale air trapping.
3. Slamming due to vertical (orthogonal) re-entry of the bottom to a wave crest with only small-scale, local air trapping.

Type 1 (typical bottom impact observed for low ship speed) can be treated by Watanabe's 3-d theory. Type 3 (typical for short waves and high ship speed) corresponds to Chuang's theory for very small deadrise angle. Type 2 (also typical for short waves and high ship speed) corresponds to Bagnolds' approach, but the air trapping and escaping mechanisms are different to simple 2-d models.

The three-dimensional treatment of slamming phenomena is still subject to research. It is reasonable to test and develop first numerical methods for two-dimensional slamming, before one progresses to computationally more challenging three-dimensional simulations. Until such methods are available with appropriate response times on engineering workstations, in practice computations will be limited to two-dimensional estimates combined with empirical corrections.

- *Hydro-elastic approaches in slamming*

  It is important to evaluate not only peak impact pressures but also structural responses to he impact, to consider the impact pressure in the design of marine structures. Whipping (large-scale, weakly dampened oscillations of the longitudinal bending moment) is a typical elastic response to impact. In the late 1960s and 1970s, slamming and whipping resulted in some spectacular ship wrecks, e.g. bulkers and container ships breaking amidships. The disasters triggered several research initiatives, especially in Japan, which eventually contributed considerably to the development of experimental and numerical techniques for the investigation of slamming and whipping.

  Let us denote the slamming impact load as $Z(t)$ and the elastic response of a ship as $S(t)$. Assuming a linear relation between them, we can write:

  $$S(t) = \int_0^\infty h(t-\tau)Z(\tau)\mathrm{d}\tau$$
  (7.111)

  $h(\tau)$ is the impulse response function of the structure. An appropriate modelling of the structure is indispensable to compute $h(\tau)$. For example, the large-scale (whipping) response can be modelled by a simple beam, whereas small-scale (local) effects can be modelled as panel responses. For complicated structures, FEM analyses determine $h(\tau)$.

  When the duration of the impact load is of the same order as the natural period of the structure, the hydro-elastic interaction is strong. The impact load on the flexible bottom can be about twice that on the rigid bottom. Various theories have been developed, some including the effect of air trapping, but these theories are not powerful enough to explain experimental data quantitatively. However, numerical methods either based on FEM or FVM could be used to analyse both fluid and structure simultaneously and should improve considerably our capability to analyse hydro-elastic slamming problems.

- *CFD for slamming*

  For most practical impact problems, the body shape is complex, the effect of gravity is considerable, or the body is elastic. In such cases, analytical solutions are very difficult or even impossible. This leaves CFD as a tool. Due to the required computer resources, CFD applications to slamming appeared only since the 1980s. While the results of boundary element methods for water entry problems agree well with analytical results, it is doubtful whether they are really suited to this problem. Real progress is more likely to be achieved with field methods like FEM, FDM, or FVM. Various researchers have approached slamming problems, usually employing surface-capturing methods, e.g. marker-and-cell methods or level-set techniques. Often the Euler equations are solved as viscosity plays a less important role than for many other problems in ship hydrodynamics. But also RANS solutions including surface tension, water surface deformation, interaction of air and water flows etc. have been presented. The numerical results agree usually well with experimental results for two-dimensional problems. Due to the large required computer resources, few really three-dimensional applications to ships have been presented.

7.3.1.3 Wetness

By wetness is meant the shipping of spray or green seas over the ship and, unless otherwise qualified, refers to wetness at the bow.

It is generally not possible to calculate wetness accurately but it may be assessed by:

(a) calculating the relative vertical movement of the bow and water surface and assuming that the probability of deck wetness is the same as that of the relative motion exceeding the freeboard at the stem head;
(b) running a model in waves and noting for each of a range of sea conditions the speed at which the model is wet and assuming that the ship will behave in a similar way. Model tests at the right F_n can represent green seas but not spray effects.

Methods based on (a) are usually adopted. If it is valid to assume that the relative motion between ship and waves is sinusoidal and that the probability of deck immersion follows a Rayleigh distribution:

$$P_r = \exp\left[-\frac{F^2}{2m_0}\right], \quad F = \text{freeboard} \qquad (7.112)$$

The average time interval between the deck being wet at a given station is

$$t_w = \frac{2\pi}{P_r \omega} = \frac{2\pi}{\omega} \exp\left[\frac{F^2}{2m_0}\right] \qquad (7.113)$$

where ω = average frequency of the relative motion

$$= \sqrt{\frac{m_2}{m_0}}$$

m_2 = variance of the relative velocity

Besides trying to reduce the incidence of wetness the naval architect should:

(a) design decks forward so that water clears quickly;

(b) avoid siting forward any equipments which may be damaged by green seas or which are adversely affected by salt water spray.

A bulwark can be fitted to increase freeboard provided it does not trap water. The sizes of freeing ports required in bulwarks are laid down in various international regulations.

A review of research on deck wetness is given in ITTC (2005).

7.3.1.4 Propeller emergence

Using an arbitrary assumption that the propeller should be regarded as having emerged when a quarter of its diameter, D, is above water, criteria corresponding to those used in wetness follow, viz:

$$P_r = \exp\left[-\left(T_p - \frac{D}{4}\right)^2 \bigg/ 2m_0\right] \qquad (7.114)$$

T_p = depth of propeller boss below the still waterline

Average time interval between emergencies

$$t_p = \frac{2\pi}{P_r \omega}, \quad \omega = \sqrt{\frac{m_2}{m_0}}$$

7.3.1.5 Degradation of human performance

Besides reducing comfort, motions can reduce the ability and willingness of humans to work and make certain tasks more difficult. Thus in controlling machinery, say, motions may degrade the operator's ability to decide what he should do and, having decided, make the execution of his decision more difficult. There is inadequate knowledge of the effects of motion on human behaviour but in broad terms it depends upon the acceleration experienced and its period. These can be combined in a concept of subjective motion. In this, the combinations of acceleration and frequency are determined at which the subjects feel the motion to have the same intensity. Denoting this level as subjective magnitude (SM) with a value of 10, other combinations of acceleration and frequency are assessed as of $SM = 10n$ when they were judged to be n times as intense as the original base SM. It is found that

$$SM = A(f)a^{1.43},$$
a = acceleration amplitude in 'g'

With the frequency, f, in Hz, $A(f) = 30 + 13.53(\log_e f)^2$

Assuming the sinusoidal results can be applied to random motions

$$a = \frac{2}{g}\sqrt{m_{4a}}, \quad f = \frac{1}{2\pi}\sqrt{\frac{m_{2a}}{m_{0a}}}$$

where m_{0a}, m_{2a} and m_{4a} are the variances of the absolute motion, velocity and acceleration respectively in SI units.
Then

$$SM = \left[3.087 + 1.392\left\{\log_e \frac{1}{2\pi}\sqrt{\frac{m_{2a}}{m_{0a}}}\right\}^2\right]m_{4a}^{0.715}$$

(7.115)

The motions experienced by any individual will depend upon their position in the ship. An overall figure for a ship can be obtained by applying a weighting curve representing the distribution of personnel in the ship. Alternatively for a localized activity (e.g. on the bridge) the *SM* for that one location can be obtained. Unfortunately no clear-cut limiting *SM* could be proposed although a figure of 15 has been suggested as an absolute maximum. Five actions a designer can take to prevent or mitigate the adverse affects of ship motion, especially sea sickness, are

(a) Locate critical activities near the effective centre of rotation.
(b) Minimize head movements.
(c) Align operator position with the ship's principal axes.
(d) Avoid combining provocative sources.
(e) Provide an external visual frame of reference.

Another approach to the effect of motions on personnel is the concept of *motion sickness incidence* (MSI). The MSI is the percentage of individuals likely to vomit when subject to the given motion for a given time. Plots can be made as in Figure 7.58. The limitations of this approach are that the data relate to unacclimatised subjects, sinusoidal motions and the fact that human performance may be degraded long before vomiting occurs.

Studies on the effects of motions on human performance, and applications to operational limits include O'Hanlon and McCauley (1974), Lawther and Griffin (1986, 1987, 1988), Bales (1980), Graham et al. (1992) and Karayannis and Molland (2003). Typical suggested limiting criteria are contained in the Nordforsk Report (1987), Smith and Thomas (1989), ISO (1985, 1997), and the Code of Safety for High Speed Craft, IMO (2000).

7.4 Overall seakeeping performance

A number of possible limiting seakeeping criteria have been discussed in Section 7.3. Their variety and the range of sea conditions expected in service mean that no single performance parameter is likely to be adequate in defining a design's overall seakeeping performance. This applies even within the restricted definition of seakeeping adopted in this Chapter. However a methodology is developing which permits a rational approach, the steps of which are now outlined.

(a) The *sea states* in which the ship is to operate are established. The need may be specific in the sense that the ship will operate on a particular route at certain seasons of the year, or it may he as general as world-wide operations all the year round. Ocean wave statistics can be used to determine the ranges of wave height, period and direction likely to be met for various percentages of time, Hogben et al. (1986). As described later, the technique of wave climate synthesis can be to improve the reliability of predictions based on observed wind and wave data. This establishes the number of days a year the ship can be expected to experience various wave conditions and these can be represented by appropriate wave spectra e.g. by adopting the formulation recommended by the ITTC, Section 1.7.

(b) The *ship responses* in the various sea states can be assessed from a knowledge of its responses in regular waves. Even in long-crested seas the ship response depends upon the severity of the sea, the ship speed and the ship's heading relative to the wave crest line. Thus motions can be represented by a polar diagram, such as Figure 7.59, in which contours are drawn for given values of response for each of a range of significant wave heights. Assuming a linear dependency the contours can be expressed as response operators.

If it is desired to compare designs on the basis of their relative motions at various speeds (or Froude numbers) the areas within the polar plot can be used. This will average out variations with heading. If it is known that the ship will transit on some headings more frequently than others the polar plots can be adjusted by means of suitable weighting factors. For vessels which are symmetrical about their centre-line plane in geometry and loading the polar plots will be symmetrical about the vertical axis.

(c) *Limiting conditions.* It is not usually the motion amplitudes *per se* which limit the ability of a ship to carry out its intended mission. More often it is a combination of motions and design features leading to an undesirable situation which can only be alleviated by reduction in

Seakeeping 539

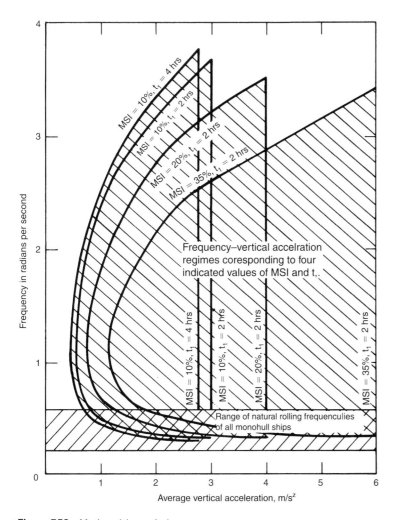

Figure 7.58 Motion sickness index.

speed or a change of heading. That is to say the ship's freedom of action is restricted.

The usual action is to reduce speed as this has the effect of avoiding synchronism with wave components other than short waves which have less effect on motions anyway. A change of course is only effective when there is a predominant wave direction and often can only be adopted for relatively short periods of time.

Various limiting seakeeping criteria have been discussed above. For any chosen criterion, the speed above which the agreed acceptable limit of the criterion is expected to be exceeded can be plotted on a polar diagram. As with motions, the area within the plot, adjusted if necessary by weighting factors for different headings, can be used as an overall measure of the design's performance in terms of the selected criterion in the given sea state. The greater the area the greater the range of speeds and headings over which the vessel can operate.

Plots, and hence areas, will be required for each wave spectrum of interest, each of which can be characterized by its significant wave height.

Another concept is that of a *stratified measure of merit*. The time the ship is expected to encounter various wave conditions has been established from ocean wave statistics. The product of this time and the area under the polar plot for the chosen seakeeping criterion is plotted against the significant wave height appropriate to the wave spectrum. The area under the resulting curve represents the overall performance of the design with respect to the selected criterion, over the period of time covered by the wave data used. This could be a single voyage, a year or the whole life of the ship.

540 Maritime engineering reference book

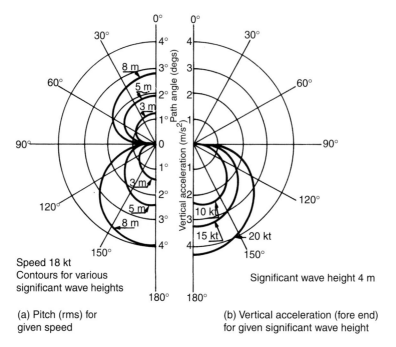

(a) Pitch (rms) for given speed

(b) Vertical acceleration (fore end) for given significant wave height

Figure 7.59 Typical polar diagrams.

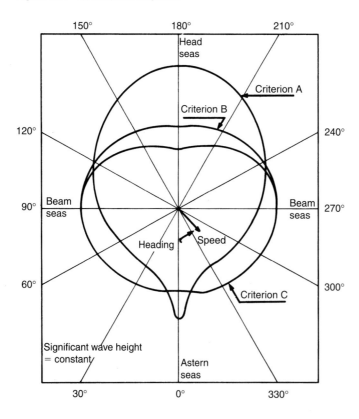

Figure 7.60 Seakeeping speed polar diagram.

(d) The *operational* ability of a design will not always be limited by the same criterion. Thus the design's overall potential must be assessed in relation to all the possible limiting criteria. If all criteria of interest are plotted on a common polar plot the area within the inner curve at each heading represents the overall limiting performance of that design in the selected sea state. The measure of merit concept can be used as for a single criterion.

This operational ability assumes a common mission throughout the life of the ship. If it is known in advance that the ship will have different missions at different times the method will need to be modified to reflect the different influence of the various criteria on the missions concerned.

In practice a captain must judge the operational importance of maintaining speed against the risk to the ship. The captain of a warship is more likely to reduce speed on a peacetime transit than in a wartime operation.

This general approach is one method of assessing the relative seagoing performance of competitive vehicle types. Other 'scoring' methods suggested are:

(a) The percentage time a given vehicle in a given condition of loading can perform its function in a specified area, in a given season at a specified speed without any of a range of chosen seakeeping criteria exceeding agreed values.
(b) The time a vehicle needs to transit between two specified locations in calm water divided by the time the vehicle would require to travel between the same locations in rough weather without any of the selected criteria value being exceeded.

7.5 Data for seakeeping assessments

It will be appreciated from the foregoing that two things are necessary to enable an assessment to be made of seakeeping performance, namely a knowledge of:

(a) wave conditions for the area to which the assessment is related and specifically how the total energy of the wave system is distributed with respect to frequency;
(b) the responses of the ship in regular sinusoidal waves covering the necessary frequency band. These responses are normally defined by the appropriate response amplitude operators in the form of response per unit wave height, Figure 7.11.

7.5.1 Selection of wave data

Chapter 1 gives information on the type of data available for sea conditions likely to be met in various parts of the world. Much of this is based on visual observations, both of waves and winds. As such they involve an element of subjective judgment and hence uncertainty. In particular visual observations of wave periods are likely to be unreliable. Care is necessary therefore in the interpretation and analysis of wave data if sound design decisions are to be derived from them.

The National Maritime Institute, (now BMT Ltd), developed a method, known as wave climate synthesis, for obtaining reliable long-term wave data from indirect or inadequate source information. This approach can be used when instrumented wave measurements are not available. Essentially relationships derived from corresponding sets of instrumented and observed data are used to improve the interpretation of observed data. Various sources of data and of methods of analysing it are available such as the Marine Information and Advisory Service of the Institute of Oceanographic Sciences and agencies of the World Meteorological Organization. Much of the data is stored on magnetic tape.

The NMI analysis used probabilistic methods based on parametric modelling of the joint probability of wave height and wind speed. Important outputs are:

(a) *Wave height*
When a large sample is available raw visual data provide reasonable probability distributions of wave height. However, comparisons of instrumented and visual data show that better distributions can be derived using best fit functional modelling to smooth the joint probability distributions of wave height and wind speed. This is illustrated in Figure 7.61 for OWS *India* in which the 'NMIMET Visual' curve has been so treated.

Analysis of joint probabilities for wave height and wind speed from measured data leads to the relationship

Mean wave height $= H_r = [(aW_r^n)^2 + H_2^2]^{\frac{1}{2}}$,
$W_r =$ wind speed.

(7.116)

Standard deviation of the scatter about the mean is

$$\sigma_r = H_2(b + cW_r)$$

The joint probability distribution is given by a gamma distribution

$$P(H_s/H_r, \sigma_r) = \frac{q^{p+1}}{\Gamma(p+1)} H_s^p \exp(-qH_s) \quad (7.117)$$

where $p = \dfrac{H_r^2}{\sigma_r^2} - 1$

$q = H_r/\sigma_r^2$

$H_s =$ significant wave height

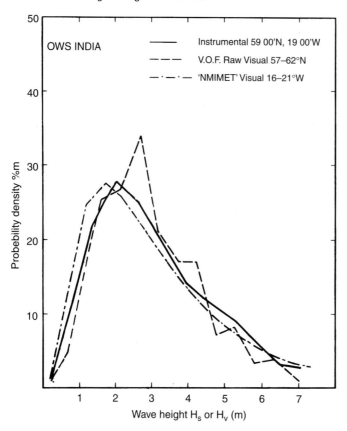

Figure 7.61 Visual and measured wave height probabilities.

H_2, a, b, c and n are the model parameters for which, in the absence of more specific data, suitable standard values may be used. The following values, Table 7.10, have been recommended on the basis of early work using instrumental data from a selection of six stations. In quoting them it should be noted that they are subject to review in the light of more recent work and meanwhile should be regarded as only valid for use with measured wind speeds up to a limit of about 50 knots. It should also be noted that the numerical values cited are to be used in association with units of metres for wave height and knots for wind speed.

Table 7.10

	H_2 (metres)	a	b	c	n
Open ocean	2.0	0.033	0.5	0.0125	1.46
Limited fetch	0.5	0.023	0.75	0.0188	1.38

The wave height probabilities follow from the wind speed probabilities using:

$$P(H_s) = \sum_r (H_s/H_r, \sigma_r) \times P(W_r) \qquad (7.118)$$

Wave directionality data can be obtained if the joint probability distributions of wave height and wind speed are augmented by corresponding joint probabilities of wave height and period and wind speed and direction.

(b) *Wave periods*

Reliability of visual observations of wave period is poor. NMI adopted a similar approach to that used for wave height but using a different functional representation. Based on analysis of instrumented wave height/period data, wave height and period statistics can be synthesized when reliable wave height data are available using:

$$P(T) = \sum_r P(T/H_r) \times P(H_r) \qquad (7.119)$$

where

$$P(T/H_r) = F_1(\mu_h, \sigma_h, \mu_t, \sigma_t, \rho)$$

$$= \left[2\pi(1-\rho^2)\sigma_t^2\right]^{-\frac{1}{2}} \exp\left\{\frac{-1}{2\left(\sqrt{(1-\rho^2)}\sigma_t\right)^2}\right.$$

$$\left.\left[(t-\mu_t) - \rho\frac{\sigma_t}{\sigma_h}(h-\mu_h)\right]^2\right\}$$

$$P(H_r) = F_2(\mu_h, \sigma_h)$$

$$= \frac{1}{\sqrt{2\pi}\sigma_h} \exp\left\{-\frac{(h-\mu_h)^2}{2\sigma_h^2}\right\}$$

$$\left\{1 - \frac{C_s}{6}\left[3\left(\frac{h-\mu_h}{\sigma_h}\right) - \left(\frac{h-\mu_h}{\sigma_h}\right)^3\right]\right\}$$

where

μ_h = mean value of h
σ_h = standard deviation of h
μ_t = mean value of t = $\ln \mu_T - \sigma_t^{3/2}$
σ_t = standard deviation of t = $0.244 - 0.0225\mu_H$
ρ = correlation coefficient = $0.415 + 0.049\mu_H$
C_s = skewness parameter = $E\left(\frac{[h-\mu_h]^3}{\sigma_h^3}\right)$

In these expressions h and t are the logarithmic values of H and T respectively. μ_h, σ_h and C_s follow from the given probability distribution of H as does μ_H the mean wave height.

$$\mu_T = 3.925 + 1.439\mu_H$$

The numerical values of the coefficients in the formulae for σ_t, ρ and μ_t were derived by regression analysis of over 20 sets of instrumental data.

(c) *Extreme wave height*
Sometimes the designer needs to estimate the most probable value of the maximum individual wave height in a given return period. After the probabilities of H_s are obtained the corresponding cumulative probabilities are computed and plotted on probability paper.

The methods used by NMI for analysing these cumulative probabilities for H_s are suitable for use when, as in the case of visual data, wave records are not available.

The data define exceedance probabilities for H_s up to a limiting level $1/m$ where m is the number of H_s values (or visual estimates of height) available. It is commonly required to extrapolate these to a level $1/M$ corresponding to an extreme storm of specified return period, R years, and duration, D hours, and in this case $M = 365 \times 24 \times R/D$.

In the NMI method this extrapolation is achieved by use of a 3-parameter Weibull distribution, the formula for the cumulative probability being:

$$P(x > H_s) = \exp-\left[\frac{(H_s - H_0)^n}{b}\right] \quad (7.120)$$

with values of the parameters n, b and H_0 determined numerically by least square fitting of the available data. The most probable maximum individual wave height H_{max} corresponding to the significant height H_{sM} for the extreme storm having exceedance probability $1/M$ is then estimated by assuming a Rayleigh distribution of heights in the storm, so that $H_{max} \doteq (\frac{1}{2} \ln N)^{\frac{1}{2}} H_{sM}$, where N is an estimate of the number of waves in the storm given by $N = 3600D/T$, where T is an estimated mean wave period.

7.5.2 Obtaining response amplitude operators

It has been shown that response amplitude operators are convenient both in presenting the results of regular motions in non-dimensional form and as a means of deducing overall motion characteristics in irregular seas. How are these RAOs to be obtained for a given ship? If it is a new design then calculation or model experiments must be used. If the ship exists then ship trials are a possibility. As with other aspects of ship performance the naval architect makes use of all three approaches.

7.5.2.1 *Theory*

Theory helps in setting up realistic model tests which in turn help to develop the theory indicating where simplifying assumptions, e.g. that of linearity of response, are acceptable. Full-scale trials provide evidence of correlation between ship and model or ship and theory.

As knowledge has built up confidence in theory, and as more powerful computers have facilitated more rigorous but lengthy calculations, theory has become the favoured approach to assessing ship motions at least in the early design stages and for conventional forms. Models can be used for the final form to look at deck wetness and rolling in quartering seas for which the theory is less reliable, or to confirm, data for unusual hull configurations. Theoretical methods are discussed in Section 7.7.

7.5.2.2 *Model experiments*

Several methods of model testing are in essence available, namely,

(a) measuring the response of a model in regular waves and deducing a set of response amplitude operators for use in the superpositioning theory to predict performance in an irregular sea. For reliable results such tests must be conducted in relatively moderate waves, typically with wave length to height ratio of 40:1. The test facilities can be relatively simple but a large number of runs will be required to cover adequately the range of speed and wavelength involved;

(b) running the model in a standard irregular pattern and analysing the data to provide the response amplitude operators. Several test runs will generally be necessary for each speed to obtain sufficient motion cycles to provide adequate confidence levels in the subsequent statistical analysis. The wavemakers must be capable of creating an irregular wave pattern with the desired spectral characteristics;

(c) *transient wave* or *impulse* wave testing. This method can be regarded as a special case of method (b). The wavemaker starts at high frequency, slows down and then stops. Thus it initially produces short waves which are gradually overtaken by the later, longer waves. The model starts in calm water, passes through a short wave sequence and finishes its run in calm water. By analysing the complete wave and motion records a full picture is obtained of the model responses with a considerable reduction in testing time. The range of frequencies present in the wave sequence must cover the range for which response operators are required, the waves must not become so steep that they break and the records must not be affected by reflections from the end of the tank; see Clauss (2000) for an example of this technique.

(d) creating a representation of an actual (recorded on ship trial, say,) irregular wave pattern and running the model in it to record motion, wetness, power, etc. In practice it would be very difficult to ensure the model experienced the same wave pattern as the ship which would be the only way to compare directly the pattern of wetness and slamming. However, if the spectral form is correct the model and ship performance can be compared on a statistical basis and thus provide some check on the adequacy of the linear super-positioning assumption.

It must be remembered that the results of a particular series of experiments can only be regarded as a sample of all possible experimental outcomes. To illustrate this, consider a single run in an irregular wave system. The actual surface shape repeats itself only after a very long period of time, if at all. The actual model data obtained from a particular run depends on when, during that period, the experiment is run. Tests for statistical significance can be applied.

7.5.2.3 *Ship trials*

In principle it is the ship trial which should provide the final check on the adequacy of theoretical and experimental predictions of ship behaviour. Unfortunately the actual ocean never exhibits a standard long-crested sinusoidal wave pattern. Thus no direct measurement of individual response operators is possible for comparison with predictions. Even truly long-crested irregular seas are never met. On occasion the sea surface may approximate to the long crested form but waiting for good conditions can be expensive of time and money. If they are met the sea and motions can be measured and response operators deduced. Comparison of ship with prediction can be made on the basis of these response operators or on the motion parameters predicted for the sea spectrum as measured.

Unfortunately, it is often the more extreme conditions that are of most concern and it is for these that the usual assumptions of linear superposition are likely to yield the greatest inaccuracies. These would be additional to those arising due to inaccuracies in recording waves and motions, sea variations over a recording period and differences in wind and tide. Thus a comparison of two ships as a result of trials conducted at different times and in different sea conditions has a number of limitations, particularly under limiting conditions for operations. Some of these are avoided, or reduced, if the two ships can go on trials in company. The sea condition will then be the same for both ships although it must be remembered that the sea condition may favour one design, particularly if the ships are of different length. It will be appreciated, therefore, that all trials data must be used intelligently with due allowance for the above factors.

The conduct of ship trials is discussed in Section 7.9.2.

7.6 Non-linear effects

Most of the remarks in this Chapter relate to simple linear equations of motion. This is because it is the simple approach, and

- it is a desirable step, to give a basic understanding, before going on to consider non-linear equations;
- it is an assumption that has served the naval architect with believable accuracy over many years when the tools were not available for more precise studies.

To some extent the success of the linear approach in the past has been due to the use of conservative design methods, or factors of safety that may have been unreasonably high. The key word here is 'may' because, in truth, the naval architect did not know. Designs to these methods had performed well in service but had they been over designed? Some would argue that because ships are still lost, they were inadequately designed, but perhaps there were other reasons for the losses. Or, the successful ships may have been lucky in that they did not meet the more extreme conditions for which they were designed; chance governs much of our lives.

Unfortunately, it is for the more extreme sea conditions and ship responses that non-linearities are most important. It is these conditions that are of greatest concern in safety. Extreme loading in a seaway and broaching-to are two examples of situations when it is desirable, indeed necessary, to take account of non-linear effects. How then do these effects arise?

- Some are due to the variation in the underwater form as the ship moves through, and responds to, the waves. This will arise even for moderate angles due to flare, but become severe when the deck edge becomes immersed or the bow leaves the water completely.
- Cross-coupling of motions will occur. Heave and roll will induce pitching, for example because the ship's shape and therefore forces upon it differ forward and aft.
- Some arise from the physics of the situation. Taking roll damping, the forces on appendages at high speed are proportional to roll velocity but viscous damping forces vary as the square of the roll velocity.

Software is available to enable the designer to solve non-linear, coupled, six degrees of freedom motion calculations. The exciting forces and moments arise from hydrostatics, from movement through the water (resistance and propulsion), from control surface movements (rudders and stabilizers), wind and wave forces. Such computations are discussed in the next Section.

7.7 Numerical prediction of seakeeping

7.7.1 Overview of computational methods

If the effect of the wave amplitude on the ship seakeeping is significantly nonlinear, there is little sense in investigating the ship in elementary waves, since these waves do not appear in nature and the non-linear reaction of the ship in natural seaways cannot be deduced from the reaction in elementary waves. In these non-linear cases, simulation in the time domain is the appropriate tool for numerical predictions.

However, if the non-linearity is weak or moderate the seakeeping properties of a ship in natural seaways can be approximated by superposition of the reactions in elementary waves of different frequency and direction. In these cases, the accuracy can be enhanced by introducing some relatively simple corrections of the purely linear computations to account for force contributions depending quadratically on the water velocity or considering the time-dependent change of position and wetted surface of the ship, for example. Even if iterative corrections are applied the basic computations of the ship seakeeping is still based on its reaction in elementary waves, expressed by complex amplitudes of the ship reactions. The time dependency is then always assumed to be harmonic, i.e. sinusoidal.

The Navier–Stokes equation (conservation of momentum) and the continuity equation (conservation of mass) suffice in principle to describe all phenomena of ship seakeeping flows. However, we neither can nor want to resolve all little turbulent fluctuations in the ship's boundary layer and wake. Therefore we average over time intervals which are long compared to the turbulent fluctuations and short compared to the wave periods. This then yields the Reynolds-averaged Navier–Stokes equations (RANS). By the late 1990s RANS computations for ship seakeeping were subject to research, but were still limited to selected simplified problems.

If viscosity is neglected the RANS turn into the Euler equations. Euler solvers do not have to resolve the boundary layers (no viscosity = no boundary layer) and allow thus coarser grids and considerably shorter computational times. By the late 1990s, Euler solvers were also still limited to simplified problems in research applications, typically highly non-linear free surface problems such as slamming of two-dimensional sections.

In practice, potential flow solvers are used almost exclusively in seakeeping predictions. The most frequent application is the computation of the linear seakeeping properties of a ship in elementary waves. In addition to the assumption for Euler solvers potential flow assumes that the flow is irrotational. This is no major loss in the physical model, because rotation is created by the water adhering to the hull and this information is already lost in the Euler flow model. Relevant for practical applications is that potential flow solvers are much faster than Euler and RANS solvers, because potential flows have to solve only one linear differential equation instead of four non-linear coupled differential equations. Also potential flow solvers are usually based on boundary element methods and need only to discretize the boundaries of the domain, not the whole fluid space. This reduces the effort in grid generation (the main cost item in most analyses) considerably. On the other hand, potential flow methods require a simple, continuous free surface. Flows involving breaking waves and splashes can hardly be analysed properly by potential flow methods.

In reality, viscosity is significant in seakeeping, especially if the boundary layer separates periodically from the hull. This is definitely the case for roll and yaw motions. In practice, empirical corrections are introduced. Also, for flow separation at sharp edges in the aftbody (e.g. vertical sterns, rudder, or transoms) a Kutta condition is usually employed to enforce a smooth detachment of the flow from the relevant edge.

The theoretical basics and boundary conditions of linear potential methods for ship seakeeping are treated extensively in the literature, e.g. by Newman (1978). Therefore, we can limit ourselves here to a short description of the fundamental results important to the naval architect.

The ship flow in elementary waves is described in a coordinate system moving with ship speed in the x direction, but not following its periodic motions. The derivatives of the potential give the velocity of water relative to such a coordinate system. The total velocity potential is decomposed:

$$\phi^t = (-Vx + \phi^s) + (\phi^w + \phi^I)$$

with ϕ^t potential of total flow

$-Vx$ potential of (downstream) uniform flow with ship speed V
ϕ^s potential of the steady flow disturbance
ϕ^w potential of the undisturbed wave
ϕ^I remaining unsteady potential

The first parenthesis describes only the steady (time-independent) flow, the second parenthesis the periodic flow due to sea waves. The potentials can be simply superimposed, since the fundamental field equation (Laplace equation, describing continuity of mass) is linear with respect to ϕ^t:

$$\Delta \phi^t = \left(\frac{\partial^2}{\partial x^2} + \frac{\partial^2}{\partial y^2} + \frac{\partial^2}{\partial z^2} \right) \phi^t = 0 \qquad (7.121)$$

Various approximations can be used for ϕ^s and ϕ^I which affect computational effort and accuracy of results. The most important linear methods can be classified as follows:

- *Strip Theory*
  Strip methods are the standard tool for ship seakeeping computations. They omit ϕ^s completely and approximate ϕ^I in each strip $x = $ constant independently of the other strips. Thus in essence the three-dimensional problem is reduced to a set (e.g. typically 10 to 30) of two-dimensional boundary value problems. This requires also a simplification of the actual free surface condition. The method originated in the late 1950s with work of Korvin-Kroukovsky and Jacobs. Most of today's strip methods are variations of the strip method proposed by Salvesen, Tuck and Faltinsen (1970). These are sometimes also called STF strip methods where the first letter of each author is taken to form the abbreviation. The two-dimensional problem for each strip can be solved analytically or by panel methods, which are the two-dimensional equivalent of the three-dimensional methods described below. The analytical approaches use conformal mapping to transform semicircles to cross-sections resembling ship sections (Lewis sections). Although this transformation is limited and, e.g., submerged bulbous bow sections cannot be represented in satisfactory approximation, this approach still yields for many ships results of similar quality as strip methods based on panel methods (close-fit approach). A close-fit approach is discussed by Bertram (2000). Strip methods are – despite inherent theoretical shortcomings – fast, cheap and for most problems sufficiently accurate. However, this depends on many details. Insufficient accuracy of strip methods often cited in the literature is often due to the particular implementation of a code and not due to the strip method in principle. But at least in their conventional form, strip methods fail (as most other computational methods) for waves shorter than perhaps $\frac{1}{3}$ of the ship length. Therefore, the added resistance in short waves (being considerable for ships with a blunt waterline) can also only be estimated by strip methods if empirical corrections are introduced. Section 7.7.2 describes a linear strip method in more detail.

- *Unified theory*
  Newman (1978) and Sclavounos developed at the MIT the 'unified theory' for slender bodies. Kashiwagi (1997) describes more recent developments of this theory. In essence, the theory uses the slenderness of the ship hull to justify a two-dimensional approach in the near field which is coupled to a three-dimensional flow in the far field. The far-field flow is generated by distributing singularities along the centreline of the ship. This approach is theoretically applicable to all frequencies, hence 'unified'. Despite its better theoretical foundation, unified theories failed to give significantly and consistently better results than strip theories for real ship geometries. The method therefore failed to be accepted by practice.

- *'High-speed strip theory' (HSST)*
  Several authors have contributed to the high-speed strip theory after the initial work of Chapman, (1975). A review of work since then can be found in Kashiwagi (1997). HSST usually computes the ship motions in an elementary wave using linear potential theory. The method is often called

$2\frac{1}{2}$ dimensional, since it considers the effect of upstream sections on the flow at a point x, but not the effect of downstream sections. Starting at the bow, the flow problem is solved for individual strips (sections) x = constant. The boundary conditions at the free surface and the hull (strip contour) are used to determine the wave elevation and the velocity potential at the free surface and the hull. Derivatives in the longitudinal direction are computed as numerical differences to the upstream strip which has been computed in the previous step. The computation marches downstream from strip to strip and ends at the stern resp. just before the transom. HSST is the appropriate tool for fast ships with Froude numbers $F_n > 0.4$. For lower Froude numbers, it is inappropriate.

- *Green function method (GFM)*
ISSC (1994) gives a literature review of these methods. GFM distribute panels on the average wetted surface (usually for calm-water floating position neglecting dynamical trim and sinkage and the steady wave profile) or on a slightly submerged surface inside the hull. The velocity potential of each panel (Green function) fulfils automatically the Laplace equation, the radiation condition (waves propagate in the right direction) and a simplified free-surface condition (omitting the ϕ^s completely). The unknown (either source strength or potential) is determined for each element by solving a linear system of equations such that for each panel at one point the no-penetration condition on the hull (zero normal velocity) is fulfilled. The various methods, e.g. Ba and Guilbaud (1995), Iwashita (1997), differ primarily in the way the Green function is computed. This involves the numerical evaluation of complicated integrals from 0 to ∞ with highly oscillating integrands. Some GFM approaches formulate the boundary conditions on the ship under consideration of the forward speed, but evaluate the Green function only at zero speed. This saves a lot of computational effort, but cannot be justified physically and it is not recommended.

As an alternative to the solution in the frequency domain (for excitation by elementary waves), GFM may also be formulated in the time domain (for impulsive excitation). This avoids the evaluation of highly oscillating integrands, but introduces other difficulties related to the proper treatment of time history of the flow in so-called convolution integrals. Both frequency and time domain solutions can be superimposed to give the response to arbitrary excitation, e.g. by natural seaway, assuming that the problem is linear.

All GFMs are fundamentally restricted to simplifications in the treatment of ϕ^s. Usually ϕ^s is completely omitted which is questionable for usual ship hulls. It will introduce, especially in the bow region, larger errors in predicting local pressures.

- *Rankine singularity method (RSM)*
Bertram and Yasukawa (1996) give an extensive overview of these methods covering both frequency and time domains. RSM, in principle, capture ϕ^s completely and also more complicated boundary conditions on the free surface and the hull. In summary, they offer the option for the best approximation of the seakeeping problem within potential theory. This comes at a price. Both ship hull and the free surface in the near field around the ship have to be discretized by panels. Capturing all waves while avoiding unphysical reflections of the waves at the outer (artificial) boundary of the computational domain poses the main problem for RSM. Since the early 1990s, various RSM for ship seakeeping have been developed. By the end of the 1990s, the time-domain SWAN code (SWAN = Ship Wave ANalysis) of MIT was the first such code to be used commercially.

- *Combined RSM–GFM approach*
GFM are fundamentally limited in the capturing the physics when the steady flow differs considerably from uniform flow, i.e. in the near field. RSM have fundamental problems in capturing the radiation condition for low τ values. Both methods can be combined to overcome the individual shortcomings and to combine their strengths. This is the idea behind combined approaches. These are described as 'Combined Boundary Integral Equation Methods' by the Japanese, and as 'hybrid methods' by Americans. Initially only hybrid methods were used which matched near-field RSM solutions directly to far-field GFM solutions by introducing vertical control surfaces at the outer boundary of the near field. The solutions are matched by requiring that the potential and its normal derivative are continuous at the control surface between near field and far field. In principle, methods with overlapping regions also appear possible.

7.7.2 Strip theory

This section presents the most important formulae for a linear frequency-domain strip method for slender ships in elementary waves. The formulae will be given without derivation. For a more extensive coverage of the theoretical background, the reader is referred to Newman (1978).

Two coordinate systems are used:

- The ship-fixed system x, y, z, with axes pointing from amidships forward, to starboard and downwards.

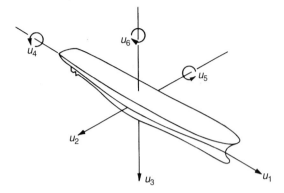

Figure 7.62 Six degrees of freedom for motions.

In this system, the ship's centre of gravity is time independent x_g, y_g, z_g.
- The inertial system ξ, η, ζ. This system follows the steady forward motion of the ship with speed V and coincides in the time average with the ship-fixed system.

The main purpose of the strip method is to compute the ship's rigid-body motions, i.e. the three translations of the origin of ship-fixed system in the ξ, η, ζ direction and the three rotations around these axes. We denote, Figure 7.62:

u_1 surge u_4 roll
u_2 sway u_5 pitch
u_3 heave u_6 yaw

The motions are combined in a six-component vector $\vec{u}$. The forces and moments acting on the ship are similarly combined in a six-component vector $\vec{F}$. $\vec{u}$ and $\vec{F}$ are harmonic functions of time t oscillating with encounter frequency ω_e:

$$\vec{F} = \mathrm{Re}(\hat{\vec{F}}e^{i\omega_e t})\vec{u} = \mathrm{Re}(\hat{\vec{u}}e^{i\omega_e t}) \quad (7.122)$$

The fundamental equation of motion is derived from $\vec{F} = M \cdot \ddot{\vec{u}}$:

$$[-\omega_e^2(M+A) + i\omega_e N + S]\hat{\vec{u}} = \hat{\vec{F}}_e \quad (7.123)$$

Here M, A, N and S are real-valued 6×6 matrices. For mass distribution symmetrical to $y = 0$ the mass matrix M is:

$$M = \begin{bmatrix} m & 0 & 0 & 0 & mz_g & 0 \\ 0 & m & 0 & -mz_g & 0 & mx_g \\ 0 & 0 & m & 0 & -mx_g & 0 \\ 0 & -mz_g & 0 & \theta_{xx} & 0 & -\theta_{xz} \\ mz_g & 0 & -mx_g & 0 & \theta_{yy} & 0 \\ 0 & mx_g & 0 & -\theta_{xz} & 0 & -\theta_{zz} \end{bmatrix}$$

$$(7.124)$$

The mass moments of inertia θ are related to the origin of the ship-fixed coordinate system:

$$\theta_{xx} = \int (y^2 + z^2)\,\mathrm{d}m; \quad \theta_{xz} = \int xz\,\mathrm{d}m; \quad \text{etc.}$$

If we neglect contributions from a dry transom stern and other hydrodynamic forces due to the forward speed of the ship, the restoring forces matrix S is:

$$S = \begin{bmatrix} 0 & 0 & 0 & 0 & 0 & 0 \\ 0 & 0 & 0 & 0 & 0 & 0 \\ 0 & 0 & \rho g A_w & 0 & -\rho g A_w x_w & 0 \\ 0 & 0 & 0 & gmGM & 0 & 0 \\ 0 & 0 & -\rho g A_w x_w & 0 & gmGM_L & 0 \\ 0 & 0 & 0 & 0 & 0 & \theta_{zz}\omega_g^2 \end{bmatrix}$$

$$(7.125)$$

Here A_w is the waterline area, x_w the x coordinate of the centre of the waterline, GM the metacentric height, GM_L the longitudinal metacentric height, ω_g the circular eigenfrequency of yaw motions. ω_g is determined by the control characteristics of the autopilot and usually has little influence on the yaw motions in seaways. In computing GM_L, the moment of inertia is taken with respect to the origin of the coordinate system (usually amidships) and not, as usual, with respect to the centre of the waterline. For corrections for dry transoms and unsymmetrical bodies reference is made to Söding (1987).

N is the damping matrix; it contains mainly the effect of the radiated waves. A is the added mass matrix. The decomposition of the force into hydrostatic (S) and hydrodynamic (A) components is somewhat arbitrary, especially for the ship with forward speed. Therefore, comparisons between computations and experiments often are based on the term $-\omega_e^2 A + S$.

$\vec{F}_e$ is the vector of exciting forces which a wave would exert on a ship fixed on its average position (diffraction problem). The exciting forces can be decomposed into a contribution due to the pressure distribution in the undisturbed incident wave (Froude-Krilov force) and the contribution due to the disturbance by the ship (diffraction force). Both contributions are of similar order of magnitude.

To determine A and N, the flow due to the harmonic ship motions $\vec{u}$ must be computed (radiation problem). For small frequency of the motion (i.e. large wave length of the radiated waves), the hydrostatic forces dominate and the hydrodynamic forces are almost negligible. Therefore large relative errors in computing A and N are acceptable. For high frequencies, the crests of the waves radiated by the ship motions are near the ship almost parallel to the ship hull, i.e. predominantly in longitudinal direction. Therefore the longitudinal velocity component of the radiated waves can be neglected. Then only the two-dimensional flow around the ship sections (strips) must be determined. This simplifies the computations a great deal.

For the diffraction problem (disturbance of the wave due to the ship hull), which determines the exciting forces $\vec{F}_e$, a similar reasoning does not hold: unlike radiation waves (due to ship motions), diffraction waves (due to partial reflection at the hull and distortion beyond the hull) form a similar angle (except for sign) with the hull as the incident wave. Therefore, for most incident waves, the diffraction flow will feature also considerable velocities in longitudinal direction. These cannot be considered in a regular strip method, i.e. if we want to consider all strips as hydrodynamically independent. This error is partially compensated by computing the diffraction flow for wave frequency ω instead of encounter frequency ω_e, but a residual error remains. To avoid also these residual errors, sometimes $\vec{F}_e$ is determined indirectly from the radiation potential following formulae of Newman (1965). However, these formulae are only valid if the waterline is also streamline. This is especially not true for ships with submerged transom sterns.

For the determination of the radiation and (usually also) diffraction (=exciting) forces, the two-dimensional flow around an infinite cylinder of the same cross-section as the ship at the considered position is solved, Figure 7.63. The flow is generated by harmonic motions of the cylinder (radiation) or an incident wave (diffraction). Classical methods used analytical solutions based on multipole methods. Today, usually two-dimensional panel methods are preferred due to their (slightly) higher accuracy for realistic ship geometries. These two-dimensional panel methods can be based on GFM or RSM.

The flow and thus the pressure distribution depends on

- for the radiation problem:
  hull shape, frequency ω_e, and direction of the motion (vertical, horizontal, rotational)
- for the diffraction problem:
  hull shape, wave frequency ω, and encounter angle μ

For the radiation problem, we compute the pressure distributions for unit amplitude motions in one degree of freedom and set all other motions to zero and omit the incident wave. For the diffraction problem, we set all motions to zero and consider only the incident wave and its diffraction. We denote the resulting pressures by:

$\hat{p}_2$ for horizontal unit motion of the cylinder
$\hat{p}_3$ for vertical unit motion of the cylinder
$\hat{p}_4$ for rotational unit motion of the cylinder around the x axis

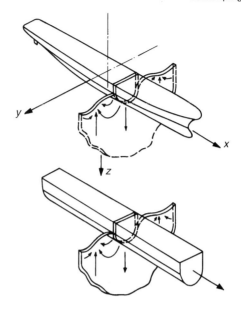

Figure 7.63 Principle of strip method.

$\hat{p}_0$ for the fixed cylinder in waves (only the pressure in the undisturbed wave)
$\hat{p}_7$ for the fixed cylinder in waves (only the disturbance of the pressure due to the body)

Let the actual motions of the cylinder in a wave of amplitude $\hat{h}_x$ be described by the complex amplitudes $\hat{u}_{2,0x}$, $\hat{u}_{3,0x}$, $\hat{u}_{4,0x}$. Then the complex amplitude of the harmonic pressure is:

$$\hat{p}_i = \hat{p}_2 \hat{u}_{2,0x} + \hat{p}_3 \hat{u}_{3,0x} + \hat{p}_4 \hat{u}_{4,0x} + (\hat{p}_0 + \hat{p}_7)\hat{h}_x \tag{7.126}$$

The amplitudes of the forces per length on the cylinder are obtained by integrating the pressure over the wetted surface of a cross-section (wetted circumference):

$$\begin{Bmatrix} \hat{f}_2 \\ \hat{f}_3 \\ \hat{f}_4 \end{Bmatrix} = \int_0^l \begin{Bmatrix} n_2 \\ n_3 \\ yn_3 - zn_2 \end{Bmatrix} \cdot \hat{p}_i \, d\ell$$

$$= \int_0^l \begin{Bmatrix} n_2 \\ n_3 \\ yn_3 - zn_2 \end{Bmatrix} \cdot (\hat{p}_2, \hat{p}_3, \hat{p}_4, \hat{p}_0 + \hat{p}_7) d\ell \cdot$$

$$\begin{Bmatrix} \hat{u}_{2,0x} \\ \hat{u}_{3,0x} \\ \hat{u}_{4,0x} \\ \hat{h}_x \end{Bmatrix} \tag{7.127}$$

$\{0, n_2, n_3\}$ is here in the inward unit normal on the cylinder surface. The index x in the last vector indicates that all quantities are taken at the longitudinal coordinate x at the ship, i.e. the position of the strip under consideration. ℓ is the circumferential length coordinate of the wetted contour. We can write the above equation in the form:

$$\vec{\hat{f}} = \hat{H} \cdot \{\hat{u}_{2,0x}, \hat{u}_{3,0x}, \hat{u}_{4,0x}, \hat{h}_x\}^T \quad (7.128)$$

The elements of the matrix $\hat{H}$, obtained by the integrals over the wetted surface in the above original equation, can be interpreted as added masses a_{ij}, damping n_{ij} and exciting forces per wave amplitude $\hat{f}_{ei}$:

$$\hat{H} = \begin{bmatrix} \omega_e^2 a_{22} - i\omega_e n_{22} & 0 & \omega_e^2 a_{24} - i\omega_e n_{24} & \hat{f}_{e2} \\ 0 & \omega_e^2 a_{33} - i\omega_e n_{33} & 0 & \hat{f}_{e3} \\ \omega_e^2 a_{42} - i\omega_e n_{42} & 0 & \omega_e^2 a_{44} - i\omega_e n_{44} & \hat{f}_{e4} \end{bmatrix}$$

(7.129)

For example, a_{22} is the added mass per cylinder length for horizontal motion.

The added mass tends towards infinity as the frequency goes to zero. However, the effect of the added mass also goes to zero for small frequencies, as the added mass is multiplied by the square of the frequency.

The forces on the total ship are obtained by integrating the forces per length (obtained for the strips) over the ship length. For forward speed, the harmonic pressure according to the linearized Bernoulli equation contains also a product of the constant ship speed $-V$ and the harmonic velocity component in the x direction. Also, the strip motions denoted by index x have to be converted to global ship motions in 6 degrees of freedom. This results in the global equation of motion:

$$[S - \omega_e^2(M + \hat{B})]\hat{u} = \hat{E}h \quad (7.130)$$

$\hat{B}$ is a complex matrix. Its real part is the added mass matrix A. Its imaginary part is the damping matrix N:

$$\omega_e^2 \hat{B} = \omega_e^2 A - i\omega_e N$$
$$= \int_L V(x) \cdot \left(1 + \frac{iV}{\omega_e} \frac{\partial}{\partial x}\right)(\hat{H}_B \cdot W(x)) \, dx$$

(7.131)

This equation can be used directly to compute $\hat{B}$, e.g. using the trapezoidal rule for the integrals and numerical difference schemes for the differentiation in x. Alternatively, partial integration can remove the x derivatives. The new quantities in the above equations are defined as:

$$\vec{\hat{E}} = \frac{\vec{\hat{F}}_E}{h} = \int_L V(x) \left(\hat{H}_E + \frac{iV}{\omega} \frac{\partial \hat{H}_{E7}}{\partial x}\right) e^{ikx \cos \mu} \, dx$$

(7.132)

$$W(x) = \begin{bmatrix} 0 & 1 & 0 & t_x & 0 & x - V/(i\omega_e) \\ 0 & 0 & 1 & 0 & -x + V/(i\omega_e) & 0 \\ 0 & 0 & 0 & 1 & 0 & 0 \end{bmatrix}$$

(7.133)

t_x is the z coordinate (in the global ship system) of the origin of the reference system for a strip. (Often a strip reference system is chosen with origin in the waterline, while the global ship coordinate system may have its origin on the keel.)

$$V(x) = \begin{bmatrix} 1 & 0 & 0 & 0 & 0 \\ 0 & 1 & 0 & 0 & 0 \\ 0 & 0 & 1 & 0 & 0 \\ 0 & t_x & 0 & 1 & 0 \\ -t_x & 0 & -x & 0 & 1 \\ 0 & x & 0 & 0 & 0 \end{bmatrix}$$

(7.134)

$$\hat{H} = \begin{bmatrix} 0 & 0 & 0 \\ \omega_e^2 a_{22} - i\omega_e n_{22} & 0 & \omega_e^2 a_{24} - i\omega_e n_{24} \\ 0 & \omega_e^2 a_{33} - i\omega_e n_{33} & 0 \\ \omega_e^2 a_{42} - i\omega_e n_{42} & 0 & \omega_e^2 a_{44} - i\omega_e n_{44} \\ 0 & 0 & 0 \end{bmatrix}$$

(7.135)

$$\hat{H}_E = \begin{Bmatrix} -i\rho g k A_x \cos \mu \\ \hat{f}_{e2} \\ \hat{f}_{e3} \\ \hat{f}_{e4} \\ -i\rho g k A_x s_x \cos \mu \end{Bmatrix}$$

(7.136)

A_x is the submerged section area at x; s_x is the vertical coordinate of the centre of the submerged section area in the global system. $\hat{H}_E$ contains both the Froude–Krilov part from the undisturbed wave (Index 0) and the diffraction part (Index 7), while $\hat{H}_{E7}$ contains only the diffraction part.

The formulae for $\hat{B}$ and $\vec{\hat{E}}$ contain x derivatives. At locations x, where the ship cross-section changes suddenly (propeller aperture, vertical stem, submerged transom stern), this would result in extremely high forces per length. To a large extent, this is actually true at the bow, but not at the stern. If the cross-sections decrease rapidly there, the streamlines separate from the ship hull. The momentum (which equals added mass of the cross-section times velocity of the cross-section) remains then in the ship's wake while the above formulae would yield in strict application zero momentum behind the ship as the added mass $\hat{H}$ is zero there. Therefore, the integration of the x derivatives over the ship length in the above formulae has to end at such locations of flow separation in the aftbody.

The global equation of motion above yields the vector of the response amplitude operators (RAOs) (=complex amplitude of reaction/wave amplitude) for the ship motions:

$$\frac{\hat{\vec{u}}}{h} = (S - \omega_e^2[M + \hat{B}])^{-1} \cdot \hat{\vec{E}} \quad (7.137)$$

The effect of rudder actions due to course deviations (yaw oscillations) was already considered in the matrix S. In addition, there are forces on the rudder (and thus the ship) due to ship motions (for centrally located rudders only due to sway, yaw, and roll) and due to the incident wave. Here it is customary to incorporate the rudder in the model of the rigid ship filling the gaps between rudder and ship. While this is sufficient for the computation of the ship motions, it is far too crude if the forces on the rudder in a seaway are to be computed.

Accurate computation of the motions, pressures, internal forces etc. requires further additions and corrections, e.g. to capture the influence of non-linear effects especially for roll motion, treatment of low encounter frequencies, influence of bilge keels, stabilizing fins etc. The special and often empirical treatment of these effects differs in various strip methods. Details can be found in the relevant literature including Lloyd (1998).

7.7.3 Rankine singularity methods

Bertram and Yasukawa (1996) give an extensive survey of these methods. A linear frequency-domain method is described briefly here to exemplify the general approach. A worked example is given by Bertram (2000).

In principle, RSM can completely consider ϕ^s. If ϕ^s is completely captured the methods are called 'fully three dimensional' to indicate that they capture both the steady and the harmonic flow three dimensionally. In this case, first the 'fully non-linear' wave resistance problem is solved to determine ϕ^s and its derivatives, including second derivatives of ϕ^s on the hull. The solution yields also all other steady flow effects, namely dynamic trim and sinkage, steady wave profile on the hull, steady and the wave pattern on the free surface. Then the actual seakeeping computations can be performed considering the interaction between steady and harmonic flow components. The boundary conditions for ϕ^I are linearized with regard to wave amplitude h and quantities proportional to h, e.g. ship motions. The Laplace equation (mass conservation) is solved subject to the boundary conditions:

1. Water does not penetrate the hull.
2. Water does not penetrate the free surface.
3. At the free surface there is atmospheric pressure.
4. Far away from the ship, the flow is undisturbed.
5. Waves generated by the ship radiate away from the ship.
6. Waves generated by the ship are not reflected at the artifical boundary of the computational domain.
7. For antisymmetric motions (sway, roll, yaw), a Kutta condition is enforced on the stern.
8. Forces (and moments) not in equilibrium result in ship motions.

For $\tau = \omega_e V/g > 0.25$ waves generated by the ship travel only downstream, similar to the steady wave pattern. Thus also the same numerical techniques as for the steady wave resistance problem can be used to enforce proper radiation, e.g. shifting source elements relative to collocation points downstream. Values $\tau < 0.25$ appear especially in following waves. Various techniques have been proposed for this case, as discussed in Bertram and Yasukawa (1996). However, there is no easy and accurate way in the frequency domain. In the time domain, proper radiation follows automatically and numerical beaches have to be introduced to avoid reflection at the outer boundary of the computational domain.

We split here the six-component motion vector of the chapter for the strip method approach into two three-component vectors, $\vec{u} = \{u_1, u_2, u_3\}^T$ describes the translations, $\vec{\alpha} = \{u_4, u_5, u_6\}^T = \{\alpha_1, \alpha_2, \alpha_3\}^T$ the rotations. The velocity potential is again decomposed as in Section 7.7.1:

$$\phi^t = (-Vx + \phi^s) + (\phi^w + \phi^I) \quad (7.138)$$

The steady potential ϕ^s is determined first. Typically, a 'fully non-linear' wave resistance code employing higher-order panels is used also to determine second derivatives of the potential on the hull. Such higher-order panels are described in the section on boundary elements. ϕ^w is the incident wave.

$$\phi^w = \text{Re}(-ic\hat{h}e^{-kz}e^{-ik(x\cos\mu - y\sin\mu)}e^{i\omega_e t}) \quad (7.139)$$

The wave amplitude is chosen to $\hat{h} = 1$.

The remaining unknown potential ϕ^I is decomposed into diffraction and radiation components:

$$\phi^I = \phi^d + \sum_{i=1}^{6} \phi^i u_i \quad (7.140)$$

The boundary conditions 1–3 and 7 are numerically enforced in a collocation scheme, i.e. at selected individual points. The remaining boundary conditions are automatically fulfilled in a Rankine singularity method. Combining 2 and 3 yields the boundary condition on the free surface, to be fulfilled by the unsteady potential $\phi^{(f)} = \phi^w + \phi^I$:

$$(-\omega_e^2 + Bi\omega_e)\hat{\phi}^{(1)} +$$
$$([2i\omega_e + B]\nabla\phi^{(0)} + \vec{a}^{(0)} + \vec{a}^g)\nabla\hat{\phi}^{(1)}$$
$$+ \nabla\phi^{(0)}(\nabla\phi^{(0)}\nabla)\nabla\hat{\phi}^{(1)} = 0 \quad (7.141)$$

With:

$\phi^{(0)} = -Vx + \phi^s$ steady potential
$\vec{a} = (\nabla\phi^{(0)}\nabla)\nabla\phi^{(0)}$ steady particle acceleration
$\vec{a}^g = \vec{a} - \{0, 0, g\}^T$
$B = -(1/a_3^g)\partial(\nabla\phi^{(0)}\vec{a}^g)/\partial z$
$\nabla = \{\partial/\partial x, \partial/\partial y, \partial/\partial z\}^T$

The boundary condition 1 yields on the ship hull

$$\vec{n}\nabla\hat{\phi}^{(1)} + \hat{\vec{u}}(\vec{m} - i\omega_e\vec{n})$$
$$+ \hat{\vec{\alpha}}[\vec{x} \times (\vec{m} - i\omega_e\vec{n}) + \vec{n} \times \nabla\phi^{(0)}] = 0 \quad (7.142)$$

Here the m terms have been introduced:

$$\vec{m} = (\vec{n}\nabla)\nabla\phi^{(0)}$$

Vectors $\vec{n}$ and $\vec{x}$ are to be taken in the ship-fixed system.

The diffraction potential ϕ^d and the six radiation potentials ϕ^i are determined in a panel method that can employ regular first-order panels. The panels are distributed on the hull and on (or above) the free surface around the ship. The Kutta condition requires the introduction of additional dipole (or alternatively vortex) elements. The preferred choice here are Thiart elements, see Bertram (2000), Chapter 6.

Test computations for a container ship (standard ITTC test case S-175) have shown a significant influence of the Kutta condition for sway, yaw and roll motions for small encounter frequencies.

To determine ϕ^d, all motions (u_i, $i = 1$ to 6) are set to zero. To determine the ϕ^i the corresponding u_i is set to 1, all other motion amplitudes, ϕ^d and ϕ^w to zero. Then the boundary conditions form a system of linear equations for the unknown element strengths which is solved, e.g., by Gauss elimination. Once the element strengths are known, all potentials and derivatives can be computed.

For the computation of the total potential ϕ^t, the motion amplitudes u_i remain to be determined. The necessary equations are supplied by the momentum equations:

$$m(\ddot{\vec{u}} + \ddot{\vec{\alpha}} \times \vec{x}_g) = -\vec{\alpha} \times \vec{G}$$
$$+ \int (p^{(1)} - \rho[\vec{u}\vec{a}^g + \vec{\alpha}(\vec{x} \times \vec{a}^g)])\vec{n}\, dS \quad (7.143)$$

$$m(\vec{x}_g \times \ddot{\vec{u}}) + I\ddot{\vec{\alpha}} = -\vec{x}_g \times (\vec{\alpha} \times \vec{G}) +$$
$$\int (p^{(1)} - \rho[\vec{u}\vec{a}^g + \vec{\alpha}(\vec{x} \times \vec{a}^g)])$$
$$\times (\vec{x} \times \vec{n})\, dS \quad (7.144)$$

$G = gm$ is the ship's weight, $\vec{x}_g$ its centre of gravity and I the matrix of the moments of inertia of the ship (without added masses) with respect to the coordinate system. I is the lower-right 3×3 sub-matrix of the 6×6 matrix M given in the section for the strip method.

The integrals extend over the average wetted surface of the ship. The harmonic pressure $p^{(1)}$ can be decomposed into parts due to the incident wave, due to diffraction, and due to radiation:

$$p^{(1)} = p^w + p^d + \sum_{i=1}^{6} p^i u_i \quad (7.145)$$

The pressures p^w, p^d and p^i, collectively denoted by p^j, are determined from the linearized Bernoulli equation as:

$$p^j = -\rho(\phi_t^j + \nabla\phi^{(0)}\nabla\phi^j) \quad (7.146)$$

The two momentum vector equations above form a linear system of equations for the six motions u_i which is easily solved.

The explicit consideration of the steady potential ϕ^s changes the results for computed heave and pitch motions for wave lengths of similar magnitude as the ship length – these are the wave lengths of predominant interest – by as much as 20–30% compared to total neglect. The results for standard test cases such as the Series-60 and the S-175 agree much better with experimental data for the 'fully three-dimensional' method. For the standard ITTC test case of the S-175 container ship, in most cases good agreement with experiments could be obtained, Figure 7.64. Only for low encounter frequencies, the antisymmetric motions are overpredicted, probably because viscous effects and autopilot were not modelled at all in the computations.

If ϕ^s is approximated by double-body flow, similar results are obtained as long as the dynamic trim and sinkage are small. However, the computational effort is nearly the same.

Japanese experiments at a tanker model indicate that for full hulls the diffraction pressures in the forebody for short head waves ($\lambda/L = 0.3$ and 0.5) are predicted with errors of up to 50% if ϕ^s is neglected (as typically in GFM or strip methods). Computations with and without consideration of ϕ^s yield large differences in the pressures in the bow region for radiation in short waves and for diffraction in long waves.

7.7.4 Problems for fast and unconventional ships

Seakeeping computations are problematic for fast and unconventional ships. Seakeeping plays a special role here, as fast ships are often passenger ferries,

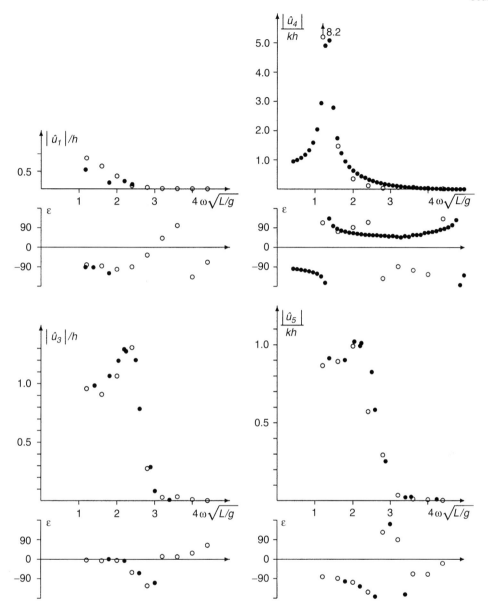

Figure 7.64 Selected response amplitude operators of motions for the container ship S-175 at $f_n = 0.275$; experiment, computation surge (top left) for $\mu = 180°$; roll (top right) for $\mu = 120°$; heave (bottom left) for $\mu = 150°$; pitch (bottom right) for $\mu = 150°$.

which need good seakeeping characteristics to attract passengers. This is the reason why, e.g., planing boats with their bad seakeeping are hardly ever used for commercial passenger transport. For fast cargo ships, the reduced speed in seaways can considerably influence transport efficiency. A hull form, which is superior in calm water, may well become inferior in moderate seaways. Warships also often require good seakeeping to supply stable platforms for weapon systems, helicopters, or planes.

Unfortunately, computational methods for conventional ships are usually not at all or only with special modifications suitable for fast and unconventional ships. The special 'High-speed strip theory', see Section 7.7.1, has been successfully applied in various forms to both fast monohulls and multihulls. Japanese validation studies showed that for a fast monohull with transom stern the HSST fared much better than both conventional strip methods and three-dimensional GFM and RSM.

However, the conventional strip methods and the three-dimensional methods did not use any special treatment of the large transom stern of the test case. This impairs the validity of the conclusions. Researchers at the MIT have shown that at least for time-domain RSM the treatment of transom sterns is possible and yields good results also for fast ships, albeit at a much higher computational effort than the HSST. In most cases, HSST should yield the best cost-benefit ratio for fast ships.

It is claimed often in the literature that conventional strip methods are only suitable for low ship speeds. However, benchmark tests show that strip methods can yield good predictions of motion RAOs up to Froude numbers $F_n \approx 0.6$, provided that proper care is taken and the dynamic trim and sinkage and the steady wave profile at the hull is included to define the average submergence of the strips. The prediction of dynamic trim and sinkage is relatively easy for fast displacement ships, but difficult for planing boats. Neglecting these effects, i.e. computing for the calm-water wetted surface, may be a significant reason why often in the literature a lower Froude number limit of $F_n \approx 0.4$ is cited.

For catamarans, the interaction between the hulls plays an important role especially for low speeds. For design speed, the interaction is usually negligible in head seas. Three-dimensional methods (RSM, GFM) capture automatically the interaction as both hulls are simultaneously modelled. The very slender form of the demihulls introduces smaller errors for GFM catamaran computations than for monohulls. Both RSM and GFM applications to catamarans can be found in the literature, usually for simplified research geometries. Strip methods require special modifications to capture, at least in good approximation, the hull interaction, namely multiple reflection of radiation and diffraction waves. Simply using the hydrodynamic coefficients for the two-dimensional flow between the two cross-sections leads to strong overestimation of the interaction for $V > 0$. Research on the motions of high-speed craft is, for example, described by Bruzzone et al. (2001) and Hudson et al. (2001).

Seakeeping computations for air-cushioned vehicles and surface effect ships are particularly difficult due to additional problems:

- The flexible skirts deform under the changing air cushion pressure and the contact with the free surface. Thus the effective cushion area and its centre of gravity change.
- The flow and the pressure in the cushion contain unsteady parts which depend strongly on the average gap between free surface and skirts.
- The dynamics of fans (and their motors) influences the ship motions.

Especially the narrow gaps between skirts and free surface result in a strongly non-linear behaviour that so far excludes accurate predictions. The motions of air cushion craft are discussed by Yun and Bliault (2000).

7.7.5 Further quantities in regular waves

Within a linear theory, the velocity and acceleration RAOs can be directly derived, once the motion RAOs are determined. The relative motion between a point on the ship and the water surface is important to evaluate the danger of slamming or water on deck. The RAOs for relative motion should incorporate the effect of diffraction and radiation, which is again quite simple once the RAOs for the ship motions are determined. However, effects of flared hull shape with outward forming spray for heave motion cannot be modelled properly within a linear theory, because these depend non-linearly on the relative motion. In practice, the section flare is important for estimating the amount of water on deck.

Internal forces on the ship hull (longitudinal, transverse, and vertical forces, torsional, transverse, and longitudinal bending moment) can also be determined relatively easily for known motions. The pressures are then only integrated up to a given cross-section instead of over the whole ship length. (Within a strip method approach, this also includes the matrix of restoring forces S, which contains implicitly many hydrostatic pressure terms.) Also, the mass forces (in matrix M) should only be considered up to the given location x of the cross-section. Stresses in the hull can then be derived from the internal forces. However, care must be taken that the moments are transformed to the neutral axis of the 'beam' ship hull. Also, stresses in the hull are of interest often for extreme loads where linear theory should no longer be applied.

The longitudinal force on the ship in a seaway is to first order within a linear theory also a harmonically oscillating quantity. The time average of this quantity is zero. However, in practice the ship experiences a significantly non-zero added resistance in seaways. This added resistance (and similarly the transverse drift force) can be estimated using linear theory. Two main contributions appear:

- Second-order pressure contributions are integrated over the average wetted surface.
- First-order pressure contributions are integrated over the difference between average and instantaneous wetted surface; this yields an integral over the contour of the waterplane.

If the steady flow contribution is completely retained (as in some three-dimensional BEM), the resulting expression for the added resistance is rather complicated and involves also second derivatives of the potential on the hull. Usually this formula is simplified assuming

- uniform flow as the steady base flow
- dropping a term involving x-derivatives of the flow
- considering only heave and pitch as main contributions to added resistance.

7.7.6 Ship responses in stationary seaway

Here the issue is how to get statistically significant properties in natural seaways from a response amplitude operator $Y_r(\omega, \mu)$ in elementary waves for an arbitrary response r depending linearly on wave amplitude. The seaway is assumed to be stationary with known spectrum $S_\zeta(\omega, \mu)$.

Since the spectrum is a representation of the distribution of the amplitude squared over ω and μ, and the RAO $\hat{Y}_r$ is the complex ratio of r_A/ζ_A, the spectrum of r is given by:

$$S_r(\omega, \mu) = |Y_r(\omega, \mu)|^2 S_\zeta(\omega, \mu) \quad (7.147)$$

Values of r, chosen at a random point in time, follow a Gaussian distribution. The average of r is zero if we assume $r \sim \zeta_A$, i.e. in calm water $r = 0$. The probability density of randomly chosen r values is:

$$f(r) = \frac{1}{\sqrt{2\pi}\sigma_r} \exp\left(-\frac{r^2}{2\sigma_r^2}\right) \quad (7.148)$$

The variance σ_r^2 is obtained by adding the variances due to the elementary waves in which the natural seaway is decomposed:

$$\sigma_r^2 = \int_0^\infty \int_0^{2\pi} S_r(\omega, \mu)\,d\mu\,d\omega \quad (7.149)$$

The sum distribution corresponding to the frequency density $f(r)$ above is:

$$F(r) = \int_{-\infty}^r f(\rho)\,d\rho = \frac{1}{2}[1 + \phi(r/\sigma_r)] \quad (7.150)$$

The probability integral ϕ is defined as:

$$\phi = 2/\sqrt{2\pi} \cdot \int_{-\infty}^x \exp(-t^2/2)\,dt$$

$F(r)$ gives the percentage of time when a response (in the long-term average) is less or equal to a given limit r. $1 - F(r)$ is then the corresponding percentage of time when the limit r is exceeded.

More often the distribution of the amplitudes of r is of interest. We define here the amplitude of r (differing from some authors) as the maximum of r between two following upward zero crossings (where $r = 0$ and $\dot{r} > 0$). The amplitudes of r are denoted by r_A. They have approximately (except for extremely 'broad' spectra) the following probability density:

$$f(r_A) = \frac{r_A}{\sigma_r^2} \exp\left(-\frac{r_A^2}{2\sigma_r^2}\right) \quad (7.151)$$

The corresponding sum distribution is:

$$F(r_A) = 1 - \exp\left(-\frac{r_A^2}{2\sigma_r^2}\right) \quad (7.152)$$

σ_r follows again the formula given above. The formula for $F(r_A)$ describes a so-called Rayleigh distribution. The probability that a randomly chosen amplitude of the response r exceeds r_A is:

$$1 - F(r_A) = \exp\left(-\frac{r_A^2}{2\sigma_r^2}\right) \quad (7.153)$$

The average frequency (occurrences/time) of upward zero crossings and also as the above definition of amplitudes of r is derived from the r spectrum to:

$$f_0 = \frac{1}{2\pi\sigma_r}\sqrt{\int_0^\infty \int_0^{2\pi} \omega_e^2 S_r(\omega, \mu)\,d\mu\,d\omega} \quad (7.154)$$

Together with the formula for $1 - F(r_A)$ this yields the average occurrence of r amplitudes which exceed a limit r_A during a period T:

$$z(r_A) = T f_0 \exp\left(-\frac{r_A^2}{2\sigma_r^2}\right) \quad (7.155)$$

Often we are interested in questions such as, 'How is the probability that during a period T a certain stress is exceeded in a structure or an opening is flooded?' Generally, the issue is then the probability $P_0(r_A)$ that during a period T the limit r_A is never exceeded. In other words, $P_0(r_A)$ is the probability that the maximum amplitude during the period T is less than r_A. This is given by the sum function of the distribution of the maximum or r during T. We make two assumptions:

- $z(r_A) \ll Tf_0$; this is sufficiently well fulfilled for $r_A \geq 2\sigma_r$.
- An amplitude r_A is statistically nearly independent of its predecessors. This is true for most seakeeping responses, but not for the weakly damped amplitudes of elastic ship vibration excited by seaway, for example.

Under these assumption we have:

$$P_0(r_A) = e^{-z(r_A)}$$

If we insert here the above expression for $z(r_A)$ we obtain the 'double' exponential distribution typical for the distribution of extreme values:

$$P_0(r_A) = e^{-T f_0 \exp(-r_A^2/(2\sigma_r^2))}$$

The probability of exceedence is then $1 - P_0(r_A)$. Under the (far more limiting) assumption that $z(r_A) \ll 1$ we obtain the approximation:

$$1 - P_0(r_A) \approx z(r_A)$$

The equations for $P_0(r_A)$ assume neither a linear correlation of the response r from the wave amplitude nor a stationary seaway. They can therefore also be applied to results of non-linear simulations or long-term distributions.

7.7.7 Simulation methods

The appropriate tool to investigate strongly non-linear ship reactions are simulations in the time domain. The seaway itself is usually linearized, i.e. computed as superposition of elementary waves. The frequencies of the individual elementary waves ω_j may not be integer multiples of a minimum frequency ω_{min}. In this case, the seaway would repeat itself after $2\pi/\omega_{min}$ unlike a real natural seaway. Appropriate methods to chose the ω_j are:

- The ω_j are chosen such that the area under the sea spectrum between ω_j and ω_{j+1} is the same for all j. This results in constant amplitudes for all elementary waves regardless of frequency.
- The frequency interval of interest for the simulation is divided into intervals. These intervals are larger where S_ζ or the important RAOs are small and vice versa. In each interval a frequency ω_j is chosen randomly (based on constant probability distribution). One should not choose the same ω_j for all the L encounter angles under consideration. Rather each combination of frequency ω_j and encounter angle μ_l should be chosen anew and randomly.

The frequencies, encounter angles, and phase angles chosen before the simulation must be kept during the whole simulation.

Starting from a realistically chosen start position and velocity of the ship, the simulation computes in each time step the forces and moments acting from the moving water on the ship. The momentum equations for translations and rotations give the translatory and rotational accelerations. Both are three-component vectors and are suitably expressed in a ship-fixed coordinate system. The momentum equations form a system of six scalar, coupled ordinary second-order differential equations. These can be transformed into a system of 12 first-order differential equations which can be solved by standard methods, e.g. fourth-order Runge–Kutta integration. This means that the ship position and velocity at the end of a small time interval, e.g. one second, are determined from the corresponding data at the beginning of this interval using the computed accelerations.

The forces and moments can be obtained by integrating the pressure distribution over the momentary wetted ship surface. Three-dimensional methods are very, and usually, too expensive for this purpose. Therefore modified strip methods are most frequently used. A problem is that the pressure distribution depends not only on the momentary position, velocity, and acceleration, but also from the history of the motion which is reflected in the wave pattern. This effect is especially strong for heave and pitch motions. In computations for the frequency domain, the historical effect is expressed in the frequency dependency of the added mass and damping. In time-domain simulations, we cannot consider a frequency dependency because there are many frequencies at the same time and the superposition principle does not hold. Therefore, the historical effect on the hydrodynamic forces and moments $\vec{F}$ is either expressed in convolution integrals ($\vec{u}$ contains here not only the ship motions, but also the incident waves):

$$\vec{F}(t) = \int_{-\infty}^{t} K(\tau)\vec{u}(\tau)\,d\tau \qquad (7.156)$$

or one considers 0 to n time derivatives of the forces $\vec{F}$ and 1 to $(n + 1)$ time derivatives of the motions $\vec{u}$:

$$B_0\vec{F}(t) + B_1\dot{\vec{F}}(t) + B_2\ddot{\vec{F}}(t) + \cdots \\ = A_0\dot{\vec{u}}(t) + A_1\ddot{\vec{u}}(t) + A_2\dddot{\vec{u}}(t)\cdots \qquad (7.157)$$

The matrix $K(\tau)$ in the first alternative and the scalars A_i, B_i in the second alternative are determined in potential flow computations for various sinkage and heel of the individual strips.

The second alternative is called state model and appears to be far superior to the first alternative. Typical values for n are 2 to 4; for larger n problems occur in the determination of the constants A_i and B_i resulting, e.g., in numerically triggered oscillations. Pereira (1988) gives details of such a simulation method. The simulation method has been extended considerably in the mean time and can also consider simultaneously the flow of water through a damaged hull, sloshing of water in the hull, or water on deck.

A far simpler and far faster approach is described, e.g., in Söding (1987). Here only the strongly non-linear surge and roll motions are determined by a direct solution of the equations of motion in the time-domain simulation. The other four degrees of freedom are linearized and then treated similarly as the incident waves, i.e. they are computed from RAOs in the time domain. This is necessary to couple the four linear motions to the two non-linear motions. (Roll motions are often simulated as independent from the other motions, but this yields totally unrealistic results.) The restriction to surge and roll much simplifies the computation, because the history effect for these degrees of freedom is negligible. Extensive validation studies for this approach with model tests gave excellent agreement for capsizing of damaged RoRo vessels drifting without forward speed in transverse waves (Chang and Blume (1998)).

Simulations often aim to predict the average occurrence $z(r_A)$ of incidents where in a given period T a seakeeping response $r(t)$ exceeds a limit r_A. A new incident is then counted when after a previous incident another zero crossing of r occurred. The average occurrence is computed by multiple simulations with the characteristic data, but other random phases ε_{jl} for the superposition of the seaway. Alternatively, the simulation time can be chosen as nT and the number of occurrences can be divided by n. Both alternatives yield the same results except for random fluctuations.

Often seldom (extremely unlikely) incidents are of interest which would require simulation times of weeks to years to determine $z(r_A)$ directly if the occurrences are determined as described above. However, these incidents are expected predominantly in the presence of one or several particularly high waves. One can then reduce the required simulation time drastically by substituting the real seaway of significant wave height H_{real} by a seaway with larger significant wave height H_{sim}. The periods of both seaways shall be the same. The following relation between the incidents in the real seaway and in the simulated seaway exists (Söding (1987)):

$$\frac{H_{sim}^2}{H_{real}^2} = \frac{\ln[z_{real}(r_A)/z(0)] + 1.25}{\ln[z_{sim}(r_A)/z(0)] + 1.25} \quad (7.158)$$

This equation is sufficiently accurate for $z_{sim}/z(0) < 0.03$. In practice, one determines in simulated seaway, e.g. with 1.5 to 2 times larger significant wave height, the occurrences $z_{sim}(r_A)$ and $z(0)$ by direct counting; then the above equation is solved for the unknown $z_{real}(r_A)$:

$$z_{real}(r_A) = z(0)$$
$$\exp\left(\frac{H_{sim}^2}{H_{real}^2}\{\ln[z_{sim}(r_A)/z(0)] + 1.25\} - 1.25\right)$$
$$(7.159)$$

7.7.8 Long-term distributions

Section 7.7.6 treated ship reactions in a stationary seaway. This section will cover probability distributions of ship reactions r during periods T with changing sea spectra. A typical example for T is the total operational time of a ship. A quantity of interest is the average occurrence $z_L(r_A)$ of cases when the reaction $r(t)$ exceeds the limit r_A. The average can be thought of as the average over many hypothetical realizations, e.g. many equivalently operated sister ships.

First, one determines the occurrence $z(r_A; H_{1/3}, T_p, \mu_0)$ of exceeding the limit in a stationary seaway with characteristics $H_{1/3}$, T_p, and μ_0 during total time T. (See Section 7.7.6 for linear ship reactions and Section 7.7.7 for non-linear ship reactions.) The weighted average of the occurrences in various seaways is formed. The weighing factor is the probability $p(H_{1/3}, T_p, \mu_0)$ that the ship encounters the specific seaway:

$$z_L(r_A) = \sum_{\text{all } H_{1/3}} \sum_{\text{all } T_p} \sum_{\text{all } \mu_0} z(r_A; H_{1/3}, T_p, \mu_0) p(H_{1/3}, T_p, \mu_0)$$
$$(7.160)$$

Usually, for simplification it is assumed that the ship encounters seaways with the same probability under n_μ encounter angles μ_0:

$$z_L(r_A) = \frac{1}{n\mu} \sum_{\text{all } H_{1/3}} \sum_{\text{all } T_p} \sum_{i=1}^{n_\mu} z(r_A; H_{1/3}, T_p, \mu_{0i}) p(H_{1/3}, T_p) \quad (7.161)$$

The probability $p(H_{1/3}, T_p)$ for encountering a specific seaway can be estimated using data such as given in Table 1.8. If the ship would operate exclusively in the ocean area for Table 1.8, the table values could be taken directly. This is not the case in practice and requires corrections. A customary correction then is to base the calculation only on 1/50 or 1/100 of the actual operating time of the ship. This correction considers, for example;

- The ship usually operates in areas with not quite so strong seaways as given in Table 1.8.
- The ship tries to avoid particularly strong seaways.
- The ship reduces speed or changes course relative to the dominant wave direction, if it cannot avoid a particularly strong seaway.
- Some exceedence of r_A is not important, e.g. for bending moments if they occur in load conditions when the ship has only a small calm-water bending moment.

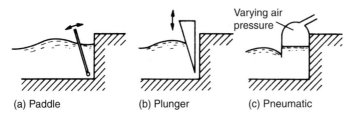

(a) Paddle   (b) Plunger   (c) Pneumatic

Figure 7.65 Types of wavemaker.

The sum distribution of the amplitudes r_A, i.e. the probability that an amplitude r is less than a limit r_A, follows from z_L:

$$P_L(r_A) = 1 - \frac{z_L(r_A)}{Z_L(0)} \qquad (7.162)$$

$z_{L(0)}$ is the number of amplitudes during the considered period T. This distribution is used for seakeeping loads in fatigue strength analyses of the ship structure. It is often only slightly different from an exponential distribution, i.e. it has approximately the sum distribution:

$$P_L(r_A) = 1 - e^{-r_A/r_0} \qquad (7.163)$$

where r_0 is a constant describing the load intensity. (In fatigue strength analyses, often the logarithm of the exceedence probability $\log(1 - P_L)$ is plotted over r_A; since for an exponential distribution the logarithm results in a straight line, this is called a log-linear distribution.)

The probability distribution of the largest loads during the period T can be determined from (see Section 7.7.6 for the underlying assumptions):

$$P_0(r_A) = e^{-z(r_A)} \qquad (7.164)$$

The long-term occurrence $z_L(r_A)$ of exceeding the limit r_A is inserted here for $z(r_A)$.

Comprehensive reviews of developments in theoretical and numerical modelling of seakeeping are given in ITTC (2002) and ITTC (2005). Developments have been reviewed by Beck and Reed (2001) and typical further developments are described by Arribas and Fernandez (2006).

7.8 Experiments and trials

7.8.1 Test facilities

Seakeeping experiments can be conducted in the conventional long, narrow ship tanks usually used for resistance tests provided they are fitted with a wavemaker at one end and a beach at the other, see Figure 5.22. Unfortunately in such tanks it is only possible to measure the response of the model when heading directly into or away from the waves.

Various forms of wavemaker have been employed, Figure 7.65. Beaches also take a number of forms but are essentially devices for absorbing the energy of the incident waves. They reduce the amplitude of the reflected waves which would otherwise modify the waves experienced by the model. There are a number of basins specially designed for seakeeping experiments. They permit the model to be run at any heading relative to, or to manoeuvre in, the waves. Short crested wave systems can be generated. The basin at QinetiQ, Haslar, UK, is depicted in Figure 7.66. It is 122 m long by 61 m wide and uses a completely free remotely-controlled model. In a number of facilities, the basin is spanned by a bridge so that models can be run under a carriage either free or constrained. This is also a feature of the 170 m by 40 m Seakeeping and Manoeuvring Facility at MARIN in which realistic short crested wave conditions can be created. Other examples of such basins include those at NSWC (Carderoc) in the USA and Marintek in Norway.

Whilst a free model has no guides that can interfere with its motion, the technique presents many difficulties. The model must contain its own propulsion system, power supplies, radio-control devices as well as being able to record much of the experimental data. It must be ballasted so as to possess the correct stability characteristics and scaled inertias in order that its response as a dynamic system will accurately simulate that of the ship. All this must be achieved in a relatively small model as too large a model restricts the effective length of each run. This is very important for experiments where data have to be presented statistically.

7.8.2 Ship seakeeping trials

Ship trials are carried out for a variety of reasons, including:

(a) to confirm that the ship meets her design intention as regards performance;

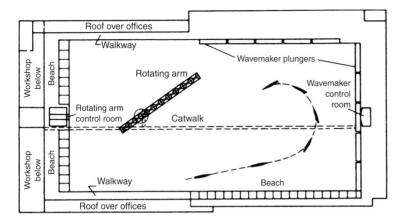

Figure 7.66 Seakeeping basin at QinetiQ, Haslar, UK.

(b) to predict performance during service;
(c) to prove that equipment can function properly in the shipboard environment;
(d) to provide data on which future ship designs can be based;
(e) to determine effect on human performance.

Whilst resistance and propulsion trials are usually carried out in calm water, those concerned with motions must by definition be carried out in rough water. Stabilizer performance is a special case and is discussed in Section 7.8.3. Two types of trial are possible:

(a) short duration trials in which the ship responses to a measured sea system are recorded;
(b) prolonged period trials in which statistical data is built up of ship response in a wide range of sea conditions.

The first type of trial is essential if it is wished to compare ship with model or calculated response operators over a range of ship speeds and headings. Then the likely long-term behaviour of the ship on voyage can be deduced as described earlier.

The second type of trial provides a comparison of actual and assessed behaviour during a voyage or over a period of time. Differences may be due either to the ship not responding to the wave systems as predicted or to the wave systems experienced not being those anticipated so the data is of limited value in assessing prediction methods. The longer the time period the better the measure of a ship's 'average' performance.

Short duration trials are expensive and the opportunity may be taken to record hull strains, motions, shaft torque and shaft thrust at the same time. Increasing attention is being paid to the performance of the people on board. The sea state itself must be recorded and this is usually by means of a wave recording buoy. For the second type of trial a simpler statistical motion recorder is used, often restricted to measurement of vertical acceleration. No wave measurements are made but sea states are observed. Statistical strain gauges may also be fitted. Satellites can be used to measure the wave system in which the trial ship is operating and GPS will normally be used to record the ship's path.

Although various methods have been proposed for measuring a multi-directional wave system it is a very difficult task. Good correlation has been achieved between calculated and measured sea loadings in some trials by applying the cosine squared spreading function to a spectrum based on buoy measurements.

In the earliest trials the waves were recorded by a shipborne wave recorder but nowadays a freely floating buoy is used. Signals are transmitted to the trials ship over a radio link or recorded in the buoy for recovering at the end of the trial. Vertical motions of the buoy are recorded by an accelerometer and movement of the wave surface relative to the buoy by resistive probes. Roll, pitch and azimuth sensors monitor the attitude of the buoy. For studying complex wave systems in detail several buoys may be used.

A typical sequence for a ship motion trial is to:

(a) carry out measured mile runs at the start of the voyage to establish the ship's smooth water performance and to calibrate the log;
(b) carry out service trials during passage to record sample ship motions and propulsive data under normal service conditions;
(c) launch the recording buoy, record conditions and recover buoy, when conditions are considered suitable, i.e. waves appear to be sufficiently long-crested;
(d) carry out a manoeuvre of the type shown in Figure 7.67 recording motions and waves for each leg.

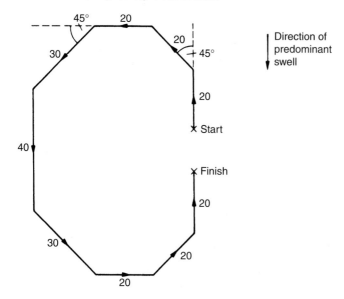

Figure 7.67 Typical seakeeping manoeuvre.

Figures denote time in minutes spent on each leg. The accuracy of the analysis depends upon the number of oscillations recorded. For this reason, the legs running with the seas are longer than those with ship running head into the waves. The overall time on the manoeuvre has to be balanced against the possibility of the sea state changing during the trial. The two sets of buoy records and a comparison of the results from the initial and final legs provides a guide to the stability of the trial conditions. The remaining steps of the sequence are:

(e) launch buoy for second recording of waves;
(f) repeat (*c*), (*d*) and (*e*) as conditions permit;
(g) carry out service trials on way back to port;
(h) carry out measured mile runs on return.

On completion of the trials a lengthy computer-based analysis of the gathered data will take place.

7.8.3 Stabilizer trials

Stabilizers, fitted to reduce rolling in a seaway, can be specified directly in terms of roll under stated conditions but such performance can never be precisely proven on trial. As an alternative a designer may relate performance to the steady angle of heel that can be generated by holding the fins over in calm water at a given speed, see Section 7.10.2.4. Heel can be measured directly to establish whether or not such contractual requirements have been met. Forced rolling trials in calm water can be used to study the performance of shipborne equipment under controlled conditions. It is often difficult to distinguish the effects of the stabilizers from the cross-coupling effect of the rudders; indeed, it is possible to build up a considerable angle of roll in calm water by the judicious use of rudder alone, see Section 8.8.

7.9 Improving seakeeping performance

7.9.1 Design and operational changes

It has been seen that overall seakeeping performance is limited not so much by motions *per se* as by the interplay between motions and other design features. Thus overall performance can be improved by such actions as:

(a) siting critical activities in less-affected areas of the ship. Examples are the siting of passenger accommodation towards the position of minimum vertical motion; placing helicopter operations aft in frigates and placing only very rugged equipment forward on the forecastle;
(b) rerouting of ships to avoid the worst sea conditions;
(c) providing local stabilization for certain equipments such as radars.

To an extent these can only be regarded as palliatives and it is necessary to consider how the motions themselves can be reduced. Care is needed to ensure the reduction is 'useful'. For example, if human performance is a limiting factor significant reductions in vertical acceleration over a wide range of higher frequencies is counter-productive if it is won at the expense of even a small increase in acceleration in the frequency band critical to humans. Bearing this reservation in mind there are a number of ways open to the naval architect:

Use can be made of radically different hull forms such as:

(a) the Small Waterplane Area Twin Hull (SWATH) ship. Essentially the use of a small waterplane area reduces the exciting forces and moments, the twin hull restoring the desired static stability qualities and weather deck area.
(b) the semi-submersible. The concept is similar to the SWATH. A major part of the vessel is well below the still waterplane so that the waves exert little force on it. This configuration is used for oil drilling rigs where a stable platform is essential and it must be held accurately in position over the seabed.
(c) hydrofoil craft. With suitable height sensors and foil incidence control systems a hydrofoil can provide a high speed, steady platform in sea states up to those in which the waves impact the hull. This depends upon foil separation from the hull.

Special ship types can be very effective in specific applications but usually there are penalties which means that the vast majority of ships are still based on a conventional monohull. In this case the designer can either improve performance by detail form changes or stabilize the whole ship. These are now considered.

7.9.2 Influence of form on seakeeping

It can be dangerous to generalize on the effect of varying form parameters on the seakeeping characteristics of a design. A change in one parameter often leads to a change in other parameters and a change may reduce motions but increase wetness. Again, the trend arising from a given variation in a full ship may not be the same as the trend from the same variation in a fine ship. This accounts for some apparently conflicting conclusions from different series of experiments. It is essential to consult data from previous similar ships and particularly any methodical series data that is available covering the range of principal form parameters applicable to the new design. Such methodical series data are those reported by Maury (2003) for Series 60 forms, Blok and Beukelman (1984) for high-speed displacement forms and Molland *et al.* (2001) for fast displacement catamarans.

With the above cautionary remarks in mind the following are some general trends based on the results of methodical series data:

Length is an important parameter in its own right. This can be appreciated by considering the response of a ship to a given wave system. If the ship is long compared with the component waves present, it will pitch and heave to a small extent only, e.g. a large passenger liner is hardly affected by waves which cause a 3000 tonnes frigate to pitch violently. With most ships there does come a time when they meet a wave system which causes resonance but the longer the ship the less likely this is.

Forward waterplane area coefficient. An increase reduces the relative motion at the bow but can lead to increased vertical wave bending moment.

Length to beam ratio has little influence on motions although lower L/B ratios are preferable.

Length to draught ratio. High values lead to resonance with shorter waves and this effect can be quite marked. Because of this, high L/T ratios lead to lower amplitudes of pitch and heave in long waves and greater amplitudes in short waves. A high L/T ratio is more conducive to slamming.

Block coefficient. Generally the higher the block coefficient the less the motions and the greater the increase in resistance, but the influence is small in both cases.

Prismatic coefficient. The higher the C_P value the less the motion amplitudes but the wetter the ship. High C_P leads to less speed loss at high speed and greater speed loss at low speed.

Beam to draught ratio. Higher values reduce vertical acceleration but may lead to greater slamming.

Longitudinal radius of gyration. In waves longer than the ship, a small radius of gyration is beneficial in reducing motions.

A bulbous bow generally reduces motions in short waves but can lead to increased motions in very long waves.

Forward sections. U-shaped sections usually give less resistance in waves and a larger longitudinal inertia. V-shaped sections usually produce lower amplitudes of heave and pitch and less vertical bow movement. Above-water flare has little effect on motion amplitudes but can reduce wetness at the expense of increased resistance and possible slamming effects.

Freeboard. The greater the freeboard the drier the ship.

7.9.3 Summary

It will be noted that a given change in form often has one effect in short waves and the opposite effect in long waves. In actual ocean conditions, waves of all lengths are present and it would not be surprising, therefore, if the motions, etc., in an irregular wave system showed less variation with form changes. Research has shown that, for conventional forms, the overall performance of a ship in waves is not materially influenced by variations in the main hull parameters. A large ship will be better than a small one.

Local form changes can assist in reducing the adverse consequences of motion, e.g. providing finer forms forward with large deadrise angle can reduce slamming forces.

Roll stabilization and pitch controls can produce a reduction in ship motions, and these are discussed in the next Section.

7.10 Ship motion control

7.10.1 Background

There is a limit to the extent to which amplitudes of motion can be reduced in conventional ship forms by changes in the basic hull shape. Fortunately, considerable reductions in roll amplitudes are possible by other means, roll being usually the most objectionable of the motions as regards comfort. In principle, the methods used to stabilize against roll can be used to stabilize against pitch but, in general, the forces or powers involved are generally too great to justify their use.

7.10.2 Roll Stabilization

7.10.2.1 *Stabilization systems*

These fall naturally under two main headings:

(a) Passive systems in which no separate source of power is required and no special control system. Such systems use the motion itself to create moments opposing or damping the motion. Some, such as the common bilge keel, Figure 7.68, are external to the main hull and with such systems there is an added resistance to ahead motion which has to be overcome by the main engines. The added resistance is offset, partially at least, by a reduction in resistance of the main hull due to the reduced roll amplitude.

Other passive systems, such as passive anti-roll tanks, are fitted internally. In such cases, there is no augment of resistance arising from the system itself.

The principal passive systems (discussed presently) fitted are:

Bilge keels
Fixed fins
Passive tank system
Passive moving weight system.

(b) Active systems in which the moment opposing roll is produced by moving masses or control surfaces by means of power. They also employ a control system which senses the rolling motion and so decides the magnitude of the correcting moment required. As with the passive systems, the active systems may be internal or external to the main hull.

The principal active systems fitted are:

Active fins
Active tank system
Active moving weight
Gyroscope.

Brief descriptions of systems
The essential requirement of any system is that the system should always generate a moment opposing the rolling moment.

(a) With active fins a sensitive gyro system senses the rolling motion of the ship and sends signals to the actuating system which, in turn, causes the fins to move in a direction such as to cause

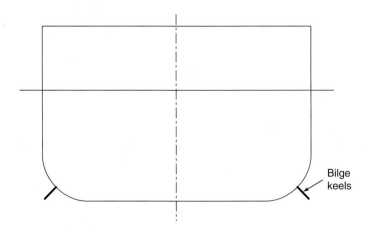

Figure 7.68 Bilge kells.

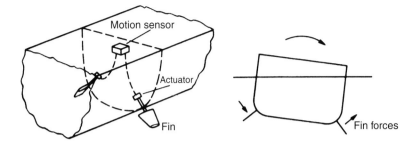

Figure 7.69 Active fin system.

forces opposing the roll. The actuating gear is usually electrohydraulic. The fins which may be capable of retraction into the hull, or may always protrude from it, are placed about the turn of bilge in order to secure maximum leverage for the forces acting upon them Figure 7.6.9. The fins are usually of the balanced spade type, but may incorporate a flap on the trailing edge to increase the lift force generated.

The capacity of a fin system is usually expressed in terms of the steady angle of heel it can cause with the ship moving ahead in still water at a given speed. Since the force on a fin varies in proportion to the square of the ship speed, whereas the GZ curve for the ship is, to a first order, independent of speed, it follows that a fin system will be more effective the higher the speed. Broadly speaking, a fin system is not likely to be very effective at speeds below about 10 knots. Details of the fin stabilizer design procedure is given in Section 7.10.2.5.

(b) Active weights systems take a number of forms, but the principle is illustrated by the scheme shown in Figure 7.70. If the weight W is attached to a rotating arm of radius R then, when the arm is at an angle α to the centre line of the ship and on the higher side,

Righting moment = $WR \sin \alpha$

Such a system has the advantage, over the fin system, that its effectiveness is independent of speed. It involves greater weight and power, however, and for these reasons is not often fitted.

(c) Active tank systems are also available in a variety of forms as illustrated in Figure 7.71. The essential, common, features are two tanks, one on each side of the ship, in which the level of water can be controlled in accord with the dictates of the sensing system. In scheme (a), water is pumped from one tank to the other so as to keep the greater quantity in the higher tank. In scheme (b), the water level is controlled

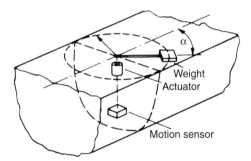

Figure 7.70 Active weight system.

indirectly by means of air pressure above the water in each tank, the tanks being open to the sea at the bottom. Scheme (b) has the advantage of requiring less power than scheme (a). In scheme (c), each tank has its own pump but otherwise is similar to scheme (a).

(d) All active stabilizing systems depend upon gyroscopes as part of their control system. If the gyroscope is massive enough, use can be made of the torque it generates when precessed to stabilize the ship. Such systems are not commonly fitted because of their large space and weight demands.

(e) Bilge keels are so simple and easy to fit that very few ships are not so fitted. They typically extend over the middle half to two-thirds of the ship's length at the turn of bilge. Compared with a ship not fitted, bilge keels can produce a reduction of roll amplitude of 35% or more. They are usually carefully aligned with the flow around the hull in calm water so as to reduce their resistance to ahead motion. Unfortunately, when the ship rolls the bilge keels are no longer in line with the flow of water and can lead to significant increases in resistance. For this reason, some large ships may be fitted with a tank stabilizing system and dispense with bilge keels.

(f) Fixed fins are similar in action to bilge keels except that they are shorter and extend further from the

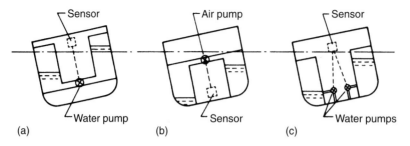

Figure 7.71 Active tank systems.

ship's side. An advantage claimed for them is that, by careful shaping of their cross-section, the lift generated at a given ahead speed can be increased compared with the drag they suffer. A disadvantage is that, projecting further from the hull, they are more susceptible to damage. They are generally less effective at low speed.

(g) Passive tank systems use the roll of the ship itself to cause water in the tanks to move in such a way as to oppose the motion, Figure 7.72. Starting from rest with water level in the two tanks, if the ship rolls to starboard water flows from port to starboard until the maximum angle of roll is reached. As the ship now tries to recover, the water will try to return but will nevertheless lag and the moment due to the water will oppose the roll velocity. Also, if the resistance of the duct is high the water will not be able to return before the ship is rolling to port, i.e. the level of water in the tanks can be made to lag the roll motion. By carefully adjusting the resistance of the duct the system can be 'tuned' to give maximum stabilizing effect. This will be when the phase lag is 90 degrees.

One limitation of such a scheme is that the system can only be 'tuned' to one frequency. This is chosen as the natural period of roll because it is at this period that the really large angles of roll can be built up. At other frequencies the passive tank system may actually lead to an increase in roll angle above the 'unstabilized' value, but this is not usually serious because the roll angles are small anyway. A more sophisticated system is one in which the resistance in the duct can be varied to suit the frequency of the exciting waves. In this way roll damping is achieved in all wave lengths.

An alternative to the side tank arrangement in Figure 7.72 is the flume or Frahm tank (Figure 7.73), where the tank is tuned by varying the constriction and depth of water (Goodrich, 1969; McGeorge, 1995).

(h) Passive moving weight systems are similar in principle to the passive tank systems but are generally less effective for a given weight of system.

7.10.2.2 *Comparison of principal systems*

Table 7.11 compares the principal ship stabilizing systems. The most commonly fitted, apart from bilge keels, are the active fin and passive tank systems.

7.10.2.3 *Performance of stabilizing systems*

The methods of predicting the performance of a given stabilizer system in reducing motion amplitudes in irregular seas are complex. A common method of specifying a system's performance is the roll amplitude it can induce in calm water, and this is more readily calculated and can be checked on trials.

When the ship rolls freely in still water, the amplitude of each successive swing decreases by an amount depending on the energy absorbed in each roll. At the end of each roll the ship is momentarily still and all its energy is stored as potential energy. If ϕ_1 is the roll angle, the potential energy is $\frac{1}{2}\Delta GM\phi_1^2$. If, on the next roll, the amplitude is ϕ_2 then the energy lost is

$$\frac{1}{2}\Delta GM(\phi_1^2 - \phi_2^2) = \Delta GM\left(\frac{\phi_1 + \phi_2}{2}\right)$$
$$(\phi_1 - \phi_2) = \Delta GM\phi\delta\phi \qquad (7.165)$$

where ϕ = mean amplitude of roll.

The reduction in amplitude, $\delta\phi$, is called the *decrement* and in the limit is equal to the slope of the curve of amplitude against number of swings at the mean amplitude concerned. That is,

$$\delta\phi = \left(-\frac{d\phi}{dn}\right)_\phi \qquad (7.166)$$

This means that when stabilizers are rolling a ship to a steady amplitude ϕ, the energy lost to damping per swing is

$$\Delta GM\phi\left(\frac{d\phi}{dn}\right)_\phi \qquad (7.167)$$

and this is the energy that must be provided by the stabilizers.

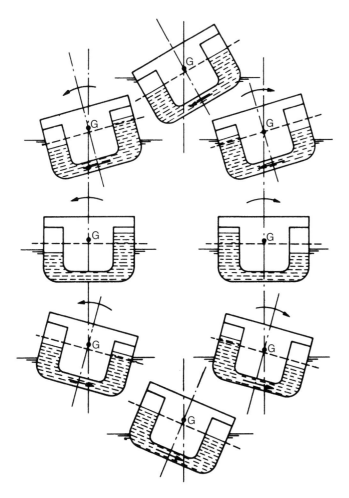

Figure 7.72 Passive tank system.

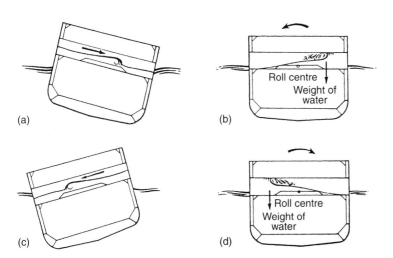

Figure 7.73 Brown-NPL passive tank stabilizer: (a) Stern view of ship with passive tank rolled to starboard. The water is moving in the direction shown, (b) Ship rolling to port. The water in the tank on the starboard side provides a moment opposing the roll velocity, (c) Ship at the end of its roll to port. The water is providing no moment to the ship, (d) Ship rolling to starboard. The water in the tank on the port side provides a moment opposing the roll velocity (McGeorge, 1995).

Table 7.10 Comparison of stabilizer systems. (Figures are for normal installations)

Type	Activated fin	Passive tank	Active tank	Massive gyro (active)	Moving weight (active)	Moving weight (passive)	Bilge keel	Fixed fin
Percentage roll reduction	90%	60–70%	No data	45%	No data	No data	35%	No data
Whether effective at very low speeds	No	Yes	Yes	Yes	Yes	Yes	Yes	No
Reduction in deadweight	1% of displacement	1–4% of displacement	Comparable with passive tank	2% of displacement	Comparable with passive tank		Negligible	
Any reduction in statical stability	No	Yes	Yes*	No	Yes*	Yes	No	No
Any increase in ship's resistance	When in operation	No	No	No	No	No	Slight	Slight
Auxiliary power requirement	Small	Nil	Large	Large	Large	Nil	Nil	Nil
Space occupied in hull	Moderate generally less than tanks	Moderate	Moderate	Large	Moderate	Less than tanks	Nil	Nil
Continuous athwartships space	No	Generally	Yes	No	Yes	Yes	No	No
Whether vulnerable to damage	Not when retracted	No	No	No	No	No	Yes	Very
First cost	High	Moderate	Probably high†	Very high	Probably high†	Probably high†	Low	Moderate†
Maintenance	Normal mechanical	Low	Normal mechanical	Probably high	Normal mechanical	Normal mechanical	Often high	Probably high

*There is an effective reduction in statical stability, since allowance must be made for the possibility of the system stalling with the weight all on one side.
†These systems have not been developed beyond the experimental stage and the cost comparison is based on general consideration.

The value of $d\phi/dn$ can be derived from model or full-scale experiments by noting successive amplitudes of roll as roll is allowed to die out naturally in otherwise still water. These amplitudes are plotted to base n (i.e. the number of swings) and the slope measured at various points to give values of $-d\phi/dn$ at various values of ϕ. (See Figure 7.74 (a), (b) and (c).)

In most cases, it is adequate to assume that $d\phi/dn$ is defined by a second order equation. That is

$$-\frac{d\phi}{dn} = a\phi + b\phi^2 \qquad (7.168)$$

or

$$-\frac{1}{\phi}\frac{d\phi}{dn} = a + b\phi \qquad (7.169)$$

By plotting $(1/\phi) \, d\phi/dn$ against ϕ as in Figure 7.74(d), a straight line can be drawn through the experimental results to give values of a and b.

Considering forcing a roll by moving weight, the maximum amplitude of roll would be built up if the weight could be transferred instantaneously from the depressed to the elevated side at the end of each swing as shown in Figure 7.75.

The change in potential energy of the weight at each transfer is $wb \sin \phi$. Hence, approximately

$$wb\phi = \Delta \text{GM} \phi \left(-\frac{d\phi}{dn} \right)$$

or

$$-\frac{d\phi}{dn} = \frac{w}{\Delta} \frac{b}{\text{GM}} \qquad (7.170)$$

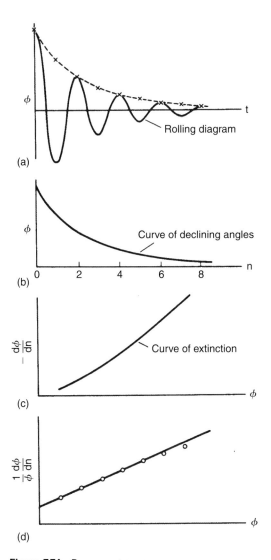

Figure 7.74 Decrement curve.

It follows that the moving weight can increase the roll amplitude up to the value appropriate to this value of $d\phi/dn$.

7.10.2.4 Fin stabilizers: Design procedure

(A) Design procedure

There are a number of references that provide a background to the use, design and operation of fin stabilizers, including Allan (1945), Conolly (1969), Gunsteren (1974), Lloyd (1975, 1977, 1998), Cox and Lloyd (1977), Fairlie-Clarke (1980) and Dallinga (1994).

The basic consideration is to provide the necessary fin force and couple to oppose the rolling moments applied by the waves to the ship. The stabilizing moment should be equal to the wave heeling moment and opposite in phase. A parameter suitable for use at the design stage for checking fin area and specifying stabilizer power is the *waveslope capacity*, Fairlie-Clarke (1980)

If a ship is heeled to an angle ϕ, the restoring moment is $\Delta gGZ = \Delta gGM \cdot \phi$, where Δ is the ship displacement and GM the metacentric height, Figure 7.76.

The stabilising moment of the fins is $2L_F R$, Figure 7.77.

Thus, if ϕ is seen as the wave slope, Figure 7.78 and $\Delta gGM \cdot \phi$ the couple applied by the wave, then the stabilising moment has to match this couple and

$$2L_F R = \Delta gGM\phi$$

and

$$\phi = 2L_F R \times 57.3/\Delta gGM \tag{7.171}$$

where ϕ, expressed in degrees, is commonly referred to as the *waveslope capacity* of the stabilizer. This criterion represents the approximate maximum sea

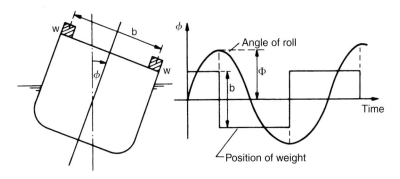

Figure 7.75 Instantaneous weight transfer.

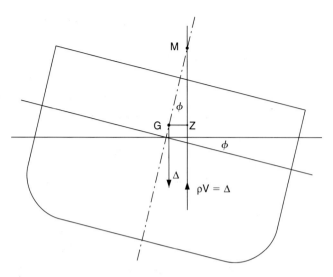

Figure 7.76 Equivalent wave couple.

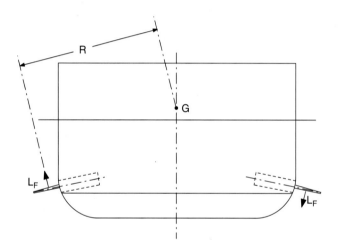

Figure 7.77 Stabilizing moment.

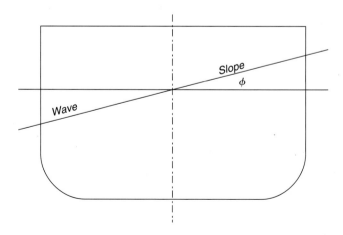

Figure 7.78 Wave slope.

state in which the stabilizer can operate effectively. It can also be considered as the angle of steady heel that the fins can create with the ship moving ahead in calm water at a given speed.

The required waveslope capacity will typically lie between 3° and 5°, depending on size and type of ship and its roll period, Fairlie-Clarke (1980). Gunsteren (1974) suggests 3°–4° for large ships and 4°–5° for small ships. Approximate waveslopes, for the Pierson–Moskowitz wave spectrum, for given ship size, are given in Table 7.12 which was assembled from data in Fairlie-Clarke (1980). Fairlie-Clarke points out that other spectra may have higher average waveslopes and that the ship will not respond in full to waveslopes such as those in Table 7.12 due to dynamic effects and entrained water which will reduce the static heeling force by about 20–25% for a ship with bilge keels.

Equation (7.171) illustrates the importance of GM in the stabilizer fin design process. For a given design waveslope, a larger GM will require a larger restoring moment and fin force. This will generally mean larger fins and fin power. Thus from the point of view of stabilizer fin design, a small GM is desirable, subject to the requirements of overall ship stability and safety.

Table 7.12 Approximate average RMS wave slopes (degrees)

Ship breadth (m)	Sea state				
	3	4	5	6	7
10	2.0	3.5	4.8	5.6	6.5
20	0.9	2.2	3.5	4.4	5.2
30	0.5	1.5	2.6	3.5	4.5

(B) Design data and process

Given the dimensions and displacement of the ship, its GM and the position of the stabilizers then, for an assumed wave slope ϕ, the required force per fin L_F can be found using Equation (7.171). The geometry and working incidence of the fin will be designed to deliver this lift. Free-stream lifting surface data suitable for fin design are discussed in Section 8.19.

(i) Location

Stabilizer fins should preferably be located near amidships to avoid breaking the surface when pitching, although within 20% L of amidships is also likely to be acceptable. The fin should preferably be at right angles to the local hull to avoid shaping the root of the stabilizer fin to clear the hull when at incidence, Figure 7.79, leading to significant root gap effect. For reasons of cost, maintenance and convenience a number of small ships have nonretractable fins. Warships tend to have nonretractable fins as they have greater immunity to damage from shock and explosion, Lloyd (1998). For safety, docking and port operation nonretractable fins need to be confined to a rectangle formed by the baseline and ship breadth, Figure 7.80. This will restrict the span of the fin and, for a given required area, lead to a smaller aspect ratio. Lloyd (1998) points out that with limited span, there is little advantage in increasing the chord to decease the effective aspect ratio to less than about 1.0. With such limited spans, the effective aspect ratio is typically from about 1.0 to 1.3. Further, lift can only be achieved by fitting more pairs of fins. Retractable fins do not suffer this restriction and larger aspect ratios with a higher hydrodynamic efficiency can be used. Retractable fins may be mounted higher and be nearly horizontal, typically at about 15° to the horizontal, without much loss in lever, Figure 7.77.

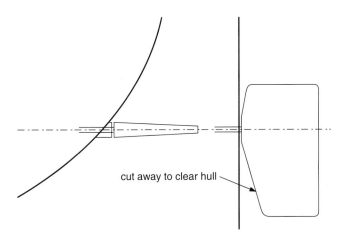

cut away to clear hull

Figure 7.79 Root gap effects with horizontal fin axis.

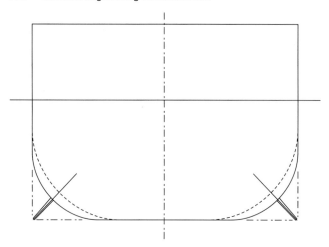

Figures 7.80 Possible stabilizer size limits (non-retractable).

In this case, hull shape and root gap losses can also be minimized.

The required level of restoring moment, or limits on lift due to aspect ratio, may be such that more than one pair of fins is necessary. It should be noted that interference effects occur between fins in line, one behind another. The downwash and upwash from the upstream fin influences the effective incidence of the downstream fin and the lift forces produced. This aspect is addressed further by Allan (1945) and Lloyd (1977, 1998).

It has been pointed out by Lloyd (1975) and Dallinga (1994) that when measured experimental data have been less than predicted, this may be due to possible effects of the free surface, cavitation, body–hull interference, boundary layer and root gap.

(ii) Design speed
Design speed will normally be the service speed. There may, however, be cases where the design speed is taken as the maximum speed achievable, or some lower cruise speed at which the ship may spend much of its time.

(iii) Influence of the ship boundary layer
The presence of the boundary layer leads to a reduction in the effective fin area. This reduction will clearly be larger with smaller aspect ratio fins.

An estimate of the boundary layer thickness at the position of the stabilizers may be made using a suitable equation for the thickness of a turbulent boundary layer:

$$\delta = x \times 0.370 \, Rn_x^{-1/5}$$

where x is the distance from the fore end and Rn_x is the Reynolds number based on x, and using a suitable equation for the displacement thickness:

$$\delta^* = x \times 0.0463 \, Rn_x^{-1/5}$$

A velocity correction can be made across the boundary layer thickness based on a power law turbulent velocity distribution. The stabilizer lift L_F (or drag) may then be calculated over the boundary layer region as

$$L_F = C_L \, \tfrac{1}{2} \rho c \int u^2 d\delta$$
$$= C_L \, \tfrac{1}{2} \rho c \int [V(y/\delta)^n]^2 d\delta \qquad (7.172)$$

An alternative approximate approach is to use the displacement thickness to correct the span, area and aspect ratio and to apply these to the database and coefficients.

(iv) Influence of hull and waves on flow speed
The water speed in the outer flow (outside boundary layer) changes as it flows around the hull with speeds up to 5–10% higher than ship speed around the bilge region where the stabilizers are located. Such speeds have been assessed for calm-water conditions, but the changes in flow speed and direction will be considerably more complicated when the ship is rolling.

In a similar manner, the sub-surface orbital motion of the waves producing the roll, Section 1.5.3, will lead to changes in the flow speed and direction. Again, these are complicated and difficult to quantify.

(v) Influence of adjacent hull shape and root gap
This can be assessed in a similar manner to the derivation of effective aspect ratio for rudders. An

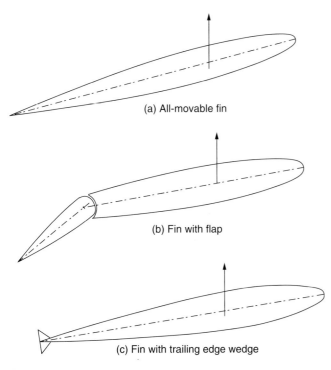

Figure 7.81 Fin stabilizer section types.

approximate equation for the aspect ratio factor k, covering a range of effective hull shapes is defined as

$$k = 2 - 0.016\,\alpha$$

where $k = \mathrm{AR}/\mathrm{AR}_G$

Hence at say 15° incidence, $k = 1.76$ and for say $AR = 4.0$, the lift curve slope is 94% of the reflection plane lift curve slope, leading to an approximate loss in lift of 6%. At 25°, the loss in lift is about 10%.

With say a gap/chord ratio of 0.01 (a gap of 10 mm for a chord of 1.0 m), $k = 1.90$ and the loss in lift is only about 2.5%.

(C) Section design

(1) Section shape

Sections used for fin stabilizers tend to be all-movable, or high lift such as a flapped foil or a foil with a wedge at the trailing edge, Figure 7.81. The all-movable section is normally used for non-retractable fins. For retractable or folding fins the high-lift devices tend to be employed in that they provide a higher lift curve slope, hence faster response to change in incidence, a delay in stall and larger stall angle and a larger maximum lift coefficient. The differences between the performance of the flapped foil and foil with trailing edge wedge tend to be marginal. The flapped foil tends to be a little more efficient in terms of lift/drag ratio, whilst the foil with trailing edge wedge has no moving parts to manufacture and maintain.

(2) Lift and drag

Lift curve slope, stall angle and maximum lift for a given fin aspect ratio and incidence can be derived from free stream data and corrections to aspect ratio carried out.

Drag on the fins may also be derived from the data. The drag of the fins is important in that it can lead to a loss in ship speed.

Minimizing the drag of non-retractable fins is important in that they produce a parasitic drag even in calm water. Typically, all-movable fins with NACA00 type sections will be employed and the thickness/chord ratio will be minimized within the limits of stock diameter and structural integrity.

As mentioned earlier, retractable fins tend to be of the high-lift type including flapped foils and sections with a wedge at the trailing edge. These both tend to have higher drag at a given incidence than the all-movable equivalent, Molland and Turnock (2007). When not in use, drag is not a problem as the fins are retracted or folded away. The actual drag on the fins can be quantified in the normal way, using drag coefficients from the database. The influence of this drag on ship speed is difficult to quantify as the stabilizers reduce roll and hence the ship resistance due to roll. Thus, the drag due to the stabilizers is offset to a certain extent by the reduction in ship resistance due to roll. These aspects of speed loss due to stabilizers are discussed in more depth by Allan (1945) and Dallinga (1994).

(3) Centre of pressure, torque and stock diameter
CPc and CPs for the relevant section and aspect ratio and incidence can be obtained from free-stream data. In order to minimize torque, but preclude negative torques, the stock will normally be located a little forward of the estimated forwardmost position of CPc. Knowing the lift, drag and centre of pressure, the torque, bending moment and stock diameter can be calculated using Equations (8.60–8.71) (see Chapter 8).

(D) Cavitation
A cavitation check can be carried out using the principles applied to rudders. Allan (1945) carried out cavitation tunnel tests on stabilizer sections and the results and limiting data are, as expected, similar to rudder data.

The cavitation check entails the calculation of the appropriate cavitation number and the use of cavitation inception curves for a given lift coefficient, Figure 5.82. Cavitation number σ is defined as

$$\sigma = \frac{(P_{AT} + \rho gh - P_v)}{0.5\rho V^2} \quad (7.173)$$

It should be noted that the depth of immersion h is likely to be less than that for rudders, particularly when taking into account the reduction in ρgh due to roll.

Cavitation is unlikely to be present at ship speeds less than about 20 knots. Overall, the maximum design and operational lift may be limited by cavitation in larger faster vessels, whilst in slower speed craft it will be limited by stall. Section shapes that delay the onset of cavitation may also be considered.

(E) Operation
(a) Control
Maximum fin angle will normally be limited by the occurrence of stall and/or cavitation. A maximum lift may be specified based on stall, cavitation, torque, or materials and structure. The fin controller will be programmed in such a way that this maximum is not exceeded. The controller, which may for example include feedback from a force measurement on the stock, will limit fin angle as necessary to avoid stall, cavitation, or some prescribed maximum lift. Some fundamentals of stabilizer control are discussed in Allan (1945) and Conolly (1969) and more recently by Perez (2005).

(b) Sway and yaw effects
It can be seen from Figure 7.77 that the stabilizer force L_F will also have vertical and horizontal components. Whilst the vertical components effectively cancel, there is a net horizontal sway force that will induce a yaw motion. This is to port in Figure 7.77 but will oscillate from port to starboard as the fin incidence is reversed. This yawing effect is generally not a problem on most ships, but will depend on the location of the fins relative to the ship LCG, a large lever increasing the yaw moment, and the directional stability of the ship, Section 8.2.

(c) Dynamic effects
The static approach to the design of fin stabilizers described neglects the dynamic behaviour of the ship and stabilizers including, for example, the roll-induced angle of attack on the fins and the influence of non-stationary motion on the lift characteristics. For a more detailed estimate of overall performance, a more rigorous and complete approach can be used which will include the application of the equation of motion for roll, with the stabilizer fins increasing the motion damping term. Such an approach is described in Conolly (1969), Gunsteren (1974) and Lloyd (1975).

(F) Roll stabilization with rudders
The heel angle produced by the rudder is described in Section 8.8. This effect can be used to produce a stabilizing moment in roll and the rudder used as a roll stabilizer. A number of investigations have been carried out to determine the efficacy of such an approach, such as those described by Cowley and Lambert (1972), Carley (1975), Cowley and Lambert (1975) and Lloyd (1975). The use of a lateral force estimator (LFE), using the lateral acceleration, to control the rudder as a stabilizer is described by Tang and Wilson (1992). A thorough review of the use of the rudder as a roll stabilizer together with suitable controllers is given by Perez (2005).

Restricted rudder angles generally have to be applied to avoid the effects of coupling roll, sway and yaw. It is suggested by Lloyd (1977) that using the rudder may be acceptable for lower-speed merchant ships but may be unsuitable for high-speed ships, as in low frequency following and quartering seas the rudder might be expected to amplify the roll motion. The use of the rudder as a roll stabilizer seems to have received only limited practical application.

(G) Roll stabilization at rest
An interesting development is stabilization by fins whilst the vessel is at rest, The Naval Architect (2006). Available fin incidence is increased to 40° from

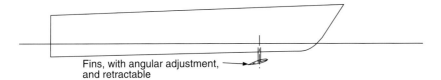

Figure 7.82 Pitch-damping fins.

the more normal maximum of 20°–25°, with fin action responding to roll sensors/controllers. Such systems are finding applications in smaller vessels such as large motor yachts, small ferries and offshore vessels.

7.10.3 Pitch damping

7.10.3.1 *Pitch damping fins*

(A) Applications

Fins are used to control the pitch of some higher speed craft, Figure 7.82. The energy in pitch is much greater than the energy in roll and the required forces and moments to stabilize pitch are much larger than those required for roll. Hence pitch damping or stabilization is generally not a practical proposition for large ships. For example, Conolly and Goodrich (1970) carried out full-scale sea trials on fixed anti-pitching fins fitted to a coastal minesweeper. They concluded that it was doubtful whether fixed fins would ever provide sufficient attenuation of motion to justify their installation in any ship. Successful stabilization in pitch tends to be limited to smaller faster semi-displacement vessels, and some success with ride control has been achieved using variable incidence lifting foils situated near the fore end, variable incidence stern flaps (trim tabs), adjustable interceptors and combinations of these various devices.

(B) Design procedure and data

Pitch-damping foils tend to take the form of a flapped foil, Figure 7.83 where the forward part can act as a fixed part and provide a suitable connection for one or two vertical supporting struts. The struts/foils may be retractable. The design procedure is similar to that for the roll stabilizer fin and, again, use can be made of the extensive data available for all-movable and flapped control surfaces, Molland and Turnock (2007).

An approximate approach is to relate the required stabilizing moment to the longitudinal hydrostatic restoring moment. The hydrostatic restoring moment for a trim angle θ is $\Delta g GM_L \theta = \Delta g GM_L \times \text{trim}/L$.

[In a similar manner, the moment to change trim 1 cm is $\Delta g GM_L / 100 L$].

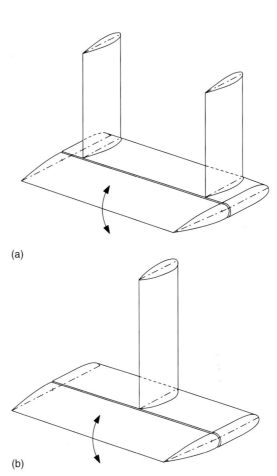

Figure 7.83 Pitch-damping foils.

The stabilizing moment of the fins is $L_{FP} \times R_P$, Figure 7.84, where $L_{FP} = C_L \times \frac{1}{2} \rho A V^2$.

If θ is considered to be the pitch angle and $\Delta g GM_L \theta$ the couple applied by the pitch motion, then the stabilizing moment must match this couple and

$$L_{FP} \times R_P = \Delta g GM_L \theta \qquad (7.174)$$

θ can be considered as the angle of steady pitch that the foil(s) can create with the ship moving ahead in

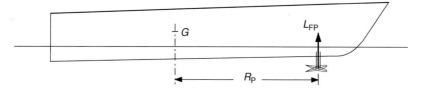

Figure 7.84 Pitch-stabilizing moment.

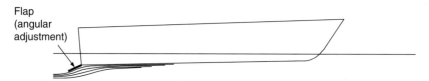

Figure 7.85 Transom flaps.

calm water. This approach tends to suggest the need for a very large control surface area.

An alternative approach is to carry out a ship motion simulation using say strip theory, Section 7.7.2, and to determine a level of damping that will have a useful effect in decreasing the pitch motion and vertical accelerations. The pitch-damping foils then need to be sized whereby the forces and moments provide an adequate level of damping.

Haywood and Benton (2005) describe the use of lifting foils and their application also to ride control of a high-speed trimaran. The lifting foils, situated at amidships, lift approximately one-third of the mass of the vessel. They are also capable of changing angle of attack due to vessel motion and form part of the ride control system. The transom also has an active transom mounted tab, linked to the ride control system. Full-scale trials demonstrated a high level of damping and good seakeeping performance. Davis et al. (2003) describe the theoretical prediction of motions and full-scale measurements on a high-speed catamaran ferry fitted with T-foils and stern flaps. Various investigations into ride control are reported in Davis and Holloway (2003), Xi and Sun (2005), Katayama et al. (2003), Folsø et al. (2003) and Doctors (2004).

(C) Operation

Like the roll stabilizer fin, the incidence of the pitch-damping foil may be limited by stall or cavitation and the fin controller will be programmed in such a way that this maximum is not exceeded. Both these characteristics may be examined in a manner similar to that for a roll stabilizer fin, Section 7.10.2.4.

7.10.3.2 *Transom flaps*

Adjustable flaps, or trim tabs, at the bottom of the transom, Figure 7.85, are employed on high-speed semi-displacement and planing craft to adjust the running trim and minimize the resistance to forward motion. They may also be used to provide pitch or ride control and may be used in conjunction with pitch-damping fins forward. Flap angle is typically changed using hydraulic actuators. Optimum flap angle will change with speed. Dissimilar flap angles port and starboard can be used to create a horizontal turning or steering moment on the hull. Further information on the design and applications of stern flaps is included in Cusanelli and Karafiath (2001), Cusanelli (2003) and Tsai et al. (2003).

7.10.3.3 *Interceptors*

These are adjustable vertical plates at the bottom of the transom, normally one port and one starboard or one in each hull in the case of a catamaran, Figure 7.86.

Vertical movement is usually carried out using hydraulic actuators. Vertical adjustment can have an effect similar to a stern flap, providing adjustment to the running trim. They are also used on faster semi-displacement craft to provide pitch or ride control and may be used in conjunction with pitch-damping fins forward. Dissimilar vertical adjustment port and starboard leads to dissimilar horizontal forces on the interceptors and a horizontal turning or steering moment on the hull. Interceptors may be used as well as, or instead of, flaps. A typical application of interceptors has been to supplement the steering of vessels propelled by waterjets. Further information

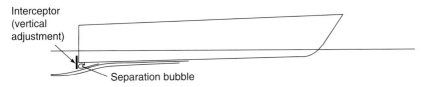

Figure 7.86 Interceptors.

on the design and applications of interceptors is included in Tsai *et al.* (2003), Katayama *et al.* (2003) and Brizzolara (2003).

References (Chapter 7)

Allan, J.F. (1945). The stabilisation of ships by active fins. *Transactions of the Royal Institution of Naval Architects*, Vol. 87, pp. 123–159.

Arndt, B., Brandl, H. and Vogt, K. (1982). 20 Years of experience – stability regulations of the West-German Navy. *Proc. of STAB'82: 2nd International Conference on Stability of Ships and Ocean Vehicles*, Tokyo, p.765.

Arribas, F.P. and Fernandez, J.A.C. (2006). Strip theories applied to vertical motions of high-speed craft. *Ocean Engineering*, Vol. 33.

Bales, N.K. (1980). Optimising the seakeeping performance of destroyer-type hulls. *Proceedings of 13th Symposium on Naval Hydrodynamics*, Tokio.

Beck, R. and Reed, A. (2001). Modern seakeeping computations for ships. *Proceedings of 23rd Symposium on Naval Hydrodynamics*. Val de Reuil, France.

Bertram, V. (2000). *Practical Ship Hydrodynamics*. Butterworth-Heinemann, Oxford, UK.

Bertram, V. and Yasukawa, H. (1996). Rankine source methods for seakeeping problems. *Jahrbuch Schiffbautechnische Gesellschaft*, pp. 411–425, Springer.

Bertram, V. and Yasukawa, H. (2001). Investigation of global and local flow details by a fully three-dimensional seakeeping method. *Proceedings of 23rd Symposium on Naval Hydrodynamics*. Val de Reuil, France.

Bhattacharyya, R. (1978). *Dynamics of marine vehicles*. Wiley, New York.

Biran, A. (2003). *Ship Hydrostatics and Stability*. Butterworth-Heinemann, Oxford, UK.

Bishop, R.E.D. and Price, W.G. (1979). *Hydroelasticity of Ships*. Cambridge University Press, London.

Blok, J.J. and Beukelman, W. (1984). The high-speed displacement ship systematic series hull forms – seakeeping characteristics. *Trans. SNAME*, Vol. 92.

Blume, P. and Hattendorf, H. G. (1982). An investigation on intact stability of fast cargo liners, *Proc. of STAB'82: 2nd International Conference on Stability of Ships and Ocean Vehicles*, Tokyo.

Blume, P. and Hattendorf, H.G. (1984). Ergebnisse von systematischen Modellversuchen zur Kentersicherheit. *Transactions STG*, Vol. 78.

Brizzolara, S. (2003). Hydrodynamic analysis of interceptors with CFD methods. Proceedings of *Seventh International Conference on Fast Sea Transportation, FAST'2003*, Ischia, Italy, October.

Bruzzone, D., Gualeni, P. and Sebastiani, L. (2001). Effects of different three-dimensional formulations on the seakeeping computations of high speed hulls. *Proceedings of 8th International Symposium on Practical Design of Ships and Other Floating Structures, PRADS'2001*. Vol. 1. Shanghai, China.

Carley, J.B. (1975). Feasibility study of steering and stabilising by rudder. *Fourth Ship Control Systems Symposium*, The Hague.

Chang, B.C. and Blume, P. (1998). Survivability of damaged roro passenger vessels. *Ship Technology Research*, Vol. 45, pp. 105–117.

Chapman, R.B. (1975). Free-surface effects for yawed surface piercing plates. *J. Ship Research*, Vol. 20, pp. 125–132.

Chuang, S.L. (1967). Experiments on slamming of wedge-shaped bodies. *J. Ship Research*, Vol. 11, pp. 190–198.

Clauss, G.F. (2000). Tailor-made transient wave groups for capsizing tests. *7th International Conference on Stability of Ships and Ocean Vehicles*. February, 2000.

Conolly, J.E. (1969). Rolling and its stabilisation by active fins. *Transactions of the Royal Institution of Naval Architects*, Vol. 111, pp. 21–48.

Conolly, J.E. and Goodrich, G.J. (1970). Sea trials of anti-pitching fins. *Transactions of the Royal Institution of Naval Architects*, Vol. 112, pp. 87–100.

Cowley, W.E. and Lambert, T.H. (1972). The use of the rudder as a roll stabiliser. *Third Ship Control Systems Symposium*, Bath, UK.

Cowley, W.E. and Lambert, T.H. (1975). Sea trials on a roll stabiliser using the ship's rudder. *Fourth Ship Control Systems Symposium*, The Hague.

Cox, G.G. and Lloyd, A.R.J.M. (1977). Hydrodynamic design basis for navy ship roll motion stabilisation. *Transaction of the Society of Naval Architects and Marine Engineers*, Vol. 85, pp. 51–93.

Cusanelli, D.S. (2003). Stern flap: An economic fuel-saving, go-faster, go-further device for the commercial vessel market. Proceedings of *Seventh International Conference on Fast Sea Transportation, FAST'2003*, Ischia, Italy, October.

Cusanelli, D.S. and Karafiath, G. (2001). Advances in stern flap design and application. Proceedings of *Sixth International Conference on Fast Sea Transportation, FAST'2001*, Southampton, UK, September.

Czichos, H. (ed.) (1991). *Hütte. Die Grundlagen der Ingenieurwissenschaften*, 29th ed. Springer-Verlag, New York, Berlin, Heidelberg.

Dallinga, R.P. (1994). Hydromechanic aspects of the design of fin stabilisers. *Transactions of the Royal Institution of Naval Architects*, Vol. 136, pp. 189–200.

Davis, M. and Holloway, D. (2003). Effect of sea, ride controls, hull form and spacing on motion and seasickness incidence for high speed catamarans. Proceedings of *Seventh International Conference on Fast Sea Transportation*, FAST 2003, Ischia, Italy, October.

Davis, M.R., Watson, N.L. and Holloway, D.S. (2003). Wave response of an 86 m high speed catamaran with active T-Foils and stern tabs. *Transactions of the Royal Institution of Naval Architects*, Vol. 145, pp. 87–106.

Doctors, L.J. (2004). Theoretical study of the tradeoff between stabiliser drag and hull motion. *Transactions of the Royal Institution of Naval Architecture*, Vol. 146, pp. 289–298.

Fairlie-Clarke, A.C. (1980). Fin stabilisation of ships. *The Naval Architect*, Published by RINA, London, January, pp. 10–11.

Folsø, R., Nielsen, U.D. and Torti, F. (2003). Ride control systems – Reduced motions on the cost of increased sectional forces. Proceedings of *Seventh International Conference on Fast Sea Transportation, FAST '2003*, Ischia, Italy, October.

Goodrich, G.J. (1969). Development and design of passive roll stabilisers. *Trans. RINA*, Vol. 111.

Graham, R., Baitis, A.E. and Meyers, W.G. (1992). On the development of seakeeping criteria. *Naval Engineering Journal*.

Gunsteren, F.F. Van (1974). Analysis of roll stabiliser performance. *International Shipbuilding Progress*, Vol. 21, No. 237, pp. 125–146.

Haywood, A.J. and Benton, H.S. (2005). The integration of lifting foils into ride control systems for fast ferries. Proceedings of *Eighth International Conference on Fast Sea Transportation, FAST '2005*, St. Petersburg, Russia, June.

Hogben, N., Dacunha, N. M. C. and Oliver, G.F. (1986). Global wave statistics. Compiled and edited by British Maritime Technology Ltd, Unwin Brothers.

Hudson, D.A., Molland, A.F., Price, W.G. and Temarel, P. (2001). Seakeeping performance of high speed catamaran vessels in head and oblique waves. *Proceedings of 6th International Conference on Fast Sea Transportation, FAST'2001*, Southampton, UK.

IMO (2000). International Code of Safety for High Speed Craft. HSC Code 2000. MSC 97/73.

ISO (1985). ISO 2631 Evaluation of exposure to whole-body z-axis vertical vibration in the frequency range 0.1 to 0.63 hz, ISO 3631/3-1985.

ISO (1997). ISO 2631 Mechanical vibration and shock-evaluation of human exposure to whole-body vibration – Part 1: General requirements.

ISSC (1994). In N.E. Jeffery and A.M. Kendrick (eds), *Int. Ship and Offshore Structures Congress*. St. John's.

ITTC (2002). Report of the Loads and Responses Committee. *Proceedings of 23rd International Towing Tank Conference*, Vol. 1. Venice, Italy. Published by INSEAN, Rome.

ITTC (2005). Report of the Seakeeping Committee. *Proceedings of 24th International Towing Tank Conference*, Vol.1. Edinburgh, UK. Published by The University of Newcastle upon Tyne, UK.

Jens, J. (1964). Stabilitäts- und Rollschwingungsuntersuchungen mit Küstenmotorschiffen. *Hansa*, Vol. 101, No. 14, p. 1419, Jahrgang, Hamburg.

Karayannis, T. and Molland, A.F. (2003). Technical and economic investigations of fast ferry operations. *Proc. of Seventh International Conference on Fast Sea Transportation, FAST'2003*, October, Ischia, Italy.

Kashiwagi, M. (1997). Numerical seakeeping calculations based on slender ship theory. *Ship Technology Research*, Vol. 44, pp. 167–192.

Kastner, S. (1969). Das Kentern von Schiffen in unregelmäßiger längslaufender See Part 1. *Schiffstechnik*, Vol. 16, Hamburg.

Kastner, S. (1970). Das Kentern von Schiffen in unregelmäßiger längslaufender See Part 2. *Schiffstechnik*, Vol. 17, Hamburg.

Kastner, S. (1970a). Hebelkurven in unregelmäßiger See. *Schiffstechnik*, Vol. 17, Hamburg.

Kastner, S. (1975). "On the statistical precision of determining the probability of capsizing in random seas" *Proc. of STAB '75: 1st International Conference on Stability of Ships and Ocean Vehicles*, Glasgow.

Katayama, T., Suzuki, K. and Ikeda, Y. (2003). A new ship motion control system for high speed craft. Proceedings of *Seventh International Conference on Fast Sea Transportation*, FAST '2003, Ischia, Italy, October.

Kobylinski, L.K. and Kastner, S. (2003). *Stability and Safety of Ships*. Elsevier, Oxford, UK.

Korobkin, A. (1996). Water impact problems in ship hydrodynamics. *Advances in Marine Hydrodynamics*, Comp. Mech. Publ. pp. 323–371.

Kwon, Y.J. (2000). Estimating the effect of wind and waves on ship speed and performance. *The Naval Architect*, September, RINA, London.

Lawther, A. and Griffin, M.J. (1986a). Prediction of the incidence of motion sickness from the magnitude, frequency and duration of vertical oscillation. *Journal of the Acoustical Society of America*, Vol. 82.

Lawther, A. and Griffin, M.J. (1986b). The motion of a ship and the consequent motion sickness amongst passengers. *Ergonomics*, Vol. 29, No. 4.

Lawther, A. and Griffin, M.J. (1988). Motion sickness and motion characteristics of vessels at sea. *Ergonomics*, Vol. 31, No. 10.

Lewis, E.V. (ed.) (1989). *Principles of Naval Architecture*, Second Revision. SNAME.

Lloyd, A.RJ.M. (1975). Roll stabilisation by rudder. *Fourth Ship Control Systems Symposium*, The Hague.

Lloyd, A.R.J.M. (1975). Roll stabiliser fins: A design procedure. *Transactions of the Royal Institution of Naval Architects*, Vol. 117, pp. 233–254.

Lloyd, A.R.J.M. (1977). Recent developments in roll stabilisation. *The Naval Architect*, published by RINA, London, March, pp. 44–45.

Lloyd, A.R.J.M. (1998). *Seakeeping: Ship Behaviour in Rough Weather.* Published by the Author, Gosport, UK.

Maury, C., Delhommeau, G., Ma, M. and Guilbaud, M. (2003). Comparison between numerical computations and experiments for seakeeping on ship models with forward speed. *Journal of Ship Research*, Vol. 47, No. 4.

McGeorge, H.D. (1995). *Marine Auxiliary Machinery*, 7th edition. Butterworth-Heinemann, Oxford, UK.

Mizoguchi, S. and Tanizawa, K. (1996). Impact wave loads due to slamming – A review. *Ship Technology Research*, Vol. 43, pp. 139–154.

Molland, A.F., Wellicome, J.F., Temarel, P., Cic, J. and Taunton, DJ. (2001). Experimental investigation of the seakeeping characteristics of fast displacement catamarans in head and oblique seas. *Trans. RINA*, Vol. 143.

Molland, A.F. and Turnock, S.R. (2007). *Marine Rudders and Control Surfaces.* Butterworth-Heinemann, Oxford, UK.

Newman, J.N. (1965). The exciting forces on a moving body in waves. *J. Ship Research*, Vol. 10, pp. 190–199.

Newman, J.N. (1978). The theory of ship motions. *Adv. Appl. Mech.*, Vol. 18, pp. 222–283.

Nordforsk Report (1987). Assessment of ship performance in a seaway. *The Nordic Co-operative Project on Seakeeping Performance of Ships*, Nordforsk.

O'Hanlon, J.F. and McCauley, M.E. (1974). Motion sickness incidence as a function of the frequency and acceleration of vertical sinusoidal motion. *Aerospace Medicine*, Vol. 45.

Pereira, R. (1988). Simulation nichtlinearer Seegangslasten. *Schiffstechnik*, Vol. 35, pp. 173–193.

Perez, T. (2005). *Ship Motion Control. Course Keeping and Roll Stabilisation using Rudder and Fins*. Springer.

Pierson, W. J. (1993). "Ship stability in heavy weather: the real situation and models thereof", *Proc. of US Coast Guard Vessel Stability Symposium*, US Coast Guard Academy, New London, Connecticut, pp. 79–93.

Rawson, K.J. and Tupper, E.C. (2001). *Basic Ship Theory*. 5th Edition, Combined Volume. Butterworth-Heinemann, Oxford, UK.

Russo, V.L. and Sullivan, E.K. (1953). Design of the Mariner type ship. Transactions of the Society of Naval Architects and Marine Engineers, Vol. 61, pp. 98–151.

Salvesen, N., Tuck, E.O. and Faltinsen, O. (1970). Ship motions and sea loads. *Trans. SNAME*, Vol. 78, pp. 250–287.

Satchwell, C.J. (1989). Windship technology and its applications to motor ships. *Trans. RINA*, Vol. 131.

Skalicky, A. (1998). "Seegangsspektrum", Assignment, Hochschule Bremen.

Smith, T.C. and Thomas, W.L. III. (1989). A survey and comparison of criteria for naval missions. David Taylor Research Centre, Ship Hydrodynamics Department, Report DTRC/SHD-1312-01.

Söding, H. (1987). Ermittlung der Kentergefahr aus Bewegungssimulationen. *Schiffstechnik*, Vol. 34, pp. 28–39.

SSC (1995). *Hydrodynamic impact on displacement ship hulls*. Ship Structure Committee, SSC-385, SSC-385A.

St. Denis, M. and Pierson, W.J. (1953). On the motions of ships in confused seas. *Transactions SNAME*, Vol. 61.

Tang, A. and Wilson, P.A. (1992). LFE Stabilisation using the rudder. Proceedings of Conference, *Manoeuvring and Control of Marine Craft*, MCMC '92, Computational Mechanics Publications, 475–492.

Tanizawa, K. and Bertram, V. (1998). Slamming. *Handbuch der Werften* XXIV, Hansa-Verlag, pp. 191–210

The Naval Architect (2006). Stabilization at rest: a new solution. Royal Institution of Naval Architects, London, September, p. 96.

Thode, H. (1965). Rollzeitmessung und Stabilität. *Hansa*, Vol. 102, No. 18, Hamburg.

Townsin, R.L. and Kwon, Y.J. (1983). Approximate formulae for the speed loss due to added resistance in wind and waves. *Trans. RINA*, Vol. 125.

Townsin, R.L., Kwon, K.J., Baree, M.S. and Kim, D.Y. (1993). Estimating the effect of weather on ship performance. *Trans. RINA*, Vol. 135.

Tsai, J.F., Hwang, J.L. and Chou, S.K. (2003). Study on the compound effects of interceptors with stern flap for two fast monohulls with transom stern. Proceedings of *Seventh International Conference on Fast Sea Transportation, FAST'2003*, Ischia, Italy, October.

Weiss, G. (1953). Erfahrungen mit der Stabilitätsprüfung durch Rollversuche. *Hansa*, Vol. 90, Hamburg.

Wendel, K. (1965). Bemessung und Überwachung der Stabilität. *Transactions STG*, Hamburg.

Xi, H. and Sun, J. (2005). Effects of actuator dynamics on stabilisation of high-speed planing vessels with controllable transom flaps. Proceedings of *Eighth International Conference on Fast Sea Transportation*, FAST '2005, St. Petersburg, Russia, June.

Yun, L. and Bliault, A. (2000). *Theory and Design of Air Cushion Craft*. Elsevier Butterworth-Heinemann, Oxford, UK.

8 Manoeuvring

Contents

8.1 General concepts
8.2 Directional stability
8.3 Stability and control of surface ships
8.4 Rudder action
8.5 Limitations of theory
8.6 Assessment of manoeuvrability
8.7 Loss of speed on a turn
8.8 Heel when turning
8.9 Turning ability
8.10 Standards for manoeuvring and directional stability
8.11 Dynamic positioning
8.12 Automatic control systems
8.13 Ship interaction
8.14 Shallow water/bank effects
8.15 Broaching
8.16 Experimental approaches
8.17 CFD for ship manoeuvring
8.18 Stability and control of submarines
8.19 Rudders and control surfaces
References (Chapter 8)

The various Sections of this Chapter have been taken from the following books, with the permission of the authors:

Barrass, C.B. and Derrett, D.R. (2006) *Ship Stability for Masters and Mates*. 6th Edition. Butterworth-Heinemann, Oxford, UK. [Section 8.13]

Bertram, V. (2000) *Practical Ship Hydrodynamics*. Butterworth-Heinemann, Oxford, UK. [Sections 8.16, 8.17]

Molland, A.F. and Turnock, S.R. (2007) *Marine Rudders and Control Surfaces*. Butterworth-Heinemann, Oxford, UK. [Sections 8.2, 8.14, 8.19 (excluding 8.19.1.3)]

Rawson K.J. and Tupper E.C. (2001) *Basic Ship Theory*. 5th Edition, Combined Volume. Butterworth-Heinemann, Oxford, UK. [Sections 8.1, 8.3–8.12, 8.15, 8.18, 8.19.1.3]

8.1 General concepts

All ships require to be controllable in direction in the horizontal plane so that they can proceed on a straight path, turn or take other avoiding action as may be dictated by the operational situation. They must further be capable of doing this consistently and reliably not only in calm water but also in waves or in conditions of strong wind. In addition, submarines require to be controllable in the vertical plane, to enable them to maintain or change depth as required whilst retaining control of fore and aft pitch angle.

Considering control in the horizontal plane, a study of a ship's manoeuvrability must embrace the following:

(a) the ease with which it can be maintained on a given course. The term steering is commonly applied to this action and the prime factor affecting the ship's performance is her directional or dynamic stability. This should not be confused with the ship's transverse stability as discussed in Chapter 3;
(b) the response of the ship to movements of her control surfaces, the rudders, either in initiating or terminating a rate of change of heading;
(c) the response to other control devices such as bow thrusters;
(d) the ability to turn completely round within a specified space.

With knowledge on these factors the designer can ensure that the ship will be controllable; calculate the size and power of control surfaces and/or thrusters to achieve the desired standards of manoeuvrability; design a suitable control system—autopilot or dynamic positioning system; provide the necessary control equations for the setting up of training simulators.

For control in the vertical plane, it is necessary to study:

(a) ability to maintain constant depth, including periscope depth under waves;
(b) ability to change depth at a controlled pitch angle.

Submarine stability and control is dealt with in more detail in Section 8.18.

It is clear from the above that all ships must possess some means of directional control. In the great majority of cases, this control is exercised through surfaces called rudders fitted at the after end of the ship. In some cases, the rudders are augmented by other lateral force devices at the bow and, in a few special applications, they are replaced by other steering devices such as the vertical axis propeller.

It is important to appreciate that it is not the rudder forces directly in themselves that cause the ship to turn. Rather, the rudder acts as a servo-system which causes the hull to take up an attitude in which the required forces and moments are generated hydrodynamically on the hull. Rudders are fitted aft in a ship because, in this position, they are most effective in causing the hull to take up the required attitude and because they benefit from the increased water velocity induced by the propellers. At low speed, when the rudder forces due to the speed of the ship alone are very small, a burst of high shaft revolutions produces a useful side force if the propellers and rudders are in line.

In the early days of man's movements on water, directional control was by paddle as in a canoe today. That is to say, the heading was controlled by applying a force either on the port or starboard side of the craft. As vessels grew in size, the course was changed by means of an oar over the after end which was used to produce a lateral force. Later again, this was replaced by a large bladed oar on each quarter of the ship and, in turn, this gave way to a single plate or rudder fitted to the transom. The form of this plate has gradually evolved into the modern rudder. This is streamlined in form to produce a large lift force with minimum drag and with leading edge sections designed to reduce the loss of lift force at higher angles of attack. In some cases the single rudder has given way to twin or multiple rudders. See Mott (1997) and Molland and Turnock (2007) for early developments in rudder design.

8.2 Directional stability

A ship is said to be directionally stable if, when deflected from its straight-line path, by say wind or waves, it returns to a new straight-line path, although this will not necessarily be in the same direction as the original path. A high measure of directional stability will result in good coursekeeping but low ability to manoeuvre, whilst a measure of directional instability will result in poor coursekeeping but good ability to manoeuvre. A compromise will normally have to be accepted, depending on the duties of the vessel under consideration. For example, a service vessel working in a port will need good stopping and manoeuvring ability whilst a large sea-going ship that covers large distances without manoeuvring will primarily require good coursekeeping. Directional instability is not a desirable property, as repeated rudder corrections generally have to be applied to maintain a course. Also, a vessel with low directional stability will readily enter a turn but may be slow to respond to reversed rudder angle in order to leave the turn.

The directional stability of a vessel will depend on several features including the fineness or fullness of the form, fine forms having higher directional stability, the operating draught and trim with an increase in trim by the stern increasing directional stability, and the amount of deadwood or fixed skeg area aft, increases in such area leading to an increase in directional stability. Increases in directional stability lead to a decrease in

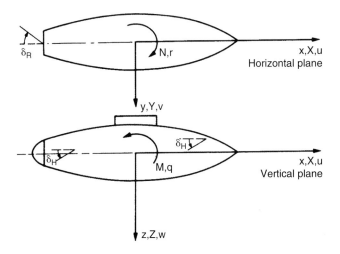

Figure 8.1 Sign convention.

manoeuvring performance and, if increases are needed, the best compromise is reached if directional stability is increased by increasing movable rudder area, if this is possible, rather than increasing fixed skeg or deadwood area, Burcher (1972).

The sign convention used in this Chapter is illustrated in Figure 8.1.

It is necessary to differentiate between 'inherent' and 'piloted' controllability. The former represents a vessel's open loop characteristics and uses the definition that when, in a given environment, a ship can attain a specified manoeuvre with some steering function, that ship is said to be manoeuvrable. This ability depends upon the environment so that some situations could arise in which the ship becomes unmanoeuvrable. Piloted manoeuvrability reflects the ability of a ship, when controlled by a human operator or an autopilot, to perform a manoeuvre such that deviations from a preset mission remain within acceptable limits. In deciding whether a ship is manoeuvrable in this sense the mission must be specified and the limits within which it is to be achieved.

8.3 Stability and control of surface ships

For a surface ship we need only consider linear motions along the x and y axes and angular motion about the z axis, the axes used being body axes. If the ship is disturbed from its straight line course in such a way that it has a small sideways velocity v it will experience a sideways force and a yawing moment which can be denoted by Y_v and N_v respectively. If this was the only disturbance, the ship would exhibit directional stability if the moment acted so as to reduce the angle of yaw and hence v. In the more general case the disturbed ship will have an angular velocity, angular and linear accelerations and will be subject to rudder actions. All these will introduce forces and moments. Considering only small deviations from a straight path so that second order terms can be neglected, the linear equations governing the motion become

$$(m - Y_{\dot{v}})\dot{v} = Y_v v + (Y_r - m)r + Y_{\delta_R}\delta_R \quad (8.1)$$

$$(I - N_{\dot{r}})\dot{r} = N_v v + N_r r + N_{\delta_R}\delta_R \quad (8.2)$$

where subscripts v, r and δ_R denote differentiation with respect to the lateral component of velocity (radial), rate of change of heading and rudder angle respectively, i.e. $Y_v = \partial Y/\partial v$, etc. Y denotes component of force on ship in y direction and N the moment of forces on ship about z-axis. m is the mass of the ship.

Equations are needed only for motion along the transverse axis and about the vertical axis as it is assumed that the ship has a steady forward speed. Put into words the equations are saying no more than that the rate of change of momentum in the y axis direction is equal to the force in that direction and that force is the sum of all such terms as (rate of change of Y with lateral velocity) × (lateral velocity).

Equations (8.1) and (8.2) can be expressed non-dimensionally by

$$(m' - Y'_{\dot{v}})\dot{v}' = Y'_v v' + (Y'_r - m')r' + Y'_{\delta_R}\delta'_R \quad (8.3)$$

$$(I' - N'_{\dot{r}})\dot{r}' = N'_v v' + N'_r r' + N'_{\delta_R}\delta'_R \quad (8.4)$$

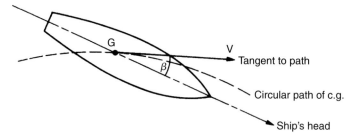

Figure. 8.2 Drift angle β.

From Figure 8.2 $v' = \dfrac{\dot{y}}{V} = -\sin\beta$ and the non-dimensional turn rate $r' = \dot{\psi}\dfrac{L}{V} = \dfrac{L}{R}$ where R is the radius of curvature of the path at that point. The coefficients Y'_v, N'_v, etc., are termed the stability derivatives.

As a typical example

$$Y'_{\delta_R} = \dfrac{1}{\tfrac{1}{2}\rho V^2 L^2} \dfrac{\partial Y}{\partial \delta_R} \tag{8.5}$$

The directional stability of a ship is related to its motion with no corrective, i.e. rudder, forces applied. In this case the equations become

$$(m' - Y'_{\dot{v}})\dot{v}' = Y'_v v' + (Y'_r - m')r' \tag{8.6}$$

$$(I' - N'_{\dot{r}})\dot{r}' = N'_v v' + N'_r r' \tag{8.7}$$

from which it follows that

$$(m' - Y'_{\dot{v}})\left[\dfrac{(I' - N'_{\dot{r}})\ddot{r}' - N'_r \dot{r}'}{N'_v}\right]$$

$$= Y'_v \left[\dfrac{(I' - N'_{\dot{r}})\dot{r}' - N'_r r'}{N'_v}\right] + (Y'_r - m')r'$$

$$(m' - Y'_{\dot{v}})(I' - N'_{\dot{r}})\ddot{r}' - [(m' - Y'_{\dot{v}})N'_r + (I' - N'_{\dot{r}})Y'_v]\dot{r}' + [N'_r Y'_v - (Y'_r - m')N'_v]r' = 0. \tag{8.8}$$

This equation is of the form

$$\left[a\dfrac{d^2}{dt^2} + b\dfrac{d}{dt} + c\right]r = 0 \tag{8.9}$$

which has as a general solution of the form

$$r' = r_1 e^{m_1 t} + r_2 e^{m_2 t} \tag{8.10}$$

where m_1 and m_2 are the roots of the equation

$$am^2 + bm + c = 0$$

$$m = \dfrac{-b \pm \sqrt{b^2 - 4ac}}{2a}$$

In a stable ship any initial oscillation must decay to zero, which requires both m_1 and m_2 to be negative.

a and b are always positive for ships and the complex solution of the differential equation does not appear to occur. The condition for stability or stability criterion then becomes $c > 0$,

i.e. $N'_r Y'_v - N'_v (Y'_r - m') > 0$

i.e. $\dfrac{N'_r}{Y'_r - m'} > \dfrac{N'_v}{Y'_v} \tag{8.11}$

Thus the condition for stability reduces to a requirement that the centre of pressure in pure yaw should be ahead of that in pure sway.

Returning to the earlier equations, for a steady turn $\dot{v}$ and $\dot{r}$ are zero giving

$$0 = Y'_v v' + (Y'_r - m')r' + Y'_{\delta_R}\delta'_R \tag{8.12}$$

$$0 = N'_v v' + N'_r r' + N'_{\delta_R}\delta'_R \tag{8.13}$$

This leads to a relationship between r' and δ'_R as follows:

$$(N'_r r' + N'_{\delta_R}\delta'_R)Y'_v = (Y'_r - m)N'_v r' + Y'_{\delta_R}\delta'_R N'_v. \tag{8.14}$$

$$[N'_r Y'_v - N'_v (Y'_r - m')]r' = [Y'_{\delta_R} N'_v - Y'_v N'_{\delta_R}]\delta'_R \tag{8.15}$$

$$\dfrac{r'}{\delta'_R} = \dfrac{Y'_{\delta_R} N'_v - Y'_v N'_{\delta_R}}{Y'_v N'_r - N'_v (Y'_r - m')} \tag{8.16}$$

It will be noted that the denominator is the stability criterion obtained above. This seems reasonable on general grounds. If the denominator in the expression for r'/δ'_R were zero then r'/δ'_R becomes infinite and the ship will turn in a circle with no rudder applied. Thus for a stable ship the denominator would be expected to be non-zero. Also by referring to Figure 8.1 it will be seen that r'/δ_R must be negative for a stable ship. Following from the sign convention and the geometry of the ship, Y'_{δ_R} is positive, N'_{δ_R}

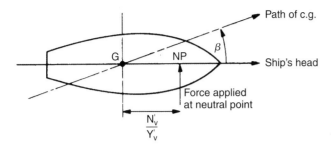

Figure 8.3 Location of neutral point.

and Y'_v are negative. It will also be seen later that Y effectively acts forward of the centre of gravity so that N'_v is also negative. Thus the denominator must be positive for a stable ship.

An important point in directional control is the so-called *Neutral Point* which is that point, along the length of the ship, at which an applied force, ignoring transient effects, does not cause the ship to deviate from a constant heading. This point is a distance ηL forward of the centre of gravity, where

$$\eta = \frac{N'_v}{Y'_v} \quad (8.17)$$

Typically, η is about one-third, so that the neutral point is about one-sixth of the length of the ship abaft the bow.

It can be readily checked, Figure 8.3, that with a force applied at the neutral point the ship is in a state of steady motion with no change of heading but with a steady lateral velocity, i.e. a steady angle of attack. When moving at an angle of attack β, lateral velocity $-v$, the non-dimensional hydrodynamic force and moment are vY'_v and vN'_v respectively, i.e. the hydrodynamic force effectively acts at a distance $(N'_v/Y'_v)L$ ahead of the centre of gravity G directly opposing the applied force, so that there is no tendency for the ship's head to change. If the applied force is of magnitude F, then the resulting lateral velocity is

$$v = \frac{F}{Y'_v} \quad (8.18)$$

Until the velocity has built up to this required value, there will be a state of imbalance and during this phase there can arise a change of heading from the initial heading.

It follows, that if the force is applied aft of the neutral point and acts towards port the ship will turn to starboard, and if applied in the same sense forward of the neutral point the ship turns to port. Clearly, the greater the distance of application of the force from the neutral point the greater the turning influence, other things being equal. This explains why rudders are more effective when placed aft. If $\eta = \frac{1}{3}$, then the 'leverage' of a stern rudder is five times that of a bow rudder. At the stern also, the rudders gain from the effect of the screw race.

If, in Equation (8.16) for r'/δ_R, $N'_{\delta_R} = x'Y'_{\delta_R}$ then

$$\frac{r'}{\delta_R} \text{ is proportional to } \frac{N'_v}{Y'_v} - x' \quad (8.19)$$

That is, for a given rudder angle, the rate of change of heading is greatest when the value of x' is large and negative. This again shows that a rudder is most effective when placed right aft.

8.4 Rudder action

The laws of dynamics demand that when a body is turning in a circle, it must be acted upon by a force acting towards the centre of the circle of sufficient magnitude to impart to the body the required radial acceleration. In the case of a ship, this force can only arise from the aerodynamic and hydrodynamic forces acting on the hull, superstructure and appendages. It is usual, in studying the turning and manoeuvring of ships, to ignore aerodynamic forces for standard manoeuvres and to consider them only as disturbing forces. That is not to say that aerodynamic forces are unimportant. On the contrary, they may prevent a ship turning into the wind if she has large windage areas forward.

In order to produce a radial force of the magnitude required, the hull itself must be held at an angle of attack to the flow of water past the ship. The rudder force must be capable of holding the ship at this angle of attack; that is, it must be able to overcome the hydrodynamic moments due to the angle of attack and the rotation of the ship. The forces acting on the ship during a steady turn are illustrated in Figure 8.4 where F_H is the force on the hull and F_R the rudder force. F_H is the resultant of the hydrodynamic forces on the hull due to the angle of attack and the rotation of the ship as it moves around the circle.

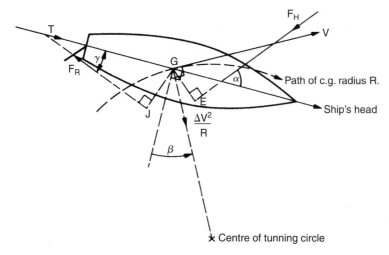

Figure 8.4 Action of rudder in turning ship.

If T is the thrust exerted by the propellers and F_H and F_R act at angles α and γ relative to the middle line plane then, for a steady turn with forces acting as shown in Figure 8.4, these forces must lead to the radial force $\Delta V^2/R$, i.e.

$$T - \frac{\Delta V^2}{R} \sin \beta = F_H \cos \alpha + F_R \cos \gamma \quad (8.20)$$

$$-\frac{\Delta V^2}{R} \cos \beta + F_H \sin \alpha = F_R \sin \gamma \quad (8.21)$$

$$F_H GE + F_R GJ = 0 \quad (8.22)$$

The radial components of the forces on the rudder and the hull, F_R and F_H, must have a resultant causing the radial acceleration.

8.5 Limitations of theory

The simple concepts of a linear theory of motion have been described in Section 8.3. This is useful in providing an insight into the manoeuvring of ships but many problems are, or appear to be, highly non-linear. This has led to the introduction of higher degrees of derivative to obtain a better representation of the way forces and moments, whose deviations from a steady state condition are other than small, can vary. Such problems concern, for example, steering in a seaway (particularly in a following sea), high-speed large-angle submarine manoeuvring when the body shape may have important effects, athwartships positioning of big ships and drilling vessels. Unfortunately, such approaches are critically dependent upon the validity of the mathematical representation adopted for the fluid forces. One limitation of these analyses is that they assume the forces and moments acting on the model to be determined by the motion obtaining at that instant and are unaffected by its history. This is not true. For instance, it has been shown that when two fins (like a ship and its rudder) are moving in tandem and the first is put to an angle of attack, there is a marked time delay before the second fin experiences a change of force. This approach uses a linear functional mathematical representation which includes a 'memory' effect and shows how the results in the frequency and time domains are related. The approach is limited to linear theory but the inclusion of memory effects provides an explanation for at least some of the effects which arise in large amplitude motions. Further theoretical approaches to manoeuvring are discussed in Section 8.17.

8.6 Assessment of manoeuvrability

Assessment of manoeuvrability is made difficult by the lack of rigorous analytical methods and of universally accepted standards for manoeuvrability. The hydrodynamic behaviour of a vessel on the interface between sea and air is inherently complex. Whilst reasonable methods exist for initial estimates of resistance and powering and motions in a seaway, the situation is less satisfying as regards manoeuvring. Much reliance is still placed upon model tests and fullscale trials using a number of common manoeuvres which are outlined below.

8.6.1 Turning circle

Figure 8.5 shows diagrammatically the path of a ship when executing a starboard turn. When the rudder is put over initially, the force acting on the rudder tends to push the ship bodily to port of its original line of advance. As the moment due to the rudder force turns the ship's head, the lateral force on the hull builds up and the ship begins to turn. The parameters at any instant of the turn are defined as:

8.6.1.1 *Drift angle*

The drift angle at any point along the length of the ship is defined as the angle between the centre line of the ship and the tangent to the path of the point concerned. When a drift angle is given for the ship without any specific point being defined, the drift angle at the centre of gravity of the ship is usually intended. Note that the bow of the ship lies within the circle and that the drift angle increases with increasing distance aft of the pivoting point which is defined below.

8.6.1.2 *Advance*

The distance travelled by the centre of gravity in a direction parallel to the original course after the instant the rudder is put over. There is a value of advance for any point on the circle, but if a Figure is quoted for advance with no other qualification the value corresponding to a 90 degree change of heading is usually intended.

8.6.1.3 *Transfer*

The distance travelled by the centre of gravity perpendicular to the original course. The transfer of the ship can be given for any point on the circle, but if a figure is quoted for transfer with no other qualification the value corresponding to a 90 degree change of heading is usually intended.

8.6.1.4 *Tactical diameter*

The value of the transfer when the ship's heading has changed by 180 degrees. It should be noted that the tactical diameter is not the maximum value of the transfer.

8.6.1.5 *Diameter of steady turning circle*

Following initial application of the rudder there is a period of transient motion, but finally the speed, drift angle and turning diameter reach steady values. This usually occurs after about 90 degrees change of heading but, in some cases, the steady state may not be achieved until after 180 degrees change of heading. The steady turning diameter is usually less than the tactical diameter.

8.6.1.6 *Pivoting point*

This point is defined as the foot of the perpendicular from the centre of the turn on to the middle line of the ship extended if necessary. This is not a fixed point, but one which varies with rudder angle and speed. It may be forward of the ship as it would be in Figure 8.5, but is typically one-third to one-sixth of the length of the ship abaft the bow. It should be

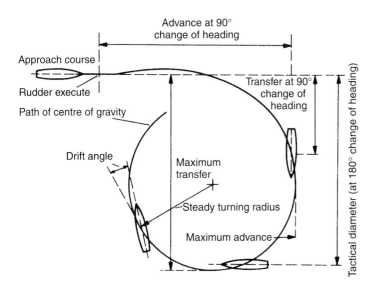

Figure 8.5 Geometry of turning circle.

586 Maritime engineering reference book

noted that the drift angle is zero at the pivoting point and increases with increasing distance from that point.

The turning circle has been a standard manoeuvre carried out by all ships as an indication of the efficiency of the rudder. Apart from what might be termed the 'geometric parameters' of the turning circle defined above, loss of speed on turn and angle of heel experienced are also studied.

8.7 Loss of speed on turn

As discussed above, the rudder holds the hull at an angle of attack, i.e. the drift angle, in order to develop the 'lift' necessary to cause the ship to accelerate towards the centre of the turn. As with any other streamlined form, hull lift can be produced only at the expense of increased drag. Unless the engine settings are changed, therefore, the ship will decelerate under the action of this increased drag. Most ships reach a new steady speed by the time the heading has changed 90 degrees but, in some cases, the slowing down process continues until about 180 degrees change of heading. Steady speed on the turn may be as low as 60% of the approach speed.

8.8 Heel when turning

When turning steadily, the forces acting on the hull and rudder are F_H and F_R. Denoting the radial components of these forces by lower case subscripts (i.e. denoting these by F_h and F_r respectively) and referring to Figure 8.6, it is seen that to produce the turn

$$F_h - F_r = \frac{\Delta V^2}{Rg} \qquad (8.23)$$

where V = speed on the turn, R = radius of turn.

$$\begin{aligned}\text{Moment causing heel} &= (F_h - F_r)KG + F_r(KH) \\ &\quad - F_h(KE) \\ &= (F_h - F_r)(KG - KE) \\ &\quad + F_r(KH - KE) \\ &= (F_h - F_r)GE - F_rEH\end{aligned}$$

$$(8.24)$$

For most ships, E, the centre of lateral resistance, and H are very close and this expression is given approximately by

$$\text{Moment causing heel} = (F_h - F_r)GE \qquad (8.25)$$

This moment causes the ship to heel outwards during the steady turn. When the rudder is initially put over, however, F_r acts before F_h has built up to any significant value and during this transient phase the ship may heel inwards. It should also be noted that the effect of F_r during the steady turn is to reduce the angle of heel, so that if the rudder angle is suddenly taken off, the ship will heel to even larger angles. If the rudder angle were to be suddenly reversed even more serious angles of heel could occur.

It will be appreciated that F_h acts at the centre of lateral resistance only if the angle of heel is small. For large heel angles, the position of E is difficult to assess. For small angles of heel

$$\begin{aligned}\Delta GM \sin \phi &= (F_h - F_r)GE \\ &= \frac{\Delta V^2}{Rg} GE\end{aligned} \qquad (8.26)$$

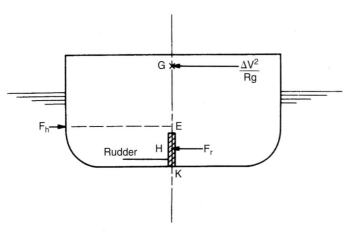

Figure 8.6 Forces producing heel when turning.

Hence

$$\frac{Rg \sin \phi}{V^2} = \frac{GE}{GM} \quad (8.27)$$

It must be emphasized, however, that the angle of heel obtained by this type of calculation should only be regarded as approximate. Apart from the difficulty of accurately locating E, some ships, particularly high speed vessels, suffer an apparent loss of stability when underway because of the other forces acting on the ship and appendages due to the flow around the ship when it is turning.

8.9 Turning ability

The turning circle characteristics are not by themselves indicators of initial response to rudder, which may be important when ships are operating in confined waters or in close company. Indeed, some factors which have a major impact on initial response have very little effect on tactical diameter. One indicator that can be used is the heading angle turned through from an initially straight course, per unit rudder angle applied, after the ship has travelled one ship length. Whilst theoretical prediction of tactical diameter is difficult because of non-linearities, linear theory can be used to calculate this initial response and it is possible to derive an expression for it in terms of the stability derivatives.

Multiple regression techniques can be used to deduce approximate empirical formulae for a design's stability derivatives from experimental data. Although they do not accurately account for all the variations in experimental data some derivatives are reproduced below as an example.

$$\begin{aligned}
-Y'_{\dot{v}}/\pi(T/L)^2 &= 1 + 0.16 C_B B/T - 5.1(B/L)^2 \\
-Y'_{\dot{r}}/\pi(T/L)^2 &= 0.67 B/L - 0.0033(B/T)^2 \\
-N'_{\dot{v}}/\pi(T/L)^2 &= 1.1 B/L - 0.041 B/T \\
-N'_{\dot{r}}/\pi(T/L)^2 &= 1/12 + 0.017 C_B B/T - 0.33 B/L \\
-Y'_v/\pi(T/L)^2 &= 1 + 0.40 C_B B/T \\
-Y'_r/\pi(T/L)^2 &= -1/2 + 2.2 B/L - 0.080 B/T \\
-N'_v/\pi(T/L)^2 &= 1/2 + 2.4 T/L \\
-N'_r/\pi(T/L)^2 &= 1/4 + 0.039 B/T - 0.56 B/L
\end{aligned} \quad (8.28)$$

8.9.1 Zig-zag manoeuvre

It can be argued that it is not often that a ship requires to execute more than say a 90 or 180 degree change of heading. On the other hand, it often has to turn through angles of 10, 20 or 30 degrees. It can also be argued that in an emergency, such as realization that a collision is imminent, it is the initial response of a ship to rudder movements that is the critical factor. Unfortunately, the standard circle manoeuvre does not adequately define this initial response and the standard values of transfer and advance for 90 degrees change of heading and tactical diameter are often affected but little by factors which have a significant influence on initial response to rudder. Such a factor is the rate at which the rudder angle is applied. This may be typically 3 degrees per second. Doubling this rate leads to only a marginally smaller tactical diameter but initial rates of turn will be increased significantly.

The zig-zag manoeuvre, sometimes called a Kempf manoeuvre after G. Kempf, is carried out to study more closely the initial response of ship to rudder movements, Figure 8.7. A typical manoeuvre would

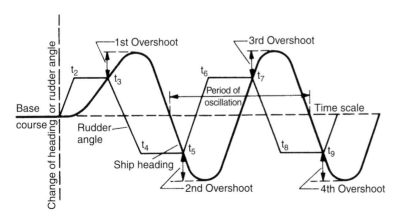

Figure 8.7 The zig-zag manoeuvre.

be as follows. With the ship proceeding at a steady speed on a straight course the rudder is put over to 20 degrees and held until the ship's heading changes by 20 degrees. The rudder angle is then changed to 20 degrees in the opposite sense and so on.

Important parameters of this manoeuvre are:

(a) the time between successive rudder movements;
(b) the *overshoot* angle which is the amount by which the ship's heading exceeds the 20 degree deviation before reducing.

The manoeuvre is repeated for a range of approach speeds and for different values of the rudder angle and heading deviation.

8.9.2 Spiral manoeuvre

This manoeuvre, sometimes referred to as the Dieudonné Spiral after J. Dieudonné who first suggested it, provides an indication of a ship's directional stability or instability.

To perform this manoeuvre, the rudder is put over to say 15 degrees starboard and the ship is allowed to turn until a steady rate of change of heading is achieved. This rate is noted and the rudder angle is reduced to 10 degrees and the new steady rate of change of heading is measured. Successive rudder angles of 5°S, 0°, 5°P, 10°P, 15°P, 10°P, 5°P, 0°, 5°S, 10°S and 15°S are then used. Thus, the steady rate of change of heading is recorded for each rudder angle when the rudder angle is approached both from above and from below. The results are plotted as in Figure 8.8, in which case (*a*) represents a stable ship and case (*b*) an unstable ship.

In the case of the stable ship, there is a unique rate of change of heading for each rudder angle but, in the case of an unstable ship, the plot exhibits a form of 'hysteresis' loop. That is to say that for small rudder angles the rate of change of heading depends upon whether the rudder angle is increasing or decreasing. That part of the curve shown dotted in the Figure cannot be determined from ship trials or free model tests as it represents an unstable condition.

It is not possible to deduce the degree of instability from the spiral manoeuvre, but the size of the loop is a qualitative guide to this. Of direct practical significance, it should be noted that it cannot be said with certainty that the ship will turn to starboard or port unless the rudder angle applied exceeds δ_S or δ_P, respectively and controlled turns are not possible at low rates of turn.

8.9.3 Pull-out manoeuvre

This manoeuvre is used to determine the directional stability of a ship. The rudder is put over to a

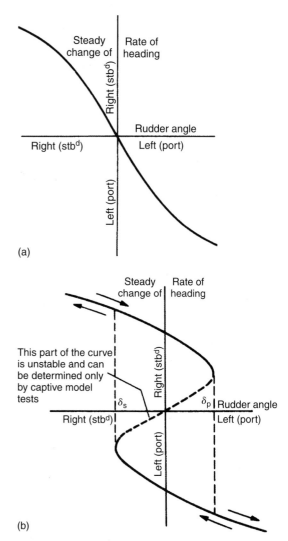

Figure 8.8 (a) Presentation of spiral manoeuvre results (stable ship) (b) Presentation of spiral manoeuvre results (unstable ship).

predetermined angle and held. When the ship is turning at a steady rate the rudder is returned to amidships and the change of rate of turn with time is noted. If the ship is directionally stable the rate of turn reduces to zero and the ship takes up a new straight path. If the ship is unstable a residual rate of turn will persist. The manoeuvre can be conveniently carried out at the end of each circle trial during ship trials.

It has been found that for a stable ship a plot of the log of rate of turn against time is a straight line after an initial transient period, Figure 8.9.

It was shown in the section on theory that the differential equation of motion had two roots m_1 and m_2 both of which had to be negative for directional

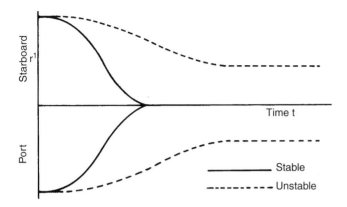

Figure 8.9 Pull-out manoeuvre, rate of turn on time base.

stability. It has been argued that the more negative root will lead to a response which dies out during the transient phase so that the straight portion of the log rate/time curve gives the root of smaller numerical value. Thus if this root is m_1,

$$r^1 = re^{m_1 t}$$
$$\log r^1 = \log r + m_1 t \qquad (8.29)$$

The area under the curve of turn rate against time gives the total heading change after the rudder is centred. Thus the less the total change the more stable the ship.

8.10 Standards for manoeuvring and directional stability

The standards required in any particular design depend upon the service for which the ship is intended but, in any case, they are not easy to define. The problem is made more difficult by the fact that good directional stability and good manoeuvrability are to some extent conflicting requirements, although they are not actually incompatible as has often been suggested. For instance, a large rudder can increase the directional stability and also improve turning performance. Also, in a long fine form increasing draught-to-length ratio can increase stability without detriment to the turning. On the other hand, increasing beam-to-length ratio improves turning but reduces the directional stability. Placing a large skeg aft will improve directional stability at the expense of poorer turning ability.

For a large ocean-going ship, it is usually possible to assume that tugs will be available to assist her when manoeuvring in the confined waters of a harbour. The emphasis in design is therefore usually placed on good directional stability for the long ocean transits. This leads to less wear on the rudder gear, especially if an automatic control system is fitted, and reduces overall average resistance. The highest degree of directional stability is demanded for ships likely to suffer disturbances in their normal service such as supply ships replenishing smaller naval units at sea.

For medium size ships which spend relatively more time in confined waters and which do not normally make use of tugs, greater emphasis has to be placed on response to rudder. Typical of these are the cross channel steamers and antisubmarine frigates.

The manoeuvring capabilities are defined using those parameters measured in the various manoeuvres described in the earlier parts of this Chapter. Typical values to be expected are discussed below.

Tactical diameter-to-length ratio (TD/L). For ships in which tight turning is desirable this may be, say, 3.25 for modern naval ships at high speed, with conventional rudders at 35 degrees. Where even smaller turning circles are required, recourse is usually made to some form of lateral thrust unit.

A TD/L value of 4.5 is suggested as a practicable criterion for merchant ship types desiring good handling performance. Values of this ratio exceeding 7 are regarded as very poor.

Turning rate. For very manoeuvrable naval ships this may be as high as 3 degrees per second. For merchant types, rates of up to 1.5 degrees per second should be achieved in ships of about 100 m at 16 knots, but generally values of 0.5–1.0 degrees per second are more typical. Turning rate is discussed further in Section 8.19.4.2.

Speed on turn. This can be appreciably lower than the approach speed, and typically is only some 60% of the latter.

Initial turning. It has been proposed that the heading change per unit rudder angle in one ship length travelled should be greater than 0.3 generally and greater than 0.2 for large tankers.

Angle of heel. A very important factor in passenger ships and one which may influence the standard of transverse stability incorporated in the design.

Directional stability. Clearly, an important factor in a well balanced design. The inequality presented earlier as the criterion for directional stability can be used as a 'stability index'. Unfortunately, this is not, by itself, very informative. A reasonable design aim is that the spiral manoeuvre should exhibit no 'loop', i.e. the design should be stable even if only marginally so. Using the pull-out manoeuvre it has been suggested that using the criterion of total heading change after the rudder is centred 15–20 degrees represents good stability, 35–40 degrees reasonable stability but that 80–90 degrees indicates marginal stability.

Time to turn through 20 degrees. This provides a measure of the initial response of the ship to the application of rudder. It is suggested that the time to reach 20 degrees might typically vary from 80 to 30 seconds for speeds of 6–20 knots for a 150 m ship. The time will vary approximately linearly with ship length.

Overshoot. The overshoot depends on the rate of turn and a ship that turns well will overshoot more than one that does not turn well. If the overshoot is excessive, it will be difficult for a helmsman to judge when to start reducing rudder to check a turn with the possible danger of damage due to collision with other ships or a jetty. The overshoot angle does not depend upon the ship size and values suggested are 5.5 degrees for 8 knots and 8.5 degrees for 16 knots, the variation being approximately linear with speed.

A thorough review of standard manoeuvring tests is presented by Burcher (1972). ITTC recommendations for ship manoeuvring trials can be found in ITTC (1975). International standards for ship manoeuvrability are contained in the IMO Standards for Ship Manoeuvrability, IMO (1993) and IMO (2002), and include requirements for turning ability, initial turning ability, yaw-checking and course-keeping ability, pull out and stopping ability. These are required to be demonstrated by means of a satisfactory performance in a turning circle, zig-zag test and a full astern stopping test. The mechanics of manoeuvring and the equations of motion may be pursued in more depth in standard naval architecture texts such as Lewis (1989).

8.11 Dynamic positioning

In ships engaged in underwater activities it may be necessary to hold the ship steady relative to some underwater datum. Typical of this situation are drilling ships and those deploying divers. If the water is shallow then it may be practical to moor the ship. In deeper water use is made of a dynamic positioning system using thrust producing devices forward and aft together with a means of detecting departures from the desired position usually using satellite navigation aids. Such a system is described by Peters *et al.* (2005).

8.12 Automatic control systems

Many ships, particularly those on long ocean voyages, travel for long periods of time on a fixed course, the only deviations in course angle being those necessitated by variations in tide, waves or wind. To use trained helmsmen for this type of work is uneconomical and boring for the people concerned. It is in these circumstances that the automatic control system or automatic 'pilot' is most valuable.

Imagine a system which can sense the difference, ψ_e, between the ordered course and the actual course and which can cause the rudder to move to an angle proportional to this error, and in such a way as to turn the ship back towards the desired course, i.e.:

$$\delta_R = \text{Const.} \times \psi_e = a\psi_e, \text{ say} \qquad (8.30)$$

Then, as the ship responds to the rudder the course error will be reduced steadily and, in consequence, the rudder angle will also reduce. Having reached the desired course, the rudder angle will reduce to zero but the ship will still be swinging so that it is bound to 'overshoot'. Thus, by repetition of this process the ship will oscillate about the desired course, the amplitude of the oscillation depending upon the value of the constant of proportionality used in the control equation. (See also Figure 6.133).

How can the oscillation be avoided or at least reduced? In a ship, a helmsman mentally makes provision for the rate of swing of the ship and applies opposite rudder before the desired course angle is reached to eliminate the swing. By introducing a rate gyro into the control system it also can sense the rate of swing and react accordingly in response to the following control equation

$$\delta_R = a\psi_e + b\left(\frac{d\psi_e}{dt}\right) \qquad (8.31)$$

By careful selection of the values of a and b, the overshoot can be eliminated although, in general, a better compromise is to allow a small overshoot on the first swing but no further oscillation as this usually results in a smaller average error. It would be possible to continue to complicate the control equation by adding higher derivative terms. A ship, however, is rather slow in its response to rudder and the introduction of higher derivatives leads to excessive rudder movement with little effect on the ship. For most applications the control equation given above is perfectly adequate.

The system can be used for course changes. By setting a new course, an 'error' is sensed and the system reacts to bring the ship to a new heading. If desired, the system can be programmed to effect a planned manoeuvre or series of course changes. For an efficient system the designer must take into account the characteristics of the hull, the control surfaces and the actuating system. The general mathematics of control theory will apply as for any other dynamic system. In some cases it may be desirable to accept a directionally unstable hull and create course stability by providing an automatic control system with the appropriate characteristics. This device is not often adopted because of the danger to the ship, should the system fail. Simulators can be used to study the relative performances of manual and automatic controls. The consequences of various modes of failure can be studied in safety using a simulator, including the ability of a human operator to take over in the event of a failure. Part-task simulators are increasingly favoured as training aids. A special example of automatic control systems is that associated with dynamic positioning of drilling or diving ships. Motion control is described and discussed by Perez (2005).

8.13 Ship interaction

8.13.1 Interaction

Interaction occurs when a ship comes too close to another ship or too close to, say, a river or canal bank. As ships have increased in size (especially in breadth moulded), interaction has become very important to consider. In February 1998, the Marine Safety Agency (MSA) issued a Marine Guidance note 'Dangers of Interaction', alerting Owners, Masters, Pilots and Tug-Masters on this topic.

Interaction can result in one or more of the following characteristics:

1. If two ships are on a passing or overtaking situation in a river the squats of both vessels could be doubled when their amidships are directly in line.
2. When they are directly in line each ship will develop an angle of heel and the smaller ship will be drawn bodily towards the larger vessel.
3. Both ships could lose steerage efficiency and alter course without change in rudder helm.
4. The smaller ship may suddenly veer off course and head into the adjacent riverbank.
5. The smaller ship could veer into the side of the larger ship or worse still be drawn across the bows of the larger vessel, bowled over and capsized.

In other words there is:

(a) a ship to ground interaction,
(b) a ship to ship interaction,
(c) a ship to shore interaction.

What causes these effects of interaction? The answer lies in the pressure bulbs that exist around the hull form of a moving ship model or a moving ship, Figure 8.10. As soon as a vessel moves from rest, hydrodynamics

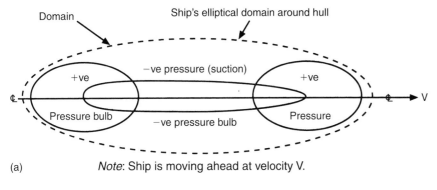

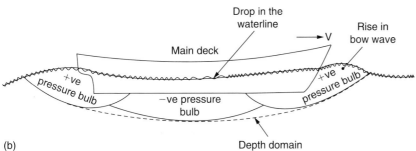

Figure 8.10 (a) Pressure distribution around ship's hull (not drawn to scale). (b) Pressure bulbs around a ship's profile when at forward speed.

produce the shown positive and negative pressure bulbs. For ships with greater parallel body such as tankers these negative bulbs will be comparatively longer in length. When a ship is stationary in water of zero current speed these bulbs disappear.

Note the elliptical domain that encloses the vessel and these pressure bulbs. This domain is very important. When the domain of one vessel interfaces with the domain of another vessel then interaction effects will occur. Effects of interaction are increased when ships are operating in shallow waters.

8.13.2 Ship to ground (squat) interaction

In a report on measured ship squats in the St Lawrence seaway, A.D. Watt stated: 'meeting and passing in a channel also has an effect on squat. It was found that when two ships were moving at the low speed of 5 knots that squat increased up to double the normal value. At higher speeds the squat when passing was in the region of one and a half times the normal value.' Unfortunately, no data relating to ship types, gaps between ships, blockage factors, etc. accompanied this statement.

Thus, at speeds of the order of 5 knots the squat increase is +100% whilst at higher speeds, say 10 knots, this increase is +50%. Figure 8.11 illustrates this passing manoeuvre. Figure 8.12 interprets the percentages given in the previous paragraph.

How may these squat increases be explained? The Ship Squat value depends on the ratio of the ship's cross-section to the cross-section of the river. This is the blockage factor 'S'. The presence of a second ship meeting and crossing will of course increase the blockage factor. Consequently the squat on each ship will increase.

Maximum squat can be calculated by using the equation:

$$\delta_{max} = \frac{C_b \times S^{0.81} \times V_k^{2.08}}{20} \text{ metres} \quad (8.32)$$

Application is best illustrated by an example:
A supertanker has a breadth of 50 m with a static even-keel draft of 12.75 m. She is proceeding along a river of 250 m and 16 m depth rectangular cross-section. If her speed is 5 kts and her C_b is 0.825, calculate her maximum squat when she is on the centreline of this river.

$$S = \frac{b \times T}{B \times H} = \frac{50 \times 12.75}{250 \times 16} = 0.159$$

$$\delta_{max} = \frac{0.825 \times 0.159^{0.81} \times 5^{2.08}}{20} = \underline{0.26 \text{ m}}$$

Assume now that this supertanker meets an oncoming container ship also travelling at 5 kts (see Figure 8.13). If this container ship has a breadth of 32 m a C_b of 0.580, and a static even-keel draft of 11.58 m, calculate the maximum squats of both vessels when they are transversely in line as shown.

$$S = \frac{(b_1 \times T_1) + (b_2 \times T_2)}{B \times H}$$

$$S = \frac{(50 \times 12.75) + (32 \times 11.58)}{250 \times 16} = 0.252$$

Supertanker:

$$\delta_{max} = \frac{0.825 \times 0.252^{0.81} \times 5^{2.08}}{20}$$
$$= \underline{0.38 \text{ m at the bow}}$$

Container ship:

$$\delta_{max} = \frac{0.580 \times 0.252^{0.81} \times 5^{2.08}}{20}$$
$$= \underline{0.27 \text{ m at the stern}}$$

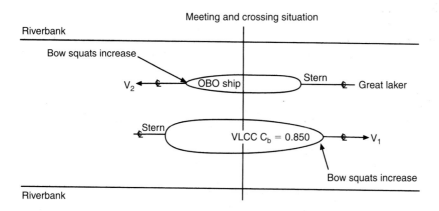

Figure 8.11 Amidships (⊕) of VLCC directly in line with amidships of OBO ship in ST Lawrence seaway.

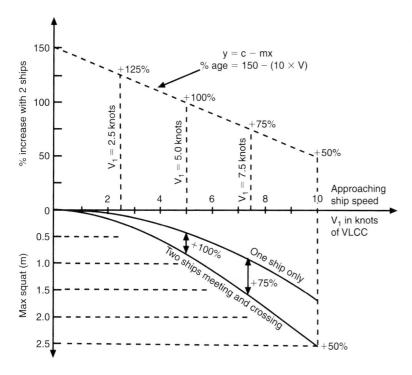

Figure 8.12 Maximum squats for one ship, and for the same ship with another ship present.

The maximum squat of 0.38 m for the supertanker will be at the bow because her C_b is greater than 0.700. Maximum squat for the container ship will be at the stern, because her C_b is less than 0.700. As shown this will be 0.27 m.

If this container ship had travelled alone on the centreline of the river then her maximum squat at the stern would have only been 0.12 m. Thus the presence of the other vessel has more than doubled her squat.

Clearly, these results show that the presence of a second ship does increase ship squat. Passing a moored vessel would also make blockage effect and squat greater. These values are not qualitative but only illustrative of this phenomenon of interaction in a ship to ground (squat) situation. Nevertheless, they are supportive of A.D. Watt's statement.

8.13.3 Ship to ship interaction

Consider Figure 8.14 where a tug is overtaking a large ship in a narrow river. Three cases have been considered:

Case 1. The tug has just come up to aft port quarter of the ship. The domains have become in contact. Interaction occurs. The positive bulb of the ship reacts with the positive bulb of the tug. Both vessels veer to port side. Rate of turn is greater on the tug. There is a possibility of the tug veering off into the adjacent riverbank as shown in Figure 8.14.

Case 2. The tug is in danger of being drawn bodily towards the ship because the negative pressure (suction) bulbs have interfaced. The bigger the differences between the two deadweights of these ships the greater will be this transverse attraction. Each ship develops an angle of heel as shown. There is a danger of the ship losing a bilge keel or indeed fracture of the bilge strakes occurring. This is 'transverse squat', the loss of underkeel clearance at forward speed. Figure 8.13 shows this happening with the tanker and the container ship.

Case 3. The tug is positioned at the ship's forward port quarter. The domains have become in contact via the positive pressure bulbs (see Figure 8.14). Both vessels veer to the starboard side. Rate of turn is greater on the tug. There is great danger of the tug being drawn across the path of the ship's heading and bowled over. This has actually occurred with resulting loss of life.

Note how in these three cases that it is the smaller vessel, be it a tug, a pleasure craft or a local ferry involved, that ends up being the casualty!!

Figures 8.15 and 8.16 give further examples of ship to ship interaction effects in a river.

594 Maritime engineering reference book

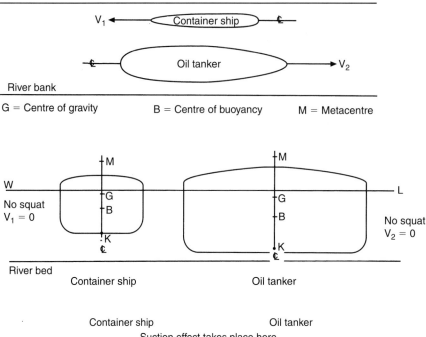

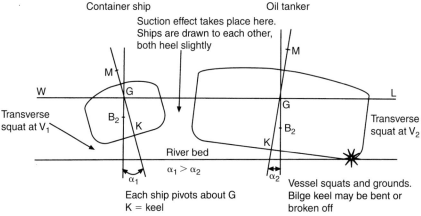

Figure 8.13 Transverse squat caused by ships crossing in a confined channel.

Methods for reducing the effects of interaction in Cases 1 to 5

Reduce speed of both ships and then if safe increase speeds after the meeting crossing manoeuvre time slot has passed. Resist the temptation to go for the order 'increase revs'. This is because the forces involved with interaction vary as the *speed squared*. However, too much a reduction in speed produces a loss of steerage because rudder effectiveness is decreased. This is even more so in shallow waters, where the propeller rpm decrease for similar input of deep water power. Care and vigilance are required.

Keep the distance between the vessels as large as practicable bearing in mind the remaining gaps between each ship side and nearby riverbank.

Keep the vessels from entering another ship's domain, for example crossing in wider parts of the river.

Cross in deeper parts of the river rather than in shallow waters, bearing in mind those increases in squat.

Make use of rudder helm. In Case 1, starboard rudder helm could be requested to counteract loss of steerage. In Case 3, port rudder helm would counteract loss of steerage.

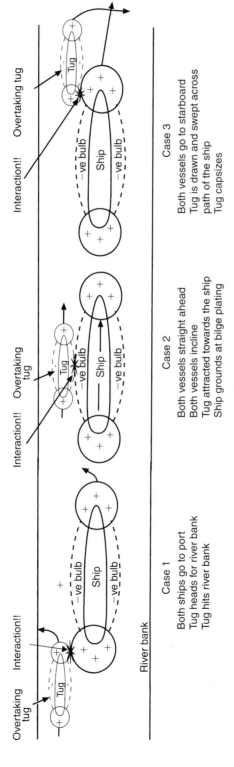

Figure 8.14 Ship to ship interaction in a narrow river during an overtaking manoeuvre.

596 Maritime engineering reference book

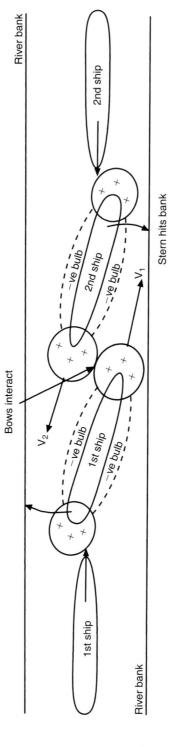

Figure 8.15 *Case 4.* Ship to ship interaction. Both sterns swing towards riverbanks. The approach situation.

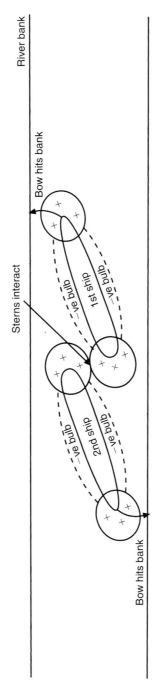

Figure 8.16 *Case 5.* Ship to ship interaction. Both bows swing towards riverbanks. The leaving situation.

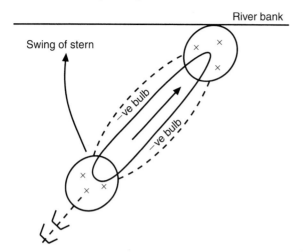

Figure 8.17 Ship to bank interaction. Ship approaches slowly and pivots on forward positive pressure bulb.

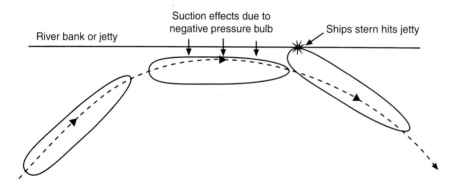

Figure 8.18 Ship to bank interaction. Ship comes in at too fast a speed. Interaction causes stern to swing towards riverbank and then hits it.

8.13.4 Ship to shore interaction

Figures 8.17 and 8.18 show the ship to shore interaction effects. Figure 8.17 shows the forward positive pressure bulb being used as a pivot to bring a ship alongside a riverbank.

Figure 8.18 shows how the positive and negative pressure bulbs have caused the ship to come alongside and then to veer away from the jetty. Interaction could in this case cause the stern to swing and collide with the wall of this jetty.

8.13.5 Summary

An understanding of the phenomenon of interaction can avert a possible marine accident. Generally a reduction in speed is the best preventive procedure. This could prevent on incident leading to loss of sea worthiness, loss of income for the shipowner, cost of repairs, compensation claims and maybe loss of life.

8.14 Shallow water/bank effects

The effectiveness of the rudder helm is influenced by the presence of the upstream propeller and the wake of the hull. The proximity of the seabed and/or banks will in turn alter these effects. Studies into these effects include those by Kijima *et al.* (1990, 1992) who use a parametric approach based on extensive model tests and theoretical analysis. Kobayashi (1995) describes a method for evaluating the manoeuvring efficiency in deep and shallow water. The MMG manoeuvring model, Kose (1982), was adapted for simulations in shallow water and was found to show satisfactory agreement with free-running and captive model tests that were carried out in shallow water. Turnock and Molland (1998) investigated shallow water effects using a surface panel code. The methodology is described in Section 8.19.

A thorough review of ship performance in confined waters is presented by Brix (1993) and

Dand (1982) reports on an extensive series of experiments involving the behaviour of steered ship models in shallow water close to a bank. Delefortrie and Vantoore (2007) model the manoeuvring behaviour of container ships in shallow water and Delefortrie et al. (2004) describe manoeuvring behaviour in shallow water with a muddy bottom.

8.15 Broaching

Broaching, or *broaching-to*, describes the loss of directional stability in waves, induced by a large yaw moment exceeding the course keeping ability of the rudders. Orbital motion of water particles in the wave can result in a zero flow past the rudders which become ineffective. This loss can cause the ship to turn beam on to the waves. The vessel might even capsize due to a large roll moment arising from the forward momentum and the large heading angle. The effect is greater because the ship's hydrostatic stability is often reduced by the presence of the waves.

Broaching is likely when the ship is running with, or being slowly overtaken by, the waves. It may be sudden, due to the action of a single wave, or be cumulative where the yaw angle builds up during a succession of waves. Although known well since man put to sea in boats, broaching is a highly non-linear phenomenon and it is only relatively recently that good mathematical simulations have been possible. Broaching is described and discussed by DuCane and Goodrich (1962), Tuite and Renilson (1998) and Vassalos et al. (1999).

When the encounter frequency of the ship with the waves approaches zero the ship can become trapped by the wave. The ship remains in the same position relative to the waves for an appreciable time. It is then said to be *surf riding*. This is a dangerous position and broaching is likely to follow. The operator can get out of this condition by a change of speed or direction, although the latter may temporarily result in large roll angles.

8.16 Experimental approaches

8.16.1 Manoeuvring tests in sea trials

The main manoeuvring characteristics are quantified in sea trials with the full-scale ship. Usually the design speed is chosen as initial speed in the manoeuvre. Trial conditions should feature deep water (water depth > 2.5 ship draught), little wind (less than Beaufort 4) and 'calm' water to ensure comparability to other ships. Trim influences the initial turning ability and yaw stability more than draught. For comparison with other ships, the results are made non-dimensional with ship length and ship length travel time (L/V).

The main manoeuvres used in sea trials follow recommendations of the Manoeuvring Trial Code of ITTC (1975) and the IMO circular MSC 389 (1985). IMO also specifies the display of some of the results in bridge posters and a manoeuvring booklet on board ships in the IMO resolution A.601(15) (1987) (Provision and display of manoeuvring information on board ships). These can also be found in Brix (1993). Typical tests include a turning circle test, a spiral manoeuvre, a pull-out manoeuvre, a zig-zag manoeuvre, a stopping trial, a hard rudder test and a man-overboard manoeuvre, Bertram (2000) and IMO (2002).

8.16.2 Model tests

Model tests to evaluate manoeuvrability are usually performed with models ranging between 2.5 m and 9 m in length, The models are usually equipped with propeller(s) and rudder(s), electrical motor and rudder gear. Small models are subject to considerable scaling errors and usually do not yield satisfactory agreement with the full-scale ship, because the too small model Reynolds number leads to different flow separation at model hull and rudder and thus different non-dimensional forces and moments, especially the stall angle (angle of maximum lift force shortly before the flow separates completely on the suction side), which will be much smaller in models (15° to 25°) than in the full-scale ship (>35°). Another scaling error also contaminates tests with larger models: the flow velocity at the rudder outside the propeller slipstream is too small (due to a too large wake fraction in model scale) and the flow velocity inside the propeller slipstream is too large (because the too large model resistance requires a larger propeller thrust). The effects cancel each other partially for single-screw ships, but usually the propeller effect is stronger. This is also the case for twin-screw twin-rudder ships, but for twin-screw midship-rudder ships the wake effect dominates for free-running models. For a captured model, propeller thrust minus thrust deduction does not have to equal resistance. Then the propeller loading may be chosen lower such that scale effects are minimized. However, the necessary propeller loading can only be estimated.

Model tests are usually performed at Froude similarity. For small Froude numbers, hardly any waves are created and the non-dimensional manoeuvring parameters become virtually independent of the Froude number. For $F_n < 0.3$, e.g., the body forces Y and N may vary with speed only by less than 10% for deep water. For higher speeds the wave resistance changes noticeably and the propeller loading increases, as does the rudder effectiveness if the rudder is placed in the propeller slipstream. Also, in shallow water, trim and sinkage change with F_n influencing Y and N. If the rudder pierces the free surface or is close enough for ventilation to occur, the Froude number is always important.

Model tests with free-running models are usually performed indoors to avoid wind effects. The track of the models is recorded either by cameras (two or more) or from a carriage following the model in longitudinal and transverse directions. Turning circle tests can only be performed in broad basins and even then usually only with rather small models. Often, turning circle tests are also performed in towing tanks with an adjacent round basin at one end. The manoeuvre is then initiated in the towing tank and ends in the round basin. Spiral tests and pull-out manoeuvres require more space than usually available in towing tanks. However, towing tanks are well suited for zigzag manoeuvres. If the ship's track is precisely measured in these tests, all necessary body force coefficients can be determined and the other manoeuvres can be numerically simulated with sufficient accuracy.

Model tests with captured models determine the body force coefficients by measuring the forces and moments for prescribed motions. The captured models are also equipped with rudders, propellers, and electric motors for propulsion.

- Oblique towing tests can be performed in a regular towing tank. For various yaw and rudder angles, resistance, transverse force, and yaw moment are measured, sometimes also the heel moment.
- Rotating arm tests are performed in a circular basin. The carriage is then typically supported by an island in the centre of the basin and at the basin edge. The carriage then rotates around the centre of the circular basin. See rotating arm in Figure 7.66. The procedure is otherwise similar to oblique towing tests. Due to the disturbance of the water by the moving ship, only the first revolution should be used to measure the desired coefficients. Large non-dimensional radii of the turning circle are only achieved for small models (inaccurate) or large basins (expensive). The technology is today largely obsolete and replaced by planar motion mechanisms which can also generate accelerations, not just velocities.
- Planar motion mechanisms (PMMs) are installed on a towing carriage. They superimpose sinusoidal transverse or yawing motions (sometimes also sinusoidal longitudinal motions) to the constant longitudinal speed of the towing carriage, Figure 8.19. The periodic motion may be produced mechanically from a circular motion via a crankshaft or by computer-controlled electric motors (computerized planar motion carriage (CPMC)). The CPMC is far more expensive and complicated, but allows the extension of model motions over the full width of the towing tank, arbitrary motions and a precise measuring of the track of a free-running model.

8.17 CFD for ship manoeuvring

For most ships, the linear system of equations determining the drift and yaw velocity in steady turning motion is nearly singular. This produces large relative errors in the predicted steady turning rate especially for small rudder angles and turning rates. For large rudder angle and turning rate, non-linear forces alleviate these problems somewhat. But non-linear hull forces depend crucially on the cross-flow resistance or the direction of the longitudinal vortices, i.e. on quantities which

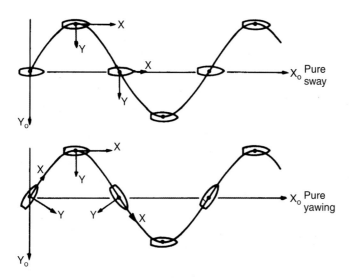

Figure 8.19 Planar motion mechanism test.

are determined empirically and which vary widely. In addition, extreme rudder forces depend strongly on the rudder stall angle which – for a rudder behind the hull and propeller – requires at least two-dimensional RANS simulations. Thus large errors are frequently made in predicting both the ship's path in hard manoeuvres and the course-keeping qualities. (The prediction of the full ship is fortunately easier as at the higher Reynolds numbers stall rarely occurs). In spite of that, published comparisons between predictions and measurements almost always indicate excellent accuracy. A notable exception is Söding (1993). The difference is that Söding avoids all information which would not be available had the respective model not been tested previously. The typical very good agreement published by others is then suspected to be either chosen as best results from a larger set of predictions or due to empirical corrections of the calculation method based on experiments which include the ship used for demonstrating the attained accuracy. Naturally, these tricks are not possible for a practical prediction where no previous test results for the ship design can be used. Thus the accuracy of manoeuvring predictions is still unsatisfactory, but differences between alternative designs and totally unacceptable designs may be easily detected using the available methods for manoeuvring prediction. With appropriate validation, it may also be possible to predict full-scale ship motions with sufficient accuracy, but the experience published so far is insufficient to establish this as state of the art.

The simplest approach to body force computations is the use of regression formulae based on slender-body theory, but with empirical coefficients found from analysing various model experiments, e.g. Clarke *et al.* (1983). The next more sophisticated approach would be to apply slender-body methods directly, deriving the added mass terms for each strip from analytical (Lewis form) or BEM computations. These approaches are still state of the art in ship design practice and have been discussed in Section 7.7.

The application of three-dimensional CFD methods, using either lift-generating boundary elements (vortex or dipole) or field methods (Euler or RANS solvers) is still predominantly a matter of research, although the boundary element methods are occasionally applied in practical design. The main individual CFD approaches are ranked in increasing complexity:

- *Lifting surface methods*
  An alternative to slender-body theory, applicable to rudder and hull (separately or in combination), is the lifting surface model. It models the inviscid flow about a plate (centre plane), satisfying the Kutta condition (smooth flow at the trailing edge) and usually the free-surface condition for zero F_n (double-body flow). The flow is determined as a superposition of horseshoe vortices which are symmetrical with respect to the water surface (mirror plane). The strength of each horseshoe vortex is determined by a collocation method from Biot–Savart's law. For stationary flow conditions, in the ship's wake there are no vertical vortex lines, whereas in instationary flow vertical vortex lines are required also in the wake. The vortex strength in the wake follows from three conditions:
  1. Vortex lines in the wake flow backwards with the surrounding fluid velocity, approximately with the ship speed u.
  2. If the sum of vertical ('bound') vortex strength increases over time within the body (due to larger angles of attack), a corresponding negative vorticity leaves the trailing edge, entering into the wake.
  3. The vertical vortex density is continuous at the trailing edge.

  Except for a ship in waves, it seems accurate enough to use the stationary vortex model for manoeuvring investigations.

  Vortex strengths within the body are determined from the condition that the flow is parallel to the midship (or rudder) plane at a number of collocation points. The vortices are located at 1/4 of the chord length from the bow, the collocation points at 3/4 of the chord length from the bow. This gives a system of linear equations to determine the vortex strengths. Transverse forces on the body may then be determined from the law of Kutta–Joukowski, i.e. the body force is the force exerted on all 'bound' (vertical) vortices by the surrounding flow.

  Alternatively one can smooth the bound vorticity over the plate length, determine the pressure difference between port and starboard of the plate, and integrate this pressure difference. For shallow water, reflections of the vortices are necessary both at the water surface and at the bottom. This produces an infinite number of reflections, a subset of which is used in numerical approximations. If the horizontal vortex lines are arranged in the ship's centre plane, only transverse forces depending linearly on v and r are generated. The equivalent to the non-linear cross-flow forces in slender-body theory is found in the vortex models if the horizontal vortex lines are oblique to the centre plane. Theoretically the position of the vortex lines could be determined iteratively to ensure that they move with the surrounding fluid flow which is influenced by all other vortices. But practically this procedure is usually not applied because of the high computing effort and convergence problems. According to classical foil theory, the direction of the horizontal vortices should be halfway between the ship longitudinal direction and the motion direction in deep water. More modern procedures arrange the vortices in longitudinal direction within the ship length, but in an oblique plane (for steady motion at a constant yaw angle) or on a circular cylinder (for steady turning motion) in the wake. The exact direction of the vortices is determined depending on water depth. Also important is the arrangement of vortex lines and collocation points on the material plate. Collocation points should be about halfway between vortex lines both in longitudinal and vertical directions. High accuracy with few vortex lines is attained if the distance between vertical vortices is smaller at both ends of the body, and if the distance between horizontal vortices is small at the keel and large at the waterline.

- *Lifting body methods*
  A body with finite thickness generates larger lift forces than a plate. This can be taken into account in different ways:

  1. by arranging horseshoe vortices (or dipoles) on the hull on both sides
  2. by arranging a source and a vortex distribution on the centre plane
  3. by arranging source distributions on the hull and a vortex distribution on the centre plane

  In the third case, the longitudinal distribution of bound vorticity can be prescribed arbitrarily, whereas the vertical distribution has to be determined from the Kutta condition along the trailing edge. The Kutta condition can be approximated in different ways. One suitable formulation is: the pressure (determined from the Bernoulli equation) should be equal at port and starboard along the trailing edge. If the ship has no sharp edge at the stern (e.g. below the propeller's axis for a stern bulb), it is not clear where the flow separation (and thus the Kutta condition) should be assumed. This may cause large errors for transverse forces for the hull alone, but when the rudder is modelled together with the hull, the uncertainty is much smaller.

  Forces can be determined by integrating the pressure over the hull surface. For a very thin body, the lifting surface and lifting body models should result in similar forces. In practice, however, large differences are found. The lifting body model with source distributions on both sides of the body has difficulties if the body has a sharp bow. Assuming a small radius at the bow waterline produces much better results. For a ship hull it seems difficult to obtain more accurate results from lifting-body theory than using slender-body theory. For the rudder and for the interaction between rudder and hull, however, lifting surface or lifting body theory is the method of choice for angles of attack where no stall is expected to occur. Beyond the stall angle, only RANS methods (or even more sophisticated viscous flow computations) may be used.

  By the early 1990s research applications for lifting body computations including free surface effects appeared for steady drift motions. The approach of Zou (1990) is typical. First the wave resistance problem is solved including dynamic trim and sinkage. Assuming small asymmetry, the difference between symmetrical and asymmetrical flow is linearized. The asymmetrical flow is then determined by a lifting body method with an additional source distribution above the free surface.

- *Field methods*
  In spite of the importance of viscosity for manoeuvring, viscous hull force calculations appeared in the 1990s only as research applications and were mostly limited to steady flow computations around a ship with a constant yaw angle. Difficulties in RANS computations for manoeuvring are:

  – The number of computational cells is much higher than for resistance computations, because both port and starboard sides must be discretized and because vortices are shed over nearly the full ship length.
  – The large-scale flow separation makes wall functions (e.g. in the standard k-ε turbulence model) dubious. But avoiding wall functions increases the necessary cell number further and deteriorates the convergence of the numerical solution methods.

  State of the art computations for ship hulls at model scale Reynolds numbers were capable of predicting transverse forces and moments reasonably well for steady flow cases with moderately constant yaw angle, but predicted the longitudinal force (resistance) with large relative errors. Flow details such as the wake in the aftbody were usually captured only qualitatively. Either insufficient grid resolutions or turbulence models were blamed for the differences with model tests. By the late 1990s, RANS results with free-surface deformation (waves) were also presented, but with the exception of Japanese research groups, none of the computations included dynamic trim and sinkage, although for shallow water these play an important role in manoeuvring.

  Despite these shortcomings, RANS computations including free-surface effects will grow in importance and eventually also drift into practical applications. They are expected to substantially improve the accuracy of manoeuvring force predictions over the next decade.

  Comprehensive reviews of developments in theoretical and numerical modelling of manoeuvring are given in ITTC (2002) and ITTC (2005). Typical further developments are described by Jacquin and Guillerm (2006) and Felli et al. (2006). Simulations of manoeuvring in waves have been developed by Bailey et al. (1998, 2002) and Lin and Zhang (2006).

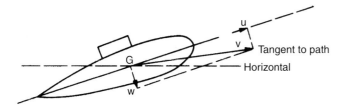

Figure 8.20 Submarine turning in vertical plane.

8.18 Stability and control of submarines

8.18.1 Control requirements and equations

The high underwater speed of some submarines makes it necessary to study their dynamic stability and control. The subject assumes great importance to both the commanding officer and the designer because of the very short time available in which to take corrective action in any emergency: many submarines are restricted to a layer of water which is of the order of at most two or three ship lengths deep. To the designer and research worker, this has meant directing attention to the change in the character of the forces governing the motion of the submarine which occurs as the speed is increased. For submarines of orthodox size and shape below about ten knots the hydrostatic forces predominate. In this case, the performance of the submarine in the vertical plane can be assessed from the buoyancy and mass distributions. Above 10 knots, however, the hydrodynamic forces and moments on the hull and control surfaces predominate.

To a certain degree, the treatment of this problem is similar to that of the directional stability of surface ships dealt with in Section 8.3. There are differences however between the two, viz.:

(a) the submarine is positively stable in the fore and aft vertical plane in that B lies above G so that having suffered a small disturbance in trim when at rest it will return to its original trim condition;
(b) the limitation in the depth of water available for vertical manoeuvres;
(c) the submarine is unstable for translations in the z direction because the hull is more compressible than water;
(d) it is not possible to maintain a precise equilibrium between weight and buoyancy as fuel and stores are being continuously consumed.

It follows, from (c) and (d) above, that the control surfaces or hydroplanes will have, in general, to exert an upward or downward force on the submarine. Also, if the submarine has to remain on a level keel or, for some reason, the submarine cannot be allowed to trim to enable the stability lever to take account of the trimming moment, the control surfaces must also exert a moment. To be able to exert a force and moment on the submarine which bear no fixed relationship one to another requires two separate sets of hydroplanes. Usually, these are mounted well forward and well aft on the submarine to provide maximum leverage, Figures 8.21 and 8.25.

Consider a submarine turning in the vertical plane, Figure 8.20.

Assume that the effective hydroplane angle is δ_H, i.e. the angle representing the combined effects of bow and stern hydroplanes.

In a steady state turn, with all velocities constant, the force in the z direction and the trimming moment are zero. Hence

$$wZ_w + qZ_q + mqV + \delta_H Z_{\delta_H} = 0 \qquad (8.33)$$

$$wM_w + qM_q + \delta_H M_{\delta_H} - mgBG\theta = 0 \qquad (8.34)$$

where subscripts w, q and δ_H denote differentiation with respect to velocity normal to submarine axis, pitching velocity and hydroplane angle respectively. Compare the equations for directional stability of surface ships:

mqV is a centrifugal force term
$mgBG\theta$ is a statical stability term

In the moment equation M_w, M_q and M_{δ_H} are all proportional to V^2, whereas $mgBG\theta$ is constant at all speeds. Hence, at high speeds, $mgBG\theta$ becomes small and can be ignored. As mentioned above, for most submarines it can be ignored at speeds above about 10 knots. By eliminating w between the two equations so simplified

$$\frac{q}{\delta_H} = \frac{M_w Z_{\delta_H} - M_{\delta_H} Z_w}{M_q Z_w - M_w(Z_q + mV)} \qquad (8.35)$$

As with the surface ship problem the necessary condition for stability is that the denominator should be positive, i.e.

$$M_q Z_w - M_w(Z_q + mV) > 0 \qquad (8.36)$$

This is commonly known as the *high speed stability criterion*.

If this condition is met and statically the submarine is stable, then it will be stable at all speeds. If it is statically stable, but the above condition is not satisfied, then the submarine will develop a diverging (i.e. unstable) oscillation in its motion at forward speeds above some critical value.

Now by definition $q = d\theta/dt = \dot{\theta}$, so that differentiating the moment equation with respect to time

$$\dot{w}M_w + \dot{q}M_q + \dot{\delta}_H M_{\delta_H} - mgBGq = 0 \quad (8.37)$$

But in a steady state condition as postulated $\dot{w} = \dot{q} = \dot{\delta}_H = 0$. Hence $q = 0$ if BG is positive as is the practical case. That is, a steady path in a circle is not possible unless BG = 0. Putting $q = 0$, the equations become

$$wZ_w + \delta_H Z_{\delta_H} = 0 \quad (8.38)$$

$$wM_w + \delta_H M_{\delta_H} - mgBG\theta = 0 \quad (8.39)$$

i.e.

$$w = -\delta_H \frac{Z_{\delta_H}}{Z_w} \quad (8.40)$$

and

$$\theta = \delta_H \left(M_{\delta_H} - \frac{Z_{\delta_H}}{Z_w} M_w \right) \Big/ mgBG \quad (8.41)$$

Now rate of change of depth $= V(\theta - w/V)$ if w is small $= V\theta - w$, i.e.

$$\frac{\text{depth rate}}{\delta_H} = \left(VM_{\delta_H} - VM_w \frac{Z_{\delta_H}}{Z_w} + mgBG \frac{Z_{\delta_H}}{Z_w} \right) \Big/ mgBG$$

$$\quad (8.42)$$

The depth rate is zero if

$$V = -\left(mgBG \frac{Z_{\delta_H}}{Z_w} \right) \Big/ \left(M_{\delta_H} - M_w \frac{Z_{\delta_H}}{Z_w} \right)$$

$$= mgBG \Big/ \left(M_w - M_{\delta_H} \frac{Z_w}{Z_{\delta_H}} \right) \quad (8.43)$$

From the equation for θ, if the hydroplanes are so situated that

$$\frac{M_{\delta_H}}{Z_{\delta_H}} = \frac{M_w}{Z_w} \quad (8.44)$$

then θ is zero. The depth rate will be $\delta_H Z_{\delta_H}/Z_w$ which is not zero. The ratio M_w/Z_w defines the position of the *neutral point*. This corresponds to the similar point used in directional stability and is usually forward of the centre of gravity. A force at the neutral point causes a depth change but no change in the angle of pitch.

The equation for depth rate can be rewritten as

$$\text{depth rate}/\delta_H = \frac{Z_{\delta_H}}{Z_w} \left(1 - \frac{V}{V_c} \right) \quad (8.45)$$

where

$$V_c = mgBG \Big/ \left(M_w - \frac{M_{\delta_H}}{Z_{\delta_H}} \cdot Z_w \right) \quad (8.46)$$

or

$$\text{depth rate}/\delta_H = \frac{Z_{\delta_H} V}{mgBG} \left\{ \frac{M_{\delta_H}}{Z_{\delta_H}} - x_c \right\} \quad (8.47)$$

where

$$x_c = \frac{M_w}{Z_w} - \frac{mgBG}{VZ_w} \quad (8.48)$$

The first of these two expressions shows that (depth rate)/δ_H is negative, zero or positive as V is greater than, equal to or less than V_c respectively. V_c is known as the *critical speed* or *reversal speed*, since at that speed the planes give zero depth change and cause reverse effects as the speed increases or decreases from this speed. Near the critical speed the value of $(1 - (V/V_c))$ is small—hence the hydroplanes' small effect in depth changing.

It will be seen that θ is not affected in this way since

$$\frac{\theta}{\delta_H} = -\frac{Z_{\delta_H}}{Z_w} \frac{1}{V_c} = -\frac{Z'_{\delta_H}}{Z'_w} \frac{V}{V_c} \quad (8.49)$$

The magnitude of θ/δ_H changes with V but not its sign. If stern hydroplanes are considered, a positive hydroplane angle produces a negative pitch angle (bow down), but depth change is downwards above the critical speed and upwards below the critical speed.

The second expression for depth change illustrates another aspect of the same phenomenon. x_c denotes a position $mgBG/VZ_w$ abaft the neutral point,

$$\frac{mgBG}{VZ_w} = \frac{mgBG}{\frac{1}{2}\rho L^2 Z'_w V^2} \quad (8.50)$$

hence x_c is abaft the neutral point by a distance which is small at high speed and large at low speed. The critical situation is given by $x_c = M_{\delta_H}/Z_{\delta_H}$, i.e. centre of pressure of the hydroplanes. The position defined by x_c is termed the *critical point*. Figure 8.21 illustrates the neutral and critical point positions. Figure 8.22 shows a typical plot of x_c/L against Froude number. The critical speed can be obtained by noting the Froude number appropriate to the hydroplane position, e.g. in the figure

$$\frac{V_c}{\sqrt{gL}} = 0.05$$

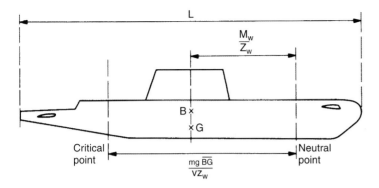

Figure 8.21 Neutral and critical points.

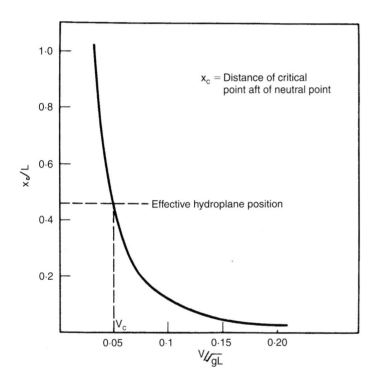

Figure 8.22 Variation of critical point with speed.

i.e.

$$V_c = 3 \text{ knots if } L = 100 \text{ m}$$

8.18.2 Experiments and trials

As in the case of the directional stability of surface ships, the derivatives needed in studying submarine performance can be obtained in conventional ship tanks using planar motion mechanisms and in rotating arm facilities. The model is run upright and on its side with and without propellers, hydroplanes and stabilizer fins to enable the separate effects of these appendages to be studied. Data so obtained are used to predict stability and fed into computers. The computer can then predict the manoeuvres the submarine will perform in response to certain control surface movements. These can be used to compare with full-scale data obtained from trials. A computer can be associated with a tilting and rotating cabin, creating a simulator for realistic training of operators and for studying the value of different display and control systems.

8.18.3 Design assessment

Modifying dynamic stability characteristics

In common with most features of ship design, it is likely that the designer will wish to modify the dynamic stability standards as defined by the initial model tests. How then can the desired standards be most effectively produced?

In most cases, the basic hull form will be determined by resistance, propulsion and seakeeping considerations. The designer can most conveniently modify the appendages to change the dynamic stability. The procedure is similar for submarines and surface ships but is illustrated below for the former.

Assuming that the hydroplanes are correctly sized, the designer concentrates on the stabilizer fins (skeg for the lateral plane). If the contributions of these fins to Z'_w and M'_w, as determined from the model results with and without fins, are $\delta Z'_w$ and $\delta M'_w$ the effective distance of the fins from the centre of gravity is X_s, say, where:

$$\frac{X_s}{L} = -\frac{\delta M'_w}{\delta Z'_w}, \quad \text{i.e. } \delta M'_w = -\frac{X_s}{L}\delta Z'_w \quad (8.51)$$

The negative sign arises because the fins are aft.

The effect of the fins on the curvature derivatives can be deduced similarly or, if not available from direct model tests, it can be argued that the rotation causes an effective change of incidence at the fin, such that:

$$\delta Z'_q = \frac{X_s}{L}\delta Z'_w \quad (8.52)$$

and

$$\delta M'_q = \left(\frac{X_s}{L}\right)^2 \delta Z'_w \quad (8.53)$$

If the derivatives, as originally determined, give rise to an unstable motion, the required increase in fin area can be deduced using the above relationships and assuming that $\delta Z'_w$ is proportional to the fin area.

Example: The stability derivatives found for a certain submarine, complete with all appendages are:

$Z'_w = -0.023, \quad Z'_q = -0.01$
$M'_w = 0.0105, \quad M'_q = -0.005$
$m' = 0.024$

The corresponding figures for Z'_w and M'_w without fins are 0 and 0.022. It is required to calculate the percentage increase in fin area required to make the submarine just stable assuming m' is effectively unaltered.

The stability criterion in non-dimensional form is

$$M'_q Z'_w - M'_w (Z'_q + m') > 0$$

Substituting the original data gives -0.000068 so that the submarine is unstable.

If the fin area is increased by $p\%$ then the derivatives become

$Z'_w = -0.02 + p(-0.02)$
$M'_w = 0.012 + p(-0.01)$, in this case $X_s = -\frac{1}{2}L$
$Z'_q = -0.01 + p(-0.01)$
$M'_q = -0.005 + p(-0.005)$

Substituting these values in the left-hand side of the stability criterion and equating to zero gives the value of p which will make the submarine just stable.

Carrying out this calculation gives $p = 14.8$, and the modified derivatives become:

$$Z'_w = -0.023, \quad Z'_q = -0.0115$$
$$M'_w = 0.0105, \quad M'_q = -0.00575 \quad (8.54)$$

8.19 Rudders and control surfaces

8.19.1 Control surfaces and applications

The purpose of a control surface is to produce a force, which is used to control the motion of the vehicle. Control surfaces may be fixed or movable but, in the marine field, they are mainly movable with the prime example being the ship rudder.

Movable control surfaces are used on most marine vessels including boats, ships of all sizes, submarines and other underwater vehicles. Typical applications may be summarized as

Rudders: used to control horizontal motion of all types of marine vehicle.
Fin stabilizers: used to reduce roll motion.
Hydroplanes (or diving planes): used to control the vertical motion of submarines and other underwater vehicles.
Fins for pitch damping: used to control pitch motion in high-speed vessels.
Transom flaps: used to control running trim and/or to provide ride control.
Interceptors: used to control running trim and/or to provide ride control.

Examples of fixed control surfaces include *anti-pitching fins* on fast vessels and *keels* on sailing yachts.

In general, lift or sideforce on a control surface may be developed by applying incidence, Figure 8.23(a), introducing asymmetry by means of fixed camber, Figure 8.23(b), or introducing variable camber by means say of a flap, Figure 8.23(c). Further increases in lift may be achieved by the application of incidence

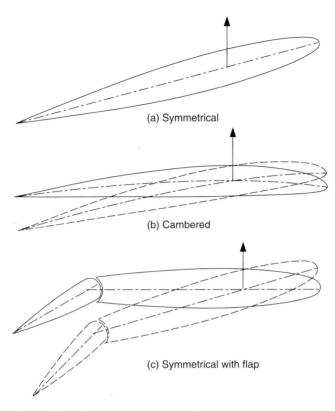

Figure 8.23 Sideforce on a control surface.

to cases (b) and (c). Since movable control surfaces generally have to act in both directions, applications in the marine field tend to be symmetrical and confined to the use of (a) or (c) in Figure 8.23, or some variants of these two basic types. The cambered shape (b) is of course used extensively for aircraft lifting surfaces as well as marine applications such as sections for propeller blades and lifting foils on hydrofoil craft.

8.19.1.1 Rudder types

The choice of the rudder type will depend on factors such as ship or boat type and size, the shape of the stern, size of rudder required and whether there is a propeller upstream of the rudder.

The principal rudder types, or concepts, are summarised in Figure 8.24 and some comments on each are as follows:

(a) *Balanced rudder*: Open sternframe with a bottom pintle, which is a support bolt or pin with a bearing. The upper bearing is inside the hull. It has been applied to vessels such as tugs and trawlers and extensively to single-screw merchant ships. Tends to have been superseded by the use of the semi-balanced skeg rudder, type (d).

(b) *Spade rudder*: A balanced rudder. Both bearings are inside the hull. Bending moments as well as torque are carried by the stock, leading to larger stock diameters and rudder thickness. Applied extensively to single and twin-screw vessels, including small powercraft, yachts, ferries, warships and some large merchant ships. Also employed as control surfaces on submarines and other underwater vehicles.

(c) *Full skeg rudder*: An unbalanced rudder. The rudder is supported by a fixed skeg with a pintle at the bottom. Applied mainly to large sailing yachts, but also applied as hydroplanes on underwater vehicles.

(d) *Semi-balanced skeg rudder*: Also known as a horn rudder or a Mariner rudder, following its early application to a ship of that type, Russo and Sullivan (1953). The movable part of the rudder is supported by a fixed skeg with a pintle at the bottom of the skeg. This pintle, at about half the rudder's vertical depth, is therefore usefully situated in the vicinity of the centre of pressure of the combined movable rudder plus skeg. Used extensively in single and twin-screw merchant ships of all sizes and some warships. In the single-screw application it is combined with an open, or Mariner, type stern arrangement.

(e) *Semi-balanced rudder, aft of skeg or deadwood*: Typically applied to twin-screw ships with a single rudder. Tends to have been superseded by the use of twin rudders of type (b) or (d).

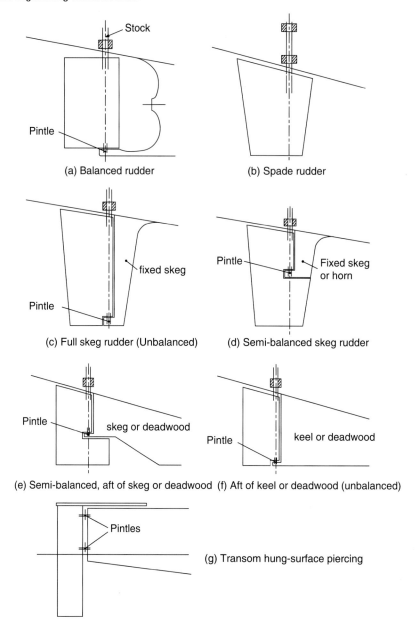

Figure 8.24 Rudder types.

(f) *Unbalanced, aft of keel or deadwood*: Typically applied to some older sailing craft.
(g) *Transom hung, surface piercing*: An unbalanced rudder. Typically applied to small sailing craft.

Other variants, such as twisted, flapped and high lift rudders may be considered as special cases of these principal rudder types.

In generic terms, rudders (a) and (b) in Figure 8.24 are balanced rudders, (d) and (e) are semi-balanced, whilst (c), (f) and (g) are unbalanced. An element of balance will reduce the rudder torque and reduce the size of the steering gear. It should, however, be noted that the centre of action of the rudder force tends to move with change in helm, or rudder angle and it is not possible to fully balance a rudder over a complete range of angles. "Balanced" is therefore only a broad generic term when used in the context of describing rudder types. Rudder balance is discussed in more detail in Section 8.19.4.2.

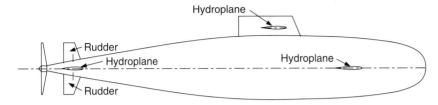

Figure 8.25 Hydroplanes: Submarine or underwater vehicle.

8.19.1.2 *Hydroplanes*

Hydroplanes are control surfaces that are used to control the vertical motion of underwater vehicles, as discussed in Section 8.18, and behave in the manner of a horizontal rudder, Figure 8.25. Aerofoil sections tend to be used for hydroplanes which will be all-movable or with a fixed skeg of 20–30% of chord from the leading edge, Figure 8.24(c). The design procedure tends to that of a control surface in a free stream allowing for boundary layer growth and interference from upstream control surfaces.

8.19.1.3 *Efficiency of control surfaces*

Ideally, the operator of any ship should define the standard of manoeuvrability required in terms of the standard manoeuvres discussed in Sections 8.6–8.10. The designer could then calculate, or measure by model tests, the various stability derivatives and the forces and moments generated by movements of the control surfaces, i.e. rudders and hydroplanes. By feeding this information to a computer simulation a prediction can be made of the ship performance, compared with the stated requirements and the design modified as necessary. By changing skeg or fin and modifying the areas of control surfaces, the desired response may be achieved.

As a simpler method of comparing ships, the effectiveness of control surfaces can be gauged by comparing the forces and moments they can generate with the forces and moments produced on the hull by movements in the appropriate plane. Strictly, the force and moment on the hull should be the combination of those due to lateral velocity and rotation, but for most purposes they can be compared separately; for example the rudder force and moment can be compared with the force and moment due to lateral velocity to provide a measure of the ability of the rudder to hold the hull at a given angle of attack and thus cause the ship to turn. The ability of the rudder to start rotating the ship can be judged by comparing the moment due to rudder with the rotational inertia of the ship. The ability of hydroplanes to cope with a lack of balance between weight and buoyancy is demonstrated by comparing the force they can generate with the displacement of the submarine. It is important that all parameters be measured in a consistent fashion and that the suitability of the figures obtained be compared with previous designs.

8.19.2 Presentation of rudder data

The notation and particulars of the control surface, forces and centre of pressure are given in Figures 8.26 and 8.27. The rudder and propeller coefficients are defined in the following manner:

Geometric definitions,

Chord (c)	length of chord from leading edge to trailing edge
Span (S)	overall length of lifting surface
Plan area (A)	span × mean chord
Root chord (C_R)	length of chord at root
Tip chord (C_T)	length of chord at tip
Taper Ratio (TR)	C_T/C_R
Mean chord ($\bar{c}$)	($C_T + C_R$)/2, or Area/Span, A/S
Aspect ratio (AR)	span/mean chord, $S/\bar{c}$, or span^2/plan area, S^2/A

Forces:

In the free stream, rudder lift (L) and drag (d) forces are non-dimensionalized using the free-stream (ship wake) speed V:

$$C_L = \frac{L}{\frac{1}{2}\rho A V^2}, \quad C_D = \frac{d}{\frac{1}{2}\rho A V^2} \quad (8.55)$$

These are presented in terms of the rudder incidence α.

Also,

$$C_N = C_L \cos\alpha + C_D \sin\alpha \quad (8.56)$$

and

$$C_N = \frac{N}{\frac{1}{2}\rho A V^2} \quad (8.57)$$

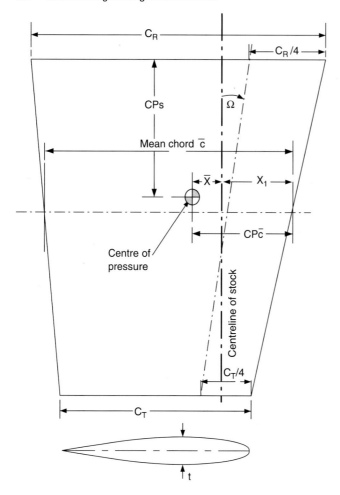

Figure 8.26 Notation for typical all-movable control surface.

In the case of the rudder downstream of a propeller, C_L and C_D are presented in terms of rudder incidence α and propeller thrust loading K_T/J^2 for particular rudder and propeller geometries, where

$$J = \frac{V}{nD} \tag{8.58}$$

and propeller thrust and torque coefficients are defined as

$$K_T = \frac{T}{\rho n^2 D^4}, \qquad K_Q = \frac{Q}{\rho n^2 D^5} \tag{8.59}$$

Different presentations of the rudder and propeller forces are used for low speed and four-quadrant operation and these are discussed in Section 8.19.5.4.

Centre of pressure, Figures 8.26 and 8.27.
Centre of pressure, chordwise,% chord from leading edge: CPc
Centre of pressure, spanwise,% span from root: CPs

For design purposes it is necessary to be able to estimate the forces and moments on a particular rudder at a given angle of attack to the flow and given inflow speed. The presentation of experimental data is usually in the form of lift, drag and moment characteristics in coefficient form over a range of rudder incidence, together with the location of the centre of pressure in the chordwise and spanwise directions, as illustrated in Figure 8.28. The presentation will also normally identify the stall angle α_{stall} and maximum lift coefficient $C_{L\max}$ together with the minimum profile drag coefficient, C_{Do}. The influence of a propeller upstream of a rudder on rudder performance characteristics will normally be depicted by curves resulting from different levels of

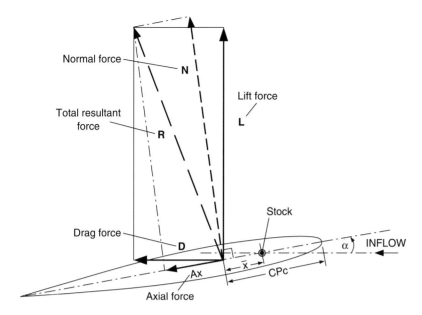

Figure 8.27 Notation of forces and angles for all-movable control surface.

propeller thrust loading, K_T/J^2, Figure 8.29, since the propeller induced velocity is a function of K_T/J^2.

8.19.3 Rudder design within the ship design process

Previous sections have described the action of the rudder and its role in course-keeping and manoeuvring. It is apparent that a fundamental requirement of the rudder is to produce sideforce in the most efficient manner, that is to produce the required lift with minimum drag. The rudder also has to fit into the practical layout of the aft end of the ship or boat with possible constraints in rudder size and shape, and operate under the influences of the upstream hull and, in many cases, propeller. It should be noted that it is advantageous to place the rudder in the slipstream of a propeller. The accelerating affect of the propeller on the flow leads to greater rudder inflow velocities and rudder forces. Also, the propeller has a straightening effect on the cross flow at the stern on a turn, increasing rudder incidence and generating more lift.

As well as a limited space for the rudder, the draught may be limited which in turn will limit the span of the rudder. A larger span and aspect ratio leads to a more efficient rudder in terms of reduced drag for a given lift. Good coursekeeping might favour a rudder with a high lift curve slope and rapid response resulting from a large aspect ratio, whilst manoeuvring performance might be enhanced with a large stall angle, which is more likely to be achieved with a small aspect ratio. Before starting the rudder design process a clear set of requirements for the rudder should be defined, but it is apparent that a compromise design is a likely outcome as a result of the various constraints.

The relative geometrical arrangement of the rudder, propeller and hull can have significant influences on both manoeuvring and propulsion performance. The effect of the hull is generally to slow down the flow into the propeller and that of the propeller is to accelerate and rotate the flow into the rudder, thus affecting its performance. The proximity of the rudder also influences the propeller upstream and the overall propulsive effect of the propeller–rudder combination. It is therefore necessary to devise an overall stern arrangement that satisfies the design requirements in terms of propulsion, speed and fuel consumption whilst ensuring the vessel is able to maintain its course and satisfy manoeuvring requirements at both low and service speeds. The position of the propeller relative to the hull is generally considered in terms of the stern tube or stern-frame design and minimum propeller to hull clearances. There is normally more freedom in the siting of the rudder relative to the propeller and hence hull.

The principal geometrical properties which affect rudder–propeller interaction relate to the relative positions of the rudder and propeller and may be summarized as the longitudinal separation (*X/D*), lateral separation (*Y/D*) and vertical position (*Z/D*), where D is the diameter of the propeller, Figure 8.30. See Molland and Turnock (2007) for discussion of separation effects.

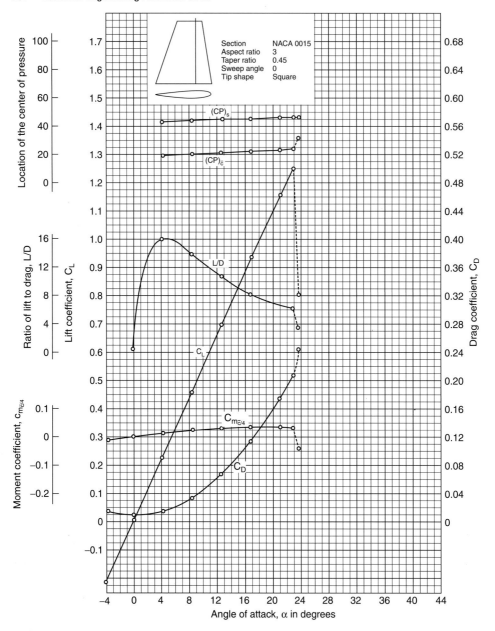

Figure 8.28 Example of test results (ahead) Whicker and Fehlner (1958).

8.19.4 Detailed rudder design

8.19.4.1 *Background*

The forces developed by a rudder or control surface depend fundamentally on its area, profile shape and aspect ratio (ratio of span to chord), section shape, the square of the inflow velocity, the density and viscosity of water and the rudder angle of attack or incidence α. For a precise estimate of the rudder forces all of the parameters have to be taken into account.

Classical approaches to rudder design have tended to concentrate mainly on the derivation of rudder torque and a suitable stock diameter. Rudder force and its centre of pressure were derived using empirical equations that were based on relatively

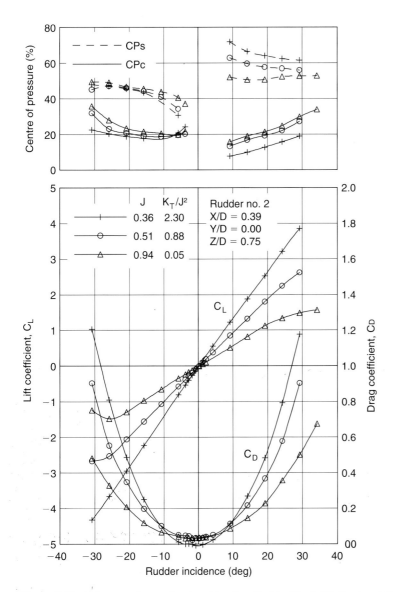

Figure 8.29 Influence of propeller thrust loading: Rudder No. 2, (Molland and Turnock, 2007).

limited model- and full-scale data. Derived equations are of the form:

Force $N = kf_1(\alpha)AV^2$
$CPc = f_2(\alpha)AV^2$
Torque $Q_R = N \times$ lever of stock position to CPc

In early formulae, $f_1(\alpha)$ is a function of the rudder angle only and does not take account of aspect ratio, although later formulae would allow for this in the constant k. There were many alternatives for $f_1(\alpha)$ and a popular choice was $\sin \alpha$. Another assumption was to take $CPc = 0.375\bar{c}$ at 35° rudder angle.

Such formulae were used extensively for many years, with different values of the coefficient developed for various ship and rudder types.

A significant change in the level of understanding of the behaviour of control surfaces, with relatively low aspect ratio suitable for marine applications, occurred with the publication of the extensive freestream tests of Whicker and Fehlner (1958). An early publication using the results of Whicker and Fehlner, and a more rigorous approach to rudder forces, torques and moments, was that of Taplin (1960). He includes example calculations to illustrate the methodology. Other publications of note through that

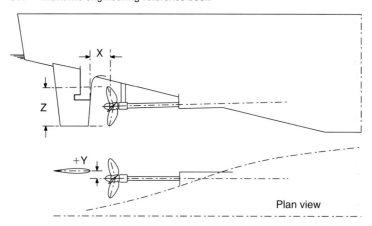

Figure 8.30 Definitions of X/D, Y/D, Z/D.

period include those of Romahn and Thieme (1957), Thieme (1965) and Okada (1966).

Harrington (1981) presents an extensive review of rudder torque prediction. He makes use of the Whicker and Fehlner data. He includes estimates of the frictional losses in the rudder bearings and compares his results with full-scale measurements.

In the field of yachts and small powercraft, relevant publications on rudder design include those of Millward (1969) and Molland (1978).

The manual of Brix (1993) presents a wide range of detailed information concerning the design of rudders of many types. It also includes a thorough review of other manoeuvring devices.

Son *et al.* (2001) and Kresic (2002) present methods for estimating the torque of semi-balanced skeg, or horn, rudders. Son *et al.* use the modified lifting line analysis of Molland (1985) for the prediction of forces and *CPc* and carried out a regression analysis of the results. Satisfactory agreement with full-scale data was obtained. Kresic developed a program for detailed estimates of torque for skeg rudders. He compared his results with model data, including the free-stream work of Goodrich and Molland (1979).

It can be seen that the publication of the free-stream characteristics of various rudder series, including all-movable, flapped and semi-balanced skeg rudders, has enabled a more rigorous analysis and physically correct approach to be used for the design of marine rudders and control surfaces. Consequently, the most common rudder design and performance prediction method currently employed entails the use of free-stream or open-water characteristics for a particular rudder or control surface. The free-stream characteristics of the rudder represent its performance in the absence of the hull, propeller, or appendages. In order to take account of the presence of the hull and/or propeller, corrections are applied to the rudder aspect ratio, the inflow velocity to the rudder and to the rudder angle of attack to yield their effective values. These effective values are then used to enter the appropriate free-stream characteristic curves, and hence to compute the rudder forces and moments. The rudder in the free-stream condition and the modifying effects on the free stream of the hull and propeller are, therefore, treated as individual components of the complete system. This approach is discussed further in the next Section.

A number of mathematical models of rudder–propeller interaction using the individual component approach have been developed over the years, generally using actuator disc theory to model the rudder axial inflow velocity. This approach, together with a relatively large number of empirical modifications, can achieve reasonable predictions in simulations. The method of using free-stream characteristics with correction factors is, however, deficient in that it does not correctly account for the actual physical interaction between the various components including, for example, the asymmetric performance of a rudder downstream of a propeller, the spill over effects when the propeller slipstream is not completely covering the rudder span, or the significant increase in stall angle when the rudder is downstream of a propeller. Consequently, test data have been derived in various investigations for the rudder–propeller combination working as a unit. In this case, the rudder plus propeller is modelled as a combination in isolation. The influences of an upstream hull and drift angle β are then applied in the form of velocity and flow straightening inputs to the basic isolated model of the rudder–propeller combination. The feasibility of this approach has been demonstrated through experimental work, Molland and Turnock (2002), which has indicated, for example, that a systematic change in drift angle applied to the rudder–propeller combination leads to an effective shift in the sideforce characteristics of the combination by an angular offset. Thus if these stages in the procedure are modelled in the manner described, with sufficient detail and adequate accuracy, then a versatile and more physically correct model of rudder action in the presence of a propeller can be established. A design methodology using the rudder–propeller interaction data of Molland and Turnock is presented by Smithwick (2000) and Molland et al. (1998).

8.19.4.2 Rudder design process

In the process of designing a rudder it is necessary to identify the performance and design requirements, choose and apply a rudder with appropriate geometric parameters and estimate its performance characteristics for the given flow conditions. The overall rudder design process may then be summarized as follows:

Input rudder parameters

(i) Number of rudders
(ii) Rudder type
(iii) Area
(iv) Aspect ratio
(v) Profile shape: taper ratio and sweep
(vi) Chordwise section shape and thickness
(vii) Position of stock, balance
(viii) Rudder location relative to hull
(ix) Rudder location relative to propeller

Input flow conditions

(i) Effective inflow velocity
(ii) Effective rudder incidence

Output data

(i) C_L over range of incidence
(ii) C_D over range of incidence
(iii) C_{Lmax}
(iv) α_{stall}
(v) Centre of pressure
(vi) Pressure (load) distribution

Outcomes

The output data are used to derive rudder torque and bending moments to size the rudder stock diameter, size the steering gear, estimate rudder scantlings from the load distributions and provide lift and drag data for coursekeeping and manoeuvring simulations. An outline of the overall rudder design flow path is shown in Figure 8.31.

The following section discusses the topics within the design process:

1. Rudder parameters

Number. The number of rudders will depend on the ship type and service, or yacht or boat size and purpose. In motor-propelled vessels, the number of rudders will generally follow the number of propellers. In sailing yachts, the number will depend on the required total rudder area and performance requirements.

Type. Typical rudder types are shown in Figure 8.24, Section 8.19.1. The rudder type chosen will often be related to the ship type and stern arrangement. There are, however, circumstances where alternatives may be available, such as the choice between an all movable spade-type rudder, a full-skeg rudder, or a semi-balanced skeg rudder. Typical reasons for choosing the alternatives include hydrodynamic performance, structural design, layout and maintenance.

Area. Rudder area would ideally be estimated using a coursekeeping and manoeuvring simulation that would indicate the size of the rudder necessary to provide a certain level of steering performance. In practice, this is generally not possible at the preliminary design stage when the stern arrangement, propeller and rudder layouts are being decided. An alternative, and often used procedure, is to estimate the area, generally based on a proportion of the immersed lateral area, from the area used for similar ships with satisfactory steering properties. This has been found to be a satisfactory procedure for existing ship types and aft end layouts. Care must however be exercised if radical changes to the aft end layout are applied, when more fundamental investigations may be necessary including model tests and simulations.

Aspect ratio. Aspect ratio may be deemed the most important parameter as far as hydrodynamic performance is concerned, with increase in aspect ratio leading to an increase in overall hydrodynamic efficiency of the control surface. In merchant ships, aspect ratio tends to evolve as a result of the rudder–propeller layout. For example, if there are any draught limitations, then maintaining a required rudder area will lead to an increase in rudder chord length and decrease in aspect ratio. Such low aspect ratios can be seen on shallow draught inland waterway vessels. The shape of the hull above the rudder is important in that it affects the aspect ratio used for the performance predictions. It should be noted that an increase in aspect ratio for a spade rudder can lead to conflicting outcomes since it will lead to an increase in rudder root bending moment, increase in root thickness for structural reasons and a consequent decrease in hydrodynamic performance.

Profile shape. Profile shape tends not to have a significant influence on hydrodynamic performance. Small amounts of taper and sweep tend to be the norm. Further adjustments to shape may occur to suit particular stern arrangements.

Section shape. The choice of chordwise section shape will follow design requirements for hydrodynamic performance. Standard aerofoil type

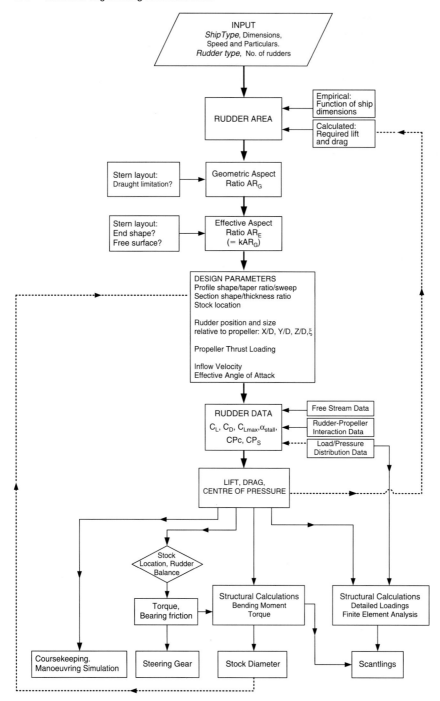

Figure 8.31 Rudder design flow path.

sections are used in most cases, but specialized sections may be employed where increased lift curve slope, delayed stall, low drag or the avoidance of cavitation is sought. Numerical methods can be usefully employed in the design of section shapes. Section thickness generally results from structural requirements. See also Figure 8.41(a–f).

Balance. Balance can be fundamental to the rudder design since it influences tiller forces and steering gear size. However, as the centre of action of the forces (centre of pressure) moves aft with increase in incidence, it is generally not possible to fully balance a rudder or control surface over a range of incidence. The location of the stock will depend on whether the centre of pressure should always be aft of the stock, which would lead to a trailing rudder in the event of a tiller or a steering gear malfunction, but with relatively high torques at large incidence, or a compromise where some negative torque is accepted at small angles in order to lower peak torques at large angles. In this case, the rudder will flop over and increase in angle of attack in the event of a tiller or steering gear malfunction. In order to limit an excessive size of steering gear, this tends to be the practice for large merchant ships. In the case where astern operation is important, such as for some ferries and warships, a compromise stock position may have to be adopted to achieve peak ahead and astern torques at broadly the same level.

Rudder-hull. The rudder location relative to the hull can be important as it may influence the end effect between the rudder and the hull, the effective aspect ratio and, consequently, hydrodynamic performance.

Rudder-propeller. The rudder location relative to the propeller influences the performance of the rudder depending on the relative longitudinal locations, the amount of asymmetry in the propeller race and the proportion of the rudder within the propeller race. The whole propeller diameter will, where possible, be within the rudder span to utilize fully the accelerated flow from the propeller. In practice, this may not always be achievable, such as the case of some small twin-screw ships.

2. *Flow conditions*

Velocity. In the case of a sailing craft, or a twin-screw motor ship with a rudder not in way of the propellers, the effective inflow velocity will be estimated by taking into account the slowing down effect of the hull. In the case where the rudder is operating downstream of a propeller, this will amount to estimating the slowing down effect of the hull on the propeller, together with the accelerating effect of the propeller on the flow into the rudder. Where performance data are available for the rudder operating downstream of a propeller, the data are entered at the appropriate propeller thrust loading, K_T/J^2. Alternatively, the propeller induced velocity is applied directly to the free-stream rudder data.

Incidence. The effective inflow incidence on a control surface is likely to be different from the set incidence. For a sailing craft, the rudder can often be operating in the downwash of the keel. For a ship or a boat just entering a turn, the hull develops a drift or leeway angle which decreases the effective rudder angle, whilst the hull and the propeller have flow straightening effects, which increase the effective angle. These factors will be taken into account in manoeuvring simulations and may be considered in the preliminary rudder design process.

3. *Output data*

For given effective inflow velocity and incidence, performance data, e.g. Figure 8.28, will be obtained and applied for the control surface type and size under consideration.

4. *Outcomes*

The output data can be used in a systematic way to estimate the forces acting on the rudder, to estimate the diameter of the rudder stock and to size the steering gear. The data will broadly be applied in the following manner:

(A) *Forces, torques, moments*: Levers of centres of centre of pressure, Figure 8.26:

$$\bar{x} = \left[\frac{CPc}{100} \times \bar{c}\right] - x_1$$
$$= \left[\frac{CPc}{100} - \frac{x_1}{100}\right] \times \bar{c} \quad (8.60)$$

$$\bar{y} = \frac{CPs}{100} \times S \quad (8.61)$$

Force data:

$$C_N = C_L \cos\alpha + C_D \sin\alpha \quad (8.62)$$

where α is the effective rudder incidence

Normal force $\quad N = C_N \times 0.5\rho A V^2 \quad (8.63)$

Rudder torque $\quad Q_R = N \times \bar{x} \quad (8.64)$

Typical curves of CPc, C_N and Q_R are shown in Figure 8.32. In the example shown, the rudder has some balance and the stock axis has been chosen whereby there is some negative torque at low angles of attack, leading to a lower maximum (positive or negative) torque.

Resultant force coefficient: $\quad C_R = \sqrt{C_L^2 + C_D^2}$
$$(8.65)$$

and resultant force $\quad R = C_R \times 0.5\rho A V^2$
$$(8.66)$$

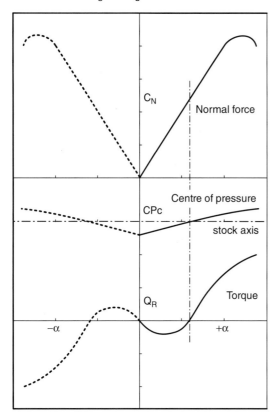

Figure 8.32 Typical curves of rudder normal force C_N, centre of pressure CPc and torque Q_R.

Root bending moment (spade rudder case)

$$M = R \times \bar{y} \qquad (8.67)$$

Equivalent bending moment

$$BM_E = \frac{M}{2} + 0.5\sqrt{M^2 + Q_R^2} \qquad (8.68)$$

Equivalent torque

$$Q_{RE} = M + \sqrt{M^2 + Q_R^2} \qquad (8.69)$$

Diameter of rudder stock

$$D = \sqrt[3]{(BM_E \times 32)/\pi \times \sigma}$$
$$\text{or} \quad = \sqrt[3]{(Q_{RE} \times 16)/\pi \times \sigma} \qquad (8.70)$$

where σ is the allowable stress in the stock material.

For an equivalent tubular rudder stock, the outside and inside diameters d_1 and d_2 have to satisfy the equation:

$$D = \sqrt[3]{(d_1^4 - d_2^4)/d_1} \qquad (8.71)$$

(B) *Load distributions and scantlings*: Examples of outline structural layouts for a spade rudder and a semi-balanced skeg rudder are shown in Figure 8.33.

The scantlings for the rudder webs and plating will normally be obtained or checked using the rules of classification societies and standards such as LR (2005), DNV (2006), ABS (1990), ABS (1994), GL (2005) and ISO (2005). Direct calculations may also be carried out. In this case, the load distributions can be established from available experimental pressure distributions, or one of the numerical methods. These load distributions may then be linked to a finite element analysis, FEA, which will determine the structural response of the rudder to the prescribed load distribution. This will allow appropriate structural scantlings for the rudder to be determined. Derived rudder scantlings, together with the derived stock diameter, will then normally be compared with the requirements of the various classification societies and standards.

(C) *Steering gear*: The rudder characteristics determine the hydrodynamic torque, Q_H, which is effectively Q_R in Equation (8.64) and Figure 8.32. The frictional torque Q_F due to friction in the rudder bearings also has to be overcome. The total torque Q_T to be provided by the steering gear is

$$Q_T = Q_F \pm Q_H \qquad (8.72)$$

The sign in Equation (8.72) depends on whether the rudder angle is being displaced or restored. The effect of friction and whether the rudder angle is being displaced or restored is shown schematically in Figure 8.34. This shows the basic hydrodynamic torque Q_H, say Q_R from Figure 8.32, together with Q_F to give the total torque Q_T. As the rudder is displaced, with increasing angle, the effect of the frictional torque Q_F is to decrease the initial negative torque and increase the positive torque at larger angles. When the rudder action is reversed, say at point A, the torque drops to point B on the restoring curve. When the rudder is restoring, with decreasing angle (rudder effectively driving the steering gear), the effect of Q_F is to decrease Q_T at positive torque and increase Q_T at negative torque. Note that, in reality, there will be a complete reversal of sign for the restoring Q_T at point B, which for schematic purposes, is not shown in Figure 8.34.

The frictional torque Q_F is derived from the reactions in the rudder stock bearings due to the total rudder normal force N. The frictional torque at each bearing is then the resultant force multiplied by the coefficient of friction multiplied by the bearing

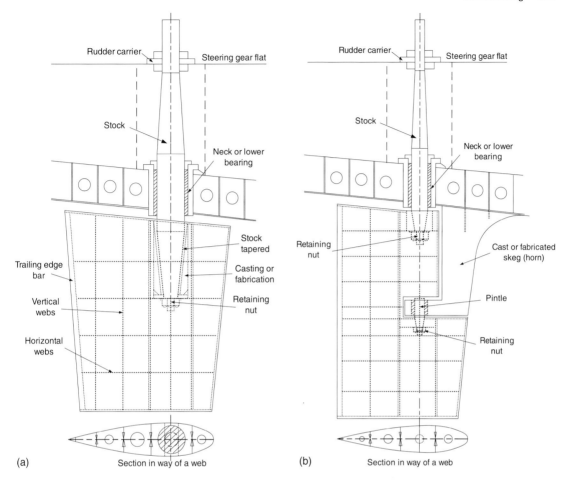

Figure 8.33 (a) Structural layout: spade rudder; (b) Structural layout: skeg rudder.

turning radius. For the case of the spade rudder, Figure 8.35.

$$Q_F = \mu_B R_1 N \left[\frac{y_1 + y_2}{y_3} \right] + \mu_B R_2 N \left[\frac{y_1 + y_2 + y_3}{y_3} \right]$$

(8.73)

where μ_B is the coefficient of friction of the bearing material and R_1 and R_2 are the radii of the bearings. Typical values of μ_B for sleeve bearings are 0.1 for metals and 0.2 for synthetic materials, Taplin (1960) and Harrington (1981).

Harrington (1981) uses a number of worked examples to illustrate the derived hydrodynamic and frictional torques for spade and semi-balanced skeg rudders. Taplin (1960) includes calculations for frictional torque. The frictional torque Q_F is assumed linear with angle, Figure 8.34, and at 30–35° may be about 5–10% of the total torque Q_T. At 10–20°, when hydrodynamic torque Q_H can be very low or tending to zero (but the normal force is still present, Figure 8.32), Q_F will represent a much higher proportion of the total torque.

(D) *Rudder rate*: The steering gear torque may be further influenced by the rate of change of rudder angle. The time to change heading and abilities such as zig–zag overshoot characteristics are affected by rudder deflection rate, although the rate has no influence on the diameter of the steady turn. Minimum rudder rates of $2\frac{1}{3}$°/s are required by regulatory bodies and classification societies. That is typically putting the rudder over from 35° one side to 35° the other side in 30 s, or 35° one side to 30° the other side in 28 s. Such rudder rates are generally acceptable for most ship types. Ideally, a fast rudder rate is called for initially, with rudder angle rising to just below stall. As drift angle develops, the rudder angle would be increased such that the effective angle remains just below stall. As pointed out by Mandel (1953), this ideal goal is generally not possible.

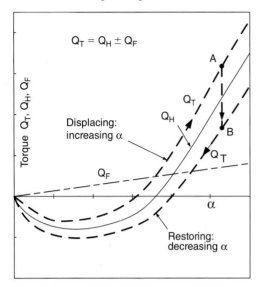

Figure 8.34 Effect of friction in bearings on rudder torque.

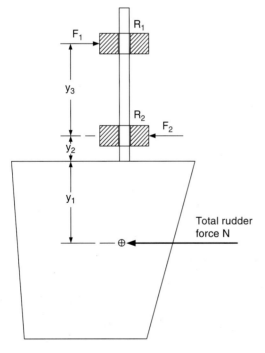

Figure 8.35 Resolution of forces at bearings.

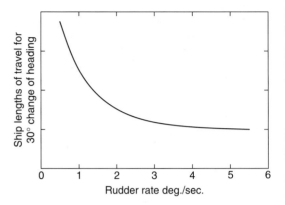

Figure 8.36 Influence of rudder rate.

than larger ones. It was also deduced that, based on the results of the investigation, a good rudder rate to select for most ships would be a rate that corresponds to about 15° rudder deflection per ¼ ship length of travel, and indicates this would amount to a rudder rate of about 5°/s for a 122 m, 20 knot ship. This is in the area where the curve in Figure 8.36 flattens and further increases in rate are not very effective. Mandel concluded that most ships will have a rate somewhat lower than 5°/s. The effect of rudder rate was examined in some detail by Eda and Crane (1965) and Eda (1983a, 1983b). Their results indicate that beyond about 3°/s, further improvements become very small. It should be finally noted that increased rudder rate will increase the torque to a certain extent and increase the load on the steering gear.

(E) *Full-scale tests*: In making predictions for full-scale rudder performance, note will be made of the relevant Reynolds number and whether any correction for scale effects should be included.

Many difficulties can arise when attempting to measure full-scale rudder forces and torques. Ideally, forces and torque would be measured using suitably sited strain gauges on the rudder and rudder stock, and such an approach was used in the trials reported by Becker and Brock (1958). Very few results of such detailed measurements have been published elsewhere. More normally, torque has to be measured indirectly using the pressures in the hydraulic rams of the steering gear. Further corrections then have to be made to allow for mechanical friction in the steering system and in the rudder stock bearings. The difficulties arising from this approach are discussed by Hagen (1972), Taplin (1975) and Harrington (1981). Son *et al.* (2001) compare, for a large tanker, the sea-trial torque from the steering gear with a regression model of rudder torque.

In making a comparison of full-scale measurements with those predicted by model, care has to be taken

Mandel (1953) investigated the influence of changes in rudder rate, based on the time to change the heading by 30°, resulting in a trend shown schematically in Figure 8.36. It was found that increases in rudder rate provided greater improvements in shorter ships

with the measurement or prediction of the actual full-scale rudder angle of attack and inflow speed, allowing for wake, propeller slipstream, hull drift angle and change in rudder angle of attack as the ship turns. With careful consideration of all the corrections, reasonable correlation between model and full-scale can be achieved. The extended discussion to Harrington's paper covers most aspects of the problems associated with full-scale predictions and tests.

A further indirect approach to checking overall rudder forces is to compare the full-scale manoeuvring trial results, such as in Clarke *et al.* (1972), with those from a mathematical manoeuvring simulation that adequately models the rudder, such as in Molland *et al.* (1996). It has to be noted that, as this approach models the hull, propeller and rudder, interaction effects need to be modelled correctly, otherwise acceptable overall manoeuvring predictions may be obtained for the wrong reasons.

8.19.5 Rudder manoeuvring forces

8.19.5.1 *Rudder forces*

The basic requirements of a rudder and the forces acting during a manoeuvre are described in Section 8.4. The manoeuvring performance of a ship is controlled by the performance of its rudder and it is therefore necessary to be able to estimate rudder forces at any stage in the manoeuvre. At a point in the manoeuvre, for a given rudder design and arrangement, it is necessary to estimate the effective rudder incidence and the effective rudder velocity. Knowing the incidence and velocity, the rudder performance data such as that in Figure 8.28, can be used to estimate the forces produced by the rudder.

In a manoeuvre, the total sideforce will be made up of:

(i) the contribution from the rudder,
(ii) the sideforce due to the propeller in oblique flow and
(iii) the sideforce developed on the hull due to the rudder–propeller combination.

The sideforce due to the propeller when the ship is in a turn can be significant and will depend on drift angle and speed. Guidance on the likely levels of propeller sideforce may be derived from research on the performance of propellers on inclined shafts, such as that of Gutsche (1964), Hadler (1966) and Peck and Moore (1973).

From the aspect of propulsion when in a manoeuvre, as well as the basic thrust and torque characteristics of the propeller, account has to be taken of:

(i) Changes in rudder drag or thrust ΔK_R due to the influence of the propeller and,
(ii) Changes in propeller thrust ΔK_T and torque due to the presence of the rudder.

An overall algorithm showing the derivation of the total manoeuvring sideforce and the net propulsive force is shown in Figure 8.37.

8.19.5.2 *Hull upstream*

The hull upstream of a rudder, or rudder–propeller combination, can have a significant influence on the rudder forces and the production of total sideforce.

(i) The hull slows down the flow speed into the rudder, or the rudder–propeller combination. The average wake speed in the vicinity of the propeller can be estimated using a suitable wake fraction. (See Chapter 5, Section 5.2).
(ii) The hull has a flow straightening effect when the ship is on a turn.
(iii) The hull contributes to the production of sideforce, due to pressure changes on the adjacent hull induced by the rudder, Figure 8.38.

The total sideforce (ship axis) may be written as

$$F_T = F_Y(1 + a_H) \quad (8.74)$$

where F_T = total sideforce and F_Y = rudder sideforce.
$(1 + a_H)$ is the contribution due to the hull, which will depend on the distance between the rudder and hull and the overall hull–propeller–rudder arrangement. Typical values of $(1 + a_H)$ are from 1.10 to 1.30. a_H may be expressed as a function of block coefficient as suggested by Hirano (1981), indicating values of $a_H = 0.15$ at $C_B = 0.55$ up to $a_H = 0.30$ at $C_B = 0.80$, and to reflect some dependence on speed. Results of tests by Gong *et al.* (1995) for a tanker with $C_B = 0.73$ and large $B/T = 3.77$ indicate values of a_H ranging from 0.12 at low $J(0.30)$ up to about 0.20 at high $J(0.90)$.

8.19.5.3 *Influence of drift angle*

When rudder angle is applied, the ship develops a drift angle β, Figure 8.2, which leads to a cross flow at the stern and a geometric drift angle at the rudder β_R, which is larger than β. The net effect is a decrease in the effective rudder incidence, although the flow-straightening effects of the propeller and hull, lessen this decrease. The final effective rudder incidence will therefore result from the effects of drift angle and flow straightening. Typical values for flow straightening are contained in Molland and Turnock (2007).

An approximate value for β for single-screw merchant ships is proposed in Lewis (1989), based on the results of Shiba (1960) and Mandel (1953), as

$$\beta = 22.5 \; L/R \; (\text{deg.}) \quad (8.75)$$

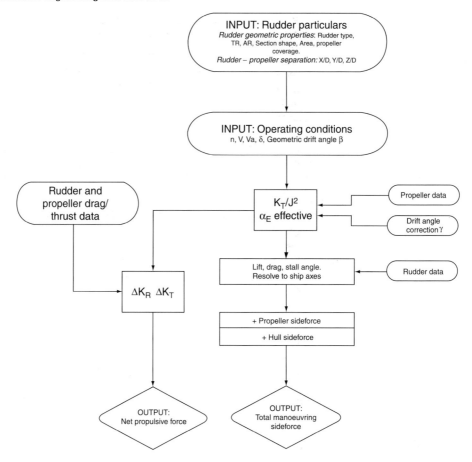

Figure 8.37 Development of total manoeuvring sideforce.

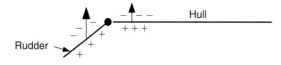

Figure 8.38 Pressures around rudder and hull.

where R is the steady turning radius.

Actual β will depend on the ship form and aft end hull–propeller–rudder arrangement and Mandel suggests β values between $18 L/R$ and $22.5 L/R + 1.4$.

For a rudder located $L/2$ aft of G, β_R at the rudder is related to the ship drift angle β at G as

$$\tan \beta_R = \tan \beta + L/2R\cos \beta \qquad (8.76)$$

As drift angle is developed on a turn, the effective incidence on the rudder decreases and the rudder helm will be increased to compensate. It is apparent that large rudder angles may need to be applied, say up to 60°, to reach the full effectiveness (e.g. stall) of the rudder. For example, for a ship on a steady turn with a diameter of 5 ship lengths, or $L/R = 0.4$, then from Equation (8.75), approximate $\beta = 9°$ and from Equation (8.76) approximate $\beta_R = 19.8°$. A typical flow-straightening angle is about 10° and a rudder with a set helm of 50° would therefore see an effective incidence of about 40°.

8.19.5.4 *Low and zero speed and four quadrants*

1. Methodology

Although most requirements for ship-manoeuvring capabilities are defined at service speed, it is crucial that a ship manoeuvres well at low speed and has a known performance in all four quadrants of operation:

(i) Ship ahead, propeller ahead
(ii) Ship ahead, propeller astern
(iii) Ship astern, propeller astern
(iv) Ship astern, propeller ahead.

The experimental database, described in Molland and Turnock (2007), include tests at zero, low speed and in all four quadrants.

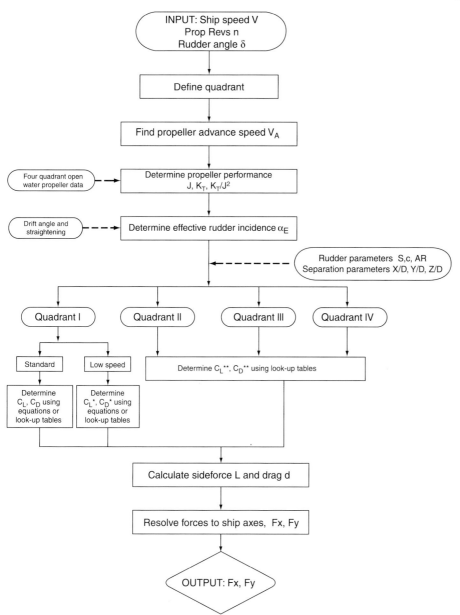

Figure 8.39 Flow chart for four-quadrant rudder force prediction algorithm.

A useful presentation of the data in the four quadrants case is

$$C_L^{**} = \frac{L}{\frac{1}{2}\rho A[V^2 + K_T n^2 D^2]}$$

$$C_D^{**} = \frac{d}{\frac{1}{2}\rho A[V^2 + K_T n^2 D^2]}$$

and are presented in terms of the propeller advance angle $\psi = \tan^{-1} J/0.7\pi$ across the four quadrants for different values of rudder incidence δ.

A flow chart showing the principal features of the approach, using curve fits and look-up tables for the database is given in Figure 8.39. In this example, for the first quadrant data, the standard and low speed presentations are used, namely:

$$C_L = \frac{L}{\frac{1}{2}\rho A V^2} \qquad C_D = \frac{d}{\frac{1}{2}\rho A V^2}$$

$$C_L^* = \frac{L}{\frac{1}{2}\rho A K_T n^2 D^2} \qquad C_D^* = \frac{d}{\frac{1}{2}\rho A K_T n^2 D^2}$$

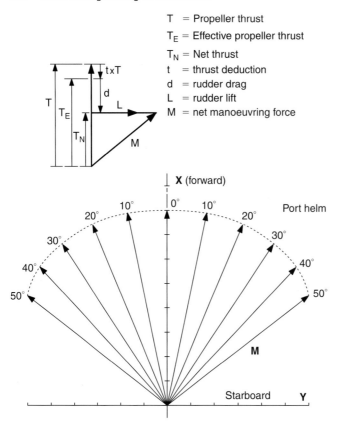

Figure 8.40 Polar plot of net manoeuvring force.

When working downstream of a propeller at low and zero speed, stall angle is delayed and significant sideforce can be generated by a rudder in the static $J = 0$ condition, an attribute utilized in the low speed handling of ships. For low speed work, it is common practice to produce polar or vector diagrams of the net manoeuvring force for various rudder angles, as shown in Figure 8.40. Such a diagram can be applied to manoeuvring simulations and can also be used to compare alternative rudder types, configurations and manoeuvring devices. In Figure 8.40 the propeller thrust has been reduced by the effect of thrust deduction, See Section 5.1.4.2.

Slow-speed manoeuvring is discussed in some detail by Brix (1993). Various simulation models of low speed and backing manoeuvres have been proposed, including those of Oltman and Sharma (1984), Abkowitz (1990) and Shouji et al. (1990). Kang et al. (1998) describe a methodology for predicting the manoeuvring of full-form ships with low speed. A mathematical model is developed and a regression analysis carried out on the results of several model tests to provide a database for manoeuvring simulations. Semi-balanced skeg, Schilling and flap rudders were investigated. It was concluded that special rudders could significantly improve the performance of ships that had poor manoeuvrability.

2. High lift rudders and control surfaces
A number of high lift rudders and control surfaces have been proposed and employed, primarily to enhance low-speed manoeuvring. The main objective of such devices is to extend the rudder angle before stall and increase the maximum lift achievable.

They often result from the modification and/or addition to the ordinary aerofoil type section. A brief review is made of some of the various devices that have been proposed for high lift purposes, as shown in Figure 8.41.

(a) *Flapped aerofoil*, Figure 8.41(a). The concept is used to increase the lift curve slope, delay stall and increase C_{Lmax}. The increased lift curve slope, giving a faster response for a given helm, can be utilized for coursekeeping and for other control surfaces requiring a fast response such as the fin stabilizer. It is the concept used in the Becker high lift rudder, Brix (1993).

(b) *Schilling rudder*, Figure 8.41(b). Has a special section designed to delay stall and increase C_{Lmax}. It is described by Bingham and Mackey (1987).

(c) *Wedge at tail*, Figure 8.41(c). Designed to increase lift curve slope and C_{Lmax}. Can be used in situations requiring a fast response, such as the fin roll stabilizer.

(d) *Gurney flap*, Figure 8.41(d). The Gurney flap amounts to a small flat plate attached at the trailing edge at right angles to the chord. Designed originally to be fitted to one side of an asymmetrical section, it could equally be applied to a symmetrical movable control surface. It induces an effective camber disproportionate to its size, increasing lift for a given incidence. Whilst increasing lift, a penalty is an increase in zero-lift drag and a reduced *L/D* at low to moderate values of lift. An investigation into this concept is included in the work of Date (2001).

(e) *Jet flap*, Figure 8.41(e). The jet flap has origins in the aircraft industry. A thin sheet of fluid is discharged from the trailing edge (either side) at an angle to the chord of the foil. The emerging jet has the effect of increasing the circulation around the foil and foil lift. English *et al.* (1972) describe the operation of the jet flap and report on the results of tests carried out in a water tunnel. With the jet flap working, they found an improvement in lift, a rearward movement in *CPc* and a reduction in drag, in spite of the increased lift. The advantage of this type of flap is that it does not have the mechanical complexities of a conventional hinged flap. Whilst meant for zero and low speed manoeuvring, it has attractions also for coursekeeping, where the rudder would be fixed amidships and the jet flap operated for coursekeeping.

(f) *Blown flap gap*, Figure 8.41(f). When a water jet is blown tangential to the suction surface of a flapped rudder, extra lift force is induced by delaying stall and increasing circulation, especially at large angles of attack A rudder employing this concept was designed and tested by Choi *et al.* (2004). Conventional rudder lift slope was improved with an ordinary flap by 35–64%, depending on flap angle/rudder angle ratio. With the addition of jet injection, there were further large increases in lift. The model test results were used in a manoeuvring simulation and it was concluded that the use of a blown flap rudder is an effective way of improving the ship's tuning ability.

(g) *Rotating cylinder in isolation*, Figure 8.41(g). The rotating cylinder in isolation produces lift due to the Magnus effect. Research has been carried out on the application of such a concept as a ship rudder/low-speed manoeuvring device, Steele and Harding (1970). The concept, using the cylinder alone, does not seem to have had many practical applications.

(h) *Rotating cylinder in association with rudder*, Figure 8.41(h). A thorough review of the design of control surfaces with rotating cylinders is carried out by Cordier (1992). With a rotating cylinder at the leading edge of the rudder, the cylinder imparts energy into the boundary layer. The boundary layer can be controlled and the flow on the back low-pressure side maintained up to very large rudder angles. For example, with this rudder type, angles up to 80° have been achieved without stalling. From the tests reported by Brix (1993), it was found that with optimum rotational speed/forward speed, increases in rudder lift of up to 100% could be achieved at large angles. Work on this concept has been carried out at NPL, Steele and Harding (1970), who also consider the use of a rotating cylinder at the leading edge of a flap. McGeough and Millwood (1981) carried out water tunnel tests on a rotating cylinder rudder (cylinder at fore end of rudder). With the cylinder rotating, the stall angle was delayed from about 20° (without rotor) to about 50°, with an increase in C_L from 0.65 to 1.46. The use of a blown trailing edge cylinder is mentioned by English (1972).

(i) *Rotating cylinder in association with flap*, Figure 8.41(i). The rotating cylinder is located at the leading edge of the flap. The concept provides flow control over the flap and was considered by Steele and Harding (1970).

(j) *End plates*, Figure 8.41(j). End plates have been used over many years to increase the effective aspect ratio of a control surface and to enhance its lift performance. In straight-line flow, there will be an increase in rudder drag due to the frictional drag on the end plates. In oblique flow, which is far more likely in practical situations, there can be significant increases in drag due to separated flow across the plates and shed vortices from the edges of the plates. For this reason, the use of end plates tends to be limited to rudders used mainly in low-speed manoeuvring situations. At low speeds, with the influences of a propeller slipstream, the use of end plates on a high lift section, such as the Schilling section, can lead to very high incidence and lift values before stall, Bingham and Mackey (1987).

(k) *Robust simple rudder*, Figure 8.41(k). It may be important to design a rudder where robustness and reliability are the key design features, such as for the rudder on a vessel working mainly in harbours and coastal waters on manoeuvring and towing duties. As drag is generally not a problem at these low speeds, a flat-plate rudder can be used and the design can concentrate on the

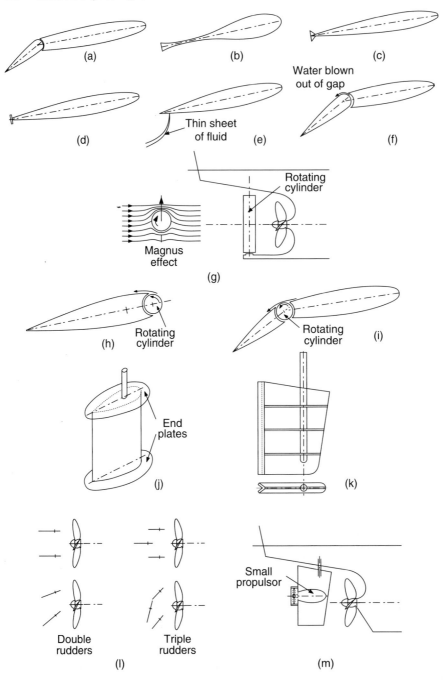

Figure 8.41 Various high-lift rudders.

method of construction and strength, rather than hydrodynamics. Plate rudders, such as that shown in Figure 8.41(k), have been employed for such vessels and construction is typically from 14 mm to 20 mm steel plate, with an oversize stock diameter. Horizontal stiffeners may be used as necessary and a wedge tail (from flanged plate) may be used at the aft end to improve the strength and increase the stall angle. Recommended scantlings for plate rudders are provided by the classification societies, such as LR (2005), DNV (2006), ABS (1990), GL (2005).

(l) *Use of double/triple rudders*, Figure 8.41(l). Twin rudders have been used to vector the slipstream from the propeller in various directions. Such a concept is used with a twin Schilling rudder installation, Bingham and Mackey (1987) and Anon. (1985), with a controller that allows differential operation of the two rudders. Guarino (1984) reports on the use of three rudders in a differential manner, Figure 8.41(l), whereby large steering forces are generated.

(m) *Active rudder*, Figure 8.41(m). This concept consists of a submerged electric motor contained in a streamlined casing, set in a normal rudder with a ducted propeller at the aft end, Brix (1993). Large rudder angles may be employed. The unit also offers some auxiliary propulsion. There tends to be little detrimental effect on the overall cruising efficiency, Anon. (1979).

8.19.6 Numerical modelling of rudder

8.19.6.1 *Available methods*

Since computational machines first became available to Naval Architects in the late 1950s progressively more complex theoretical methods have been developed to analyse the performance of rudders and control surfaces. The more elementary theoretical approaches such as Glauert's (1983) lifting-line theory were the first to be used. It comes as a surprise, especially to the modern student brought up in a world of virtual reality and commercial computational fluid dynamics (CFD) packages, how accurate such an approach can be. In essence this is because a large ship rudder is operating at a relatively high Reynolds number. Neglecting viscous effects will only cause small errors in sideforce and steering torque. Likewise, for large ships, rudder drag is typically 2–3% of total resistance so, again, accurate prediction of rudder drag is less important. However, for high-performance systems, as say typified by the appendages on America's cup yachts, where margins of victory can be a few seconds over many hours of racing, such details do become essential and the computational power of CFD becomes essential.

CFD is the use of computational techniques to solve numerically the equations defining fluid flow around, within and between bodies. Principally, the equations solved are numerical approximations to mathematical models describing the physics of fluid flow. There is, therefore, always an inherent level of approximation to reality. It is the level of abstraction of the CFD analysis from physical reality which determines the amount and form of interpretation of analysis required.

Historically, the origins of our ability to mathematically describe the detailed flow around moving objects such as ships came through the work of such luminaries as Newton, Euler, Laplace, Navier and Stokes. For those interested, a number of detailed fluid texts can be recommended, Anderson (1995), Ferziger and Peric (2004), Cebeci *et al.* (2005) and Katz and Plotkin (1991), that more fully describe the detailed theoretical background to CFD.

A hierarchy of four CFD methods are considered in order of increasing complexity:

- lifting-line methods,
- surface panel or boundary element methods,
- Reynolds-averaged Navier–Stokes (RANS) methods and
- Large Eddy Simulations (LES) and Direct Numerical Simulation (DNS) methods.

However of the four methods listed above only the first three are as yet commonly applied to rudder and control surface design. All four methods vary in complexity, and have particular advantages and disadvantages associated with their physical realism and computational cost. The most challenging aspect of numerical analysis is to capture the interaction between a propeller and rudder as well as the secondary effects of hull and free surface.

8.19.6.2 *Potential flow methods*

The methods described are based on the properties of Laplace's equation. A much fuller description of all these methods is given by Katz and Plotkin (1991).

(a) Lifting-line formulation
Glauert (1983) used a system of bound vorticity to represent the lift (or sideforce) generated by a wing or control surface. Lifting-line theory is a simple method of quickly establishing the performance of a control surface. The three-dimensional (3-D) lifting surface and wake are modelled as a series of horseshoe vortices, known as lifting lines, whose strength is initially unknown.

This method has the advantage that it provides good estimates of spanwise loading, and induced drag, whilst remaining simple to implement and computationally inexpensive. However, it has a number of specific limitations:

- The theory is limited to lifting surfaces of relatively high aspect ratio, and assumes that the wake is aligned in the local flow direction.
- It assumes that the lifting surface is of zero thickness, and therefore neglects any flow effects resulting from section camber and thickness.
- As it is based on potential flow theory, this method neglects viscous flow effects. Hence, it is unable to model directly frictional drag, flow separation and stall.

Molland (1981) describes how the standard method can be enhanced when considering the typically low aspect ratio of ship rudders.

(b) Boundary element (surface panel) methods

In the early 1960s, as a consequence of increased computing power, a new numerical approach to the lifting surface problem, known as the panel method or boundary element method began to emerge. This method promised to overcome many of the problems of the early lifting line and the later vortex lattice approach. The technique allowed the treatment of more complex geometries, and actually models the lifting surface itself, allowing the effects of thickness and camber to be calculated. The basic principle of the panel method is based on the linear superposition of source/sinks, vortices and/or doublet elements over the lifting surface, such that the boundary conditions are satisfied on the body, across the wake and in the far field. See also Section 8.17.

Extensive research has been carried out in this field. A detailed overview of this method is provided by one of the pioneers of panel methods, Hess (1990).

The advantage of this approach is that it can be used to model actual geometries without requiring further simplification of the geometry. Although panel methods are more complex than lifting-line methods, the computational effort required is still less than that needed for RANS methods. A panel code used to solve a 3-D rudder flow required only 1% of the computational effort needed by a RANS code to solve the same flow problem, Turnock and Wright (2000). The advantage of the panel method is that computations are carried out to determine unknowns only on the body, wake, and far field boundary surfaces, and not throughout whole fluid domain.

Panel methods allow considerable freedom in their numerical application so that complex flow features (rotational and viscous) such as wake roll up, separation zones and unsteadiness can be incorporated. However, as a potential flow method they, like the lifting-line method, cannot account directly for frictional drag, separation and stall effects.

(c) Coupled boundary layer

The surface panel method can be enhanced to include the effect of viscosity through coupling a method for evaluating a solution of the thin boundary layer approximations to the full Navier–Stokes (N–S) equations along a series of surface streamlines.

There are two approaches:

(1) The geometry of the body in question is altered by increasing its size in the surface normal direction by an amount equal to the displacement thickness of the local boundary layer.
(2) Rather than the imposed zero normal relative velocity condition on the body surface, a flux of momentum is applied. The magnitude of the flux is proportional to the rate of momentum exchange at the edge of the boundary layer.

The three steps necessary to include the boundary layer growth in the flow solution are:

1. solve the potential flow over the body and obtain the surface pressure distribution;
2. using the pressure distribution, calculate boundary-layer characteristics and
3. modify the surface boundary conditions for the potential flow, and solve for the next iteration.

Considerable effort has been expended in developing accurate methods for solving the thin boundary-layer equations. Particular attention has been given to the prediction of laminar-turbulent transition, attached small separation bubbles and the capture of large zones of separation. In many cases, these methods and their associated mathematical complexity and significant computational effort, are in the process of being superseded by the complete N–S solvers.

Theoretical methods that have the limited objective of predicting overall characteristics of the boundary layer, for example momentum thickness, displacement thickness and skin friction, rather than details of the actual flow, are more straightforward to apply. For the purpose of improving rudder performance prediction, such methods are all that are needed to modify the potential flow.

8.19.6.3 *Navier–Stokes methods*

It is only in the last 10 years that it has become practical to solve the N–S equations, based on Reynolds averaging around 3-D free-stream rudders, with any degree of confidence. Previous work was always limited by the availability of sufficient computational power and memory to define the 3-D computational mesh (or grid) around the rudder and a sufficiently large surrounding domain.

The development of the finite volume method, used by most commercial and research flow solvers, results from the surface integration of the conservative form of the complete N–S equations over a 3-D control volume, as for example, explained by Versteeg and Malalasekera (1995). The resulting equations express the exact conservation of the relevant flow properties within the control volume. This relationship between physical conservation and the governing equations forms one of the main attractions of the finite volume method.

1. RANS equations

Although the complete N–S equations govern both laminar and turbulent flows, they are not suitable for the direct computation of turbulent flows. To do so would require computers estimated to be of the order of at least 10^4 times faster than today's (2006) fastest supercomputer. This requires use of extremely fine grids and over a large number of time steps, in order to capture the turbulent motion at the smallest time and length scales. The statements above assume that accessible computer power has improved by a factor of 10–100 in the 15 years since the values stated by Speziale (1991).

2. DNS and LES

DNS and LES computations of marine flows are as yet uncommon. DNS involves the direct solution of the unsteady N–S equations, and are thought to be capable of resolving even the smallest eddies and time scales of turbulence within a flow. Although the DNS method does not require any additional closure equations (as in the case of the RANS method), very fine grids and extremely small time steps need to be used, in order to obtain accurate solutions. This method is currently confined to simple flow problems at relatively low Reynolds numbers. DNS computations of the fully turbulent high Reynolds number flows associated with ship flows await major advances in computational hardware. Although DNS solvers are limited to solving low Reynolds flows, they are seen as playing a role in further RANS code turbulence model development.

Like DNS codes, the use of LES solvers is still mainly as a research tool. LES, Wilcox (1998), is a method that can be used to predict accurately the large scale turbulent structures within a flow, requiring a subgrid scale model to represent the smaller scale eddies. Although only the large scale eddies are resolved individually, this still requires the use of extremely fine grids, making solutions expensive and demanding on present computer resources. For example, this method has been successfully utilized in solving numerous high Reynolds number problems, like the turbulent flow over a NACA0012 aerofoil, carried out by Creismeas (1999).

8.19.6.4 Rudder–propeller interaction

The core problem associated with investigating the influence of an upstream propeller on the behaviour of a control surface is the induction of significant swirl velocity and axial acceleration. Prediction of the magnitude of the swirl and local velocity is the key to the successful analysis. The induced flow across he rudder is unsteady but dominated by the circumferential mean flow, Tsakonas et al. (1970), Cho and Williams (1990).

(a) Lifting line/BEM
The circumferential mean flow influence can be captured by modifying the inflow to the lifting-line method. Molland and Turnock (1996). In this work the propeller race flow is estimated using blade element-momentum theory. The rudder both blocks and diverts the flow through the propeller. This effect is included through use of theoretical estimates based on surface panel calculations.

(b) Surface panel
It is possible to incorporate the effect of the propeller within a panel or lifting surface method, Li (1996), Guo and Huang (2006), Willis et al. (1994), Turnock et al. (1994), Tamashima et al. (1993), Söding (1998) and Laurens and Grosjean (2002), and recently a team at INSEAN, Felli et al. (2006), have all used a variety of boundary element methods to model ship rudders operating under both free stream and propeller flow conditions. The methods vary in their detail and complexity which they use to represent the structure of the propeller race. The ongoing work at INSEAN includes a complete, unsteady calculation of the wake shape evolution even in the presence of the rudder. The challenge in such approaches is in how to deal with the difficulty of controlling the numerical instabilities associated with wake roll-up while at the same time allowing the vortex filaments to stretch correctly as they arrive at the rudder leading edge and sweep along either side of the rudder.

The simpler approach is to start by considering just the circumferential mean influence of the propeller race. The problem can then be separated into two separate calculations, for example, Cho and Williams (1990) use this approach for wing/propeller interaction. The interaction effects are then captured through a modification of the respective inflow conditions.

Turnock (Turnock et al. (1994), Turnock (1993, 1994)) developed a surface panel method based on the perturbation potential formulation with the aim of capturing the interaction between the rudder and propeller. This program, Palisupan, captures multiple body interaction by splitting the flow solution into multiple domains within which a number of bodies are placed. The interaction between each domain is captured through the flowfield modifications induced by the bodies within one domain on all others.

Using this interaction velocity field (IVF) method the bodies are not all modelled in one numerical pass thus creating a further iteration loop around the solution method described earlier. Take the example of rudder–propeller interaction, first the propeller flow model is solved to get a velocity influence upon the rudder. The rudder flow is then solved with the modified inflow velocity field to get a subsequent field on the propeller. This process is repeated with the starting point on the propeller being the velocity influence solved in the previous numerical pass. The procedure repeats until the difference in the results of body forces have iterated down to a minimum required value.

(c) RANS

At the current time, it is not possible routinely to use RANS methods to predict the full unsteady interaction of the hull, propeller and rudder, to the level required for rudder designers, Laurens (2003), Simonsen and Stern (2005), Abdel-Maksoud and Karsten (2000). Propeller effects can be incorporated within the RANS method using one of two approaches. The simplest way is to model the propeller as an actuator disk. This method involves applying body forces, i.e., the source terms in the momentum equations, to the cells located within the propeller disk, such that the flow is accelerated in the same way as a propeller with an infinite number of blades, with the required thrust and torque. This actuator disk approach was proposed by Schetz and Favin (1977). However, this method only accounts for the axial and tangential forces, and neglects any radial force components, which would be present in the real flow. For simplicity, the effect of the propeller is usually represented as circumferentially averaged body forces, input into the steady RANS momentum equations, hence neglecting any unsteady effects. It is perfectly feasible for unsteady body forces to be included in unsteady RANS momentum equations. However, due to the high computing overheads associated with time accurate simulations, these computations are uncommon. Examples of investigations using the circumferentially averaged body forces, and time varying body force approaches are found in Turnock and Wright (2000), Laurens (2003), Simonsen and Stern (2005), Stern *et al.* (1994) and Tzabiras (1997). Various degrees of success have been obtained using this body force method, with qualitative results comparing more favourably than the quantitative results.

The second, and more complex, way of incorporating propellers within the RANS model, is to compute the actual unsteady flow over the real rotating propeller geometry. This method is complicated, requiring the generation of complex non-matching grids around the hull, propeller and rudder geometries, with fixed and rotating frames of reference. In addition, this approach needs to take into account the different time scales in the flow, as the propeller flow requires a much smaller time step than the ship flow. This approach requires extremely large computing resources. Limited examples of such calculations are available, McDonald and Whitfield (1996). Improvements in the mesh generation capabilities and more suitable turbulence models are required before the full potential of this method can be exploited. A good overview of capabilities in application of CFD to hull–propeller–rudder systems is given in Hino (2005) and ITTC (2005).

Notwithstanding the lack of progress in this area, it can be expected that progressively more success will be achieved as computing power reduces in cost and increases in availability.

8.19.6.5 *Unsteady behaviour*

A ship rudder can be located in the race of a propeller, which in turn is located within the wake of a hull; all of these are subject to a greater or lesser extent by the presence of the free surface and the motion of the ship. Such unsteady flows as seen by the rudder are still a particularly challenging area for the application of time accurate RANS equations. One of the difficulties is whether the assumption that the time period of unsteady turbulent fluctuations is sufficiently distinct from that of the variations due to the propeller race.

The highly turbulent, periodic and interactive wake produced by a propeller, gives rise to rudder performance characteristics which differ significantly from those experienced in a free stream. Date (2001), imposed periodic flow conditions, representative of the flow produced in the wake of a propeller, to both 2-D NACA0020 and high lift sections. It was expected that a better understanding of the performance of rudders operating in propeller wakes could be achieved. The response prediction of 2-D rudder sections subjected to periodic flow conditions can be regarded as a necessary first step towards understanding the requirements for full periodic 3-D rudder computations.

8.19.7 Guidelines for rudder design

When considering the overall design process and in deciding the most appropriate rudder or control surface for a particular task, three complementary areas of knowledge can be used:

(1) Empirical knowledge derived from in-service experience but, more usually these days, from the results of model-scale experimentation of varying levels of complexity and expense, Section 8.19.2.
(2) Theoretical investigation, using dimensional analysis, that allows the categorization of the appropriate flow regime and then adoption of a suitable mathematical approximation, Section 8.19.6.
(3) The use of numerical methods to solve the many fluid dynamic equations required to discretize a complete domain, Section 8.19.6.

These three areas are discussed in some detail by Molland and Turnock (2007).

The most effective design strategy will be one that permits an appropriate synthesis of the three areas in a suitable blend.

It is difficult to develop a universal guideline. As in the majority of designs, the quality and sophistication of the final product will depend on the resources made available, both in terms of expenditure associated with model testing (computational or experimental) and, most importantly, the amount of time available for work on the project.

References (Chapter 8)

Abdel-Maksoud, M., and Karsten, R. (2000). Unsteady numerical investigation of the turbulent flow around the container ship model (KCS) with and without propeller, *Proc of Gothenburg 2000, A Workshop on Numerical Ship Hydrodynamics*, Chalmers University of Technology.

Abkowitz, M.A. (1990). A manoeuvring simulation model for large angles of attack and backing propellers. *Proceedings of 19th ITTC*, Vol.2, Madrid.

ABS (1990). American Bureau of Shipping. *Guide for building and classing Motor Pleasure Yachts*.

ABS (1994). American Bureau of Shipping. *Guide for building and classing Offshore Racing Yachts*.

Anderson, J.D. (1995). *Computational Fluid Dynamics: The basics with applications*. McGraw-Hill International Editions.

Anon. (1979). High performance rudders for improved shiphandling. *The Naval Architect*, published by The Royal Institution of Naval Architects, March, pp. 53–54.

Anon. (1985). The Schilling rudder – 10 years on. *The Naval Architect*, published by The Royal Institution of Naval Architects, April.

Bailey, P.A., Price, W.G. and Temarel, P. (1998). A unified mathematical model describing the manoeuvring of a ship travelling in a seaway. *Trans. RINA*, Vol. 140.

Bailey, P.A., Hudson, D.A., Price, W.G. and Temarel, P. (2002). Time simulation of manoeuvring and seakeeping assessments using a unified mathematical model. *Trans. RINA*, Vol. 144.

Barrass, C.B. and Derrett, D.R. (2006). *Ship Stability for Masters and Mates*, 6th Edition. Butterworth-Heinemann, Oxford, UK.

Becker, L.A. and Brock, J.S. (1958). The experimental determination of rudder forces during trials of USS Norfolk. *Trans. SNAME*, Vol. 66.

Bertram, V. (2000). *Practical Ship Hydrodynamics*. Butterworth-Heinemann, Oxford, UK.

Bingham, V.P. and Mackey, T.P. (1987). High performance rudders with particular reference to the Schilling rudder. *Marine Technology*, Vol. 24, No. 4, pp. 312–320, October.

Boussinesq, J. (1877). Essai Sur La Theorie Des Eaux Courantes. *Mem. Present'es Acad. Sci*, Vol. 25.

Brix, J. (ed.) (1993). *Manoeuvring Technical Manual*. Seehafen Verlag.

Burcher, R.K. (1972). Developments in ship manoeuvrability. *Transactions of The Royal Institution of Naval Architects*, Vol. 114.

Cebeci, T., Shao, J.P., Kafyeke, F. and Laurendeau, E. (2005). *Computational Fluid Dynamics for Engineers: From Panel to Navier-stokes Methods with Computer Programs*. Springer.

Cho, J. and Williams, M. (1990). Propeller-wing interaction using a frequency domain panel method. *Journal of Aircraft*, Vol. 27, No. 3.

Choi, B., Park, H., Kim, H. and Lee, S-H. (2004). An experimental evaluation on the performance of high lifting rudder under Coanda effect. *9th Symposium on Practical Design of Ships and Other Floating Structures, PRADS'2004*. Lubeck-Travemunde, Germany.

Clarke, D., Patterson, D.R. and Woooderson, R.K. (1972). Manoeuvring trials with the 193,000 Tonne deadweight tanker 'Esso Bernicia'. *Trans. RINA*, Vol. 114.

Clarke, D., Gedling, P. and Hine, G. (1983). The application of manoeuvring criteria in hull design using linear theory. *The Naval Architect*. RINA, London.

Cordier, S. (1992). Design of control surfaces with rotating cylinders. Proc. of Conference, *Manoeuvring and Control of Marine Craft*, MCMC '92, Computational Mechanics Publications, 475–492.

Creismeas, P. (1999). *Application de la LES Implicite a Letude d'un Ecoulement Autour D'un Profil NACA0012, 7th Journees De L'Hydrodynamique*. Marseille, France, 8–10 March.

Dand, I.W. (1982). On ship-bank interaction. *Transactions of the Royal Institution of Naval Architects*, Vol. 124.

Date, J.C. (2001). Performance prediction of high lift rudders operating under steady and periodic flow conditions. Ph.D. Thesis, University of Southampton, UK.

Delefortrie, G., Vantorre, M. and Eloot, E. (2004). Linear manoeuvring derivatives in muddy navigation areas. *Trans. RINA*, Vol. 146.

Delefortrie, G. and Vantorre, M. (2007). Modelling the manoeuvring behaviour of container carriers in shallow water. *Journal of Ship Research*, Vol. 51, No. 4, December. SNAME.

DNV (2006). Det Norske Veritas. *Rules for Classification of Ships*, Part 3, Chapter 3, Hull Equipment and Safety, January.

DuCane, P. and Goodrich, G.J. (1962). The following sea, broaching and surging. *Trans. RINA*, Vol. 104.

Eda, H. and Crane, C.L. (1965). Steering characteristics of ships in calm water and in waves. *Trans. SNAME*, Vol. 73.

Eda, H. (1983a). Notes on ship controllability. *SNAME Bulletin No. 1–41*, April.

Eda, H. (1983b) Shiphandling simulation study during preliminary ship design. *Proceedings of Fifth CAORF Symposium*, Kings Point, New York, May.

English, J.W., Rowe, S. and Bain, D. (1972). Some Manoeuvring Devices for Use at Zero and Low Ship Speed. *Trans. NECIES*, Vol. 88.

Felli, M., Greco, L., Colombo, C, Salvatore, F., Di Felice, F. and Soave, M. (2006). Experimental and theoretical investigations of propeller-rudder interaction phenomena, *Proc. of 26th Symposium on Naval Hydrodynamics*, Rome, Sept.

Ferziger, J.H. and Peric, M. (2004). *Computational Methods for Fluid Dynamics*. Springer-Verlag.

GL (2005). Germanischer Lloyd. *Rules and Regulations for Pleasure Craft*.

Glauert, H. (1983). *The Elements of Aerofoil and Airscrew theory*, 2nd Edition. Cambridge University Press.

Gong, I-Y, et al. (1995). The influence of rudder area on the manoeuvrability of a ship with large beam-to-draught ratio. Proc. of *The Sixth International Symposium on Practical Design of Ships and Mobile Units*, PRADS'95, SNAK, Soeul, Korea.

Goodrich, G.J. and Molland, A.F. (1979). Wind Tunnel Investigation of Semi-Balanced Ship Skeg-Rudders. *Transactions of the Royal Institution of Naval Architects*, Vol. 121, pp. 285–307.

Guarino, S.J. (1984). *The slot augmented, flap effect rudder. 8th International Tug Convention*. Organised by Thomas Reed Publications Ltd., Singapore.

Guo, C. and Huang, S. (2006). Study of the rudder with additional thrust fins used by non-linear vortex lattice method. *Journal of Huazhong University of Science and Technology*, Vol. 34, No. 6, pp. 87–89.

Gutsche, F. (1964). The study of ships' propellers in oblique flow. *Schiffbauforschung*, Vol. 3, No. 3/4, pp. 97–122, Translation from German, DRIC Translation No. 4306.

Hadler, J.B. (1966). The prediction of power performance of planing craft. *Trans SNAME*, Vol. 74.

Hagen, G.R. (1972). A Contribution to the Hydrodynamic Design of Rudders. Ministry of Defence, *Third Ship Control Systems Symposium*, Bath, September.

Harrington, R.L. (1981). Rudder torque prediction. *Trans. SNAME*, Vol. 89.

Hess, J.L. (1990). Panel methods in computational fluid dynamics. *Annual Review of Fluid Mechanics*, Vol. 22, pp. 255–274.

Hino (2005). *Proc. of CFD Workshop*, Ed. T. Hino, Tokyo.

Hirano, M. (1981). A practical calculation method of ship manoeuvring motion at initial design stage. *The Society of Naval Architects of Japan, Naval Architecture and Ocean Engineering*, Vol. 19, pp. 68–80.

IMO (1993). IMO Resolution A.751(18). *Interim standards for ship manoeuvrability*.

IMO (2002). IMO Resolution MSC137(76). *Standards for Ship Manoeuvrability*.

Inoue, S. and Kijima, K. (1978). The hydrodynamic derivatives on ship manoeuvrability in the trimmed condition. *Proceedings of ITTC*, Vol.2, pp. 87–92.

ISO (2005). ISO Draft International Standard: Hull Construction – Scantlings – Rudders. ISO/DIS 12215-8.

ITTC (1975). ITTC 1975 Manoeuvring trial code. *14th International Towing Tank Conference*.

ITTC (2002). Report of the Manoeuvring Committee. *Proceedings of 23rd International Towing Tank Conference*, Vol.1. Venice, Italy. Published by INSEAN, Rome.

ITTC (2005). Report of Resistance Committee, *Proceedings of 24th International Towing Tank Conference*, Vol.1. Edinburgh, UK. Published by The University of Newcastle upon Tyne, UK.

ITTC (2005). Report of the Manoeuvring Committee. *Proceedings of 24th International Towing Tank Conference*, Vol.1. Edinburgh, UK. Published by The University of Newcastle upon Tyne, UK.

Jacquin, E., Guillerm, P-E. et al. (2006). Simulation of unsteady ship manoeuvring using free-surface RANS solver. *26th Symposium on Naval Hydrodynamics*, September. ONR/INSEAN, Rome, Italy.

Kang, G-G., et al. (1998). The Manoeuvrability of Full Form Ships with Low Speed. Proc. of *International Symposium on Forces Acting on a Manoeuvring Vessel*, MAN'98, Val de Reuil, France.

Katz, J. and Plotkin, A. (1991). *Low-speed aerodynamics: from wing theory to panel methods*. McGraw-Hill.

Kijima, K., Katsuno, T., Nakiri, Y. and Furakawa, Y. (1990). On the manoeuvring performance of a ship with the parameter of loading condition. *Journal of the Society of Naval Architects of Japan*, Vol. 168, pp. 141–148.

Kijima, K., Tanaka, S., Matsunga, M. and Hori, T. (1992). Manoeuvring characteristics of a ship in deep and shallow waters as a function of loading condition. Proc. of Conference, *Manoeuvring and control of marine craft*, MCMC'92, Computational Mechanics, pp. 73–86.

Kobayashi, E. (1995). The development of practical simulation system to evaluate ship manoeuvrability in shallow water. Proc. of *The Sixth International Symposium on Practical Design of Ships and Mobile Units*, PRADS'95, SNAK, Seoul, Korea.

Kose, K. (1982). On a new mathematical model of manoeuvring motions of a ship and its applications. *International Shipbuilding Progress*, Vol. 29, No. 336, pp. 205–220.

Kracht, A. (1982). Kavitation an Rudern. *Jarbuch Schiffbautechnische*. Springer.

Kresic, M. (2002). Estimating hydrodynamic force and torque acting on a horn-type rudder. *Marine Technology*, Vol. 39, No. 2, April.

Laurens, J.-M. (2003). Unsteady hydrodynamie behaviour of a rudder operating in the propeller slipstream. *Ship Technology Research*, Vol. 50, No. 3, July.

Laurens, J-M. and Grosjean, F. (2002). Numerical simulation of the propeller-rudder interaction. *Ship Technology Research*, Vol. 49, No. 1, Feb.

Lewis, E.V. (ed.) (1989). *Principles of Naval Architecture*. Published by The Society of Naval Architects and Marine Engineers, New York.

Li, D.-Q. (1996). Non-linear method for the propeller-rudder interaction with the slipstream deformation taken into account. *Computer Methods in Applied Mechanics and Engineering*, Vol. 130, No. 1–2, pp. 115–132.

Lin, W-M., Zhang, S. *et al.* (2006). Numerical simulation of ship manoeuvring in waves. *26th Symposium on Naval Hydrodynamics*, September. ONR/INSEAN, Rome, Italy.

LR (2005). Lloyds Register of Shipping. *Rules and Regulations for the Classification of Ships*, Ship Control Systems, Part 3, Chapter 13, July.

Mandel, P. (1953). Some hydrodynamic aspects of appendage design. *Trans. S.N.A.M.E.*, Vol. 61.

McDonald, H. and Whitfield, D.L. (1996) Self-Propelled Manoeuvring Underwater Vehicles, *21st Symposium on Naval Hydrodynamics*, Trondheim, Norway, June 24–28, pp 478–489.

McGeough, F.G. and Millward, A. (1981). The effect of cavitation on the rotating cylinder rudder. *International Shipbuilding Progress*, Vol. 28, No. 317.

Millward, A. (1969). The design of spade rudders for yachts. *Southampton University Yacht Research Report*, S.U.Y.R No.28.

Molland, A.F. (1978). Rudder Design Data for Small Craft. University of Southampton, *Ship Science Report No. 1/78*.

Molland, A.F. (1981). *The free-stream characteristics of ship skeg-rudders*, PhD Thesis, University of Southampton.

Molland, A.F. (1985). A Method for Determining the Free-Stream Characteristics of Ship Skeg-Rudders. *International Shipbuilding Progress*, Vol. 32, No. 370, June.

Molland, A.F. and Turnock, S.R. (1996). A compact computational method for predicting forces on a rudder in a slipstream. *Transactions of The Royal Institution of Naval Architects*, Vol. 138, pp. 227–244.

Molland, A.F., Turnock, S.R. and Wilson, P.A. (1996). Performance of an Enhanced Rudder Force Prediction Model in a Ship Manoeuvring Simulator. *Proc. of International Conference on Marine Simulation and Ship Manoeuvrability, MARSIM '96*, Copenhagen, Denmark, pp. 425–434.

Molland, A.F., Turnock, S.R. and Smithwick, J.E.T. (1998). Design Studies of the Manoeuvring Performance of Rudder-Propeller Systems. *Proc. 7th Int. Symp. on Practical Design of Ships and Mobile Units, PRADS '98*, The Hague, The Netherlands, September, 807–816.

Molland, A.F. and Turnock, S.R. (2002). Flow Straightening Effects on a Ship Rudder due to Upstream Propeller and Hull. *International Shipbuilding Progress*, Vol. 49, No. 3, pp. 195–214.

Molland, A.F. and Turnock, S.R. (2007). *Marine Rudders and Control Surfaces*. Butterworth-Heinemann, Oxford, UK.

Mott, L.V. (1997). *The development of the rudder: a technological tale*. Chatham Publishing, London.

Okada, S. (1966). On the performance of rudders and their designs. *The Society of Naval Architects of Japan*, 60th Anniversary Series, Vol. 11.

Oltman, P. and Sharma, S.D. (1984). Simulation of combined engine and rudder manoeuvres using an improved model of hull-propeller interactions. *15th ONR Symposium on Naval Hydrodynamics*, Hamburg.

Peck, J.G. and Moore, D.H. (1973). Inclined-shaft propeller performance characteristics. Paper G, *SNAME Spring Meeting*, April.

Perez, T. (2005). *Ship Motion Control. Course Keeping and Roll Stabilisation using Rudder and Fins*. Springer.

Peters, O.A.J., Kramers, C.H.M. and De Keizer, D. (2005). Improving trailing suction hopper dredger design with fully integrated dynamic positioning/dynamic tracking system. *Trans. RINA*, Vol. 147.

Rawson, K.J. and Tupper, E.C. (2001). *Basic Ship Theory*, Combined Volume. 5th Edition. Butterworth-Heinemann, Oxford, UK.

Romahn, K. and Thieme, H. (1957). On the selection of balance area for rudders working in the slipstream (in German). *Schiffstechnic*, No. 21.

Russo, V.L. and Sullivan, E.K. (1953). Design of the Mariner-Type ship. *Transactions of the Society of Naval Architects and Marine Engineers*, Vol. 61.

Schetz, J.A. and Favin, S. (1977). A Numerical solution for the near wake of a body with propeller. *Journal of Hydronautics*, Vol. 11, No. 10, pp. 136–141.

Shiba, H. (1960). Model Experiments about the Manoeuvrability and Turning of Ships. *First Symposium on Ship Manoeuvrability*, DTMB Report 1461, October.

Shouji, K., Ishiguru, T. and Mizoguchi, S. (1990). Hydrodynamic forces by propeller and rudder interaction at low speed. *Proceedings of MARSIM and ICSM*, Society of Naval Architects of Japan.

Simonsen, C.D. and Stern, F. (2005). RANS manoeuvring simulation of ESSO OSAKA with rudder and a body-force propeller. *Journal of Ship Research*, Vol. 49, No. 2, pp. 98–121.

Smithwick, J.E.T. (2000). Enhanced design performance prediction method for rudders operating downstream of a propeller. Ph.D. Thesis, University of Southampton, UK.

Söding, H. (1982). Prediction of ship steering capabilities. *Schiffstechnik*, pp. 3–29.

Söding, H. (1984). Bewertung der Manövriereigenschaften im entwurfsstadium. *Jarbuch Schiffbautechnische Gesellschaft*, Springer.

Söding, H. (1986). Kräfte am Ruder. *Handbuch der Werften XVIII*. Hansa-Verlag.

Söding, H. (1993). CFD for manoeuvring of ships. *19th WEGEMT School*, Nantes.

Söding, H. (1998a). Limits of potential theory in rudder flow prediction. *Ship Technology Research*, Vol. 45, No. 3, pp 141–155.

Söding, H. (1998b). Limits of potential theory in rudder flow prediction. *22nd Symposium on Naval Hydrodynamics*, ONR Washington.

Son, D.I., Ahn, J.H. and Rhee, K.P. (2001). An empirical formula for steering gear torque of tankers with a horn rudder. *Proc. 8th Int. Symp. on Practical Design of Ships and Mobile Units, PRADS'2001*.

Speziale, C.G. (1991). Analytical Methods for the Development of Reynolds-stress Closures in Turbulence. *Annual review of fluid mechanics*, Vol. 23, pp. 107–157.

Steele, B. and Harding, M. (1970). The application of rotating cylinders to ship manoeuvring. *National Physical Laboratory*, Ship Report 148.

Stern, F., Kim, H.T., Zhang, D.H., Toda, Y., Kerwin, J. and Jessup, S.L. (1994). Computation of viscous flow around propeller-body configurations: Series 60 C_B = 0.6 Ship Model. *Journal of Ship Research*, Vol. 38, No. 2, pp. 137–157.

Tamashima, M., Matsui, S., Yang, J., Mori, K. and Yamazaki, R. (1993). The method for predicting the performance of propeller-rudder system with rudder angle and its application to rudder design. *Transactions of the West-Japan Society of Naval Architects*, Vol. 85.

Taplin, A. (1960). Notes on rudder design practice. *D.T.M.B. Report 1461*, October.

Taplin, A. (1975). Sea trials for measuring rudder torque and force. *Proc. of Fourth Ship Control Systems Symposium*. Royal Netherlands Naval College, The Hague. Vol. 5, pp. 99–115.

Thieme, H. (1965). Design of ship rudders. *Jarbuch der Schiffbautechnischen Gesellschaft*, Vol. 56, 1962 or DTMB Translation No. 321.

Tsakonas, S., Jacobs, W.R. and Ali, M.R. (1970). Application of the unsteady lifting-lifting-surface theory to the study of propeller-rudder interaction. *Journal of Ship Research*, Vol. 14, No. 3, pp. 181–194.

Tuite, A.J. and Renilson, M.R. (1998). The effect of principal design parameters on broaching-to of a fishing vessel in following seas. *Trans RINA*, Vol. 140.

Turnock, S.R. (1993). Prediction of ship rudder-propeller interaction using parallel computations and wind tunnel measurements, PhD Thesis, University of Southampton.

Turnock, S.R. (1994). A transputer based parallel algorithm for surface panel analysis. *Ship Technology Research Schiffstechnik*, Vol. 41, No. 2, pp. 93–104, May.

Turnock, S.R., Molland, A.F. and Wellicome, J.F. (1994). Interaction velocity field method for predicting ship rudder-propeller interaction, *Proc. of SNAME Propeller/Shafting Symposium*, Paper 18, 1–14, Virginia Beach USA, Sept.

Turnock, S.R. and Molland, A.F. (1998). The effects of shallow water and channel walls on the low-speed manoeuvring performance of a mariner hull with rudder. *International Symposium and Workshop on Forces Acting on a Manoeuvring Vessel*, MAN'98, Val de Reuil, France.

Turnock, S.R. and Wright, A.M. (2000). Directly coupled fluid structural model of a ship rudder behind a propeller. *Marine Structures*, Vol. 13, pp. 53–72.

Tzabiras, G.D. (1997). A numerical study of additive bulb effects on the resistance and self-propulsion characteristics of a full ship form. *Ship Technology Research*, Vol. 44, pp. 98–108.

Vassalos, D., Umeda, N., Hamamoto, M. and Tsangaris, M. (1999). Modelling extreme behaviour in astern seas. *Trans. RINA*, Vol. 141.

Versteeg, H.K. and Malalasekera, W. (1995). *An Introduction to Computational Fluid Dynamics*. Longman Scientific and Technical.

Whicker, L.F. and Fehlner, L.F. (1958). Free Stream Characteristics of a Family of Low Aspect Ratio Control Surfaces for Application to Ship Design. *DTMB Report 933*, December.

Wilcox, D.C. (1998). *Turbulence Modelling for CFD*. D.C.W. Industries.

Willis, C.J., Crapper, G.D. and Millward, A. (1994). A numerical study of the hydrodynamic forces developed by a marine rudder. *Journal of Ship Research*, Vol. 38, No. 3, pp. 182–192.

Wolff, K. (1981). Ermittlung der manövriereigenschaften fünf repräsentativer schiffstypen mit hilfe von CPMC-modellversuchen. *IfS Report 412*, University of Hamburg.

Zou, Z.J. (1990). Hydrodynamische kräfte am manövrierenden schiff auf flachem wasser bei endlicher froudezahl. *IfS Report 503*, University of Hamburg.

9 Ship design, construction and operation

Contents

9.1 Introduction
9.2 Ship design
9.3 Materials
9.4 Ship construction
9.5 Ship economics
9.6 Optimization in design and operation
References (Chapter 9)

The various Sections of this Chapter have been taken from the following books and sources, with the permission of the authors:

Eyres, D.J. (2007) *Ship Construction*. 6th Edition. Butterworth-Heinemann, Oxford, UK. [Sections 9.3.2, 9.3.3, 9.3.5, 9.4]

Molland, A.F. (2005) *Ship Design and Economics*. Lecture Notes. School of Engineering Sciences, University of Southampton, UK. [Sections 9.1, 9.2]

Schneekluth, H. and Bertram, V. (1998) *Ship Design for Efficiency and Economy*. 2nd Edition. Butterworth-Heinemann, Oxford, UK. [Section 9.6]

Shenoi, R.A. and Dodkins, A.R. (2000) Design of Ships and Marine Structures made from FRP Composite Materials, in Kelly. A. and Zweben, C. (eds), *Comprehensive Composite Materials*, Vol. 6, Elsevier Science Ltd, Oxford, UK. [Section 9.3.4]

Watson, D.G.M. (1998) *Practical Ship Design. Elsevier Science*, Oxford, UK. [Section 9.5]

9.1 Introduction

This Chapter provides a broad overview of ship design, construction and operation. In the first sections, the basic practical aspects of deriving the technical ship dimensions, masses, stability and body plan are described. The next sections describe the materials used in ship construction, the effects of corrosion, and a brief description of the ship construction process. The economics of ship operation are described, indicating the interactions with technical design. Finally, optimization applied to ship design and operation is described and how optimization may be used to achieve the most suitable design.

9.2 Ship design

9.2.1 Overview

9.2.1.1 General

A ship is a complex vehicle. Its production requires the involvement of a wide range of engineering disciplines. Ship design is not an exact science but embraces a mixture of theoretical analysis and empirical data accumulated from previous successful designs. Due to the complex interrelationships between features of the technical design, and the construction of the ship and its operation, the final ship design will often represent a compromise between conflicting ship requirements.

The development of the overall ship design and its production cannot normally be treated in technical isolation as operational requirements have to be considered. For example, the ship will often form part of a through transport system; this may range from sophisticated container systems with dedicated ships operating between specified ports, or ferries and RO/RO vessels relying on a regular wheeled through cargo, to tramp vessels on non-regular schedules which rely on carrying various types of cargo between various ports. A review of some of these ship types is given in Chapter 2.

The route and its environment, type of cargo, quantity to be moved, value of the cargo and port facilities are typical features which will be considered when evolving the size, speed and specification of a suitable ship (or ships). Specific service requirements will be similarly considered when evolving vessels such as warships, passenger ships or fishing vessels.

Shipowners operate ships to make a satisfactory profit on their investment. The evolution of a technical design can therefore be considered as a component part of an overall economic model. In evolving a ship design it is therefore necessary to assess the operating requirements and the environment in which the vessel is to operate, to evolve the feasible technical design and to economically justify the viability of the proposal.

In an overall final design process the design objectives have to be clearly identified and constraints in the process incorporated. The following discusses some of the alternative objectives:

Design for functionability, or capability: this is a pre-requisite without which the ship does not fulfil its role, whether it be a warship or a large tanker.

Design for efficiency and economy: this is normally also a pre-requisite and might take several forms including designing to minimize running costs, maintenance costs, cranage/turnround time for container ships, or turnround time for ferries (e.g. manoeuvring), all with a view to improving the overall efficiency of the operation.

Design for production: In this case producibility is important, and savings in construction costs may be assessed, Kuo et al. (1984), Andrews et al. (2005). In this case, the analysis may, for example, be trading increases in steel mass (and hence decrease in deadweight) against decreases in production costs.

Design for maintenance: this will often amount to increase in space and improved access for maintenance of tanks or machines. This might entail accepting surplus volume and an increase in ship first cost.

Design for the environment: aspects may include pollution, emissions, noise and wave wash. These objectives are becoming increasingly important. Some of these aspects are covered in MARPOL.

Design for disposal, or scrap: this is becoming more important in the design process, whereby ease of disposal (e.g. cutting up hull, or removing machinery) is taken into account.

Each objective is important in its own right. Whilst achievement of all the objectives is desirable, but unlikely, some weighting as to the relative importance of the various objectives will normally be necessary.

The following Sections consider the practical aspects of evolving the technical design model, bearing in mind operational patterns and requirements, and its extension in Section 9.5 to include economic considerations and evaluations. Assessments of alternative design methodologies and philosophies are not carried out; these can be studied further in texts and references such as Gillmer (1977), Watson (1998), Schneekluth and Bertram (1998), Eames and Drummond (1977) and Andrews (1981, 1998, 2007).

9.2.1.2 *Ship design process*

The ship design process may be broken down broadly into two stages:

(i) Conceptual and/or preliminary design
(ii) Detailed or tender or contract design

The principal ship dimensions and power to meet the intended service will be evolved at stage (i). If the results of stage (i) are technically and economically viable then stage (ii) will follow. The development of the detailed requirements up to contract stage should not normally have a significant effect on the basic particulars evolved at stage (i).

This section is concerned, in the main, with the evolution of the preliminary ship design and its evaluation. Involvement in detailed and constructional aspects is limited to a brief overview in Section 9.4.

The preliminary design process will normally take the form of a techno-economic appraisal, using a fundamental engineering economy approach.

The increase in effort to improve efficiency has led to an increasing use of economic investigation. Whilst a primary and traditional function of the ship designer or naval architect is to derive a feasible technical design, it is unlikely that this will be achieved in technical isolation without taking account of economic considerations, either directly or indirectly.

The application of engineering economics to ship design is basically the conversion of the marine transport requirements into a range of feasible ship designs which must then be evaluated for their technical and economic performance.

The overall flow path shown in Figure 9.1 can thus be established.

Many of the techno-economic evaluations amount to an investigation of the trade-off between first and operating costs; it is important to note that the 'best' design need not necessarily be of lowest first cost, but that which shows the most profitable combination of first and operating costs over the life cycle.

9.2.2 Technical ship design

9.2.2.1 *Principal requirements*

The principal requirements of a technical ship design may be summarized as follows:

1. Is adequate in size and arrangement for intended service — implies ability to carry a specified volume of cargo and have adequate space for machinery, fuel and crew etc.
2. Floats at correct draught — implies sum of weights of lightship and deadweight equals force due to buoyancy (function of ship form)
3. Floats upright — implies adequate stability
4. Achieves correct speed — implies satisfactory estimates of resistance and propulsive power (plus margins) and installation of suitable engine(s).
5. Is structurally safe/sound — implies structural design with the ability to withstand forces in the marine environment; typically built to the requirements of a classification society
6. Meets requirements for manoeuvring, coursekeeping and seakeeping — implies choice of suitable hull form
7. Meets international standards of safety and reliability — meets requirements of IMO

The derivation of a feasible technical design will take the form of an 'iterative process of analysis and synthesis'; i.e. is a repetitive process whereby the design is resolved into simple elements and relevant calculations made, after which the elements are combined into the total ship design.

For example, for a deadweight determined design, items 1 to 4 might be modelled as in Figure 9.2.

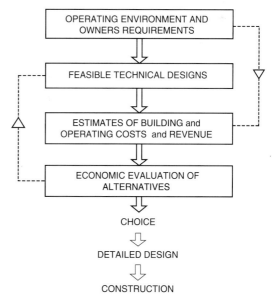

Figure 9.1 Overall flow path.

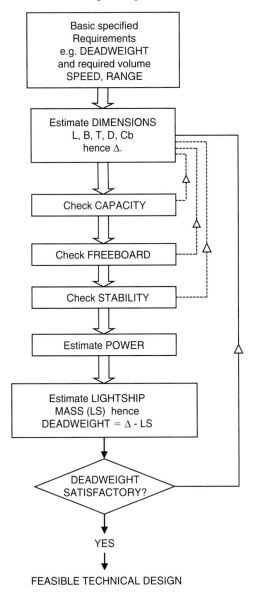

Figure 9.2 Preliminary design path.

- Development of a new service or carrying a different kind of cargo on an existing route aimed at capturing an increased percentage of the trade
- Development of a new service on a new route

In each situation, the owner is faced with decisions concerning the number of ships required, their type, size and speed.

Before the design process can be initiated, the basic technical data relating to the operational requirements have to be defined and specified, or derived or assumed if several alternatives are to be investigated.

For a deadweight carrier, the basic specification requirements would typically be as follows:

Deadweight:	Will become a variable if alternative ship sizes are to be investigated.
Speed:	Possible 'hydrodynamic optimum' speed for particular ship length, or may be dependent on nature/value of cargo, e.g. perishable fruit or passengers. Will be specified, or alternatives may be investigated.
Capacity:	Specified to suit the design cargo stowage rate (m^3/tonne).
Range:	Range and route, length of voyage – fuel capacity, route – weather conditions/power margins.
Stability:	Minimum requirements usually specified for most onerous conditions (e.g. loaded arrival is not uncommon).
Strength:	Minimum requirements usually specified to be those of one of the classification societies.
Manoeuvrability:	Specification may relate to rudders, etc. lateral thrusters, tug dues, etc. mooring arrangements.
Trim:	Normally required by stern.
Dimensional Constraints	e.g. specific breadth or width or draught limitations, such as canal or port limits in breadth and/or draught.
General Arrangement:	To meet the needs of specified crew, cargo handling, passenger accommodation.
Statutory Regulations:	To meet requirements.

9.2.2.2 *Specification*

The owner's operational requirements need to be established, which then allows the development of a basic specification.

An owner might typically be seeking a new ship design to suit one or a combination of the following alternatives:

- Replacement or conversion of old vessels
- Expansion or modification of services on an existing route in an effort to enlarge the participation

At the *preliminary* design stage, for a deadweight carrier, it is often suitable to treat the following as primary requirements:

- DEADWEIGHT
- SPEED
- RANGE

And to treat the following as 'checking' or constraint requirements:

CAPACITY
STABILITY
FREEBOARD
Plus others if necessary

It can be noted that this procedure covers a large proportion of merchant ship types, but an alternative known as a *capacity approach* is necessary in the case of *capacity carriers* such as passenger ships, ferries, warships and container ships. In such cases the pre-requisite is to contain a certain capacity or volume rather than to lift a particular deadweight. The *CAPACITY* or *SPACE DESIGN* approach is discussed later in Section 9.2.4.

9.2.3 Deadweight determined designs

A *deadweight design approach* is based on equating the sum of the component masses of the vessel to its displacement. It is applicable to the majority of ship types including tankers, bulk and ore carriers, and most cargo vessels.

9.2.3.1 *Deadweight and dimensions*

(a) *Deadweight (DW):*
Includes cargo, fuel, FW, stores crew and effects. Cargo is the only component of deadweight, which will earn revenue, hence other items of deadweight should be kept to a minimum.

(b) *Lightship mass (LS):*
Condition is that of a ship when ready to put to sea, but without cargo, fuel, stores and provisions. The primary components of lightship mass are steel, outfit and machinery.

(c) *Displacement (Δ):*
Total ship mass: equals mass of water displaced, equals 1.025 L.B.T.C_B, and Deadweight = Displacement − Lightship
A primary aim is to design a ship with minimum Δ to meet the requirements of the owner, hence obtaining the most economical ship in respect of the machinery, fuel consumption and initial cost.

(d) *Deadweight coefficient (C_D):*

is defined as C_D = Total Deadweight/Displacement

and can be treated as a very approximate criterion or measure of 'efficiency' of the vessel.
A preliminary value of displacement can be determined from C_D, when the DW has been defined.
Typical values of C_D are as follows:

Cargo ships	0.65–0.75
Large tankers/Bulk	0.79–0.85
Ore	0.82
*Container	0.60
*Refrigerated cargo	0.55–0.60
*Passenger	0.35

*For these the predominant factor is that of space, hence C_D of little significance.

Note that C_D will vary with cargo type since bulky cargoes require greater volume (hence steel), hence C_D will be lower. Similarly a higher speed (for same DW) will involve increases in machinery mass hence in LS and reduction in C_D. Hence special care is needed in the use of this coefficient.

Derivation of dimensions:
(1) *Length (L):*
Usually a minimum consistent with speed and form; length is generally the most expensive dimension. A preliminary estimate of length may be made using:
$L = f(\nabla^{1/3})$, where the function depends on ship type.
$L/\nabla^{1/3}$ lies typically in the range 5.5 to 6.5 for cargo vessels and tankers and 6.5 to 8.5 for higher speed vessels and passenger ships.

(2) *Breadth (B):*
Has direct influence on stability. L/B has influence on hull resistance, hence power. L/B tends to be larger for faster ships.
L/B ratios for cargo ships lie within range 6–7 and for passenger ships with range 6.5–7.5.
Typical empirical values (from Watson and Gilfillian (1977)) are as follows:

L/B = 4 .... (L < 30m)
L/B = 4 + 0.025 (L−30)... (L = 30–130m)
L/B = 6.5 .... (L > 130m)

(3) *Draught (T):*
B/T ratio related to hydrodynamic performance and stability

B/T for cargo ships typically varies between 2 and 2.5
B/T for pass. ships typically varies between 3 and 5
T related to D in respect of freeboard and T/D is typically 0.7–0.8 for cargo vessels, bulk carriers and tankers

(4) Depth (D):
Can be approximated from L/D ratio (related to strength of ship), or B/D ratio (related to stability)
L/D for cargo, tankers bulk carriers typically 12–13
B/D $\cong$ 1.9 for DW carriers such as tankers and bulk carriers
B/D $\cong$ 1.7 for stability limited capacity carriers

(5) Block coefficient (C_B):
For economical propulsion from a hydrodynamic point of view, length and fullness at a given speed are closely related. Typical approximate formulae relate C_B with $V/\sqrt{L}$ as follows.

$$C_B = a - b\, V/\sqrt{L} \qquad [\text{V knots, L metres}]$$

Where typical values of a and b are 1.23 and 0.395.

(6) Derived dimensions:
A simple model to derive the principal dimensions for a given deadweight (DW) and speed (V), using functional relationships between the dimensions might be developed as shown in Figure 9.3.
L may be assumed, or approximated using $L = f(\nabla^{1/3})$ in first cycle.

9.2.3.2 Cargo capacity check

A deadweight design approach ensures that the correct *mass* of cargo can be carried. The volumetric *capacity* of the vessel must be such that the volume of the required mass can be contained.
Typical cargo stowage rates are as follows:

General cargo	1.4–1.7 m³/tonne
Refrigerated cargo	1.8–2.0 m³/tonne
Crude Oil	1.05 m³/tonne (approx.)

In the design process, the required volumetric cargo capacity V_C can be estimated from a knowledge of the cargo deadweight and the stowage rate for the cargo type.

Definitions:
Moulded volume
Total internal hull volume to inside of shell plating.

Grain capacity
Taken to top of beams, inside shell plating and tank top and is moulded volume less actual volume of all obstructions such as structure. Grain capacity is typically about 1.5–2% less than moulded volume in the holds and about 3% less in double bottoms.

Bale capacity
Capacity taken to underside of beams, inside of frames, inside of beam knees etc. Bale capacity typically about 10–12% less than the grain capacity.

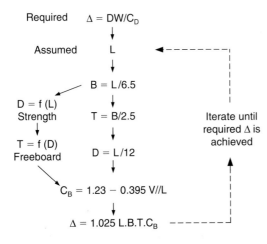

Figure 9.3 Flow chart for dimensions (deadweight approach).

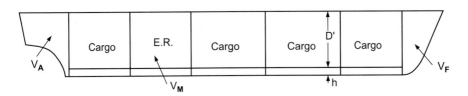

Figure 9.4 Capacity check.

Capacity check

At the detailed design stage the capacity check presents no problems as it can readily be estimated from the ship hull form and general arrangement.

At the preliminary stage approximate relationships have to be applied.

If the capacity of a suitable basis vessel is available, then a preliminary estimate for the new design can be made by scaling the dimensions. The method is more accurate if all the underdeck volume is used, i.e. including other non-cargo spaces such as machinery. Hence if G_1 is the total underdeck grain capacity for the basis vessel then the corrected total underdeck volume for the new design will be:

$$G_2 = G_1 \times L_2/L_1 \times B_2/B_1 \times D_2/D_1 \\ \times C_{B2}\,@\,0.85D/C_{B1}\,@\,0.85D$$

(Depth used should take account of shear and double bottom)

The requirements of non-cargo spaces, such as machinery and accommodation etc. underdeck, for the new design proposal are then subtracted from G_2 to give the estimated cargo capacity.

If a suitable particular basis vessel is not available at the preliminary stage, a preliminary check may be made using a data base from similar ships. This may be carried out as follows, Figure 9.4:

Total Underdeck Volume $V_T = LBD'C_B'$

where D' allows for double bottom, hence $D' = D-h$ and $h = f(D)$; C_B' = block coefficient at 80% D and is a criterion of fullness up to main deck with C_B' derived from C_B @ design draught T.

From the total volume V_T will be deducted:

(i) the volume of the machinery space V_M:
   An approximate assumption is that the volume required by the main engine and auxillary machinery is a function of power, hence $V_M = f(\text{Power}) = f(\Delta^{2/3}\,V^3)$, where the function will depend on ship type, size and position of the machinery space.
(ii) the non-cargo volumes within the length but forward and aft of the cargo space $(V_F + V_A)$; these will be typically expressed as a percentage of the total volume V_T for a particular ship type

i.e. $(V_F + V_A) = f(LBD'C_B')$

Thus the cargo capacity may be expressed approximately in terms of the variables already determined at that stage in the design path i.e.

$$V_C = V_T - V_M - (V_F + V_A) = f_1(LBD'C_B') \\ - f_2(\Delta^{2/3}V^3) - f_3(LBD'C_B')$$

where f_1, f_2 and f_3 would be obtained from similar basis ships.

9.2.3.3 Summary of overall model: Deadweight approach

The derivation of the dimensions can now be incorporated into an overall technical design model, as shown in Figure 9.5.

The model illustrated is simple, but lends itself to systematic variations in components of the design, e.g. methodical variation of deadweight, speed, dimensions, etc. The model shows functional relationships between the principal dimensions. It should be noted that all dimensions could be 'free floating' in the design procedure provided adequate constraints confine particulars to physical limitations, the power estimate is adequate to predict changes due to distorted dimensional relationships and data requirements are within the range of any empirical relationships used.

It should be noted that at this stage in the design process a design has been derived which is *feasible*, although it may not be the best for its intended purpose. It will be seen later that the model providing the derivation of the alternative feasible technical designs can be incorporated in a larger model in which economic evaluations of the alternatives can be carried out see Section 9.6.

9.2.4 Capacity (or space) determined designs

A *capacity design approach* is used where the dimensions are required to be determined (primarily) by the need to provide a requisite space. Examples include passenger ships, most naval vessels, cargo ships with high stowage rates (such as for meat, bananas and cars) passenger/car ferries and container ships.

9.2.4.1 Cargo ships

The capacity design approach for this vessel type is based on the equality: $V_C = V_H - V_S$, where V_C = vol. available for cargo, V_H = total underdeck volume and V_S = volume for machinery and other essentials. V_C is known and V_S may be estimated from basis vessels and/or known power requirements. Hence dimensions based on V_H may be derived as shown in Figure 9.6.

9.2.4.2 Passenger ships

In this case it is necessary to consider the volume of the whole ship including erections. The main

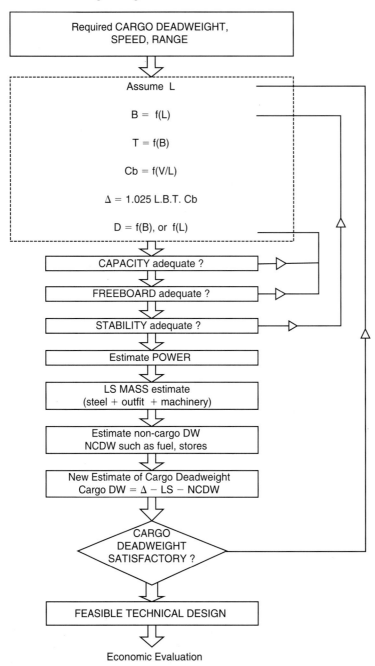

Figure 9.5 Preliminary design path.

problem is essentially one of calculating the volume required to permit the arrangement of the required passenger and crew spaces, machinery spaces etc., rather than the process of obtaining the actual dimensions to give the required volume, as described below.

The total volume requirement (V_T) is derived from a summation of the volumes of passenger accommodation (α pass. no.), public areas, crew accommodation (α crew no.), machinery (α power requirements) and volumes for fuel, FW, stores, etc. Values may be analysed from past ship data or

Ship design, construction and operation 645

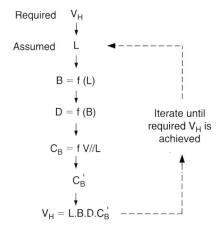

Figure 9.6 Flow chart for dimensions (capacity approach).

from sources such as Watson (1962) and Watson and Gilfillian (1977).

Total volume (V_T) = Hull volume (V_H)
     + superstructure volume

The superstructure is typically 30%–40% of total volume for passenger ships and current practice for a particular vessel type can be readily investigated.

Hence with total volume known, volume of hull (V_H) can be deduced.

In most cases, main hull volume V_H can be estimated on the basis:

$V_H = L.B.D.C_B'$, where $C_B' = C_B$ @ 0.8D
e.g. $C_B' = C_B + (1 - C_B)(0.8D - T)/3T$

Hence the required volume may be modelled as shown in Figure 9.6.

At this stage, the main hull profile and erections can be drawn to give the required volume and general arrangement. A *STABILITY* check (see Section 9.2.5) is necessary before any further detailed arrangements are developed. This is important in the case of the passenger ship since late changes (say in beam) may have a significant influence on the internal arrangement/layout of cabins, etc. – or on car lane widths in the case of large car ferries. Similarly, *FREEBOARD* (see Section 9.2.8) would be checked at an early stage in the case of large car ferries in order to site the freeboard deck and deck heights etc.

9.2.4.3 Container ships

These may be defined as 'linear dimensions' ships (e.g. see Watson (1962) and Watson and Gilfillian (1977)), further examples being St Laurence Seaway/Panama with breadth restrictions and some car ferries with B as a function of car lanes, etc.

Classical references on container ship design are Henry and Karsch (1966), Meek (1970) and Meek et al. (1972) and these provide good accounts of the basic design procedures and problems associated with the design of container ships. See also Section 9.6.6.2.

Container ships may be classified as capacity or space determined designs and their size is generally defined by their container capacity – e.g. 1500, 3000, 6000 or 10000 TEU. (Container sizes $20' \times 8' \times 8.5'$ height or $40' \times 8' \times 8.5'$). TEU = Twenty Foot Equivalent Unit (e.g. $40'$ container = 2 TEU's).

Stowage rates for containers are typically max 20 tonnes/20′ container, but actual stowage rate is about 12 tonnes/container. For example, if the cargo deadweight is specified, then number of containers (n) can be derived.

Due to the high stowage rate of cargo in containers and containers in holds, this leads to the requirement for a large quantity of containers on deck. A decision is first made on number of containers on deck, e.g. say 6 deep in holds, 3 high on deck – hence approximately 2/3 of total n containers to be stowed below deck.

Hence ship hull dimensions may be designed around a capacity to contain say 2/3 n containers + machinery volume + double bottoms + peaks, etc.

Containers are best stowed in a rectangular space (say about midships), consequently machinery/accommodation is ideally placed, although not always feasible or desirable, see Meek (1970), and current designs tend to have machinery/accommodation about ¾ aft.

Beam:
B = f [container breadth + clearances + sufficient deck width outside line of hatches for required longitudinal and torsional strength]

Clearance between containers 9″–12″ for preliminary design.

Minimum strength width each side of hatches may be assumed to be about 10% B, i.e. about 20% overall.

Length:
Can be adjusted to give suitable dimensional ratios based on B. Length enclosing containers = f[containers length + clearances + bulkheads/stiffs].

In detail, Meek (1970) quotes length (or breadth) as being = f [container length + tolerance + structure to support cell guides + cell guide clearance + cell guide tolerance + adequate ship structure, etc].

Depth:
D = f [container depth + double bottom depth].

DαB is very important in view of stability requirements with deck cargo, i.e. static stability and dynamical re-wind effects – possible use of water ballast.

Note also development of hatchless container ships (cell guides running up above deck level) facilitating faster turn around time. (Removal/replacement of container ship pontoon hatch covers time consuming.)

9.2.4.4 *High speed passenger/vehicle ferries*

The derivation of the dimensions for these vessel types is usually based on areas (for given tween deck height) for given number of passengers and/or vehicles.

A description of the concept design and derivation of dimensions for these vessel types (currently for monohulls and catamarans) is given in Karayannis *et al.*(1999) and Molland *et al.*(2003), together with regression equations relating areas to passengers and/or vehicle numbers and (L × B) to required areas. Figures 9.7 and 9.8 and Table 9.1 from Molland *et al.* (2003), describe typical design flow paths and regression formulae for high speed ferries.

In this approach, described by Molland *et al.* (2003), the initial derivation of the dimensions is based on suitable values of the L/B ratio and L × B product, and hence a solution for L and B.

Hydrostatics/Hydrodynamics:
L/B is based on hull hydrostatic and hydrodynamic requirements and suitable assumptions for $L/\nabla^{1/3}$, C_B and B/T:

$$\frac{L}{B} = \left[\left(\frac{L}{\nabla^{1/3}}\right)^3 \cdot C_B \cdot \frac{T}{B}\right]^{1/2} \quad (9.1)$$

Areas:
The product L × B is based on required passenger and vehicle areas:

$$L \times B = f(A_p, A_V), \quad (9.2)$$

where $A_S = f(N_P)$, $A_P = f(A_S)$ and $A_V = f(N_V)$.

Suitable forms of these relationships, as well as ranges of the design parameters, are given in Table 9.1. The L × B product is derived using a three-step procedure as shown in Figures 9.7 and 9.8. This offers more flexibility in selecting the desired level of seating comfort and overall accommodation quality, which is achieved by appropriate adjustment of the passenger area relationships.

The solution for L becomes:

$$L = \left[(L \times B) \times \left(\frac{L}{B}\right)\right]^{1/2} \quad (9.3)$$

and B can then be derived from L/B, T from B/T and $\Delta = \rho.L.B.T.C_B$

For catamarans, L/B is derived as:

$$\frac{L}{B} = 1/\left[\frac{S}{L} + \frac{b}{L}\right] \quad (9.4)$$

where b is the breadth of a demihull and S the separation of the demihull centrelines.

In the case of catamarans, L/b is derived as:

$$\frac{L}{b} = \left[\left(\frac{L}{\nabla^{1/3}}\right)^3 \times C_B \times \frac{T}{b}\right]^{1/2} \quad (9.5)$$

in this case, B in Equation (9.1) is replaced by b and displaced volume ∇ refers to one of the hulls; the catamaran displacement then becomes $\Delta = 2.\rho.L.b.T.C_B$. The estimate of the overall depth D_O (including superstructure) in Table 9.1 is only approximate, and is provided primarily for use in the equipment numeral E for the hull and superstructure mass estimate.

As the principal hull parameters did not show any reliable trends with speed, the first estimate of dimensions in the iterative cycle is based only on passenger and vehicle requirements, together with appropriate values of hydrodynamic parameters as starting points. This creates an anomaly in the design procedure. For example, a change in speed for a particular design, whilst retaining the same passenger and vehicle requirements, results in a change in propulsive power and machinery mass and hence overall mass balance. This problem is overcome by incorporating a mass balance directly within the procedures for the derivation of dimensions, Figures 9.7 and 9.8.

In the design path, Figures 9.7 and 9.8, suitable values for $L/\nabla^{1/3}$, C_B and B/T are chosen and used in Equation (9.1). These may then be modified in further design iterations in order to achieve a satisfactory balance of masses, generally by adjusting the displacement. There are several ways in which the parameters may be modified, but an approach which has been found to be effective and efficient is to retain overall constancy of L/B, hence constant L from Equation (9.3), which results in constancy of Equation (9.1). Hence for constant L/B, combinations of $L/\nabla^{1/3}$, C_B and B/T within Equation (9.1) may be chosen depending on any other design constraints. For example, (i) for fixed ∇ and $L/\nabla^{1/3}$, C_B can be increased and B/T increased to retain constant ∇; (ii) if a change in ∇ is accepted, C_B and $L/\nabla^{1/3}$ may be changed with B/T constant, or B/T and $L/\nabla^{1/3}$ changed with C_B constant or suitable changes made to both B/T and C_B. The procedure for catamarans is similar, but using Equations (9.4) and (9.5).

It is seen that the proposed approach truly integrates areas and masses into the initial design

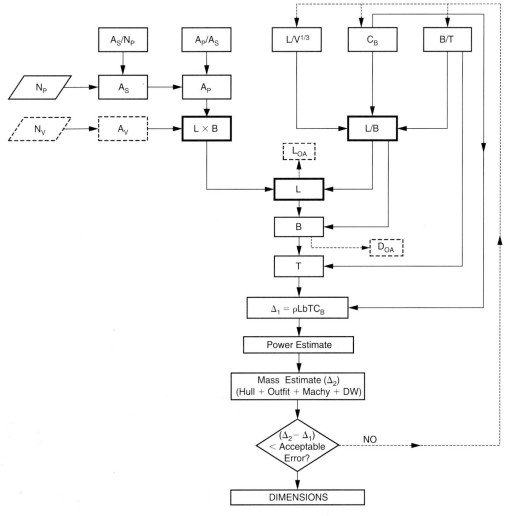

Figure 9.7 Estimation of main dimensions – monohulls.

cycle for the derivation of the dimensions. It is quite different from the more traditional approaches where, when establishing preliminary dimensions for a capacity carrier, the emphasis might be placed on volumes or areas followed by a mass check or, for displacement vessels, using a balance of masses followed by a capacity check.

A starting point in the design process can be established by using the mid values of the various parameters given in Table 9.1.

9.2.5 Stability check

Criterion for transverse stability: The transverse metacentric height GM may be used as a measure of ship stability, where GM may be calculated as:

$$GM = (KB + BM) - KG \qquad (9.6)$$

(see also Chapter 3 for a more detailed discussion of stability)

KG Depends on disposition of structure and contents; can be calculated in detail at advanced stage of design if time allows, or by inclining experiment after launch.

KB Calculated from a knowledge of the underwater form.

BM J_T/∇
   J_T = transverse 2nd moment of area of the waterplane about the centreline.
   ∇ = immersed volume.

Estimation of GM at the preliminary design stage e.g. for use in the model described in Figure 9.5.

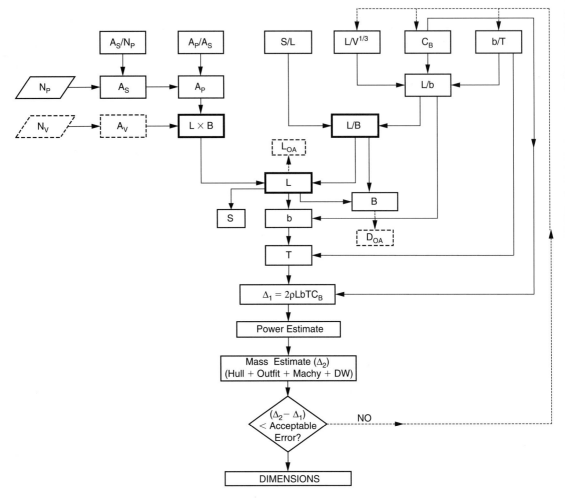

Figure 9.8 Estimation of main dimensions – catamarans.

Table 9.1 Design equations and range of parameters.

Item	Pass. Only Monos	Pass. Only Cats	Car/Pass. Monos	Car/Pass. Cats
$L \times B$ (m²)	$146 + 1.86 \times 10^{-3} A_p^2$	$138 + 0.91 A_p$	$121 + 0.27 A_p + 0.60 A_v$	$471 + 0.55 A_p + 0.28 A_v$
A_s/N_p (m²)	0.55–0.75	0.55–0.85	0.85–1.25	0.80–1.40
A_p/A_s	1.10–1.30	1.10–1.30	1.15–1.45	1.30–1.70
A_v (m²)	–	–	$156 + 10.2 N_v$	$12.4 N_v$
S/L	–	0.20–0.25	–	0.20–0.25
$L/\nabla^{1/3}$	5.0–7.5	8.0–10.5	6.5–9.0	8.5–11.0
(majority)	(5.5–6.5)	(8.5–9.5)	(7.0–8.5)	(9.5–10.5)
B/T	3.5–8.5	b/T 1.5–3.0	3.5–7.5	b/T 1.5–3.0
(majority)	(4.0–6.5)		(4.5–6.5)	
D_O	$4 + 0.6B$	$4 + 0.44B$	$4 + 0.6B$	$4 + 0.44B$
C_B	0.35–0.45	0.40–0.55	0.35–0.45	0.40–0.55
L_O/L	1.13–1.15	1.13–1.15	1.13–1.15	1.13–1.15

where A_P = Total pass. area (m²), A_S = Seating area (m²), A_V = Vehicle area, N_P = No. of pass. N_V = Number of vehicles

KG, KB and J_T will not be known accurately at the preliminary stage. In this case, empirical relationships can be used to provide satisfactory approximate estimates.

e.g. $KB = T/6[(5C_W - 2C_B)/C_W]$ or
$KB = T[C_W/(C_W + C_B)]$

C_W = waterplane area coefficient = waterplane area/L × B

i.e. $KB = f_i(T)$
$BM = J_T/\nabla$
$\nabla = L.B.T.C_B$ hence known
$J_T = iLB^3$

i for waterplane shape can be related to C_W or C_B
Hence $BM = J_T/\nabla = f_2[LB^3/LBTC_B] = f_2[B^2/T\,C_B]$
KG is often defined as a function of depth D, i.e. $KG = f_3(D)$. For example, typical values for various ship types are as follows:

KG lightship = 0.63–0.7 D
= 0.69–0.66 D Tankers
= 0.63–0.66 D Bulk carrier
= 0.66–0.68 D Cargo
= 0.71–0.75 D Cargo Insulated
= 0.90 D Tug
= 0.84 D Trawler

KG loaded = 0.53 D Tankers
= 0.57 D Bulk carriers
= 0.65 D Container ships

At the preliminary stage it is normally satisfactory to assume f_1, f_2 and f_3 constant for a particular ship type. Assuming a value of GM > 0.5 m may be a typical criterion at the preliminary design stage. A maximum GM to preclude the possibility of a very small roll period (which may be undesirable, leading to high accelerations) may also be incorporated. Roll period for general cargo vessels can be approximated as:

$$P = \frac{0.43\,B}{\sqrt{GM}}\ secs \quad (9.7)$$

(i.e. large GM implies short periods and high accelerations) hence, if a minimum period of say 10 secs. is required, then $GM < 0.0018\,B^2$ is required.

Therefore a possible overall stability criterion/constraint for use in the model in Figure 9.5 might be:

$$0.0018\,B^2 > f_1(T) + f_2\left[\frac{B^2}{TC_B}\right] - f_3(D) > 0.5 \quad (9.8)$$

Note: There may be practical variations in GM due to distortion of dimensions, e.g. B limited for some Δ hence high L/B, low BM; shallow draught implies low KB; container ships with large deck cargo implies high KG etc.

9.2.6 Lightship mass estimates

The estimation of the masses of the various items which make up the lightship mass is an important factor in the design process. Masses have a bearing on the technical characteristics of the ship (such as draught and deadweight), and are often used as the basis for cost estimation see Section 9.6.4.2.

The lightship mass is normally summarized under three main headings:

1. STEEL (W_s): Steel hull and superstructure.
2. OUTFIT (W_o): Accommodation, deck fittings, piping, lifeboats etc.
3. MACHINERY (W_m): Main propulsion and auxiliaries such as generators, compressors, boilers etc.

A MARGIN will also be incorporated, depending on the level of uncertainty of the lightship estimate.

Mass estimates, both at the preliminary and detailed design stages are usually grouped under these headings.

It is possible at the detailed design stage, and particularly during and just after construction, to derive reasonably accurate estimates of these masses, although a lot of effort and time will usually be involved.

At the preliminary design stage rapid estimates are required and empirical approximations which relate the component masses to the principal ship particulars (such as dimensions and power) have to be used. The variables in the relationships usually have a physical justification, and the relationships will be 'calibrated' for different ship types. It is the duty of a naval architect to update such empirical relationships whenever possible.

9.2.6.1 *Steel mass*

This normally forms a significant part of the hull mass. Since total ship mass must equal displacement, fixed for a given vessel, a change in steel mass leads to a change in deadweight.

Steel mass/Δ_{mt} should be as low as possible with typical values as follows:

Cargo Ship	Steel/Δ%	20
Cargo & Passenger		28
Passenger		30
Cross Channel		35
Oil Tankers		18

Factors affecting steel mass include: draught (in relation to dimensions), proportions (e.g. L/D), fineness or fullness of form, number of decks and bulkheads, extent of deckhouses and erections, type of construction (structural design).

Use of recorded steel mass for a basis ship forms the most common method of making a preliminary estimate for a new proposal.

Care is required as to what basis steel mass includes: i.e. some yards include hull forgings and castings. Further, whilst generalized data gives a good guide, care must be exercised in interpretation as even for similar ships (or even sisters) steel masses can differ due to alternative methods of construction, owners extras, classification requirements for special vessels/conditions etc.

For *special ship types*, or *novel designs*, detailed mass calculations based on preliminary plans may have to be resorted to.

Steel *ordered* is subject to a rolling margin of $\pm 2\frac{1}{2}$%. Steel purchased is *invoiced mass* and hence steel received varies to within $\pm 2\frac{1}{2}$% of that ordered.

Steel built into the ship is known as *Net Steel Mass*; net steel is about 8–10% less than invoiced mass, i.e. 8–10% scrap.

Methods of estimating steel mass:
(a) Cubic number:
Used for preliminary estimates only.

$$C_N = \frac{LBD}{1000} \text{ where D is to uppermost deck}$$

$$\text{and steel mass } (W_S) = C \cdot \frac{LBD}{1000},$$

where C is some constant derived from basis vessel(s). Method attaches no importance to draught or erections. It also assumes L, B and D to influence the steel mass by the same amount which is not true. Such an approach has to be used with caution. Only if the basis ship is similar and there is little difference in L, B and D can good results be obtained.

(b) Dimensional corrections and differences:
This approach is normally based on data for a basis vessel. Dimensional corrections can be made for length, breadth and depth separately, with subsequent allowances for any other differences between basis and design.

For the dimensional correction, it is required to have the mass/unit change in length, breadth and depth. Also, since steel mass is more sensitive to some dimensions than others it is assumed, for example, that of steel mass:

85% is affected by L
55% is affected by B
30% is affected by D.

This increase in steel mass can be written as follows:

$$\text{increase} \quad \delta W_S = 0.85 w_1 \left[\frac{L_2 - L_1}{L_1}\right] + 0.55 w_1 \left[\frac{B_2 - B_1}{B_1}\right] + 0.30 w_1 \left[\frac{D_2 - D_1}{D_1}\right] \quad (9.9)$$

where w_1 = steel mass of basis.

As well as the basic *dimensional* correction, *difference* corrections will also be made for changes in scantlings due to change in dimensions, change in *form, sheer*, and any other changes such as in erections, superstructures, bulkheads etc., Munro-Smith (1950).

Corrections for changes in sheer (which will normally be small) erections, superstructures and watertight bulkheads etc. would then be carried out as required.

It should be noted that, if sufficient mass data are available for vessels of similar type, the 'weightings' or importance of the various dimensions can be derived as follows:

$$\text{assume} \quad W_s = k L^a B^b D^c \quad (9.10)$$

taking logs

$$\log W_s = \log k + a \log L + b \log B + c \log D \quad (9.11)$$

$$\text{differentiate:} \quad \frac{dW_S}{W_S} = a\frac{dL}{L} + b\frac{dB}{B} + c\frac{dD}{D}$$

i.e. in same format as Equation (9.9), and where coefficients a, b and c may be obtained from multiple linear regression of Equation (9.11). Alternatively, once the coefficients are determined, Equation (9.10) may be used directly.

C_B, T and other variables may be added to Equation (9.10) provided adequate parametric data for similar vessels are available.

(b) Mass/Unit length:
(i) Method uses the midship section for both basis *and* new design. Steel mass for new design

proportioned on the change in mass/m and change in length.
i.e. if W = mass of steel for basis, W_1 = mass/m for basis and W_2 = mass/m for new design, then:

$$\text{Steel mass for new design} = W \times \frac{W_2}{W_1} \times \frac{L_2}{L_1}. \quad (9.12)$$

The method assumes the mass for each ship to be distributed in the same proportion to each other throughout length as they do at amidships. Further corrections may be made, as necessary, for changes in erections, bulkheads etc.

(ii) Use may also be made of mass/m at the preliminary design stage (without a basis vessel) by estimating the mass/m amidships (from midship section) using Classification Society rules, and distributing mass through ship say according to Sectional Area Curve (see proposal by Watson and Gilfillian (1977)). Integration of the mass distribution will give total mass. The method has seen more applications in recent years, since the Classification Rules are available on computer. Parametric variation of dimensions allows a database to be established (for a particular vessel type) and regression equations may be fitted to the data for design purposes.

(c) Group mass method:
Steel mass of known ships analysed into suitable subdivisions or groups. For each group a parameter proportional to say volume or area is derived which can be applied to new designs. Typical groups will include: shell plating, framing, bulkheads, deck plating, erections etc.

The method is best suited to shipyards who have detailed data, and who have established computer data bases, normally also including the hours to work the materials in a particular group. Such an approach allows total masses, building costs and scheduling to be estimated.

(d) Steel mass as function of Lloyds equipment numeral:
Proposed by Watson (1962), updated by Watson and Gilfillian (1977).
Net steel mass plotted against Lloyds equipment numeral:

$$E = L(B + T) + 0.85\,L\,(D - T) + 0.85\,\Sigma l_1 h_1 + 0.75\,\Sigma l_2 h_2$$

where l_1 and h_1 = length and height of full width erections
l_2 and h_2 = length and height of houses.

If extent of houses/erections not known at design stage, for ordinary cargo ships an allowance of 200–300 (metric units) can be used.

(Numeral shown is in fact 'old' numeral, and in 1962 paper Watson plotted invoiced steel – 1977 paper retains 'old' numeral but plots net steel).

Steel masses plotted by Watson were corrected to standard fullness $C_B = 0.7$, measured at 0.8D.

Corrections to steel mass for variation in C_B from 0.7 are made using the following relationship:

$$W_S = W_{S_{0.7}}(1 + 0.5(C_B' - 0.7)) \quad (9.13)$$
W_S = steel mass for actual C_B' at 0.8D

$W_{S_{0.7}}$ = steel mass at $C_B' = 0.7$ as lifted from graph (or following equation).

Watson found following formula to give satisfactory fit to data:

$$W_{S_{0.7}} = kE^{1.36} \quad (9.14)$$

with k for different ship types as shown in Table 9.2:

Table 9.2 k values for steel mass.

Type	k
Tankers/bulk carriers	0.029–0.035
Containers	0.033–0.040
Cargo	0.029–0.037
Tugs	0.044
Trawlers	0.041–0.042
Ferries	0.024–0.037
Passenger	0.037–0.038

Hence combining block coefficient correction with above formula:

$$\text{Net steel mass} = kE^{1.36}(1 + 0.5(C_B' - 0.7)) \quad (9.15)$$

This method offers a good approach at the preliminary design stage for the relevant ship types.

(e) Detailed (direct) calculations:
These are lengthy and laborious. They are required for unusual design proposals. Method does yield LCG, VCG. Ship has to be fairly well defined with approximate body plan, position of DB, DKs, 1/2 girths, frame spacing and outline of section and scantlings etc.

9.2.6.2 Outfit mass

(i) Dimensional corrections:
This approach is normally based on data for a basis vessel. It assumes that part of the outfit mass is constant between basis and new design and the

remainder to vary as length and breadth. Hence the corrected mass for new design W_{O2} is given by:

$$W_{O2} = xW_{O1} + (1-x)\left[W_{O1} \times \frac{L_2}{L_1} \times \frac{B_2}{B_1}\right] \quad (9.16)$$

where the value of x will depend on ship type and size, say 0.5 in the absence of better information.

The approach will normally be too inaccurate for vessels such as ferries and passenger vessels etc.

(ii) Empirical approach:
Uses typical empirical formulae for outfit mass based on L × B for various ship types. Values proposed in the Watson and Gilfillian (1977) paper are as follows:

$$W_O = k' \times L \times B \text{ tonnes}$$

With typical values for k' shown in Table 9.3:

Table 9.3 k' values for outfit mass.

Passenger ships	$K' = 0.7$–1.55 (L = 100–250m)	
Trawlers	$k' = 0.3$–0.5 (L = 25–80m)	Intermediate values by linear interpolation
Cargo vessels	$k' = 0.4$	
Container	$k' = 0.32$	
Tankers/Bulk Carriers	$k' = 0.25$–0.18 ….(L = 150–300m)	

Such formulae and plotting of empirical data can be very approximate due to wide variations in outfit mass that can occur for a particular ship type. They must be used with care.

9.2.6.3 Machinery mass

It is important to note that the *TOTAL* machinery mass is made up of the main propulsion machinery *together with* the remaining machinery such as auxiliaries, compressors, boilers, piping etc.

A preliminary power estimate is required prior to carrying out the machinery mass estimate. This may be made using standard series or suitable regression data or simple relationships based on displacement and speed (e.g. Power $\alpha \, \Delta^{2/3} V^3$, see Chapter 5).

In estimating the machinery mass the most effective approach is to break down the total machinery mass into the propulsion machinery mass and the remainder. Data is readily available for main engine(s), allowing data fits to be made and updated. Such an approach, and the equations proposed by Watson (1977) is as follows:

For diesel installation:

$$\text{Main engine(s) mass} = 9.38\left[\frac{P}{N}\right]^{0.84} \quad (9.17)$$

where P = installed power (HP) and N = engine rpm (not prop) (a typical assumption at the preliminary design stage is N = 110 rpm for low speed diesel and 500 rpm for medium speed diesel).

Remaining mass (tonnes) = $k'' \, P^{0.7}$

where $k'' = 0.56$ for bulk carriers and general cargo
$= 0.59$ for tankers
$= 0.65$ for passenger vessels and ferries

Hence, TOTAL Machinery mass W_m (tonnes)

$$= 9.38\left[\frac{P}{N}\right]^{0.84} + k'' \, P^{0.7} \quad (9.18)$$

For a steam turbine installation, Watson proposes:

$$\text{Total machinery Mass tonnes} = 0.16 \, (\text{SHP})^{0.89}$$

where SHP = installed maximum power.

9.2.6.4 Margin

The sum of the net steel and outfit constitute the hull mass; any underestimation can only be made up for in loss of deadweight, DW. Also any departures from design causing increase in hull mass will influence DW. Thus a margin is normally allowed. Amount of margin allowed will depend on the degree of uncertainty of the lightship estimate, and penalty clauses regarding non-compliance with the specified deadweight.

Typical values are:

1% of Lightship mass + 0.1% load displacement	Watson (1962)
2% of DW	Munro-Smith (1950)
2% Lightweight	Watson and Gilfillian (1977)

Such margins normally adjusted to give a round figure for the lightship mass.

9.2.6.5 Masses of fast ferries

Estimation procedures for the masses of the aluminium alloy hull, outfit and machinery for fast ferries (monohulls and catamarans) have been proposed by Karayannis *et al.* (1999), Karayannis and Molland (2001) and Molland *et al.* (2003) using methods similar to those already described. Satisfactory values were obtained when compared with data from basis ships, and the methods are particularly suitable for use at the preliminary design stage.

9.2.6.6 *Vertical centre of gravity (KG)*

A *detailed* mass check will normally incorporate VCG and LCG information.
Actual ship KG can be derived from an inclining experiment, see Section 3.6.

For approximate and preliminary purposes (such as for stability check in Section 9.2.5), KG normally expressed as a function of depth (D) for a particular ship type/size. This function may be derived from a basis vessel (correcting for depth and changes in machinery mass KG etc and any other significant changes) or a database of similar vessels, and applied to a new design. For cargo ships, bulk carriers etc lightship KG is typically 0.68 to 0.72D.

If more detailed data are available, different levels of breakdown of the components of KG may be applied.

9.2.7 Design of ship lines

The design of the ship's lines is fundamental to the ship design process. This Section considers the design of the ship lines and body plan (see Figures 3.2 and 9.44), and the modification to form.

There are several ways to establish the lines and body plan but four fundamentals must be achieved:

1. Correct displacement Δ on selected principal dimensions.
2. Correct LCB – determined partly by disposition of structure/cargo/machinery, etc (LCG) and also by best form for resistance.
3. Position of metacentre in worst condition of loading of vessel – dependent on beam and shape of waterline.
4. Shape of Sectional Area Curve (SAC) for satisfactory propulsion; Cp, entrance, run, LCB and maximum slope of SAC.

Two distinct phases in designing the lines:

(a) Achievement of form characteristics, per 1–4 above.
(b) Ensuring form determined corresponds to a fair body.

- The body plan is normally drawn to moulded lines. Thus form coefficients for steel ships are normally for moulded displacement; a typical allowance for the shell is 5 tonnes/1000 tonnes displacement (0.5%) to give *extreme* displacement.
- A body plan can be developed from first principles. However, a similar basis ship, or suitable published body plan is frequently chosen and modified to the correct C_B, C_p, LCB, L, B, etc for the new design.
- Changes to the basis for the proposed design, such as to C_p and LCB, are most conveniently carried out by modifying the basis Sectional Area Curve (SAC). It is a very robust method and allows complete control over any changes. The Sectional Area Curve will also be employed when developing a body plan/lines from first principles.

9.2.7.1 *Sectional area curve (SAC) – definitions:*

$\bar{x}$ = LCB = longitudinal centre of area of curve

$$\frac{V}{L.A_m} = Cp_T = \frac{\text{Area Under Curve}}{\text{Enclosing Rectangle}}$$

$$Cp_F = \frac{\text{Immersed vol. for 'd Amidships}}{L/2 \times A_m}$$

$$Cp_E = \frac{\text{Immersed vol. of Entrance}}{E \times \text{Largest section area}}$$

Cp_A, Cp_R obtained similarly

It is usual to normalise the length in terms of stations 0~10 (or 0~1) and nondimensionalise areas in terms of Area amidships A_m, Figure 9.10.

Using this notation, C_P equals the area under the sectional area curve

9.2.7.2 *Modifications to sectional area curve*

(a) To change Cp:

$$\frac{r_1}{r_2} = \frac{1 - Cp'}{1 - Cp}$$

Cp' = desired Cp for new design (correction applied both ends), Figure 9.11.

Modifying Cp in this way also changes LCB, hence logical procedure is to

(a) correct for Cp, and recalculate LCB
(b) correct to LCB of new form.

(b) To change LCB:

Find the centre of area of SAC (longitudinal and about base), and modify as shown in Figure 9.12.

$$\delta x(\%L) = \frac{BB'(\%L)}{\bar{y}} \times y$$

The simple $(1 - Cp)$ change as described above has a number of disadvantages:

(1) There is no control over Parallel mid body, PMB, in the derived form
(2) It cannot be used to reduce Cp in a form with no PMB

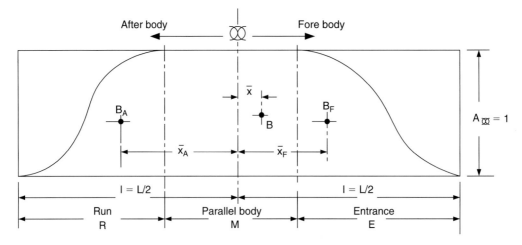

Figure 9.9 Sectional area curve.

(3) The basic form without PMB cannot be increased in fullness without inserting PMB
(4) The prismatic of entrance and run cannot be adjusted.

The above simple methods are however suitable for many applications.

These deficiencies can be overcome using the Lackenby Transformations, described in the next section.

9.2.7.3 Sectional area curve transformations

Useful procedures are the Lackenby transformations, Lackenby (1950).

The deficiencies in the $(1 - Cp)$ and LCB curve swinging methods can be overcome by the more detailed numerical calculations described by Lackenby.

Lackenby includes the basic $(1 - Cp)$ and curve swinging methods and makes a comprehensive review of alternatives including the cases of:

$(1 - Cp)$ method $\}$ either holding LCB constant or changing LCB

Change in Cp for case of no parallel mid body $\}$ either holding LCB constant or changing LCB

- *Summary of formulae*: which are applied to forward and aft *separately*

  ϕ = original Cp, $\delta\phi$ = change in Cp see Figure 9.13

- One-minus Prismatic method

$$\delta x = \frac{\delta\phi}{(1-\phi)}(1-x), \quad \text{i.e. same as } r_2, r_1 \text{ method.}$$

also $\delta p = \dfrac{\delta\phi}{(1-\phi)}(1-x)$, where P = original PMB.

$h = \dfrac{\phi(1 - 2\bar{x})}{(1 - \phi)}$ $\bar{x}$ = centroid of original area — forward or aft

h = centroid of added 'sliver'

(h is used in derivation of $\delta\phi_f$ or $\delta\phi_a$)

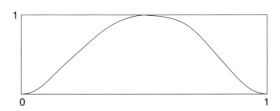

Figure 9.10 Sectional area curve.

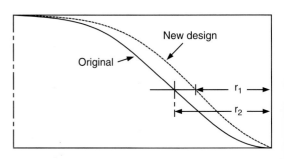

Figure 9.11 SAC: Change in Cp.

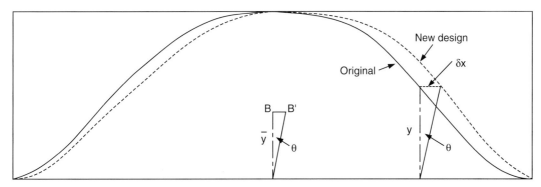

Figure 9.12 SAC: Change in LCB.

- Case with no PMB:
  δx variation assumed as $\delta x = c.x(1 - x)$
  i.e. maximum change at $x = 0.5$. (max. shift restricted to shoulders in the case of $(1 - Cp)$ variation).

whence $\delta x = \dfrac{\delta \phi x(1-x)}{\phi(1-2\bar{x})}$ and $h = \dfrac{2\bar{x} - 3k^2}{(1-2\bar{x})}$

where k = lever of second moment of original curve, Figure 9.13.
- Derivation of $\delta\phi_f$ and $\delta\phi_a$ to suit required LCB change (or need to keep constant).

[Suffixes f and a represent forward or aft].

$$\delta\phi_f = \dfrac{2[\delta\phi_t(h_a + \bar{z}) + \delta\bar{z}(\phi_t + \delta\phi_t)]}{(h_f + h_a)}$$

and $\delta\phi_a = \dfrac{2[\delta\phi_t(h_f - \bar{z}) - \delta\bar{z}(\phi_t + \delta\phi_t)]}{(h_f + h_a)}$

ϕ_t = total Cp
$\bar{z}$ = distance of LCB of Basis ship from amidships, as a fraction of half-length (+ve forward)
$\delta\bar{z}$ = required fractional shift of LCB.

also: $\phi_t = (\phi_f + \phi_a)/2$ and $\delta\phi_t = (\delta\phi_f + \delta\phi_a)/2$

Examples of special cases:
e.g. if LCB is to remain unchanged, $\delta\bar{z} = 0$

and $\delta\phi_f = \dfrac{2\delta\phi_t(h_a + \bar{z})}{(h_f + h_a)}$; $\delta\phi_a = \dfrac{2\delta\phi_t(h_f - \bar{z})}{(h_f + h_a)}$

or if ϕ_t is to remain unchanged, $\delta\phi_t = 0$

and $\delta\phi_f = \dfrac{2\delta\bar{z}.\phi_t}{(h_f + h_a)}$ and $\delta\phi_a = \dfrac{-2\delta\bar{z}\phi_t}{(h_f + h_a)}$

- Proofs of the above formulae are given in Lackenby's paper and applications are illustrated by worked examples.

9.2.7.4 Preparation of body plan

(A) From 'FIRST PRINCIPLES': Using Sectional Area and load water line curves:

Midship Area:
There is not much freedom with midship area coefficient (C_m). There are high values in cargo ships/tankers leading to large cargo carrying capacity.

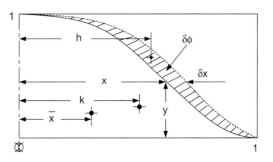

Figure 9.13 Notation for Lackenby transformation.

Typical values:

| C_B | .75 | .70 | .65 | .6 | .55 |
| C_m | .987 | .984 | .980 | .975 | .960 |

$C_B = C_m \cdot C_p$

i.e. as C_B reduces, C_m reduces.

In fine form ships low C_m implies large rise of floor and large bilge radius.

Choice of rise of floor may also depend on directional stability and drainage from double bottom tanks. Bilge radius varies with fullness of section C_m.

Section Design: Assuming Sectional Area Curve and Waterline are available, e.g. using a polynominal approach, or from standard series data.

For a particular station, area A from SAC, B/2 from LWL curve hence proceed as shown in Figure 9.14. Repeat at other sections, and fair using waterlines.

(B) Body plan 'FROM BASIS': with change in Cp and/or LCB.

Correct the basis SAC and lift offsets for new design from basis ship lines at a revised station spacing from ends, Figure 9.15. New lines are automatically fair.

Trans immersed area at $9A$ = area at $9'E$

At position 9′ on basis lines (1/2 breadth) plan, lift waterline offsets and plot at station 9 for new design; repeat for other stations. This procedure results in fair new lines and body plan.

Correct for $\dfrac{T_1}{T}$ and $\dfrac{B_1}{B}$ if necessary (maintains C_B and LCB).

Note: this method (B) lends itself to a computer based approach.

i.e.

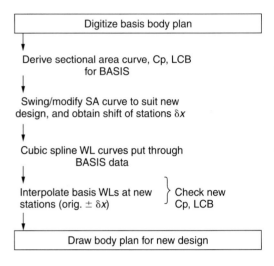

Alternatively, the required shift of stations may be applied to commercial ship lines packages.

(C) *STANDARD SERIES DATA*:

A number of standard series have been published which provide a useful source of hull forms, as well as providing resistance/powering data.

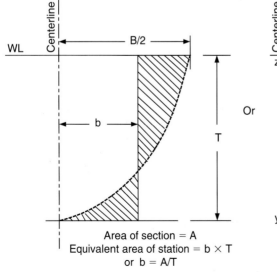
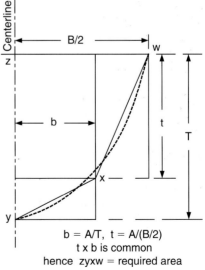

Figure 9.14 Section design.

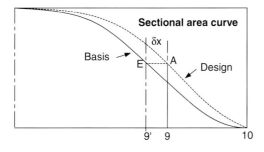
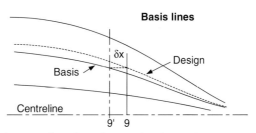

Plot waterline offsets from station 9' at station 9 on new body plan

Figure 9.15 Modifying ship lines.

BSRA Series:	*Trans. RINA* 1961, 1966 BSRA Report No. NS 333 gives refaired lines + bulbous bows	⎫ ⎬ SS Merchant Forms ⎭
Dawson Coasters:	*IESS* 1958/59 & earlier	
Series 60:	Todd, *SNAME* Vol. 61, 1953	
Taylor Series:	Taylor/Gertler revised DTMB Rep. 806	⎫ ⎬ TS ⎭
Linblad Series:	*Trans RINA* 1946/49	
NPL Round Bilge Series:	Smaller Semi-Displacement Craft RINA Monograph No. 4	
Series 64 Round Bilge Series:	*Marine Technology*, No. 2, July 1965, SNAME	
Series 62 Planing Hulls:		
NTUA Series (double chine):	Semi-displacement craft. Radojcic *et al.* (2001)	

See also Section 5.1, Resistance and propulsion.

9.2.8 Statutory regulations

Legislation exists which is concerned with the safety of ships and the well being of all who sail in them. International legislation with regard to shipping is now dealt with by the I.M.O. (International Maritime Organization, formerly I.M.C.O.) which was set up in 1959 by the U.N. Its various committees meet periodically and I.M.O. arranges various conferences such as SOLAS 1960, 1974, International Load Line Conference 1966 and the Tonnage Conference 1969.

Further description and discussion of the role of IMO is included in Chapter 11.

Implementation of the legislation is the responsibility of the government of the country concerned. In the UK it is administered by the Maritime and Coastguard Agency (MCA) and the rules are drawn up by virtue of the Merchant Shipping Acts.

Surveyors verify that ships are built and operated in accordance with the regulations.

Typical matters (and hence legislation) with which the rules are concerned are: Stability; Load lines (freeboard); Subdivision; Tonnage; Life Saving Appliances; Crew Accommodation Regulations; Fire Appliances and Protection; Carriage of Grain Cargoes; Dangerous Cargoes.

Since the regulations are statutory they are of fundamental importance in the design and operation of ships and, consequently, have to be integrated in the design process from the early conceptual stages to the detailed final stages. Stability, freeboard and subdivision, in particular, are fundamental to the initial design process. Stability and subdivision are described and discussed in Chapter 3. Information on freeboard and tonnage may be obtained from Eyres (2007), Tupper (2004), IMO (1966) and IMO (1969).

9.2.9 Concept design content: example

The typical content which may be covered in the design process at concept stage is shown in Figure 9.16, and applies much of the content in the earlier Sections of this Chapter. Aspects such as powering, structures, seakeeping and manoeuvring and safety are dealt with in other Chapters.

9.3 Materials

9.3.1 Introduction

A description is given of the principal materials used in the construction of the main components of a

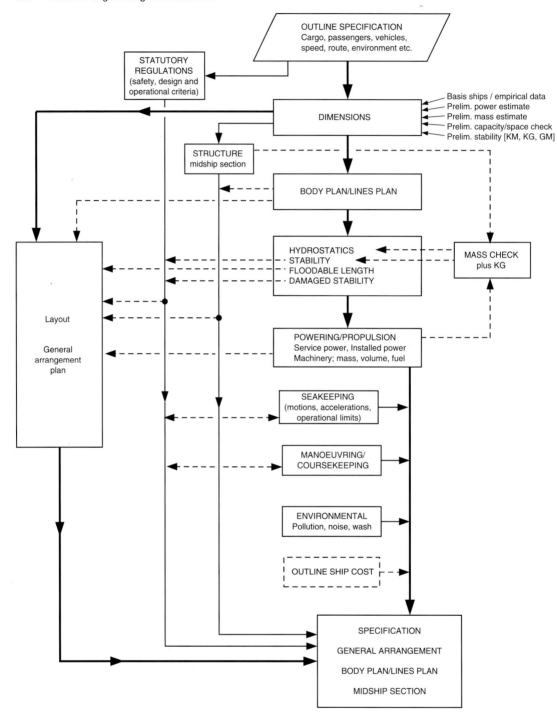

Figure 9.16 Typical content of concept design model.

ship or marine structure. These amount basically to steels, aluminium alloys and composites. An outline of corrosion, corrosion control and anti-fouling is included.

9.3.2 Steel

The production of all steels used for shipbuilding purposes starts with the smelting of iron ore and the making of pig-iron. Normally the iron ore is smelted in a blast furnace, which is a large, slightly conical structure lined with a refractory material. To provide the heat for smelting, coke is used and limestone is also added. This makes the slag formed by the incombustible impurities in the iron ore fluid, so that it can be drawn off. Air necessary for combustion is blown in through a ring of holes near the bottom, and the coke, ore, and limestone are charged into the top of the furnace in rotation. Molten metal may be drawn off at intervals from a hole or spout at the bottom of the furnace and run into moulds formed in a bed of sand or into metal moulds.

The resultant pig-iron is from 92 to 97% iron, the remainder being carbon, silicon, manganese, sulphur, and phosphorus. In the subsequent manufacture of steels the pig iron is refined, in other words the impurities are reduced.

9.3.2.1 *Manufacture of steel*

Steels may be broadly considered as alloys of iron and carbon, the carbon percentage varying from about 0.1% in mild steels to about 1.8% in some hardened steels. These may be produced by one of four different processes, the open hearth process, the Bessemer converter process, the electric furnace process, or an oxygen process. Processes may be either an acid or basic process according to the chemical nature of the slag produced. Acid processes are used to refine pig-iron low in phosphorus and sulphur which are rich in silicon and therefore produce an acid slag. The furnace lining is constructed of an acid material so that it will prevent a reaction with the slag. A basic process is used to refine pig-iron that is rich in phosphorus and low in silicon. Phosphorus can be removed only by introducing a large amount of lime, which produces a basic slag. The furnace lining must then be of a basic refractory to prevent a reaction with the slag. About 85% of all steel produced in Britain is of the *basic* type, and with modern techniques is almost as good as the *acid* steels produced with superior ores.

Only the open hearth, electric furnace, and oxygen processes are described here as the Bessemer converter process is not used for shipbuilding steels.

Open hearth process. The open hearth furnace is capable of producing large quantities of steel, handling 150 to 300 tonnes in a single melt. It consists of a shallow bath, roofed in, and set above two brick-lined heating chambers. At the ends are openings for heated air and fuel (gas or oil) to be introduced into the furnace. Also these permit the escape of the burned gas which is used for heating the air and fuel. Every twenty minutes or so the flow of air and fuel is reversed.

In this process a mixture of pig-iron and steel scrap is melted in the furnace, carbon and the impurities being oxidized. Oxidization is produced by the oxygen present in the iron oxide of the pig-iron. Subsequently carbon, manganese, and other elements are added to eliminate iron oxides and give the required chemical composition.

Electric furnaces. Electric furnaces are generally of two types, the arc furnace and the high-frequency induction furnace. The former is used for refining a charge to give the required composition, whereas the latter may only be used for melting down a charge whose composition is similar to that finally required. For this reason only the arc furnace is considered in any detail. In an arc furnace melting is produced by striking an arc between electrodes suspended from the roof of the furnace and the charge itself in the hearth of the furnace. A charge consists of pig-iron and steel scrap and the process enables consistent results to be obtained and the final composition of the steel can be accurately controlled.

Electric furnace processes are often used for the production of high-grade alloy steels.

Oxygen process. This is a modern steelmaking process by which a molten charge of pig-iron and steel scrap with alloying elements is contained in a basic lined converter. A jet of high purity gaseous oxygen is then directed onto the surface of the liquid metal in order to refine it.

Steel from the open hearth or electric furnace is tapped into large ladles and poured into ingot moulds. It is allowed to cool in these moulds, until it becomes reasonably solidified permitting it to be transferred to 'soaking pit' where the ingot is reheated to the required temperature for rolling.

Chemical additions to steels. Additions of chemical elements to steels during the above processes serve several purposes. They may be used to deoxidize the metal, to remove impurities and bring them out into the slag, and finally to bring about the desired composition.

The amount of deoxidizing elements added determines whether the steels are 'rimmed steels' or 'killed steels'. Rimmed steels are produced when only small additions of deoxidizing material are added to the molten metal. Only those steels having less than 0.2% carbon and less than 0.6% manganese can be rimmed. Owing to the absence of deoxidizing material, the oxygen in the steel combines with the carbon and other gases present and a large volume of gas is liberated. So long as the metal is molten the gas passes upwards through the molten metal. When solidification takes place in ingot form, initially from the sides and bottom and then across the top, the gasses can no longer leave the metal. In the central portion of the ingot a large quantity of gas is trapped with the result that the core of the rimmed ingot is a mass of blow holes. Normally the hot rolling of the ingot into thin sheet is sufficient to weld the surfaces of the blow holes together, but this material is unsuitable for thicker plate.

The term 'killed' steel indicates that the metal has solidified in the ingot mould with little or no evolution of gas. This has been prevented by the addition of sufficient quantities of deoxidizing material, normally silicon or aluminium. Steel of this type has a high degree of chemical homogeneity, and killed steels are superior to rimmed steels. Where the process of deoxidation is only partially carried out by restricting the amount of deoxidizing material a 'semi-killed' steel is produced.

In the ingot mould the steel gradually solidifies from the sides and base as mentioned previously. The melting points of impurities like sulphides and phosphides in the steel are lower than that of the pure metal and these will tend to separate out and collect towards the centre and top of the ingot which is the last to solidify. This forms what is known as the 'segregate' in way of the noticeable contraction at the top of the ingot. Owing to the high concentration of impurities at this point this portion of the ingot is often discarded prior to rolling plate and sections.

9.3.2.2 Heat treatment of steels

The properties of steels may be altered greatly by the heat treatment to which the steel is subsequently subjected. These heat treatments bring about a change in the mechanical properties principally by modifying the steel's structure. Those heat treatments which concern shipbuilding materials are described.

Annealing. This consists of heating the steel at a slow rate to a temperature of say 850°C to 950°C, and then cooling it in the furnace at a very slow rate. The objects of annealing are to relieve any internal stresses, to soften the steel, or to bring the steel to a condition suitable for a subsequent heat treatment.

Normalizing. This is carried out by heating the steel slowly to a temperature similar to that for annealing and allowing it to cool in air. The resulting faster cooling rate produces a harder stronger steel than annealing, and also refines the grain size.

Quenching (or hardening). Steel is heated to temperatures similar to that for annealing and normalizing, and then quenched in water or oil. The fast cooling rate produces a very hard structure with a higher tensile strength.

Tempering. Quenched steels may be further heated to a temperature somewhat between atmospheric and 680°C, and some alloy steels are then cooled fairly rapidly by quenching in oil or water. The object of this treatment is to relieve the severe internal stresses produced by the original hardening process and to make the material less brittle but retain the higher tensile stress.

Stress relieving. To relieve internal stresses the temperature of the steel may be raised so that no structural change of the material occurs and then it may be slowly cooled.

9.3.2.3 Steel sections

A range of steel sections are rolled hot from ingots. The more common types associated with shipbuilding are shown in Figure 9.17. It is preferable to limit the sections required for shipbuilding to those readily available, that is the standard types; otherwise the steel mill is required to set up rolls for a small amount of material which is not very economic.

9.3.2.4 Shipbuilding steels

Steel for hull construction purposes is usually mild steel containing 0.15 to 0.23% carbon, and a reasonably high manganese content. Both sulphur and phosphorus in the mild steel are kept to a minimum (less than 0.05%). Higher concentrations of both are detrimental to the welding properties of the steel, and cracks can develop during the rolling process if the sulphur content is high.

Steel for a ship classed with Lloyds Register is produced by an approved manufacturer, and inspection and prescribed tests are carried out at the steel mill before dispatch. All certified materials are

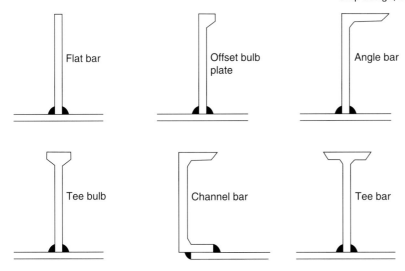

Figure 9.17 Steel sections of shipbuilding.

marked with the Society's brand and other particulars as required by the rules.

Ship classification societies originally had varying specifications for steel: but in 1959, the major societies agreed to standardize their requirements in order to reduce the required grades of steel to a minimum. There are now five different qualities of steel employed in merchant ship construction and now often referred to as IACS steels. These are graded A, B, C, D and E, Grade A being an ordinary mild steel to Lloyds Register requirements and generally used in shipbuilding. Grade B is a better quality mild steel than Grade A and specified where thicker plates are required in the more critical regions, Grades C, D and E possess increasing notch-tough characteristics, Grade C being to American Bureau of Shipping requirements. Lloyds Register requirements for Grades A, B, D and E steels may be found in Chapter 3 of Lloyds Rules for the Manufacture, Testing and Certification of Materials, Lloyds Register (2004).

9.3.2.5 *High tensile steels*

Steels having a higher strength than that of mild steel are employed in the more highly stressed regions of large tankers, container ships and bulk carriers. Use of higher strength steels allows reductions in thickness of deck, bottom shell, and framing where fitted in the midships portion of larger vessels; it does, however, lead to larger deflections. The weldability of higher tensile steels is an important consideration in their application in ship structures and the question of reduced fatigue life with these steels has been suggested. Also, the effects of corrosion with lesser thicknesses of plate and section may require more vigilant inspection.

Higher tensile steels used for hull construction purposes are manufactured and tested in accordance with Lloyds Register requirements. Full specifications of the methods of manufacture, chemical composition, heat treatment, and mechanical properties required for the higher tensile steels are given in Chapter 3 of Lloyds Rules for the Manufacture, Testing and Certification of Materials. The higher strength steels are available in three strength levels, 32, 36, and 40 (kg/mm^2) when supplied in the as rolled or normalized condition. Provision is also made for material with six higher strength levels, 42, 46, 50, 55, 62 and 69 (kg/mm^2) when supplied in the quenched and tempered condition. Each strength level is subdivided into four grades, AH, DH, EH and FH depending on the required level of notch-toughness.

9.3.2.6 *Corrosion resistant steels*

Steels with alloying elements, that give them good corrosion resistance and colloquially referred to as stainless steels are not commonly used in ship structures, primarily because of their higher initial and fabrication costs. Only in the fabrication of cargo tanks containing highly corrosive cargoes might such steels be found.

For oil tankers the inner surfaces, particularly the deckhead and bottom, are generally protected by high cost corrosion resistant coatings that require vigilant inspection and maintenance. A recent development in the manufacture of an alloyed shipbuilding steel with claimed improved corrosion resistance properties and its approval by Lloyds Register for use in certain cargo tanks of a 105 000 dwt tanker indicate that in the future the need to coat oil cargo tanks might be dispensed with.

9.3.2.7 *Steel sandwich panels*

As an alternative to conventional shipyard fabricated stiffened steel plate structures, proprietary manufactured steel sandwich panels have become available and used on ships where their lighter weight was important. Such panels consist of a steel core in the form of a honeycomb with flanges to which the external steel sheets are resistance (spot) or laser (stake) welded. Early use of these bought in steel sandwich panels was primarily for non-hull structures in naval construction where their light weight was important. Also when fabricated using stainless steel their corrosion-resistance and low maintenance properties have been utilized.

A proprietary steel sandwich plate system (SPS) has been developed which consists of an elastomer core between steel face plates. Elastomers are a specific class of polyurethane that has a high tolerance to mechanical stress i.e. it rapidly recovers from deformation. The SPS elastomer also has a high resistance to most common chemical species. Initial application of SPS in shipbuilding has been in passenger ship superstructures where the absence of stiffening has increased the space available and provided factory finished surfaces with built in vibration damping, acoustic insulation and fire protection. SPS structures have been approved with an A 60 fire-resistance rating. Also SPS overlays have been applied to repair existing work deck areas. SPS structures can be fabricated using joining technologies presently used in the shipbuilding industry, but the design of all joints must take into account the structural and material characteristics of the metal-elastomer composite. The manufacturer envisages the use of SPS panels throughout the hull and superstructure of ships providing a simpler construction with greater carrying capacity and less corrosion, maintenance and inspection. In association with the manufacturer Lloyds Register in early 2006 published provisional Rules for the use of this sandwich plate system for new construction and ship repair. The Rules cover construction procedures, scantling determination for primary supporting structures, framing arrangements and methods of scantling determination for steel sandwich panels.

The Norwegian classification society, Det Norske Veritas (DNV), have proposed for bulk carrier hulls the use of a lightweight concrete/steel sandwich. They envisage a steel/concrete/steel composite structure for the cargo hold area of say 600 mm width for the side shell but somewhat greater width for the double bottom area. This sandwich would be much narrower than for a comparable steel-only double skin bulk carrier thus increasing the potential carrying capacity although water ballast may have to be carried in some designated holds as the double skin would not be available for this purpose. DNV consider the other advantages of the concrete/steel sandwich to be reduced stress concentrations with less cracking in critical areas, considerable elimination of corrosion and elimination of local buckling. At the time of writing DNV were undertaking a two-year investigation programme in association with a shipyard to study the practicalities of their sandwich proposal.

9.3.2.8 *Steel castings*

Molten steel produced by the open hearth, electric furnace, or oxygen process is pored into a carefully constructed mould and allowed to solidify to the shape required. After removal from the mould a heat treatment is required, for example annealing, or normalizing and tempering to reduce brittleness. Stern frames, rudder frames, spectacle frames for bossings, and other structural components may be produced as castings.

9.3.2.9 *Steel forgings*

Forging is simply a method of shaping a metal by heating it to a temperature where it becomes more or less plastic and then hammering or squeezing it to the required form. Forgings are manufactured from killed steel made by the open hearth, electric furnace, or oxygen process, the steel being in the form of ingots cast in chill moulds. Adequate top and bottom discards are made to ensure no harmful segregations in the finished forgings and the sound ingot is gradually and uniformly hot worked. Where possible the working of the metal is such that metal flow is in the most favourable direction with regard to the mode of stressing in service. Subsequent heat treatment is required, preferably annealing or normalizing and tempering to remove effects of working and non-uniform cooling.

9.3.3 Aluminium alloy

9.3.3.1 *General*

There are three advantages which aluminium alloys have over mild steel in the construction of ships. Firstly aluminium is lighter than mild steel (approximate weight being aluminium 2.723 tonnes/m^3, mild steel 7.84 tonnes/m^3), and with an aluminium structure it has been suggested that up to 60% of the weight of a steel structure may be saved. This is in fact the principal advantage as far as merchant ships are concerned, the other two advantages of aluminium being a high resistance to corrosion and its non-magnetic properties. The non-magnetic properties can have advantages in warships and locally in way of the magnetic compass, but they are generally of little importance in merchant vessels. Good corrosion properties can be utilized, but correct maintenance procedures and careful insulation from the adjoining steel structure are necessary. A major disadvantage of the use of aluminium alloys is their higher initial and fabrication costs. The higher costs must be offset by an increased earning capacity of the vessel, resulting from a reduced lightship weight or increased passenger accommodation on the same ship dimensions. Experience with large passenger liners on the North Atlantic service has indicated that maintenance costs of aluminium alloy structures can be higher for this type of ship and service.

A significant number of larger ships have been fitted with superstructures of aluminium alloy and, apart from the resulting reduction in displacement, benefits have been obtained in improving the transverse stability. Since the reduced weight of superstructure is at a position above the ship's centre of gravity this ensures a lower centre of gravity than that obtained with a comparable steel structure. For example on the Queen Elizabeth 2, with a limited beam to transit the Panama Canal, the top five decks constructed of aluminium alloy enabled the ship to support one more deck than would have been possible with an all steel construction.

Only in those vessels having a fairly high speed and hence power, also ships where the deadweight/lightweight ratio is low, are appreciable savings to be expected. Such ships are moderate – and high – speed passenger liners having a low deadweight. It is interesting to note however that for the Queen Mary 2, not having a beam limitation, the owners decided to avoid aluminium alloy as far as possible to ensure ease of maintenance over a life cycle of 40 years. A very small number of cargo liners have been fitted with an aluminium alloy superstructure, principally to clear a fixed draught over a river bar with maximum cargo.

The total construction in aluminium alloy of a large ship is not considered an economic proposition and it is only in the construction of smaller multihull and other high speed craft where aluminium alloys higher strength to weight ratio are fully used to good advantage.

9.3.3.2 *Production of aluminium*

For aluminium production at the present time the ore, bauxite, is mined containing roughly 56% aluminium. The actual extraction of the aluminium from the ore is a complicated and costly process involving two distinct stages. Firstly the bauxite is purified to obtain pure aluminium oxide known as alumina; the alumina is then reduced to a metallic aluminium. The metal is cast in pig or ingot forms and alloys are added where required before the metal is cast into billets or slabs for subsequent rolling, extrusion, or other forming operations.

Sectional material is mostly produced by the extrusion process. This involves forcing a billet of the hot material through a die of the desired shape. More intricate shapes are produced by this method than are possible with steel where the sections are rolled. However, the range of thickness of section may be limited since each thickness requires a different die. Typical sections are shown in Figure 9.18.

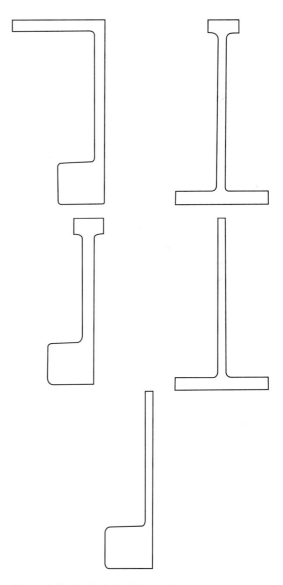

Figure 9.18 Typical aluminium alloy sections.

Table 9.4 Alloying elements.

Element	5083	5086	6061	6082
Copper	0.10 max	0.10 max	0.15–0.40	0.10 max
Magnesium	4.0–4.9	3.5–4.5	0.8–1.2	0.6–1.2
Silicon	0.40 max	0.40 max	0.4–0.8	0.7–1.3
Iron	0.40 max	0.50 max	0.70 max	0.50 max
Manganese	0.4–1.0	0.2–0.7	0.15 max	0.4–1.0
Zinc	0.25 max	0.25 max	0.25 max	0.20 max
Chromium	0.05–0.25	0.05–0.25	0.04–0.35	0.25 max
Titanium	0.15 max	0.15 max	0.15 max	0.10 max
Other elements				
each	0.05 max	0.05 max	0.05 max	0.05 max
total	0.15 max	0.15 max	0.15 max	0.15 max

Aluminium alloys. Pure aluminium has a low tensile strength and is of little use for structural purposes; therefore the pure metal is alloyed with small percentages of other materials to give greater tensile strengths, Table 9.4. There are a number of aluminium alloys in use, but these may be separated into two distinct groups, non-heat treated alloys and heat treated alloys. The latter as implied are subjected to a carefully controlled heating and cooling cycle in order to improve the tensile strength.

Cold working of the non-heat treated plate has the effect of strengthening the material and this can be employed to advantage. However, at the same time the plate becomes less ductile, and if cold working is considerable the material may crack; this places a limit on the amount of cold forming possible in shipbuilding. Cold worked alloys may be subsequently subjected to a slow heating and cooling annealing or stabilizing process to improve their ductility.

With aluminium alloys a suitable heat treatment is necessary to obtain a high tensile strength. A heat treated aluminium alloy which is suitable for shipbuilding purposes is one having as its main alloying constituents magnesium and silicon. These form a compound Mg_2Si and the resulting alloy has very good resistance to corrosion and a higher ultimate tensile strength than that of the non-heat treated alloys. Since the material is heat treated to achieve this increased strength, subsequent heating, for example welding or hot forming, may destroy the improved properties locally, Kecsmar and Shenoi (2004).

Aluminium alloys are generally identified by their Aluminium Association numeric designation. The 5000 alloys being non-heat treated and the 6000 alloys being heat treated. The nature of any treatment is indicated by additional lettering and numbering.

Lloyds Register prescribe the following commonly used alloys in shipbuilding:

5083-0	annealed
5083-F	as fabricated
5083-H321	strain hardened and stabilized
5086-0	annealed
5086-F	as fabricated
5086-H321	strain hardened and stabilized
6061-T6	solution heat treated and artificially aged
6082-T6	solution heat treated and artificially aged

Riveting. Riveting may be used to attach stiffening members to light aluminium alloy plated structures where appearance is important and distortion from the heat input of welding is to be avoided.

The commonest stock for forging rivets for shipbuilding purposes is a non-heat treatable alloy NR5 (R for rivet material) which contains 3–4% magnesium. Non-heat treated alloy rivets may be driven cold or hot. In driving the rivets cold relatively few heavy blows are applied and the rivet is quickly closed to avoid too much cold work, i.e. becoming work hardened so that it cannot be driven home. Where rivets are driven hot the temperature must be carefully controlled to avoid metallurgical damage. The shear strength of hot driven rivets is slightly less than that of cold driven rivets.

9.3.3.3 *Aluminium alloy sandwich panels*

As with steel construction, proprietary aluminium alloy honeycomb sandwich panels are now available to replace fabricated plate and stiffener structures and can offer extremely low weight options for the superstructures of high speed craft.

9.3.3.4 *Fire protection*

It is considered necessary to mention when discussing aluminium alloys that fire protection is more critical in ships in which this material is used because of the low melting point of aluminium alloys. During a fire the temperatures reached may be sufficient to cause a collapse of the structure unless protection is provided. The insulation on the main bulkheads in passenger ships will have to be sufficient to make the aluminium bulkhead equivalent to a steel bulkhead for fire purposes.

For the same reason it is general practice to fit steel machinery casings through an aluminium superstructure on cargo ships.

9.3.4 Composite materials

9.3.4.1 *Overview*

In this section a summary is made of the design of marine structures made from composite materials. Attention is focused on fibre-reinforced plastics (FRP), but it should be noted that the term 'composites' can include materials such as fibre-reinforced metals, fibre-reinforced cement and combinations of FRP, wood, metal and concrete see also Section 9.3.2.7. This section on the marine applications of FRP composite materials has been taken from Shenoi and Dodkins (2000). Further developments in the properties and applications of composites can be found in references such as Shenoi and Wellicome (1993), Clarke *et al.* (1998), Jeong and Shenoi (2001), Kelly and Zweben (2000), Backman (2005) and Vasiliev and Morozov (2007).

9.3.4.2 *Introduction*

Polymeric composite materials have been used in ships, boats, and other marine structures for over 50 years, Smith (1990), Shenoi and Wellicome (1993). The motivation for their use has varied from application to application. In naval minehunters, for instance, the main driver for their usage is the non-magnetic and non-conducting capability of glass reinforced plastics (GRP). In the case of dinghies, canoes, and small harbour craft, GRP is preferred because of competitive first cost and the ease with which complex shapes required for such craft can be fabricated. Yet another factor leading to increased use is the good fire resistance of fibre reinforced plastics (FRP) – this is so with regard to applications in offshore structures. Other issues encouraging the increased use of FRP are: (i) low operating (maintenance) cost; (ii) good fatigue resistance; (iii) high specific strength; (iv) good corrosion resistance; (v) good thermal resistance; and (vi) reduced parts count.

The purpose of this section is to provide a broad overview to the design of ship structures made from composite materials. As a precursor to this, it is essential to understand certain key features distinguishing ships and marine structures. Ships are products of a one-off variety. There are very few series of similar ships. The largest production run could be of the order of about 10–12 ships of one kind. This is a very rare occurrence, though. What this means is that the design effort has to be dedicated for each ship order. This contrasts with a large-volume production of an aircraft type (e.g., Boeing 747, Airbus A320, etc.) or a motor car model (e.g., Ford Fiesta, Volvo 460 series, etc.), where the design effort can be focused to a greater degree.

The lead time for ships, from order to delivery, is very short. For large tankers and bulk carriers this can be as short as a few months. For naval ships, such as minehunters, this can be about three to four years. For smaller craft, the time span can vary from a few months to a couple of years. This places a tremendous pressure on marine designers and production engineers to produce practical and cost-effective solutions rapidly.

Ships and other marine craft, with a few exceptions, are generally low-cost modes of transport. The cargo freight rates or passenger ticket prices for the marine mode are several orders of magnitude smaller than for aircraft. This implies that ships have to have a much lower life cycle cost. A significant component of this economic balance is primary (or production) cost. Thus marine designers need to search for solutions using relatively inexpensive materials in production processes which do not require substantial tooling or other forms of high-cost infrastructural investment.

The implication of the above-listed three features is that marine design has to be done rapidly, using technology that is well proven and with relatively large factors of safety to account for uncertainties in a variety of production and operational areas. This section seeks to provide an overview of marine structural design in composite materials, particularly with regard to materials selection, design procedures, structural synthesis, and external influences on design.

9.3.4.3 *Materials selection*

(a) Materials selection

1. Reinforcements

For marine applications, generally the choice of reinforcement is simplified because cost constraints render the more expensive high-performance reinforcements such as carbon and aramid unattractive. The emphasis for bulk use is strongly on glass fibre. This has been used in a variety of forms including unidirectional tows, woven and stitched fabrics, and chopped random mats. There are some areas in high-performance craft where combinations of carbon and aramid fibres are being considered, Serter (1997), Maccari and Farolfi (1992). However, glass still accounts for over 95% of the usage in marine applications.

Some key property parameters influencing the selection of structural materials for marine use, Gibson (1993), are shown in Figures 9.19 and 9.20. Figure 9.19 compares various materials in terms of strength per unit weight and stiffness per unit weight. It is evident that composites have better characteristics than metals with regard to specific strength. However, in terms of specific stiffness, only carbon and aramid composites outperform metals. Glass-based composites are more flexible. Apart from mechanical performance, structural materials also have cost implications. In Figure 9.20 it is clear that none of the composites is competitive with metals in stiffness-critical areas. Furthermore, in strength-critical cases, only glass-based composites can compete with metals. This is the underlying reason for the large usage of glass in large-volume applications such as ships, offshore structures, and other marine artifacts.

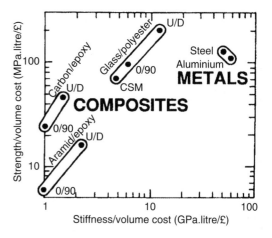

Figure 9.20 Strength and stiffness per volume cost for typical engineering materials, Gibson (1993).

The mechanical properties are engineered by the appropriate use of different forms of the reinforcement, (see Figure 9.21). High fibre volume fractions are desirable in some applications and certain regions of ships. Unidirectional rovings give the highest fibre volume fraction, usually in the range 0.5–0.65. In woven fabrics, the volume fraction is generally 0.4–0.55, while with random mats only 0.25–0.33 is achievable.

2. Matrix resins

The matrix plays a critical role in determining off-axis strength, damage tolerance, corrosion resistance, and thermal stability. Current technology and cost constraints limit the selection to thermosets and there are three widely used candidates, as shown in Table 9.5, each with particular strengths and drawbacks.

Unsaturated polyesters are the most widely used resins in the marine industry. Their principal advantage, apart from low cost, lies in their cure chemistry. The free radical cure reaction, triggered by the addition of a peroxide catalyst, offers a rapid but controllable cure, while the resins themselves have a long shelf life. For this reason, polyesters are easily fabricated. Among the various types of polyester resins, the isophthalic variety offers the most attractive combination of mechanical strength and resistance to the marine environment. However, from a cost viewpoint, the orthophthalic variety holds attractions for the small boat industry.

Vinyl ester resins lie midway between polyesters and epoxies. While retaining some of the fabricability of the free radical cure, they offer better mechanical properties and are often preferred in demanding

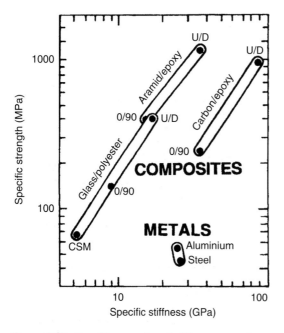

Figure 9.19 Specific strength and stiffness properties for typical engineering materials, Gibson (1993).

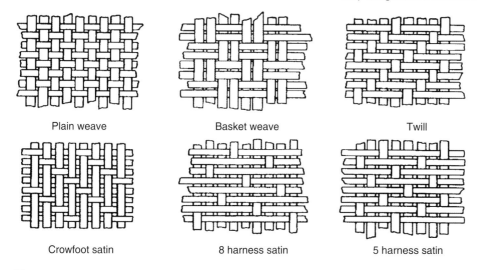

Figure 9.21 Types of fibre reinforcement with potential use in marine applications.

Table 9.5 Candidate resins for use in marine applications.

Resin	Cost (£/tonne)	Mechanical strength	Corrosion resistance	Fire performance
Polyester	1200–1600	xx	xx	x
Vinyl ester	2200–2600	xxx	xxx	x
Epoxy	>4000	xxxxx	xxxxx	x

applications, particularly where chemical or environmental resistance is needed.

Epoxy resins, of which there are several variants, offer the most outstanding combination of strength, toughness, and corrosion resistance. They are, however, expensive. Fabrication can also be more difficult and hazardous compared with polyesters and vinyl esters. They are most widely used with higher performance fibres in vessels where high strength, toughness, and damage tolerance are prime requirements.

3. Core materials

The choice in this context is mainly between PVC foams, balsa wood, and honeycomb materials. Expanded closed-cell polyvinyl chloride (PVC) foam has been widely used in many marine applications. It is available in a range of densities, varying from 45 to over 200 kg m^{-3}. There are also several varieties of these, including linear PVC which has high ductility but low mechanical properties and cross-linked PVC which has high strength and stiffness but is relatively brittle. PVC foams offer good resistance to water penetration, good thermal and electrical insulation, and effective vibration and damping characteristics. Their main deficiencies are reduction of strength and stiffness at modestly elevated temperatures (typically a loss of 50% of compressive and shear moduli and strengths at temperatures in the range 40–60°C), outgassing at temperatures up to 100°C and chemical breakdown, with emission of HCl vapour at temperatures of over 200°C.

End grain balsa is one of the most efficient, Hearmon (1948), and moderately priced sandwich core materials. Its main deficiency is susceptibility to water penetration and consequential swelling, debonding, and rot. Although some success has been claimed for the balsa core sandwich construction in boats, Lippay and Levine (1968), a number of disastrous instances of water penetration and subsequent deterioration of balsa core have also occurred. For these reasons, use of balsa core in the primary hull and deck structure of ships and boats is not normally advisable.

Sandwich panels and shells with ultralight honeycomb cores in aluminium, FRP, or resin-impregnated paper, developed in many cases for aerospace structures, are generally too expensive for marine construction. However, they have a limited application in decks and bulkheads of weight-critical craft such as hydrofoils, hovercraft, and high-performance sailing yachts and in specialized components such as radomes. They are unlikely to be suitable for the primary hull structure of high-speed craft because of the risk of water penetration and core-skin debonding under impact loads, though they have recently found extensive application in racing yachts where robustness and durability are sacrificed in favour of performance.

With all forms of sandwich construction, regardless of core materials used, a sound and consistent bond between core and the skins is of paramount importance. Production techniques, quality control and inspection methods need to be applied with great care on ship scale sandwich fabrication.

(b) Mechanical properties

1. Static properties

These may be carried out using BS, ISO, European, ASTM or naval standards, Sims (1993). A selection of these are listed below.

Component fractions	BS2782, ISO 1172/7822, ASTM 2584/2374/D3171
Tensile	BS2782, ISO3268, EN2597, ASTM 2585
Compressive	ISO 8515/604, ASTM D3410, EN 2850
Flexural	ISO 178, ASTM D790, BS 2782, EN 2561
In-plane shear	ASTM D3518/4255/3846
Interlaminar shear	ISO 4585, BS 2782, EN2563, ASTM D2344
Sandwich materials	ASTM C273/297/364/365/393/394/408

Laminates incorporating woven reinforcement are tested in both the warp and weft directions. It is important to emphasize that variability in properties may be high especially for marine laminates. The values of strength and modulus to be used in design calculations should correspond to 'mean minus two standard deviations' limit derived from mechanical test data. This implies a 97.5% probability that the design figure will be exceeded in the actual structure.

2. Long-term properties

All resins absorb a certain small percentage of moisture. The general pattern of degradation in a marine laminate is an initial fall of 10–20% in mechanical properties in the first 12–18 months followed by a very slow fall over the rest of the period. However, if a saturated laminate is subjected to continuous tensile stress, then the degradation may be more severe. Laboratory experiments have shown that sustained stress levels should not exceed about 20% of the ultimate values, Dodkins (1993). This is generally not a problem in ships' structures designed for extreme load cases due to wave action etc., but special attention should be paid to structure supporting dead loads. It is advisable to keep strain levels below those at which resin microcracking occurs, say 0.3–0.5%, thus avoiding moisture ingress.

Accelerated aging tests are also now coming into more prominence. The acceleration is usually achieved by increasing the temperature above ambient, though well below the heat distortion temperature. This is typically 60–70°C for a non-post cured polyester laminate. Quicker results may be obtained by heating to 80–90°C and comparing the performance to that of a laminate with known resistance to aging.

In specifying the lay-up of a shell laminate of the ship's hull, an all-woven roving configuration is acceptable, though the outer surface must contain a layer of chopped strand mat. The mat takes up a greater proportion of resin than the roving and short fibres ensure that if the glass becomes exposed on the surfaces, water cannot wick far along the fibres thus ensuring that any minor surface damage as a result of impact and abrasion remains localized. The mat layer therefore forms a protective barrier to the underlying woven layers. The use of gel coats on yachts perform the same function, while giving a good aesthetic finish at the same time.

3. Fire resistance

The fire resistance of a structure is difficult to characterize accurately, the requirements being generally: (i) maintenance of strength and stiffness until a fire is extinguished; (ii) limitation of temperature and prevention of spread of flames to adjacent compartments; and (iii) minimization of smoke and toxic fumes.

As with all polymeric materials, FRP is combustible. However, FRP, although flash-igniting in air temperatures of 350–400°C, burns slowly, is readily extinguished by water sprinkling or oxygen exclusion, and providing woven reinforcement is used, provides a partially effective barrier in the form of exposed glass fibres. Because of its low conductivity,

Table 9.6 Fire-related properties of metals and FRP.

Material	Melting temperature (°C)	Thermal conductivity w/(m.°C)	Heat distortion temperature (°C) (BS2752)	Self-ignition temperature (°C)	Flash ignition temperature (°C)	Oxygen index (%) (ASTM D2863)	Smoke density D_m (ASTM E662)
Aluminium	660	240	–	–	–	–	–
Steel	1430	50	–	–	–	–	–
E-Glass	840	1.0	–	–	–	–	–
Polyester resin	–	0.2	70	–	–	20–30	–
Phenolic resin	–	0.2	120	–	–	35–60	–
GRP (polyester based)	–	0.4	120	480	370	25–35	750
GRP (phenolic based)	–	0.4	200	570	530	45–80	75

GRP meets requirement (ii) effectively. Emission of smoke and fumes by burning polyester resin presents a serious problem which requires special ventilation and fire-fighting facilities. An increased effectiveness can be provided by the use of intumescent and other fire-retardant coatings which can be incorporated in the surface of the laminate as a gel coat.

Phenolic-based FRP, which offers a high level of strength and stiffness retention at temperatures of up to 250°C, high flash-ignition temperature (about 530°C), and oxygen index (>45%), together with low smoke and toxic fume emission, should be considered carefully for fire-critical structures such as bulkheads in accommodation areas and machinery compartments.

Some of the fire-related characteristics of marine structural materials are listed in Table 9.6, Dow and Bird (1994), Don and Bird (1994). Fire is an important issue which is currently affecting the increased use of composites in marine vessels; legislative issues covering this aspect are discussed in more detail in Section 9.3.4.6.

(c) Production considerations

1. Production processes
(i) Open mould wet lay-up (hand lay-up)
Until very recently, production of marine structures was achieved using one of two techniques. Small, 'low-tech' dinghies, yachts, and similar craft have been produced by the spray-gun technique, where reinforcement strands and resin are injected together on to the surface of a mould. This results in a randomly oriented reinforcement in a resin-rich form. Larger ships have been built using the hand lay-up techniques. Some automation in resin impregnation was achieved even in the early days of GRP shipbuilding, Smith (1990), though this was limited to some shipyards and certain parts of the structure.

(ii) Vacuum-assisted resin infusion moulding
The most significant breakthrough in FRP fabrication in a marine context occurred in the early 1990s with the introduction of vacuum-assisted resin infusion moulding. This is now gradually replacing the wet lay-up process in a comprehensive manner. Vosper Thornycroft (now UT), for instance, used SCRIMP (or Seeman Composites Resin Infusion Moulding Process) for the production of large ship scale mouldings. Bulkheads and plate panels up to 10 m × 10 m in size and 18 mm thickness are currently in production. Hull and superstructure mouldings up to 30 m long have also been successfully produced.

Such resin infusion techniques have distinct advantages over the wet lay-up process.

(i) High compaction under vacuum results in laminates of high quality and fibre content and enhanced mechanical properties with improved uniformity.
(ii) Air voids are virtually eliminated.
(iii) A cleaner production process is achieved with very low styrene emissions.
(iv) Electromagnetic screening in the form of a metallic mesh can be embedded in the lay-up prior to resin infusion so that it can become an integral part of the structure.
(v) A weight saving of 15% has been achieved on single skin parts.
(vi) For sandwich construction, both skins can be wetted-out and bonded to the core in a single infusion process.

2. Practical considerations

(i) Wet lay-up

FRP laminating should take place inside climate controlled buildings. Polyester resin cures at room temperature (above 16°C). A styrene fume extraction system maintains shop airborne styrene levels in compliance with regulatory requirements. This covers the requirements imposed by the wet lay-up process, where significant quantities of styrene are released through evaporation from large areas of exposed wet laminate. It is noteworthy that in the vacuum-assisted resin infusion process, the laminate is sealed under a nylon film and nearly all styrene is cross-linked during the curing process.

The characteristics of laminating materials affect both the quality of the final laminate and the production time. These features may be evaluated realistically only by large-scale production trials (which are discussed in the next section). Important aspects to consider in materials selection are discussed below.

(i) Ease of cutting the reinforcement cloth is largely dependent on the cloth weight. If it is intended to use a range of standard widths, then it is advantageous to have the material supplied ready cut to those widths with the edges stitched in order to prevent fraying. Woven roving is more likely to fray than combination cloth where the stitching and mat layer hold the cut edges.

(ii) While most cloths will wrap easily around a cylindrically shaped mould surface, not all will form around a corner or a shape with double curvature without some tailoring. Examples are the snapped ends of stiffeners, tapered stiffener sections, and the bow section of the hull. Cloths with poor drapeability should be avoided as they lead to excessive tailoring which results in cloth joints that are too close together, necessitating additional material to compensate for the loss in strength. Conversely, a cloth that is very drapeable is too easily distorted such that the rovings are pulled out of a straight line. Coloured threads may be incorporated with the warp rovings to help maintain the straightness during lay-up.

(iii) The use of a heavy cloth implying the need to have a reduced number of plies does not always lead to reduced production time. Each of the heavier cloth layers will take longer to set up and consolidate. In any case, there is a limit on the weight of cloth and resin that can be laid wet-on-wet at one time (see the issue below on resin curing).

(iv) A low viscosity resin reduces the time taken for consolidation (wet-out of the cloth and removal of air bubbles by rolling). The resin should be thixotropic to reduce drainage on vertical surfaces. However, this tends to conflict with the ease of wetting out. This feature is particularly important in resin selection.

(v) The curing reaction of the polyester resin is exothermic. In the wet lay-up process, if the laminate thickness build-up is too fast, then the later layers tend to insulate the earlier layers and prevent dissipation of the heat. The laminate then begins to heat up more, which further accelerates the curing reaction until a runaway situation develops and the temperature may rise to a point where the laminate becomes permanently heat damaged.

Before finalization of a lay-up for any structure, it is desirable to carry out production trials. Initial trials may be carried out on small panels measuring about $1\,m \times 2\,m$. These should have any envisaged stiffening to be bonded to the surface to check drapeability over the sides and ends of the stiffening. Handling may be evaluated to a limited extent and the resin ratio and ply thicknesses can also be checked. At this early stage in the materials selection process, a large number of materials can be evaluated economically.

Having short-listed the materials with adequate handling characteristics, panels of about $3\,m \times 3\,m$ could be fabricated; these samples should be used to cut samples for testing to determine mechanical properties. Panels of this size are important because they reflect the level of difficulty involved in large-scale production; laminates will normally include a realistic void content and butted and staggered cloth edges.

Once the mechanical tests have been completed, the number of fibre/resin combinations can be reduced to perhaps two or three. These materials may now be tested again in more realistic production trials. Short sections of the hull, perhaps three to four frame spaces long as a minimum, should be laid up in the hull mould. These sections should include examples of all principal structural features of the proposed design such as frames, bulkheads, stiffeners, tee joints, beam knee joints, stiffener-to-shell connections, etc.

(ii) Vacuum-assisted resin infusion moulding

Mould surfaces used for this process need to be completely airtight as full vacuum is applied to the entire moulding area. However, unlike conventional resin transfer moulding, no positive resin injection pressure is applied, so that moulds need not be heavily reinforced. A drawback though is the higher cost of consumable materials such as tubing and flow medium and resin lost in feed and vacuum tubes. However, reusable bags have been developed to suit series production of small to medium sized mouldings.

Good control of the workshop environment, resin, and mould temperatures must be maintained for SCRIMP to produce consistent results. The process very much depends on knowing gel times, which are largely temperature dependent.

When producing large mouldings, thought must be given to the means of catalyzing and infusing large quantities (perhaps over 1 tonne) of resin in timescales of 40 min or less. This requires a high level of shop floor planning and teamwork.

Many of the comments made earlier with regard to selection of constituent materials also apply to SCRIMP. However, compared with wet lay-up, low viscosity resins are preferred in order to maximize flow rate and as all plies of fabric representing the full lay-up are applied dry and infused together, the weight of individual fabric layers is not important, thereby allowing the specification of fewer plies of heavier fabrics in order to reduce the manhours required for mould preparation.

For any new lay-up or materials, production trials are required to finalize the infusion set-up and procedure, measure flow rates across the mould, and select the gel time required. This is best carried out on a glass-topped moulding table such that the resin flow front can be viewed and timed from above and below the laminate. The results of such trials are normally sufficient to scale up to full-sized production mouldings, provided the effects of all features of the production mouldings have been checked.

9.3.4.4 *Design concepts*

(a) Design spiral
Ship design and, by inference, ship structural design, is iterative in nature (Figure 9.22). This is specifically derived for preliminary design purposes when the ship framing on the decks and bottom is longitudinal in nature (see Section 9.3.4.5(c)); it typifies one aspect of the more all-inclusive process of overall ship design and is therefore a spiral within spirals. Inferences to be drawn from the illustration are that among the interlocking constraints which must be satisfied, albeit in harmony with each other, the web frame and longitudinal spacings are tentatively set as initial conditions on which the final, optimized design is to be based. Other optimized designs for varying frame spacings could also be investigated.

First estimates of plate thicknesses, section details, joint specifications, longitudinal scantlings, and materials choices will by necessity be rough. However, as the process proceeds towards convergence, more characteristics of the design become known and, hence, more refined methods can be used. The progression from approximate analytical expressions to more refined finite element analysis based techniques is discussed in more detail later in this section. With each iterative cycle, necessary modifications from the one previous become smaller. A designer has to make the decision to finalize the design, after a requisite number of iterations, having met all performance-based requirements, based on its ability to be produced in a cost-efficient manner and to be maintainable at reasonable costs during the operational life of the ship.

(b) Design loads
1. General
The first step in any structural design is to define the loads that will act on the structures. For ship and boat design, this exercise can be exhaustive and tedious. Primary loads from the operation of the vessel in the seaway must ideally allow for the variability of the ocean environment itself. Secondary and tertiary loads resulting from locally-induced sources such as the main engines, heavy cargo in one compartment, etc. may also be critical in some cases. In addition to the magnitude and direction of the loads, it is also important to know the frequency with which the force systems acts in order that fatigue calculations may be adequately carried out. Details of such calculations may be learnt from Chapter 4 or any standard naval architectural textbook, Lewis (1988). Table 9.7 lists the principal loads to be considered in warship design, for instance.

There are a variety of guides available from regulatory authorities in various countries such as Lloyd's Register of Shipping (1998a), American Bureau of Shipping (1998). These are increasingly based on first-principles mechanics concepts, though because of the very nature of uncertainties associated with seaways, there is still a significant reliance on empiricism and operational experience.

The purpose of this section is to outline the principal types of loads and their characteristics. Loads to be considered here include:

(i) hull girder bending loads that act over the entire length of the ship;
(ii) wave slamming loads on ships and high-speed craft;
(iii) deck and bulkhead loads;
(iv) point loads.

These aspects are described in more detail in Chapter 4.

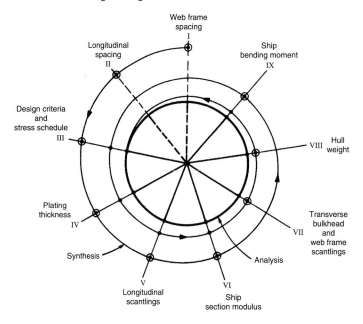

Figure 9.22 Structural design spiral.

Table 9.7 Loads imposed on warship structures.

Basic loads	Sea loads	Operational loads	Combat loads
Live loads	Hull bending	Flooding	Primary Shockwave
Structure self-weight	Wave slamming	Helicopter landing	Gun blast pressures
Tank pressures	Roll/pitch/heave inertia	Replenishment at sea	Explosion-induced whipping
Equipment weights	Wind loads	Docking	Fragmentation
		Anchoring	Gun recoil
		Berthing	Missile efflux pressures

2. The hull as a longitudinal girder

Classical approaches to ship structural design treat the hull structure as a beam for purposes of analysis. The validity of this approach is related to the vessel's length-to-beam (L/B) and length-to-depth (L/D) ratios. Hull girder methods are applied to L/D values greater than 12. From practical considerations this refers to vessels greater than about 50 m in length.

(i) Still water bending moment

Before a ship even goes out to sea, some stress distribution profile exists within the structure. Figure 9.23 shows how the summation of buoyancy and weight distribution curves of an idealized rectangular barge lead to shear force and bending moment distribution diagrams. Stresses apparent in the still water condition generally become extreme only in cases where concentrated loads are applied to the structure, which can be the case when the holds of a cargo vessel are selectively filled.

(ii) Wave bending moment

A quasistatic approach to predicting stresses in a seaway involves the superposition of a trochoidal wave with a wavelength equal to ship length in the hogging and sagging conditions (see Figure 9.24). The wave height is usually taken as $L/10$ ($L < 60$ m), $L/15$ (60 m $< L <$ 90 m), $L/20$ (90 m $< L <$ 150 m), and $0.6\ L^{0.6}$ ($L > 150$ m). Except for very slender craft, this will not apply to smaller vessels. See also Section 4.1.2.14.

3. Dynamic forces on large ships
(i) Ship oscillation forces

The dynamic response of a vessel operating in a given sea spectrum is very difficult to predict

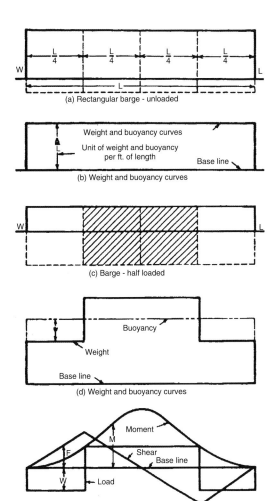

Figure 9.23 Still water bending moment distribution in an idealized rectangular barge.

analytically. Accelerations experienced in the vessel vary as a function of the vertical, longitudinal, and transverse location. These accelerations produce virtual increases of the weight of the concentrated masses, resulting in consequential increases in stress. The designer should have a feel for the worst locations and the type of dynamic behavior that can produce extreme load scenarios. It is generally assumed that combined roll and pitch forces near the deck edge forward represents a (worst case) condition for the extreme accelerations for the ship. There are a number of two- and three-dimensional codes that are used for determining the dynamic response of ship hull girders, Bishop and Price (1979), Faltinsen (1992). In the main, these are used for analytical purposes: for design synthesis, reliance is still placed to a large degree on classification society rules, Lloyd's Register of Shipping (1998a), American Bureau of Shipping (1998).

(ii) Dynamic phenomena
This is principally related to high-frequency loading such as vibrations. Such loading can be either steady state, as with propulsion system induced phenomena, or transient, such as slamming through waves. In the former case, load amplitudes are generally within the design limits of the hull structural material choice. However, repetitive loading implies that fatigue can be a significant issue. Further, a preliminary vibration analysis of major structural elements (such as the hull girder, engine foundations, deck houses, masts, etc.) is generally prudent to ensure that the natural frequencies are not near the propeller shaft or propeller blade rotation rate for normal operating modes. See Section 4.3.

4. Wave slamming on small craft
Slamming is defined in the classical sense as 'high impulsive water pressures at certain speeds in severe seas when the ship motions become large enough to result in forefoot emergence', Ochi and Motter (1973). The timescale during which a slamming pressure acts on a panel is very short, of the order of hundredths of a second. At a given point on a ship panel, however, the maximum pressure will be present only for a few thousandths of a second. This is of particular relevance in the context of composite ships because FRP composite panels have eigenfrequencies of the same order of magnitude as the frequency of the peak-slamming load. Although Ochi and Motter (1973) estimated that more than 300 papers on slamming had already been published by the 1970s, the complexity of the phenomenon is such that universally applicable analytical solutions are still beyond the designers' capabilities.

On a practical level, slam load prediction is still done on the basis of the pioneering work by Heller and Jasper (1960). The method is based on relating strain in a structure from a static load to the corresponding value from dynamic conditions. The ratio of the dynamic to static strains is the so-called dynamic response factor. Such work was extended in recent times by seeking the response of composite single skin and sandwich panels to drop load tests, which sought to simulate slam conditions, Hayman *et al.* (1991). This showed that slam pressure is not uniform on a panel; the pressure pulse typically starts at one edge of a panel and works its way to the other edge – as shown in Figure 9.25. Such work has now been incorporated into design guidelines from classification societies, Lloyd's Register of Shipping (1988b).

A further description of slamming is contained in Section 4.1.2.22.

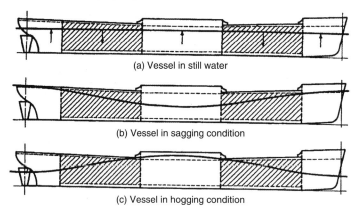

Figure 9.24 Superposition of the static wave profile.

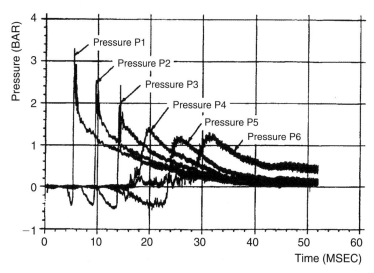

Figure 9.25 Slam pressure variation on an FRP plate panel (after Hayman *et al.* 1991).

(c) Design margins

Typical design margins based on current practice are listed in Table 9.8. The margins for short-term static loads are those which, if applied to the ultimate strength of the laminate, will give the resin microcracking stress. Stresses higher than the microcracking stress are deemed to cause significant permanent damage to the laminate, although the structure would still be able to take further load up to the ultimate value.

In case of local buckling of panels between stiffeners, the low margin of 1.5 is only justified if positive measures are taken to prevent premature detachment of stiffeners from the panel. This may involve bolting of flanges to the panel or using resilient adhesive which prevents peeling of the flanges.

Regarding fatigue, the margin of 5.0 applied to the ultimate stress only relates to high strain rate applications such as slamming and whipping in the forward regions of the ship. Static short-term margins may be used for the overall structure if the maximum operating strains are less than about 20% of the ultimate limit of the matrix.

Bearing in mind the low modulus of FRP, it is important to evaluate the structural deformations carefully. When designing tanks to withstand internal pressure, a limit of $L/200$ (where L is the tank length) is typically imposed. This avoids excessive

Table 9.8 Design margins.

Load action	Margin
Static short-term loads (tension)	3.0
Static short-term loads (compression)	2.0
Static long-term loads (dry)	4.0
Static long-term loads (immersed)	6.0
Load reversal	5.0
Local buckling (stiffeners parallel to load)	1.5
Column buckling of plate/stiffener combinations	2.0
Buckling (stiffeners perpendicular to load)	3.5

deformation and possible subsequent damage to boundary joints. Panels between stiffeners on lightweight decks, for example in the superstructure, should be limited in deflection to $B/80$ (where B is the panel width) to avoid them feeling springy to walk on. Careful consideration should be given to the selection of coatings (e.g., non-skid deck paint) as materials formulated for application to steel may not be sufficiently flexible for use on an FRP substrate.

9.3.4.5 *Design synthesis*

(a) Choice of topology

There are four radically different styles to choose from – top hat stiffened single skin, monocoque single skin, sandwich, and corrugated construction. The advantages and disadvantages of each of these are given in Table 9.9. From this it may be concluded that:

(i) Sandwich construction offers a fairly low cost (at least for one-offs or small production runs) and high stiffness-to-weight at the potential expense of service durability. However, these aspects are being addressed in the context of small craft and the experience should no doubt filter through to the applications in larger ships.
(ii) With a large capital investment, monocoque construction may be mechanized to a very large extent, thereby minimizing labour cost. However, the result is a heavy structure. It is best suited for long production runs and where the vessel weight is not of particular consequence. Quality assurance during build and operation can also be problematical, thereby potentially restricting use to more sophisticated customers (e.g., navies).
(iii) Stiffened single skin construction offers the lowest technical risk in that design, build inspection, maintenance, and repair are all straightforward. Thus, where weight and durability under a variety of load conditions need to be good, where weight has a slightly lower emphasis, single skin stiffened construction is suitable; this is particularly so for displacement vessels (as opposed to high-speed craft, where dynamic lift implies weight criticality). Cost is higher than for sandwich, especially for one-offs; this difference though is reduced when production runs of five or more vessels are planned.
(iv) Corrugated construction offers lighter weight than stiffened single skin, but is unlikely to be considered for hull structures without considerable further development. It is relatively expensive, particularly in terms of tooling cost and lay-up complexity.

It is possible to mix these different forms of construction to combine the advantages and obtain the best compromise for a particular application (see Table 9.10). For example, it may be attractive to specify a single skin hull and main deck, corrugated watertight bulkheads, and sandwich construction for secondary structure such as internal decks, minor bulkheads, and superstructure. The UK Sandown minehunters and the RNLI Severn class lifeboat are examples where two or more construction styles have been used.

(b) Structural elements

1. General

As mentioned in Section 9.3.4.2, ship design is characterized by the need to have a workable set of plans at very short notice. Structural design consequently suffers from constraints. The tendency of designers is to start from a known case, modify it slightly to suit changed circumstances for the new design, and then test key elements of the design in more detailed studies.

This approach is being questioned now in view of the fact that there is a growing tendency among ship owners to ask for much higher performances from the new ships. Designers are having less and less past material to base their empiricism on; increasingly therefore, designs are being based on confirmed first principles.

Structural design is based on three principal levels of load–response estimation, namely the primary (hull bending), secondary (plate bending), and tertiary (or stiffener) stresses. In general, the primary stresses are the dominant stresses for larger ships, e.g., $L > 60\,\text{m}$. Even so, detail calculations are required to ensure that suitable margins exist in the structure to be able to cope with the variety and potential severity of loads.

Table 9.9 Comparison of structural styles.

Configuration	Advantages	Disadvantages
Top hat stiffened single skin	Properties and responses well known Automation possible Easy to fit equipment Costs reduce with number of hulls Quality control is easy Survey in service is straightforward	Fairly expensive to build Care is needed to provide good impact resistance
Monocoque single skin	Easily automated Low labour cost Few secondary bonds below waterline Good shock resistance	Very heavy High material cost Survey methods difficult Attachments and support to machinery difficult Quality control difficult
Sandwich	High specific bending stiffness Can be built without a mould Secondary bonding can be minimized Construction/maintenance costs low Easy to fit equipment	Survey methods need refinement Long-term durability is potential problem Precautions needed to protect core from fire
Corrugated	Relatively lightweight Low labour and material cost Automation is possible	Lower transverse strength Internal fitting may prove to be difficult Awkward mould Strange appearance

Table 9.10 Comparison of weights and costs for different structural styles.

Configuration	Relative weight	Relative cost
Single skin—longitudinal stiffening	1.00	1.00 (0.75)[a]
Corrugations with 0.16 m depth	1.24	1.55
PVC foam core sandwich	0.73	0.62
Monocoque thick GRP	3.04	1.92

[a]Compliant resin used instead of bolts.

2. Plating design

The plating thickness is determined by the requirement for it to resist a combination of lateral pressures and in-plane loading. The magnitudes and proportions of the loads vary from location to location in the ship. For instance in larger ships, the bottom shell will be subject lateral loading owing to local water from the outside, payload or cargo weight from the inside, and in-plane loading owing to global bending of the hull girder. Bulkheads are generally designed on the basis of linearly varying water pressure arising from one of the two compartments that the bulkhead separates being flooded. The deck, for instance, requires a particularly critical examination. This is because, under a sagging condition, compressive stresses in the deck plating can be significant enough to warrant exhaustive stability checks. If, in addition, there are transverse loads on the deck, then the situation becomes even more severe.

The basis of design is orthotropic plate analysis. There are several versions. At the simplest, designers use fundamental equations from isotropic theory with orthotropy incorporated by lumping the section properties on to the plate thickness to give artificially contrived panel stiffnesses, Smith (1968). A quick check can of course be made using cylindrical bending equations of the type derived by Pagano (1968). More refined approaches which incorporate shear inertia effects, Ochoa and Reddy (1992), are now being used more and more in order to specify scantlings for sandwich plates.

Failure limits in strength need to be identified explicitly. They are based on phenomenological issues:

(i) matrix cracking
(ii) fibre breakage
(iii) fibre–matrix debonding
(iv) interfacial cracking
(v) delamination
(vi) core shear cracking
(vii) skin wrinkling
(viii) skin-core debond.

This does require an explicit definition on a structural level and is done using one of the many different macroscopic criteria, e.g., maximum stress/strain, Tsai–Wu, Tsai–Hill, Hoffman, Hart-Smith, etc.

3. Stiffener design
For top hat stiffeners such as that illustrated in Figure 9.26, the designer has to select from a range of variables:

(i) section height
(ii) section width
(iii) web angle
(iv) flange width
(v) web lay-up
(vi) table lay-up.

The almost infinite freedom that this represents is tempered by the need to standardize as much as possible. The penalty for not doing so is to give the production department an almost impossible task in shaping foam former sections, tapering from one size to another, and tailoring cloth widths to suit varying section sizes. Too many changes in the number of lay-up plies along the run of one stiffener can cause many problems to the laminator. Overall the result can be a most significant reduction in productivity.

The best approach, Dodkins *et al.* (1994), is to devise a range of standard section sizes, preferably 10 or less for a ship, and try to restrict the choice to one section size for the length of each run of a stiffener. For example, the longitudinal hull bottom stiffener may run through several compartments, being supported at different, possibly unevenly spaced, locations. By selecting from the standard range, it should be possible to cope with the different spans by varying the lay-up from one compartment to the next and achieve this without incurring significant weight penalty.

The failure modes that are of interest here are:

(i) shear failure of the webs
(ii) tensile/compressive failure of the table
(iii) tensile/compressive failure of the base panel
(iv) local buckling of the table
(v) shear buckling of the webs
(vi) interlaminar shear/tensile failure of connection between flange and base plate.

4. Joints
Joints become necessary in a structure for three main reasons. These relate to production or processing restrictions, the need to gain access within the structure during its working life, and repair of the original structure.

The production-related feature arises because large structures cannot be formed in one process, thereby needing components to be joined to produce

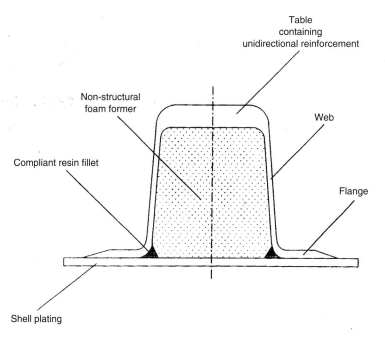

Figure 9.26 Top hat stiffener configuration.

the completed product. Considerations that limit process size include exotherm, resin working time, cloth size and drapeability, mould accessibility, and release limitations. Considering access and repair, if the components within the structure require regular servicing, then the structural elements that obstruct access need to be joined to the remaining structure in such a way as to allow them to be removed with reasonable ease. If the hidden components require only very occasional treatment, then the structure can be cut out as necessary and treated as a repair. Here the jointing method can be treated as permanent.

There are two main classes of joints, namely those that effect in-plane load transfer and those that connect two structural elements orthogonal to each other. The latter can refer either to frame-to-shell connections or bulkhead-to-shell connections.

(i) In-plane joints

These can be either bonded or bolted; the choice depends very much on the application being considered. Typical examples of bonded joints are shown in Figure 9.27. A bonded connection provides a greater area to transmit load. This ensures that all the fibres at the joint interface are used to carry load so that stress concentrations are reduced. They are cheaper and easier to produce and can be formed from one side of the panel. However, some environmental control is usually necessary during the construction process. One shortcoming is that when initial failure occurs in a purely bonded joint, it can propagate easily since there are no fibres across the joint to act as crack arrestors. Therefore special attention is devoted to the design to ensure that such events are minimized. Such joints are permanent and cannot be easily removed.

Bolted connections provide a strong link across the joint interface; they are easily removed and can usually be formed under adverse conditions. When used in conjunction with an adhesive, the bolts can act as crack arrestors in the event of final failure. However, since the load is transmitted through a small area, stress concentrations occur that can lead to early failure. They require access from both sides of the plate panels, are heavy, and can be expensive to build.

The literature on in-plane connections is extensive, Godwin and Matthews (1980), Greene (1997), and the reader can refer to such work for a clearer exposition of the subject. In the marine context, most in-plane connections between two panels are done using primarily bonded connections, Smith (1990).

(ii) Frame-to-shell connections

Some typical arrangements of such connections are shown in Figure 9.28. Frames are normally laid up over a foam former and bonded to the shell when the latter is fully cured. The main purpose of this connection is to transmit shear stresses between the shell and frame flanges under local bending caused by lateral pressure or concentrated lateral loads. Design of the connection, Greene (1997), requires an evaluation of the envelope of the maximum shear forces in each frame. Another development, Dodkins *et al.* (1994), is that of preforming top hat sections and bonding these cured sections to the shell. The major benefits of this approach are reduced production time and cost and also greater flexibility in the design of the joint.

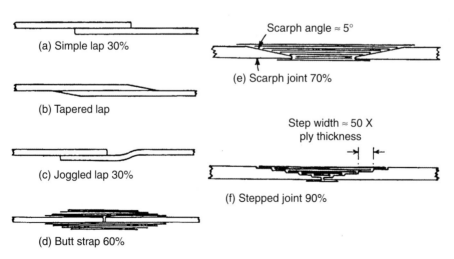

Figure 9.27 Typical arrangements and efficiencies of in-plane joints.

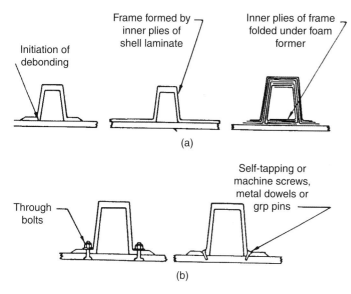

Figure 9.28 Typical frame-to-shell connections: (a) types of attachment; (b) reinforcement of joint.

(iii) Bulkhead-to-shell connections
Most ship and boat hulls rely critically on transverse bulkheads to provide rigidity and strength under transverse loads; this involves the transmission of direct and membrane shear stresses across the bulkhead-to-shell connection. An effective arrangement is provided by a double-angle arrangement; examples of such arrangements in sandwich construction are shown in Figure 9.29. Design of the boundary angle has principally been based on equating its stiffness with those of the two plates being connected, i.e., the bulkhead and shell plates. Since, in most cases, the material used in the boundary angle is the same as that in the parent plates, the thickness of the overlaminate is usually specified as a function of the thickness of the two plates. However, more recent work has shown the importance of designing joints to be flexible, Shenoi and Hawkins (1992). This is in order to avoid the effect of a 'hard point' created by the very presence of the bulkhead plate and avoid stress concentrations.

5. Finite element analysis
To model a ship's hull, or even a section, using layered finite elements would be an extremely laborious task and would require a great deal of computing resources. The preferred approach is to conduct a multilevel numerical modelling exercise. The global response of the hull structure can be modelled with sufficient accuracy using general-purpose codes and isotropic elements. This gives a reasonably realistic distribution of strains around the hull section and deformation of the hull and deck panels between bulkheads. Figure 9.30 shows a typical stress output from a study of a minehunter.

To examine stress distribution at a detailed level, local models of stiffened panels can be created and boundary conditions can be determined from the global model. Layered orthotropic elements can be used at this stage. Then all the pertinent failure mechanisms such as those listed in the previous sections can be examined. These detailed results can be used for design optimization purposes, where the lay-up can be verified and altered to yield the correct response modes without deficiency. Figure 9.31 illustrates the detail that can be obtained through modelling with this level of care.

(c) Arrangement and layout issues
1. Influence of the general arrangement
Certain features of the ship's general arrangement can have a marked influence on the complexity, and hence cost, of the structure. In particular, the following points should be noted in order to keep the structural arrangement as simple as possible.

(i) Major bulkheads should be placed in positions of multiples of frame spacing. This avoids the complication of varying frame spacing along the ship's length or landing bulkheads on the shell in positions too close to existing frames.
(ii) Bulkhead positions should lead to approximately equal compartment lengths along the length of

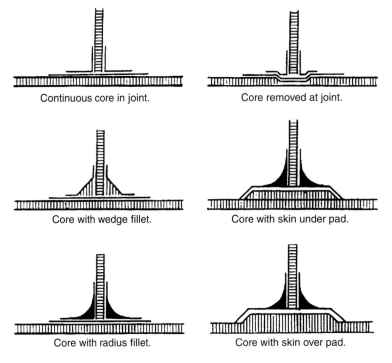

Figure 9.29 Typical tee connections in marine sandwich construction.

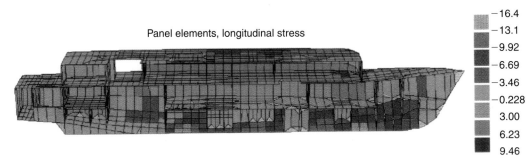

Figure 9.30 Finite element modelling of a whole ship.

the ship. This is not always practical to achieve, but in extreme cases of long compartments adjacent to short compartments, it may be necessary to taper longitudinal stiffeners, resulting in a high labour effort to shape the foam formers and tailor the lay-up cloths. It is preferable to maintain a constant former section and accommodate reasonable variations in spans by varying lay-up alone.

One exception is likely to be in the engine room space, where longitudinals have large spans, but in any case need to be shaped to provide engine and gear box foundations.

(iii) It is not essential to position main transverse bulkheads at either end of the lower tier of the superstructure. The flexibility of the FRP material will ensure that there are no significant stress concentrations at these locations.

(iv) In optimizing transverse and longitudinal frame spacing, it is important to consider the space between stiffeners required for bolted skin fittings as well as ensuring good access to all stiffener surfaces for laminators. This means a frame spacing of about 1.0–1.5 m for hull and main deck and 0.6–1.0 m for superstructure and internal structure.

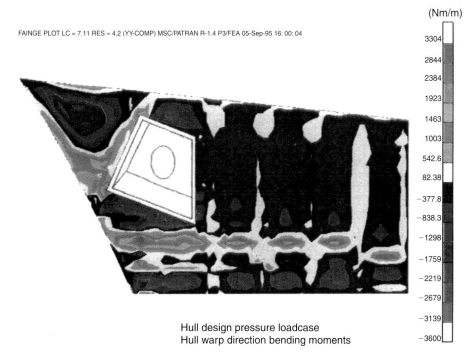

Figure 9.31 Finite element modelling of a stiffened panel.

(v) The main deck is required to have a number of hatches and shipping openings. These should be confined to the centre of the ship and kept as far apart as possible. Thus longitudinals can run straight and parallel to the centreline outside the line of openings, with transverse beams running between inner longitudinals to provide local support to the edges of the openings. This maximizes the longitudinal section modulus and avoids cranking of the longitudinals around openings (which adds to the complexity and reduces labour productivity).

(vi) In positioning the deck and bulkhead penetrations, allowance should be made for tee joints at bulkhead-to-shell and bulkhead-to-deck connections. Penetrations should be kept clear of these joints, although bonding angles can be through-bolted to provide a strong attachment point and also serve to clamp the bonding angle to the plating.

(vii) A unique property of composite materials is that almost any shape of structure may be produced by the use of an appropriately shaped mould. However, as far as possible and where feasible, efforts should be made to maximize the use of flat panels and assemblies. This is particularly so for superstructures, deck houses, and other secondary structural regions.

2. Structural arrangement

Figure 9.32 shows the structural elements in a midship section of a modern FRP vessel of predominantly stiffened single skin construction. A key feature of modern design, Dodkins (1993), is the adoption of longitudinal framing. Advantages of this form of stiffening (over transverse framing) are that:

(i) more of the structure is effective in resisting hull girder bending;
(ii) stiffener intersections are greatly reduced;
(iii) instability problems, especially in the deck structure, are minimized.

These have had to be weighted against the perceived drawbacks, which are:

(i) stiffener bases must be shaped to land upright on the varying deadrise angle of the ship's bottom;
(ii) laminating longitudinals on the side shell is difficult;
(iii) the main transverse bulkheads need to be stronger and heavier in order to support the longitudinals.

The lower ends of the side frames are simply butted onto the outermost bottom longitudinals at the turn of the bilge. This part of the hull structure is inherently rigid and external pressures do not place excessive

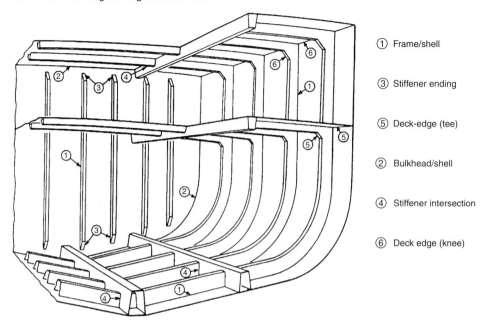

Figure 9.32 Midship section of a Sandown class minehunter.

load on these joints. The upper ends of the joints are terminated alternately by a snape or a beam knee connection to the main deck beams. The arrangement ensures good continuity and transverse strength between the hull and the main deck structure.

Deck plating thickness ranges from 15 to 25 mm, reflecting the variation of longitudinal bending and demands of local loading. Shell plating is about 20 mm thick in most parts, with extra reinforcement placed locally by way of highly loaded regions such as the forward end which is prone to slamming loads, tanks where there is local fluid loading, and in the engine room where extra stiffness has to be provided for machinery supports.

9.3.4.6 *External issues*

(a) Regulatory issues

Two features that characterize ships, which have been alluded to in Section 9.3.4.2, and which influence design practice, are the very short lead time from tender/order to delivery of the ships and the fact that they are generally made-to-order, one-off products. The effect of these is to place tremendous pressure on designers to produce designs that are both practical and optimal. An ideal approach to adopt would be one based entirely on first principles. In such practice, all possible combinations of requirements would need to be assessed thoroughly and hypotheses tested rigorously. Such assessment would be time-consuming and expensive – two luxuries that are ill-affordable by the marine community. Primarily because of this set of constraints, designers place a great deal of reliance on 'rules and regulations' of respected independent regulatory bodies – classification societies such as Lloyd's Register of Shipping, the American Bureau of Shipping, Det Norske Veritas, Germanischer Lloyd, Bureau Veritas. See also Section 11.3.

Design codes are the instruments through which classification societies exercise a partial control over the design activity. Codes specify minimum requirements to be satisfied by any designer. The adoption of optimal solutions is a natural attitude of a rational designer, which leads very often to a design based on minimum code requirements. Thus codes govern the main features of design and they also represent the existing practice in a sector of the marine industry. They result from the experience of applying evolving guidance principles and they shape the new ships to be produced. Because the codes need to be universal in application, both geographically and in terms of the product range, they have to be simple to use. This, in turn, implies that the expressions used to calculate the design variables and parameters have to be simple to understand and apply. The simplicity sometimes conflicts with the need to assure adequate safety margins. This forces the classification societies to be quite conservative to balance the lack of accuracy in the design formulations.

Very recent developments in information technology and the proper harnessing of computing power however are encouraging. This is allowing an integration of hydrodynamics and loading calculations, definition of ship geometry, synthesis of structural elements, materials characterization, and production modelling capabilities. Designers are thus being able to assess the global effects of the change of a structural design parameter on whole ship performance fairly quickly. The capabilities in this context are still in their infancy. Regulatory bodies and the insurance industry that underwrites the financing of ships and shipping need to be convinced of the validity and correctness of such tools. The validation process is underway in a number of different ways and forums. The entirely first principles based process should therefore be a reality soon.

(b) Statutory issues

Apart from the issues discussed above, all ships have to conform to statutes of the country in which they are registered. These laws are, in the main, derived from resolutions of the International Maritime Organization (IMO), see Section 11.2.2.

The principal relevance of the IMO and the statutory implications in ship structural design is that there is a requirement for the main structure in ships to be built of non-combustible materials. Steel is a non-combustible material; aluminium alloys (even though they melt at relatively low temperatures) also do not burn. Both these are acceptable structural materials for ships. FRP composites, however, are combustible. Therefore they are subject to stringent checks under various clauses. The most recent example of this is the adoption of a code for the design of high-speed craft, IMO (2000), HSC code.

The HSC code applies to vessels of high speed which are engaged in international voyages, covering passenger craft which do not proceed more than four hours from a port of refuge, and cargo craft of 500 gross tonnes and upwards, which do not proceed for more than eight hours from a port of refuge. The HSC code includes requirements of 'fire restrictive' (or combustible) materials with respect to their use in primary, secondary, and tertiary structures and components. The requirements of the HSC code are principally aimed at:

(i) *fire prevention* – the use of non-combustible or fire-restricting materials, such that fire prevention is controlled by low flame spread materials, limited heat flux and limited heat release, together with the control of harmful gases and smoke;
(ii) *structural performance* – controlling the structural integrity at elevated temperatures;
(iii) *fire containment* – controlling fires developed in major and moderate fire hazard areas by the use of fire resisting divisions.

These place tremendous burdens on the designer to demonstrate conformance of the structure with the HSC requirements. Current attributes of the candidate materials are such that FRP composites will rarely be allowed for use in structural applications in ships. However, the code and requirements are being re-examined in a more fundamental manner both in terms of evaluation of the safety case for ships where the whole picture of passenger safety and evacuation following a fire is considered (rather than one of just the candidate materials) and in terms of prescribing the correct tests for checking conformance. In a curious and paradoxical context, FRP composites are being used in offshore structures following the disastrous Piper Alpha fire precisely because of their fire-resistive capabilities, Gibson (1993). The future therefore looks promising for the application of polymeric composites in major ship structural applications.

9.3.5 Corrosion

9.3.5.1 *Nature and forms of corrosion*

There is a natural tendency for nearly all metals to react with their environment. The result of this reaction is the creation of a corrosion product which is generally a substance of very similar chemical composition to the original mineral from which the metal was produced.

Atmospheric corrosion. Protection against atmospheric corrosion is important during the construction of a ship, both on the building berth and in the shops. Serious rusting may occur where the relative humidity is above about 70%; the atmosphere in British shipyards is unfortunately sufficiently humid to permit atmospheric corrosion throughout most of the year. But even in humid atmospheres the rate of rusting is determined mainly by the pollution of the air through smoke and/or sea salts.

Corrosion due to immersion. When a ship is in service the bottom area is completely immersed and the waterline or boot topping region may be intermittently immersed in sea water. Under normal operating conditions a great deal of care is required to prevent excessive corrosion of these portions of the hull. A steel hull in this environment can provide ideal conditions for the formation of electro-chemical corrosion cells.

Electro-chemical nature of corrosion. Any metal in tending to revert to its original mineral state releases energy. At ordinary temperatures in aqueous solutions the transformation of a metal atom into a mineral molecule occurs by the metal passing into solution. During this process the atom loses one or more electrons and becomes an ion, i.e. an electrically charged atom, with the production of an electric current (the released energy). This reaction may only occur if an electron acceptor is present in the aqueous solution. Thus any corrosion reaction is always accompanied by a flow of electricity from one metallic area to another through a solution in which the conduction of an electric current occurs by the passage of ions. Such a solution is referred to as an electrolyte solution; and because of its high salt content sea water is a good electrolyte solution.

A simple corrosion cell is formed by two different metals in an electrolyte solution (a galvanic cell) as illustrated in Figure 9.33. It is not essential to have two different metals as we shall see later. As illustrated a pure iron plate and a similar pure copper plate are immersed in a sodium chloride solution which is in contact with oxygen at the surface. Without any connection the corrosion reaction on each plate would be small. Once the two plates are connected externally to form an electrical path then the corrosion rate of the iron will increase considerably, and the corrosion on the copper will cease. The iron electrode by means of which the electrons leave the cell and by way of which the conventional current enters the cell is the anode. This is the electrode at which the oxidation or corrosion normally takes place. The copper electrode by means of which the electrons enter the cell and by way of which the conventional current leaves the cell is the cathode, at which no corrosion occurs. A passage of current through the electrolyte solution is by means of a flow of negative ions to the anode and a flow of positive ions to the cathode.

Electro-chemical corrosion in aqueous solutions will result from any anodic and cathodic areas coupled in the solution whether they are metals of different potential in the environment or they possess different potentials as the result of physical differences on the metal surface. The latter is typified by steel plate carrying broken millscale in sea water (Figure 9.33) or corrosion currents flowing between areas of well painted plate and areas of defective paintwork.

In atmospheric corrosion and corrosion involving immersion both oxygen and an electrolyte play an important part. Plates freely exposed to the atmosphere will receive plenty of oxygen but little moisture, and the moisture present therefore becomes the controlling factor. Under conditions of total immersion it is the presence of oxygen which becomes the controlling factor.

Bimetallic (galvanic) corrosion. Although it is true to say that all corrosion is basically galvanic, the term 'galvanic corrosion' is usually applied when two different metals form a corrosion cell.

Many ship corrosion problems are associated with the coupling of metallic parts of different potential which consequently form corrosion cells under service conditions. The corrosion rates of metals and alloys in sea water have been extensively investigated and as a result galvanic series of metals and alloys in sea water have been obtained.

A typical galvanic series in sea water is shown in Table 9.11.

The positions of the metals in the table apply only in a sea water environment; and where metals are grouped together they have no strong tendency to form couples with each other. Some metals appear twice because they are capable of having both a passive

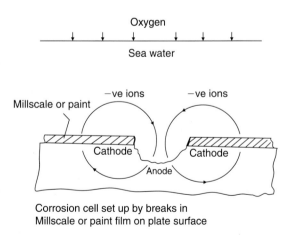

Figure 9.33 Corrosion cell.

Table 9.11 Galvanic series of metals and alloys in sea water.

Noble (cathodic or protected) end
Platinum, gold
Silver
Titanium
Stainless steels, passive
Nickel, passive
High duty bronzes
Copper
Nickel, active
Millscale
Naval brass
Lead, tin
Stainless steels, active
Iron, steel, cast iron
Aluminium alloys
Aluminium
Zinc
Magnesium
Ignoble (anodic or corroding) end

with the attachment of bronze and aluminium alloy fittings. Where aluminium superstructures are introduced, the attachment to the steel hull and the fitting of steel equipment to the superstructure require special attention. This latter problem is overcome by insulating the two metals and preventing the ingress of water as illustrated in Figure 9.34. A further development is the use of explosion-bonded aluminium/steel transition joints also illustrated. These joints are free of any crevices, the exposed aluminium to steel interface being readily protected by paint.

Stress corrosion. Corrosion and subsequent failure associated with varying forms of applied stress is not uncommon in marine structures. Internal stresses produced by non-uniform cold working are often more dangerous than applied stresses. For example, localized corrosion is often evident at cold flanged brackets.

and an active state. A metal is said to be passive when the surface is exposed to an electrolyte solution and a reaction is expected but the metal shows no sign of corrosion. It is generally agreed that passivation results from the formation of a current barrier on the metal surface, usually in the form of an oxide film. This thin protective film forms, and a change in the overall potential of the metal occurs when a critical current density is exceeded at the anodes of the local corrosion cells on the metal surface.

Among the more common bimetallic corrosion cell problems in ship hulls are those formed by the mild steel hull with the bronze or nickel alloy propeller. Also above the waterline problems exist

Corrosion/erosion. Erosion is essentially a mechanical action but it is associated with electro-chemical corrosion in producing two forms of metal deterioration. Firstly, in what is known as 'impingement attack' the action is mainly electro-chemical but it is initiated by erosion. Air bubbles entrained in the flow of water and striking a metal surface may erode away any protective film that may be present locally. The eroded surface becomes anodic to the surrounding surface and corrosion occurs. This type of attack can occur in most places where there is water flow, but particularly where features give rise to turbulent flow. Sea water

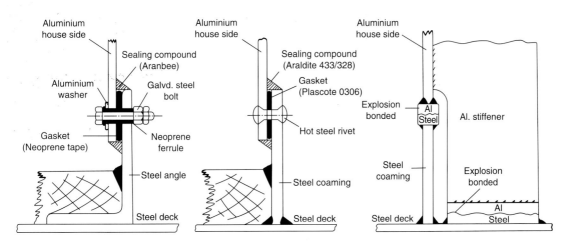

Figure 9.34 Aluminium to steel connections.

discharges from the hull are a particular case, the effects being worse if warm water is discharged.

Cavitation damage is also associated with a rapidly flowing liquid environment. At certain regions in the flow (often associated with a velocity increase resulting from a contraction of the flow stream) the local pressures drop below that of the absolute vapour pressure. Vapour cavities, that is areas of partial vacuum, are formed locally, but when the pressure increases clear of this region the vapour cavities collapse or 'implode'. This collapse occurs with the release of considerable energy, and if it occurs adjacent to a metal surface damage results. The damage shows itself as pitting which is thought to be predominantly due to the effects of the mechanical damage. However it is also considered that electro-chemical action may play some part in the damage after the initial erosion. See also Section 5.5.

Corrosion allowance. Plate and section scantlings specified for ships in the rules of classification societies include corrosion additions to the thickness generally based on a 25 year service life. The corrosion allowance is based on the concept that corrosion occurs on the exposed surface of the material at a constant rate, no matter how much material lies behind it. That is if a plate is 8 mm or 80 mm thick, corrosion will take place at the same rate, not at a faster rate in the thicker plate.

9.3.5.2 *Corrosion control*

The control of corrosion may be broadly considered in two forms, cathodic protection and the application of protective coatings, i.e. paints.

Cathodic protection. Only where metals are immersed in an electrolyte can the possible onset of corrosion be prevented by cathodic protection. The fundamental principle of cathodic protection is that the anodic corrosion reactions are suppressed by the application of an opposing current. This superimposed direct electric current enters the metal at every point lowering the potential of the anode metal of the local corrosion cells so that they become cathodes.

There are two main types of cathodic protection installation, sacrificial anode systems and impressed current systems.

(1) *Sacrificial anode systems* – Sacrificial anodes are metals or alloys attached to the hull which have a more anodic, i.e. less noble, potential than steel when immersed in sea water. These anodes supply the cathodic protection current, but will be consumed in doing so and therefore require replacement for the protection to be maintained.

This system has been used for many years, the fitting of zinc plates in way of bronze propellers and other immersed fittings being common practice. Initially results with zinc anodes were not always very effective owing to the use of unsuitable zinc alloys. Modern anodes are based on alloys of zinc, aluminium, or magnesium which have undergone many tests to examine their suitability; high purity zinc anodes are also used. The cost, with various other practical considerations, may decide which type is to be fitted.

Sacrificial anodes may be fitted within the hull, and are often fitted in ballast tanks. However, magnesium anodes are not used in the cargo-ballast tanks of oil carriers owing to the 'spark hazard'. Should any part of the anode fall and strike the tank structure when gaseous conditions exist an explosion could result. Aluminium anode systems may be employed in tankers provided they are only fitted in locations where the potential energy is less than 28 kg.m.

(2) *Impressed current systems* – These systems are applicable to the protection of the immersed external hull only. The principle of the systems is that a voltage difference is maintained between the hull and fitted anodes, which will protect the hull against corrosion, but not overprotect it thus wasting current. For normal operating conditions the potential difference is maintained by means of an externally mounted silver/silver chloride reference cell detecting the voltage difference between itself and the hull. An amplifier controller is used to amplify the micro-range reference cell current, and it compares this with the preset protective potential value which is to be maintained. Using the amplified DC signal from the controller a saturable reactor controls a larger current from the ship's electrical system which is supplied to the hull anodes. An AC current from the electrical system would be rectified before distribution to the anodes. Figure 9.35 shows such a system.

Originally, consumable anodes were employed but in recent systems non-consumable relatively noble metals are used; these include lead/silver and platinum/palladium alloys, and platinized titanium anodes are also used.

A similar impressed current system employs a consumable anode in the form of an aluminium wire up to 45 metres long which is trailed behind the ship whilst at sea. No protection is provided in port.

Although the initial cost is high, these systems are claimed to be more flexible, to have a longer life, to reduce significantly hull maintenance, and to weigh less than the sacrificial anode systems.

Care is required in their use in port alongside ships or other unprotected steel structures.

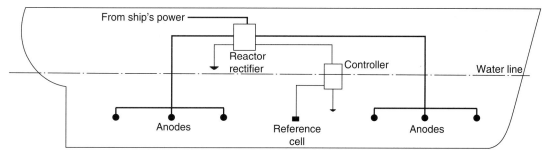

Figure 9.35 Impressed current cathode protection system.

Protective coatings (paints). Paints intended to protect against corrosion consist of pigment dispersed in a liquid referred to as the 'vehicle'. When spread out thinly the vehicle changes in time to an adherent dry film. The drying may take place through one of the following processes.

(a) When the vehicle consists of solid resinous material dissolved in a volatile solvent, the latter evaporates after application of the paint, leaving a dry film.
(b) A liquid like linseed oil as a constituent of the vehicle may produce a dry paint film by reacting chemically with the surrounding air.
(c) A chemical reaction may occur between the constituents of the vehicle after application, to produce a dry paint film. The reactive ingredients may be separated in two containers ('two-pack paints') and mixed before application. Alternatively ingredients which only react at higher temperatures may be selected, or the reactants may be diluted with a solvent so that the reaction occurs only slowly in the can.

Corrosion-inhibiting paints for application to steel have the following vehicle types:

(a) *Bitumen or pitch* Simple solutions of bitumen or pitch are available in solvent naphtha or white spirit. The bitumen or pitch may also be blended by heat with other materials to form a vehicle.
(b) *Oil based* These consist mainly of vegetable drying oils, such as linseed oil and tung oil. To accelerate the drying by the natural reaction with oxygen, driers are added.
(c) *Oleo-resinous* The vehicle incorporates natural or artificial resins into drying oils. Some of these resins may react with the oil to give a faster drying vehicle. Other resins do not react with the oil but heat is applied to dissolve the resin and cause the oil to body.
(d) *Alkyd resin* These vehicles provide a further improvement in the drying time and film forming properties of drying oils. The name alkyd arises from the ingredients, alcohols and acids. Alkyds need not be made from oil, as an oil-fatty acid or an oil-free acid may be used.

(*Note.* Vehicle types (b) and (d) are not suitable for underwater service, and only certain kinds of (c) are suitable for such service.)

(e) *Chemical-resistant* Vehicles of this type show extremely good resistance to severe conditions of exposure. As any number of important vehicle types come under this general heading these are dealt with individually.

  (i) *Epoxy resins* Chemicals which may be produced from petroleum and natural gas are the source of epoxy resins. These paints have very good adhesion, apart from their excellent chemical resistance. They may also have good flexibility and toughness where co-reacting resins are introduced. Epoxy resins are expensive owing to the removal of unwanted side products during their manufacture, and the gloss finish may tend to 'chalk' making it unsuitable for many external decorative finishes. These paints often consist of a 'two-pack' formulation, a solution of epoxy resin together with a solution of cold curing agent, such as an amine or a polyamide resin, being mixed prior to application. The mixed paint has a relatively slow curing rate at temperatures below 10°C. Epoxy resin paints should not be confused with epoxy-ester paints which are unsuitable for underwater use. Epoxy-ester paints can be considered as alkyd equivalents, as they are usually made with epoxy resins and oil-fatty acids.

  (ii) *Coal tar/epoxy resin* This vehicle type is similar to the epoxy resin vehicle except that, as a two-pack product, a grade of coal tar pitch is blended with the resin. A formulation of this type combines to some extent the chemical resistance of the epoxy resin with the impermeability of coal tar.

  (iii) *Chlorinated rubber and isomerized rubber* The vehicle in this case consists of a solution of plasticized chlorinated rubber, or isomerized rubber. Isomerized rubber is produced chemically from natural rubber, and it has the same chemical composition but a different molecular structure. Both these derivatives of natural rubber have a wide range of solubility in organic solvents, and so allow a vehicle of higher solid content. On drying, the film thickness is greater than would be obtained if natural rubber were used. High build coatings of this type are available, thickening or thixotropic agents being added to produce a paint which can be applied in much thicker coats. Coats of this type are particularly resistant to attack from acids and alkalis.

(iv) *Polyurethane resins* A reaction between isocyanates and hydroxyl-containing compounds produces 'urethane' and this reaction has been adapted to produce polymeric compounds from which paint film, fibres, and adhesives may be obtained. Paint films so produced have received considerable attention in recent years, and since there is a variety of isocyanate reactions, both one-pack and two-pack polyurethane paints are available. These paints have many good properties; toughness, hardness, gloss, abrasion resistance, as well as chemical and weather resistance. Polyurethanes are not used under water on steel ships, only on superstructures, etc., but they are very popular on yachts where their good gloss is appreciated.

(v) *Vinyl resins* Vinyl resins are obtained by the polymerization of organic compounds containing the vinyl group. The solids content of these paints is low; therefore the dry film is thin, and more coats are required than for most paints. As vinyl resin paints have poor adhesion to bare steel surfaces they are generally applied over a pretreatment primer. Vinyl paint systems are among the most effective for the underwater protection of steel.

(f) *Zinc-rich paints* Paints containing metallic zinc as a pigment in sufficient quantity to ensure electrical conductivity through the dry paint film to the steel are capable of protecting the steel cathodically. The pigment content of the dry paint film should be greater than 90%, the vehicle being an epoxy resin, chlorinated rubber, or similar medium.

Corrosion protection by means of paints. It is often assumed that all paint coatings prevent attack on the metal covered simply by excluding the corrosive agency, whether air or water. This is often the main and sometimes only form of protection; however there are many paints which afford protection even though they present a porous surface or contain various discontinuities.

For example certain pigments in paints confer protection on steel even where it is exposed at a discontinuity. If the reactions at the anode and cathode of the corrosion cell which form positive and negative ions respectively, are inhibited, protection is afforded. Good examples of pigments of this type are red lead and zinc chromate, red lead being an anodic inhibitor, and zinc chromate a cathodic inhibitor. A second mode of protection occurs at gaps where the paint is richly pigmented with a metal anodic to the basis metal. Zinc dust is a commercially available pigment which fulfils this requirement for coating steel in a salt water environment. The zinc dust is the sacrificial anode with respect to the steel.

9.3.5.3 *Anti-fouling systems*

The immersed hull and fittings of a ship at sea, particularly in coastal waters, are subject to algae, barnacle, mussel and other shellfish growth that can impair its hydrodynamic performance and adversely affect the service of the immersed fittings.

Fittings such as cooling water intake systems are often protected by impressed current anti-fouling systems and immersed hulls today are finished with very effective self polishing anti-fouling paints.

Impressed current anti-fouling systems. The functional principle of these systems is the establishment of an artificially triggered voltage difference between copper anodes and the integrated steel plate cathodes. This causes a minor electrical current to flow from the copper anodes, so that they are dissolved to a certain degree. A control unit makes sure that the anodes add the required minimum amount of copper particles to the sea water, thus ensuring the formation of copper oxide that creates ambient conditions precluding local fouling. A control unit can be connected to the management system of the vessel. Using information from the management system the impressed current anti-fouling system can determine the amount of copper that needs to be dissolved to give optimum performance with minimum wastage of the anodes.

Anti-fouling paints. Anti-fouling paints consist of a vehicle with pigments which give body and colour together with materials toxic to marine vegetable and animal growth. Copper is the best known toxin used in traditional anti-fouling paints.

To prolong the useful life of the paint the toxic compounds must dissolve slowly in sea water. Once the release rate falls below a level necessary to prevent settlement of marine organisms the anti-fouling composition is no longer effective. On merchant ships the effective period for traditional compositions was about 12 months. Demands in particular from large tanker owners wishing to reduce very high docking costs led to specially developed anti-fouling compositions with an effective life up to 24 months in the early 1970s. Subsequent developments of constant emission organic toxin antifoulings having a leaching rate independent of exposure time saw the paint technologists by chance discover coatings which also tended to become smoother in service. These so called self-polishing antifoulings with a lifetime that is proportional to applied thickness and therefore theoretically unlimited, smooth rather than roughen with time and result in reduced friction drag. Though more expensive than their traditional counterparts, given the claim that each 10 micron (10^{-3} mm) increase in hull roughness can result in a 1% increase in fuel consumption, their self polishing characteristic as well as their longer effective life, up to 5 years protection between drydockings, made them attractive to the shipowner.

The benefits of the first widely used SPC (self polishing copolymer) anti-fouling paints could be traced to the properties of their prime ingredients the tributylen compounds or TBT's. TBT's were extremely active against a wide range of fouling organisms, also they were able to be chemically bonded to the acrylic backbone of the paint system. When immersed in sea water a specific chemical reaction took place which cleaved the TBT from the paint backbone, resulting in both controlled release of the TBT and controlled disappearance or polishing of the paint film. Unfortunately, it was found that the small concentrations of TBT's released, particularly in enclosed coastal waters, had a harmful effect on certain marine organisms. This led to he banning of TBT anti-fouling paints for pleasure boats and smaller commercial ships in many developed countries and the introduction of regulations limiting the release rate of TBT for antifouling paints on larger ships. The International Convention On The Control Of Harmful Anti-Fouling On Ships, 2001 subsequently required that

(a) ships shall not apply or reapply organotin compounds which act as biocides in anti-fouling systems on or after 1 January 2003; and
(b) no ship shall have organotin compounds which act as biocides in anti-fouling systems (except floating platforms, FSU's and FPSO's built before 2003 and not docked since before 2003)

(Note! Organotin means an organic compound with one or more tin atoms in its molecules used as a pesticide, hitherto considered to decompose safely, now found to be toxic in the food chain. A biocide is a chemical capable of killing living organisms.)

Anti-fouling paints subsequently applied have generally focused on either the use of copper-based self polishing anti-fouling products, which operate in a similar manner to the banned TBT products, or the use of the so-called low-surface-energy coatings. The latter coatings do not polish or contain booster biocides, instead they offer a very smooth, low-surface-energy surface to which it is difficult for fouling to adhere. When the vessel is at rest some fouling may occur but once it is underway and reaches a critical speed the fouling is released.

9.3.5.4 *Painting ships*

To obtain the optimum performance from paints it is important that the metal surfaces are properly prepared before application of paints and subsequently maintained as such throughout the fabrication and erection process. Paints tailored for the service conditions of the structure to which they apply, and recommended as such by the manufacturer, only should be applied.

Surface preparation. Good surface preparation is essential to successful painting, the primary cause of many paint failures being the inadequacy of the initial material preparation.

It is particularly important before painting new steel that any millscale should be removed. Millscale is a thin layer of iron oxides which forms on the steel surface during hot rolling of the plates and sections. Not only does the non-uniform millscale set up corrosion cells as illustrated previously, but it may also come away from the surface removing any paint film applied over it.

The most common methods employed to prepare steel surfaces for painting are:

Blast cleaning
Pickling
Flame cleaning
Preparation by hand

(a) *Blast cleaning* is the most efficient method for preparing the surface and is in common use in all large shipyards. Following the blast cleaning it is desirable to brush the surface, and apply a coat of priming paint as soon as possible since the metal is liable to rust rapidly.

There are two main types of blasting equipment available, an impeller wheel plant where the abrasive is thrown at high velocity against the metal surface, and a nozzle type where a jet of abrasive impinges on the metal surface. The latter type should preferably be fitted with vacuum recovery equipment, rather than allow the spent abrasive and dust to be discharged to atmosphere, as is often the case in ship repair work. Impeller wheel plants which are self-contained and collect the dust and re-circulate the clean abrasive are generally fitted within the shipbuilding shops.

Cast iron and steel grit, or steel shot which is preferred, may be used for the abrasive, but non-metallic abrasives are also available. The use of sand is prohibited in the United Kingdom because the fine dust produced may cause silicosis.

(b) *Pickling* involves the immersion of the metal in an acid solution, usually hydrochloric or sulphuric acid in order to remove the millscale and rust from the surface. After immersion in these acids the metal will require a thorough hot water rinse. It is preferable that the treatment is followed by application of a priming coat.

(c) Using an *oxy-acetylene flame* the millscale and rust may be removed from a steel surface. The process does not entirely remove the millscale and rust, but it can be quite useful for cleaning plates under inclement weather conditions, the flame drying out the plate.

(d) *Hand cleaning* by various forms of wire brush is often not very satisfactory, and would only be used where the millscale has been loosened by weathering, i.e. exposure to atmosphere over a long period.

Blast cleaning is preferred for best results and economy in shipbuilding, and it is essential prior to application of high performance paint systems used today. Pickling which also gives good results can be expensive and less applicable to production schemes; flame cleaning is much less effective; and hand cleaning gives the worst results.

Temporary paint protection during building. After the steel is blast cleaned it may be several months before it is built into the ship and finally painted. It is desirable to protect the material against rusting in this period as the final paint will offer the best protection when applied over perfectly clean steel.

The formulation of a prefabrication primer for immediate application after blasting must meet a number of requirements. It should dry rapidly to permit handling of the plates within a few minutes and working the plates within a day or so. It should be non-toxic, and it should not produce harmful porosity in welds nor give off obnoxious fumes during welding or cutting. It must also be compatible with any subsequent paint finishes to be applied. Satisfactory formulations are available, for example a primer consisting of zinc dust in an epoxy resin.

Paint systems on ships. The paint system applied to any part of a ship will be dictated by the environment to which that part of the structure is exposed. Traditionally the painting of the external ship structure was divided into three regions.

(a) Below the water-line where the plates are continually immersed in sea water.
(b) The water-line or boot topping region where immersion is intermittent and a lot of abrasion occurs.
(c) The topsides and superstructure exposed to an atmosphere laden with salt spray, and subject to damage through cargo handling.

However, now that tougher paints are used for the ship's bottom the distinction between regions need not be so well defined, one scheme covering the bottom and water-line regions.

Internally by far the greatest problem is the provision of coatings for various liquid cargo and salt water ballast tanks.

(a) *Below the Water-line* The ship's bottom has priming coats of corrosion-inhibiting paint applied which are followed by an anti-fouling paint. Paints used for steels immersed in sea water are required to resist alkaline conditions. The reason for this is that an iron alloy immersed in a sodium chloride solution having the necessary supply of dissolved oxygen gives rise to corrosion cells with caustic soda produced at the cathodes. Further the paint should have a good electrical resistance so that the flow of corrosion currents between the steel and sea water is limited. These requirements make the standard non-marine structural steel primer red lead in linseed oil unsuitable for ship use below the water-line. Suitable corrosion-inhibiting paints for ships' bottoms are pitch or bitumen types, chlorinated rubber, coal tar/epoxy resin, or vinyl resin paints. The anti-fouling paints may be applied after the corrosion-inhibiting coatings and should not come into direct contact with the steel hull, since the toxic compounds present may cause corrosion.

(b) *Water-line or boot topping region* Generally modern practice requires a complete paint system for the hull above the water-line. This may be based on vinyl and alkyd resins or on polyurethane resin paints.

(c) *Superstructures* Red lead or zinc chromate based primers are commonly used. White finishing paints are then used extensively for superstructures. These are usually oleo-resinous or alkyd paints which may be based on 'non-yellowing' oils, linseed oil-based paints which yellow on exposure being avoided on modern ships.

Where aluminium alloy superstructures are fitted, under no circumstance should lead based paints be applied; zinc chromate paints are generally supplied for application to aluminium.

Cargo and ballast tanks. Severe corrosion may occur in a ship's cargo tanks as the combined result of carrying liquid cargoes and sea water ballast, with warm or cold sea water cleaning between voyages. This is particularly true of oil tankers. Tankers carrying 'white oil' cargoes suffer more general corrosion than those carrying crude oils which deposit a film on the tank surface providing some protection against corrosion. The latter type may however experience severe local pitting corrosion due to the non-uniformity of the deposited film, and subsequent corrosion of any bare plate when sea water ballast is carried. Epoxy resin paints are used extensively within these tanks, and vinyl resins and zinc rich coatings may also be used.

Further useful information on paints and anti-fouling systems is given in Anon. (2003, 2005), IMO (2005) and Swain et al. (2007).

9.4 Ship construction

9.4.1 Introduction

This section outlines typical examples of ship structure, and the complexity of stiffening arrangements. An outline of shipyard layout and shipbuilding process is given, together with a description of the links between the design, drawing and manufacturing process.

9.4.2 Typical examples of structure

Figures 9.36 to 9.41 illustrate some typical components of structure. Figure 9.36 shows a typical transom stern, stern frame and the stiffening arrangement in the aft peak. Figure 9.37 shows a typical midship section for a transversely framed cargo ship and Figure 9.38 the midship section for a container ship, showing side shell, bottom shell and tank top plating and stiffening arrangements. Figure 9.39 shows the midship section for a longitudinally stiffened high-speed catamaran ferry, using aluminium alloy. Double bottom construction is illustrated in Figures 9.40(a) and 9.40(b), (a) showing a transversely framed double bottom and (b) a longitudinally framed version. Figure 9.41 illustrates a fore end layout, showing the bulbous bow and fore peak structure.

9.4.3 Shipyard layout

The past two or three decades have seen the emergence of a substantial number of new shipyards, primarily in Asia and Eastern Europe, that have been specifically planned to construct the larger ships being ordered today, using contemporary shipbuilding practices and production methods. A number of traditional shipbuilders have also established new yards where they can also build larger ships and/or exploit the new technology and production methods. In general the remaining shipbuilders will have had to re-configure their site in order to utilize new technology and improve production, whilst continuing to build ships. In many cases the latter will still be restrained as to the size and type of ship that can be built.

An ideal layout for a modern shipyard is based on a production flow basis, with the yard extending back from the river or shore at which the berths or building dock are located. The furthest area from the berths is reserved for the material stockyard, and between the two are arranged in sequence the consecutive work and shop processes. Too often existing shipyards follow the river bank, and are restricted by their location in a built up area or the physical river bank slope from extending back from the river, so that modified production flow lines are required.

Planning a new shipyard, or re-planning an existing one, will involve decisions to be made on the following:

Size and type of ship to be built.
Material production per year to be achieved.
Material handling equipment to be supplied.
Machining processes to be installed.
Unit size and weight to be fabricated and erected.
Amount of outfit and engine installation to be undertaken.
Control services to be supplied.
Administration facilities required.

Shipyards usually have a fitting out basin or berth where the virtually completed ship is tied up after launching and the finishing off work and static trails may be carried out.

Before considering the actual layout of the shipyard it is as well to consider the relationship of the work processes involved in building a ship as illustrated in Figure 9.42.

An idealized layout of a new shipyard is indicated in Figure 9.43 which might be appropriate for a smaller yard specializing in one or two standard type ships with a fairly high throughput so that one covered building dock or berth was sufficient.

At this point it may be convenient to mention the advantages and disadvantages of building docks as opposed to building berths. Building docks can be of advantage in the building of large vessels where launching costs are high, and there is a possibility of structural damage owing to the large stresses imposed by a conventional launch. They also give good crane clearance for positioning units. The greatest disadvantage of the building dock is its high initial cost.

Many yard re-constructions have incorporated undercover construction facilities in the form of docks or slipways within building halls. Others have building halls at the head of the slipway with advanced transfer systems installed so that the hull can be extruded out of the hall onto the slipway for

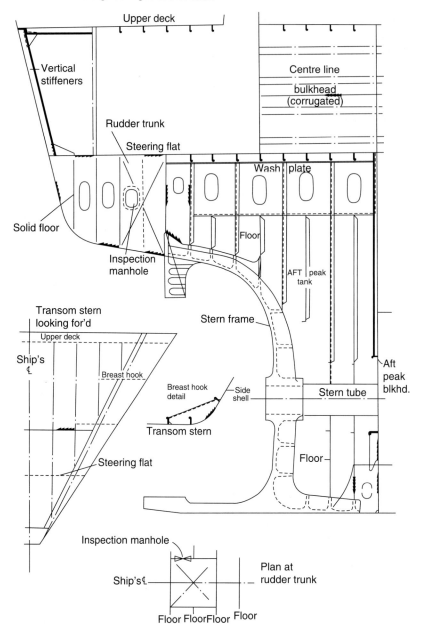

Figure 9.36 Transom stern.

launching. Such facilities permit ship construction in a factory type environment providing protection from the worst effects of weather and darkness.

9.4.4 Ship drawing office, Loftwork and CAD/CAM

This section describes the original functions of the ship drawing office and subsequent full or 10/1 scale lofting of the hull and its structural components and the current use by shipyards of computer aided design (CAD) for these purposes. The subsequent introduction and extensive use of computer aided manufacturing (CAM) in shipbuilding is also covered.

9.4.4.1 *Ship drawing office*

The ship drawing office was traditionally responsible for producing detailed working structural, general

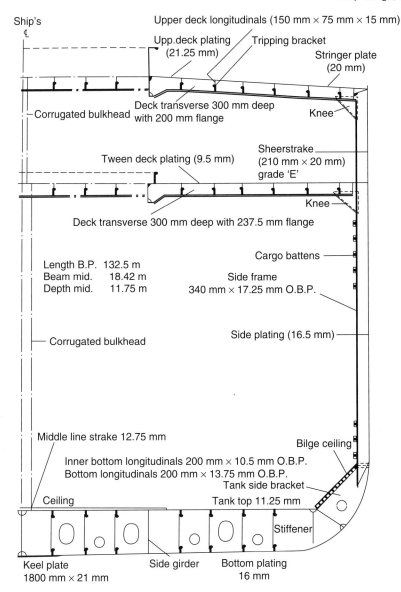

Figure 9.37 General cargo ship – midship section.

arrangement and outfit drawings for a new ship. It was also common practice for the drawing office to contain a material ordering department that would lift the necessary requirements from the drawings and progress them.

Structural drawings prepared by the drawing office would be in accordance with Lloyd's or other classification society rules and subject to their approval; also owner's additional requirements and standard shipyard practices would be incorporated in the drawings. General arrangements of all the accommodation and cargo spaces and stores would also be prepared, incorporating statutory requirements as well as any shipowner's requirements and standards. Outfit plans including piping arrangements, ventilation and air conditioning (which may be done by an outside contractor), rigging arrangements, furniture plans, etc. were also prepared. Two plans of particular significance were the ships 'lines plan' and 'shell expansion'.

Lines plan. A preliminary version of this was, in effect, prepared at the time of the conceptual design to give the required capacity, displacement and propulsive characteristics. It was subsequently

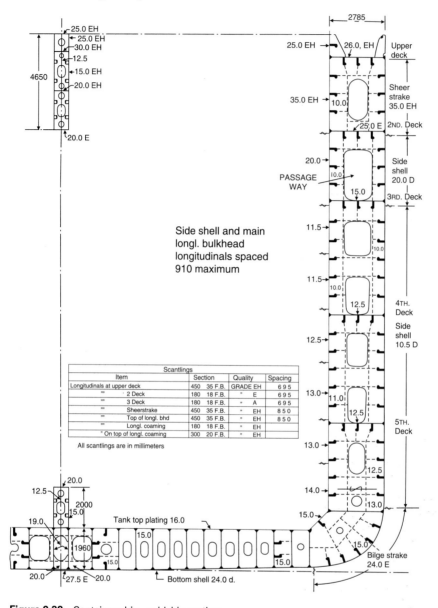

Figure 9.38 Container ship – midship section.

refined during the preliminary design stage and following any tank testing or other method of assessing the hull's propulsive and seakeeping characteristics. The lines plan is a drawing, to a suitable scale, of the moulded lines of the vessel in plan, profile, and section. Transverse sections of the vessel at equally spaced stations between the after and forward perpendiculars are drawn to form what is known as the body plan. Usually ten equally spaced sections are selected with half ordinates at the ends where a greater change of shape occurs. A half transverse section only is drawn since the vessel is symmetrical about the centre line, and forward half sections are drawn to the right of the centre line with aft half sections to the left. Preliminary body plans are drawn initially to give the correct displacement, trim, capacity, etc., and must be laid off in plan and elevation to ensure fairness of the hull form. When the final faired body plan is available the full lines plan is completed showing also the profile or sheer plan of the vessel and the plan of the water-line shapes at different heights above the base.

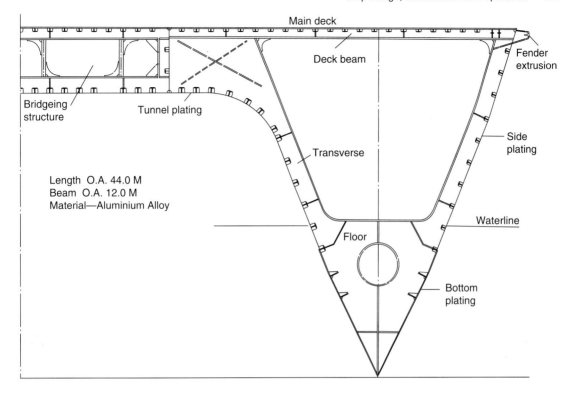

Figure 9.39 High-speed craft (catamaran) – section.

A lines plan is illustrated in Figure 9.44. The lines of the lateral sections in the sheer plan as indicated are referred to as 'bow lines' forward and 'buttock lines' aft. Bilge diagonals would be drawn with 'offsets' taken along the bilge diagonal to check fairness.

When the lines plan was completed manually the draughtsmen would compile a 'table of offsets', that is a list of half breadths, heights of decks and stringer, etc., at each of the drawn stations. These 'offsets' and the lines plan were then passed to loftsmen for full size or 10 to 1 scale fairing. Since the original lines plan was of necessity to a small scale which varied with the size of ship, the offsets tabulated from widely spaced stations and the fairing were not satisfactory for building purposes. The offsets used for building the ship would subsequently be lifted by the loftsman from the full size or 10 to 1 scale lines for each frame.

3-dimensional representation of shell plating. When preparing the layout and arrangement of the shell plating at the drawing stage it was often difficult to judge the line of seams and plate shapes with a conventional 2-dimensional drawing. Shipyards used to therefore make use of a 'half block model' which was in effect a scale model of half the ship's hull from the centre line outboard, mounted on a base board. The model was either made up of solid wooden sections with faired wood battens to form the exterior, or of laminated planes of wood faired as a whole. Finished with a white lacquer the model was used to draw on the frame lines, plate seams, and butts, lines of decks, stringers, girders, bulkheads, flats, stem and stern rabbets, openings in shell, bossings etc.

Shell expansion. The arrangement of the shell plating taken from a 3-dimensional model may be represented on a 2-dimensional drawing referred to as a shell expansion plan. All vertical dimensions in this drawing are taken around the girth of the vessel rather than their being a direct vertical projection. This technique illustrates both the side and bottom plating as a continuous whole. In Figure 9.45 a typical shell expansion for a tanker is illustrated. This also shows the numbering of plates, and lettering of plate strakes for reference purposes and illustrates the system where strakes 'run out' as the girth decreases forward and aft. This drawing was often subsequently retained by the shipowner to identify plates damaged in service. However a word of caution is necessary at this point because since prefabrication became the accepted practice any shell expansion drawing produced will generally have a numbering system related to the erection

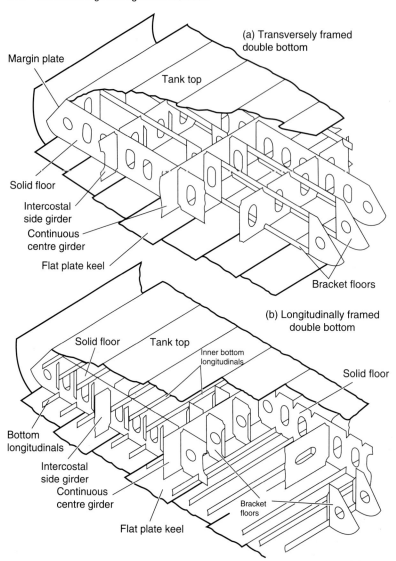

Figure 9.40 Double bottom construction.

of fabrication units rather than individual plates. However single plates were often marked in sequence to aid ordering and production identification.

9.4.4.2 *Loftwork following drawing office*

The mould loft in a shipyard was traditionally a large covered wooden floor area suitable for laying off ship details at full size.

When the loftsmen received the scale lines plan, and offsets from the drawing office, the lines would be laid off full size and faired. This would mean using a great length of floor even though a contracted sheer and plan were normally drawn, and aft and forward body lines were laid over one another. Body sections were laid out full size as they were faired to form what was known as a 'scrieve board'.

The scrieve board was used for preparing 'set bars' (curvature to match plate) and bevels (maintain web of frame perpendicular to ships centre line) for bending frames and for making templates and mouldings for plates which required cutting and shaping.

Shell plates were developed full size on the loft floor and wooden templates made so that these plates

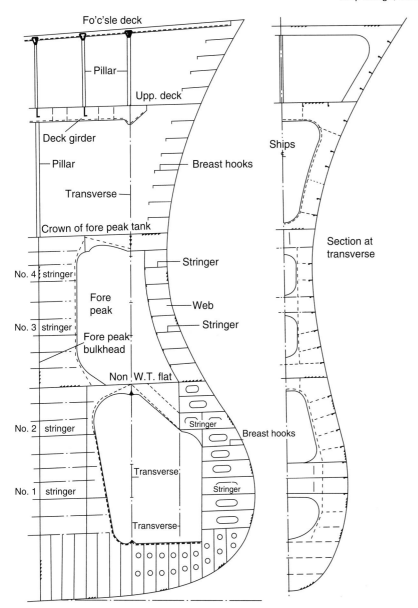

Figure 9.41 Bulbous bow.

could be marked and cut to the right shape before fitting to the framing on the berth.

10/1 scale lofting. In the late 1950s the 10/1 lofting system was introduced and was eventually widely adopted. This reduced the mould loft to a virtual drawing office and assisted in the introduction of production engineering methods. Lines could be faired on a 10/1 scale and a 10/1 scale scrieve board created. Many yards operated a flame profiling machine which used 10/1 template drawings to control the cutting operation. In preparing these template drawings the developed or regular shape of the plates was drawn in pencil on to special white paper or plywood sheet painted white, and then the outline was traced in ink on to a special transparent material. The material used was critical, having to remain constant in size under different temperature and humidity conditions and having a surface which would take ink without 'furring'. Many of the outlines of plates to be cut by the profiler could be traced directly from the scrieve board, for example floors and transverses.

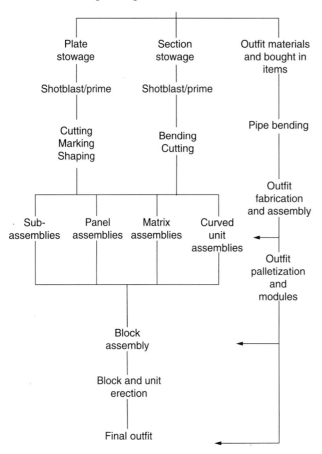

Figure 9.42 Shipbuilding process.

9.4.4.3 Computer Aided Design (CAD)/Computer Aided Manufacturing (CAM)

The first use of computers in the shipbuilding industry probably occurred in the 1960s and because of the high costs involved were only used by the largest shipbuilders running programs developed in-house on a mainframe or mini computer for hull lines fairing, hydrostatics, powering calculations etc. The hull design would have been drawn by hand and stored on the computer as tables of offsets.

In the late 1970s the graphics terminal and the Engineering Workstation became readily available and could be linked to a mini computer. These computers cost considerably less than the earlier mainframes and commercial ship design and construction software became available for them. The larger shipyards quickly adopted these systems. They developed further in the following two decades to run on UNIX Workstations and Windows NT machines and have expanded to cover virtually all the computing needs of a large shipyard.

The early 1980s saw the appearance of the Personal Computer (PC) and several low-cost software packages that performed simple hull design, hydrostatics and powering estimate tasks. These were popular with small shipyards and also reportedly with some larger shipyards for preliminary design work. They were however somewhat limited and incompatible so that it was difficult to build a system that covered all the shipyards CAD/CAM requirements. During the 1990s the available PC software standardized on hardware, operating systems, programming languages, data interchange file formats and hull geometry and are now widely used by naval architects and the ship and boat building industry in general.

Ship product model. Software systems for large shipbuilders is based on the concept of the 'Ship Product Model' in which the geometry and the attributes of all elements of the ship derived from the contract design and classification society structural requirements are stored. This model can be visualized at all stages and can be exploited to obtain information for production of the ship. See Figure 9.46.

At the heart of the 'Ship Product Model' is the conceptual creation of the hull form and its subsequent

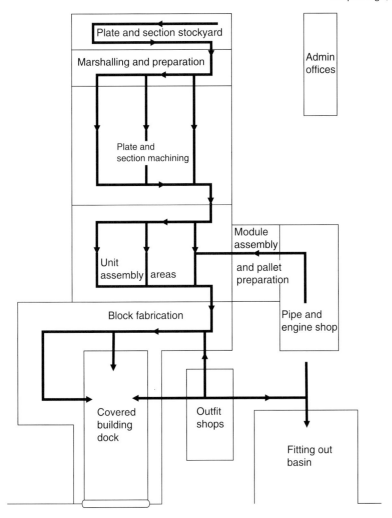

Figure 9.43 Shipyard layout.

fairing for production purposes which is accomplished without committing any plan to paper. This faired hull form is generally held in the computer system as a 'wire model' which typically defines the moulded lines of all structural items so that any structural section of the ship can be generated automatically from the 'wire model'. The model can be worked on interactively with other stored shipyard standards and practices to produce detailed arrangement and working drawings. The precision of the structural drawings generated enables them to be used with greater confidence than was possible with manual drawings and the materials requisitioning information can be stored on the computer to be interfaced with the shipyards commercial systems for purchasing and material control. Sub-assembly, assembly and block drawings can be created in 2-dimensional and 3-dimensional form and a library of standard production sequences and production facilities can be called up so that the draughtsman can ensure that the structural design uses the shipyards resources efficiently and follows established and cost effective practices. Weld lengths and types, steel weights and detailed parts lists can be processed from the information on the drawing and passed to the production control systems. A 3-dimensional steel assembly can be rotated by the draughtsman on screen to assess the best orientation for maximum downhand welding.

The use of 3-dimensional drawings is particularly valuable in the area of outfit drawings where items like pipework and ventilation/air-conditioning trunking can be 'sighted' in the 3-dimensional mode and more accurately measured before being created in the 2-dimensional drawing.

Stored information can be accessed so that lofting functions such as preparing information for bending

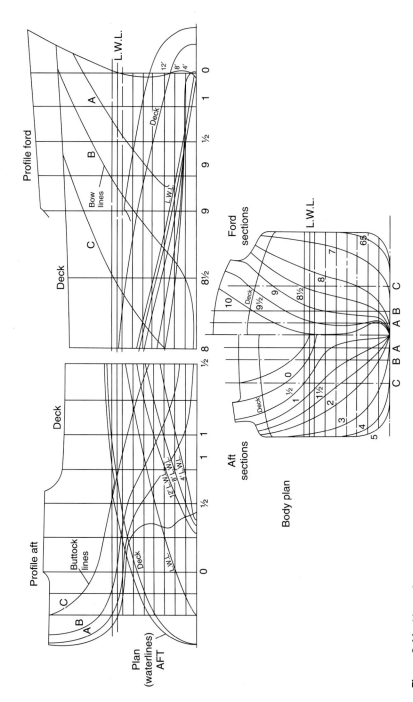

Figure 9.44 Lines plan.

Ship design, construction and operation 701

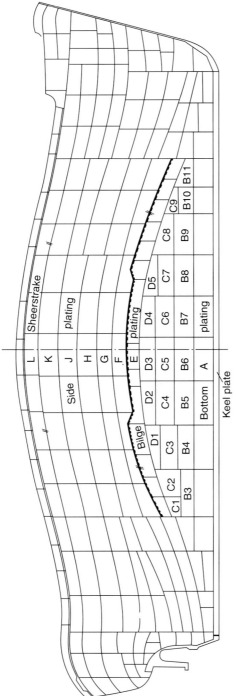

Figure 9.45 Shell expansion.

Framing, stringers, decks and openings in side shell are also shown on the shell expansion but have been omitted for clarity

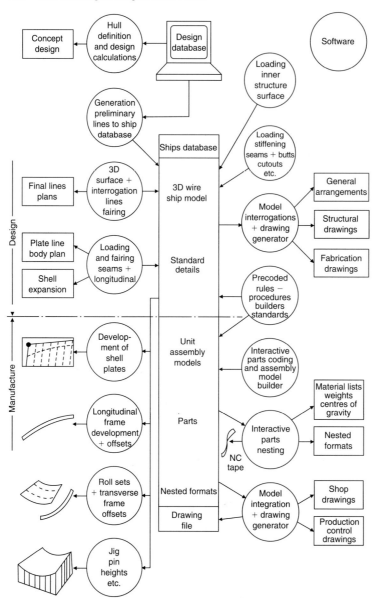

Figure 9.46 Integration of design and manufacture – ship product model.

frames and longitudinals, developing shell plates, and providing shell frame sets and rolling lines or heat line bending information for plates can be done via the interactive visual display unit.

For a numerically controlled profiling machine the piece parts to be cut can be 'nested', i.e. fitted into the most economic plate which can be handled by the machine with minimum wastage (see Figure 9.47). This can be done at the drawing stage when individual piece parts are abstracted for steel requisitioning and stored later being brought back to the screen for interactive nesting. The order in which parts are to be marked and cut can be defined by drawing the tool head around the parts on the graphics screen. When the burning instructions are complete the cutting sequence may be replayed and checked for errors. A check of the NC data can be carried out with a plotter. Instructions for cutting flame planed plates and subsequently joining them into panel assemblies and pin heights of jigs for setting up curved shell plates for welding framing and other members to them at the assembly stage can also be determined (see Figure 9.46).

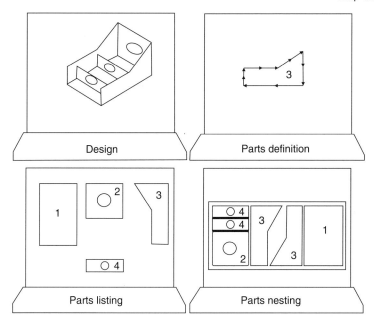

Figure 9.47 Assembly plate parts listing and nesting.

The basic Ship Product Model also contains software packages for the ships outfit including piping, electrical and heating, ventilating and air conditioning (HVAC) systems for a ship.

Further useful information on shipyard processes and production can be found in references such as Taggart (1980), Kuo *et al.* (1984), Torroja and Alonso (2000), Whitfield *et al.* (2003), Lamb *et al.* (2006) and Eyres (2007).

9.5 Ship economics

The following section is taken from Watson (1998). Further information on ship operational economics may be obtained from references such as Benford (1963), Goss (1965), Gilfillian (1969), Buxton (1972), Fisher (1972), Carreyette (1978), Erichsen (1989), Stopford (1997), Karayannis and Molland (2003) and Cullinane (2005).

9.5.1 Shipowners and operators

The operational economics of a ship can be looked at in a number of different ways depending on the type of trade in which it is used and how it is employed.

9.5.1.1 *Types of trade*

Whilst there is an enormous diversity in the type and size of ships, all are generally employed in one of five principal ways, namely as liners, cruise ships, industrial carriers, service vessels or as tramps. The first four of these categories can be classed as owner-operated ships, whilst the last category consists mainly of ships let out on charter.

(i) Liners
To be designated as a liner, a vessel must ply on a regular advertised service; examples are container ships and ferries, see Sections 2.2.2 and 2.2.3. Because ships providing this sort of service sail on scheduled dates and, when passengers are carried, at scheduled times, departing whether the ships are fully loaded or not, the cost of running a service of this type can be high. Freight rates and ticket prices must be set to achieve a satisfactory return over a period of time against the anticipated demand.

(ii) Cruise ships
The first cruises were offered by passenger liner companies using their liners either in their normal country to country service or on special voyages. These cruises were usually arranged at a time of year when passenger numbers in their normal services were likely to be on the low side.

With the decline of passenger services caused by the growth of air travel, passenger liners ceased to be available for use in this way and purpose built cruise liners started to make their appearance. These are now becoming more like floating hotels or holiday camps and the cruise business is currently one of the fastest growing areas of shipping. See also Section 2.2.6.

Typically, cruise ships undertake trips of one or two weeks duration generally steaming at night and with arrangements made for passengers to go ashore and see the sights and enjoy a new locality each day.

Although each cruise is a scheduled service, the fact that cruise schedules and itineraries can be changed at relatively short notice gives these ships an operational flexibility which liner services do not have.

(iii) Industrial carriers

A number of large companies with a substantial shipping requirement either for the import of their raw materials or for the export of their finished products or both own a number of ships to cover at least a baseload part of their shipping requirement.

Typical examples of this are the tanker fleets owned by oil companies; ships specially designed to carry iron ore and/or coal owned by steelmakers; and ships designed to carry cars in bulk owned by major car manufacturers, see Sections 2.2.4 and 2.2.5.

The owners of these ships generally assume total responsibility for all aspects of cost when the vessel is employed in their own trade. The object of such an ownership is to minimize the costs of an overall industrial process, but the lack of flexibility which has often been a characteristic of such operators has sometimes been found to do the opposite and this type of shipowner has been diminishing in recent years.

The U.S. anti-pollution laws have had a severe impact on some of the major oil companies who now refuse to trade with their own vessels in U.S. waters because of the virtually unlimited liability that applies there and instead charter in from traditional shipowners.

(iv) Service vessels

Very few, if any, service vessels carry cargo, their function being to supply services to other vessels or installations at sea. Examples of service vessels are tugs, dredgers, navigational service vessels, offshore safety vessels, etc. These services may be paid for directly as in the case of tugs or indirectly through port dues or taxation in some other cases. But the owners of all these ships need to calculate ship operating expenses on an owner operator basis.

(v) Tramps

A ship can be said to be tramping when it is prepared to go wherever a suitable cargo is available. Tramp ships can be employed in various ways under different types of charter which are explained in 9.5.1.2. Most bulk carriers and oil tankers, together with many small container ships and coasters operate as tramps, making this the method of employment of the majority of ships.

9.5.1.2 *Methods of employment*

An owner will generally employ a ship in one of four ways, namely: in his own trade, in tramp trades as an operator, or in tramp trades by time chartering or bareboat chartering the ship to another party. The extent to which an owner bears the costs of operations under each of these situations is discussed in the following paragraphs and is illustrated in Figure 9.48 which is a slightly modified version of a figure originally given in Dr. Buxton's 1972 R.I.N.A. paper 'Engineering economics applied to ship design', Buxton (1972) – a paper which, along with Dr. Buxton's earlier B.S.R.A. report 'Engineering economics and ship design', contributed substantially to this section.

(i) Ships used by an owner in his own trade

The types of trade in which ships are used by owners in their own trade have been outlined in 9.5.1.1. When ships are used in this way, the owner will generally assume total responsibility for all aspects of cost incurred.

(ii) Ships used by an owner as operator

An owner operator can arrange for the employment of a ship in a number of different ways, viz:

(i) by taking on Contracts of Affreightment to move a large volume of cargo in regular shipments of a set size, based on a set rate per tonne moved;

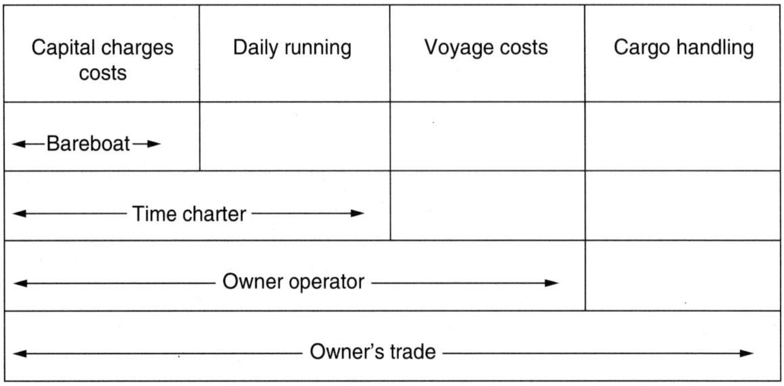

Capital charges costs	Daily running	Voyage costs	Cargo handling
←Bareboat→			
←——— Time charter ———→			
←——————— Owner operator ———————→		
←——————————— Owner's trade ———————————→

Figure 9.48 Changing responsibilities of the owner from bareboat to owner's trade.

(ii) by letting the ship on Voyage Charter to carry a single cargo on a set rate per tonne; or
(iii) by letting the ship for a single voyage on Time Charter for a set rate per day.

Under Contracts of Affreightment and Voyage Charters the owner will meet the capital cost, running costs and voyage costs (comprising port charges and bunkers). The terms of the charter will determine who pays the cargo handling costs as follows:

Gross terms (Gross)   Shipowner pays for loading and discharge
Free on board (FOB)   Charterer pays for loading
Free discharge (FD)   Charterer pays for discharge
Free in and out (FIO) Charterer pays for loading and discharge

Under a single voyage time charter the charterer will meet the voyage costs as well as the cargo handling costs.

(iii) Tramping – let out on time charter
In a time charter, the shipowner undertakes to provide a ship for the charterer to use either for a fixed time of anything from a few months to 20 years or for a single round voyage.

The charterer is responsible for arranging cargoes and voyages during the charter and also for paying all voyage expenses including fuel, port and canal dues, cargo handling charges.

The shipowner provides the ship and crew and is responsible for the capital charges and daily running costs. Hire is only payable for time in service and ceases during breakdown and repair, although it continues if the ship is delayed in port or sails empty for reasons not attributable to the ship.

(iv) Tramping – let out on bareboat charter
In this case the charterer provides the crew and is responsible for maintenance with the shipowner's sole responsibility being the provision of the ship and meeting the capital charges. In effect the charterer uses the ship as if he owned it.

9.5.2 Economic criteria

9.5.2.1 *The basis of these criteria*

There are a number of different economic criteria which may be used to assess the likely success of a shipping investment or to compare the profitability of alternatives. These criteria should take account of:

– the time value of money,
– the full life of the investment,
– changes in items of income and expenditure which can be expected over the life,
– the economic facts of life such as interest rates; taxes; loans and investment grants.

The time value of money represents the fact that a sum of money available now is of much more value than the same sum not available for a number of years.

Interest is fundamental to the calculations whether there is a need to borrow or not. This takes account of the fact that if available cash is used the interest it might have earned is being foregone.

9.5.2.2 *Interest*

This may be simple or compound and the following relationships apply:

– Simple interest
  Total repayment after N years: $F = P(1 + N \cdot i)$
– Compound interest
  Total repayment after N years: $F = P(1 + i)^N$
  In this case the factor $(1 + i)^N$ is called the compound amount factor (CA), and P = original investment.

9.5.2.3 *Present worth*

The reciprocal of CA is called the present worth (PW) factor.

$$PW = 1/(CA) = (1 + i)^{-N}$$
$$P = (PW)F$$

The present worth of F, which includes all the accumulated interest is the same as the present sum of money P.

9.5.2.4 *Repayment of principal*

If the loan is repaid by annual instalments of principal plus interest, this may take two forms:

(i) principal repaid in equal instalments with interest being paid on the reducing balance; or
(ii) equal annual payments with interest predominating in the early years and capital repayments in the later years.

The concept of equal annual payments enables a present sum of money to be converted into an annual repayment sum spread over a number of years with the annual sum A being linked to the sum invested – the 'present sum P' by the capital recovery factor (CRF)

$$A = (CRF)P; \text{ and } CRF = \frac{i(1+i)^N}{(1+i)^N - (1)}$$

$$\text{or } \frac{i}{1 - (1+i)^{-N}}$$

The reciprocal of (CRF) is Series Present Worth factor (SPW). This is the multiplier required to convert a number of regular annual payments into a present sum.

9.5.2.5 Sinking fund factor

To find the annual sum (A) which accumulates to provide a future sum (F), this is multiplied by the sinking fund factor (SF)

$$A = F(SF); \text{ and } (SF) = \frac{i}{(1+i)^N - (1)}$$

The reciprocal of (SF) is the series compound amount factor (SCA)

$$SCA = 1/SF \text{ and } F = (SCA)A$$

With this brief introduction to, or refresher on, economics, the economic criteria commonly used in shipping can now be introduced.

9.5.2.6 Net present value

In this type of calculation the net present values (NPV) of income and expenditure are calculated over the assumed life of the ship (N) years. The final sum should be positive for the investment to be profitable at the assumed discount rate – or where alternatives are being compared it should be the larger sum.

$$NPV = \sum_{1}^{N} [PW \text{ (cargo tonnage} \times \text{ freight rate)} \\ - PW(\text{operating costs}) - PW \\ (\text{ship acquisition costs})]$$

9.5.2.7 Required freight rate

The required freight rate (RFR) is that which will produce a zero NPV, i.e. the break-even rate. Transposing the equation above gives:

$$RFR = \sum_{1}^{N} \left[\frac{(PW(\text{Operating costs}) + PW}{\text{Cargo tonnage}} \right]$$

9.5.2.8 Yield

In the above calculations a rate of interest must be assumed. If the freight rate is known or at least assumed, the rate at which money can be borrowed with NPV = 0, can be made the criterion.

9.5.2.9 Inflation and exchange rates

It is perhaps worth pointing out that economic forecasts of the sort described in the foregoing paragraphs are made on fixed money values. Inflation and the consequent reduction in the future value of money together with changes in exchange rates do not enter into these calculations although both of these must be estimated and taken into account in more detailed projections. This might be when fixing rates which are intended to apply over more than a limited period of time and/or when payments are to be made in a currency other than that in which the costs are incurred.

9.5.3 Operating costs

The next three sections as well as describing the components of operating costs try to suggest some ways of minimising these.

9.5.3.1 Capital charges

As Figure 9.48 shows, capital charges are included in the costing of all the different modes of ship operation and are in fact the only cost component in Bareboat chartering. Included in capital charges are:

– loan repayment
– loan interest
– profit
– taxes

9.5.3.2 Capital amortization

Loan interest and loan repayment can conveniently be taken together as capital amortisation.

The biggest component of capital charges is the repayment of the loan used to pay the shipbuilder. Payments to shipbuilders are almost invariably made in a number of instalments during the building period with a final instalment at the end of the guarantee period (usually a year after delivery).

Before the ship starts earning, its total cost will have increased above the tender price due to the interest payments on the sums paid out together with such other costs as those incurred in supervising construction, engaging the crew and in providing owners' supply items and initial stores.

Moreover, it will be an exceptional contract that does not result in some additional payments for changes in specification during building.

One obvious way to minimize capital charges is to keep the capital cost low, which may be achieved by good buying in relation to shipbuilding prices.

The initial building cost can, in principle, be kept down by building to a lower standard, although if this involves accepting that the ship will have a shorter than normal life this may not be a cost effective thing to do.

When considering capital economy measures, care must be taken to ensure that any lower standards adopted do not lead to higher operating costs that will negate any savings made.

The second largest component of capital charges is the sum paid in interest on the money borrowed to meet the costs incurred in building the ship and getting it into service.

Consequently, another way – and probably in the long term one of the most important ways – of minimizing capital charges, is by obtaining the most advantageous interest rates available.

Finally, at the end of whatever operating life is being assumed in the financial costing, the ship will still have a value, even if this is only as scrap, and an allowance for this should be made when assessing the cost of capital amortization.

The general assumption made in most financial assessments is that ships will have an operating life of 20 years. Although many continue in service for much longer periods, others become obsolete much earlier either as a result of changes in technology and/or in trading patterns and a 20 year period is probably a reasonable compromise.

9.5.3.3 *Profit and taxes*

The profit which the shipowner plans to make together with the taxes which this profit will incur forms the second part of capital charges.

9.5.3.4 *Depreciation*

Although depreciation does not enter into operating cost calculations, it seems desirable to include a short paragraph on the subject at this point as it does have a very significant effect on shipping company accounts, the tax paid and the profit made in particular years.

Depreciation is the process of writing off capital costs in company accounts. There are two classical methods of treating depreciation, namely:

(i) Straight line depreciation. If a 20-year life is assumed, the depreciation would be 5% per annum.
(ii) Declining balance depreciation. If a 15% per annum basis is assumed, then:

Year 1: $15\% \times 100$ $= 15\%$
Year 2: $15\% \times (100-15)$ $= 12.75\%$
Year 3: $15\% \times (100-15-12.75) = 10.84\%$
Year 10: 3.52%
Year 20: 0.94%

In most countries there are special provisions for the treatment of shipping depreciation from a taxation point of view. These treatments vary from country to country as do the rates of tax imposed.

Most of these treatments permit the writing off of a ship's capital cost at rather faster rates than the classical treatments. In general it pays a shipowner to depreciate as fast as the profits permit thus reducing or at least deferring tax payments.

9.5.3.5 *Ship values*

Although the book value of a ship at any time will be its original cost plus the cost of any repairs or alterations and minus the accumulated depreciation, the value of a ship as measured by its possible selling price is likely to fluctuate dramatically during its lifetime. This does not enter into operating cost calculations, although some owners significantly improve their profits by playing the market in this way.

9.5.4 Daily running costs

Included in daily running costs are:

– crew costs
– provisions and stores
– maintenance and repairs
– insurance
– administration and general charges

These costs are added for time charter calculations and of course also apply to voyage charters and owner operation. These are costs incurred whether the ship is at sea or in port.

9.5.4.1 *Crew costs*

The two major factors which determine crew costs today are crew numbers and the nationality of different sections of the officers and crew.

The effect of numbers is offset to some extent by the fact that a smaller crew will generally tend to have more 'chiefs' and fewer 'indians' and the fact that all the members of a reduced crew will (or certainly ought to) have a higher standard of training and as a consequence will (or ought to) be paid more *per capita*.

The automation and higher quality materials required to reduce watch-keeping and maintenance and thus enable the reduced crew to work the ship satisfactorily will increase the capital cost, whilst there is also likely to be a demand for higher class accommodation although this will be offset by the reduced number of cabins required.

9.5.4.2 *Provisions and stores*

Provisions are usually bought locally at the ship's trading ports and the annual cost is calculated on a per person per day basis.

Ships consume an extraordinary variety and quite considerable quantity of miscellaneous stores with the three most important items being chandlery, paint, chemicals and gases but with smaller sums being expended on such items as fresh water, laundry and charts.

Lubricating oil is sometimes included with this item, but it seems more logical to include it with bunkers.

9.5.4.3 *Maintenance and repair*

With today's small crews, maintenance at sea is necessarily limited, but careful planning by the ship's staff whilst at sea can greatly speed work carried out when in port and minimize its cost.

A big item under this heading is drydocking, but this is no longer an annual event with three or even five year intervals becoming usual.

Budgets for maintenance will generally include sums for work on the hull and superstructure, cargo spaces and systems, the main and auxiliary machinery, the electrical installation and the safety equipment plus survey fees.

Also included under this heading is the cost of riding squads which are now used to carry out maintenance and repairs which would have formerly been done by the crew but which is beyond the capability of the reduced crews of today.

9.5.4.4 *Insurance*

Insurance can be subdivided into Hull and P & I. The cost of Hull insurance is directly related to the capital cost of the ship with the insurance history of the managing company exercising a secondary effect. Costs have escalated significantly in recent years due to the number of major casualties and a generally ageing tonnage. Policies now provide for more deductibles and in the event of a claim these can increase running costs considerably.

P & I premiums have also increased greatly because of the U.S. Oil Pollution act and worries about crew standards.

9.5.4.5 *Administration and general charges*

Administration costs are a contribution to the office expenses of a shipping company or the fees payable to a management company plus a not inconsiderable sum for communications and sundries, together with flag charges.

Amongst the items included in general charges can be the cost of hiring items of ship's equipment such as the radio installation which are sometimes hired rather than bought as part of the ship.

The charge for the hire can be reduced by making a bulk deal for several ships with one company. The decision between buying and hiring demands reconsideration from time to time as prices, interest rates and tax measures change. At present the use of hired equipment is reducing.

It is also wise to allow in this heading a sum for exceptional items when preparing a cost estimate as regrettably only too often there will be something which cannot be foreseen.

9.5.5 Voyage costs

Included in voyage costs are:

– bunkers
– port and canal dues
– tugs, pilotage
– miscellaneous port expenses

These items are added when moving from a time charter to a voyage charter calculation and of course apply to owner operation.

9.5.5.1 *Bunkers*

(i) Oil fuel
The factors affecting oil fuel costs are the distance travelled, the average power used, the specific fuel consumption and the cost per tonne of fuel. The first of these can be minimized by good navigation which must also take into account favourable and adverse currents.

The second can be minimized by steaming at as slow a speed as enables the required schedule to be kept; by keeping the hull finish to a high standard of smoothness (a task that is much easier than it used to be with the latest long life and self polishing anti-fouling paints); and at an earlier stage, by good design of the ship's lines and the propeller.

Specific fuel consumption can be minimized at the design stage by a good choice of machinery and at the operating stage by keeping the engine well maintained.

The cost of fuel can be minimized by a careful choice of bunkering port, although any cost saving thus obtained must first meet any additional costs if a diversion is required or there is any reduction in cargo carrying capacity or increase in average voyage displacement increasing power and consumption. The fuel cost can also be reduced by the use of a poorer quality of fuel, although any saving must be assessed against any extra costs for purifiers, etc. needed for the fuel to be used and any increases in maintenance and repair costs that may result from its use. Bulk buying is yet another way of getting fuel at an advantageous price.

(ii) Diesel oil

Here the factors involved are the number of days, as generators are kept running in port as well as at sea, and the average electrical load. Because the cost of diesel oil is much higher than that of oil fuel it is advantageous to meet as much as possible of the electrical load by the use of shaft driven alternators.

(iii) Lubricating oil

Although the quantity of lubricating oil consumed is relatively small its high, unit cost results in it being a considerable item of expenditure. This item is sometimes included with stores, but as the usage depends on the distance travelled it seems better grouped with bunkers.

9.5.5.2 *Port and canal dues, pilotage, towage etc.*

(a) Port and canal dues

Port and canal dues depend on the tonnage of the vessel and on the trading pattern. Low gross and/or net tonnages are particularly important on some routes, such as those using the Suez or Panama canals or The St. Lawrence Seaway.

Booklets giving canal dues can be obtained from:

- Panama Canal Commission, Balboa, Republic of Panama (Fax: 507-272-2122)
- Suez Canal Authority, Ismailia, Arab Republic of Egypt (Fax: 064-320-784)
- St. Lawrence Seaway Authority, 360 Albert St, Ottawa, Canada (Fax: 613-598-4620)

(b) Pilotage costs

Pilotage costs are usually also assessed on gross tonnage but can be reduced in certain trades by having a ship's officer with a pilotage certificate where this procedure is followed.

(c) Towage and mooring costs

Tug charges can be eliminated or reduced if the ship is fitted with a bow thruster or approved high performance steering equipment.

The time spent in mooring can be reduced by fitting special deck machinery such as self-tensioning winches.

9.5.6 Cargo handling costs

Cargo handling costs include the costs arising from both loading and unloading cargo together with any claims that may arise relating to the cargo.

Cargo handling costs are excluded from voyage charter costs but have to be met in owner operation.

Cargo handling time can be reduced and with it the costs of this operation, by the provision of good cargo handling features such as:

1. large hatches giving good access;
2. shipside doors where appropriate;
3. hatch covers which can be speedily opened and closed;
4. fork lift trucks to speed stowage;
5. cargo handling cranes or derricks on the ship with a lift capacity optimized to the cargo carried and a speedy cycle time;
6. in appropriate cases by providing the ships with self discharging facilities.

Where the trade is based on a small number of specific ports there is the alternative of minimizing the ship cost and using shoreside cargo handling facilities.

Containerization or palletization of the cargo can make a step change in cargo handling time and cost.

9.6 Optimization in design and operation

9.6.1 Overview

Most design problems may be formulated as follows: determine a set of design variables (e.g. number of ships, individual ship size and speed in fleet optimization; main dimensions and interior subdivision of ship; scantlings of a construction; characteristic values of pipes and pumps in a pipe net) subject to certain relations between and restrictions of these variables (e.g. by physical, technical, legal, economical laws). If more than one combination of design variables satisfies all these conditions, we would like to determine that combination of design variables which optimizes some measure of merit (e.g. weight, cost, or yield).

9.6.2 Introduction to methodology of optimization

Optimization means finding the best solution from a limited or unlimited number of choices. Even if the number of choices is finite, it is often so large that it is impossible to evaluate each possible solution and then determine the best choice. There are, in principle, two methods of approaching optimization problems:

1. Direct search approach
   Solutions are generated by varying parameters either systematically in certain steps or randomly. The best of these solutions is then taken as the estimated optimum. Systematic variation soon

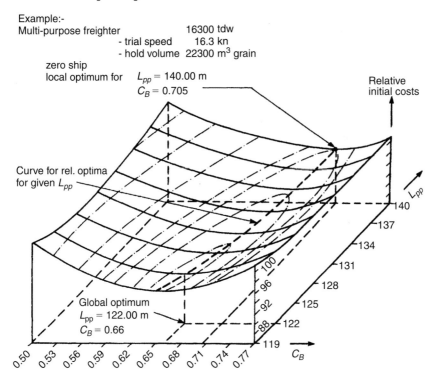

Figure 9.49 Example of overall costs dependent on length and block coefficient.

becomes prohibitively time consuming as the number of varied variables increases. Random searches are then employed, but these are still inefficient for problems with many design variables.
2. Steepness approach
   The solutions are generated using some information on the local steepness (in various directions) of the function to be optimized. When the steepness in all directions is (nearly) zero, the estimate for the optimum is found. This approach is more efficient in many cases. However, if several local optima exist, the method will 'get stuck' at the nearest local optimum instead of finding the global optimum, i.e. the best of *all* possible solutions. Discontinuities (steps) are problematic; even functions that vary steeply in one direction, but very little in another direction make this approach slow and often unreliable.

Most optimization methods in ship design are based on steepness approaches because they are so efficient for smooth functions. As an example consider the cost function varied over length L and block coefficient C_B (Figure 9.49). A steepness approach method will find quickly the lowest point on the cost function, if the function $K = f(C_B, L)$ has only one minimum. This is often the case.

Repeating the optimization with various starting points may circumvent the problem of 'getting stuck' at local optima. One option is to combine both approaches with a quick direct search using a few points to determine the starting point of the steepness approach. Also repeatedly alternating both methods – with the direct approach using a smaller grid scale and range of variation each time – has been proposed.

A pragmatic approach to treating discontinuities (steps) assumes first a continuous function, then repeats the optimization with lower and upper next values as fixed constraints and taking the better of the two optima thus obtained. Although, in theory, cases can be constructed where such a procedure will not give the overall optimum, in practice this procedure apparently works well.

The target of optimization is the objective function or criterion of the optimization. It is subject to boundary conditions or constraints. Constraints may be formulated as equations or inequalities. All technical and economical relationships to be considered in the optimization model must be known and expressed as functions. Some relationships will be exact, e.g. $\nabla = C_B \cdot L \cdot B \cdot T$; others will only be approximate, such as all empirical formulae, e.g. regarding resistance or weight estimates.

Procedures must be sufficiently precise, yet may not consume too much time or require highly detailed inputs. Ideally all variants should be evaluated with the same procedures. If a change of procedure is necessary, for example, because the area of validity is exceeded, the results of the two procedures must be correlated or blended if the approximated quantity is continuous in reality.

A problem often encountered in optimization is having to use unknown or uncertain values, e.g. future prices. Here plausible assumptions must be made. Where these assumptions are highly uncertain, it is common to optimize for several assumptions ('sensitivity study'). If a variation in certain input values only slightly affects the result, these may be assumed rather arbitrarily.

The main difficulty in most optimization problems does not lie in the mathematics or methods involved, i.e. whether a certain algorithm is more efficient or robust than others. The main difficulty lies in formulating the objective and all the constraints. If the human is not clear about his objective, the computer cannot perform the optimization. The designer has to decide first what he really wants. This is not easy for complex problems. Often the designer will list many objectives which a design shall achieve (e.g. see Section 9.2.1). This is then referred to in the literature as 'multi-criteria optimization', e.g. Sen (1992), Ray and Sha (1994). The expression is nonsense if taken literally. Optimization is only possible for one criterion, e.g. it is nonsense to ask for the best and cheapest solution. The best solution will not come cheaply, the cheapest solution will not be so good. There are two principle ways to handle 'multi-criteria' problems, both leading to one-criterion optimization:

1. One criterion is selected and the other criteria are formulated as constraints.
2. A weighted sum of all criteria forms the optimization objective. This abstract criterion can be interpreted as an 'optimum compromise'. However, the rather arbitrary choice of weight factors makes the optimization model obscure and we prefer the first option.

Throughout optimization, design requirements (constraints), e.g. cargo weight, deadweight, speed and hold capacity, must be satisfied. The starting point is called the 'basis design' or 'zero variant'. The optimization process generates alternatives or variants differing, for example, in main dimensions, form parameters, displacement, main propulsion power, tonnage, fuel consumption and initial costs. The constraints influence, usually, the result of the optimization. Figure 9.50 demonstrates, as an example, the effects of different optimization constraints on the sectional area curve.

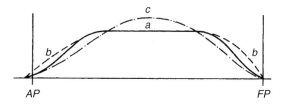

Figure 9.50 Changes produced in sectional area curve by various optimization constraints:
a is the basis form;
b is a fuller form with more displacement; optimization of carrying capacity with maximum main dimensions and variable displacement;
c is a finer form with the displacement of the basis form *a*, with variable main dimensions.

Optimized main dimensions often differ from the values found in built ships. There are several reasons for these discrepancies:

1. *Some built ships are suboptimal*
   The usual design process relies on statistics and comparisons with existing ships, rather than analytical approaches and formal optimization. Designs found this way satisfy the owner's requirements, but better solutions, both for the shipyard and the owner, may exist. Technological advances, changes in legislation and in economical factors (e.g. the price of fuel) are reflected immediately in an appropriate optimization model, but not when relying on partially outdated experience. Modern design approaches increasingly incorporate analyses in the design and compare more variants generated with the help of the computer. This should decrease the differences between optimization and built ships.
2. *The optimization model is insufficient*
   The optimization model may have neglected factors that are important in practice, but difficult to quantify in an optimization procedure, e.g. seakeeping behaviour, manoeuvrability, vibrational characteristics, easy cargo-handling. Even for directly incorporated quantities, often important relationships are overlooked, leading to wrong optima, e.g.:
   (a) A faster ship usually attracts more cargo, or can charge higher freight rates, but often income is assumed as speed independent.
   (b) A larger ship will generally have lower quay-to-quay transport costs per cargo unit, but time for cargo-handling in port may increase. Often, the time in port is assumed to be size independent.
   (c) In reefers the design of the refrigerated hold with regard to insulation and temperature requirements affects the optimum main dimensions. The additional investment and

annual costs have to be included in the model to obtain realistic results.

(d) The performance of a ship will often deteriorate over time. Operating costs will correspondingly increase, Malone et al. (1980), Townsin et al. (1981), but are usually assumed time independent.

The economic model may use an inappropriate objective function. Often there is confusion over the treatment of depreciation. This is not an item of expenditure, i.e. cash flow, but a book-keeping and tax calculation device, see Sections 9.5.3.4 and 9.6.4. The optimization model may also be based on too simplified technical relationships. Most of the practical difficulties boil down to obtaining realistic data to include in the analysis, rather than the mechanics of making the analysis. For example, the procedures for weight estimation, power prediction and building costs are quite inaccurate, which becomes obvious when the results of different published formulae are compared. The optimization process may now just maximize the error in the formulae rather than minimize the objective.

The result of the optimization model should be compared against built ships. Consistent differences may help to identify important factors so far neglected in the model. A sensitivity analysis concerning the underlying estimation formulae will give a bandwidth of 'optimal' solutions and any design within this bandwidth must be considered as equivalent. If the bandwidth is too large, the optimization is insignificant.

A critical view on the results of optimization is recommended. But properly used optimization may guide us to better designs than merely reciprocating traditional designs. The ship main dimensions should be appropriately selected by a naval architect who understands the relationships of various variables and the pitfalls of optimization. An automatic optimization does not absolve the designer of his responsibility. It only supports him in his decisions.

9.6.3 Scope of application in ship design

Formal optimization of the lines including the bulbous bow even for fixed main dimensions is beyond our current computational capabilities. Although such formal optimization has been attempted using CFD methods, the results were not convincing despite high computational effort, Janson (1997). Instead, we will focus here on ship design optimization problems involving only a few (less than 10) independent variables and rather simple functions. A typical application would be the optimization of the main dimensions. However, optimization may be applied to a wide variety of ship design problems ranging from fleet optimization to details of structural design.

In fleet optimization, the objective is often to find the optimum number of ships, ship speed and capacity without going into further details of main dimensions, etc. A ship's economic efficiency is usually improved by increasing its size, as specific cost (cost per unit load, e.g. per TEU or per ton of cargo) for initial cost, fuel, crew, etc., decrease. However, dimensional limitations restrict size. The draught (and thus indirectly the depth) is limited by channels and harbours. However, for draught restrictions one should keep in mind that a ship is not always fully loaded and harbours may be dredged to greater draughts during the ship's life. The width of tankers is limited by building and repair docks. The width of containerships is limited by the span of container bridges. Locks restrict all the dimensions of inland vessels. In addition, there are less obvious aspects limiting the optimum ship size:

1. The limited availability of cargo coupled to certain expectations concerning frequency of departure limits the size on certain routes.
2. Port time increases with size, reducing the number of voyages per year and thus the income.
3. The shipping company loses flexibility. Several small ships can service more frequently various routes/harbours and will thus usually attract more cargo. It is also easier to respond to seasonal fluctuations.
4. Port duties increase with tonnage. A large ship calling on many harbours may have to pay more port dues than several smaller ships servicing the same harbours in various routes, thus calling each in fewer harbours.
5. In container line shipping, the shipping companies offer door-to-door transport. The costs for feeder and hinterland traffic increase if large ships only service a few 'hub' harbours and distribute the cargo from there to the individual customer. Costs for cargo-handling and land transport then often exceed savings in shipping costs.

These considerations largely concern shipping companies in optimizing the ship size. Factors favouring larger ship size are, Buxton (1976):

- Increased annual flow of cargo.
- Faster cargo-handling.
- Cargo available one way only.
- Long-term availability of cargo.
- Longer voyage distance.
- Reduced cargo-handling and stock-piling costs.
- Anticipated port improvements.
- Reduced unit costs of building ships.
- Reduced frequency of service.

We refer to Benford (1963, 1965) for more details on selecting ship size.

After the optimum size, speed, and number of ships has been determined along with some other specifications, the design engineer at the shipyard is usually tasked to perform an optimization of the main dimensions as a start of the design. Further stages of the design will involve local hull shape, e.g. design of the bulbous bow lines, structural design, etc. Optimization of structural details often involves only a few variables and rather exact functions. Söding (1977) presents as an example the weight optimization of a corrugated bulkhead. Other examples are found in Liu et al. (1981) and Winkle and Baird (1985).

For the remainder of the Section we will discuss only the optimization of main dimensions for a single ship. Pioneering work in introducing optimization to ship conceptual design in Germany has been performed by the Technical University of Aachen (Schneekluth, 1957, 1967; Malzahn et al., 1978). Such an optimization varies technical aspects and evaluates the result from an economic viewpoint. Fundamental equations (e.g. $\nabla = C_B \cdot L \cdot B \cdot T$), technical specifications/constraints, and equations describing the economical criteria form a more or less complicated system of coupled equations, which usually involve nonlinearities. Gudenschwager (1988) gives an extensive optimization model for Ro-Ro ships with 57 unknowns, 44 equations, and 34 constraints.

To establish such complicated design models, it is recommended to start with a few relations and design variables, and then to improve the model step by step, always comparing the results with the designer's experience and understanding the changes relative to the previous, simpler model. This is necessary in a complicated design model to avoid errors or inaccuracies which cannot be clarified or which may even remain unnoticed without applying this stepwise procedure. Design variables which involve step functions (number of propeller blades, power of installed engines, number of containers over the width of a ship, etc.) may then be determined at an early stage and can be kept constant in a more sophisticated model, thus reducing the complexity and computational effort. Weakly variation-dependent variables or variables of secondary importance (e.g. displacement, underdeck volume, stability) should only be introduced at a late stage of the development procedure. The most economic solution often lies at the border of the search space defined by constraints, e.g. the maximum permissible draught or Panamax width for large ships. If this is realized in the early cycles, the relevant variables should be set constant in the optimization model in further cycles. Keane et al. (1991) discuss solution strategies of optimization problems in more detail.

Simplifications can be retained if the associated error is sufficiently small. They can also be given subsequent consideration.

9.6.4 Economic basics for optimization

9.6.4.1 *Discounting*

An outline of the economic criteria has been given in Section 9.5.2. For purposes of optimization, all payments are discounted, i.e. converted by taking account of the interest, to the time when the vessel is commissioned. The rate of interest used in discounting is usually the market rate for long-term loans. Discounting decreases the value of future payments and increases the value of past payments. Individual payments thus discounted are, for example, instalments for the new building costs and the re-sale price or scrap value of the ship. The present value (discounted value) K_{pv} of an individual payment K paid N years later—e.g. scrap or re-sale value—is:

$$K_{pv} = K \cdot \frac{1}{(1+i)^N} = K \cdot \text{PW}$$

where i is the interest rate. PW is the present worth factor. For an interest rate of 8%, the PWF is 0.2145 for an investment life of 20 years, and 0.9259 for 1 year. If the scrap value of a ship after 20 years is 5% of the initial cost, the discounted value is about 1%. Thus the error in neglecting it for simplification is relatively small.

A series of constant payments k is similarly discounted to present value K_{pv} by:

$$K_{pv} = k \cdot \frac{(1+i)^N \cdot i}{(1+i)^N - 1} = k \cdot \text{CRF}$$

CRF is the capital recovery factor. The shorter the investment life, the greater is the CRF at the same rate of interest. For an interest rate of 8%, the CRF is 0.1018 for 20 years and 1.08 for 1 year of investment life.

The above formulae assume payment of interest at the end of each year. This is the norm in economic calculations. However, other payment cycles can easily be converted to this norm. For example, for quarterly payments divide i by 4 and multiply N by 4 in the above formulae.

For costs incurred at greater intervals than years, or on a highly irregular basis, e.g. large-scale repair work, an annual average is used. Where changes in costs are anticipated, future costs should be entered at the average annual level as expected. Evaluation of individual costs is based on present values which may be corrected if recognizable longer-term trends exist. Problems are:

1. The useful life of the ship can only be estimated.
2. During the useful life, costs can change with the result that cost components may change in absolute terms and in relation to each other. After the oil crisis of 1973, for example, fuel costs rose dramatically.

All expenditure and income in a ship's life can thus be discounted to a total 'net present value' (NPV). Only the cash flow (expenditure and income) should be considered, not costs which are used only for accounting purposes.

Yield is the interest rate i that gives zero NPV for a given cash flow. Yield is also called Discounted Cash Flow Rate of Return, or Internal Rate of Return. It allows comparisons between widely different alternatives differing also in capital invested. In principle, yield should be used as the economic criterion to evaluate various ship alternatives, just as it is used predominantly in business administration as the benchmark for investments of all kinds. The operating life should be identical for various investments then. Unfortunately, yield depends on uncertain quantities like future freight rates, future operating costs, and operating life of a ship. It also requires the highest computational effort as building costs, operating costs and income must all be estimated.

Other economic criteria which consider the time value of money include NPV, NPV/investment, or Required Freight Rate (= the freight rate that gives zero NPV); they are discussed in more detail by Buxton (1972, 1976). The literature is full of long and rather academic discussions on what is the best criterion. But the choice of the economic criterion is actually of secondary importance in view of the possible errors in the optimization model (such as overlooking important factors or using inaccurate relationships).

Discounting decreases the influence of future payments. The initial costs, not discounted, represent the single most important payment and are the least afflicted by uncertainty. (Strictly speaking, the individual instalments of the initial costs should be discounted, but these are due over the short building period of the ship.) The criterion 'initial costs' simplifies the optimization model, as several variation-independent quantities can be omitted. Initial costs have often been recommended as the best criterion for shipyard as this maximises the shipyard's profit. This is only true if the price for various alternatives is constant. However, in modern business practice the shipyard has to convince the shipowner of its design. Then price will be coupled to expected cash flow.

In summary, the criterion for optimization should usually be yield. For a simpler approach, which may often suffice or serve in developing the optimization model, initial costs may be minimized.

9.6.4.2 Initial costs (building costs)

Building costs can be roughly classified into:

- Direct labour costs.
- Direct material costs (including services bought).
- Overhead costs.

Overhead costs are related to individual ships by some appropriate key, for example equally among all ships built at the accounting period, proportional to direct costs, etc. See also Carreyette (1978) for a discussion of costs.

For optimization, the production costs are divided into (Figure 9.51):

1. *Variation-dependent costs*
   Costs which depend on the ship's form:
   (a) Cost of hull.
   (b) Cost of propulsion unit (main engine).
   (c) Other variation-dependent costs, e.g. hatchways, pipes, etc.
2. *Variation-independent costs*
   Costs which are the same for every variant, e.g. navigation equipment, living quarters, etc.

Buxton (1976) gives some simple empirical estimates for these costs.

Building costs are covered by own capital and loans. The source of the capital may be disregarded. Then also interest on loans need not be considered in the cash flow. The yield on the capital should then be larger than alternative forms of investment, especially the interest rate of long-term loans. This approach is too simple for an investment decision, but suffices for optimizing the main dimensions.

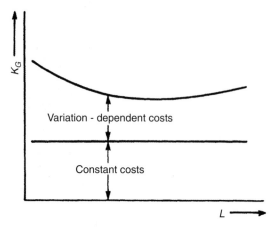

Figure 9.51 Division of costs into length-dependent and length-independent.

Typically 15–45% of the initial costs are attributable to the shipyard, the rest to outside suppliers. The tendency is towards increased outsourcing. Of the wages paid by the shipyard, typically 20% are allotted to design and 80% to production for one-of-a-kind cargo ships, while warships feature typically a 50:50 proportion.

Determining the variation-dependent costs
Superstructure and deckhouses are usually assumed to be variation-independent when considering variations of main dimensions. The variation-dependent costs are:

1. The hull steel costs.
2. The variation-dependent propulsion unit costs.
3. Those components of equipment and outfit which change with main dimensions.

The steel costs
The yards usually determine the costs of the processed steel in two separate groups:

1. The cost of the unprocessed rolled steel. The costs of plates and rolled sections are determined separately using prices per ton. The overall weight is determined by the steel weight calculation. The cost of wastage must be added to this.
2. Other costs. These comprise mainly wages. This cost group depends on the number of man-hours spent working on the ship within the yard. The numbers differ widely, depending on the production methods and complexity of construction. As a rough estimate, 25–35 man-hours/t for containerships are cited in older literature. There are around 30–40% more man-hours/t needed for constructing the superstructure and deckhouses than for the hull, and likewise for building the ship's ends as compared with the parallel middlebody. The amount of work related to steel weight is greater on smaller ships. For example, a ship with $70\,000\,\text{m}^3$ underdeck volume needs 15% less manufacturing time per ton than a ship with $20\,000\,\text{m}^3$, Kerlen (1985).

For optimization, it is more practical to form 'unit costs per ton of steel installed', and then multiply these unit costs by the steel weight. These unit costs can be estimated as the calculated production costs of the steel hull divided by the net steel weight. Kerlen (1985) gives the specific hull steel costs as:

$$k_{St}[\text{MU/t}] = k_0 \cdot \left(\frac{4}{\sqrt[3]{L/m}} + \frac{3}{L/m} + 0.2082 \right)$$
$$\cdot \left(\frac{3}{2.58 + C_B^2} - 0.07 \cdot \frac{0.65 - C_B}{0.65} \right)$$

k_0 represents the production costs of a ship 140 m in length with $C_B = 0.65$. The formula is applicable for ships with $0.5 \leq C_B \leq 0.8$ and $80\,\text{m} \leq L \leq 200\,\text{m}$. The formula may be modified, depending on the material costs and changes in work content.

Propulsion unit costs
For optimization of main dimensions, the costs of the propulsion plant may be assumed to vary continuously with propulsion power. They can then be obtained by multiplying propulsion power by unit costs per unit of power. A further possibility is to use the catalogue prices for engines, gears and other large plant components in the calculation and to take account of other parts of the machinery by multiplying by an empirical factor. Only those parts which are functions of the propulsion power should be considered. The electrical plant, counted as part of the engine plant in design – including the generators, ballast water pipes, valves and pumps – is largely variation-independent.

The costs of the weight group 'equipment and outfit'
Whether certain parts are so variation-dependent as to justify their being considered depends on the ship type. For optimization of initial costs, the equipment can be divided into three groups:

1. Totally variation-independent equipment, e.g. electronic units on board.
2. Marginally variation-dependent equipment, e.g. anchors, chains and hawsers which can change if in the variation the classification numeral changes. If variation-dependence is not pronounced, the equipment in question can be omitted.
3. Strongly variation-dependent equipment, e.g. the cost of hatchways rises roughly in proportion to the hatch length and the 1.6th power of the hatch width, i.e. broad hatchways are more expensive than long, narrow ones.

Relationship of unit costs
Unit costs relating to steel weight and machinery may change with time. However, if their ratio remains constant, the result of the calculation will remain unchanged. If, for example, a design calculation for future application assumes the same rates of increase compared with the present for all the costs entered in the calculation, the result will give the same main dimensions as a calculation using only current data.

9.6.4.3 Annual income and expenditure

The income of cargo ships depends on the amount of cargo and the freight rates. Both should be a function of speed in a free market. At least the interest of the tied-up capital cost of the cargo should be included as a lower estimate for this speed dependence. The issue will be discussed again in Section 9.6.5. for the effect of speed.

Expenditure over the lifetime of a ship includes:

1. *Risk costs*
   Risk costs relating to the ship consist mainly of the following insurance premiums:
   - Insurance on hull and associated equipment.
   - Insurance against loss or damage by the sea.
   - Third-party (indemnity) insurance.

   Annual risk costs are typically 0.5% of the production costs.

2. *Repair and maintenance costs*
   The repair and maintenance costs can be determined using operating cost statistics from suitable basis ships, usually available in shipping companies.

3. *Fuel and lubricating costs*
   These costs depend on engine output and operating time.

4. *Crew costs*
   Crew costs include wages and salaries including overtime, catering costs, and social contributions (health insurance, accident and pension insurance, company pensions). Crewing requirements depend on the engine power, but remain unchanged for a wide range of outputs for the same system. Thus crew costs are usually variation-independent. If the optimization result shows a different crewing requirement from the basis ship, crew cost differences can be included in the model and the calculation repeated.

5. *Overhead costs*
   - Port duties, lock duties, pilot charges, towage costs, haulage fees.
   - Overheads for shipping company and broker.
   - Hazard costs for cargo (e.g. insurance, typically 0.2–0.4% of cargo value).

   Port duties, lock duties, pilot charges and towage costs depend on the tonnage. The proportion of overheads and broker fees depend on turnover and state of employment. All overheads listed here are variation-independent for constant ship size.

6. *Costs of working stock and extra equipment*
   These costs depend on ship size, size of engine plant, number of crew, etc. The variation-dependence is difficult to calculate, but the costs are small in relation to other cost types mentioned. For this reason, differences in working-stock costs may be neglected.

7. *Cargo-handling costs*
   Cargo-handling costs are affected by ship type and the cargo-handling equipment both on board and on land. They are largely variation-independent for constant ship size.

Taxes, interest on loans covering the initial building costs and inflation have only negligible effects on the optimization of main dimensions and can be ignored.

9.6.4.4 The 'cost-difference' method

Cash flow and initial costs can be optimized by considering only the differences with respect to the 'basis ship'. This simplifies the calculation as only variation-dependent items remain. The difference costs often give more reliable figures.

Objective function for initial costs optimization
The initial difference costs consist of the sum of hull steel difference costs and propulsion unit difference costs:

$$\Delta K_G[\text{MU}] = W_{St_0} \cdot k_{St_0} - W_{St_n} \cdot k_{St_n} + \Delta K_M \cdot C_M$$
$$= W_{St_0} \cdot k_{St_0} - W_{St_n} \cdot k_{St_n} + \Delta P_B \cdot k_M \cdot C_M$$

ΔK_G	[MU]	difference costs for the initial costs
W_{St_0}	[t]	hull steel weight for basis variant
W_{St_n}	[t]	hull steel weight for variant n
k_{St}	[MU/t]	specific costs of installed steel
ΔK_M	[MU]	difference costs for the main engine
C_M		factor accounting for the difference costs of the 'remaining parts' of the propulsion unit
ΔP_B	[kW]	difference in the required propulsion power
k_M	[MU/kW]	specific costs of engine power

In some cases the sum of the initial difference costs should be supplemented further by the equipment difference costs.

Objective function for yield optimization
The yield itself is not required, only the variant which maximizes yield. Again, only the variation-dependent cash flow needs to be considered. The most important items are the differences in:

1. Initial costs
2. Fuel and lubricant costs
3. Repair and insurance costs
4. Net income if variation-dependent

The power requirements are a function of trial speed, therefore the initial costs of the propulsion unit depend on the engine requirements under trial speed conditions. The fuel costs should be related to the service speed. The annual fuel and lubricant costs then become:

$$k_{f+l}[\text{MU/yr}] = P_{B,D} \cdot F \cdot (k_f \cdot s_f + k_l \cdot s_l)$$

$P_{B,D}$ [kW]   brake power at service speed
F [h]   annual operating time
k_f [MU/t]   cost of 1 t of fuel (or heavy oil)
s_f [t/kWh]   specific fuel consumption
k_l [MU/t]   cost of 1 t of lubricating oil
s_l [t/kWh]   specific lubricant consumption

9.6.4.5 *Discontinuities in propulsion unit costs*

Standardised propulsion unit elements such as engines, gears, etc. introduce steps in the cost curves (Figures 9.52 and 9.53). The stepped curve can have a minimum on the faired section or at the lower point of a break. With the initial costs, the optimum is always situated at the beginning of the curve to the right of the break. Changing from a smaller to a larger engine reduces the engine loading and thus repair costs. The fuel costs are also stepped where the number of cylinders changes (Figure 9.54). At one side of the break point the smaller engine is largely fully loaded. On the other side, the engine with one more cylinder has a reduced loading, i.e. lower fuel consumption. Thus when both initial costs and annual costs are considered the discounted cash flow is quasi-continuous.

The assumption of constant speed when propulsion power is changed in steps is only an assumption for

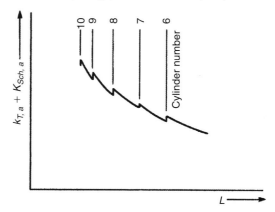

Figure 9.54 Annual fuel and lubricant costs ($k_f + k_l$) as a function of number of engine cylinders and ship's length.

comparison when determining the optimum main dimensions. In practice, if the propulsion plant is not fully employed, a higher speed is adopted.

9.6.5 Discussion of some important parameters

9.6.5.1 *Width*

A lower limit for B comes from requiring a minimum metacentric height GM and, indirectly, a maximum possible draught. The GM requirement is formulated in an inequality requiring a minimum value, but allowing larger values which are frequently obtained for tankers and bulkers.

9.6.5.2 *Length*

Suppose the length of a ship is varied while cargo weight, deadweight and hold size, but also $A_M \cdot L$, B/T, B/D and C_B are kept constant (Figure 9.55). (Constant displacement and underdeck volume, approximate constant cargo weight and hold capacity.) Then a 10% increase in length will reduce A_M by 10%. D, B and T are each reduced by around 5%. L/B and L/D are each increased by around 16%.

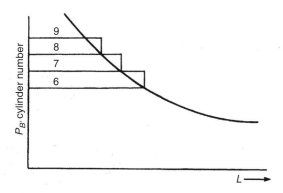

Figure 9.52 Propulsion power P_B and corresponding engine cylinder number as a function of ship's length.

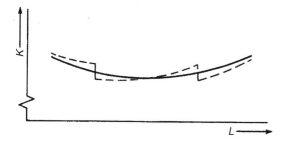

Figure 9.53 Effect of a change in number of engine cylinders on the cost of the ship.

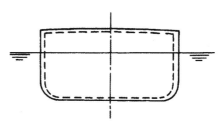

Figure 9.55 Variation of midship section area A_M with proportions unchanged.

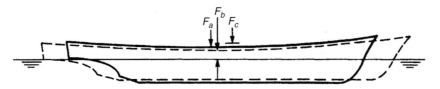

Figure 9.56 Effect of length variation on the freeboard. F_a = freeboard of basis form, F_b = freeboard of distorted ship, F_c = desired freeboard after lengthening.

For this kind of variation, increasing length has these consequences:

1. Increase in required regulation freeboard with decrease in existing freeboard.
2. Decrease in initial stability.
3. Better course-keeping ability and poorer course-changing ability.
4. Increase in steel weight.
5. Decrease in engine output and weight—irrespective of the range of Froude number.
6. Decrease in fuel consumption over the same operational distance.

Increase in the regulation freeboard
The existing freeboard is decreased, while the required freeboard is increased (Figure 9.56). These opposing tendencies can easily lead to conflicts. The freeboard regulations never conflict with a shortening of the ship, if C_B is kept constant.

Reduction in initial stability
The optimization often requires constant initial stability to meet the prescribed requirements and maintain comparability. A decrease in GM is then, if necessary, compensated by a slight increase of B/T, reducing T and D somewhat. This increases steel weight and decreases power savings.

Course-keeping and course-changing abilities
These characteristics are in inverse ratio to each other. A large rudder area improves both.

Increase in steel weight, decrease in engine output and weight, decrease in fuel consumption
These changes strongly affect the economics of the ship, see Section 9.6.4.

9.6.5.3 Block coefficient

Changes in characteristics resulting from reducing C_B:

1. Decrease in regulation freeboard for $C_B < 0.68$ (referred to 85% D).
2. Decrease in area below the righting arm curve if the same initial stability is used.
3. Slight increase in hull steel weight.
4. Decrease in required propulsion power, weight of the engine plant and fuel consumption.
5. Better seakeeping, less added resistance in seaway, less slamming.
6. Less conducive to port operation as parallel middlebody is shorter and flare of ship ends greater.
7. Larger hatches, if the hatch width increases with ship width. Hatch covers therefore are heavier and more expensive. The upper deck area increases.
8. Less favourable hold geometry profiles. Greater flare of sides, fewer rectangular floor spaces.
9. The dimensional limits imposed by slipways, docks and locks are reached earlier.
10. Long derrick and crane booms, if the length of these is determined by the ship's width and not the hatch length.

Initial stability
GM remains approximately constant if B/T is kept constant. However, the prescribed GM is most effectively maintained by varying the width using Mühlbradt's formula:

$$B = \frac{B_0}{C[(C_B/C_{B0})^2 - 1] + 1}$$

$C = 0.12$ for passenger and containerships
$C = 0.16$ for dry cargo vessels and tankers.

Seakeeping
A small C_B usually improves seakeeping. Since the power requirement is calculated for trial conditions, no correction for the influence of seastate is included. Accordingly, the optimum C_B for service speed should be somewhat smaller than that for trial speed. There is no sufficiently simple and accurate way to determine the power requirement in a seastate as a function of the main dimensions. Constraints or the inclusion of some kind of consideration of the

seakeeping are in the interest of the ship owner. If not specified, the shipyard designer will base his optimization on trial conditions.

Size of hold
For general cargo ships, the required hold size is roughly constant in proportion to underdeck volume. For container and Ro-Ro ships, reducing C_B increases the 'noxious spaces' and more hold volume is required.

Usually the underdeck volume $\nabla_D = L \cdot B \cdot D \cdot C_{BD}$ is kept constant. Any differences due to camber and sheer are either disregarded or taken as constant over the range of variation. C_{BD} can be determined with reasonable accuracy by empirical equations:

$$C_{BD} = C_B + c \cdot \left(\frac{D}{T} - 1\right) \cdot (1 - C_B)$$

with $c = 0.3$ for U-shaped sections and $c = 0.4$ for V-shaped sections. See also Section 9.2.4.2.

With the initial assumption of constant underdeck volume, the change in the required engine room size, and any consequent variations in the unusable spaces at the ship's ends and the volume of the double bottom are all initially disregarded. A change in engine room size can result from changes in propulsion power and in the structure of the inner bottom accommodating the engine seatings.

The effect on cost
A C_B variation changes the hull steel and propulsion system costs. Not only the steel weight, but also the price of the processed tonne of steel is variation-dependent. A tonne of processed steel of a ship with full C_B is relatively cheaper than that of a vessel with fine C_B. See also Carreyette formula, Carreyette (1978).

The specific costs of hull steel differ widely over the extent of the hull. We distinguish roughly the following categories of difficulty:

1. Flat areas with straight sections in the parallel middlebody.
2. Flat areas with straight sections not situated in the parallel middlebody, e.g. a piece of deck without sheer or camber at the ship's ends. More work results from providing an outline contour adapted to the outer shell and because the shortening causes the sections to change cross-section also.
3. Slightly curved areas with straight or curved sections. The plates are shaped locally using forming devices, not pre-bent. The curved sections are pre-formed.
4. Areas with a more pronounced curvature curved only in one direction, e.g. bilge strake in middlebody. The plates are rolled cold.
5. Medium-curved plates curved multidimensionally, e.g. some of those in the vicinity of the propeller aperture. These plates are pressed and rolled in various directions when cold.
6. Highly curved plates curved multidimensionally, e.g. the forward pieces of bulbous bows. These plates are pressed or formed when hot, or heat line bending used.

Decreasing C_B complicates design and construction, thus increasing costs:

1. More curved plates and sections, fewer flat plates with rectangular boundaries.
2. Greater expenditure on construction details.
3. Greater expenditure on wooden templates, fairing aids, gauges, etc.
4. More scrap.
5. More variety in plates and section with associated costs for storekeeping and management.

An increase in C_B by $\Delta C_B = 0.1$ will usually increase the share of the weight attributable to the flat areas of the hull (group (1) of the above groups) by 3%. About 3% of the overall hull steel will move from groups (3)–(5) to groups (1) and (2). The number of highly curved plates formed multidimensionally (group (6)) is hardly affected by a change in C_B. The change in weight of all curved plates and sections of the hull depends on many factors. It is approximately $0.33\Delta C_B \cdot$ hull steel weight.

9.6.5.4 Speed

The speed can be decisive for the economic efficiency of a ship and influences the main dimensions in turn. Since speed specifications are normally part of the shipping company requirements, the shipyard need not give the subject much consideration. Since only the agreement on trial speed, related to smooth water and full draught, provides both shipyard and shipping company with a clear contractual basis, the trial speed will be the normal basis for optimization. However, the service speed could be included in the optimization as an additional condition. If the service speed is to be attained on reduced propulsion power, the trial speed on reduced power will normally also be stated in the contract. Ships with two clearly defined load conditions can have both conditions considered separately, i.e. fully loaded and ballast.

Economic efficiency calculations for the purpose of optimizing speed are difficult to formulate due to many complex boundary conditions. Schedules in a transport chain or food preservation times introduce constraints for speed. (For both fish and bananas, for example, a preservation period of around 17 days is assumed.)

Speed variation may proceed on two possible assumptions:

1. Each ship in the variation series has *constant transportation capacity*, i.e. the faster variant has smaller carrying capacity.
2. Each ship in the variation series has a *constant carrying capacity*, i.e. the faster variant has a greater transportation capacity than the slower one and fewer ships are needed.

Since speed increase with constant carrying capacity increases the transportation capacity, and a constant transportation capacity leads to a change of ship size, it is better to compare the transport costs of 1 tonne of cargo for various ships on one route than to compare costs of several ships directly.

Essentially there are two situations from which an optimization calculation can proceed:

1. Uncompetitive situation. Here, speed does not affect income, e.g. when producer, shipping company and selling organizations are under the same ownership as in some areas of the banana and oil business.
2. Competitive situation. Higher speed may attract more cargo or justify higher freight rates. This is the prime reason for shipowners wanting faster ships. Both available cargo quantity and freight rate as a functions of speed are difficult to estimate.

In any case, all variants should be burdened with the interest on the tied-up capital of the cargo. For the uncompetitive situation where the shipowner transports his own goods, this case represents the real situation. In the competitive case, it should be a lower limit for attractiveness of the service. If the interest on cargo costs are not included, optimizations for dry cargo vessels usually produce speeds some 2 knots or more below normal.

Closely related with the question of optimum speed is that of port turnaround times. Shortening these by technical or organizational changes can improve the ship's profitability to a greater extent than by optimizing the speed.

Some general factors which encourage higher ship speeds are, Buxton (1972, 1976)

- High-value cargo.
- High freight rates.
- Competition, especially when freight rates are fixed as in Conferences.
- Short turn-around time.
- High interest rates.
- High daily operating costs, e.g. crew.
- Reduced cost of machinery.
- Improved hull form design, reduced power requirements.
- Smoother hulls, both new and in service, e.g. by better coatings.
- Cheap fuel.
- Lower specific fuel consumption.

9.6.6 Special cases of optimization
9.6.6.1 *Optimization of repeat ships*

Conditions for series shipbuilding are different from those for single-ship designs. Some of the advantages of series shipbuilding can also be used in repeat ships. For a ship to be built varying only slightly in size and output from a basis ship, the question arises: 'Should an existing design be modified or a new design developed?' The size can be changed by varying the parallel middlebody. The speed can be changed by changing the propulsion unit. The economic efficiency (e.g. yield) or the initial costs have to be examined for an optimum new design and for modification of an existing design.

The advantages of a repeat design (and even of modified designs where the length of the parallel middlebody is changed) are:

1. Reduced design and detailed construction work can save considerable time, a potentially crucial bargaining point when delivery schedules are tight.
2. Reduced need for jigs for processing complicated components constructed from plates and sections.
3. Greater reliability in estimating speed, deadweight and hold size from a basis ship, allowing smaller margins.
4. Greater accuracy in calculating the initial costs using a 'cost difference' method.

Where no smaller basis ship exists to fit the size of the new design, the objective can still be reached by shortening a larger basis ship. This reduces C_B. It may be necessary to re-define the midship area if more than the length of the parallel middlebody is removed. Deriving a new design from a basis ship of the same speed by varying the parallel middlebody is often preferable to developing a new design. In contrast, transforming a basis ship into a faster ship merely by increasing the propulsion power is economical only within very narrow limits.

Simplified construction of steel hull
Efforts to reduce production costs by simplifying the construction process have given birth to several types of development. The normal procedure employed in cargo shipbuilding is to keep C_B far higher than optimum for resistance. This increases the portion of the most easily manufactured parallel middlebody.

Blohm and Voss adopted a different method of simplifying ship forms. In 1967 they developed and built the *Pioneer* form which, apart from bow and stern bulbs, consisted entirely of flat surfaces. Despite 3–10% lower building costs, increased power requirement and problems with fatigue strength in the structural elements at the knuckles proved this approach to be a dead end.

Another simple construction method commonly used in inland vessels is to build them primarily or entirely with straight frames. With the exception of the parallel middlebody, the outer shell is usually curved only in one direction. This also increases the power requirement considerably.

Ships with low C_B can be simplified in construction – with only little increase in power requirement – by transforming the normally slightly curved surfaces of the outer shell into a series of curved and flat surfaces. The curved surfaces should be made as developable as possible. The flat surfaces can be welded fairly cheaply on panel lines. Also, there is less bending work involved. The difference between this and the *Pioneer* form is that the knuckles are avoided. C_B is lower than in the *Pioneer* class and conventional ships. Optimization calculations for simple forms are more difficult than for normal forms since often little is known about the hydrodynamic characteristics and building costs of simplified ship forms.

There are no special methods to determine the resistance of simplified ships, but CFD methods may bring considerable progress within the next decade. Far more serious is the lack of methods to predict the building costs by consideration of details of construction, Kaeding (1997).

9.6.6.2 Optimizing the dimensions of containerships

The width

The effective hold width of containerships corresponds to the hatch width. The area on either side of the hatch which cannot be used for cargo is often used as a wing tank. Naturally, the container stowage coefficient of the hold, i.e. the ratio of the total underdeck container volume to the hold volume, is kept as high as possible. The ratio of container volume to gross hold volume (including wing tanks) is usually 0.50–0.70. These coefficients do not take into account any partial increase in height of the double bottom. The larger ratio value applies to full ships with small side strip width and the smaller to fine vessels and greater side strip widths.

For constant C_B, a high container stowage coefficient can best be attained by keeping the side strip of deck abreast of the hatches as narrow as possible. Typical values for the width of this side strip on containerships are:

For small ships:	≈ 0.8–1.0 m
For medium-sized ships:	≈ 1.0–1.5 m
For larger ships:	≈ 1.2–2.0 m

The calculated width of the deck strip adjacent to the hatches decreases relative to the ship's width with increasing ship size. The variation in the figure also decreases with size.

If the ship's width were to be varied only in steps as a multiple of the container width, the statistics of the containership's width would indicate a stepped or discontinuous relationship. However, the widths are statistically distributed fairly evenly. The widths can be different for a certain container number stowed across the ship width, and ships of roughly the same width may even have a different container number stowed across the ship. The reason is that besides container stowage other design considerations (e.g. stability, carrying capacity, favourable proportions) influence the width of containerships. The difference between the continuous variation of width B and that indicated by the number and size of containers is indicated by the statistically determined variation in the wing tank width, typically around half a container width. The practical compromise between strength and construction considerations on the one hand and the requirement for good utilization on the other hand is apparently within this variation.

The length

The length of containerships depends on the hold lengths. The hold length is a 'stepped' function. However, the length of a containership depends not only on the hold lengths. The length of the fore peak may be varied to achieve the desired ship length. Whether the fore end of the hold is made longer or shorter is of little consequence to the container capacity, since the fore end of the hatch has, usually, smaller width than midships, and the hold width decreases rapidly downwards.

The depth

Similarly the depth of the ship is not closely correlated to the container height, since differences can be made up by the hatchway coaming height. The double bottom height is minimized because wing tanks, often installed to improve torsional rigidity, ensure enough tank space for all purposes.

Optimization of the main dimensions

The procedure is the same as for other ships. Container stowage (and thus hold space not occupied by containers) are included at a late stage of refining the optimization model. This subsequent variation is subject to, for example, stability constraints.

The basis variant is usually selected such that the stowage coefficient is optimized, i.e. the deck strips alongside the hatches are kept as narrow as possible.

If the main dimensions of the ship are now varied, given constant underdeck capacity and hold size, the number of containers to be stowed below deck will no longer be constant. So the main dimensions must be corrected. This correction is usually only marginal.

Since in slender ships the maximum hold width can only be fully utilized for a short portion of the length, a reduction in the number of containers to be stowed across the width of the midship section would only slightly decrease the number of containers. So the ratio of container volume to hold volume will change less when the main dimensions are varied on slender containerships than on fuller ships.

9.6.7 Developments of the 1980s and 1990s

9.6.7.1 *Concept exploration models*

Concept exploration models (CEMs) have been proposed as an alternative to 'automatic' optimization. The basic principle of CEMs is that of a direct search optimization: a large set of candidate solutions is generated by varying design variables. Each of these solutions is evaluated and the most promising solution is selected. However, usually all solutions are stored and graphically displayed so that the designer gets a feeling for how certain variables influence the performance of the design. It thus may offer more insight to the design process. However, this approach can quickly become impractical due to efficiency problems. Erikstad (1996) gives the following illustrating example: given ten independent design variables, each to be evaluated at ten different values, the total number of combinations becomes 10^{10}. If we assume that each design evaluation takes 1 millisecond, the total computer time needed will be 10^7 seconds – more than 3 months.

CEM applications have resorted to various techniques to cope with this efficiency problem:

- Early rejection of solutions not complying with basic requirements, Georgescu et al. (1990).
- Multiple steps methods where batches of design variables are investigated serially, Nethercote et al. (1981).
- Reducing the number of design variables, Erikstad, (1994).
- Increasing the step length.

Erikstad (1994) offers the most promising approach, which is also attractive for steepness search optimization. He presents a method to identify the most important variables in a given design problem. From this, the most influential set of variables for a particular problem can be chosen for further exploration in a CEM. The benefit of such a reduction in problem dimension, while keeping the focus on the important part of the problem, naturally increases rapidly with the dimension of the initial problem. Experience of the designer may serve as a short cut, i.e. select the proper variables without a systematic analysis, as proposed by Erikstad.

Among the applications of CEM for ship design are:

- A CEM for small warship design, Eames and Drummond (1977), based on six independent variables. Of the 82 944 investigated combinations, 278 were acceptable and the best 18 were fully analysed.
- A CEM for naval SWATH design, Nethercote *et al.* (1981), based on seven independent variables.
- A CEM for cargoship design, Georgescu *et al.* (1990), Wijnholst (1995), based on six independent variables.

CEM incorporating knowledge-based techniques have been proposed by Hees (1992) and Erikstad (1996), who also discuss CEM in more detail.

9.6.7.2 *Optimization shells*

Design problems differ from most other problems in that from case to case different quantities are specified or unknown, and the applicable relations may change. This concerns both economic and technical parts of the optimization model. In designing scantlings for example, web height and flange width may be variables to be determined or they may be given if the scantling continues other structural members. There may be upper bounds due to spatial limitations, or lower bounds because crossing stiffeners, air ducts, etc. require a structural member to be a certain height. Cut-outs, varying plate thickness, and other structural details create a multitude of alternatives which have to be handled. Naturally most design problems for whole ships are far more complex than the sketched 'simple' design problem for scantlings.

Design optimization problems require in most cases tailor-made models, but the effort of modifying existing programs is too tedious and complex for designers. This is one of the reasons why optimization in ship design has been largely restricted to academic applications. Here, methods of 'machine intelligence' may help to create a suitable algorithm for each individual design problem. The designer's task is then basically reduced to supplying:

- a list of specified quantities;
- a list of unknowns including upper and lower bounds and desired accuracy;
- the applicable relations (equations and inequalities).

In conventional programming, it is necessary to arrange relations such that the right-hand sides contain only known quantities and the left-hand side only one unknown quantity. This is not necessary in modern optimization shells. The relations may be given in arbitrary order and may be written in the most convenient way, e.g. $\nabla = C_B \cdot L \cdot B \cdot T$, irrespective of which of the variables are unknown and which are given. This 'knowledge base' is flexible in handling diverse problems, yet easy to use.

Such optimization shells include CHWARISMI, Söding (1977), and DELPHI, Gudenschwager (1988). These shells work in two steps. In the first step the designer compiles all relevant 'knowledge' in the form of relations. The shell checks if the problem can be solved at all with the given relations and which of the relations are actually needed. Furthermore, the shell checks if the system of relations may be decomposed into several smaller systems which can be solved independently. After this process, the modified problem is converted into a Fortran program, compiled and linked. The second step is then the actual numerical computation using the Fortran program.

The following example illustrates the concept of such an optimization shell. The problem concerns the optimization of a containership and is formulated for the shell in a quasi-Fortran language:

```
      PROGRAM CONT2
C Declaration of variables to be read from file
C TDW      t           deadweight
C VORR     t           provisions
C VDIEN    m/s         service speed
C TEU      -           required TEU capacity
C TUDMIN               share of
                       container capacity underdeck (<1.)
C NHUD                 number of bays under deck
C NHOD                 number of bays on deck
C NNUD                 number of stacks under deck
C NNOD                 number of stacks on deck
C NUEUD                number of tiers under deck
C MDHAUS   t           mass of deckhouse
C ETAD     -           propulsive efficiency
C BMST     t/m**3      weight coefficient for hull
C BMAUE    t/m**2      weight coefficient for E&O
C BMMA     t/kW        weight coefficient for engine
C BCST     DM/t        cost per ton steel hull
C BCAUE    DM/t        cost per ton E&O (initial)
C BCMA     DM/t        cost per ton engine (initial)
C
C Declaration of other variables
C LPP      m           length between perpendiculars
C BREIT    m           width
C TIEF     m           draft
C CB                   block coefficient
C VOL      m**3        displacement volume
C CBD                  block coefficient related to
                       main deck
C DEPTH    m           depth
C LR       m**3        hold volume
C TEUU                 number of containers under deck
C TEUO                 number of containers on deck
C NUEOD                number of tiers on deck
C GM       m           metacentric height
C PD       kW          delivered power
C MSTAHL   t           weight of steel hull
C MAUE     t           weight of E&O
C MMASCH   t           machinery weight
C CSCHIF   DM          initial cost of ship
C CZUTEU   DM/TEU      initial cost/carrying capacity

C
C Declare type of variables
      REAL BCAUE, BCMA, BCST, BMAUE, BMMA, BMST, ETAD,
      MDHAUS, REAL TEU, TDW, TUDMIN, VDIEN, VORR REAL
      NHOD, NHUD, NNOD, NNUD, NUEUD
C Input from file of required values
      CALL INPUT (BCAUE, BCMA, BCST, BMAUE, BMMA, BMST,
      ETAD, MDHAUS, & TDW, TEU, TUDMIN, VDIEN, VORR, NHOD,
      NHUD, NNOD, NNUD, NUEUD)
C unknowns      start    initial    lower    upper
C               value    stepsize   limit    limit
      UNKNOWNS LPP    (120.,     20.0,      50.0,     150.0),
      &        BREIT  (20.,       4.0,      10.0,      32.2),
      &        TIEF   (5.,        2.0,       4.0,       6.4),
      &        CB     (0.6,       0.1,       0.4,      0.85),
      &        VOL    (7200.,   500.0,    1000.0,   30000.0),
      &        CBD    (0.66,      0.1,        .5,      0.90),
      &        DEPTH  (11.,       2.0,       5.0,      28.0),
      &        LR     (12000.,  500.0,   10000.0,   50000.0),
      &        TEUU   (.5*TEU,   20.0,       0.0,      TEU ),
      &        TEUO   (.5*TEU,   20.0,       0.0,      TEU ),
      &        NUEOD  (2.,         .1,       1.0,       4.0),
      &        GM     (1.0,       0.1,       0.4,       2.0),
      &        PD     (3000.,   100.0,     200.0,   10000.0),
      &        MSTAHL (1440.,   100.0,     200.0,   10000.0),
      &        MAUE   (360.,     50.0,      50.0,    2000.0),
      &        MMASCH (360.,     50.0,      50.0,    2000.0).
      &        CSCHIF (60.E6,    1.E6,      2.E6,     80.E6),
      &        CZUTEU (30000.   5000.,    10000.,   150000.)
C **** Relations decribing the problem ****
C mass and displacement
      VOL        = LPP*BREIT*TIEF*CB
      VOL*1.03   = MSTAHL + MAUE + MMASCH 1 TDW
      MSTAHL     = STARUM (BMST,LPP,BREIT,TIEF,DEPTH,CBD)
      MAUE       = BMAUE*LPP*BREIT
      MMASCH     = BMMA*(PD/0.85)**0.89
C stability
      GM         = 0.43*BREIT - (MSTAHL*0.6*DEPTH
      &                         +MDHAUS*(DEPTH+6.0)
      &                         +MAUE*1.05*DEPTH
      &                         +MMASCH*0.5*DEPTH
      &                         +VORR*0.4*DEPTH
      &                         +TEUU*MCONT*
      &                         (0.743-0.188*CB)
      &                         +TEUO*MCONT*(DEPTH +
      &                         2.1+0.5*NUEOD*HCONT))
      &                         /VOL/1.03
C hold
      CBD    = CB+0.3*(DEPTH-TIEF)/TIEF*(1.2 CB)
      LR     = LPP*BREIT*DEPTH*CBD*0.75
C container stowing/main dimensions
      LPP    .GE. (0.03786+0.0016/CB**5)*LPP
      &           +0.747*PD**0.385
      &           +NHUD*(LCONT+1.0)
      &           +0.07*LPP
      LPP    .GE. 0.126*LPP+13.8
      &           +(NHOD-2.)*(LCONT+1.0)
      &           +0.07*LPP
      BREIT .GE. 2.*2.0+BCONT*NNUD+(NNUD+1.)*0.25
      BREIT .GE. 0.4+BCONT*NNOD+(NNUD-1)*0.04
      DEPTH .GE. (350+45*BREIT)/1000.+NUEUD*HCONT
      &           -1.5
      TEU    = TEUU+TEUO
      TEUU .GE. TUDMIN*TEU
      TEUU   = (0.9*CB+0.26) *NHUD*NNUD*NUEUD
      TEUO   = (0.5*CB+.55) *NHOD*NNOD*NUEOD
C propulsion
      PD     = VOL**0.567*VDIEN**3.6/(153.*ETAD)
C building cost
      CSCHIF    = BCST*MSTAHL*SQRT(.7/CB)
                  +BCAUE*MAUE + BCMA*MMASCH
      CZUTEU    = CSCHIF/(TEUU+TEUO)
C freeboard approximation
      DEPTH - TIEF . GE. 0.025*LPP
C L/D ratio
      LPP/DEPTH.GE.8.
      LPP/DEPTH.LE.14.
C Criterion: minimize initial cost/carried container
      MINIMIZE CZUTEU
      SOLVE
```

```
C Output
    CALL OUTPUT (LPP, BREIT, TIEF, CB, VOL, CBD, DEPTH,
    LR, TEUU, TEUO,NUEOD,
    &           GM,PD,MSTAHL, MAUE,MMASCH, CSCHIF,
                CZUTEU)
    END

    REAL FUNCTION STARUM (BMST, LPP,B,T,D,CBD)
C weight of steel hull following SCHNEEKLUTH, 1985
    REAL B, BMST, CBD, C1, D, LPP, T, VOLU
    VOLU = LPP*B*D*CBD
    C1 = BMST* (1.+0.2E-5*(LPP-120.)**2)
    . STARUM = VOLU*C1
    &           *(1.+0.057*(MAX(10.,LPP/D)-12.))
    &           *SQRT(30./(D-14.))
    &           *(1.+0.1*(B/D-2.1)**2)
    &           *(0.92+(1.-CBD)**2)
    END
```

The example shows that the actual formulation of the problem is relatively easy, especially since it can be based on existing Fortran procedures (steel weight in this example).

Even an optimization shell is not foolproof and errors occur frequently when beginners start using the shell. Not the least of the problems is that users formulate problems which allow no solution as improper constraints are imposed.

Another problem is that, in reality, many design problems are not so clearly defined. While there are, in principle, techniques to include uncertainty in the optimization (other than through sensitivity analyses), e.g. Schmidt (1996), extended functionality always comes at the price of added complexity for the user, which in our experience at present prevents acceptance.

Optimization shells of the future should try to extend functionality without sacrificing user-friendliness. Perhaps further incorporation of knowledge-based techniques, namely in formulating and interpreting results, could be the path to a solution. But even the most 'intelligent' system will not relieve the designer of the task to think and to decide.

References (Chapter 9)

Allen, H.G. (1969). *Analysis and Design of Structural Sandwich Panels*. Pergamon Press, Oxford.

American Bureau of Shipping (1998). Rules and Regulations for the Classification of Ships.

Andrews, D.J. (1981). Creative ship design. *Trans. RINA*, Vol. 123.

Andrews, D.J. (1998). A comprehensive methodology for the design of ships (and other complex systems). *Proceedings of The Royal Society*, Series A (1998) 454, January.

Andrews, D.J. (2007). The art and science of ship design. *Trans. RINA*, Vol. 149.

Andrews, D.J., Burger, D. and Zhang, J. (2005). Design for production using the design building block approach. *Trans. RINA*, Vol. 147.

Anon. (2003). *Paint terminology explained*. The Naval Architect, *RINA*, London.

Anon. (2005). *Paints and coatings technology*. The Naval Architect, *RINA*, London.

Backman, B. (2005). *Composite Structures, Design, Safety and Innovation*. Elsevier, Oxford, UK.

Benford, H. (1963). Principles of engineering economy in ship design. *Trans. SNAME*, Vol. 71.

Benford, H. (1965). *Fundamentals of ship design economics. Department of Naval Architects and Marine Engineers, Lecture Notes*. University of Michigan.

Bishop, R.E.D. and Price, W.G. (1979). *Hydroelasticity of Ships*. Cambridge University Press, Cambridge, UK.

Buxton, I.L. (1972). Engineering economics applied to ship design. *Trans. RINA*, Vol. 114.

Buxton, I.L. (1976). Engineering economics and ship design. British Ship Research Association report, 2nd edn.

Carreyette, J. (1978). Preliminary ship cost estimation. *Trans. RINA*, Vol. 120.

Clarke, S.D., Shenoi, R.A., Hicks, I.A. and Cripps, R.M. (1998). Fatigue characteristics for FRP sandwich structures of RNLI lifeboats. *Trans. RINA*, Vol. 140.

Cullinane, K. (ed.) (2005). *Shipping Economics: Research in Shipping Economics*, Vol. 12. Elsevier Oxford, UK.

Dodkins, A.R. (1993). In Composite Materials in Maritime Structures, eds. R. A. Shenoi and J. F. Wellicome, Cambridge, Ocean Technology Series, Cambridge University Press, Cambridge, UK, Vol. II, pp. 3–25.

Dodkins, A.R., Shenoi, R.A. and Hawkins, G.L. (1994). *Journal of Marine Structures*, Vol. 7, pp. 365–398.

Dow, R.S. and Bird, J. (1994). In Proceedings of the Conference on Structural Materials in Marine Environments, London, pp. 1–34.

Eames, M.C. and Drummond, T.G. (1977). Concept exploration – An approach to small warship design. *Trans. RINA*, Vol. 119.

Erichsen, S. (1989). *Management of Marine Design*. Butterworths.

Erikstad, S.O. (1994). Improving concept exploration in the early stages of the ship design process. *5th International Marine Design Conference*, Delft, p. 491.

Erikstad, S.O. (1996). A Decision Support Model for Preliminary Ship Design. Ph.D. thesis, University of Trondheim.

Eyres, D.J. (2007). *Ship Construction*, 6th edition. Butterworth-Heinemann, Oxford, UK.

Faltinsen, O.M. (1992). *Sea Loads on Ships and Offshore Structures, Cambridge, Ocean Technology Series*. Cambridge University Press, Cambridge, UK.

Fisher, K.W. (1972). Economic optimisation procedures in preliminary ship design (Applied to the Australian Ore Trade). *Trans. RINA*, Vol. 114.

Fried, N. (1967). *The potential of filament wound materials for the construction of deep submergent pressure hulls.* Conf. on Filament Winding, Plastics Institute, London.

Georgescu, C., Verbaas, F. and Boonstra, H. (1990). *Concept exploration models for merchant ships. CFD and CAD in Ship Design.* Elsevier Science Publishers, p. 49.

Gibson, A.G., Shenoi, R.A. and Wellicome, J.F. (eds) (1993). *Composite Materials in Maritime Structures, Cambridge, Ocean Technology Series*, Vol. II. Cambridge University Press, Cambridge, UK, pp. 199–228.

Gilfillian, A.W. (1969). The economic design of bulk cargo carriers. *Trans. RINA*, Vol. 111.

Gillmer, T.C. (1977). *Modern Ship Design.* Naval Institute Press, Annapolis, Maryland.

Glenn, D. (1985). Playing TAG. *Yachting World*, Vol. 137, Nov., p. 71.

Godwin, E.W. and Matthews, F.L. (1980). *Journal of Composites*, Vol. 11, No. 3, pp. 155–160.

Goss, R.O. (1965). Economic criteria for optimal ship design. *Trans. RINA*, Vol. 107.

Greene, E. (1997). Design Guide for Marine Applications of Composites, Ship Structure Committee Report SSC403, US Coast Guard, NTIS #PB98-111-651.

Gudenschwager, H. (1988). Optimierungscompiler und Formberechnungsverfahren: Entwicklung und Anwendung im Vorentwurf von RO/RO-Schiffen. IfS-Report 482, University of Hamburg.

Hayman, B., Haug, T. and Valsgôrd, S. (1991). In Proceedings of the 1st International Conference on Fast Sea Transportation, Trondheim, Norwegian Institute of Technology, Trondheim, Norway, pp. 55–67.

Hearmon, R.F. (1948). Elasticity of Wood and Plywood, Forest Products Special Report No. 7, HMSO, London.

Hees, M. Van. (1992). Quaestor: A knowledge-based system for computations in preliminary, ship design. PRADS'92, Newcastle, p. 21284.

Heller, S.R. and Jasper, N.H. (1960). *Transactions of The Royal Institution of Naval Architects*, Vol. 102, pp. 49–65.

Henry, J.J. and Karsch, H.J. (1966). Container ships. *Trans. SNAME*, Vol. 74.

IMO (1966). *International Conference on Load Lines, 1966* (2005 edition). IMO Publication (IMO-701E).

IMO (1969). International Conference on Tonnage Measurement of Ships, 1969. IMO Publication (IMO-713E).

IMO (2000). International code of safety for High Speed Craft. HSC code 2000. MSC 97/93.

IMO (2005). *Anti-fouling systems – International Convention on the Control of Harmful Anti-fouling Systems on Ships.* IMO Publication.

Janson, C.E. (1997). Potential Flow Panel Methods for the Calculation of Free-surface Flows with Lift; Ph.D. thesis, Gothenborg.

Jeong, H.K. and Shenoi, R.A. (2001). Structural reliability of fibre reinforced composite plates. *Trans. RINA*, Vol. 143.

Kaeding, P. (1997). Ein Ansatz zum Abgleich von Fertigungs- und Widerstandsaspekten beim Formentwurf. *Jahrbuch Schiffbautechn. Gesellschaft.*

Karayannis, T. and Molland, A.F. (2001). A decision making model for alternative high-speed ferries. *Proc. of Sixth International Conference on Fast Sea Transportation, FAST '2001*, Southampton, September 2001.

Karayannis, T. and Molland, A.F. (2003). Technical and economic investigations of fast ferry operations. *Proc. of Seventh International Conference on Fast Sea Transportation, FAST '2003*, Ischia, Italy, October.

Karayannis, T., Molland, A.F. and Williams, Y. Sarac. (1999). Design data for high speed vessels. *Proc. of Fifth International Conference on Fast Sea Transportation, FAST '99*, Seattle.

Keane, A.J., Price, W.G. and Schachter, R.D. (1991). Optimization techniques in ship concept design. *Trans. RINA*, Vol. 133, p. 123.

Kecsmar, J. and Shenoi, R.A. (2004). Some notes on the influence of manufacturing on the fatigue life of welded aluminium marine structures. *Journal of Ship Production*, Vol. 20, No. 3, August.

Kelly, A. and Zweben, C. (eds) (2000). *Comprehensive Composite Materials*, Vol. 6. Elsevier, Oxford, UK.

Kerlen, H. (1985). Über den Einfluß der Völligkeit auf die Rumpfstahlkosten von Frachtschiffen. IfS Rep. 456, University of Hamburg.

Kuo, C., MacCallum, K.J. and Shenoi, R.A. (1984). An effective approach to structural design for production. *Trans. RINA*, Vol. 126.

Lackenby, H. (1950). On the systematic variation of ship forms. *Trans. RINA.*

Lamb, T., Chung, H., Spicknall, M., Shin, J.G., Woo, J.H. and Koenig, P. (2006). Simulation-based performance improvement for shipbuilding processes. *Journal of Ship Production*, Vol. 22, No. 2, May. SNAME.

Lewis, E.V. (ed.) (1988). *Principles of Naval Architecture*, Vols I and II. The Society of Naval Architects and Marine Engineers, New York.

Lippay, A. and Levine, R.S. (1998a). In Proceedings of the Conference on Fishing Vessel Construction Materials, Montreal, pp. 41–50.

Liu, D., Hughes, O. and Mahowald, J. (1981). Applications of a computer-aided, optimal preliminary ship structural design method. *Trans. SNAME*, Vol. 89, p. 275.

Lloyd's Register of Shipping (1998a). Rules and Regulations for the Building and Classification of Ships.

Lloyd's Register of Shipping (1998b). Rules and Regulations for the Classification of Special Service Craft.

Lloyds Register (2004). *Rules and Regulations for the Classification of Ships*. Part 2, Rules for the Manufacture, testing and Certification of Materials.

Maccari, A. and Farolfi, F. (1992). In Proceedings of the International Conference on Nautical Construction with Composite Materials, Paris, pp. 39–47.

Malone, J.A., Little, D.E. and Allman, M. (1980). Effects of hull foulants and cleaning/coating practices on ship performance and economics. *Trans. SNAME*, Vol. 88, p. 75.

Malzahn, H., Schneekluth, H. and Kerlen, H. (1978). OPTIMA, Ein EDV-Programm für Probleme des Vorentwurfs von Frachtschiffen. Report 81, Forschungszentrum des Deutschen Schiffbaus, Hamburg.

Mandel, P. and Leopold, R. (1966). Optimisation methods applied to ship design. *Trans. SNAME*, Vol. 74.

Marshall, A. (1982). Sandwich construction. In G. Lubin (ed.), *Handbook of Composites*. Van Nostrand Reinhold, New York.

Matthews, F.L., Kilty, P.F. and Godwin, E.W. (1982). *Journal of Composites*, Vol. 13, No. 1, pp. 29–37.

Meek, M. (1970). The First OCL container ships. *Trans. RINA*, Vol. 112.

Meek, M., Adams, R., Chapman, J.C., Reibel, H. and Wieske, P. (1972). The structural design of the OCL container ships. *Trans. RINA*, Vol. 114.

Molland, A.F. (2005). *Ship Design and Economics*. Lecture Notes. School of Engineering Sciences, University of Southampton, UK.

Molland, A.F. and Karayannis, T. (1997). Concept Exploration and Assessment of Alternative High Speed Ferry Types. *Proc. of Fourth International Conference on Fast Sea Transportation, FAST '97*, Sydney.

Molland, A.F., Karayannis, T., Taunton, D.J. and Sarac-Williams, Y. (2003). Preliminary estimates of the dimensions, powering and seakeeping characteristics of fast ferries. *Eighth International Marine Design Conference, IMDC 2003*, Athens, Greece, May.

Munro-Smith, R. (1950). *Elements of Ship Design*. Marine Media Management Ltd.

Murphy, R.D., Sabat, D.J. and Taylor, R.J. (1965). Least cost ship characteristics by computer techniques. *Marine Technology* (SNAME), Vol. 2, No. 2.

Ness, D. and Whiley, D. (1991). Advanced Composites for High-Performance Marine Craft. *Marine Structures*.

Nethercote, W.C.E., Eng, P. and Schmitke, R.T. (1981). A concept exploration model for SWATH ships. *The Naval Architect*, p. 113.

Ochi, M.K. and Motter, L.E. (1973). *Transactions of the Society of Naval Architects and Marine Engineers*, Vol. 81, pp. 144–176.

Ochoa, O.O. and Reddy, J.N. (1992). *Finite Element Analysis of Composite Laminates*. Kluwer/Academic Press.

Pagano, N.J. (1968). *Journal of Composite Materials*, Vol. 3, pp. 398–409.

Papanikolaou, A. and Kariambas, E. (1994). Optimization of the preliminary design and cost evaluation of fishing vessel. *Schiffstechnik*, Vol. 41, p. 46.

Pegg, R.L. and Reyes, H. (1986). Composites promise Navy weight, tactical advantages. *Sea Technology*, July, p. 31.

Radojcic, D., Grigoropoulos, G.J., Rodic, T., Kuvelic, T. and Damala, D.P. (2001). The resistance and trim of semi-displacement, double chine, transom-stern hull series. *Proc. of Sixth International Conference on Fast Sea Transportation, FAST '2001*, Southampton, September.

Ray, T. and Sha, O.P. (1994). Multicriteria optimization model for containership design. *Marine Technology*, Vol. 31, No. 4, p. 258.

Reichard, R.P. (1986). Structural Design of Multihull Sailboats. *Proc. Internat. Conf. on Marine Applications of Composite Materials*, SNAME, Melbourne, Florida.

Roberts, M.L. and Smith, C.S. (1988). Design of submarine structures. *Proc. Int. Conf. on Undersea Defence Technology*, London, October.

Schmidt, D. (1996). Programm-Generatoren für Optimierung unter Berücksichtigung von Unsicherheiten in schiffstechnischen Berechnungen. IfS Rep. 567, University of Hamburg.

Schneekluth, H. (1957). Die wirtschaftliche Länge von Seefrachtschiffen und ihre Einfluß faktoren. *Schiffstechnik*, Vol. 13, p. 576.

Schneekluth, H. (1967). Die Bestimmung von Schiffslänge und Blockkoeffizienten nach Kostengesichtspunkten. *Hansa*, p. 367.

Schneekluth, H. and Bertram, V. (1998). *Ship Design for Efficiency and Economy*, 2nd edition. Butterworth-Heinemann, Oxford, UK.

Sen, P. (1992). Marine Design: The Multiple Criteria Approach. *Trans. of The Royal Institution of Naval Architects*, Vol. 134.

Serter, E. (1997). *Jane's International Defence Review*, Vol. 10, pp. 60–63.

Shenoi, R.A. and Hawkins, G.L. (1992). *Journal of Composites*, Vol. 23, No. 5, pp. 335–345.

Shenoi, R.A. and Wellicome, J.F. (eds) (1993). *Composite Materials in Marine Structures*, Vol. 1 and 2, Cambridge University Press, Cambridge, UK.

Shenoi, R.A. and Dodkins, A.R. (2000). Design of ships and marine structures made from FRP composite materials. In A. Kelly and C. Zweben (eds), *Comprehensive Composite Materials*, Vol. 6. Elsevier Science Ltd., Oxford, UK.

Shibley, A.M. (1982). Filament winding. In G. Lubin (ed), *Handbook of Composites*. Van Nostrand Reinhold, New York.

Sims, G.D. (1993). In Composite Materials in Maritime Structures, eds. R. A. Shenoi and J. F. Wellicome, Cambridge, Ocean Technology Series, Cambridge University Press, Cambridge, UK, Vol. 1, pp. 316–340.

Slobodzinsky, A. (1982). Bag molding processes. In G. Lubin (ed.), *Handbook of Composites*. Van Nostrand Reinhold, New York.

Smith, C.S. (1968). *Journal of Ship Research*, Vol. 12, pp. 249–270.

Smith, C.S. and Chalmers, D.W. (1987). Design of ship superstructures in fibre reinforced plastic. *Trans. RINA*, Vol. 129.

Smith, C.S. (1990). *Design of Marine Structures in Composite Materials*. Elsevier Science, Oxford, UK.

Söding, H. (1977). Ship design and construction programs (2). *New Ships*, Vol. 22/8, p. 272.

Stopford, M. (1997). *Maritime Economics*, 2nd edition. Routledge, London.

Swain, G.W., Kovach, B., Touzot, A., Casse, F. and Kavanagh, C.J. (2007). Measuring the performance of today's antifouling coatings. *Journal of Ship Production*, Vol. 23, No. 3, August. SNAME.

Taggart, R. (ed.) (1980). *Ship Design and Construction*. Publ. by SNAME, New York.

Torroja, J. and Alonso, F. (2000). Developments in computer aided ship design and production. *Trans. RINA*, Vol. 142.

Townsin, R.L., Byrne, D., Svensen, T.E. and Milne, A. (1981). Estimating the technical and economic penalties of hull and propeller roughness. *Trans. SNAME*, Vol. 89, p. 295.

Tucker, J.S. (1979). Glass reinforced plastic submersibles. *Trans. NEC Inst. Engrs & Shipbuilders*, Vol. 95, February No. 2.

Tupper, E.C. (2004). *Introduction to Naval Architecture*. Butterworth-Heinemann, Oxford, UK.

Vasiliev, V.V. and Morozov, E. (2007). *Advanced Mechanics of Composite Materials*, 2nd edition. Elsevier, Oxford, UK.

Watson, D.G.M. (1962). Estimating preliminary dimensions in ship design. *Trans. I.E.S.S.*

Watson, D.G.M. (1998). *Practical Ship Design*. Elsevier Science, Oxford, UK.

Watson, D.G.M. and Gilfillian, A.W. (1977). Some ship design methods. *Trans. RINA*, Vol. 119.

Whitfield, R.I., Duffy, A.H.B., Meehan, J. and Wu, Z. (2003). Ship product modelling. *Journal of Ship Production*, Vol. 19, No. 4, November. SNAME.

Wijnholst, N. (1995). *Design Innovation in Shipping*. Delft University Press.

Wilhelmi, G.F. and Schab, H.W. (1977). Glass reinforced plastic piping for shipboard applications. *Naval Engineers J.*, April.

Winkle, I.E. and Baird, D. (1986). Towards more effective structural design through synthesis and optimisation of relative fabrication costs. *Trans. RINA*, Vol. 128.

Wittman, C. and Shook, G.D. (1982). Hand lay-up techniques. In G. Lubin (ed.), *Handbook of Composites*. Van Nostrand Reinhold, New York.

Young, P.R. (1982). Thermoset matched die moulding. In G. Lubin (ed.), *Handbook of Composites*. Van Nostrand Reinhold, New York.

10 Underwater vehicles

Contents

10.1 Introduction
10.2 A bit of history
10.3 ROV design
10.4 ROV components
References (Chapter 10)

The various Sections of this Chapter have been taken from the following book, with the permission of the authors:

Christ, R.D. and Wernli, R.L. (2007). *The ROV Manual*. Butterworth-Heinemann, Oxford, UK.

10.1 Introduction

Underwater vehicles, in the broadest sense, cover manned and unmanned vehicles, Figure 10.1, with the unmanned vehicles being divided into autonomous underwater vehicles (AUVs) which are non-tethered, and remotely operated vehicles (ROVs) which are tethered. The manned versions include submarines and passenger carrying submersibles. The roles of the unmanned vehicles include the use of AUVs by oceanographers to map the features of the ocean and operators such as the oil and gas industry to map the seabed. ROVs are used for many purposes including underwater observation, exploration of the seabed, underwater construction and maintenance of subsea projects and underwater inspection and cleaning of ships' hulls.

Griffiths (2003) describes developments in AUV design, construction and operation, whilst Curtin *et al.* (2005) provide a review of trends in AUV development. Burcher and Rydill (1994) cover many of the design and performance characteristics of submarines, including propulsion and control which are also applicable to AUVs and a wider range of underwater vehicles.

This Chapter is extracted from Christ and Wernli (2007) and focuses on the design of ROVs, although much of what is said is often applicable to other classes of underwater vehicles. The properties of the oceans in which the underwater vehicles operate are included in Chapter 1, which addresses the properties of the marine environment as a whole.

10.2 A bit of history

10.2.1 Introduction

The strange thing about history is that it never ends. In the case of Remotely Operated Vehicles (ROVs), that history is a short one, but very important nonetheless, especially for the observation-class ROVs.

Two critical groups of people have driven ROV history: (1) Dedicated visionaries and (2) exploiters of technology. Those who drove the development of ROVs had a problem to solve and a vision, and they did not give up the quest until success was achieved. There were observation-class vehicles early in this history, but they were far from efficient. In time, however, the technology caught up with the smaller vehicles, and those who waited to exploit this technology have led the pack in fielding smaller, state-of-the-art ROVs.

This section will discuss what an ROV is, address some of the key events in the development of ROV technology, and address the breakthroughs that brought observation-class ROVs to maturity.

10.2.2 What is an ROV?

Currently, underwater vehicles fall into two basic categories (Figure 10.1): Manned Underwater Vehicles and Unmanned Underwater Vehicles (UUVs). The US Navy often uses the definition of UUV as synonymous with Autonomous Underwater Vehicles (AUVs), although that definition is not a standard across industry.

According to the US Navy's UUV Master Plan (2004 edition, Section 1.3), an 'unmanned undersea vehicle' is defined as a:

Self-propelled submersible whose operation is either fully autonomous (pre-programmed or real-time adaptive mission control) or under minimal supervisory control and is untethered except, possibly, for data links such as a fiber-optic cable.

The civilian moniker for an untethered underwater vehicle is the AUV, which is free from a tether and can run either a pre-programmed or logic-driven course, Griffiths (2003). The difference between the AUV and the ROV is the presence (or absence) of direct hard-wire communication between the vehicle and the surface. However, AUVs can also be linked to the surface for direct communication through an acoustic modem, or (while on the surface) via an RF (radio frequency) and/or optical link. But in this Chapter, we are concerned primarily with the ROV.

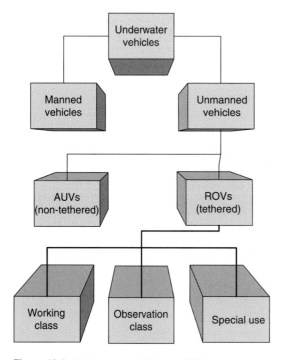

Figure 10.1 Underwater vehicles to ROVs.

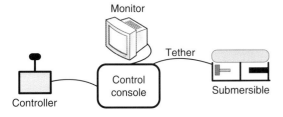

Figure 10.2 Basic ROV system components.

Simplistically, an ROV is a camera mounted in a waterproof enclosure, with thrusters for manoeuvring, attached to a cable to the surface over which a video signal is transmitted (Figure 10.2). Practically all of today's vehicles use common consumer industry standards for commercial off-the-shelf (COTS) components. The following section will provide a better understanding of the scope of this definition.

10.2.3 In the beginning

One way to discuss the historical development of ROVs is to consider them in terms of the cycle of life – from infancy to maturity. Anyone who has raised a child will quickly understand such a categorization. In the beginning the ROV child was 'nothing but a problem: Their bottles leaked, their hydraulics failed, sunlight damaged them, they were too noisy and unreliable, were hard to control and needed constant maintenance. Beginning to sound familiar?' (Wernli 1998).

Some have credited Dimitri Rebikoff with developing the first ROV – the *POODLE* – in 1953. However, the vehicle was used primarily for archaeological research and its impact on ROV history was minimal – but it was a start.

Although entrepreneurs like Rebikoff were making technology breakthroughs, it took the US Navy to take the first real step to an operational system. The Navy's problem was the recovery of torpedoes that were lost on the seafloor. Replacing a system that essentially grappled for the torpedo, the Navy (under a contract awarded to VARE Industries, Roselle, New Jersey) developed a manoeuvrable underwater camera system – a Mobile Underwater Vehicle System. The original VARE vehicle, the *XN-3*, was delivered to the Naval Ordnance Test Station (NOTS) in Pasadena, California, in 1961. This design eventually became the Cable-Controlled Underwater Research Vehicle (*CURV*).

The Navy's *CURV* (and its successor – *CURV III*) made national headlines twice:

- The *CURV* retrieved a lost atomic bomb off the coast of Palomares, Spain in 1966, from 2850 feet (869 metres) of water, even though working beyond its maximum depth. The *CURV*'s sister vehicle, *CURV II*, is shown in Figure 10.3.
- *CURV III*, which had become a 'flyaway' system, was sent on an emergency recovery mission from San Diego to a point offshore, near Cork, Ireland in 1973. With little air left for the two pilots of the *PISCES III* manned submersible, which was trapped on the bottom in 1575 feet (480 metres) of water, the *CURV III* attached a recovery line that successfully pulled the doomed crew to safety.

Figure 10.3 The Navy's *CURV II* vehicle.

With such successes under its belt, the Navy expanded into more complex vehicles, such as the massive Pontoon Implacement Vehicle (PIV), which was developed to aid in the recovery of sunken submarines, shown with the integrated Work Systems Package (WSP) (Figure 10.4).

At the other end of the scale, the US Navy developed one of the very first small-size observation ROVs. The *SNOOPY* vehicle, which was hydraulically operated from the surface, was one of the first portable vehicles (Figure 10.5(a)).

This version was followed by the *Electric SNOOPY*, which extended the vehicle's reach by going with a fully electric vehicle. Eventually sonars and other sensors were added and the childhood of the small vehicles had begun.

Navy-funded programs helped Hydro Products (San Diego, CA) get a jump on the ROV field through the development of the *TORTUGA*, a system dedicated to investigating the utility of a submarine-deployed ROV. These developments led to Hydro Products' RCV line of 'flying eyeball' vehicles (Figure 10.5(b)).

These new intruders, albeit successful in their design goals, still could not shake that lock on the market by the manned submersibles and saturation divers. In 1974, only 20 vehicles had been constructed, with 17 of those funded by various governments. Some of those included:

- France – *ERIC* and *Telenaute* and ECA with their *PAP* mine countermeasure vehicles.
- Finland – *PHOCAS* and Norway – the *SNURRE*
- UK – British Aircraft Corporation (*BAC-1*) soon to be the *CONSUB 01*; *SUB-2*, *CUTLET*
- Heriot-Watt University, Edinburgh – *ANGUS* (001, 002, and 003)
- Soviet Union – *CRAB-4000* and *MANTA* vehicles.

It could be said that ROVs reached adolescence, which is generally tied to a growth spurt, accented by bouts of unexplained or irrational behaviour, around 1975. With an exponential upturn, the number of vehicles grew to 500 by the end of 1982. And the funding line also changed during this period. From 1953 to 1974, 85% of the vehicles built were government funded. From 1974 to 1982, 96% of the 350 vehicles produced were funded, constructed, and/or bought by private industry.

The technological advancements necessary to take ROVs from adolescence to maturity had begun. This was especially true in the electronics industry, with the miniaturization of the onboard systems and their increased reliability. With the ROV beginning to be accepted by the offshore industry, other developers and vehicles began to emerge:

- USA – Hydro Products – the *RCV 125*, *TORTUGA*, *ANTHRO*, *AMUVS* were soon followed by the *RCV 225*, and eventually the *RCV 150*; AMETEK, Straza Division, San Diego – turned their Navy funded *Deep Drone* into their *SCORPIO* line; Perry Offshore, Florida – started their RECON line of vehicles based on the US Navy's *NAVFAC SNOOPY* design.

Figure 10.4 Co-author R. Wernli (right) directs the launch of the US Navy's WSP/PIV.

Underwater vehicles 733

Figure 10.5(a) US Navy's hydraulic *SNOOPY*.

Figure 10.5(b) Hydro Products' RCV 225 and RCV 150 vehicles.

- Canada – International Submarine Engineering (ISE) started in Canada (*DART*, *TREC*, and *TROV*).
- France – Comex Industries added the *TOM-300*, C.G. Doris produced the *OBSERVER* and *DL-1*.
- Italy – Gay Underwater Instruments unveiled their spherical *FILIPPO*.
- The Netherlands – Skadoc Submersible Systems' *SMIT SUB* and *SOP*.
- Norway – Myers Verksted's *SPIDER*.
- Sweden added SUTEC's *SEA OWL* and Saab-Scandia's *SAAB-SUB*.
- UK – Design Diving Systems' *SEA-VEYOR*, Sub Sea Offshore's *MMIM*, Underwater Maintenance

Co.'s *SCAN*, Underwater and Marine Equipment Ltd.'s *SEA SPY*, *AMPHORA*, and *SEA PUP*, Sub Sea Surveys Ltd.'s *IZE*, and Winn Technology Ltd.'s *UFO-300*, *BOCTOPUS*, *SMARTIE*, and *CETUS*.
- Japan – Mitsui Ocean Development and Engineering Co., Ltd. had the *MURS-100*, *MURS-300* and *ROV*.
- Germany – Preussag Meerestechnik's *FUGE*, and VFW-Fokker GmbH's *PIN-GUIN B3* and *B6*.
- Other US – Kraft Tank Co. (*EV-1*), Rebikoff Underwater Products (*SEA INSPECTOR*), Remote Ocean Systems (*TELESUB-1000*), Exxon Production Research Co. (*TMV*), and Harbor Branch Foundation (*CORD*).

From 1982 to 1989 the ROV industry grew rapidly. The first ROV conference, ROV '83, was held with the theme 'A Technology Whose Time Has Come!' Things had moved rapidly from 1970, when there was only one commercial ROV manufacturer. By 1984 there were 27. North American firms (Hydro Products, AMETEK, and Perry Offshore) accounted for 229 of the 340 industrial vehicles produced since 1975.

Not to be outdone, Canadian entrepreneur Jim McFarlane bought into the business with a series of low-cost vehicles – *DART*, *TREC*, and *TROV* – developed by International Submarine Engineering (ISE) in Vancouver, British Columbia. But the market was cut-throat; the dollar to pound exchange rate caused the ROV technology base to transfer to the UK in support of the Oil and Gas Operations in the North Sea. Once the dollar/pound exchange rate reached parity, it was cheaper to manufacture vehicles in the UK. Slingsby Engineering, Sub Sea Offshore, and the OSEL Group cornered the North Sea market and the once dominant North American ROV industry was soon decimated. The only North American survivors were ISE (due to their diverse line of systems and the can-do attitude of their owner) and Perry, which wisely teamed with their European competitors to get a foothold in the North Sea.

However, as the oil patch companies were fighting for their share of the market, a few companies took the advancements in technology and used them to shrink the ROV to a new class of small, reliable, observation-class vehicles. These vehicles, which were easily portable when compared to their larger offshore ancestors, were produced at a cost that civil organizations and academic institutions could afford.

The *MiniRover*, developed by Chris Nicholson, was the first real low-cost, observation-type ROV. This was soon followed by Deep Ocean Engineering's (DOE's) *Phantom* vehicles. Benthos (now Teledyne Benthos) eventually picked up the *MiniRover* line and, along with DOE, cornered the market in areas that included civil engineering, dam and tunnel inspection, police and security operations, fisheries, oceanography, nuclear plant inspection, and many others.

The 1990s saw the ROV industry reach maturity. Testosterone-filled ROVs worked the world's oceans; No job was too hard or too deep to be completed. The US Navy, now able to buy vehicles off the shelf as needed, turned its eyes toward the next milestone – reaching the 20 000-foot (6279 metres) barrier. This was accomplished in 1990, not once, but twice:

- *CURV III*, operated by Eastport International for the US Navy's Supervisor of Salvage, reached a depth of 20 105 feet (6128 metres).
- The Advanced Tethered Vehicle (*ATV*), developed by the Space and Naval Warfare Systems Center, San Diego broke the record less than a week later with a record dive to 20 600 feet (6279 metres).

It didn't take long and this record was not only beaten by Japan, but obliterated. Using JAMSTEC's *Kaiko* ROV (Figure 10.6) Japan reached the deepest point in the Mariana Trench – 35 791 feet (10 909 metres) – a record that can be tied, but never exceeded.

Figure 10.6 *Kaiko* – the world's deepest diving ROV.

Figure 10.7 Schilling Robotics UHD.

The upturn in the offshore oil industry is increasing the requirement for advanced undersea vehicles. Underwater drilling and subsea complexes are now well beyond diver depth, some exceeding 3000 metres deep. Due to necessity, the offshore industry has teamed with the ROV developers to ensure integrated systems are being designed that can be installed, operated, and maintained through the use of remotely operated vehicles. ROVs, such as Schilling Robotics UHD ROV (Figure 10.7), are taking underwater intervention to a higher technological level.

In the late 1990s, Wernli (1998), it was estimated that there were over 100 vehicle manufacturers, and over 100 operators using approximately 3000 vehicles of various sizes and capabilities. According to the 2006 edition of *Remotely Operated Vehicles of the World*, there are over 450 builders and developers of ROVs (including AUVs), and over 175 operators. As far as the number of actual vehicles in the field … well, tracking that number will be left to the statisticians. However, the number of small, observation-class vehicles being used today probably comes close to the all-inclusive 3000 vehicles worldwide in 1998, which provides a perfect segue into the next section.

critical missions. Whether performing dam inspections, body recoveries, fish assessment or treasure hunting, technology has allowed the development of the advanced systems necessary to complete the job – and stay dry at the same time.

Technology has moved from vacuum tubes, gear trains and copper/steel cables to microprocessors, magnetic drives and fibre-optic/Kevlar cables. That droopy-drawered infant has now graduated from college and can work reliably without constant maintenance. As computers have moved from trunk-sized 'portable' systems to those that fit in a pants pocket, observation-class ROVs have moved from portable, i.e. a team of divers can carry one, to handheld vehicles that can complete the same task. Examples of observation-class vehicles in use today are shown in Figure 10.8. Table 10.1 provides a summary of those vehicles that weigh in less than 70 kg (150 lb) and have 25 or more in the field.

The following sections will provide an overview of observation-class ROVs, related technologies, and words to the wise about how to – and how not to – use them to complete an underwater task. How well this is done will show up in the History Chapter of future publications.

10.2.4 Today's observation-class vehicles

One can spend a lifetime running numbers and statistics; However, it can be easily said that the technology is here to reach into the shallower depths in a cost-effective manner and complete a series of

10.3 ROV design

This Chapter describes the different types of underwater systems, the basic theory behind vehicle design/communication/propulsion/integration,

Figure 10.8 Authors Bob Christ (L) and Robert Wernli (R) with examples of today's observation class vehicles.

Table 10.1 Observation-class vehicles – small vehicles that weigh less than approx. 70 kg (150 lb) with over 25 sold.

Name	Company	Wt. (kg) in air	Depth (m)	Built
AC-ROV	AC-CESS CO, UK	3	75	75
Firefly	Deep Ocean Engineering, USA	5.4	46	30
H300	ECA Hytec, France	65	300	24
Hyball	SMD Hydrovision Ltd., UK	41	300	185
Little Benthic Vehicle (LBV)	SeaBotix, Inc., USA	10–15	150–1500	300
Navaho	Sub-Atlantic (SSA alliance), UK	42	300	35
Offshore Hyball	SMD Hydrovision Ltd., UK	60	300	50
Outland 1000	Outland Technology Inc., USA	17.7	152	39
Phantom 150	Deep Ocean Engineering, USA	14	46	27
Phantom XTL	Deep Ocean Engineering, USA	50	150	81
Prometeo	Elettronica Enne, Italy	48–55	–	36
RTVD-100MKIIEX	Mitsui, Japan	42	150	310*
Seaeye 600 DT	Seaeye Marine Ltd., UK	65	300	63
Seaeye Falcon	Seaeye Marine Ltd., UK	50	300–1000	72
Stealth	Shark Marine Technologies Inc., Canada	40	300	50+
VideoRay[†]	VideoRay LLC, US	4–4.85	0–305	550+

*Includes RTV-100.
[†]Includes Deep Blue, Explorer, Pro 3XE, and Pro 3XE GTO.
Information from: *Remotely Operated Vehicles of the World*, 7th edition, Clarkson Research Services Ltd., 2006/2007, ISBN 1-902157-75-3.

and explains the means by which a typical ROV gets everyday underwater tasks performed.

10.3.1 Underwater vehicles to ROVs

The previous section provided an introduction to what an ROV is and is not, along with a brief history of how these underwater robots arrived at their present level of worldwide usage. Since these vehicles are an extension of the operator's senses, the communication with them is probably the most critical aspect of vehicle design.

The communication and control of underwater vehicles is a complex issue and sometimes occludes the lines between ROV and AUV. Before addressing the focus of this Chapter, the issues involved will be investigated further.

The basic issues involved with underwater vehicle power and control can be divided into the following categories:

- Power source for the vehicle
- Degree of autonomy (operator controlled or program controlled)
- Communications linkage to the vehicle.

10.3.1.1 Power source for the vehicle

Vehicles can be powered in any of the following three categories: Surface-powered, vehicle-powered, or a hybrid system.

- *Surface-powered* vehicles must, by practicality, be tethered, since the power source is from the surface to the vehicle. The actual power protocol is discussed more fully later in this Chapter, but no vehicle-based power storage is defined within this power category.
- *Vehicle-powered* vehicles store all of their power-producing capacity on the vehicle in the form of a battery, fuel cell, or some other means of power storage needed for vehicle propulsion and operation.
- *Hybrid system* involves a mixture of surface and submersible supplied power. Examples of the hybrid system include the battery-powered submersible with a surface-supplied charger (through a tether) for recharging during times of less-than-maximum power draw; a surface-powered vehicle with an on-board power source for a transition from ROV to AUV (some advanced capability torpedo designs allow for swim-out under ship's power to transition to vehicle power after clearing the area) and other variations to this mix.

10.3.1.2 Degree of autonomy

According to the National Institute of Standards and Technology, Huang (2004), unmanned vehicles may be operated under several modes of operation, including fully autonomous, semi-autonomous, tele-operation, and remote control.

- *Fully autonomous* – A mode of operation of an unmanned system (UMS) wherein the UMS is expected to accomplish its mission, within a defined scope, without human intervention. Note that a team of UMSs may be fully autonomous while the individual team members may not be, due to the need to coordinate during the execution of team missions.
- *Semi-autonomous* – A mode of operation of a UMS wherein the human operator and/or the UMS plan(s) and conduct(s) a mission and requires various levels of human–robot interaction (HRI).
- *Tele-operation* – A mode of operation of a UMS wherein the human operator, using video feedback and/or other sensory feedback, either directly controls the motors/actuators or assigns incremental goals, waypoints in mobility situations, on a continuous basis, from off the vehicle and via a tethered or radio/acoustic/optic/other linked control device. In this mode, the UMS may take limited initiative in reaching the assigned incremental goals.
- *Remote control* – A mode of operation of a UMS wherein the human operator, without benefit of video or other sensory feedback, directly controls the actuators of the UMS on a continuous basis, from off the vehicle and via a tethered or radio-linked control device using visual line-of-sight cues. In this mode, the UMS takes no initiative and relies on continuous or nearly continuous input from the user.

10.3.1.3 Communications linkage to the vehicle

The linkage to the vehicle can come in several forms or methods depending upon the distance and medium through which the communication must take place. Such linkages include:

- Hard-wire communication (either electrical or fibre optic)
- Acoustic communication (via underwater analogue or digital modem)
- Optical communication (while on the surface)
- Radio frequency (RF) communication (while on or near the surface).

What is communicated between the vehicle and the operator can be any of the following:

- *Telemetry* – The measurement and transmission of data or video through the vehicle via tether, RF, optical, acoustic, or other means.
- *Tele-presence* – The capability of an unmanned system to provide the human operator with some amount of sensory feedback similar to that which the operator would receive if inside the vehicle.
- *Control* – The upload/download of operational instructions (for autonomous operations) or full tele-operation.
- *Records* – The upload/download of mission records and files.

ROVs receive their power, their data transmission or their control (or all three) directly from the surface through direct hard-wire communication (i.e. the tether). In short, the difference between an ROV and an AUV is the tether (although some would argue that the divide is not that simple).

10.3.1.4 *Special-use ROVs*

Some of the special-use remotely operated vehicles come in even more discriminating packages:

- *Rail cameras* – Work on the drilling string of oil and gas platforms/drilling rigs (a pan and tilt camera moving up and down a leg of the platform to observe operations at the drill head with or without intervention tooling).
- *Bottom crawlers* – Lay pipe as well as communications cables and such while heavily weighted in the water column and being either towed or on tracks for locomotion.
- *Towed cameras* – Can have movable fins that allow 'sailing' up or down (or side to side) in the water column behind the towing vehicle.
- *Swim-out ROVs* – Smaller free-swimming systems that launch from larger ROV systems.

Although this Chapter covers many of the technologies associated with all underwater vehicles, the subject matter will focus on the free-swimming, surface-powered, teleoperated (or semi-autonomous) observation-class remotely operated vehicle with submersible weights of less than 200 pounds (91 kg).

10.3.2 Autonomy plus: 'Why the tether?'

In order to illustrate where ROVs fit into the world of technology, an aircraft analogy will be discussed first, then the vehicle in its water environment.

Autonomy with regard to aerial vehicles runs the full gamut from man occupying the vehicle while operating it (e.g. a pilot sitting in the aircraft manipulating the controls for positive navigation) to artificial intelligence on an unmanned aerial vehicle making unsupervised decisions on navigation and operation from start to finish (Figure 10.9). However, where the human sits (in the vehicle or on a separate platform) is irrelevant to the autonomy discussion, since it does not effect how the artificial brain (i.e. the controller) thinks and controls.

10.3.2.1 *An aircraft analogy*

To set the stage with an area most are familiar with, the control variations of an aircraft will be defined as follows:

- *Man in vehicle* – Pilot sitting aboard the aircraft in seat manning controls.
- *Man in vehicle with AutoPilot* – Pilot sitting aboard aircraft in seat with AutoPilot controlling the aircraft's navigation (pilot supervising the systems).
- *Man in remote location with tele-operation* – Technician sitting in front of control console on the ground (or another aerial platform) with radio frequency link to the unmanned aircraft while the technician is manipulating the controls remotely.
- *Vehicle operating with artificial intelligence and full autonomy* – No human supervisor directly controlling the vehicle. The vehicle controls are pre-programmed with the vehicle making objective decisions as to the conduct of that flight from inception to termination.

Predator UAVs (unmanned aerial vehicles) have recently been retrofitted with weapons, producing the new designation UCAV (unmanned combat aerial vehicle). The most efficient technology would allow that UCAV (without human intervention) to find, detect, classify, and deliver a lethal weapon upon the target, thus eliminating the threat. And here is the crux – is a responsible commander in the field comfortable enough with the technology to allow a machine the decision of life and death? This may be an extreme example, but (for now) a human must remain in the decision loop. To continue this example, would any passenger (as a passenger) fly in a commercial airliner without a pilot physically present in the cockpit? This may not happen soon, but one can safely predict that unpiloted airliners are in our future. Unattended trolleys are currently used in many airports worldwide.

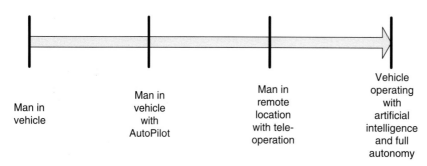

Figure 10.9 Degrees of autonomy.

10.3.2.2 Underwater vehicle variations

Now, the aircraft analogy will be reconsidered with underwater vehicle control in mind.

- *Man in vehicle* – Manned submersible pilot sitting aboard the vehicle underwater in the pilot's seat manning the controls.
- *Man in vehicle with AutoPilot* – Same situation with AutoPilot controlling the submersible's navigation (pilot supervising the systems).
- *Man in remote location with tele-operation* – Technician sitting in front of control console on the surface (or other submerged platform) with tether or other data link to the submersible while the technician is manipulating the controls remotely.
- *Vehicle operating with artificial intelligence and full autonomy* – No human supervisor directly controlling the vehicle. The vehicle controls of the Autonomous Underwater Vehicle (AUV) are pre-programmed, with the vehicle making objective decisions as to the conduct of that dive from inception to termination.

During operation Iraqi Freedom, Mine Countermeasure (MCM) AUVs were used for mine clearance operations. The AUV swam a pre-programmed course over a designated area to search and detect mine-like objects on the bottom. Other vehicles (or marine mammals or divers) were then sent to these locations to classify and (if necessary) neutralize the targets.

The new small UUVs are going through a two-stage process where they Search (or survey), Classify, and Map. The Explosive Ordinance Disposal (EOD) personnel then return with another vehicle (or marine mammal or human divers) to Reacquire, Identify and Neutralize the target. Essentially, the process is to locate mine-like targets, classify them as mines if applicable, then neutralize them. What if the whole process can be done with one autonomous vehicle? And again the crux – is the field commander comfortable enough with the vehicle's programming to allow it to distinguish between a Russian KMD-1000 Bottom-type influence mine and a manned undersea laboratory before destroying the target? For the near term, man will remain in the decision loop for the important operational decisions. But again, one can safely predict that full autonomy is in our future.

10.3.2.3 Why the tether?

Radio frequency (RF) waves penetrate only a few wavelengths into water due to water's high attenuation of its energy. If the RF is of a low frequency, the waves will penetrate further into water due to longer wavelengths. But with decreasing RF frequencies, data transmission rates suffer. In order to perform remote inspection tasks, live video is needed at the surface so that decisions by humans can be made on navigating the vehicle and inspecting the target. Full tele-operation (under current technology) is possible only through a high-bandwidth data link.

With the UAV example above, full tele-operation was available via the RF link (through air) between the vehicle and the remote operator. In water, this full telemetry is not possible (with current technologies) through an RF link. Acoustic in-water data transmission (as of 2006) is limited to less than 100 kilobytes per second (insufficient for high-resolution video images). A hard-wire link to the operating platform is needed to have a full tele-operational in-water link to the vehicle. Thus, the need exists for a hard-wire link of some type, in the foreseeable future, for real-time underwater inspection tasks.

10.3.2.4 Tele-operation versus remote control

An ROV pilot will often operate a vehicle remotely with his/her eyes directly viewing the vehicle while guiding the vehicle on the surface to the inspection target. This navigation of the vehicle through line of sight (as with the remote control airplane) is termed Remote Control (RC) mode. Once the inspection target is observed through the vehicle's camera or sensors, the transition is made from RC operational mode to tele-operation mode. This transition is important because it changes navigation and operation of the vehicle from the operator's point of view to the vehicle's point of view. Successful management of the transition between these modes of operation during field tasks will certainly assist in obtaining a positive mission completion.

Going back to the UAV analogy, many kids have built and used RC model aircraft. The difference between an RC aircraft and a UAV is the ability to navigate solely by use of onboard sensors. A UAV can certainly be operated in an RC mode while the vehicle is within line-of-sight of the operator's platform, but once line-of-sight is lost, navigation and control are only available through tele-operation or pre-programming.

The following is an example of this transition while performing a typical observation-class ROV inspection of a ship's hull: The operator swims the vehicle on the surface (Figure 10.10) via RC to the hull of the vessel until the inspection starting point is gained with the vehicle's camera, then transitions to navigation via the vehicle's camera.

10.3.2.5 Degrees of autonomy

An open-loop control system is simply a condition on a functioning machine whereby the system has

Figure 10.10 Surface swim in RC mode.

two basic states: 'On' or 'Off'. The machine will stay On/Off for as long as the operator leaves it in that mode. The term 'open-loop' (or essentially 'no loop') refers to the lack of sensor feedback to control the operation of the machine. An example of an open-loop feedback would be a simple light switch that, upon activation, remains in the 'On' or 'Off' condition until manually changed.

Beginning with pure tele-operation (which is no autonomy), the first step toward full autonomy is the point at which the vehicle begins navigation autonomously within given parameters. This is navigation through 'closed-loop feedback'.

Closed-loop feedback is simply control of an operation through sensor feedback to the controller. A simple example of a closed-loop feedback system is the home air-conditioning thermostat. At a given temperature, the air-conditioner turns on, thus lowering the temperature of the air surrounding the thermostat (if the air-conditioning is ducted into that room). Once the air temperature reaches a certain pre-set value, the thermostat sends a signal to the air-conditioner (closing the control signal and response loop) to 'turn off', completing this simple closed-loop feedback system.

The most common first step along this line for the ROV system is the auto heading and auto depth functions. Any closed-loop feedback control system can operate on an ROV system, manipulating control functions based upon sensor output. Operation of the vertical thruster as a function of constant depth (as measured by the variable water pressure transducer) is easily accomplished in software to provide auto depth capability. For example, consider an auto depth activation system on an ROV at 100 feet (30 metres) of seawater. The approximate (gauge) pressure is 3 atmospheres or 45 psig (3 bar). As the submersible sinks below that pressure (as read by the pressure transducer on the submersible), the controller switches on the vertical thruster to propel the vehicle back toward the surface until the 45 psig (3 bar) reading is reacquired (the reverse is also applicable).

The same applies to auto altitude, where variation of the vertical thruster maintains a constant height off the bottom based upon echo soundings from the vehicle's altimeter. Similarly, auto standoff from the side of a ship for hull inspections can be based upon a side-looking acoustic sensor, where variation of the sounder timing can be used to vary the function of the lateral thruster.

Any number of closed-loop variables can be programmed. The submersible can then be given a set of operating instructions based upon a matrix of 'if/then' commands to accomplish a given mission. The autonomy function is a separate issue from communications. A tethered ROV can be operated in full autonomy mode just as an untethered AUV may be operated in full autonomy mode. The only difference between a fully autonomous ROV and a fully autonomous AUV is generally considered to be the presence/absence of a hard-wire communications link, i.e. a tethered AUV is actually an ROV.

10.3.3 The ROV

The following sections will address the most critical areas of ROV design and operational considerations.

10.3.3.1 *What is the perfect ROV?*

This section, and those to follow, will investigate the premise that vehicle geometry does not affect the motive performance of an ROV (over any appreciable tether length) nearly as much as the dimensions of the tether.

Accordingly, the perfect ROV would have the following characteristics:

- Minimal tether diameter (for instance, a single strand of unshielded optical fibre)
- Powered from the surface having unlimited endurance (as opposed to battery operated with limited power available)

- Very small in size (to work around and within structures)
- Have an extremely high data pipeline for sensor throughput.

ROV systems are a trade-off of a number of factors, including cost, size, deployment resources/platform, and operational requirement. But the bottom line, which will become obvious in the following sections, is that the tether design can help create, or destroy, the perfect ROV.

10.3.3.2 ROV classifications

ROV systems come in three basic categories:

- *Observation class* – Observation-class ROVs are normally a 'flying eye' designed specifically for lighter usage with propulsion systems to deliver a camera and sensor package to a place where it can provide a meaningful picture or gather data. The newer observation-class ROVs enable these systems to do more than just see. With its tooling package and many accessories, the observation-class ROV is able to deliver payload packages of instrumentation, intervention equipment, and underwater navigational aids, enabling them to perform as a full-function underwater vehicle.
- *Work class* – Work-class systems generally have large frames (measured in multiple yards/metres) with multi-function manipulators, hydraulic propulsion/actuation, and heavy tooling meant for larger underwater construction projects where heavy equipment underwater needs movement.
- *Special use* – Special-use ROV systems describe tethered underwater vehicles designed for specific purposes. An example of a special-use vehicle is a cable burial ROV system designed to plow the sea floor to bury telecommunications cables.

10.3.3.3 Size considerations

(a) Vehicle size versus task suitability
As discussed in the previous Section, ROVs range from the small observation-class to large, complex work-class vehicles. The focus of this Chapter is on the smaller observation-class systems. The various sizes of observation-class ROV systems have within them certain inherent performance capabilities (Figure 10.11). The larger systems have a higher payload and thruster capability, allowing better open-water operations. The smaller systems are much more agile in getting into tight places in and around underwater structures, making them more suitable for enclosed structure penetrations. The challenge for the ROV technician is to find the right system for the job.

Table 10.2 provides examples of tasks along with the 'best' system size selection.

(b) The ROV crew
For a larger work-class system, the ROV crew consists of a supervisor as well as two or more team members possessing specialized knowledge in areas such as mechanical or electrical systems. On the smaller systems, the supervisor does not normally have the range of personnel due to costing issues. For deployment of an observation-class ROV system, the ROV crew should comprise, at the very least, an operator as well as a tether tender. For performing an inspection task, it is quite helpful to also have a third person to take prolific notes as well as a second set of eyes to view the job for content and completeness.

(c) Platform or vessel of opportunity
The operations platform for larger ROV systems could range from a drilling rig deck to the moon pool of a specialized, dynamically positioned Diver Support Vessel (DSV) outfitted specifically for ROV operations. With observation-class systems, any number of work platforms may be used depending upon the work environment and the equipment being deployed. As a minimum, the work platform will require the following characteristics:

1. A water ingress point footprint large enough to deploy the submersible safely.
2. A comfortable platform from which to operate the video and electronics console.
3. A direct line of communication from the ROV pilot to the operator of the mobile deployment platform.
4. A sufficient power supply to run all equipment for the duration of the operation.

10.3.3.4 Buoyancy and stability

As discussed further in Section 10.4, any vehicle has movement about six degrees of freedom (Figure 10.12). Three translations (surge, heave, and sway along the longitudinal, vertical, and transverse (lateral) axes respectively) and three rotations (roll, yaw, and pitch about these same respective axes). This Section will address the interaction between vehicle static and dynamic stability and these degrees of freedom.

ROVs are not normally equipped to pitch and roll. The system is constructed with a high centre of buoyancy and a low centre of gravity to give the camera platform maximum stability about the longitudinal and lateral axes (Figure 10.13). Most small ROV systems have fixed ballast with variable positioning to allow trimming of the system nose-up/nose-down or for roll adjustment/trim. In

Figure 10.11 Various system sizes of observation-class ROV systems.

Table 10.2 Task versus ROV system size.

Tasking	Best size
External pipeline inspection	Large
External hull inspection	Medium/large
Internal wreck survey	Small
Open-water scientific transect	Medium/large
Calm-water operations	All sizes

the observation class, the lead (or heavy metal) ballast is normally located on tracks attached to the bottom frame to allow movement of ballast along the vehicle to achieve the desired trim.

(A) Hydrostatic equilibrium

According to Archimedes' principle, any body partially or totally immersed in a fluid is buoyed up by a force equal to the weight of the displaced fluid. If somehow one could remove the body and instantly fill the resulting cavity with fluid identical to that surrounding it, no motion would take place: The body weight would exactly equal that of the displaced fluid, see Sections 3.1 and 3.2.

The resultant of all of the weight forces on this displaced fluid (Figure 10.14) is centred at a point within the body termed the 'centre of gravity' (CG) (see Figures 10.13 and 10.15). This is the sum of all the gravitational forces acting upon

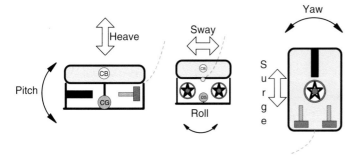

Figure 10.12 Vehicle degrees of freedom.

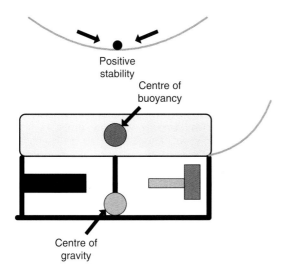

Figure 10.13 Positive ROV stability.

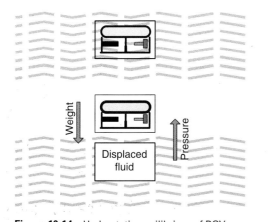

Figure 10.14 Hydrostatic equilibrium of ROV.

the body by gravity. The resultant of the buoyant forces countring the gravitational pull acting upward through the CG of the displaced fluid is termed the 'centre of buoyancy' (CB). There is one variable in the stability equation that is valid for surface vessels with non-wetted area that is not considered for submerged vehicles. The point where the CB intersects the hull centreline is termed the metacentre and its distance from the CG is termed the metacentric height (usually written as GM). For ROV considerations, all operations are with the vehicle submerged and ballasted very close to neutral buoyancy, making only the separation of the CB and the CG the applicable reference metric for horizontal stability.

Per Van Dorn (1993):

> *The equilibrium attitude of the buoyant body floating in calm water is determined solely by interaction between the weight of the body, acting downward through its CG, and the resultant of the buoyant forces, which is equal in magnitude to the weight of the body and acts upward through the CB of the displaced water. If these two forces do not pass through the same vertical axis, the body is not in equilibrium, and will rotate so as to bring them into vertical alignment. The body is then said to be in static equilibrium.*

(B) Transverse stability

Paraphrasing Van Dorn (1993) to account for ROVs, having located the positions of the CG and the upright CB of the vehicle, one can now investigate the transverse (lateral) stability. This is done without regard to external forces, merely by considering the hull of the vehicle to be inclined through several angles and calculating the respective moments exerted by the vertically opposing forces of gravity and buoyancy. These moments are generated by horizontal displacements (in the vehicle's reference frame) of the CB relative to the CG, as the vehicle inclines, such that these forces are no longer collinear, but are separate by same distance d, which is a function of the angle of inclination, Θ. The magnitude of both forces remains always the same, and equal to the vehicle's weight W,

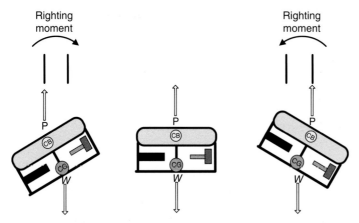

Figure 10.15 Righting forces acting on an ROV.

but their moment ($W \times d$) is similarly a function of Θ. If the moment of the buoyancy (or any other) force acts to rotate the vehicle about its CG opposite to the direction of inclination, it is called a righting moment; If in the same direction, it is called a heeling moment.

Referring to Figure 10.16, as the BG becomes smaller, the righting moment decreases in a logarithmic fashion until static stability is lost.

See also Chapter 3 for a description and discussion of stability and Section 8.18 for the stability of submarines.

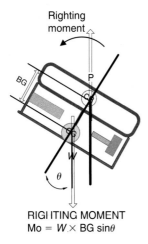

RIGHTING MOMENT
$Mo = W \times BG \sin\theta$

Figure 10.16 ROV righting moment.

(C) Water density and buoyancy

It is conventional operating procedure to have vehicles positively buoyant when operating to ensure they will return to the surface if a power failure occurs. This positive buoyancy would be in the range of 1 lb (450 grams) for small vehicles and 11–15 lb (5–7 kg) for larger vehicles, and in some cases, work-class vehicles will be as much as 50 lb (23 kg) positive. Another reason for this is to allow near-bottom manoeuvring without thrusting up, forcing water down, thus stirring up sediment. It also obviates the need for continual thrust reversal. Very large vehicles with variable ballast systems that allow for subsurface buoyancy adjustments are an exception. The vehicles operated by most observation-class operators will predominantly have fixed ballast.

As more fully described in Chapter 1, the makeup of the water in which the submersible operates will determine the level of ballasting needed to properly operate the vehicle. The three major water variables affecting this are temperature, salinity, and pressure.

More than 97% of the world's water is located in the oceans. Many of the properties of water are modified by the presence of dissolved salts. The level of dissolved salts in sea water is normally expressed in grams of dissolved salts per kilogram of water (historically expressed in imperial units as parts per thousand or 'PPT' with the newer accepted unit as the practical salinity unit or 'PSU'). Open ocean seawater contains about 35 PSU of dissolved salt. In fact, 99% of all ocean water has salinity of between 33 and 37 PSU.

Pure water has a specific gravity of 1.00 at maximum density temperature of about 4°C (approximately 39°F). Above 4°C, water density decreases due to molecular agitation. Below 4°C, ice crystals begin to form in the lattice structure, thereby decreasing density until the freezing point. It is well known that ice floats, demonstrating the fact that its density is lower than water.

At a salt content of 24.7 PSU, the freezing point and the maximum density temperature of sea water coincide at −1.332°C. In other words, with salt content above 24.7 PSU, there is no maximum density of sea water above the freezing point.

Most ROVs have a fixed volume. When transferring a submersible from a freshwater environment (where the system was neutrally ballasted) to a higher density salt-water environment, the ROV pilot will notice that the system demonstrates a more positive buoyancy, much like an ice cube placed into a glass of water. In order to neutralize the buoyancy of the system, ballast weights will need to be added to the submersible until neutral buoyancy is re-achieved. The converse is also true by going from salt water into fresh water or between differing temperature/salinity combinations with the submersible (Figure 10.17).

Water is effectively (for our purposes) incompressible. At deeper depths, water will be at a higher density, slightly affecting the buoyancy of the submersible. The water density buoyancy shift at the deeper operating depths is partially offset due to the compression of the air-filled spaces of the submersible. This balance is more or less dependent upon the system design and the amount of air-filled space within the submersible.

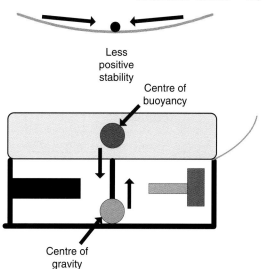

Figure 10.18 ROV with ballast moved up.

10.3.3.5 Dynamic stability

As with a child's seesaw, the further a weight is placed from the fulcrum point, the higher the mechanical force, or moment, needed to 'upset' that weight (the term 'moment' is computed by the product of the weight times the arm or distance from the fulcrum). It is called 'positive stability' when an upset object inherently rights itself to a steady state. When adapting this to a submersible, positive longitudinal and lateral stability can be readily achieved by having weight low and buoyancy high on the vehicle. This technique produces an intrinsically stable vehicle on the pitch and roll axis.

In most observation-class ROV systems, the higher the stability the easier it is to control the vehicle. With lower static stability, expect control problems (Figure 10.18).

External forces, however, do act upon the vehicle when it is in the water, which can produce apparent reductions in stability. For example, the force of the vertical thruster when thrusting down appears to the vehicle as an added weight high on the vehicle and, in turn, makes the centre of gravity appear to rise, which destabilizes the vehicle in pitch and roll. The centre of buoyancy and centre of gravity can be calculated by taking moments about some arbitrarily selected point.

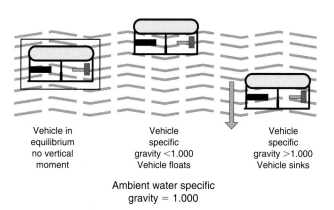

Figure 10.17 Effect of water specific gravity on ROV buoyancy.

746 Maritime engineering reference book

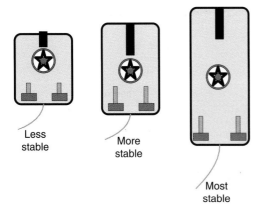

Figure 10.19 Vehicle geometry and stability.

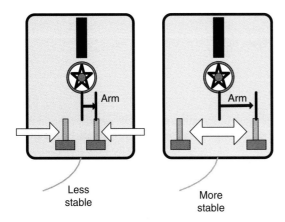

Figure 10.20 Thruster placement and stability.

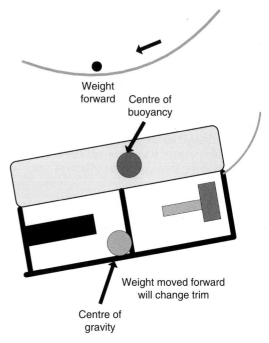

Figure 10.21 Vehicle trim with weight forward.

Other design characteristics also affect the stability of the vehicle along the varying axis. The so-called 'aspect ratio' (total mean length of the vehicle versus total mean width of the vehicle) will determine the vehicle's hull stability (Figure 10.19), as will thruster placement (Figure 10.20).

Most attack submarine designers specify a 7:1 aspect ratio as the optimum for the manoeuvring-to-stability ratio, see Moore and Compton-Hall (1987). For ROVs, the optimal aspect ratio and thruster placement will be dependent upon the anticipated top speed of the vehicle, along with the need to manoeuvre in confined spaces.

(1) Mission-related vehicle trim
Two examples of operational situations where observation-class ROV trim could be adjusted to assist in the completion of the mission follow:

1. If an ROV pilot requires the vertical viewing of a standpipe with a camera tilt that will not rotate through 90°, the vehicle may be trimmed to counter the lack in camera mobility (Figure 10.21).
2. If the vehicle is trimmed in a bow-low condition while performing a transect or a pipeline survey (Figure 10.22), when the thrusters are operated the vehicle will tend to drive into the bottom, requiring vertical thrust (and stirring up silt in the process). The vehicle ballast could be moved aft to counter this condition.

(2) Point of thrust/drag
Another critical variable in the vehicle control equation is the joint effect of both the point of net thrust (about the various axis) and the point of effective total drag.

The drag perspective will be considered first. One can start with the perfect drag for a hydrodynamic body (like an attack submarine) then work toward some practical issue of manufacturing a small ROV.

As stated best in Burcher and Rydill (1994), there are two basic types of drag with regard to all underwater bodies:

1. *Skin friction drag* – Friction drag is created by the frictional forces acting between the skin and the water. The viscous shear drag of water flowing tangentially over the surface of

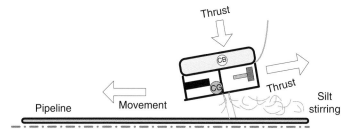

Figure 10.22 Movement down a pipeline with vehicle out of trim.

the skin contributes to the resistance of the vehicle. Essentially this is related to the exposed surface area and the velocities over the skin. Hence, for a given volume of vehicle hull, it is desirable to reduce the surface area as much as possible. However, it is also important to retain a smooth surface, to avoid roughness and sharp discontinuities, and to have a slowly varying form so that no adverse pressure gradients are built up, which cause increased drag through separation of the flow from the vehicle's hull.

2. *Form drag* – A second effect of the viscous action of the vehicle's hull is to reduce the pressure recovery associated with non-viscous flow over a body in motion. Form drag is created as the water is moved outward to make room for the body and is a function of cross-sectional area and shape. In an ideal non-viscous flow there is no resistance since, although there are pressure differences between the bow and stern of the vehicle, the net result is a zero force in the direction of motion. Due to the action of viscosity there is reduction in the momentum of the flow and, whilst there is a pressure build up over the bow of the submersible, the corresponding pressure recovery at the stern is reduced, resulting in a net resistance in the direction of motion. This form drag can be minimized by slowly varying the sections over a long body, i.e. tending toward a needle-shaped body even though it would have a high surface-to-volume ratio.

Floating bodies will also have wave resistance. A full description of the components of ship resistance is given in Chapter 5.

As shown in Figure 10.23, there is an optimum aspect ratio whereby the total drag formed from both form drag and skin friction is minimized. Assuming a smoothly shaped contour forming a cylindrical hull, that aspect is somewhere in the range of a 6:1 aspect ratio (length-to-diameter ratio). The practicalities of building a cost-effective underwater vehicle (including the engineering headaches of procuring and forming constantly changing form factors) always get in the

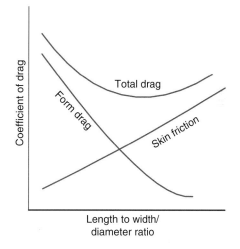

Figure 10.23 Vehicle drag curves.

way of obtaining the perfect underwater design. Figure 10.24 shows the ideal submarine form, then a slightly modified form factor popular in the defence industry. From this perfect form, the various aspects of the drag computation can be analysed.

Skin friction drag

The drag dynamics of submerged vehicles were worked out during manned submersible research done in the 1970s by the office of the Oceanographer of the Navy. According to Busby (1976):

Skin friction is a function of the viscosity of the water. Its effects are exhibited in the adjacent, thin layers of fluid in contact with the vehicle's surface, i.e. the boundary layer (Figure 10.25). The boundary layer begins at the surface of the submersible where the water is at zero velocity relative to the surface. The outer edge of the boundary layer is at water stream velocity. Consequently, within this layer is the velocity gradient and shearing stresses produced between the thin

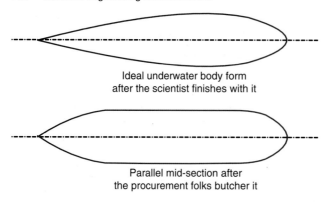

Figure 10.24 Underwater vehicle body forms.

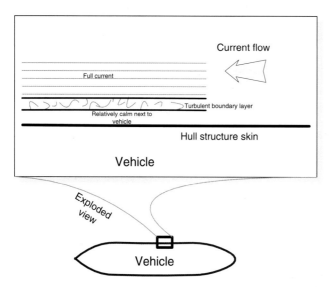

Figure 10.25 Ideal form with skin surface detail.

layers adjacent to each other. The skin friction drag is the result of stresses produced within the boundary layer. Initial flow within the boundary layer is laminar (regular, continuous movement of individual water particles in a specific direction) and then abruptly terminates into a transition region where the flow is turbulent and the layer increases in thickness. To obtain high vehicle speed, the design must be toward retaining laminar flow as long as possible, for the drag in the laminar layer is much less than that within the turbulent layer (see Figure 5.13).

An important factor determining the condition of flow about a body and the relative effect of fluid viscosity is the 'Reynolds number'. This number was evolved from the work of Englishman Osborne Reynolds in the 1880s. Reynolds observed laminar flow become abruptly turbulent when a particular value of the product of the distance along a tube and the velocity, divided by the viscosity, was reached. The Reynolds number expresses in non-dimensional form a ratio between inertia forces and viscous forces on a particle, and the transition from the laminar to the turbulent area occurs at a certain critical Reynolds number value. This critical Reynolds number value is lowered by the effects of surface imperfections and regions of increasing pressure. In some circumstances, sufficient kinetic energy of the flow may be lost from the boundary layer such that the flow separates from the body and produces large pressure or form drag.

The Reynolds number can be calculated by the following formula:

$$Re = \rho Vl/\mu = Vl/\nu,$$

where:
ρ = density of fluid (slugs/ft³) [kg/m³]
V = velocity of flow (ft/s) [m/s]
μ = coefficient of viscosity (lb-s/ft²) [kg/ms]
$\nu = m/\rho$ = kinematic viscosity (ft²/s) [m²/s]
l = a characteristic length of the body (ft) [m].

An additional factor is roughness of the body surface, which will increase frictional drag. Naval architects generally add a roughness-drag coefficient to the friction-drag coefficient value for average conditions, see Chapter 5, Sections 5.1 and 5.7.

Form drag
A variation on a standard dynamics equation can be used for ROV drag curve simulation. With an ROV, the two components causing typical drag to counter the vehicle's thruster output are the tether drag and the vehicle drag (Figure 10.26). The function of an ROV submersible is to push its hull and pull its tether to the work site in order to deliver whatever payload may be required at the work site. The only significant metric that matters in the motive performance of an ROV is the net thrust to net drag ratio. If that ratio is positive (i.e. net thrust exceeds net drag), the vehicle will make headway to the work site. If that ratio is negative, the vehicle becomes a very high-tech and very expensive boat anchor.

ROV thrusters must produce enough thrust to overcome the drag produced by the tether and the vehicle. The drag on the ROV system is a measurable quantity derived by hydrodynamic factors that include both vehicle and tether drag. The drag produced by the ROV is based upon the following formula:

$$\text{Vehicle drag} = 1/2 \times \rho A V^2 C_d,$$

where
ρ = (density of sea water)/(gravitational acceleration), where density of sea water = 64 lb/ft³ (1025 kg/m³) and gravitational acceleration = 32.2 ft/s² (9.8 m/s²).
A = Characteristic area on which C_d (the drag coefficient) is non-dimensionalised. For an ROV, A is defined as the cross-sectional area of the front of the vehicle. In some cases, the ROV volume raised to the 2/3 power is used.
V = Velocity in feet per second − (1 knot) = 1.689 feet/second = 0.51 metres/second.
C_d = Non-dimensional drag coefficient. This ranges from 0.8 to 1 when based on the cross-sectional area of the vehicle.

Total drag of the system is equal to the vehicle drag plus the tether drag (Figure 10.27).

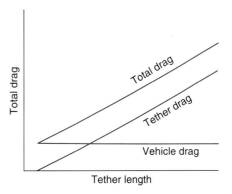

Figure 10.27 Component drag at constant speed.

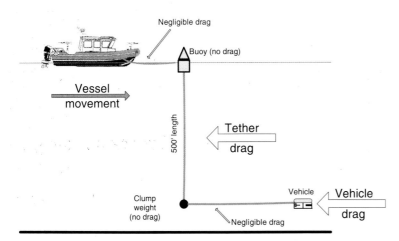

Figure 10.26 System drag components.

In the case of cables, the characteristic area, A, is the cable diameter in inches divided by 12, times the length perpendicular to the flow.

The C_d for cables ranges from 1.2 for unfaired cables; 0.5–0.6 for hair-faired cable; and 0.1–0.2 for faired cables. Since the cylindrical form has the highest coefficient of drag, the use of cable fairings to aid in drag reduction can have a significant impact.

Accordingly, the total drag of the system is defined as:

Total drag = $1/2 \rho A_v V^2 C_{dv} + 1/2 \rho A_u V_u^2 C_{du}$
(where v = vehicle; u = umbilical).

A simple calculation can be performed if it is assumed that the umbilical cable is hanging straight down and that the tether from the end of the umbilical (via a clump or TMS) to the vehicle is horizontal with little drag (Figure 10.26). For this calculation, it will be assumed that the ship is station keeping in a 1-knot current (1.9 km/h) and the vehicle is working at a depth of 500 feet (152 m). The following system parameters will be used:

Unfaired umbilical diameter = 0.75 inch (1.9 cm)
A, the characteristic area of the vehicle = 10 square feet (0.93 square metres)

Based on the above, the following is obtained:

Vehicle drag = $1/2 \times 64/32.2 \times 10 \times (1.689)^2 \times 0.9$ = 25.5 pounds (11.6 kg)

Umbilical drag = $1/2 \times 64/32.2 \times (0.75/12 \times 500) \times (1.689)^2 \times 1.2$

= 106.3 pounds (48.2 kg)

Note: Computations will be the same in both imperial and metric if the units are kept consistent. This simple example shows why improvements in vehicle geometry do not make significant changes to system performance. The highest factor affecting ROV performance is tether drag.

The following discussion will consider the drag of individual components.

Drag computations for the vehicle assume a perfectly closed frame box. Drag computations for the tether are in the range of a cross-section of ROV systems sampled during recent field trials of small observation-class systems.

By varying the tether diameter, the relationships in Figure 10.28 can be developed.

Figure 10.29 shows that by varying the speed with a constant length of tether, the vehicle will display a similar curve, producing a drag curve that is proportional to velocity squared.

The power required to propel an ROV is calculated by multiplying the drag and the velocity as follows:

Power = Drag $\times V/550$

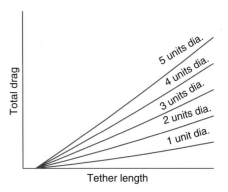

Figure 10.28 Linear tether drag at constant speed with varying diameter.

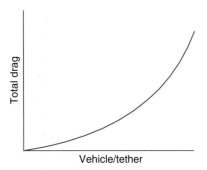

Figure 10.29 Linear tether drag at varying speed with constant diameter.

The constant 550 is a conversion factor that changes feet/pounds/seconds to horsepower. As discussed previously, the drag of a vehicle is proportional to the velocity of the vehicle squared. Accordingly, the propulsion power used is proportional to the velocity cubed. To increase the forward velocity by 50%, for example from 2 knots to 3 knots, the power increases by $(3/2)^3$, or $(1.5)^3$, which is 3.4 times more power. To double the speed, the power increases by $(2)^3$, or eight times. Increased speed requirements have a severe impact on vehicle design.

Table 10.3 lists some observation-class systems tested during United States Coast Guard (USCG) procedures trials (without specific names and using figures within each vehicle manufacturer's sales literature) with their accompanying dimensions.

At a given current velocity (i.e. 1 knot), the drag can be varied (by increasing the tether length) until the maximum thrust is equal to the total system drag. That point is the maximum tether length for that speed that the vehicle will remain on station in the current. Any more tether in the water (i.e. more drag) will result in the vehicle losing way against the current. Eventually (when the end of the tether is

Table 10.3 Specifications of ROVs evaluated.

System and parameter	Large ROV A	Small ROV A	Small ROV B	Large ROV B	Small ROV C	Medium ROV A
Depth rating (ft)	500	330	500	1150	500	1000
Length (in)	24	10	14	39	21	18.6
Width (in)	15	7	9	18	9.65	14
Height (in)	10	6	8	18	10	14
Weight in air (lb)	39	4	8	70	24	40
Number of thrusters	4	3	3	4	4	4
Lateral thruster	Yes	No	No	Yes	Yes	No*
Approx. thrust (lb)	25	2	5	23	9	12
Tether diameter (in)	0.52	0.12	0.44	0.65	0.30	0.35
Rear camera	No	No	Yes	No	No	No
Side camera	No	No	No	Yes	No	No
Generator req. (kW)	3	1	1	3	1	3

*Medium ROV A possesses lateral thrusting capabilities due to offset of vertical thrusters.

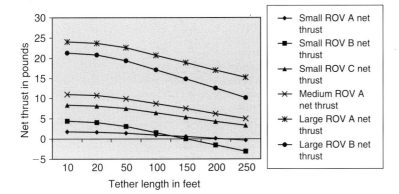

Figure 10.30 Drag curves of systems tested at 0.5 knot.

reached) the form drag will turn the vehicle around, causing the submersible to become the high-tech equivalent of a sea anchor.

The charts in Figures 10.30–10.33 show the approximate net thrust (positive forward thrust versus total system drag) curves at 0.5, 1, 1.5, and 2.0 knots for the ROVs described in Table 10.3.

The net thrust is shown with the horizontal line representing zero net thrust. All points below the zero thrust line are negative net thrust, causing the vehicle to lose headway against an oncoming current. Note: These tether lengths represent theoretical cross-section drag for a length of tether perfectly perpendicular to the oncoming water (see the example, Figure 10.26). Vehicle drag assumes a perfectly closed box frame with the dimensions from Table 10.3 for the respective system.

With a tether in a perfectly streamlined configuration (i.e. tether following directly behind the vehicle) the tether drag profile changes and is significantly reduced (Figure 10.34).

The obvious message from this data is that the tether drag on the vehicle is the largest factor in ROV deployment and usage. The higher the thrust-to-drag ratio and power available, the better the submersible pulls its tether to the work site.

(3) Tether effects
Tether pull point
Stability testing was performed on a small ROV system at Penn State University's Advanced Research Lab in their water tunnel. The water flow was slowly brought up while observing the vehicle's handling characteristics as well as its computed, versus actual, zero net thrust point.

This particular vehicle had a tether pull point significantly above the line of thrust (see Figure 10.35), resulting in a 'bow up' turning moment. As the speed

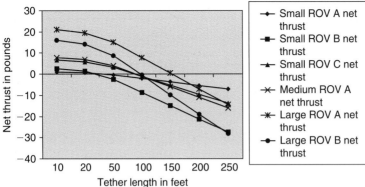

Figure 10.31 Drag curves of systems tested at 1.0 knot.

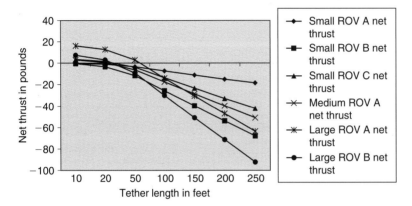

Figure 10.32 Drag curves of systems tested at 1.5 knots.

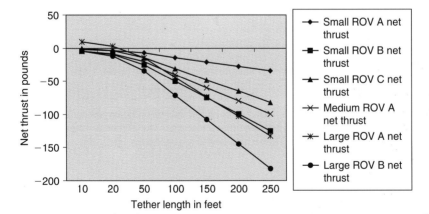

Figure 10.33 Drag curves of systems tested at 2.0 knots.

ramped up during the tests, with little tether in the water, the vehicle was still able to maintain control about the vertical plane by counteracting the 'bow up' tendency with vertical thrust-down. However, at a constant speed with the tether being lengthened, the tether drag produced an increasingly higher tether turning moment, eventually overpowering the vertical thruster and shooting the submersible to the surface in an uncontrolled fashion.

If the tether is placed in close proximity to the thruster, parasitic drag will occur due to the skin friction and form drag from the thruster discharge flow across the tether. When selecting the tether placement, it is best to design the tether pull point (Figure 10.36) as close to the centre point of thrust to balance any turning moment due to the tether pull point.

Tether pull/lay
Some vehicle manufacturers place the tether pull point atop the vehicle. The benefit to this placement is the tether will not lay as easily in the debris located on the bottom, allowing a cleaner tether channel from the vehicle to the surface. This is beneficial if the vehicle is operated in minimal currents with little or no horizontal offset. If either a horizontal offset or a current (or both) is encountered, the vehicle may experience difficulty through partial (if not total) loss of longitudinal and/or lateral stability.

Hydrodynamics of vehicle and tether
The most typical arrangement for an observation-class system involves a clean tether (i.e. without clump weight) following the vehicle to the work site. The tether naturally settles behind the vehicle and slopes in the current as it feeds toward the surface. As the vehicle speed ramps up, the flow drag on the tether correspondingly melds the tether into its wake, forming a 'sail' of sorts behind the vehicle (Figure 10.37). A small reduction in the drag due to reduced angle of incidence to the oncoming water flow is more than offset with the additional form and flow drag of the excessive tether in the water. There is an old technique used by surface-supplied commercial divers to counter the excessive umbilical in the water – grab hold of something on the bottom while the tender takes up the slack. The same technique can be applied to ROVs by placing the vehicle on a stationary item on the bottom then having the tender pull the excessive tether back on deck.

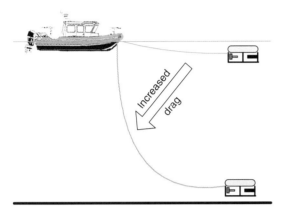

Figure 10.34 Tether profile drag increases to the perpendicular point.

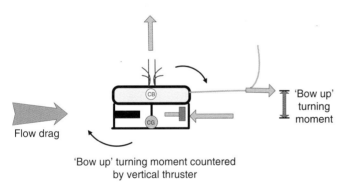

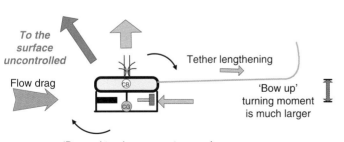

Figure 10.35 Vehicle stability considerations.

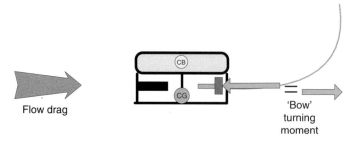

Figure 10.36 Minimal bow turning moment with tether on thrust line.

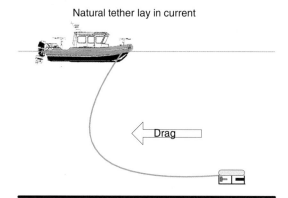

Figure 10.37 Natural tether lay behind the vehicle as the speed ramps up.

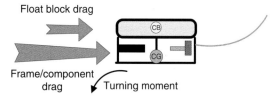

Figure 10.38 Bow turning moment due to asymmetrical drag as speed ramps up.

(4) Thrusters and speed
As the speed of the vehicle ramps up, the low speed stability due to static stability is overtaken by the placement of the thrusters with regard to the centre of total drag. In short, for slow-speed vehicles the designer can get away with improper placement of thrusters. For higher speed systems, thruster placement becomes a more important consideration in vehicle control (Figure 10.38).

Propeller efficiency and placement
Propellers come in all shapes and sizes based upon the load and usage. Again using the aircraft correlation, on fixed-pitch aircraft propellers for small aircraft there are 'climb props' and 'cruise props'. The climb prop is optimized for slower speeds, allowing better climb performance while sacrificing on cruise speeds. Conversely, a cruise prop has better range and speed during cruise, but climb performance suffers. Likewise, a tugboat propeller would not be best suited for a high-speed passenger liner.

Propellers have an optimum operational speed. Some propellers are optimized for thrust in one direction over another. A common small ROV thruster on the market today uses such a propeller for forward and downward thrusting. The advantage to this propeller arrangement is better thrust performance in the forward direction while sacrificing turning reversal and upward thrust performance. Other propellers have equal thrusting capabilities in both directions. Both have their strengths and weaknesses. All ROVs are slow systems and should make use of propellers' maximizing power at slow speeds in order to counter the combined tether/vehicle drag. Remember, an ROV is a tugboat and not a speedboat.

Propellers also produce both cavitation and propeller-tip vortices, causing substantial amounts of drag. As the spinning propeller moves water across the blade, the centrifugal force (instead of moving aft to produce a forward thrust vector) throws the water toward the tips of the blade, spilling over the end of the blade in turbulent flow. Kort nozzles form a basic hub around the propeller to substantially reduce the instance of tip vortices. The Kort nozzle then maintains the water volume within the thruster unit, allowing for more efficient movement of the water mass in the desired vectored direction. Propeller cavitations are a lesser problem due to the speed at which the small ROV thruster propeller turns and are inconsequential to this analysis. A detailed discussion of propeller design, ducted propellers and cavitation is included in Chapter 5.

Thrust to drag and bollard pull
The following factors come into play when calculating vehicle speed and ability to operate in current:

- *Bollard pull* is a direct measurement of the ability of the vehicle to pull on a cable. Values provided by manufacturers can vary due to lack of standards for testing: 'Actual bollard pull can only be measured in full scale, and is performed during so-called bollard pull trials. Unfortunately the test results are not only dependent on the performance of the [vehicle] itself, but also on test method and set-up, on trial site and on environmental conditions ...' (Jukola and Skogman 2002).
- *Hydrodynamics* is another aspect of ROV design that must be considered holistically. Although a vehicle shape and size may make it very hydrodynamic, i.e. certain smaller enclosed systems, there is often a trade-off in stability. Some manufacturers seem to spend considerable effort making their ROVs more hydrodynamic in the horizontal plane, but in deep sea operations diving to depth may consume considerable time.

It is bollard pull, vehicle hydrodynamics, and tether drag together that determine most limitations on vehicle performance. The smaller the tether cable diameter, the better – in all respects (except, of course, power delivery). Stiffer tethers can be difficult to handle, but they typically provide less drag in the water than their more flexible counterparts. Flexible tethers are much nicer for storage and handling, but they tend to get tangled or hang up more often than those that are slightly stiffer.

The use of ROVs in current is an issue that is constantly debated among users, designers, and manufacturers. This is not a topic that can be settled by comparing specifications of one vehicle to another. One of the most common misconceptions is that maximum speed equates to an ability to deal with current. When operating at depth (versus at the surface), the greatest influence of current is on the tether cable. It is the ability of the vehicle to pull this cable that allows it to operate in stronger currents. A vehicle with more power, but not necessarily more speed, will be better able to handle the tether (an example of which would be bollard pull of a tugboat versus that of a speedboat). The most effective way to determine a vehicle's ability to operate in current is to test the vehicle in current. Experience of the operator can have a significant effect on how the vehicle performs in higher current situations. Realistically, no small surfaced-powered ROV can be considered effective in any current over three knots.

10.3.3.6 *Vehicle control*

Vehicle manufacturers use a variety of techniques for gross versus fine control of vehicle movement to conform to the operating environment. One small vehicle manufacturer makes use of a horizontal and vertical gain setting to allow for varying power versus joystick position combinations. Another vehicle manufacturer allows for variable power delivery scaling via software on the controller. The reason this power scaling is necessary is that when towing the tether and vehicle combination to the work site, the full power complement is needed for the muscling operation. Once at the work site, finer adjustments are needed to ease the ROV into and out of tight places. If the power were set to full gain in a confined area, a quarter joystick movement could over-ramp the power so quickly that the vehicle could ram into a wall, damaging the equipment and causing some embarrassing personnel reviews.

The joystick control matrix can also significantly affect the ease of control over the vehicle (Figures 10.39 and 10.40). For example, if during a small turning adjustment, such as 20°, the thrusters may ramp up power so quickly that the operator cannot stop the turn until after reaching the 90° rotation point. Effective control of the vehicle would be lost.

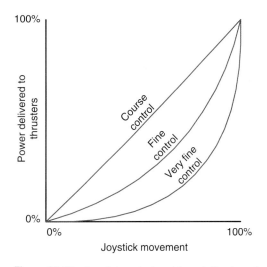

Figure 10.39 Joystick control matrix variation based upon varying power delivery versus joystick position.

(a) Control versus speed

Unlike underwater vehicles built for high speed (examples of high-speed underwater vehicles are a torpedo or a nuclear attack submarine), most observation-class ROV submersibles are designed for speeds no greater than 3 knots. In fact, somewhere in the speed range of 6–8 knots for underwater vehicles, interesting hydrodynamic forces act upon the system, which require strong design and engineering considerations that address drag and control issues. At higher speeds, small imperfections in vehicle ballasting and trim propagate to larger

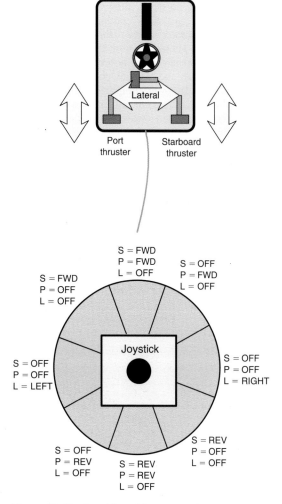

Figure 10.40 Joystick position versus thruster activation on a four-thruster configuration.

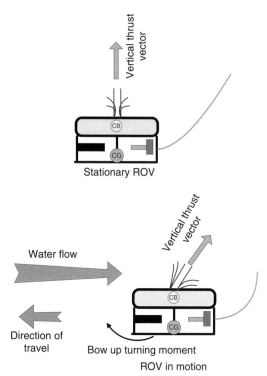

Figure 10.41 Apparent thrust vector change due to water flow across vehicle.

forces that simple thruster input may not overcome. As an anecdote to unexpected consequences for high-speed underwater travel, during trials for the USS *Albacore* (AGSS 569) it was noted with some surprise (especially by the Commanding Officer) that the submarine snap rolled in the direction of the turn during high-speed manoeuvring!

For the ROV at higher speeds, any thruster that thrusts on a plane perpendicular to the relative water flow will have the net vector of the thrust reaction move in the direction of the water flow (Figure 10.41). Also, one current vehicle manufacturer makes use of vectored thrust for vehicle control, which mitigates the thrust vector problem at higher speeds.

At what point is control over the vehicle lost? The answer to that is quite simple. The loss of control happens when the vehicle's thrusters can no longer counter the forces acting upon the vehicle while performing a given task. Once the hydrodynamic forces exceed the thruster's ability to counter these forces (on any given plane), control is lost. One of the variables must be changed in order to regain control.

(b) Auto stabilization

With sensor feedback fed into the vehicle control module, any number of parameters may be used in vehicle control through a system of closed-loop control routines. Just as dogs follow a scent to its source, ROVs can use sensor input for positive navigation. Advances are currently being made for tracking chemical plumes from environmental hazards or chemical spills. A much simpler version of this technique is the rudimentary auto depth/altitude/heading.

Auto depth is easily maintained through input from the vehicle's pressure-sensitive depth transducer. Auto altitude is equally simple, but the vehicle manufacturer is seldom the same company as the sensor manufacturer (causing some issues with communication standards and protocols between sensor and vehicle). The most common compass modules used in observation-class ROV systems are the inexpensive flux gate-type. These flux gate-type compasses have a sampling rate (while accurate) slower than the yaw swing rate of most small vehicles, which cause the vehicle to 'chase the heading'. Flux gate auto heading is better than no auto heading, but several manufacturers of small systems have countered this 'heading chase' problem by using a gyro.

Gyros for small ROVs come in two basic types, the slaved gyro and the rate gyro. The slaved gyro samples the magnetic compass to slave the gyro periodically to correspond with its magnetic counterpart. Since the auto heading function of an ROV is simply a heading hold function, some manufacturers have gotten away with using a simple rate gyro for 'heading stabilization'. When the heading hold function is slaved to a gyro only, sensing a turn away from the initial setting and a rate at which the turn is progressing (i.e. the rate gyro has no reference to any magnetic heading), the vehicle is then only referenced to a given direction. Hence the term 'heading stabilization' due to the lack of any reference to a specific compass direction.

10.3.3.7 *Deployment techniques*

Deployment methods vary, but there are a few common methods that have proven successful. The deployment methods can be divided into two main categories: Direct deployed and TMS deployed.

(1) Directly deployed
For station-keeping operations with smaller ROVs, the vehicle can be directly deployed from the deck of the boat. Larger vehicles can be directly deployed, but the risk of damage increases as the weight of the vehicle increases. This is due to the vehicle's momentum building through vessel sway while the vehicle is suspended in air (i.e. the vehicle becomes a 'wrecker's ball'). Directly deployed vehicles are more vulnerable to any currents prevalent from the surface to the operating depth.

(2) Tether management system (TMS)
The tether management system (TMS) can be part of a cage deployment system (Figure 10.43) or can simply be attached to the so-called clump weight (Figure 10.42). The main function for the TMS is to manage a soft tether cable – the link from the TMS to the ROV for electrical power and sensors, including video and telemetry. The tether cable allows the ROV to make excursions at depth for a distance of 500 feet (150 m) or more from the point of the clump weight. Some refer to the TMS as the entire system of cage or 'top hat' deployment, tether management, vehicle protection, and junction point

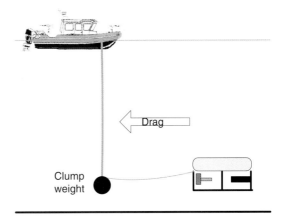

Figure 10.42 Clump weight deployed ROV.

for the surface/vehicle link. Technically, the TMS is the tether handling machinery only.

Clump deployed
The use of a clump weight has become prevalent in the observation-class category. If working on or near the sea bottom, clump weights enable the ROV operator to easily manage the tether 'lay' from the insertion point next to the vessel all the way to the clump weight location next to the bottom.

This allows the weight to absorb the cross-section drag of the current, relieving the submersible of the tether drag from the surface to the working depth. The vessel can be manoeuvred to a point directly above the work site, thus locating the centre point of an operational circle at the clump weight. In short, the vehicle only needs to drag the tether length between the clump weight and the vehicle for operations on the bottom.

Cage deployed
Cages are used for larger ROVs to protect the vehicle against abrasions and deployment damage due to the instability of most vessels of opportunity while underway.

Cages also function as a negatively buoyant anchor to overcome the drag imposed from the cross-section of the cable presented to the current (between the platform and the cage) at shallower depths (Figure 10.43). This allows the weight of the cage to fight the current instead of the vehicle fighting the current. The cage further provides room for a tether management system to meter the softer tether in small amounts, thus lowering the risk of tether entanglement. The cage umbilical is normally

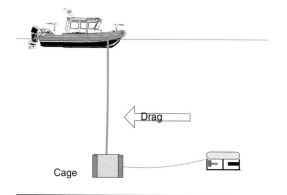

Figure 10.43 Cage deployed ROV.

made of durable material (steel, Kevlar, etc.) with the conductors for the vehicle buried within the core of the umbilical. For deeper diving submersibles, the umbilical encases fibre-optic data links to the cage, requiring digital modem feeds from the cage to the control unit at the surface.

10.4 ROV components

10.4.1 Introduction

This Section discusses the major components (Figure 10.44) of a typical ROV system along with everyday underwater tasks ROVs perform.

As discussed in the prior section on ROV design, the design team must consider the overall system. To reinforce the importance of this point, a few additional comments about the design process are warranted.

An ROV is essentially a robot. What differentiates a robot from its immovable counterparts is its ability to move under its own power. Along with that power of locomotion comes the ability to navigate the robot, with ever increasing levels of autonomy to achieve some set goal. While the ROV system, by its nature, is one of the simplest robotic designs, complex assignments can be accomplished with a variety of closed-loop aids to navigation. Some ROV manufacturers are aggressively embracing the open source computer-based control models, allowing users to design their own navigation and control matrix. This is an exciting development in the field of subsea robotics and will allow development of new techniques, which will only be limited by the user's imagination. This concept takes the control of the development of navigation capabilities (which is the mission) from the hands of the design engineer (who may or may not understand the user's needs) into the hands of the end user (who does understand the needs). Designing efficient and cost-effective systems with the user in mind is critical to the success of the product and ultimately the mission.

Do not over-design the system. The old saying goes that a chain is only as strong as its weakest link. Accordingly, all components of an ROV system should be rated to the maximum operating depth of the underwater environment anticipated, including safety factors. However, they should not be over-designed. As the operating depth proceeds into deeper water, larger component wall thicknesses will be required for the air-filled spaces (pressure-resistant housings) on the vehicle. This increased wall thickness results in an increased vehicle weight, which requires a larger flotation system to counter the additional weight. This causes an increase in drag due to a larger cross-section, which requires more power. More power drives the cable to become larger, which increases drag, etc. It quickly becomes a vicious design spiral.

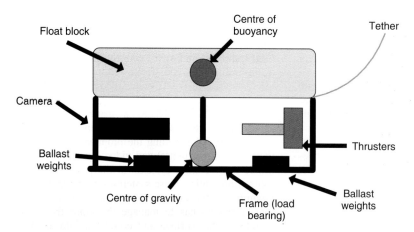

Figure 10.44 ROV submersible components.

Careful consideration should be given during the design phase of any ROV system to avoid over-engineering the vehicle. By saving weight and cost during the design process, the user will receive an ROV that has the capability of providing a cost-effective operation. This is easier said than done, as 'bells and whistles' are often added during the process, or the 'latest and greatest' components are chosen without regard to the impact on the overall system. Keep these ideas in mind as the various component choices are presented in the remainder of this Section.

10.4.2 Mechanical and electro/mechanical systems

Since weight is one of the most critical design factors, the components/subsystems having a significant impact in this area are discussed first.

10.4.2.1 *Frame*

The frame of the ROV provides a firm platform for mounting, or attaching, the necessary mechanical, electrical, and propulsion components. This includes special tooling/instruments such as sonar, cameras, lighting, manipulator, scientific sensor, and sampling equipment. ROV frames have been made of materials ranging from plastic composites to aluminium tubing. In general, the materials used are chosen to give the maximum strength with the minimum weight. Since weight has to be offset with buoyancy, this is critical.

The ROV frame must also comply with regulations concerning load and lift path strength. The frame can range in size from 6 in × 6 in to 20 ft × 20 ft. The size of the frame is dependent upon the following criteria:

- Weight of the complete ROV unit in air
- Volume of the onboard equipment
- Volume of the sensors and tooling
- Volume of the buoyancy
- Load-bearing criteria of the frame.

10.4.2.2 *Buoyancy*

Archimedes' principle states: An object immersed in a fluid experiences a buoyant force that is equal in magnitude to the force of gravity on the displaced fluid. Thus, the objective of underwater vehicle flotation systems is to counteract the negative buoyancy effect of heavier than water materials on the submersible (frame, pressure housings, etc.) with lighter than water materials; A near neutrally buoyant state is the goal. The flotation foam should maintain its form and resistance to water pressure at the anticipated operating depth. The most common underwater vehicle flotation materials encompass two broad categories: Rigid polyurethane foam and syntactic foam.

The term 'rigid polyurethane foam' comprises two polymer types: Polyisocyanurate formulations and polyurethane formulas. There are distinct differences between the two, both in the manner in which they are produced and in their ultimate performance.

Polyisocyanurate foams (or 'trimer foams') are generally low-density, insulation-grade foams, usually made in large blocks via a continuous extrusion process. These blocks are then put through cutting machines to make sheets and other shapes. ROV manufacturers generally cut, shape, and sand these inexpensive foams, then coat them with either a fibreglass covering or a thick layer of paint to help with abrasion and water intrusion resistance. These resilient foam blocks have been tested to depths of 1000 feet of seawater (fsw) (305 m) and have proven to be an inexpensive and effective flotation system for shallow water applications (Figure 10.45).

Polyisocyanurate foams have excellent insulating value, good compressive-strength properties, and temperature resistance up to 300°F. They are made in high volumes at densities between 1.8 and 6 lb per cubic foot, and are reasonably inexpensive. Their stiff, brittle consistency and their propensity to shed dust (friability) when abraded can serve to identify these foams.

For deep-water applications, syntactic foam has been the foam of choice. Syntactic foam is simply an air/microballoon structure encased within a resin body. The amount of trapped air within the resin structure will determine the density as well as the durability of the foam at deeper depths. The technology, however, is quite costly and is normally saved for the larger deep-diving ROV systems.

10.4.2.3 *Propulsion and thrust*

The propulsion system significantly impacts the vehicle design. The type of thrusters, their configuration, and the power source to drive them usually take priority over many of the other components.

(a) Propulsion systems

ROV propulsion systems come in three different types: Electrical, hydraulic, and ducted jet propulsion. These different types have been developed to suit the size of vehicle and anticipated type of work. In some cases, the actual location of the work task has dictated the type of propulsion used. For example, if the vehicle is operated in the

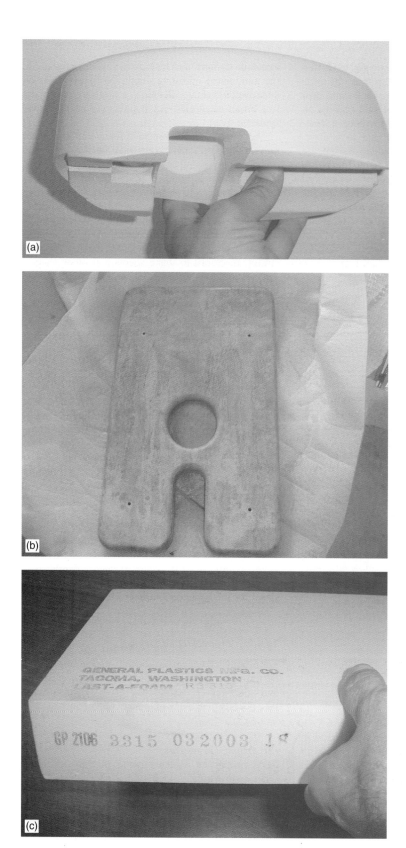

Figure 10.45 Polyurethane fibreglass encased and simple painted float blocks.

vicinity of loosely consolidated debris, which could be pulled into rotating thrusters, ducted jet thruster systems could be used. If the vehicle requires heavy duty tooling for intervention, the vehicle could be operated with hydraulics (including thruster power). Hydraulic pump systems are driven by an electrical motor on the vehicle, requiring a change in energy from electrical to mechanical to hydraulic – a process that is quite energy inefficient. A definite need for high mechanical force is required to justify such an energy loss and corresponding costs.

The main goal for the design of ROV propulsion systems is to have high thrust-to-physical size/drag and power-input ratios. The driving force in the area of propulsion systems is the desire of ROV operators to extend the equipment's operating envelope. The more powerful the propulsion of the ROV, the stronger the sea current in which the vehicle can operate. Consequently, this extends the system's performance envelope.

Another concern is the reliability of the propulsion system and its associated sub-components. In the early development of the ROV, a general practice was to replace and refit electric motor units every 50–100 hours of operation. This increased the inventory of parts required and the possibilities of errors by the technicians in reassembling the motors. Thus, investing in a reliable design from the beginning can save both time and money.

The propulsion system has to be a trade-off between what the ROV requires for the performance of a work task and the practical dimensions of the ROV. Typically, the more thruster power required, the heavier the equipment on the ROV. All parts of the ROV system will grow exponentially larger with the power requirement continuing to increase. Thus, observation ROVs are normally restricted to a few minor work tasks without major modifications that would move them to the next heavier class.

(b) Thruster basics

The ROV's propulsion system is made up of two or more thrusters that propel the vehicle in a manner that allows navigation to the work site. Thrusters must be positioned on the vehicle so that the moment arm of their thrust force, relative to the central mass of the vehicle, allows a proper amount of manoeuvrability and controllability.

Thrust vectoring is the only means of locomotion for an ROV. There are numerous placement options for thrusters to allow varying degrees of manoeuvrability. Manoeuvring is achieved through asymmetrical thrusting based upon thruster placement as well as varying thruster output.

The three-thruster arrangement (Figure 10.46) allows only fore/aft/yaw, while the fourth thruster also allows lateral translation. The five-thruster variation allows all four horizontal thrusters to thrust in any horizontal direction simultaneously.

Also, placing the thruster off the longitudinal axis of the vehicle (Figure 10.47) will allow a better turning moment, while still providing the vehicle with strong longitudinal stability.

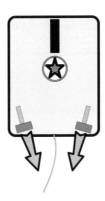

Figure 10.47 Thruster aligned off the longitudinal axis.

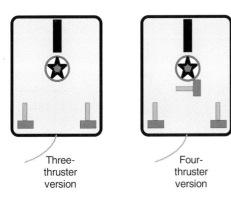 

Figure 10.46 Thruster arrangement.

One problem with multiple horizontal thrusters along the same axis, without counter-rotating propellers, is the torque steer issue (Figure 10.48). With two or more thrusters operating on the same plane of motion, a counter-reaction to this turning moment will result. Just as the propeller of a helicopter must be countered by the tail rotor or a counter-rotating main rotor, the ROV must have counter-rotating thruster propellers in order to avoid the torque of the thrusters rolling the vehicle counter to the direction of propeller rotation. If this roll does occur, the resulting asymmetrical thrust and drag loading could give rise to course deviations – the effect of which is known as 'torque steering'.

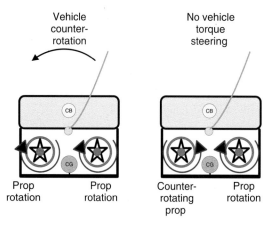

Figure 10.48 Thruster rotational effect upon vehicle.

(c) Thruster design
Underwater electrical thrusters are composed of the following major components:

- Power source
- Electric motor
- Motor controller (this may be part of the thruster or may be part of a separate driver board)
- Thruster housing and attachment to vehicle frame
- Gearing mechanism (if thruster is geared)
- Drive shafts, seals, and couplings
- Propeller
- Kort nozzle and stators.

The most critical of these components will be discussed in more detail in the following sections.

Power source
On a surface-powered ROV system, power arrives to the vehicle from a surface power source. The power can be in any form from basic shore power (e.g. 110 VAC 60 Hz or 220 VAC 50 Hz – which is standard for most consumer electrical power delivery worldwide) to a DC battery source. For an observation-class ROV system running on DC power, the AC source is first rectified to DC on the surface, then sent to the submersible for distribution to the thrusters. The driver and distribution system location will vary between manufacturers and may be anywhere from on the surface control station, within the electronics bottle of the submersible, to within the actual thruster unit. The purpose of this power source is the delivery of sufficient power to drive the thruster through its work task.

Electric motor
Electric motors come in many shapes, sizes, and technologies, each designed for different functions. By far the most common thruster motor on observation-class ROV systems is the DC motor, due to its power, availability, variety, reliability, and ease of interface. The DC motor, however, has some difficult design and operational characteristics. Factors that make it less than perfect for this application include:

- The optimum motor speed is much higher than the normal in-water propeller rotation speed, thus requiring gearing to gain the most efficient speed of operation
- DC motors consume a high amount of current
- They require a rather complex pulse width modulation (PWM) motor control scheme to obtain precise operations.

Permanent magnet DC motors. Per Clark and Owings (2003) the permanent magnet DC motor has, within the mechanism, two permanent magnets that provide a magnetic field within which the armature rotates. The rotating centre portion of the motor (the armature) has an odd number of poles – each of which has its own winding (Figure 10.49). The winding is connected to a contact pad on the centre shaft called the commutator. Brushes attached to the (+) and (−) wires of the motor provide power to

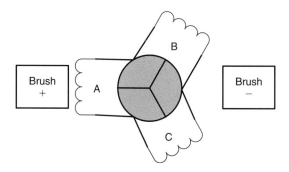

Figure 10.49 Commutator and brushes.

the windings in such a fashion that one pole will be repelled from the permanent magnet nearest it and another winding will be attracted to another pole. As the armature rotates, the commutator changes, determining which winding gets which polarity of the magnetic field. An armature always has an odd number of poles, and this ensures that the poles of the armature can never line up with their opposite magnet in the motor, which would stop all motion.

Near the centre shaft of the armature are three plates attached to their respective windings (A, B, and C) around the poles. The brushes that feed power to the motor will be exactly opposite each other, which enables the magnetic fields in the armature to forever trail the static magnetic fields of the magnets. This causes the motor to turn. The more current that flows in the windings, the stronger the magnetic field in the armature, and the faster the motor turns.

Even as the current flowing in the windings creates an electromagnetic field that causes the motor to turn, the act of the windings moving through the static magnetic field of the motor causes a current in the windings. This current is opposite in polarity to the current the motor is drawing from the power source. The end result of this current, and the countercurrent (CEMF – counter electromotive force), is that as the motor turns faster it actually draws less current.

This is important because the armature will eventually reach a point where the CEMF and the draw current balance out at the load placed on the motor and the motor attains a steady state. The point where the motor has no load is the point where it is most efficient. It is also the point where the motor is the weakest in its working range. The point where the motor is strongest is when there is no CEMF; All current flowing is causing the motor to try to move. This state is when the armature is not turning at all. This is called the stall or startup current and is when the motor's torque will be the strongest. This point is at the opposite end of the motor speed range from the steady-state velocity (Figure 10.50).

While the maximum efficiency of an electric motor is at the no-load point, the purpose of having the motor in the first place is to do work (i.e. produce mechanical rotary motion that may be converted into linear motion or some other type of work). The degree of turning force delivered to the drive shaft is known as motor torque.

Motor torque is defined as the angular force the motor can deliver at a given distance from the shaft. If a motor can lift 1 kilogram from a pulley with a radius of 1 m it would have a torque of 1 newton-metre. One newton equals 1 kilogram-metre/second^2, which is equal to 0.225 pounds (1 inch equals 2.54 centimetres, 100 centimetres are in a metre, and there are 2π radians in one revolution).

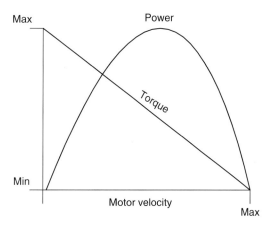

Figure 10.50 Motor velocity versus torque and power.

The formula for mechanical power in watts is equal to torque times the angular velocity in radians/second. This formula is used to describe the power of a motor at any point in its working range. A DC motor's maximum power is at half its maximum torque and half its maximum rotational velocity (also known as the no-load velocity). This is simple to visualize from the discussion of the motor working range: Where the maximum angular velocity (highest revolutions per minute (RPM)) has the lowest torque, and where the torque is the highest, the angular velocity is zero (i.e. motor stall or start torque).

Note that as motor velocity (RPM) increases, the torque decreases; At some point the power stops rising and starts to fall, which is the point of maximum power.

When sizing a DC motor for an ROV, the motor should be running near its highest efficiency speed, rather than its highest power, in order to get the longest running time. In most DC motors this will be at about 10% of its stall torque, which will be less than its torque at maximum power. So, if the maximum power needed for the operation is determined, the motor can be properly sized. A measure of efficiency and operational life can then be obtained by oversizing the motor for the task at hand.

Brushless DC motors. For brushless DC motors, a sensor is used to determine the armature position. The input from the sensor triggers an external circuit that reverses the feed current polarity appropriately. Brushless motors have a number of advantages that include longer service life, less operating noise (from an electrical standpoint), and, in some cases, greater efficiency.

Gearing
DC motors can run from 8000 to 20 000 RPM and higher. Clearly, this is far too fast for ROV

Figure 10.51 Fluid sealing of direct dive thruster coupling.

applications if vehicle control is to be maintained. Thus, to match the efficient operational speed of the motor with the efficient speed of the thruster's propeller, the motor will require gearing. Gearing allows two distinctive benefits – the power delivered to the propeller is both slower and more powerful. Further, with the proper selection of a gearbox with a proper reduction ratio, the maximum efficiency speed of the motor can match the maximum efficiency of the thruster's propeller/Kort nozzle combination.

Drive shafts, seals, and couplings
The shafts, seals, and couplings for an ROV thruster are much like those for a motorboat. The shaft is designed to provide torque to the propeller while the seal maintains a watertight barrier that prevents water ingress into the motor mechanism.

Drive shafts and couplings vary with the type of propeller driving mechanism. Direct drive shafts, magnetic couplings, and mechanical (i.e. geared) couplings are all used to drive the propeller.

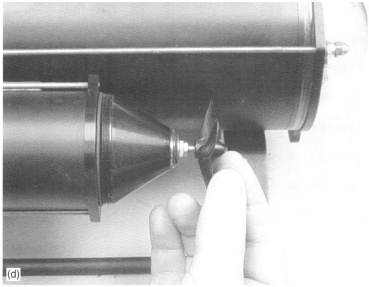

Figure 10.51 Continued.

Technology advances are being exploited in attempts to miniaturize thrusters. In one case, a new type of thruster housing places the drive mechanism on the hub of the propeller instead of at the drive shaft, allowing better torque and more efficient propeller tip flow management. Others are developing miniature electric ring thrusters, where the propeller, which can be hubless, is driven by an external 'ring' motor built into the surrounding nozzle. Such a design eliminates the need for shafts, and sometimes seals, altogether.

There are various methods for sealing underwater thrusters. Some manufactures use fluid-filled thruster housings to lower the difference in pressure between the sea water and the internal thruster housing pressure by simply matching the two pressures (internal and external). Still others use a lubricant bath between the air-filled spaces and the outside water (Figure 10.51). A common and highly reliable technique is the use of a magnetically coupled shaft, which allows the air-filled housing to remain sealed (Figure 10.52).

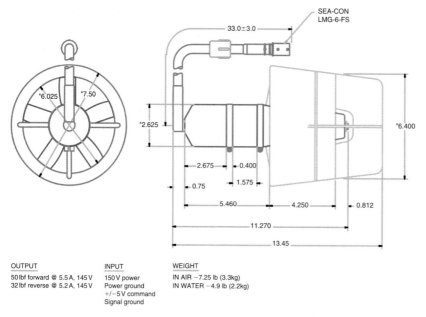

Figure 10.52 Magnetically coupled thruster diagram.

Propeller design

The propeller is a turning lifting body designed to move and vector water opposite to the direction of motion. Many thruster propellers are designed so their efficiency is much higher in one direction (most often in the forward and the down directions) than in the other. Propellers have a nominal speed of maximum efficiency, which is hopefully near the vehicle's normal operating speed. Some propellers are designed for speed and others are designed for power. When selecting a propeller, choose one with the ROV's operating envelope in mind. The desired operating objectives should be achieved through efficient propeller output below the normal operating envelope. A detailed discussion of propeller design is included in Chapter 5.

Kort nozzle

A Kort nozzle is common on most underwater thruster models. The efficacy of a Kort nozzle is the mechanism's help in reducing the amount of propeller vortices generated as the propeller turns at high speeds. The nozzle, which surrounds the propeller blades, also helps with reducing the incidence of foreign object ingestion into the thruster propeller. Also, stators help reduce the tendency of rotating propellers' swirling discharge, which tends to lower propeller efficiency and cause unwanted thruster torque acting upon the entire vehicle. Bi-directional tip-driven thrusters have been developed for underwater vehicles and ROVs, such as that described by Abu Sharkh *et al.* (2003).

10.4.3 Primary subsystems

The ability to sense the environment, either visually or through other means, and perform work at the desired location, is the mission of the ROV. The subsystems necessary for this task are discussed in the following sections.

10.4.3.1 *Lighting*

This explanation of lighting comes courtesy of Ronan Gray of Deep Sea Power & Light. The need for underwater lighting becomes apparent below a few feet from the surface. Ambient visible light is quickly attenuated by a combination of scattering and absorption, thus requiring artificial lighting to view items underwater with any degree of clarity. We see things in colour because objects reflect wavelengths of light that represent the colours of the visible spectrum. Artificial lighting is therefore necessary near the illuminated object to view it in true colour with intensity. Underwater lamps provide this capability.

Lamps convert electrical energy into light. The main types or classes of artificial lamps/light sources used in underwater lighting are incandescent, fluorescent, high-intensity gas discharge, and

Table 10.4 Light source characteristics (table and lighting description courtesy of Deep Sea Power and Light).

Source	Lumens/watt	Life(h)	Colour	Size	Ballast
Incandescent	15–25	50–2500	Reddish	M–L	No
Tungsten–halogen	18–33	25–4000	Reddish	S–M	No
Fluorescent	40–90	10 000	Varies	L	Yes
Green fluor.	125	10 000	Green	L	Yes
Mercury	20–58	20 000	Bluish	M	Yes
Metal halide	70–125	10 000	Varies	M	Yes
High-press. sodium	65–140	24 000	Pink	M	Yes/I
Xenon arc	20–40	400–2000	Daylight	VS	Yes/I
HMI/CID	70–100	200–2000	Daylight	S	Yes/I
Low-press. sodium	100–185	18 000	Yellow	L	Yes
Xenon flash	30–60	NA	Daylight	M	NA

V, very; S, small; M, medium; L, large; I, ignitor required; NA, not applicable.

light-emitting diode (LED) – each with its strengths and weaknesses. All types of light are meant to augment the natural light present in the environment. Table 10.4 shows the major types of artificial lighting systems, as well as their respective characteristics.

- *Incandescent* – The incandescent lamp was the first artificial light bulb invented. Electricity is passed through a thin metal element, heating it to a high enough temperature to glow (thus producing light). It is inefficient as a lighting source with approximately 90% of the energy wasted as heat. Halogen bulbs are an improved incandescent. Light energy output is about 15% of energy input, instead of 10%, allowing them to produce about 50% more light from the same amount of electrical power. However, the halogen bulb capsule is under high pressure instead of a vacuum or low-pressure noble gas (as with regular incandescent lamps) and, although much smaller, its hotter filament temperature causes the bulbs to have a very hot surface. This means that such glass bulbs can explode if broken, or if operated with residue (such as fingerprints) on them. The risk of burns or fire is also greater than other bulbs, leading to their prohibition in some underwater applications. Halogen capsules can be put inside regular bulbs or dichroic reflectors, either for aesthetics or for safety. Good halogen bulbs produce a sunshine-like white light, while regular incandescent bulbs produce a light between sunlight and candlelight.
- *Fluorescent* – A fluorescent lamp is a type of lamp that uses electricity to excite mercury vapour in argon or neon gas, producing short-wave ultraviolet light. This light then causes a phosphor coating on the light tube to fluoresce, producing visible light. Fluorescent bulbs are about 40% efficient, meaning that for the same amount of light they use one-fourth the power and produce one-sixth the heat of a regular incandescent. Fluorescents typically do not have the luminescent output capacity per unit volume of other types of lighting, making them (in many underwater applications) a poor choice for underwater artificial light sources.
- *High-intensity discharge* – High-intensity discharge (HID) lamps include the following types of electrical lights: Mercury vapour, metal halide, high-pressure sodium and, less common, xenon short-arc lamps. The light-producing element of these lamp types is a well-stabilized arc discharge contained within a refractory envelope (arc tube) with wall loading (power intensity per unit area of the arc tube) in excess of $3\,W/cm^2$ ($19.4\,W/in^2$). Compared to fluorescent and incandescent lamps, HID lamps produce a large quantity of light in a small package, making them well suited for mounting on underwater vehicles. The most common HID lights used in underwater work are of the metal halide type.
- *LED* – A light-emitting diode (LED) is a semiconductor device that emits incoherent narrow-spectrum light when electrically biased in the forward direction. This effect is a form of electroluminescence. The colour of the emitted light depends on the chemical composition of the semiconducting material used, and can be near-ultraviolet, visible, or infrared. LED technology is useful for underwater lighting because of its low power consumption, low heat generation, instantaneous on/off control, continuity of colour throughout the life of the diode, extremely long life, and relatively low cost of manufacture. LED lighting is a rapidly evolving technology – look for more usage of LEDs in the underwater lighting field soon.

Most observation-class ROV systems use the smaller lighting systems, including halogen and metal halide HID lighting.

The efficiency metric for lamps is efficacy, which is defined as light output in lumens divided by energy input in watts, with units of lumens per watt (LPW). Lamp efficacy refers to the lamp's rated light output per nominal lamp watts. System efficacy refers to the lamp's rated light output per system watts, which include the ballast losses (if applicable). Efficacy may be expressed as 'initial efficacy', using rated initial lumens at the beginning of lamp life. Alternatively, efficacy may be expressed as 'mean efficacy', using rated mean lumens over the lamp's lifetime; Mean lumens are usually given at 40% of the lamp's rated life and indicate the degree of lumen depreciation as the lamp ages.

An efficient reflector will not only maximize the light output that falls on the target, but will also direct heat forward and away from the lamp. The shape of the reflector will be the main determinant in how the light output is directed. Most are parabolic, but ellipsoidal reflectors are often used in underwater applications to focus light through a small opening in a pressure housing. The surface condition of a reflector will determine how the light output will be dispersed and diffused. The majority of reflectors are made of pure, highly polished aluminium that will reflect light back at roughly the same angle to the normal at which it was incident. By adding dimples or peens to the surface, the reflected light is dispersed or spread out. When a plain white surface is used, the reflected light is diffused in all directions.

10.4.3.2 Cameras

Currently, most small ROV systems use inexpensive charge-coupled device (CCD) cameras as their main viewing device. These camera systems are mounted on small circuit boards and produce a video signal transmitted in a format sent up the tether to the video capture device on the surface. The actual protocol of the signal emanating from the camera and control box (after transmission through the tether) is manufacturer-specific, but usually falls under either composite or RF (radio frequency) video. The protocol of the video signal will determine the receiving adapter on the viewing device. Refer to the manufacturer's instructions for the specific protocol and/or adapter for the system.

Production of an ROV camera assembly can be accomplished on any electronics bench with rudimentary equipment. A simple chip camera system sold through any surveillance camera manufacturing company is mounted on a block along with a motor and gearing system for panning and/or tilting, plus focusing (if manual focusing is desired). Once the camera is mounted (Figure 10.53), simple wiring and switching will accomplish both control of the individual camera as well as switching between various camera systems aboard the vehicle.

Various regions of the world use different video formats. In the USA, as well as a few other countries, NTSC (National Television Standards Committee) is the standard format, while most of Europe, Africa, and Asia use PAL (Phased Array Lines) format. The SECAM format used predominately in France has been declining in recent years and will in all likelihood eventually be eliminated.

Camera technology is evolving rapidly. The High Definition format is quickly being adapted to ROVs as it trickles down to the smaller vehicles due to decreasing size and lower cost structure. Digital still camera technology is also being adapted for high-resolution image capture of underwater items. Look for major improvements in size, functionality, and cost in the near future.

10.4.3.3 Sensors

As stated earlier, most industrial ROV systems provide the capability to transmit data from the submersible to the surface. This allows the ROV system to deliver a suite of instruments to the work site, powered by the vehicle, with data transmitted through the tether to the surface. Any combination of sensor and instrument (heading/gyro/depth, etc.) is available as payload to the modern ROV system, assuming proper data protocol transmissions and power delivery are available. Figure 10.54 shows (left to right) a sonar, hydrology sensor, and manipulator configured for attachment to an ROV.

Major issues regarding the integration of sensors involves the data transmission protocol and the method through which this transmission takes place. The ROV manufacturer must provide a throughput within the ROV system to allow for sensor integration. Some examples of common sensors packages placed aboard ROV systems in industrial applications include:

- Radiation sensors
- CTD (conductivity/temperature/depth) sensors
- Pressure-sensitive depth transducer
- Magnetic flux gate compass module
- Slaved or rate gyro for heading stabilization
- Ultrasonic thickness gauges for measuring metal thickness and quality
- Imaging sonar
- Acoustic positioning
- Digital cameras
- Multi-parameter environmental sensors (e.g. turbidity, chlorophyll, DO, pH, and ORP sensors, which are discussed in Chapter 1).

Underwater vehicles 769

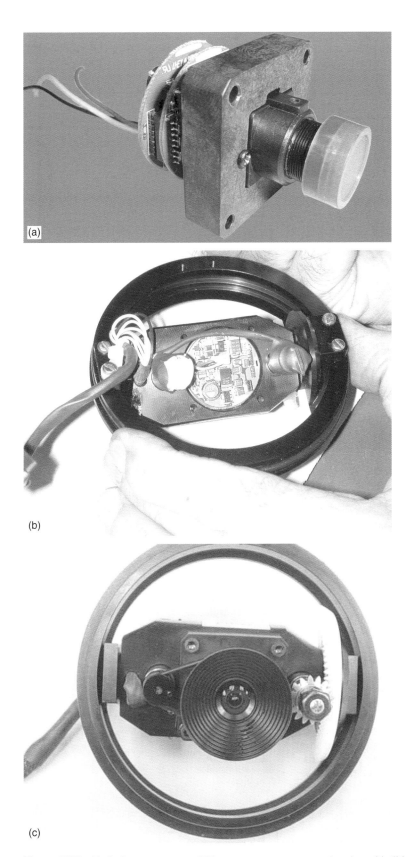

Figure 10.53 Typical arrangement – CCD camera system mounted to ring with tilting mechanism.

Figure 10.54 Various sensors for mounting to ROV.

10.4.3.4 *Manipulator and tool pack*

Most professional ROV systems allow for a simple A/B power source to provide the locomotion needs of intervention tooling packs. On the larger hydraulic ROV systems, a simple independent A and B connector is provided to power turning or cutting equipment for subsea work. On observation-class systems, a simple 12 or 24 VDC source is provided to run manipulators or other small tools needed for specific jobs. Power can also be redirected from main thruster power for more demanding mechanical work.

A basic single-function manipulator package, common on many small ROV systems, consists of a 24 VDC electric motor running a worm gear to open and close small grabber arms for light intervention duties (Figures 10.55(a) and 10.55(b)). A common problem with small manipulators without a limit switch at the end of the travel of the worm gear is that the ROV operator will continue activating the manipulator. When the worm gear reaches the end of travel, the motor can become stalled at the full close or full open point of the jaw. If the jaw is left in this position (i.e. with the worm gear stalled at the end of its travel and torqued against the end of the worm gear stop) and allowed to sit for a length of time, oxidation can seize the screw/teeth, preventing it from moving the worm gear away from the stop. If this happens, the jaw must be either squeezed/pulled to take the pressure off the worm gear, and the motor then activated, or the mechanism must be disassembled to unfreeze the worm gear.

10.4.4 Electrical considerations

The following sections discuss specific issues and relationships regarding the tether, power, data, and the connectors that bring it all together.

10.4.4.1 *The tether*

The tether and the umbilical are essentially the same item. The cable linking the surface to the cage or tether management system (TMS) is termed the 'Umbilical', while the cable from the TMS to the submersible is termed the 'Tether'. Any combination of electrical junctions is possible in order to achieve power transmission and/or data relay. For instance, AC power may be transmitted from the surface through the umbilical to the cage, where it is then changed to DC to power the submersible's thrusters and electronics. Further, video and data may be transmitted from the surface to the cage via fibre-optics (to lessen the noise due to AC power transmission), then changed to copper for the portion from the cage to the submersible, thus eliminating the AC noise problem. Figure 10.56 is an example of the neutrally buoyant tether for the Outland 1000 observation-class ROV system (courtesy of Outland Technology).

The umbilical/tether can be made up of a number of components:

- Conductors for transmitting power from the surface to the submersible

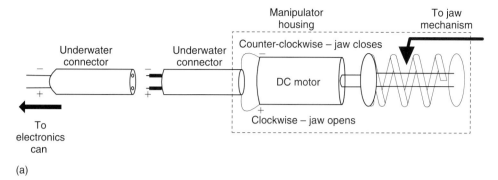

(a)

Figure 10.55(a) Typical setup for small ROV manipulator.

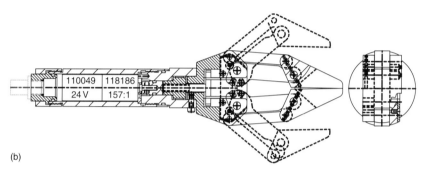

(b)

Figure 10.55(b) Small manipulator CAD drawing (courtesy of Inuktun Services Ltd.).

- Control throughput for telemetry (conducting metal or fibre optic)
- Video/data transmission throughput (conducting metal or fibre optic)
- Strength member allowing for higher tensile strength of cable structure
- Lighter-than-water filler that helps the cable assembly achieve neutral buoyancy
- Protective outer jacket for tear and abrasion resistance.

Most observation-class ROV systems use direct current power for transmission along the tether to power the submersible. The tether length is critical in determining the power available for use at the vehicle. The power available to the vehicle must be sufficient to operate all of the electrical equipment on the submersible. The electrical resistance of the conductors within the tether, especially over longer lengths, could reduce the vehicle power sufficiently during high-load conditions to affect operations.

The maximum tether length for a given power requirement is a function of the size of the conductor, the voltage, and the resistance. For example, using a water pipe analogy, there is only a certain amount of water that will flow through a pipeline at a given pressure. The longer the pipe, the higher the internal resistance to movement of the water. As long as the water requirements at the receiving end do not exceed the delivery capacity of the pipe (at a given pressure), the system delivery of water will be adequate. If there were to be a sudden increase in the water requirement (a fire requiring water, everyone watering their lawn simultaneously, etc.), the only way to get adequate water to the delivery end would be to increase the pressure or to decrease the resistance (i.e. shorten the pipe length or increase the diameter) of the pipe. The same holds true in electrical terms between tether length, total power required, voltage, and resistance (Figure 10.57).

Ohm's law deals with the relationship between voltage and current in an ideal conductor. This relationship states that the potential difference (voltage) across an ideal conductor is proportional to the current through it. So, the voltage (V, or universally as E) is equal to the current (I) times the resistance (R). This is stated mathematically as $V = IR$. Further, power (measured in watts) delivered to a circuit is a product of the voltage and the current.

Thus, based on Ohm's law, the voltage drop over a length of cable can be calculated by using the formula, $V = IR$, where V is the voltage drop, I is

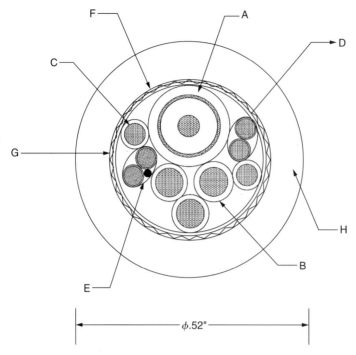

A – 75-ohm mini coax. Cap. 16.6 pf/ft. 22-7 TC, XLPO-Foam, Alum,1 spiral T.C. shield
B – 3×3 #18 (19/30) T.C. (OR/BL/PUR)
C – 2×3 #22 (19/34) T.C. (WH)
D – 1 3× #24 AWG TP, (19/36) T.C.
E – 1×3 #24 AWG TSP, (19/36) T.C. polypropylene insulation 0.010 wall
F – MYLAR tape
G – KEVLAR weave
H – 0.045 Green cellular foam polyurethane

Figure 10.56 Cross-section of neutrally buoyant tether.

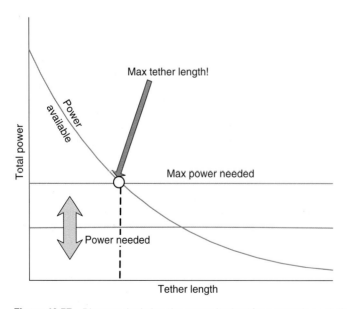

Figure 10.57 Diagram depicting the power budget for power through the tether.

the current draw of the vehicle in amps, and R is the total electrical resistance of the power conductor within the tether in ohms. The current draw of a particular component (light, thruster, camera, etc.) can be calculated if the wattage and voltage of the component are known. The current draw is equal to the component wattage divided by the component voltage (or amps = watts/volts).

For example, referring to the table of electrical resistances for various wire gauges (Table 10.5), the voltage required to operate a 24-volt/300-watt light at 24 volts over 250 feet of 16-gauge cable can be calculated as follows: The current draw, I, of a 24-volt/300-watt lamp operating at 24 volts is 300 watts/24 volts = 12.5 amps. The resistance of 16-gauge wire is approximately 4 ohms/1000 feet (Table 10.5). Since the total path of the circuit is from the power supply to the light and back to the power supply, the total resistance of the cable is twice the length of the cable times the linear resistance, or for this example, $R = (2 \times 250\,ft) \times (4\,ohms/1000\,ft) = 2.0\,ohms$. Since $V = IR$, the voltage drop, V, is equal to 12.5 amps × 2.0 ohms = 25 volts. This means that 25 volts is lost due to resistance, so the power supply will need to provide at least 49 volts (the 24 volts necessary to operate the light plus the additional voltage loss of 25 volts) to power this 24-volt/300-watt light over a 250-foot cable.

Table 10.5 Standard copper wire gauge resistance over nominal lengths (example and table courtesy of Deep Sea Power and Light).

Wire gauge	Ohms/1000 ft (approx.)
20	10
18	6
16	4
14	2.5
12	1.5

10.4.4.2 *Power source*

The ROV system is made up of a series of compromises. The type of power delivered to the submersible is a trade-off of cost, safety, and needed performance. Direct current (DC) allows for lower cost and weight of tether components; Since inductance noise is minimal, it allows for less shielding of conductors in close proximity to the power line. Alternating current (AC) allows longer transmission distances than that available to DC while using smaller conductors.

Most operators of ROV systems specify a power source independent of the vessel of opportunity. The reason for this separation of supply is that the time the vessel is in most need of its power is normally the time when the submersible is most in need of its power. Submersible systems attempting to escape a hazardous bottom condition have been known to lose power at critical moments while the vessel is making power-draining repositioning thrusts on its engines. This can cause entanglement of the vehicle. With a separate power source, submersible manoeuvring power is separated from the power needs of the vessel.

With the advent of the lightweight microgenerators for use with small ROVs, the portability of the ROV system is significantly enhanced. Some operators prefer usage of the battery/inverter combination for systems requiring AC power. Also, some smaller systems use only DC as their power source. Either method should have the power source capable of supplying uninterrupted power to the system at its maximum sustained current draw for the length of the anticipated operation.

10.4.4.3 *AC versus DC considerations*

Electrical power transmission techniques are an important factor in ROV system design due to their effect upon component weights, electrical noise propagation and safety considerations. The DC method of power transmission predominates the observation-class ROV systems due to the lack of need for shielding of components, weight considerations for portability, and the expense of power transmission devices. On larger ROV systems, AC power is used for the umbilical due to its long power transmission distances, which are not seen by the smaller systems. AC power in close proximity to video conductors could cause electrical noise to propagate due to EMF (electromotive force) conditions. The shielding necessary to lower this EMF effect could cause the otherwise neutrally buoyant tether to become negatively buoyant, resulting in vehicle control problems. And the heavy and bulky transformers are a nuisance during travel to a job site or as checked baggage aboard aircraft.

Larger work-class systems normally use AC power transmission from the surface down the umbilical to the cage (the umbilical normally uses fibre-optic transmission, lowering the EMF noise through the video) since the umbilical does not require neutral buoyancy. At the cage, the AC power is then rectified to DC to run the submersible through the neutrally buoyant tether that runs between the cage and vehicle.

10.4.4.4 *Data throughput*

The wider the data pipeline from the submersible to the surface, the greater the ability for the vehicle to deliver to the operator the necessary job-specific data as well as sensory feedback needed to properly control the vehicle. With the cost of broadband fibre-optic transmission equipment dropping into the range of most small ROV equipment manufacturers' budgets, more applications and sensors should soon become available to the ROV marketplace. The ROV is simply a delivery platform for transporting the sensor package to the work location. The only limitation to full sensor feedback to the operator will remain one of lack of funding and imagination. The Human-Robot Interface (the intuitive interaction protocol between the human operator and the robotic vehicle) is still in its infancy; However, sensors are still outstretching the human's ability to interpret this data fast enough to react to the feedback in a timely fashion. This subject is probably the most exciting field of development for the future of robotics and will be of considerable interest to the next generation of ROV pilots.

10.4.4.5 *Data transmission and protocol*

Most small ROV manufacturers simply provide a spare twisted pair of conductors for hard-wire communication of sensors from the vehicle to the surface. The strength of this method is that the sensor vendor does not need engineering support from the ROV manufacturer in order to design these sensor interfaces. The weakness is that unless the sensor manufacturers collude to form a set of transmission standards, each sensor connected to the system 'hogs' the data transmission line to the detriment of other sensors needed for the task. A specific example of this problem is the need for concurrent use of an imaging sonar system and an acoustic positioning system. Unless the manufacturers of each sensor package agree upon a transmission protocol to share the single data line, only one instrument may use the line at a time. A few manufacturers have adapted industry standard protocols for such transmissions, including TCP/IP, RS-485, and other standard protocols. The most common protocol, RS-232, while useful and seemingly ubiquitous in the computer industry, is distance limited through conductors, thus causing transmission problems over longer lengths of tether.

The move toward open source PC-based sensor data processing has led to the production of data protocol converters for use in ROV sensor interpretation. Most small ROV sensor manufacturers transmit data with the RS-485 protocol, requiring a converter at the surface to both isolate the signal and to convert it to USB (or RS-232) protocol for easy processing with a standard laptop computer. Standards for these protocol converters are slow in evolving (due to the size of the customer base). Thus, the ROV system integrator must become familiar with the wiring and pin arrangement for these converters to assure data transmission from the sensor, through the vehicle and tether to the software at the surface, is achieved.

10.4.4.6 *Underwater connectors*

The underwater connector is said to be the bane of the ROV business. Salt water is highly conductive, causing any exposed electrical component submerged in salt water to short to ground. The result is the 'Ubiquitous ground fault'. The purpose of an underwater connector is to conduct needed electrical currents through the connector while at the same time squeezing the water path and sealing the connection to lower the risk of electrical leakage to ground.

The underwater connector is lined with synthetic rubber that blocks the ingress path of water while allowing a positive electrical connection. Connectors sometimes experience cathodic delamination, causing rubber peeling and flaking from the connector walls. Connector maintenance (Figure 10.58) should include:

- Use small amounts of silicone grease to lubricate the connector, thus allowing easier slide on and off. Using too much grease, a widespread problem, can interfere with sealing.
- Always pull the connector by its body instead of its tail (cable), since the wire splice is located in the connection. Pulling on the tail could part the solder joint and ruin the electrical continuity within the connector.
- Keep the connectors as clean as possible through regularly scheduled maintenance tasks that include cleaning the contacts and lubricating the rubber lining.
- Spray the connector body with silicone spray to keep the housing from drying out, which could result in flaking and rubber degradation.

Even when the contacts are right and the connector has good design features, the connector must be appropriate for the intended use and environment. The connector materials must be able to withstand the environmental conditions without degradation. For example, extended exposure to sunlight (ultraviolet energy) will cause damage to neoprene, and many steels will corrode in sea water. Check that the connector will fully withstand the environment.

The connector must not adversely affect the application. For example, all ferrous materials (steel, etc.) should be avoided in cases where the connector's magnetic signature might affect the system. In extreme cases, even the nickel used under gold plating could have an effect and should be reviewed.

The physical size of the connector, its weight, ease of use (and appropriateness for the application),

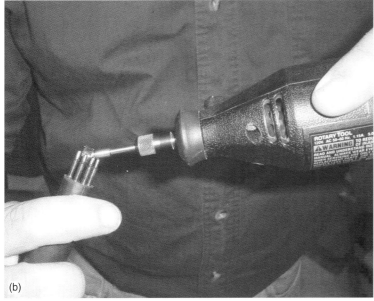

Figure 10.58 Underwater connectors must be serviced to assure proper electrical connectivity.

durability, submergence (depth) rating, field repairability, etc. should all be assessed. The use of oil-filled cables or connectors should be considered.

Ease of installation and use is especially important, so realistically appraise the technical ability of those personnel who will actually install or use the equipment. If they are inexperienced, a more 'user-friendly' connector may be a better choice. And, if possible, train operators in the basics of proper connector use: Use only a little lubricant, avoid over-tightening, note acceptable cable bending radii, provide grounding wires for steel connectors in aluminium bulkheads, etc.

Splicing and repairing underwater cables and connectors, while quite simple, requires some basic precautions to avoid water ingress into the electrical spaces, thus grounding the connection. Examples are shown in Figures 10.59–10.64.

10.4.5 Control systems

The control system controls the different functions of the ROV, from controlling the propulsion system to switching of the light(s) and video camera(s). From simple relay control systems in the past to today's digital fibre optics, these systems are equipped with a computer and subsystem control interface. The control system has to manage the input from the operator at the surface and convert it into actions subsea. The data required by the operator on the

Figure 10.59 Internal workings of male/female underwater connectors.

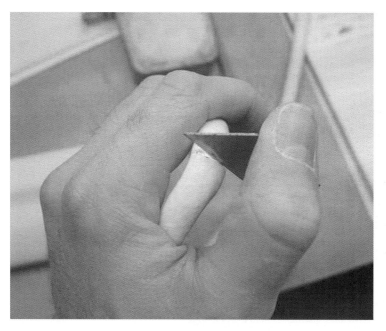

Figure 10.60 Slicing is conducted by peeling back the conductors of the tether.

surface to accurately determine the position in the water is collected by sensors (sonar and acoustic positioning) and transmitted to the operator.

Over the last 10–15 years, computers utilized for these purposes have been designer computers with sophisticated computer programs and control sequences. Today, one can find standard computers in the heart of these systems. There has been a shift back to simpler control systems recently with the commercial advent of the PLC (Programmable Logic Computer). This is used in numerous manufacturing processes since it consists of easily assembled modular building blocks of switches, analogue in/outputs, and digital in/outputs.

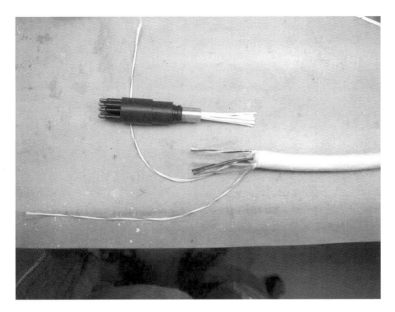

Figure 10.61 Pin out the conductors on the tether to correspond to the connector.

Figure 10.62 The electrical connection is made through standard bench techniques.

10.4.5.1 *The control station*

Control stations vary from large containers, with their spacious enclosed working area for work class systems, to simple PC gaming joysticks with PHDs (personal head-mounted displays) for some micro-ROV systems. All have in common a video display and some form of controlling mechanism (normally a joystick, such as at Figure 10.65). On older analogue systems, a simple rheostat controls the variable power to the electric motors, while newer digital controls are necessary for more advanced vehicle movements.

With the rise of robotics as a sub-discipline within electronics, further focus highlighted the need to control robotic systems based upon intuitive interaction through emulation of human sensory inputs. Under older analogue systems, a command of

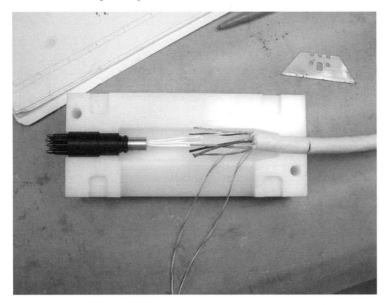

Figure 10.63 The connected conductors are laid into a potting mould then sealed with potting compound.

Figure 10.64 Finished connector along with plastic guard.

'look left/go left' was a complex control command requiring the operation of several rheostats to gain vector thrusting to achieve the desired motion. As digital control systems arose, more complex control matrices could be implemented much more easily through allowing the circuit to proportionally control a thruster based upon the simple position of a joystick control. The advent of the modern industrial joystick coupled with programmable logic circuits has allowed easier control of the vehicle while operating through a much simpler and more intuitive interface. The more sensors available to the 'human' that allow intuitive interaction with the 'robot', the easier it is for the operator to figuratively operate the vehicle from the vehicle's point of view. This interaction protocol between operator and vehicle has become known as the Human–Robot Interface and is the subject of intensive current research. Look for major developments within this area of robotics over both the short and long terms.

10.4.5.2 *Motor control electronics*

Since observation-class ROV systems use mainly electronic motors for thruster-based locomotion, a study of basic motor control is in order.

A basic control of direction and proportional scaling of electrical motors is necessary to finely control the motion of the submersible. If only 'On' or 'Off were the choices of motor control via switches, the operator would quickly lose control of the vehicle due to the inability to make the fine corrections needed for accurate navigation. In the early days of ROVs, the simple analogue rheostat was used for motor control. It was quite a difficult task to control a vehicle with the operation of three or four independent rheostat knobs while attempting to fly a straight line. Later came digital control of electric motors and the finer science of robotics took a great leap forward.

The basic electronic circuit that made the control of electronic motors used in robotics and industrial

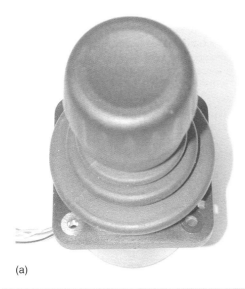

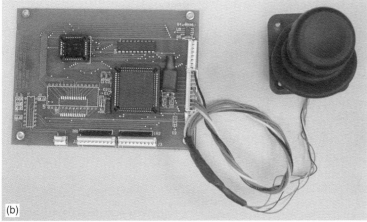

Figure 10.65 Industrial joystick and circuit board used in vehicle control.

components so incredibly useful is known as the 'H-bridge'. An understanding of the H-bridge (discussed later in this section) and the digital control of that H-bridge will help significantly with the understanding of robotic locomotion.

Consider the analysis of a simple electric circuit (Figure 10.66). As discussed earlier in this Chapter, Ohm's law gives a relationship between voltage (V), current (I) and resistance (R) that is stated as $V = IR$. In Figure 10.66, the current and voltage are known, thus the resistance can be calculated to be 3.0 ohms.

(a) Inductors
An inductor is an energy storage device that can be as simple as a single loop of wire or consist of many

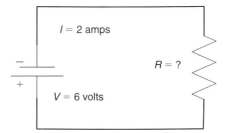

Question:	Answer
$I = 2$ amps	$V = IR$
$V = 6$ volts	$R = V/I$
$R = ?$ ohms	$R = 3.0$

Figure 10.66 Simple electrical circuit.

turns of wire wound around a core. Energy is stored in the form of a magnetic field in or around the inductor. Whenever current flows through a wire, it creates a magnetic field around the wire. By placing multiple turns of wire around a loop, the magnetic field is concentrated into a smaller space, where it can be more useful. When applying a voltage across an inductor, current starts to flow. It does not instantly rise to some level, but rather increases gradually over time (Figure 10.67). The relationship of voltage to current vs. time gives rise to what is known as inductance. The higher the inductance, the longer it takes for a given voltage to produce a given current – sort of a 'shock absorber' for electronics. Whenever there is a moving or changing magnetic field in the presence of an inductor, that change attempts to generate a current in the inductor. An externally applied current produces an increasing magnetic field, which in turn produces a current opposing that applied externally, hence the inability to create an instantaneous current change in an inductor. This property makes inductors useful as filters in power supplies. The basic mathematical expression for inductance (L) is $V = L \times dI/dT$.

The ideal world and the real world depart, since real inductors have resistance. In this world, the current eventually levels out, leaving the strength of the magnetic field plus the level of the stored energy as proportional to the current (Figure 10.68).

So, what happens when the switch is opened? The current dissipates quickly in the arc (Figure 10.69).

Diodes are used to suppress arcing, allowing the recirculation currents to dissipate more slowly. The current continues to flow through the inductor, but power is dissipated across the diode through the inductor's internal resistance. This is the basis of robotic locomotion electronic dampening (sort of a shock absorber for your thruster's drive train).

Permanent magnet DC motors can be modelled as an inductor, a voltage source, and a resistor. In this case, the torque of the motor is proportional to the current and the internal voltage source is proportional to the RPM (Figure 10.70) or back EMF (when the current is released and the motor turns into a generator as the motor spools down).

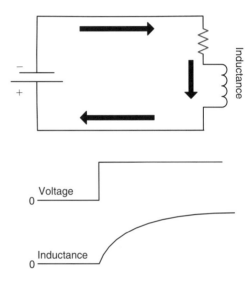

Figure 10.68 Inductor/voltage interaction with resistance.

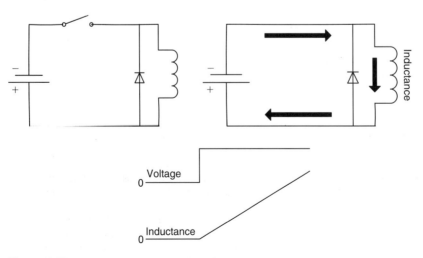

Figure 10.67 Inductor/voltage interaction without resistance.

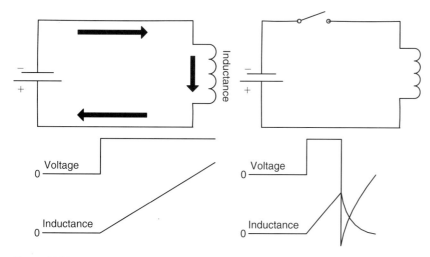

Figure 10.69 Inductor/voltage interaction as circuit is opened.

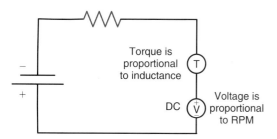

Figure 10.70 Diagram depicting inductance/torque interaction.

The stall point of an electric motor is at the point of highest torque as well as the point of highest current, and is proportional to the internal resistance of the motor.

That leads to the concern over back EMF within thruster control electronics design – once the DC motor is disengaged, the turning motor mass continues to rotate the coils within the armature. This rotating mass converts the electric motor into a generator, rapidly reversing and spiking the voltage in the reverse direction. Unless there is some circuit protection within the driver board circuit, damage to the control electronics will often result (Figure 10.71).

(b) The H-bridge
The circuit that controls the electrical motor is known as the 'H-bridge' due to its resemblance to the letter 'H' (Figure 10.72). Through variation of the switching, as well as inductance filtering (described earlier), the direction and the ramp-up speed are controlled through this circuit in an elegant and simplistic fashion.

(c) PWM control
Pulse width modulation (PWM) is a modulation technique that generates variable-width pulses to represent the amplitude of an analogue input signal. The output switching transistor is on more of the time for a high-amplitude signal and off more of the time for a low-amplitude signal. The digital nature (fully on or off) of the PWM circuit is less costly to fabricate than an analogue circuit that does not drift over time.

PWM is widely used in ROV applications to control the speed of a DC motor and/or the brightness of a light bulb. For example, if the line were closed for 1 μs, opened for 1 μs, and continuously repeated, the target would receive an average of 50% of the voltage and run at half speed or the bulb at half brightness. If the line were closed for 1 μs and open for 3 μs, the target would receive an average of 25%.

There are other methods by which analogue signals are modulated for motor control, but observation-class ROV systems predominate with the PWM mode due to cost and simplicity of design.

Figure 10.71 Damaged driver board due to back EMF.

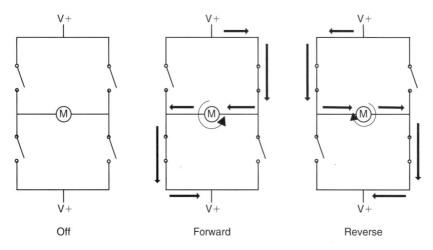

Figure 10.72 H-bridge diagram depicting basic operation.

References (Chapter 10)

Abu Sharkh, S.M., Turnock, S.R. and Hughes, A.W. (2003). Design and performance of an electric tip-driven thruster. *Proceedings of the Institution of Mechanical Engineers, Part M: Journal of Engineering for the Maritime Environment*, Vol. 217, No. 3.

Benthos, Inc. *StingRay Mk II Operations and Maintenance Manual*. Website for Benthos, Inc. is located at: http://www.benthos.com/.

Bohm H. and Jensen V. (1997). *Build Your Own Underwater Robot*. West Coast Words, ISBN 0-9681610-0-6.

Bowditch, N. (2002). *The American Practical Navigator*. National Imagery and Mapping Agency, ISBN 1-57785-271-0.

Burcher, R. and Rydill, L. (1994). *Concepts in Submarine Design*. Cambridge University Press.

Busby, R.F. (1976). *Manned Submersibles*. Office of the Oceanographer of the Navy.

Christ, R.D. and Wernli, R.L. (2007). *The ROV Manual*. Butterworth-Heinemann, Oxford, UK.

Curtin, T.B., Crimmins, D.M., Curcio, J., Benjamin, M. and Roper, C. (2005). Autonomous underwater vehicles: trends and transformations. *Marine Technology Society Journal*, Vol. 39, No. 3.

Deep Sea Power and Light. *Frequently Asked Questions* website page at: http://www.deepsea.com/faq.html.

Department of the Army, Corps of Engineers (2002). *Coastal Engineering Manual*. Publication number EM 1110-2-1100.

Desert Star Systems LLC. Various Operational Manuals for the Dive Tracker line of acoustical positioning systems. Reprinted with permission. Located on the web at: http://www.desertstar.com/.

Duxbury, A.C. and Alison, B. (1997). *An Introduction to the World's Oceans*, 5th edition. William C. Brown, ISBN 0-697-28273-2.

Edge, M. (1999). *The Underwater Photographer*, 2nd edition. Butterworth-Heinemann, ISBN 0-240-51581-1.

Everest, F.A. (1994). *The Master Handbook of Acoustics*, 3rd edition. TAB Books (a division of McGraw-Hill), ISBN 0-8306-4438-5.

Fondriest Environmental. Informational website page on environmental sensors at: http://www.fondriest.comlparameter.htm. Reprinted with permission.

Giancoli, D.C. (1991). *Physics*, 3rd edition. Prentice-Hall, ISBN 0-13-672510-4.

Gray, R. (2004). Light sources, lamps and luminaries. *See Technology*, Vol. 45, No. 12, pp. 39–43, December.

Griffiths, G. (2003). Technology and applications of autonomous underwater vehicles. *Ocean Science Technology*, Vol. 2, p. 342.

Huang, H.-M. (2004). *Autonomy Levels for Unmanned Systems (ALFUS) Framework*, Volume I: Terminology, Version 1.1 National Institute of Standards and Technology Special Publication 1011, September.

Imagenex Technology Corporation. Sonar theory from their website located at: http://www.imagenex.comlsonar theory.pdf. Reprinted with Permission.

Inuktun Services, Ltd. *ROV Seamor Operations Manual*. Website for Inuktun Services, Ltd. is located at: http://www.seamor.com/.

Joiner, J.T. (ed.) (2001). *NOM Diving Manual*, 4th edition. Best, ISBN 0-941332-70-5.

Jukola, H. and Skogman, A. (2002). *Bollard Pull*. Paper presented at the 17th International Tug and Salvage Convention, ITS 2002, 13–17 May, Bilbao, Spain.

Kongsberg, S.A.S. (2002). *Introduction to Underwater Acoustics*.

Linton, S.J. *et al.* (1986). *Dive Rescue Specialist Training Manual*. Concept Systems, Inc., ISBN 0-943717-42-6.

Loeser, H.T. (1992). *Sonar Engineering Handbook*. Peninsula, ISBN 0-932146-02-3.

Medwin, H. and Clay, C.S. (1998). *Fundamentals of Acoustical Oceanography*. Academic Press, ISBN 0-12-487570-X.

Milne, P.B. (1983). *Underwater Acoustic Positioning Systems*. E. & F. N. Spon Ltd, ISBN 0-419-12100-5.

Moore, J.E. and Compton-Hall, R. (1987). *Submarine Warfare: Today and Tomorrow*. Adler & Adler, ISBN 091756121X.

Olsson M.S. *et al.* (2000). ROV lighting with metal halide. White Paper on Metal Halide Technology displayed on the Deep Sea Power and Light Web Site at: http://www.deepsea.com. Reprinted with permission. Revision 1 dated 27 January 2000.

Outland Technology, Inc. *ROV Model Outland 1000 Operations Manual*. Website for Outland Technology, Inc. is located at: http://www.outlandtech.com/.

Remotely Operated Vehicle Subcommittee of the Marine Technology Society. Educational website page at: http://www.rov.org/student/education.cfm.

Remotely Operated Vehicles of the World (2006/2007). 7th edition, Clarkson Research Services Ltd., ISBN 1-902157-75-3.

Seafriends.org. *Underwater Photography – Water and Light*. From their educational web page located at: http://www.seafriends.org.nz/phgraph/water.htm.

Segar, D.A. (1998). *Introduction to Ocean Science*. Wadsworth, ISBN 0-314-09705-8.

Sonardyne International Ltd. *Acoustic Theory*. Website page at: http://www.sonardyne.co.uk/theory.htm.

Teather, R.G. (1994). *Royal Canadian Mounted Police Encyclopedia of Underwater Investigations*. Best, ISBN 0-941332-26-8.

Thurman, H.V. (1994). *Introductory Oceanography*, 7th edition. Macmillan, ISBN 0-02-420811-6.

Urick, R.J. (1975). *Principles of Underwater Sound*. McGraw-Hill, ISBN 0-07-066086-7.

US Geological Survey. Educational website page on water science at: http://ga.water.usgs.gov/edu/earthhowmuch.html.

Van Dorn, W.G. (1993). *Oceanography and Seamanship*, 2nd edition. Cornell Maritime Press, ISBN 0-87033-434-4.

Waite, A.D. (2002). *Sonar for Practicing Engineers*, 3rd edition. John Wiley, ISBN 0-471-49750-9.

Wernli, R. (1998). *Operational Effectiveness of Unmanned Underwater Systems*. Marine Technology Society, ISBN 0-933957-22-X.

Wilson, W.D. (1960). Equation for the speed of sound in sea water. *Journal of the Acoustic Society of America*.

11 Marine safety

Contents

11.1 Background
11.2 Regulatory authorities
11.3 Classification societies
11.4 Safety of marine systems
11.5 Safety management of ship stability
References (Chapter 11)

The various Sections of this Chapter have been taken from the following books, with the permission of the authors:

Eyres, D.J. (2007) *Ship Construction.* Sixth Edition. Butterworth-Heinemann, Oxford, UK. [Section 11.3]

Kobylinski, L.K. and Kastner, S. (2003) *Stability and Safety of Ships.* Elsevier, Oxford, UK. [Section 11.5]

Kristiansen, S. (2005) *Maritime Transportaion.* Elsevier Butterworth-Heinemann, Oxford, UK. [Sections 11.1, 11.2]

Pillay, A. and Wang, J. (2003) *Technology and Safety of Marine Systems.* Elsevier Science Ltd., Oxford, UK. [Section 11.4]

11.1 Background

This Chapter reviews current safety practices employed in the marine industry. The role of regulatory bodies and classification societies is outlined. Typical safety assessment techniques are described, together with a description of formal safety assessment (FSA). A description of the safety management of ship stability and relevant operational requirements is included, providing examples of practical applications.

11.1.1 International trade and shipping

Waterborne transport of materials and goods has for centuries been the main prerequisite for trade between nations and regions, and has without doubt played an important role in creating economic development and prosperity. The cost of maritime transport is very competitive compared with land and airborne transport, and the increase to the total product cost incurred by shipping represents only a few percent. Negative aspects of waterborne transport include longer transport time as a result of relatively low ship speed, congestion in harbours resulting in time delays, as well as less efficient integration with other forms of transport and distribution.

Shipping has from time to time been under attack for unacceptable safety and environmental performance, and this will be discussed in the next Sections. At this point we only make the following remark: in view of the relatively low cost of transport, it is a paradox that some areas of shipping have a relatively low standard of safety. Efficient transport should be able to pay for acceptable safety.

It has been discussed for some time whether basic economic mechanisms could ensure safe shipping. In this context the following questions are relevant:

- Is there any economic motivation for high levels of safety?
- Who should pay for increased safety?
- Are there any trade-offs between safety and efficiency?

These questions will be addressed briefly in this Chapter.

11.1.2 The actors in shipping

In shipping there are a number of actors that have an influence on safety, and the most important of these are presented in Table 11.1.

It should be evident that the different actors within the shipping domain to some degree have competing interests that may complicate the issue of safety, and this is a result of various factors such as the following:

- Who is controlling whom?
- Who sets the quality standards?
- What is the motivation for safe operation?
- Who is picking up the bill after an accident?

We will return to the questions of safety management and the regulation of shipping in later Sections.

11.1.3 The shipowner

In the case of severe shipping accidents and losses, the shipowner and/or ship management company will be

Table 11.1 Actors in shipping that influence safety.

Actor	Influence on safety
Shipbuilder	• Technical standard of vessel
Shipowner	• Decides whether technical standards will be above minimum requirements • Selects crew or management company for crew and operation • Make decisions regarding operational and organizational safety policies
Cargo owner	• Pays for the transport service and thereby also the quality and safety of the vessel operation • May undertake independent assessments of the quality of the shipper
Insurer	• Takes the main part of the risk on behalf of the shipper and cargo owner (i.e. vessel, cargo, third party – P&I) • May undertake independent assessment of the quality of the shipper
Management company	• Responsible for crewing, operation and upkeep (i.e. maintenance) of the vessel on behalf of the shipowner
Flag state	• Control of vessels, crew standards and management standards
Classification society	• Control of technical standards on behalf of insurer • Undertakes some control functions on behalf of the flag state
Port administration	• Responsible for safety in port and harbour approaches • May control safety standard of vessels, and in extreme cases deny access for substandard vessels

subjected to particular attention. This is natural given the fact that the shipowner owns the damaged/lost vessel, as well as manning, maintaining and operating it. Questions that are always raised in the context of maritime accidents are whether the shipowner has demonstrated a genuine concern for safety, and whether the standards of the vessel and its crew have been sacrificed for profit. The shipowner may counter such questions by claiming that the standards will not be better than what the market is willing to pay for.

With regard to vessel safety standards there has recently been an increasing focus on the cargo owner, as this is the party that decides which ship to charter and at what price. The charter party (i.e. the contract) gives the cargo owner considerable authority to instruct the Master with respect to the operation of the vessel. Given this important role in terms of safety, it may be seen as a paradox that the cargo owner has minimal, if any, liability in the case of shipping accidents.

Shipowners take some key decisions that have profound consequence for safety. The choice of flag state for registration of vessels, choice of classification society and arrangements for insurance are some key decisions. There exist international markets for these services in which different standards and corresponding fees can be found. The safety standard will therefore to a large degree be a result of what the owner is willing to pay for these services. A much discussed and fairly controversial topic is the increasing practice of 'flagging out', in which the shipowner registers a vessel in a country other than where it operates. Flagging out is mainly done for economic reasons, as shown in Figure 11.1. Availability of cheap labour, and the costs and strictness of safety control, seem to be key concerns for the owner. Based on this it must be asked whether shipowners, through their choice of flag, sacrifice safety.

The safety aspect of ship operation should also be seen in a wider context as shipowners have a number of different objectives that need to be balanced. These objectives include the following:

- Stay in business: return on investment.
- Marketing: win well-paying freight contracts.
- Service: minimize damage on cargo, keep on schedule.
- Efficiency: operate and maintain vessel.
- Employer: attract competent personnel.
- Subcontracting: select efficient service providers.
- Availability: minimize unplanned off-hire.

It is not necessarily obvious that these objectives and priorities for a given company are consistent with a high safety standard at all times. In this view it is of great importance that shipowners have clearly defined policies that never compromise on safety. An alternative view is that there is no conflict between cost, efficiency and safety. The main argument for this position is that in order to stay in business and thrive in the long term, shipowners have to operate safely and keep their fleets well maintained and up to standard. This view may, however, be a little naïve, as substandard shipping companies may not necessarily have a long-term perspective of their business. An OECD (Organization for Economic Cooperation and Development) study has shown that such substandard shipping companies are competitive on price and take their fair share of the market.

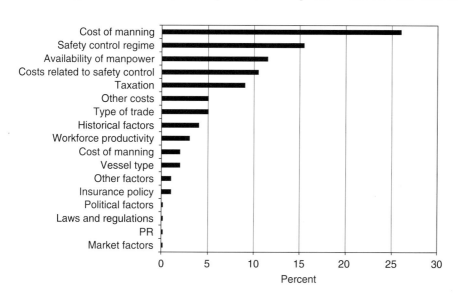

Figure 11.1 Reasons for flagging out: distribution of answers from questionnaire study. (Adapted from Bergantino and Marlow, 1998.)

11.1.4 Safety and economy

It is reasonable to assume that one of the prerequisites for achieving an acceptable safety standard for a vessel is that the company it belongs to has a sound economy and thereby is able to work systematically and continuously with safety-related matters such as training of personnel, developing better technical standards and improving management routines. This is made difficult by the fact that in the business of shipping one usually finds that income and revenues are fluctuating dramatically over time, which creates a rather uncertain business environment. This can be illustrated by Figure 11.2, which shows how the charter rate for the transport of crude oil has varied in the period from 1980 to 2002. As can be seen from this figure, charter rates for oil are heavily influenced by political development such as wars and economic crises. From a top quotation of approximately 80,000 USD/day in the autumn of 2000, charter rates fell to approximately 10,000 USD/day within a period of one year, i.e. a fall in charter rates by a factor of 8. In light of the fact that the minimum rate to result in a profit is around 22,150 USD/day (Front Line Ltd., Investor presentation, 3rd Quarter 2002.), it is clear that the economic basis for continuous and systematic safety work is not the best. The volatility of tanker shipping is also reflected in tanker vessel prices, as shown in Figure 11.3.

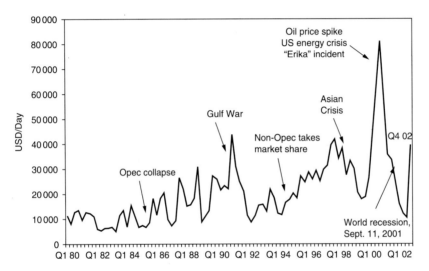

Figure 11.2 VLCC T/C equivalent rate for key routes, 1980 to 2002, given quarterly (Source: P. F. Bassøe AS & Co., *Tanker Fundamentals*, Nov. 2002.)

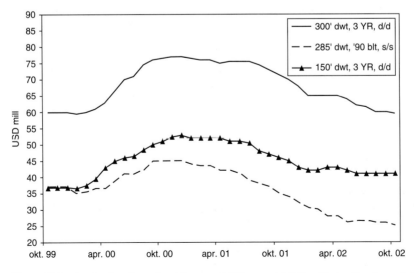

Figure 11.3 Crude oil tanker prices. (Source: P. F. Bassøe AS & Co., *Tanker Fundamentals*, Nov. 2002.)

It is also possible to argue for the opposite view, namely that the low cost of sea transport should put shipping in a favourable position compared with more expensive transport modes. As can be seen in Table 11.2, the relative cost of sea transport (i.e. freight cost in percent of sales price) is in the order of 2–6%. If we assume that the safety of shipping could be improved significantly by a 50% increase in the freight rate, this would have resulted only in a 1–3% increase in the sales price.

An argument against sea transport in these kinds of discussions is the longer transport time due to relatively low speeds and delays in ports. However, an increasing competitiveness of shipping on shorter routes can now be seen. The increase in road transport is currently representing a great environmental problem in central Europe and other densely populated areas due to exhaust emissions, and the road system is becoming more and more congested. A study by Abeille et al. (1999), illustrated by Table 11.3, shows that seaborne transport may be economically competitive even on shorter distances.

There is currently a growing national, regional and international concern for the emissions related to the burning of fossil fuels, and this also affects shipping. However, according to Kristensen (2002), some ship types perform environmentally better than road transport. Some results from Kristensen's study of the environmental cost of road and sea transport can be found in Table 11.4. In studying Table 11.4 it must, however, be recognized that estimating environmental consequences and economic aspects of transport is a highly uncertain and controversial exercise. Nevertheless, Kristensen's study indicates that container and bulk carriers are far better than road transport. Ro-Ro vessels, on the other hand, are less favourable due to large motor installations (resulting in higher speed) and lower cargo capacity. The author also points out that these figures are subject to change due to the continuous toughening of emission standards.

Table 11.2 Relative price of seaborne transport, Far East–Europe/US (distance: 9000 nm)[a].

Product/vessel/ capacity utilization (%)	Sales price/ unit (USD)	Freight cost/ unit (USD)	Relative freight cost
1 barrel of crude oil/VLCC/50%	30	1.5	5.0%
1 tonne of wheat/Bulk 52,000 DW/100%	220	14	6.4%
1 car/Multipurpose Ro-Ro/50%	21,000	558	2.7%
1 refrigerator/ Container 6600 TEU/80%	550	9	1.6%

[a]Market prices in fall 2002.

Table 11.3 Comparison of costs for different modes of transport on the Barcelona–Genoa route (costs expressed in Euros).

Transport mode	Trailer 16.6 m	40 feet container
Railway	—	1300–1500
Road	900	—
Short sea, Ro-Ro	1300	1200

Source: Abeille et al. (1999).

Table 11.4 Environmental cost of road and sea transport, in euros per 1000 tonne-km.

Transport mode/source of cost		Environmental cost, €/1000 tonne-km
Truck transport	Emissions	7–12
	Accidents	4–7
	Noise	5–15
	Congestion	5–12
	Total cost	21–46
Container vessel, 3000 TEU		6–8
Bulk carrier, 40,000 dwt		2–3
Ro-Ro cargo ship, 3000 lane-meters		33–48

Source: Kristensen (2002).

11.1.5 Maritime safety regime

Given the factors pointed out earlier in this Section, it is not completely obvious that safety is an important issue to companies in the maritime transport domain. One may, however, argue that advanced modern ship design achieves high levels of safety, that training of crew is now of a fairly high standard, and that shipping companies are relatively advanced when compared with similar types of businesses. In addition to this, shipping is subject to rigorous control and continuously has the attention of both governments and the public. Table 11.5 shows that seaborne transport today is strictly regulated as a result of a series of internationally ratified safety conventions.

The average loss rate for the world fleet, measured in annual percent relative to the fleet at risk, has been reduced significantly during the same period studied in Table 11.5. In 1900 the average loss rate was 3%. This had been reduced to 0.5% in 1960, and further down to 0.25% in 2000.

Table 11.5 Milestones in maritime safety[a].

Year	Initiative or regulation
1914	Safety of Life at Sea (SOLAS): Ship design and lifesaving equipment
1929	First international conference to consider hull subdivision regulations
1948	The International Maritime (Consultative) Organization (IMO) is set up as a United Nations agency
1966	Load Line Convention: Maximum loading and hull strength
	Rules of the road
	The International Association of Classification Societies (IACS): Harmonization of classification rules and regulations
1969	Tonnage Convention
1972	International Convention on the International Regulations for Preventing Collisions at Sea (COLREG)
1974	IMO resolution on probabilistic analysis of hull subdivision
1973	Marine Pollution Convention (MARPOL 73)
1978	International Convention on Standards for Training, Certification and Watchkeeping for Seafarers (STCW)
1979	International Convention on Maritime Search and Rescue (SAR)
1988	The Global Maritime Distress and Safety System (GMDSS)

[a]An excellent summary is given by Vassalos (1999).

Table 11.6 Recent maritime accidents and responses.

Background	Response
Need to increase maritime safety, protection of the marine environment, and improve working conditions on board vessels. Flag state control is not regarded as efficient enough	Declaration adopted in 1980 by the Regional European Conference on Maritime Safety that introduced Port state control of vessels, known as the *Paris Memorandum of Understanding (MOU)*[a]
The loss of Ro-Ro passenger ferry *Herald of Free Enterprise* (Dover, 1987), and the loss of passenger ferry *Scandinavian Star* (Skagerak, 1990)	IMO adopts the *International Management Code for the Safe Operation of Ships and for Pollution Prevention (ISM Code)*: Ship operators shall apply quality management principles throughout their organization
Grounding of oil tanker *Exxon Valdez* in Alaska 1989, resulting in oil spill and considerable environmental damages	US Congress passes the *Oil Pollution Act (OPA '90)*: Ship operators have unlimited liability for the removal of spilled oil and compensation for damages[b]
The flooding, capsize and sinking of the Ro-Ro passenger vessel *Estonia*	*Stockholm agreement (1995)*: NW European countries agree to strengthen design requirements that account for water on deck
A need for greater consistency and cost-effectiveness in future revisions of safety regulations	*Interim Guidelines for the Application of Formal Safety Assessment (FSA) to the IMO Rule-Making Process*, 1997
Hull failure and sinking of the oil tanker *Erika* off the coast of France, 1999	European Commission approves a directive calling for tighter inspection of vessels, monitoring of classification societies, and elimination of single-hull tankers[c]
Oil tanker *Prestige* sinks off the coast of Spain, 2002	The European Commission speeds up the implementation of ERIKA packages 1 and 2
Spreading of exotic organisms through dumping of ballast water has resulted in widespread ecosystem changes	Increased focus on research on these issues, and introduction of new regulation and control measures[d]

[a]http://www.parismou.org/
[b]OPA (full text): http://www.epa.gov/region09/waste/sfund/oilpp/opa.html
[c]Erika Package 1: http://www.nee.gr/Files/erikal.pdf
[d]Australian initiative: http://www.ea.gov.au/coasts/pollution/

It is too early at this stage of the Chapter to discuss whether the safety level in maritime transport is acceptable. However, it can on the other hand be argued that there is a case for increased safety efforts unless it can be shown that this cannot be defended with regard to the resources spent. Another way of thinking is that safety should be on the agenda as long as accidents are rooted in trivial errors or failures (very often human errors). Thirdly, ship accidents should have our attention as long as they lead to fatal outcomes and the consequences for the environment are unknown. The examples given in Table 11.6 show that maritime safety still is on the agenda and will continue to be so in the foreseeable future.

11.1.6 Why safety improvement is difficult

Despite the fact that safety is at the top of the agenda both in the shipping business itself and by regulators, it may appear that the pace of safety improvements is rather low. The degree to which this general observation is true will not be discussed in any depth here.

However, some explanations for such a view are presented below:

- *Short memory*: When safety work is successful, few accidents tend to happen. This lack of feedback can make people believe, both on a conscious and unconscious level, that they are too cautious and therefore can relax on the strict requirements they normally adhere to. An even simpler explanation is complacency, i.e. that people tend to forget about the challenges related to safety if no accidents or incidents give them a 'wake-up call'. This weakness seems to degrade the safety work effectiveness of both companies and governments.
- *Focus on consequences*: People have a tendency to focus on the consequences of an accident rather than its root causes. There is, for instance, great uncertainty attached to whether oil pollution is reduced in the best way by double-hull tankers or heavy investment in containment and clean-up equipment. Doing something about consequences is generally much more expensive compared to averting, or reducing the probability and the initiating causes of an accident.
- *Complexity*: Safety involves technological, human and organizational factors, and it can be very difficult to identify the most cost-effective set of safety-enhancing measures across all potential alternatives. There is also a tendency among companies, organizations and governments to go for technical fixes, whereas the root causes in a majority of cases are related to human and organizational factors. It seems to be easier to upgrade vessels than to change people's behaviour.
- *Unwillingness to change*: Humans have a tendency to avoid changing their behaviour, also when it comes to safety critical tasks. People sometimes express their understanding of the need for change, but in practice use all means to sabotage new procedures. In some companies, 'cutting corners' is unfortunately a natural way of behaving.
- *Selective focus*: Formal safety assessment (i.e. a risk analysis and assessment methodology described in a later Section) is in general seen as a promise for more efficient control of risk. However, such methods may be criticized in a number of ways: they oversimplify the systems studied, a number of failure combinations are overlooked due to the sheer magnitude of the problem, and operator omissions (e.g. forgetting or overlooking something) are not addressed in such models.

11.1.7 The risk concept

The concept of risk stands central in any discussion of safety. With reference to a given system or activity, the term 'safety' is normally used to describe the degree of freedom from danger, and the risk concept is a way of evaluating this. The term 'risk' is, however, not only used in relation to evaluating the degree of safety and, as outlined in Table 11.7, the risk concept can be viewed differently depending on the context.

Engineers tend to view risk in an objective way in relation to safety, and as such use the concept of risk as an objective safety criteria. Among engineers the following definition of risk is normally applied:

$$R = P \cdot C \qquad (11.1)$$

where P = the probability of occurrence of an undesired event (e.g. a ship collision) and C = the

Table 11.7 Different aspects of the risk concept.

Aspect	Comments
Psychological	People often relate to risk in a subjective and sometimes irrational way. Some people are even attracted to risk
Values/ethics	Risk can be perceived in the light of fundamental human values: life is sacred, one should not experiment with nature, and every individual has a responsibility for ensuring safety
Legal	Risks and safety are to a large degree controlled by laws and regulations, and people might therefore be liable for accidents they cause
Complexity	The nature of accidents is difficult to understand because so many different influencing factors and elements are involved: machines, people, environment, physical processes and organizations
Randomness	There is often a fine line between safe and unsafe operation. A lack of system understanding may lead to a feeling that accidents happen at random
Delayed feedback	It is difficult to see the cause and effect mechanisms and thereby whether introduced safety measures have a positive effect on safety. Some measures even have to be applied for a considerable period of time before they have a real effect on system safety

Table 11.8 Different perception of risk.

Factor	Negative → higher perceived risk	Positive → lower perceived risk
Is the hazard confronted as a result of a personal choice or decision?	Involuntary	Voluntary
Is the consequence (effect) of the act evident?	Immediate	Delayed
Is the cause and effect mechanism clear to the decision-maker?	Uncertain	Certain
Is the individual in a 'pressed' situation that leaves no alternatives?	No alternatives	Alternatives
Does the decision-maker experience some degree of control?	No control	Have control
Risk at work is not the same as risk in your spare time	Occupational	Hobby, sport
Is the risk unknown in the sense that it is seldom or still not experienced?	'Dread' hazard	Common
Are the consequences given once and for all?	Irreversible	Reversible

expected consequence in terms of human, economic and/or environmental loss.

Equation (11.1) shows that objective risk has two equally important components, one of probability and one of consequence. Risk is often calculated for all relevant hazards, hazards being the possible events and conditions that may result in severity. For example, a hazard with a high probability of occurrence and a high consequence has a high level of risk, and a high level of risk corresponds to a low level safety for the system under consideration. The opposite will be the case for a hazard with a low probability and a low consequence. Safety is evaluated by summing up all the relevant risks for a specific system. This objective risk concept will be studied in greater detail in later Sections.

An important question is how people relate to and understand the concept of risk. Table 11.8 gives a brief overview of some of the factors that determine the subjectively experienced and perceived risk. In performing risk analyses, engineers should always keep these subjective aspects in mind so as to improve communication with different individuals/groups and be able to achieve mutual understanding of complicated safety issues, etc.

11.1.8 Acceptable risk

Some might argue that any risk is unacceptable. This view is questioned by Rowe (1983), who gives the following reasons for the opposite standpoint (i.e. that some risks are indeed acceptable):

- Threshold condition: a risk is perceived to be so small that it can be ignored.
- Status quo condition: a risk is uncontrollable or unavoidable without major disruption in lifestyle.
- Regulatory condition: a credible organization with responsibility for health and safety has, through due process, established an acceptable risk level.
- De facto condition: a historic level of risk continues to be acceptable.
- Voluntary balance condition: a risk is deemed by a risk-taker as worth the benefits.

Rowe (1983) further outlines three different models for how an acceptable level of risk is established in society:

1. *Revealed preferences:*
   - By trial and error society has arrived at a near optimum balance between risks and benefits.
   - Reflect a political process where opposing interests compete.
2. *Expressed preferences:*
   - Determine what people find acceptable through a political process.
   - The drawback is that people may have an inconsistent behaviour with respect to risk and don't see the consequences of their choices.
3. *Implied preferences:*
   - Might be seen as a compromise between revealed and expressed preferences.
   - Are a reflection of what people want and what current economic conditions allow.

Our attitude to risk is also reflected by our views on why accidents happen. Art, literature and oral tradition all reflect a number of beliefs that are rooted in culture and religion (see also Kouabenan, 1998):

- Act of God: an extreme interruption with a natural cause (e.g. earthquake, storm, etc.).
- Punishment: people are punished for their sins by accidents.
- Conspiracy: someone wants to hurt you.
- Accident proneness: some people are pursued by bad luck all the time. They will be involved in more than their fair share of accidents.
- Fate: "What is written in the stars, will happen." You cannot escape your fate.
- Mascots: an amulet may protect you against hazards.
- Black cat: you should sharpen your attention.

11.1.9 Conflict of interest

As already pointed out earlier in this Section, in any maritime business activity, and almost any other activity for that matter, there will be conflicts of interests

Table 11.9 Conflicts of interests between the involved parties in the case of an oil terminal/refinery located in a coastal area.

Actor	Potential positive effects	Potential negative effects
Oil company	Earn money Enter new markets Increases their market share	Production stops – lost income Some liability for accidents Bad reputation because of pollution
Shipowners	New/alternative trades Earn money	Limited liability for accidents Loose contract Be grey/blacklisted
Employees	Employment Income Improved standard of living	Exposed to accidents Few other job alternatives
Population	Infrastructure improvements More jobs	Air pollution Restriction on land-use and outdoor life
Local economy	Tax income Improved service to the public Increased population	Local economy becomes highly dependent on the terminal Increased wage cost Traditional businesses unable to attract competent personnel Pollution may affect primary industries (fishing, fish farming)
Society at large	Contribution to national economy	Must take most of the cost in case of major accidents and oil spill

between the different parties involved and affected by that activity. For example, consider an oil company that is planning to establish an oil terminal with refinery capacity in a local community somewhere on the coast. For the different parties affected by this establishment there will be both positive and negative effects, as illustrated by Table 11.9. For the community population the project may, for example, result in infrastructure improvements and an increased number of jobs, both of which are obvious benefits. On the other hand the project may result in negative effects such as air pollution and restrictions on land-use and outdoor life. A problem related to activities such as the oil terminal is that the positive and the negative effects, including the income and costs, may be unevenly distributed between the affected parties.

11.1.10 Expertise and rationality

It should be clear from the preceding observations in this Section that decisions related to safety are difficult for a number of reasons. As engineers we are inclined to perform rational analyses using computational methods. On the other hand it has been commented that as humans we often have an irrational attitude to risks. In addition, people view risks differently, and there may be conflicts of interests in safety-related issues. The dominant view in the field of safety assessment is that formal risk analysis methods should be used on many types of activities, and that the public in general is often ill informed and therefore should have little influence on such complicated matters. Perrow (1999) has questioned this view. Some of his observations of contemporary practices in risk analyses are as follows:

- Expected number of fatalities: whether you die from diabetes or murder is irrelevant.
- Who is at risk? Whether 50 unrelated persons from many communities or 50 persons from a small community of 100 inhabitants die in an accident is also irrelevant.
- People are to a large degree sceptical about nuclear power plants but still continue to smoke. It is irrational to dread the nuclear plants that have shown excellent safety performance, whereas smoking is an undisputed factor of risk in relation to lung cancer. However, this argument totally neglects the fact that smoking is a result of intense marketing and advertisement.
- Risk assessment ignores social class distribution of risk, as may be illustrated by the corporate vice-president's dilemma: By investing USD 50 million in a proposed safety measure, the life of one extra worker can be saved. However, by rejecting the proposal the company will avoid USD 20 million on price increases and be able to give USD 30 million in dividends. The last option is chosen as the price is very high given the depressed labour market.

Perrow (1999) also gives an interesting discussion of three basic forms of rationality: 'absolute', enjoyed by economist and engineers, 'bounded or limited' rationality proposed in cognitive science and organizational psychology, and 'cultural or social' rationality mainly practised by the public. The author delivers many interesting arguments for these alternatives to absolute rationality.

11.2 Regulatory authorities

11.2.1 Introduction

This Section provides an outline of the regulation of safety in seaborne transport. The control of safety is primarily based on the rules (conventions and resolutions) given by the United Nations agency the International Maritime Organization (IMO). These rules have international application but some reference will also be made to national regulations by taking the Norwegian legal regime as an example.

When we use the term *safety*, it will encompass:

- Safety and health of persons
- Safety of vessel
- Environmental aspects

Safety is regulated on the basis of different *legal sources*, the key ones of which are the following:

- International laws and regulations
  - UN Law of the Seas (UNCLOS)
  - European Union (EU) Directives
- National laws and regulations
- Case law (court rulings)
- National territorial zones
- IMO conventions and resolutions
- Classification construction rules
- Port State control MOU guidelines

It should also be kept in mind that there are a number of *actors* that have an impact on safety. The primary ones are:

- Flag and Port State control (Maritime Directorate)
- International Maritime Organization (IMO)
- Classification Societies
- Insurance companies
- Charterer, cargo owner

11.2.1.1 *The structure of control*

Seen from a national point of view, the regulation of safety is based on a set of international rules that is adopted by the legislative assembly (Parliament) (see Figure 11.4). The concrete rules and regulations are written or translated by the responsible government branch (Foreign and International Trade Department). The role of the Maritime Administration is to ensure that regulations are followed by the shipowners through proper control and certification. This is what is termed Flag State control (FSC). The figure also shows that the Classification Society has a role in the certification process, although this is primarily

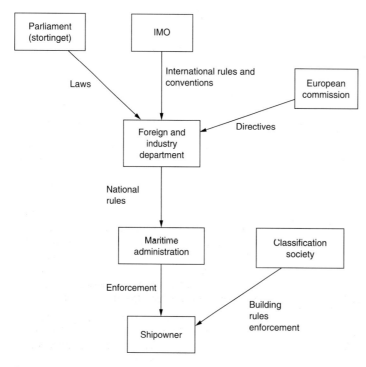

Figure 11.4 Regulation of maritime safety.

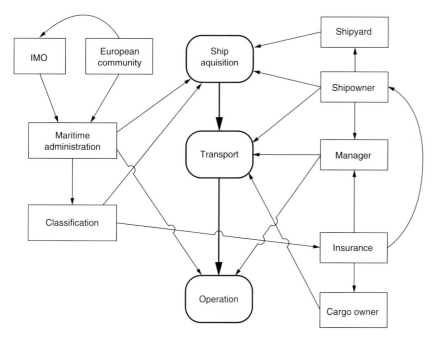

Figure 11.5 Actors and interactions in safety control.

related to the insurance of the vessel, cargo and third-party interests.

The control of safety in shipping is complex for a number of reasons:

- International, regional and national laws and regulations.
- Control is exercised by a number of agencies.
- Control affects the various life-cycles of the vessel.

A simple outline of the number of actors and interactions is shown in Figure 11.5. It should also be kept in mind that shipping as an internationally oriented business is highly competitive and is also influenced by dramatic economic cycles. Seen from the shipowner's perspective, the safety standard is a result of the cross-pressure between control and commercial competition (see Figure 11.6).

11.2.2 International maritime organization (IMO)

The main principle in the regulation of shipping are harmonized national rules based on international conventions and resolutions given by the IMO. This is an organization under the United Nations system. Its prime function is to establish rules based on participation by the member states. IMO has a complex set of committees that draft and revise regulations which are adopted by the General Assembly. A new regulation has to be ratified by a minimum number of states before it enters into force. IMO has no power to enforce the international safety regulations. This is the task of the member states in their role as so-called Flag States.

11.2.2.1 SOLAS

The main objective of the SOLAS Convention (Safety of Life at Sea) is to specify minimum standards for the construction, equipment and operation of ships (SOLAS, 2001). The present version of the SOLAS Convention was adopted in 1974 and was later revised and supplemented with so-called Protocols. It entered into force in 1980. SOLAS-74 has 12 articles and 12 chapters with the following specific requirements:

I. General provisions
II.
  1. Construction – Subdivision and stability, machinery and electrical installations
  2. Construction – Fire protection, detection and fire extinction
III. Life-saving appliances and arrangements
IV. Radiotelegraphy and radiotelephony
V. Safety of navigation
VI. Carriage of grain
VII. Carriage of dangerous goods
VIII. Nuclear ships
IX. Management for the safe operation of ships (ISM Code)

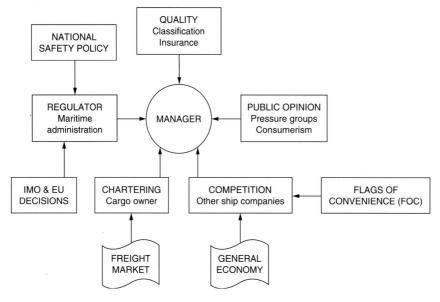

Figure 11.6 Safety subject to regulation and competition.

X. Safety measures for high speed craft
XI. Special measures to enhance safety
XII. Additional safety measures for bulk carriers.

The convention has been amended a number of times since its adoption in order to be in accordance with the development of new technology and new safety knowledge. The regulation is to a large degree prescriptive by specifying solutions in minute technical detail. Performance criteria are only applied to a limited degree. This has two main drawbacks: technical solutions specified in SOLAS may become obsolete even before it enters into force, and the lack of focus on performance criteria does not stimulate the designer to find or invent better solutions.

The SOLAS-74 Convention has been ratified by most nations. In order to become effective, the convention has to be translated into the official national language and be formally adopted by the government branch. The implementation of SOLAS-74 by the Norwegian Flag State is given in:

- Regulation of 15 June 1987 No. 506 on inspection for issuing of certificates for passenger and cargo vessels and barges, etc.
- Regulation of 15 September 1992 No. 695 on building of passenger and cargo vessels and barges.

11.2.2.2 International Convention on Load Lines, 1966

It has long been recognized that limitations on the draught to which a ship may be loaded make a significant contribution to her safety. These limits are given in the form of freeboard, which, besides external weather-tightness and watertight integrity, constitute the main requirement of the Convention.

The first International Convention on Load Lines (ILLC, 2002), adopted in 1930, was based on the principle of reserve buoyancy. It was also recognized then that the freeboard should ensure adequate stability and avoid excessive stress on the ship's hull as a result of overloading.

The regulations take into account the potential hazards present in different geographical zones and different yearly seasons. The technical annex contains several additional safety requirements concerning doors, freeing ports, hatchways and other items. The Convention includes Annex I with the following four chapters:

I. General
II. Conditions of assignment of freeboard
III. Freeboards
IV. Special requirements for ships assigned timber freeboards

Annex II covers zones, areas and seasonal periods, and Annex III certificates, including the International Load Line Certificate.

11.2.2.3 STCW Convention

The International Convention on Standards of Training, Certification and Watchkeeping for Seafarers (STCW) was the first to establish basic requirements on training, certification and watchkeeping for seafarers at an international level. The technical provisions of the Convention are given in an Annex containing six chapters:

1. General provisions.
2. Master-deck department: This chapter outlines basic principles to be observed in keeping a navigational watch. It also lays down mandatory minimum requirements for the certification of masters, chief mates and officers in charge of navigational watches on ships of 200 grt or more.
3. Engine Department: Outlines basic principles to be observed in keeping an engineering watch. It includes mandatory minimum requirements for certification of officers of ships with main propulsion machinery of 3000 kW.
4. Radio Department.
5. Special requirements for tankers.
6. Proficiency in survival craft.

The 1995 amendments represented a major revision of the 1978 Convention (STCW, 1996). The original Convention had been criticized on many counts. It referred to vague phrases such as 'to the satisfaction of the Administration', which admitted quite different interpretations of minimum manning standards. Others criticized that the Convention was never uniformly applied and did not impose strict obligations on the Flag States regarding its implementation.

11.2.2.4 MARPOL

Both SOLAS and ICCL have an indirect effect on preventing pollution from ships. However, there was a dramatic development of specialized tankers after the Second World War in terms of ship size and complexity of operation. The International Convention for the Prevention of Pollution from Ships (MARPOL) seeks to address the environmental aspects related to design and operation of these ships more directly (MARPOL, 2002).

The Convention prohibits the deliberate discharge of oil or oily mixtures for all seagoing vessels, except tankers less than 150 gross tons and other ships less than 500 gross tons, in areas denoted 'prohibited zones'. In general these zones extend at least 50 n. miles from the coastal areas, although zones of 100 miles and more were established in areas which included the Mediterranean and Adriatic Seas, the Gulf and Red Sea, the coasts of Australia and Madagascar, and some others.

MARPOL introduces a number of measures:

- Segregated ballast tanks (SBT): ballast tanks only used for ballast as cargo oil is prohibited. Reduces cleaning problem.
- Protective location of SBT: SBT arranged in bottom or sides to protect cargo tanks against impact or penetration.
- Draft and trim requirements: to ensure safe operation in ballast condition.
- Tank size limitation to limit potential oil outflow.
- Subdivision and stability in damaged condition.
- Crude oil washing (COW).
- Inert gas system (IGS) for empty cargo tanks.
- Slop tanks for containing slop, sludge and washings.

The implementation of MARPOL is based on a complex scheme where ship size and whether it is an existing or a new building determine which requirements apply. An interesting outline of the tankship technology has developed, and the present environmental challenges are given in NAS (1991).

11.2.2.5 The ISM Code

The introduction of the International Management Code for Safe Operation and Pollution Prevention (ISM, 2002) represented a dramatic departure in regulatory thinking by the IMO. It acknowledges that detailed prescriptive rules for design and manning have serious limitations. Inspired by principles from quality management and internal control, the ISM Code will stimulate safety consciousness and a systematic approach in every part of the organization both ashore and onboard.

The ISM Code itself is a fairly short document of about 9 pages. The main intention with ISM is to induce the shipping companies to create a safety management system that works. The Code does not prescribe in detail how the company should undertake this, but just states some basic principles and controls that should be applied. The philosophy behind ISM is commitment from the top management, verification of positive attitudes and competence, clear placement of responsibility and quality control of work processes.

The Code states the following objectives for the adoption of a management system:

1. To provide for safe practices in ship operation and a safe working environment;
2. To establish safeguards against all identified risks; and
3. To continuously improve the safety management skills of personnel ashore and aboard, including preparing for emergencies related both to safety and environmental protection.

The Code has 13 chapters, which are listed in Table 11.10. When addressing the effect of ISM on safety, there are two key aspects: the material content of the regulation, and what is an acceptable

Table 11.10 Organization of the ISM Code elements.

Management function	Chapter	ISM element
Objective, policy	1.2.2	Provide safe practices, establish safeguards and continuously improve skills
	2	Safety and environmental protection policy
Requirements	1.2.3	Compliance with rules and regulations
		Other IMO Conventions: SOLAS, STCW, MARPOL, COLREG, Load lines, etc.
	1.3	Functional requirements: policy, instructions, authority, communication, accident reporting, emergency preparedness, audits
Controls	3	Company responsibilities and authority
	4	Designated persons
	6	Resources and personnel
	7	Development of plans for shipboard operations
	8	Emergency preparedness
	10	Maintenance of the ship and equipment
Safety management system	11	Documentation
Implementation of controls	5	Master's responsibility and authority
Monitoring of the system	9	Reports and analysis of non-conformities, accidents and hazardous occurrences
	12	Company verification, review and evaluation
The periodic system review	13	Certification, verification and control

compliance with the Code. In order to implement ISM correctly, certain elements are required:

- Documentation of how the ISM Code will be implemented.
- External verification and certification.
- Reporting (logging) of the safety management processes.
- Internal (company) verification.

Apart from this, the *Guidelines on Implementation of ISM* (ISM, 2002) is fairly vague on how to verify that a *safety management system* (*SMS*) conforms with the Code. It admits that certain criteria for assessment are necessary, but also warns against the emergence of prescriptive requirements and solutions prepared by external consultants. The obvious philosophy behind this attitude is that the SMS should be an integral part of the management thinking of the company. In that sense the SMS should reflect the objectives of the Code but otherwise be implemented in such a way that it is viewed as an element of the culture, organization and decision-making processes of the company.

The ISM Code specifies certain requirements for the safety management system (SMS) of the operating company. In order for the SMS to work, certain distinct functions have to be in place. The core of the SMS is made up of certain *controls* which are defined in terms of, see ISM (2002):

- Responsibility and authority.
- Provision of resources and support.
- Procedures for checking of competence and operational readiness, training, and shipboard operations.
- Establishing minimum standards for the maintenance system.

Another key feature of the ISM concept is the definition of a *monitoring function*, which is based on audits and reporting of events. The audit will ensure that errors and shortcoming in the SMS are corrected and that the system is updated in view of new requirements and experience gained. The auditing and event reporting will also address operational errors and failures directly and thereby lead to corrective action in terms of modified systems and improved procedures.

Chapter 13 states that the company should have a *certificate of approval* which documents that the SMS is in accordance with the intentions and specific requirements of the ISM Code. It should be kept in mind that ISM has a relation to existing or traditional regulatory approaches for design, equipment, training and emergency preparedness. The Code should be understood in the context of existing safety regulations that have already been mentioned: SOLAS, ILLC, MARPOL, COLREG and STCW. ISM does not address any of the specific requirements in these conventions, but just assumes that the management system should ensure that they are met.

The ISM Code is discussed further in Section 11.5 on safety management.

11.2.3 Flag State Control

As already pointed out, the set of internationally accepted safety rules and regulations are not enforced by the IMO but by the so-called Flag States. The national maritime administration is acting as Flag State on behalf of the country in question. Based on plans, technical documentation and inspections, a ship is subject to registration and awarded the necessary safety-related certificates.

11.2.3.1 The Seaworthiness Act

Each country has to give a legal basis for exercising this role as Flag State. In Norway the competence of the Maritime Administration is laid down in the Seaworthiness Act (Falkanger *et al.*, 1998). The law regulates shipping activity in relation to the public sphere and also defines the role of the national Maritime Administration (in Norway, the Maritime Directorate, in the UK The Maritime and Coastguard Agency). The key functions specified by the law are:

- Safety control activity in general
- The competence of the Maritime Directorate
- Investigation of accidents (Sea Court)
- Inspection and detention (withholding a vessel)
- Certificates
- Safety and occupational health-related activities onboard
- Equipment standard
- Cargo condition and safety
- Manning and working hours
- Control of passenger vessels
- Responsibility of Master and Owner

Section 2 of the Seaworthiness Act defines *seaworthiness* as follows:

> *A ship is considered unseaworthy when, because of defects in hull, equipment, machinery or crewing or due to overloading or deficient loading or other grounds, it is in such a condition, that in consideration of the vessel's trade, the risk to human life associated with going to sea exceeds what is customary.*

The law basically applies to vessels greater than 50 gross register tons, but the Administration (Flag State) may decide that other vessels also have to be built in accordance with the rules under the law.

The jurisdiction of this law is in principle limited to Norwegian vessels. The maritime administration acts in this manner as *Flag State*. However, international law has developed during the last decades and today accepts that a nation may exercise some control and, if necessary, detain a foreign vessel viewed as a risk to human life (passenger transport) and coastal environment (oil pollution). The maritime administration in that sense acts as a *Port State*. We will return to this role later.

Shipping activity in Norway or more precisely Norwegian national register vessels (NOR) are subject to both private and public law. The international register in Norway (NIS) is regulated through a separate act. Some of the key laws are (Sjøfartsdirektoratet, 1988):

1. The Maritime Code (Sjøloven av 24. juni 1994 nr. 39).
2. The Seaworthiness Act (Sjødyktighetsloven av 9. juni 1903 nr. 7).
3. The Seaman's Act (Sjømannsloven av 30. Mai 1975 nr. 18).
4. Norwegian International Register Act (NIS-loven av 12. juni 1987 nr. 48).

11.2.3.2 Delegation of Flag State Control

Some Flag States accept foreign vessels and have become what is commonly termed international or offshore registers. The standard of some of these registers has been questioned and they have been branded as *Flags of Convenience* (FOC). They are suspected to offer registration to foreign owners mainly for economic reasons and are viewed as having a lenient enforcement of safety regulations. Another characteristic is the lack of or minimal maritime administration. A common practice is to delegate the control to an independent certifying authority, primarily classification societies and even consultants.

11.2.3.3 Effectiveness of Flag State Control

Flag State control (FSC) has for years been a key principle in the safety control of shipping. Based on internationally accepted rules, the safety is to be ensured by the maritime authority of the nation of registration of the vessel. It has, however, become evident that different Flag States have varying competence and motivation to undertake their role. This was clearly demonstrated in a small survey of the *SAFECO I* project (Kristiansen and Olofsson, 1997). Table 11.11 shows the loss rate for some selected Flag States. It is clear that the annual loss rate may vary by a factor of more than 10. This great variation can even be observed among European Flag States, as shown in Figure 11.7.

11.2.3.4 The Flag State Audit Project

The Seafarers International Research Centre (SIRC) at Cardiff University has undertaken an assessment of the performance of the main Flag States (Alderton and Winchester, 2001). Some shipowners prefer to register their fleet under a flag other than the national one. This has been a practice for years but has gained renewed importance during the present trend toward globalization and deregulation of industry

Table 11.11 Total loss rate by flag for vessels greater than 100 grt.

Flag	Fleet size 1993	Loss rate per 1000 shipyears 1994–95
Denmark	599	3.1
France	769	1.2
Germany	1234	1.3
Netherlands	1006	0.84
Norway	1691	1.3
United Kingdom	1532	2.7
North Europe selected	6831	1.8[a]
Cyprus	1591	5.3
Greece	1929	1.8
Italy	1548	1.7
Malta	1037	5.5
Portugal	307	—
Spain	2111	3.0
Mediterranean selected	8523	3.2[a]
Japan	9950	1.3
Korea (South)	2085	4.2
Philippines	1469	2.3
Singapore	1129	0.4
USA	5646	2.4
Bahamas	1121	2.6
Liberia	1611	1.9
Panama	5564	3.9
Worldwide	80655	2.4

[a]Weighted estimation on the basis of the selected countries.
Source: *World Fleet Statistics* and *Casualty Return*, Lloyd's Register, London.

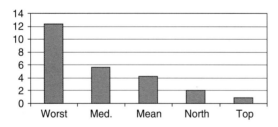

Figure 11.7 Loss rate of European fleet segments, 1994–95: loss rate per 1000 ship-years. (Med. – Mediterranean countries, North – Northern and Central Europe.)

and trade. Some of these flags lack both motivation and competence to enforce the international safety standards set by IMO. These flags have been termed Flags of Convenience (FOC). However, today it seems too simple to distinguish between national flags and FOCs. The International Transport Workers' Federation therefore commissioned a study of the performance of the various flags operating today.

The first step in the study was to define a set of criteria for ranking flags. It was decided to create an index (FLASCI) based on a weighted ranking of following factors:

1. The nature of the maritime administration
2. Administrative capacity
3. Maritime law
4. Seafarers' safety and welfare
5. Trade union law
6. Corruption
7. Corporate practice

The relative weighting and detailed factors assessed are summarized in Table 11.12. Data were retrieved from a literature search, and review of Internet sources and other available information on the Flag States such as Port State control statistics. The FLASCI scores are summarized in Table 11.13.

The Flag States got scores of between 19 and 84 and inspection of the findings suggested that the Flag States might be grouped into five categories, as shown in Table 11.13. The study clearly shows that flags show greater variation in performance than has generally been accepted. Some of the main findings were:

- Some of the so-called second registers perform as well as the best national registers: Norway (NIS), Denmark (DIS), Germany (GIS) and France (Kerguelen Islands).
- A few of the established FOCs are performing relatively well: Bermuda (63). Other FOCs such as Bahamas (43) and Liberia (43) are ranked lower but are still better than the worst performing.
- There seems to be a clear correlation between low performance and short operation as flag (new entrants). Port State control of these flags shows a quite high detention rate as shown in Table 11.14.

The last point can be explained by the apparent dynamics in the 'market' of Flag States. FOCs will, after some time when they are more established, be under pressure to improve their performances. As they eventually do this, it will open a market for new flags that will offer a more lenient safety regime. The SIRC study also showed that the fleets of the new entrants have a much higher growth rate than the average rate for the world fleet.

The SIRC study also analysed the working conditions on board and it was confirmed to be a less attractive climate on new entrant flag vessels. This is discussed further by Kristiansen (2005).

11.2.4 Port state control

11.2.4.1 UNCLOS

The basis for international shipping is the principle of freedom of the seas. The international legal basis is defined in the *United Nations Convention on the*

Table 11.12 Flag State Conformance Index (FLASCI).

Flag State fleet 15%	Port State control rates	Own-citizen beneficial ownership
	Casualty rates	
	Pollution incidence	Abandonment of crews
	Own-citizen labour force participation	Appearance in crew complaints DB
FS administrative capacity 30%	Death records	Casualty investigation capacity
	Crew records of service	Statistics of ships, owners and labour force
	Health screening procedures and records	Certification of seafarers
	Accessibility of consular services	Involvement in training and education
	Enforcement of IMO and ILO Conventions	
FS maritime law 20%	Ratification of IMO and ILO Conventions	Specialist law practitioners
	Provisions of maritime legal code	Location of registry
	Publication of relevant law reports	'Ownership' of registry
Miscellaneous maritime 5%	Maritime welfare support and maritime charities	Government ministries with maritime remit
		Stock exchange maritime listings
	Maritime interest groups	State-owned shipping
Trade union law 10%	Legal rights for migrant labour	Provision for trade union recognition
	Independent trade unions	Enforcement of trade union recognition procedures
	Mediation/arbitration procedures	
Corruption 10%	Probity of public officials	Integrity of political institutions and legal process
	Misapplication of public funds	Corporate integrity
Corporate practice 10%	Regulation of financial institutions	Regulation of accounting standards
	Regulation of non-resident companies	Legal definition of corporate public responsibility

Source: Alderton and Winchester (2001).

Table 11.13 Ranking of selected Flag States.

Category	Selected flags (score)	Score range
Traditional maritime nations	NOR (84), UK (80), DIS (77), NIS (77),	84–72
Centrally operated second registers	Netherlands (76), GIS (75), Kerguelen Islands (72)	
Semi-autonomous second registers	Hong Kong (64), Bermuda (63), Latvia (60), Cayman Islands (62), Estonia (58)	64–58
Established open registers (seeking EU membership)	Cyprus (50), Malta (49), Russia (48), Bahamas (43), Liberia (43), Panama (41)	50–41
National registers		
New open registers	Marshall Islands (36), Ukraine (36), Honduras (35), Lebanon (35)	36–35
New entrants to the open register markets	St. Vincent and Grenadines (30), Bolivia (30), Belize (27), Equatorial Guinea (24), Cambodia (19)	30–19

Source: Alderton and Winchester (2001).

Table 11.14 Detention rate for 'new entrant' flags.

	Belize	*Bolivia*	*Cambodia*	*Equatorial Guinea*
Asia–Pacific MOU (average 7%)	24.7%	No data	30%	11.1%
Paris MOU (average 9%)	31.4% Blacklisted	70%	24.8% Blacklisted	14.3%
USCG (average 5%)	50.6% Targeted	No data	Too few inspections	28.6% Targeted

Source: Alderton and Winchester (2001).

Law of the Sea or *UNCLOS* (AMLG, 2004). The principle has the following key elements:

- Ships may sail without restriction in all waters on innocent passage (Article 17).
- The country of registration (Flag State) has the sole jurisdiction over the ship (Article 91).
- Other countries have limited jurisdiction even in own territorial sea.

The coastal state has at the outset the following rights:

- The outer limit of the territorial sea is 12 nm from the coast (baseline) within which it has full jurisdiction.
- The exclusive economic zone stretches out to 200 nm:
  - Very limited control jurisdiction.
  - Certain rights to take measures to preserve the marine environment.
  - However, the control should be exercised in accordance with international practice or non-discrimination against foreign vessels (Article 227).

The above means that the coastal states have to exercise their rights with respect to pollution hazards with delicacy. This becomes even more complicated when a state has both a substantial international trading fleet and a threatened coast. A good example is one of the initiatives of Spain and France in the aftermath of the *Prestige* accident. In an EU communication the following is stated:

> ... INVITES Member States to adopt measures, in compliance with international law of the sea, which would permit coastal States to control and possibly to limit, in a non-discriminatory way, the traffic of vessels carrying dangerous and polluting goods, within 200 miles of their coastline ...

This position has been strongly opposed by INTERTANKO, which stresses that any measure in this area must adhere to international law and more specifically UNCLOS.

11.2.4.2 MOU PSC

The basic principle is that under the international safety conventions a certificate issued by Flag State A is equivalent to a certificate issued by state B. However, a Port State may challenge a certificate if there are indications that the condition of the foreign vessel is not in accordance with the particulars of the certificate.

The legal basis for Port State control (PSC) in Europe is found in the so-called Paris MOU (MOU, 2004), the 'Memorandum of Understanding on Port State Control' signed in 1982 by 19 European states and Canada.

The introduction of PSC was initially heavily opposed by shipping interests who feared that it would have a negative impact on the principle of equal market access and free competition. But in the end all involved parties acknowledged the shortcomings of Flag State control and the necessity of giving Port States authority to control shipping in their own waters.

The MOU has been given legal basis in national and international law, for instance in Norway by Regulation of 1 July 1996, No. 774, regarding control of foreign vessels, and similarly in Europe by Council Directive 95/21/EC of 19 June 1995 (Directives of the European Commission have status as law).

The objective for each Port State is to control 25% of the foreign flag ships calling at their ports on an annual basis. An inspection may result in:

- *Deficiency:* a non-conformity, technical failure or lack of function. A deadline for correction will be given.
- *Detention:* a serious deficiency or multitude of deficiencies that must be corrected before the vessel is allowed to leave the port.
- *Banning:* ships having a multitude of detentions or lacking an ISM certificate may be banned from European waters.

Since the Paris MOU was established, a number of similar MOUs have been set up in other parts of the world. More information is available on the EQUASIS homepage (EQUASIS, 2004).

The findings and actions of Paris MOU are published in yearbooks (MOU, 2004). A summary of the number of inspections and relative frequency of deficiencies and detentions is shown in Figure 11.8.

Table 11.15 gives a summary of the relative deficiency rate for specific inspection areas. The following areas have a relative high frequency:

- Life-saving appliances
- Safety in general
- Safety of navigation

The Flag States show quite different performance in terms of deficiencies and detentions. The worst performing states are shown in Figure 11.9. Both 'classical' FOCs and new entrants have a quite high detention rate: Honduras, Belize and St. Vincent & Grenadines.

11.3 Classification societies

11.3.1 Background

Classification societies are independent bodies which set standards for the design, maintenance and repair of ships.

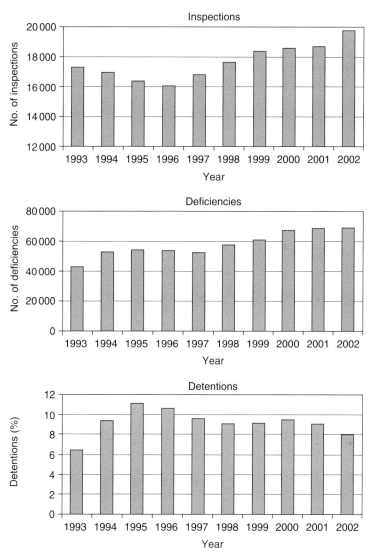

Figure 11.8 Port State control findings by Paris MOU. (Source: MOU, 2004.)

A cargo shipper and the underwriter who are requested to insure a maritime risk require some assurance that any particular vessel is structurally fit to undertake a proposed voyage. To enable the shipper and underwriter to distinguish the good risk from the bad a system of classification has been formulated over a period of more than 200 years. During this period reliable organizations have been created for the initial and continuing inspection of ships so that classification may be assessed and maintained. The class in this sense is a form of quality check for the insurance company. The Classification Society has otherwise no official role relative to international and national regulation, although there are a few special exceptions.

Recent amendment to the requirements of the International Convention for the Safety of Life at Sea (SOLAS), see Section 11.2.2.1, have required ships to which that convention applies to be designed, constructed and maintained in compliance with the structural, mechanical and electrical requirements of a classification society which is recognised by the flag administration or with applicable national standards of that administration which provide an equivalent level of safety. In general flag administrations recognise specific classification societies for this purpose rather than maintaining such national standards.

Whilst there are reported to be more than 50 ship classification organizations worldwide the 10 major

Table 11.15 Deficiency rate in % for inspection areas

	No. of deficiencies			Def. in % of total number			Ratio of def. to inspections × 100			Ratio of def. to indiv. Ships × 100		
	2000	2001	2002	2000	2001	2002	2000	2001	2002	2000	2001	2002
Ship's certificates and documents	3465	3581	3369	5.1	5.2	4.88	18.8	19.2	17.04	30.8	30.7	28.50
Training certification and watchkeeping for seafarers	1179	1302	5522	1.7	1.9	7.99	6.4	7.0	27.94	10.5	11.2	46.71
Crew and Accommodation (ILO 147)	1963	2113	1853	2.9	3.1	2.68	10.7	11.3	9.37	17.5	18.1	15.67
Food and catering (ILO 147)	1031	876	664	1.5	1.3	0.96	5.6	4.7	3.36	9.2	7.5	5.62
Working space (ILO 147)	678	703	602	1.0	1.0	0.87	3.7	3.8	3.05	6.0	6.0	5.09
Life-saving appliances	10942	10516	9009	16.2	15.3	13.04	59.5	56.3	45.58	97.3	90.2	76.20
Fire safety measures	8789	8547	8158	13.0	12.4	11.81	47.8	45.8	41.27	78.1	73.3	69.00
Accident prevention (ILO 147)	1506	1586	1429	2.2	2.3	2.07	8.2	8.5	7.23	13.4	13.6	12.09
Safety in general	9243	8951	9306	13.7	13.0	13.47	50.2	47.9	47.08	82.2	76.8	78.71
Alarm, signals	330	326	301	0.5	0.5	0.44	1.8	1.7	1.52	2.9	2.8	2.55
Carrtage of cargo and dangerous goods	836	1323	1028	1.2	1.9	1.49	4.5	7.1	5.20	7.4	11.3	8.69
Load lines	3816	3906	3507	5.6	5.7	5.08	20.7	20.9	17.74	33.9	33.5	29.66
Mooring arrangements (ILO 147)	878	1109	1060	1.3	1.6	1.53	4.8	5.9	5.36	7.8	9.5	8.97
Propulsion and aux. machinery	3671	3713	3606	5.4	5.4	5.22	20.0	19.9	18.24	32.6	31.8	30.50
Safety of navigation	8055	8315	6769	11.9	12.1	9.80	43.8	44.5	34.25	71.6	71.3	57.25
Radio communication	2638	2703	2421	3.9	3.9	3.50	14.3	14.5	12.25	23.5	23.2	20.48
MARPOL, annex I	4875	5116	4421	7.2	7.4	6.40	26.5	27.4	22.37	43.3	43.9	37.39
Oil tankers, chemical tankers and gas carriers	212	151	502	0.3	0.2	0.29	1.2	0.8	1.02	1.9	1.3	1.71
MARPOL, annex II	71	43	64	0.1	0.1	0.09	0.4	0.2	0.32	0.6	0.4	0.54
SOLAS-related operational deficiencies	1132	1262	1353	1.7	1.8	1.96	6.2	6.8	6.85	10.1	10.8	11.44
MARPOL-related operational deficiencies	618	456	341	0.9	0.7	0.49	3.4	2.45	1.73	5.5	3.9	2.88
MARPOL, annex III	31	13	21	0.0	0.0	0.03	0.2	0.1	0.11	0.3	0.1	0.18
MARPOL, annex V	742	758	701	1.1	1.1	1.01	4.0	4.1	3.55	6.6	6.5	5.93
ISM	929	1239	3210	1.4	1.8	4.65	5.0	6.6	16.24	8.3	10.6	27.15
Bulk carriers, additional safety measures	9	50	51	0.0	0.1	0.07	0.0	0.3	0.26	0.1	0.4	0.43
Other def. clearly hazardous to safety	44	33	4	0.1	0.1	0.07	0.2	0.2	0.24	0.4	0.3	0.41
Other def. not clearly hazardous	52	65	63	0.1	0.1	0.09	0.3	0.3	0.32	0.5	0.6	0.53
Total	67735	68756	69079									

Source: MOU (2004).

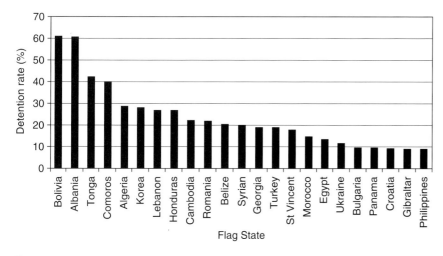

Figure 11.9 Detention rate in % for Flag States above the average rate. (Source: MOU, 2004.)

classification societies that claim to class some ninety-four per cent of all commercial tonnage involved in international trade worldwide are members of the International Association of Classification Societies (IACS). These members of IACS are –

American Bureau of Shipping (ABS)	USA
Bureau Veritas (BV)	France
China Classification Society (CCS)	China
Det Norske Veritas (DNV)	Norway
Germanischer Lloyd (GL)	Germany
Korean Register (KR)	Korea
Lloyds Register (LR)	Great Britain
Nippon Kaiji Kyokai (Class NK)	Japan
Registro Italiano Navale (RINA)	Italy
Russian Maritime Register of Shipping (RS)	Russia

11.3.2 Rules and regulations

The classification societies each publish rules and regulations which are principally concerned with the strength and structural integrity of the ship, the provision of adequate equipment, and the reliability of the machinery. Ships may be built in any country to a particular classification society's rules and they are not restricted to classification by the relevant society of the country where they are built. A typical example of such rules is Lloyds Register (2007).

In recent years, under the auspices of IACS, member societies have been engaged in the development of common structural rules for ships. The first two of these common structural rules, for bulk carriers of 90 metres or more in length and for oil tankers of 150 metres or more in length came into force on 1 April 2006. These common rules will be incorporated into each member societies rule book.

These and other common rules to be developed by IAC members anticipate the nature of future standards to be made under the International Maritime Organization's proposed Goal Based New Ship Construction Standards.

11.3.3 Lloyds Register

Only the requirements of Lloyds Register, which is the oldest of the classification societies are dealt with in detail in this Chapter. The requirements of other classification societies that are members of IACS are not greatly different.

Founded in 1760 and reconstituted in 1834, Lloyds Register was amalgamated with the British Corporation, the only other British classification society in existence at that time, in 1949. Ships built in accordance with Lloyds Register rules or equivalent standards, are assigned a class in the Register Book, and continue to be classed so long as they are maintained in accordance with the Rules.

11.3.4 Lloyds Register classification symbols

All ships classed by Lloyds Register are assigned one or more character symbols. The majority of ships are assigned the characters 100A1 or ✠100A1.

The character Figure 100 is assigned to all ships considered suitable for sea-going service. The character letter A is assigned to all ships which are built in accordance with or accepted into class as complying with the Society's Rules and Regulations. The character Figure 1 is assigned to ships carrying on board anchor and/or mooring equipment complying with the Society's Rules and Regulations. Ships which the Society agree need not be fitted with anchor and mooring equipment may be assigned the character letter N in lieu of the character Figure 1. The Maltese cross mark is assigned to new ships constructed under the Society's Special Survey, i.e. a surveyor has been in attendance during the construction period to inspect the materials and workmanship.

There may be appended to the character symbols, when considered necessary by the Society or requested by the owner, a number of class notations. These class notations may consist of one or a combination of the following. Type notation, cargo notation, special duties notation, special features notation, service restriction notation. Type notation indicates that the ship has been constructed in compliance with particular rules applying to that type of ship, e.g. 100A1 'Bulk Carrier'. Cargo notation indicates the ship has been designed to carry one or more specific cargoes, e.g. 'Sulphuric acid'. This does not preclude it from carrying other cargoes for which it might be suitable. Special duties notation indicate the ship has been designed for special duties other than those implied by type or cargo notation, e.g. 'research'. Special features notation indicates the ship incorporates special features which significantly affect the design, e.g. 'movable decks'. Service restriction notation indicates the ship has been classed on the understanding it is operated only in a specified area and/or under specified conditions, e.g. 'Great Lakes and St. Lawrence'.

The class notation ✠LMC indicates that the machinery has been constructed, installed and tested under the Society's Special Survey and in accordance with the Society's Rules and Regulations. Various other notations relating to the main and auxiliary machinery may also be assigned.

Vessels with a refrigerated cargo installation constructed, installed and tested under the Society's Special Survey and in accordance with its Rules and Regulations may be assigned the notation ✠Lloyds RMC. A classed liquefied gas carrier or tanker in which the cargo reliquefaction or cargo refrigeration equipment is approved, installed and tested in accordance with the Society's Rules and regulations may be assigned the notation ✠Lloyds RMC (LG).

Where additional strengthening is fitted for navigation in ice conditions an appropriate notation may be assigned.

11.3.5 Classification of ships operating in ice

Classification societies such as Lloyds Register and a number of Administrations whose waters experience icing have for many years had regulations defining and categorizing ice conditions and specifying design and standard requirements for ships operating in ice. Lloyds Register have assigned special features notations to many existing ships for operation in first-year ice and for operation in multi-year ice. First year ice notations are for additional strengthening where waters ice up in winter only and multi-year ice for service in Artic and Antarctic waters.

The increasing maritime trading within Arctic waters in the past decade and the desire to ship oil, gas and other commodities from there all year round appears to have resulted in the class societies adopting to some extent the ice strengthening requirements of the 'Finnish-Swedish Ice Class Rules 1985' developed for vessels trading in winter and for which the keel was laid after 1 November 1986. These requirements intended primarily for vessels operating in the Northern Baltic in winter are given for four different ice classes.

Ice Class 1AA
Ice Class 1A
Ice Class IB
Ice Class 1C

The hull scantling requirements determined under these rules are based on certain assumptions concerning the nature of the ice load the ship's structure may be subjected to. These assumptions having been determined from full scale observations made in the Northern Baltic.

This increased trading in Artic waters has also created particular interest in the establishment of universal requirements for ships operating in ice.

Both IMO and IACS have been involved in this work with IMO producing guidelines in December 2002 for ships operating in Arctic ice-covered waters for which they prescribe seven 'Polar Class' descriptions. These range from PC 1 for year-round operation in all Arctic ice covered waters to PC 7 for summer/autumn operation in thin first-year ice which may include old ice inclusions. Subsequently IACS set up a working group to develop Unified Requirements for Polar Ships that would cover –

(a) Polar class descriptions and applications;
(b) Structural requirements for Polar class ships
(c) Machinery requirements for Polar class ships

It was intended that with the completion of these Uniform Requirements for Polar Ships and their adoption by the IACS Council, the IACS member societies will have one year in which to implement these common standards for ships operating in ice.

11.3.6 Structural design programs

In recent years the principal classification societies have developed software packages for use by shipyards which incorporate dynamic-based criteria for the scantlings, structural arrangements and details of ship structures. This was a response to a perception that the traditional semi-empirical published classification rules based on experience could be inadequate for new and larger vessel trends. The computer programs made available to shipyards incorporate a realistic representation of the dynamic loads likely to be experienced by the ship and are used to determine the scantlings and investigate the structural responses of critical areas of the ship's structure.

Lloyds Register 'Ship Right Procedures for the Design, Construction and Lifetime Care of Ships' incorporates programs for structural design assessment (SDA) and fatigue design assessment (FDA). Also incorporated are Construction Monitoring (CM) procedures which ensure that the identified critical locations on the ship are built to acceptable standards and approved construction procedures. (These provisions are mandatory for classification of tankers of more than 190 metres in length and for other ships where the type, size and structural configuration demand).

11.3.7 Periodical Surveys

To maintain the assigned class the vessel has to be examined by the Society surveyors at regular periods.

The major hull items to be examined at these surveys only are indicated below.

(i) *Annual survey* All steel ships are required to be surveyed at intervals of approximately one year. These annual surveys are where practicable held concurrently with statutory annual or other load line surveys. At the survey the surveyor is to examine the condition of all closing appliances covered by the conditions of assignment of minimum freeboard, the freeboard marks, and auxiliary steering gear. Watertight doors and other penetrations of watertight bulkheads are also examined and the structural fire protection verified. The general condition of the vessel is assessed, and anchors and cables are inspected where possible at these annual surveys. Dry bulk cargo ships are subject to an inspection of a forward and after cargo hold.

(ii) *Intermediate surveys* Instead of the second or third annual survey after building or special survey an intermediate survey is undertaken. In addition to the requirements for annual survey particular attention is paid to cargo holds in vessels over 15 years of age and the operatings systems of tankers, chemical carriers and liquefied gas carriers.

(iii) *Docking surveys* Ships are to be examined in dry dock at intervals not exceeding $2\frac{1}{2}$ years. At the drydocking survey particular attention is paid to the shell plating, stern frame and rudder, external and through hull fittings, and all parts of the hull particularly liable to corrosion and chafing, and any unfairness of bottom.

(iv) *In-water surveys* The society may accept in-water surveys in lieu of any one of the two dockings required in a five-year period. The in-water survey is to provide the information normally obtained for the docking survey. Generally consideration is only given to an in-water survey where a suitable high resistance paint has been applied to the underwater hull.

(v) *Special surveys* All steel ships classed with Lloyds Register are subject to special surveys. These surveys become due at five yearly intervals, the first five years from the date of build or date of special survey for classification and thereafter five years from the date of the previous special survey. Special surveys may be carried out over an extended period commencing not before the fourth anniversary after building or previous special survey, but must be completed by the fifth anniversary.

The hull requirements at a special survey, the details of the compartments to be opened up, and the material to be inspected at any special survey are listed in detail in the Rules and regulations (Part 1, Chapter 3). Special survey hull requirements are divided into four ship age groups as follows:

1. Special survey of ships – five years old
2. Special survey of ships – ten years old
3. Special survey of ships – fifteen years old
4. Special survey of ships – twenty years old and at every special survey thereafter

In each case the amount of inspection required increases and more material is removed so that the condition of the bare steel may be assessed. It should be noted that where the surveyor is allowed to ascertain by drilling or other approved means the thickness of material, non-destructive methods such as ultrasonics are available in contemporary practice for this purpose. Additional special survey requirements are prescribed for oil tankers, dry bulk carriers, chemical carriers and liquefied gas carriers.

When classification is required for a ship not built under the supervision of the Society's surveyors, details of the main scantlings and arrangements of the actual ship are submitted to the Society for approval. Also supplied are particulars of manufacture and testing of the materials of construction together with full details of the equipment. Where details are not available, the Society's surveyors are allowed to lift the relevant information from the ship. At the special survey for classification all the hull requirements for special surveys (1), (2), and (3) are to be carried out. Ships over 20 years old are also to comply with the hull requirements of special survey (4), and oil tankers must comply with the additional requirements stipulated in the Rules and Regulations. During this survey the surveyor assesses the standard of the workmanship, and verifies the scantlings and arrangements submitted for approval. It should be noted that the special survey for classification will receive special consideration from Lloyds Register in the case of a vessel transferred from another recognised Classification Society. Periodical surveys where the vessel is classed are subsequently held as in the case of ships built under survey, being dated from the date of special survey for classification.

11.3.8 Hull Planned Maintenance Scheme

Lloyds Register offers a Hull Planned Maintenance Scheme (HPMS) that may significantly reduce the scope of the periodical surveys of the hull. The classification society works closely with the shipowner to set up an inspection programme that integrates classification requirements with the shipowners own planned maintenance programme. Ship staff trained and accredited by Lloyds Register are authorised to inspect selected structural items according to an approved schedule. Compliance is verified by Lloyds Register surveyors at an annual audit.

11.3.9 Damage repairs

When a vessel requires repairs to damaged equipment or to the hull it is necessary for the work to be carried out to the satisfaction of Lloyds Register surveyors. In order that the ship maintains its class, approval of the repairs undertaken must be obtained from the surveyors either at the time of the repair or at the earliest opportunity.

11.4 Safety of marine systems

11.4.1 Introduction

11.4.1.1 *Background*

Safety was not considered to be a matter of public concern in ancient times, when accidents were regarded as inevitable or as the will of the gods. Modern notions of safety were developed only in the 19th century as an outgrowth of the industrial revolution, when a terrible toll of factory accidents aroused humanitarian concern for their prevention. Today the concern for safety is worldwide and is the province of numerous governmental and private agencies at the local, national and international levels.

The frequency and severity rates of accidents vary from country to country and from industry to industry. A number of accidents in the chemical, oil and gas, marine and nuclear industries over the years have increased the public and political pressure to improve the safety which protects people and the environment. In the evolution of the approach to safety, there has been an increasing move towards risk management in conjunction with more technical solutions. Hazardous industries have developed approaches for dealing with safety and loss prevention, from design standards to plaint inspections and technical safety, through to safety auditing and human factors (Trbojevic and Soares (2000)).

As far as the marine industry is concerned, tragic accidents such as the *Herald of Free Enterprise* and *Derbyshire*, together with environmental disasters such as *Exxon Valdez* and *Amoco Cadiz*, have focused world opinion on ship safety and operation (Wang (2002)). This demand for improved safety requires comprehensive safety analyses to be developed. Such safety analyses will ensure efficient, economic and safe ship design and operation.

11.4.1.2 *Safety and reliability development in the maritime industry*

Reliability and safety methods saw a rapid development after the Second World War. These methods were mainly concerned with military use for electronics and rocketry studies. The first predictive reliability models appeared in Germany on the V1 missile project where a reliability level was successfully defined from reliability requirements and experimentally verified on components during their development stages (Bazovsky (1961)).

The first formal approach to shipboard reliability was the Buships specification, MIL-R-22732 of July 31, 1960, prepared by the United States of America's Department of Defence and addressed ground and shipboard electronic equipment (MIL (1960)). Subsequently in 1961 the Bureau of Weapons issued the MIL standards concerning reliability models for avionics equipment and procedures for the prediction and reporting of the reliability of weapon systems. This was due to the fact that the growing complexities of electronic systems were responsible for the failure rates leading to a significantly reduced availability on demand of the equipment.

In February 1963 the first symposium on advanced marine engineering concepts for increased reliability was held at the office of Naval Research at the University of Michigan. In December 1963 a paper entitled *'Reliability Engineering Applied to the Marine Industry'*

(Harrington and Riddick (1963)) was presented at the Society of Naval Architects and Marine Engineers (SNAME) and in June the following year another paper, entitled *'Reliability in Shipbuilding'* (Dunn (1964)), was presented. Following the presentation of these two papers, SNAME in 1965 established Panel M-22 to investigate the new discipline as applied to marine machinery and make it of use to the commercial marine industry.

In the last three decades, stimulated by public reaction and health and safety legislation, the use of risk and reliability assessment methods has spread from the higher risk industries to an even wider range of applications. The Reactor Safety Study undertaken by the U.S.A (U.S Nuclear Regulatory Commission (1975)) and the Canvey studies performed by the UK Health & Safety Executive (HSE (1978, 1981a,b)) resulted from a desire to demonstrate safety to a doubtful public. Both these studies made considerable use of quantitative methods, for assessing the likelihood of failures and for determining consequence models.

11.4.1.3 *Present status*

There is a long history in the United Kingdom (UK) of research, development and successful practical application of safety and reliability technology. There is a continuing programme of fundamental research in areas such as software reliability and human error in addition to further development of the general methodology. Much of the development work was carried out by the nuclear industry.

Based on the considerable expertise gained in the assessment of nuclear plants, a National Centre for System Reliability (NCSR) was established by the UK Atomic Energy Authority (UKAEA) to promote the use of reliability technology. This organization plays a leading role in research, training, consultancy and data collection. The NCSR is part of the safety and reliability directorate of the UKAEA, which has played a major role in formulating legislation on major hazards, and has carried out major safety studies on industrial plants. It is noted that some of the major hazard studies commissioned at the national level in the UK have included the evaluation of the risks involved as a result of marine transportation of hazardous materials such as liquefied gases and radioactive substances. It is expected that the recent legislation in relation to the control of major hazards will result in a wider use of quantitative safety assessment methods and this will inevitably involve the marine industry.

Most chemical and petrochemical companies in the UK have made use of safety and reliability assessment techniques for plant evaluation and planning. Similar methods are regularly employed in relation to offshore production and exploration installations.

The Royal Navy has introduced reliability and maintainability engineering concepts in order to ensure that modern warships are capable of a high combat availability at optimum cost (Gosden and Galpin (1999)). The application of these methods has been progressively extended from consideration of the operational phase and maintenance planning to the design phase.

To date, comparatively little use of safety and reliability assessment methods has been made in connection with merchant shipping. Lloyd's Register of Shipping has for a long period, collected information relating to failures and has carried out development work to investigate the application of such methods to the classification of ships. Apart from this, some consultancy work has also been carried out on behalf of ship owners. One example is the *P&O Grand Princess*, for which a comprehensive safety and availability assurance study was carried out at the concept design stage of this cruise ship. Established risk assessment techniques were used including Failure Mode and Effects Analysis (FMEA), flooding risk analysis and fire risk analysis. The resultant ship was believed to be better and safer than it would have been otherwise (Best and Davies (1999)). P&O has now developed an in-house safety management system which is designed to capture any operational feedback, so as to improve the safety and efficiency of its cruise fleet operation and to use it for better design in the future.

The merchant ship-building yards in the UK, having seen the success of the warship yards in applying Availability, Reliability and Maintainability (ARM) studies at the design stage, are actively seeking benefits from adopting a similar approach. Some joint industry-university research projects are being undertaken to explore this area.

11.4.1.4 *Databases*

The early reliability studies, particularly on electronics, made use of failure data obtained by testing a large number of components. As the techniques found more widespread applications, the methods for statistically analysing data from real life experience became more advanced and large communal databases of reliability data were created.

In the 1980's, the maritime classification societies, commercial institutions and other authorities realised the importance of statistical data collection on failure or repair data and eventually, data on general accident statistics were provided (HSE (1992)). These data give general trends and are not directly useable in quantitative assessments. By far the most useful sets of statistics on marine accidents are presented in the publications of the UK Protection and Indemnity (P&I) Club of insurers (P & I Club (1992)).

Accident investigation is a common method used by many organizations in attempt to enhance safety. Discovering the causes of casualties may allow steps to be taken to preclude similar accidents in the future. Since 1981 the United States Coast Guard (USCG) has maintained a computer database summarising the causes of investigated marine casualties. In 1992 the USCG implemented a new computer casualty database, the Marine Investigation Module (MINMOD), which changed the way marine casualty investigations were reported (Hill *et al.* (1994)). The new system implemented several improvements that were expected to enhance the validity and completeness of the casualty data reported. One of the most important changes made was the adoption of a chain-of-events analysis of accident causes, enabling a more complete description of all accident-related events and their associated causes.

In the past, accident statistics were not gathered systematically and the data type was not consistent. This led to the analyst not knowing if the set of data is applicable to the analysis under consideration. Some commercial institutions have focused on developing databases of maritime accidents. The accident information is presented systematically and in some cases correlation is available. Typical examples include:

- OREDA (Offshore Reliability Data) – A database of offshore accidents which was first published in 1982 and has been updated annually ever since (OREDA (1982)).
- Marine Incident Database System (MIDS) – A database maintained by the Marine Accident Investigation Branch (MAIB).
- World Casualty Statistics – A collection of data published annually by Lloyds Register of Shipping.
- The Institute of London Underwriters.
- CASMAIN – A database maintained by the United States Coast Guard.
- SEAREM – A British Isles database developed and refined under the stewardship of the Royal National Lifeboat Institution (RNLI).

Over the last several years, progressive maritime organizations around the world have been cooperating to form a worldwide information network, called RAM/SHIPNET, to support the optimisation of safety, reliability, and cost effectiveness in vessel operations. The mission of RAM/SHIPNET is to form an efficient information network for vessel operators and other industry participants to collect and share sanitised performance information on vessel equipment. It consists of distributed and partially shared Reliability, Availability, and Maintainability (RAM) databases. RAM/SHIPNET was established to collect equipment performance data and to share this data at different levels by linking chief engineers, ship operators/managers, regulatory agencies, equipment manufacturers and shipyards/designers. First generation stand-alone data collection and processing tools were developed and the system became ready for implementation. The roll-out period is in progress for full validation, demonstration, and implementation of RAM/SHIPNET (Inozu and Radovic (1999)).

The databases that are described in this Section, are still lacking specific information of equipment and component failures. Novel methods have to be developed to handle this shortcoming. These novel techniques should integrate expert judgement with available data in a formal manner to ensure the accuracy and the applicability of the safety assessment carried out.

In the following Sections, fishing vessels are chosen as a major test case while other types of ships are also used. Fishing vessels are generally smaller with a unique operating nature and the accidents concerning these vessels have been overlooked in the past.

11.4.2 Ship safety and accident statistics

11.4.2.1 *Background*

Recognising the need for attention to safety of commercial fishing vessels, the IMO organized an international conference, which culminated in the Torremolinos International Convention for the safety of fishing vessels in 1977 (IMO (1977)). It established uniform principles and rules regarding design, construction and equipment for fishing vessels 24 m (79 feet) in length and over. This Convention was a major milestone. It provided benchmarks for improving safety, and many fishing nations have adopted its measures into their marine safety programmes.

The IMO convention on Standard of Training, Certification and Watch keeping for seafarers (STCW) 1978 is another important influence on fishing vessel safety. Although the STCW 1978 specifically exempts fishing vessels, it has inspired efforts to develop personnel qualification standards (STCW 95 also exempts fishing vessels). Notable among these efforts is the Document for Guidance on Fishermen's Training and Certification (IMO (1988)) and the Code of Safety for Fishermen and Fishing Vessels (IMO (1975a)). Other IMO codes and guidelines include the Voluntary Guidelines for the Design, Construction and Equipment of Small Fishing Vessels (IMO (1980)) and the Code of Safety for Fishermen and Vessel Design and Construction (IMO (1975b)). These standards are jointly prepared by the IMO and two other United

Nations subsidiaries, the Food and Agricultural Organization (FAO) and the International Labour Organization (ILO). They provide guidance on training and education and detailed curriculum development.

There are strong safety programmes among the IMO member states that include equipment standards, inspection requirements and certification or licensing of vessel operators and crew. These programmes vary in each country. For example, Canada, Norway and the UK have extensive requirements, while other countries are less stringent. Generally fishing vessels in length 15 m or longer are addressed; however some countries address vessels as small as 9 m, such as New Zealand and 12 m as in the UK.

In the UK, comprehensive regulations have come into force since 1975. Surveys and certification of fishing vessels with the length of 12 m or longer are required; they apply to about 2000 vessels. For vessels with the length of over 16.5 m, deck officers and engineers have comprehensive entry level professional training, certification, manning and watch keeping requirements.

Studies on the effect of compulsory programmes have been conducted in Norway, The Netherlands, UK and Spain, but they have tended to focus on training, statistics and causes of accidents rather than performance of technical systems in relation to compulsory programmes. It appears that fatalities have generally been reduced, while the rates of incidence for injuries related to vessel casualties and workplace accidents appear unchanged. The lack of apparent change in injury rates may be related to working conditions and methods, vessel design, training deficiencies and changes in the number of fishing vessels and fishermen (Carbajosa (1989), Dahle and Weerasekara (1989), Hoefnagal and Bouwman (1989), Stoop (1989)). The number of vessel casualties over the years has changed. For example, in the UK, since safety rules were applied to all vessels over 12 m during the mid 1980's, the number of losses of these vessels has significantly reduced. However, losses of vessels under 12 m have more than doubled, perhaps partly because of a large increase in the number of vessels under 12 m, to which only life saving and fire safety government regulations apply, Hopper and Dean (1992).

11.4.2.2 Code of practice for the safety of small fishing vessels

The development of a Code of Practice for small fishing vessels marked the beginning of the first major review of fishing vessel safety regulations since 1975. The principal aim in developing the Code was to update the safety equipment requirements for small fishing vessels. Its secondary aim was to build on the concept of hazard identification and risk assessment, and to introduce an assessment by owners of the fitness of their vessels (House of Commons (2000)).

The Code of Practice for the safety of small fishing vessels has been effective since the 1st of April 2001. The aim of this Code of Practice is to improve safety in the under 12 metre sector of the fishing industry and to raise the safety awareness of all those involved with the construction, operation and maintenance of fishing vessels with a registered length of less than 12 metres.

(a) Development
In 1992 the National Audit Office, in its report entitled 'Department of Transport: Ship Safety', noted an increase in the fishing vessel accident rate from 1978 to 1989, due in part to an increase in the numbers of smaller vessels (National Audit Office (1992)). It observed the absence, until 1990, of any programme of inspection of fishing vessels with a registered length of less than 12 metres. At about the same time, the House of Lords Select Committee on Science and Technology recommended that fishing vessels down to 7 m in length should be brought within the licensing, crew certification and structural safety regimes.

In response, the Surveyor General's Organization of the Department of Transport (now the Maritime & Coastguard Agency (MCA)), in consultation with industry members of the Fishing Industry Safety Group (FISG), decided to develop a Code of Practice for fishing vessels with a registered length of less than 12 metres. The content of the Code has been the subject of extensive discussion with representatives of the under 12 metre sector of the fishing industry within a Steering Committee set up by FISG to oversee the Code's development. The Code has been applied from the 1st of April 2001 to all United Kingdom registered fishing vessels with a registered length of less than 12 metres.

(b) Code requirements
To comply with the Code of Practice, a vessel owner is required:

- To carry safety equipment on the vessel appropriate to its length and construction (i.e. decked or open).
- To complete or arrange completion of an assessment of the health and safety risks arising in the normal course of work activities or duties on the vessel in accordance with the provisions of the Merchant Shipping and Fishing Vessels (Health and Safety at Work) Regulations 1997 and MGN (Marine Guidance Note) 20 (M + F) (MSA (1998)).
- To certify annually that the vessel complies with the Code, by declaring that the safety equipment

has been properly maintained and serviced in accordance with manufacturers' recommendations and that an appropriate, up-to-date health and safety risk assessment has been completed.
- To present the vessel for inspection either voluntarily or as requested by the MCA.

The checklist of requirements for the Code of Practice for the safety of small fishing vessels is in 4 categories. The vessels addressed in this Code of Practice include:

1. Decked vessels 10 m and above registered length to less than 12 m registered length.
2. All decked vessels up to 10 m registered length.
3. Open vessels 7 m and above to less than 12 m registered length.
4. Open vessels less than 7 m registered length.

11.4.2.3 *The Fishing Vessels (Safety Provisions) Safety Rules 1975*

In 1968, three vessels were tragically lost off the coast of Iceland. The investigation of these three vessel accidents determines the loss as 'capsizing due to ice accumulation'. Following the official inquiry into these losses, a rule regime was investigated which eventually arrived on the statute as 'The Fishing Vessels (Safety Provisions) Safety Rules 1975'. Unfortunately the formulation of the rules did not result in an analysis of the organizational or human failing, present in many safety tragedies within the fishing community. The rules are primarily concerned with vessels of over 12 metres registered length. Smaller vessels are addressed, but only life saving appliances and firefighting measures are included. The 1975 rules do not concern themselves with the whole vessel, but may be noted to consider the vessel from the deck and accommodation line downwards. The winches, wires and fishing equipment are not covered by the rules.

Following the introduction of the 1975 Rule, the European Common Fisheries policy brought in a licensing scheme for vessels over 10 metres. This coupled with a de-commissioning scheme for larger vessels, resulted in a huge increase in the number of under 10 metre vessels. These vessels did not need licenses to fish and need not comply with the majority of the 1975 Rules. However, in 1996, the Ministry of Agriculture Fisheries and Food introduced fishing licenses for vessels of under 10 metres overall length. The introduction of this law has reduced the size of the fleet. The greatest incidence of risk has now moved to vessels in the 7 to 20 metre range, with particular safety concern for those vessels under 12 metres. Following concern emanating from the Parliament, inspections on these under 12 metre vessels have been requested. Since 1993, under 12 metre vessels have been subjected to safety inspection.

11.4.2.4 *Accident data for fishing vessels*

Comparisons of the safety record of the fishing industry with other industries indicate that the industry continues to be the most dangerous by a significant margin. In 1995/96 there were 77 fatal injuries per 100 000 fishermen as opposed to 23.2 per 100 000 employees in the mining and quarrying industry (the next highest category in that year) (MAIB (1995)). In 1992 there were 494 reported fishing vessel accidents from a fleet of 10 953 vessels. In 1997, figures indicate 485 reported fishing vessel accidents from a significantly reduced fleet of 7 779 vessels. These statistics do not include personal accidents to fishermen while at sea; it is believed that these are under-reported (MAIB (1997)).

The accident data presented in this section are predominantly gathered from the Marine Accident Investigation Branch (MAIB). The MAIB is a totally independent unit within the Department of the Environment Transport and the Regions (DETR) and reports directly to the Secretary of State. The MAIB received 1 418 accident and incident reports in 1999. Accidents to ships accounted for 641 of those reports.

The data presented here is collected from 1992 to 1999 and reflects all the reported incidents and accidents relating to fishing vessels. It is thought that the actual accident and incident figures are higher than what is presented here, as many accidents are not reported to the coastguard authorities.

Figure 11.10 shows the total number of vessels lost (primary y-axis) and total number of vessels registered (secondary y-axis) from 1992 to 1999. These figures include all vessel sizes ranging from under 12 metre to over 24 metres. From this graph, it is evident that the percentage of vessels lost increased from 1992 to 1994 and then reduced from 1994 to 1998. From 1998 onwards, it is noted that there was a sharp increase in the percentage of vessels lost. Overall, the percentage of vessels lost was between 0.27% (minimum in 1997/98) and 0.45% (maximum in 1999) of the total registered vessels, as seen in Figure 11.11.

There were approximately 7,460 UK-registered fishing vessels in 1999 (end December 1999 figure). During the year 370 accidents and incidents involving these vessels were reported to the MAIB. 33 fishing vessels were lost which at 0.45% of the total fleet represent the highest rate since 1994. Machinery damage is noted as the main contributor to the high number of accidents as seen in the pie chart of Figure 11.12.

An analysis of the data from previous years shows that machinery damage has contributed to over 50% of all accidents. This could be attributed to several

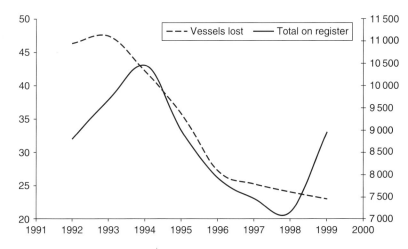

Figure 11.10 Vessels registered and lost (1992–1999).

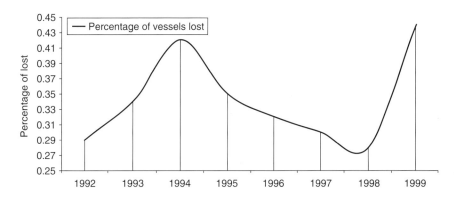

Figure 11.11 Proportion of vessels lost.

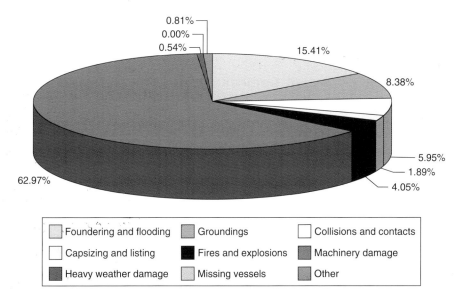

Figure 11.12 Accidents to vessels by accident type in 1999.

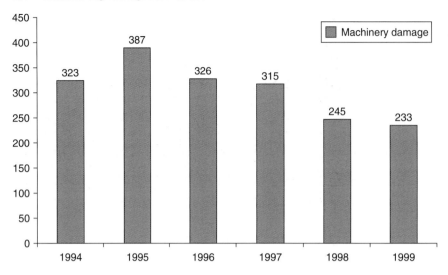

Figure 11.13 Accidents caused by machinery damage.

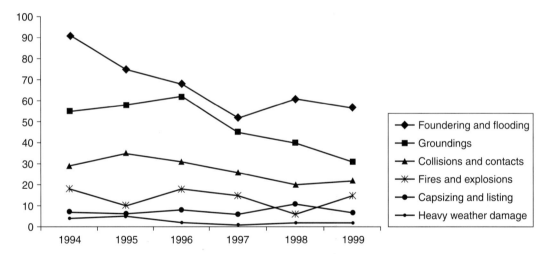

Figure 11.14 Accidents to fishing vessels by accident type.

factors including poorly maintained equipment, incorrect operation, age, lack of automation, etc. The graph in Figure 11.13 shows the number of accidents caused by machinery damage from 1994 to 1999. Although the figures indicate a decreasing trend, the number of accidents related to this category is still high and certainly unacceptable from a safety perspective.

The next highest contributor to accidents is found to be flooding and foundering followed by grounding and then collision and contact. A comparison of all accident types is made as seen in Figure 11.14. Flooding and foundering is estimated to cause almost 15% to 20% of accidents on fishing vessels.

These data are cumulated and presented as a pie chart in Figure 11.15 to reveal the contribution of each accident type for the sampling period. As revealed earlier, machinery damage is found to be the most common cause of accidents on fishing vessels, contributing 64.4% of all accidents. Foundering and flooding (14.2%), grounding (10.2%), collision and contacts (5.7%), and fires and explosions (2.9%) follow.

To determine the severity of the accidents on fishing vessels, data reflecting the accidents to vessel crew together with the number of deaths are gathered and presented in Figures 11.16 and 11.17. These bar charts show that almost 30% of accidents to crew

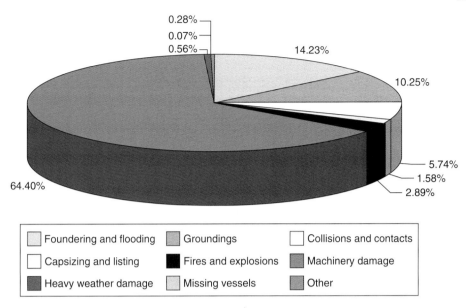

Figure 11.15 Accidents by nature (1994–1999).

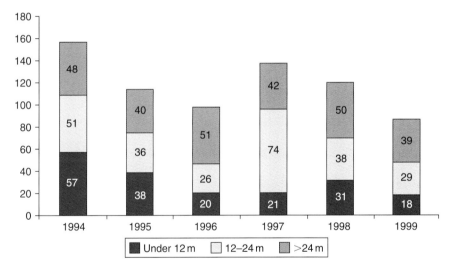

Figure 11.16 Accidents to crew.

on vessels that are under 12 metre result in deaths and for vessels that are 12–24 metres and more than 24 metres in length, these figures are calculated to be 13% and 15%, respectively. The results indicate that vessels under 12 metres have the highest casualty rates and suffer severe consequences when an accident happens. This could be attributed to the size and stability of these vessels when sailing in bad weather conditions. The number of under 12 metre vessels that were lost is much higher than the other vessels as seen in Figure 11.18. The trend in the number of vessels lost is difficult to determine, as it does not follow any specific mathematical rule. However, by comparing the graphs in Figures 11.10, 11.11 and 11.18, it can be concluded that from 1997, the number of vessels lost increased as the percentage of registered vessels decreased.

Table 11.16 gives the detailed breakdown of accidents by vessel length and accident cause for 1999 (MAIB (1999a)). From this table, it is noted that a great proportion of fishing vessel accidents (20%) is caused by negligence/carelessness of the crew. This could

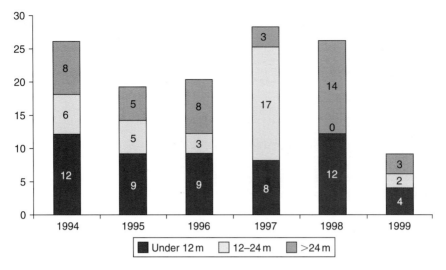

Figure 11.17 Deaths to crew.

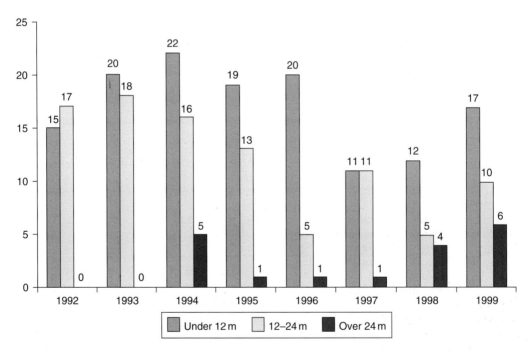

Figure 11.18 Vessels lost.

be summarised as human error attributed by several factors including competency of the crew, fatigue, poor manning of vessel and difficult operating conditions. A method assessing human error and means to reduce these errors are described in Pillay and Wang (2003). Accidents caused by the lifting gear (15%) and other fishing gear equipment (12%) are also high compared to the other accident causes.

11.4.2.5 *Data analysis*

In many cases of fishing vessel accidents, information is incomplete or totally lacking. This makes it difficult to analyse the events that lead to the accident. Accurate historical and current data on vessels, fishermen, professional experience, hours and nature of exposure and safety performance of personnel

Table 11.16 Accidents by Vessel Length and Accident Cause.

Accidents by Vessel Length and Accident Cause (more than one cause may be applicable to a particular accident)

Accident	Under 12 metres	12–24 metres	Over 24 metres	Total
Negligence/carelessness of injured person	4	10	9	23
Ship movement	1	3	3	7
Lifting gear	2	8	7	17
Miscellaneous fishing gear and equipment	3	5	6	14
Failure of deck machinery and equipment	–	2	5	7
Sea washing inboard	3	1	4	8
No known cause	2	–	2	4
Trawl boards	–	1	4	5
Door or hatch not secured	1	1	–	2
Failure to comply with warnings/orders	–	1	–	1
Unsecured non-fishing gear on deck	–	–	–	–
Unfenced opening	1	–	–	1
Fatigue	–	–	–	–
Failure to use protective clothing or equipment	3	1	1	5
Slippery surface	–	–	3	3
Lifting/carrying by hand incorrectly	–	–	3	3
Failure of engine room and workshop equipment	–	–	–	–
Others	2	5	4	11

and equipment are fundamental to assessing safety problems, monitoring results of safety programmes and measuring the effectiveness of safety improvement strategies (Loughran et al. (2002)). Very few data are regularly collected or published on these parameters. The limited data make it difficult to quantify safety problems, determine casual relations and assess safety improvement strategies. However, the data that are available indicate that significant safety problems exist and that human error, vessels and equipment inadequacies and environmental conditions all contribute to them.

Marine accidents that have occurred could have been prevented with greater attention to safety. This is particularly true for fishing vessels. Recent inquiries into the losses of fishing vessels 'Pescado' (MAIB (1998)) and 'Magaretha Maria' (MAIB (1999b)) have raised concerns as to how similar accidents may be prevented in the future. The data analysis in Section 11.4.2.4 shows that there is a rise of fishing vessel accidents and the trend seems to be continuing in an upward fashion. From the literature survey, it is found that safety assessment of fishing vessels has been limited to stability consideration and very little work has been carried out on the operational and equipment safety assessment. From the data given in Section 11.4.2.4, it can be deduced that fishing vessel safety needs to be addressed and the number of accidents and incidents related to the operation and equipment is to be reduced. In order to direct the attention of the safety assessment on fishing vessels, the probable causes of each accident category have been investigated and are summarised as follows:

1. Machinery damage

The highest number of incidents reported in the official statistics relates to machinery damage. Although most machinery failures do not threaten the vessel or lives of the crew, given other factors such as bad weather or being in a tideway, the consequences could be disastrous. Upon investigation of several fishing vessels in the UK, it was found that maintenance activities on board these vessels were almost non-existent. This is thought to lead to the high number of machinery failures. The present situation concerning maintenance on fishing vessels is discussed in detail in Pillay and Wang (2003) where a method for improving the current status is described.

2. Foundering/Flooding

Typically these incidents are caused by burst pipes, fittings working loose, leaking glands and sprung planks. Flooding is a particular problem with smaller wooden vessels. Smaller vessels are often of clinker construction where the strakes are lapped against each other and clenched. They are reliant upon the swelling nature of the wood when soaked for making a good seal. This method of construction is particularly vulnerable in heavy sea conditions. These types of accidents can also happen on vessels that are of metal construction. Sometimes incompatible metals become rapidly corroded in a seawater environment; examples are copper piping adjacent to steel or aluminium structures, which resulted in a relatively new vessel suffering a major flooding incident, Hopper and Dean (1992).

3. Grounding

These incidents are associated with all classes of fishing vessels and can be due to various causes. Engine or gearbox failures and propellers fouled by ropes or fishing nets are common causes. However, many cases have been associated with navigational error. This may be a failure to plot a proper course, failure to keep a check on vessel position with wind and tidal drift, reliance on auto-pilots and electronic plotters or a failure to keep a proper lookout. There are no requirements to carry on board a certified navigator (especially for vessels under 12 metres registered length), hence the navigators on these vessels rely heavily upon experience and 'gut feeling', which in turn could increase the level of navigator error.

4. Collisions and contacts

Almost all collision and contact incidents involve a fishing vessel and a merchant vessel and almost without exception they are due to human error. Large merchant vessels may have a poor line of sight from the wheelhouse and small fishing vessels are not easily seen under the bow. Apart from that, skippers on fishing vessels are too involved in the fishing operation. The fishing operation itself requires sudden stopping or course changing which could lead to unavoidable collisions. Collisions and contacts could also occur involving two or more fishing vessels. This is especially true when pair trawling is in progress. However, the consequences are less severe and the incident normally occurs due to errors of judgement by one or both parties involved.

5. Fires and explosions

The investigation of these accidents has shown that in most cases the fire originated from the engine room and was caused by oil or fuel coming into contact with hot exhausts. Other causes are heating and cooking stoves and electrical faults. There have been several cases where the fire had started in the accommodation area due to the crew smoking cigarettes in the sleeping bunk. The number of accidents caused by fire has been relatively low compared to other categories. However, due to the limited fire fighting resources on board fishing vessels, it has the potential to cause severe damage and even loss of life.

6. Capsizing

From the MAIB reports, it is evident that the majority of capsizing incidents occurred during the fishing and recovery of gear operations. This shows that for the vessels that did capsize, there was an insufficient factor of safety in the present stability criteria. This insufficient factor is introduced by the act of fishing and the associated moment lever introduced by the gear along with the wind lever in the dynamic situation at sea (Loughran et al. (2002)). This is perhaps the most lethal type of incident in terms of loss of life. The capsizing of small fishing vessels happens in a matter of minutes and this leaves little chance for the crew to escape. Extreme sea conditions are one of the many factors that lead to a capsize. As most skippers and crew depend on the catch for their daily income, skippers have been known to put their vessel through extreme sea conditions to get to a fishing ground and sometimes drift within the fishing grounds waiting for the sea to calm in order to resume fishing operations. However, the most common cause of capsizing is when the fishing gear becomes snagged. Trawl gear fouled on some sea bed obstruction is common for a fishing skipper. Attempts to free badly fouled gear by heaving on the winch can result in forces that are large enough to roll the vessel over. Heaving on both warps at the same time will produce a balanced situation but if one side suddenly becomes free, the force on the opposite side may be sufficient to capsize the vessel.

7. Heavy weather damage

The number of vessels suffering weather damage is comparatively low as seen in the graph in Figure 11.14. Small vessels are particularly vulnerable to these accidents, especially when they go out further away from the coastline for their fishing operation (due to the reduced fishing opportunities in British waters). These small vessels will be working far offshore where they cannot withstand the severe weather and wave conditions that can occur unexpectedly. Heavy weather can weaken the hull structure of the vessel and at the same time, cause deck fittings to come loose and lead to an accident.

11.4.2.6 Containership accident statistics

1. Introduction to containerships

Due to a rapidly expanding world trade, the traditional multi-purpose general-cargo liner became increasingly labour and cost intensive. A system was required to accommodate the needs of physical distribution, a system that would offer convenience, speed and above all low cost. By this system, goods should be able to be moved from manufacturer to final distribution using a common carrying unit, compatible with both sea and land legs of transportation. The result was expected to be that all costly and complicated transhipment operations at seaports would be eliminated. The whole process resulted in the development and introduction of the 'freight container', a standard box, filled with commodities, detachable from its carrying vehicle,

Table 11.17 World Fully Cellular Containerships in TEUs.

	Under 1000 TEU		1000–1999 TEU		2000–2999 TEU	
	VSL	TEU	VSL	TEU	VSL	TEU
1999	1,836	765,922	851	1,177,368	426	1,060,460
(1998)	1,751	714,155	807	1,117,310	381	956,349

	3000–3999 TEU		4000 + TEU		Total	
	VSL	TEU	VSL	TEU	VSL	TEU
1999	205	711,498	188	889,982	3,506	4,605,230
(1998)	189	653,444	152	704,559	3280	4,145,817

and as easy to carry by sea as by air, road and rail. The beginning of the container era was marked with the sailing of the 'container tanker' 'MAXTON' on 26th April 1956 from Newark N.J. to Houston, loaded with 58 containers (Chadwin et al. (1999), Stopford (1997)).

During the first years of containerisation, transportation was carried out with modified tankers or dry cargo vessels, broadly accepted as the first generation of containerships (Containerisation International (1996), Stopford (1997)). It was not until 1965 that the first orders for purpose built cellular vessels were placed, forming the second generation of container vessels. These were the 'Bay Class' ships of 1 600 TEUs (twenty-foot-long equivalent units) capacity. In the late 1970's the third generation appeared increasing the sizes up to Panamax and capacities up to 3 000 TEUs. Following the increasing demand for tonnage but without being prepared to lose the Panama Canal flexibility the industry moved to the development of the fourth generation of container vessels, keeping the Panamax dimensions and increasing the capacity up to 4 200 TEUs represented by the 'Econ Class' ships (Containerisation International (1996), Stopford (1997)).

Further development in the shipbuilding industry and the need for the creation of 'economies of scale' resulted in the appearance of the fifth generation of container ships, the Post-Panamax in the 1980's (Stopford (1997)). A recent research in the container sector of the shipping industry indicates that the world fully cellular containership fleet increased to more than 3,500 vessels with a total carrying capacity exceeding 4.6 million TEUs in 1999 and with an average annual growth rate up to 11.1% as shown in Table 11.17 (Nippon Yusen Kaisha Research Group (1999)). It is also noteworthy that the growth rate of post-Panamax containerships is the largest of all the containership sizes, amounting up to 26.3%. See also Section 2.2.2.

Although there were not many major casualties in terms of loss of lives, resulting from accidents involving containerships, this particular ship type has more of its fair share of losses due to incidents involving cargo damage, personal injury, collision, ship structural failure and pollution. Major accidents in the last decade include the total loss of the 'C/V Pioneer Container' in 1994 due to a collision in the South China Sea, the loss of the 'C/V River Gurara' in 1996, the extensive damages suffered by the 'C/V Toyama Maersk' in 1997 due to a collision with a Gas Carrier in the Singapore Strait, the loss of the 'C/V MSC Carla' in 1998 which broke into two in bad weather conditions, and the extensive damages suffered by the 'M/V APL China' in 1999 due to severe bad weather conditions. Statistics indicate that incidents involving containerships account up to about 7% of the total (Wang and Foinikis (2001)).

In terms of incident categories, containerships differ from most other ship types in that shore error accounts for a high percentage of all major incidents. The result is an equally high percentage of cargo damage. Although containerships follow the same pattern as the majority of cargo vessels, as far as the types of damages are concerned, they do differentiate in various aspects. The relative statistics available show that the percentage of incidents is higher in newer containerships, decreasing as they age, while in other cargo ship types, higher incident rates occur at their middle age. The same statistics show that a high percentage of all incidents caused by human error, was due to shore based personnel error, which is far higher than other cargo ship types. As far as ship size is concerned the smaller ships of this type are better placed with fewer incidents.

Other operational characteristics of containerships, such as the fact that they very rarely travel in ballast condition and that there are few opportunities for overnight stay at ports, contribute to the overall performance of these vessels and their operators. It should be stressed that although a relatively large amount of detailed data exists, organizations such as classification societies, as well as private shipping companies are reluctant to release it. This is mainly

820 Maritime engineering reference book

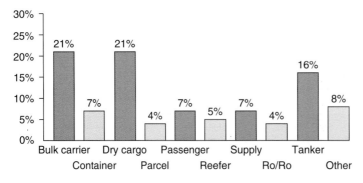

Figure 11.19 Distribution of incidents per ship type.

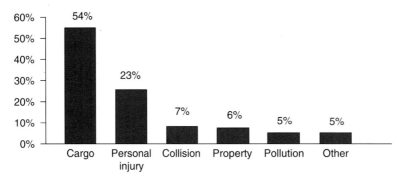

Figure 11.20 Incident categories involving containerships.

attributable to the high level of competition in the market. On the other hand, government agencies are either not ready yet to dedicate the necessary resources for data collection, or the time period for which relevant government projects are run is not sufficient to produce reliable data.

2. Containership accident statistics
Classification societies and P&I Clubs can be a very useful source of failure data mainly because of the large amount of vessels each one represents. However, data from these organizations should be critically evaluated before used or combined with others. Classification societies tend to look into safety, mainly from the viewpoint of compliance with the various sets of rules in force (Wang and Foinikis (2001)). On the other hand, P&I Clubs tend to deal with the matter from the viewpoint of financial losses due to lack of safety and are not immediately interested in the regulatory aspect of loss prevention. A recent research carried out by one of the world's leading P&I Clubs, the UK P&I Club shows that for the ten-year period from 1989 to 1999 incidents involving containerships account up to 7% of the total as shown in Figure 11.19, UK P&I Club (1999).

In terms of incident categories, containerships differ from most other ship types in that shore error accounts for up to 21% of all major incidents. The result is a fairly high percentage of cargo damage, 54%. All the values of incident categories are shown in Figure 11.20 while the total number of incidents is 273 for the period 1989–1999, UK P&I Club (1999).

In terms of ship size and age, the 10-year study shows that the smaller ships of this type are better placed. 87% of the major incidents have occurred on containerships above 10,000 grt as shown in Figure 11.21. Equally interesting is the fact that 44% of incidents involving containerships have occurred on ships of less than 10 years of age as shown in Figure 11.22 (UK P&I Club (1999)). The human error factor in incidents involving containerships is shown to be in decline, following two peak periods in 1988 and 1991 as shown in Figure 11.23, UK P&I Club (1999).

Administrations tend to look into marine casualties from the viewpoint of 'reportable incidents' within their jurisdiction which results to a differentiation in the relevant numbers, as the sample of vessels considered is smaller than that of P&I Clubs and classification societies. Furthermore, due to their

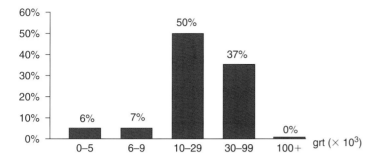

Figure 11.21 Distribution of incidents as per ship size (in grt).

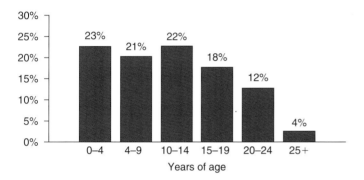

Figure 11.22 Distribution of incidents as per ships age.

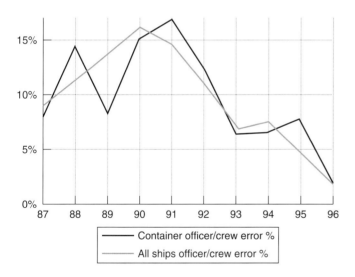

Figure 11.23 Containership-officer/crew error-frequency trend.

orientation towards ship safety and environmental protection, areas such as cargo damage and third party liability (i.e. fines) may not be considered. Nevertheless, results of such data are equally useful for the identification of major problematic areas of the various ship types.

11.4.2.7 Conclusion

A review has been performed on available incident data relevant to fishing vessels. It was found that the amount of data relating to this type of vessel is limited. The only data source that compiles fishing vessel accident/incident data has been identified to be the MAIB. Over the years, the database maintained by the MAIB has considerably improved in terms of its format. However, the database still lacks information about the casual relationship between the causes and effects of the accidents/incidents.

Data interpretation should be carried out with caution, as it is highly likely that there is some degree of underreporting of incidents. This would entail that the actual number of deaths, accidents and vessel losses, would be much higher than the figures presented here. However, the data gathered and analysed in this Section show that there is a real problem in the fishing vessel industry. The likelihood of accidents and the associated severity are still high for maritime standards, and the number of accidents/incidents has to be reduced.

A statistical analysis on containership accidents is also carried out. The result indicates that containership accident categories differ from other types of ships. Like fishing vessels, there is also a lack of proper reporting of accidents/incidents for containerships.

11.4.3 Safety analysis techniques

11.4.3.1 Background

Reliability and safety analyses are different concepts that have a certain amount of overlapping between them. Reliability analysis of an item involves studying its characteristics expressed by the probability that it will perform a required function under stated conditions for a stated period of time. If such an analysis is extended to involve the study of the consequences of the failures in terms of possible damage to property and the environment or injuries/deaths of people, the study is referred to as safety analysis.

Risk is a combination of the probability and the degree of the possible injury or damage to health in a hazardous situation (British Standard (1991)). Safety is the ability of an entity not to cause, under given conditions, critical or catastrophic consequences. It is generally measured by the probability that an entity, under given conditions, will not cause critical or catastrophic consequences (Villemuer (1992)).

Safety assessment is a logical and systematic way to seek answers to a number of questions about the system under consideration. The assessment of the risk associated with an engineering system or a product may be summarised to answer the following three questions:

1. What can go wrong?
2. What are the effects and consequences?
3. How often will they happen?

The answer obtained from these questions will provide the information about the safety of the system.

Such information is interesting but is of no practical significance unless there is a method for controlling and managing the risks associated with specific hazards to tolerable levels. Hence, a complete safety assessment will require a fourth question to be answered:

4. What measures need to be undertaken to reduce the risks and how can this be achieved?

Safety analysis can be generally divided into two broad categories, namely, quantitative and qualitative analysis, Wang and Ruxton (1997). Depending on the safety data available to the analyst, either a quantitative or a qualitative safety analysis can be carried out to study the risk of a system in terms of the occurrence probability of each hazard and its possible consequences.

11.4.3.2 Qualitative safety analysis

Qualitative safety analysis is used to locate possible hazards and to identify proper precautions that will reduce the frequencies or consequences of such hazards. Generally this technique aims to generate a list of potential failures of the system under consideration. Since this method does not require failure data as an input to the analysis, it relies heavily on engineering judgement and past experience.

A common method employed in qualitative safety analysis is the use of a risk matrix method (Halebsky (1989), Tummala and Leung (1995)). The two parameters that are considered are the occurrence likelihood of the failure event and the severity of its possible consequences. Upon identifying all the hazards within the system under consideration, each hazard is evaluated in terms of these two parameters. The severity of all the failure events could be assessed in terms of the four categories (i.e. Negligible, Marginal, Critical and Catastrophic) as shown in Table 11.18.

The occurrence likelihood of an event is assessed qualitatively as frequent, probable, occasional, remote or improbable as depicted in Table 11.19 (Military Standard (1993)). Each of these categories can be represented quantitatively by a range of probabilities. For example, such a range of probabilities can be seen in column three of Table 11.19. This is to provide a rough guideline for the experts or analysts who are providing the information or carrying out the analysis.

It is reasonable to assign a high priority if the hazard has a catastrophic consequence and a frequent probability. On the other hand, it is also reasonable to assign a low priority if the hazard has a negligible consequence and an improbable probability. Based on this logic, certain acceptable criteria can be developed. All identified hazards can be prioritised corresponding to safety and reliability objectives by appropriate hazard indexes using the hazard severity and the corresponding hazard probabilities as shown in Table 11.20 (Military Standard (1980)). The hazard probabilities shown in this table are used to carry out qualitative analysis for a military defence system. These probabilities can be assigned appropriately when different systems are considered. If an identified hazard is assigned with a hazard index of 4C, 3D, 4D, 2E, 3E or 4E, it needs an immediate corrective action. A hazard with an index 3B, 4B, 2C, 2D or 3C would require a possible corrective action. Similarly, a hazard with index 3A, 4A, 2B, 1D or 1E would be tracked for a corrective action with low priority; or it may not warrant any corrective action. On the other hand, a hazard with index 1A, 2A, 1B or 1C might not even require a review for action.

All the identified hazards within the system under study can be evaluated using this method to produce a risk ranking based on the highest priority down to the lowest priority. A variation of this qualitative risk matrix approach is presented in Pillay and Wang (2003) with its application to the safety analysis of a ship.

Table 11.18 Assessment of Hazard Severity and Categories.

Hazard consequences	Hazard severity	Category
Less than minor injury or less than minor system or environmental damage, etc	Negligible	1
Minor injury or minor system or environmental damage, etc	Marginal	2
Severe injury or major system or environmental damage, etc	Critical	3
Death, system loss or severe environmental damage, etc	Catastrophic	4

Table 11.19 Assessment of Hazard Probabilities and Levels.

Hazard categories	Qualitative	Quantitative	Level
Improbable	So unlikely, it can be assumed occurrence may not be experienced	The probability is less than 10^{-6}	A
Remote	Unlikely but possible to occur in the lifetime of an item	The probability is between 10^{-6} and 10^{-3}	B
Occasional	Likely to occur sometime in the life of an item	The probability is between 10^{-3} and 10^{-2}	C
Probable	Will occur several times in the life time of an item	The probability is between 10^{-2} and 10^{-1}	D
Frequent	Likely to occur frequently	The probability is greater than 10^{-1}	E

11.4.3.3 Quantitative safety analysis

Quantitative safety analysis utilises what is known and assumed about the failure characteristics of each individual component to build a mathematical model that is associated with some or all of the following information:

- Failure rates.
- Repair rates.
- Mission time.
- System logic.
- Maintenance schedules.
- Human error.

Similar to the qualitative analysis, the occurrence probability of each system failure event and the magnitude of possible consequences are to be obtained. However, these parameters are to be quantified.

(A) Event probabilities

There are predominantly three methods that could be used to determine the occurrence probability of an event, namely (Preyssl (1995)):

1. Statistical method.
2. Extrapolation method.
3. Expert judgement method.

Table 11.20 Priority Matrix Based on Hazard Severity and Hazard Probability.

Hazard probability	Hazard Severity			
	(1) Negligible	(2) Marginal	(3) Critical	(4) Catastrophic
(A) Improbable ($x < 10^{-6}$)	1A	2A	3A	4A
(B) Remote ($10^{-3} > x > 10^{-6}$)	1B	2B	3B	4B
(C) Occasional ($10^{-2} > x > 10^{-3}$)	1C	2C	3C	4C
(D) Probable ($10^{-1} > x > 10^{-2}$)	1D	2D	3D	4D
(E) Frequent ($x > 10^{-1}$)	1E	2E	3E	4E

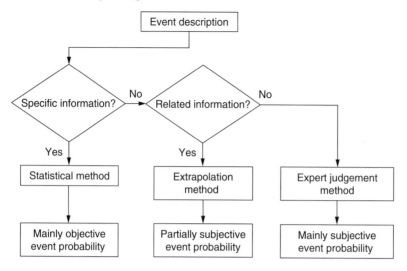

Figure 11.24 Event probability determination.

The statistical method involves the treatment of directly relevant test of experience data and the calculation of the probabilities. The extrapolation method involves the use of model prediction, similarity considerations and Bayesian concepts. Limited use of expert judgement is made to estimate unknown values as input to the extrapolation method. The expert judgement method involves direct estimation of probabilities by specialists.

These methods can be used together in an effective way to produce a reasonable estimate of the probability of an event occurring. The flowchart in Figure 11.24 shows the type of event probability produced depending on the available data.

(B) Failure probability distributions

There are a number of probability distributions to model failures. The distribution types can be found in various sources (Henley and Kumamoto (1992), Hoover (1989), Law and Kelton (1982), Rubinstein (1981), Savic (1989)). The typical ones are listed as follows:

- Beta.
- Exponential.
- Gamma.
- Lognormal.
- Normal.
- Triangular.
- Uniform.
- Weibull.

In this Section, only two particular types of distributions (i.e. Exponential and Normal distributions) are briefly described.

For many items, the relationship of failure rate versus time can be commonly referred to as the 'bathtub' curve. The idealised 'bathtub' curve shown in Figure 11.25 has the following three stages:

1. Initial period
   The item failure rate is relatively high. Such failure is usually due to factors such as defective manufacture, incorrect installation, learning curve of equipment user, etc. Design should also aim at having a short 'initial period'.
2. Useful life.
   In this period of an item, the failure rate is constant. Failures appear to occur purely by chance. This period is known as the 'useful life' of the item.
3. Wear-out period
   In this period of an item, the item failure rate rises again. Failures are often described as wear-out failures.

(i) Exponential distribution

A risk assessment mainly concentrates on the useful life in the 'bathtub' curve in Figure 11.25. In the useful life region, the failure rate is constant over the period of time. In other words, a failure could occur randomly regardless of when a previous failure occurred. This results in a negative exponential distribution for the failure frequency. The failure density function of an exponential distribution is as follows:

$$f(t) = \lambda e^{-\lambda t} \qquad (11.2)$$

where failure rate $\lambda = 1/MTBF$ and $t =$ time of interest.

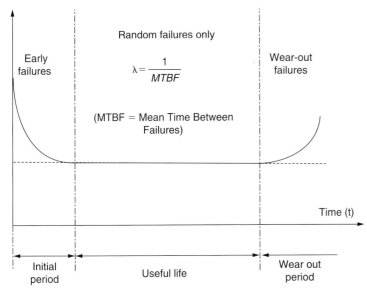

Figure 11.25 The 'bathtub' curve.

(*MTBF*: Mean Time Between Failure)
Failure probability of an item at time t is:

$$P(t) = 1 - e^{-\lambda t} \tag{11.3}$$

Example
Given that the Mean Time Between Failure for an item is 10,000 hours, calculate the failure probabilities of the item at $t = 0$, 10,000 and 100,000 hours if failures follow an exponential distribution.

Solution
  $\lambda = 1/MTBF = 0.00001$ per hour
  When $t = 0$, $P(0) = 1 - e^{-\lambda t} = 1 - e^0 = 0$
  When $t = 10,000$, $P(10,000) = 1 - e^{-\lambda t} = 1 - e^{-0.00001 \times 10,000} = 0.632$
  When $t = 100,000$, $P(100,000) = 1 - e^{-\lambda t} = 1 - e^{-0.00001 \times 10,000} = 1$
From the above, it can be seen that at $t = 0$ the item does not fail and after a considerable time it fails.

(ii) Normal distribution
Normal distributions are widely used in modelling repair activities. The failure density function of a normal distribution is:

$$f(t) = \frac{1}{\sqrt{2\pi}\sigma} e^{-(t-\mu)^2/2\sigma^2} \tag{11.4}$$

where μ = mean and σ^2 = standard deviation of t.

(c) Event consequences
The possible consequences of a system failure event can be quantified in terms of the possible loss of lives and property damage, and the degradation of the environment caused by the occurrence of the failure event (Smith (1985, 1992)). Experts of the particular operating situation normally quantify these elements in monetary terms. Quantifying human life in monetary terms could be difficult as it involves several moral issues that are constantly debated. Hence, it is normally expressed in terms of the number of fatalities, Henley and Kumamoto (1992).

The process of risk assessment is initially performed qualitatively and later extended quantitatively to include data when it becomes available. The interactions and outcomes of both these methods are seen in Figure 11.26. Using the quantified method, risk evaluation can be carried out to determine the major risk contributors and the analysis can be attenuated to include cost benefit assessment of the risk control options.

11.4.3.4 Cause and effect relationship

As discussed in the previous two sections, safety analysis techniques can be initially categorised either as qualitative or quantitative methods. However, the way each analysis explores the relationship between causes and effects can be categorised further into four different categories, namely,

1. Deductive techniques.
2. Inductive techniques.

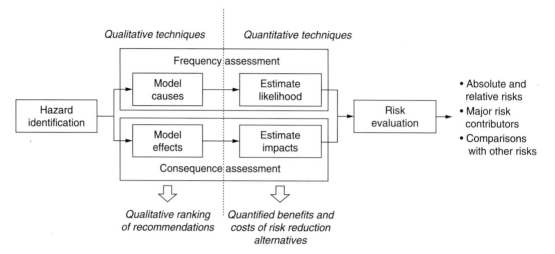

Figure 11.26 Qualitative and quantitative analysis.

Table 11.21 Ways to Investigate Cause-Effect Relationship.

		Effects	
		Known	*Unknown*
Cause	Known	Descriptive techniques	Inductive techniques
	Unknown	Deductive techniques	Exploratory techniques

3. Exploratory techniques.
4. Descriptive techniques.

Deductive techniques start from known effects to seek unknown causes, whereas inductive techniques start from known causes to forecast unknown effects. Exploratory techniques establish a link between unknown causes to unknown effects while descriptive techniques link known causes to known effects. These four ways to investigate the relationship between causes and effects are illustrated in Table 11.21, Pillay (2001).

11.4.3.5 *Preliminary hazard analysis (PHA)*

Preliminary Hazard Analysis (PHA) was introduced in 1966 after the Department of Defence of the United States of America requested safety studies to be performed at all stages of product development. The Department of Defence issued the guidelines that came into force in 1969 (Military Standard (1969, 1999)).

Preliminary Hazard Analysis is performed to identify areas of the system, which will have an effect on safety by evaluating the major hazards associated with the system. It provides an initial assessment of the identified hazards. PHA typically involves:

1. Determining hazards that might exist and possible effects.
2. Determining a clear set of guidelines and objectives to be used during a design.
3. Creating plans to deal with critical hazards.
4. Assigning responsibility for hazard control (management and technical).
5. Allocating time and resources to deal with hazards.

'Brainstorming' techniques are used during which the design or operation of the system is discussed on the basis of the experience of the people involved in the brainstorming activity. Checklists are commonly used to assist in identifying hazards.

The results of the PHA are often presented in tabular form, which would typically include information such as but not limited to Henley and Kumamoto (1992), Smith (1992), Villemuer (1992):

1. A brief description of the system and its domain.
2. A brief description of any sub-systems identified at this phase and the boundaries between them.
3. A list of identified hazards applicable to the system, including a description and unique reference.
4. A list of identified accidents applicable to the system including a description, a unique reference and a description of the associated hazards and accident sequences.
5. The accident risk classification.
6. Preliminary probability targets for each accident.
7. Preliminary predicted probabilities for each accident sequence.
8. Preliminary probability targets for each hazard.

9. A description of the system functions and safety features.
10. A description of human error which could create or contribute to accidents.

The advantages of using the PHA method include:

1. It identifies the potential for major hazards at a very early stage of project development.
2. It provides basis for design decisions.
3. It helps to ensure plant to plant and plant to environment compatibility.
4. It facilitates a full hazard analysis later.

The disadvantage of PHA is that it is not comprehensive and must be followed by a full HAZard and OPerability (HAZOP) study.

(i) Subsystem Hazard Analysis/System Hazard Analysis

Subsystem Hazard Analysis (SSHA) or System Hazard Analysis (SHA) is one requiring detailed studies of hazards, identified in the PHA, at the subsystem and system levels, including the interface between subsystems and the environment, or by the system operating as a whole. Results of this analysis include design recommendations, changes or controls when required, and evaluation of design compliance to contracted requirements. Often subsystem and system hazards are easily recognised and remedied by design and procedural measures or controls. These hazards are often handled by updating and expanding the PHA, with timing of the SSHA/SHA normally determined by the availability of subsystem and system design data (usually begins after the preliminary design review and completed before the critical design review).

(ii) Operating and Support Hazard Analysis

Operating and Support Hazard Analysis (OSHA) is an analysis performed to identify those operating functions that may be inherently dangerous to test, maintenance, handling, transportation or operating personnel or in which human error could be hazardous to equipment or people. The information for this analysis is normally obtained from the PHA. The OSHA should be performed at the point in system development when sufficient data is available, after procedures have been developed. It documents and evaluates hazards resulting from the implementation of operations performed by personnel. It also considers:

1. The planned system configuration at each phase of activity.
2. The facility interfaces.
3. The planned environments.
4. The support tools or other equipment specified for use.
5. The operation or task sequence.
6. Concurrent task effects and limitations.
7. Regulatory or contractually specified personnel safety and health requirements.
8. The potential for unplanned events including hazards introduced by human error.

OSHA identifies the safety requirements (or alternatives) needed to eliminate identified hazards or to reduce the associated risk to an acceptable level.

11.4.3.6 *What-if analysis*

What-If analysis uses a creative team brainstorming 'what if' questioning approach to the examination of a process to identify potential hazards and their consequences. Hazards are identified, existing safeguards noted, and qualitative severity and likelihood ratings are assigned to aid in risk management decision making. Questions that begin with 'what-if' are formulated by engineering personnel experienced in the process or operation preferably in advance.

There are several advantages and disadvantages of using the What-If technique. The advantages include:

1. Team of relevant experts extends knowledge and creativity pool.
2. Easy to use.
3. Ability to focus on specific element (i.e. human error or environmental issues).

The disadvantages include:

1. Quality is dependent on knowledge, thoroughness and experience of team.
2. Loose structure that can let hazards slip through.
3. It does not directly address operability problems.

11.4.3.7 *Hazard and Operability (HAZOP) studies*

A HAZard and OPerability (HAZOP) study is an inductive technique, which is an extended Failure Mode, Effects and Criticality Assessment (FMECA). The HAZOP process is based on the principle that a team-approach to hazard analysis will identify more problems than when individuals working separately combine results.

The HAZOP team is made up of individuals with varying backgrounds and expertise. The expertise is brought together during HAZOP sessions and through a collective brainstorming effort that stimulates creativity and new ideas, a thorough review of the process under consideration is made. In short it can be applied by a multidisciplinary team using a checklist to stimulate systematic thinking for identifying potential hazards and operability problems, particularly in the process industries (Bendixen et al. (1984)).

The HAZOP team focuses on specific portions of the process called 'nodes'. A process parameter (e.g. flow) is identified and an intention is created for the node under consideration. Then a series of guidewords is combined with the parameter 'flow' to create a deviation. For example, the guideword 'no' is combined with the parameter 'flow' to give the deviation 'no flow'. The team then focuses on listing all the credible causes of a 'no flow' deviation beginning with the cause that can result in the worst possible consequences the team can think of at the time. Once the causes are recorded, the team lists the consequences, safeguards and any recommendations deemed appropriate. The process is repeated for the next deviation until completion of the node. The team moves on to the next node and repeats the process.

(i) Guidewords, Selection of parameters and deviations

The HAZOP process creates deviations from the process design intent by combining guidewords (no, more, less, etc.) with process parameters resulting in a possible deviation from the design intent. It should be pointed out that not all guideword/parameter combinations would be meaningful. A sample list of guidewords is given below:

- No
- More
- Less
- As Well As
- Reverse
- Other Than

The application of parameters will depend on the type of process being considered, the equipment in the process and the process intent. The most common specific parameters that should be considered are flow, temperature, pressure, and where appropriate, level. In almost all instances, these parameters should be evaluated for every node. The scribe shall document, without exception, the team's comments concerning these parameters. Additionally, the node should be screened for application of the remaining specific parameters and for the list of applicable general parameters. These should be recorded only if there is a hazard or an operability problem associated with the parameter. A sample set of parameters includes the following:

- Flow
- Temperature
- Pressure
- Composition
- Phase
- Level
- Relief
- Instrumentation

(ii) HAZOP process

A HAZOP study can be broken down into the following steps (McKelvey (1988)):

1. Define the scope of the study.
2. Select the correct analysis team.
3. Gather the information necessary to conduct a thorough and detailed study.
4. Review the normal functioning of the process.
5. Subdivide the process into logical, manageable sub-units for efficient study and confirm that the scope of the study has been correctly set.
6. Conduct a systematic review according to the established rules for the procedure being used and ensure that the study is within the special scope.
7. Document the review proceedings.
8. Follow up to ensure that all recommendations from the study are adequately addressed.

The detailed description of the methodology can be found in Bendixen et al. (1984), McKelvey (1988), Kletz (1992), Wells (1980).

(iii) HAZOP application to fishing vessels

In order to apply the HAZOP process for the study of a fishing vessel system, the conventional method given in the previous sub-section is modified and can be summarised as follows (Pillay (2001)):

1. Define the system scope and team selection
   - Firstly define the scope of the study and then accordingly select the appropriate team to be involved in the study
2. Describe the system
   - Describe the system in some detail. This description should clarify the intention of the system as a whole from an operational viewpoint.
   - The information generated here will help the analyst understand the system and its criticality to the safe operation of the vessel. The data will later prove to be useful when used to determine the consequences of component failure in Step 6 of the approach.
3. Break it down into smaller operations for consideration and identify each component within the considered system.
   - Having attained the overall picture, break it down into its sub-operations/routines. It is difficult to see all the problems in a complex process but when each individual process is analysed on its own, the chances are that little will be missed

Table 11.22 Examples of Guidewords.

Guide words	Examples
No	No flow, no signal
Less	Less flow, less cooling
More	Excess temperature, excess pressure
Opposite	Cooling instead of heating
Also	Water as well as lubricating oil
Other	Heating instead of pumping
Early	Opening the drain valve too soon
Late	Opening the drain valve too late
Part of	Incomplete drainage

out. Ideally, each operation should be singled out, but it is frequently more convenient to consider more than one operation at a time due to its inter-relationship and dependency.
- The identification of each component can be achieved by first looking at historical failure data that is available and then complementing it with components identified from equipment drawings. Component failure data can be obtained from logbooks, docking reports, Chief engineer's reports and maintenance reports.

4. Determine design intention for each component that is identified.
   - At this stage, the purpose or intention of each component is ascertained. This helps to determine the functional purpose of the specific operation and shows how it relates/interacts to achieve the process intentions.

5. Apply a series of guidewords to see how that intention may be frustrated.
   - This is the heart of HAZOP. Having decided the intention of a process, this stage analyses the ways in which it can go wrong.
   - Examples of guide words are as illustrated in Table 11.22.

6. For meaningful deviations from the intention, look for possible causes and likely consequences.
   - At this stage, the root of the problem is identified and the possible consequences are predicted and complemented with any historical data available. The consequences are considered for four major categories (personnel, environment, equipment and operation). At this point, it is determined how the failure of a component will affect the safety and integrity in terms of these four categories.

7. Consider possible action to remove the cause or reduce the consequences.
   - A HAZOP team usually provides ideas to remove a cause or deal with the possible consequences. This could be suggestion of improvements in design, operational procedure, maintenance periods and redundancy arrangements. It would be very unusual for every single one of these actions to be put into practice, but at least a rational choice could be made.

8. Reiteration
   - Consider how the improvements will affect the operation of the system and re-evaluate what can go wrong (with the improvements incorporated).

These steps can be illustrated in the flowchart in Figure 11.27. There are several advantages of using HAZOP to assess the safety of fishing vessels. These include:

1. It is the most systematic and comprehensive PHA methodology.
2. It provides greatest safety assurance.
3. It can be used in conjunction with Human Error Analysis (HEA).
4. It is the only PHA to address both safety/operability problems and environmental hazards.

The HAZOP process can be time consuming and costly if it is not well prepared in advance and can be tedious if it is not well facilitated. A comprehensive HAZOP study will require many experts and a considerable duration.

11.4.3.8 *Fault tree analysis (FTA)*

Fault Tree Analysis (FTA) is a formal deductive procedure for determining combinations of component failures and human errors that could result in the occurrence of specified undesired events at the system level (Ang and Tang (1984)). It is a diagrammatic method used to evaluate the probability of an accident resulting from sequences and combinations of faults and failure events. This method can be used to analyse the vast majority of industrial system reliability problems. FTA is based on the idea that:

1. A failure in a system can trigger other consequent failures.
2. A problem might be traced backwards to its root causes.

The identified failures can be arranged in a tree structure in such a way that their relationships can be characterised and evaluated.

(i) Benefits to be gained from FTA
There are several benefits of employing FTA for use as a safety assessment tool. These include:

1. The Fault Tree (FT) construction focuses the attention of the analyst on one particular undesired system failure mode, which is usually identified as the most critical with respect to the desired function, Andrews and Moss (2002).

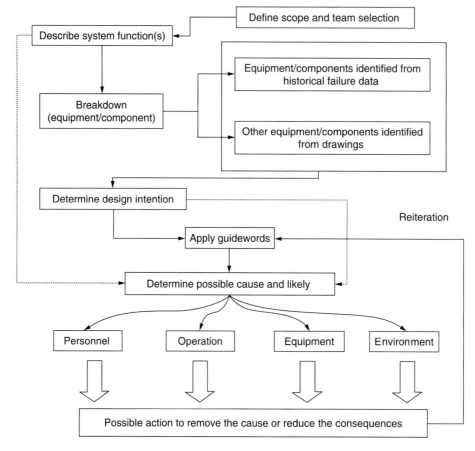

Figure 11.27 Flowchart of HAZOP process applied to fishing vessels.

2. The FT diagram can be used to help communicate the results of the analysis to peers, supervisors and subordinates. It is particularly useful in multi-disciplinary teams with the numerical performance measures.
3. Qualitative analysis often reveals the most important system features.
4. Using component failure data, the FT can be quantified.
5. The qualitative and quantitative results together provide the decision-maker with an objective means of measuring the adequacy of the system design.

An FT describes an accident model, which interprets the relation between malfunction of components and observed symptoms. Thus the FT is useful for understanding logically the mode of occurrence of an accident. Furthermore, given the failure probabilities of the corresponding components, the probability of a top event occurring can be calculated. A typical FTA consists of the following steps:

1. System description.
2. Fault tree construction.
3. Qualitative analysis.
4. Quantitative analysis.

These steps are illustrated in Figure 11.28.

(ii) System definition
FTA begins with the statement of an undesired event, that is, failed state of a system. To perform a meaningful analysis, the following three basic types of system information are usually needed:

1. Component operating characteristics and failure modes: A description of how the output states of each component are influenced by the input states and internal operational modes of the component.
2. System chart: A description of how the components are interconnected. A functional layout diagram of the system must show all functional interconnections of the components.
3. System boundary conditions: These define the situation for which the fault tree is to be drawn.

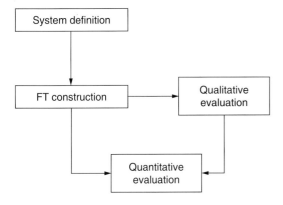

Figure 11.28 FTA method.

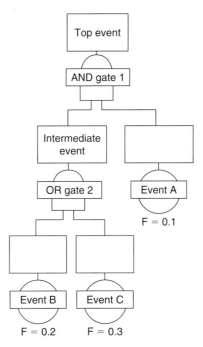

Figure 11.29 Fault tree example.

(iii) Fault Tree construction

FT construction, which is the first step for a failure analysis of a technical system, is generally a complicated and time-consuming task. An FT is a logical diagram constructed by deductively developing a specific system failure, through branching intermediate fault events until a primary event is reached. Two categories of graphic symbols are used in an FT construction, logic symbols and event symbols.

The logic symbols or logic gates are necessary to interconnect the events. The most frequently used logic gates in the fault tree are **AND** and **OR** gates. The **AND** gate produces an output if all input events occur simultaneously. The **OR** gate yields output events if one or more of the input events are present.

The event symbols are rectangle, circle, diamond and triangle. The rectangle represents a fault output event, which results from combination of basic faults, and/or intermediate events acting through the logic gates. The circle is used to designate a primary or basic fault event. The diamond describes fault inputs that are not a basic event but considered as a basic fault input since the cause of the fault has not been further developed due to lack of information. The triangle is not strictly an event symbol but traditionally classified as such to indicate a transfer from one part of an FT to another. Figure 11.29 gives an example of a fault tree. The fault tree in Figure 11.29 is constructed using Fault Tree+ (Isograph Limited (1995)). In the fault tree in Figure 11.29, it can be seen that the occurrence probabilities of basic events *A, B* and *C* are assumed to be 0.1, 0.2 and 0.3 under certain conditions for a given period of time, respectively.

To complete the construction of a fault tree for a complicated system, it is necessary first to understand how the system works. This can be achieved by studying the blue prints of the system (which will reflect the interconnections of components within the system). In practice, all basic events are taken to be statistically independent unless they are common cause failures. Construction of an FT is very susceptible to the subjectivity of the analyst. Some analysts may perceive the logical relationships between the top event and the basic events of a system differently. Therefore, once the construction of the tree has been completed, it should be reviewed for accuracy, completeness and checked for omission and oversight. This validation process is essential to produce a more useful FT by which system weakness and strength can be identified.

(iv) Qualitative Fault Tree evaluation

Qualitative FTA consists of determining the minimal cut sets and common cause failures. The qualitative analysis reduces the FT to a logically equivalent form, by using the Boolean algebra, in terms of the specific combination of basic events sufficient for the undesired top event to occur (Henley and Kumamoto (1992)). In this case each combination would be a critical set for the undesired event. The relevance of these sets must be carefully weighted and major emphasis placed on those of greatest significance.

(v) Quantitative Fault Tree evaluation

In an FT containing independent basic events, which appear only once in the tree structure, then the top event probability can be obtained by working the basic event probabilities up through the tree.

In doing so, the intermediate gate event probabilities are calculated starting at the base of the tree and working upwards until the top event probability is obtained.

When trees with repeated events are to be analysed, this method is not appropriate since intermediate gate events will no longer occur independently. If this method is used, it is entirely dependent upon the tree structure whether an overestimate or an underestimate of the top event probability is obtained. Hence, it is better to use the minimal cut-set method.

In Boolean algebra, binary states 1 and 0 are used to represent the two states of each event (i.e. occurrence and non-occurrence). Any event has an associated Boolean variable. Events A and B can be described as follows using Boolean algebra:

$$A = \begin{cases} 1 & \text{event occurs} \\ 0 & \text{event does not occur} \end{cases}$$

$$B = \begin{cases} 1 & \text{event occurs} \\ 0 & \text{event does not occur} \end{cases} \quad (11.5)$$

Suppose '+' stands for 'OR' and '·' for 'AND'. Suppose '$\bar{A}$' stands for 'not A'. Then the typical Boolean algebra rules are described as follows:

Identity laws

$A + 0 = A$

$A + 1 = 1$

$A \cdot 0 = 0$

$A \cdot 1 = A$

Indempotent laws

$A + A = A$

$A \cdot A = A$

Complementative laws

$A \cdot \bar{A} = 0$

$A + \bar{A} = 1$

Commutative laws

$A + B = B + A$

$A \cdot B = B \cdot A$

Associative laws

$(A + B) + C = A + (B + C)$

$(A \cdot B) \cdot C = A \cdot (B \cdot C)$

Distributive laws

$A \cdot (B + C) = A \cdot B + A \cdot C$

$A + (B \cdot C) = (A + B) \cdot (A + C)$

Absorption laws

$A + A \cdot B = A$

$A \cdot (A + B) = A$

De Morgan's laws

$\overline{A \bullet B} = \bar{A} + \bar{B}$

$\overline{A + B} = \bar{A} \bullet \bar{B}$

The above rules can be used to obtain the minimum cut sets leading to a top event in a fault tree. The occurrence probability of a top event can then be obtained from the associated minimum cut sets. The following two mini-trees are used to demonstrate how the occurrence probability of a top event can be obtained:

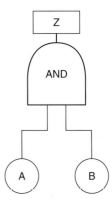

Obviously the minimum cut set for the mini-tree below is $A \cdot B$.

If one event is independent from the other, the occurrence probability of top event Z is

$$P(Z) = P(A \cdot B) = P(A) \times P(B)$$

where $P(A)$ and $P(B)$ are the occurrence probabilities of events A and B.

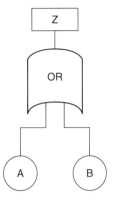

Obviously the minimum cut set for the mini-tree above is $A + B$.

If one event is independent from the other, the occurrence probability of top event Z is

$P(Z) = P(A + B)$
$= P(A) + P(B) - P(A \cdot B)$
$= P(A) + P(B) - P(A) \times P(B)$

where $P(A)$ and $P(B)$ are the occurrence probabilities of events A and B.

FTA may be carried out in the hazard identification and risk estimation phases of the safety assessment of ships to identify the causes associated with serious system failure events and to assess the occurrence likelihood of them. It is worth noting that in situations where there is a lack of the data available, the conventional FTA method may not be well suited for such an application. As such, a new modified method incorporating FTA and Fuzzy Set Theory (FST) may be used which is presented and discussed in detail in Pillay and Wang (2003).

(vi) FTA example
An example
The risk assessment of a marine system is carried out at the early design stages. It has been identified that a serious hazardous event (top event) arises if

events $X1$ and $X2$ happen; or
event $X3$ occurs.

$X1$ occurs when events A and B happen.
$X2$ occurs when

event B happens; or
events B and C occur.

Event $X3$ occurs when

events C and D happen; or
events A, C and D happen.

Events A, B, C and D are basic events. It is assumed that events A, B, C and D follow an exponential distribution. The failure rates (1/hour) for events A, B, C and D are 0.0001, 0.0002, 0.0003 and 0.0004, respectively.

(i) Draw the fault tree for the above problem.
(ii) Find the minimum cut sets.
(iii) Discuss how the likelihood of occurrence of the top event can be reduced/eliminated.
(iv) Calculate the occurrence likelihood of the top event at time $t = 10,000$ hours assuming that events A, B, C and D are independent of each other.

Solution

(i) The fault tree is built as shown in Figure 11.30.
(ii) Top event $= X1 \cdot X2 + X3$

$= A \cdot B \cdot (B + B \cdot C) + C \cdot D + A \cdot C \cdot D$
$= A \cdot B \cdot B + C \cdot D$
$= A \cdot B + C \cdot D$

(iii) When events A and B or events C and D happen, the top event happens. Therefore, to avoid the occurrence of the top event, it is required to make sure that events A and B do not happen simultaneously and events C and D do not happen simultaneously. To reduce the occurrence likelihood of the top event, it is required to reduce the occurrence likelihood of four basic events A, B, C and D.

(iv) At $t = 10,000$ hours

$P(A) = 1 - e^{-\lambda t} = 1 - e^{-0.0001 \times 10,000} = 0.632$
$P(B) = 1 - e^{-\lambda t} = 1 - e^{-0.0002 \times 10,000} = 0.865$
$P(C) = 1 - e^{-\lambda t} = 1 - e^{-0.0003 \times 10,000} = 0.95$
$P(D) = 1 - e^{-\lambda t} = 1 - e^{-0.0004 \times 10,000} = 0.982$
$P(\text{Top event}) = P(A \cdot B + C \cdot D) = P(A \cdot B)$
$\quad + P(C \cdot D) - P(A \cdot B \cdot C \cdot D)$
$= P(A) \times P(B) + P(C) \times P(D)$
$\quad - P(A) \times P(B) \times P(C) \times P(D)$
$= 0.97$

The likelihood of occurrence of the top event at time $t = 10,000$ hours is 0.97.

It should be noted that when calculating the failure probability of the top event, the application of the simplification rules may be required. This is demonstrated by the following example:

Example
Given that $P(A) = P(B) = P(C) = P(D) = 0.5$ and also that basic events A, B, C and D are independent, calculate $P(A \cdot B + B \cdot C + A \cdot C)$.

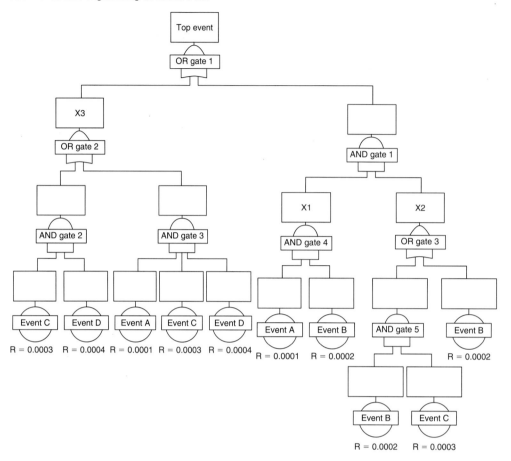

Figure 11.30 A fault tree.

Solution

$$P(A \cdot B + B \cdot C + A \cdot C)$$
$$= P(A \cdot B) + P(B \cdot C + A \cdot C)$$
$$\quad - P(A \cdot B \cdot (B \cdot C + A \cdot C))$$
$$= P(A) \times P(B) + P(B \cdot C) + P(A \cdot C)$$
$$\quad - P(B \cdot C \cdot A \cdot C)$$
$$\quad - P(A \cdot B \cdot B \cdot C + A \cdot B \cdot A \cdot C)$$
$$= P(A) \times P(B) + P(B) \times P(C)$$
$$\quad + P(A) \times P(C) - P(A \cdot B \cdot C)$$
$$\quad - P(A \cdot B \cdot C + A \cdot B \cdot C)$$
$$= P(A) \times P(B) + P(B) \times P(C) + P(A) \times P(C)$$
$$\quad - P(A \cdot B \cdot C) - P(A \cdot B \cdot C)$$
$$= P(A) \times P(B) + P(B) \times P(C)$$
$$\quad + P(A) \times P(C) - 2 \times P(A \cdot B \cdot C)$$
$$= P(A) \times P(B) + P(B) \times P(C)$$
$$\quad + P(A) \times P(C) - 2 \times P(A) \times P(B) \times P(C)$$
$$= 0.5$$

The top events of a system to be investigated in FTA may also be identified through a PHA or may correspond to a branch of an event tree or a system Boolean representation table (Wang *et al.* (1995)). The information produced from FMECA may be used in construction of fault trees. Detailed description of FTA and its applications can be found in various published documents such as Andrews and Moss (2002), Ang and Tang (1984), Halebsky (1989), Henley and Kumamoto (1992).

11.4.3.9 *Event tree analysis*

In the case of standby systems and in particular, safety and mission-oriented systems, the Event Tree Analysis (ETA) is used to identify the various possible outcomes of the system following a given initiating event which is generally an unsatisfactory operating event or situation. In the case of continuously operated systems, these events can occur (i.e. components can fail) in any arbitrary order. In the ETA, the components can be considered in any order since they do not operate

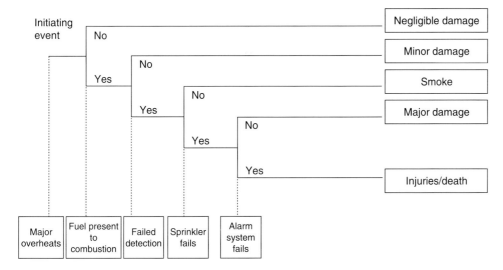

Figure 11.31 Example of an event tree.

chronologically with respect to each other. ETA provides a systematic and logical approach to identify possible consequences and to assess the occurrence probability of each possible resulting sequence caused by the initiating failure event (Henley and Kumamoto (1992), Villemuer (1992)).

Event Tree Example

A simple example of an event tree is shown in Figure 11.31. In the event tree, the initiating event is 'major overheats' in an engine room of a ship. It can be seen that when the initiating event 'major overheats' takes place and if there is no fuel present, the consequences will be negligible in terms of fire risks. If there is fuel present, then it is required to look at if the detection fails. If the answer is no, then the consequences are minor damage, otherwise it is required to investigate if the sprinkler fails. If the sprinkler works, then the consequences will be smoke, otherwise it is required to see if the alarm system works. If the alarm system works, then the consequences will be major damage, otherwise injuries/deaths will be caused.

ETA has proved to be a useful tool for major accident risk assessments. Such an analysis can be effectively integrated into the hazard identification and estimation phases of a safety assessment programme. However, an event tree grows in width exponentially and as a result it can only be applied effectively to small sets of components.

11.4.3.10 *Markov chains*

Markov methods are useful for evaluating components with multiple states, for example, normal, degraded

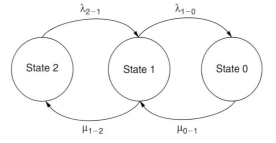

Figure 11.32 Markovian model for a system with three states.

and critical states (Norris (1998)). Consider the system in Figure 11.32 with three possible states, 0, 1 and 2 with failure rate λ and repair rate μ. In the Markovian model, each transition between states is characterised by a transition rate, which could be expressed as failure rate, repair rate, etc. If it is defined that:

$P_i(t)$ = probability that the system is in state i at time t.
$\rho_{ij}(t)$ = the transition rate from state i to state j.

and if it is assumed that $P_i(t)$ is differentiable, it can be shown that:

$$\frac{dP_i(t)}{dt} = \left(\sum_j \rho_{ij}(t)\right) \bullet P_i(t) + \sum_j (\rho_{ij}(t) \bullet P_j(t))$$

(11.6)

If a differential equation is written for each state and the resulting set of differential equations is solved, the time dependent probability of the system being in each state is obtained (Modarres (1993)). Markov chains are mainly a quantitative technique, however, using the state and transition diagrams, qualitative information about the system can be gathered.

11.4.3.11 *Failure mode, effects and critical analysis (FMECA)*

The process of conducting a Failure Mode, Effects and Critical Analysis (FMECA) can be examined in two levels of detail. Failure Mode and Effects Analysis (FMEA) is the first level of analysis, which consists of the identification of potential failure modes of the constituent items (components or subsystems) and the effects on system performance by identifying the potential severity of the effect. The second level of analysis is Criticality Analysis for criticality ranking of the items under investigation. Both of these methods are intended to provide information for making risk management decisions.

FMEA is an inductive process that examines the effect of a single point failure on the overall performance of a system through a 'bottom-up approach' (Andrews and Moss (2002)). This analysis should be performed iteratively in all stages of design and operation of a system.

The first step in performing an FMEA is to organize as much information as possible about the system concept, design and operational requirements. By organizing the system model, a rational, repeatable, and systematic means to analyse the system can be achieved. One method of system modelling is the system breakdown structure model – a top down division of a system (e.g. ship, submarine, propulsion control) into functions, subsystems and components. Block diagrams and fault-tree diagrams provide additional modelling techniques for describing the component/function relationships.

A failure mode is a manner that a failure is observed in a function, subsystem, or component (Henley and Kumamoto (1992), Villemuer (1992)). Failure modes of concern depend on the specific system, component, and operating environment. Failure modes are sometimes described as categories of failure. A potential failure mode describes the way in which a product or process could fail to perform its desired function (design intent or performance requirements) as described by the needs, wants, and expectations of the internal and external customers/users. Examples of failure modes are: fatigue, collapse, cracked, performance deterioration, deformed, stripped, worn (prematurely), corroded, binding, seized, buckled, sag, loose, misalign, leaking, falls off, vibrating, burnt, etc. The past history of a component/system is used in addition to understanding the functional requirements to determine relevant failure modes. For example, several common failure modes include complete loss of function, uncontrolled output, and premature/late operation (IMO (1995)).

The causes of a failure mode (potential causes of failure) are the physical or chemical processes, design defects, quality defects, part misapplication, or others, which are the reasons for failure (Military Standard (1980)). The causes listed should be concise and as complete as possible. Typical causes of failure are: incorrect material used, poor weld, corrosion, assembly error, error in dimension, over stressing, too hot, too cold, bad maintenance, damage, error in heat treat, material impure, forming of cracks, out of balance, tooling marks, eccentric, etc. It is important to note that more than one failure cause is possible for a failure mode; all potential causes of failure modes should be identified, including human error.

The possible effects are generally classified into three levels of propagation: local, next higher level, and end effect. An effect is an adverse consequence that the customer/user might experience. The customer/user could be the next operation, subsequent operations, or the end user. The effects should be examined at different levels in order to determine possible corrective measures for the failure (Military Standard (1980)). The consequences of the failure mode can be assessed by a severity index indicating the relative importance of the effects due to a failure mode. Some common severity classifications include (1) Negligible, (2) Marginal, (3) Critical and (4) Catastrophic.

Criticality analysis allows a qualitative or a quantitative ranking of the criticality of the failure modes of items as a function of the severity classification and a measure of the frequency of occurrence. If the occurrence probability of each failure mode of an item can be obtained from a reliable source, the criticality number of the item under a particular severity class may be quantitatively calculated as follows:

$$C = \sum_{i=1}^{N} E_i L_i t$$

where:
E_i = failure consequence probability of failure mode i (the probability that the possible effects will occur, given that failure mode i has taken place.
L_i = occurrence likelihood of failure mode i.
N = number of the failure modes of the item, which fall under a particular severity classification.
t = duration of applicable mission phase.

Once all criticality numbers of the item under all severity classes have been obtained, a criticality matrix can be constructed which provides a means of comparing the item to all others. Such a matrix

Table 11.23 An Example of FMEA.

Name	Control system				
Function	Controlling the servo hydraulic transmission system				
Failure rate	36 (failures per million hours)				
Failure mode no.	Failure mode rate	Failure mode	Effects on system	Detecting method	Severity
1	0.015	Major leak	Loss of hoisting pressure in lowering motion. Load could fall.	Self-annunciation	Critical (3)
2	0.31	Minor leak	None.	Self-annunciation	Negligible (1)
3	0.365	No output when required.	Loss of production ability.	Self-annunciation & by maintenance	Marginal (2)
4	0.155	Control output for lowering motion cannot be stopped when required.	Possibility of fall or damage of load. Possibility of killing and/or injuring personnel.	Self-annunciation & by maintenance	Catastrophic (4)
5	0.155	Control output for hoisting up motion cannot be stopped when required.	Possibility of fall or damage of load. Possibility of killing and/or injuring personnel.	Self-annunciation & by maintenance	Catastrophic (4)

display shows the distributions of criticality of the failure modes of the item and provides a tool for assigning priority for corrective action. Criticality analysis can be performed at different indenture levels. Information produced at low indenture levels may be used for criticality analysis at a higher indenture level. Failure modes can also be prioritised for possible corrective action. This can be achieved by calculating the Risk Priority Number (RPN) associated with each failure mode. This is studied in detail in Pillay and Wang (2003).

Part of the risk management portion of the FMEA is the determination of failure detection sensing methods and possible corrective actions (Modarres (1993)). There are many possible sensing device alternatives such as alarms, gauges and inspections. An attempt should be made to correct a failure or provide a backup system (redundancy) to reduce the effects propagation to rest of system. If this is not possible, procedures should be developed for reducing the effect of the failure mode through operator actions, maintenance, and/or inspection.

FMEA/FMECA is an effective approach for risk assessment, risk management, and risk communication concerns. This analysis provides information that can be used in risk management decisions for system safety. FMEA has been used successfully within many different industries and has recently been applied in maritime regulations to address safety concerns with relatively new designs. While FMEA/FMECA is a useful tool for risk management, it also has qualities that limit its application as a complete system safety approach. This technique provides risk analysis for comparison of single component failures only.

FMECA Example
Example
Table 11.23 shows an FMEA for a control system of a marine crane hoisting system (Wang (1994), Wang et al. (1995)). It can be seen that for the control system there are five failure modes. Failure mode rate is the ratio of the failure rate of the failure mode to the failure rate of the item. From Table 11.23 it can be seen that the sum of the five failure mode rates is equal to 1.

Suppose the failure consequence probabilities for the failure modes in Table 11.23 are 20%, 100%, 20%, 10% and 30%, respectively. The duration of interest is 10,000 hours. Formulate the criticality matrix of the above system.

Solution
From Table 11.23, it can be seen that failure mode 2 is classified as severity class 1, failure mode 3 as severity class 2 and failure mode 1 as severity class 2 while failure modes 4 and 5 are classified as severity class 4.

Severity class 1: Criticality number
$$= E_2 \times L_2 \times t$$
$$= 1 \times 0.31 \times 0.000036 \times 10\,000$$
$$= 0.1116$$

Severity class 2: Criticality number
$$= E_3 \times L_3 \times t$$
$$= 0.2 \times 0.365$$
$$\times 0.000036 \times 10\,000$$
$$= 0.02628$$

Severity class 3: Criticality number
$$= E_2 \times L_2 \times t$$
$$= 0.2 \times 0.015$$
$$\times 0.000036 \times 10\,000$$
$$= 0.00108$$

Severity class 4: Criticality number
$$= E_4 \times L_4 \times t + E_5 \times L_5 \times t$$
$$= 0.1 \times 0.155 \times 0.000036$$
$$\times 10\,000 + 0.3 \times 0.155$$
$$\times 0.000036 \times 10\,000$$
$$= 0.02233$$

The criticality matrix can be formulated as follows:

Severity class	Criticality number
1	0.1116
2	0.02628
3	0.00108
4	0.2232

If the criticality matrices for other systems are produced, comparisons can be made to determine which system needs more attention in the design stages.

11.4.3.12 Other analysis methods

Apart from the methods described above, several other methods have gained popularity in the industry. Many of these methods have been developed to a very advanced stage and have been integrated with other analysis tools to enhance their applicability.

(a) Diagraph-based Analysis (DA)
Diagraph-based Analysis (DA) is a bottom up, event-based, qualitative technique. It is commonly used in the process industry, because relatively little information is needed to set up the diagraph (Kramer and Palowitch (1987)). In a DA, the nodes correspond to the state variables, alarm conditions or failure origins and the edges represent the casual influences between the nodes. From the constructed diagraph, the causes of a state change and the manner of the associated propagation can be found out (Umeda (1980)). Diagraph representation provides explicit casual relationships among variables and events of a system with feedback loops. The DA method is effective when used together with HAZOP (Vaidhyanathan and Venkatasubramanian (1996)).

(b) Decision table method
Decision table analysis uses a logical approach that reduces the possibility of omission, which could easily occur in a fault tree construction (Dixon (1964)). A decision table can be regarded as a Boolean representation model, where an engineering system is described in terms of components and their interactions (Wang et al. (1995)). Given sufficient information about the system to be analysed, this approach can allow rapid and systematic construction of the Boolean representation models. The final system Boolean representation table contains all the possible system top events and the associated cut sets. This method is extremely useful for analysing systems with a comparatively high degree of innovation since their associated top events are usually difficult to obtain by experience, from previous accident and incident reports of similar products, or by other means. A more detailed discussion on the use of this method for safety assessment can be found in Wang (1994), Wang et al. (1995).

(c) Limit state analysis
Limit state analysis is readily applicable to failure conditions, which occur when the demand imposed on the component, or system exceeds its capability. The probability of failure is the probability that the limit state functions are violated. These probabilities are estimated by the statistical analysis of the uncertainty or variability associated with the functions' variables. In most cases, the analytical solution of the probability of failure is very difficult and sometimes almost practically impossible. However, by incorporating the Monte Carlo simulation method, this setback can be addressed. This method is normally used in structural reliability predictions and represents only half of a safety assessment (as it does not consider the severity of the failure), Bangash (1983), Damkilde and Krenk (1997).

11.4.3.13 Conclusion

In this Section, typical safety analysis methods are outlined in terms of their requirements, advantages and limitations. Some of these techniques have been successfully used in the industry and still continue to be used. However, the application of these conventional techniques to ship safety assessment may not be as straightforward as it may seem. Certain modifications are needed to enhance the application of such methods to the maritime industry. These modifications include the ability of the analysis methods to handle data that is associated with a high degree of uncertainty and the integration of expert opinion in a formal manner, where there is no bias of opinion.

The conventional methods can be used together within the framework of a formal safety assessment process. The formal safety assessment process will be described and discussed in the following Sections, detailing how the analysis methods identified here can be used effectively.

11.4.4 Formal safety assessment of ships and its relation to offshore safety case approach

11.4.4.1 Offshore safety assessment

Following the public inquiry into the Piper Alpha accident (Department of Energy (1990)), the responsibilities for offshore safety regulations were transferred from the Department of Energy to the Health & Safety Commission (HSC) through the Health & Safety Executive (HSE) as the single regulatory body for offshore safety. In response to the accepted findings of the Piper Alpha enquiry the HSE Offshore Safety Division launched a review of all offshore safety legislation and implemented changes. The changes sought to replace legislation which was seen as prescriptive with a more 'goal setting' regime. The mainstay of the regulations is the health and safety at work act. Under that act, a draft of the offshore installations (safety case) regulations (SCR-1992) was produced in 1992 (HSE (1992)). It was then modified, taking into account the comments arising from public consultation. The regulations came into force in two phases – at the end of May 1993 for new installations and November 1993 for existing installations. The regulations require operational safety cases to be prepared for all offshore installations. Both fixed and mobile installations are included. Additionally all new fixed installations require a design safety case. For mobile installations the duty holder is the owner whereas for fixed installations the duty holder is the operator.

A safety case covers all aspects of the safety of the plant or process in question, and determines how the risks involved are to be minimised. It should include sufficient data to demonstrate that (HSE (1992)):

- Hazards with the potential to cause major accidents have been identified.
- Risks have been evaluated and measures have been taken to reduce them to a As Low As Reasonably Practicable level (ALARP).

A safety case should be prepared demonstrating safety by design, describing operational requirements, providing for continuing safety assurance by means of regular review, and setting out the arrangements for emergency response. It should also include identification of a representative sample of major accident scenarios and assessments of the consequences of each scenario together with an assessment in general terms of the likelihood of its happening. The report suggests that innovative safety analysis methods and cost-benefit analysis may be beneficially used for the prediction and control of safety.

The report on the public inquiry into the Piper Alpha accident (the Cullen Report) recommends Quantitative Risk Assessment (QRA) to be used in the process of hazard identification and risk assessment in preparing a safety case. QRA can help to provide a structured objective approach to the assessment of risks, provided that it relies on and is supplemented by good engineering judgement and the limitation of the data used is roughly understood. The significant pathway leading to serious failure conditions can be systematically identified using QRA and hence all reasonably practicable steps can be taken to reduce them.

The HSE framework for decisions on the tolerability of risk is shown in Figure 11.33 where there are three regions: intolerable, ALARP and broadly acceptable. Offshore operators must submit operational safety cases for all existing and new offshore installations to the HSE Offshore Safety Division for acceptance. An installation cannot legally operate without an accepted

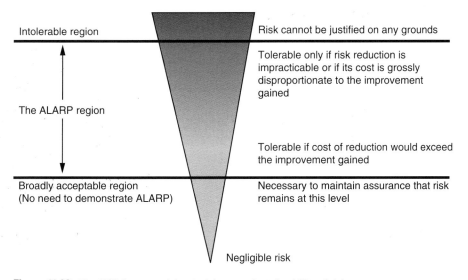

Figure 11.33 The HSE framework for decisions on the tolerability of risk.

operational safety case. To be acceptable a safety case must show that hazards with the potential to produce a serious accident have been identified and that associated risks are below a tolerability limit and have been reduced as low as is reasonably practicable. For example, the occurrence likelihood of events causing a loss of integrity of the safety refuge should be less than 10^{-3} per platform year (Spouge (1997)) and associated risks should be reduced to an ALARP level.

Management and administration regulations (MAR-1995) were introduced to cover areas such as notification to the HSE of changes of owner or operator, functions and powers of offshore installation managers (HSE (1995b)). MAR-1995 is applied to both fixed and mobile offshore installations excluding subsea offshore installations. The importance of safety of offshore pipelines has also been recognised. As a result, pipeline safety regulations (PSR-1996) were introduced to embody a single integrated, goal setting, risk based approach to regulations covering both onshore and offshore pipelines (HSE, 1996d)). Fires and explosions may be the most significant hazards with potential to cause disastrous consequences in offshore installations. Prevention of fire and explosion, and emergency response regulations (PFEER-1995) were therefore developed in order to manage fire and explosion hazards and emergency response from protecting persons from their effects (HSC (1997)). A risk-based approach is promoted to be used to deal with problems involving fire and explosion, and emergency response. PFEER-1995 supports the general requirements by specifying goals for preventive and protective measures to manage fire and explosive hazards and to secure effective emergency response and ensure compliance with regulations by the duty holder.

After several years' experience of employing the safety case approach in the UK offshore industry, the safety case regulations were amended in 1996 to include verification of safety-critical elements and the offshore installations and wells (design and construction, etc.) regulations 1996 (DCR-1996) were introduced to deal with various stages of the life cycle of the installation (HSE (1996b)). From the earliest stages of the life cycle of the installation the duty holder must ensure that all safety-critical elements be assessed. Safety-critical elements are such parts of an installation and such of its plant (including computer programs), or any part thereof the failure of which could cause or contribute substantially to; or a purpose of which is to prevent, or limit the effect of a major accident (HSE (1996c)). In DCR-1996, a verification scheme is also introduced to ensure that a record is made of the safety-critical elements; comment on the record by an independent and competent person is invited; a verification scheme is drawn up by or in consultation with such person; a note is made of any reservation expressed by such person; and such scheme is put into effect (HSE (1996c)). All such records are subject to the scrutiny of the HSE at any time. More detailed information about the DCR-1996 can be found in (HSE (1996a, b, c)). DCR-1996 allows offshore operators to have more flexibility to tackle their own offshore safety problems. Offshore duty holders may use various safety assessment approaches and safety based decision making tools to study all safety-critical elements of offshore installations and wells to optimise safety. This may encourage offshore safety analysts to develop and employ novel safety assessment and decision making approaches and to make more efforts to deal with offshore safety problems.

The relationships between such typical offshore safety regulations can be seen in Figure 11.34 where the core is the safety case regulations and others are closely related to them.

Compliance with the current offshore safety regulations is achieved by applying an integrated risk-based approach, starting from feasibility studies and extending through the life cycle of the installation. This is achieved through stages of hazard identification for the life cycle of installation from concept design to

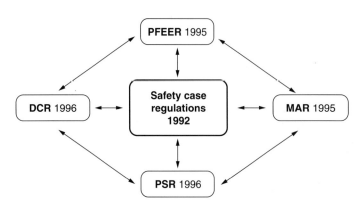

Figure 11.34 Relationships between offshore safety regulations.

decommissioning and the use of state-of-the-art risk assessment methods (Janardhanan and Grillo (1998)). In a risk-based approach, early considerations are given to those hazards which are not foreseeable to design out by progressively providing adequate measures for prevention, detection, control and mitigation and further integration of emergency response.

The main feature of the new offshore safety regulations in the UK is the absence of a prescriptive regime, defining specific duties of the operator and definition as regard to what are adequate means. The regulations set forth high level safety objective while leaving the selection of particular arrangements to deal with hazards in the hands of the operator. This is in recognition of the fact that hazards related to an installation are specific to its function and site conditions.

Recently, the industrial guidelines on a framework for risk related decision support have been produced by the UK Offshore Operators Association (UKOOA) (UKOOA (1999)). In general, the framework could be usefully applied to a wide range of situations. Its aim is to support major decisions made during the design, operation and abandonment of offshore installations. In particular, it provides a sound basis for evaluating the various options that need to be considered at the feasibility and concept selection stages of a project, especially with respect to 'major accidents hazards' such as fire, explosion, impact, loss of stability, etc. It can also be combined with other formal decision making aids such as Multi-Attribute Utility Analysis (MAUA), Analytical Hierarchy Process (AHP) or decision trees if a more detailed or quantitative analysis of the various decision alternatives is desired.

It should be noted that there can be significant uncertainties in the information and factors that are used in the decision making process. These may include uncertainties in estimates of the costs, time-scales, risks, safety benefits, the assessment of stakeholder views and perceptions, etc. There is a need to apply common sense and ensure that any uncertainties are recognised and addressed.

The format of safety case regulations was advocated by Lord Robens in 1972 when he laid emphasis on the need for self regulation, and at the same time he pointed out the drawbacks of a rule book approach to safety (Sii (2001)). The concept of the safety case has been derived and developed from the application of the principles of system engineering for dealing with the safety of systems or installations for which little or no previous operational experience exists (Kuo (1998)). The five key elements of the safety case concepts are illustrated in Figure 11.35. These elements are discussed as follows:

1. Hazard identification
   This step is to identify all hazards with the potential to cause a major accident.
2. Risk estimation
   Once the hazards have been identified, the next step is to determine the associated risks. Hazards can generally be grouped into three risk regions known as the intolerable, tolerable and negligible risk regions as shown in Figure 11.33.
3. Risk reduction
   Following risk assessment, it is required to reduce the risks associated with significant hazards that deserve attention.
4. Emergency preparedness
   The goal of the emergency preparedness is to be prepared to take the most appropriate action in the event that a hazard becomes a reality so as to

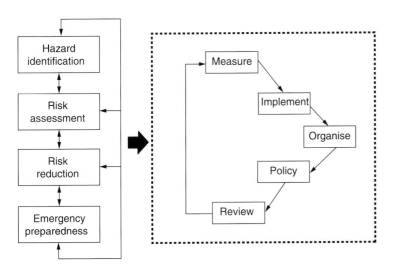

Figure 11.35 The five key elements of the safety case concepts.

minimise its effects and, if necessary, to transfer personnel from a location with a higher risk level to another one with a lower risk level.

5. Safety management system
   The purpose of a safety management system (SMS) is to ensure that the organization is achieving the goals safely, efficiently and without damaging the environment. One of the most important factors of the safety case is an explanation of how the operator's management system will be adapted to ensure that safety objectives are actually achieved.

A safety case is a written submission prepared by the operation of an offshore installation. It is a standalone document which can be evaluated on its own but has cross-references to other supporting studies and calculations. The amount of detail contained in the document is a matter of agreement between the operator and the regulating authority. In general, the following elements of an offshore installation are common for many safety cases:

1. A comprehensive description of the installation.
2. Details of hazards arising from the operation installation.
3. Demonstrations that risks from these hazards have been properly addressed and reduced to an ALARP level.
4. Description of the safety management system, including plans and procedures in place for normal and emergency operations.
5. Appropriate supporting references.

The following activities characterise the development of a safety case:

1. Establish acceptance criteria for safety, including environment and asset loss.
2. Consider both internal and external hazards, using formal and rigorous hazard identification techniques.
3. Estimate the frequency or probability of occurrence of each hazard.
4. Analyse the consequences of occurrence of each hazard.
5. Estimate the risk and compare with criteria.
6. Demonstrate ALARP.
7. Identify remedial measures for design, modification or procedure to reduce the frequency of occurrence or to mitigate the consequences.
8. Prepare the detailed description of the installation including information on protective systems and measures in place to control and manage risk.
9. Prepare a description of the safety management system and ensure that the procedures, which are appropriate for the hazards, are identified.

The following seven parts drawn from a safety case (Sii (2001), Wang (2002)) are subjects that can be found in a typical safety case for the operations of an offshore installation:

Part I Introduction and management summary
Part I of an operational safety case is an introduction and management summary. It will:

- Describe the scope and structure of the safety case.
- Describe the ownership and operatorship of the installation.
- Provide brief summaries of Part II to VII, highlighting major conclusions.

A summary of all key features contained in the safety case is outlined, including:

- Definition of the safety case.
- Objectives.
- Scope and structure of the seven parts of the safety case.
- Usage of the safety case.
- Custodian of the safety case.
- Review periods and updates.
- Application of the hazard management process to the operation.
- Hazard analysis of the operation.
- Remedial work.
- Conclusions drawn concerning the safety of the operation.

Part II: Operations safety management system
Part II is a concise description of the safety management system at the installation. It summarises both the corporate and installation specific policies, organizational structures, responsibilities, standards, procedures, processes, controls and resources which are in place to manage safety on safety case.

The six main sections of Part II cover the following:

- Policies and objectives.
- Organization, responsibilities and resources.
- Standards and procedures.
- Performance monitoring.
- Audits and audit compliance.
- Management review and improvement.

Part III: Activities catalogue
Part III contains the activities catalogue which lists all safety activities applicable to the operation in the activity specification sheet. The activity specification sheet describes the activity and the hazard management objectives of that activity, safety related inputs and outputs, methods used to achieve the hazard management objectives along with management

controls applied and the accountability for meeting the stated objectives. Any areas of concern arising from these sheets are noted as deficiencies.

Part IV: Description of operations
Part IV describes the essential features of the installation in sufficient detail to allow the effectiveness of safety systems to be appreciated. As such it describes the purpose of the installation and the processes performed there and its relationship to the location, reservoir and other facilities. Operational modes and manning for the installation are described, e.g. normal operation, shut down configurations, maintenance modes, etc.

The essence of Part IV is not to give a detailed physical description but to explain how the various systems relate to the safety of the installation and how their use can affect safety.

Part V: Hazard analysis, hazard register and manual of permitted operations (MOPO)
Part V provides a description of the hazards, their identification, ranking and assessment, the means by which they are to be controlled, and the recovery mechanisms. The design reviews and audits carried out to identify and assess hazards are also described.
Part V contains the following four sections.

- Hazard assessment.
- Hazard register (including the hazard/activity matrix).
- Safety critical operational procedures (SCOP).
- Manual of permitted operations (MOPO).

The sections are constructed as follows:

(a) A summary of all hazard investigations, design reviews and audits carried out, stating the major findings and recommendations from those investigations and the follow up of recommended action items.
(b) The hazard register, which describes each hazard in terms of
  - The way it was identified.
  - The methods used to assess the possible dangers presented by the hazard.
  - The measures in place to control the hazard.
  - The methods used to recover from any effects of the hazard.

Part V also contains the hazard/activity matrix, which cross-refers the activities identified in Part III with their effects on the identified hazards.

(c) The MOPO defines the limits of safety operation permitted when the defences are reduced, operating conditions are unusually severe or during accidental activities.

(d) A list of all safety critical operations procedures (SCOP), identifying the key hazard controls and recovery procedures required for the installation.

Part VI: Remedial action plan
Part VI records any deficiencies identified during the studies which lead to Parts II, III, IV and V, that require action to be taken. The record known as the 'remedial action plan' includes:

- A statement of each identified deficiency.
- The proposed modifications to address the problem.
- An execution plan to show action parties and planned completion dates.

This remedial action plan will be used as the basis of the improvement plan, and as such the plan will be regularly reviewed and updated annually.

Part VII: Conclusion and statement of fitness
Part VII includes summaries of the major contributors to risk, the acceptance criteria for such risks, deficiencies identified and planned remedial actions.

Part VII ends with a 'statement of fitness' which is the asset owner's statement that he appreciates and understands the hazards of the operation and considers that sufficient hazard control mechanisms are in place for the operation to continue. This statement is signed by the asset owner and approved by the signature of the operations directors.

In offshore safety analysis, it is expected to make safety based design/operation decisions at the earliest stages in order to reduce unexpected costs and time delays regarding safety due to late modifications. It should be stressed that a risk reduction measure that is cost-effective at the early design stages may not be ALARP at the late stage. HSE's regulations aim to have risk reduction measures identified and in place as early as possible when the cost of making any necessary changes is low. Traditionally, when making safety based design/operation decisions for offshore systems, the cost of a risk reduction measure is compared with the benefit resulting from reduced risks. If the benefit is larger than the cost, then it is cost-effective, otherwise it is not. This kind of cost benefit analysis based on simple comparisons has been widely used as a general principle in offshore safety analysis.

Conventional safety assessment methods and cost benefit analysis approaches can be used to prepare a safety case. As the safety culture in the offshore industry changes, more flexible and convenient risk assessment methods and decision making approaches can be employed to facilitate the preparation of a safety case. The UKOOA framework for risk related decision support can provide an umbrella under which various risk assessment and decision making tools are employed.

The guidelines in the UKOOA framework set out what is generally regarded in the offshore industry as good practice. These guidelines are a living document. The experience in the application of the framework changes in working practices, the business and social environment and new technology may cause them to need to be reviewed and updated to ensure that they continue to set out good practice. It should be noted that the framework produced by the UKOOA is only applicable to risks falling within the ALARP region shown in Figure 11.33.

Life-cycle approach is required to manage the hazards that affect offshore installations. It should be noted that offshore safety study has to deal with the boundaries of other industries such as marine operations and aviation. In offshore safety study, it is desirable to obtain the optimum risk reduction solution for the total life cycle of the operation or installation, irrespective of the regulatory boundaries (UKOOA (1999)). The basic idea is to minimise/eliminate the source of hazard rather than place too high reliance on control and mitigatory measures. To reduce risks to an ALARP level, the following hierarchical structure of risk control measures should follow:

1. Elimination and minimisation of hazards by 'inherently safer' design.
2. Prevention.
3. Detection.
4. Control.
5. Mitigation of consequences.

Decisions evolve around the need to make choices, either to do something or not to do something, or to select one option from a range of options. These can either take the form of rigid criteria, which must be achieved, or take the form of goals or targets which should be aimed for, but which may not be met. The UK offshore oil and gas industry operates in an environment where safety and environmental performances are key aspects of successful business. The harsh marine environment and the remoteness of many of the installations also provide many technical, logistic and operational challenges. Decision making can be particularly challenging during the early stages of design and sanction of new installations where the level of uncertainty is usually high.

In many situations, there may be several options which all satisfy the requirements. It may also be difficult to choose a particular option, which is not obviously the best. If this is the case, then there is a need to consider what is or may be 'reasonably practicable' from a variety of perspectives and to identify and assess more than just the basic costs and benefits. The decision making process can be set up to (UKOOA (1999)):

1. Define the issue.
2. Examine the options.
3. Take the decision.
4. Implement, communicate and review the decision.

Making risk based decisions may be very challenging because it may be difficult to:

1. Ensure the choices have been properly selected and defined.
2. Find ways to set out criteria and objectives.
3. Identify risk issues and perceptions.
4. Assess the performance of options against aspects that may not be quantifiable or which may involve judgements and perceptions that vary or are open to interpretation.
5. Establish the relative importance of often widely different types of objectives and factors.
6. Deal with uncertainties in estimates, data and analyses.
7. Deal with conflicting objectives and aspects of performance.
8. Deal with differences in resolution of estimates, data and analyses – these may not be able to provide a fair reflection of the actual differences between the options being considered.
9. Deal with or avoid hidden assumptions or biases.

A narrow view in the decision making process may result in decisions creating problems in other areas later on. For example, in a lifecycle view of the project or installation, decisions made during design to cut engineering and installation costs may lead to higher operating costs, reducing the overall profitability of the venue.

Safety and risk factors in the decision making process include risk transfer, risk quantification, cost benefit analysis, risk levels and gross disproportion, risk aversion, perception, risk communication, stakeholders and uncertainties. In general, decision making can be carried out on a 'technological' basis or based on value judgements and perception. It can be difficult to determine which basis is most appropriate to a given decision, especially where the different bases may indicate conflicting best outcomes. It is therefore important to understand the decision context and use this to identify the importance of the various decision bases in any given decision situation, UKOOA (1999).

The factors that affect offshore safety based decision making include degree of novelty vs. well understood situation or practice; degree of risk trade-offs and uncertainties; strength of stakeholder views and risk perceptions; and degree of business and economic implications, UKOOA (1999). The study of such factors will determine the basis on which

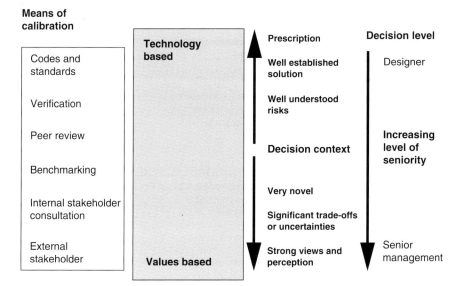

Figure 11.36 Decision levels.

decision making can be best conducted. Different means of calibrating or checking the basis of the decision need to be matched to the type of decision context as shown in Figure 11.36, UKOOA (1999). Calibration should be used to ensure that the basis of decision making has been properly assessed and is still appropriate. Decision calibration changes with decision context. As the design context moves from prescription to strong views and perceptions, means of calibration change from codes and standards to external stakeholder consultation through verification, peer review, benchmarking and internal stakeholder consultation.

The framework proposed by the UKOOA is also capable of reflecting the differences between the design for safety approach for fixed offshore installations operating in the UK continental shelf and mobile offshore installation operating in an international market. Fixed offshore installations in the UK continental shelf are usually uniquely designed and specified for the particular duty and environment, and their design basis can be set against very specific hazards and specific processing and operation requirements. Many of the more complex design decisions therefore often fall into the 'Type B' context in the detailed framework shown in Figure 11.37. Mobile offshore installations have to operate in very different environments and tackle a wide range of operational activities and reservoir conditions (UKOOA (1999)). The design cannot be based on specific hazards or duties but needs to address more global duties and operating envelopes. It is less likely to make specific risk based approaches to the design. It is usually the case that the installation is designed around a generic operating envelope. Because many of the hazards are generally well understood, common solutions incorporated into code, standards and ship classification society rules have been developed. Therefore, many mobile offshore installation design decisions fall into the 'Type A' context. Where neither codes and rules can be effectively applied nor traditional analysis can be carried with confidence, such installations may be categorised as the 'Type C' context, UKOOA (1999).

11.4.4.2 Formal ship safety assessment

As serious concern is raised over the safety of ships all over the world, the International Maritime Organization (IMO) has continuously dealt with safety problems in the context of operation, management, survey, ship registration and the role of the administration. The improvement of safety at sea has been highly stressed. The international safety-related marine regulations have been governed by serious marine accidents that have happened. The lessons were first learnt from the accidents and then the regulations and rules were produced to prevent similar accidents to occur. For example, the capsize of the *Herald of Free Enterprise* in 1987 greatly affected the rule developing activities of the IMO (Cowley (1995)), Sekimizu (1997)). The accident certainly raised serious questions on operation requirements and the role of management, and stimulated discussions in those areas at the IMO. This finally resulted in the adoption of the International Management System (ISM) Code. The *Exxon Valdes* accident in 1989 seriously damaged the environment by the large scale oil spill. It facilitated the implementation

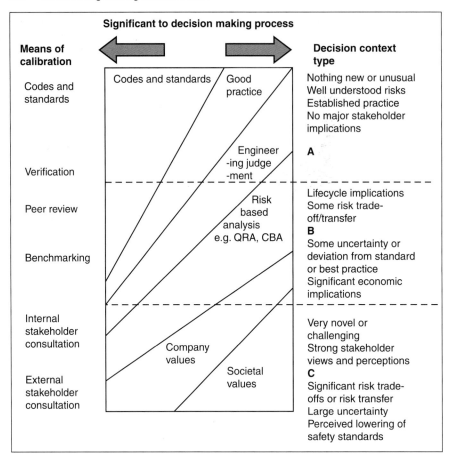

Figure 11.37 The detailed UKOOA framework.

of the international convention on Oil Pollution Preparedness, Response and Co-operation (OPRC) in 1990. Double hull or mid-deck structural requirements for new and existing oil tankers were subsequently applied (Sekimizu (1997)). The *Scandinavian Star* disaster in 1990 resulted in the loss of 158 lives. Furthermore, the catastrophic disaster of the *Estonia*, which capsized in the Baltic Sea in September 1994, caused more than 900 people to lose their lives. Those accidents highlighted the role of human error in marine casualties, and as a result, the new Standards for Training, Certificates and Watchkeeping (STCW) for seafarers were subsequently introduced.

After Lord Carver's report on the investigation of the capsize of the *Herald of Free Enterprise* was published in 1992, the UK Maritime & Coastguard Agency (MCA, previously named as Marine Safety Agency (MSA)) quickly responded and in 1993 proposed that the IMO should explore the concept of formal safety assessment and introduce it in relation to ship design and operation. This proposal was submitted to the 62nd session of the Maritime Safety Committee (MSC) held from 24–28 May 1993 (MSA (1993)). The IMO reacted favourably to the UK's formal safety assessment submission. At the 65th meeting of the MSC in May 1995, strong support was received from the member countries and a decision was taken to make formal safety assessment a high priority item on the MSC's agenda. Accordingly, the UK decided to embark on a major series of research projects to further develop an appropriate framework and conduct a trial application on the chosen subject of high speed passenger catamaran ferries. The framework produced was delivered to MSC 66 in May 1996, with the trial application programmed for delivery to MSC 68 in May 1997. An international formal safety assessment working group was formulated at MSC 66 and MSC 67 where draft international guidelines were generated, including all key elements of the formal safety assessment framework developed by the UK.

Several applications of formal safety assessment have been attempted by the IMO on various vessels and systems. These include the application to the transportation of dangerous goods on passenger/ ro-ro cargo vessels (IMO (1998a)), the effects of introducing Helicopter Landing Areas (HLA) on cruise ships (IMO (1998b)), high speed catamaran passenger vessels (IMO (1997b)), novel emergency propulsion and steering devices for oil tankers (IMO (1998d)) and the trial application is on a bulk carrier (IMO (1998c), MCA (1998)). The IMO has approved the application of formal safety assessment for supporting rule making process, IMO (1997a), Wang (2001, 2002).

Formal safety assessment in ship design and operation may offer great potential incentives. The application of it may:

1. Improve the performance of the current fleet, be able to measure the performance change and ensure that new ships are good designs.
2. Ensure that experience from the field is used in the current fleet and that any lessons learned are incorporated into new ships.
3. Provide a mechanism for predicting and controlling the most likely scenarios that could result in incidents.

The possible benefits have already been realised by many shipping companies. For example, the P & O Cruises Ltd in the UK has reviewed the implementation of risk assurance methods as a strategic project and proposed short-term/medium-term and long-term objectives (Vie and Stemp (1997)). Its short-term/ medium-term objectives are to provide a reference point for all future risk assurance work; to develop a structure chart that completely describes vessel operation; to complete a meaningful hazard identification as the foundation of the data set; to enable identification of realistic options for vessel improvement; to be a justified record of modifications adopted or rejected; and to be capable of incorporating and recording field experience to ensure that the knowledge is not lost. The idea of formal safety assessment may well be fitted to the above objectives in order to improve the company's performance.

Formal safety assessment is a new approach to maritime safety which involves using the techniques of risk and cost-benefit assessment to assist in the decision making process. It should be noted that there is a significant difference between the safety case approach and formal safety assessment in terms of their application to design and operations (Wang (2002)). A safety case approach is applied to a particular ship, whereas formal safety assessment is designed to be applied to safety issues common to a ship type (such as high-speed passenger vessel) or to a particular hazard (such as fire). The philosophy of formal safety assessment is essentially same as the one of the safety case approach. Many ship owners have begun to develop their own ship safety cases. The major difference between such ship specific applications of the approach and its generic application by regulators is that whilst features specific to a particular ship cannot be taken into account in a generic application, the commonalities and common factors which influence risk and its reduction can be identified and reflected in the regulator's approach for all ships of that type (IMarE (1998)). This should result in a more rational and transparent regulatory regime being developed. Use of formal safety assessment by an individual owner for an individual ship on the one hand and by the regulator for deriving the appropriate regulatory requirements on the other hand, are entirely consistent, IMarE (1998).

It has been noted that many leading classification societies including Lloyds Register of Shipping and American Bureau of Shipping are moving towards a risk based regime. It is believed that the framework of formal safety assessment can facilitate such a move.

A formal ship safety assessment framework proposed by the UK MCA consists of the following five steps:

1. The identification of hazards.
2. The assessment of risks associated with those hazards.
3. Ways of managing the risks identified.
4. Cost benefit assessment of the options.
5. Decisions on which options to select.

Formal safety assessment involves much more scientific aspects than previous conventions. The benefits of adopting formal safety assessment as a regulatory tool include (MSA (1993)):

1. A consistent regulatory regime which addresses all aspects of safety in an integrated way.
2. Cost effectiveness, whereby safety investment is targeted where it will achieve the greatest benefit.
3. A pro-active approach, enabling hazards that have not yet given rise to accidents to be properly considered.
4. Confidence that regulatory requirements are in proportion to the severity of the risks.
5. A rational basis for addressing new risks posed by ever changing marine technology.

11.4.4.3 *Risk criteria*

Risk criteria are standards which represent a view, usually that of a regulator, of how much risk is acceptable/tolerable (HSE (1995a)). In the decision making process, criteria may be used to determine if risks are acceptable, unacceptable or need to reduce to an ALARP level. When Quantitative Risk Assessment

(QRA) is performed, it is required to use numerical risk criteria. The offshore industry has extensively used QRA and significant experience has been gained. The shipping industry has functioned reasonably well for a long time without consciously making use of risk criteria. Recently QRA has been used extensively for ships carrying hazardous cargoes in port areas and for ships operating in the offshore industry (Spouge (1997)). It is noted that in general there is no quantitative criteria in formal safety assessment even for a particular type of ship although the MCA trial applications have used QRA to a certain extent. As time goes on, it is believed that more QRA will be conducted in marine safety assessment. Therefore, numerical risk criteria in the shipping industry need to be dealt with in more detail.

As described previously in this Section, risk assessment involves uncertainties. Therefore it may not be suitable to use risk criteria as inflexible rules. The application of numerical risk criteria may not always be appropriate because of uncertainties in inputs. Accordingly, acceptance of a safety case is unlikely to be based solely on a numerical assessment of risk.

Risk criteria may be different for different individuals. They would also vary between societies and alter with time, accident experience and changing expectation of life. Risk criteria can therefore only assist judgements and be used as guidelines for decision making.

In different industries, risk criteria are also different. For example, in the aviation industry, a failure with catastrophic effects must have a frequency less than 10^{-9} per aircraft flying hour. In the nuclear industry, the basic principles of the safety policy recommended by the International Commission Radiological Protection (ICRP) are that no practice shall be adopted unless it has a positive net benefit; that all exposures shall be kept As Low As Reasonably Achievable (ALARA), taking economic and social factors into account; and that individual radiation doses shall not exceed specific criteria (ICRP (1977)). There are no explicit criteria used by ICRP.

As far as risk criteria for ships are concerned, the general criteria may include: (1) the activity should not impose any risks which can reasonably be avoided; (2) the risks should not be disproportionate to the benefits; (3) the risks should not be unduly concentrated on particular individuals; and (4) the risks of catastrophic accidents should be a small proportion of the total (Spouge (1997)). More specifically, individual risk criteria and social risk criteria need to be defined. For example, the maximum tolerable risk for workers may be 10^{-6} per year according to the HSE industrial risk criteria. In the regions between the maximum tolerable and broadly acceptable levels, risks should be reduced to an ALARP level, taking costs and benefits of any further risk reduction into account (Wang (2001, 2002)).

11.4.4.4 *Discussion and conclusion*

An offshore installation/a ship is a complex and expensive engineering structure composed of many systems and is usually unique with its own design/operational characteristics, Wang and Ruxton (1997). Offshore installations/ships need to constantly adopt new approaches, new technology, new hazardous cargoes, etc. and each element brings with it a new hazard in one form or another. Therefore, safety assessment should cover all possible areas including those where it is difficult to apply traditional safety assessment techniques, some of which are described in Section 11.4.3. Such traditional safety assessment techniques are considered to be mature in many application areas. Depending on the uncertainty level/the availability of failure data, appropriate methods can be applied individually or in combination to deal with the situation. All such techniques can be integrated in a sense that they formulate a general structure to facilitate risk assessment.

Lack of reliable safety data and lack of confidence in safety assessment have been two major problems in safety analysis of various engineering activities. To solve such problems, further development may be required to develop novel and flexible safety assessment techniques for dealing with uncertainty properly and also to use decision making techniques on a rational basis.

In offshore safety assessment, a high level of uncertainty in failure data has been a major concern that is highlighted in the UKOOA's framework for risk related decision support. Different approaches need to be applied with respect to different levels of uncertainty.

Software safety analysis is another area where further study is required. In recent years, advances in computer technology have been increasingly used to fulfil control tasks to reduce human error and to provide operators with a better working environment in ships. This has resulted in the development of more and more software intensive systems. However, the utilisation of software in control system has introduced new failure modes and created problems in the development of safety-critical systems. The DCR-1996 has dealt with this issue in the UK offshore industry. In formal ship safety assessment, every safety-critical system also needs to be investigated to make sure that it is impossible or extremely unlikely that its behaviour will lead to a catastrophic failure of the system and also to provide evidence for both the developers and the assessment authorities that the risk associated with the software is acceptable within the overall system risks, Wang (1997).

The formal safety assessment philosophy has been approved by the IMO for reviewing the current safety and environmental protection regulations and studying any new element proposal by the IMO; and justifying and demonstrating a new element proposal to the IMO by an individual administration. Further applications may include the use of formal safety assessment for granting exemptions or accepting equivalent solutions for specific ships under the provisions of SOLAS Chapter 1 by an individual administration; for demonstrating the safety of a specific ship and its operation in compliance with mandatory requirements to the acceptance of the Flag Administration by an individual owner; as a management tool to facilitate the identification and control of risks as a part of the Safety Management System in compliance with the ISM Code by an individual owner. Several possible options regarding the application of formal safety assessment have been under investigation at the IMO. Among the possible application options, the individual ship approach may have the great impact on marine safety and change the nature of the safety regulations at sea since it may lead to deviation from traditional prescriptive requirements in the conventions towards performance-based criteria. This may be supported by ship type specific information. However, this would raise concerns due to the difficulty in the safety evaluation process by other administrations particularly when acting as port states although the merits of it may also be very significant. At the moment, unlike in the UK offshore industry, there is no intention to put in place a requirement for individual ship safety cases.

It is also very important to take into account human error problems in formal safety assessment. Factors such as language, education and training, that affect human error, need to be taken into account. The application of formal safety assessment may also encourage the Flag States to collect operation data. Another important aspect that needs to be considered is the data problem. The confidence of formal safety assessment greatly depends on the reliability of failure data. If formal safety assessment is applied, it may facilitate the collection of useful data on operational experience which can be used for effective pro-active safety assessment.

More test case studies also need to be carried out to evaluate and modify formal ship safety assessment and associated techniques and to provide more detailed guidelines for the employment of them. This would enable validation of them and can also direct the further development of flexible risk modelling and decision making techniques and facilitate the technology transfer to industries.

It is clear that it would be possible to reduce marine accidents by good design, training, and operation in an appropriate systematic management system. As the public concern regarding maritime safety increases, more and more attention has been directed to the wide application of formal safety assessment of ships as a regulatory tool. It is believed that the adoption of such a tool in ship design and operation will reduce maritime risks to a minimum level.

11.4.5 Formal safety assessment (FSA)

11.4.5.1 *Formal safety assessment*

Formal Safety Assessment (FSA) has as its objective the development of a framework of safety requirements for shipping in which risks are addressed in a comprehensive and cost effective manner. The adoption of FSA for shipping represents a fundamental cultural change, from a largely reactive approach, to one, which is integrated, proactive and soundly based upon the evaluation of risk.

As described in Section 11.4.4, the FSA framework consists of five steps. The interaction between the five steps can be illustrated in a process flowchart as shown in Figure 11.38. As it can be seen, there are repeated iterations between the steps, which makes FSA effective as it constantly checks itself for changes within the analysis. Each step within the FSA can be further broken down into individual tasks and is represented in Figure 11.39. The execution and documentation of each task is vital, as it will enable the preceding tasks/steps to be carried out with ease. In order for the assessment to be accurate, the analyst must understand and appreciate the objectives of each step and execute it without any 'short-cuts'.

Depending on the requirement of the safety analysts and the safety data available, either a qualitative or a quantitative safety analysis can be carried out to study the risks of a system in terms of the occurrence probability of each hazard and possible consequences. As described in Section 11.4.3, qualitative safety analysis is used to locate possible hazards and to identify proper precautions (design changes, administrative policies, maintenance strategies, operational procedures, etc.) that will reduce the frequencies or/and consequences of such hazards.

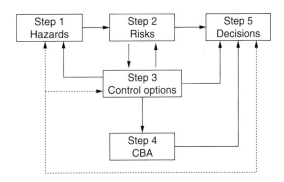

Figure 11.38 Flowchart of FSA process.

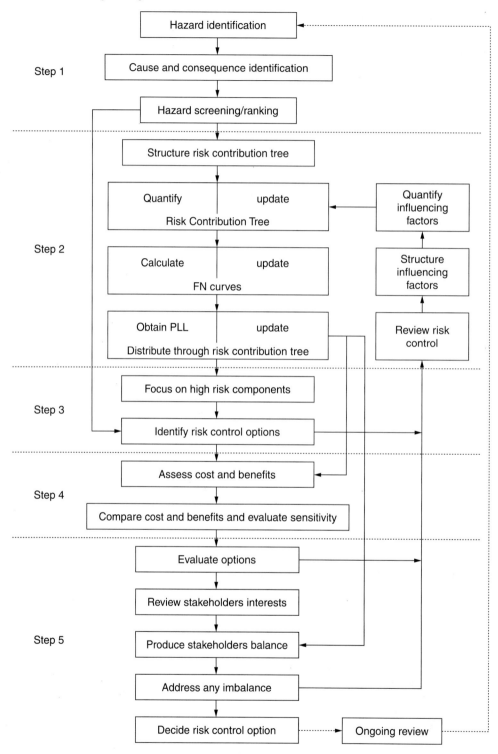

Figure 11.39 Detailed breakdown of FSA process.

1. Step 1 – Hazard Identification
Various methods may be used individually or in a combination to carry out Step 1 of the FSA approach. Such typical methods include: Preliminary Hazard Analysis (PHA), Fault Tree Analysis (FTA), Event Tree Analysis (ETA), Cause-Consequence Analysis (CCA), Failure Mode, Effects and Criticality Analysis (FMECA), HAZard and OPerability analysis (HAZOP), Boolean Representation Method (BRM) and Simulation analysis (Henley and Kumamoto (1996), Smith (1992), Villemeur (1992), Wang (1994)). The use of these methods as safety analysis techniques has been reviewed in Section 11.4.3.

In the hazard identification phase, the combined experience and insight of engineers is required to systematically identify all potential failure events at each required indenture level with a view to assessing their influences on system safety and performance. This is achieved using 'brainstorming' techniques. The hazard identification phase can be further broken down into several steps as follows:

Problem definition – Define the bounds of study, generic vessel and generic stakeholder for the vessel.

Problem identification – The problem boundaries of a formal safety assessment study can be developed in the following manner:

- Range of the vessel.
- Geographic boundaries.
- Risks to be considered.
- Vessel systems.
- Relevant regulations.
- Measures of risk.

In addition, the following factors specifically related to the vessel are defined:

- The generic vessel.
- Vessel accident category.
- Vessel stakeholders.
- Vessel operational stages.

Hazard identification – The HAZard IDentification (HAZID) consists of determining which hazards affect the vessel's activities under consideration using 'brainstorming' techniques. At the HAZID session the following information is gathered:

- Operational stage.
- Vessel systems.
- Hazards, causes and consequences.

Structuring HAZID output – The approach to structuring the HAZID output is to convert the information gathered at the HAZID meeting into hazard worksheets which record the causes, accident sub-categories, consequences and the source of information. These hazard worksheets provide a means for recording the output from the HAZID meeting and other hazards identified during the analysis period, that is, from incident databases or interviews with the vessel personnel.

Risk exposure groups – The next step is to group the causes into risk exposure groups. This can be achieved by using the guidewords taken from the risk exposure source given in MSC 68/14 (IMO (1993)). The groups are then further sub-divided, during the hazard-structuring phase into risk exposure sub-groups. An example of this can be found in (MSC (1997b)). In order to sort the large amount of information collected at the HAZID meeting, accident sub-categories are established for each accident category and all the identified consequences are grouped according to the contributing factors.

Hazard screening – The purpose of hazard screening during Step 1 is to provide a quick and simple way of ranking hazards. It is a process of establishing, in broad terms, the risks of all identified accident categories and accident sub-categories, prior to the more detailed quantification, which will be conducted in Step 2. Risk is a combination of the frequency of occurrence of an accident type with the severity of its consequences. Possible consequences may be loss of lives, environmental pollution or damage to ship/cargo or financial loss. Accordingly, risk can also be read as the estimated loss in a given period of time. Two approaches can be used for the assignment of screening risk level in order to check the robustness of the resulting hazard rankings and to assist in the resolution of the rankings in cases where several hazards have similar ranking levels. These are:

1. Risk matrix approach (Loughran *et al.* (2002)).
2. Cumulative loss approach (MSC (1997a)).

2. Step 2 – Risk Estimation
Information produced from the hazard identification phase will be processed to estimate risk. In the risk estimation phase, the likelihood and possible consequences of each System Failure Event (SFE) will be estimated either on a qualitative basis or a quantitative basis (if the events are readily quantified). The risk estimation phase can be further broken down into several steps as follows:

Structuring of Risk Contribution Tree (RCT) – The causes and outcomes that were identified in Step 1 are structured in Step 2 for its employment in various parts of the Risk Contribution Tree (RCT). An RCT is structured in two distinct ways. Below the accident category, the structure is a graphical representation of the accident sub-categories and of the combinations of contributory factors relevant to each accident sub-category. Its structure is similar to a Fault Tree in its use of logical symbols, and the term 'Contribution Fault Tree' has therefore been employed. Above the accident category level,

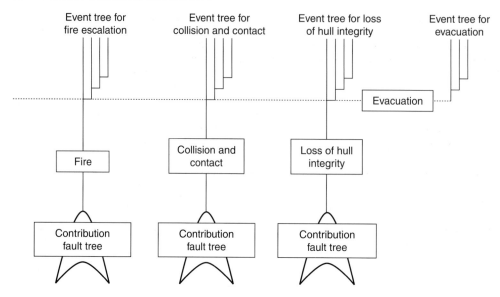

Figure 11.40 An example of Risk Contribution Tree (RCT).

the structure is an event tree representation of the development of each category of accident into its final outcome. An example of an RCT is shown in Figure 11.40.

Structuring and quantification of influence diagrams – The purpose of influence diagrams is to identify the influences, which determine the likelihood of an accident, and to enable those influences to be quantified. It also provides information for use in Step 3 of the FSA process. An example of an influence diagram for a fire accident is given in Pillay and Wang (2003). An influence diagram takes into account three different types of influence, which are due to:

1. Hardware failure.
2. Human failure.
3. External event.

Additionally, each influence diagram incorporates dimensions of design, operation and recovery. Here, recovery refers to taking remedial action to recover from an error or failure before the accident occurs.

Quantification of RCT – The quantification of the RCT is accomplished by using available historical data from the incident database and where such data is absent, expert judgement is used to complement the quantification. The level of potential consequences of a System Failure Event (SFE) may be quantified in economic terms with regard to loss of lives/cargo/property and the degradation of the environment caused by the occurrence of the SFE. Finally, the calculation of FN (i.e. frequency (F) –fatality (N)) curves and Potential Loss of Life (PLL) through the RCT is carried out.

3. Step 3 – Risk Control Options (RCOs)
The next step aims to propose effective and practical Risk Control Options (RCOs). Focusing on areas of the risk profile needing control, several RCOs are developed and recorded in a Risk Control Measure Log (RCML). Upon identifying all possible RCOs from the estimated risks, the RCOs in the RCML are used to generate a Risk Control Option Log (RCOL). The information in the RCOL will be used in Step 4 of the FSA process.

In general, RCO measures have a range of following attributes (MSA (1993)):

1. Those relating to the fundamental type of risk reduction (preventative or mitigating).
2. Those relating to the type of action required and therefore to the costs of the action (engineering or procedural).
3. Those relating to the confidence that can be placed in the measure (active or passive, single or redundant).

The main objective of an RCO is to reduce frequencies of failures and/or mitigate their possible consequences.

4. Step 4 – Cost-Benefit Analysis (CBA)
Upon gathering the various control options, the next step is to carry out a Cost-Benefit Analysis (CBA) on each option. CBA aims at identifying the benefits from reduced risks and costs associated with the implementation of each risk control option

for comparison. The evaluation of costs and benefits may be conducted using various techniques (IMO (1993)). It should be initially carried out for the overall situation and then for those interested entities influenced by the problem consideration.

5. Step 5 – Decision-making

The final step is the decision-making phase, which aims at making decisions and giving recommendations for safety improvement. At this point, the various stakeholders' interest in the vessel under study is considered. The cost and benefit applicable to each stakeholder have to be determined in order to decide the best risk control option – each RCO will have a different impact on the identified stakeholders, as such, the most effective RCO should strike a balance between the cost and benefit for each stakeholder. In reality, this is not always possible, hence, any imbalance has to be addressed and justified before the selected RCO is accepted as being the best option. The information generated in Step 4 of the FSA process can be used to assist in the choice of a cost-effective RCO. However, the cost factor may not be the only criterion that should be considered. As such, at this stage, certain multi criteria decision-making techniques should be employed to select the most favourable RCO (Wang et al. (1996), Pillay and Wang (2001), Wang et al. (2002)).

6. A Brief Discussion

There are many different types of ships such as fishing vessels, cruising ships, bulk carriers and containerships. It has been noted that different types of ships have different characteristics in terms of available failure data, the corresponding safety regulations, etc. As a result, the formal safety assessment framework described above should be applied on a flexible basis. For example, for fishing vessels, due to the poor safety culture in the fishing industry and lack of reliable failure data, the FSA framework described above may not be entirely applied. A modified FSA framework with a more qualitative nature may be more useful as described by Pillay and Wang (2003). In contrast, the FSA framework described above may be relatively easily applied to containerships.

11.5 Safety management of ship stability

11.5.1 Introduction

11.5.1.1 *Need to introduce a ship stability management system*

Sufficient stability is most important for operating a ship safely. Stability of a ship at sea depends very much on the actions taken by the master. The safe operation of the ship needs a thorough knowledge of the current loading status of the ship, of the ship behaviour in extreme seas, and of the best ways to cope with dangerous situations, Kaps and Kastner (1984). There may even be rare situations, where relying on the minimum required stability during ship operations is not sufficient at all.

Intact stability of the ship is a vital element in the shipboard consideration of safety and environmental protection. Proper management of ship's stability is a key shipboard operation. To prevent capsizing in severe seas, there are a number of problems:

- Capsizing is a rare extreme event, so direct experience can rarely be available.
- The risk of capsizing is hidden and can hardly be foreseen.
- A quick reaction is needed in case of danger from capsizing.
- Decisions under pressure with a lack of time cause human errors with severe consequences.

Hoffman (1976) pointed out the state of the art. Quote: *The navigation of a ship in rough seas demands comprehensive judgement from the navigator to assume full control of the ship's operation, usually this is done by the master. The response of the ship to the environment encountered is often a question of trial and error, and the limited guessing exercised by the master is usually based on visual observation of the sea state and the meteorological conditions.*

Special shipboard guidance is needed for the operation of a ship in rough weather. There is a need for objective information on wave height and period, which the ship encounters. Furthermore, depending on size, loading and design, ships behave quite differently in the same seaway. The ship capabilities to cope with the seaway in her specific loading condition must be made available to the master.

New computing capabilities, coupled with risk assessment techniques, can be utilised in defining the chain of events that end up in extreme roll and even capsize. These techniques can provide improvements in safety through design, operator guidance, training, and life-cycle management, Alman, et al, (1999).

11.5.1.2 *Tools of efficient stability management*

The methods of assessing stability of the ship in service can be subdivided into preparations before departure and into control while at sea. For sailing in heavy weather, special precautions must be taken. The probability of failure and the operability of the ship are subject to the favourable influence of seamanship actions. It is logical to subdivide actions taken before severe conditions are encountered, and those once severe conditions are encountered. Prior action is avoidance of severe weather, proper ballasting of the ship, lashing of cargo, and making sure what the current stability of the ship is.

Once the ship is in heavy weather, short-term measures are avoidance of resonance and of reduced stability in a wave crest, see Section 11.5.2.4 on guidance to the master. Decisions can be based on current data of ship and environment when supported by online measurements.

To cope with rare extreme situations, guidelines for stability management and the development of typical scenarios can assist the master. Severe situations actually experienced can be evaluated accordingly.

It must be kept in mind, that ship stability in still water generally serves as a reference. Dynamic effects and environmental factors are implemented in modifying the simple *GZ* criteria. This is a debatable method. New forms of criteria reflecting the ship dynamics in waves may be encouraged in the future, see the discussion at STAB 96 and at IMO, Francescutto (2002). To relate every influence on stability in terms of the familiar *GZ* curve in still water, simply does not show the operator the corresponding situation at sea.

In addition to the mandatory ISM code (International Safety management), discussed in Sections 11.2.2.5 and 11.5.2.5 more detailed guidelines on managing ship stability can be useful. In order to cope with the problems of ensuring safety at sea, a strict regime on control of all stability aspects while in service must be practised. IMO has discussed a proposal on the management of ship's stability, IMO (2000a). This proposal was to serve as an aid for the operator to focus on problems on shipboard stability management. However, the IMO sub-committee could not agree on a final document, as it does not want to regulate too much and feels this must be left to the practice of the ship operator. In Section 11.5.2.3 the main topics of this proposal are discussed.

11.5.1.3 *The master's range of judgement for operational stability assessment*

The minimum requirements on ship stability are deterministic, while the real sea behaviour follows probabilistic laws. The acceptance criteria of stability are based on the righting capability of the ship, and it is often taken for granted that the demand from environmental forces and moments will be less. Wendel (1965) introduced the moment balancing of heeling and righting moments. Wendel's method was applied to set up the stability requirements of the German Navy. Arndt, *et al* (1982) gave a detailed report. Although there is a good Navy experience with balancing, requirements in merchant shipping adhere to the still water *GZ* curve alone.

For the future, the best way is seen in defining so-called rational stability criteria, based on the actual conditions of the ship in the particular environment, Kobylinski (2000). This is a field under discussion among experts, and certainly not to be resolved completely in the near future. In a workshop on information to the master at the International Conference STAB 94, Rakhmanin stated:

- *Stability is a key factor ensuring ship safety.*
- *The shipmaster bears the ultimate responsibility for the ship's fate and lives aboard.*
- *Any system of standard requirements is far from ideal and cannot guarantee 100% level of safety. This stands for the master's knowledge, experience and skill, and may play a decisive role for ship survival.*

The shipowner has to provide the ship with the Loading and Stability Manual, where the hydrostatic parameters of typical loading conditions are compiled, see Section 11.5.2.2. It seems natural, that masters take the *GZ*-curve the ship was designed for by naval architects as measures for their operational condition of stability. However, the minimum requirements do not show the assumptions at which the minimum set values were derived. This has a dangerous drawback, when the ship encounters extremely severe environmental conditions at sea. The minimum *GZ*-curve allowed does not guarantee safe ship at any rate, see also Kobylinski (2000).

The inadequacy of common design criteria on ship stability to include the role of human factors in marine regulations was expressed by Cleary (1994), quote:

1. *Regulations assume knowledgeable humans, but humans often assume that any marine craft that has been 'approved' is so well protected that the operator can test it to the limit.*
2. *The output of a naval architect is in a form, which shows how well the design meets minimum rules. But that output is not intelligent information for operators.*
3. *The trend to avoid 'stopping the ship' at any cost affects operation. It affects ship systems safety, and it affects ship's actual level of safety.*
4. *The IMO trend is to be driven by consensus.*
5. *Video training is proposed as a tool, which includes the limits of safety.*

It is dangerous to rely on the mandatory minimum when more is needed in an extreme situation. Kastner (1986) proposed to define a *safety margin of operational stability*. This margin is needed additionally to the required minimum to cope with extreme situations. Running the ship in an extreme situation with an additional safety margin depends

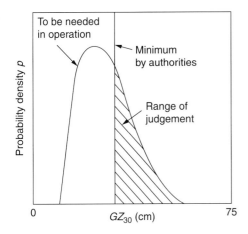

Figure 11.41 Master's range of judgement in operational ship stability.

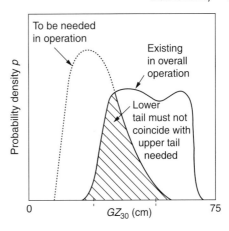

Figure 11.42 Comparing needed and existing GZ_{30} showing master's range of judgement.

on the master. We call it the *master's range of judgement*, see Figure 11.41. This expression points out the need to act responsibly in severe seas, rather than relying on given standards.

Figure 11.41 shows the probability density of the GZ at 30 degrees of heel as might be needed during a voyage, compared to a deterministic single value required by authorities. The larger values of GZ may be needed very rarely. Extreme conditions need thoughtful navigation by the master. Provisions include avoiding severe situations at all times by e.g. ship routeing, by changing ballast in advance, by changing speed and heading. Figure 11.42 compares the needed and existing GZ_{30} of a container ship to demonstrate the importance of the master's range of judgement.

A workshop at STAB 94 concluded, that although economics drives companies to operate at minimum safety levels imposed by law, those levels should be exceeded for certain, although rare but severe, situations.

11.5.1.4 Seakeeping guidance and survivability criteria

Stability information available to the ship operator is generally still based on ship hydrostatics. Although dynamic ship characteristics are part of the background leading to improved stability criteria, IMO (1993), no specific reference is made. Guidance on stability safety management, see Section 11.5.2, and the development of typical scenarios for the ship are recommended, Hoffman, (1976), Kastner, (1986).

Development of scenarios with options for the master cannot give easy answers. The process is essentially one of trial and error, Hoffman (1976). The master expects correcting an unsatisfactory condition at sea, without leading to a substantial worsening of another condition. Guidance as to the relative merits is useful, such as on resonance avoidance. The direction of travel of the storm, the available power, the ability to steer the ship at reduced speeds can affect the decision.

The appropriate reaction to potential danger will always need the know-how, experience and logic of the master faced with a decision to minimise adverse heavy weather conditions. Guidance charts on the expected ship response have been proposed. The charts must be simple to understand, and be based on a theoretical calculation for specific ships. Kuo in his discussion to McTaggart and de Kat (2000) expressed fears that use of graphs and tables in adverse weather may be counterproductive. In his view, any guidance must act as a 'second opinion' to the experienced ship captain. Figure 11.43 shows the safety region of a container ship for all combinations of ship speed and heading at constant Beaufort number, Hoffman (1976). The safety region is defined by a number of different ship motion criteria, such as vertical acceleration at the F.P., slamming, deck wetness at the F.P., vertical-bending moment at the main section, propeller top emersion, lateral acceleration, rolling amplitude. Comparing different wind velocities, Figure 11.43 shows the extreme reduction of the safe zones from Beaufort 8 to Bft. 10.

Hutchison (1981) proposed a diagram which gives the operability domain of the ship showing downtime regions in terms of the significant wave height and the mean wave period. The hatched downtime regions depend on multiple criteria, in this example (A) deck inclination, (B) local acceleration, (C) critical motion transverse displacement. The partial areas have been numbered from 1 to 7 to show the different downtime causes (e.g. no. 1 is due to A and not B or C, no. 7 is due to A and B and C).

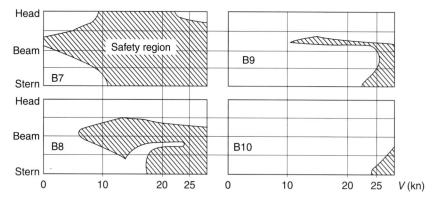

Figure 11.43 Safety region of container ship.

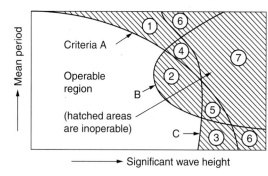

Figure 11.44 Operability domain showing downtime regions.

To each of the downtime regions, a probability can be associated, which is based on the probability of the significant wave height and mean period combinations within the subject sea area. Independence of statistically distributed processes for all influencing parameters is assumed. The probability of a response value greater than x during an individual voyage through independent contiguous spatial domains and during the seasons is derived from the sea state distribution with seamanship actions incorporated in the following linear probabilistic approach, Hutchison (1981):

$$P(H_s, T_m, V, \chi \mid Y, \Lambda) \\ = P(V, \chi \mid H_s, T_m, \chi_0) \cdot P(H_s, T_m, V, \chi_0 \mid Y, \Lambda) d\chi_0 \tag{11.7}$$

χ_0 Original heading with seamanship action
Y Spatial domain of sea area
Λ Season domain of sea area

Action parameters are the change of the relative heading χ and of the ship speed V. For every set we

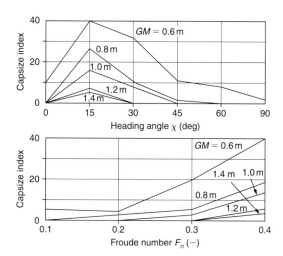

Figure 11.45 Calculated capsize index for GM, heading angle, and speed.

can associate a response statistic, for example the expected maximum roll. The equation combines multiple trials of independent distributed processes. This method does not include the nonlinear extreme ship motion such as large roll and capsize correctly. However, it can be a reasonable tool to identify risk related downtime losses, and the improvement by good scamanship.

The progress of numerical simulation allows calculating extreme ship motions including capsize. Figure 11.45 shows the results of a capsize index from simulation by de Kat and Thomas (1994). The upper diagram shows quartering seas to be most dangerous. The lower diagram shows the increase of danger from capsizing with increasing ship speed (expressed by the Froude number $F_n = V / \sqrt{gL}$).

Rainey and Thompson (1991) proposed the Transient Capsize Diagram, see Figure 11.46. It depicts the wave

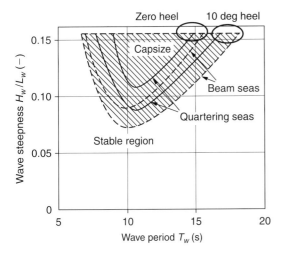

Figure 11.46 Typical transient capsize diagram.

Table 11.24 Safety modes of ship operation in rough seas by Russian Register of Shipping.

Calculation	Gauges and monitoring
Maximum speed	Speed and heading
Intensive roll at resonance	Draught, RPM/pitch, rudder angle
Extreme ship motion accelerations	Acceleration
Slamming	Ship motion
Deck wetness with or without deck cargo	Statistical and spectral analysis
Self-surging or broaching to	Short-term and long-term characteristics
Propeller air rush/racing	Control of accuracy and reliability
Calculated sailing modes due to criteria	Correction of calculated data
Safe sailing modes	
Heading and speed	Trim

height above which a vessel will capsize, as a function of the wave period. The diagram can be determined by physical model tests or computer simulation. (However, a conclusion from the results shown in Figure 11.46 that beam seas are generally more dangerous than quartering seas must be questioned!). Rainey, et al (1990) proposed to describe the 'capsizability' by the diagram Figure 11.46 using simulation in socalled transient conditions of the rolling ship. Instead of simulating the long runs of a vessel in irregular seas, they suggest using only groups of regular waves, which are preceded by relatively calm conditions (see the grouping phenomenon in Chapter 7). In using equivalent regular wave conditions of a wave group, the calculation effort reduces considerably, when reaching some critical wave height. For the critical wave height, the range of initial roll angles and velocities does not effect the outcome anymore, so the number of simulations can be reduced. The authors suggest using the transient capsize diagram as the basis for stability regulation which includes the ship dynamics, and as an operational aid.

Kuteynikov and Lipis (2000) report on the index method of ship operational seaworthiness, as it was developed at the Russian Maritime Register of Shipping. They calculate criteria for operational restrictions reflecting the influence of individual ship parameters, the loading condition with draught and stability, the sea region, and the season of navigation. Their calculation scheme can be used for operational decisions on a quantitative basis with regard to weather restrictions.

The authors point at the operational stability of a ship in various real life situations. Operational stability is significantly defined by the ability of the master to load and control the vessel, to observe the characteristics of the ship design, and to act in dangerous sea conditions.

What seems to be a very basic principle in the Russian regulations is the possibility to assign restrictions to ships not satisfying an unrestricted region of navigation. There are five categories with two control parameters: the permissible distance from a place of refuge and the allowable height of waves with a 3% probability exceeding the level of 8.5 m to 3.5 m.

Seaworthiness is defined by various characteristics such as loss of speed, deck wetness, slamming, propeller racing, decrease of manoeuvrability. Rolling is calculated in connection with possible shifting of cargo. To support stability management and to account for the human element, a supplement to the stability booklet is provided aboard. This supplement contains information on the safe modes of ship operation in the form of polar diagrams. The software SAFESEA is provided. Polar diagrams indicate the safe and the dangerous zones on the screen. The calculation procedures are supported by measurements. Table 11.24 compiles the set-up of the onboard system.

Kuteynikov and Lipis (2000) define a factor of possibility for choosing safe ship navigation modes as the ratio, W, of all safe areas, Q_s, and of the total area, $Q_0 = \pi V_0^2$, in the directional diagram, with V_0, the speed in still water. Then W is split up into factor W_s, accounting for the safe mode selection in the possible area due to the ship characteristics, and W_v accounting for the integral average loss of speed for different wind and wave directions:

$$W = \frac{Q_s}{Q_0} \qquad (11.8)$$

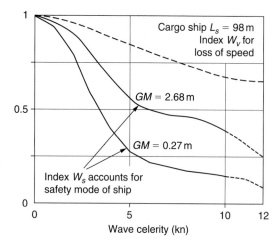

Figure 11.47 Short-term index W_s and W_v for safe ship operation.

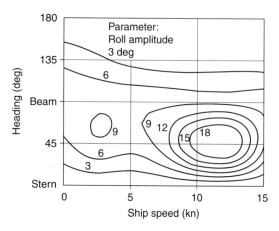

Figure 11.48 Roll response versus speed and heading.

Figure 11.47 shows an example for a cargo ship at fully loaded and ballast conditions in the North Atlantic.

In a systematic study on the ship response in severe seas by Papanikolaou, *et al* (2000), the authors point at the improvement of bulk-carrier safety by operational measures. The calculations proved that stern quartering seas are most dangerous with respect to the ship's stability and might lead to capsize. For four types of bulk-carriers ranging from 150 m to 280 m at about 15 kn maximum speed, roll response peaks up at higher speeds. Figure 11.48 depicts the significant amplitudes of the calculated roll for the handysize bulker with 152.40 m length. A 'seakeeping information booklet' can serve as guidance for avoiding dangerous situations in extreme weather conditions and was proposed by the authors to IMO in 1998. Operational measures such as weather routeing, change of speed and heading appear to be most effective.

Of particular concern is the stability safety of fishing vessels, see Section 11.4.2. Statistics show a drastic increase with human error as the main cause due to excessive ship motion reducing the level of attention of the crew, Boccodamo, *et al* (2000). Boccodamo, *et al* calculated the limiting wave height versus the wave period to be critical for the incidence of motion induced interruptions for the work on deck in fishing operations. Therefore, too large a GM leading to bad seakeeping characteristics is not advised.

Nabergoj, *et al* (2000) investigated loss of stability, parametric roll resonance, and broaching to for fishing vessels with a series of experiments. It is interesting to note that the amplitudes of parametric roll are sensitive to the rolling motion excitation threshold. According to Basin (1969), the threshold

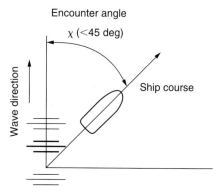

Figure 11.49 Definition of encounter angle in IMO resolution.

is given by the intensity of stability modulation as the ratio of GM variation in waves versus the still water GM:

$$a_0 = \frac{\delta GM}{GM_0} \geq 4\nu_{0\phi} \qquad (11.9)$$

Here $\nu_{0\phi}$ is the non-dimensional roll damping (parameter D in Chapter 7, Section 7.2.11). The authors conclude that limitations in a_0 by ship hull design can improve the roll behaviour, when limiting parametric roll to about 17 degrees.

To control operational stability, interactive computer software has been developed. An example is given by He, *et al* (2000) with their program CapeBoat. Important is the user-friendly graphical interface. The program includes a stability safety database with storing the maximum roll angles and time histories, computed under different sea and operating conditions.

Alman, et al (1999) reported new computing abilities coupled to risk assessment techniques in defining the chain of events that can lead to capsize. These techniques provide improvements in safety through design, operator guidance, training, and life-cycle management.

Results of German federally funded co-operative research were presented at the STG summer meeting in Gdansk 2001. Cramer and Krüger (2001) reported on numerical simulation methods to predict large roll angles. This becomes important for Ro/Ro and Ro/Pax vessels, characterised by high values of initial metacentric heights, so the seakeeping behaviour becomes more and more important. Pereira, et al (2001) reported on nonlinear simulation techniques for ship motions in six degrees of freedom. They use a strip method converting the hydrodynamic coefficients obtained in the frequency domain to be converted for the use in the time simulation. In the Hamburg model basin, Blume and Brink (2001) set up model testing for validating the prediction of extreme roll up to capsizing. Clauss, et al (2002) report on deterministic seakeeping model tests for the analysis of extreme roll and capsizes. The federally funded program ROLL-S further develops numerical methods for simulating ship motions in extreme seas. The target is to design safer ships with reduced risk from capsizing.

Computer assistance with numerical methods of seakeeping guidance and survivability criteria can give the ship operator an objective tool for making decisions. When risks and the expected downtime of avoiding severe conditions are identified, the improved safety and the expected savings will allow operating the ship both safely and economically.

11.5.2 Guidelines on in–service ship stability

11.5.2.1 *Purpose of guidelines for operational stability*

Basic requirements for stability curves are described in Chapter 3. Ship operational stability can be divided into four parts of shipboard action:

- To ensure compliance with the minimum standards on hydrostatic stability during loading and unloading before departure,
- To keep control of free surfaces, fuel consumption and ballast at sea,
- To ensure proper lashing and securing of cargo,
- To limit excessive ship motion at sea and to prevent capsizing.

The code of intact stability for all types of ships, IMO Resolution A.749 (18) was adopted in 1993, with later Amendments, IMO (1999). The preparation took many years of research and discussion. The code contains some requirements on the ship operation too. In chapter 2.3 on general precautions against capsizing, it reads: *Compliance with the stability criteria does not ensure immunity against capsizing, regardless of the circumstances, or absolve the master from his responsibilities. Masters should therefore exercise prudent and good seamanship, having regard to the season of the year, weather forecasts and the navigational zone, and should take the appropriate action as to speed and course warranted by the prevailing circumstances.*

Chapter 2.5 of IMO Res. A.749 (18) refers to operational procedures related to weather conditions. Quote: *Special attention should be paid when a ship is sailing in following or quartering seas because dangerous phenomena such as parametric resonance, broaching to, reduction of stability on the wave crest, and excessive rolling may occur singularly, in sequence or simultaneously in a multiple combination, creating a threat of capsize. Particularly dangerous is the situation when the wavelength is in the order of 1.0 to 1.5 ship's length. A ship's speed and/or course should be altered appropriately to avoid the above mentioned phenomena.*

Ensuring sufficient ship stability is a key task in ship operation, not just a matter of regulations for minimum standards. The safe handling of ships with respect to stability must cover so many topics and different situations, that it requires an educated and experienced ship officer. It is certainly a large step forward for improving stability safety by IMO acknowledging the importance of operational stability in resolution A.749(18).

Guidelines as proposed by Kaps in a contribution to IMO compiled the state of the art. The aspects common to all types should be addressed in a specific guidance to be provided by the shipping company:

1. Care for and monitoring of the ship's water tight and weather tight integrity,
2. Control of KG_c in the course of cargo, ballast or bunker operations (index c includes corrections for free surfaces),
3. Avoidance of transverse shifting of masses, such as cargo or heavy items of equipment,
4. Avoidance of large liquid surfaces,
5. Preparedness for corrective measures to mitigate detrimental effects on stability,
6. Keeping of records for control and reference (e.g. light ship characteristics).

11.5.2.2 *Loading and stability manual*

Each ship should be provided with a Loading and Stability Manual (stability booklet), generally developed by the shipyard, and approved by the administration.

IMO has discussed standardising this manual. It must contain sufficient information to enable the master to operate the ship safely. This manual should comply with the model developed by IMO in MSC/Circular 920. The general purpose is to provide all necessary information to the master for the proper loading and ballasting of the ship for the control of stability, draught, and structural integrity.

For evaluating displacement, draughts and stability, a recommended form sheet for manual moment calculation of stability and trim is provided, together with a form for calculating and plotting of the righting lever curve. However, care should be taken on the scale of the GZ-axis to be comparable for different loading conditions.

The Model Loading and Stability Manual by IMO includes chapters on:

- Operation of the ship;
- Typical approved loading conditions;
- Control of stability, trim and longitudinal strength;
- Technical information with respect to:
  Capacity plan,
  Cargo space information,
  Tank space information,
  Hydrostatic particulars,
  Light ship particulars,
  Load line particulars,
  Stability limits,
  Longitudinal strength criteria,
  Other operating restrictions.

Stability limits are generally given in the form of a limiting KG curve for the maximum permissible KG depending on the draught.

The latter operating restrictions are extremely important. They include:

- Restrictions to high initial stability with regard to securing of deck cargoes, in particular containers (reference to the cargo-securing manual),
- Stability and trim requirements with regard to damage control (reference to the damage control plan),
- Restrictions with regard to shiphandling in heavy weather, (see the next Subsection 11.5.2.3).

Finally, the Model Loading and Stability Manual should contain useful reference information such as:

- Inclining test report.
- Intact stability criteria.

See also Chapter 3 for the basic properties and regulatory requirements of stability curves.

11.5.2.3 Guidelines on the management of ship stability

The management of stability may differ considerably among ship types and trade patterns.

The above mentioned (in Subsection 11.5.1.2) presentation of guidelines on the management of ship's stability is meant as an aid to focus on the problems involved. The proposal comprises the following topics:

- Assessment of stability before departure (section 3), see above Subsection 11.5.2.2,
- Control of stability while at sea (section 4),
- Measures before and during heavy weather (section 5),
- Training requirements (section 10).

In appendices, practical recommendations are given on:

- Simplified draught survey,
- Measurement of stability by in-service inclining test.
- Measurement of stability by observing natural periods of roll (see Subsection 7.2.11.6).
- Adaptation of test results for assessing final conditions.

Methods of assessment of stability before departure
The methods of stability assessment refer to the minimum stability requirements. They include additional criteria for certain types of cargo or modes of operation, as given in the Intact Stability Code (Resolution IMO A.749 (18)). Deviations of the ship status from the required values can lead to severe danger for the ship. It is important to reduce uncertainties in the assessment before leaving port. We have three levels of assessment, at increasing accuracy:

Comparing the intended loading plan with similar conditions where stability is known.

Individual calculation of masses and moments of cargo distribution and tank filling (mass and centre of gravity of cargo and tank fillings must be available with sufficient accuracy, and computer programs will facilitate the method).

Measurement of stability by an in-service inclining test, or by observation via the observed natural period of roll is applied. Measurement is the most accurate method. When carried out during the loading process, it needs further adaptation to the final state at departure, and to the worst condition during the voyage.

Sailing in heavy weather
Fuel consumption, ballast, water absorption of cargo, and icing, must of course be controlled while at sea. All cargo should be properly stowed and secured, in accordance with the code of safe practice for cargo stowage and securing.

In severe weather, the speed of the ship should be reduced, and/or the course changed, if excessive rolling, propeller emergence, shipping of water on deck, or heavy slamming occurs. Water trapping in deck wells should be avoided. Special attention should be paid when the ship is travelling in following or stern quartering seas. Dangerous dynamic phenomena such as parametric resonance, broaching to, reduction of stability on the wave crest can occur and lead to danger from capsizing, see the next section.

11.5.2.4 Guidance to the master for avoiding dangerous situations in following and quartering seas

First, national authorities started to include basic recommendations on avoidance of resonance in stability requirements, BMV (1984). Takaishi (1994), Takaishi *et al* (1997) developed the technical background and objectives of new international IMO guidelines approved in 1995. This MSC 707 guidance to the master for avoiding dangerous situations in following and quartering seas, consists of four parts:

- **Explanation of dangerous ship response in following and quartering seas**
  (a) surf–riding and broaching to;
  (b) reduction of intact stability on the wave crest amidships;
  (c) synchronous rolling motion;
  (d) parametric rolling motion;

- **Dangerous conditions with respect to navigation**
  (a) the ship approaches to the phase velocity of the waves, causing mainly:
  – surf–riding and broaching to,
  – reduction of intact stability on the wave crest amidships.
  (b) ship speed approaches group velocity of the waves, causing mainly:
  – reduction of intact stability,
  – synchronous rolling,
  – parametric rolling.

- **Procedure to avoid dangerous situations**
  Ship speed and heading must be changed to avoid the dangerous zone for
  (c) surf–riding, Figure 11.50
  (d) high wave group, Figure 11.51: $0.8 < V/T$ (knot/s) < 2.0;
  (e) synchronous and parametric rolling, Figure 11.51: T_E or $2\,T_E$ not equal T_r.

- **Necessary data for the operation**
  (a) significant wave height;
  (b) mean wave period, or wave length;
  (c) wave direction relative to ship course;
  (d) ship speed;
  (e) ship course with respect to the waves (heading angle); Figure 11.49.

The following figures give the graphs as contained in MSC Circular 707.

The operation diagram, (Figures 11.51 and 11.53), shows on the right-hand side the ratio $V\cos\chi/T_w$ (also known as the so-called V/T parameter, see the circle diagram on the left-hand side), versus the ratio wave period to encounter period T_W/T_E.

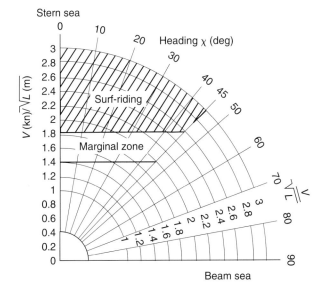

Figure 11.50 Operation diagram indicating dangerous zone due to surf-riding according to IMO.

This linear relationship is derived from Equation (7.71) as follows:

$$T_E = \frac{T_w^2}{T_w - \frac{2\pi}{g} V \cos\chi}$$

$$\frac{T_w}{T_E} = 1 - \frac{2\pi}{g} \cdot \frac{V \cos\chi}{T_w} \quad (11.10)$$

This is a linear equation of the form $y = b - \frac{1}{K} x$

with $b = 1$ and

$$\frac{1}{K} = \frac{2\pi}{g} \cong 0.64 \text{ s}^2/\text{m} \quad \text{or} \quad K \cong 1.56 \text{ m/s}^2$$

$$y = \frac{T_w}{T_E} \text{ and } x = \frac{V \cos\chi}{T_w}$$

For V in knots, the transformation factor to m/s is: $1 \text{ kn} = 1 \text{ nm/h} = 1852 \text{ m}/3600 \text{ s} = 0.514 \text{ m/s}$.

Equation (11.10) reduces to (for V in knots and T_w in seconds):

$$\frac{T_w}{T_E} = 1 - \frac{1}{K} \cdot \frac{V_{kn} \cos\chi}{T_w} \cong 1 - 0.33 \cdot \left\langle \frac{V}{T} \right\rangle \quad (11.11)$$

However, we want to have the V/T parameter, so we write Equation (11.11) into:

$$\left\langle \frac{V}{T} \right\rangle = \frac{V_{kn} \cos\chi}{T_w} = K \cdot \left(1 - \frac{T_w}{T_{E'}}\right) \cong K \cdot \left(1 - \frac{T_w}{T_E}\right) \quad (11.12)$$

This equation is depicted in Figures 11.51 and 11.52 using the 'wave factor' $K = 3.03$ kn/s, valid for regular sine waves, as used in the MSC 707 guidance. In fact, this is a very surprising result, when compared with all the detailed graphs of Section 7.2.11. Using the dimensionless ratio T_W/T_E and the V/T parameter allows compressing formula (7.71) for all partial waves into one straight line, no

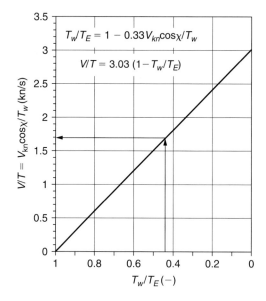

Figure 11.52 Ratio of wave period to encounter period versus V/T parameter.

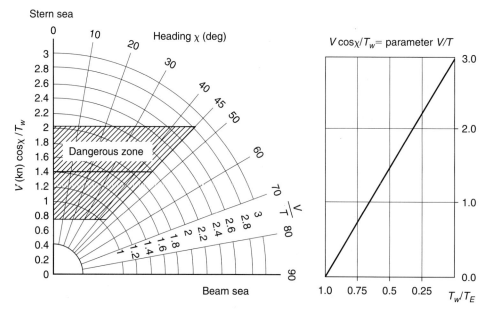

Figure 11.51 Operation diagram indicating dangerous zone due to resonance.

Operation diagram for the master

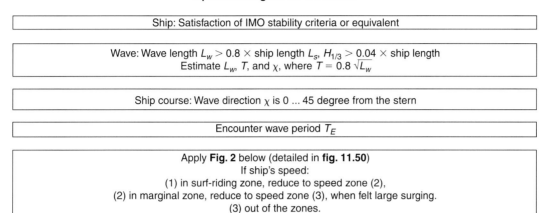

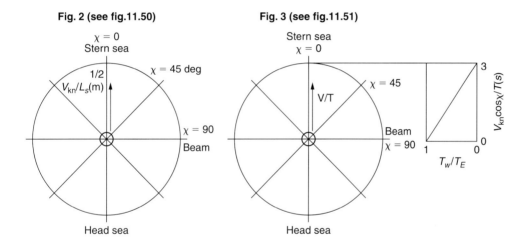

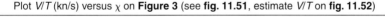

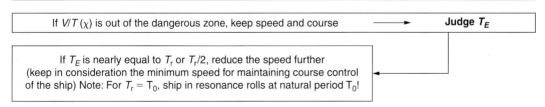

Figure 11.53 Operation diagram for the master, according to IMO MSC/Circ. 707 (1995).

matter what wave period we look at. This is quite an achievement for practical use by the master. The reciprocal action of ship and surrounding water, however, has not been taken into account. This might be of particular concern when ship and wave progress is almost the same, $T_E \to \infty$.

For determining the ship speed and heading (wave direction to the ship course), Grochowalski, et al (1994) suggested using three methods:

1. At the moment of wave impact at stern,
2. Averaged over one wave encounter,

3. The maximum during passage of a high wave group.

It is sufficient to obtain heading χ by visual observation. The period of the wave, T_w, is taken as the period of heaving motion of foam on the sea surface generated by breaking waves and is measured with the use of a stop watch. When the wave length is estimated (by visual observation in comparison with the ship length or by reading the mean distance between successive wave crests on the radar images of the waves), wave period can be calculated by $T_w = 0.8\sqrt{L_w}$. The encounter period, T_E, is equal to the period of pitching.

The natural roll period, T_0, must be measured in a calm sea. Alternatively, this value can be roughly estimated by using the Weiss formula, $T_0 = C_r B / \sqrt{GM}$ (Equation (7.43)), which gives T_0 as a function of GM, ship breadth B and the rolling coefficient, C_r. (Equ. (7.45)). Nonlinear effects due to the shape of the GZ curve and variation of GZ in following and quartering seas (see (Section 7.2.10)) should be considered.

The advantage of any such operational measures of improving ship stability is the imminent response the master can observe by his measures. Since in random seas the vessel experiences excitation by groups of waves, survival of the first severe wave does not mean surviving the next ones. Once the ship is in a dangerous situation, the master must take appropriate action. This is mainly a reduction of speed but also a change of heading, to prevent low cycle resonance.

IMO has decided to review the guidance MSC Circular 707 with a view to improving it, in particular with respect to large ships, on the basis of new technical developments and in the light of experience gained from the Circular's application. Inclusion of head sea parametric rolling is considered, Francescutto (2002).

Umeda (1994) validated the new guidance to the master by comparing results of model experiments on fishing vessels and container ships, and he gave some assumed scenarios to show the application. The IMO guidance complies with ratings such as:

- Practical for ship operation,
- Ease of use by mariners,
- Considers stability requirements by existing criteria,
- Corresponds to physical phenomena,
- Excludes capsizing as observed by model experiments,
- Is applicable to all conventional types of ships without limitations in size.

Umeda (1994) also proposed to control the severe rolling condition aboard a ship by the dangerous zone as given in Figure 11.54. This criterion can be derived again from Equation (11.12). For resonance at $T_E = T_0$, we replace in Equation (11.12) T_E by T_0, and we assume a wide range within $0.7T_0$ to $1.3T_0$ to be dangerous. Furthermore, the natural roll period, T_0, valid for still water at small roll is assumed to increase by the factor 1.25 due to the 'softening spring effect'. This factor accounts for the reduction of the GZ on a wave crest. The combined zone coefficients are $0.7 \cdot 1.25 = 0.875$ and $1.3 \cdot 1.25 = 1.63$. Thus, the criterion for the dangerous range of the V/T parameter is as follows:

$$3.03 \left(1 - \frac{1}{0.875} \cdot \frac{T_w}{T_0} \right) < \frac{V \cos\chi}{T_w}$$
$$< 3.03 \left(1 - \frac{1}{1.63} \cdot \frac{T_w}{T_0} \right) \text{ for } T_E = T_0 \quad (11.13)$$

In extending to the parametric resonance at $T_E = 0.5 \cdot T_0$, we add the range from $0.3T_0$ to $0.7T_0$, see the second dangerous zone as shown in Figure 11.54.

Equation (11.13) can be applied to all types of conventional ships without any limitations in size.

An insufficient stability margin together with unfavourable environmental conditions call have serious stability consequences with a final capsize. The knowledge of ship dynamics is important for both ship operators and ship designers. Ship proportions and hull form, loading of the ship, environmental conditions and ship operation all contribute to ship safety. To reproduce the capsizing mechanism theoretically, forces and moments on captive ship models and the behaviour of free running models must be correlated. Many authors have done studies and experiments, let us here cite Crudu et al (1994). From experiments at 1/30 scale with a ship of 2700 dwt and 86 m length the authors stress the point that both main and parametric resonance ($T_E = T_0$ and $T_E = 0.5T_0$) can result in

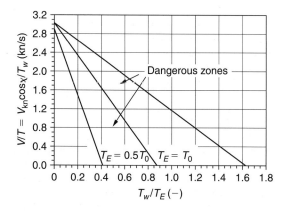

Figure 11.54 Dangerous zones for severe rolling.

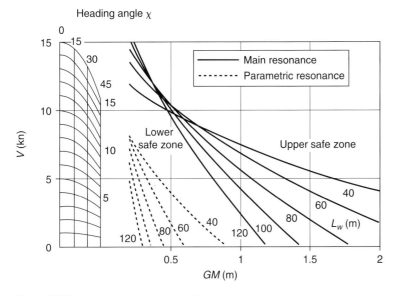

Figure 11.55 Safe zones for ship running in following waves.

large roll amplitudes, see their graph on safe zones in Figure 11.55.

In comparing Figure 11.55 with the other diagrams shown, the reader will observe the same background to identify resonance regions. Future application aboard ships will hopefully develop into some widespread recognition of a good guidance in ship operation to avoid severe roll resonance.

In addition to just identifying the resonance zones, it seems appropriate to want information on the maximum roll to be expected. In irregular sea, this can only be given as a probability measure, depending on the particular situation the ship is in. Taggart and de Kat (2000) presented results from computer simulation. Figure 11.56 depicts an example on the probability of the hourly exceedance versus the maximum roll angle for most likely conditions given capsize. The data of this example are: speed 10 kn, heading 45 deg (stern quartering), T_v 12.4 s, H_v 9.5 m, average wind speed 49 kn, thick line from a number of simulations, thin line curve is fitted Gumbel distribution.

In a circle diagram with both the parameters ship speed V and heading χ, areas of V-χ combinations with large risk level can be identified.

Measurements in service can give direct exact data. Takaishi (2000) reported a new development for an on-board system to execute the operational guidance automatically. The system consists of a wave directional analysis. It evaluates measurements of wave probes and accelerometers installed on the ship.

As a definition of severe weather, i. e. when the operator is actually using the IMO guidance,

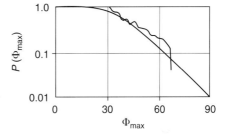

Figure 11.56 Hourly exceedance of maximum roll from simulation.

Renilson (1994, 1997) suggested to plot the ratio GM/L (metacentric height – ship length ratio) as a function of H_W/L (wave height ship length ratio). From capsizing statistics of ship models, a demarcation line results between ships that are safe and unsafe, see Figure 11.57. If $GM < 0.2\,H_W$, caution is required and the guide must be consulted.

At the STAB Conferences, the operational factors in stability safety of ships in heavy seas were discussed in detail. The guidelines of IMO MSC Circular 707 are seen as a reasonable step forward. For the future, more detailed criteria for each individual ship might be developed. Grochowalski, et al (1994) stressed that adequate guidelines require careful and detailed examinations of various operational conditions and physical phenomena. Detailed criteria are mainly based on theoretical results. Francescutto and Bulian (2002)

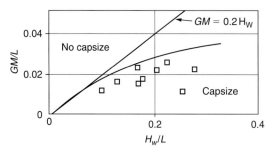

Figure 11.57 Capsize region to adopt IMO guideline.

describe the non-linear and stochastic aspects of parametric rolling modelling. A computerised advice to the master will eventually provide the solution in developing a set of detailed criteria for each potentially dangerous situation. However, any attempt to offer precise guidance may fail, due to the complexity of real situations at sea. Guidance must provide operators with a more general understanding, e.g. where risky combinations of speed and heading exist (see paper and discussion by Alman, et al (1999).

11.5.2.5 *International safety management code (ISM)*

IMO adopted Resolution A.741 (18) on the International management code for the safe operation of ships and for pollution prevention in 1993 [IMO, ISM Code (1994) and ISM (2002). For ferries, the code was implemented in July 1996, for passenger ships, tankers and chemical carriers in July 1998, and for general cargo ships and drilling rigs by July 2002. The ISM code has also been adopted by the International Convention for the Safety of Life at Sea (SOLAS), with the new Chapter IX: Management for the Safe Operation of Ships. Further comments on the ISM code were given earlier in Section 11.2.2.5. In most cases, more than one problem contributes to a disaster. Dölling [2000] compiles examples of operational negligence:

- Seamen on duty do not act properly,
- Decision competence not clear,
- Loss of time due to delay of action,
- Shiphandling not adequate,
- Control of ship condition not satisfactory,
- Crew is not aware of the technical limitations of the ship,
- The owner is reluctant to fix known deficiencies in the ship operation,
- Technical equipment for improving safety is not installed, unless it is mandatory.

The purpose of the ISM code is to provide an international standard for the safe management and operation of ships and for pollution prevention. The theoretical framework is formed by three elements, Zharen and Duncan (1994):

1. Definitions, objectives, and functional requirements,
2. Safety and environmental protection policy,
3. Company responsibilities and authority.

The shipping company must establish procedures for key shipboard operations concerning the safety of the ship and the prevention of pollution. Safety audits must be carried out and certified by independent surveyors, and a certificate for the ship must be issued by the administration of the flag states or the respective classification societies. Maintenance of the ship and equipment should also be assured.

The main topics in the ISM code are the emergency preparedness by establishing procedures and programs for drill and exercises (section 8), to ensure that the master is properly qualified for command, and is given the necessary support to perform his duties safely (Section 6). Responsibilities and authority aboard the ship must be defined, and a person ashore having direct access to the highest level of company management must be designated to provide a link between the company and those on board, to ensure the safe operation of the ship (Section 4). The master's responsibility and authority is particularly defined (Section 5), in particular implementing the policy, motivating the crew, issuing instructions, verifying that requirements are observed, and reviewing the safety management system (SMS) for the ship.

In the short term, legislators expect the IMO code to give a significant contribution to establishing an effective compliance regime for the mandatory requirements and the non-mandatory codes, guides, and standards in shipping, Knudsen (1993). At a later stage, safety management in shipping may develop to levels where all factors to safety and protection of the environment are subject to adequate management control.

The ISM code is certainly a large step forward towards improving safety procedure in shipping operation. Now, the shipping company must set up a safety and environmental protection policy, which describes how the objectives will be achieved. A main advantage of the ISM code for safer shipping can be seen in the requirement of identifying risks and establishing safeguards. It is mandatory to continuously improve safety management skills of personnel. However, systematic identification of hazards and specific risk analysis methodology are not required by the Code, Knudsen (1993).

11.5.3 The human factor. Maritime education and training

Assurance on the quality of ship and equipment, and setting up management schemes, will only work when the responsible persons involved in shipping contribute with knowledge and experience. IMO has put a lot of effort into regulations and guidelines supporting personnel development. In 1997, IMO adopted Resolution A.850 (20) (IMO 1997b) on human element vision, principles and goals for the organization. IMO acknowledges the close relationship between the human elements and safety. Quote: *The human element is a complex multi-dimensional issue that affects maritime safety and marine environmental protection. It involves the entire spectrum of human activities performed by ships crews, shore-based management, regulatory bodies, recognised organizations, shipyards, legislators, and other relevant parties, all of whom need to co-operate to address the human element effectively.*

One of the goals with respect to maritime education and training (MET) is, quote: *to provide material to educate seafarers so as to increase their knowledge and awareness of the impact of human element issues on safe ship operations, to help them to do the right thing.*

Statistical analysis of marine accidents demonstrates that the human element is involved considerably. Boniface and Bea (1996) cite the UK P&I Club analysis of the claims in 1993 and estimated human errors to be the primary cause of 62% of those accidents. Around 60% of accident claims are directly due to human error, a further 30% are indirectly related to human error too, by failure of components designed and operated by human beings, Kuo, (2000).

There are mainly two sources of malfunction due to wrong operation of complicated technical systems. One is negligence and laxness in handling the system, the other is overload and pressure. Figure 11.58 shows the probability density of human errors versus the intensity of stressors, Salvendy, (1987). The human errors have a flat minimum in a wide range of stressors, where persons can cope with all demands. However, when the culture of operating the system is too lax, the probability of human errors increases. The same reduced ability to find the right decisions results in overload and pressure on persons.

Personnel selection, training, experience and support have a positive effect. They decrease the probability of human errors considerably.

There is no clear–cut definition of the term 'human factors'. An evident definition by Gregory (1987) is: 'Human factors concern the efficiency of persons in their working environment.'

Kuo (2000) presented a thorough study on the impact of human factors on ship stability. For practical applications, he defines as follows: 'The

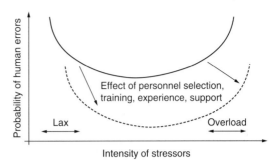

Figure 11.58 Effect of stress on probability of human errors.

term human factors is concerned with the interfacing of a set of personal capabilities and characteristics with a combination of hardware, software, working environment and organizational culture in the effective performance of a task.'

In both interpretations the main aspect of human factors is the connection of two distinct groups. They connect persons, described by their capabilities and their performance, with a complicated technical system.

The impact of human factors on safe ship operation is considerable. This is particularly true with respect to stability. Kuo (2000) gives some important points to be looked at:

- Standard procedures may sometimes not be appropriate and may be difficult to implement, thus leading to accidents.
- Information may not be communicated correctly among multilingual and multicultural crews.
- Emphasis on training with practical demonstrations, group exercises and training software is advised.
- Assigning specific responsibilities to clearly designated individuals is a prerequisite.

In identifying human factors contributing to accidents, we speak of 'human errors'. Kuo (2000) cites two opposing views. 'Nothing can be done about it because it involves human beings, or, human errors must and can be eliminated.' A realistic option is certainly to recognise the different natures of human beings, and trying to minimise the probability of the occurrence of human errors and impact by appropriate methods.

Human errors are associated with human actions, omissions and neglect. Their impact can be extremely serious. Kuo (2000) ends up with three levels of human error:

1. Skill-based
   Tasks are carried out on a routine basis, but concentration may fail.
2. Rule-based
   Here procedures are often carried out from experience of training. A good rule can be

misused, or a poor rule is implemented, or the appropriate rule not applied at all.
3. Knowledge-based
   Here the solution must be worked out, as previously applied methods are inappropriate. This level is often associated with unexpected emergency situations.

Furthermore, Kuo identifies twelve factors that can lead to human errors, such as: Attitude, communication, concentration, confidence, experience, human limitations, information, management, procedure, time constraints, tiredness, training. He suggests selecting only three of the most significant factors to apply an analytical decision analysis method. This will overcome most of the deficiencies. His so-called *safety case concept* includes a methodology to address the most relevant factors. The first step is to analyse the problem, then to set up a safety management system incorporating the best way to manage safety, Kuo, (1998). This concept is particularly good for considering situations or systems for which there is little previous experience, and for problems associated with human factors. See also Section 11.4.3. Part of the safety case concept analysis is:

- Hazard identification (what aspects of a system can go wrong),
- Risk assessment (what are the chances and effects of the hazards),
- Risk reduction,
- Emergency preparedness (what to do if a severe situation arises).

Improved methods in management, engineering and operation are needed to reduce the probability of human errors. The reduction of human errors by personnel training, experience and support is shown in Figure 11.58.

Boniface and Bea (1996) reviewed the key principles in human performance methodologies. They developed a human error modelling system, named Human Error Risk Reduction Operating System (HERROS), to provide insights and solutions at all stages of the life cycle of a ship.

IMO (1995a) revised the International Convention on Standards of Training, Certification and Watchkeeping for Seafarers (STCW). It is to be seen as an international maritime training guide. Still, operational decisions that have influence upon capsizing are left to the discretion of the captain, as Dahle (1993) complained. STCW quotes specifically, that training of ship and engine officers should give them knowledge on the calculation of intact stability, and on the safe stowage and handling of cargo. Instructions on how to handle the ship in severe conditions with minimum stability are considered important.

Miller and Paitl (2001) report on the U.S. fishing vessel-training suite. It includes a training document 'Best practices guide to vessel stability', supported by an interactive stability trainer. This trainer is a free floating, scaled fishing vessel which replicates actual operating conditions such as sloshing liquid, loading catch, lifting operations, and icing conditions.

The backbone of proper shiphandling is Maritime education and training (MET). An important role in international enhancement is played by the World Maritime University (WMU) in Malmö/Sweden, founded on behalf of IMO in 1983, Weinstein, (1996).

11.5.4 Operational stability in the future – A wishful forecast

There is widespread agreement on the need to improve stability and safety of ships. It has been recognised that better information to the master is important. Improvements in the area of human factors must be co-ordinated with research on maritime safety regulations.

Grochowalski, *et al* (1994) suggested, making a distinction between 'stability' and 'stability safety'. Stability safety depends on the ship characteristics and on ship operation, which must be optimised together, while stability alone, is just a feature of the ship to resist capsizing.

To stress the importance of ship operation on the ability of the ship in service to prevent capsize and to reduce damage from rolling and roll acceleration, the term *operational stability* has been used throughout this Section. The main concern of safe operation with respect to intact stability is to avoid severe situations such as synchronous and parametric rolling, surf riding, and broaching, and to give provisions for survival.

The inclusion of ship dynamics to assess operational stability beyond the conventional hydrostatic approach will improve taking vital action in advance. General application and further development of the tools now available in operational stability of ships and ocean vehicles will contribute substantially to safer shipping in the near future.

The main topics of operational stability are currently at different stages of development and implementation and are compiled as follows:

1. Management of stability and safety with reference to the ship type and mission, supported by performance based intact stability criteria,
2. Automatic estimate of ship static stability status continuously,
3. Ship-borne measurement equipment on behaviour at sea,
4. Numerical time domain simulation of dynamic ship behaviour with reliable computer programs allowing study, training and forecast on adverse seakeeping scenarios,

5. Short-term routeing assisted by sea data evaluation from ship-borne gauges and from satellites,
6. Storage of black box data on extreme events and decisions taken (voyage data recorder VDR),
7. Risk analysis and development of risk oriented scenarios for survival in heavy seas,
8. Reduction of human errors by mandatory assistance using onboard computer systems, but responsibility stays with the master,
9. Maritime education and training (MET) using modern tools such as video, virtual reality systems, numerical simulators, computer feedback trainers, and physical models for demonstration of risk scenarios,
10. Steady transfer of research and feedback among operators, designers, administrators and scientists.

References (Chapter 11)

Abeille, M. *et al.* (1999). *Preliminary Study on the Feasibility of Substituting Land Transport of Chemical Products by Shortsea Shipping.* Centre d'Estudis del Risc Tecnològic, Universitat Politècnica de Catalunya, Barcelona.

Alderton, T. and Winchester, N. (2001). The Flag State Audit. *Proceedings of SIRC's Second Symposium*, Cardiff University, 29 June. Seafarers International Research Centre, Cardiff, UK. http://www.sirc.cf.ac.uk/pubs.html

Alman, P.R., Minnick, P.V., Sheinberg, R. and Thomas, W.L., III (1999). Dynamic capsize vulnerability: reducing the hidden operational risk. *SNAME Transactions*, Vol. 107.

AMLG (2004). *Admirality and Maritime Law Guide – International Conventions.* http://www.admiraltylawguide.com/conven/unclospart2.html

Andrews, J.D. and Moss, T.R. (2002). *Reliability and Risk Assessment.* Professional Engineering Publishing Ltd., London and Bury St Edmunds, UK.

Ang, A.H.S. and Tang, W.H. (1984). *Probability Concepts in Engineering Planning and Design.* John Wiley & Sons, UK.

Arndt, B., Brandl H. and Vogt K. (1982). '20 Years of experience – stability regulations of the West-German Navy. *Proc. of STAB'82: 2nd International Conference on Stability of Ships and Ocean Vehicles*, Tokyo, p.765.

Bangash, Y. (1983). Containment Vessel Design and Practice. *Progress-in-Nuclear-Energy*, Vol. 11, No. 2, pp. 107–181.

Basin, A.M. (1969). Ship motions, Transport, Moscow (in Russian).

Bazovsky, I. (1961). *Reliability Theory and Practice.* Prentice Hall, Englewood Cliffs New Jersey.

Bendixen, L.M., O'Neil, J.K. and Little, A.D. (1984). Chemical Plant Risk Assessment Using HAZOP and Fault Tree Methods. *Plant/Operations Progress*, Vol. 3, No. 3, pp. 179–184.

Bergantino, A. and Marlow, P. (1998). Factors influencing the choice of flag: Empirical evidence. *Maritime Policy and Management*, Vol. 25, No. 2, pp. 157–174.

Best, P.J. and Davies, W.B. (1999). The Assessment of Safety for Vessels in Service: Practical Examples of the Application of FSA Techniques from Inland Vessels to Ocean Going Ferries. *Transactions of IMarE*, Vol. 111, No. 2, pp. 51–58.

Blume, P. and Brink, K. (2001). Model tests for the validation of extreme roll motion prediction, *STG summer meeting*, Gdansk.

BMV (1984). (Bundesminister für Verkehr). Bekanntmachung über die Anwendung der Stabilitätsvorschriften für Frachtschiffe, Fahrgastschiffe und Sonderfahrzeuge vom 24. Oktober 1984, Hamburg.

Boccodamo, G., Cassella, P. and Scamardella, A. (2000). Stability, operability and working conditions onboard fishing vessels, *Proc. of STAB'2000: 7th International Conference on Stability of Ships and Ocean Vehicles*, Vol. 1, Launceston, Tasmania, pp. 32–51.

Boniface, D.E. and Bea, R.G. (1996). Assessing the risks of and countermeasures for human and organisational error. *SNAME Transactions*, Vol. 104, pp. 157–177.

British Standard (1991). BS EN 292: Safety of Machinery – Basic Concepts, General Principles for Design, Part 1: Basic Terminology Methodology; Part 2: Technical Principles and Specification.

Carbajosa J.M. (1989). Report on Work Accident in the Sea Fishing Sector, International Symposium of Safety and Working Conditions Aboard Fishing Vessels, L'Université du Québec á Rimouski, Quebec, Canada, Aug 22–24.

Chadwin, M.L., Pope, J.A. and Talley, W.K. (1999). *Ocean Container Transportation, an Operational Perspective.* Taylor & Francis, New York.

Clauss, G., Hennig, J., Kühnlein, W., Brink, K.E. and Cramer, H. (2002). *Analysis of extreme roll motions in severe seas with embedded rough wave sequences for the development of safer ships with reduced capsizing risk.* Transactions STG, Hamburg.

Cleary, W.A. (1994). Human factors in marine regulation, Workshop on human factors, synopsis by the moderator S. Allen STAB'94: 5th International Conference on Stability of Ships and Ocean Vehicles, Melbourne, Florida.

Containerisation International, (1996). Year Book on the 40 Years of Containerisation, 1996, pp. v–x.

Cowley J. (1995). The Concept of the ISM Code, Proceeding of Management and Operation of Ships: Practical Techniques for Today and

Tomorrow, *The Institute of Marine Engineers*, 24–25 May, London, Paper 1.

Cramer, H. and Krueger, S. (2001). Numerical capsize simulations and consequences for ship design, *STG summer meeting*, Gdansk.

Crudu, L., Nabergoj, R., Obreja, D.C. and Trincas, G. (1994). Ship stability in following waves. Theoretical and experimental investigations, *Proc. of STAB'94: 5th International Conference on Stability of Ships and Ocean Vehicles*, Vol. 1. Melbourne, Florida.

Dahle E.A. and Weerasekara J.C.D. (1989). Safe Design Construction of Fishing Vessel, International Symposium of Safety and Working Conditions Aboard Fishing Vessels, L'Université du Quēbec á Rimouski, Quebec, Canada, Aug 22–24.

Dahle, E.A. (1993). Qualification, education and training *Proc. of US Cost Guard Vessel Stability Symposium*, US Coast Guard Academy, New London, Connecticut, pp. 242–249.

Damkilde, L. and Krenk, S. (1997). Limits – a System for Limit State Analysis and Optimal Material Layout. *Computers & Structures*, Vol. 64, No. 1, pp. 709–718.

de Kat, J.O. (2000). Dynamics of a ship with partially flooded compartment. In D. Vassalos, M. Hamamoto, A. Papanikolaou and D. Moulyneux (eds), *Contemporary Ideas of Ship Stability*. Elsevier Science, pp. 249–264.

de Kat, J.O. and Thomas, W.L. (1994). The use of numerical simulation tools in the derivation of operational guidelines for severe weather conditions, Workshop on Manoeuvring and survival in storm seas, synopsis by the moderator J.O. de Kat *STAB'94: 5th International Conference on Stability of Ships and Ocean Vehicles*, Melbourne, Florida.

Department of Energy (1990). The Public Inquiry into the Piper Alpha Disaster, (Cullen Report), ISBN 0 10 113102, London.

Dixon, P. (1964). Decision Tables and Their Applications. *Computer and Automation*, Vol. 13, No. 4, pp. 376–386.

Dölling, W. (2000). *Anspruch Technik contra Factor Mensch*. Transactions STG, Hamburg.

Dunn T.W. (1964). Reliability in Shipbuilding, SNAME, Spring Meeting, June 4–5, Boston, Massachusetts, USA.

EPA (2004). Oil Pollution Act, US Environmental Protection Agency. http://www.cpa.gov/rcgion5/defs/html/opa.htm

EQUASIS (2004). EQUASIS – Public web-site promoting quality shipping http://www.equasis.org/

Eyres, D.J. (2007). *Ship Construction*. Sixth Edition. Butterworth-Heinemann, Oxford, UK.

Falkanger, T., Bull, H.J. and Brautaset, L. (1998). *Introduction to Maritime Law: The Scandinavian Perspective*. Tano Aschehoug, Oslo.

Francescutto, A. (2002). Intact stability – the way ahead, *Proceeding of 6th International Ship Stability Workshop*, Webb Institute, New York.

Francescutto, A. and Bulian, G. (2002). Nonlinear and stochastic aspects of parametric rolling modelling, *Proceeding of 6th International Ship Stability Workshop*, Webb Institute, New York.

Gosden, S.R. and Galpin, L.K. (1999). Breaking the Chains of Preventive Maintenance: The Royal Navy Experience. *Maintenance & Asset Management Journal*, Vol. 14, No. 5, pp. 11–15.

Gregory, R.L. (1987). *Oxford companion to the mind*. Oxford University Press.

Grochowalski, S., Archibald, J.B., Connolly, F. J. and Lee, C. K. (1994). Operational factors in stability safety of ships in heavy seas *Proc. of STAB'94: 5th International Conference on Stability of Ships and Ocean Vehicles*, Vol. 4, Melbourne, Florida.

Halebsky, M. (1989). System Safety Analysis Techniques as Applied to Ship Design. *Marine Technology*, Vol. 26, No. 3, pp. 245–251.

Harrington R.L. and Riddick R.P. (1963). Reliability Engineering Applied to the Marine Industry, SNAME, Hampton Road Section, December 1963.

Henley, E.J. and Kumamoto, H. (1992). *Probabilistic Risk Assessment*. IEEE Press, NY, USA.

Hill S.G., Byres J.C., Rothblum A.M. and Booth R.L. (1994). Gathering and Recording Human Related Casual Data in Marine and Other Accident Investigation, Proceedings of the Human Factors and Ergonomics Society 38th Annual Meeting, Nashville, Tennessee, USA, October 24–28, pp. 863–867.

Hoefnagal W.A.M. and Bouwman K. (1989). Safety Aboard Dutch Fishing Vessels, International Symposium of Safety and Working Conditions Aboard Fishing Vessels, L'Université du Quēbec á Rimouski, Quebec, Canada, Aug 22–24.

Hoffman, D. (1976). The impact of seakeeping on ship operations. *Marine Technology*, Vol. 13, No. 3, pp. 241–262.

Hoover, S. and Perry, R. (1989). *Simulation, a Problem-solving Approach*. Addison-Wesley Publishing Company, USA.

Hopper, A.G. and Dean, A.J. (1992). Safety in Fishing-Learning from Experience. *Safety Science*, Vol. 15, pp. 249–271.

House of Commons (2000). *Fishing Safety, Hansard Debates for 16 Nov 2000 (Part – 1)*. Westminster Hall, United Kingdom.

IISE (1997). Prevention of Fire and Explosion, and Emergency Response on Offshore Installations, Health & Safety Executive, HSE Books, Suffolk, UK, ISBN: 0717613860.

HSE (1978). *Canvey: An Investigation of Potential Hazards from Operations in the Canvey Island/Thurrock Area*. Health & Safety Executive, HMSO, United Kingdom.

HSE (1981a). *Canvey: A Second Report. A Review of Potential Hazards from Operations in the Canvey Island/Thurrock Area*. Health & Safety Executive, HMSO, United Kingdom.

HSE (1981b). *Canvey: A Second Report. A Summary of a Review of Potential Hazards from Operations in the Canvey Island/Thurrock Area*. Health & Safety Executive, HMSO, United Kingdom.

HSE (1992a). *The Offshore Installation (Safety Case) Regulation 1992*. Health & Safety Executive, ISBN 00118820559.

HSE (1992b). *The Tolerability of Risk from Nuclear Power Stations*, 1992 Edition. Health & Safety Executive, ISBN 00118863681.

HSE (1995a). *Generic Terms and Concepts in the Assessment and Regulation of Industrial Risks*. Health & Safety Executive, Discussion Document DDE2, HSE Books, Suffolk, UK.

HSE (1995b). *A Guide to the Offshore Installations and Pipeline Works (Management and Administration) Regulations 1995*. Health & Safety Executive, HSE Books, Suffolk, UK, ISBN: 0717609383.

HSE (1996a). *A Guide to the Integrity, Workplace Environment and Miscellaneous Aspects of the Offshore Installations and Wells (Design and Construction, Etc.) Regulations 1996*. Health & Safety Executive, HSE Books, Suffolk, UK, ISBN 0717611647.

HSE (1996b). *A Guide to the Installation Verification and Miscellaneous Aspects of Amendments by the Offshore Installations and Wells (Design and Construction, etc.) Regulations 1996 to the Offshore Installations (Safety Case) Regulations 1992*. Health & Safety Executive, HSE Books, Suffolk, UK, ISBN 0717611930.

HSE, (1996c). The Offshore Installations and Wells (Design and Construction, etc.) Regulations 1996, ISBN 0-11-054451-X, No. 913.

HSE (1996d). *A Guide to the Pipelines Safety Regulations 1996*. HSE Books, Suffolk, UK, ISBN: 0717611825.

Hutchison, B.L. (1981). Risk and operability analysis in the marine environment. *Transactions SNAME*, Vol. 89, pp. 127–154.

ICRP (1977). *Recommendations of the ICRP, ICRP Publication 26, International Commission on Radiological Protection*. Pergamon Press, Oxford.

IEE (1999). Health and Safety Information, Quantified Risks Assessment Techniques (Part 2) Event Tree Analysis – ETA, No. 26(b), *Institution of Electrical Engineers*, UK.

ILLC (2002). *International Conference on Load Lines 1966, 2002 Edition*. International Maritime Organization, London.

IMarE, (1998) Proceeding of New Safety Culture, Organised by the Institute of Marine Engineers (IMarE) and the MCA, London, 4th December.

IMO (1975a). *Code of Safety for Fishermen and Fishing Vessels, Part A – Safety & Health Practices for Skippers and Crew*. International Maritime Organization, London.

IMO (1975b). *Code of Safety for Fishermen and Fishing Vessels, Part B – Safety & Health Requirements for the Construction and Equipment of Fishing Vessels*. International Maritime Organization, London.

IMO (1977). *Final Act of the Conference with Attachments Including the Torremolinos International Convention for Safety of Fishing Vessels, International Conference On Safety Of Fishing Vessels*. International Maritime Organization, London.

IMO (1980). *Voluntary Guidelines for the Design, Construction and Equipment of Small Fishing Vessels, An International Maritime Training Guide*. International Maritime Organization, London.

IMO (1988). *Document for Guidance on Fishermen's Training and Certification, An International Maritime Training Guide*. International Maritime Organization, London.

IMO (1993). *Draft Guidelines for FSA Application to the IMO Rule-making Process, MSC 68/14*. IMO Maritime Safety Committee, London.

IMO (1995). *International Code of Safety for High-Speed Craft*. International Maritime Organisation, London, pp. 175–185.

IMO (1997a). *IMO/MSC Circular 829, Interim Guidelines for the Application of Formal Safety Assessment to the IMO Rule-Making Process*. International Maritime Organization, London, 17th November.

IMO (1997b). *FSA Trial Application to High Speed Passenger Catamaran Vessel*. International Maritime Organization, UK, MSC 68/14/2 & 68/INF.

IMO (1998a). Trial Application of FSA to the Dangerous Goods on Passenger/Ro-Ro Vessels, MSC 69/INF.24, International Maritime Organization, Submitted by IMO Finland.

IMO (1998b). FSA Study on the Effects of Introducing Helicopter Landing Areas (HLA) on Cruise Ships, MSC 69/INF.31, International Maritime Organization, Submitted by IMO Italy.

IMO (1998c). Report of the Working Group on FSA, IMO MSC 66/WP 1.4, International Maritime Organization, United Kingdom.

IMO (1998d). Novel Emergency Propulsion and Steering Devices for Oil Tanker Analysed with the FSA Method, MSC 69/14/1, International Maritime Organization, Submitted by IMO Germany.

IMO (1999). Amendments to the code on intact stability for all types of ships covered by IMO instruments, IMO Res. MSC.75(69), London.

Inozu, B. and Radovic, I. (1999). Practical Implementations of Shared Reliability and Maintainability Databases on Ship Machinery: Challenges and Rewards. *Transactions of IMarE*, Vol. 111, No. 3, pp. 121–133.

ISM (2002). *International Safety Management Code (ISM Code) and Guidelines on Implementation of the ISM Code*, 2002 Edition. International Maritime Organization, London.

Isograph Limited (1995). *Fault Tree + Version 6.0.* Manchester, United Kingdom.

Janardhanan, K. and Grillo, P. (1998). Latest UK Offshore Safety Legislation. *The Journal of Offshore Technology*, May, pp. 37–39.

Kaps, H. and Kastner, S. (1984). On the physical and operational background of cargo securing aboard ships, report BMV (Bundesminister für Verkehr), Hochschule Bremen.

Kastner, S. (1986). Operational stability of ships and safe transport of cargo, *Proc. of STAB'86: 3rd International Conference on Stability of Ships and Ocean Vehicles*, Vol. 1, Gdansk pp. 207–215.

Kletz, T.A. (1992). HAZOP and HAZAN: Identifying and Assessing Process Industry Hazards. *AIChE.*, 3rd Edition.

Knudsen, R. (1993). Safety management. *Proc. of US Cost Guard Vessel Stability Symposium*, New London, Connecticut, pp. 208–220.

Kobylinski, L. (2000). Stability standards – future outlook, *Proc. of STAB'2000: 7th International Conference on Stability of Ships and Ocean Vehicles*, Vol. 1 Launceston, Tasmania, pp. 52–61.

Kouabenan, D.R. (1998). Beliefs and the perception of risks and accidents. *Risk Analysis*, Vol. 18, No. 3, pp. 243–252.

Kobylinski, L.K. and Kastner, S. (2003). *Stability and Safety of Ships.* Elsevier, Oxford, UK.

Kramer, M.A. and Palowitch, B.L. (1987). A Rule Based Approach to Fault Diagnosis Using the Signed Directed Graph. *AIChE Journal*, Vol. 33, No. 7, pp. 1067–1077.

Kristensen, H.O. (2002). Cargo transport by sea and road – technical and economic environmental factors. *Marine Technology*, Vol. 39, No. 4, pp. 239–249.

Kristiansen, S. and Olofsson, M. (1997). *SAFECO – Safety of Shipping in Coastal Waters.* Marintek As, Trondheim, Norway, Operational Safety and Ship Management – WPII.5.1: Criteria for Management Assessment. Report No. MT23 F97-0175.

Kristiansen, S. (2005). *Maritime Transportation.* Elsevier Butterworth-Heinemann, Oxford, UK.

Kuo, C. (1998). *Managing ship safety.* LLP Ltd., ISBN 1-85978-841-6.

Kuo, C. (2000). The impact of human factors on ship stability, *Proc. of STAB'2000: 7th International Conference on Stability of Ships and Ocean Vehicles*, Vol. 1, Launceston, Tasmania, pp. 4–22.

Kuteynikov, M.A. and Lipis, V.B. (2000). On indices of ships operational seaworthiness, *Proc. of STAB'2000: 7th International Conference on Stability of Ships and Ocean Vehicles*, Vol. 1, Launceston, Tasmania, pp. 176–189.

Law, A.M. and Kelton, W.D. (1982). *Simulation Modelling and Analysis.* McGraw-Hill Book Company.

Lloyds Register (2007). *Rules and Regulations for the Classification of Ships.* Lloyds Register, London.

Loughran, C., Pillay, A., Wang, J., Wall, A. and Ruxton, T. (2002). A Preliminary Study of Fishing Vessel Safety. *Journal of Risk Research*, Vol. 5, No. 1, pp. 3–21.

Loughran, C.G., Pillay, A., Wang, J., Wall, A.D. and Ruxton, T. (2002). Formal Fishing Vessel Safety Assessment. *Journal of Risk Research*, Vol. 5, No. 1, pp. 3–21.

MAIB (1995). Marine Accident Investigation Branch – Safety Digest Summary, No. 2/95, United Kingdom.

MAIB (1997). MAIB Annual Report, Department of the Environment, Transport and the Regions, United Kingdom.

MAIB (1998). Marine Accident Investigation Branch: Marine Accident Report 1/98, Department of the Environment, Transport and the Regions, United Kingdom, September, ISBN 1 85 112101 3.

MAIB (1999a). Marine Accident Investigation Branch: Annual Report 1999, Department of the Environment, Transport and the Regions, United Kingdom.

MAIB (1999b). Marine Accident Investigation Branch: Marine Accident Report 3/99, Department of the Environment, Transport and the Regions, United Kingdom, July, ISBN 1 85112 109 9.

MARPOL (2002). *MARPOL 73/78*, Consolidated Edition 2002. International Maritime Organization, London.

Mc Taggart, K. and De Kat, J.O. (2000). Capsize risk of intact frigates in irregular seas. *Transcations SNAME*, Vol. 108, pp. 147–177.

MCA (1998). Formal Safety Assessment for Bulk Carriers (Including Annexes A-I), Informal Paper submitted by UK to IMO/MSC, 70th session, London, 27th November, (IMO/MSC 70/INF PAPER).

McKelvey, T.C. (1988). How to Improve the Effectiveness of Hazard and Operability Analysis. *IEEE Transaction on Reliability*, Vol. 37, No. 1, pp. 167–170.

MIL (1960). *Buships Specification.* Department of Defence, United States of America, MIL-R-22732.

Military Standard (1969). Department of Defence; System Safety Program Requirements, MIL-STD-882, USA.

Military Standard (1980). Procedures for Performing a Failure Mode, Effects and Criticality Analysis, MIL-STD-1629, November, AMSC Number N3074.

Military Standard (1993). System Safety Program Requirements, MIL-STD-882c, January, AMSC Number F6861.

Military Standard (1999). Department of Defence, Military Standards; System Safety Program Requirements, MIL-STD-882D, USA America.

Miller, T.C. and Paitl, G.J. (2001). A vessel is its own best lifeboat: Prevention of casualties through education. *Marine Technology*, Vol. 38, No. 1, pp. 26–30.

Modarres, M. (1993). *What Every Engineer Should Know about Reliability and Risk Analysis*. Marcel Dekker Inc., NY, USA.

MOU (2004). The Paris Memorandum of Understanding on Port State Control, Yearbook 2002. http://www.parismou.org/

MSA (1993). Formal Safety Assessment, MSC 66/14, Submitted by the United Kingdom Marine Safety Agency to IMO Maritime Safety Committee.

MSA (1998). *Implementation of EC Directive 89/391 – Merchant Shipping and Fishing Vessels (Health and Safety at Work) Regulations 1997*. Marine Safety Agency, Department of the Environment, Transport and the Regions, Southampton, UK.

MSC (1997a). Formal Safety Assessment: Trial Application to High-speed Passenger Catamaran Vessels, 68/Informal paper, 29th May, International Maritime Organization, London.

MSC (1997b). FSA Trial Application to High Speed Passenger Catamaran Vessel, MSC 68/14/2 & 68/INF., Maritime Safety Committee, International Maritime Organization, United Kingdom.

Nabergoj, R., Rakhmanin, N.N., Trincas, G. and Messina, G. (2000). A methodology to provide guidelines for increasing operating safety of fishing vessels, Proc. of STAB'2000: *7th International Conference on Stability of Ships and Ocean Vehicles*, Vol. 1, Launceston, Tasmania, pp. 160–175.

NAS (1991). *Tanker Spills: Prevention by Design*. Committee on Tank Vessel Design, Marine Board, National Research Council, National Academy of Sciences. The National Academies Press, Washington, DC.

National Audit Office. (1992). Department of Transport: Ship Safety, HC 186, Parliamentary Session 1991–92, United Kingdom, ISBN 010 218692 8.

Nippon Yusen Kaisha Research Group (1999). World Containership Fleet and Its Operations 1999, Internal Report, Tokyo, 11th June 1999.

Norris J.R. (1998). Markov Chains, Statistical and Probabilistic Mathematics: Series 2, Cambridge University, ISBN. 0521633966253.

NOU (1991). The Scandinavian Star Disaster of 7 April 1990. Norwegian Official Reports NOR 1991: 1 E, Government Administration Services, Government Printing Service, Oslo.

OREDA (1982). *Offshore Reliability Data Handbook (OREDA)*. PennWell Books, 1982 onwards.

P & I Club (1992). *Analysis of Major Claims 1992*. UK Mutual Steam Ship Association Ltd. Publication, Bermuda.

Papanikolaou, A., Gratsos, G., Boulougouris, E. and Eliopoulou, E. (2000). Operational measures for avoiding dangerous situations in extreme weather conditions, Proc. of STAB'2000: *7th International Conference on Stability of Ships and Ocean Vehicles* Vol. 1, pp. 137–148.

Pereira, R., Puntigliano, F. and Bertram, V. (2001). Nonlinear simulation of capsizing of undamaged ships in seaways. *STG summer meeting*, Gdansk.

Perrow, C. (1999). *Normal Accidents*. Princeton University Press, Princeton, NJ.

Pillay A. and Wang J. (2001). Human Error Assessment and Decision-making Using Analytical Hierarchy Processing, Proceeding of the International Mechanical Engineering Congress and Exposition – ASME, November 11–16, New York.

Pillay A. (2001). Formal Safety Assessment of Fishing Vessels, PhD Thesis, School of Engineering, Liverpool John Moores University, UK.

Pillay, A. and Wang, J. (2003). *Technology and Safety of Marine Systems*. Elsevier Science Ltd., Oxford, UK.

Preyssl, C. (1995). Safety Risk Assessment and Management- the ESA Approach. *Reliability Engineering and System Safety*, Vol. 49, No. 3, pp. 303–309.

Rainey, R.C.T., Thompson, J.M.T., Tam, G.W. and Noble, P.G. (1990). The transient capsize diagram – a route to soundly-based new stability regulations, Proc. of STAB'90: 4th International Conference on Stability of Ships and Ocean Vehicles, Naples.

Rainey, R.C.T. and Thompson, J.M.T. (1991). The transient capsize diagram – a new method of quantifying stability in waves. *Journal of Ship Research*, Vol. 35, No. l, pp. 58–62.

Renilson, M. (1994). Contribution to workshop on manoeuvring and survival in storm seas, *Proc. of STAB'94: 5th International Conference on Stability of Ships and Ocean Vehicles*, Melbourne, Florida.

Renilson, M. (1997). *A note on the capsizing of Vessels in Following and Quartering Seas*. Oceanic Engineering International, Vol. 1. No. 1, St. John's, Newfoundland.

Rowe, W.D. (1983) Acceptable levels of risk for undertakings. Colloquium *Ship Collisions with Bridges and Offshore Structures*, Copenhagen. IABSE Reports Vol. 41, International Association for Bridge and Structural Engineering, Zürich.

Rubinstein, R. (1981). *Simulation and the Monte Carlo Method*. John Wiley & Sons, NY, USA.

Salvendy (ed.) (1987). *Handbook of human factors*. John Wiley and Sons, New York.

Savic, D. (1989). *Basic Technical Systems Simulation*. Butterworths Basic Series.

Sekimizu K. (1997). Current Work at IMO on Formal Safety Assessment, Proceeding of Marine Risk Assessment: A Better Way to Manage Your Business, *The Institute of Marine Engineers*, London, 8–9 April.

Sii H.S. (2001). Marine and Offshore Safety Assessment, PhD Thesis, Staffordshire University/Liverpool John Moores University, UK.

Sjøfartsdirektoratet (1998). *Rules of the Norwegian Ship Control [In Norwegian: Den Norske Skipskontrolls Regler]*. Erlanders Forlag, Oslo.

Smith, D.J. (1985). *Reliability and Maintainability and Perspective*, 2nd Edition. Macmillan Publishers Ltd., London, UK.

Smith, D.J. (1992). *Reliability, Maintainability and Risk*, 4th Edition. Butterworths-Heineman Limited, Oxford, United Kingdom.

SOLAS (2001). *SOLAS*, Consolidated Edition 2001. International Maritime Organization, London.

Spouge, J. (1997). Risk Criteria for Use in Ship Safety Assessment, Proceeding of Marine Risk Assessment: A Better Way to Manage Your Business, *The Institute of Marine Engineers*, London, 8–9 April.

STCW (1996). *Convention on Standards of Training, Certification and Watchkeeping for Seafarers*, 1996 Edition. International Maritime Organization, London.

Stoop, J. (1989). *Safety and Bridge Design, International Symposium of Safety and Working Conditions Aboard Fishing Vessels*. L'Université du Quèbec á Rimouski, Quebec, Canada, Aug 22–24.

Stopford, M. (1997). *Maritime Economics*, 2nd Edition. Routledge, London.

Takaishi, Y. (1994). IMO guidance to the masters for avoiding dangerous situations in following and quartering seas, Workshop on Manoeuvring and survival in storm seas, synopsis by the moderator J. O. de Kat *STAB'94: 5th International Conference on Stability of Ships and Ocean Vehicles*, Melbourne, Florida.

Takaishi, Y. (2000). Development of an onboard system to execute the operational guidance in following and quartering seas, Workshop D, STAB'2000: *7th International Conference on Stability of Ships and Ocean Vehicles Conference*, Launceston, Tasmania.

Takaishi, Y., Matsuda, K. and Watanabe, K. (1997). Probability to encounter high run of waves in the dangerous zone shown on the operational guidance/IMO for following and quartering seas, Proc. of STAB'97: *6th International Conference on Stability of Ships and Ocean Vehicles Conference*, Vol. 1 Varna, pp. 173–179.

Trbojevic V.M. and Soares C.G. (2000). Risk Based Methodology for a Vessel Safety Management System, Proceeding ESREL 2000 and SRA-Europe Annual Conference, ISBN: 90 5809 141 4, Scotland, UK, 15–17 May, Vol. 1, pp. 483–488.

Tummala, V.M.R. and Leung, Y.H. (1995). A Risk Management Model to Assess Safety and Reliability Risks. *International Journal of Quality and Reliability Management*, Vol. 13, No. 8, pp. 53–62.

U.S Nuclear Regulatory Commission (1975). Reactor Safety Study: An Assessment of Accident Risks in U.S. Commercial Nuclear Power Plants, October 1975, WASH-1400 (NUREG 75/014).

UK P&I CLUB, (1999) Analysis of Major Claims – Ten Years Trends in Maritime Risk, London, 1999.

UKOOA (1999). Industry Guidelines on a Framework for Risk Related Decision Making, Published by the UK Offshore Operators Association, April.

Umeda, T., Kuryama, T.E., O'Shima, E. and Matsuyama, H. (1980). A Graphical Approach to Cause and Effect Analysis of Chemical Processing Systems. *Chem. Eng. Sci.*, Vol. 35, pp. 2379–2386.

Umeda, N. (1994). Operational stability in following and quartering seas – a proposed guidance and its validation, *Proc. of STAB'94: 5th International Conference on Stability of Ships and Ocean Vehicles*, Vol. 2, Melbourne, Florida.

Vaidhyanathan, R. and Venkatasubramanian, V. (1996). *Digraph-based Models for Automated HAZOP Analysis*. Computer Integrated Process Operations Consortium (CIPAC), School of Chemical Engineering Prudue University, USA.

Vassalos, D. (1999). Shaping ship safety: the face of the future. *Marine Technology*, Vol. 36, No. 2, pp. 61–76.

Vie R.H. and Stemp J.B. (1997). The Practical Application of Risk Assurance Technology Techniques to Cruise Vessel Design and Operation, Proceeding of Marine Risk Assessment: A Better Way to Manage Your Business, *The Institute of Marine Engineers*, London, 8–9 April.

Villemeur, A. (1992). *Reliability, Availability, Maintainability and Safety Assessment, Methods and Techniques*, Vol. 1. John Wiley & Sons, Chichester, UK.

Wang, J. and Foinikis, P. (2001). Formal Safety Assessment of Containerships. *International Journal of Marine Policy*, Vol. 25, pp. 143–157.

Wang, J. and Foinikis, P. (2001). Formal Safety Assessment of Containerships. *Marine Policy*, Vol. 21, pp. 143–157.

Wang, J. and Ruxton, T. (1997). A Review of Safety Analysis Methods Applied to the Design Process of Large Engineering Products. *Journal of Engineering Design*, Vol. 8, No. 2, pp. 131–152.

Wang J. (1994). Formal Safety Assessment Approaches and Their Application to the Design Process, PhD Thesis, Engineering Design Centre, University of Newcastle upon Tyne, July.

Wang, J. (1997). A Subjective Methodology for Safety Analysis of Safety Requirements

Specifications. *IEEE Transactions on Fuzzy Systems*, Vol. 5, No. 3, pp. 418–430.

Wang, J. (2001). Current Status of Future Aspects of Formal Safety Assessment of Ships. *Safety Science*, Vol. 38, pp. 19–30.

Wang J. (2002). A Brief Review of Marine and Offshore Safety Assessment, *Marine Technology*, SNAME, Vol. 39, No. 2, April, pp. 77–85.

Wang, J. (2002). Offshore Safety Case Approach and Formal Safety Assessment of Ships. *Journal of Safety Research*, Vol. 33, No. 1, pp. 81–115.

Wang, J., Ruxton, T. and Labrie, C.R. (1995). Design for Safety of Marine Engineering Systems with Multiple Failure State Variables. *Reliability Engineering & System Safety*, Vol. 50, No. 3, pp. 271–284.

Wang J., Sii H.S., Pillay A. and Lee J.A. (2002). Formal Safety Assessment and Novel Supporting Techniques in Marine Applications, Proceeding of the International Conference on Formal Safety Assessment Conference, ISBN: 0 903055805, RINA, London, September 18–19, pp. 123–142.

Wang, J., Yang, J.B., Sen, P. and Ruxton, T. (1996). Safety Based Design and Maintenance Optimisation of Large Marine Engineering Systems. *Applied Ocean Research*, Vol. 18, No. 1, pp. 13–27.

Wang, J., Yang, J.B. and Sen, P. (1996). Multi-person and Multi-attribute Design Evaluations Using Evidential Reasoning based on Subjective Safety and Cost Analyses. *Reliability Engineering & System Safety*, Vol. 52, No. 2, pp. 113–129.

Weinstein, K. (1996). Education and training for maritime management: The role of the World Maritime University, *Symposium by Carl Duisberg Gesellschaft*, Bremen.

Wells, G.L. (1980). *Safety in Process Plants Design*. John Wiley & Sons, NY, USA.

Wendel, K. (1965). *Bemessung und Überwachung der Stabilität*. Transactions STG, Hamburg.

Yang, J.B. and Sen, P. (1994). A General Multi-level Evaluation Process for Hybrid MADM with Uncertainty. *IEEE Transactions on Systems, Man and Cybernetics*, Vol. 24.

Zharen, W.M. and Duncan, W. (1994). Environmental risk assessment and management in the maritime industry. The Interaction with ISO 9000, ISM, and ISM management systems, *Transactions SNAME*, Vol. 102, pp. 137–164.

12 Glossary of terms and definitions

Contents

12.1 Abbreviations
12.2 Symbols
12.3 Terms and definitions

The various parts of this Chapter have been taken from the following books, with the permission of the authors:

Barrass, C.B. (2004) *Ship Design and Performance*. Butterworth-Heinemann, Oxford, UK. [Parts of Section 12.3]
Eyres, D.J. (2007) *Ship Construction*. 6th Edition. Butterworth-Heinemann, Oxford, UK. [Parts of Section 12.3]
Rawson K.J. and Tupper E.C. (2001) *Basic Ship Theory*. 5th Edition, Combined Volume. Butterworth-Heinemann, Oxford, UK. [Section 12.2]
Tupper, E.C. (2004) *Introduction to Naval Architecture*. Butterworth-Heinemann, Oxford, UK. [Parts of Section 12.3]

12.1 Abbreviations

ABS	American Bureau of Shipping
ATTC	American Towing Tank Conference
AUV	Autonomous underwater vehicle
Bn	Beaufort number
BEM	Boundary element method, or blade element-momentum theory
BHP	Brake horsepower
BSRA	British Ship Research Association
BV	Bureau Veritas
CAD	Computer aided design
CAM	Computer aided manufacture
CB	Centre of buoyancy
CCS	China Classification Society
CF	Centre of flotation
CFD	Computational fluid dynamics
CLR	Centre of lateral resistance
CG	Centre of gravity
DHP	Delivered horsepower
DNS	Direct numerical simulation
DNV	Det Norske Veritas
DOF	Degrees of freedom
DTMB	David Taylor Model Basin
EHP	Effective horsepower
EPSRC	Engineering and Physical Sciences Research Council
FDM	Finite difference method
FEA	Finite element analysis
FEM	Finite element method
FEU	Forty foot equivalent unit [container]
FRP	Fibre reinforced plastic
FVM	Finite volume method
GL	Germanischer Lloyd
GPS	Global positioning system
hp	Horsepower
HSVA	Hamburg Ship Model Basin
IACC	International Americas Cup Class
IACS	International Association of Classification Societies
IESS	Institution of Engineers and Shipbuilders in Scotland
IMarE	Institute of Marine Engineers (became IMarEST from October 2001)
IMarEST	Institute of Marine Engineering, Science and Technology
IMO	International Maritime Organization
INSEAN	Instituto di Architectura Navale (Rome)
ISO	International Standards Organization
ISSC	International Ship and Offshore Structures Congress
ITTC	International Towing Tank Conference
JASNAOE	Japan Society of Naval Architects and Ocean Engineers
KRS	Korean Register of Shipping
LCG	Longitudinal centre of gravity
LE	Leading edge of foil or fin
LES	Large eddy simulation
LNG	Liquified natural gas
LPG	Liquified petroleum gas
LR	Lloyd's Register of Shipping
MAIB	Marine Accident Investigation Branch (UK)
MARIN	Maritime Research Institute of the Netherlands
MARPOL	Marine Pollution (IMO)
MCR	Maximum continuous rating
MER	Marine Engineers Review
NACA	National Advisory Council for Aeronautics (USA)
NC	Numerical control
NECIES	North East Coast Institution of Engineers and Shipbuilders (UK)
NI	The Nautical Institute
NKK	Nippon Kaiji Kyokai
NPL	National Physical Laboratory
N-S	Navier Stokes
ONR	Office of Naval Research (USA)
P&I	Protection and indemnity
PPT	Parts per thousand
PSU	Practical salinity units
RANS	Reynolds Averaged Navier Stokes
RINA	Registro Italiano Navale
RINA	Royal Institution of Naval Architects
RMRS	Russian Maritime Register of Shipping
RNLI	Royal National Lifeboat Institution
ROV	Remotely operated underwater vehicle
rps	Revolutions per second
rpm	Revolutions per minute
SHP	Shaft horsepower
SNAME	Society of Naval Architects and Marine Engineers (USA)
SNAJ	Society of Naval Architects of Japan (later to become JASNAOE)
SNAK	Society of Naval Architects of Korea
SOLAS	Safety of Life at Sea (IMO)
SSPA	Statens Skeppsprovningsansalt, Götoborg
SWATH	Small waterplane area twin hull
TE	Trailing edge of foil or fin
TEU	Twenty foot equivalent unit [container]
THP	Thrust horsepower
TRINA	Transactions of the Royal Institution of Naval Architects
ULCC	Ultra large crude carrier
UUV	Unmanned underwater vehicle
VLCC	Very large crude carrier
WEGEMT	West European Graduate Education in Marine Technology (now, European Association of Universities in Marine Technology)
WHOI	Woods Hole Oceanographic Institution

12.2 Symbols

The principal symbols used in the various sections of the text are shown below. Other symbols are defined locally within the text.

General

a	Linear acceleration
A	Area in general
B	Breadth
D, d	Diameter
E	Energy
F	Force
g	Acceleration due to gravity
h	Depth or pressure head
h_w, ζ_w	Height of wave, crest to trough
H	Total head, Bernoulli
L	Length in general
L_w, λ	Wave-length
m	Mass
n	Rate of revolution
p	Pressure intensity
P_v	Vapour pressure of water
P_∞	Ambient pressure at infinity
P	Power in general
q	Stagnation pressure
Q	Rate of flow
r, R	Radius in general
s	Length along path
t	Time in general
$t°$	Temperature
T	Period of time for a complete cycle
u	Reciprocal weight density, specific volume
u, v, w	Velocity components in direction of x-, y-, z-axes
U, V	Linear velocity
w	Weight density
W	Weight in general, or weight of ship (Chapter 3).
x, y, z	Body axes and Cartesian co-ordinates Right-hand system fixed in the body, z-axis vertically down, x-axis forward. Origin at c.g.
x_0, y_0, z_0	Fixed axes Right-hand orthogonal system nominally fixed in space, z_0-axis vertically down, x_0-axis in the general direction of the initial motion.
α	Angular acceleration
γ	Specific gravity
Γ	Circulation
δ	Thickness of boundary layer
θ	Angle of pitch or trim
μ	Coefficient of dynamic viscosity
ν	Coefficient of kinematic viscosity
ρ	Mass density
ϕ	Angle of roll, heel or list
χ	Angle of yaw
ω	Angular velocity or circular frequency
∇	Volume of displacement

Geometry of ship

A_M	Midship section area
A_W	Waterplane area
A_x	Maximum transverse section area
B	Beam or moulded breadth
BM	Metacentre above centre of buoyancy
C_B	Block coefficient
C_M	Midship section coefficient
C_P	Prismatic coefficient
C_{VP}	Vertical prismatic coefficient
C_W	Waterplane coefficient
D	Depth of ship
F	Freeboard
GM	Transverse metacentric height
GM_L	Longitudinal metacentric height
I	Transverse second moment of area about centreline (Chapter 3)
I_L	Longitudinal moment of inertia of waterplane about CF
I_P	Polar moment of inertia
I_T	Transverse moment of inertia
J_T	Transverse second moment of area about centreline (Chapter 9).
KB	Height of centre of buoyancy above keel (or baseline)
KM	Height of metacentre above keel (or baseline)
L	Length of ship
LCB	Longitudinal centre of buoyancy
LCG	Longitudinal centre of gravity
L_{OA}	Length overall
L_{PP}	Length between perpendiculars (or L_{BP})
L_{WL}	Length of waterline
M	Metacentre
S	Wetted surface
T	Draught
Δ	Displacement mass (tonnes)
λ	Scale ratio—ship/model dimension
∇	Displacement volume (m^3)

Ship structures

a	Length of plate
b	Breadth of plate
C	Modulus of rigidity
ε	Linear strain
E	Modulus of elasticity, Young's modulus
σ	Direct stress

σ_y	Yield stress		C_F	Frictional resistance coefficient
g	Acceleration due to gravity		C_L	Lift coefficient
I	Planar second moment of area		C_R	Residuary resistance coefficient
J	Polar second moment of area		C_T	Specific total resistance coefficient
j	Stress concentration factor		C_W	Specific wave-making resistance coefficient
k	Radius of gyration		D	Drag force
K	Bulk modulus		F_n	Froude number
l	Length of member		F_{nh}	Depth Froude number
L	Length		I	Idle resistance
M	Bending moment		J	Advance coefficient
M_p	Plastic moment		K_Q	Torque coefficient
M_{AB}	Bending moment at A in member AB		K_T	Thrust coefficient
m	Mass		L	Lift force
P	Direct load, externally applied		n	Propeller rate of revolution (rps)
P_E	Euler collapse load		N	Propeller rate of revolution (rpm)
p	Distributed direct load (area distribution), pressure		P_D	Delivered power at propeller
			P_E	Effective power
p'	Distributed direct load (line distribution)		P_I	Indicated power
τ	Shear stress		P_S	Shaft power
r	Radius		P_T	Thrust power
S	Internal shear force		Q	Torque
s	Distance along a curve		R	Resistance in general
T	Applied torque		R_n	Reynolds number
t	Thickness, time		R_F	Frictional resistance
U	Strain energy		R_R	Residuary resistance
W	Weight, external load		R_T	Total resistance
y	Lever in bending		R_W	Wave-making resistance
δ	Deflection, permanent set, elemental (when associated with element of breadth, e.g. δb)		s_A	Apparent slip ratio
			t	Thrust deduction fraction
ρ	Mass density		T	Thrust
ν	Poisson's ratio		U	Velocity of a fluid
θ	Slope		U_∞	Velocity of an undisturbed flow
			V	Speed of ship
			V_A	Speed of advance of propeller
			w_T	Taylor wake fraction

Propeller geometry

A_D	Developed blade area		w_F	Froude wake fraction
A_E	Expanded area		W_n	Weber number
A_O	Disc area		β	Appendage scale effect factor
A_P	Projected blade area		β	Advance angle of a propeller blade section
BAR	Blade area ratio		δ	Taylor's advance coefficient
b	Span of aerofoil or hydrofoil		η	Efficiency in general
c	Chord length		η_b	Propeller efficiency behind ship
d	Boss or hub diameter		η_D	Quasi propulsive coefficient (QPC)
D	Diameter of propeller		η_h	Hull efficiency
f_M	Camber		η_O	Propeller efficiency in open water
P	Propeller pitch		η_r	Relative rotative efficiency
R	Propeller radius		σ	Cavitation number
t	Thickness of aerofoil		ρ	Density
Z	Number of blades of propeller			
α	Angle of attack			
ϕ	Pitch angle of screw propeller		**Seakeeping**	

Resistance and propulsion

			c	Wave velocity
			f	Frequency
			f_E	Frequency of encounter
a	Resistance augment fraction		I_{xx}, I_{yy}, I_{zz}	Real moments of inertia
C_D	Drag coefficient		I_{xy}, I_{xz}, I_{yz}	Real products of inertia

k	Radius of gyration	C_L	Lift coefficient
m_n	Spectrum moment where n is an integer	C_N	Normal force coefficient
M_L	Horizontal wave bending moment	d	Drag force
		K, M, N	Moment components on body relative to body axes
M_T	Torsional wave bending moment	L	Lift force
M_V	Vertical wave bending moment	N	Normal force
		O	Origin of body axes
s	Relative vertical motion of bow with respect to wave surface	p, q, r	Components of angular velocity relative to body axes
		X, Y, Z	Force components on body
$S_\zeta(\omega), S_\theta(\omega)$, etc.	One-dimensional spectral density	α	Angle of attack
		β	Drift angle
$S_\zeta(\omega,\mu), S_\theta(\omega,\mu)$, etc.	Two-dimensional spectral density	δ	Rudder angle
		ϕ	Heel angle
T	Wave period	χ	Heading angle
T_E	Period of encounter	ω_C	Steady rate of turn
T_z	Natural period in smooth water for heaving		
T_θ	Natural period in smooth water for pitching		
T_ϕ	Natural period in smooth water for rolling		
$Y_{\theta\zeta}(\omega)$	Response amplitude operator—pitch		
$Y_{\phi\zeta}(\omega)$	Response amplitude operator—roll		
$Y_{\chi\zeta}(\omega)$	Response amplitude operator—yaw		
β	Leeway or drift angle		
δ_R	Rudder angle		
ε	Phase angle between any two harmonic motions		
ζ	Instantaneous wave elevation		
ζ_A	Wave amplitude		
ζ_w	Wave height, crest to trough		
θ	Pitch angle		
θ_A	Pitch amplitude		
κ	Wave number		
ω_E	Frequency of encounter		
Λ	Tuning factor		

Conversion of units

1 m = 3.28 ft
1 ins = 25.4 mm
1 kg = 2.205 lb
1 ton = 2240 lb
1 lbf/ins^2 = 6895 N/m^2
1 mile = 5280 ft
1 mile/hr = 1.61 km/hr
1 hp = 0.7457 kW

1 ft = 12 ins
1 km = 1000 m
1 tonne = 1000 kg
1 lbf = 4.45 N
1 bar = 14.7 lbf/ins^2
1 Nautical mile (Nm) = 6078 ft
1 knot = 1 Nm/hr
1 knot = 0.5144 m/s
1 UK gal. = 4.546 litres

12.3 Terms and definitions

In many cases the fuller definition of these terms, and the context in which they are used, will be found in the main text. They can be found by reference to the index. Where appropriate, the usually accepted abbreviation is given. Terms shown in bold in the explanations are defined elsewhere in the glossary.

Added mass. The effective increase in mass of a hull, due to the entrained water, when in motion.

Added weight method. One method used in the calculation of a ship's damaged **stability** when it is partially flooded. It regards the water which has entered as an added weight, the basic hull envelope remaining. The other approach uses the concept of **lost buoyancy**.

Aframax. A term used for the largest dry bulk carriers.

Aft. At or near the **stern**.

After perpendicular (AP). See **perpendiculars**.

Air draught. The vertical distance from the summer waterline to the highest point in the ship, usually the top of a mast.

Manoeuvring

A_C	Area under cut-up
A	Area of rudder
AR	Effective aspect ratio of a rudder or control surface.
AR_G	Geometric aspect ratio of a rudder or control surface.
s	Span of rudder or control surface
c	Chord of rudder or control surface
C_D	Drag coefficient

Amidships. The point that lies midway between the fore and after **perpendiculars.** It is also used to refer more generally to the central section of a ship.

Appendage. A small addition to the main part or main structure.

Approximate integration. Simple, approximate, methods and rules used for calculating the areas of plane shapes and the volumes of three-dimensional shapes. The most commonly used in naval architecture are the **Simpson's rules**.

Aspect ratio. The ratio of the span to the chord length of a lifting surface, for example, rudder or stabilizer fin.

Athwartships. Across the ship, at right angles to the **centreline.**

Attributes. Those features and characteristics of a ship that give it the capabilities demanded of it.

Auxiliary machinery. Machinery other than the ship's main engines.

Availability. The likelihood that a given function will be available when needed.

Bale capacity. Capacity in hold to edge of frames and stiffeners; reflects the stowage of bales or boxes.

Beam. The transverse width of the ship at any point. Unless otherwise specified the term applies to the maximum width or breadth of the hull.

Bending moments. The moments due to forces acting on a ship which are trying to distort the hull. The most important are those causing the ship to bend the hull in its vertical plane.

Bilge keel. A longitudinal member fitted to the **bilge** to help damp rolling and to provide some protection to the **bilge strake.**

Bilge. Rounded portion of hull between side and bottom; also known as turn of bilge.

Bilge strake. Strake (or line) of plating at the bilge.

Block coefficient (C_B). One of the **coefficients of fineness**. It is the ratio of the volume of a ship's underwater form to the volume of the circumscribing rectangular solid. $[C_B = \nabla/(L \times B \times T)]$

Body plan. A figure showing the cross sections of a ship's hull.

Bollard pull. Test carried out on a tug to measure the pulling force.

Bonjean curve. A curve drawn for a transverse section of a ship the ordinate of which, at any given height, represents the area of the section up to that height.

Bow and buttock lines. Lines which mark the intersections of a ship's hull by vertical planes parallel to its **centreline plane**. The bowlines relate to the forward part of the hull and the buttock lines to the after part. They are used in the fairing of hulls and also help visualize the flow past the hull.

Breadth. see **beam**

Brittle fracture. A mode of failure of a material under load. It is associated with steels of low **toughness** at low temperature.

Built-in stresses. Stresses which are generated in a structure due to shaping and securing the plates and stiffeners and which remain there. Often called **residual stresses**. Many arise from the welding of the structure.

Bulkhead deck. The uppermost weathertight deck to which transverse watertight bulkheads are taken. The concept is important to a study of the ability of a hull to sustain damage.

Bulkhead. A partition between compartments providing a subdivision of the ship. Bulkheads provide means of compartmentalizing the ship to separate out different activities. If water-tight they will limit the extent of flooding following damage. Bulkheads may be transverse or longitudinal.

Bulwark. Strake of shell plating extending above the top deck.

Buoyancy. The upward force acting on a floating, or submerged, body due to the water pressures on its boundary. It is equal in value to the weight of water displaced if the body is in equilibrium.

Camber. The amount by which a deck is higher at the centre than at the sides. Weather decks are cambered to assist in the shedding of water. Other decks usually have no camber in order to facilitate construction.

Capesize. A term applied to large cargo vessels that cannot transit either the Panama or Suez Canals. They are usually of the order of 120 000–180 000 DWT.

Capsize. A ship is said to capsize when it loses **transverse stability** and rolls over and sinks.

Cavitation. The formation of bubbles on an aerofoil section in areas of reduced pressure. Can occur on heavily loaded ship propellers.

Centre of buoyancy (CB). That point through which the **buoyancy** force acts. It is defined in space by its longitudinal, vertical and transverse (respectively, LCB, VCB and TCB) position relative to a set of orthogonal axes. It is also the centroid of volume of the displaced water.

Centre of flotation (CF). The centroid of area of a **waterplane**. A small weight added, or removed, from the ship vertically in line with the CF will cause a change of **draught** without **heel** or trim. For a symmetrical ship the CF will be on the centreline and its position is given relative to **amidships**.

Centre of gravity (CG). The point through which the force due to gravity, that is the weight of the body, acts. Its position is defined in a similar way to the **centre of buoyancy** and is very important in calculations of **stability**.

Centreline plane. The forward and aft vertical plane splitting the ship into two. Most ships are symmetrical about their centreline planes. The hull form is defined by the distances of its outer surface from this plane.

Centreline. The fore and aft line at the middle of the ship.

Chain locker. Space or compartment forward of collision bulkhead in which anchor chain is stored.

Charpy test. A simple test indicating the **toughness** of a metal.

Classification societies. Bodies concerned with the construction and classification of ships ensuring that they meet all the international and national standards. A new vessel will be classified by one of the societies and will be subsequently checked by that society to ensure that those standards are maintained.

Coefficients of fineness. These relate to the underwater form and give a broad indication of the hull shape. They are the ratios of certain areas and volumes to their circumscribing rectangles or prisms.

Cofferdam. A narrow void compartment between two adjoining tanks to insulate against the leakage of liquids.

Collision bulkhead. First watertight bulkhead from bow of ship.

Computer-aided design (CAD). Computer-based systems assisting in the design of ships and other products.

Computer-aided manufacture (CAM). Computer-based systems assisting in the building of ships or the manufacture of other products.

Containers. Boxes of standard dimension for the carriage of goods by road, rail and ship. See also TEU and FEU.

Coursekeeping. To steer a prescribed course.

Cross curves of stability. A series of curves showing how a ship's **transverse stability** varies, with displacement, for a range of **heel** angles.

Curve of statical stability. A plot showing how the righting lever experienced by a ship varies with angle as the ship is rotated about a fore and aft axis. It defines a ship's **stability** at large angles. Also known as the **GZ curve**.

Damping. The dissipation of energy experienced by a moving or vibrating body.

Deadweight. The weight of cargo, fuel, water, crew and effects. It equals in value the **displacement** less the **lightship weight** for the ship. Deadweight mass (tonne) is usually used for ship design purposes.

Deck. The part(s) of a ship that correspond to a floor in a building.

Depth. Vertical distance from underside of bottom of ship to a particular deck (e.g. main deck).

Directional stability. A measure of a ship's stability in course keeping. See also **dynamic stability**.

Discounted cash flow. A means of assessing the net present worth of any artefact.

Displacement sheet. A convenient tabular means of calculating a ship's displacement and the position of its **centre of buoyancy**.

Displacement. The weight of water displaced by a freely floating body. It is equal to the weight of the ship if the ship is in equilibrium. Displacement mass (tonnes) or displacement volume (m^3) is usually used for ship design purposes.

Double bottom. The double bottom is created with a watertight inner (or double) skin **(tank top)** forming a double hull at the bottom the ship. The double skin or hull **(inner bottom)** provides protection in the event of bottom damage. Double bottom tanks may be used for ballast water.

Draught. The depth of any point on the ship's hull below the waterline. Unless otherwise specified it is usually the draught at mid length.

Drift angle. The angle between a ship's head and the direction in which it is moving.

Dynamic stability. Another term for the **directional stability**.

Dynamical stability. A measure of a ship's ability to absorb the energy of waves. For any angle of **heel** it represents the energy needed to heel the ship to that angle.

Energy spectrum. A convenient way of showing how energy is distributed between frequencies. Used in defining waves and resulting ship motions.

Expert systems. Computer-based systems developed to help in the control and management of a ship.

Fatigue. Refers to the failure of material or a structure under repeated stress cycles.

Flare. The extra width of the deck compared with the **beam** at the waterline. Usually refers to the hull in the vicinity of the bow where the flare is likely to be most pronounced.

Floodable length. The length of the hull, at any point, that can flood without immersing the **margin line**. Important in studying the safety of ships.

Fore peak. A compartment just abaft the bow and forward of the **collision bulkhead**. Also used as ballast tank for trimming ship.

Forecastle. Deckhouse on main deck at forward end of ship.

Forefoot. Bottom of the curve of the **stem** where it meets the keel.

Foreward (For'd). At, near or towards the bow of a ship.

Formal safety assessment (FSA). A process for assessing the safety of a ship by studying the risks, their likelihood and consequences.

Forward perpendicular (FP). See **perpendiculars**.

Foundering. A ship is said to founder when it sinks bodily into the water. That is it does not roll over **(capsize)** or sink by the bow or stern **(plunge)**.

Fracture. The breaking of material under load. The nature of the fracture will vary with the toughness of the material.

Free surface effects. Any free liquid surfaces inside the ship will reduce the effective **stability**.

Freeboard. The height of the deck at side above the waterline. See also **loadline markings**.
Freeing port. An opening in the **bulwark** plating to allow water to run overboard.
Geosim. Abbreviation for 'geometrically similar' ship or model.
Grain capacity. The grain, or liquid volume; is the moulded internal volume of a hold or space less the volume of structure.
Gross tonnage. See **tonnage**.
GZ curve. See **curve of statical stability.**
GZ. The distance from the **centre of gravity** to the line of action of the buoyancy force. It is a measure of a ship's ability to resist heeling moments.
Handysize. A term applied to bulk carriers of 40 000–65 000 DWT.
Hawse pipe. A tube extending from the **forecastle** deck to the side of ship through which the anchor chain passes.
Heave. The vertical movement of a ship, as a rigid body, in a seaway.
Heel. The slow angular movement of a ship about a fore and aft axis. Angular movements as a result of waves are referred to as **rolling**.
Hogging. A ship is said to hog when the hull is bent concave downwards by the forces acting on it. Hogging is the opposite of **sagging**.
Hold. That part of a ship where cargo or supplies are carried.
Human factors. The study of human behaviour and needs.
Hydrostatic curves. A set of curves showing, for a range of draughts, the **displacement,** the **TPC** and **MCT**, and the positions of the **CF, CB** and **metacentre**.
Inboard. In a direction towards the **centreline** of the ship.
Inclining experiment. A test carried out as a ship is completed to confirm the **displacement** and centre of gravity position of the ship.
Inner bottom. Plating forming the upper boundary of the **double bottom**.
International Maritime Organization (IMO). An international body concerned with the safety of ships.
Keel. Bar or plate on centreline at bottom of ship.
Laminar flow. The smooth flow of water past a body before turbulence sets in.
Length. A distance measured in the fore and aft direction. Three lengths are of importance in defining a ship's hull:

(1) **Length between perpendiculars (L_{BP}).** The length between the forward and aft **perpendiculars**.
(2) **Length overall (L_{OA}).** The distance between the forward most and after most points of the hull.
(3) **Length on the waterline (L_{WL}).** This will vary with draught. Unless otherwise specified it is taken as the length on the design waterline.

Lightship. In simple terms the weight of hull and machinery. See **deadweight**.
Load line markings. Markings on the ship's side defining the minimum **freeboard** allowable in different ocean areas and different seasons of the year. Also known as Plimsol mark.
Loll. A ship which is slightly unstable in the vertical position will **heel** until the **GZ curve** becomes zero. It is said to loll and the angle it takes up is the angle of loll.
Longitudinal A line in the fore and aft direction parallel to the **centreline**. Also refers to a longitudinal stiffener running parallel (or nearly parallel) to the **centreline**.
Longitudinal centre of buoyancy (LCB). The fore and aft location of the **centre of buoyancy**.
Longitudinal centre of gravity (LCG). The fore and aft location of the **centre of gravity**.
Longitudinal stability. The **stability** of a ship for rotation (**trim**) about a transverse axis.
Lost buoyancy method. One approach to the calculation of a ship's damaged **stability** when it is partially flooded. The other approach uses the concept of **added weight**.
Manoeuvrability. The ability of a ship to respond to its rudder and to steer a course.
Margin line. A line, at least 76 mm below the upper surface of the **bulkhead deck**. This line helps to define the amount of flooding a ship can safely take at any point along its length.
MARPOL. A statutory regulation of **IMO** dealing with the prevention of pollution.
Mean time to failure (MTF). A statistical representation of the failure rates of elements in a system.
Mean time to repair (MTR). As **MTF** but relating to the time needed to repair, or replace, the failed component.
Metacentre. The intersection of successive vertical lines through the **centre of buoyancy** as a ship is heeled progressively. For small inclinations the metacentre is on the centreline of the ship.
Metacentric diagram. A plot showing how the **metacentre** and **centre of buoyancy** change as draught increases.
Metacentric height (GM). The vertical separation of the **metacentre** and the **centre of gravity** as projected on to a transverse plane.
Midship area coefficient (C_M). One of the **coefficients of fineness**. It is the ratio of the underwater area of the midship section to that of the circumscribing rectangle [$C_M = A_M/(B \times T)$].
Midship section. A cross section of the ship amidships, showing details of plating, frames and stiffening.
Midships. The midpoint between the two **perpendiculars**.
Moment to change trim (MCT). The moment needed to change **trim** between **perpendiculars** by a unit amount, usually 1 m.

Moulded dimensions. Dimensions of the hull taken to the inside of the plating. Used in calculating the volumes of internal spaces.
Net tonnage. See **tonnage**.
Neutral stability. See **stability**.
Offsets. The distances of the outer hull surface from the **centreline plane**. The offsets define the hull shape and are usually presented in a table showing offsets for each waterline at each transverse section.
Outboard In a direction towards the side of the ship.
Panamax. A term applied to cargo vessels that are just able to pass through the Panama Canal with maximum breadth 32.2 m and maximum length 275 m. They are generally about 65 000–80 000 DWT.
Parallel middle body. A length of the ship, near amidships, which has a constant cross section.
Permeability. A measure of the free volume in a compartment defining the maximum amount of water that can enter as a result of damage. It will be less than unity because of stiffeners and equipment in the space.
Perpendiculars. Two vertical lines used in the definition of the **length** of a ship. They are:

(1) the **forward perpendicular** which is through the intersection of the forward side of stern with the design waterline;
(2) the **after perpendicular** which is through some convenient point aft, often the rudder pintle.

Pitching. The angular motion about a transverse axis experienced by a ship in a seaway.
Plummer blocks. Supports for a shaft (such as the propeller shaft).
Plunging. A ship is said to plunge when it sinks bow or stern first through loss of longitudinal **stability**.
Port Left hand side of ship when looking forward.
Port State Control (PSC). Controls ensuring that a foreign ship using a port meets requirements laid down by the country concerned for the ship, its crew and the way it is operated.
Prismatic coefficient (C_P). One of the **coefficients of fineness**. It is the ratio of the underwater volume to the product of the area of the midship section and length. [$C_p = \nabla/(A_M \times L)$].
Pull-out manoeuvre. A manoeuvre used to demonstrate the **directional stability** of a ship.
Rake. A feature inclined to the vertical in a transverse view is said to have rake. For instance, a bow, mast or the end of a superstructure block.
Range of stability. The range of **heel** angles over which a ship is **stable** that is over which the **GZ** is positive. It is shown on the **curve of statical stability**.
Reefer. Abbreviation for a refrigerated cargo ship.
Reserve of buoyancy. The buoyancy that could be provided by the watertight volume of the ship above the design waterline.

Residual stresses. See **built-in stresses**.
Resistance. The retarding force a ship experiences when moving through the water. The two main components are the frictional resistance due to the passage of water over the hull and the resistance due to wavemaking.
Rise of floor. The amount by which the bottom of the hull rises from the keel to the turn of bilge.
Rolling. The rotation about a fore and aft axis experienced by a ship in a seaway.
RO-RO. Roll on Roll off vessel; designed to carry cars, lorries etc.
Safety of Life at Sea (SOLAS). A statutory regulation of **IMO** dealing with the safety of life at sea.
Sagging. A ship is said to sag if the forces acting on it make it bend longitudinally concave up. Sagging is the opposite of **hogging**.
Scantlings. The dimensions of the structural elements making up a ship's structure.
Seakeeping. Strictly a study of all aspects of a ship which render it fit to operate safely at sea. Usually, in naval architectural texts it is restricted to a study of a ship's motions in a seaway and other effects attributable to the waves in which it is operating.
Shearing force. The resultant vertical force acting at a transverse section of a ship.
Sheer. The rise of the deck from amidships towards the bow and stern.
Sheer strake. The upper **strake** of shell plating, just below the bulwarks (at edge of deck).
Shell. The outside plating of a ship.
Ship routing. An attempt to guide a ship into areas where it will experience less severe weather and so reduce passage times.
Simpson's rules. Rules for calculating approximate areas and volumes.
Slamming. The impact of the hull, usually the bow area, with the sea surface when in waves.
SOLAS. See **Safety of Life at Sea**.
Spiral manoeuvre. A manoeuvre conducted to indicate whether or not a ship is directionally stable.
Squat. When a ship travels fast in shallow water it tends to sink lower in the water, or squat. The effect is due to the reduced pressure under the ship.
Stability. Stability is a measure of what happens when a ship is moved away from a position of equilibrium. If, after the disturbance is removed, it tends to return to its original position it is said to have positive stability. If it remains in the disturbed position it is said to have neutral stability. If the movement increases it is said to have negative stability.
Stabilization. The reduction of rolling motions by the use of a passive or active system leading to roll reduction.
Starboard. Right hand side of ship when looking forward.

Stem. Solid or rolled metal shape forming the extreme fore end of a ship.
Stern. Back end of ship
Stern tube. Tube in stern through which the **tailshaft** passes.
Strake. A fore and aft course of shell or deck plating.
Stress concentration. A localized area in which the stress level is greater than that around it due to some discontinuity in the structure.
Strip theory. A simplified theory for calculating ship motions.
Suezmax. A term applied to cargo ships which are just able to transit the Suez Canal.
Superstructure. Enclosed space above the level of the main (or upper) deck.
Table of offsets. A table setting out the hull **offsets**.
Tailshaft. Aftermost section of the propeller shafting, carrying propeller.
Tank top. Plating at top of **double bottom**.
TEU. A measure used to define the size of a container ship. It is the number of 20-foot equivalent units it could carry (FEU = 40-foot equivalent).
Thrust block. A bearing arrangement, aft of the engine(s), by which the thrust of the propeller is transmitted to the ship.
Tonnage. A measure of the volume of a ship. In simple terms the **gross tonnage (GRT)** represents the total enclosed volume of the ship and the **net tonnage (NT)** represents the volume of cargo and passenger spaces. Tonnage is defined by internationally agreed formulae, and is used for dues for drydocking and pilotage and port and harbour dues etc. It should be noted that tonnage represents a function of volume and should not be confused with deadweight mass (tonnes), Lightship mass (tonnes) or displacement mass (tonnes).
Tonnes per centimetre immersion (TPC). The extra buoyancy experienced due to increasing the draught by 1 cm.
Torsional strength. The strength of the hull in resisting twisting about a longitudinal axis.

Toughness. An attribute of a material which determines how it will **fracture**.
Transverse. Across the ship, at right angles to the **centreline**.
Transverse planes. Vertical planes normal to the centreline plane of the ship.
Transverse sections. The intersections of **transverse planes** with the envelope of the ship's hull.
Transverse stability. A measure of a ship's **stability** in relation to rotation about a longitudinal axis.
Trim. The slow rotation of a ship about a transverse axis, leading to changes in draught forward and aft.
Tumble home. The amount by which the **beam** reduces in going from the waterline to the deck.
Turning circle. A manoeuvre to help define how a ship responds to rudder movements.
Tween deck height. Vertical distance between adjacent decks, measured from the tops of deck beams at shipside.
Vanishing stability. The point at which the ordinate of **GZ curve** becomes zero.
Vertical prismatic coefficient (C_{VP}). One of the **coefficients of fineness.** It is the ratio of the volume of a ship's underwater form to the product of the waterplane area and draught.
Waterline. The horizontal line which is the projection of a **waterplane** on to the ship's **centreline plane**.
Waterplane. A horizontal plane intersecting the ship's hull.
Waterplane coefficient (C_w). One of the **coefficients of fineness**. It is the ratio of the area of the waterplane to the circumscribing rectangle [$C_w = A_w/(L \times B)$]
Wetness. A ship is said to be wet if it ships water over the deck when in a seaway.
Zig-zag manoeuvre. A manoeuvre to study how a ship responds to rudder movements.

Author Biographies

Bryan Barrass is a marine consultant, author and Principal Lecturer at Liverpool John Moores University, UK.

Volker Bertram is Project Manager at Hamburg Ship Model Basin. He was formerly based at the Technical University of Hamburg as Professor for Ship Design.

Adrian Biran is an Adjunct Associate Professor at the Technion (Israel Institute of Technology, Haifa). He has been with the Faculty of Mechanical Engineering at Technion since 1973. He has many years' experience as a Senior Design Engineer, project leader, researcher and research engineer, and has written several books on ships and on MATLAB.

John Carlton is Senior Principal Surveyor and Global Head of Marine Technology and Investigations at Lloyd's Register, and is Visiting Lecturer at City University London.

Robert D. Christ is Principal in SeaTrepid, a marine consulting firm, and Co-Founder and former Senior VP of VideoRay, a US manufacturer of small observation class ROVs.

Alan Dodkins was the lead structural design engineer for the First of Class design of the all-FRP construction Sandown Class Minehunter in the 1980s and later became VT's Chief Structural Engineer. He is currently on secondment from BVT to Thales Naval Limited, part of the Aircraft Carrier Alliance, and is responsible for the hull structural design of the Future Carrier.

D.J. Eyres is a former lecturer in Naval Architecture at Plymouth University. He was employed as a Naval Architect and Principal Surveyor by the New Zealand Government. He was also Manager of Policy and Standards Development with the Maritime Safety Authority of New Zealand, and was on the IMO Maritime Safety Committee from 1989 to 2000.

Jørgen Juncher Jensen is Professor in Marine Engineering at the Department of Mechanical Engineering, Technical University of Denmark. He has been at DTU since 1973 and has published more than 150 papers. He is a member of the Danish Scientific Research Council for Technology and Production and the Danish Academy of Technical Sciences.

Sigismund Kastner is a lecturer at Bremen University of Applied Sciences, Germany.

Lech K. Kobylinksi is a lecturer at the Technical University of Gdansk, Poland and a member of the Foundation for Safety of Navigation and Environment Protection.

Svein Kristiansen is Professor of Marine Systems Design at the Norwegian University of Science and Technology. He is a member of the Executive Committee of West European Graduate Education in Marine Technology, and is Scientific Advisor for MARINTEK. He is also Chairman of the Association of Scientific Employees at the Norwegian Marine Technology Research Institute.

David McGeorge is a senior lecturer in marine engineering at Warsash, and London Metropolitan Universities. He is a senior examiner for Department of Transport's Certificates of Competency in Marine Engineering. He is also the author of many authoritative books on marine engineering.

A.F. Molland is Emeritus Professor of Ship Design at the University of Southampton, UK. He has lectured ship design and operation for many years. He has carried out extensive research and published widely on ship design and various aspects of ship hydrodynamics.

Dr. Anand Pillay is the Managing Director of Safetec's Asian operations. He has previously served as a Senior Safety Engineer with Lloyds Register and also on board merchant vessels up to the rank of Chief Engineer, working predominantly on large foreign going crude oil carriers. He was a lecturer at a Liverpool John Moores University and has published over 50 peer reviewed International Conference and Journal papers.

Kenneth Rawson, was Professor of Design and Technology at Brunel University and Chief Naval Architect for the UK Ministry of Defence.

H. Schneekluth was Project Manager at Hamburg Ship Model Basin. He was formerly based at the Technical University of Hamburg as Professor for Ship Design.

R.A. Shenoi is Research Professor in Lightweight Structures in the School of Engineering Sciences (SES) at the University of Southampton, UK. He has published over 250 papers on subjects including the mechanics of composite materials/structures, naval architecture and ship design.

D.A. Taylor was a Marine Consultant for Harbour Craft services Ltd, Hong Kong and was formerly Senior Lecturer in Marine Technology at Hong Kong Polytechnic University.

Eric Tupper is a member of the Royal Corps of Naval Constructors and Honorary Vice-President of the Royal Institution of Naval Architects. His career has included ship design, hydrodynamics and structural research, and ship production.

Stephen R. Turnock is Reader in Maritime Fluid Dynamics at the University of Southampton, UK. He has published widely on computational and experimental fluid dynamics in areas such as underwater vehicles, rudders and propulsion systems and renewable energy.

Professor Jin Wang leads the Liverpool Logistics, Offshore and Marine (LOOM) Centre at Liverpool John Moores University. He has authored or co-authored more than 200 technical publications. He has won several awards for his research work including two Denny Medals from the Institute of Marine Engineering, Science and Technology (IMarEST) in 2004 and 2007.

D. G. M. Watson gained extensive experience in shipyards and was for many years a senior naval architect for YARD, Glasgow.

Robert Wernli is a mechanical engineer and president of First Centurion Enterprises. He has 35 years experience in the development of undersea vehicles and work systems. In addition to his work as an independent consultant, he has begun a second career as a writer and has several fiction and non-fiction books published.

Doug Woodyard, after seagoing experience as an engineer officer in the British Merchant Navy, held editorial positions at the Institution of Mechanical Engineers and the Institute of Marine Engineers before editing *The Motor Ship*. He has been editor of *Marine Propulsion and Auxiliary Machinery* for the past five years, and contributes to other marine journals.

Index

A

Abnormal waves
   damage due to, 20–21
   progressive, 20
Absorption of electro-magnetic energy, 31
AC power, *vs.* DC power, 773
Acoustic seabed classification, 28
Agulhas wave, 21
Air resistance, 220
Aircraft analogy, for ROV, 738
Alkalinity of water, 35–36
Ambient air, ships, 38–39
Ammonium (NH_4^+), in water, 30
Amplitudes of ship vibration, 175
Analysis techniques, for safety, *see* Safety, analysis techniques
Angle of heel, 94
Appendage skin friction, 192
Archimedes, 77, 742
Atwood's formula, 89
Auf'm Keller, 196
Automatic control systems, 590–591
AUV *see* Autonomous underwater vehicle
Autonomous underwater vehicles, 730
Autonomy, ROV, 737, 739–740
Auxiliary machinery and equipment
   bilge systems, 418
      regulations, 418–420
   ballast, 425–426
   domestic water systems, *see* Domestic water systems
   incinerators, 439
   oil/water separators, 420
      oil content monitoring, 424–425
      principle of operation, 420–423
      pumping, 423
      Simplex-Turbulo oil/water separator, 423–424
      tanker ballast, 425
   sewage systems, 436–439
Azimuthing propulsors, 350–351

B

Beaufort scale, 3–4
Bending moment
   formulae for wave induced, 133
   hull strength, 123
Bending stresses, hull strength, 125–126
Bio-chemical oxygen demand (BOD), sewage systems, 436
Biological sewage treatment, 437–439
Body plan, 77, 653, 700
Bollard pull, 755, 233
Bonjean curves, 79, 122
Breaking wave, 7
Brittle fracture, hull strength, 144–146
Broaching, 599
Brushless DC motor, 763
Buckling of panels, 159
Built-in stress concentrations, structural material, 143–144
Bulbous bows, 188–190
Bulk cargo carriers, 51–58
   dry bulk carriers, 55–58
   tankers, 52–55
Bulk cargoes, weight movement, 92–93
Buoyancy
   hull strength, 122–123
   of ROV, 759
      hydrostatic equilibrium, 742–743
      water density and, 744–745

C

Calcium (Ca^{2+}), in water, 30
Calm water resistance
   appendage skin friction, 192
   bulbous bows, 188–190
   naked hull skin friction, 191–192
   transom immersion resistance, 190
   viscous form, 190–191
   viscous resistance, 192–195
   wave making resistance, 184–188

Cameras, 668
Capillary waves, 5
Car carriers, 51
Cargo ships, 45
    IMO intact stability code, 98–101
Cause and effect relationship, in safety analysis, 825–826
Cavitation, 259–260, 239
    and propellers, 265–272
        in design, 272–282
            bucket diagram, 278
            physics of, 260–265
Celerity, 8
Centre of gravity, ships, 85–86, 653
CFD, *see* Computational Fluid Dynamics (CFD)
Charge-coupled device (CCD) cameras, 668
Chemical carriers, 56
Chloride (Cl^-), in water, 30
Chlorophyll, 22–23
Circulation, water, 23–24
Classification societies, 802, 803–805
    damage repairs, 808
    HPMS, 808
    Lloyds Register, 805
        classification symbols, 805–806
        planned maintenance, 808
    rules and regulations, 805
    ships operating in ice, 806
    structural design programs, 806–807
    surveys, 807–808
Climatic extremes, 39–40
Coastal zone classification, 38
Code of practice, for small fishing vessels, 811–812
Communications linkage, ROV, 737
Composite materials, 170, 665–683
Compressibility factor, 24–25
Computational Fluid Dynamics (CFD), 195, 600–602
    for propellers, 258–259
    for rudders, 627–630
    viscous flow computations, 210
    wave resistance computations, 206–210
Conductance, water specific, 36
Conductivity, water, 25
Conflict of interest, 792–793
Container ships, 20, 45–49
    accident statistics, 818–821
    IMO intact stability code for, 102
Contra-rotating propeller, 351–352
Control
    bridge control, 475–479
        electrical supply control, 480
    centralized control, 475
    controller actions, 465–466
        derivative action, 466
        integral action, 466
        multiple-term controllers, 467
        two-step or on-off, 466
    controllers, 467–469
    correcting unit, 469–470
    integrated control, 480–481
    systems, 470–475
    theory, 462–463
    transmitters
        electrical, 464–465
        hydraulic devices, 465
        pneumatic devices, 463–464
    unattended machinery spaces, 475
Control surfaces, 606–607
    efficiency of, 609
    hydroplanes, 609
    see also Rudders
Control systems, of ROV, 775, 776–777
    control station, 777–778
    motor control electronics, 778–779
        H-bridge, 781, 782
        inductors, 779–781
        PWM control, 781
Coupling, ship vibrations, 172
Crack extension, hull stremgth, 144–146
Cross curves of stability, 91
Crude oil carriers, 52–53
Cruise ships, 58
Cummins, 403
Currents, 20, 23
Cycloidal propeller, 357

D

Damage repairs, Classification societies, 808
Damage stability, 94–98
    regulations, 110–114
        German Navy, 113–114
        probabilistic, 111–112
        SOLAS, 110–111
        UK Navy, 113
        US Navy, 112–113
Data, 773–774
    transmission and protocol, 774
Data assessment, 541–544
    response amplitude operators, 543–544
        model experiments, 543–544
        ship trials, 544
        theory, 543
    wave data selection, 541–543
DC power, AC power *vs.*, 773
Deadweight, 641
Deckhouses, structural material, 147–149
Deep water, sinusoidal waves describe regular waves in, 7–8
Deflection, hull strength, 129–130

Density profiles, sonic velocity, 34–35
Density, sea water, 25–27
Deployment techniques, of ROV
    auto stabilization, 757–758
Depth, water, 27–28
Destroyers, 70
Diagraph-based Analysis (DA), 838
Diesel engines
    admiralty coefficient, 366
    astern running, 368–370
    bmep, 365
    derating, 365
    exhaust temperatures, 364–365
    fuel coefficient, 366
    high-speed engines, 399–403
    intelligent, 387–390
    low speed, 381–387
    maximum rating, 362–364
    mean effective pressures, 365
    medium speed, 390–399
    propeller build-up, 367
    propeller performance, 367
    rating, 362
Directional stability, 580–581
Dispersion, of wave field, 8–9
Dissolved gases, 28–30
Dissolved oxygen, 29–30
Dissolved salts, in water, 32
Domestic water systems
    corrosion, 431
    evaporators, 432–434
    fresh water, 426–427
    low pressure evaporators, 428
    reverse osmosis, 434
    salinometer, 428
        electric, 428–431
    sanitary system, 427
    sterilization, 431
        chlorine sterilization, 431–432
        Electro-Katadyn method, 432
        ultra-violet sterilizer, 432
    tanks, 436
    water production, 427–428
    water treatment, 434–435
Dry bulk carriers, 55–58
Dust and sand, 39–40
Dynamic stability, of ROV, 745–746
    drag, 746
        form drag, 747, 749–751
        skin friction drag, 746–747, 747–749
    mission-related vehicle trim, 746
    point of net thrust, 746
    tether effects, 751, 752–755
Dynamical stability, ships, 93–94
Dynamically supported craft (DSC), IMO code, 101–102

E
Echo sounder, 27–28
Ekman spiral, 24
Elastic stability, frame structures of ship, 166–167
Electric motors, for ROV propulsion, 762
    brushless DC motor, 763
    gearing, 763, 764
    permanent magnet DC motor, 762–763
Electro-magnetic transmission, through water, 30–31
Elsan type plant, sewage systems, 437
Energy method, ship framework, 163–165
Engine selection, 370–373
    electric drives, 375–380
    mechanical drives, 373–375
Equilibrium, of floating body, 77–79
    in still water, 77
    underwater volume, 77–79
ETA, *see* Event tree analysis, for safety
European Commission regulation
    damge stability of internal water vessel, 114
    intact stability of internal water vessel, 109
Event tree analysis, for safety, 834–835
Experiments and trials, seakeeping
    ship seakeeping trials, 558–560
    stabilizer trials, 560
    test facilities, 558
Extreme climatic conditions, 39–40
Extreme waves, in normal stationary sea, 20

F
Factor of subdivision, 110
Failure
    ship structural, 134–135
    structural design, 149–150
Failure mode, effects and critical analysis (FMECA),
        for safety, 836–838
Fatigue, structural material, 146–147, 149
Fault Tree Analysis (FTA), for safety, 829–834
Ferries, large, 58–59
Finite element analysis (FEA), structural design,
        168–169
Fishing vessels, 60–62
    IMO intact stability code for, 101
Fixed pitch propellers, 346–348
Flag States, 798–799, 801
    audit, 799–800
    control, 798
    delegation of, 799
    effectiveness of, 799
    Seaworthiness Act, 799
Flexural vibrations, ship, 172
Flexure, hull strength, 130
Floating body
    equilibrium of, 77–79
    stability

Floating body (*Continued*)
    at large angles, 89–92
    at small angles, 79–84
    see also Ships
Floodable length, 96–98
Flooding, 94–98
    asymmetrical, 96
    floodable length, 96–98
Flow visualization tests, for resistance, 203
Fluid free surface, 87–88
Fluorescent lighting, 767
FMECA, see Failure mode, effects and critical analysis (FMECA), for safety
Formal safety assessment (FSA), 849–853
Frame, of ROV, 759
Frameworks, 159–168
Freak waves, 20–21
Free surface, fluid, 87–88
Frequency spectrum, sea surafce level, 5
Fresh water, 30
Frigates, 70
Froude's analysis, of resistance, 183–184
Froude, William, on ship resistance, 183–184
FSA, see Formal safety assessment (FSA)
FTA, see Fault Tree Analysis (FTA), for safety
Full scale trials, hull strength, 134

G
Gas carriers, 53–54
Gas turbines, 403–404
    air filtration, 411–412
    cycles and efficiency, 406–409
    emissions, 409–410
    lubrication, 410–411
    plant configurations, 404–406
    turbine designs, 412–414
Geometrical discontinuties, hull strength, 142–143
German Navy, damage stability regulations, 113–114
GM see metacentric height
GM see longitudinal stability
Green sea loading, 40
Grillages, ship design, 151–155
Ground interactions, to ships, 592–593
Gust, 3
GZ curve, 90, 501–502

H
H-bridge control, 781, 782
Hail, exposure of equipments, 40
Harvald method, 196–197
Hazard and Operability (HAZOP) studies, for safety, 827–829
HAZOP, see Hazard and Operability (HAZOP) studies, for safety

High speed craft, 62–66
    hydrofoil craft, 64–65
    IMO intact stability code for, 102–103
    monohulls, 64
    multi-hulled vessels, 65–66
    RIB, 66
    SES, 64
High winds, exposure of equipments, 40
high-intensity discharge (HID) lighting, 767
High-speed hull, 215–216
    fast and unconventional ships, 217–220
    model test data, 216–217
    standard series data, 216
Higher order waves, Stokes and Airy theory, 5–7
Hogging, 118–119
Hollow box girder, hull strength, 130–131
HPMS, see Hull Planned Maintenance Scheme (HPMS)
Hull Planned Maintenance Scheme (HPMS), 808
Hull strength, 118–149
    material considerations, 142–149
    standard calculation, 119–142
Humps, and hollows, 187
Hydrodynamics, of ROV, 755
Hydroelastic analysis, hull strength, 142
Hydrofoil craft, 64–65
Hydroplanes, 603, 609
Hydrostatic curves, 84–85

I
Ice strengthened ships, 60–62
Icebreakers and ice strengthened ships, 60
Icing
    exposure of equipments, 39
    IMO code on intact stability, 102
ILLC, see International Convention on Load Lines (ILLC)
IMO, see International Maritime Organization (IMO)
Improving performance, seakeeping, 560–562
    design and operational changes, 560–561
    influence of form, 561
Incandescent lighting, 767
Incinerators, 439
Inclining experiment, 88–89
    IMO code for, 102
Influence lines, hull strength, 126–129
Instrumentation, 461–462
    see also Control
Intact stability regulations, 98–110
    criteria for sail vessels, 107–108
    IMO code, 98–103
    internal water vessels, 109
    small workboats and pilot boats, 108–109
    UK Navy, 106–107
    US Navy, 103–106

International Association of Classification Societies (IACS), 132, 805
International Convention on Load Lines (ILLC), 796
International Convention on Standards of Training, Certification and Watchkeeping for Seafarers (STCW), 796–797
International Maritime Organization (IMO), 795
  ILLC, 796
  ISM code, 797–798
  MARPOL, 797
  SOLAS, 795–796
  STCW Convention, 796–797
International Maritime Organization (IMO),code on intact stability, 98–103
International safety management code (ISM) code, 797–798, 866
International trade and shipping, 786
Ionic concentration, water, 30
Irregular seaway, statistics, 16–17
Irregular waves, 10–12
ISM, *see* International safety management code (ISM) code

J

Joint North Sea Wave Program (JONSWAP) seaway spectrum, 13
JONSWAP seaway spectrum, *see* Joint North Sea Wave Program (JONSWAP) seaway spectrum

K

Kelvin, Lord, on wave making resistance, 184–188
Kort nozzle, 348, 766

L

Lifting surface models, 248–249
Lifting-line-lifting-surface, 249–250
Light refraction, 31
Light transmission, through water, 30–31
Light-emitting diode (LED), 767–768
Lighting, 766–767
  fluorescent, 767
  high-intensity discharge (HID), 767
  incandescent, 767
  light-emitting diode (LED), 767–668
Limit design method, ship framework, 165–166
Limit state analysis, for safety, 838
Limiting criteria, seakeeping, 529–538
  degradation of human performance, 537–538
  propeller emergence, 537
  slamming, 531–537
  speed and power in waves, 529–531
  wetness, 537
Lines plan, 78, 656, 693, 700

Liquefied gas carriers, 53
Lloyds Register, 805
  classification symbols, 805–806
Loading
  hull strength, 123
    realistic assessment of, 136–137
  panels under lateral, 155–156
  structural design, 149–150
Longitudinal stability, 83–84
Longitudinal strength, 118–134
  realistic assessment, 135

M

Magnetohydrodynamic propulsion systems, 359–361
Manoeuvring
  automatic control systems, 590–591
  broaching, 599
  CFD, 600–602
  directional stability, 580–581
  dynamic positioning, 590
  experimental approaches, 599–600
    manoeuvring tests in sea trials, 599
    model tests, 599–600
  heel, during turning, 586–587
  limitations of theory, 584
  manoeuvrability assessment, 584–586
    turning circle, 584–586
    pull-out manoeuvre, 588
    spiral manoeuvre, 588
    zig-zag manoeuvre, 587
  rudders and control surfaces, 606–631
    control surfaces and applications, 606–609
    detailed rudder design, 612–621
    numerical modelling of rudder, 627–630
  rudder action, 583–584
    rudder data presentation, 609–611
    rudder design guidelines, 630–631
    rudder design, in ship design process, 611
    rudder manoeuvring forces, 621–627
  shallow water/bank effects, 598–599
  ships interaction
    to ground, 592–593
    to ship, 593–594
    to shore, 598
  speed loss, on turn, 586
  stability and control, of surface ships, 581–583
  standards, 589–590
  submarine stability and control, 603–606
    control requirements and equations, 603–604
    design assessment, 606
    experiments and trials, 605
  turning ability, 587–589
    pull-out, 588–589
    spiral, 588
    zig-zag, 587–588

Marine pollution, 40–41
Markov methods, for safety, 835–836
MARPOL, 40, 638, 797
Materials, ship construction, 657–691
   aluminium alloy, 662–665
      fire protection, 665
      production, 663–664
      sandwich panels, 664–665
   composite materials, 665–683
      design concepts, 671–675
      design synthesis, 675–682
      external issues, 682–683
      introduction, 665
      materials selection, 665–671
      mechanical properties, 668
      production considerations, 669
      overview, 665
   corrosion, 683–691
      anti-fouling systems, 688–689
      control, 686–688
      nature and forms, 683–686
      painting ships, 689–691
   steel, 659–662
      castings, 662
      corrosion resistant, 661–662
      forgings, 662
      heat treatment, 660
      high tensile, 661
      manufacturing, 659–660
      sandwich panels, 662
      sections, 660
      shipbuilding, 660–661
Merchant ships, 45–62
   bulk cargo carriers, 51–58
   car carriers, 51
   container ships, 45–49
   fishing vessels, 62
   general cargo ships, 45
   icebreakers and ice strengthened ships, 60–62
   passenger ships, 58–60
   roll-on roll-off ships, 49–51
   tugs, 60
Metacentric diagrams, 83
metacentric height, 81, 504–505, 858
Metacentric height, lolled condition, 91
Mine countermeasures vessels, 70
Mobile offshore drilling units
   IMO code for, 101
Model test facilities, for resistance, 203
Moment distribution process
   for ship framework, 161–162
Monohulls, high speed craft, 62
Mould growth, exposure of equipments, 39

Multi-hulled vessels, 64–65
MV Derbyshire, 20, 72

N
Naked hull skin friction, 191–192
Natural gas carriers, 53
Navier–stokes methods, for rudders, 628–629
Nitrate (NO_3-), in water, 30
non-linear effects, 544–545
Numerical modelling, rudders, 627–630
   available methods, 627
   Navier–stokes methods, 628–629
   potential flow methods, 627–628
   rudder–propeller interaction, 629–630
numerical prediction, of seakeeping, 545–556
   overview of computational methods, 545–547
   quantities, in regular waves, 554–555
   Rankine singularity methods, 551–552
   ship responses, in stationary seaway, 555–556
   strip theory, 547–551

O
Observation-class vehicles, 735, 741
Ocean currents, 23–24
Ocean coverage, 38
Oceanography, 21–38
   coastal zone and sea bottom, 38
   distribution of water, 22
   water properties, 22–38
Offshore safety assessment, 839–845
Opposing waves, 20
Optimisation, in design and operation, 709–724
   block coefficient, 718–719
   concept exploration, 722
   developments of 1980s and 1990s, 722–724
      concept exploration models, 722
      optimisation shells, 722–724
   economic basics, 713–717
      annual income and expenditure, 715–716
      cost difference method, 716–717
      discontinuities in propulsion unit costs, 717
      discounting, 713–714
      initial costs, 714–715
   length, 717–718
   methodology, 709–712
   scope of application, 712–713
   special cases, 720–724
      dimensions of containerships, 721–722
      repeat ships, 720–721
   speed, 719–720
   width, 717
Orbital motion, of water particle in wave, 9–10

ORP, *see* Oxidation reduction potential (ORP)
Oxidation reduction potential (ORP), 35

P

Paddle wheels, 357–359
Panamax, 51
Panels of plating, 155–159
    behaviour under lateral loading, 155–156
    buckling, 159
    flat plates under lateral pressure, 156–159
Passenger ships, 56–58
    IMO intact stability code, 98–101
Permanent magnet DC motor, 762–763
Permeability of ships, 95
Permissible length, 110
Petroleum gas carriers, 53–55
pH, water, 35–36
PHA, *see* Preliminary hazard analysis (PHA)
Pierson/Moskowitz and Bretschneider,
        spectrum formulae, 12–13
Pilot boats
    damage stability code, 114
    intact stability code for, 108–109
Plating
    panels of, 155–159
    stiffened, 151–155
    swedged, 155
Podded propellers, 350–351, 379
Pollution, marine, 40
Port State control, 800, 802
    MOU PSC, 802
Preliminary hazard analysis (PHA), 826–827
Pressure see water pressure
Product carriers, 53
Progressive abnormal waves, 20
Propeller
    controllable-pitch propeller, 445, 353–356
    law, 366
    maintenance, 445–446
    mounting, 445
    roughness, 317
    slip, 365–366
    trailing and locking, 368
    see also specific entries
Propeller design
    and analysis loop, 282–283
    choice of propeller type, 284–287
    considerations
        balance between propulsion and cavitation,
            300–301
        basic design process, 303
        blade area ratio, 298–299
        blade number, 297–298
        cavitation, 295, 299, 445
        composite blades, 302–303
        duct form, 300
        for partial hull tunnels, 302
        hub form, 299–300
        pitch–diameter ratio, 298
        propeller tip, 301–302
        rotation, 295–297
        section form, 299
        shaft inclination, 300
        skew, 299
    constraints, 283–284
    process, 303–309
    propeller design basis, 287–292
    standard series data in, 292
        cavitation effects, 295
        diameter, 292–293
        mean pitch ratio, 293
        open water efficiency, 293
        propeller thrust determination, 295
        rpm for power, 293–295
Propeller theories, 247–259
    boundary element methods, 254–255
    CFD, 258–259
    lifting surface models, 248–249
    lifting-line-lifting-surface, 249–250
    specialist propulsors, methods for, 256–258
    vortex lattice method, 250–254
Propeller, performance characteristics, 233
    open water, 233–239, 202
        cavitation on, 239, 259–282
    scale effects, 239–243
    specific propeller, 243
        controllable pitch propeller, 243–244
        ducted propellers, 245–246
        fixed pitch propellers, 243
        high-speed propellers, 246
        standard series data, 246–247
Propeller-induced forces, ship vibrations, 175–178
Propulsion systems, 346
    azimuthing propulsors, 350–351
    contra-rotating propeller, 351–352
    controllable pitch propellers, 353–356, 445
    cycloidal propeller, 357
    ducted propellers, 348–350
    fixed pitch propellers, 346–348
    magnetohydrodynamic, 359–361
    overlapping propellers, 352–353
    paddle wheels, 357–359
    podded propellers, 350–351, 379
    superconducting motors, 362
    tandem propellers, 353
    waterjet, 356–357

Propulsion systems, of ROV, 759, 761
  drive shafts, 764–765
  electric motors, 762
    brushless DC motor, 763
    gearing, 763, 764
    permanent magnet DC motor, 762–763
  gearing, 763, 764
  permanent magnet DC motor, 762–763
  propeller design, 766
  sealing, 765
  thrusters, 761–762
    design, 762
Propulsion tests, 203
Propulsive coefficients, 210–211
  hull efficiency, 212–213
  QPC, *see* Quasi-propulsive coefficient
  relative rotative efficiency, 211
  thrust deduction factor, 211–212

Q

Qualitative safety analysis, 822–823
Qualities, seakeeping, *see* Seakeeping qualities
Quantitative safety analysis, 823–825
Quasi-Propulsive Coefficient (QPC), 196–197, 213, 367

R

Rain, exposure of equipments, 40
RANS method, 628–629, 630
Reefers, *see* Refrigerated cargo ships
Refraction, light, 31
Refrigerated cargo ships, 45
Regulatory authorities
  Flag States, *see* Flag States
  IMO, 795
    ILLC, 796
    ISM Code, 797–798, 866
    MARPOL, 40, 638, 797
    SOLAS, 795–796, 110–111, 803
    STCW Convention, 796–797
  Port State control, 800, 802
  structure of control, 794–795
Remotely operated vehicles (ROVs), 730–731, 735–737
  aircraft analogy, 738
  autonomy, 737, 739–740
  buoyancy, 741–743
    and water density, 744–745
  characteristics for perfect ROV, 740–741
  communications linkage, 737
  components, 758–759
  control, 755, 775
  crew, 741
  deployment techniques, 757–758
  design, 735
  dynamic stability, *see* Dynamic stability, of ROV
  frame, 759
  historical developments of, 730–735
  hydrodynamics, 755
  lighting, 766–768
  manipulators, 770
  observation-class vehicles, 735, 741
  operations platform, 741
  power sources, 737, 773
  propulsion systems of, *see* Propulsion systems, of ROV
  sensors, 768, 769
  special use, 738, 741
  stability
    and water density, 743–744
  tele operation *vs.* remote control, 739
  vehicle size *vs.* task suitability, 741
  work class, 741
Resistance
  air resistance, 220
  calm water resistance
    appendage skin friction, 192
    bulbous bows, 188–190
    high speed hulls, 215–218
    naked hull skin friction, 191–192
    transom immersion resistance, 190
    viscous form, 190–191
    viscous resistance, 192–195
    wave making resistance, 184–188
  evaluation, *see* Resistance evaluation methods
  Froude's analysis, 183–184
Resistance evaluation methods
  CFD, 206–210
  direct model tests, 200–206
  propulsion tests, 203
  regression-based, 197–200
  resistance test, 200
  traditional and standard analysis
    Auf'm Keller, 196
    Ayre's method, 196
    Harvald method, 196–197
    standard series data, 197
    Taylor's method, 195–196
Restricted water effects, 214–215
Response amplitude operator (RAO), 136, 490–496, 543, 553–554
Rhodamine, in water, 36
RIB, *see* Rigid inflatable boats (RIB)
Rigid inflatable boats (RIB), 66
Roll-on roll-off ships, 49–51
Rolling tests, IMO code, 102
Rough water, influence, 213–214
Rudders
  action, 583–584

data presentation, 609–611
design within ship design, 611–612
detailed design, 612–615
  process, 615–621
drift angle, 621–622
forces, 621
guidelines, 630–631
high lift, 624–627
hull upstream of, 621
low and zero speed, 622–627
numerical modelling, 627–630
Navier-stokes methods, 628
Potential flow methods, 627
rudder-propeller interaction, 629
types, 607–608

S

Safety management, of ship stability, 853
  educationa and training, 867–868
  futuristic aspects, 868–869
  guidelines on, 859–866
    loading and stability manual, 859–860
  ISM, 797–798, 866
  operational stability assessment, 854–855
  seakeeping guidance and survivability criteria, 855–859
  stability manual, 859
  tools for, 853–854
Safety of Life at Sea *see* SOLAS
Safety, maritime
  acceptable risk, 792
  accident statistics, 810–811
  ALARP, 839
  analysis techniques, 822
    cause and effect relationship, 825–826
    decision table method, 838
    Diagraph-based Analysis (DA), 838
    ETA, 834–835
    FMECA, 836–838
    FTA, 829–834
    HAZOP, 827–829
    limit state analysis, 838
    Markov methods, 835–836
    PHA, 826–827
    qualitative safety analysis, 822–823
    quantitative safety analysis, 823–825
    What-If analysis, 827
  and economy, 788–789
  cause and effect relationship, 825–826
  code of practice for small fishing vessels, 811–812
  conflict of interest, 792–793
  containership accident statistics, 818–821
  current status, 809
  databases, 809–810
  expertise and rationality, 793–794
  fishing vessels
    accident data for, 812–816
    data analysis, 816–818
    Safety Rules 1975, 812
  formal ship safety assessment, 845–847
  FSA, 849–853
  improvement difficulties, 791
  maritime domain, 789–790
  offshore safety assessment, 839–845
  PHA, 826–827
  regulatory authorities, *see* Regulatory authorities
  reliability and safety methods, 808–809
  risk concept, 791–792
  risk criteria, 847–848
Safety, structural, 140
Sagging, 118–119
Sailing vessels
  damage stability regulations, 114
  intact stability regulation criteria for, 107–108
  principal forces, 66–67
Salinity, 32–34
  profiles, sonic velocity, 34–35
Salt water, 32–34
Sand and dust, 39–40
Scattering, light, 31
Sea surface level, variation, 4–5
Sea water density, 25–27
Seabed classification, acoustic, 28
Seakeeping
  data assessment, 541–544
  experiments and trials, 558–560
  human performance, 537
  improving performance, 560–562
  limiting criteria, 529–538
  non-linear effects, 544–545
  numerical prediction, 545–558
  fast craft, 552
  Rankine singularity, 551
  simulation methods, 556
  strip theory, 547
  overall performance, 538–541
  performance, 538–541
  propeller emergence, 537
  qualities, 485–486
    importance, 485–486
    motions, 485
    ship routing, 485
    slamming, 485, 531
    speed in waves, 485, 530
    wetness, 485, 537
  ship motion control, 562–575
  ship motions, 486–529
Seaway
  spectrum, 11
  statistics of irregular, 16–17

Seaworthiness Act, 799
Second moment of area, hull strength, 123–125
Section modulus, hull strength, 129
Sensors, ROV, 768, 769
Sensors, warships, 68
Service analysis
  equations for the roughness induced power penalties, 321–324
  fouling, 309–317, 317–321
  hull drag reduction, 317
  hull roughness, 309–317
  performance monitoring, 324–333
  propeller roughness, 317–321
  weather effects, 309
SES, *see* Surface effect ships (SES)
Sewage systems, 436–439
  bio-chemical oxygen demand, 436
  biological sewage treatment, 437–439
  coliform count, 437
  effluent quality standards, 436
  Elsan type plant, 437
  holding tanks, 437
  suspended solids, 436–437
Shaft bearings, 443
Shafting, 439–440
  shaft bearings, 443
  sterntube bearing, 443
  sterntube seals, 444
  thrust block, 440–443
Shallow water/bank effects, 214–215, 598–599
Shear force, 123
Shear stresses, hull strength, 126
Ship construction, 691–703
  CAD/CAM, 698–703
  examples of structure, 691
  lines plan, 693
  loftwork, 696–697
  shipyard layout, 691
  ship drawing office, 692–696
Ship design, 638–657
  capacity determined design, 643–647
    cargo ships, 643
    container ships, 645–646
    high speed passenger/vehicle ferries, 646–647
    passenger ships, 643–645
  concept design content, 657
  deadweight determined design, 641–643
    cargo capacity check, 642–643
    deadweight and dimensions, 641–642
    summary of overall model, 643
  lightship mass estimates, 649–653
    machinery mass, 652
    margin, 652
    masses of fast ferries, 652
    outfit mass, 651–652
    steel mass, 649–651
    verticle centre of gravity, 653
  overview, 638–639
    general, 638
    ship design process, 639
  ship lines design, 653–657
    preparation of body plan, 655–657
    sectional area curve, 653
    sectional area curve modifications, 653–654
    sectional area curve transformations, 654–655
  stability check, 647–649
  statutory regulations, 657, 795
  technical ship design, 639–641
    principal requirements, 639
    specification, 640–641
Ship economics
  cargo handling costs, 709
  criteria, 705–706
    basis, 705
    inflation and exchange rates, 706
    interest, 705
    net present value, 706
    present worth, 705
    repayment of principal, 705–706
    required freight rate, 706
    sinking fund factor, 706
    yield, 706
  daily running costs, 707–708
    administration and general charges, 708
    crew costs, 707
    insurance, 708
    maintenance and repair, 708
    provisions and stores, 707–708
  operating costs, 706–707
    capital amortisation, 706–707
    capital charges, 706
    depreciation, 707
    profit and taxes, 707
    ship values, 707
  shipowners and operators, 703, 786
    methods of employment, 704–705
    types of trade, 703–704
  voyage costs, 708–709
    bunkers, 708–709
    pilotage, 709
    port and canal dues, 709
    towage, 709
Ship girder, 118
  structural failure of, 135
Ship management company, 786–787
ship motion control, 562–575
  pitch damping, 573–575

interceptors, 574–575
  transom flaps, 574
 roll stabilization, 562–573
  fin stabilizers, 567–573
  performance of stabilizing systems, 564–567
  principal systems, comparison, 564
  stabilization systems, 562–564
  with rudders, 572
Ship motions, 486–529
 approximate period of roll, 488
 damped, in still water, 488
 data presentation, 490–492
 degrees of freedom, 486
 encounter period, 513–515
 heaving, 487
 in irregular seas, 492–495
 in oblique seas, 496
 in regular waves, 489–490
  assumptions, 489
  pitching and heaving, 490
  rolling in beam sea, 489–490
 large amplitude rolling, 497–499, 508–509
 rolling, 487
 roll excitation, 499–529
  at large amplitudes, 508–509
  bandwidth, of transformed sea spectrum, 526–527
  encounter frequency, 515
  encounter period, 513–515
  GM, 504–505
  GZ-variation, 509–513
  irregular time series, of wave encounter, 527–529
  linear restoring moment, 501–502
  mass moment of inertia, 501
  modes, 505–506
  motion direction, of rigid body, 499–501
  natural roll period, 502
  relevant frequencies, of encounter spectra, 523–526
  roll damping, 503–504
  ship roll, in beam seas, 489–490, 506–508
  wave encounter, in irregular seas, 521–523
  wave energy and encounter spectra, 523
  wave group, 515–521
 surge, 497
 sway, 497
 undamped, in still water, 486–488
  heaving, 487–488
  rolling, 487
  yaw, 497
Ship owner, 786–787
ship routing, 485
Ship vibrations, 171–178
 amplitude, 175

calculating levels of, 175
 direct calculation, 173–174
 flexural, 172
 formulae, 173
  approximate, 174–175
 propeller-induced forces, 175–178
 reduction, 175
 testing of equipment, 178
 torsional, 172
Shipping
 actors in, 786
 international trade and, 786
Ships
 ambient air, 38–39
 angle of heel due to turning of, 94
 centre of gravity of, 85–86
 icebreakers and ice strengthened, 60–62
 longitudinal strength standards, 132–134
 main hull strength, 118–149
 marine environment, 3
 realistic structural response, 137–140
 stability, *see* Stability, of floating bodies
 structural safety assessment, 140–142
 structural units of, 150–151
 structures, 118–178
  design and analysis, 149–170
  main hull strength, 118–149
  vibrations, 171–178
 trim changes, 83
 wall-sided, 86–87
 see also Vessels
Ships interaction, 591–592
 to ground, 592–593
 to ship, 593–598
 to shore, 598
Shipyard layout, 691–692
Shore interaction, to ships, 598
Simulation method, seakeeping, 556–558
 long-term distribution, 557–558
Sinkage, compartment open to sea, 94–95
Sinusoidal waves, 7–10
Slamming, 485, 531
Slamming, hull strength, 142
Slope deflection analysis
 ship framework, 162–163
Slope, hull strength, 129–130
Smith correction, 131
Solar radiation
 exposure of equipment, 39
 underwater, 33
SOLAS, 110–111, 795–796, 803
Sonic velocity in water, 34–35
Sound channels in water, 34–35

Special use ROV, 738, 741
Specific conductance, water, 36
Spectrum formulae, Pierson/Moskowitz, Bretschneider, JONSWAP, 12–13
Stability, of floating bodies
  at large angles, 89–92
  at small angles, 79–84
    concept of, 79–81
    longitudinal, 83–84
    metacentric diagrams, 83
    problems, 85–87
    transverse metacentre, 81–82
      for simple geometrical forms, 82–83
  damaged, 94–98
    regulations, 110–114
  dynamical, 93–94
  influence of wind on, 93–94
  manual, 859
  regulations of intact, *see* Intact stability regulations
Stability, of ROV, 743–744
Stabilizers *see* ship motion control
Standard series data, propellers, 292
  cavitation effects, 295
  diameter, 292–293
  mean pitch ratio, 293
  open water efficiency, 293
  propeller thrust determination, 295
  rpm for power, 293–295
Standing waves, 20
Statical stability curves, 89–91
  features of, 91–92
  from cross curves, 91
Stationary random sea, wave height in, 13–16
Stationary sea
  extreme waves in normal, 20
  wave height in random, 13–16
Statistics of irregular seaway, 16–17
Stealth warship, 68
Steam turbines, 414
  astern arrangements, 416
  construction, 416–417
  types, 414–416
Steering gear, 446–448
  electric steering, 458
  electrical remote control, 450, 452, 453
  power units, 452
    ram type, 453, 455
    rotary vane type, 455–456, 459–460
  telemotor control, 448, 450, 452
  testing, 461
  twin-system steering gears, 458, 461
  variable delivery pumps, 448, 449–450, 451

Sterntube bearing, 443
Sterntube seals, 444
Stiffened plating, 151–155
  treatment of, 155
Stokes and Airy theory, higher order waves, 5–7
Storm classes, 18–20
Stress concentration, ship, 142
  structural material and built-in, 143–144
Structural design
  analysis, 149–170
  composite materials, 170
  discontinuities, 147
  finite element analysis, 168–169
  loading and failure, 149–150
  panels of plating, 155–159
  realistic assessment of structural elements, 169–170
  stiffened plating, 151–155
  structural units of ship, 150–151
  three dimensional framework, 159–168
Structural elements assessment, ship design, 169–170
Structural failure, hull strength, 134–135
Structural material, hull strength, 142–149
  built-in stress concentrations, 143–144
  crack extension and brittle fracture, 144–146
  discontinuities in structural design, 147
  fatigue, 146–147
  geometrical discontinuities, 142–143
  superstructures and deckhouses, 147–149
Structural response, ships, 137–140
Structural safety assessment, ships, 140–142
Submarines, 70–72
  stability and control, 603–606, 743
Submerged body, hydrostatic curves of fully, 84–85
Suezmax, 51
Superstructures, structural material, 147–149
Surf beat, 4
Surface currents, 20
Surface effect ships (SES), 62
Surface ships, hydrostatic curves, 84
Surge, 4
Suspended solids, sewage systems, 436–437
SWATH, 65
Swedged plating, ship design, 155
Swell, 4
Swiss regulations for internal navigation, 109

T
Tandem propellers, 353
Tankers, 52–55
Taylor's method, 195–196
Temperature profiles, sonic velocity, 34–35
Tether, 770–773
Tether management system (TMS), 770

Three-dimensional extrapolation method,
  for resistance, 205–206
Three-dimensional frameworks, ship, 159–168
  elastic stability, 166–167
  end constraint, 167–168
  methods of analysis, 161–166
Thrust block, 440–443
Thrusters, of ROV, 761–762
Tides, 24
Torsional vibrations, ship, 172
Total dissolved solids (TDS), in water, 36
Towing tank facilities, around the world, 204
Tractor tugs, see Tugs
Transom immersion resistance, 190
Transverse metacentre, 81–82
  for simple geometrical forms, 82–83
Trim, 83
  compartment open to sea, 94–95
  problems in, 85–87
Trochoid, 5, 120–121
Tugs, 58–60
Turbidity, of water, 35
Trimaran, 65
Tween decks, 45
Two-dimensional extrapolation method,
  for resistance, 203–205

U
UK Navy
  damage stability regulations, 113
  intact stability regulations, 106–107
Underwater connectors, 774–775
Underwater volume, 77–79
US Navy
  damage stability regulations, 112–113
  intact stability regulations, 103–106

V
Vehicle control, of ROV, 755–756
  auto stabilization, 756–757
Vessels
  circular cross section, 83
  constant triangular section, 82–83
  fishing, 62–63
  intact stability regulations for internal water, 109
  mine countermeasures, 70
  multi-hulled, 65–66
  rectangular cross section, 82
  see also Ships
Vibration levels, ship, 175
Vibration testing of equipments, 178
Vibrations, ship, 171–178
  propeller excited, 175
Viscosity, 35

Viscous form resistance, 190–191
Viscous resistance, 192–195
Vortex lattice method, 250–254

W
Wake field
  characteristics, 220–222
  definition, 222–223
  effective, 228–230
  normal field, 223–225
  parameters' estimation, 225–228
  scaling, 230–233
Wall-sided formula, 86–87
Warships, 66–72
  sensors of, 68
  stealth, 66
  sustaining damage, 68
  types, 69–72
  vibration response and endurance test levels for
    surface, 178
  vulnerability, 69–70
  weapons, 68–69
Water circulation, 23–24
Water depth, 27–28
Water distribution on earth, 22
Water pressure, 31–32
Water properties, oceanography, 22–38
Water quality, 35–36
Water temperature, 36–38
Water vessels
  damage stability regulations for internal, 114
  intact stability regulations for internal, 109
Waterjet propulsion systems, 356–357
Wave climate, 18–20
Wave field, normal dispersion, 8–9
Wave pressure correction, hull strength, 131–132
Waves
  breaking, 7
  capillary, 5
  classes, 18–20
  data, 17–18
  freak, 20–21
  height
    in stationary random sea, 13–16
    Weibull distribution of extreme, 16
  irregular, 10–12, 16
  main hull strength and calculation for,
    120–121
  making resistance, 184–188
  orbital motion, 9–10
  progressive abnormal, 20
  regular, 5–7
    higher order waves, 5–7
    trochoid, 5, 120

Waves (*Continued*)
  significant wave height, 12
  sinusoidal, 7–10
  standing, 20
Weapons, warship, 68–69
Weibull distribution, extreme wave heights, 16
Weight distribution, hull stregnth, 121–122
Weight movements, 92–93
What-If analysis, for safety, 827
Wilson's formula, 34
Wind sea, 5
Wind velocity, marine environment, 3–4
WMO, *see* World Meteorological Organization (WMO)
Work class, ROV, 741
Workboats, small
  damage stability code, 114
  intact stability code for, 108–109
World Meteorological Organization (WMO), 17

Y

Yachts, 66–67